The Central Nervous System of Vertebrates   Volume 3

«Le système nerveux est au fond tout l'animal;
les autres systèmes ne sont là que pour le servir
ou pour l'entretenir...»
GEORGES CUVIER (1812)

»Hirnanatomie allein getrieben wäre eine sterile Wissenschaft.
Erst in dem Momente, wo man die Frage nach dem Verhältnis
der anatomischen Struktur zu der Funktion aufwirft,
gewinnt sie Leben.«
LUDWIG EDINGER (1908)

"Our primary interest is in the behavior of the living body,
and we study brains because these organs are the chief instruments
which regulate behavior."
CHARLES JUDSON HERRICK (1948)

Springer-Verlag Berlin Heidelberg GmbH

R. Nieuwenhuys
H. J. ten Donkelaar
C. Nicholson

The Central Nervous System of Vertebrates

Volume 3

With Chapters
in Cooperation with:

J.L. Dubbeldam
P.A.M. van Dongen
and J. Voogd

With 405 Figures and 23 Tables

Springer

Additional material to this book can be downloaded from http://extras.springer.com

ISBN 978-3-642-62127-7   ISBN 978-3-642-18262-4 (eBook)
DOI 10.1007/978-3-642-18262-4

Libary of Congress Cataloging-in-Publication-Data
Nieuwenhuys, R., 1927–
   The central nervous system of vertebrates / R. Nieuwenhuys, H.J. ten Donkelaar, C. Nicholson :
with contributions by J.L. Dubbeldam ... [et al.]. p.   cm.
   Includes bibliographical references.

1. Central nervous system – Anatomy. 2. Vertebrates – Anatomy. 3. Anatomy, Comparative.
4. Neuroanatomy. I. Donkelaar, H.J. ten (Hendrik Jan), 1946–, II. Nicholson, Charles.
III. Title. [DNLM: 1. Central Nervous System – anatomy & histology. 2. Anatomy, Comparative.
3. Vertebrates – anatomy & histology. WL
300 NB82ca 1997]   QM451.N498 1997   573.8'616 – dc21   DNLM/DLC
  for Library of Congress

This work is subject to copyright. All rights are reserved, whether the whole or part of the material is concerned, specifically the rights of translation, reprinting, reuse of illustrations, recitation, broadcasting, reproduction on microfilm or in any other way, and storage in data banks. Duplication of this publication or parts thereof is permitted only under the provisions of the German Copyright Law of September 9, 1965, in its current version, and permission for use must always be obtained from Springer-Verlag. Violations are liable for prosecution under the German Copyright Law.

© Springer-Verlag Berlin Heidelberg 1998

The use of general descriptive names, registered names, trademarks, etc., in this publication does not imply, even in the absence of a specific statement, that such names are exempt from the relevant protective laws and regulations and therefore free for general use.
Product liability: The publishers cannot guarantee the accuracy of any information about dosage and application contained in this book. In every individual case the user must check such information by consulting the relevant literature.

Production: PRO EDIT GmbH, Heidelberg
Typesetting: Mitterweger Werksatz GmbH, Plankstadt
Cover Design: Erich Kirchner, Heidelberg
SPIN: 10087711   27/3136 – 4 3 2 1 0 – Printed on acid-free paper

R. Nieuwenhuys
H. J. ten Donkelaar
C. Nicholson

The Central Nervous System of Vertebrates

Volume 3

With Chapters
in Cooperation with:

J.L. Dubbeldam
P.A.M. van Dongen
and J. Voogd

With 405 Figures and 23 Tables

Additional material to this book can be downloaded from http://extras.springer.com

ISBN 978-3-642-62127-7   ISBN 978-3-642-18262-4 (eBook)
DOI 10.1007/978-3-642-18262-4

Libary of Congress Cataloging-in-Publication-Data
Nieuwenhuys, R., 1927–
   The central nervous system of vertebrates / R. Nieuwenhuys, H.J. ten Donkelaar, C. Nicholson :
with contributions by J.L. Dubbeldam ... [et al.]. p.   cm.
   Includes bibliographical references.

1. Central nervous system – Anatomy. 2. Vertebrates – Anatomy. 3. Anatomy, Comparative.
4. Neuroanatomy. I. Donkelaar, H.J. ten (Hendrik Jan), 1946–, II. Nicholson, Charles.
III. Title. [DNLM: 1. Central Nervous System – anatomy & histology. 2. Anatomy, Comparative.
3. Vertebrates – anatomy & histology. WL
300 NB82ca 1997]   QM451.N498 1997   573.8'616 – dc21   DNLM/DLC
for Library of Congress

This work is subject to copyright. All rights are reserved, whether the whole or part of the material is concerned, specifically the rights of translation, reprinting, reuse of illustrations, recitation, broadcasting, reproduction on microfilm or in any other way, and storage in data banks. Duplication of this publication or parts thereof is permitted only under the provisions of the German Copyright Law of September 9, 1965, in its current version, and permission for use must always be obtained from Springer-Verlag. Violations are liable for prosecution under the German Copyright Law.

© Springer-Verlag Berlin Heidelberg 1998

The use of general descriptive names, registered names, trademarks, etc., in this publication does not imply, even in the absence of a specific statement, that such names are exempt from the relevant protective laws and regulations and therefore free for general use.
Product liability: The publishers cannot guarantee the accuracy of any information about dosage and application contained in this book. In every individual case the user must check such information by consulting the relevant literature.

Production: PRO EDIT GmbH, Heidelberg
Typesetting: Mitterweger Werksatz GmbH, Plankstadt
Cover Design: Erich Kirchner, Heidelberg
SPIN: 10087711   27/3136 – 4 3 2 1 0 – Printed on acid-free paper

List of Contributors

DUBBELDAM, J. L.
Van der Klaauw Laboratory, University of Leiden,
P. O. Box 9516, 2300 RA Leiden,
The Netherlands

MEEK, J.
Department of Anatomy & Embryology,
University of Nijmegen, P. O. Box 9101,
6500 HB Nijmegen, The Netherlands

NICHOLSON, C.
Department of Physiology & Biophysics, New York
University Medical Center, 550 First Avenue,
New York, 10016 NY, USA

NIEUWENHUYS, R.
Papehof 25, 1391 BD Abcoude,
The Netherlands

SMEETS, W. J. A. J.
Department of Anatomy & Embryology,
Free University, Van der Boechorststraat 7,
1081 BT Amsterdam, The Netherlands

TEN DONKELAAR, H. J.
Department of Anatomy & Embryology,
University of Nijmegen, P.O. Box 9101,
6500 HB Nijmegen, The Netherlands

VAN DONGEN, P. A. M.
Linge 9, 5032 EV Tilburg, The Netherlands

VOOGD, J.
Department of Anatomy, Erasmus University
Rotterdam, P. O. Box 1738, 3000 DR Rotterdam,
The Netherlands

WICHT, H.
Klinikum der J.-W.-Goethe-Universität, Zentrum
der Morphologie, University of Frankfurt,
Theodor-Stern-Kai 7, 60590 Frankfurt am Main,
Germany

Contents

VOLUME 1

Preface V
Purpose and Plan IX
Acknowledgements for Figures XI

I. GENERAL INTRODUCTORY PART

Chapter 1	Structure and Function of the Cellular Elements in the Central Nervous System C. NICHOLSON	1
Chapter 2	Structure and Organisation of Centres R. NIEUWENHUYS	25
Chapter 3	Structure and Organisation of Fibre Systems R. NIEUWENHUYS	113
Chapter 4	Morphogenesis and General Structure R. NIEUWENHUYS	158
Chapter 5	Histogenesis R. NIEUWENHUYS	229
Chapter 6	Comparative Neuroanatomy Place, Principles and Programme R. NIEUWENHUYS	273
Chapter 7	Notes on Techniques H. J. TEN DONKELAAR and C. NICHOLSON	327

II. SPECIALISED PART

Chapter 8	Introduction R. NIEUWENHUYS	357
Chapter 9	Amphioxus R. NIEUWENHUYS	365
Chapter 10	Lampreys, Petromyzontoidea R. NIEUWENHUYS and C. NICHOLSON	397
Chapter 11	Hagfishes, Myxinoidea H. WICHT and R. NIEUWENHUYS	497
Chapter 12	Cartilaginous Fishes W. J. A. J. SMEETS	551
Chapter 13	Brachiopterygian Fishes R. NIEUWENHUYS	655
Chapter 14	Chondrostean Fishes R. NIEUWENHUYS	701

VOLUME 2

II. SPECIALISED PART (Contd.)

Chapter 15	Holosteans and Teleosts J. Meek and R. Nieuwenhuys	759
Chapter 16	Lungfishes R. Nieuwenhuys	939
Chapter 17	The Coelacanth, *Latimeria chalumnae* R. Nieuwenhuys	1007
Chapter 18	Urodeles H. J. ten Donkelaar	1045
Chapter 19	Anurans H. J. ten Donkelaar	1151
Chapter 20	Reptiles H. J. ten Donkelaar	1315

VOLUME 3

II. SPECIALISED PART (Contd.)

Chapter 21	Birds J. L. Dubbeldam	1525
Chapter 22	Mammals J. Voogd, R. Nieuwenhuys, P. A. M. van Dongen and H. J. ten Donkelaar	1637

III. GENERAL CONCLUDING PART

Chapter 23	Brain Size in Vertebrates P. A. M. van Dongen	2099
Chapter 24	The Meaning of It All R. Nieuwenhuys, H. J. ten Donkelaar and C. Nicholson	2135
Subject Index		2197

II. SPECIALISED PART (Contd.)

21 Birds 1525

22 Mammals 1637

CHAPTER 21

Birds

J.L. Dubbeldam

21.1	Introduction	1525
21.2	Nomenclature	1528
21.3	Gross Features of the CNS	1528
21.4	Morphogenesis and Histogenesis	1532
21.4.1	Morphogenetic Events	1532
21.4.2	Histogenesis	1535
21.5	Spinal Cord	1542
21.5.1	Introductory Remarks	1542
21.5.2	Spinal Grey	1542
21.5.3	Marginal Nuclei and Corpus Gelatinosum	1549
21.5.4	Dorsal Roots	1549
21.5.5	Funiculi	1552
21.6	Rhombencephalon	1554
21.6.1	Introductory Remarks	1554
21.6.2	Cranial Nerves	1554
21.6.3	Somatosensory Complex	1555
21.6.4	Vestibulo-cochlear Complex	1557
21.6.5	Viscerosensory Complex	1559
21.6.6	Motor Nuclei	1560
21.6.7	Oliva Inferior and Nuclei of the Pontine Region	1562
21.6.8	Reticular Formation	1562
21.6.9	Fibre Systems of the Brain Stem	1564
21.7	Cerebellum	1564
21.7.1	Introductory Remarks	1564
21.7.2	Cerebellar Cortex	1564
21.7.3	Histochemistry of the Cerebellar Cortex	1567
21.7.4	Central Nuclei	1567
21.7.5	Large Cerebellar Fibre Systems	1567
21.8	Mesencephalon	1568
21.8.1	Introductory Remarks	1568
21.8.2	Mesencephalic Trigeminal Nucleus	1568
21.8.3	Tegmental Cell Groups	1573
21.8.4	Reticular Formation	1573
21.8.5	Isthmic Cell Groups	1573
21.8.6	Colliculus Mesencephali and Tectum Mesencephali	1575
21.8.7	Nuclei of the Lobus Mesencephali	1579
21.8.8	Mesencephalic Fibre Systems	1580
21.9	Diencephalon	1581
21.9.1	Introductory Remarks	1581
21.9.2	Nervus Opticus and Chiasma Opticum	1581
21.9.3	Synencephalon or Pretectal Region	1581
21.9.4	Epithalamus	1583
21.9.5	Dorsal Thalamus	1583
21.9.6	Ventral Thalamus	1585
21.9.7	Thalamofrontal Fibre Systems	1586
21.9.8	Hypothalamus	1586
21.10	Telencephalon	1589
21.10.1	Introductory Remarks	1589
21.10.2	Paleostriatal Complex	1591
21.10.3	Neostriatum, Hyperstriatum Ventrale and Related Structures	1596
21.10.4	Vocalisation Circuit: A Specific Sensorimotor Circuit	1604
21.10.5	Neostriatum and Hyperstriatum Ventrale: Associative Functions and the Role in Learning	1606
21.10.6	Archistriatum	1607
21.10.7	Cortical Regions	1609
21.10.8	Septal Area	1615
21.10.9	Olfactory System and Terminal Nerve	1615
21.10.10	Ascending, Descending and Intrinsic Telencephalic Fibre Systems	1616
21.11	Concluding Remarks	1620
	References	1620

21.1
Introduction

The modern birds display a great diversity of adaptations to many different ecological niches and a great behavioural variability. However, the most striking and unifying characteristic of all birds is their ability to fly. Almost every avian feature is an adaptation to flight. This is probably also true for the high and constant body temperature associated with the need for a high metabolic rate to sustain flight. In this respect birds differ radically from the various groups of reptiles, though they still seem to share many other features, particularly with the crocodilian line, and even though it is generally believed that birds are descendants of the dinosaurian stock (e.g. Ostrom 1974). Many avian species exhibit a well-developed social behaviour and with it an elaborate pattern of vocalisations.

The recent class Aves can be subdivided into two main groups. Most birds belong to the superorder Neognatha, comprising more than 20 ordines and representing the advanced stage of bird evolution. Several flightless birds – the Ratites: ostrich, emu, cassowary, rhea and kiwi, but also the extinct moa – are placed with the tinamous in the superorder Paleognatha. There is no unanimity, either about

Fig. 21.1. The pigeon, *Columba livia* L.

the phylogeny of or about the relationships between the orders of neognaths. The earliest avian forms – *Archaeopteryx* and *Protoavis*, representing the subclass Archaeornithes – appeared late in the Jurassic. Descendants of these early birds form the group of Ornithurae living in the Cretaceous. According to Cracraft (1988) this group can be split into the Hesperornithiformes and the Carinatae (Fig. 21.2). One of the characters of the latter group is the presence of a keel (carina) on the sternum; this group is at the base of the modern birds, the Neornithes comprising both the paleognath and neognath birds. According to the cladogram presented in Fig. 21.2 the taxa of galliform and anseriform birds are monophyletic and form together the sister group of the complex of the other neognath orders. Feduccia (1995), however, sketches a different picture of the evolution: he assumes a bottleneck transition around the Cretaceous-Tertiary boundary (with a small number of species during this transition), followed by an explosive evolution within a period of 5–10 million years. In his opinion the divergence of the neognath orders within such an evolutionary short period of time complicates the ascertainment of 'higher level relationships by DNA-DNA hybridisation or cladistic methodology'. This short survey may suffice to demonstrate the existence of diverging interpretations.

The pigeon, *Columba livia* L. (Fig. 21.1), has been chosen as the representative of the modern birds. It belongs to the order Columbiformes, comprising approximately 305 species. The genus *Columba* contains 51 species, which are found all over the world and show considerable adaptive radiations (Martin 1979). 'The' pigeon does not exist. Many domestic strains have been bred, leading to a great variety of pigeon characteristics. The original form is the rock pigeon, thought to have evolved in arid or semi-arid regions. Its present wide distribution may be an effect of the human agricultural activity providing the bird with new suitable feeding grounds (Martin 1979). Feral pigeons inhabit many cities all over the world and are derived from the domesticated form and share part of its variability. Some wild-type characteristics may even have been bred out in the domesticated and feral forms. This may pertain to morphological features, but possibly also behavioural changes – e.g. in the patterns of vocalisation – have occurred (Goodwin 1983). This variability is also found for the shape and structure of the brain (e.g. Ebinger and Löhmer 1984). The well-known atlas of Karten and Hodos (1967) is perfect when

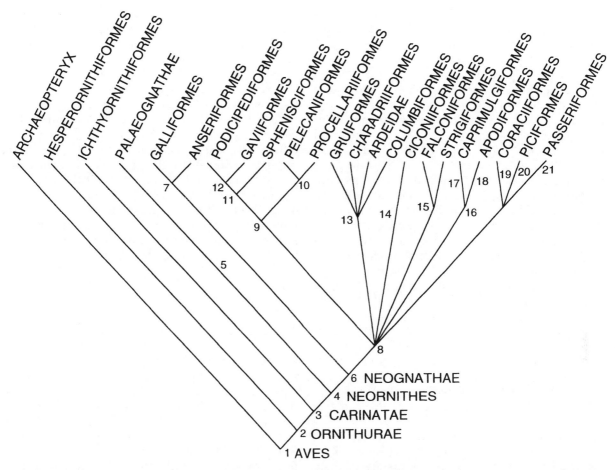

Fig. 21.2. Cladogram of major recent and extinct avian taxa based upon morphological characters and partly supported by molecular data. In this cladogram the Palaeognathae are considered a monophyletic group derived from the Carinatae, birds in possession of a keel on the sternum. Note also the assumed monophyly of the clade of Galliformes and Anseriformes. (From Cracraft 1988)

using the white carneaux pigeon; however, the use of other, even closely related breeds requires adjustments of the stereotactic atlas coordinates.

Rock pigeons are primarily seed-eating birds, but occasionally also take other types of food, such as berries, leaves or even snails. Feral and domestic pigeons take a still greater variety of food, but always show the typical pecking behaviour. Several studies have been devoted to the pecking and drinking behaviour in pigeons (e.g. Zeigler et al. 1980; Zweers 1982a,b). These studies prove that, although the pigeon is often considered to represent a 'basic avian form', it possesses a highly specialised type of pecking and drinking behaviour. Many other avian species have developed new feeding habits, such as probing in charadriiform birds, straining in many anseriform birds and flamingoes, fishing and grazing. Some birds, most notably the psittaciforms, possess a highly developed manipulative ability. This variability in feeding habits and other 'beak activities' is associated with a great diversity of sensory and motor specialisations. The same is true for many other aspects of avian behaviour.

The pigeon has been chosen as the core species, as it has been – and still is – used in many neurobiological and behavioural studies. Probably one of the best-studied areas is that of vision and the organisation of the visual system (recent reviews in Karten 1979; Güntürkün 1991; Zeigler and Bischof 1993). Also another specific ability, homing behaviour, has received considerable attention and still is the subject of much debate. Many aspects of the pigeon's sensory abilities and behaviour have been discussed in the book edited by Abs (1983). Surprisingly enough, so far the neurobiology of flight has received only scant attention. One of the consequences of flight is the dissociation of the activity of the forelimbs/wings and hindlimbs or legs. This should have consequences for the organisation of the locomotory system in the spinal cord. The same is true for the synchronisation of left and right wing activity. The possession of a well-developed cerebellum and vestibular system is considered a prerequisite for flying behaviour.

Birds are typical 'eye-animals', but exceptions exist. Night-hunting owls and oilbirds rely heavily upon the auditory system for orientation (e.g. Knudsen 1980), whereas many probing and straining birds depend upon tactile sense to detect and recognise food (e.g. Bolze 1969; Berkhoudt 1980). The mallard, *Anas platyrhynchos* L., is included in this chapter as a typical 'tactile' bird. Until recently, birds were supposed to have a poorly developed sense of taste, but recent work has changed this picture (see Berkhoudt 1985). Also the significance of the sense of smell has been poorly understood for a long time and only slowly is more knowledge about the role and importance of the sense of smell becoming available. Most intriguing is the suggestion that olfaction is important in the homing behaviour of pigeon, but strong differences of opinion exist (e.g. Papi et al. 1978; Waldvogel and Phillips 1991). There is increasing evidence that pigeons may use different senses in this behaviour depending upon the area where they hatch and grow up. So far, the possible role and biological basis of magnetic sense are little understood (e.g. Able and Able 1993).

Several other birds have been used in studies of specific aspects of behaviour. The chicken, *Gallus gallus forma domestica*, is used in many studies on learning and imprinting (recent survey in Andrew 1991). A large body of work exists about the neural basis of vocalisations, most notably in songbirds and psittaciforms (in particular the budgerigar). A few sections of the forebrains of representatives of these two groups (*Corvus monedula* L., jackdaw, and *Melopsittacus undulatus* Shaw, budgerigar) have been included, not only to illustrate this aspect of vocalisation, but also because these birds possess brains with the most highly developed telencephala.

For a long time the organisation of the avian telencephalon has intrigued many investigators. The most prominent feature is its 'striated' appearance and for a long time it has been considered to consist of a strongly proliferated corpus striatum with a rudimentary cortex (cf. Ariëns Kappers et al. 1936). Lately this picture has changed drastically after recognition of the nature of the so-called dorsal ventricular ridge. This appears to be a characteristic of the whole sauropsid stock; its development and importance have been discussed extensively by Ulinski (1983, 1990) and Karten (1991). Considerable differences in telencephalic size and organisation exist (e.g. Stingelin 1958). Portmann (1946, 1947) expressed the relative size of the brain of a specific bird in a 'cerebralisation index', defined as the cerebral weight of this brain/weight of the brain stem of a gallinaceous bird of the same body weight. Cerebralisation is lowest in Gallinaceae (about 2.5–3.2), highest in passeriform birds (7–18.9) – particularly in corvids – and in psittaciform birds (7.4–27.60, highest in the genus *Ara*). Recent studies (Rehkämper et al. 1991) confirm this picture. The pigeon has a relatively low index, but differences exist between different breeds, homing pigeons having relatively larger brains than non-homing breeds (Rehkämper et al. 1988).

In conclusion, basically the organisation of spinal cord and brain stem has many features in common with that in both reptiles and mammals. The most impressive differences are found in the organisation of thalamus and telencephalon.

21.2
Nomenclature

The most commonly used nomenclature is that of the atlas of Karten and Hodos (1967). Particularly for the forebrain, however, a different nomenclature is sometimes used (e.g. Kuhlenbeck 1977). In principle, the nomenclature of the Nomina Anatomica Avium (NAA) (Breazile and Kuenzle 1993) will be used in this chapter; it corresponds largely to those of Karten and Hodos (1967) and Breazile and Hartwig (1989). When the nomenclature of the NAA differs from the nomenclature used elsewhere in this book, the latter will be used and the names of the NAA will be mentioned between square brackets.

Several stereotactic atlases are available. The most commonly used is that of the pigeon (Karten and Hodos 1967). Other atlases are those by van Tienhoven and Juhasz (1962: chicken), Zweers (1971: brain stem of the mallard), Baylé et al. (1974: *Coturnix coturnix*), Stokes et al. (1974: *Serinus canaria*), Vowles et al. (1975: *Streptopelia risoria*), Youngren and Phillips (1978: chick), Matochick et al. (1991: *Fulmarus glacialis*) and Kuenzel and Masson (1988: chicken). A complication in the comparison of the various atlases is the sometimes strongly differing orientation of the brain in the stereotactic plane. All authors use more or less the same nomenclature.

21.3
Gross Features of the CNS

The central nervous system (CNS) comprises brain and spinal cord, the latter forming the larger part in the pigeon (Fig. 21.3). Different from the situation in mammals, the number of cervical vertebrae, and with it the number of cervical spinal nerves, is

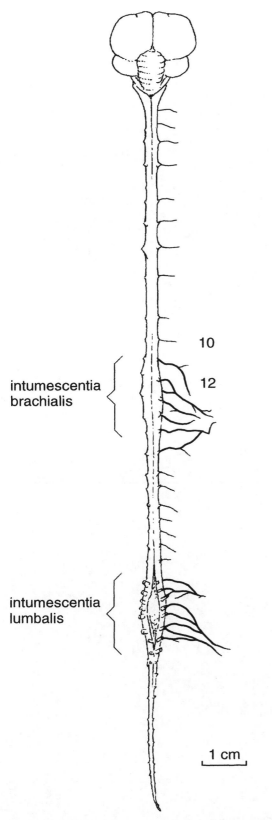

Fig. 21.3. Central nervous system of the pigeon in dorsal view

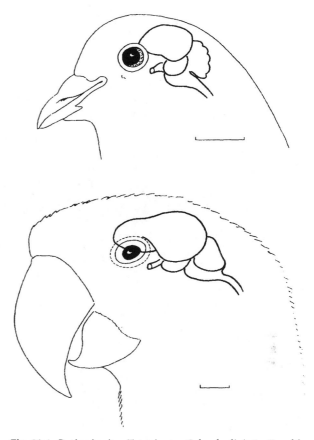

Fig. 21.4. Brains in situ. *Top*, pigeon, *Columba livia* L.; *B*, gold and blue macaw, *Ara ararauna* L. *Bar*, 1 cm

highly variable among the avian species (Bubien-Waluszewska 1985). Also a certain variability occurs within a species. Generally, 14 cervical, 6 thoracic and 6 lumbal nerves are found in the pigeon (Kaiser 1923). The spinal nerves 12–15 form the plexus brachialis (Fig. 21.3), but in about 50% of cases spinal nerve 11 also participates. The plexus lumbalis is formed by spinal nerves 21–26, sometimes with a contribution from spinal nerve 20. The most caudal, sacrococcygeal part is highly variable. The spinal cord of the pigeon reaches the entire length of the vertebral canal, there being no cauda equina. Characteristic of the avian spinal cord is the presence of a slight cervical enlargement at the level of the brachial plexus and the existence of a conspicuous lumbosacral enlargement (Fig. 21.3). The latter is further characterised by the presence of the rhomboid sinus filled with the so-called glycoid body.

The shape of the brain differs greatly among the avian species (Fig. 21.4). Apart from the differences in size of the brain parts, the differences in shape are largely connected to the position of the brain in the head (Dubbeldam 1968, 1989). Extreme situ-

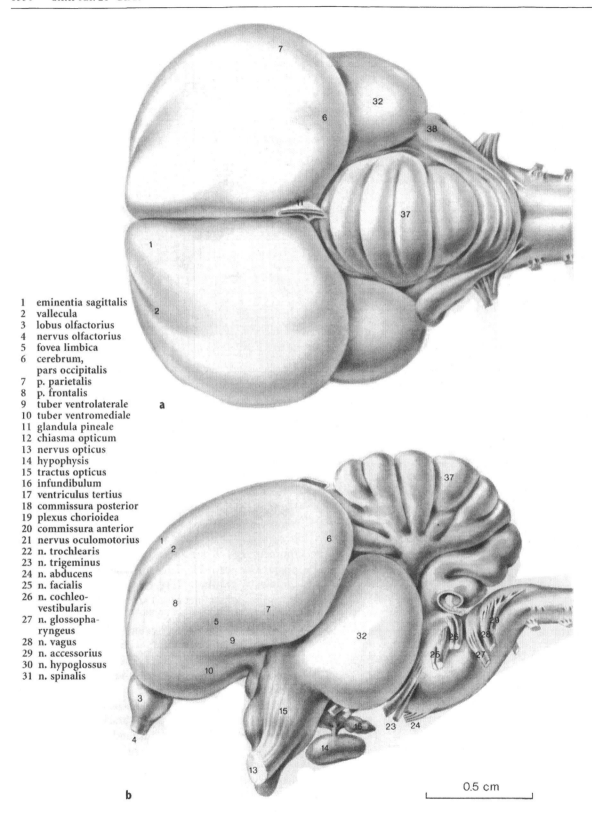

Fig. 21.5a,b. Brain of the pigeon in dorsal (a) and lateral (b) view.

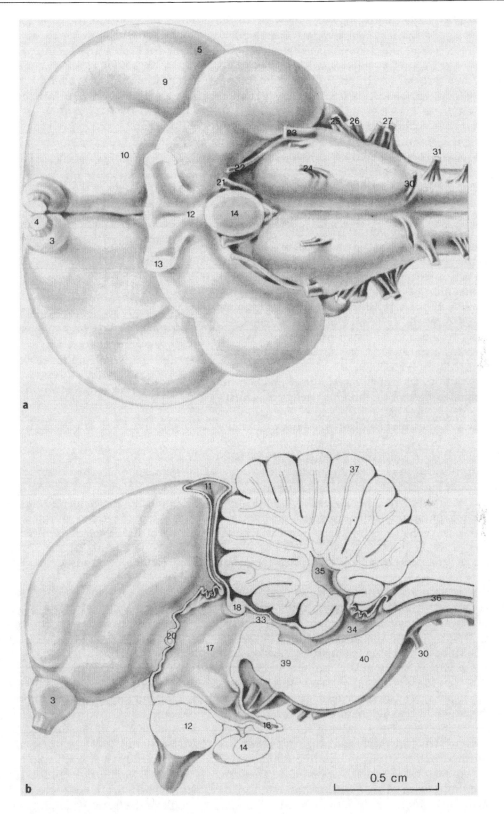

Fig. 21.6a,b. Brain of the pigeon in ventral view (a) and midsagittal section through the brain (b). For abbreviations see Fig. 21.5

ations are found in, for example, the cormorant (*Phalacrocorax carbo sinenses* Shaw and Nodder), with a horizontal and approximately straight brain axis, and the common snipe (*Capella gallinago* L.), with a vertically directed and strongly curved brain axis. The situation in most birds is somewhere between these extremes; this is also true for the pigeon, with a brain axis under an angle of about 45° to the length axis of the skull (Fig. 21.4). Its brain stem comprising myelencephalon, mesencephalon and diencephalon (the *reste du tronque* of Portmann 1946) is ventrally curved in the myelen-mesencephalic region with a second curvature in the transition of mesencephalon-diencephalon (Fig. 21.6b). The longest axis of the telencephalic hemispheres again makes an angle of about 40° with the length axis of the whole brain. The pigeon possesses a well-developed cerebellum consisting of ten folia, most of which can be subdivided into two or three subfolia (Fig. 21.6b). Laterally it bears the auricula cerebelli. It is not possible to delineate a distinct vermis and hemisphaerae cerebelli.

Another avian feature is the lateral position of the optic lobes (Figs. 21.5a, b). In the early embryo the tectum mesencephali has a dorsal position not different from that in other vertebrates. During ontogeny, however, the diencephalic region shortens and the cerebral hemispheres expand caudally, 'pushing' the optic lobes laterally (Huber 1949). Depending on the size of the hemispheres the position of the optic lobes varies from dorsolateral (e.g. owl, duck, Fig. 21.48) to ventrolateral (e.g. jackdaw, Fig. 21.45) to the brain stem. Again, the optic lobes in the pigeon occupy an intermediate position.

The telencephalon is the most conspicuous part of the brain. It comprises two large hemispheres that cover the dorsal aspect of the diencephalon and most of the mesencephalon (Fig. 21.5a). Caudally, the rostral part of the cerebellum is wedged between the hemispheres. In the pigeon each hemisphere bears a small, but distinct bulbus olfactorius (Fig. 21.5b). Overall, the surface of the telencephalon is smooth, but each hemisphere has a slight dorsal bump – the dorsal Wulst or eminentia sagittalis – bordered by a groove, the vallecula (Figs. 21.5a, 21.41). A superficial fissura, the fovea limbica, delineates the occipital lobe (polus occipitalis). Stingelin (1958), studying the telencephalic shape and structure in a broad range of species, distinguished brains with a *basale Frontbildungstendenz* and with a dorsal *Frontbildungstendenz* depending upon the position of the Wulst (Fig. 21.56). Moreover, he attempted to interpret differences in shape and size in terms of increasing complexity of brain structures. Without doubt the rich variety of shapes and sizes of avian forebrain reflect differences in functional specialisation (e.g. Dubbeldam 1990a) as well as an increasing complexity of brains among the species (e.g. Rehkämper et al. 1991). However, our knowledge of the organisation of the avian forebrain is still too incomplete to provide a satisfactory functional or phylogenetic interpretation of the various avian brain types.

21.4
Morphogenesis and Histogenesis

The study of the morphogenesis and histogenesis of the avian brain has contributed greatly to a better understanding of its organisation and to the recognition of homologies in the brains of various vertebrates. Particularly, the theory of neuromery has great importance in this respect; much of the research on neuromeres has been done in the chicken (e.g. Bergquist and Källén 1954; Vaage 1969; Puelles et al. 1987; Lumsden 1990). The neuromeres were recognised as early as the nineteenth century (cf. von Kupffer 1906; review in Vaage 1969). The following account of the development of the avian brain is based mainly upon work in the chicken. For a fuller discussion of the concepts, see Chap. 3.

21.4.1
Morphogenetic Events

An important aspect of the early morphogenesis in birds is that the early embryo is no more than a disc on top of the large mass of yolk. As early as 18 h after the beginning of the incubation, the neural plate can be recognised in cross-sections of the embryo (Patten 1957). It is difficult to use age in the comparison of young embryos, as specimens of the 'same' age may look different. For this reason Hamburger and Hamilton (1951) described a series of stages using morphological criteria: the number of somites to define the early stages, after stage 22 (possessing 22 somites), the development of the limbs, etc. Use of the stages instead of age facilitates the comparison between different studies. Therefore, the stages will be used to describe further development as far as possible.

The neural folds begin to develop after about 22 h, i.e. in the two-somite stage, corresponding to stage 7 of Hamburger and Hamilton (1951). The neural folds begin to approach each other in the middle of the head region, this process extending in the rostral and caudal direction. Closure of the neural tube begins after 27 h (stage 9); the rostral neuropore (Fig. 21.7) is open until about stage HH11.

21.4 Morphogenesis and Histogenesis 1533

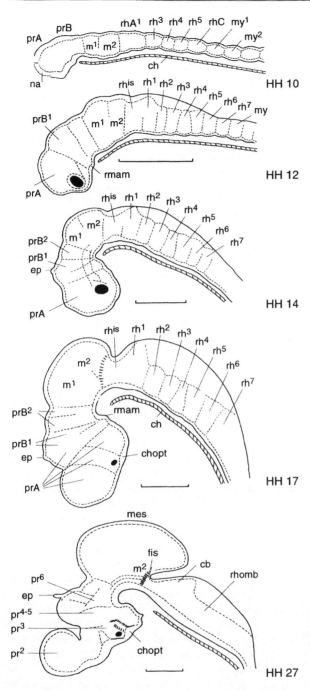

Fig. 21.7. Early developmental stages of the brain of the chicken. For each the stage according to the table of Hamburger-Hamilton is indicated. *cb*, cerebellar plate; *ch*, chorda; *chopt*, chiasma opticum; *ep*, epiphysis; *fis*, fovea isthmi; m^1, m^2, mesencephalic neuromeres 1, 2; *mes*, mesencephalon; my^1, my^2, etc., myelomeres 1, 2, etc.; *na*, neuroporus anterior; *prA*, etc., prosomeres A, etc.; *rhA*, rh^2, etc., rhombomeres A, 2, etc.; *rhom*, rhombencephalon; *rmam*, recessus mammilaris; *st*, septum transversum. *Bar in HH 10–17* represents 0.5 mm, *in HH 27* 1 mm. (Redrawn after Vaage 1969)

Often there is a transient occlusion of the spinal canal after closure of the tube (Schoenwolf and Desmond 1984; Desmond and Schoenwolf 1985). The enlargement of the cephalic part of the neural tube begins around stage 9–10, and furrows bordering two bulges – archencephalon and deuterencephalon – can soon be recognised. The latter subdivides into the mesencephalon and rhombencephalic bulge. At the same time the optic vesicles evaginating from the archencephalon become visible. From this stage on a pattern of neuromeres develops in the prospective brain and spinal cord (Fig. 21.7), but diverging and sometimes opposing interpretations of the observations exist. It is beyond the scope of this chapter to review all the literature on this topic, as most of it has been discussed extensively by Vaage (1969; see also Chap. 3). However, the current interpretation of the avian brain, and in particular of the forebrain, relies heavily upon embryological evidence. A few aspects will therefore be considered.

Neuromeres are transient areas in the brain; slight furrows in the neural tube indicate the boundaries between successive neuromeres (Fig. 21.7). Bergquist and Källén (1954) argue that neuromere-like structures appear and disappear in three successive waves, called the proneuromeres, neuromeres and migration areas, respectively. These structures coincide topographically with proliferation zones in the neural tube. According to these authors no neuromeric structures are present between these waves. Vaage (1969), however, demonstrated that the number of neuromeres increases by splitting without a temporary disappearance of the neuromeres. Successive periods of cell proliferation produce waves of migrating cells from the ventricular surface to the outer surface, the so-called migration zones. The following criteria can be used for homologisation of cell groups: (a) which migration area (last generation neuromeres), (b) which migration zone and (c) which part of this migration zone (Bergquist and Källén 1954). Bergquist and Källén describe four longitudinal columns of migration areas: a dorsal one, the source of somatosensory cell groups; a dorsolateral and a ventrolateral column, the origin of viscerosensory and visceromotor cells, respectively; and a ventral one, the source of the somatomotor nuclei. In the interpretation of these authors, the last column extends to the rostral margin of the mesencephalon, the dorsal one not further than the rostral margin of the rhombencephalon, whereas the two lateral columns reach the optic chiasm, thus including the mesencephalon and part of the prosencephalon. The importance of these columns had been recognised long ago, but in particular their importance in the prosencephalic region has been disputed (see below).

A certain consensus exists about the interpretation of the rhombencephalic neuromeres, but diverging opinions have been put forward about the mesencephalon and archencephalon. Many authors recognised the three-vesicle stage consisting of archencephalon, mesencephalon and rhombencephalon. Generally, the archencephalon is considered a prechordal structure and the other two vesicles epichordal structures (Fig. 21.7: HH10), the plica encephali indicating the border between the pre- and epichordal areas. However, Bergquist and Källén, referring to Kingsbury (1920), state that the floor plate of the neural tube is coextensive with the notochord and sulcus dorsalis. Using this criterion the pre-epichordal limit would be at the fovea isthmi, i.e. caudal to the mesencephalon. These authors stress that the differences in interpretation may be caused (at least partly) by the use of different criteria:

1. A morphological criterion, i.e. the plica encephali ventralis as border archen-deuterencephalon
2. An embryological criterion: distinction of a metameric zone and a premetameric zone, i.e. deuterencephalon and archencephalon, respectively
3. A morphogenetic criterion: archencephalon and deuterencephalon develop under the influence of two different inducing systems (Nieuwkoop 1952)

In the opinion of Vaage (1969) the first and third criteria lead to the same conclusion about the border archen-deuterencephalon. Whereas Bergquist and Källén (1954) contend that no neuromeres are present in the archen- or prosencephalic region, Vaage described the development of three to eight prosomeres in this region (Fig. 21.7: HH14–21). Puelles et al. (1987) again concluded that the par- and synencephalic neuromeres are still epichordal and that only the secondary prosencephalon is prechordal and non-neuromeric. The discrepancies in interpretation may be partly caused by the existence of other morphogenetic processes.

From about stage 10 the prosencephalon begins to bend ventrally – the cranial flexure – with the result that the mesencephalic bulge seems to occupy the most rostral position (Fig. 21.7: HH14). Around stage 13 the rostral part of the embryo tordates to the left and the prosencephalon becomes visible again, now in lateral view. At this stage three prosomeres can be recognised, in the terminology of Vaage PrA, B^1 and B^2, the boundary between PrA and B possibly being the di-telencephalic border (Fig. 21.7: HH15). Finally, the early prosomeres differentiate into eight prosomeres. Prosomeres B^1 and B^2 form the prospective parencephalic and synencephalic regions of the diencephalon (cf. von Kupffer 1906). The parencephalon carries a small dorsal evagination, the epiphysis (Fig. 21.7), that will develop into the pineal organ (Patten 1957). Rendahl (1924) distinguished an anterior and a posterior parencephalon, and therefore three diencephalic segments. Several other interpretations of the numbers of prosomeres exist (reviewed in Vaage) that cannot easily be harmonised. Figdor and Stern (1993) using scanning electron microscopy describe four diencephalic neuromeres (D1-D4). The borders between these neuromeres and with the mesencephalon and telencephalon appear as ridges in the wall of the third ventricle during stages 12–17. Neuromere D1 differentiates into ventral thalamus and hypothalamus, D2 into dorsal thalamus with habenula and stria medullaris, D3 into the anterior pretectum (without the commissura posterior and its tract) and D4 into the caudal pretectum including the posterior commissure. The anterior border of the tectal commissure and exit point of the oculomotor nerve form the border with the mesencephalon. Figdor and Stern (1993) also demonstrated that during development cells do not cross the neuromeral borders. Their neuromeres 1 and 2 correspond more or less with the parencephalon anterius and posterius of Rendahl (1924), respectively, whereas his synencephalon comprises the neuromeres D3 and 4. Martinez et al. (1993), however, dispute this interpretation and using another marking technique describe not more than three diencephalic segments.

At the end of day 3 (around stage 17) the anterolateral walls of the prosencephalon begin to evaginate and the telencephalic vesicles or prospective telencephalic hemispheres become visible (Fig. 21.7: HH17, 27). Initially the expansion of these vesicles is in the dorsal direction, but from day 4 also in the rostral direction gradually narrowing the fissura interhemispherica. The expansion in the diencephalic area is mainly in the dorsoventral direction; indeed a certain rostrocaudal shortening has been described (Huber 1949), resulting in a caudal displacement of the telencephalon (Fig. 21.8). This displacement is coupled with a lateral displacement of the developing mesencephalic tectum, resulting in a lateral instead of a dorsal position of the prospective optic lobes. The pontine flexure is visible from about stage 27 and increases considerably from stage 31. At the same time the brain begins to widen (Rogers 1960). The cerebellum can be recognised from day 10 (stage 36), the first fissures appearing on day 11

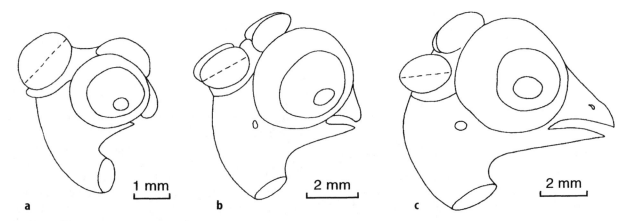

Fig. 21.8a-c. Lateral view of the head of chick embryos showing the caudal displacement of the telencephalon and change of position of the tectum mesencephali during development. a embryo day 7; b day 8; c day 9. (Redrawn after Huber 1949)

(stage 37). The essential features of the brain are present by day 12 (Romanoff 1960).

21.4.2
Histogenesis

A few things remain to be said about the histogenesis and development of the neuronal centres. The first neuroblasts appear as early as in stage 5; these are precursors of primarily medium-sized cells of the reticular formation in the medulla, the di-mesencephalic junction, as well as the intermediate zone of the spinal cord (McConnell and Sechrist 1980). The first prospective motoneurons and sensory neurons appear after 26–28 h and 32 h, respectively. In the chicken 12 myelomeres have been recognised in the spinal cord. More caudally, no myelomeres develop; the rostral myelomeres fuse into an uninterrupted longitudinal cell column. There is no dispute about the existence of rhombomeres, except about the exact number. Vaage (1969) mentions a maximum of seven rhombomeres in the chicken (Fig. 21.7) adding an isthmic rhombomere of uncertain status (Lumsden 1990). Lumsden distinguished eight rhombomeres, the rhombospinal border lying medial to the most rostral somite (i.e. first rudimentary somite) but not marked by a morphological discontinuity. He demonstrated that the motoneurons contributing to a specific cranial nerve are derived from two adjoining rhombomeres: the trigeminal nucleus from rh^2 and rh^3, the facial from rh^4 and rh^5, etc. (Fig. 21.9). The motoneurons of the trochlear nerve arise from the 'isthmic rhombomere' of Vaage. One rhombomere can contribute to more than one nerve, e.g. rh^5 to N. VI and N. VII, rh^6 to N. VI and IX (Lumsden 1990). The oculomotor complex has its origin in the mesencephalic neuromeres or mesomeres. It is disputed whether one or two mesomeres occur. It is interesting to note that the segmental organisation of the spinal cord may be caused by mesoderm/somatic induction (e.g. Keynes and Stearns 1984), whereas the development of cranial nerves seems to be closely related to the transient existence of rhombomeres (Vaage 1969; Lumsden 1990), a segmentation resulting from the existence of proliferation zones (Bergquist and Källén 1954) that may arise as the result of 'differential activation of regulating genes' (Lumsden 1990).

Puelles and Medina (1994) studied the early development of tyrosine hydroxylase and dopamine immunoreactivity in the chicken embryo. In particular the distribution, alar or basal origin and segmental origin were analysed. These authors point out that limits of catecholaminergic cell groups generally fit within the morphological pattern of the segmental scheme. Furthermore, separate groups of catecholaminergic cells seem to emerge from the basal and alar regions (Puelles and Medina 1994). Such a parcellation may help the identification of specific cell groups in different classes of vertebrates. In a recent study in the pigeon, Medina and Reiner (1994) using choline acetyltransferase immunoreactivity used the same approach to identify corresponding avian and mammalian cell areas. Such an approach should be done very carefully, as there is the tendency to use histochemical data to better define segmental borders as well as to use these borders to identify cells groups with similar histochemical characteristics.

A few things should be said about the other brain parts. The adult tectum mesencephali is a multi-layered structure. Three slightly overlapping phases can be recognised in its development (LaVail and

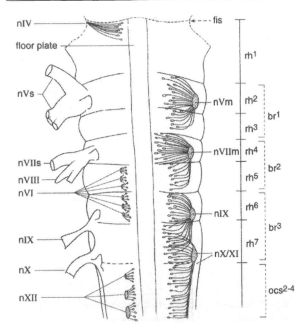

Fig. 21.9. Diagram of the ventral aspect showing the relationship of rhombomeres (rh^{1-8}) and the cranial nerve motor centres (*right side*) in stage 21 chick embryo. In addition the relationship of rhombomeres and the branchiomeres (br^{1-3}) and occipital somites (ocs^{2-4}) are indicated. (Redrawn after Lumsden 1990)

Cowan 1971). During phase 1 (days 3–6) cell proliferation causes an increase in the surface, but not in the thickness of the tectum. Phase 2 is characterised by migration and differentiation of neuroblasts; three migration waves can be recognised (LaVail and Cowan 1971). Recent studies revealed that one matrix cell gives rise to both neuroblasts and glia cells (Galileo et al. 1990). Clonically related neurons migrate radially to form 'ontogenetic columns' throughout the layers of the tectum, thus contributing different cell types (Gray et al. 1988). The first retinal fibres reach the tectum by day 6, cover it by day 9 and penetrate the tectum on day 10 (Crossland et al. 1974); according to Rager and von Oeynhausen (1979) the first fibres may even penetrate the tectum at around day 8. The gradual arrival of retinal fibres and gradual maturation of tectal cells result in the development of a retinotopical organisation of the tectal afferents (Rager 1980). Most retinal fibres reach the contralateral tectum, but a small number has also been found ipsilaterally at around day 9. These have disappeared again on day 15 (Thanos and Bonhoeffer 1984). The orderly ingrowth of retinal fibres may depend at least partly upon positional cues on the tectum (Thanos et al. 1984). Rager (1980) suggests that synchronisation of these fibres and local maturation of tectal cells may be important. The third phase of tectal development comprises the differentiation of tectal cells and the final segregation of cells and fibres giving rise to the outer laminae of the stratum griseum et fibrosum superficiale. In the pigeon ipsi- and contralateral retinal afferents are present at hatching; degeneration of the ipsilateral fibres, ingrowth of the contralateral projection and maturation occur during the first 6 days after hatching (Bagnoli et al. 1987).

As for the cerebellum: in stage HH17 the roof of the fourth ventricle is mainly occupied by neural epithelium. Rüdeberg (1961) distinguished two migration layers during further development, the first appearing around stage HH19, the second becoming visible around stage HH25 (Fig. 21.10a). The medial part of the first migration (A2) and the second migration (B) form a distinct cell mass (A2B) giving rise to the cerebellar nuclei (Fig. 21.10b–e). These nuclei develop independently of the vestibular region (Rüdeberg 1964). At the same time the remaining A layer enters the early external granular layer (Fig. 21.10c), an external proliferating cell layer giving birth to the cells of the future granular layer and to glioblasts (Hanaway 1967; review in Feirabend 1990). Slightly after this external layer a second, inner cortical cell layer becomes visible (Fig. 21.10d); the cells of this layer develop from the ventricular epithelium. At least the Purkinje cells, but possibly also Golgi cells, develop from this internal layer (Hanaway 1967; Feirabend 1990). Feirabend et al. (1985) further demonstrated that both rostrocaudally and mediolaterally directed gradients exist during the differentiation resulting in a parcellation of the early cerebellar cortex. Around the 8th day of incubation clusters of prospective Purkinje cells can be recognised (Fig. 21.10e) that form distinct longitudinal bands; from day 9 on granular cells seem to migrate gradually through granular raphes between these clusters of Purkinje cells from the external granular layer to the (inner) granular layer of the adult cerebellar cortex. Later on these raphes as well as the external granular layer disappear and an uninterrupted monolayer of Purkinje cells becomes visible. It is tempting to conclude that these transient granular raphes and longitudinal bands of Purkinje cells are the origin of the longitudinal modules described in the adult cerebellum (Feirabend et al. 1985). There is no relation between the pattern of glial fibres in the early cerebellum and this adult pattern of longitudinal zones (Roeling and Feirabend 1988). Details about the differentiation of the Purkinje cells are provided by Bertossi et al. (1986). The first spinocerebellar fibres reach the cerebellar plate on embryonic day 8 (Okado et al. 1987a).

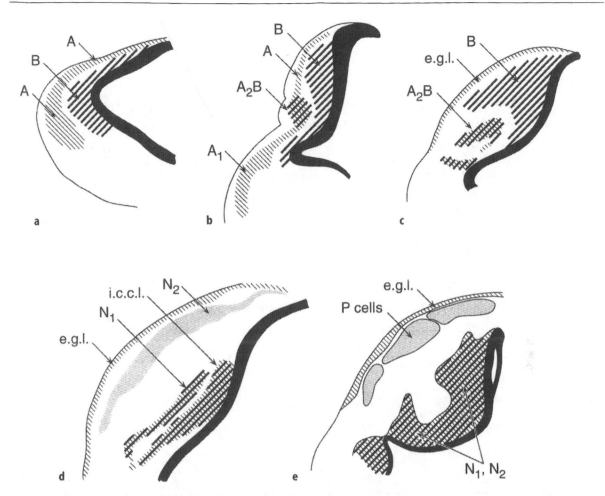

Fig. 21.10a–e. Early steps in cerebellar development. A,B, first and second cell migration; A_2B, early anlage of central nuclei; *e.g.l.*, external granular layer; *i.c.c.l.*, inner cortical cell layer; N_1, N_2, rudiments of intermediate and medial cerebellar nuclei; *P cells*, clusters of Purkinje cells. (Adapted from Feirabend 1983)

In the period of cortical differentiation the cerebellar fissures also develop. Sætersdal (1959) reviewed much of the older literature and described the gradual appearance of the respective fissures in several species of birds. The first to appear is the posterolateral fissure subdividing the cerebellar disc into a corpus cerebelli and a flocculonodular lobe. The order of appearance of the transverse fissures is more or less comparable in all species of birds. In the chicken the fissura prima appears late in stage 35, the following fissures becoming visible between this stage and stage 40. The first sign of fissure formation (their anlage) appears in the external granular layer, before an indentation in the cerebellar surface can be seen (Melian et al. 1986; cf. also Feirabend 1983).

It has been mentioned before that the role of prosomeres during the development of the telencephalic hemispheres is not clear. Prospective cell areas develop from specific periventricular proliferation zones (Tsai et al. 1981). Generally, neuroblasts forming lateral areas are born earlier than those forming the more medial areas. Also a rostrocaudal gradient and a ventrodorsal gradient can be recognised (Tsai et al. 1981). In contrast to the migration waves there is a certain overlap in the time of origin of cell groups. Another interesting aspect is that glia seems to develop after day 9. The consequence is that glia is not instrumental in the origin of the cell free zones, where later laminae appear, most notably the lamina medullaris dorsalis (Tsai et al. 1981).

The use of embryological data has been of great importance for the interpretation of the avian telencephalic organisation. Källén (1953) used the pattern of proliferation zones and migration waves

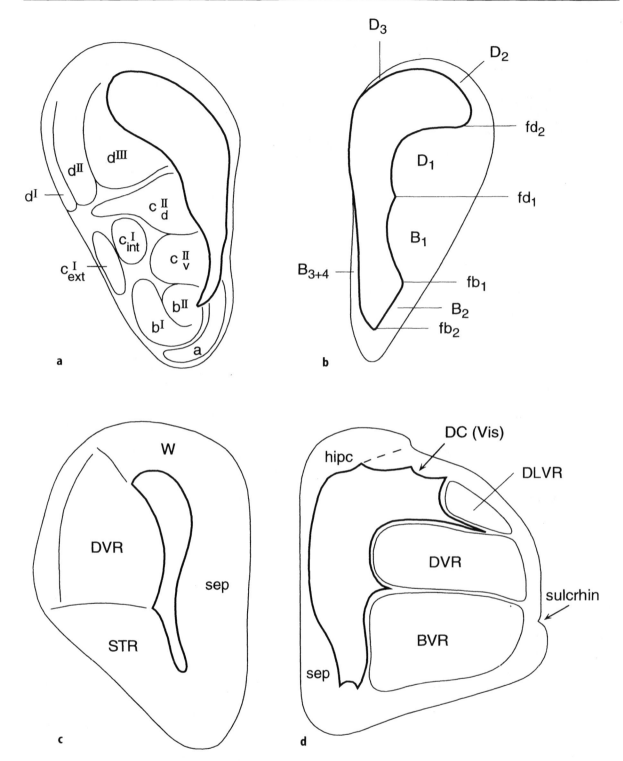

Fig. 21.11a–d. Four interpretations of the ontogenesis of the telencephalon. **a** Redrawn after Källen (1962); *letters and numbers* indicate successive migration waves. **b** Redrawn after Kuhlenbeck (1973); *capitals* indicate dorsal (D) and basal (B) primordia separated by ventricular fissures (fb_1, fb_2, fd_1, fd_2). **c** Redrawn after Ulinski (1983). Dorsal ventricular ridge (*DVR*) and striatum (*STR*) corresponding with the dorsal (*d*) and ventral (*c*) migrations of Källen, respectively. **d** Redrawn after Karten (1991): generalised plan of non-mammalian, amniote telencephalon showing three neuromeres: dorsal ventricular ridge (*DVR*), basal ventricular ridge (*BVR*) and dorsolateral ventricular ridge (*DLVR*). *DC*, dorsolateral cortex; *sulcrhin*, sulcus rhinalis

to identify the origin of the respective telencephalic regions and to homologise these with cell areas in reptiles and mammals. In particular, the development of the so-called dorsal ventricular ridge (DVR) in reptiles and birds has been discussed by several authors (Källén 1953; Ulinski 1983; Fig. 21.11a,c). In reptiles the cortical zones extend farther laterally than in birds. In the latter DVR retains a lateral position, whereas in reptiles it forms a more ventral bulge protruding dorsally into the lateral ventricle (Ulinski and Margoliash 1990). Källén considers the DVR in the chicken a pallial structure containing populations of cells that in mammals form part of the cerebral cortex. In a recent discussion Karten (1991) suggests the existence of three prosomeres in the lateral telencephalic wall of the avian embryo corresponding to the migration zones of Källén: a basal ventricular ridge containing the future elements of the basal ganglion, a dorsal and a dorsolateral ventricular ridge containing the cell populations that are homologous to those in the mammalian neocortex (Fig. 21.11d).

A divergent interpretation is given by Kuhlenbeck (1973) and followed by Haefelfinger (1957) and Jones and Levi-Montalcini (1958). Whereas the interpretations of Bergquist and Källén are based on the development of migration zones (Källén 1962), Kuhlenbeck uses intraventricular sulci to define a longitudinal pattern in embryos and adults. He distinguishes a dorsal or pallial region – dorsal to the ventricle – and a basal region (Fig. 21.11b). The latter is the origin of his nucleus basalis; only a small part of the dorsal region is considered 'pallial', viz. the nuclei diffusi dorsalis and laterodorsalis and the lateral corticoid area. A large part of the dorsal region is called the hypopallial ridge and gives rise to the epibasal nuclei (Kuhlenbeck 1973; for a comparison of the terminology, see Tables 21.3, 21.4). Källén (1962) states that the cell columns, the basis for his interpretation, do not correspond to the intraventricular ridges of Kuhlenbeck: the bordering sulci often lie in the middle of these cell columns. Recently, Puelles and Rubenstein (1993) demonstrated that the neuromeric model fits well with the expression patterns of putative regulatory genes.

In contrast to these interpretations, Ariëns Kappers et al. (1936) consider the neostriatum as an addition to the basal part representing the putamen. According to their interpretation the dorsolateral area (= DVR) corresponds to the reptilian hypopallium and gives rise to the ventral hyperstriatum; this represents (part of) the n. caudatus and – caudally – the amygdala (archistriatum). These authors distinguish only a dorsal corticoid region. Their interpretation is based both upon embryological and comparative adult data.

The consequences of the various interpretations of the embryological phenomena will be considered in Sect. 21.10.

Abbreviations Used in the Figures

acc	nucleus accumbens
acdl	area corticoidalis dorsolateralis
al	ansa lenticularis
aph	area parahippocampalis
arch	archistriatum
atv	area tegmentalis ventralis (Tsai)
aur	auricula cerebelli
bcm	brachium colliculi mesencephali
bol	bulbus olfactorius
blv	blood vessel
ca	commissura anterior
ccb	commissura cerebelli
cdm	columna dorsalis magnocellularis medullae spinalis
cg	corpus gelatinosum
chab	commissura habenularum
chopt	chiasma opticum
cinf	commissura infima
cio	capsula interna occipitalis
cmod	nucleus centralis dorsalis medullae oblongatae
cmov	nucleus centralis ventralis medullae oblongatae
cmspl	columna lateralis (of the columna motoria)
cmspm	columna medialis (of the columna motoria)
cnd	cornu dorsale
cnv	cornu ventrale
cp	commissura posterior
cpp	cortex prepiriformis
ct	corpus trapezoideum
ctm	commissura tecti mesencephali
dcd	decussatio cochlearis dorsalis
dip	nucleus intercalatus posterior thalami
dla	nucleus dorsolateralis anterior thalami
dlamc	nucleus dorsolateralis anterior magnocellularis
dlp	nucleus dorsolateralis posterior thalami
dma	nucleus dorsomedialis anterior thalami
dmp	nucleus dorsomedialis posterior thalami
dnIV	decussatio nervi trochlearis
dso	decussatio supraoptica
ect	ectostriatum
emmed	eminentia mediana

ems	eminentia sagittalis (Wulst)	lsp	lemniscus spinalis
EW	nucleus of Edinger-Westphal	maml	nucleus mammillaris lateralis
fae	fibrae arcuatae externae	mamm	nucleus mammillaris medialis
fai	fibrae arcuatae internae	mcl	mitral cell layer
fd	funiculus dorsalis	meV	nucleus mesencephalicus nervi trigemini
fipl	fissura posterolateralis cerebelli		
fipp	fissura prepyramidalis cerebelli	mld	nucleus mesencephalicus lateralis, pars dorsalis
fis	fissura secunda cerebelli		
fpr	fissura prima cerebelli	mpv	nucleus mesencephali profundus, pars ventralis
fL	field L		
fl	funiculus lateralis	nII	nervus opticus
fli	fovea limbica	nIII	nervus oculomotorius
flm	fasciculus longitudinalis medialis	nIV	nervus trochlearis
flt	fasciculus lateralis telencephali	nV	nervus trigeminus
fmt	fasciculus medialis telencephali	nVIIIc	nervus octavus, pars cochlearis
fol	fila olfactoria	nVIIIv	nervus octavus, pars vestibularis
ftqf	fibrae of tractus quintofrontalis	nIX	nervus glossopharyngeus
fu	fasciculus uncinatus	nX	nervus vagus
fv	funiculus ventralis	nXII	nervus hypoglossus
gcm	griseum centrale mesencephali	nstrc	neostriatum caudale
gcms	griseum centrale medullae spinalis	nstrf	neostriatum frontale
gld	nucleus geniculatus lateralis, pars dorsalis	nstri	neostriatum intermedium
		nua	nucleus angularis
gll	lamina glomerulosa	nubas	nucleus basalis
glv	nucleus geniculatus lateralis, pars ventralis	nucbi	nucleus cerebellaris internus
		nucbivm	nucleus cerebellaris internus, pars ventromedialis
habl	nucleus habenularis lateralis		
habm	nucleus habenularis medialis	nucbl	nucleus cerebellaris lateralis
hipc	cortex hippocampi	nucbm	nucleus cerebellaris intermedius
his	hyperstriatum intercalatum supremum	nucl	nucleus cervicalis lateralis
hypoth	hypothalamus	nucoe	nucleus coeruleus
hyperstra	hyperstriatum accessorium	nucs	nucleus commissuralis septi
hyperstrd	hyperstriatum dorsale	nudl	nucleus dorsolateralis of cornu dorsale
hyperstrv	hyperstriatum ventrale	nuem	nucleus ectomammillaris
iha	nucleus intercalatus hyperstriati accessorii	nugc	nuclei gracilis et cuneatus
		nuhyma	nucleus medialis hypothalami anterioris
inf	infundibulum		
inp	nucleus intrapeduncularis	nuhymp	nucleus medialis hypothalami posterioris
ip	nucleus interpeduncularis		
ismc	nucleus magnocellularis isthmi	nuic	nucleus intercollicularis
ispc	nucleus parvocellularis isthmi	nuichy	nucleus intercalatus hypothalami
lam	nucleus laminaris cochlearis	nuicp	nucleus interstitialis commissurae posterioris
ld	ligamentum denticulatum		
lfs	lamina frontalis superior	nuif	nucleus interfacialis
lfsp	lamina frontalis suprema	nuinfhy	nucleus inferior hypothalami
lge	lamina granularis externa	nuio	nucleus isthmo-opticus
lgi	lamina granularis interna	nuips	nucleus interstitio-pretecto-subpretectalis
lh	lamina hyperstriatica		
licb	lingula cerebelli	nulath	nucleus lateralis anterior thalami
ll	lemniscus lateralis	nulc	nucleus linearis caudalis
lm	lemniscus medialis	nulhy	nucleus lateralis hypothalami
lmd	lamina medullaris dorsalis	null	nucleus lemnisci lateralis
lme	lamina molecularis externa	nulld	nucleus dorsalis lemnisci lateralis
lmi	lamina molecularis interna	nulm	nucleus lentiformis mesencephali
lmv	lamina medullaris ventralis	numa	nucleus marginalis
lpo	lobus parolfactorius	numc	nucleus magnocellularis cochlearis

nuppc	nucleus principalis precommissuralis	rdV	radix descendens nervi trigemini
nupl	nucleus pontis lateralis	rgc	nucleus reticularis gigantocellularis
nupm	nucleus pontis medialis	rl	nucleus reticularis lateralis
nupr	nucleus proprius	rlm	formatio reticularis lateralis mesencephali
nuprmc	nucleus preopticus magnocellularis		
nupt	nucleus pretectalis	rnVI	radix nervi abducentis
nupvhy	nucleus paraventricularis hypothalami	rnXII	radix nervi hypoglossi
nurdVcd	nucleus radicis descendentis nervi trigemini, subn. caudalis	rot	nucleus rotundus
		rp	nucleus reticularis pontis
		rpc	nucleus reticularis parvocellularis
nurdVip	nucleus radicis descendentis nervi trigemini, subn. interpolaris	rpg	nucleus reticularis paragigantocellularis
nurdVor	nucleus radicis descendentis nervi trigemini, subn. oralis	rpgc	nucleus reticularis pontis gigantocellularis
nurob	nucleus robustus archistriati		
nus	nucleus semilunaris	rpm	nucleus (reticularis) paramedianus
nusc	nucleus suprachiasmaticus	rpo	nucleus reticularis pontis rostralis
nusl	nucleus septi lateralis	rsp	tractus rubrospinalis
nusm	nucleus septi medialis	rub	nucleus ruber
nuso	nucleus supraopticus	rv	radix ventralis nervi spinalis
nusov	nucleus supraopticus ventralis	sac	stratum album centrale
nuspt	nucleus subpretectalis	sce	stratum cellulare externum
nuspth	nucleus superficialis parvocellularis thalami	sci	stratum cellulare internum
		sfgs	stratum fibrosum et griseum superficiale
nuss	nucleus supraspinalis (cervicalis medialis)		
		sgc	stratum griseum centrale
nut	nucleus tuberis	sgd	nucleus substantiae gelatinosae
nuts	nucleus tractus solitarii	sgp	stratum griseum periventriculare
nuvmhy	nucleus ventromedialis hypothalami	sgr	stratum granulare
oli	oliva inferior [nucleus olivaris caudalis]	sm	stria medullaris
		smol	stratum moleculare
ols	oliva superior [nucleus olivaris rostralis]	sop	stratum opticum
		sP	stratum Purkinje
om	tractus occipitomesencephalicus	spcbd	tractus spinocerebellaris dorsalis
ov	nucleus ovoidalis	spcbv	tractus spinocerebellaris ventralis
pap	nucleus papilioformis (Arends and Zeigler 1991b)	spl	nucleus spiriformis lateralis
		spm	nucleus spiriformis medialis
pcbi	pedunculus cerebellaris inferior	spo	nucleus semilunaris parovoidalis
pcbm	pedunculus cerebellaris medius	st	nucleus subtrigeminalis
pcbs	pedunculus cerebellaris superior	subcd	nucleus subcoeruleus dorsalis
pcbsd	pedunculus cerebellaris superior, pars descendens	subcv	nucleus subcoeruleus ventralis
		subrot	nucleus subrotundus
pcc	pedunculus cerebellaris caudalis	t	nucleus teniae
pci	pedunculus cerebellaris intermedius	tda	tractus dorsoarchistriatalis
pcr	pedunculus cerebellaris rostralis	tect	tectum mesencephali
pic	cortex piriformis	tegmv	nucleus tegmentalis ventralis
plcvl	plexus chorioideus ventriculi lateralis	tegpp	nucleus tegmentalis pedunculopontinus
podl	nucleus preopticus dorsolateralis		
pra	nucleus preopticus anterior	tel	telencephalon
pst	tractus pretectosubpretectalis	tem	tractus nuclei ectomammillaris
pstra	paleostriatum augmentatum	tfa	tractus frontoarchistriaticus
pstrp	paleostriatum primitivum	tinf	tractus infundibuli
pvm	nucleus periventricularis magnocellularis	tio	tractus isthmo-opticus
		tol	tuberculum olfactorium
ra	nucleus raphes	tpo	area temporoparieto-occipitalis
rap	nucleus raphes pontis	tqf	tractus quintofrontalis
rd	radix dorsalis nervi spinalis	tri	nucleus triangularis

ts	tractus solitarius
tsc	torus semicircularis
tsm	tractus septomesencephalicus
tth	tractus tectothalamicus
tthf	tractus thalamofrontalis
vall	vallecula telencephali
vcb	ventriculus cerebelli
ved	nucleus vestibularis descendens
vel	nucleus vestibularis lateralis
vemd	nucleus vestibularis dorsomedialis
ves	nucleus vestibularis superior
vet	nucleus vestibularis tangentialis
vl	ventriculus lateralis
volf	ventriculus olfactorius
vq	ventriculus quartus
vt	ventriculus tertius
vtmes	ventriculus tecti mesencephali
IIIdl	nucleus nervi oculomotorii dorsolateralis
IIIdm	nucleus nervi oculomotorii dorsomedialis
IIIvm	nucleus nervi oculomotorii ventromedialis
Vm	nucleus motorius nervi trigemini
Vpr	nucleus sensorius principalis nervi trigemini
VI	nucleus nervi abducentis
VIImd	nucleus motorius nervi facialis, pars dorsalis
Xm	nucleus motorius nervi vagi
XIIm	nucleus motorius nervi hypoglossi

21.5
Spinal Cord

21.5.1
Introductory Remarks

The spinal cord of the pigeon forms the largest part of the CNS. It consists of at least 37 segments; the numbers of segments vary in the respective regions (e.g. Huber 1936), particularly in the coccygeosacral region. Shape, diameter and organisation of the cord show distinct regional differences (Figs. 21.12–21.15). The largest diameters are found in the lumbal and cervical intumescences. Even then, the surface of the largest cross-section in the lumbal region of, for example, the ostrich is about four times that in the cervical region (Streeter 1903). Both the dorsal and the ventral horn contribute to this difference.

21.5.2
Spinal Grey

The figures in the atlas section (Figs. 21.12–12.33) do not give an adequate impression of the rich variety of cell types in the respective regions. This variety has led to a series of slightly different interpretations of the subdivisions of the grey substance. Several authors attempted to apply the nomenclature of Rexed (1952, 1954), subdividing the grey matter into laminae (Leonard and Cohen 1975: pigeon; Brinkman and Martin 1973: chicken). However, it has proved difficult to recognise all layers at all levels of the spinal cord. Other subdivisions are proposed by Matsushita (1968) and Van den Akker (1970) and in the reviews by Nieuwenhuys (1964) and Breazile and Hartwig (1989). Figure 21.34b and Table 21.1 serve to harmonise the different interpretations and nomenclatures.

The dorsal horn or cornu dorsale is quite large in cervical regions, but relatively small at more caudal levels. It comprises an outer monolayer of rather large cells – the nucleus dorsolateralis (nudl, Fig. 21.14) – and the central substantia gelatinosa dorsalis. Leonard and Cohen (1975) distinguish two layers within this region (LII–III, Fig. 21.34b). These layers are also obvious in the caudal nucleus of the descending trigeminal tract (Figs. 21.16, 21.17; Dubbeldam and Karten 1978). The ventral continuation of the dorsal horn is the nucleus proprius that consists of several cell groups, viz. a dorsal, a magnocellular and a ventral subdivision or layers IV and V of Leonard and Cohen (Fig. 21.34b). The arrangement of the dorsal layers in the chicken (Brinkman and Martin 1973) looks slightly different from that in the pigeon (also see p. 1552).

The cornu ventrale is most prominent at the levels of the cervical and lumbal intumescences, where a medial and a lateral column (Fig. 21.12) can be easily recognised (Nieuwenhuys 1964; Breazile and Hartwig 1989). Matsushita (1968) distinguished three motor cell groups in the chicken – a ventrolateral, a ventromedial and a dorsomedial – that can be recognised specifically in the lumbosacral intumescence. These cell groups may correspond to the three motor cell areas indicated by Van den Akker (1970). From a functional point of view a subdivision into two cell groups seems more appropriate, a (ventro-)lateral one innervating the muscles deriving from the dorsal muscular mass, and a (dorso-)medial cell group innervating the ventral muscle mass (Landmesser 1978). Motoneurons innervating a specific muscle are arranged in a longitudinal column, the axons leaving the spinal cord via two or more ventral roots (e.g. leg muscles: Lance-Jones and Landmesser 1981; wings: Strasznicky and Tay

21.5 Spinal Cord 1543

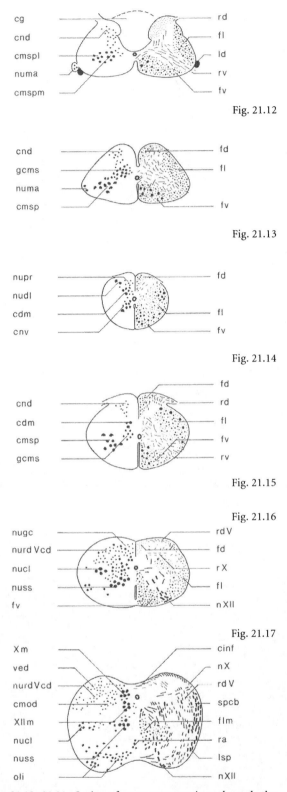

Figs. 21.12–21.31. Series of transverse sections through the spinal cord and brain of the pigeon. *The left half of each figure* shows the cell picture, based on Nissl-stained sections; *the right half* the fibre systems, based on Klüver-Barrera and Häggquist-stained sections

1983; neck muscles: Watanabe and Ohmori 1988; Hörster et al. 1990; Zijlstra and Dubbeldam 1994).

It is interesting to note that striking differences exist in the density of serotonergic fibres surrounding the motor neurons (Homma et al. 1988): a dense aggregation of fibres was found around ventral horn cells in the cervical cord of the pigeon. In the brachial region, however, the dense innervation was present mainly in the medial motor column and only in small parts of the lateral column, whereas in the lumbosacral region no serotonergic fibres were observed in the medial motor column but more prominently in the lateral column. It has been suggested that serotonergic fibres may innervate specifically those motor neurons that have to do with the erect position of the body (Homma et al. 1988). In the lumbosacral region of the chicken substance P and somatostatin immunoreactive fibres were also observed in close relation with presumed motoneurons (Lavalley and Ho 1983).

Rostrally the motor column continues as the n. supraspinalis (Figs. 21.16–21.18). Originally, this cell group was considered the hypoglossal motor nucleus (e.g. Ariëns Kappers et al. 1936); it appears to innervate neck muscles, the axons leaving the medulla through the caudal hypoglossal roots (Wild 1981; Watanabe and Ohmori 1988; Zijlstra and Dubbeldam 1994), but it has no part in the innervation of the muscles of the syrinx and tongue (e.g. Manogue and Nottebohm 1982; Dubbeldam and Bout 1990). As it does contribute to the hypoglossal nerve, it should be considered the ventral motor nucleus of N. XII and the n. intermedius (the 'new' hypoglossal centre) the dorsal motor nucleus of N. XII (see p. 1560).

The intermediate zone or griseum centrale of the spinal cord (Fig. 21.15) is little differentiated and most authors consider it to be one area (Table 21.1); Matsushita is the only one to describe a zona intermedia, a n. posterolateralis and a n. dorsolateralis ("often difficult to recognise" and not to be confused with our n. dorsolateralis of cornu dorsale). More obvious is the columna dorsalis magnocellularis (Clarke's column), consisting of large (35–40 μm) cells. Again, Matsushita distinguishes a n. posteromedialis at the cervical intumescence and a n. dorsalis corresponding with Clarke's column; it extends caudally from the cervical intumescence, initially increasing but becoming small again at the sacrolumbal level. According to Matsushita the two cell groups merge partially. In the transition of spinal cord to myelencephalon a group of large cells appears in the lateral region, the n. cervicalis lateralis (Figs. 21.16, 21.17). The homology and connections of this cell group are not known. The substantia gelatinosa centralis surrounds the central canal.

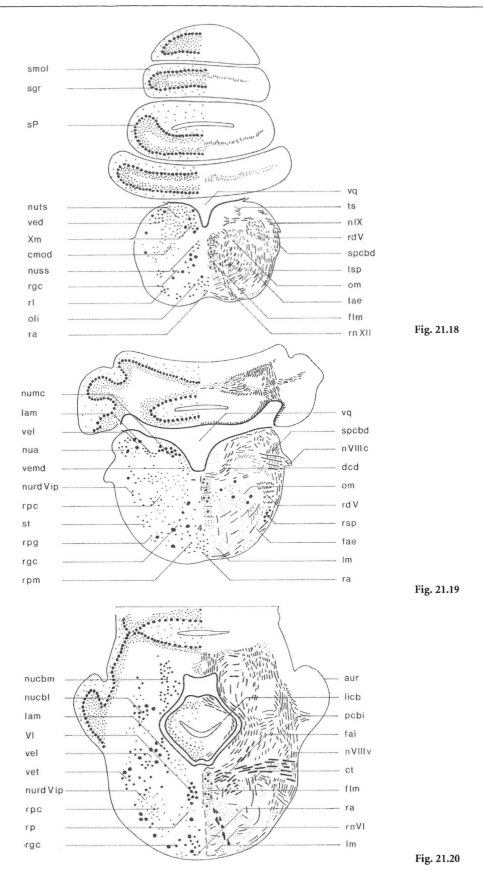

Fig. 21.18

Fig. 21.19

Fig. 21.20

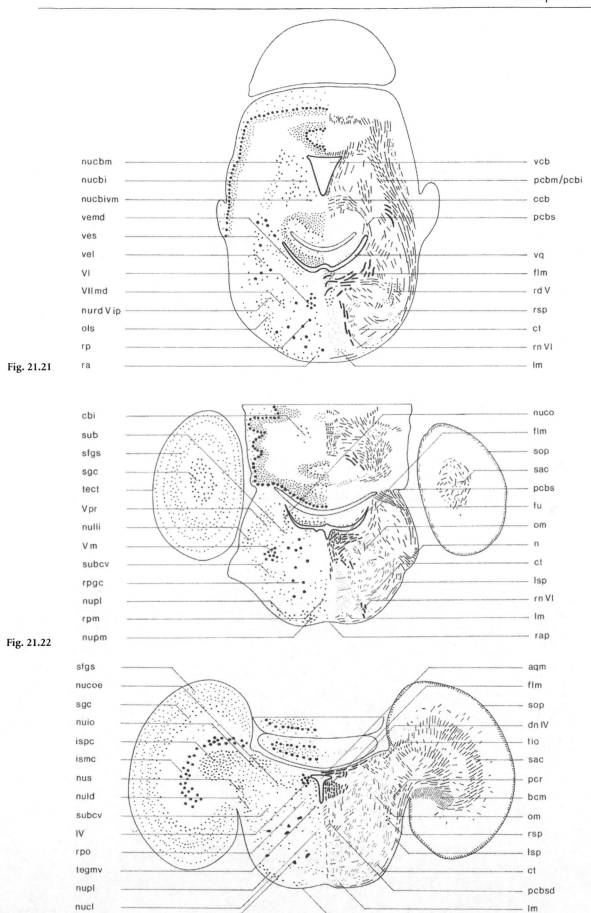

Fig. 21.21

Fig. 21.22

Fig. 21.23

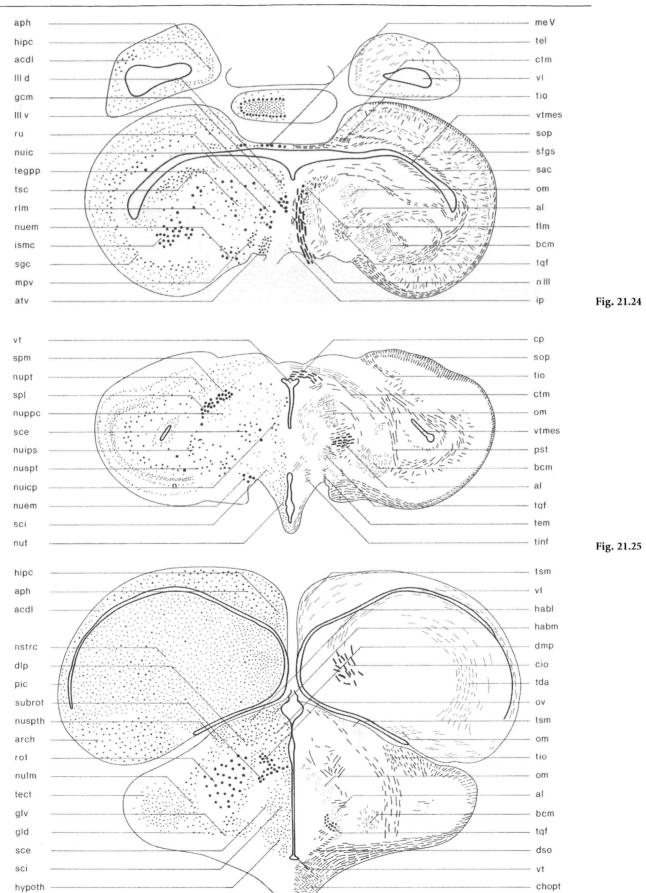

Fig. 21.24

Fig. 21.25

Fig. 21.26

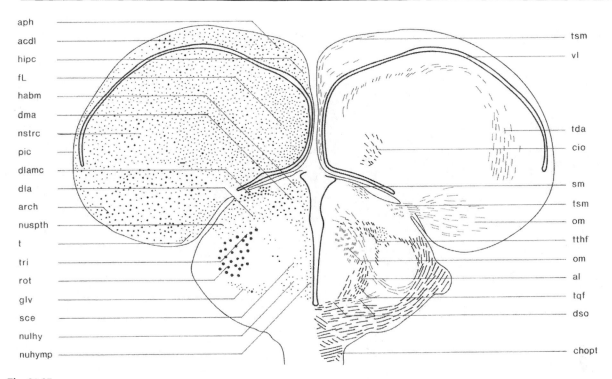

Fig. 21.27

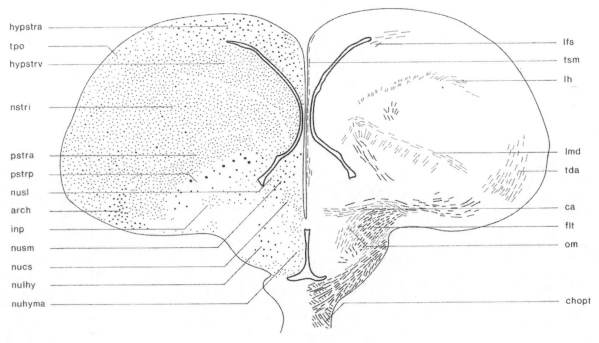

Fig. 21.28

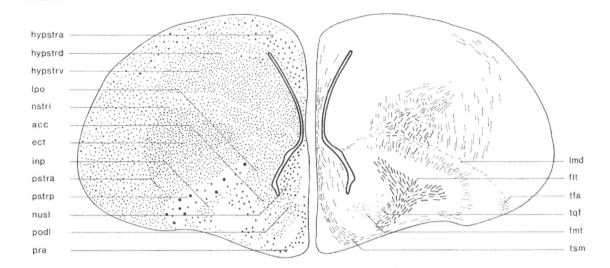

Fig. 21.29

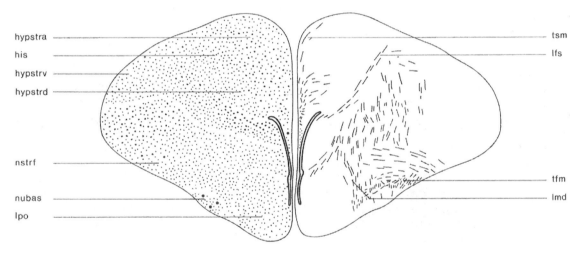

Fig. 21.30

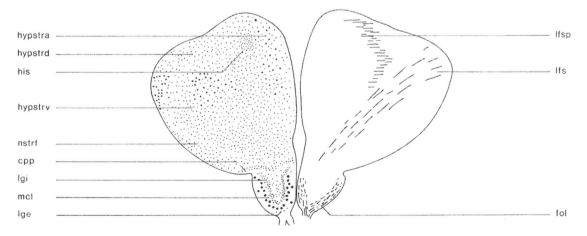

Fig. 21.31

It contains ventral and dorsal commissural cell groups. The n. intermediomedialis (i.e. the column of Terni) and – more sparsely – n. intermediolateralis and n. intercalatus contain autonomic preganglionic neurons (e.g. Cabot and Bogan 1987; Ohmori et al. 1992); the thoracolumbal and sacral subdivisions represent the sympathetic and parasympathetic neurons, respectively (Ohmori et al. 1984).

21.5.3
Marginal Nuclei and Corpus Gelatinosum

Another category of spinal neurons form the marginal or paragriseal nuclei; these consist of small- to medium-sized fusiform cells lying in the white matter near the ventrolateral surface. Sections at thoracic and cervical levels contain only a few scattered cells, the minor nuclei marginales or nuclei of Hoffmann-Kölliker (Ariëns Kappers et al. 1936). No cells occur at the levels of the ventral roots, but cells in the lumbosacral region form small lateral lobes (Fig. 21.12) that are known as nn. marginales majores, in the older literature generally called the accessory lobes of Lachi (Ariëns Kappers et al. 1936). The nature and function of these cells are not well understood. One suggestion is that the paragriseal cells belong to the neuroglia – possibly of the astrocyt type, with a capacity for glycogen storage (de Gennero and Benzo 1978) and involved with the synthesis of myelin. Other suggestions are that these cells have a mechanoreceptor function (Schroeder and Murray 1987) or a secretory function (Bodega et al. 1989). There is recent evidence that cells of the major marginal nuclei have projections to both the ipsilateral and (sparsely) contralateral ventral horns as well as to the contralateral major marginal nuclei (Antal et al. 1994; Necker 1994). These features support the suggested role in mechanoreceptive function.

A specific avian feature is the sinus rhomboidalis with corpus gelatinosum at the lumbosacral level (Figs. 21.12, 21.34a). Though this has been a well-known feature for a long time, no satisfactory explanation for its existence has been offered (review in de Gennaro 1982). The corpus gelatinosum embodies rather large cells of neural origin containing much intracellular glycogen. It has been suggested that these cells are modified glial cells (Welsch and Wächtler 1969), perhaps a special type of astrocyte (Lyser 1973) with a secretory function in connection with the glucose homeostasis of the cerebrospinal fluid (Welsch and Wächtler 1969; Azcoitia et al. 1985). Some of the cells interrupt the ependymal lining of the central canal. According to Sansone (1977) the glycogen body is not restricted to the sacrolumbal region: a zone of scattered glycogen-rich cells extends over the whole length of the spinal cord dorsal to the central canal, possibly even continuing into the brain stem as far rostrally as the oculomotor nuclei as a population of scattered glycogen-rich cells in the floor of the ventricles.

21.5.4
Dorsal Roots

Experimental studies show that the dorsal ganglia project to the whole dorsal horn including Clarke's column, but also to the intermediate zone (layer VI) as well as to the ventral lamina IX (Leonard and Cohen 1975b). There is anatomical and physiological evidence that layers I–IV are the somatosensory centres (Wild 1985; Necker 1985), layer I containing the nociceptive elements. Probably the projections to layer IX are proprioceptive, rather than somatosensory. Woodbury (1992) using electrophysiology describes a comparable distribution of afferents in the chicken; this author confirms the observation of Brinkman and Martin (1973) that layer III occupies a medial, rather than a ventral, position with respect to layers I and II. Honig (1982) demonstrated that a specific area can receive afferents via more than one dorsal root (see also Necker and Schermuly 1985). The projections to the dorsal horn are somatotopically organised (e.g. Schulte and Necker 1994). It also seems possible to differentiate the terminal fields of the somatic fibres (L II and III) and of the visceral fibres. The latter project to the lateral edge and to the base of the dorsal horn and to the dorsal grey commissure (Ohmori et al. 1987). Afferents from muscle spindles may project to Clarke's column (Necker, 1990). Holloway et al. (1976) suggest that there is a certain convergence of various modalities on dorsal horn cells; the activity of these cells is modified by supraspinal influences. This control may be exerted by fibres in the fasciculus dorsolateralis, but the anatomical origin of this influence has not yet been specified (Holloway et al. 1978). Lesion experiments showed that a dense substance-P-like projection to layers I and II does not derive from supraspinal sources, but from the dorsal root ganglia (Davis and Cabot 1984). Serotonergic fibres in these layers represent bulbospinal projections from the raphe nuclei. This is also true for projections to the sympathetic preganglionic cells (column of Terni) and to the ventral horn (Davis and Cabot 1984).

The presence of dorsal root fibres in the dorsal funiculus and their projections to the spinal grey and dorsal column nuclei were demonstrated long ago (Kühn and Trendelenburg 1911; Necker and Schermuly 1985); the majority of these fibres repre-

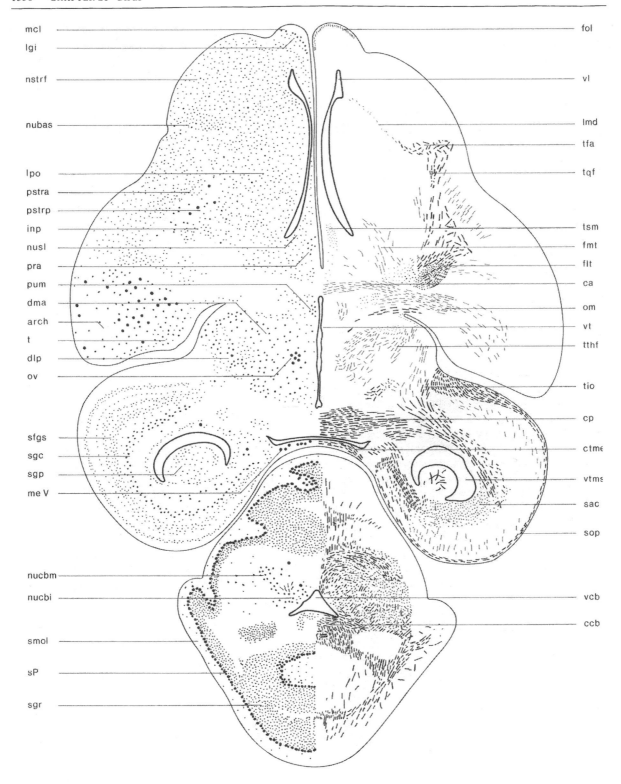

Figs. 21.32, 21.33. Horizontal sections through the brain of the pigeon. *The left half* shows the cell picture, based on Nissl-stained sections, *the right half* fibre systems based on Klüver-Barrera-stained sections

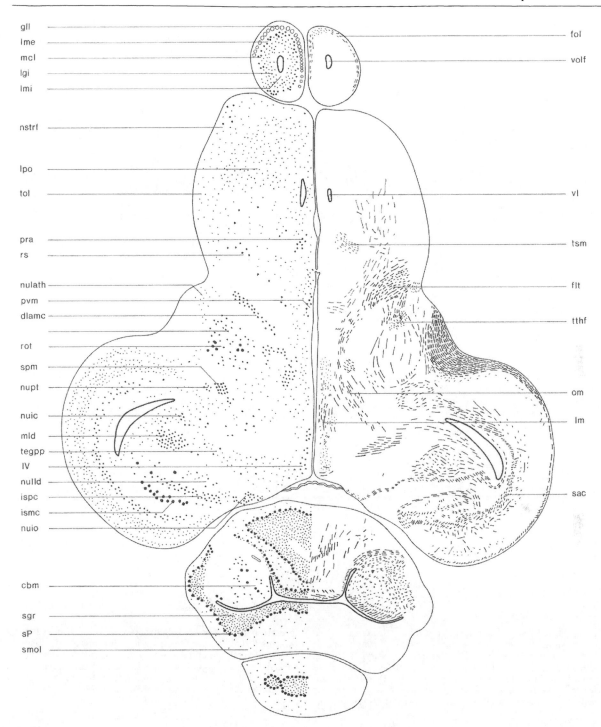

Fig. 21.33

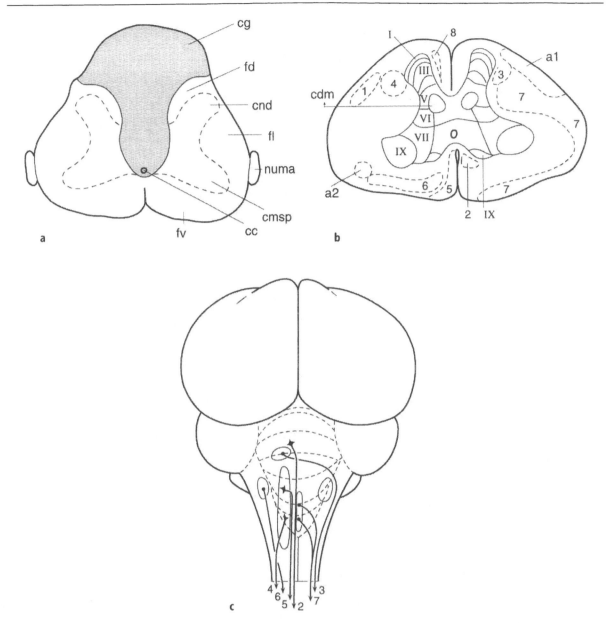

Fig. 21.34. a Cross-section through lumbar cord with corpus gelatinosum. b Generalised cross-section through spinal cord. Within the grey matter laminae are indicated; cf. Table 21.1. Within the funiculi a number of descending fibre systems are indicated (after Cabot et al. 1982). c Horizontal view of descending spinal fibre systems, same numbers as used in b. *1*, hypothalamo-spinal fibres; *2*, (periaquaductal) interstitiospinal fibres; *3*, rubrospinal fibres; *4*, dorsolateral pontine reticulospinal fibres; *5*, medial pontine reticulospinal fibres; *6*, vestibulospinal fibres; *7*, raphe (medial medullary/caudal pontine) -spinal fibres; *8*, dorsal column-spinal fibres; *a1*, dorsolateral and lateral ascending fibres; *a2*, ventral ascending fibres. (After van den Akker 1970)

sent cutaneous afferents. In addition, terminal fields of dorsal root ganglion fibres have been found in n. cuneatus externus, more rostrally in an area dorsolateral to the descending trigeminal root that may be comparable to the mammalian group (Necker 1991a) and even in the princeps nucleus of the trigeminal nerve (Wild 1985).

21.5.5
Funiculi

Dorsal, lateral and ventral funiculi are present over the whole length of the spinal cord. The dorsal funiculus contains the highest percentage of thin fibres (0–3 µm) and hardly any fibres thicker than 6 µm (Yamada 1966). The ventral funiculus has the

Table 21.1. Comparison of nomenclatures of the substantia grisea of the avian spinal cord

Breazile and Hartwig (1989)	Leonard and Cohen (1975a)	Brinkman and Martin (1973)	Van den Akker (1970)	Matsushita (1968)
Nucleus dorsolateralis	I	1	1	cpm
Nucleus substantia gelatinosa	II III	2	2	sg
Nucleus proprius:				
Dorsal part	IV	3/4	3	
Nucleus magnocellularis				
Ventral part	V	5	4	
Nucleus intermedius medullae spinalis	Column of Clarke	6	5	ZI/DL
Nucleus intermediomedialis (Terni)			6 (part)	
Substantia gelatinosa centralis	I/VII/VIII	7	4	sgc cca/ccp
Nucleus motorius	IX		7	
Columna medialis[a]		8		DM/VM
Columna lateralis[a]		9		VL

[a] Huber (1936), Nieuwenhuys (1964).

highest percentage of thick fibres (9 µm or larger), viz. up to nearly 10% at the level of the cervical cord and perhaps even slightly more at more caudal levels (Figs. 21.12–21.16). The lateral and laterodorsal funiculi, again, contain more thinner fibres with a high percentage in the classes of 3–9 µm (Yamada 1966). The dorsal funiculus contains ascending primary fibres projecting to the nuclei of the cuneatus-gracilis complex as well as a lateral contingent of descending fibres that sprout from the dorsal column nuclei (Cabot et al. 1982). Descending fibres in the ventral funiculus (Fig. 21.34b) have their origin in the peri-aquaductal grey (interstitiospinal tract), the medial pontine reticular formation (reticulospinal tract), the vestibular nuclei and the raphe nuclei (e.g. Janzik and Glees 1967; Cabot et al. 1982; Glover and Petursdottir 1988). Rabin (1975) demonstrated that vestibular nuclei influence the activity of motoneurons in the cervical cord, but not those innervating muscles of the wings and legs. In contrast to earlier reports, there is no experimental evidence of a tectospinal tract (cf. Breazile and Hartwig 1989; Dubbeldam, unpublished observations). Intercollicular cells may project to the rostral cervical cord (Webster and Steeves 1988); most vestibulospinal fibres, too, may not reach farther than the cervical cord projecting to the motor cells of the neck muscles (Arends et al. 1991). Rubrospinal fibres and those from the lateral pontine area are concentrated in the medial dorsolateral funiculus, whereas hypothalamospinal fibres occupy a more lateral position (Fig. 21.12b). Raphe fibres are dispersed throughout the dorsolateral funiculus and in parts of the lateral funiculus. Catecholaminergic neurons in the lateral medullary reticular formation and in the locus coeruleus are assumed to project to the preganglionic sympathetic neurons of the cervical cord (Smolen et al. 1979). Webster and Steeves (1991) provide evidence that reticulospinal fibres in the lateral funiculus are essential to initiate walking.

Gross and Oppenheim (1985) found some additional sources of supraspinal afferents in early embryos, but it is not clear whether these fibres persist in the adult animal. In any case, these cell groups could not be labelled from the spinal cord in hatched chicken (Okado and Oppenheim 1985). Another point of debate is the existence of an avian equivalent of the mammalian corticospinal tract. The avian archistriatum projects through the occipito-mesencephalic tract to the first segments of the cervical cord (Zeier and Karten 1971; Dubbeldam et al. 1997). In psittaciform birds the basal branch of the septomesencephalic tract, sprouting from the telencephalic Wulst, reaches the cervical cord (Zecha 1962). In other birds this tract is said to extend not farther caudally than the mesencephalon (cf Verhaart 1971), but in the mallard a few fibres reach at least the spino-myelencephalic transitory region (Dubbeldam 1976). However, the evidence of an avian 'pyramidal tract' descending into the spinal cord does not seem very convincing (Webster et al. 1990).

Oscarsson et al. (1963) present physiological evidence of the presence of crossed as well as uncrossed ascending fibres in the lateral funiculus. These authors suggest that the crossed fibres are related to leg nerves and may correspond with the mammalian ventral spinocerebellar tract. A dorsal tract either does not exist or has an organisation different from that in mammals. The two fibre compartments have been identified anatomically (e.g. van den Akker 1970; Karten 1963) and both contain spinocerebellar fibres. Part of these enter the cerebellum through the caudal cerebellar peduncle (corpus restiforme), the remaining fibres through the rostral peduncle (brachium conjunctivum; Vielvoye 1977). Ascending fibres in the lateral funiculus reach many other targets in the brain stem. The contours of the respective fibre groups are not sharply delineated, but roughly the same distribution has been found in experimental studies (Van den Akker 1970; Karten 1963; Cabot et al. 1982).

The precise origins of several of the ascending tracts have been specified by Necker (1989). Lamina I cells, but also cells in layers V–VIII, project to the reticular formation of the brain stem. Spinocerebellar fibres have their origin in Clarke's column, but also arise from dorsolateral cells in the ventral horn. Spinothalamic projections arise from layer V–VI in the lumbal enlargement. A comparable projection from the cervical enlargement has not been found (Schneider and Necker 1989). The spinothalamic fibres form part of the spinal lemniscus of the brain stem (see below).

Hardly anything is known about the organisation of the propriospinal system and its relation with, for example, the independent organisation of the locomotory systems (walking, swimming, flying, etc.). Jacobson and Hollyday (1982) claim that the spinal cord possesses a central generator for locomotory movements and suggest that the components for the different types of locomotion each have a specific descending control; locomotory activities can be evoked by stimulation in the lateral funiculus. Webster and Steeves (1988) could differentiate the supraspinal input of the sacrolumbal and cervical/brachial regions, the latter being the main recipient of cerebellar and mesencephalic input. Three areas may have a prominent role in locomotion: ipsilaterally the rhombencephalic gigantocellular reticular formation, and bilaterally parts of the parvocellular and the lateral mesencephalic reticular formation.

21.6
Rhombencephalon

21.6.1
Introductory Remarks

The avian brain stem contains well-differentiated cell groups and fibre systems that can also be easily distinguished in non-experimental material. Basically, the organisation of the brain stem is similar to that in other vertebrates. The four longitudinal cell columns are differentiated into series of distinct nuclei. The cranial nerves, sensory nuclei, motor nuclei and reticular centres will be discussed in turn.

21.6.2
Cranial Nerves

Birds possess 12 pairs of cranial nerves. An extensive survey of all nerves can be found in Bubien-Waluszewska (1981); in this review a short résumé will be given of nerves V–XII (Figs. 21.5, 21.6). The nervus hypoglossus (N. XII) leaves the brain stem ventrally through a series of radices; branches of this nerve innervate muscles of the tongue, trachea and syrinx and some of the rostral cervical muscles. According to Wild (1990) the XIIth nerve also carries sensory fibres innervating mechanoreceptors in the tongue. Such fibres may have their cell bodies in the jugular ganglion and terminate in nuclei of the trigeminal system. The accessory nerve exits laterally close to the vagus nerve and joining this nerve; it innervates the cucullaris capitis muscle, but some fibres may be distributed elsewhere with the vagus fibres. The vagus has two sensory ganglia, the nodose ganglion (ganglion distale) lying in the thoracic inlet and a proximal ganglion, part of the ganglion jugulare complex. The first contains mainly viscerosensory elements, the second also somatic sensory cells (Dubbeldam et al. 1979). The vagus arises from the brain stem by a series of rootlets that are continuous with those of the glossopharyngeal nerve. The latter also possesses a proximal and a distal ganglion, the proximal ganglion forming the other part of the jugular ganglion. The distal or petrosal ganglion is located just outside the bony skull; cells of this ganglion convey somatosensory and gustatory information from the tongue. The efferent fibres of the two nerves form an important component of the sympathetic system. The lingual nerve innervates the tongue; the laryngopharyngeal nerve innervates laryngeal and pharyngeal muscles, but also glands of the larynx and trachea. The eighth nerve consists of a pars vestibularis and a pars cochlearis. The latter part contains a cochlear and a lagenar ganglion. The vestibular ganglion may also contain cells originating from the ganglion radicis of the facial nerve. The ganglion geniculatum of the facial nerve is not very conspicuous and contains mainly visceral afferents. In contrast to older ideas (Ariëns Kappers et al. 1936), there is strong evidence that the facial nerve innervates taste buds in the upper and lower beak (Krol and Dubbeldam 1979; Gentle 1984). The bulk of the facial nerve consists of visceral efferents with a ramus palatinus carrying parasympathetic fibres to the ganglia ethmoidale and sphenopalatinum and a hyomandibular branch carrying (among others) the fibres innervating the depressor mandibulae muscle and several tongue muscles (e.g. Dubbeldam and Bout 1990). The nervus abducens is one of the oculomotor nerves, leaving the pontine region of the brain stem ventrally (Fig. 21.5b).

The trigeminal nerve is large in most birds. The sensory component innervates mechanoreceptors in the head region; large numbers of tactile sense organs are present in the upper and lower beak, often with high concentrations in the tips (bill-tip organ, Gottschaldt 1985) innervated by the ophthalmic and mandibular branches. The third branch, the maxillary nerve, innervates the orbital and nasopalatine region. The trigeminal ganglion consists of a more or less separated ophthalmic part and partly overlapping maxillary and mandibular parts (e.g. Dubbeldam and Veenman 1979). The motor branch innervates in particular the jaw muscles with the exception of the depressor mandibulae. The motor component also carries afferents innervating the muscle spindles.

21.6.3
Somatosensory Complex

The nuclei of the trigeminal system form the somatosensory complex of the brain stem. Part of the sensory fibres of the trigeminal nerve split, forming an ascending and a descending tract (Dubbeldam and Karten 1978), other fibres joining either of the two tracts. The ascending fibres end in the principal sensory nucleus of N. V (Fig. 21.22). The descending tract (Figs. 21.16–21.21) extends into the spinal cord and is accompanied by a series of cell groups, from rostral to caudal the subnn. oralis and interstitialis, subn. interpolaris and subn. caudalis (Fig. 21.35a). The latter forms the dorsal horn in the first cervical segments. There is some evidence that particularly small ganglion cells send their fibres into the descending tract (Kishida et al. 1985). Zeigler and Witkovsky (1968) suggest that the trigeminal nuclei receive somatosensory as well as proprioceptive information from the head region. Another group of proprioceptive fibres belong to the mesencephalic trigeminal nucleus (see Sect. 21.8.2) and have collaterals projecting to the trigeminal motor nuclei (e.g. Arends and Dubbeldam 1982).

The princeps nucleus is a well-defined cell group; the size of this cell group differs considerably among the avian species depending upon the numbers of mechanoreceptors in the beak and tongue (Stingelin 1965; Dubbeldam 1990a). It is the origin of the quintofrontal tract (Figs. 21.24–21.29, 21.35a) that, partly decussating partly ipsilaterally, ascends to project to the n. basalis or n. trigeminalis prosencephali (Cohen and Karten 1974). In the mallard a small cell group, seemingly a medial part of the princeps, does not receive afferents from the trigeminal ganglion but from the mesencephalic trigeminal nucleus (Dubbeldam 1980). Apparently, it does not contribute to the quintofrontal tract and so far the character of this cell group remains unclear; it has been named the supratrigeminal nucleus, but may have nothing to do with the centre with the same name in mammals.

The subn. oralis projects to the cerebellum (Arends et al. 1984), and the subnn. interpolaris and caudalis are sources of reticular projections and intrinsic trigeminal connections (Fig. 21.35a). An additional lateral system of fine fibres has been recognised in the pigeon; along its caudal course it seems to project to small cells lateral to the radix descendens nervi trigemini and ends in the n. cuneatus externus [accessorius]; it has been suggested that this system may correspond to the lateral descending trigeminal system of the pit-organ-possessing snakes (Dubbeldam and Karten 1978). There is recent evidence that birds possess a modest trigeminal projection to the solitary complex (Wild and Zeigler 1996).

In some birds fibres from the distal glossopharyngeal (*Anas*, Dubbeldam et al. 1979) and proximal jugular ganglia (*Cacatua*, Wild 1981; finches, Wild 1990) join the trigeminal system with projections to the principal sensory nucleus of N. V (PrV) and subnuclei oralis and interpolaris of the descending trigeminal tract (TTD) (Fig. 21.35b); there is solid evidence that these fibres innervate mechanoreceptive elements in the tongue. It has also been suggested that fibres innervating the syrinx in the zebra finch join the trigeminal system, but there is only indirect evidence (Bottjer and Arnold 1982). This stresses the importance of the trigeminal system as a system for the processing of exteroceptive and propioceptive mechanoreceptive information (Dubbeldam 1991).

The glossopharyngeal and vagal nerves also carry fibres that enter the n. cuneatus externus (Dubbeldam et al. 1979; Dubbeldam 1984); possibly these fibres convey somatosensory information from the neck region. The n. cuneatus externus is a small cell group lateral to the subn. caudalis.

The nuclei gracilis and cuneatus (Fig. 21.16) are rather small nuclei receiving primary ascending somatosensory fibres as well as secondary afferents originating in cervical lamina IV and lumbar medial lamina V (Necker 1991b). Reinke and Necker (1992) showed using electrophysiology that several types of both cutaneous and deep mechanoreceptors, but no muscle spindles, project directly to these nuclei. The nuclei gracilis and cuneatus form the main source of the lemniscus medialis.

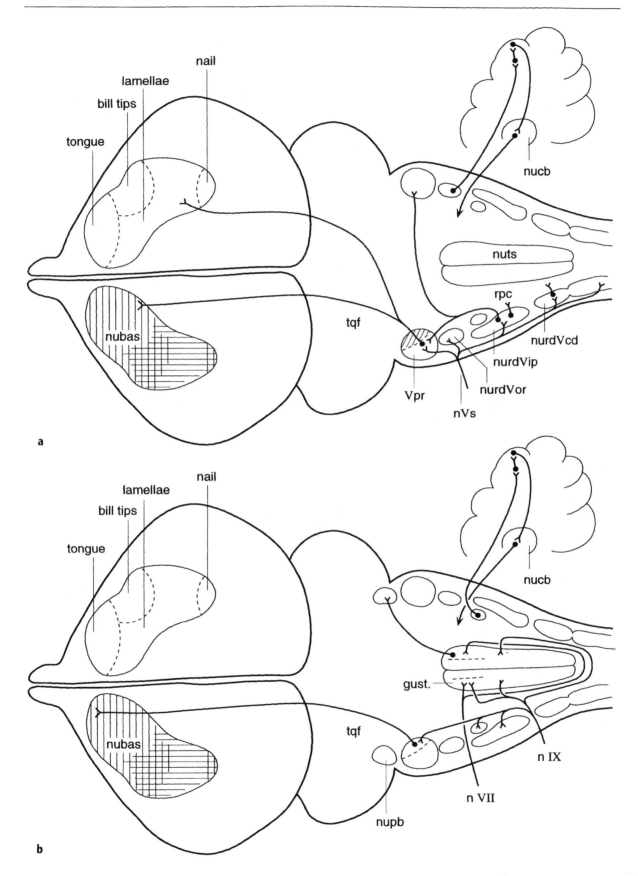

Other centres contributing to this fibre system may be the external cuneate nucleus and, possibly, cells in layer I of the subn. caudalis (e.g. Zecha 1966; Arends et al. 1984). The fibres collect into a bundle that traverses the caudal medulla to cross the midline and then ascends along the ventral surface of the brain stem (Figs. 21.19-21.22). More rostrally the fibres swing through the deep layers of the tectum mesencephali and enter the diencephalon; its main terminal fields are the n. dorsolateralis caudalis and n. dorsointermedius ventralis anterior thalami (Wild 1989), with another terminal field in the ventral thalamus (Zecha 1966; Arends et al. 1984).

21.6.4
Vestibulo-cochlear Complex

The second group of somatic sensory nuclei are those of the eighth nerve. The cochlear component of this nerve has two targets, the n. angularis (Fig. 21.19) sitting on top of the cochlear root, and the n. cochlearis magnocellularis (Boord and Rasmussen 1963). The latter cell group contains neurons without, or with one or two dendrites that are arranged in dorsoventral columns (Jhaveri and Morest 1982). This is particularly visible in birds with large auditory nuclei such as night-hunting owls. The magnocellular nucleus has a tonotopic organisation: high frequencies elicit responses in the rostromedial part of this nucleus, low frequencies in the caudolateral part (Rubel and Parks 1975). The n. laminaris (Figs. 21.19, 21.20) lies between, and slightly rostral to, these two nuclei; it is a monolayer of cells with a distinct dorsal and a ventral neuropil. Each laminaris cell receives bilateral afferents from the magnocellular nucleus (Parks and Rubel 1975), an ipsilateral projection upon the dorsal dendrite, and a contralateral projection upon the ventral one (Fig. 21.36b,c). The fibres from the contralateral magnocellular nucleus reach the n. laminaris through the dorsal cochlear decussation (Fig. 21.16). Wold and Hall (1975) describe a direct, bilateral cochlear projection to the nucleus laminaris in the chicken; these authors suggest that the reported presence of a magnocellular projection to this nucleus in the pigeon (Boord and Rasmussen 1963) may in fact be caused by inadvertently interrupted cochlear fibres; different parts of the cochlear ganglion could send fibres to the different nuclei. The observations of Parks and Rubel (1975), however, support those of Boord. There is experimental evidence that the development of the dendritic symmetry of the laminaris cells depends upon an undisturbed bilateral use of the ears (Gray et al. 1982).

Hotta (1971) also described a tonotopical organisation of the n. angularis. It has been suggested that in some species of birds it may consist of a lateral cochlear part and a medial vestibular part (Wallenberg 1964). These parts should be partly separated; in most species, however, this is not obvious (Dubbeldam 1968) and there is no further experimental evidence to support this assumption. Cochlear efferents arise from the n. reticularis paragigantocellularis lateralis (Fig. 21.19) medial to the ventral n. facialis (Schwarz et al. 1981; Strutz and Schmidt 1982; Cole and Gummer 1990).

Efferents from the angular and laminar nuclei cross through the corpus trapezoideum (Figs. 21.20-21.23) to form the contralateral lemniscus lateralis (Boord 1968; Conlee and Parks 1986). On account of its structure and input the laminar nucleus has been compared to the mammalian medial superior olive (Whitehead and Morest 1981). Von Bartheld et al. (1989) found many GABA-ergic cells in the neuropil of magnocellular nuclei (Mc) and laminar nuclei (Lam) with the neurites of these cells branching within the nuclei. The oliva superior [n. olivaris rostralis] of the pigeon could correspond to the lateral oliva superior of the mammals; it is a round group of rather small cells in the ventrolateral part of the brain stem (Fig. 21.21), close to the ventral facial motor nucleus and accessory nucleus of N. VI. It receives collaterals from the ipsilateral and contralateral ascending auditory tracts and contributes to the projection to the n. mesencephalicus lateralis, pars dorsalis (Conlee and Parks 1986).

The lateral lemniscus is the first link of the ascending pathway (Fig. 21.36a). In addition to the projections to the oliva superior, it sends fibres to the contralateral nuclei of the lateral lemniscus and to the lateral mesencephalic nucleus. Conlee and Parks (1986) also describe a modest projection from the nuclei angularis laminaris cochlearis to the ipsilateral nucleus mesencephalicus lateralis, pars dorsalis. The latter cell group is the source of a projection to the thalamic n. ovoidalis (Fig. 21.36a); this, again, sends fibres to field L of the telencephalon. Details will be discussed in the following sections.

The vestibular system is more complex. Different subdivisions of this complex have been described depending upon the species studied. Renggli (1967)

◄ **Fig. 21.35a,b.** Horizontal projection of trigeminal (**a**), facial and glossopharyngeal (**b**) afferents in the brain stem of the mallard with some of the secondary connections. *On one side* regions in the nucleus basalis (*nubas*) receiving information from specific peripheral areas are indicated. Glossopharyngeal and facial afferents converge in an anterolateral part of the nucleus of the solitary tract (*nuts*); this region represents the gustatory centre (*gust*)

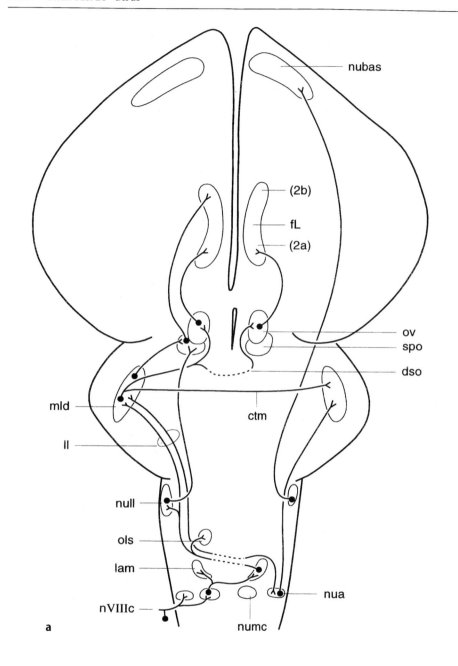
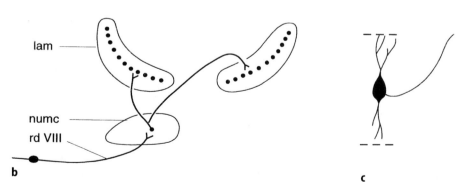

Fig. 21.36. a Scheme of avian ascending auditory pathways, one via the n. mesencephalicus lateralis, pars dorsalis (*mld*) and n. ovoidalis (*ov*) to field 2a of the auditory neostriatum (field L, *fL*), a second one via the nuclei of the lateral lemniscus and n. semilunaris parovoidalis to field 2b. The third pathway runs from the ventral nucleus lemnisci lateralis to the n. basalis (*right half of figure*). *Broken lines* indicate ventrally crossing fibres. **b** Diagram showing pattern of projections from n. magnocellularis cochlearis (*numc*) cells to the ventral neuropil of the ipsilateral n. laminaris cochlearis (*lam*) and dorsal neuropil of the contralateral lam. *Dots in lam* indicate the monolayer of cells. **c** Diagram of n. laminaris cochlearis cell with dorsal and ventral dendrites. *Broken lines* indicate border of the nucleus

provides an overview of the terminology used by the various earlier authors. Wold (1976) distinguished only six main nuclei. Using the terminology of the NAA we find from rostral to caudal the n. vestibularis superior (Fig. 21.21), a compact collection of deeply staining cells that can be subdivided into a lateral and a quadrangular part; also the n. vestibulocerebellaris may be part of this cell group (Wold 1976). Ventral and caudal to the superior vestibular nucleus lies the lateral vestibular complex (Figs. 21.19–21.21) with a gigantocellular n. vestibularis lateralis dorsalis and a magnocellular n. lateralis ventralis. Often the name Deiters' nuclei is used for these two cell groups. The small-celled n. vestibularis (dorso-)medialis (Figs. 21.10, 21.21) or n. triangularis lies ventral to the ventricle. Medium-sized interstitial cells in the vestibular root form the n. vestibularis tangentialis (Fig. 21.20); the n. vestibularis descendens extends from this level into the caudal myelencephalon (Figs. 21.17–21.20). Wold (1976) considered two more cell groups to be part of the vestibular complex, viz. cell group A (medial to the superior vestibular nucleus, but not part of it) and cell group B (ventral to the tangential nucleus); the latter may be a tail of the ventral vestibularis lateralis. The vestibular ganglion sends afferents to the central part of the n. vestibularis superior, the ventral part of the vestibularis lateralis, the dorsomedial part of the vestibularis mediodorsalis, to the n. vestibularis descendens and to cell groups A and B (Wold 1975). The tangential nucleus receives spoon-shaped en passant terminals from large vestibular fibres (Peusner and Morest 1977) as well as fine afferents (Cox and Peusner 1990). Boord and Karten (1974) also mention a lagenar projection to the external cuneate nucleus.

Vestibular efferents derive bilaterally from the n. reticularis pontis caudalis (Schwarz et al. 1978, 1981) and from the parvocellular reticular formation (Eden and Correia 1982a); the latter authors suggest that the lateral, tangential and medial nuclei may also contain perikarya of vestibular efferents. These efferents appear to be acetylcholinergic, but there is no direct adrenergic influence on the labyrinth (Eden and Correia 1981).

Ascending and descending vestibular pathways: Wold (1988) describes ascending projections from the quadrangular part of the nucleus vestibularis superior, from the ventral nucleus vestibularis lateralis and (slightly) from the nucleus vestibularis dorsomedialis to the n. dorsolateralis posterior thalami, p. rostralis. Other targets are the nucleus of the fasciculus longitudinalis medialis (Darkevitch), n. interstitialis (Cajal), n. pretectalis medialis, n. spiriformis medialis (Fig. 21.25), n. intercollicularis (Figs. 21.24, 21.33) and the oculomotor nuclei. Several previous experimental studies (e.g. Wold 1978a; Correia et al. 1983; cf. also Cox and Peusner 1990) specifically mention the vestibulo-oculomotor connections, the ascending fibres following both the ipsilateral and the contralateral fasciculus longitudinalis medialis.

Descending vestibular fibres are found in the ipsilateral and contralateral fasciculus longitudinalis medialis ('medial tract') as well as ipsilaterally through the reticular formation ('lateral tract'; Eden and Correia 1982b). An ipsilateral descending projection derives from the dorsal and ventral lateral nuclei; Wold (1978b) suggests that these nuclei correspond to the rostroventral and caudodorsal parts of the lateral vestibular nucleus in the cat. A bilateral vestibular projection has its origin in the medial and descending nuclei; the rostral parts of these nuclei project to the cervical region, the caudal parts to the thoracic part of the spinal cord. Furthermore a complex pattern of commissural vestibular connections is present (Wold 1979). Glover (1994) studying vestibulo-ocular and vestibulospinal projections in the chicken embryo distinguished hodologically rather than cytoarchitectonically defined cell groups, i.e. one cell group projects exclusively through one of the ascending or descending fibre systems. Such a cell group exhibits characteristics consistent with functional homogeneity, but is not restricted to one of the cytoarchitectonically defined nuclei. It is still an open question whether each hodologically defined cell pool represents a separate sensorimotor channel (Glover 1994).

21.6.5
Viscerosensory Complex

The main component of the viscerosensory system is the nucleus of the solitary tract. In birds this tract consists of vagal, glossopharyngeal and facial afferents; after entering the brain stem fascicles run medially and slightly rostrally before turning to caudal; some of the fibres descend to the commissura infima (Fig. 21.17) and cross to ascend contralaterally. The nucleus of the solitary tract can be subdivided into many subnuclei (Dubbeldam et al. 1976, 1979; Katz and Karten 1979, 1983a; Dubbeldam 1984), but there is no unanimity about the nomenclature. In Fig. 21.37 the nomenclature of Katz and Karten (1983a) is used. The cell groups consist of small or medium-sized perikarya. The medial tier nuclei receive gastrointestinal afferents, the lateral tier afferents from the pulmonary and cardiovascular systems (Katz and Karten 1983a). The subn. medioventralis anterior, p. anterolateralis

(Fig. 21.37a) receives convergent facial and glossopharyngeal input and represents the gustatory centre (Fig. 21.35b; Dubbeldam et al. 1979; Ganchrow et al. 1986). For a long time the sense of taste in birds was considered to be of little importance, the scarce taste buds in the tongue being innervated by branches of the glossopharyngeal nerve. The rediscovery (!) of taste buds in the tips of the upper and lower beak (cf. Berkhoudt 1985) and the demonstration of the innervation of these organs by facial branches (Krol and Dubbeldam 1979; Gentle 1979) have changed this picture. Efferents from the solitary complex reach reticular tegmental, parabrachial as well as hypothalamic, dorsal thalamic and basal telencephalic nuclei; there are also intrinsic solitary connections as well as projections to the dorsal motor nucleus of N. X (Arends et al. 1988).

21.6.6
Motor Nuclei

The visceromotor column comprises two groups of motorpools, the special visceral motor nuclei innervating the branchial musculature and the motor nuclei of the autonomic system. The most important component of the latter consists of the dorsal nuclei of N. IX and X (Figs. 21.17, 21.18). Katz and Karten (1983b) used cytological criteria to divide this complex into a series of subnuclei. Eleven cytoarchitectonically distinct nuclei were distinguished contributing to the midcervical vagal trunk and presumably each innervating a specific peripheral target. Other components of the autonomous nervous system are possibly the lateroventral n. salivatorius of the facial nerve (Bout and Dubbeldam 1985) and the n. oculomotorius accessorius (Edinger-Westphal) that sends its fibres to the ciliary ganglion (see Sect. 21.8).

The most important somatic motor nucleus in the medulla is the n. motorius N. hypoglossi (XIIm, Fig. 21.17). In recent publications the n. intermedius has been considered the 'real' hypoglossal motor centre; the rostral part of this nucleus innervates tongue muscles (Wild 1981; Dubbeldam and Bout 1990), the caudal part innervating muscles of the larynx, pharynx and syrinx (Manogue and Nottebohm 1982; Youngren and Phillips 1983; Vicario and Nottebohm 1988). The n. hypoglossus as defined in the older literature (cf Ariëns Kappers et al. 1936) innervates neck muscles (Wild 1981; Zijlstra and Dubbeldam 1994), but more recently has been called the n. cervicalis medialis (Watanabe and Ohmori 1988) or n. supraspinalis (Figs. 21.16, 21.17; Wild 1981). However, it contributes to the most caudal rootlets of the hypoglossal nerve (Watanabe and Ohmori 1988; Zijlstra and Dubbeldam 1994) and thus should be considered a ventral part of the n. hypoglossus. The n. intermedius should then be considered the dorsal part, an interpretation that links up with that in Ariëns Kappers et al. (1936). The supraspinal nucleus is continuous with the ventral horn of the cervical cord. A minor motor group is the n. retrofacialis; it innervates the musculus geniohyoideus of the tongue and was described for the first time in the galah (Wild 1981). It may be particularly well developed in birds with fleshy tongues.

The avian n. ambiguus lies ventral to the descending trigeminal tract and consists of slightly elongated cells (30×15 µm; pigeon: Grabatin and Abs 1986). It innervates laryngeal muscles. Recently, Wild (1993a) described a n. retroambigualis, situated in the transition area of lower medulla and cervical cord, caudal to the n. ambiguus and ventral to the lateral n. cervicalis. The n. retroambigualis receives afferents from the n. robustus archistriati, from the dorsomedial nucleus of the n. intercollicularis and from the parabrachial complex. It appears to be a relay in the pathway for respiration and vocalisation, its efferents projecting close to abdominal motorneurons (Wild 1993a; cf. also Vicario 1993).

The motor nuclei of the trigeminal and facial nerves form one complex. The facial motor nucleus in the pigeon consists of a dorsal (Fig. 21.21) and a ventral cell group, the former innervating the depressor muscle of the jaw, the latter innervating several tongue muscles (Wild and Zeigler 1980; Den Boer et al. 1986) and the m. dermotemporalis (Wild and Zeigler 1980). Earlier, Goodman and Fisher (1962) suggested that the latter muscle is innervated by cervical nerves; this aspect still needs confirmation. In a number of birds the ventral cell group is split into an intermediate and a ventral cell group (Knechtl 1954), innervating the tongue muscles and the neck muscle, respectively (e.g. duck, Dubbeldam and Bout 1990). The trigeminal motor nucleus (Fig. 21.22) also consists of several subnuclei innervating jaw muscles, the intermandibular muscle and the retractor muscle of the lower eyelid (Wild and Zeigler 1980; Den Boer et al. 1986).

The motor nuclei of Nn. III, IV and VI innervating the extrinsic eye muscles form the other important group of somatic motor centres. The n. nervi abducentis (VI, Figs. 21.20, 21.21) lies lateral to the fasciculus longitudinalis medialis; it consists of two populations of cells, one innervating the ipsilateral m. rectus lateralis, the other one sending input to the contralateral oculomotor complex (Cabrera et al. 1989). The accessory nucleus of N. VI lies more or less between the oliva superior and descending

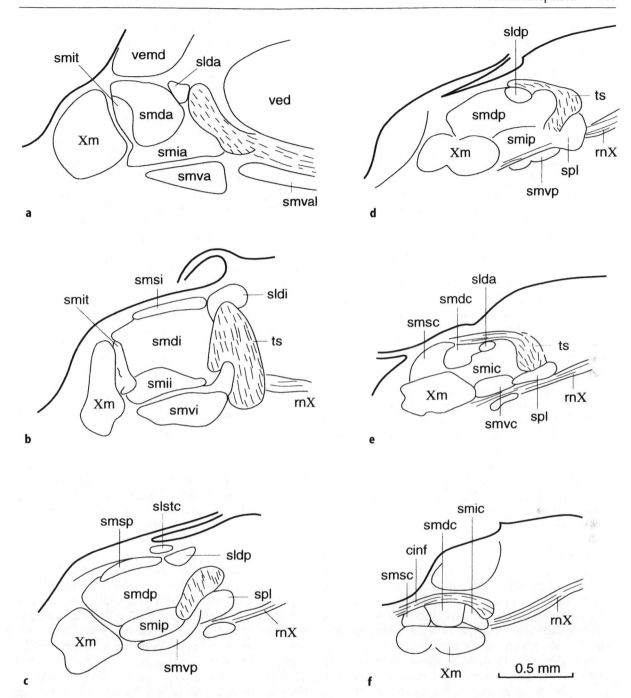

Fig. 21.37a–f. Six sections from rostral (**a**) to caudal (**f**) through the nuclear complex of the solitary tract to show the various subnuclei within this complex. The following nuclei solitarii are shown: *slda, sldc, sldi, sldp,* lateralis dorsalis, pars anterior, p. caudalis, p. intereminentialis, p. posterior; *slstc,* lateralis dorsalis taenia chorioidea; *smda, smdc, smdi, smdp,* medialis dorsalis, p. anterior, p. caudalis, p. intereminentialis, p. posterior; *smia, smic, smii, smip,* medialis intermedius, p. anterior, p. caudalis, p. intereminentialis, p. posterior; *smit,* medialis interpositus; *smsc, smsi, smsp,* medialis superficialis, p. caudalis, p. intereminentialis, p. posterior; *smva, smval, smvc, smvi, smvp,* medialis ventralis, p. anterior, p. anterolateralis, p. caudalis, p. eminentialis, p. posterior; *spl,* parasolitarius lateralis. (Adapted from Katz and Karten 1983a)

trigeminal tract; it innervates the m. quadratus and m. pyramidalis of the membrana nicticans (Labandeira-Garcia et al. 1987), though Bravo and Inzunza (1985) claim that these muscles are innervated by the oculomotor complex sensu stricto. This complex, however, is part of the mesencephalon and will be discussed in Sect. 21.8.

21.6.7
Oliva Inferior and Nuclei of the Pontine Region

The avian inferior olive [complexus olivaris caudalis] is a distinct complex of cell groups (Figs. 21.17, 21.18). Its cells are rather large, the caudal cells being slightly smaller than those in the other parts. A dorsal and a ventral lamina can easily be distinguished (Vogt-Nilsen 1954). Each can be further subdivided into several compartments. The medial part of the ventral lamina forms the most rostral pole of the oliva, the dorsal lamina the most caudal one. The oliva inferior receives afferents from several cell groups and is the origin of the climbing fibres to the cerebellum (Freedman et al. 1977). Furber (1983) using embryological and hodological data suggested that the dorsal lamina contains from lateral to medial the equivalents of the dorsal and principal olivary nucleus of the mammals as well as the rostral and a small part of the caudal nucleus, whereas the ventral lamina would correspond to most of the caudal medial nucleus. This grouping, however, is primarily based upon Larsell's interpretation of the cerebellum, which no longer seems tenable. Using the concept of a zonal organisation of the cerebellar cortex and on the basis of the mediolateral zonal pattern of olivocerebellar connections, Arends and Voogd (1989) distinguish a medial column and a ventral and a dorsal lamella, but also conclude that homologisation with the mammalian situation is at best tentative at this moment. From this point of view the use of the terms n. olivaris principalis, accessorius medialis and dorsalis (NAA) seems less appropriate.

Brodal et al. (1950) were the first to establish experimentally the character of the pontine nuclei. These are modest cell groups, separated from the ventral surface of the brain stem by a thin layer of fibres (commissure of Wallenberg) (Figs. 21.22, 21.23) and dorsally bordered by the trapezoid body. The medial pontine nucleus sends fibres to the contralateral half of the cerebellum, the lateral nucleus to the ipsilateral half. Brodal et al. reported that pontine fibres end in folia VI–VIII,b and in the paraflocculus, but other authors found a wider distribution (e.g. Freedman et al. 1975b). Several sources of pontine input have been identified. Münzer and Wiener (in Ariëns Kappers et al. 1936) described the tr. tectobulbaris to the lateral pons nucleus. Furthermore this cell group receives an archistriatal input via the occipitomesencephalic tract (Zeier and Karten, 1971), thus forming a telencephalic-cerebellar link.

The parabrachial nuclei are a number of cell groups lateral to the pedunculus cerebellaris rostralis and medial to the fasciculus uncinatus (Fig. 21.22). These nuclei are assumed to be part of the gustatory pathway receiving afferents from the solitary complex (Arends et al. 1988). Wild (1990) distinguished a number of subnuclei with multiple connections in the brain stem (n. hypoglossus, ambiguus), midbrain, hypothalamus, dorsal thalamus and basal telencephalon.

21.6.8
Reticular Formation

The medullary and pontine parts of the reticular formation form a continuous system and for reasons of convenience will be discussed as a separate entity. Petrovicky (1966) provided the first extensive description of the reticular cell groups in the pigeon, but also reported that these cell groups cannot always be recognised in other avian species (Petrovicky 1973). His nomenclature has largely been followed in the atlas of Karten and Hodos (1967) as well as in the NAA. The most caudal of these cell groups are the nn. centrales medullae oblongatae dorsalis and ventralis. The latter centre lies in the ventral half of the medulla and dorsal to the inferior olive and consists of medium-size and small cells. The n. centralis dorsalis medullae oblongatae (Figs. 21.17, 21.18) is more or less the rostral continuation of the grey of the spinal lateral horn and also consists of small- to medium-size cells. A series of raphe nuclei (Figs. 21.17–21.21, Fig. 21.22) including the n. centralis superior and nn. lineares caudalis (Fig. 21.23), intermedius and rostralis extend from the caudal medulla into the rostral mesencephalic tegmentum. Many of the raphe cells contain serotonin (Dubé and Parent 1981). Okado et al. (1992) describe the early appearance of serotonergic cells in embryos of the chick and demonstrate that the raphe cells are the main source of descending spinal serotonergic fibres.

The ventral-ventromedial region of the medulla is occupied by the n. reticularis gigantocellularis (Figs. 21.18–21.20). It extends from the level of the inferior olive to the pons nuclei and contains medium-size to large cells. At the pontine level it is replaced by the pontine reticular gigantocellular nucleus (Figs. 21.22, 21.23) with large and giant cells. The n. pontis rostralis (Fig. 21.23) with mainly medium-size and occasional large cells is again the rostral continuation of the nucleus reticularis pontis gigantocellularis.

The n. reticularis parvocellularis (Figs. 21.19, 21.20) is more or less the rostral continuation of the n. centralis dorsalis medullae oblongatae. It lies dorsolateral to the n. reticularis paragigantocellularis, consists of small cells and extends from the level of XIIm to that of IV. It can be subdivided into a dorsolateral zone (including the plexus of Horsley) and a ventromedial zone (e.g. Bout 1987). The two zones have reciprocal connections and both receive afferents from the archistriatum; the dorsolateral zone receives further input from the subnn. caudalis and interpolaris of the descending trigeminal system, whereas a rostral and a caudal compartment of it receives projections from the mesencephalic trigeminal nucleus. These compartments are premotor areas with ipsilateral projections to the motor nuclei of N.V (jaw closing mus cles) and of Nn. V+VII (innervating jaw opening muscles), respectively (Bout and Dubbeldam 1994). The caudal compartment coincides more or less with the n. centralis medullae oblongatae dorsalis. So far, it is unknown where the intermediate compartment sends its efferents. The medioventral zone of the n. reticularis parvocellularis sends bilateral projections to the motor nuclei of the jaw muscles (Bout 1987). Other connections of the two zones are projections to the area intertrigeminalis - a reticular zone between the sensory and motor nuclei of N.V -, parabrachial nuclei and oliva inferior (Arends and Dubbeldam 1982; Bout 1987). The intertrigeminal area is another source of input of the trigeminal motor nuclei (Berkhoudt et al. 1982: pigeon; Arends and Dubbeldam 1982: mallard).

It is interesting to note that some of the n. reticularis parvocellularis cells appear to be the origin of efferent connections: labelled n. reticularis parvocellularis cells were found between the n. ambiguus and the dorsal motor nucleus of N.X after applying horseradish peroxidase (HRP) to N.X (Katz and Karten 1983b), in a corresponding area in the pontine region after HRP injections in N.VIII (Whitehead and Morest 1981), as well as around the trigeminal motor nuclei after HRP injections in jaw muscles or the trigeminal nerve (Den Boer et al. 1986). Recently, Code and Carr (1994) provided more data describing two groups of cochlear efferents in the chicken, a dorsomedial group between the dorsal facial motor centre and root of the abducens nerve and a ventrolateral group around the oliva superior and medial to the ventral facial motor nucleus. In labelling experiments cells in these groups could be labelled bilaterally. About 70% of these cells are cholinergic (Code and Carr 1994). There is no good interpretation of the nature of these cells.

In addition to these large reticular fields several other reticular nuclei have been identified (Petrovicky 1966): a diffuse n. reticularis lateralis (Fig. 21.18), a n. paramedianus (Figs. 21.19, 21.22), mediodorsal to the nucleus reticularis gigantocellularis but without the large cells, the n. reticularis paragigantocellularis (Fig. 21.19). The parabrachial nuclei have already been mentioned; Petrovicky and Kolesárová (1989) described this complex in many species of birds and mammals.

Not indicated in the figures, but described by Petrovicky (1966), is a group of annularis nuclei surrounding the medial fasciculus longitudinalis. Lateral to it lies the n. tegmenti dorsalis; this is the coeruleus complex of the atlas of Karten and Hodos. This complex comprises the n. coeruleus proper (Figs. 21.22, 21.23) with medium-size, spindle-shaped cells and the subcoeruleus cell groups (Figs. 21.22, 21.23). The n. coeruleus as well as the subcoerulean nuclei contain catecholamine neurons (Dubé and Parent 1981; Takatsuki et al. 1981; also Tohyama et al. 1974: budgerigar), but more abundantly serotonin-containing cells (Dubé and Parent 1981; Kitt and Brauth 1986a). They send bilateral projections to the paleostriatum complex, septum, tuberculum olfactorium, dorsomedial cortex or hippocampus+area parahippocampalis, to the dorsal archistriatum and n. taeniae and parts of the lateral neostriatum, to the hyperstriatum dorsale and ventrale and preoptic area. An important difference from the mammalian situation is the projection to the paleostriatal complex, i.e. avian 'striatum'; Kitt and Brauth suggest that only cells in the caudal nucleus coeruleus represent 'real' coerulean cells, whereas the rostral part as well as the dorsal and ventral subcoerulean nuclei possessing dopaminergic cells may be comparable to the mammalian lateral tegmental cell group (see Sect. 21.8.4). The caudal locus coeruleus and subcoerulean nuclei do contain noradrenergic cells (Reiner et al. 1994: pigeon; Dubé and Parent, 1981); these can be considered the 'real coerulean' cells. In the chicken locus coeruleus is the area with the highest density of β-adrenergic receptors (Dermon and Kouvelas 1989).

Ascending fibres from this region follow several pathways, such as the occipitomesencephalic tract, quintofrontal tract, ansa lenticularis and lateral and medial forebrain bundles, to reach the forebrain areas. A separate fibre-tract reaches a number of dorsal thalamic nuclei (Kitt and Brauth 1986a).

Some of the coerulean cells send descending fibres to the thoracic spinal cord (Chikazawa et al. 1983). Serotonergic cells are found in the nn. coeruleus and subcoeruleus, further predominantly medially in the brain stem: n. linearis caudalis extending into a dorsolateral triangular cell group, several raphe nuclei and ventral reticular formation (Ikeda and Gotoh 1971; Dubé and Parent 1981). The characteristic serotonergic terminals are present in the various motor nuclei of the brain stem (e.g. of nerves V and VII in the chicken, Dubbeldam and den Boer-Visser 1993).

In an immunocytochemical study in the chicken, Blähser and Dubois (1980) found scattered Met-enkephalin-immunoreactive cells in various brain regions. More striking, however, is their description of three enkephalinergic networks, one closely related to the cells and axons of the vagal, glossopharyngeal and sensory trigeminal nerves; a second, medial network around medial reticular cell groups (nn. linearis caudalis, paramedianus, raphes); and a third network between the commissura posterior and area septalis. There is a partial overlap with somatostatin-positive elements, e.g. in the substantia intermedia centralis of the cervical spinal cord, parts of the descending trigeminal system, pontine and medial reticular formation, area ventralis and mammillary nuclei (Blähser and Dubois 1980). These observations were largely confirmed by de Lanerolle et al. (1981).

21.6.9
Fibre Systems of the Brain Stem

The fasciculus longitudinalis medialis (Figs. 21.17–21.24) is the most conspicuous fibre system of the rhombencephalon. It is the rostral continuation of the spinal ventral funiculus and extends into the mesencephalon. The fasciculus longitudinalis medialis is composed of a mixture of large and small axons and carries fibres from several reticular cell groups continuing into the ventral funiculus of the spinal cord. In the rhombencephalon vestibular fibres join the fasciculus longitudinalis medialis, descending fibres projecting to the intermediate zone of the spinal cord and ascending fibres to the nuclei of the oculomotor complex. For further details, see Sect. 21.8.8.

Several fibre systems have already been mentioned such as the medial lemniscus and the lateral lemniscus. Another lemniscal system is the lemniscus spinalis carrying spinotectal and spinothalamic fibres, whereas the quintofrontal tract is the ascending trigeminal system. A large descending system originates from the n. ruber (tractus rubrospinalis), from the archistriatum (occipitomesencephalic tract) and from the eminentia sagittalis (septomesencephalic tract). The rubrospinal tract runs laterally through the brain stem (Figs. 21.19, 21.21), the occipitomecenphalic tract dorsally through the parvocellular reticular formation (Figs. 21.18, 21.19, 21.22), whereas the small basal branch of the septomesencephalic branch occupies a position along the ventral periphery (Verhaart 1971: parrot; Dubbeldam 1976: mallard). Details of these fibre systems will be discussed together with their source of origin.

21.7
Cerebellum

21.7.1
Introductory Remarks

Birds possess a large cerebellum. It has a uniform cerebellar cortex and a distinct group of deep nuclei. It has been suggested in the older literature that most of the cerebellum corresponds to the mammalian vermis (review in Nieuwenhuys 1967b), whereas the auricle would comprise the flocculus and parafloccular parts. The corpus cerebelli consists of a foliated medial portion and a basal part with a flattened cortex. Larsell (1948, Larsell and Whitlock 1952) compared the cerebellar structure in various avian species and concluded that essentially ten folia can be recognised, while some of these can be subdivided in secondary folia (Fig. 21.38a). These folia differ considerably among the species, but some intraspecific variation also exists. Larsell (1948) assumed that the unfolded part of the cortex represents the cerebellar hemispheres. This assumption, however, is not undisputed.

21.7.2
Cerebellar Cortex

The cerebellar cortex comprises an outer molecular layer, a monolayer of Purkinje cells and an inner granular layer (Figs. 21.18, 21.32). Essentially, the structure of the cortex is not different from that in the mammals (Hummel and Goller 1979); only the thickness is smaller. Several interpretations of the cortical subdivisions exist. Goodman et al. (1964) recognised four regions: three longitudinal, viz. a vermal, a paravermal and a lateral zone, and the floccular region (Fig. 21.39a). The lateral zone consisting of the lateral parts of folia V through VIII was suggested to correspond to the hemispheres, as it receives the most abundant input from the pontine nuclei (Brodal et al. 1950). The paravermal and vermal zones receive modest and small numbers of pontine fibres, respectively. An additional argument was that stimulation in the respective zones results in postural movements comparable to those in mammals after stimulation in either the vermis or the hemispheres. Raymond (1958) demonstrated that wing movements can be evoked by stimulation in folium VI, tail movements in folia II–IV and IX, beak and tongue movement in folia II and III, as well as in VI–VIII, etc.

Feirabend and Voogd (1986) stress that subdivision of the cerebellar cortex should be based on the pattern of afferentation. These authors distinguish eight longitudinal zones, zones 1–5 corresponding to the vermis, zones 6–8 to the hemispheres (Fig. 21.39b). Each parasagittal zone consists of a medial compartment containing large and small fibres and a lateral compartment with predominantly thin fibres (Fig. 21.18). Longitudinal patterns of spinocerebellar (Vielvoye 1977), ponto-cerebellar (Freedman et al. 1975a), trigemino- and vestibulo-cerebellar (Arends 1981; Arends and Zeigler 1989) fibres have been described. Basically, comparable patterns have been found in mammals (Feirabend 1983).

The oliva inferior projects throughout the cortex, each subnucleus having its own terminal region (Arends and Voogd 1989). These projections are mainly contralateral (Armstrong and Clarke 1979). Stimulation experiments suggest that the dorsal and ventral spinocerebellar tracts project to folia II–VIb and VIII+IX (Whitlock 1952). More recent tracer studies show that the lumbal region projects largely to folia I–V and lightly to folia VI and IX, the thoracic region to VI and a little bit to IX (Fig. 21.38b), whereas the cervical region projects to folia II–IX (Okado et al. 1987b). Necker (1992) suggests that folium IX receives somatosensory input from dorsal horn cells and folia III–VI proprioceptive input from Clarke's column and from the dorsolateral ventral horn (his 'spinal border cells'). The folia IX and X+auriculum (flocculonodular node) form the vestibulo-cerebellum receiving both direct and indirect vestibular input (Wilson et al. 1972, 1974; Schwarz and Schwarz 1983). Folia V–VII receive an indirect trigeminal input from the subn. interpolaris (Fig. 21.38b; Arends et al. 1984; Arends and Zeigler 1989) and not a direct one as suggested by, e.g. Sanders (1929), Woodburne (1936) and others; neither receives the cerebellum afferents from other trigeminal nuclei, including the mesencephalic trigeminal nucleus (as suggested by Bortolami et al. 1969). Clarke (1974) recorded visual responses in folia VI–IX, but particularly in VII and VIII; he reports the existence of a retinotopic projection to folium VII. However, there is no evidence of a direct tectocerebellar tract as has been claimed in earlier studies. Tectal input may reach the cerebellum through the lateral pons nucleus (Clarke 1977). IXc,d and X

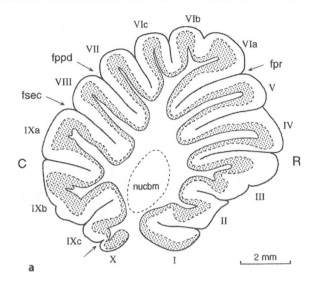

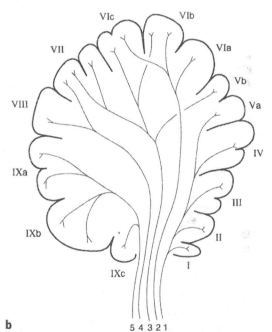

Fig. 21.38. a Parasagittal section through cerebellum of the pigeon showing the cerebellar lobuli with cortex and central nucleus (*nucbm*). **b** Same section with some afferent systems. *1,* (lumbo-)spinocerebellar projection; *2,* (thoraco-)spinocerebellar projection (Okado et al. 1987b); *3,* trigeminocerebellar projection (Arends and Zeigler, 1989); *4,* tectopontine visuocerebellar projection (Clarke 1977); *5,* ectomammillary optomotor cerebellar projection. (Brauth and Karten 1977)

are the target of axons from the ectomammillary nucleus (Fig. 21.38b; Brauth and Karten 1977), from their nucleus of the basal optic tract (Sect. 21.9.3) and to a lesser degree from the n. lentiformis mesencephali, p. magnocellularis. The latter also has an input into folia VI–IXa. Extracellular recordings reveal that units in the flocculus and in

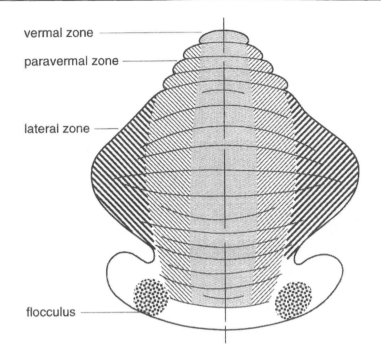

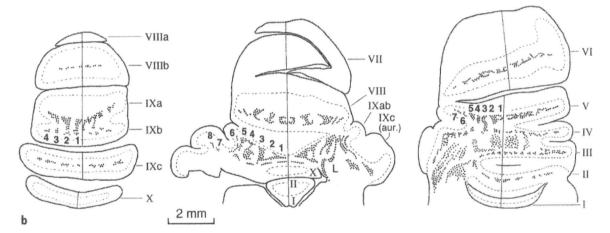

Fig. 21.39a,b. Longitudinal zones in the cerebellum. **a** Three longitudinal zones as defined by Goodman et al. (1964). This interpretation rests on electrophysiological stimulation experiments. **b** Eight zones as distinguished by Feirabend (1983). This interpretation rests on the recognition of compartments, each containing small and large fibres. *Dots* indicate the groups of large fibres in the respective compartments

folia IXc,d and X react to optokinetic stimuli; mossy fibres have a monocular field, but most Purkinje cells react to stimuli in the binocular field and also show a clear direction-selectivity (Wylie et al. 1993).

The medial spiriform nucleus (Fig. 21.25) is the origin of a thalamo-cerebellar projection reaching the contralateral cerebellar cortex through the pedunculus cerebellaris superior (rostralis) ('brachium conjunctivum cerebellopetale', Karten and Finger 1976) and crossing dorsally to the cerebellar ventricle (Clarke 1977). Other sources of cerebellar input are parts of the reticular formation in the trapezoid body, contralateral nucleus ruber and bilateral gigantocellular nucleus (Clarke 1977).

The cerebellar cortex itself projects in an orderly way to the central cerebellar nuclei and to nuclei of the vestibular complex. Lobuli II–IXa project to the ipsilateral central cerebellar and vestibular nuclei, lobuli IXb and X mainly to the processus vestibulocerebellaris and vestibular nuclei (Arends and Zeigler 1991a; Sect. 21.7.4). There is also a rostrocaudal topographic order of corticonuclear projections.

21.7.3
Histochemistry of the Cerebellar Cortex

Several recent studies provide data about the distribution of neurotransmitter receptors in the cerebellar cortex. A high density of muscarinic cholinergic receptors is found in the molecular region (Dietl et al. 1988a), whereas a moderate number of GABA-benzodiazepine receptors was reported for the molecular and granular layers (Dietl et al. 1988b). Albin et al. (1991) found a higher density of GABA-A-binding sites in the granular layer than in the molecular layer, a higher density of GABA-B sites in the molecular layer, whereas benzodiazepine-binding sites had a similar density in the two layers. N-Methyl-D-aspartate (NMDA)-binding sites were more frequent again in the granular layer (Albin et al. 1991). As for dopamine, only a low concentration of D^1 receptors was found (Dietl and Palacios 1988). Aminergic fibres deriving from the locus coeruleus enter the cortex, with a netlike pattern in the granular layer and a parallel fibre-like pattern in the molecular layer (Mugnaini and Dahl 1975). Again, a pattern of sagittal zones can be recognised.

21.7.4
Central Nuclei

There is no unanimity about the number of central cerebellar nuclei that can be distinguished (for the older literature, see Nieuwenhuys 1967b). Sometimes two nuclei are recognised; other studies mention three or even four (see Ariëns Kappers et al. 1936). Doty (1946) described the central nuclei in the sparrow as a single grey band that is greatly folded and convoluted. This may be the case in some birds, other species possessing simple, unfolded nuclei (Renggli 1967). This is also true for the pigeon. According to the atlas of Karten and Hodos (1967) a n. cerebellaris internus [medialis] (Figs. 21.21, 21.22), a n. intermedius (Figs. 21.20, 21.21) and a lateral cerebellar nucleus (Fig. 21.20) can be recognised (cf. also Kuenzel and Masson 1988). The comparability of these nuclei to those of the mammalian cerebellum is still an open question. Arends and Zeigler (1991a,b) distinguish only a lateral and a medial nucleus; the latter may be subdivided into an internal, intermediate ventromedial and an intercalate portion. These authors use the connections as an important criterion. The medial zone of the cerebellar cortex projects to the medial cerebellar nucleus; most extracerebellar connections from this nucleus leave the cerebellum through the fasciculus uncinatus (Figs. 21.22, 21.40). The lateral nucleus receives afferents from the more lateral cortical zones and sends its efferents mainly through the rostral cerebellar peduncle.

Our n. intermedius corresponds to the intermediate portion, and our n. internus to the internal and intercalate portions of the medial nucleus as distinguished by Arends and Zeigler. The processus vestibulocerebellaris is positioned as a parvocellular extension of the lateral cerebellar nucleus; Wold (1976) seems to include it in the superior vestibular nucleus. It receives afferents from the medial folia XIb and X and can be compared to the parvocellular cerebellar nucleus of the mammals (Arends and Zeigler 1991a). The lateral zones of folia IXb and X project to the lateral cerebellar nucleus and the auricula to the n. infracerebellaris and medial parts of nn. vestibularis superior and tangentialis.

21.7.5
Large Cerebellar Fibre Systems

There are three major afferent cerebellar fibre systems, the pedunculus cerebellaris caudalis (Figs. 21.20, 21.21) or corpus restiforme, the pedunculus cerebellaris intermedius (Fig. 21.21) or brachium pontis and the pedunculus cerebellaris rostralis (Figs. 21.21–21.24). The fibres of the dorsal spinocerebellar tract, the vestibulocerebellar tract and the olivocerebellar fibres (after crossing as external arcuate fibres: Figs. 21.19, 21.20) enter the cerebellum via the caudal peduncle. The intermediate peduncle carries pontocerebellar fibres, the rostral peduncle fibres from the ventral spinocerebellar tract. The latter also carries cerebellar fibres to the upper part of the brain stem; most of these derive from the lateral cerebellar nucleus and decussate through the ventral commissura cerebelli (Fig. 21.21). Targets are n. ruber and n. interstitialis (Cajal) and midbrain central grey, n. spiriformis medialis and n. precommissuralis principalis, several dorsal thalamic nuclei, stratum cellulare externum, n. ectomammilaris and n. geniculatus lateralis, pars ventralis (Fig. 21.40; Arends and Zeigler 1991b). A small descending portion of the rostral peduncle projects to the n. pontis medialis, the reticular n. papilioformis and gigantocellular and paramedian cell groups. A small component reaches the inferior olivary complex (medial column and ventral layer).

The fasciculus uncinatus is the other efferent cerebellar system with predominantly descending projections from the intermediate/medial nuclear complex. Fibres terminate mainly in the vestibular nuclei, medial reticular nuclei, the dorsolateral zone of the parvocellular formation, the lateral and the paragigantocellular reticular nuclei and the dorsal lamella of the oliva inferior (Arends and Zeigler 1991b). A small rostral component projects to the n. coeruleus and n. subcoeruleus dorsalis. Arends and Zeigler also mention modest projections to the trigeminal motor nucleus and n. oralis of the descending trigeminal system.

21.8
Mesencephalon

21.8.1
Introductory Remarks

The most striking feature of the avian mesencephalon is the development of the two large tectal bulges ('optic lobes') that have a lateral rather than a dorsal position. This position also differs considerably between species (Figs. 21.23, 21.24, 21.45). The mesencephalic ventricles (Figs. 21.24, 21.25, 21.32) expand far laterally leaving only a thin sheet as roof of the aquaduct (Figs. 21.23, 21.24). Large numbers of fibres of the trochlear decussation and intertectal commissure cross through this roof. Large cell bodies of the mesencephalic trigeminal nucleus (Fig. 21.24) occur interspersed between these fibres. Grossly, the mesencephalic cell groups can be subdivided into tegmental nuclei, isthmic nuclei and the cell groups of the optic lobe including the tectal cortex (tectum mesencephali). In a series of papers Graef (1973a-e) described the cyto- and myeloarchitectonics of the mesencephalon of the chicken.

21.8.2
Mesencephalic Trigeminal Nucleus

First a few remarks are necessary about the mesencephalic nucleus of the trigeminal nerve. This nucleus consists of a medial part (Fig. 21.24) in the commissura tecti mesencephali and a lateral part, the latter consisting of dispersed cells in the roofs of the ventricles of the mesencephalic lobes

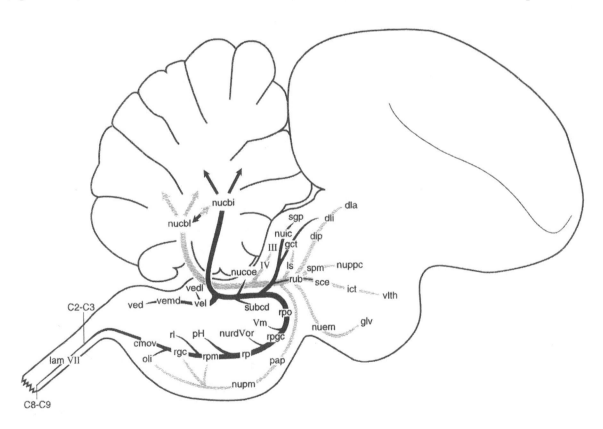

Fig. 21.40. Scheme of the two main pathways of the efferents from the central cerebellar nuclei. *Black*, fasciculus uncinatus; *grey*, efferents via the rostral cerebellar peduncle. (Adapted from Arends and Zeigler 1991b)

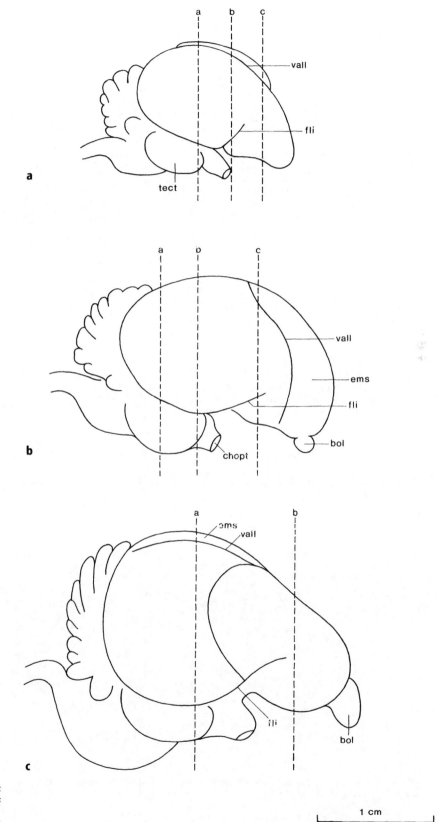

Fig. 21.41a–c. Lateral view of the brains of **a** *Melopsittacus undulatus*; **b** *Corvus monedula*; **c** *Anas platyrhynchos*. Broken lines indicate levels of the sections depicted in the following figures

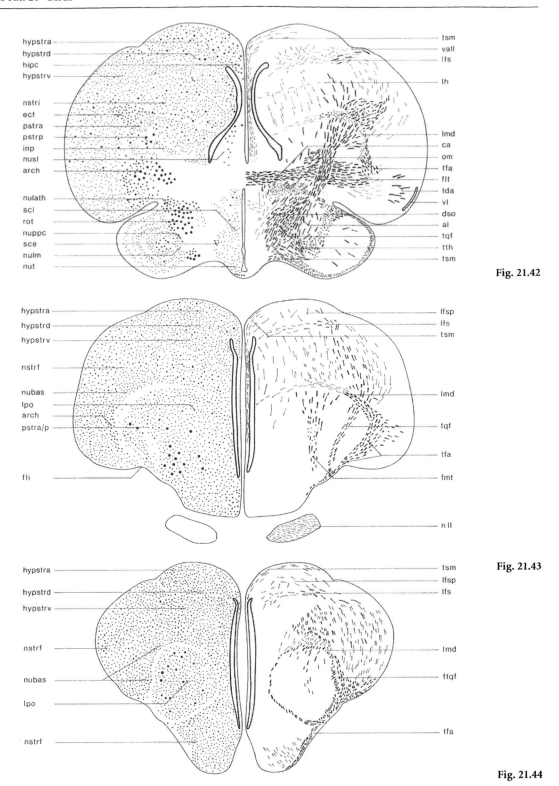

Figs. 21.42–21.44. Three cross-sections through the brain of *Melopsittacus undulatus*. *Left side* cells, *right side* fibre systems. Compare Figs. 21.12–21.31

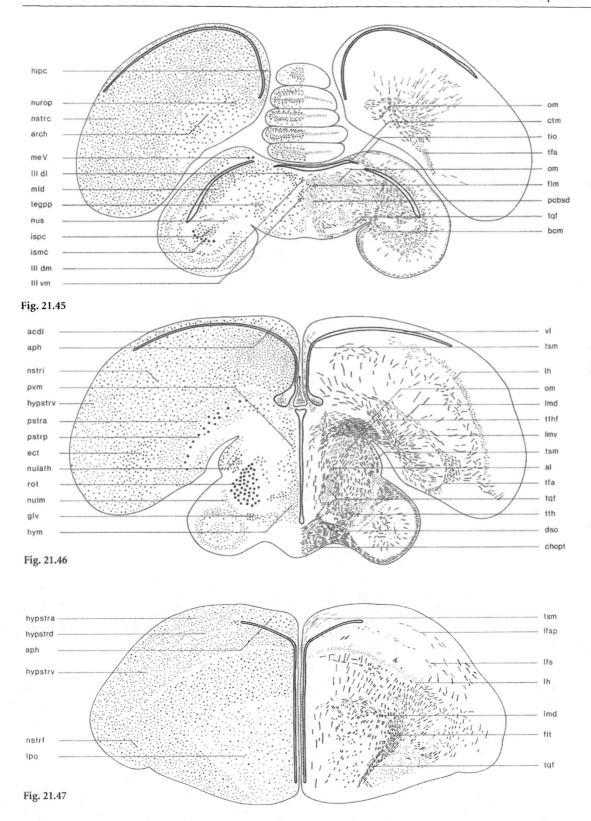

Figs. 21.45–21.47. Three cross-sections through the brain of *Corvus monedula*. *Left side* cells, *right side* fibre systems. Compare Figs. 21.12–21.31

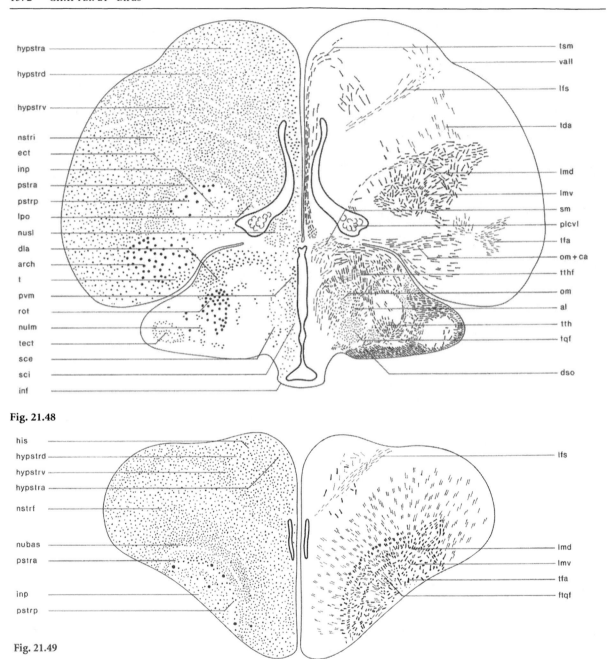

Figs. 21.48, 21.49. Two cross-sections through the brain of *Anas platyrhynchos*. *Left side* cells, *right side* fibre systems. Compare Figs. 21.12–21.31

(Fig. 21.45); generally, these cells are considered displaced ganglion cells. Bortolami et al. (1969) described two populations, i.e. unipolar and multipolar cells. Recently, Scott et al. (1994) found a heterogeneity in histochemical characteristics of cells in the different parts of this nucleus in hatchling chickens: e.g. cells expressing substance P reactivity were most abundant in the medial part, parvalbumin-reactive cells in the lateral parts. It can be questioned whether such differences reflect different functions for the different cells. Previously, Manni et al. (1965) demonstrated electrophysiologically that these cells innervate the muscle spindles of the jaw muscles. Further these cells have connections with trigeminal motor nuclei, forming a monosynaptic reflex arch and with the rostral and caudal compartments of the dorsolateral zone of the parvocellular reticular formation in the medulla (Bout 1987; see Sect. 21.6.8). A projection to the cerebellar lobi VIc and VII, suggested by Bortolami et al. (1972), was not found in other studies.

21.8.3
Tegmental Cell Groups

Several somatic motor nuclei are present in the dorsomedial part of the tegmentum, viz. caudally the nucleus of the trochlear nerve (IV, Fig. 21.23) and rostral to it the group of oculomotor nuclei (IIId, IIIv, Fig. 21.24). The trochlear nucleus is characterised by its densely packed large cells, its fibres converging at the dorsolateral edge of the nucleus and crossing through the ventricular roof. It innervates the superior oblique muscle. The subdivisions of the oculomotor complex, i.e. the dorsolateral, dorsomedial and the ventromedial cell group (IIIdl, IIIdm, IIIvm), are shown best in Fig. 21.45. The dorsolateral group innervates the ipsilateral m. rectus inferior, the dorsomedial group the m. rectus medialis. The ventromedial nucleus can be subdivided into a distinct lateral part innervating the ipsilateral m. obliquus inferior and a medial part innervating the contralateral m. rectus superior (Heaton and Wayne 1983; see also Isomura 1973). Bravo and Inzunza (1985) claim that these nuclei also innervate bilaterally the muscles of the nictating membrane, but this is denied by Labandeira-Garcia et al. (1987; see description of accessory abducens nucleus, Sect. 21.6.6). Hillebrand (1972) found evidence for a central oculomotor nucleus in galliform birds, but not in other groups of birds; he suggests that this may be connected with the possibility of convergence of the eyes as in mammals. The oculomotor nuclei receive input from the vestibular complex (e.g. Correia et al. 1983), from the n. nervi abducentis, the ipsilateral n. campi Foreli and the contralateral n. interstitialis of Cajal (Labandeira-Garcia et al. 1989).

The accessory oculomotor nucleus is considered the avian nucleus of Edinger-Westphal (Narayanan and Narayanan 1976); it receives input from the pretectal area (recipient of retinal fibres, Gamlin et al. 1984) and from the suprachiasmatic nucleus (Gamlin et al. 1982). A lateral and a medial population of cells with different properties have been described; the lateral group sends fibres to the ciliary ganglion and has a role in the control of accommodation, pupil reflex and choriocapillary blood flow (Gamlin and Reiner 1991; Fujii 1992; review in Reiner et al. 1983).

21.8.4
Reticular Formation

The reticular formation of the rhombencephalon continues into the tegmentum mesencephali. The nn. lineares have already been mentioned. The dorsolateral zone between the levels of n. trochlearis and n. ruber is occupied by the n. laterodorsalis, the ventrolateral area by the n. tegmenti pedunculopontinus (Fig. 21.24) with a pars compacta and a pars dissipata reticularis (Petrovicky 1966; Karten and Dubbeldam 1973); the n. tegmentalis pedunculopontinus is one of the end stations of the ansa lenticularis. The pars compacta contains catecholaminergic cells, i.e. mainly dopaminergic cells (Brauth et al. 1978) and can be regarded part of the avian substantia nigra. Its cells project to the paleostriatum augmentatum (Figs. 21.28, 21.29, 21.57), in particular to its caudal, ventral and lateral parts; the rostral and medial pstra and lobus parolfactorius receive input primarily from the area ventralis of Tsai, possibly another part of the substantia nigra (Reiner 1986b). The pars compacta is also the source of fibres projecting to the deep layers of the tectum. The pars dissipata may represent the substantia nigra reticulata; it has substance P positive, but no catecholinergic cells (Reiner 1986a,b).

Histochemical data are indispensable to define more precisely the avian equivalent of the mammalian substantia nigra. On account of hodology and histochemical data the n. tegmenti pedunculopontinus had already been recognised as an important component. Comparable cell groups have been identified in several reptiles (Smeets 1991). In addition, the rostral part of the locus coeruleus in the pigeon (Karten and Hodos 1967) contains dopaminergic and no noradrenergic cells and thus could be regarded a caudal extension of the avian substantia nigra (Reiner et al. 1994). These authors stress that the n. tegmenti pedunculopontinus as defined by

Karten and Dubbeldam (1973) should be named substantia nigra to avoid confusion with the mammalian nucleus tegmenti pudenculopontinus, pars compacta (tpc).

The n. ventralis tegmenti (Fig. 21.24) lies medioventrally in the tegmentum, lateral to the root of the oculomotor nerve (nIII, Fig. 21.24). Particularly its lateral zone contains medium-sized cells with rather coarse Nissl bodies. It has many catecholinergic cells projecting to the paleostriatal complex (Reiner 1986b) and receives input from the lobus parolfactorius (Kitt and Brauth 1981). The n. interpeduncularis lies medial to the root of nerve III. Rostrodorsal to Tsai's area lies the n. ruber (Fig. 21.24); it consists partly of large cells (40–50 µm), partly of small and medium-size cells (15–35 µm); the latter dominate rostrally, the larger cells more caudally. Wild et al. (1979) demonstrated that this cell group is the origin of a rubrospinal tract. This fibre system decussates and descends all the way down to the lumbar level of the spinal cord with terminations in layers V, VI and – more sparsely – VII. A very dense terminal field occurs rostrally in the ventrolateral medulla oblongata with a few fibres and terminals in the dorsolateral parvocellular reticular zone (plexus of Horsley).

21.8.5
Isthmic Cell Groups

There are four distinct isthmic nuclei: the nn. isthmi parvocellularis and magnocellularis, the n. semilunaris and the isthmo-optic nucleus. Clarke (1982) includes the nuclei of the lateral lemniscus in the isthmic complex and states that this whole complex has developed along a single spatiotemporal gradient of proliferation.

The isthmo-optic nucleus (Fig. 21.23) is a small but conspicuous cell group. Its bipolar cells form a double layer that is wrapped together; the dendrites of these cells form one plexus receiving tectal afferents (Crossland 1979; Güntürkün 1987). This projection is topographically organised (Holden and Powell 1972); other projections derive from the oculomotor nuclei (Angaut and Repérant 1978). Miceli et al. (1995) describe the GABA-ergic input of the neuropil of the isthmo-optic nucleus; part of the GABA-ergic afferents have a tectal origin; others may derive from interneurons lying in the neuropil of n. isthmo-opticus. A third group of GABA-ergic projections may still have another, not yet specified origin (Miceli et al. 1995). The isthmo-optic nucleus is the origin of the isthmo-optic tract (Figs. 21.24, 21.32, 21.45), an efferent fibre system projecting to the amacrine cell layer of the retina (Cowan and Powell 1963; Crossland and Hughes 1978). This projection, too, has a topological organisation: a specific locus of n. isthmo-opticus receives an exclusive input from a tectal region that receives afferents from that part of the retina that is recipient of efferents from this specific locus of n. isthmo-opticus (McGill et al. 1966a,b). In addition, an ectopic population of multipolar 'isthmo-optic' cells projecting to the retina has been described (Hayes and Webster 1981). The n. geniculatus lateralis, p. ventralis may be another target of tractus isthmo-opticus fibres (Galifret et al. 1971). The meaning of the efferent retinal projection is not clear. Pearlman and Hughes (1976) found an inhibition of retinal ganglion cell output when cooling n. isthmo-opticus. These authors assume an indirect effect through the amacrine cells: these cells inhibit ganglion cells but are inhibited by activity of the isthmo-optic cells. There appears to be a relation between the number of isthmo-optic cells and the of feeding behaviour: the nucleus is particularly well developed in birds that search for food on the ground (Feyerabend et al. 1994).

The magnocellular and parvocellular isthmic nuclei (Figs. 21.23, 21.33, 21.45) lie deep in the optic lobe, the magnocellular cell group bordering the other one laterally and rostrally. The parvocellular isthmic nucleus has reciprocal, topographically organised connections with the tectum (Hunt et al. 1977; Güntürkün and Remy 1990). Yan and Wang (1986) demonstrated that the magnocellular part, too, receives visual input instead of auditory, as has been suggested before. Both the parvocellular and magnocellular parts have a retinotopic organisation; receptive fields of the cells are large and oval shaped (Wang and Frost, 1991). The magnocellular part contains GABA-ergic cells (Granda and Crossland 1989). The parvocellular nucleus is richer in GABA-ergic fibres and benzodiazepine receptors, while both nuclei contain moderate numbers of GABA-receptors (Veenman and Reiner 1994; Veenman et al. 1994). In addition to the isthmotectal projections, also isthmocerebellar, isthmo-oculomotor and isthmostriatal projections have been described (Showers and Lyons 1968).

The n. semilunaris (Fig. 21.23) appears as a ventromedial, but separate portion of the parvocellular isthmic nucleus. It also has reciprocal connections with the tectum, but projects to the deep layers (L8–13), parvocellular isthmic nucleus to more superficial layers (L3 and 5; Hunt et al. 1977). A small n. isthmi ventralis is wedged between the rostroventral poles of parvocellular isthmic nucleus and n. semilunaris and sends fibres to the contralateral tectum (Martinez and Puelles 1989).

The nuclei of the lemniscus lateralis receive input from this tract. Three nuclei can be distinguished: a dorsal one (Fig. 21.23) with a dorsal and a ventral part ventromedial to the n. semilunaris, a ventral one – ventrally and more caudally extending to the oliva superior – and an intermediate n. lemnisci lateralis, lateral to the principal sensory nucleus of the trigeminal nerve and medial to the parabrachial complex. The latter is the n. paraprincipalis of Dubbeldam and Karten (1978) and the source of a direct projection to the telencephalic n. basalis (Fig. 21.36a; Arends and Zeigler 1986). The other two nuclei project to the n. mesencephalicus lateralis, pars dorsalis (see below). In addition, the ventral nucleus and ventral part of the dorsal nucleus project directly to the n. ovoidalis and, more in particular, to the n. semilunaris parovoidalis (Fig. 21.36a; Wild 1987c). Finally, the ventral nucleus as well as the oliva superior send recurrent ipsilateral fibres to the cochlear nuclei (Fig. 21.67; Arends and Zeigler 1986; also Arends 1981). Manley et al. (1988) ascribe part of the ventral nucleus a role in binaural intensity processing.

21.8.6
Colliculus Mesencephali and Tectum Mesencephali

In birds the optic lobe is a more or less a lateral structure. Its outer part, the colliculum or tectum mesencephali, is a multilayered structure (Figs. 21.50a,b, 21.51). A large number of studies have been devoted to the structural and functional organisation of the tectum. Recent reviews by Hunt and Brecha (1984) and Jassik-Gerschenfeld and Hardy (1984) summarise many of these studies. The tectal layers have been numbered in different ways (Table 21.2, Fig. 21.41). In this chapter we use the classification of Cajal distinguishing 15 layers [cf. Reiner and Karten 1982; a slightly different numbering is used by O'Flaherty (1970) in the mallard]. A complication is that not all layers are equally pronounced throughout the tectal cortex. Hayes and Webster (1985) distinguish two fields, a dorsorostral field with a wide L3, a wide and trilaminate L4 and a narrow L5, and a ventrocaudal field with a narrow L3 and L4 but a wide L5. These fields are separated by a rather narrow transition zone. The outer layer (L1, stratum opticum in Figs. 21.24, 21.25, 21.32) contains the retinal axons; layers 2–7 are the recipients of these fibres. The retinal projection is entirely contralateral; it is retinotopically organised. Distribution of afferents over the tectum is not uniform, whereas different types of retinal afferents reach different tectal layers (Hunt and Webster 1975). Electrophysiological recordings show the existence of monosynaptic, bisynaptic and polysynaptic responses. Only the latter two are also found in the deeper tectal layers (Leresche et al.

1986). The intrinsic organisation of the tectum is predominantly vertical, i.e. perpendicular to its surface (Fig. 21.51b). Layers 2/3, 5 and 7 contain horizontal dendritic plexuses, whereas cell bodies are concentrated in layers 4 and 6 (Fig. 21.50b). The deeper lying cells have their dendrites and axons oriented perpendicularly to the surface of the tectum (Stone and Freeman 1971). Many of these cells are intrinsic and responsible for the strong radial organisation of the tectum. Particularly radially oriented cells in deeper layers (layer 10 and 13) are the source of efferents projecting to a wide range of cell groups. L10/11 contain large bipolar cells, and layer 13 is characterised by large ganglionic cells with far extending dendrites (O'Flaherty 1970; Hardy et al. 1985). Cells in the superficial layers are sensitive to both moving and stationary stimuli and have small receptive fields (Bilge 1971). Jassik-Gerschenfeld and Guichard (1972) distinguish in the pigeon four classes of units, cells with concentrically organised fields (6%), with 'on' or 'off' fields (1.3%), movement-sensitive cells (72%) – often with a central activating and a surrounding suppressing region – and directionally selective cells (21%). Small receptive field cells are found in the outer 1000 µm; in deeper layers only large field, movement selective cells were found (Jassik-Gerschenfeld and Guichard 1972; Hughes and Pearlman 1974). Cells can also differ, having overlapping or spatially separated light and dark excitatory regions in their receptive fields (Hardy et al. 1982). In addition, Cotter (1976) recorded auditory and somatosensory responses in the tectum of the chicken, whereas a spatiotopic organised auditory projection has been described in the barn owl (Knudsen and Knudsen 1983). Puelles et al. (1987) found in the tectum of 1-week-old chicks an elongate strip devoid of retinal terminals that represents the locus of the optic nerve+pecten.

In addition to the retinal and other sensory inputs the tectum receives afferents from a great variety of other cell groups, e.g. from the hyperstriatum accessorium (L2–4, 6–7, 12; Bagnoli et al. 1980; Miceli et al. 1987) and archistriatum (L13, Zeier and Karten 1971), from the n. ventrolateralis thalami (L11–14) and from the n. spiriformis lateralis (L11–13) and n. semilunaris (L8–13). A few other nuclei project to the retinorecipient layers, viz. parvocellular isthmic nucleus (L3 and 5; Hunt et al. 1977), contralateral pretectal nuclei (L5) and locus coeruleus (L2–4, 7). Still other cell groups send their fibres to superficial as well as to deep tectal layers. Mestres and Delius (1982) also suggest an input from the lateral septum. In the owl *Athene noctua* the Wulst projects to the caudodorsal half of the ipsilateral, and rostroventral part of the contra-

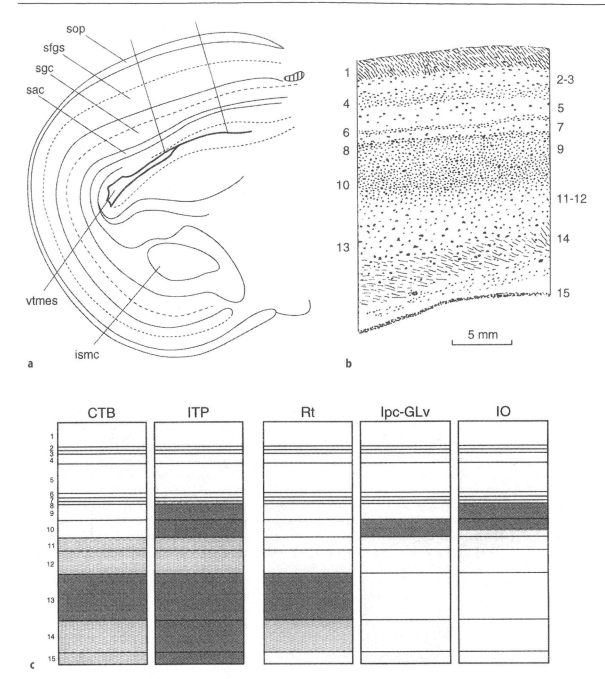

Fig. 21.50. a Cross-section through tectum mesencephali of pigeon showing main layers. **b** Detail of layers numbered according to Cajal, cf. Table 21.2. The two *lines* in **a** indicate the place of this detail. **c** Layers of origin of tectofugal fibre systems: *CTB*, decussating tectobulbar system; *ITP*, ipsilateral tectopontine system; *Rt*, tectorotundal projection; *Ipc-GLv*, projections to isthmic and lateral geniculate nuclei; *IO*, projection to isthmo-optic nucleus. (**c** adapted from Reiner and Karten 1982)

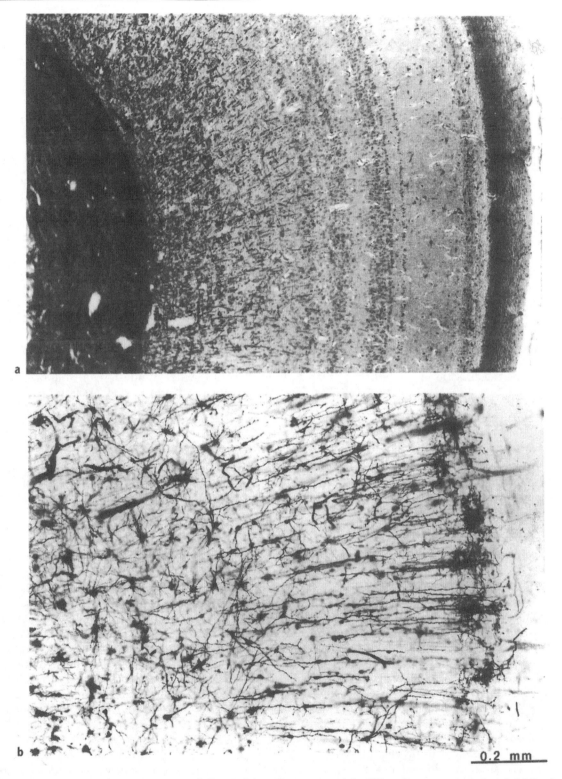

Fig. 21.51a,b. Cross-sections of tectum mesencephali. **a** Jackdaw, *Corvus monedula*, Klüver-Barrera preparation. **b** Mallard, *Anas platyrhynchos*, Golgi preparation

Table 21.2. Nomenclature of tectal layers

Abbreviations	Nomenclature of this chapter	Numbering layers according to Cowan et al. (1961)	Cajal (1909–1911)
sop	Stratum opticum	I	1
sfgs	Stratum griseum et fibrosum superficiale	IIa–j	2–12
sgc	Stratum griseum centrale	III	13
sac	Stratum album centrale	IV	14
–	Stratum griseum ventriculare	V	15

lateral tectum (Casini et al. 1992). In the pigeon the visual Wulst has been shown to influence the activity of tectal units (Bagnoli et al. 1979; Leresche et al. 1983).

A few large fibre systems derive from the tectum. Layer 13 is the main source of the most prominent ascending system, the brachium colliculi (Figs. 21.23–21.25, 21.50c: Rt) with bilateral projections to the nn. subpretectalis and interstitiopretectalissubpretectalis and -continuing as the tectothalamic tract (Figs. 21.42, 21.46) – with bilateral terminations in the n. rotundus; the fibres cross via the decussatio supraoptica, pars dorsalis (Figs. 21.26, 21.27, 21.42, 21.46, 21.48; Voneida and Mello 1975; Hunt and Künzle 1976a). Although the contralateral projection may be modest in the pigeon, in other species it can be quite substantial (e.g. zebra finch, Bischof and Niemann 1990) Even in the pigeon, however, this decussation is important for interhemispheric transfer of visual information (Meier 1971). Ipsilateral projections through the tectothalamic tract also reach the nn. pretectalis, triangularis thalami, subrotundus, dorsolateralis anterior and other thalamic nuclei (Hunt and Künzle 1976a). L10 cells send their axons through the stratum opticum to the nn. lentiformis mesencephali, opticus principalis thalami and ventrolateralis thalami.

Many fibres cross through the intertectal and posterior commissures (Figs. 21.24, 21.25, 21.32). The majority of these fibres belong to the categories of unmyelinated and small myelinated fibres (Ehrlich and Saleh 1982). Many of the crossing fibres terminate in the substantia grisea periventricularis, lateral reticular formation, n. geniculatus, pars ventralis, area pretectalis, n. linearis caudalis and n. posteroventralis (Voneida and Mello 1975). Robert and Cuénod (1969) using electrophysiology found that visual stimuli indeed reach the contralateral tectum via the intertectal and posterior commissures. Mello (1968) demonstrated that monocularly learned visual discrimination persisted after unilateral tectal lesions, whereas Güntürkün and Böhringer (1987) suggested that visual lateralisation may depend upon asymmetrical interaction between the tecta through the commissure. Parsons and Rogers (1993) found a loss of lateralisation of pecking responses in chicken after sectioning the tectal and posterior commissures. Bilateral destruction of the tectal layers cause severe intensity and pattern discrimination defects (Hodos and Karten 1974).

A crossed tectobulbar pathway descends contralaterally close to the midline to terminate in the paramedial reticular formation of the brain stem (Figs. 21.18–21.20; Reiner and Karten 1982; unpublished observations in mallard); its main source is layer 13 with modest contributions from layers 11–12, 14 and 15 (Fig. 21.50c). An ipsilateral tectopontine and -reticular tract follows the lateral periphery of the brain stem to project to the lateral pontine nucleus (Clarke 1977; Dubbeldam, unpublished) and reticular cell groups (Hunt and Künzle 1976a). Most of these fibres arise from layers 9–10, 13, 14 and 15 (Fig. 21.50c); layer 10 sends efferents to the parvocellular isthmic nucleus (Hunt et al. 1977) and n. geniculatus lateralis, p. ventralis (Fig. 21.26), layers 8–10 to the isthmo-optic nucleus (Fig. 21.50c; Reiner and Karten 1982).

Histochemical analysis has contributed greatly to a better understanding of the tectal organisation. Hunt et al. (1976) mention the activity of many neurotransmitters in the tectum: glutamate, aspartate (excitatory effects), acetylcholine (either excitatory or inhibitory effects), GABA, glycine and proline (inhibitory effects). Ehrlich et al. (1987) found substance P activity throughout the tectum with terminals particularly in layers 2–4 and 7 – those in layers 2 and 3 probably deriving from the retina – and in layers 9–13. Layers 10–12 and 13 contain numerous SP-positive cells. It has been suggested that glutamate and acetylcholine characterise different populations of retinal afferents, whereas glycine-reactive fibres may derive from the nucleus parvocellularis isthmi (Hunt et al. 1976). GABA-ergic populations of cells in layer 10 send fibres through the stratum opticum to the pretectum and ventral thalamus, whereas stellate GABA-ergic cells in layer 4 may project to the deeper layers and be instrumental in the columnar organised inhibition of – among others – layer 14 cells projecting to the contralateral tectum (Hunt and Künzle 1976b). Layer 5 contains a chiefly horizontally organised system of GABA-ergic elements. Dietl et al. (1988b) report a high density of GABA-benzodiazepine receptors in the tectum, in particular in layers 2–12. Layers 11–13 receive an enkephalinergic input from the lateral spiriform nucleus. There is a cholinergic tectal input to the n. parvocellularis isthmi.

21.8.7
Nuclei of the Lobus Mesencephali

Several cell groups are present within the optic lobe, ventral to the lateral ventricle. Most obvious is the n. mesencephalicus lateralis, pars dorsalis (Figs. 21.33, 21.45). It is an important relay of the ascending auditory pathway receiving afferents through the lemniscus lateralis (Fig. 21.36; e.g. Harman and Phillips 1967; Boord 1968) and sending ascending fibres bilaterally to the thalamic n. ovoidalis, fibres decussating through the decussatio supraoptica, pars dorsalis (Karten 1967). The medial border contains met-enkephalinergic cells projecting to the shell region of the nucleus ovoidalis (Durand et al. 1992, 1993). Ariëns Kappers et al. (1936) considered the nucleus mesencephalicus lateralis, pars dorsalis homologous to the reptilian torus semicircularis. Both the atlases of Karten and Hodos (1967) and of Kuenzel and Masson (1988) indicate the nucleus mesencephalicus lateralis, pars dorsalis as part of the torus semicircularis (Fig. 21.24). Sometimes it is called the colliculus inferior, suggesting homology with this structure in mammals. The remaining, ventral part of the torus is part of the reticular formation.

Many details about its organisation result from studies in the owl *Tyto alba*; the nucleus mesencephalicus lateralis, pars dorsalis is a very large centre in these night-hunting birds. Knudsen (1982) subdivided the nucleus into a central nucleus, a lateral external nucleus and a dorsal superficial nucleus. The core receives a tonotopically arranged projection from the contralateral n. laminaris, the shell region from the n. angularis (Takahashi and Konishi 1988). Commissural fibres from the core region project to the contralateral shell region (Takahashi et al. 1989). In this centre tonotopic information is translated into spatiotopic information (Knudsen 1982) that is transferred to the tectum mesencephali (Knudsen and Knudsen 1983). This auditory map develops during ontogeny and is attuned to the visual map of the tectum during a sensitive period (Knudsen 1982). Whereas initially the projections from the nn. angularis and laminaris were assumed to be entirely contralateral (e.g. Boord 1968: pigeon), later studies showed the presence of a modest ipsilateral ascending component (Correia et al. 1982; Conlee and Parks 1986). There is also physiological evidence for the existence of units sensitive to binauricular stimuli (Coles and Aitkin 1979).

Medially to the nucleus mesencephalicus lateralis, pars dorsalis and torus lies the n. intercollicularis (Figs. 21.24, 21.33). Part of this region, i.e. the n. dorsomedialis intercollicularis, appears to be of importance for vocalisation, or, more in particular, for respiration (Wild and Arends 1987). It may be the same area called the torus externus by Newman (1970). In oscines (Passeriformes) it is the target of a projection from the n. robustus archistriati (Sect. 21.10.6). Ball et al. (1989) using neurochemistry state that an equivalent of the dorsomedial nucleus can also be identified in non-passeriform birds. In male ring doves the intercollicular nucleus is assumed to be involved in vocal courtship behaviour. Cells appear to concentrate testosterone (Zigmond et al. 1980; Cohen 1981), whereas cell size depends upon the presence of testosterone (Panzica et al. 1991a). In female ring doves nest-cooing has a follicle-stimulating effect (self-stimulation); here, the n. intercollicularis is a relay between the auditory system (with many reciprocal connections with the nucleus mesencephalicus lateralis, pars dorsalis) and the hypothalamus (Akesson et al. 1987). Ball et al. (1989) described a sexual dimorphic distribution of α-adrenergic receptors in the nucleus intercollicularis of the Japanese quail (see also Sect. 21.9.8). The intercollicular nucleus also appears to receive somatosensory input (Wild 1989).

Puelles et al. (1994) dispute the current interpretation of this region pointing to several inconsistencies in the literature, e.g. concerning the precise definition of the dorsomedial nucleus. Using histochemical criteria these authors propose a new scheme for the torus semicircularis in the chicken distinguishing six regions (Fig. 21.52): (a) a toral periventricular lamina corresponding to the stratum griseum et fibrosum periventriculare (Karten 1967) and superficial nucleus of the inferior colliculus (Knudsen and Knudsen 1983); (b) a toral external nucleus corresponding to Knudsen's external nucleus containing a spatial map; (c) a toral central nucleus consisting of a core and a shell region (cf. Takahashi and Konishi, 1988); (d) a caudomedial shell nucleus, distinguished mainly on the "basis of its exclusive plexus of NPY+ fibres and sparse population of NPY+ neurons"; (e) a paracentral nucleus; probably in most other studies it is included in the rostral medioventral part of the

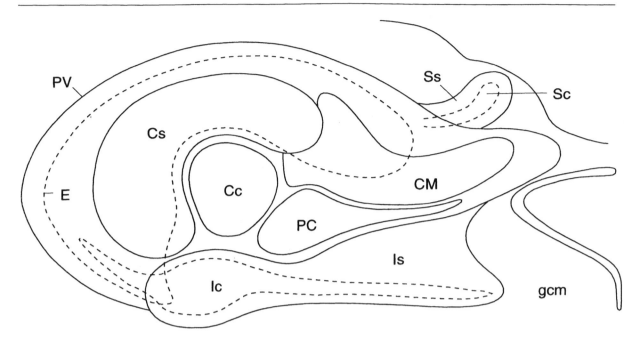

Fig. 21.52. Diagram of cross-section through the torus semicircularis showing the subnuclei as defined by Puelles et al. (1994): central nucleus with a core (*Cc*) and a shell (*Cs*) region; caudomedial shell nucleus (*CM*); paracentral torus nucleus (*PC*); intercollicular area with a core (*Ic*) and a shell (*Is*) region; external toral nucleus (*E*). *PV*, toral periventricular layer; *Sc, Ss*, pre-isthmic area with a core and a shell region. (Redrawn after Puelles et al. 1994)

intercollicular nucleus; and (f) a hilus nucleus, caudal to the central core nucleus where the lateral lemniscus enters the torus. For a more detailed discussion of the the literature concerning these cell groups refer to the study of Puelles et al. (1994).

De Lanerolle et al. (1981) mention the presence of Met-enkephalin-immunoreactive cells in many cell groups, such as the nn. intercollicularis, mesencephalicus lateralis pars dorsalis, spiriformis lateralis (Sect. 21.9.3) and others and ascribe this peptide a possible role in vocalisation. For this suggestion they have indirect evidence; other functions, however, cannot be excluded (de Lanerolle et al. 1981).

The tectal grey (griseum tectale) is a more rostral cell mass bordering the pretectum; its rostral part receives a retinal input (Gamlin and Cohen 1988a).

21.8.8
Mesencephalic Fibre Systems

Many fibre systems traverse the mesencephalon. We have already mentioned the afferents and efferents of the tectum mesencephali and nuclei of the optic lobe. The ansa lenticularis (Figs. 21.24–21.27) derives from the paleostriatal complex and has a terminal field in the n. tegmenti pedunculopontinus or substantia nigra. Other extra-telencephalic fibre systems with projections to mesencephalic cell groups are the occipito-mesencephalic tract (Figs. 21.24, 21.25) and the dorsal branch of the septomesencephalic tract. The latter enters the tectum along the dorsal margin of the n. pretectalis (Verhaart 1971, 1974). The occipito-mesencephalic tract runs dorsal to the n. tegmenti pedunculopontinus and close to the isthmo-optic nucleus; it consists of rather fine fibres. In the mesencephalon the occipito-mesencephalic tract has many projections to various regions, such as the lateral reticular formation, n. intercollicularis and stratum griseum centrale of the tectum (Fig. 21.69; Zeier and Karten 1971).

The avian fasciculus longitudinalis medialis (Figs. 21.17–21.24) is a large fibre system reaching from the spinal cord as continuation of the ventral funiculus into the mesencephalon. It is composed of a mixture of small to large fibres; as for its trajectory and composition it is comparable to the fasciculus longitudinalis medialis in other classes of vertebrates. The fasciculus longitudinalis medialis contains ipsilateral ascending vestibular fibres – some of these projecting to the oculomotor nuclei – and descending fibres from vestibular nuclei and from the nucleus of the fasciculus longitudinalis medialis (Darkewitsch) with projections to the hypoglossal motor nuclei and ventral horn of the spinal cord.

21.9
Diencephalon

21.9.1
Introductory Remarks

The avian diencephalon is well differentiated; sensory centres occupy a large part of the diencephalon and can be clearly delineated. There is no sharp border with the mesencephalon. In the section on morphogenesis it has already been mentioned that according to the interpretation of Figdor and Stern (1993) four diencephalic neuromeres might be distinguished during early ontogeny, the line between the anterior end of the tectal commissure (Fig. 21.24) and the exit point of N. III being the border between mes- and diencephalon. According to this interpretation four diencephalic subregions can be recognised: a posterior pretectum extending between the rostral and caudal border of the posterior commissure (Fig. 21.25) and tract, an anterior pretectum between the rostral border of the posterior commissure and habenular commissure and habenulo-interpeduncular tract; also the dorsal thalamus+habenula and stria medullaris and the ventral thalamus+hypothalamus. This interpretation is disputed by Martinez et al. (1993), who studying the proliferation of the presumptive boundary cells found not more than four boundary zones dividing the diencephalon in not more than three subdivisions (see also Sect. 21.4.1). The latter interpretation supports that of Rendahl (1924); his synencephalon comprises the last two neuromeres of Figdor and Stern (1993). In the following discussion we will use this tripartition of Rendahl and of Martinez et al. for the description of the diencephalic centres. More recently the cyto- and myeloarchitecture of the diencephalon have been described in a series of papers by Völker (1971a,b; 1972a,b). Before these areas are discussed a few remarks are first necessary about the visual system, as this is an important component of the diencephalon.

21.9.2
Nervus Opticus and Chiasma Opticum

According to most authors the nervus opticus crosses through the optic chiasm completely (Figs. 21.26–21.28, 21.46) and projects onto many centres in the mesencephalon and diencephalon. However, Bons (1969, 1976) using the degeneration technique found small numbers of ipsilateral projections in the mallard, whereas Repérant (1973) does not exclude the possibility of a very modest ipsilateral component in the pigeon. The terminal fields of the retinal afferents have been described in many species such as the pigeon (Cowan et al. 1961; Repérant 1973), white-crowned sparrow (Hartwig 1970), owl and several other species (Hirschberger 1967, 1971) and quail (Norgren and Silver 1989). There is no uniform nomenclature for the various terminal fields; for a complete survey of the visual centres and an overview of the various nomenclatures we refer to Ehrlich and Mark (1984). In a study using autoradiography Streit et al. (1980) also describe the transneuronal projections of the visual system. Here the most important primary terminal fields are summarised.

The projections to the tectum mesencephali and tectal grey have already been mentioned. Important diencephalic targets are the pretectal optic area, n. lateralis anterior and anterior dorsolateral complex, n. geniculatus lateralis, p. ventralis, cell groups of the accessory optic system and the hypothalamus. Details on the various cell groups will be given in the following discussion. An overview of the ascending visual system is given in Fig. 21.53.

21.9.3
Synencephalon or Pretectal Region

Several clusters of cell groups occupy the pretectal region. A first group lines the border between the tectum mesencephali and the synencephalon. Dorsally lies the pretectal nucleus, a spherical cell group of multipolar cells surrounded by a layer of neuropil (Fig. 21.25). The n. subpretectalis lies ventrally and is connected with the nucleus pretectalis by the tractus pretectosubpretectalis (Fig. 21.25). The cells of the n. interstitio-pretectosubpretectalis are dispersed between the fibres of this tract. The area pretectalis – between the nucleus pretectalis and isthmooptic tract – and n. pretectalis diffusus receive retinal input (Gamlin and Cohen 1988a); efferents of the pretectal area reach the tectum, the accessory oculomotor nucleus, interstitial nucleus of Cajal and lateral pons (Gamlin and Cohen 1988b).

The *spiriform complex*, medial to nucleus pretectalis consists of several nuclei that may correspond to part of the ventral complex of nuclei of the mammalian thalamus (Karten and Dubbeldam 1973). The n. spiriformis lateralis (Fig. 21.25) receives afferents mainly from the paleostriatum primitivum, n. ansae lenticularis anterior and n. tegmenti pedunculopontinus (substantia nigra) and has an enkephalinergic projection to the deeper layers of the tectum, in particular layers 11–13 (Reiner et al. 1982a,b). It is considered an important relay in the pathway by which the basal ganglion influences motor functions. The nucleus spiriformis medialis

(Fig. 21.25) is a relay receiving telencephalic afferents from the archistriatum and with projections to the cerebellar cortex (Karten and Finger 1976). The n. spiriformis lateralis contains many catecholaminergic fibres and terminals, the n. spiriformis medialis only a few (Fuxe and Ljunggren 1965).

Medial to the spiriform nuclei lies the n. principalis precommissuralis (Fig. 21.25). It has an input from the occipitomesencephalic tract; both n. spiriformis medialis and n. principalis precommissuralis also receive afferents from the central cerebellar nuclei (Arends and Zeigler 1991b). The n. dorsointermedius posterior [caudalis] is one of the recipients of an input from the ansa lenticularis (Karten and Dubbeldam 1973); it also receives archistriatal and cerebellar afferents. A dorsal zone of it may project to the paleostriatal complex, but its main target is the intermediate neostriatum (Kitt and Brauth 1982; Wild 1987b). Kitt and Brauth suggest that this pathway may be comparable to the ventral thalamo-motor/premotor pathway in mammals; however, there is no evidence for a direct projection to motor areas in the avian forebrain. The n. dorsolateralis posterior, although considered part of the

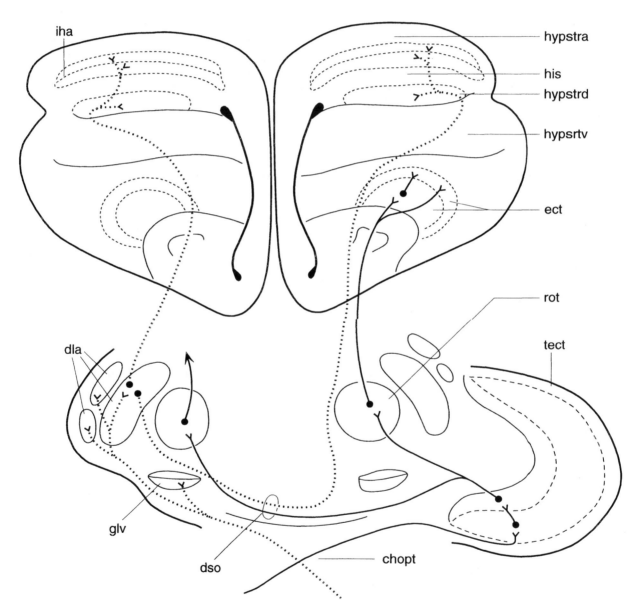

Fig. 21.53. Scheme of ascending visual pathways based on work in the owl *Speotyto cunicularia* (Karten et al. 1973). *Right half (solid line)*: retinal projection+tecto-ectostriatal pathway; note projection to contralateral rot. *Left half (broken line)*: retinal projection+thalamo-Wulst pathway. (Adapted from Karten 1979)

spiriform complex, will be discussed in the section on the dorsal thalamus (under the somatosensory system).

Nuclei optici accessorii: The n. lentiformis mesencephali (Fig. 21.26) comprises a parvocellular and a magnocellular part and receives a retinotopic projection (Bodnarenko et al. 1988). It is also called the n. superficialis synencephali (Rendahl 1924; Völker 1971a) or n. externus (Ehrlich and Mark 1984). The n. lentiformis mesencephali merges ventrally with the n. ectomammillaris (Figs. 21.24, 21.25) or nucleus of the basal optic root. Together the two cell groups form the accessory optic system. The n. lentiformis mesencephali is responsible for the horizontal optokinetic nystagmus (OKN); its units are mainly sensitive to horizontal movements and are responsive to a wide range of velocities (Gioanni et al. 1983a, Winterson and Brauth 1985). The n. ectomammillaris recipient of afferents from 'displaced' retinal cells (Karten et al. 1977; Reiner et al. 1979); it is rich in neurotransmitters and neuropeptides, several of which may derive from the retina (Britto et al. 1989). This centre is highly sensitive to moving targets (Britto et al. 1981) and is involved with the vertical OKN (McKenna and Wallman 1985), though a horizontal component cannot be excluded (Gioanni et al. 1983b). Wylie and Frost (1990a) discern four subdivisions containing cells sensitive to upward, downward, forward and backward movements, respectively. Part of the cells are responsive to binocular stimulation (Wylie and Frost 1990b). GABA-ergic innervation appears to be important in the control of horizontal and vertical OKN (Bonaventure et al. 1992).

The two nuclei have reciprocal connections; their efferents reach the vestibulocerebellum, i.e. folia IXc,d and the paraflocculus (Fig. 21.68). The n. ectomammillaris has further projections to the oculomotor nuclei, to the n. interstitialis of Cajal (relay to neck motor neurons), to the inferior olivary complex and to the contralateral n. ectomammillaris (Brecha et al. 1980); the lateral pons receives input from the n. lentiformis mesencephal (McKenna and Wallman 1985). On the other hand the visual Wulst (Sect. 21.10.7.1) has projections to the n. ectomammillaris and modulates the directional selectivity of its neurons (Britto et al. 1990).

21.9.4
Epithalamus

The epithalamus consists of a medial and a lateral habenular nucleus (Fig. 21.26). The cells of the latter are slightly smaller than those of the medial habenular nucleus; cells in the medial habenular nucleus are often round with evenly distributed gross Nissl bodies (Völker 1971b). The habenular region receives input through the stria medullaris (Fig. 21.27) from septal, preoptic and archistriatal regions; its output is through the tractus habenulo-interpeduncularis ('fasciculus retroflexus'). Just as in mammals this tract is cholinergic (Sorenson et al. 1989). Semm and Demaine (1984) report excitatory and inhibitory responses in the habenular nuclei after photic as well as after pineal stimulation. Sometimes, the medial and lateral subhabenular nuclei are also considered part of the epithalamus. These cell groups do not seem to have a mammalian equivalent (Breazile and Hartwig, 1989).

Birds possess a glandula pinealis (Fig. 21.6b) generally consisting of a corpus and a pedunculus, but its form is highly variable among species (Hodges 1981). In some species a persistent connection exists between the lumen in the stalk and the recessus pinealis of the third ventricle (e.g. in *Larus* and *Phalacrocorax*); other species possess a pineal organ with a solid stalk. It receives catecholaminergic fibres from the superior ganglion cervicale (e.g. Sato and Wake 1983). Zimmermann and Menaker (1975) concluded that the pineal body is not neurally coupled to other parts of the circadian system, but may primarily have a secretory function, possibly related to the control of reproductive functions and/or regulation of the circadian rhythm. In some species, however, it may also have a photoreceptive function (Menaker and Oksche 1974). Korf and Vigh-Teichmann (1984) describe in pigeon, mallard and fowl pinealocytes that may be rudimentary photoreceptive elements. The pineal organ of these, but also of other species, contains neurons (Korf et al. 1982; Korf and Vigh-Teichmann 1984); in some species (not in the pigeon) these cells can be acetylcholinesterase positive (Ueck 1982). There is evidence of pineofugal nerve fibres (tractus pinealis) projecting to the habenular nuclei and possibly to the periventricular layer of the hypothalamus (Korf et al. 1982; Sato and Wake 1983). Secretory cells release melatonin and are involved in the maintenance of the circadian rhythm (e.g. Julliard et al. 1977; Ueck and Umar 1983; Hasegawa and Ebihara 1992). In the Japanese quail the periodic production of melatonin is regulated by light via the eyes as well as directly via the pineal organ (Barrett and Underwood 1991).

21.9.5
Dorsal Thalamus

The dorsal thalamus comprises a group of neural centres many of which form relays for the ascending sensory systems; more or less all sensory modalities can be found here.

21.9.5.1
Visual Centres

Two groups of elements can be recognised, one receiving direct retinal input, the other one comprising relays of the ascending tectofugal pathway. The n. dorsolateralis anterior (rostralis) thalami (Fig. 21.27) is the most important thalamic recipient of retinal input. It comprises a pars magnocellularis, a p. lateralis and a p. medialis. Sometimes this complex is called the nucleus opticus principalis thalami (Karten et al. 1973); it may be an equivalent of the dorsal lateral geniculate body. Its retinal input lacks a clear retinotopic organisation (Remy and Güntürkün 1991) and the cells have small receptive fields with a directional sensitivity. This centre may have a role in pattern analysis (De Britto et al. 1975), but not in pattern or intensity discrimination (Hodos et al. 1973). The principal optic nucleus is the source of the second important ascending visual pathway (Fig. 21.53) with bilateral projections – fibres crossing through the ventral part of the decussatio supraoptica – to the visual Wulst (Hunt and Webster 1972; Karten et al. 1973; Mihailovic et al. 1974; Bagnoli et al. 1990). A ventral cell group in the lateral part of the n. dorsolateralis anterior appears to be the source of a GABA-ergic pathway to the ipsilateral Wulst, whereas the n. dorsolateralis anterior magnocellularis may have a cholinergic projection to the same area (Bagnoli et al. 1983).

The n. rotundus (Figs. 21.26, 21.27) is a well-delineated cell group consisting of large cells. It is a relay in the other large ascending visual pathway, viz. that from the optic tectum to the ectostriatum (Fig. 21.44; Karten and Revzin 1966; Revzin and Karten 1966). A modest number of tectal fibres decussate through supraoptic decussation to reach the contralateral n. rotundus (Ngo et al. 1994); this input may consist of collaterals of the ipsilateral tectorotundal fibres. Much work has done on the functional characteristics of this cell group. Bilateral lesions of n. rotundus caused severe impairment of visual discrimination performance (Hodos and Karten 1966). The tectorotundal pathway is also important for interhemispheric transfer of visual information (Watanabe et al. 1986). Single unit recordings during stimulation with diffuse monochromatic light showed that units sensitive for different wavelengths are not evenly distributed throughout this centre; generally, excitatory responses were found in the dorsal half, inhibitory responses in the ventral half (Granda and Yazulla 1971). This ventral half receives afferents from the nn. subpretectalis and interstitiopretectalis-subpretectalis (Benowitz and Karten 1976b). These authors further showed that different parts of n. rotundus receive input from different parts of the tectum mesencephali, the superficial tectal layers projecting mainly to the rostral n. rotundus, deeper layers to medial and caudal parts. It is interesting to note that Martinez-de-la-Torre et al. (1990) report that the rostromedial part of the n. rotundus in the chicken is AChE-negative, the ventrolateral part AChE-positive. GABA-ergic innervation, however, was found throughout the n. rotundus, but fibre-profiles are suggestive of a different origin, e.g. ventrally from the n. subpretectalis and dorsally from the n. postero-ventralis thalami (Ngo et al. 1992).

A second ascending tectofugal pathway reaches the n. dorsolateralis posterior (Gamlin and Cohen 1986); this nucleus will be discussed later. The n. subrotundus (Fig. 21.26) lies lateroventral to the n. ovoidalis and medial to the n. rotundus (Völker 1972b) and possibly has connections with these nuclei (Ariëns Kappers et al. 1936).

21.9.5.2
Auditory Centres

The n. ovoidalis (Fig. 21.26) is the most important auditory relay in the thalamus; it receives bilateral input from the mesencephalic auditory centres (Karten 1967) and projects to the neostriatal field L (Fig. 21.36). It consists of a uniform population of cells, but has a tonotopic organisation (Bigalke-Kunz et al. 1987). This nucleus has been compared to the mammalian medial geniculate nucleus; the presence of many perikarya positive for calcitonin-gene-related peptide supports this notion (Brauth and Reiner 1991). Ströhmann et al. (1994), however, found that different from the situation in mammals the ovoidalis cells do not show the ability to switch from a tonic mode of firing to Ca^{2+}-bursting, different states that may have to do with mechanisms of sleep and wakefulness. A second cell group, the n. semilunaris parovoidalis (Fig. 21.36), receives an ipsilateral input from the n. mesencephalicus lateralis, pars dorsalis (Karten 1967) as well as from the nuclei of the lateral lemniscus. The surroundings of these two nuclei form an additional subdivision of the auditory thalamus called the ovoidal shell by Durand et al. (1992). It receives afferents from a cell group situated between the mesencephalic auditory centre (n. mesencephalicus lateralis, pars dorsalis) and the n. intercollicularis (Sect. 8.7); it has projections to the hypothalamus, ventral paleostriatum and caudal neostriatum and hyperstriatum ventrale Durand et al., 1992, 1993). There appears also to be an auditory projection to the n. dorsolateralis posterior (see below).

21.9.5.3
Somatosensory Centres

The ascending somatosensory/trigeminal system of the head, i.e. the quintofrontal tract, bypasses the thalamus. That of the body, however, has several thalamic relays. Delius and Bennetto (1972) were the first to record somatosensory responses from the n. dorsolateralis posterior (caudalis) (Fig. 21.26); this is a distinct cell group with rather large cells (15–25 µm, Schneider 1991). There is anatomical evidence of additional somatosensory input (Wild 1987a), as well as of visual input (e.g. Gamlin and Cohen 1986). Furthermore, Korzeniewska and Güntürkün (1990) describe afferents from the nucleus of the descending trigeminal tract (sparse), from vestibular nuclei (also Wild 1988) and parts of the brain stem reticular formation, from several dorsal thalamic nuclei and from the telencephalic Wulst (Sect. 21.10.7.1). Korzeniewska (1987) recorded somatosensory, visual as well as auditory responses from this cell group, about 30% of the neurons showing polysensory properties. In this respect the n. dorsolateralis posterior thalami is comparable to the posterior thalamic complex in mammals (Korzeniewska and Güntürkün, 1990). The n. dorsolateralis posterior thalami projects to the paleostriatal complex – but sparsely to the ventral paleostriatum augmentatum and medial parolfactory lobe – as well as to the neostriatum (Wild 1987a,b). In the canary and other oscines a n. uvaeformis has been described, a centre projecting to the n. interfacialis and 'high vocalisation centre' of the telencephalon (Nottebohm et al. 1982; see also Sect. 21.10.4). In view of its position this centre could be a specialised part of the n. dorsolateralis posterior thalami.

The n. dorsalis intermedius ventralis anterior, dorsomedial to the n. rotundus and n. triangularis, is another recipient of somatosensory input (Wild 1987a; Korzenievska and Güntürkün 1990). Its cells are medium sized (10–20 µm, Schneider 1991); its efferents project to the hyperstriatum accessorium (Funke 1989; Wild 1987a). A third somatosensory centre could be the n. superficialis parvocellularis thalami (Fig. 21.27; synonym n. tractus septomesencephali) with bilateral projections to the accessory hyperstriatum (Karten and Revzin 1966). It has been suggested that the n. superficialis parvocellularis thalami receives a direct spinal projection (Karten and Revzin 1966). There is anatomical and physiological evidence that a substantial direct thalamic projection derives from the lumbal intumescence (legs), but not from the brachial intumescence (wings) (Schneider and Necker 1989). The largest portion of somatosensory (93%) cells was found to be driven by stimulation of the skin or feathers, a small number (4%–5%) showing proprioceptive responses (Schneider 1991).

The medial edge of the anterior dorsomedial nucleus (Fig. 21.27) receives solitarius input (Arends et al. 1988); the anterior and posterior nuclei also receive afferents from the parabrachial nuclei (Wild et al. 1990). The cell groups project in particular to the medial part of the parolfactory lobe and ventral paleostriatum (Wild 1987b) as well as to the medial zone of the dorsal and accessory hyperstriatum (Karten et al. 1973), i.e. the equivalent of the limbic cortex.

21.9.6
Ventral Thalamus

The ventral geniculate nucleus (Figs. 21.26, 21.27, 21.46) is a layered cell group consisting of an outer parvocellular layer, a central neuropil and an internal magnocellular layer. It receives a direct topographically organised retinal input in its outer layer (e.g. Crossland and Uchwat 1979; Ehrlich and Mark 1984) – at least partly collaterals of fibres projecting to the tectum (Britto et al. 1989) – and an input from the hyperstriatum accessorium (Ehrlich et al. 1989). In addition it has reciprocal connections with the tectum mesencephali (Guiloff et al. 1987) and may be important for colour vision (Maturana and Varela 1982). Gioanni et al. (1991) ascribe this centre a role in the optokinetic reflex, but have no solid evidence about a possible pathway for this function. The n. lateralis anterior [rostralis] (Figs. 21.33, 21.42) is another recipient of retinal afferents, but so far there is no solid evidence about the target of its efferents (Meier et al. 1974).

The entopeduncular nuclear group consists of small to medium-sized cells scattered along the fasciculus prosencephali lateralis (Völker 1972b); it may correspond to the mammalian subthalamic nucleus (Breazile and Kuenzel 1993). The n. intercalatus thalami, lateral to the n. entopeduncularis, receives direct input from the spinal cord and has projections to the n. dorsolateralis posterior (Schneider 1991). Völker (1972b) describes a n. decussationis supraopticae ventralis and a n. posterior decussationis supraopticae ventralis; these may correspond to the nn. ansae lenticularis anterior and posterior as defined by Karten and Dubbeldam (1973). Notwithstanding its position there is no good reason to compare it to the mammalian subthalamic region.

The decussatio supraoptica (Figs. 21.26, 21.27, 21.42, 21.46) has different groups of fibres passing through the dorsal and a ventral part. Saleh and Ehrlich (1984) also mention a subventral part in the chicken; the three parts differ in fibre composition, the dorsal decussatio supraoptica and subventral decussatio supraoptica containing about 24% and 20% myelinated fibres, respectively, whereas this is about 0.5% for the ventral decussatio supraoptica.

21.9.7
Thalamofrontal Fibre Systems

The fasciculus lateralis telencephali (fasciculus prosencephali lateralis) (Fig. 21.28) is a large fibre system entering the telencephalon from the thalamus. It lies lateral to the occipitomesencephalic tract and contains both ascending fibres (e.g. quintofrontal tract, Figs. 21.24–21.27, 21.42–21.44, 21.48, 21.49) and descending components (e.g. ansa lenticularis, Figs. 21.24–21.27). The tractus thalamofrontalis (Figs. 21.27, 21.46, 21.48) is another important component. This is a fibre system containing medium-size fibres that have their origin in various thalamic sensory relay nuclei and ascend into the telencephalon. In the older literature a lateral, intermediate and medial thalamofrontal tract have been distinguished (discussion in Verhaart 1971).

21.9.8
Hypothalamus

Kuenzel and Van Tienhoven (1982) describing the hypothalamus of the chicken provide a survey of the various nomenclatures used for the hypothalamic cell groups and propose a partly new nomenclature that is for the most part used in the NAA (Breazile and Kuenzel 1993). A slightly different nomenclature is used by Yamauchi and Yasuda (1985, 1988), who also provide data about the cytology and fibres of this region. A third study in the chicken not cited in Kuenzel and van Tienhoven is the cyto- and myeloarchitectonic analysis by Völker (1971b). The medial hypothalamus of the goose is described by Rehák et al. (1985). Following the NAA three main regions with 19 nuclei and two areas can be distinguished. Many cell groups are depicted in Fig. 21.54; see also Figs. 21.26–21.29.

21.9.8.1
Regio Caudalis (Mammillaris) Hypothalami

The most caudal area comprises the mammillary and supramammillary cell groups consisting of medium-sized to rather large cells (Fig. 21.54a). Berk and Hawkin (1985) demonstrate ascending projections reaching to the area parahippocampalis, to the areas corticoidea dorsolateralis and temporoparieto-occipitalis and to the cortex piriformis, to the caudal striatum, to the ventral paleostriatum and to basal parts of the olfactory lobe. These projections have their origin in the lateral mammillary nucleus and other posterolateral cells around this nucleus; it may correspond to the mammalian supramammillary nucleus. The tractus infundibuli (Figs. 21.25, 21.54a) decussating via the commissura supramammillaris is part of this fibre system (Benowitz and Karten 1976a; cf. also Verhaart 1971). The medial mammillary nucleus is a gonadotrophin-secreting centre; Ball et al. (1989) found a sexual dimorphy expressed by the density of α-adrenergic receptors in this centre in the Japanese quail. A comparable sexual dimorphy was found in the n. intercollicularis, but not in other centres containing α-adrenergic receptors, such as the infundibulum, preoptic area and supra-optica area (see below).

21.9.8.2
Regio Medialis (Tuberalis)

The nucleus tuberis of Karten and Hodos (1967) (Figs. 21.25, 21.54a,b) corresponds to the n. infundibuli of Kuenzel and van Tienhoven. The organisation of the nucleus tuberis is highly variable among avian species (e.g. Oehmke 1971). The stratum cellulare internum [n. periventricularis hypothalami] (Figs. 21.25, 21.26) receives an input from the solitary complex (Arends et al. 1988); it is also the source of a descending pathway with projections to some subnuclei of the solitary complex and part of the vagal motor nucleus (Berk 1987). Other cell groups belonging to this region are the nn. ventromedialis (Fig. 21.54c) and dorsomedialis hypothalami (n. medialis posterior, Figs. 21.27, 21.54b). The ventromedial (posterior) nucleus receives afferents from the medial septum and hippocampus (Bouillé et al. 1977).

Another cell group receiving solitary input is the n. paraventricularis (Fig. 21.54c,d) or n. periventricularis magnocellularis (Arends et al. 1988). Additional afferents derive from the contralateral paraventricular nucleus, area lateralis hypothalami (Figs. 21.27, 21.28, 21.54b–d) and n. coeruleus, as well as from the lateral septum (Korf et al. 1982). The paraventricular nucleus consist of large and small cells. The latter may represent (at least partly) CSF-contacting cells, whereas the large cells send vasotocin-containing axons to the puititary lobe (Korf 1984; cf. also Panzica and Viglietti-Panzica 1983), making this cell group part of the neuroendocrine system (Fig. 21.55). In the chaffinch (*Fringilla coelebs*) Zigmond et al. (1980) found testosterone-positive cells here.

Other cell groups in this region are the n. medialis anterior [rostralis] (Figs. 21.24, 21.54d) and the

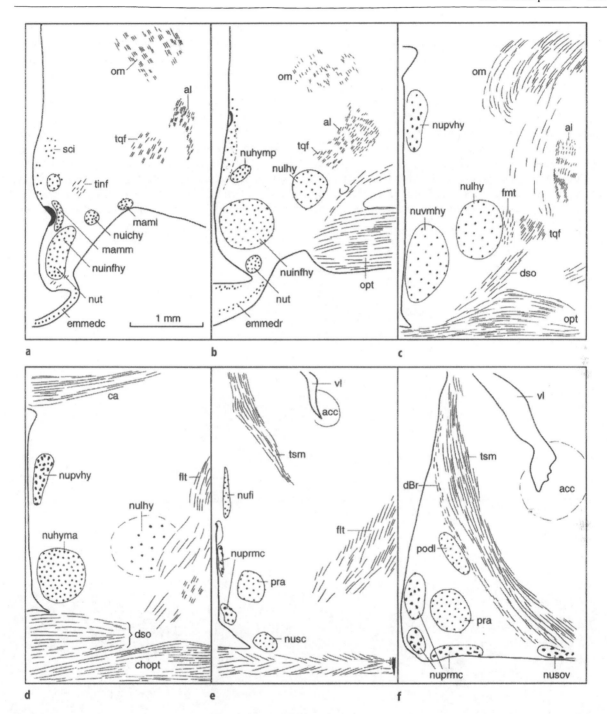

Fig. 21.54a–f. Cell groups in the hypothalamus and preoptic area of the chicken. **a** level A 5.5 (atlas of Van Tienhoven and Juhasz 1962); **b** A 6; **c** A 7.5; **d** A 8.5; **e** A 9; **f** A 9.5. (Adapted from Kuenzel and Van Tienhoven 1982)

n. suprachiasmaticus. The nuclei medialis hypothalami anterioris and medialis hypothalami posterioris may have a role in the control of plasma corticosteroids (Bouillé and Baylé 1973). In this context it is interesting to note that the 'amygdaloid' part of the archistriatum projects particularly to the medial hypothalamus (Zeier and Karten 1971), whereas occipitomesencephalic tract interruptions raise plasma corticosteroid levels (Martin et al. 1979). The nuclei medialis hypothalami anterioris and medialis hypothalami posterioris have intrinsic hypothalamic connections, but also projections to, e.g. the medial septum and, in particular, the n. intercollicularis (Berk and Butler 1981). There are two cell groups receiving visual input, a medial and a lateral one (e.g. Norgren and Silver 1989). In most studies the latter is assumed to represent the n. suprachiasmaticus (Fig. 21.54e; e.g. Blümcke 1961: chicken; Meier 1973: jackdaw; Bons 1974, 1976: mallard; Cooper et al. 1983: dove; Cassone and Moore 1987: sparrow). Takahashi and Menaker (1982) showed in the house sparrow that this nucleus is crucial for the generation of the circadian rhythmicity. Norgren and Silver (1990), however, found that the cells in the n. suprachiasmaticus of the ring dove differ histochemically from those in

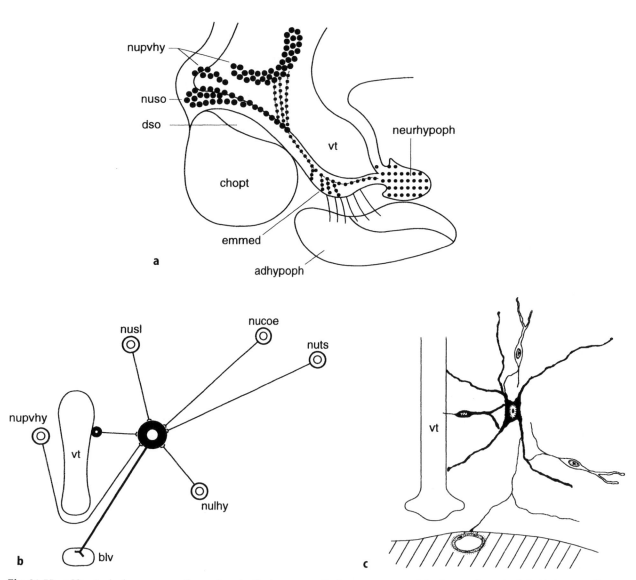

Fig. 21.55. a Vacotocin-immunoreactive system in the hypothalamus and hypophysis of the mallard (redrawn after Bons et al. 1978). b Nucleus paraventricularis (*black cells*) with some sources of afferent connections. Small paraventricular cells have contacts with the cerebrospinal fluid, large cells send axons to pituitary capillaries. c Drawing of large and small cell types in the paraventricular nucleus. (b,c redrawn after Korff 1984)

mammals, viz. by the lack of vasoactive intestinal polypeptide (VIP) and neurophysin (NP)-immunoreactive cells. Wallman et al. (1994) found that the nuclei do not express deoxyglucose labelling due to diurnal rhythms, but that the lateral nucleus does respond to 'novel motion'. These authors argue that the diurnal rhythm in birds may be gated by retinal elements; in their opinion this does not exclude the possibility of comparability of the lateral cell group to the mammalian suprachiasmatic nucleus; therefore we maintain this name for the lateral cell group. Cassone and Moore (1987) report glutamic acid decarboxylase (GAD)-positive and bombesin-positive cells as well as 5-HT-, substance P (SP)- and neuropeptide Y (NPY)-positive axons in this nucleus in the sparrow.

Vasotocin-neurophysin cells in the pigeon have been found in the periventricular hypothalamus, viz. n. paraventricularis, and lateral preoptic area (Bons 1980: mallard, quail; Berk et al. 1982: pigeon; Kiss et al. 1987: canary), but Panzica (1985) also reports modest numbers of vasotocin-immunoreactive cells in a small portion of the suprachiasmatic nucleus of chicken and quail; this is the medial cell group receiving retinal input. It lies more rostrally and medially at the border of the third ventricle. Cell nuclei containing progesterone receptors were found in all parts of the hypothalamus, but most abundantly in the preoptic area (Sterling et al. 1987).

21.9.8.3
Regio Preoptica

Several preoptic nuclei have been distinguished. The nn. preopticus magnocellularis (Fig. 21.54e,f; synonym n. paraventricularis preopticus) and filiformis (Fig. 21.54e) contain luteinising hormone releasing hormone (LHRH)-positive cells (Bons et al. 1978; Mikami et al. 1988). The n. preopticus medialis (preopticus anterior, Figs. 21.29, 21.54f) is sexually dimorphic and testosterone sensitive (Panzica et al. 1991b). It has reciprocal connections with the septal nuclei, nucleus tuberis, area ventralis of Tsai and substantia grisea centralis; the last centre may be part of a pathway controlling male copulatory behaviour (Balthazart et al. 1994). Péczely and Antoni (1984) found CRF-like reactivity in the nn. supraopticus and preopticus magnocellularis and neurophysin-like reactivity in these cell groups as well as in the nn preopticus medialis and supraopticus lateralis pars externus (lateralis). Blähser et al. (1978) report the presence of somatostatin-containing cells in the median and optic tract parts of the supraoptic nucleus. Berk and Butler (1981) described in the pigeon among others projections from the medial preoptic nucleus to the medial septum, the medial premammillary area, and the dorsomedial thalamus. Takei et al. (1979) and Thornton (1986) concluded that the preoptic area is involved in the regulation of drinking without specifying which nuclei might be involved; cells in this area send efferents to the subfornical organ (Takei et al. 1979). The supraoptic nucleus is the second area with neurosecretory cells with axons to the hypophysis (Oksche et al. 1964).

In addition to the preoptic and hypothalamic centres, Yamauchi and Yasuda (1981) described several small cell groups in the chicken surrounding the commissura anterior; the functional characteristics of these cell groups are unknown.

21.10
Telencephalon

21.10.1
Introductory Remarks

The avian telencephalon is well differentiated. The hemispheres are built up of thick layers of cells stacked one upon the other. The lateral ventricles are thin slits that extend far dorsolaterally and ventrolaterally in the caudal half of the telencephalon (Fig. 21.26). Dorsally and laterally it separates the mediodorsal, dorsal and the dorsolateral corticoid areas and the piriform cortex from the large bulk of the forebrain (Fig. 21.27). Medioventrally the septal area with distinct nn. septales medialis and lateralis (Figs. 21.28, 21.29) can easily be recognised. Rostrally, the telencephalon bears small, but distinct olfactory bulbs.

The comparison of the avian, reptilian and mammalian forebrains has led to divergent interpretations of the homology of the respective telencephalic components and to the introduction of different nomenclatures. The nomenclature of the NAA as used in this chapter corresponds largely with that of Karten and Hodos (1967) and was used in the book by Ariëns Kappers et al. (1936). The introduction of this nomenclature reflected the belief that the brain of birds possesses not more than a rudimentary neocortex and consists mainly of a hypertrophied corpus striatum (Edinger et al. 1903; Ariëns Kappers et al. 1936). Kuhlenbeck (1973) introduced an entirely different nomenclature. He distinguished a 'secondary pallium' comprising four areas corresponding to the area corticoidea lateralis, Wulst, hippocampus and area hippocampalis; all other areas are considered basal (nucleus basalis) and epibasal grisea (the neostriatum, hyperstriatum ventrale and archistriatum). This has been discussed more extensively in

Table 21.3. Different nomenclatures of telencephalic structures

NAA, Breazile and Kuenzel (1993)	Ariëns Kappers et al. (1936)[a]	Karten and Hodos (1967)	Kuhlenbeck (1977)[b]	Rose (1914)	Edinger et al. (1903)
Eminentia sagittalis					
Hyperstriatum accessorium	id	id	N. diffusus dorsalis	B	Cortex frontalis
Hyperstriatum intercalatum supremum	N. intercalatus hyperstriati[c]	H. intercalatus superior	N. diffusus dorsolateralis	A	Frontalmark
Hyperstriatum dorsale	id	id	N. epibasalis dorsalis pars superior	C	
DVR					
Hyperstriatum ventrale	id	id	N. epibasalis dorsalis pars inferior	D	Hyperstriatum (upper part)
Neostriatum[d] pars rostralis	Neostriatum frontale	id	N. epibasalis centralis pars medialis	G₁	Hyperstriatum (lower part)
pars intermedia	Neostriatum intermediale	Neostriatum intermedium	N. epibasalis centralis pars posterior	G	
pars caudalis	Neostriatum caudale	id	N. epibasalis centralis accessorium	G₂	
Ectostriatum	id	id		S	Ektostriatum
Nucleus basalis	id	id	N. epibasalis ventrolateralis	R	Mesostriatum laterale
Archistriatum	id	id	N. epibasalis caudalis	K	Epistriatum
Complexus paleostriaticus					
Paleostriatum augmentatum	id	id	N. basalis	H	Mesostriatum
Paleostriatum primitivum	id	id	N. entopeduncularis	J	N. entopeduncularis
Paleostriatum ventrale	–	id	–	–	–
Lobus parolfactorius	–	–			N. parolfactorius
Nucleus intrapeduncularis	–	?			

[a] Also Huber and Crosby (1929).
[b] Also Kuhlenbeck (1938), Jones and Levi-Montalcini (1958).
[c] Ariëns Kappers et al. (1936) distinguish a n. intercalatus laminae supremae and a n. intercalatus laminaris superioris; these cell groups fuse caudally and compose together the n. intercalatus hyperstriati.
[d] Field L (Rose) is part of the neostriatum caudale.

Sect. 21.4.2. Rose (1914) used letters to indicate the different cell areas. Table 21.3 presents an overview of the most commonly used nomenclatures.

The current interpretation of telencephalic organisation is based largely upon embryological as well as hodological data. In a series of comparative embryological studies, Bergquist and Källén (see Sect. 21.4.1) tried to demonstrate that large parts of the avian dorsal ventricular ridge (DVR) are pallial in origin and correspond to the areas giving birth to mammalian neocortex (Table 21.4). The regions deriving from the DVR, i.e. neostriatum, hyperstriatum ventrale and archistriatum, contain cell populations homologous to those in the mammalian neocortex, whereas the paleostriatal complex represents the whole basal ganglion. Nauta and Karten (1970) using the patterns of afferent and efferent connections reached the same conclusion; these authors distinguish an internal and an external striatum corresponding to the corpus striatum and pallium, respectively. Their conclusion is that the external striatum contains the same cell groups as are found in the mammalian cortex, be it with a different arrangement of cells and fibres. Webster (1979) and Ulinski (1983; 1990) have elaborated this concept, but at the same time warn that ambiguities about the homology of specific cell groups still exist (see also discussion in Dubbeldam 1991). In a more recent study Karten (1991) stresses the necessity to compare the fate of presumed prosomeres in birds and mammals. Both the developmental origin of cells groups, and the pattern of afferent and efferent connections as well as of the intrinsic connections within the dorsal ventricular ridge are used to stress similarities in mammalian and non-mammalian forebrains (Karten and Shimizu 1989). Finally, Veenman et al. (1995) propose a further subdivision of the pallium in birds distinguishing sensory regions (e.g. ectostriatum: core), neostriatal belt regions and an external pallium (Fig. 21.62; Sect. 21.10.2). These authors suggest that the core may contain 'layer IV' cells, belt regions, 'layer II/III' cells and the external pallium 'layer V/VI' cells.

Before discussing the respective areas two general remarks are necessary. The first is that distinct regional differences in cell size and density can be found in several telencephalic regions, but particularly in the neostriatum, hyperstriatum ventrale and archistriatum (Rehkämper et al. 1984, 1985; see Sect. 21.10.3). This hinders an easy characterisation of these areas. In the second place considerable differences in size and cell types can be found in corresponding regions of different birds; moreover, considerable quantitative regional differences between species are expressed in the outer shape of the forebrain (Fig. 21.56, brain types after Stingelin 1958). There is no easy way to connect such quantitative differences to specific aspects of behaviour (e.g. Dubbeldam 1990a) or evolutionary development (discussions in Stingelin 1958; Rehkämper and Zilles 1991). In general, neostriatum and hyperstriatum ventrale appear to form one complex with integrative functions (Dubbeldam 1991) and differences in cerebralisation between avian groups may be reflected in particular in the relative size of these areas. For example, passeriform birds have considerably larger brains than galliform birds of the same body size, but, in addition, neostriatum+hyperstriatum ventrale comprise about 68 % of the telencephalon in passeriforms vs. 63 % in galliforms (Rehkämper et al. 1991).

The following description is based upon the telencephalon of the pigeon with comments on the organisation in other species. Recent discussions of the telencephalic structure can be found in Pearson (1972), Nauta and Karten (1970), Cohen and Karten (1974), Benowitz (1980), Dubbeldam (1991) and Veenman et al. (1995).

21.10.2
Paleostriatal Complex

The internal striatum or paleostriatal complex (Figs. 21.28–21.30, 21.32) forms the basal part of the hemisphere. It is rostrally, dorsally and caudally bordered by the lamina medullaris dorsalis and consists of several cell groups. In the pigeon the paleostriatum augmentatum with medium-size cells and the paleostriatum primitivum with its large cells (Figs. 21.28, 21.29, 21.58c) can easily be distinguished in the more caudal sections. The fasciculus lateralis telencephali [prosencephali lateralis] enters this area ventrally; its fibres fan out in the paleostriatum primitivum (Fig. 21.29), thus forming the ventral medulla laminaris (Fig. 21.48). Another group of medium-size cells – the n. intrapeduncularis – sits 'on top' of the fasciculus lateralis telencephali. In more rostral sections a third component with medium-sized cells, the lobus parolfactorius, becomes more prominent (Figs. 21.29, 21.30, 21.32, 21.33). This is also true for *Corvus* (Fig. 21.47) and *Melopsittacus* (Figs. 21.43, 21.44). In anseriform birds, however, paleostriatum augmentatum and primitivum, and apparently also n. intrapeduncularis, extend far more rostrally than the lobus parolfactorius (Fig. 21.49); here, the latter cell area is clearly separated from the paleostriatum augmentatum (Fig. 21.48; Dubbeldam and Visser 1987).

Embryological (Källén 1953), histochemical and hodological data (Karten 1969; Parent and Olivier

Table 21.4. Embryological origin and mammalian homologies

Avian regions, NAA, Breazile and Kuenzel (1993)	Embryological origin		Mammalian equivalent region
	After Kuhlenbeck (1973)	After Källén (1962)	
Eminentia sagittalis			Isocortex
Hyperstriatum accessorium	D2	d^I+d^{II}	
Hyperstriatum intercalatum supremum	D2	d_{dII}^{III}	
Hyperstriatum dorsale	D1	id	
Hippocampus	D2	d^I+d^{II}	Hippocampus
Area parahippocampalis	D3	id	
DVR			Isocortex
Hyperstriatum ventrale	D1	dd_I^{III}	
Neostriatum	D1	d_v^{III}	
Ectostriatum	D1	d_I^{III}	
Nucleus basalis	D1	$c^I ext$	
Archistriatum (sensorimotor)	D1	d_I^{III}	
(amygdaloid)	–	$c^I ext + c^{II} d$	Amygdala
Complexus paleostriaticus			Corpus striatum
Paleostriatum augmentatum	B1+B2	id	Caudatoputamen
Paleostriatum primitivum	B2	$c^I int$	Globus pallidus
Paleostriatum ventrale	–	?	
Lobus parolfactorius (n. accumbens)	B2	$b^I + b^{II} + c_v^{II}$	Caudatoputamen (n. accumbens)
Nucleus intrapeduncularis	–		N. basalis (Meynert) ?
Nucleus lateralis septi	B3+B4	a	Septal nuclei
Nucleus medialis septi	B3+B4	$b_m + c_m$	id

1970; Karten and Dubbeldam 1973) led to the conclusion that the paleostriatum complex corresponds to the mammalian corpus striatum. Originally, the paleostriatum augmentatum was compared to the caudate-putamen, the paleostriatum primitivum to the globus pallidus externus and the n. intrapeduncularis to the globus pallidus internus (Karten and Dubbeldam 1973). Even though a 'simple' homologisation of brain parts should be avoided, the basic idea is supported by new data, albeit with some modifications. One important addition is that the lateral zone of the lobus parolfactorius should be considered part of the caudate-putamen complex, whereas the medial lobus parolfactorius may represent another entity. Moreover, the n. intrapeduncularis appears not to be a pallidal element. Such conclusions are based upon new and partly more detailed anatomical and histochemical data; many of these are reviewed by Webster (1979), Reiner et al. (1984a) and Parent (1986) and summarised in the following paragraphs.

In the pigeon topographically organised projections from the neostriatal area temporo-parieto-occipitalis and area corticoidea lateralis have been described (Brauth et al. 1978). Recently, Veenman et al. (1995) describe the external pallium comprising all lateral parts of the neostriatum and hyperstriastum between the Wulst and archistriatum as the source of these projections. In the goose and duck other parts of neostriatum also contribute (Veenman and Gottschaldt 1986; Dubbeldam and Visser 1987), but more in particular its dorsal layer projects to the paleostriatum (Dubbeldam 1990b). Initially it had not been recognised that the lobus parolfactorius is not a homogeneous structure, but can be subdivided into a lateral and a medial zone. This has led to seemingly different descriptions of the connections of the lobus parolfactorius by different authors. Bons and Oliver (1986) describe in the quail intratelencephalic afferents from the overlying parts of the medial hyperstriatum ventrale, from the n. taeniae and ventral archistriatum. In the duck more lateral parts of the frontal neostriatum also contribute afferents to the lobus parolfactorius, but mainly to its lateral half (Dubbeldam and Visser 1987). Non-telencephalic lobus parolfactorius afferents have been described from locus coeruleus and the subcoerulean nucleus, the ventral tegmental area of Tsai, the n. dorsomedialis anterior thalami and from the hypothalamus (Bons and Oliver 1986; Kitt and Brauth 1986a). Wild (1987b) also mentions the n. dorsolateralis anterior and posterior and n. subhabenularis lateralis and describes a mediolateral organised projection from these nuclei to the paleostriatal complex, including the ventral paleostriatum. The n. tegmenti pedunculopontinus has been mentioned as a source of ascending projections to the paleostriatum augmentatum and lobus parolfactorius (Kitt and Brauth 1986b), whereas the n. ansae lenticularis anterior sends fibres to the paleostriatum primitivum (e.g. Brauth et al. 1978). Reiner et al. (e.g.

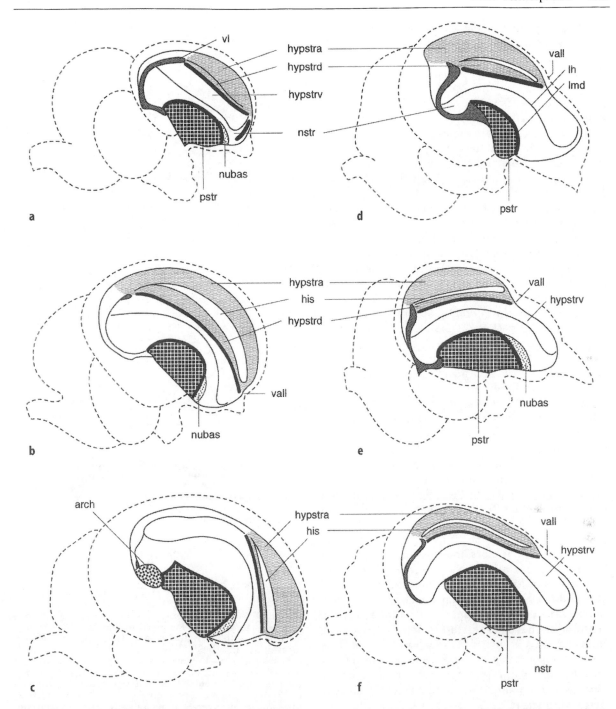

Fig. 21.56a–f. View of a rather far medial section through the telencephalon showing the main regions of the hemisphere. The shape of the brain is indicated in *broken lines*. Note in particular the different positions of the Wulst in the various birds. **a** pigeon, *Columba livia*; **b** little owl, *Athene noctua*; **c** rook, *Corvus frugilegus*; **d** spoonbill, *Platalea leucorodia*; **e** mallard, *Anas platyrhynchos*; **f** yellow-headed amazone, *Amazona ochrocephala*. **a, b** and **c** represent examples of the *dorsale Frontbildungstendenz*; **d, e** and **f** of the *basale Frontbildungstendenz*. (Redrawn after Stingelin 1958)

1994) have proposed the names somatic striatum and visceral striatum – the latter including the n. accumbens, bed nucleus of stria terminalis and olfactory tubercle – and somatic and visceral pallidum, i.e. paleostriatum primitivum and ventral paleostriatum, respectively. Using these subdivisions Veenman et al. (1995) describe the external pallium, dorsal archistriatum, hyperstriatum accessorium and lateral part of hyperstriatum dorsale as sources of projections to the somatic paleostriatum, the area prehippocampalis, medial part of the dorsal hyperstriatum, the ventral and medial archistriatum including the nucleus taeniae and piriform cortex as sources of projections to the visceral striatum. The central part of the archistriatum projects to both areas (Veenman et al. 1995).

Efferents from the paleostriatum augmentatum project upon the paleostriatum primitivum; those from the paleostriatum primitivum form together with fibres from the lateral lobus parolfactorius the ansa lenticularis (Figs. 21.25–21.27, 21.42, 21.45, 21.48) with projections to the n. dorsointermedius posterior thalami, n. ansae lenticularis p. anterior and p. posterior, n. spiriformis lateralis and n. tegmenti pedunculopontinus (Fig. 21.57; Karten and Dubbeldam 1973). Kitt and Brauth (1981) provide more details confirming that the paleostriatum augmentatum projects to the paleostriatum primitivum, i.e. the equivalent of the globus pallidus and source of the pallidotegmental projections. Both paleostriatum augmentatum and the lateral part of lobus parolfactorius are the origin of a striatotegmental projection to the n. tegmenti pedunculopontinus. Furthermore, Reiner et al. (1984) distinguish a pallidotegmental pathway to the n. spiriformis lateralis and n. dorsointermedius posterior and a striatotegmental pathway to the n. tegmenti pedunculopontinus. In addition, Székely et al. (1994) describe an intratelencephalic projection to parts of the hyperstriatum, neostriatum and lateral septum in the chicken; their pictures suggesting that the medial lobus parolfactorius is at least partly the source of these projections. Finally, Kitt and Brauth claim that the n. intrapeduncularis (inp) is not a 'pallidal' component; so far, its character is not clear.

In addition to the pattern of connections, the presence of specific neurotransmitters and neuropeptides strongly supports the comparability of the paleostriatum complex and the mammalian n. caudatus/putamen+globus pallidus. The study by Fuxe and Ljunggren (1965) is one of the first addressing the question of the distribution of monoamine-containing cells and fibres in the pigeon. Parent and Olivier (1970) and Karten and Dubbeldam (1973) demonstrate that the paleostriatal complex is rich

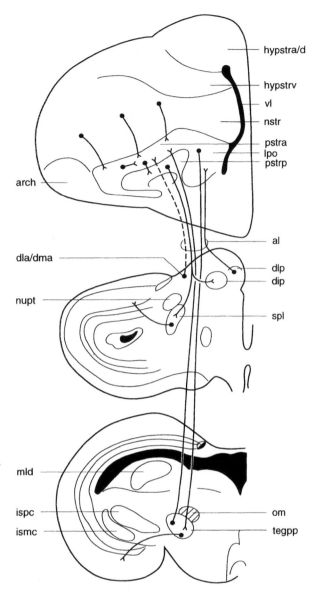

Fig. 21.57. Connections of the avian paleostriatal complex composed from data in the literature (see text): striatal projections from the lateral lobus parolfactorius (*lpo*) to the n. tegmenti pedunculopontinus (*tegpp*, substantia nigra); pallidal projections from the paleostriatum primitivum (*pstrp*) to n. dorsointermedius posterior thalami (*dip*) and n. spiriformis lateralis (*SpL*); nigral (dopaminergic) projection from the n. tegmenti pedunculopontinus to the paleostriatum augmentatum. This receives a topologically arranged projection from the neostriatum, has a number of intrinsic connections and projects to the paleostriatum primitivum. The paleostriatal complex receives afferents from several dorsal thalamus nuclei, but not from the dorsointermedius posterior thalami (Wild 1987). The intrinsic paleostriatum augmentatum cells are characterised by acetylcholine, somatostatin or pancreatic polypeptide like immunoreactivity; paleostriatum augmentatum-paleostriatum primitivum projection cells by dynorphin, enkephalin or substance P, striato-nigral cells by GABA, dynorphin and substance P. (Reiner et al. 1984)

in AChE (paleostriatum augmentatum, lobus parolfactorius and n. intrapeduncularis), whereas the paleostriatum augmentatum and lateral lobus parolfactorius contain a rich plexus of catecholamine-positive fibres and terminals. Many studies about the distribution of catecholinergic elements (e.g. Ikeda and Gotoh 1971: chicken; Yamamoto et al. 1977; Takatsuki et al. 1981; Shiosaka et al. 1981: budgerigar; Chikazawa et al. 1982, 1983: chicken; Bagnoli and Casini 1985: pigeon) have been reviewed by Reiner et al. (1994). Hardly any dopaminergic cells are found in the telencephalon, but the paleostriatal complex contains a dense dopaminergic neuropil. Most important is the existence of the ascending projection of dopaminergic fibres from the substantia nigra (Fig. 21.57). Moons et al. (1994) found that L-DOPA and dopamine have a similar distribution in the chicken. Particularly the paleostriatum augmentatum and lobus parolfactorius contain abundant dopamine D^1 and D^2 receptors (Dietl and Palacios 1988) as well as muscarinic cholinergic receptors (Dietl et al. 1988). In other parts of the telencephalon mainly D^1 receptors are found (Dietl and Palacios 1988), whereas muscarinic receptors dominate in the hyperstriatum ventrale, part of the archistriatum, n. basalis and parts of the eminentia sagittalis (Wächtler 1985; Wächtler and Ebinger 1989).

As for GABA: the density of GABA-positive cells is higher in the paleostriatum augmentatum and lobus parolfactorius than in other parts of the telencephalon; the density of neuropil and fibres is very high in the paleostriatum primitivum and lobus parolfactorius (Domenici et al. 1988) and extremely high in the ventral paleostriatum (Veenman and Reiner 1994). GABA-ergic projections upon substantia nigra and lateral spiriform (and other regions) appear to have their origin in these paleostriatal regions (Veenman and Reiner 1994). The density of neurotensin-binding sites (Brauth et al. 1986) is high in hyperstriatum ventrale, still quite high in the intermediate neostriatum, archistriatum and hyperstriatum accessorium, but low in the paleostriatum augmentatum and lobus parolfactorius and very low in the paleostriatum primitivum. On the other hand cells immunoreactive for Lys^8-Asn^9-neurotensin$^{8-13}$ are restricted to the paleostriatum complex with the highest density in the paleostriatum primitivum; again, a dense projection from here to the tegmentum is found. Expression of preproenkephalin mRNA was high in the lobus parolfactorius, paleostriatum augmentatum, as well as in the nu. accumbens and septum (Molnar et al. 1994); in other regions the distribution is different in different species. Distribution of tyrosine-hydroxylase-positive fibres in the pallial parts of the telencephalon is relatively uniform, but sensory end stations are devoid of such fibres, whereas other regions – dorsal hyperstriatum ventrale, medial frontal neostriatum, dorsal archistriatum and lateral corticoid areas – receive a relatively dense input. Reiner (1986) demonstrated the existence of substance-P-like and dynorphin-like immunoreactivity in the striatopallidal and striatotegmental projections. The density of substance-P-positive cells is greater in the medial part of the lobus parolfactorius than in its lateral half and in the paleostriatum augmentatum (Reiner et al. 1983). Furthermore many small enkephalinergic cells occur within paleostriatum augmentatum and lobus parolfactorius with a dense positive plexus in paleostriatum primitivum; in contrast, little reactivity was found in the components of the dorsal ventricular ridge and Wulst with the exception of a band of enkephalinergic cells within the hyperstriatum dorsale (Reiner et al. 1984). These data all support the notion of the comparability of the paleostriatum complex and mammalian basal ganglion and of the comparability of the cell populations of avian and mammalian pallium.

Reiner et al. (1984) suggest that the avian 'striatum' serves as a 'neural locus of the control circuitry for stereotyped species-specific behaviour routines'. It has no major influence on motor functions via a thalamotelencephalic circuit, but may do so via projections to the n. spiriformis lateralis and tectum mesencephali (Reiner et al. 1984). Rieke (1980) reports that kainic acid injections in the paleostriatum of the pigeon cause behavioural disturbances comparable to those found after experiments in mammals, e.g. circling. In studies on synaptic changes in the paleostriatal complex and, in particular, in lobus parolfactorius connected with passive avoidance learning (Stewart et al. 1987; Hunter and Stewart 1993), it has been suggested that lobus parolfactorius may be involved with the motor aspect of learning rather than with the sensory aspect. Another locus in lobus parolfactorius, area X, may have a comparable role in learning of vocalisations (Sect. 21.10.4). Finally, it has been suggested that the paleostriatum has a role in spatial orientation, i.e. visual orientation (Bugby in Reiner et al. 1984) and tactile orientation (Dubbeldam and Visser 1987). Recently, Cohen and Knudsen (1994) found space-specific auditory units in the paleostriatum augmentatum and also suggested a role in 'spatially guided behaviour'.

The character of the n. accumbens in birds is more problematic. It is indicated in the atlas of Karten and Hodos (1967), but there is some doubt about the precise status of this cell group. The projections to the habenular nuclei and ventral area of

Tsai do suggest that part of the medial lobus parolfactorius corresponds to the n. accumbens (Kitt and Brauth 1981; see also Veenman et al. 1995). In addition, Székely et al. (1994) depict projections to the preoptic area and hypothalamus in the chicken. In contrast to the rest of the lobus parolfactorius the presumptive accumbens seems to be devoid of substance-P-positive cells (Reiner et al. 1983). Lewis et al. (1981) suggest that the lobus parolfactorius-area X in songbirds (part of the vocalisation circuitry, Sect. 21.10.4) could be the avian homologue of the n. accumbens, mainly on account of the catecholaminergic input from the area of Tsai. However, these observations are not conclusive, as the mammalian medial caudate and olfactory tubercle receive the same input as the n. accumbens. It has further been suggested that a medial zone of the lobus parolfactorius represents the nucleus striae terminalis (Ramirez and Delius 1979,b); it receives an input from the amygdaloid part of the archistriatum and may have a role in agonistic behaviour.

Finally, a ventral paleostriatum has also been described (Kitt and Brauth 1981; Dubbeldam and Visser 1987). This is an ill-defined area in the ventral telencephalon, bordered dorsally by the lobus parolfactorius and laterally by the fasciculus lateralis telencephali. Part of it could correspond to the n. accumbens as indicated in the atlas of Karten and Hodos. Veenman et al. (1995) coined the term 'ventral pallidum' for this area. It consists of small cells giving rise to fibres that follow the trajectory of the ansa lenticularis but extend farther caudally with projections to the locus coeruleus and subcoerulean nuclei (Brauth et al. 1978), the reticular formation of the myelencephalon (plexus of Horsley), as well as subnuclei of the solitary complex and vagus motor complex (Berk 1987; Bout 1987). Fibres from the bed nucleus of the stria terminalis and hypothalamic fibres follow the same trajectory (Berk 1987). Hall et al. (1984) describe a GABA-ergic projection from the n. accumbens, ventral paleostriatum and ventral lobus parolfactorius to the substantia nigra. The ventral paleostriatum itself is recipient of afferents from lobus parolfactorius and possibly other telencephalic areas, in particular the lateral part of the caudal neostriatum and the archistriatum (Veenman et al. 1995).

21.10.3
Neostriatum, Hyperstriatum Ventrale and Related Structures

The neostriatum lies on top of the paleostriatal complex, segregated from it by the dorsal medullar lamina (Figs. 21.28, 21.29), and also dorsal to the archistriatum (Fig. 21.27). It extends over the whole length of the hemisphere, but generally a frontal (Figs. 21.30, 21.31), an intermediate (Figs. 21.28, 21.29) and a caudal part (Figs. 21.26, 21.27) are distinguished. Dorsally, the lamina hyperstriatica (Figs. 21.27, 21.42, 21.46) separates the neostriatum and hyperstriatum ventrale. The lamina hyperstriatica appears as a cell-poor zone in Nissl-stained sections (Figs. 21.28–21.30). The neostriatum contains small- to medium-sized cells. Many cells are clustered in groups of two to five (or even more) cells (Fig. 21.58a). The distribution of cells is not uniform. Rehkämper et al. (1985) measuring the packing density of perikarya (Fig. 21.59) distinguished 16 areas (Ne 1–16). At least several of these areas appear to correspond to cell groups described in previous studies and will be mentioned in the following discussion. Within the boundaries of neostriatum two distinct cell areas can be recognised that are often considered separate entities: rostrally and dorsal to the lamina medullaris dorsalis the n. basalis (Ne 1) or n. trigeminalis prosencephali (Cohen and Karten 1974), and in the caudolateral region the ectostriatum (Ne 7+Ne 5?).

The *n. basalis* (Fig. 21.30) is the destination of the ascending quintofrontal tract carrying trigeminal fibres from the princeps nucleus of N. V (Zeigler and Karten 1973; Dubbeldam et al. 1981). Essentially, this nucleus is a somatosensory cell area processing tactile information from the oral region (Berkhoudt et al. 1981; Veenman and Gottschaldt 1986; Félix and Roesch 1986). A dorsal and a ventral layer can be recognised in the n. basalis; this is more evident in *Anas* (Fig. 21.49; Dubbeldam et al. 1981) than in the pigeon or budgerigar (Figs. 21.43, 21.44). Moreover, the n. basalis of *Anas* also receives a glossopharyngeal input through the trigeminal princeps nucleus (Figs. 21.35, 21.60). The same may be true in psittaciform birds (cf. Wild 1981), but not in the pigeon (Witkovski et al. 1973). These differences in input reflect differences in the numbers and distribution of mechanoreceptors in the oral region (Berkhoudt et al. 1981) and are connected with differences in feeding behaviour. Recently, Arends and Zeigler (1986) demonstrated that the lateral part of the n. basalis of the pigeon receives an auditory input from the intermediate nucleus of the lemniscus lateralis; a comparable projection has been described in the budgerigar (Hall et al. 1993). Such a projection had been suggested previously, but the origin of this projection was ascribed to vestibular rather than to auditory nuclei (Schall et al. 1986: pigeon). Earlier studies reported auditory short-latency responses from the frontal neostriatum surrounding the n. basalis (Delius et al. 1979: pigeon; Kirsch et al. 1980: starling).

The trigeminal projection to the n. basalis is

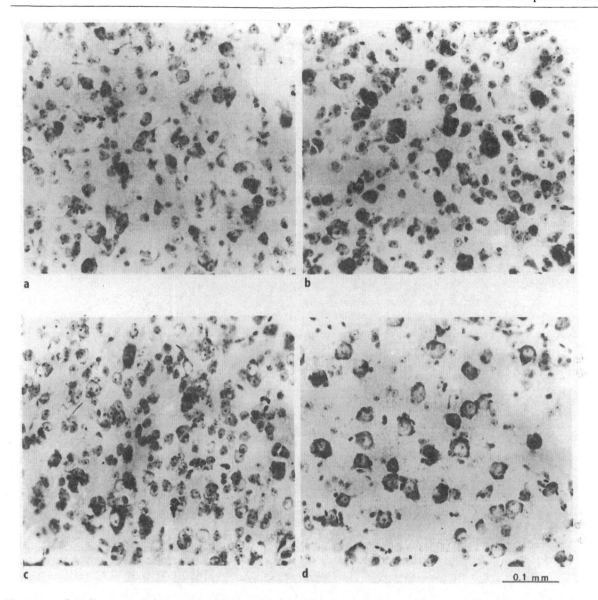

Fig. 21.58a-d. Cell types and clustering in the telencephalon of the pigeon. a Intermediate neostriatum; b hyperstriatum ventrale; c paleostriatum augmentatum; d archistriatum

topologically organised (Witkovski et al. 1973; Berkhoudt et al. 1981). This is particularly well visible in anseriform birds: the basalis cells in the dorsal layer are arranged in vertical columns, the fascicles of afferents ascending between these columns (Dubbeldam et al. 1981). These cells are small and possess three to five dendritic trees each with only a few branches. There is no direct basalis-archistriatum projection through the tractus frontoarchistriaticus, as previously had been assumed, but a radially organised pattern of projections from the n. basalis to the overlying layers of the frontal neostriatum and hyperstriatum ventrale (Veenman and Gottschaldt 1986; Dubbeldam and Visser 1987); from here projections reach the lateral neostriatum and archistriatum through the frontoarchistriatic tract, Fig. 21.61). Several authors ascribe the n. basalis an important role in the maintenance of feeding behaviour (review in Zeigler 1986) or more particularly in pecking control (e.g. Wild et al. 1985; Schall and Delius 1986).

The *ectostriatum* (Figs. 21.29, 21.42, 21.46, 21.48) is a distinct cell area, composed of a core (ectostriatum centrale, Ne 7; Fig. 21.59) and a periectostriatal belt (ectostriatum periphericum = Ne 5?). The core is characterised by round perikarya and a plexus of rather coarse axons (Fig. 21.29). It is the main recipient of rotundal fibres (Revzin and

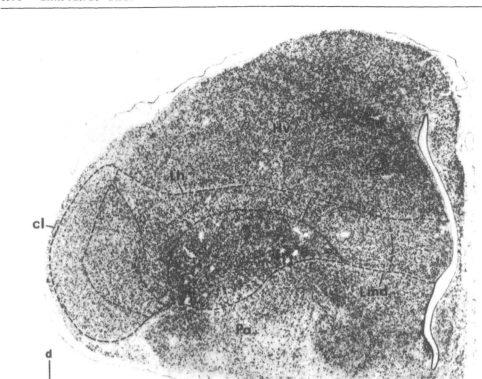

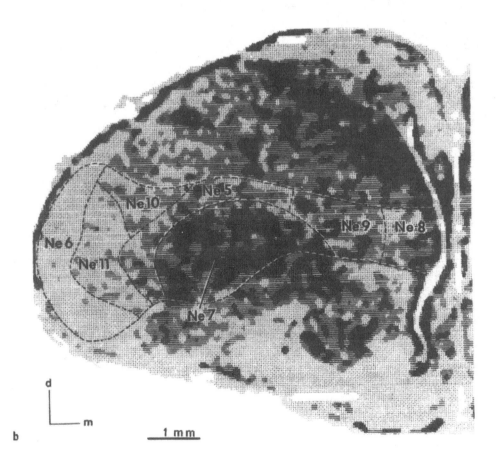

Karten 1966; Karten and Hodos 1970). In the chicken afferents enter the core at the ventromedial border and and branch two or three times before making synaptic contact with the cells of the core (Tömböl 1991). Benowitz and Karten (1976b) described the topological relation between the n. rotundus and the ectostriatal core. The rotundal input predominantly has a contralateral origin, but with a small ipsilateral component (Sect. 21.9.5); Engelage and Bischof (1988) demonstrated electrophysiologically that the ectostriatum in the zebra finch indeed receives contralateral and ipsilateral excitatory projections. Other sources of ectostriatal afferents are the thalamic n. triangularis (Fig. 21.27), the n. dorsolateralis posterior thalami and (possibly) the hyperstriatum ventrale (Watanabe et al. 1985). The cells of the belt are smaller than those in the core and slightly elongated; this area contains rather fine axons receiving its main input from the core. A few rotundal fibres penetrate the belt as well. It is difficult to delineate the belt region sharply from the overlying neostriatum in Nissl-stained preparations. In Golgi preparations of the chicken the core cells appear to possess four to six dendrites, each with two or three branches, whereas the belt cells carry seven to ten main dendrites with numerous bifurcations (Tömböl 1991, cf. also Watanabe et al. 1985). It is interesting to note the parallel with n. basalis cells with sparse dendrites and the overlying part of neostriatum containing cells with many more and more often branching dendrites (Dubbeldam et al. 1981). Several efferent ectostriatal pathways can be recognised. Caudal and intermediate belt cells have connections with the lateral neostriatum caudale and from there with the sensorimotor part of the archistriatum (Ritchie 1979; cf. also Karten 1979). A second ectostriatal pathway derives from rostral belt cells sending efferents to the hyperstriatum ventrale; from here projections via a relay in the medial neostriatum caudale reach the amygdaloid part of the archistriatum (Ritchie 1979).

There is a tendency to compare the ectostriatum to the mammalian peristriate cortex and the visual part of the eminentia sagittalis (Sect. 21.10.7.1) to the striate cortex. In both cases the assumed homology of the regions in mammals and birds is based upon the pattern of afferent and efferent connections, but there is no apparent cortical organisation in the ectostriatum (cf. Karten 1991); at best the ectostriatal core can be compared to layers III and IV of the peristriate cortex (Dubbeldam 1991). Veenman et al. (1995) elaborate the comparison with the mammalian situation, further suggesting that the belt region contains 'layer I–III' cells and the pallium externum 'layer IV' cells (Fig. 21.62). This interpretation rests upon both hodological and histochemical evidence; even though their pictures are very suggestive, their most interesting suggestion seems to be true primarily for the ectostriatal system, whereas the situation is not the same for other sensory systems.

Hellmann et al. (1995) describe the distribution of cytochrome oxidase activity in the ectostriatum of the pigeon and suggest that the differences in activity may reflect functional subdivisions within this cell area.

The neostriatum proper receives a few more ascending inputs from the thalamus. Field L (terminology of Rose 1914) is a medial region with densely packed granular cells in the caudal neostriatum (Fig. 21.27). Field L is the target of fibres from the thalamic n. ovoidalis and is considered the major telencephalic auditory centre (Karten 1968). Probably, Erulkar (1955) recorded auditory responses from this area. Recent studies in the guinea fowl (Bonke et al. 1979a,b) suggest that field L can be subdivided into three zones (L1–3), the intermediate one (L2) receiving the main input from the n. ovoidalis. So far such zones have not been recognised in the pigeon, but Rehkämper et al. (1985) suggested that the three zones may correspond to their areas Ne 13, Ne 12 and Ne 14, respectively. In a recent experimental study Wild et al. (1993) confirmed the existence of L1–3 in the pigeon; here, too, L2 is recipient of thalamic efferents. The authors further subdivided this area into subfields L2a (ventromedial) and b (dorsolateral) receiving input mainly from the n. ovoidalis and n. semilunaris parovoidalis, respectively. Earlier Fortune and Margoliash (1992) had described five parts of the field L complex in the zebra finch (L1, L2a and b, L3 and L). These authors state that field L sensu Rose is more or less separate from the other parts of the complex and does not correspond to Karten's field L. Therefore, it may be preferable to avoid the term 'field L' and use the name 'auditory neostriatum' (neostriatum auditivum).

Scheich (1983) using 2-deoxyglucose demonstrated the existence of three bands – a rostral, an intermediate and a caudal one – in L2 of the chicken. This author suggests that the intermediate one may consist of alternating ipsilateral and contralateral dominance columns parallel to isofre-

◀ Fig. 21.59. a Nissl-stained section through forebrain of the pigeon. b Computer plot of same section using grey level to distinguish areas within the neostriatum. *Stippled lines* indicate areal borders. *cl*, corticoid layer; *Hv*, hyperstriatum ventrale; *Lh*, lamina hyperstriatica; *Lmd*, lamina dorsalis medullaris; *Pa*, paleostriatum. (From Rehkämper et al. 1985)

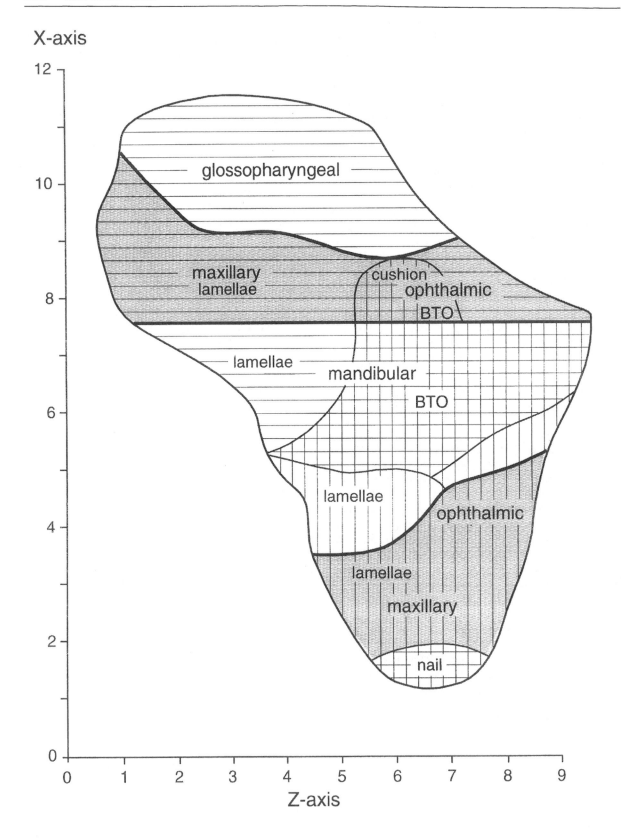

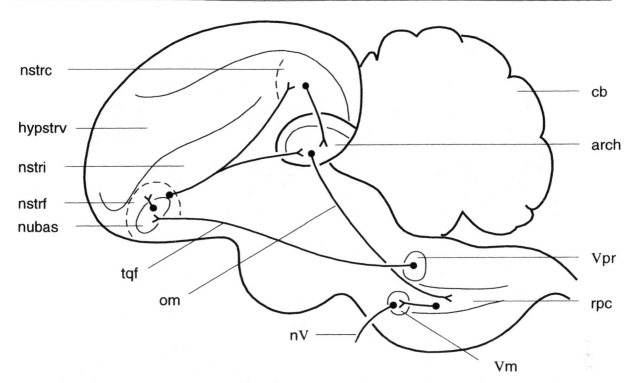

Fig. 21.61. Trigeminal sensorimotor circuit of the pigeon. Nucleus basalis (*nubas*), the overlying part of the frontal neostriatum (*nstrf*) and the lateral zone of the caudal neostriatum (*nstrc*) participate in this circuit with the intermediate archistriatum as source of the occipitomesencephalic tract (*om*). (Redrawn after Wild et al. 1985)

quency contours. Earlier, Leppelsack (1974) did not electrophysiologically recognise a tonotopic organisation in field L of the starling, but Zaretsky and Konishi (1976) reported such an organisation in the zebra finch, Rübsamen and Dorscheidt (1986) in the European starling. This has been confirmed in the pigeon (Wild et al. 1993). L2 of the pigeon sends its efferents to L1 and possibly also to L3 (Wild et al. 1993). Efferents of field L in the guinea fowl reach the ventrocaudal hyperstriatum ventrale, a dorsal zone in the caudal neostriatum and part of the paleostriatal complex (Bonke et al. 1979a). Brauth and McHale (1988) describe in the budgerigar projections to neostriatum, pars ventralis and to the rostromedial archistriatum. Particularly the neostriatal projection received a lot of attention, as in some passeriform birds it lies close to another region, the 'high vocalisation centre' (HVC, originally considered part of the hyperstriatum ventrale; Kelley and Nottebohm 1979); HVC may receive an auditory input from field L through this secondary neostriatal auditory centre. The circuits for vocalisation will be discussed separately.

In the transition of intermediate and caudal neostriatum, between ectostriatum and field L, lies a somatosensory area receiving an input from the thalamic n. dorsolateralis posterior (Wild 1987a; Funke 1989). It may correspond to (part of) Ne 9; the cytology of this region is not homogeneous (Rehkämper et al. 1985). It is not clear whether this is a genuine somatosensory region. Güntürkün (1984) reported short-latency visual responses from the caudolateral telencephalon suggesting the existence of another visual region. Possibly, this region corresponds to the end station of a second tectofugal pathway (Gamlin and Cohen 1986). Brauth et al. (1987) mention the neostriatum, pars dorsolateralis in the budgerigar as the second target of ovoidalis fibres; possibly the source of these fibres corresponds to the ovoidal shell of Durand et al. (1992). In the pigeon this caudolateral region does receive afferents from the n. dorsolateralis posterior thalami, but this nucleus seems to be a relay for several

◀ Fig. 21.60. Horizontal projection of the right nucleus basalis in the mallard (cf. Fig. 21.35A). *X-axis* and *Z-axis* correspond to those of the atlas of Zweers (1971). *Horizontal lines* mark regions receiving ipsilateral projections, *vertical lines* regions receiving contralateral projections; a central part receives bilateral projections. This central part represents the bill tip organs (*BTO*), i.e. the rostral tips of upper and lower beak, each containing high densities of mechanoreceptors. *Shaded areas* represent the two sides of the upper beak; the glossopharyngeal part receives input from the tongue. (From Berkhoudt et al. 1981)

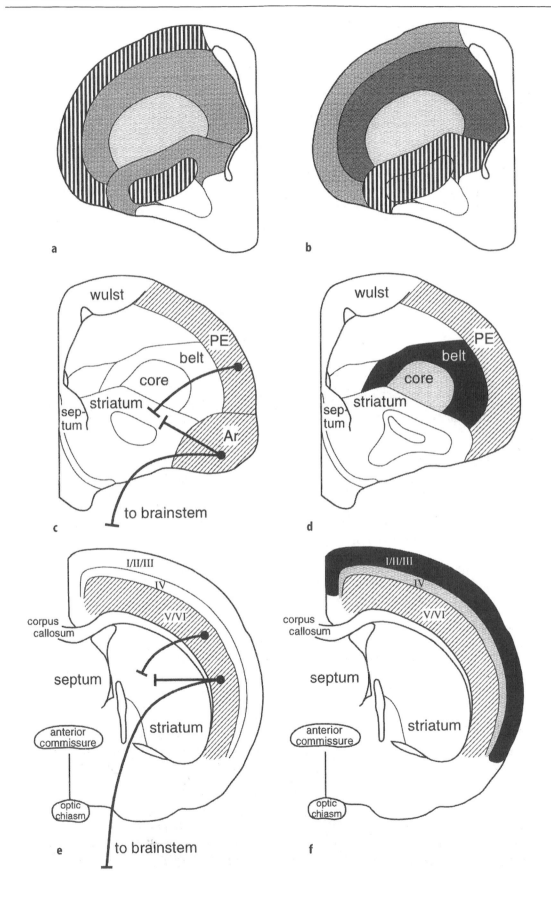

sensory modalities to this neostriatal area; see discussion in Wild (1987a) and in the section on dorsal thalamic nuclei. Finally, there may also be some overlap of somatosensory and auditory afferents in the intermediate/caudal neostriatum (Wild 1987a; electrophysiological data of Erulkar 1955). Efferents from this area reach the rostral n. intercalatus of the hyperstriatum accessorium (see Sect. 21.10.7.1) and – according to Funke (1989) – the overlying part of the hyperstriatum ventrale.

So, in summary four sensory end stations can be recognised within the neostriatum, from rostral to caudal n. basalis (tactile+possibly an auditory component), ectostriatum (visual), the intermediate/caudal 'somatosensory' (several modalities) and field L (auditory). A fifth region is the n. interfacialis (Fig. 21.63) that for the first time has been described in the canary (Nottebohm et al. 1982). It receives afferents from the thalamic n. uvaeformis, possibly a relay of efference copies from the syringeal part of the motor n. hypoglossus (Okuhuta and Nottebohm 1992). Part of the fibres from the n. uvaeformis pass the interface nucleus to project directly to the more dorsal region, often called the HVC, in the neostriatum (Fig. 21.62). Other regions of the neostriatum do not receive thalamic sensory input but are closely connected to the sensory areas and possibly involved with the processing of sensory information. Still other parts may have 'associative' functions.

The 'posterodorsolateral' neostriatum ('PDL', area 16 of Rehkämper et al. 1985) is rich in catecholaminergic (possibly dopaminergic) fibres (Mogenson and Divac 1982). Divac and Mogenson (1985; Divac et al. 1985), citing histochemical, biochemical and behavioural evidence, suggested that this region may be the pigeon's 'prefrontal cortex'. Waldmann and Güntürkün (1993) cite projections from the n. tegmenti pedunculopontinus, pars compacta and the area of Tsai as additional evidence for this notion; these centres may be the source of the dopaminergic projections to this area. It further receives tertiary visual and auditory information. A complication is the variability of content and distribution of catecholamines in different strains of pigeons (Divac et al. 1988); furthermore, Bissoli et al. (1988) report significant differences in content of ChAT and AChE between the mammalian prefrontal cortex and its possible avian equivalent. Apart from these complications it can be questioned whether the existence of certain histochemical and eventual hodological similarities allow such far-reaching conclusions (discussion in Reiner 1986). We will return to the question of comparability in general, but first we will have a closer look at the hyperstriatum ventrale.

The hyperstriatum ventrale lies on top of a large part of the neostriatum (Figs. 21.27, 21.29, 21.30). It is bordered ventrally by the lamina hyperstriatica (Figs. 21.28, 21.42, 21.46, 21.47) and separated dorsally from the hyperstriatum dorsale by the lamina frontalis superior (Figs. 21.28, 21.29, 21.30, 21.42–21.44, 21.47). Again distinct regional differences in cell size, shape and density are obvious. Clusters of small to medium-sized cells are more numerous than in the neostriatum, particularly in the medial parts of hyperstriatum ventrale (Fig. 21.58b). Rehkämper et al. (1984) distinguish a medial and a lateral zone differing in grey level index. In particular in the lateral zone a dorsal and a ventral layer can also be recognised. Both the medial-lateral difference and the two layers in lateral hyperstriatum ventrale are very clear in the forebrain of the mallard (Figs. 21.48, 21.49; Dubbeldam and Visser 1987). Tömböl et al. (1988) using Golgi technique describe two major classes of neurons: Golgi I type, i.e. putative projection neurons with long axons, and Golgi II type or putative local circuit neurons with short axons; for each several subtypes can be distinguished. The distribution of the (sub-)types is different in the various regions of hyperstriatum ventrale. These authors also suggest the existence of intraregional connections; this supports the earlier observation of Veenman and Gottschaldt (1986) about the 'horizontal' connections.

Together, the hyperstriatum ventrale and neostriatum form more than 50% of the telencephalon (Ebinger and Löhmer 1984; 1987; Dubbeldam 1990a). An interesting exception is the emu, where the two brain parts including the ectostriatum form 43% of the telencephalon (Cobb 1966). The hyperstriatum ventrale and neostriatum form a complex with many intrinsic connections. Rostrally, a pattern of reciprocal connections of n. basalis with specific layers in the overlying parts of the neostriatum frontale and hyperstriatum ventrale has been described in the goose (Veenman and Gottschaldt 1986) and the mallard (Dubbeldam and Visser 1987). A comparable pattern may exist in the pigeon (Fig. 21.61). These parts of the neostriatum frontale and (possibly) hyperstriatum ventrale send efferents to the archistriatum via a relay in the lateral neostriatum (possibly Ne 6 of Rehkämper et

◀ Fig. 21.62a–f. Interpretation of the organisation of the avian forebrain according to Veenman et al. (1994, 1995). This interpretation rests on histochemical and hodological data. a,b Scheme of densities of GABA-ergic terminals (a) and receptors (b) in the pigeon forebrain. c,e Summary of projections from the external pallium (*PE*) and archistriatum (*AR*) to the striatum (paleostriatal complex) and the brain stem in the pigeon (c) as compared to the situation in the rat (e). d,f Comparability of cell regions in pigeon (d) and rat (f). [a,b from Veenman et al. (1994), c–f from Veenman et al. (1995)]

al., 1985; cf. Wild et al. 1985: feeding circuit). This lateral area is again part of the external pallium as discussed in Veenman et al. (1995). Furthermore there are projections from the frontal neostriatum to the paleostriatal complex (at least in anseriform birds) and from the hyperstriatum ventrale to the lateral zone of the lobus parolfactorius. More caudally, a more or less comparable pattern of connections has been suggested for the ectostriatum (Dubbeldam 1991, based upon Ritchie 1979). Lesions in the lateral telencephalon – i.e. dorsolateral neostriatum, area corticoidea dorsolateralis and area temporoparieto-occipitalis – appear to disturb visuomotor activity (Jäger 1990). Lateral hyperstriatal lesions may impair certain aspects of conditional discrimination learning (MacPhail et al. 1993).

The situation is somewhat different for field L. Efferents reach a narrow zone in the caudal hyperstriatum ventrale and the caudal neostriatum, with a modest additional projection to the medial paleostriatal complex (Bonke et al. 1979a; Kelley and Nottebohm 1979).

21.10.4
Vocalisation Circuit: A Specific Sensorimotor Circuit

In some passeriform birds the situation is complicated by the existence of a vocalisation circuit including the HVC in the caudal neostriatum and the n. robustus in the archistriatum (Fig. 21.45) Apparently, the HVC receives indirect auditory input from field L. In addition, it receives input from the thalamic n. uvaeformis, partly directly, partly indirectly through the n. interfacialis in the neostriatum; possibly, an efference copy from the hypoglossal (syringeal) motor nucleus may reach the HVC through this pathway (Okuhata and Nottebohm 1992). The HVC has a direct projection to the n. robustus; the two centres have a crucial role in sustaining learned vocalisations (Simpson and Vicario 1990). Recently, this part of the vocalisation system (Fig. 21.63) has been called the 'telencephalic pathway of learned song' (Nottebohm 1993). There is a substantial sexual dimorphy, the HVC and nucleus robustus archistriati being much larger in males than in females in species where only males sing (Fig. 21.64a; e.g. Brenowitz and Arnold 1986). DeVoogd and Nottebohm (1981) found that the HVC contains the same cell types in males and females, but dendrites are longer and branch more often in males. Recently, Vu et al. (1994) provided evidence that the HVC in zebra finches is part of a programming network, whereas the n. robustus is merely part of the premotor pathway.

A second component of the vocalisation circuitry is the pathway for acquisition of learned song (Fig. 21.63): a circuit from the HVC to area X of the lobus parolfactorius, from here to the n. dorsolateralis thalami, pars medialis, then to the lateral n. magnocellularis in the frontal neostriatum (LMAN); this again sends fibres to the n. robustus (Nottebohm et al. 1982; Nottebohm 1993). LMAN and nucleus X have an important role in the development of song (learning), but not in sustaining song in the adult situation (Nottebohm 1991). In both components acetylcholine appears to be an important transmitter (Ryan and Arnold 1981).

An intriguing observation is the seasonal change of size of the HVC in the canary that may be connected with replacement of part of the song in successive breeding seasons (Nottebohm 1981); a comparable phenomenon has been described in *Euplectus franciscanus* (Fig. 21.64b; Arai et al. 1989). Such changes in size may partly be due to the increase and decrease in the size of cells, but also new neurons are born and become part of the HVC (Alvarez-Buylla et al. 1988; Nordeen and Nordeen 1988; Kim et al. 1991). These processes are under hormonal control (review in Konishi and Gurney 1982). Gahr and Konishi (1988) demonstrated the presence of oestrogen-binding neurons in the HVC of the canary, but not in that of the zebra finch. Earlier Zigmond et al. (1980) demonstrated that the HVC as well as the LMAN contain testosterone-positive cells; the only other positive telencephalic region is the lateral septum. In consequence of the preceding remarks, the seasonal changes cannot be ascribed to learning, but possibly to the storage of a 'learned skill'.

Little is known about the organisation of the neural substrate for vocalisations in non-songbirds. In a lesion study in zebra finches Simpson and Vicario (1990) found that the so-called long call was affected in males, but not in females. These authors suggested a separate pathway for the control of this non-learned part of the song, proposing a role for the intercollicular nucleus. At least part of this nucleus is part of the respiratory pathway, but a role in vocalisation has been suggested as well (see Sect. 21.8.7). Another surprise has been the description of the 'vocal control pathways' in the budgerigar (Striedter 1994). There are many parallels with the pathways in songbirds (Fig. 21.65a). However, auditory input does not come from field L, but from the auditory part of the n. basalis; efferents from here project to overlying parts of the frontal neostriatum, then to the lateral neostriatum and from here to the central nucleus of the anterior archistriatum. This cell group sends direct projections to the hypoglossal motor nucleus innervating the

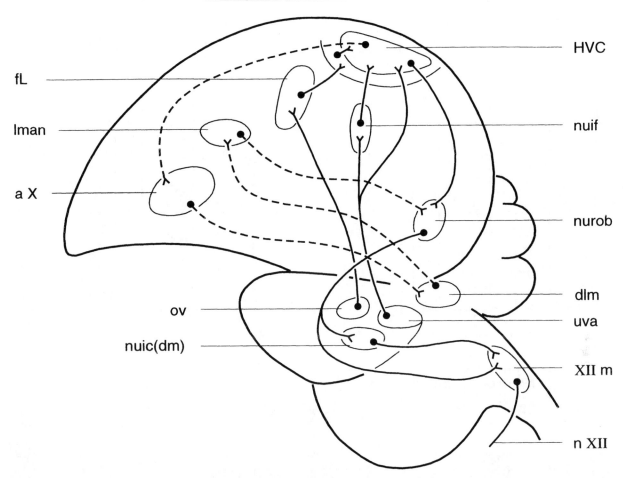

Fig. 21.63. Vocalisation circuits of a songbird. The high vocalisation centre (*HVC*) receives indirect auditory input from field L(*fL*) and direct input from the thalamic n. uvaeformis (*uva*) and the neostriatal n. interfacialis (*nif*) and has a projection to the n. robustus (*nurob*) of the archistriatum. This is a sensorimotor circuit for the production of learned song; n. robustus has projections to a dorsomedial subnucleus of the n. intercollicularis [*nuic(dm)*, part of a respiratory pathway] and motor nucleus of N. XII innervating the syrinx. The other part (*broken lines*) includes a lateral n. magnocellularis of the frontal neostriatum (*lman*) and area X (*aX*) of the lobus parolfactorius and is essential for the acquisition of song. (Redrawn after Nottebohm 1993)

tongue and trachea. Also a dorsal thalamic cell group and parts of the lobus parolfactorius and a more medial part of the frontal neostriatum are involved. Consequences of these observations are (a) the notion that highly developed vocal control systems have developed in different avian groups independently and (b) the recognition of many parallels in the organisation of these pathways in the two groups of birds. This supports the assumption of the existence of a basic 'generalised' pattern of sensorimotor pathways in the neostriatum-hyperstriatum ventrale complex (cf. Sect. 21.10.3). Finally, it is interesting to note that in 1905 Kalischer indicated a 'phonation *Bereich*' in the telencephalon of the parrot (Fig. 21.65b); this region corresponds to the lateral neostriatal area in the budgerigar that is part of the vocalisation pathway.

Comparing the avian vocalisation system to that in mammals, it can be questioned whether there is a lateralisation of function: it has been concluded that both sides make equal contributions (McCasland 1987), but differences between species may exist (e.g. Paton and Manogue 1982). Possibly, the presence of the field L-neostriatum/hyperstriatum ventrale-archistriatum circuit is connected with the role of auditory information in other aspects of the behaviour, e.g. recognition of conspecifics, of the sounds caused by prey (as in owls), etc. (e.g. Knudsen et al. 1993; Dubbeldam 1993).

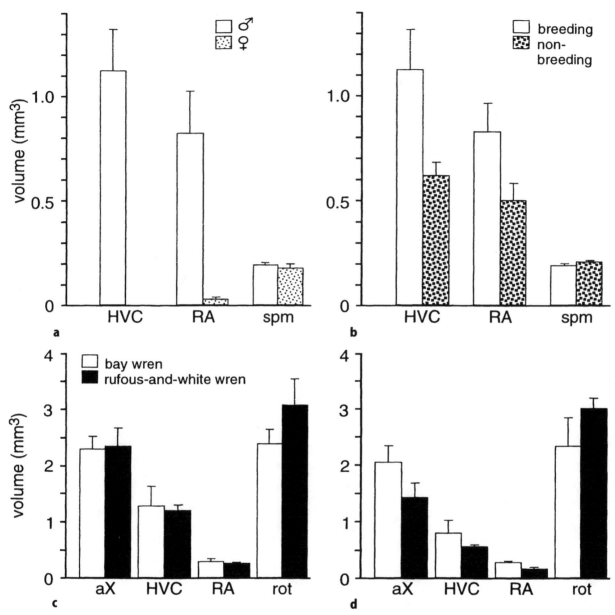

Fig. 21.64. a Quantitative comparison of the high vocalisation centre (*HVC*) and nucleus robustus (*RA*) in male and female orange bishop birds (*Euplectes franciscanus*). The medial spiriform nucleus (*spm*) serves as a reference. **b** Quantitative comparison of HVC and RA in male orange bishop birds in the breeding and the non-breeding season. **c** Size of three vocal centres in males of two species of wrens having songs of similar complexity. Nucleus rotundus (*rot*) serves as a reference; *aX*, area X of lobus parolfactorius. **d** Size of three vocal centres in females of two species of wrens. The song of female bay wrens is about as complex as that of males; the song of female rufous-and-white wrens is considerably simpler. [a and b from Arai et al. (1989), c and d from Brenowitz and Arnold (1986)]

21.10.5
Neostriatum and Hyperstriatum Ventrale: Associative Functions and the Role in Learning

An important question is that of the possible associative functions of the neostriatum and hyperstriatum ventrale. In the preceding sections it was explained that large parts of these regions are part of specific sensorimotor circuits. Impressive quantitative differences between avian species may reflect differences in sensory specialisation and at the same time differences in the ecology (Stingelin 1958; discussion in Dubbeldam 1990a). Moreover, different parts of this complex appear to be involved with different aspects of behaviour or, more in particular, with specific aspects of learning or of the storage of memory. In the preceding section LMAN and its role in learning of vocalisation have been mentioned, whereas HVC, too, has a role in a 'learned' behaviour.

An example of hyperstriatal involvement is the role of the medial intermediate region of the hyperstriatum ventrale ('IMHV') in filial imprinting (cf. Horn 1985; Johnson and Horn 1987). This IMHV receives afferents from the Wulst, from the neostriatum and from other parts of the hyperstriatum ventrale (Bradley et al. 1985) and may have a (direct or indirect) projection to part of the lobus parolfactorius suggestive of a role of this region in learning as well (Stewart et al. 1987). Here, again, a parallel exists with area X as part of the substrate for the learning of song. However, there is no unanimity of opinion about the importance of different neostriatal and/or ventral hyperstriatal regions as substrate for learning.

In a series of studies Salzen (e.g. Salzen 1975; Salzen et al. 1975, 1979) stresses the importance of the lateral neostriatum in learning or imprinting, whereas this author could not find evidence for the involvement of the IMHV. We already mentioned the study of MacPhail et al. (1993) stressing the importance of the lateral hyperstriatum ventrale. These lateral regions are part of the external pallium as defined by Veenman et al. (1995; Sect. 21.10.2). What can be the explanation of these seemingly contradictory results? It may be helpful to return for a moment to the sensorimotor circuits: for example, several pathways seem to exist relaying 'sensory' information from the nucleus basalis to sources of extratelencephalic fibre systems, viz. via the archistriatum, via the paleostriatum augmentatum/primitivum complex and via the lobus parolfactorius (Dubbeldam and Visser 1987). The same is true for the visual/ectostriatal system, as explained previously. This means that different channels exist through which sensory input can be used to generate 'motor' activity – at least a medial one via the medial neostriatum and amygdaloid ('limbic') archistriatum (possibly also the lobus parolfactorius is part of it) and a lateral one via the lateral neostriatum and the somatic sensorimotor archistriatum. Both pathways may be part of a distributed system underlying different aspects of learning behaviour. I would like to suggest that depending on the type of learning behaviour one of the two 'pathways' may have more importance. The lateral 'pathway' may be in particular important for activities with a strong sensorimotor component ('skill'), whereas the medial 'pathway' may be involved in types of behaviour with a strong motivational component. The occurrence of visuomotor disturbances (Jäger 1990) after lateral neostriatum lesions may support the first suggestion, the involvement of the medial ectostriatal/visual pathway in heart rate conditioning (Ritchie 1979) the second suggestion.

As for the vocalisation system, it seems reasonable to identify the components HVC and nucleus robustus as elements of the lateral sensorimotor pathway ('song is a learned motor skill') and area X and LMAN as elements of a medial, limbic (motivational) pathway; the observations in the budgerigar support this distinction (Sect. 21.10.4). As for the latter two centres: their presence is essential in song learning (acquisition pathway, Nottebohm 1993). It is interesting to note that no direct connection seems to exist between these two centres. The same may be true for the IMHV and lobus parolfactorius in the model of filial imprinting: the existence of more than one 'sensorimotor' output may support this assumption. Another consequence of this interpretation of the vocalisation circuits is that it is not possible to compare the HVC to IMHV as has sometimes been done in the literature.

21.10.6
Archistriatum

The archistriatum (arch) of the pigeon occupies the parieto-occipital pole of the hemisphere (Figs. 21.26–21.28, 21.32). In other species it lies deeper and is dorsally, laterally and caudally surrounded by the neostriatum (Figs. 21.42, 21.43, 21.45, 21.48). It stands out by its large cells (Fig. 21.58d) and the distinct fascicles converging into the occipitomesencephalic tract (Fig. 21.28) and anterior commissure [commissura rostralis] (Figs. 21.28, 21.32). The abundant presence of frontoarchistriatic fibres gives the most dorsal (in pigeon) or lateral (e.g. in mallard) zone a belt-like appearance: the lamina archistriatalis dorsalis (Zeier and Karten 1971; Dubbeldam and Visser 1987). Regional differences in cell size, packing density and distribution of fibres suggest that the archistriatum can be subdivided into many subnuclei (Zeier and Karten 1971, 1973). Initially, a subdivision in an anterior part or pars rostralis, an intermediate part with a dorsal and a ventral region, a medial part including the n. taeniae (Fig. 21.27) and a caudal or posterior archistriatum has been used. The rostral and intermediate arch were assumed to represent a sensorimotor region, the medial and caudal parts the amygdaloid regions (Fig. 21.69; Zeier and Karten 1971). Recent studies provide additional details. For example, observations in the mallard reveal that the most rostral part is the source of fibres projecting to the contralateral telencephalon (Dubbeldam et al. 1997). The composite figure in Zeier and Karten

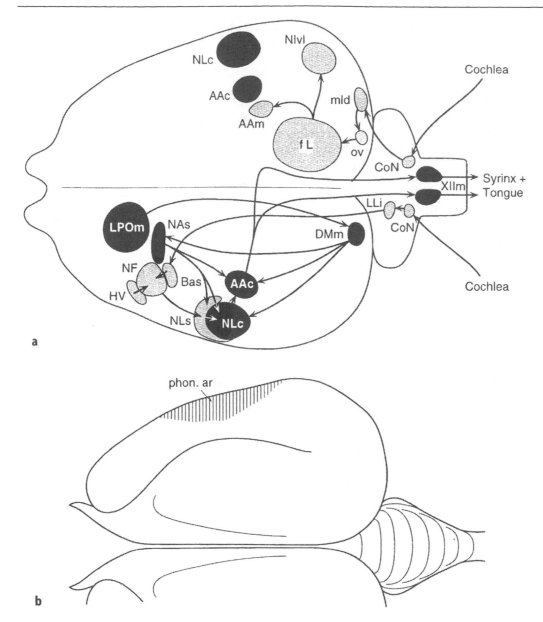

Fig. 21.65. a Vocalisation circuit of the budgerigar described by Striedter (1994). *Upper half* shows ascending auditory pathway to field F with connections to other intratelencephalic centres. *Lower half* shows auditory pathway to nucleus basalis (*Bas*) and centres of two vocalisation pathways; cf. with Fig. 21.62. *AAc, AAm,* central, medial nucleus of anterior archistriatum; *CoN,* cochlear nuclei; *DMm,* magnocellular nucleus of dorsomedial thalamus; *HV,* hyperstriatum ventrale; *LLi,* nucleus intermedius lemnisci lateralis; *LPOm,* magnocellular nucleus of lobus parolfactorius; *NAs,* supralaminar area of frontal neostriatum; *NIvl,* ventrolateral neostriatum intermedium; *NLc, NLs,* central, supracentral nucleus of lateral neostriatum. b Dorsal view of brain of parrot showing the 'phonation area' (*phon. ar.*) as defined by Kalischer (1905)

(1973) suggests that the same is true in the pigeon. More in general, an important criterion for subdividing the archistriatum is the pattern of afferent and efferent connections, another one the histochemical diversity. It is not possible to subdivide the archistriatum on the basis of Golgi cytoarchitecture (Tömböl and Davies 1994).

The archistriatum receives afferents through three major fibre systems, i.e. through the anterior commissure (Figs. 21.25, 21.42) from the contralateral archistriatum, through the tractus frontoarchistriaticus (Figs. 21.29, 21.32, 21.42, 21.43, 21.45, 21.46, 21.48, 21.49), from the frontal neostriatum and through the tractus dorsoarchistriaticus (Figs. 21.26-21.28) from the hyperstriatum ventrale and lateral neostriatum. The commissural fibres project upon the most rostral part (Zeier and Karten 1973; Dubbeldam and Visser 1987); many frontoarchistriatic fibres terminate laterally in the rostral and intermediate archistriatum, but also - and predominantly - in the lateral and caudal neostriatum; these regions, too, have projections to the intermediate archistriatum (Wild et al. 1985; Dubbeldam and Visser 1987). The dorsoarchistriatic system appears to be the most important input system of the archistriatum with projections mainly upon the intermediate part (Zeier and Karten 1971, 1973). In the duck the ventral intermediate and caudal arch receives afferents from more medial parts of the caudal neostriatum (Dubbeldam and Visser 1987). Some songbirds possess a n. robustus (Fig. 21.49) receiving specifically input from the HVC; it has direct projections to nuclei involved in vocalisation and respiration, such as the n. dorsalis medialis of the intercollicular complex and the (dorsal) n. motorius hypoglossi (e.g. Wild 1993b). Previously, the existence of a nucleus robustus has also been suggested in the budgerigar (Paton et al. 1981), but neither in this species nor in other birds can it be recognised in normal material. The recent study by Striedter changes this picture (Sect. 21.10.4). Finally, the lateral archistriatum receives afferents from the ventral paleostriatum (Dubbeldam and Visser 1987).

The sensorimotor part of the archistriatum is the main origin of the tractus occipitomesencephalicus (Figs. 21.18-21.28, 21.42, 21.45, 21.46, 21.48); otherwise than suggested by its name this tract descends the whole length of the brain stem reaching the first segments of the cervical cord. It has terminations in many cell groups throughout the brain stem (Fig. 21.69): the nn. dorsolateralis and dorsointermedius posterior thalami, medial spiriform nucleus, n. subrotundus, n. principalis precommissuralis, the mesencephalic reticular formation, n. intercollicularis, stratum griseum centrale of the tectum mesencephali, locus coeruleus and nn. subcoeruleus dorsalis and ventralis, n. pontis lateralis, n. parvocellularis reticularis of the myelencephalon, the subnn. oralis and interpolaris of the descending trigeminal root, n. subtrigeminalis (Zeier and Karten 1971; Dubbeldam et al. 1997) and the n. intermedius or hypoglossus dorsalis. This fibre system appears to have a major role in the control of premotor and motor regions of the brain stem (Arends and Dubbeldam 1982; Tellegen and Dubbeldam 1994). Through the projection to the medial spiriform nucleus the archistriatum can also influence the cerebellar cortex. The medial and caudal, amygdaloid parts of the archistriatum are the source of a hypothalamic projection (Zeier and Karten 1971; Schriber 1978), but also contribute a modest component to the tractus occipitomesencephalicus proper with projections to the mesencephalic reticular formation of the brain stem (Dubbeldam et al. 1997). Intratelencephalic projections derive from all parts of the archistriatum. In the mallard, the dorsal half of the rostral arch projects to the contralateral arch, the dorsal intermediate arch to the contralateral neostriatum and hyperstriatum ventrale, the dorsal and ventral intermediate arch to the ipsilateral neostriatum, hyperstriatum ventrale and hyperstriatum intercalatum supremum, whereas the lateral belt and laterocaudal zone contain cells projecting to the septum. The neuropil of the last-mentioned part of the neostriatum is positive for acetylcholine (Dubbeldam et al. 1997).

In addition to the anatomical there is also behavioural evidence for the subdivision into a sensorimotor and a amygdaloid part. For example, archistriatal lesions in ducks have a 'taming' effect (Phillips 1964), i.e. suppress arousal. For example, animals no longer try to avoid the experimenter. Interruption of the occipitomesencephalic tract has the same effect (e.g. Wright and Spence 1976) and at the same time raises the plasma corticostereoid level (Martin et al. 1979). Ramirez and Delius (1979a) provide evidence that such effects are mainly due to damage of the amygdaloid part.

21.10.7
Cortical Regions

Even though a typical neocortex is absent in birds, a few cortical areas can be recognised. The most evident regions are the dorsal eminentia sagittalis or Wulst (Figs. 21.28, 21.29) and adjacent mediodorsal hippocampal areas (Figs. 21.26, 21.27).

21.10.7.1
Eminentia Sagittalis

Originally the Wulst was considered the sole equivalent of the mammalian and reptilian cortices (see introduction to Sect. 21.10). It is a multilayered structure, but not over its whole length. Caudally, the hyperstriatum accessorium (Fig. 21.28) lies directly upon the hyperstriatum ventrale and lateral to the area parahippocampalis; more rostrally it expands farther medially and at the most rostral level it forms the most dorsal and medial zone of the Wulst. From caudal to rostral, first the hyperstriatum dorsale (Fig. 21.29) appears between the hyperstriatum ventrale and the hyperstriatum accessorium. Then the other components, i.e. hyperstriatum intercalatum supremum (Figs. 21.30,

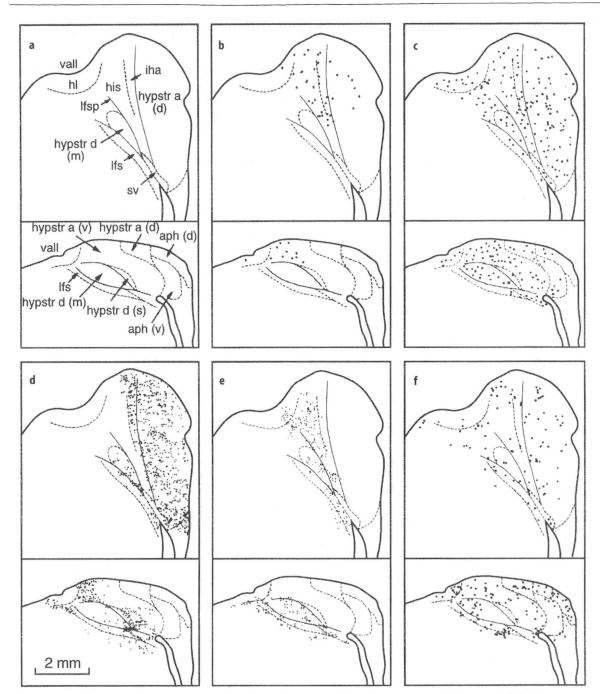

Fig. 21.66a–f. Distribution of immunoreactive cells in the subdivisions of the eminentia sagittalis. *Each panel* represents a rostral and a caudal section through the Wulst. **a** Subdivisions used in this study: hyperstriatum accessorium (*hyperstra*) and area parahippocampalis (*aph*) with a dorsal (*d*) and a ventral (*v*) part; hyperstriatum dorsale (*hyperstrd*) with a medial (*m*) and a shell (*s*) region; a lateral hyperstriatum (*hl*) and a supraventricular area (*sv*). **b–i** immunoreactive cells; **b** nicotinic acetylcholine receptors; **c** glutamic acid decarboxylase; **d** substance P; **e** cholecystokinine; **f** leucine-enkephaline; **g** neuropeptide Y; **h** somatostatin release-inhibiting factor; **i** corticotropin-releasing factor. Compare with Table 21.5. (From Shimizu and Karten 1990)

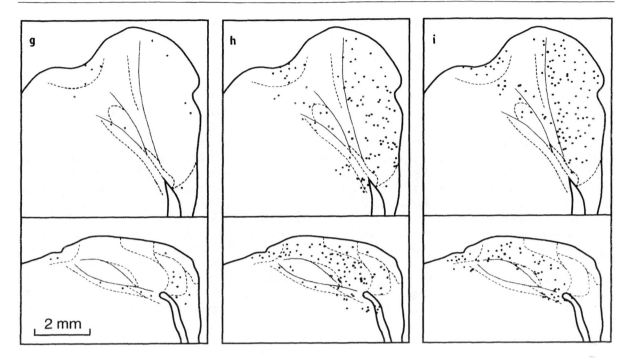

Fig. 21.66g–i.

21.31) with a lamina externa and a lamina interna, appear. Ariëns Kappers et al. (1936) distinguish a n. intercalatus laminae supremae and a n. intercalatus laminae superioris. The latter consists of scattered cells within the lamina frontalis superior, the other cell group corresponding to our hyperstriatum intercalatum supremum (Tables 21.3, 21.4). According to Ariëns Kappers et al. the two cell groups fuse caudally and form together the n. intercalatus hyperstriati. Karten et al. (1973) distinguish a hyperstriatum intercalatum supremum – as a layer of dispersed cells – and a n. intercalatus hyperstriati accessorii, i.e. a granular cell group wedged between the hyperstriatum intercalatum supremum and hyperstriatum accessorium. This n. intercalatus (Figs. 21.53, 21.66) is the specific target of bilateral projections from thalamic nuclei.

Shimizu and Karten (1990) analysed the Wulst for the presence of a large number of neurotransmitters and neuropeptides and found the distribution of immunohistochemically characterised fields generally fitting well in its laminar organisation (Fig. 21.66). Within each of the laminae several subregions with specific histochemical characteristics could be recognised; this is true for the distribution of specific types of cells and fibres (Table 21.5).

The thalamic fibres enter the forebrain through the lateral forebrain bundle (fasciculus lateralis telencephali), then pass through the neostriatum en hyperstriatum ventrale to reach the Wulst; fibres to the contralateral Wulst pass through the decussatio supraoptica to join the fasciculus lateralis telencephali (Meier et al. 1974). It is possible to subdivide the Wulst into a rostral, an intermediate and a caudal region, each with a specific input from a combination of thalamic centres (Miceli et al. 1990). Delius and Bennetto (1972) were the first to record somatosensory responses from the Wulst of the pigeon. More recently, Wild (1987a) and Funke (1989) showed that the rostral part receives a mainly ipsilateral somatosensory input from the n. dorsalis intermedius ventralis anterior. Visual projections derive from the 'nucleus opticus' (Karten et al. 1973) or, more specifically the n. dorsolateralis anterior thalami with its subdivisions; possibly, the same region receives input from the n. superficialis parvocellularis (Miceli and Repérant 1982: pigeon; Nixdorf and Bischof 1982: zebra finch; Bagnoli et al. 1990: owl). The visual projections are bilateral and retinotopically organised. In a study using evoked potentials more than 50 % of the units appeared to respond to contralateral stimulation, about 25 % to ipsilateral stimulation, whereas about 20 % could be driven by bilateral stimulation (Perisic et al. 1971).

The intercalated nucleus of the hyperstriatum accessorium is very distinct in the burrowing owl (Fig. 21.53; Karten et al. 1973), but less conspicuous in the pigeon. In the zebra finch – a bird hardly with a binocular field – the hyperstriatum intercalatum supremum is not differentiated at all; here, the hyperstriatum accessorium seems the recipient of visual afferents (Bredenkötter and Bischof 1990). Wilson (1980a) could not distinguish an intercalated nucleus in the chicken and found visual responses in the hyperstriatum accessorium and a narrow medial strip of hyperstriatum intercalatum supremum. Denton (1981), however, recorded 'early' field potentials in a region he calls the n. intercalatus of hyperstriatum accessorium and 'late' responses in the hyperstriatum accessorium itself after stimulation of the contralateral eye. Apparently, the differentiation of layers is more advanced in birds with large binocular fields, such as the owl (e.g. Pettigrew and Konishi 1976).

An intriguing aspect of the visual Wulst in the chicken is the occurrence of a transient asymmetry of the thalamohyperstriatal projection in cocks (Boxer and Stanford 1985; review in Rogers 1991). In young animals there is a bilateral projection from the left dorsal thalamus, but an ipsilateral projection from the right side. In adult animals the asymmetry has disappeared: now, both sides have bilateral projections – the fibres crossing through the supra-optic decussation – but a functional lateralisation, i.e. hemispheric specialisation, may persist. This asymmetry is determined by light exposure of the embryo (Rogers and Sink 1988); it can also be reversed by testosterone treatment (Zappia and Rogers 1987). The difference between juvenile males and females is reflected by a difference in visual discrimination learning using either the left or the right eye (Rogers 1991). This asymmetry could be related to asymmetry of the memory system for visual imprinting. It is tempting to assume that this difference between males and females is connected with differences in behaviour.

Deng and Wang (1992) report a certain overlap of somatosensory and visual input in the pigeon: somatosensory responses were found in the border zone of the hyperstriatum accessorium and hyperstriatum intercalatum supremum, visual responses in more dorsal parts of the hyperstriatum accessorium. We have already mentioned the suggestion of Korzeniewska that the multisensory convergence in the n. dorsolateralis posterior thalami could be comparable to that in the mammalian n. posterior thalami (Korzeniewska 1987). On the other hand, Funke (1989) found no visual responses in the somatosensory part of the Wulst. As for the existence of an auditory input, hardly any hard evidence exists. Wild (1987a) suggests the possibility of an auditory input in the hyperstriatum accessorium deriving from the multisensory caudal neostriatum. The somatosensory Wulst receives additional input from the frontal neostriatum and hyperstriatum dorsale (Funke 1989). Wild (1987a) described a reciprocal connection between this part of the Wulst and the somatosensory (caudal/intermediate) neostriatum; Funke does mention the neostriatal-Wulst projection but then a Wulst-(caudal) hyperstriatum ventrale projection instead of a Wulst-neostriatum projection.

The hyperstriatum dorsale (Figs. 21.29–21.31) is a layer of relatively large cells. It seems to receive other than visual or somatosensory input, e.g. from more medial dorsal thalamic nuclei, as well as from the n. coeruleus and n. tegmenti pedunculopontinus, pars compacta (Bagnoli and Burkhalter 1983). Little is known about the intrinsic organisation of the Wulst.

The Wulst is source of the septomesencephalic tract (Figs. 21.25–21.30, 21.42–21.44, 21.46–21.48). At least part of this fibre system has its origin in the hyperstriatum accessorium (Karten et al. 1973; Miceli and Repérant 1983). Bagnoli et al. (1980) suggest that a small part of the hyperstriatum intercalatum supremum may contribute fibres to this bundle. The tractus septomesencephalicus descends through the septum and consists of two extratelencephalic parts, a dorsal branch and a basal branch. The latter has its origin in the rostral, somatosensory part of the Wulst (Kalischer 1905: parrot; Karten 1971: the owl *Speotyto cunicularia*) and reaches the prerubral area – and in the owl also the n. ruber –, the medial spiriform nucleus, the medial reticular formation and medial pontine nucleus. In spite of its name more caudal projections than in the mesencephalon have been described, most notably to the cuneate-gracilis complex (owl, Karten 1971; parrot, Zecha 1962). Zecha was the first to suggest that these fibres might be the equivalent of part of the pyramidal tract. There is no evidence of more caudal projections or a role of this system in prehensile functions (Webster et al. 1990). The presence of this caudal projection, however, is not a general avian feature (Verhaart 1971; Dubbeldam 1976).

The dorsal branch of the septomesencephalic tract (Fig. 21.28) bends laterally after leaving the septum; in the pigeon it has projections to the ipsilateral retinorecipient thalamic cell groups, the ventral geniculate nucleus, pretectal nuclei and tectum (Karten et al. 1973; Bagnoli et al. 1980; Miceli and Repérant 1983; Miceli et al. 1987) as well as to the n. ectomammilaris (Rio et al. 1983). A few fibres cross through the decussatio supraoptica and commissura posterior with terminals in the contralateral ventral geniculate body. In owls, a significant projection to the contralateral tectum mesencephali also exists (Karten et al. 1973; Casini et al. 1992). There is physiological evidence that the Wulst influences the excitability of visual units in the tec-

Table 21.5. Relative densities of immunostained cells bodies and neuropil in the Wulst[a] (from Shimizu and Karten 1990)

Putative neurotransmitters	hyperstra(d)	hyperstra(v)	iha	his	hyperstrd(m)	vhyperstrd(l)	hypstr(l)
Cells							
ChAT	−	−	−	−	−	−	−
nAChr	+	−	+	+	−	−	+
TH	−	−	−	−	−	−	−
5-HT	−	−	−	−	−	−	−
GAD-1440	++	++	++	++	++	++	++
GAD-2	−	−	−	−	−	−	−
GABA$_A$	−	−	−	−	−	−	−
SP	++	++	++	−	+	+	++
CCK	−	−	−	++	+++	+++	++
L-ENK	+	+	+	+	+	+	+
NPY	−	−	−	−	−	−	+
NT	−	−	−	−	−	−	−
SRIF	++	++	−	+	+	+	++
CRF	++	+	+	+	−	−	++
VIP	−	−	−	−	−	−	−
Neuropil							
ChAT	+	+	++	+	−	+	++
nAChr	+	+	+	+	+	+	+
TH	++	+	+	+	+	+	+
5-HT	+++	+++	++	++(+)	+	+	+
GAD-1440	+	+	+	+	+	+	+
GAD-2	+	+	++	+	+(+)	+(+)	++
GABA$_A$	+	+	+	+	++	+	+
SP	+	++	++	−	+	+	+
CCK	+	+	−	++	+++	++	++
L-ENK	+	+	+	−	+	−	+
NPY	+	+	+	+	+	+	+
NT	+	+	−	+	++	+	(+)
SRIF	+(+)	+(+)	+	+	+	+	+
CRF	+	+	+	(+)	(+)	(+)	+
VIP	+	+	−	−	−	−	−

[a] Density: −, absent; +, low; ++, moderate; +++, high. Parentheses indicate a density that is slightly less than that indicated.

tum (Bagnoli et al. 1979; Leresche et al. 1983). Nau and Delius (1981) suggest that the Wulst may also modulate the intertectal inhibitory pathway. Furthermore it influences the directional sensitivity of neurons in the ectomammillary nucleus and may thus be involved in the control of the optokinetic nystagmus (Britto et al. 1990).

Intratelencephalic connections have also been described. Those of the somatosensory Wulst have already been mentioned. The visual Wulst sends projections to the ectostriatal belt and the overlying part of neostriatum (Karten et al. 1973). These fibres form a lateral fibre-system following the same route as the visual afferents. The Wulst may facilitate ectostriatal activity (Engelage and Bischof 1994). Hunt and Webster (1972) also found degenerated fibres following this route and reaching the dorsal thalamus, but assume that these are retrogradely degenerated afferents of the Wulst. Adamo (1967) described in the raven efferents from the Wulst leaving the telencephalon through the lateral forebrain bundle and projecting to the thalamus. This difference may due to the fact that in the latter study other parts of the Wulst than the intercalated nucleus were damaged by the lesions (Karten et al. 1973). The intermediate archistriatum is an intratelencephalic source of afferents to the Wulst.

Karten et al. (1973) stressed the similarities in organisation between the avian thalamohyperstriatal system and the mammalian lateral geniculate, p. dorsalis-striate cortex; particularly in the owl *Speotyto cunicularis* the visual Wulst looks like an 'upside-down' cortex. Interesting in this context is that Revzin (1970) reports the existence of a rough columnar organisation in the visual Wulst of the pigeon. He also demonstrated that the visual part of the Wulst contains units with relatively small visual fields with roughly a topographic organisation (Revzin 1969). Wilson (1980b) found in the chicken many units (62%) that responded to visual targets within a circumscribed part of the visual field, whereas a smaller proportion could only be activated by diffuse flashes. This view about the similarity of the two systems in birds and mammals is shared by many authors, though Pettigrew and Konishi (1976) caution that the specific organisation of visual Wulst and striate cortex may be correlated to the specific feature of binocular vision. Furthermore, it is generally accepted that at least the visual part of the Wulst corresponds to the reptilian dorsal cortex (e.g. Karten et al. 1973).

Finally a few remarks are necessary about the functional aspects. A major point has been the difference in function of the tecto-thalamo-ectostriatal and the thalamo-hyperstriatal systems. It has been suggested that the latter functions in brightness and pattern discrimination (Pritz et al. 1970). The effects of Wulst lesions, however, are not only related to visual defects. In particular lesions of the n. intercalatus hyperstriati affect the reversal learning procedures in pigeons (Powers 1989). Neither the size nor the location of the lesion affects the performance in normal learning, but the animals seem to suffer from an inability to learn in a new context without the interference of previous experience (Shimizu and Hodos 1989).

21.10.7.2
Hippocampus and Area Parahippocampalis

The dorsomedial cortex is designated hippocampus and area parahippocampalis (Figs. 21.26–21.29, 21.45–21.47). This is not a well-defined area. Ariëns Kappers et al. (1936) discuss this area at length. In a comparison with the reptilian situation these authors distinguish a four-layered dorsal hippocampus or entorhinal area and a dorsomedial hippocampus; these correspond to our parahippocampal area and hippocampus, respectively. Van Tienhoven and Juhasz (1962) use the names area para-entorhinalis and area entorhinalis. There is no simple way to homologise subdivisions of this area with the mammalian hippocampal structures. Several criteria have been used, firstly, the arrangement in layers. In the emu – a paleognath bird – Craigie (1935, in Ariëns Kappers et al. 1936) distinguished a complex of layers in the parahippocampal area and suggested that these may be homologous to those in the mammalian cortex. The situation in neognath birds is not clear. Mollà et al. (1986) describe a superficial plexiform layer, a granular layer and a periventricular layer in the dorsomedial cortex of the chicken; the medial, hippocampal, part contains pyramidal and bipyramidal cells that are absent in the area parahippocampalis. According to these authors the absence of a compact granular layer and of mossy fibres frustrate an easy comparison; moreover, periventricular neurons have not been described in the mammalian hippocampus.

A second criterion could be the pattern of afferent and efferent connections. Afferent connections of the hippocampus and parahippocampalis derive from the hyperstriatum accessorium, the n. tractus diagonalis, n. taeniae, and area corticoidea dorsolateralis (Fig. 21.27); further reciprocal connections between hippocampus and parahippocampalis as well as between left and right hippocampus exist (Casini et al. 1986). The parahippocampal area also receives afferents from the medial septal nucleus, from the lateral hypothalamus, n. superficialis parvocellularis and locus coeruleus (Benowitz and Karten 1976a). Projections from the dorsomedial cortex reach the septal nuclei, the n. tractus diagonalis, n. taeniae and cortex dorsolateralis, a modest projection reaching approximately the n. mammillaris (Krayniak and Siegel 1978a; Casini et al. 1986). Krayniak and Siegel did not find the hypothalamic projection and compared the efferent hippocampal tract with the precommissural part of the mammalian fornix. Bons et al. (1976) described a hippocampal projection to the n. posterior lateralis hypothalami in the pigeon; this may be the same as that indicated by Casini et al. (1986). The existence of a projection to the medial part of the intermediate hyperstriatum ventrale (Bradley et al. 1985) has not been observed by the other authors. Several parallels between this pattern of connections and that in mammals support the notion that the dorsomedial cortex represents the hippocampal and parahippocampal areas.

Additional evidence comes from histochemical studies. Krebs et al. (1991) studied the distribution of serotonin-like, choline acetyltransferase (ChAT)-like, thyroid hormone (TH)-like and GAD-like immunoreactivity in the dorsomedial cortex. All four substances are found in the neuropil and in terminals, particularly in the more dorsal region. The first three are also present in a fibre tract passing the septo-hippocampal junction, the fourth one being present in cell bodies throughout the dorsomedial region. Also the distribution of six neuropeptides in this region has been described (Erichsen et al. 1991; cf. also Erichsen et al. 1994). Again, these observations support the notion of an avian hippocampus and parahippocampal area. Nevertheless, a detailed comparison of the subdivisions of this system in birds and mammals can at best be speculative.

From a functional point of view it is interesting to note that evidence of an important role of the hippocampus in spatial memory and cognition has been found for both mammals and birds (Sherry et al. 1992). One line of evidence is its role in spatial navigation (review in Bingman 1990): hippocampus ablation in pigeons results specifically in impairment of landmark navigation (use of familiar landmarks) during homing. Other evidence comes from studies of food-storing birds (Krebs et al 1989;

Sherry et al. 1989; Hampton et al. 1995): food-storing species of passerines appear to have a relatively larger hippocampal complex than related non-food-storing species. In addition, Clayton and Krebs (1994) proved that in marsh tits the size of the hippocampus was influenced by experience; this influence is independent of age. Apparently, experience prevents cell loss.

21.10.7.3
Other Corticoid Regions

The area corticoidea dorsolateralis (Figs. 21.26, 21.27) is a lateral corticoid area that is separated from the neostriatum by the lateral ventricle. It has a connection with the hippocampal region. The piriform cortex (Figs. 21.26, 21.27) is discussed in Sect. 21.10.9. The area temporoparieto-occipitalis (Fig. 21.28) is part of the neostriatum-hyperstriatum ventrale complex and has already been discussed.

21.10.8
Septal Area

Relatively little is known about the organisation of the septum in birds. Breazile and Kuenzle (1993) subdivide the area septalis into a n. septalis lateralis (Figs. 21.29, 21.32, 21.42, 21.48), a n. septalis medialis (Fig. 21.28), a n. accumbens and a n. interstitialis. Krayniak and Siegel (1978b) distinguish a dorsal and a ventral septal region; the dorsal part consists of the lateral and medial septal nuclei, the most important element of the ventral part being the nucleus of the diagonal band [n. tractus diagonalis]. The latter cell group lies lateral to the anterior preoptic nucleus (Fig. 21.29). The medial and lateral nuclei are also depicted in the atlas of Karten and Hodos (1967); this also mentions a n. commissuralis septi, a cell group ventral to the lateral nucleus and pallial commissure. It is not mentioned in the atlas of the chicken (Kuenzel and Masson 1988), neither in the NAA. Generally, the n. accumbens is considered part of the paleostriatal complex (Sect. 21.10.2).

Even though it can be questioned whether or not the n. tractus diagonalis should be considered part of the septum, this nucleus has a few things in common with the dorsal septal area. Both regions receive input from the hippocampus and parahippocampal area. The dorsal septum sends a projection to the lateral hypothalamus, dorsomedial thalamus, periaqueductal grey and mesencephalic reticular formation (Krayniak and Siegel 1978b). The n. tractus diagonalis, too, sends fibres to the lateral hypothalamus, but also to the n. tuberis, medial hypothalamus, ventral mesencephalic tegmental area, lateral habenula and hippocampus and parahippocampal area. According to Benowitz and Karten (1976a) the area parahippocampalis is also recipient of an input from the medial septum.

21.10.9
Olfactory System and Terminal Nerve

The olfactory system of the pigeon, as judged by the size of the olfactory bulbs, is of intermediate size compared to other avian species (Bang and Cobb 1968; Matochik et al. 1991). The olfactory organ of the pigeon seems rather well developed (Müller et al. 1979). The olfactory bulb is a multilaminated structure with an outer layer of fila olfactoria, distinct glomerular layer, external plexiform layer, mitral cell layer, internal plexiform layer and granular layer (Fig. 21.31); it possesses a central ventricle (volf, Fig. 21.33). Efferents of the olfactory bulb project to several areas (Fig. 21.67; Rieke and Wenzel 1978; Reiner and Karten 1985): through a medial olfactory tract to the medial septal nucleus and to more dorsal parts of the medial telencephalon, through a lateral olfactory tract to the piriform cortex, i.e. a caudolateral corticoid area separated from the archistriatum by the lateral ventricle. An intermediate projection reaches the olfactory tubercle and rostral part of the lobus parolfactorius. Fibres seem to cross the diagonal band of Broca and enter the diencephalon through the stria medullaris (Figs. 21.27, 21.48) to cross through the commissura habenularis and to ascend into the contralateral hemisphere projecting to corresponding regions. The anterior commissure may have little importance for the interhemispheric transfer of olfactory information (Gagliardo and Teyssèdre 1988). Reiner and Karten mention a projection to the n. taeniae and Rieke and Wenzel a projection to part of the ventral hyperstriatum. The latter anatomical observation is supported by electrophysiological recordings (Macadar et al. 1980).

Ebinger et al. (1992) describe the olfactory projections in the mallard; these authors distinguish distinct medial projections to an area prepiriformis, an intermediate and retrobulbar region and also to the medial septum, medial part of the parolfactory lobe and minor parts of neostriatum and hyperstriatum ventrale and a lateral projection to the cortex piriformis of Reiner and Karten (but not to the taenia). Ebinger et al. propose on topological grounds designating the latter area the 'regio periamygdalaris', suggesting an homology with the mammalian area with the same name, whereas the term 'piriform cortex' is based on a comparison with reptiles (Reiner and Karten 1985). Bingman et al. (1994)

found projections from the piriform cortex to the parolfactory lobe, septum, parahippocampal area and dorsal hyperstriatum, as well as to dorsomedial thalamic nuclei and parts of the hypothalamus. These authors suggest that the piriform cortex is part of a system that regulates navigational map learning using olfactory stimuli. Papi and Casini (1990) previously suggested that the piriform cortex may be important for orientation in unfamiliar sites.

The different configurations of the olfactory bulb and rostral hemisphere in pigeon (Figs. 21.31, 21.33) and chicken (Kuenzel and Masson 1988) on the one hand, and in the duck on the other hand hamper an easy comparison. However, following the interpretation of this area by Nieuwenhuys (1967a), a cautious conclusion – based on the figures in Reiner and Karten and in Ebinger et al. – can be that the cortex prepiriformis (i.e. n. olfactorius anterior, pars lateralis) as well as the n. olfactorius anterior, p. medialis (i.e. area retrobulbaris) are recipients of bulbar efferents. The olfactory tubercle in the duck seems to be a minor structure. The prepiriform cortex is caudally followed by the n. olfactorius anterior sensu Karten and Hodos (1967; see also Kuenzel and Masson 1988). This interpretation differs slightly from that of Breazile and Hartwig (1989). Teuchert et al. (1986) using horseradish peroxidase and degeneration preparations suggest the existence of additional, but transient efferent as well as afferent olfactory connections in duck embryos and young ducklings: a number of myelinated fibres enter the lamina frontalis superior (Figs. 21.48, 21.49) from the medial olfactory pathway and the lamina medullaris dorsalis from the lateral pathway, but disappear in the adult birds.

For a long time the sense of smell was assumed to be of no or little importance in birds. One of the reasons was that it is difficult to train birds using conventional olfactory stimuli (e.g. Neuhaus 1963, Fink 1965). Even though later evidence shows that physiological responses (changes in respiration, heart rate) can be recorded, it is still questionable whether smell in birds is important for recognition and the guidance of behaviour, whether it may have an affective value or whether it is simply vestigial and of no significance for their behaviour (Wenzel 1971). There seems to be little evidence that the absence of olfactory input deprives the pigeon of significant environmental clues, but it may modulate the affective behaviour (Wenzel and Rausch 1977). At least in some species, however, smell is important, e.g. in feeding (e.g. Grubb 1972; Würdinger 1979) and in the discrimination of plant volatiles (Clark and Mason 1987).

A major point of debate has been – and still is – the importance of smell in homing behaviour. Conflicting observations led to a paper with the title 'Do American and Italian pigeons rely on different homing mechanisms' (Papi et al. 1978). Much evidence in favour of an important role of the olfactory system is provided by Italian researchers (e.g. Papi et al. 1974; Meschini 1983; Wiltschko and Wiltschko 1987; Wallraff et al. 1992). However, there is increasing evidence that olfactory cues are not always essential for the formation of navigational maps (Waldvogel and Phillips 1991) and that pigeons may rely upon different senses (visual cues, geomagnetic cues, e.g. Lednor and Walcott 1988; Beason et al. 1995) depending upon the area where they are raised (e.g. Wallraff 1988). In this context a distinction should be made between 'true navigation' and 'landmark navigation'; visual, but also olfactory cues, may be important particularly in landmark navigation (Bingman 1990).

For a long time a *nervus terminalis* had not been recognised in birds, but there is some recent evidence that fibres and cells of a terminal nerve are incorporated in the olfactory nerve and bulb (e.g. von Bartheld et al. 1987). These elements are immunoreactive for a molluscan cardioexcitatory peptide (FMRF-amide) (Wirsig-Wiechmann 1990) and may contain LHRH (Norgren et al. 1992).

21.10.10
Ascending, Descending and Intrinsic Telencephalic Fibre Systems

Many fibre systems have been described in the preceding sections. These will be shortly summarised with some additional comments. Most *ascending sensory systems* enter the telencephalon through the fasciculus lateralis telencephali: the quintofrontal tract from the n. sensorius principalis N. V, the thalamofrontal tracts from several thalamic sensory nuclei, e.g. nn. rotundus, ovoidalis and dorsolateralis anterior. Some of these fibres may fan out and form (part of) the lamina medullaris ventralis (Figs. 21.46, 21.48, 21.49; e.g. Dubbeldam et al. 1981). Distinct streams of fibres can be seen to reach the various sensory regions in the telencephalon, in particular the ectostriatum (Figs. 21.42, 21.48) and n. basalis (Fig. 21.43). The most important ascending pathways are summarised in Fig. 21.68.

Fig. 21.67a–h. Distribution of silver grains after an injection of tritiated proline/tritiated leucine mixture in the olfactory bulb (*OB*) of the pigeon. A medial pathway to the septum (*sep*) and a lateral pathway to the piriform cortex (*cpir*) and dorsolateral cortex (*acdl*) can be distinguished. Some fibres enter the stria medullaris (*sm*) to decussate through the habenular commissure (*chab*). (Redrawn after Reiner and Karten 1985)

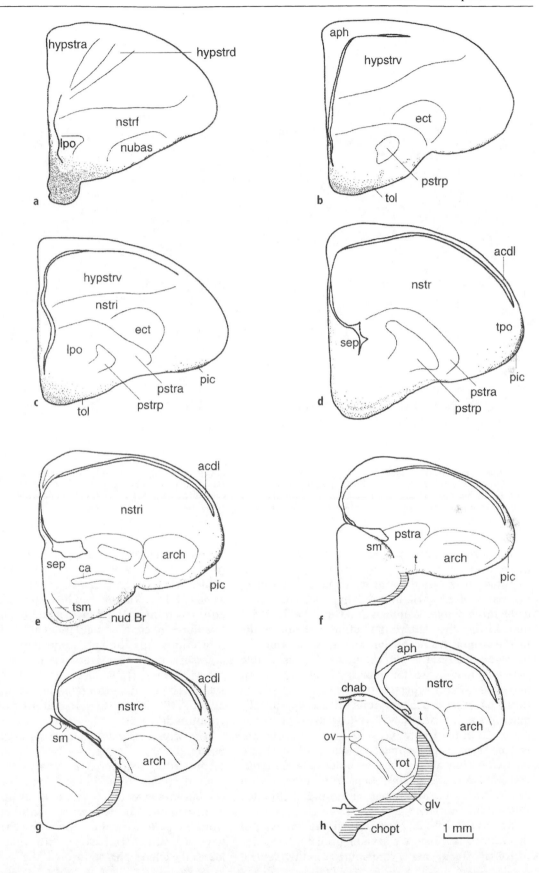

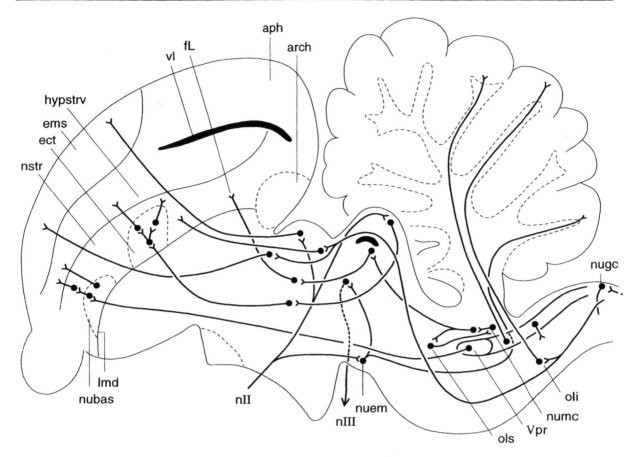

Fig. 21.68. Diagram of some of the sensory pathways in lateral view: the somatosensory pathway+projection to cerebellum via the oliva inferior; the trigeminal pathways with cerebellar projection from the n. radicis descendentis nervi trigemini, subn. oralis and reticular projections from the n. radicis descendentis nervi trigemini, subn. interpolaris and caudalis; auditory pathway with feedback via oliva superior; visual pathways including accessory optic-oculomotor loop

The intrinsic fibre systems: a predominantly vertically organised pattern of ascending and descending fibres traverses the sensory regions and overlying parts of neostriatum and hyperstriatum ventrale; for example, Veenman and Gottschaldt (1986) described such a pattern in relation to the n. basalis in the frontal telencephalon of the goose, Dubbeldam and Visser (1987) in the mallard. Comparable patterns may exist for more caudal regions of neostriatum and hyperstriatum ventrale in connection with other sensory regions (discussion in Dubbeldam 1991). The tractus frontoarchistriaticus has its origin in the frontal neostriatum and terminates in the sensorimotor part of the archistriatum, partly possibly also in the lateral neostriatum intermedium. More caudally fibres from more dorsal regions reach the archistriatum through the tractus dorsoarchistriaticus (see Sect. 21.10.6).

Relatively little is known about the contents of the various laminae. We have mentioned that quintofrontal fibres may contribute to the ventral medullary lamina. Some neostriatal fibres may reach the archistriatum by way of the lamina medullaris dorsalis (Figs. 21.28, 21.29, 21.42–21.44, 21.46–21.49), whereas archistriatal efferents may also reach medial telencephalic regions through this fibre system (Dubbeldam et al. 1997). Thalamic projections to the n. intercalatus hyperstriati accessorii reach this area via the lamina frontalis superior (Figs. 21.28, 21.30, 21.31; Watanabe et al. 1983). The lamina frontalis suprema (Figs. 21.43, 21.44, 21.47) forms the dorsal border of the hyperstriatum dorsale.

The large extratelencephalic systems: three large descending fibre systems have been described in the preceding sections, viz. the septomesencephalic tract deriving from the eminentia sagittalis, the occipitomesencephalic tract deriving from the archistriatum and the ansa lenticularis, arising from the paleostriatal complex. The first two tracts have a distinct trajectory; the ansa lenticularis leaves the telencephalon through the fasciculus later-

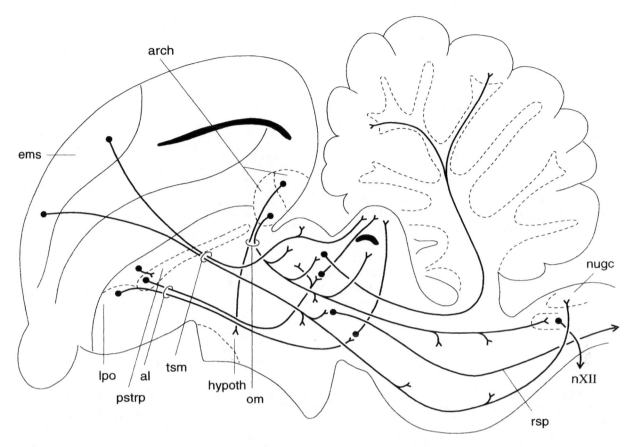

Fig. 21.69. Diagram of some of the extratelencephalic pathways in lateral view: occipitomesencephalic tract (*om*) with hypothalamic component and "motor" component, the latter with a relay in n. spiriformis medialis projecting to the cerebellum; septomesencephalic tract (*tsm*) with a dorsal and a basal component, the latter with a projection to the n. ruber, source of the rubrospinal tract (*rsp*); ansa lenticularis (*al*) with a striatal component via the n. tegmenti pedunculopontinus (substantia nigra) and tectum mesencephali and a pallidal component to the n. dorsointermedius posterior thalami and via the n. spiriformis lateralis to the tectum

alis telencephali (Fig. 21.69). Several ascending groups of fibres use these pathways to reach the forebrain, e.g. those arising from the locus coeruleus and avian substantia nigra. Huber and Crosby (1929) described a tractus striomesencephalicus et striocerebellaris (see Ariëns Kappers et al. 1936). Verhaart (1971) found no evidence for the existence of the striocerebellar component. This author doubted whether the striomesencephalic tract could be compared to the mammalian ansa lenticularis, but the experimental study by Karten and Dubbeldam (1973) concluded that the striomesencephalic tract does indeed correspond to the mammalian ansa lenticularis. Verhaart (1971) described a system of fine fibres running medially with his tractus and apparently having its origin in the lobus parolfactorius, but he declined to specify the character of this system.

A fourth descending system is the fasciculus medialis telencephali (Figs. 21.29, 21.32). Karten and Dubbeldam (1973) traced this tract after lesions in the rostral lobus parolfactorius of the pigeon; it is a system of fine fibres that could easily be followed into the hypothalamus. Later studies showed that this system has a terminal field in the dorsal thalamus (Kitt and Brauth 1981). In the section on the paleostriatum it was pointed out that the lateral part of the lobus parolfactorius contributes to the ansa lenticularis; the medial forebrain bundle probably therefore derives from the rostral and medial lobus parolfactorius. Reiner et al. (1983) showed that substance-P-positive fibres of the medial lobus parolfactorius course caudally through the diencephalon and project to the ventral tegmental area (Tsai), n. tegmenti pedunculopontinus and part of the locus coeruleus.

21.11
Concluding Remarks

The overview presented makes it clear that birds possess relatively large and well-developed brains. Striking differences with on one hand the reptiles and on the other hand the mammals are found, in particular in the organisation of the telencephalon. Notwithstanding recent attempts to clarify the origin of the neocortex in mammals and in spite of the recognition of common features of systems within the avian and mammalian forebrain – as described in the preceding sections – there is still no satisfactory explanation, either about the precise origin, or about the functional implications of the different organisations. On the other hand, the avian forebrain is reminiscent of that of the reptiles, but it lacks some of the distinct cortical structures and the differentiation of the dorsal ventricular ridge seems to be different in birds and reptiles. Apparently, the avian brain represents its own evolutionary line of development.

As for its size, the avian telencephalon is comparable to that of many mammals. It is possible to identify groups with low and high 'cerebralisation degrees'. Again, the functional interpretation of such differences can at best be speculative. An important aspect is that birds exhibit a wide array of ecological specialisations occupying all kinds of niches and rich repertoires of behaviours. This ecological and behavioural diversity is reflected in a fascinating diversity of brain structures among the avian species. This diversity is not only a matter of quantitative differences, but also impressive qualitative differences have been found between species as shown in, for example, Stingelin's studies. This rich diversity makes birds an excellent subject for the study of the relation between brain and behaviour as exemplified in the many studies on learning (e.g. Andrew 1991) and on vocalisation (e.g. Nottebohm 1993).

Studies on avian brains are largely neglected in the mammalian literature. Probably, this is caused by the difficulties in comparing the avian and mammalian brain structure. However, repeatedly studies of the avian brain have contributed to a better understanding of brain structure and function in general, e.g. the studies on the avian visual system. As the knowledge of the basic organisation of the bird brain, and in particular that of the forebrain, increases, new contributions can be expected to the fundamental knowledge of brain function and specialisation and its role in specific aspects of behaviour and ecology.

Acknowledgements. Mr. Martin Brittijn prepared many of the drawings for this chapter and Ms. Trudy Hamerling-van Delft prepared the list of references. I thank Dr. Hans ten Donkelaar for his comments on a previous version of the manuscript.

References

Able KP, Able MA (1993) Daytime calibration of magnetic orientation in a migratory bird requires a view of skylight polarization. Nature 364:323

Abs M (1983) Physiology and behaviour of the pigeon. Academic Press, London

Adamo NJ (1967) Connections of efferent fibers from hyperstriatal areas in chicken, raven, and African lovebird. J Comp Neurol 131:337–356

Adamo NJ, King RL (1967) Evoked responses in the chicken telencephalon to auditory, visual and tactile stimulation. Exp Neurol 17:498–504

Akesson TR, De Lanerolle NC, Cheng MF (1987) Ascending vocalization pathways in the female ring dove: projections of the nucleus intercollicularis. Exp Neurol 95:34–43

Albin RL, Sakurai SY, Makowiec RL, Gilman S (1991) Excitatory and inhibitory amino acid neurotransmitter binding sites in the cerebellar cortex of the pigeon (Columba livia). J Chem Neuroanat 4:429–437

Alvarez-Buylla A, Theelen M, Nottebohm F (1988) Birth of projection neurons in the higher vocal center of the canary forebrain before, during, and after song learning. Proc Natl Acad Sci USA 85:8722–2726

Andrew RJ (1991) Neural and behavioural plasticity. The use of domestic chick as a model. Oxford University Press, Oxford

Angaut P, Repérant J (1978) A light and electron microscopic study of the nucleus isthmo-opticus in the pigeon. Arch d'Anat Micr Morphol Exp 67:63–78

Antal M, Polgár E, Berki Á, Birinyi A, Puskár Z (1994) Development of specific populations of interneurons in the ventral horn of the embryonic chick lumbosacral spinal cord. Eur J Morphol 32:201–206

Arai O, Taniguchi I, Saito N (1989) Correlation between the size of song control nuclei and plumage color change in orange bishop birds. Neurosci Lett 98:144–148

Arends JJA (1981) Sensory and motor aspects of the trigeminal system in the mallard (Anas platyrhynchos L.). Thesis, Leiden University

Arends JJA, Dubbeldam JL (1982) Exteroceptive and proprioceptive afferents of the trigeminal and facial motor nuclei in the mallard (Anas platyrhynchos L.). J Comp Neurol 209:313–329

Arends JJA, Dubbeldam JL (1984) The subnuclei and primary afferents of the descending trigeminal system in the mallard (Anas platyrhynchos L.). Neuroscience 13:781–795

Arends JJA, Voogd J (1989) Topographical aspects of the olivo-cerebellar system in the pigeon. Exp Brain Res Series 17:52–57

Arends JJA, Zeigler HP (1986) Anatomical identification of an auditory pathway from a nucleus of the lateral lemniscal system to the frontal telencephalon nucleus basalis of the pigeon. Brain Res 398:375–381

Arends JJA, Zeigler HP (1989) Cerebellar connections of the trigeminal system in the pigeon (Columba livia). Brain Res 487:69–78

Arends JJA, Zeigler HP (1991a) Organization of the cerebellum in the pigeon (Columba livia): I. Corticonuclear and corticovestibular connections. J Comp Neurol 306:221–244

Arends JJA, Zeigler HP (1991b) Organization of the cerebellum in the pigeon (Columba livia): II. projections of the cerebellar nuclei. J Comp Neurol 306:245–272

Arends JJA, Woelders-Blok A, Dubbeldam JL (1984) The efferent connections of the nuclei of the descending trigeminal tract in the mallard (Anas platyrhynchos L.). Neuroscience 13:797–817

Arends JJA, Wild JM, Zeigler HP (1988) Projections of the nucleus of the tractus solitarius in the pigeon (Columba livia). J Comp Neurol 278:405–429

Arends JJA, Allan RW, Zeigler HP (1991) Organization of the cerebellum in the pigeon (Columba livia): III Corticovestibular connections with eye and neck premotor areas. J Comp Neurol 306:273–289

Ariëns Kappers CU, Huber GC, Crosby EC (1936). The comparative anatomy of the nervous system of vertebrates, including man, 2 vols. MacMillan, New York (reprint in 3 vols, 1967, Hafner, New York)

Armstrong RC, Clarke PHG (1979) Neuronal death and the development of the pontine nuclei and the inferior olive in the chick. Neuroscience 4:1635–1645

Azcoitia I, Fernandez-Soriano J, Fernandez-Ruiz B (1985) Is the avian glucogen body a secretory organ? J Hirnforsch 26:651–567

Bagnoli P, Burkhalter A (1983) Organization of the afferent projections to the Wulst in the pigeon. J Comp Neurol 214:103–113

Bagnoli P, Casini G (1985) Regional distribution of catecholaminergic terminals in the pigeon visual system. Brain Res 337:272–286

Bagnoli P, Francesconi W, Magni F (1979) Interaction of optic tract and visual Wulst impulses on single units of the pigeons optic tectum. Brain Behav Evol 16:19–37

Bagnoli P, Grassi S, Magni F (1980) A direct connection between visual Wulst and tectum opticum in the pigeon (Columba livia) demonstrated by horseradish peroxidase. Arch Ital Biol 118:72–88

Bagnoli P, Burkhalter A, Streit P, Cuénod M (1983) [^{3}H]-GABA selective retrograde labeling of neurons in the pigeon thalamo-Wulst pathway. Arch Ital Biol 121:47–53

Bagnoli P, Porciatti V, Fontanesi G, Sebastiani L (1987) Morphological and functional changes in the retinotectal system of the pigeon during the early posthatching period. J Comp Neurol 256:400–411

Bagnoli P, Fontanesi G, Casini G, Porciatti V (1990) Binocularity in the little owl, Athene noctua. I. Anatomical investigation of the thalamo-Wulst pathway. Brain Behav Evol 35:31–39

Ball GF, Foidart A, Balthazart J (1989a) A dorsomedial subdivision within the nucleus intercollicularis identified in the Japanese quail (Coturnix coturnix japonica) by means of α_2-adrenergic receptor autoradiography and estrogen receptor immunohistochemistry. Cell Tissue Res 257:123–128

Ball GF, Nock B, McEwen BS, Balthazart J (1989b) Distribution of the α_2-adrenergic receptors in the brain of the Japanese quail as determined by quantitative autoradiography: implications for the control of sexually dimorphic reproductive processes. Brain Res 491:68–79

Ballam GO (1982) Bilateral and multimodal sensory interactions of single cells in the pigeon's midbrain. Brain Res 245:27–34

Balthazart J, Dupiereux V, Aste N, Viglietti-Panzica C, Barrese M, Panzica GC (1994) Afferent and efferent connections of the sexually dimorphic medial preoptic nucleus of the male quail revealed by in vitro transport of DiI. Cell Tissue Res 276:455–475

Bang BG, Cobb S (1968) The size of the olfactory bulb in 108 species of birds. Auk 85:55–61

Barakat-Walter I, Riederer BM (1991) Brain spectrins 240/235 and 240/235E: differential expression during development of chicken dorsal root ganglia in vivo and in vitro. Eur J Neurosci 3:431–440

Barrett RK, Underwood H (1991) Retinally perceived light can entrain the pineal melatonin rhythm in Japanese quail. Brain Res 563:87–93

Baylé JD, Ramade F, Olivier J (1974) Stereotaxic topography of the brain of the quail, Coturnix coturnix japonica. J Physiol [Paris] 68:219–241

Beason RC, Dussourd N, Deutschlander ME (1995) Behavioural evidence for the use of magnetic material in magnetoreception by a migratory bird. J Exp Biol 198: 141–146

Beaudet A, Burkhalter A, Reubé JC, Cuénod M (1981) Selective bidirectional transport of [^{3}H]D-aspartate in the pigeon retinotectal pathway. Neuroscience 6:2021–2034

Benowitz LI (1980) Functional organization of the avian brain. In: Ebbeson SOE (ed) Comparative neurology of the telencephalon. Plenum, New York, pp 389–421

Benowitz LI, Karten HJ (1976a) The tractus infundibuli and other afferents to the parahippocampal region of the pigeon. Brain Res 102:174–180

Benowitz LI, Karten HJ (1976b) Organization of the tectofugal visual pathway in the pigeon: a retrograde transport study. J Comp Neurol 167:503–520

Bergquist H, Källén B (1954) Notes on the early histogenesis and morphogenesis of the central system nervous system in vertebrates. J Comp Neurol 100:627–660

Berk ML (1987) Projections of the lateral hypothalamus and bed nucleus of the stria terminalis to the dorsal vagal complex in the pigeon. J Comp Neurol 260:140–156

Berk ML, Butler AB (1981). Efferent projections of the medial preoptic nucleus and medial hypothalamus in the pigeon. J Comp Neurol 203:379–399

Berk ML, Hawkin RF (1985) Ascending projections of the mammilary region in the pigeon: emphasis on telencephalic connections. J Comp Neurol 239:330–340

Berk ML, Reaves TA, Hayward JN, Finkelstein J (1982) The localization of vasotocin and neurophysin neurons in the diencephalon of the pigeon, Columba livia. J Comp Neurol 204:392–406

Berkhoudt H (1980) The morphology and distribution of cutaneous mechano receptors (Herbst and Grandy corpuscles) in the bill and the tongue of the mallard (Anas platyrhynchos L.). Neth J Zool 30:1–34

Berkhoudt H (1985) Structure and function of avian taste receptors. In: King AS, McLelland J (eds). Form and function in birds, vol 3. Academic Press, London, pp 463–496

Berkhoudt H, Dubbeldam JL, Zeilstra S (1981) Studies on the somatotopy of the trigeminal system in the mallard, Anas platyrhynchos L. IV. Tactile representation in the nucleus basalis. J Comp Neurol 196:407–420

Berkhoudt H, Klein BG, Zeigler HP (1982) Afferents to the trigeminal and facial motor nuclei in pigeon (Columba livia) L.: central connections of jaw motoneurons. J Comp Neurol 209:301–312

Bertossi ML, Roncali L, Mancini L, Ribatti D, Nico B (1986) Process of differentiation of cerebellar Purkinje neurons in the chick embryo. Anat Embryol 175:25–34

Bigalke-Kunz B, Rübsamen R, Dorscheidt GJ (1987) Tonotopic organization and functional characterization of the auditory thalamus in a songbird; the European starling. J Comp Physiol A 161:255–265

Bilge M (1971) Electrophysiological investigations on the pigeon's optic tectum. Q J Exp Physiol 56:242–249

Bingman VP (1990) Spatial navigation in birds. In: Kesner RP, Olten DS (eds) Neurobiology of comparative cognition. L Erlbaum Ass, Hilsdale NJ, pp 423–447

Bingman VP, Casini G, Nocjar C, Jones TJ (1994) Connections of the piriform cortex in homing pigeons (Columba livia) studied with fast-blue and WGA-HRP. Brain Behav Evol 43:206–218

Bischof HJ, Niemann J (1990) Contralateral projections of the optic tectum of the zebra finch (Taenopygia guttata castanotis). Cell Tissue Res 262:307–313

Bissoli R, Battistini S, Guarnieri T, Contestabile A (1988) Regional levels of neurotransmitter markers in the pigeon telencephalon: a comparison with possible homologous areas of the rat telecephalon. J Neurochem 50:1731–1737

Blähser S, Dubois MP (1980) Immunocytochemical demonstration of met-enkephalin in the central nervous system of the domestic fowl. Cell Tissue Res 213:53–68

Blähser S, Fellman D, Bugnon C (1978) Immunocytochemical demonstration of somatostatin-containing neurons in the hypothalamus of the domestic mallard. Cell Tissue Res 195:183-187

Blümcke S (1961) Vergleichend experimentell-morphologische Untersuchungen zur Frage einer retinohypothalamischen Bahn bei Huhn, Meerschweinchen und Katze. Z Mikrosk-Anat Forsch 67:469-513

Bodega V, Suarez I, Fernandez B (1989) The possible secretory function of the marginal nuclei of the avian spinal cord. J Anat 165:19-28

Bodnarenko SR, Rojas X, McKenna OC (1988) Spatial organization of the retinal projection to the avian lentiform nucleus of the mesencephalon. J Comp Neurol 269:431-447

Bolze G (1969) Anordnung und Bau der Herbstchen Körperchen in Limicolen Schnabeln im Zusammenhang mit der Nahrungsfindung. Zool Anz 181:313-355

Bonaventure N, Kim MS, Jardon B (1992) Effects on the chicken monocular OKN of unilateral microinjections of GABA-A antagonist into the mesencephalic structures responsible for OKN. Exp Brain Res 90:63-71

Bonke BA, Bonke D, Scheich H (1979a) Connectivity of the auditory forebrain nuclei in the guinea fowl (Numida meleagris). Cell Tissue Res 200:101-121

Bonke D, Scheich H, Langner G (1979b) Responsiveness of units in the auditory neostriatum of the guinea fowl (Numida meleagris) to species-specific calls and synthetic stimuli. J Comp Physiol 132:243-255

Bons N (1969) Neuro-anatomie. Mise en évidence du croissement incomplet des nerfs optiques au niveau du chiasma chez le canard. C R Acad Sci Paris 268:2186-2188

Bons N (1974) Mise en évidence au microscope électronique, de terminaisons nerveuses d'origine rétinienne dans l'hypothalamus antérieur du Canard. C R Acad Sc Paris 278:319-321

Bons N (1976) Retinohypothalamic pathway in the duck (Anas platyrhynchos). Cell Tissue Res 168:343-360

Bons N (1980) The topography of mesotocin and vasotocin systems in the brain of the domestic mallard and Japanese quail: immunocytochemical identification. Cell Tissue Res 213:37-51

Bons N, Oliver J (1986) Origin of the afferent connections to the parolfactory lobe in quail shown by retrograde labelling with a fluorescent neuron tracer. Exp Brain Res 63:125-134

Bons N, Bouillé C, Baylé JD, Assenmacher I (1976) Light and electron microscopic evidence of hypothalamic afferences originating from the hippocampus. Experientia 32:1443-1445

Bons N, Kerdelhué B, Assenmacher I (1978) Immunocytochemical identification of an LHRH-producing system originating in the preoptic nucleus of the duck. Cell Tissue Res 188:99-106

Boord RL (1968) Ascending projections of the primary cochlear nuclei and nucleus laminaris in the pigeon. J Comp Neurol 133:523-542

Boord RL, Karten HJ (1974) The distribution of primary lagenar fibers within the vestibular nuclear complex of the pigeon. Brain Behav Evol 10:228-235

Boord RL, Rasmussen GL (1963) Projection of cochlear and lagenar nerves on the cochlear nuclei of the pigeon. J Comp Neurol 120:463-475

Bortolami R, Callegari E, Lucchi ML (1972) Anatomical relationship between mesencephalic trigeminal nucleus and cerebellum in the duck. Brain Res 47:317-329

Bortolami R, Veggetti A, Ciampoli A (1969) Electron microscopic observations on the mesencephalic trigeminal nucleus MTN of duck. J Submicr Cytol 1:235-245

Bottjer SW, Arnold AP (1982) Afferent neurons in the hypoglossal nerve of the zebrafinch (Poephila guttata): localization with horseradish peroxidase. J Comp Neurol 210:190-197

Bouillé C, Baylé JD (1973) Experimental studies on the adrenocorticotrope area of the pigeon hypothalamus. Neuroendocrinology 11:73-91

Bouillé C, Raymond J, Baylé JD (1977) Retrograde transport of horseradish peroxidase from the n. posterior medialis hypothalami to the hippocampus and the medial septum in the pigeon. Neuroscience 2:435-439

Bout RG (1987) Neuroanatomical circuits for proprioceptive and motor control of feeding movements in the mallard (Anas platyrhynchos L.). Thesis, Leiden University

Bout RG, Dubbeldam JL (1985) An HRP study of the central connections of the facial nerve in the mallard (Anas platyrhynchos L.). Acta Morphol Neerl-Scand 23:181-193

Bout RG, Dubbeldam JL (1994) The reticular premotor neurons of the jaw muscle motor nuclei in the mallard (Anas platyrhynchos L.). Eur J Morph 32:134-137

Boxer MI, Stanford D (1985) Projections to the posterior visual hyperstriatal region in the chick: an HRP study. Exp Brain Res 57:494-498

Bradley P, Davies DC, Horn G (1985) Connections of the hyperstriatum ventrale of the domestic chick (Gallus domesticus). J Anat 140:577-589

Brauth SE, Karten HJ (1977) Direct accessory optic projections to the vestibulo-cerebellum: a possible channel for oculomotor control systems. Exp Brain Res 28:73-84

Brauth SE, MacHale CM (1988) Auditory pathways in the budgerigar. II Intratelencephalic pathways. Brain Behav Evol 32:193-207

Brauth SE, Reiner A (1991) Calcitonin-gene related peptide is an evolutionarily conserved marker within the amniote thalamo-telencephalic auditory pathway. J Comp Neurol 313:227-239

Brauth SE, Ferguson JL, Kitt CA (1978) Prosencephalic pathways related to the paleostriatum of the pigeon (Columba livia). Brain Res 147:205-221

Brauth SE, Kitt CA, Reiner A, Quirion R (1986) Neurotensin binding sites in the forebrain and midbrain of the pigeon. J Comp Neurol 253:358-373

Brauth SE, McHale CM, Brasher CA, Dooling RJ (1987) Auditory pathways in the budgerigar. I Thalamotelencephalic pathways. Brain Behav Evol 30:174-199

Bravo H, Inzunza O (1985) The oculomotor nucleus, not the abducent, innervates the muscles which advance the nictating membrane in birds. Acta Anat 122:99-104

Breazile JE, Hartwig H-G (1989) Central nervous system. In: King AS, McLelland J (eds) Form and function in birds. Academic Press, London, pp 485-566

Breazile JE, Kuenzel WJ (1993) Systema nervosum centrale. In: Baumel JJ, King AS, Breazile JE, Evans HE, Vanden Berge JC (eds) Handbook of avian anatomy: Nomina Anatomica Avium, 2nd edn. Nuttall Ornithol Club 23, Cambridge, Mass, pp 493-554

Brecha N, Karten HJ, Hunt SP (1980) Projections of the nucleus of the basal optic root in the pigeon: an autoradiographic and horseradish peroxidase study. J Comp Neurol 189:615-670

Bredenkötter M, Bischof H-J (1990) Differences between ipsilaterally and contralaterally evoked potentials in the visual Wulst of the zebrafinch. Vis Neurosci 5:155-163

Brenowitz EA, Arnold AP (1986) Interspecific comparisons of the size of neural song complexity in duetting birds: evolutionary implications. J Neurosci 6:2875-2879

Brinkman R, Martin AH (1973) A cytoarchitectonic study of the spinal cord of the domestic fowl (Gallus gallus domesticus). I. Brachial region. Brain Res 56:43-62

Britto LRG, Natal CL, Marcondes AM (1981) The accessory optic system in pigeons: receptive field properties of identified neurons. Brain Res 206:149-154

Britto LRG, Hamassaki DE, Keyser KT, Karten HJ (1989a) Neurotransmitters, receptors, and neuropeptides in the accessory optic system: an immunohistochemical survey in the pigeon (Columba livia). Vis Neurosci 3:463-475

Britto LRG, Keyser KT, Hamassaki DE, Shimizu T, Karten HJ (1989b) Chemically specific retinal ganglion cells collateralize to the pars ventralis of the lateral geniculate nucleus and optic tectum in the pigeon (Columba livia). Vis Neurosci 3:477-482

Britto LRG, Gasparotto OC, Hamassaki DE (1990) Visual telencephalon modulates directional selectivity of accessory optic neurons in pigeons. Vis Neurosci 4:3–10

Brodal A, Kristiansen K, Jansen J (1950) Experimental demonstration of a pontine homologue in birds. J Comp Neurol 92:23–69

Bubien-Waluszewska A (1981) The cranial nerves. In: King AS, McLelland J (eds) Form and function in birds, vol 2. Academic Press, London, pp 385–438

Bubien-Waluszewska A (1985) Somatic peripheral nerves. In: King AS, McLelland J (eds) Form and function in birds, vol 3. Academic Press, London, pp 149–193

Cabot JB, Bogan N (1987) Light microscopic observations on the morphology of sympathic preganglionic neurons in the pigeon, Columba livia. Neuroscience 20:467–486

Cabot JB, Reiner A, Bogan N (1982) Avian bulbospinal pathways: anterograde and retrograde studies of cells of origin, funicular trajectories and laminar terminations. Prog Brain Res 57:79–106

Cabrera B, Pásaro R, Delgado-Garcia JM (1989) Cytoarchitectonic organisation of the abducens nucleus in the pigeon (Columba livia). J Anat 166:203–211

Cajal S Ramón y (1909–1911) Histologie du système nerveux. T II. Reprint (1972) Instituto Ramon y Cajal, Madrid, pp 196–212

Casini G, Bingman VP, Bagnoli P (1986) Connections of the pigeon dorsomedial forebrain – study with WGA-HRP and ^{3}H-proline. J Comp Neurol 245:454–470

Casini G, Porciatti V, Fontanesi G, Bagnoli P (1992) Wulst efferents in the little owl, Athene noctua. An investigation of projections to the optic tectum. Brain Behav Evol 39:101–115

Cassone VM, Moore RY (1987) Retinohypothalamic projection and suprachiasmatic nucleus of the house sparrow, Passer domesticus. J Comp Neurol 266:171–182

Chikazawa H, Fujioka T, Watanabe T (1982) Catecholamine-containing neurons in the mesencephalic tegmentum of the chicken. Anat Embryol 164:303–313

Chikazawa H, Fujioka T, Watanabe T (1983) Bulbar catecholaminergic neurons projecting to the thoracic spinal cord of the chicken. Anat Embryol 167:411–423

Clark L, Mason JR (1987) Olfactory discrimination of plant volatiles by the European starling. Anim Behav 35:227–235

Clarke PGH (1974) The organization of visual processing in the pigeon cerebellum. J Physiol 242:267–285

Clarke PGH (1977) Some visual and other connections to the cerebellum of the pigeon. J Comp Neurol 174:535–552

Clarke PGH (1982) The generation and migration of the chick's isthmic complex. J Comp Neurol 207:208–222

Clayton NS, Krebs JR (1994) Hippocampal growth and attrition in birds affected by experience. Proc Natl Acad Sci USA 91:7410–7414

Cobb S (1966) The brain of the emu (Dromaeus novaehollandiae). II. Anatomy of the principal nerve cell ganglia and tracts. Breviora 250:1–27

Code RA, Carr CE (1994) Choline acetyltransferase-immunoreactive cochlear efferent neurons in the chick auditory brain stem. J Comp Neurol 340:161–173

Cohen DH, Karten HJ (1974) The structural organization of avian brain: an overview. In: Goodman IJ, Schein MW (eds) Birds. Brain and behavior. Academic Press, New York, pp 29–76

Cohen DH, Pitts LH (1967) The hyperstriatal region of the avian forebrain: somatic and autonomic responses to electrical stimulation. J Comp Neurol 131:323–336

Cohen J (1981) Hormones and midbrain mediation of courtship behavior in the male ring dove (Streptopelia risoria). J Comp Physiol 95:512–528

Cohen YE, Knudsen EI (1994) Auditory tuning for spatial cues in the barn owl basal ganglia. J Neurophysiol 72:285–298

Cole KS, Gummer AW (1990) A double-label study of efferent projections to the cochlea of the chicken, Gallus domesticus. Exp Brain Res 82:585–588

Coles RB, Aitkin LM (1979) The response properties of auditory neurones in the midbrain of the domestic fowl (Gallus gallus) to monaural and binaural stimuli. J Comp Physiol A 134:241–251

Conlee JW, Parks TN (1986) Origin of ascending auditory projections to the nucleus mesencephalicus lateralis pars dorsalis in the chicken. Brain Res 367:96–113

Cooper ML, Pickard GE, Silver R (1983) Retinohypothalamic pathway in the dove demonstrated by anterograde HRP. Brain Res Bull 10:715–718

Correia MJ, Eden AR, Westlund KN, Coulter JD (1982) Organization of ascending auditory pathways in the pigeon (Columba livia) as determined by autoradiographic methods. Brain Res 234:205–212

Correia MJ, Eden AR, Westlund KN, Coulter JD (1983) A study of some of the ascending and descending vestibular pathways in the pigeon (Columba livia) using anterograde transneuronal autoradiography. Brain Res 278:53–61

Cotter JR (1976) Visual and nonvisual units recorded from the optic tectum of Gallus domesticus. Brain Behav Evol 13:1–21

Cowan WM, Powell TPS (1963) Centrifugal fibres in the avian visual system. Proc R Soc London 158:232–252

Cowan WM, Adamson L, Powell TPS (1961) An experimental study of the avian visual system. J Anat 95:545–563

Cox RG, Peusner KD (1990) Horseradish peroxidase labeling of the efferent and afferent pathways of the avian tangential vestibular nucleus. J Comp Neurol 296:324–341

Cracraft J (1988) The major clades of birds. In: Benton MJ (ed) The phylogeny and classification of the tetrapods, vol 1 Amphibians, reptiles, birds. Clarendon Press, Oxford, pp 339–363

Craigie EH (1935) The hippocampal and parahippocampal cortex of the emu (Dromiceius). J Comp Neurol 61:563–591

Crossland WJ (1979) Identification of tectal synaptic terminals in the avian isthmooptic nucleus. In: Granda AM, Maxwell HJ (eds). Neural mechanisms of behavior in the pigeon. Plenum, New York, pp 267–286

Crossland WJ, Hughes CP (1978) Observations on the afferent and efferent connections of the avian isthmo-optic nucleus. Brain Res 145:239–256

Crossland WJ, Uchwat CJ (1979) Topographic projections of the retina and optic tectum upon the ventral lateral geniculate nucleus in the chick. J Comp Neurol 185:87–106

Crossland WJ, Cowan WM, Rogers LA, Kelley JP (1974) The specification of the retino-tectal projection in the chick. J Comp Neurol 155:127–164

Crossland WJ, Cowan WM, Rogers LA (1975) Studies on the development of the chick optic tectum. IV An autoradiographic study of the development of retino-tectal connections. Brain Res 91:1–23

Davis BM, Cabot JB (1984) Substance P-containing pathways to avian sympathetic preganglionic neurons: evidence for major spinal-spinal circuitry. J Neurosci 4:2145–2159

De Britto LRG, Brunelli M, Francesconi W, Magni F (1975) Visual response pattern of thalamic neurons in the pigeon. Brain Res 97:337–343

De Gennaro LD (1982) The glycogen body. In: Farner DS, King JR, Parkes KC (eds) Avian biology. Academic Press, New York, pp 341–371

De Gennaro LD, Benzo CA (1978) Ultrastructural characterization of the accessory lobes of Lachi Hofmann's nuclei in the nerve cord of the chick. II. Scanning and transmission electron microscopy with observations on the glycogen body. J Exp Zool 206:229–240

De Lanerolle NC, Elde RP, Sparber SB, Frick M (1981) Distribution of methionine-enkephalin immunoreactivity in the chick brain: an immunohistochemical study. J Comp Neurol 199:513–533

Delius JD, Bennetto K (1972) Cutaneous sensory projections to the avian forebrain. Brain Res 37:205–221

Delius JD, Runge TE, Oeckinghaus H (1979) Short-latency auditory projection to the frontal telencephalon of the pigeon. Exp Neurol 63:594–609

Den Boer PJ, Bout RG, Dubbeldam JL (1986) Topographical representation of the jaw muscles within the trigeminal motor nucleus: horseradish peroxidase study in the mallard Anas platyrhynchos. Acta Morphol Neerl-Scand 24:1–18

Deng C, Wang B (1992) Overlap of somatic and visual response areas in the Wulst of pigeon. Brain Res 582:320–322

Denton CJ (1981) Topography of the hyperstriatal visual projection area in the young domestic chicken. Exp Neurol 74:482–498

Dermon CR, Kouvelas ED (1989) Quantitative analysis of the localization of adrenergic binding sites in chick brain. J Neurosci Res 23:297–303

Desmond ME, Schoenwolf GC (1985) Timing and positioning of occlusion of the spinal neurocele in the chick embryo. J Comp Neurol 235:479–487

DeVoogd TJ, Nottebohm F (1981) Sex differences in dendritic morphology of a song control nucleus in the canary: a quantitative Golgi study. J Comp Neurol 196:309–316

Dietl MM, Palacios JM (1988) Neurotransmitter receptor in the avian brain. I. Dopamine receptors. Brain Res 439:354–359

Dietl MM, Cortés R, Palacios JM (1988a) Neurotransmitter receptors in the avian brain. II. Muscarinic cholinergic receptors. Brain Res 439:360–365

Dietl MM, Cortés R, Palacios JM (1988b). Neurotransmitter receptors in the avian brain. III. GABA-benzodiazepine receptors. Brain Res 439:366–371

Divac I, Mogensen J (1985) The prefrontal "cortex" in the pigeon catecholamine histofluorescence. Neuroscience 15:677–682

Divac I, Mogensen J, Björklund A (1985) The prefrontal 'cortex' in the pigeon. Biochemical evidence. Brain Res 332:365–368

Divac I, Mogenson J, Björkland A (1988) Strain differences in catecholamine content of pigeon brains. Brain Res 444:371–373

Domenici L, Waldvogel HJ, Matute C, Streit P (1988) Distribution of GABA-like immunorectivity in the pigeon brain. Neuroscience 25:931–950

Doty EJ (1946) The cerebellar nuclear gray in the sparrow (Passer domesticus). J Comp Neurol 84:17–30

Dubbeldam JL (1968) On the shape and the structure of the brainstem in some species of birds. Thesis, Leiden University

Dubbeldam JL (1976) The basal branch of the septomesencephalic tract in the mallard (Anas platyrhynchos L.). Acta Morphol Neerl-Scand 14:48

Dubbeldam JL (1980) Studies on the somatotopy of the trigeminal system in the mallard, Anas platyrhynchos L. II. Morphology of the principal sensory nucleus. J Comp Neurol 191:557–571

Dubbeldam JL (1984) Afferent connections of N. facialis and N. glossopharyngeus in the pigeon (Columba livia) and their role in feeding behavior. Brain Behav Evol 24:47–57

Dubbeldam JL (1989) Shape and structure of the avian brain, an old problem revisited. Acta Morphol Neerl-Scand 27:33–43

Dubbeldam JL (1990a) On the functional interpretation of quantitative differences in forebrain organization – the trigeminal and visual system in birds. Neth J Zool 119:241–253

Dubbeldam JL (1990b) Origin of projections upon the paleostriatal complex in the mallard, Anas platyrhynchos L. Soc Neurosci 16:245

Dubbeldam JL (1991) The avian and mammalian forebrain: correspondences and differences. In: Andrew RJ (ed). Neural and behavioral plasticity. The use of the chick as a model. Oxford University Press, Oxford, pp 65–91

Dubbeldam JL (1993) Brain organization and behaviour. A discussion of neuronal systems in birds. Acta Biotheor 41:469–479

Dubbeldam JL, Bout RG (1990) The identification of the motornuclei innervating the tongue muscles in the mallard (Anas platyrhynchos); an HRP study. Neurosci Lett 119:223–227

Dubbeldam JL, den Boer-Visser AM (1993) Immunohistochemical localization of substance P, enkephalin and serotonin in the brainstem of the chicken, with emphasis on the sensory trigeminal system. Verh Anat Gesellsch 88:7–8

Dubbeldam JL, Karten HJ (1978) The trigeminal system in the pigeon (Columba livia). I. Projections of the Gasserian ganglion. J Comp Neurol 180:661–678

Dubbeldam JL, Veenman CL (1978) Studies on the somatotopy of the trigeminal system in the mallard, Anas platyrhynchos L.: I The ganglion trigeminale. Neth J Zool 28:150–160

Dubbeldam JL, Visser AM (1987) The organization of the nucleus basalis-neostriatum complex of the mallard (Anas platyrhynchos L.) and its connections with the archistriatum and the paleostriatum complex. Neuroscience 21:487–517

Dubbeldam JL, Karten HJ, Menken SBJ (1976) Central projections of the chorda tympani nerve in the mallard, Anas platyrhynchos. J Comp Neurol 170:415–420

Dubbeldam JL, Brus ER, Menken SBJ, Zeilstra S (1979) The central projections of the glossopharyngeal and vagus ganglia in the mallard, Anas platyrhynchos L. J Comp Neurol 183:149–168

Dubbeldam JL, Brauch CSM, Don A (1981) Studies on the somatotopy of the trigeminal system in the mallard, Anas platyrhynchos L. III. Afferents and organization of the nucleus basalis. J Comp Neurol 196:391–405

Dubbeldam JL, den Boer-Visser AM, Bout RG (1997) Organization and efferent connections of the archistriatum of the mallard, Anas platyrhynchos L: an anterograde and retrograde tracing study. J Comp Neurol (in press)

Dubé L, Parent A (1981) The monoamine-containing neurons in avian brain:I A study of the brain stem of the chicken (Gallus domesticus) by means of fluorescence and acetylcholinesterase histochemistry. J Comp Neurol 196:695–708

Durand SE, Tepper JM, Cheng M-F (1992) The shell region of the nucleus ovoidalis: a subdivision of the avian auditory thalamus. J Comp Neurol 323:495–518

Durand SE, Zuo MX, Zhou SL, Cheng M-F (1993) Avian auditory pathways show met-enkephalin-like immunoreactivity. Neuroreport 4:727–730

Ebinger P, Löhmer R (1984) Comparative quantitative investigations on brains of rock doves, domestic and urban pigeons (Columba l. livia). Z Zool System Evol-Forsch 22:136–145

Ebinger P, Löhmer R (1987) A volumetric comparison of brains between greylag geese Anser anser L. and domestic geese. J Hirnforsch 28:291–299

Ebinger P, Rehkämper G, Schröder H (1992) Forebrain specialization and the olfactory system in anseriform birds. An architectonic and tracing study. Cell Tissue Res 268:81–90

Eden AR, Correia MJ (1981) Vestibular efferent neurons and catecholamine cell groups in the reticular formation of the pigeon. Neurosci Lett 25:239–242

Eden AR, Correia MJ (1982a) Identification of multiple groups of efferent vestibular neurons in the adult pigeon using horseradish peroxidase and DAPI. Brain Res 248:201–208

Eden AR, Correia MJ (1982b) An autoradiographic and HRP study of vestibulocollic pathways in the pigeon. J Comp Neurol 211:432–440

Edinger L, Wallenberg A, Holmes G (1903) Untersuchungen über die vergleichende Anatomie des Gehirns. 5. Das Vorderhirn der Vögel. Abhandl Senckenb Ges 20(4):343–426

Ehrlich D, Mark R (1984a) An atlas of the primary visual projections in the brain of the chick (Gallus gallus). J Comp Neurol 223:592–610

Ehrlich D, Mark R (1984b) Topography of primary visual centres in the brain of the chick (Gallus gallus). J Comp Neurol 223:611–625

Ehrlich D, Saleh CN (1982) Composition of the tectal and posterior commissures of the chick (Gallus domesticus). Neurosci Lett 33:115–121

Ehrlich D, Keyser KT, Karten HJ (1987) Distribution of substance P-like immunoreactive retinal ganglion cells and their pattern of termination in the optic tectum of chick (Gallus gallus). J Comp Neurol 266:220–233

Ehrlich D, Stuchberry J, Zappia J (1989) Organization of the hyperstriatal projection to the ventral lateral geniculate nucleus in the chick (Gallus gallus). Neurosci Lett 104:1–6

Engelage J, Bischof HJ (1988) Enucleation enhances ipsilateral flash evoked responses in the ectostriatum of the zebrafinch (Taeniopygia guttata castanotis Gould). Exp Brain Res 70:79–89

Engelage J, Bischof HJ (1994) Visual Wulst influences on flash evoked responses in the ectostriatum of the zebra finch. Brain Res 652:17–27

Erichsen JT, Bingman VP, Krebs JR (1991) The distribution of neuropeptides in the dorsomedial telencephalon of the pigeon (Columba livia): a basis for regional subdivisions. J Comp Neurol 314:487–492

Erichsen JT, Ciocchetti A, Fontanesi G, Bagnoli P (1994) Neuroactive substances in the developing dorsomedial telencephalon of the pigeon (*Columba livia*): differential distribution and time course of maturation. J Comp Neurol 345:537–561

Erulkar SD (1955) Tactile and auditory areas in the brain of the pigeon. An experimental study by means of evoked potentials. J Comp Neurol 103:421–458

Feduccia A (1995) Explosive evolution in tertiary birds and mammals. Science 267:637–638

Feirabend HKP (1983) Anatomy and development of longitudinal patterns in the architecture of the cerebellum of the white leghorn. (Gallus domesticus). Thesis, Leiden University

Feirabend HKP (1990) Development of longitudinal patterns in the cerebellum of the chicken (Gallus domesticus): a cytoarchitectural study on the genesis of cerebellar modules. Eur J Morphol 28:169–223

Feirabend HKP, Voogd J (1986) Myeloarchitecture of the cerebellum of the chicken (Gallus domesticus): an atlas of the compartmental subdivision of the cerebellar white matter. J Comp Neurol 251:44–66

Feirabend HKP, van Luxemburg EH, van Denderen-van Dorp H, Voogd J (1985) A ^{3}H-thymidine autoradiography study of the development of the cerebellum of the white leghorn (Gallus domesticus): "evidence for longitudinal neuroblast generation patterns". Acta Morphol Neerl-Scand 23:115–126

Félix B, Roesch T (1986) Telencephalic bill projections in the Landes goose. Somatosens Res 4:141–152

Feyerabend B Malz CR, Meyer DL (1994) Birds that feed on the wing have few istmo-optic neurons. Neurosci Lett 182:66–68

Figdor MC, Stern CD (1993) Segmental organization of embryonic diencephalon. Nature 363:630–634

Fink E (1965) Geruchsorgan und Ruchvermögen bei Vögeln. Zool Jb Physiol 71:429–450

Fortune ES, Margoliash D (1992) Cytoarchitectonic organization and morphology of cells of the field L complex in male zebra finches (Taeniopygia guttata). J Comp Neurol 325:388–404

Freedman SL, Feirabend HKP, Vielvoye GJ, Voogd J (1975a) Re-examination of the pontocerebellar projection in the adult white leghorn (Gallus domesticus). Acta Morphol Neerl-Scand 13:236–238

Freedman SL, Voogd J, Vielvoye GJ (1975b) Microscopic evidence for climbing fibers in the avian cerebellar cortex following inferior olivary lesions. Anat Rec 181:358

Freedman SL, Voogd J, Vielvoye GJ (1977) Experimental evidence for climbing fibers in the avian cerebellum. J Comp Neurol 175:243–252

Fujii JT (1992) Repetitive firing properties in subpopulations of the chick Edinger-Westphal nucleus. J Comp Neurol 316:279–268

Funke K (1989) Somatosensory areas in the telencephalon of the pigeon. 1. Response characteristics. 2. Spinal pathways and afferent connections. Exp Brain Res 76:603–638

Furber SE (1983) The organization of the olivo-cerebellar projection in the chicken. Brain Behav Evol 22:198–211

Fuxe K, Ljunggren L (1965) Cellular localization of monoamines in the upper brainstem of the pigeon. J Comp Neurol 125:355–382

Gagliardo A, Teyssèdre A (1988) Interhemispheric transfer of olfactory information in homing pigeon. Behav Brain Res 27:173–178

Gahr M, Flügge G, Güttinger H-R (1987) Immunocytochemical localization of estrogen-binding neurons in the songbird brain. Brain Res 402:173–177

Gahr M, Konishi M (1988) Developmental changes in estrogen-sensitive neurons in the forebrain of the zebra finch. Proc Natl Acad Sci USA 85:7380–7383

Galifret Y, Condé-Courtine F, Repérant J, Servière J (1971) Centrifugal control in the visual system of the pigeon. Vision Res Suppl 3:185–200

Galileo DS, Gray GE, Owens GC, Majors J, Sanes RJ (1990) Neurons and glia arise from a common progenitor in chicken optic tectum: demonstration with two retroviruses and cell-specific antibodies. Proc Natl Acad Sci USA 87:458–462

Gamlin PDR, Cohen DH (1986) A second ascending visual pathway from the optic tectum to the telencephalon in the pigeon (Columba livia). J Comp Neurol 250:296–310

Gamlin PDR, Cohen DH (1988a) Retinal projections to the pretectum in the pigeon (Columba livia). J Comp Neurol 269:1–17

Gamlin PDR, Cohen DH (1988b) Projections of the retinorecipient pretectal nuclei in the pigeon (Columba livia). J Comp Neurol 269:18–46

Gamlin PDR, Reiner A (1991) The Edinger-Westphal nucleus: sources of input influencing accommodation, pupilloconstriction, and choroidal blood flow. J Comp Neurol 306:425–438

Gamlin PDR, Reiner A, Karten HJ (1982) The avian suprachiasmatic nucleus has a substance P-positive projection to the nucleus of Edinger-Westphal. Anat Rec 202:61A

Gamlin PDR, Reiner A, Erichsen JT, Karten HJ, Cohen DH (1984) The neural substrate for the pupillary light reflex in the pigeon (Columba livia). J Comp Neurol 226:523–543

Ganchrow D, Ganchrow JR, Gentle MJ (1986) Central afferent connections and origin of efferent projections of the facial nerve in the chicken (Gallus gallus domesticus). J Comp Neurol 248:455–463

Gentle MJ (1979) Single unit responses from the solitary complex following oral stimulation in the chicken. J Comp Physiol A 130:259–264

Gentle MJ (1984) The chorda tympani and taste in chicken. Experientia 40:1253–1255

Gioanni H, Rey J, Villalobos J, Richard D, Dalbera A (1983a) Optokinetic nystagmus in the pigeon (Columba livia). II. Role of the pretectal nucleus of the accessory optic system (AOS). Exp Brain Res 50:237–247

Gioanni H, Villalobos J, Rey J, Dalbera A (1983b) Optokinetic nystagmus in the pigeon (Columba livia). III. Role of the nucleus ectomamillaris (nEM): interactions in the accessory optic system (AOS). Exp Brain Res 50:248–258

Gioanni H, Palacios A, Sansonetti A, Varela F (1991) Role of the nucleus geniculatus lateralis ventralis (GLv) in the optokinetic reflex: a lesion study in the pigeon. Exp Brain Res 86:601–607

Glover JC (1994) The organization of vestibulo-ocular and vestibulospinal projections in the chicken embryo. Eur J Morphol 32:193–200

Glover JC, Petursdottir G (1988) Pathway specificity of reticulospinal and vestibulospinal projections in the 11-day chicken embryo. J Comp Neurol 270:25–38

Goodman DC, Fisher HJ (1962) Functional anatomy of the feeding apparatus in the waterfowl. Aves: Anatidae. South Illinois University Press, Carbondale

Goodman DC, Horel JA, Freemon FR (1964) Functional localization in the cerebellum of the bird and its bearing on the evolution of cerebellar function. J Comp Neurol 123:45–54

Goodwin D (1983) Behaviour. In: Abs M (ed) Physiology and behaviour of the pigeon. Academic Press, London, pp 285–308

Gottschaldt K-M (1985) Structure and function of avian somatosensory receptors. In: King AS, McLelland J (eds) Form and function in birds, vol 3. Academic Press, London, pp 375–461

Grabatin O, Abs M (1986) On efferent innervation of the larynx in the pigeon (Columba livia domestica L.). Anat Anz 162:101–108

Graef W (1973a) Zyto- und Myeloarchitektonik des Mesencephalon beim Haushuhn (Gallus domesticus L.). I. Zytoarchitektonik im kaudalen Abschnitt. Anat Anz 133:144–152

Graef W (1973b) Zyto- und Myeloarchitektonik des Mesencephalon beim Haushuhn (Gallus domesticus L.). II. Zytoarchitektonik im mittleren Abschnitt. Anat Anz 133:503–510

Graef W (1973c) Zyto- und Myeloarchitektonik des Mesencephalon beim Haushuhn (Gallus domesticus L.). III. Zytoarchitektonik im rostralen Abschnitt. Anat Anz 133:511–522

Graef W (1973d) Zyto- und Myeloarchitektonik des Mesencephalon beim Haushuhn (Gallus domesticus L.). IV. Zytoarchitektonik in Horizontalansichten. Anat Anz 134:38–44

Graef W (1973e) Zyto- und Myeloarchitektonik des Mesencephalon beim Haushuhn (Gallus domesticus L.). VI. Zytoarchitektonik in Frontalansichten. Anat Anz 134:433–444

Granda RH, Crossland WJ (1989) GABA-like immunoreactivity of neurons in the chick diencephalon and mesencephalon. J Comp Neurol 287:455–469

Granda AM, Yazulla S (1971) The spectral sensitivity of single units in the nucleus rotundus of pigeon, Columba livia. J Gen Physiol 57:363–384

Gray GE, Glover JC, Majors J, Sanes JR (1988) Clonically related cells migrate radially to form "ontogenetic columns" in laminated structures. Proc Natl Acad Sci USA 85:7356–7360

Gray L, Smith Z, Rubel EW (1982) Developmental and experimental changes in dendritic symmetry in n. laminaris of the chick. Brain Res 244:360–364

Gross GH, Oppenheim RW (1985) Novel sources of descending input to the spinal cord of the hatchling chick. J Comp Neurol 232:162–179

Grubb TC (1972) Smell and foraging in shearwaters and petrels. Nature 237:404–405

Guiloff GD, Maturana HR, Varella FJ (1987) Cytoarchitecture of the avian ventral lateral geniculate nucleus. J Comp Neurol 264:509–526

Güntürkün O (1984) Evidence for a third primary visual area in the telencephalon of the pigeon. Brain Res 294:247–254

Güntürkün O (1987) A Golgi study of the isthmic nuclei in the pigeon (Columba livia). Cell Tissue Res 248:439–448

Güntürkün O (1990) The topographical projection of the nucleus isthmi pars parvocellularis (Ipc) onto the tectum opticum in the pigeon. Neurosci Lett 111:18–22

Güntürkün O (1991) The functional organization of the avian visual system. In: Andrew RJ (ed) Neural and behavioural plasticity. The use of the chick as a model. Oxford University Press, Oxford, pp 92–105

Güntürkün O, Böhringer PG (1987) Lateralization reversal after intertectal commissurotomy in the pigeon. Brain Res 408:1–5

Güntürkün O, Remy M (1990) The topographical projection of the nucleus isthmi pars parvocellularis (Ipc) into the tectum opticum in the pigeon. Neurosci Lett 111:18–22

Haefelfinger HR (1957) Beiträge zur Vergleichenden Ontogenese des Vorderhirns bei Vögeln. Thesis, University of Basel

Hall K, Brauth SE, Kitt CA (1984) Retrograde transport of [^{3}H]GABA in the striatotegmental system of the pigeon. Brain Res 310:157–163

Hall WS, Cohen PL, Brauth SE (1993) Auditory projections to the anterior telencephalon in the budgerigar (Melopsittacus undulatus). Brain Behav Evol 41:97–116

Hamburger V, Hamilton HL (1951) A series of normal stages in the development of the chick embryo. J Morphol 88:49–92

Hampton RR, Sherry DF, Shettleworth SJ, Khurgel M, Ny G (1995) Hippocampal volume and food-storing behavior are related in Parids. Brain Behav Evol 45:54–61

Hanaway J (1967) Formation and differentiation of the external granular layer of the chick cerebellum. J Comp Neurol 131:1–14

Hardy O, Leresche N, Jassik-Gerschenfeld D (1982) The spatial organization of the excitatory regions in the visual receptive fields of the pigeon's optic tectum. Exp Brain Res 46:59–68

Hardy O, Leresche N, Jassik-Gerschenfeld D (1985) Morphology and laminar distribution of electrophysiologically identified cells in the pigeon's optic tectum: an intracellular study. J Comp Neurol 233:390–404

Harman AL, Phillips RE (1967) Responses in the avian midbrain, thalamus and forebrain evoked by click stimuli. Exp Neurol 18:276–286

Hartwig H-G (1970) Das visuelle System von Zonotrichia leucophrys gambelii. Neurohistologische Studien auf experimenteller Grundlage. Z Zellforsch 106:556–583

Hasegawa M, Ebihara S (1992) Circadian rhythms of pineal melatonin release in the pigeon measured by in vivo microdialysis. Neurosci Lett 148:89–92

Hassler R (1971) Vergleichend Experimentell-histologische Untersuchung zur Retinale Repräsentation in den Primären Visuellen Zentren einiger Vogelarten. Thesis, Johann Wolfgang Goethe-Universität, Frankfurt am Main

Hayes BP, Webster KE (1981) Neurones situated outside the isthmo-optic nucleus and projecting to the eye in adult birds. Neurosci Lett 26:107–112

Hayes BP, Webster KE (1985) Cytoarchitectural fields and retinal termination: an axonal transport study of laminar organization in the avian optic tectum. Neuroscience 16:641–657

Heaton MB, Wayne DB (1983) Patterns of extraocular innervation by the oculomotor complex in the chick. J Comp Neurol 216:245–252

Hellmann B, Waldmann C, Güntürkün O (1995) Cytochrome oxydase activity reveals parcellations of the pigeon's ectostriatum. Neuroreport 6:881–885

Hillebrand A (1972) Is there a homologon of the central oculomotor nucleus in the chicken and turkey? Anat Anz 132:24–31

Hirschberger W (1967) Histologische Untersuchungen an den primären visuellen Zentren des Eulengehirnes und der retinalen Repräsentation in ihnen. J Ornithol 108:187–202

Hirschberger W (1971) Vergleichend Experimentellhistologische Untersuchung zur Retinalen Repräsentation in den Primären Visuellen Zentren einiger Vogelarten. Thesis, Frankfurt am Main.

Hodges RD (1981) Endocrine glands. In: King AS, McLelland J (eds) Form and function in Birds. Academic Press, London, pp 149–234

Hodos W, Karten HJ (1966) Brightness and pattern discrimination deficits in the pigeon after lesions of nucleus rotundus. Exp Brain Res 2:151–167

Hodos W, Karten HJ (1974) Visual intensity and pattern discrimination deficits after lesions of the optic lobe in pigeons. Brain Behav Evol 9:165–194

Hodos W, Karten HJ, Bonbright Jr JC (1973) Visual intensity and pattern discrimination after lesions of the thalamofugal visual pathway in pigeons. J Comp Neurol 148:447–468

Holden AL, Powell TPS (1972) The functional organization of the isthmo-optic nucleus in the pigeon. J Physiol 223:419–447

Holloway JA, Trouth CO, Moolenaar GM, Wright LE (1976) Responses of dorsal horn cells of Gallus domesticus to cutaneous and peroneal nerve stimuli. Exp Neurol 53:756–767

Holloway JA, Keyser GF, Wright LE, Trouth CO (1978) Supraspinal inhibition of dorsal horn cell activity and location of descending pathways in the chicken, Gallus domesticus. Brain Res 145:380–384

Homma S, Sako H, Kohno K, Okado N (1988) The pattern of distribution of serotonergic fibers in the anterior horn of the chick spinal cord. Anat Embryol 179:25–31

Honig MG (1982) The development of sensory projection patterns in embryonic chick hind limb. J Physiol 330:175–202

Horn G (1985) Memory, imprinting, and the brain. Clarendon Press, Oxford

Hörster W, Franchini A, Daniel S (1990) Organization of neck muscle motorneurons in the cervical spinal cord of the pigeon. Neuroreport 1:93–96

Hotta T (1971) Unit responses from the nucleus angularis in the pigeon's medulla. Comp Biochem Physiol 40A:415–424

Huber GC, Crosby EC (1929) The nuclei and fiber paths of the avian diencephalon, with consideration of the telencephalic and certain mesencephalic centers and connections. J Comp Neurol 48:1–225

Huber JF (1936) Nerve roots and nuclear groups in the spinal cord of the pigeon. J Comp Neurol 65:43–91

Huber W (1949) Analyse métrique du redressement de la tête chez l'embryo de poulet. Rev Suisse Zool 56:286–291

Hughes CP, Pearlman AL (1974) Single unit receptive fields and the cellular layers of the pigeon optic tectum. Brain Res 80:365–377

Hummel G, Goller H (1979) Feinstruktur der Kleinhirnrinde vom Haushuhn. Tierärztl Wschr 92:269–273

Hunt SP (1972) Thalamo-hyperstriate interrelations in the pigeon. Brain Res 44:647–651

Hunt SP, Brecha N (1984). The avian optic tectum: a synthesis of morphology and biochemistry. In: Vanegas H (ed) Comparative neurology of the optic tectum. Plenum Press, New York, pp 619–648

Hunt SP, Künzle H (1976a) Observations on the projections and intrinsic organization of the pigeon optic tectum: an autoradiographic study based on anterograde and retrograde, axonal and dendritic flow. J Comp Neurol 170:153–172

Hunt SP, Künzle H (1976b) Selective uptake and transport of label within three identified neuronal systems after injection of ³H-GABA into the pigeon optic tectum: an autoradiographic and Golgi study. J Comp Neurol 170:173–190

Hunt SP, Webster KE (1972) Thalamo-hyperstriate interrelations in the pigeon. Brain Res 44:647–651

Hunt SP, Webster KE (1975) The projection of the retina upon the optic tectum of the pigeon. J Comp Neurol 162:433–446

Hunt SP, Henke H, Künzle H, Ruebi JC, Schenker T, Streit P, Felix D, Cuénod M (1976) Biochemical neuroanatomy of the pigeon optic tectum. Exp Brain Res Suppl 1:521–525

Hunt SP, Streit P, Künzle H, Cuénod M (1977) Characterization of the pigeon isthmo-tectal pathway by selective uptake and retrograde movement of radioactive compounds and by golgi-like horseradish peroxidase labeling. Brain Res 129:197–212

Hunter A, Stewart MG (1993) Long term increases in the numerical density of synapses in the chick lobus parolfactorius after passive avoidance training. Brain Res 605:251–255

Ikeda H, Gotoh J (1971) Distribution of monoamine-containing cells in the central nervous system of the chicken. Jpn J Pharmacol 21:763–784

Isomura G (1973) On the localization of nerve centers of the extrinsic ocular muscles in the fowl. Anat Anz 133:33–50

Jacobson RD, Hollyday M (1982) Electrically evoked walking and fictive locomotion in the chick. J Neurophysiol 48:238–256

Jäger R (1990) Visuomotor feeding perturbations after lateral telencephalic lesions in pigeons. Behav Brain Res 40:73–80

Jäger R (1993) Lateral forebrain lesions affect pecking accuracy in the pigeon. Behav Processes 28:181–188

Janzik HH, Glees P (1967) The origin of the spinal ventromedial tract in the chick an experimental study. J Hirnforsch 9:91–96

Jassik-Gerschenfeld D, Guichard J (1972) Visual receptive fields of single cells in the pigeon's optic tectum. Brain Res 40:303–317

Jassik-Gerschenfeld D, Hardy O (1984) The avian optic tectic: neurophysiology and behavioral correlations. In: Vanegas H (ed) Comparative neurology of the optic tectum. Plenum Press, New York, pp 649–686

Jhaveri S, Morest DK (1982) Neuronal architecture in nucleus magnocellularis of the chicken auditory system with observations on nucleus laminaris: a light and electron microscope study. Neuroscience 7:809–836

Johnson MH, Horn G (1987) The role of a restricted region of the chick forebrain in the recognition of individual conspecifics. Behav Brain Res 23:269–275

Jones AE, Levi-Montalcini R (1958) Patterns of differentiation of the nerve centers and fiber tracts in the avian cerebral hemispheres. Arch Ital Biol 96:231–284

Juilliard MT, Hartwig HG, Collin JP (1977) The avian pineal organ. Distribution of endogenous monoamines; a fluorescence microscopic, microspectrofluorimetric and pharmacological study in the parakeet. J Neural Transm 40:269–287

Kaiser L (1923). De segmentale innervatie bij de duif (Columba livia, var. domestica). Nwe Verh Bataafs Genootschap Proefonderv Wijsbegeerte 2e R 9, pp 1–104.

Kalischer O (1905) Das Grosshirn der Papageien in anatomischer und physiologischer Beziehung. Abh Preuss Akad 4:1–105

Källén B (1953) On the nuclear differentiation during ontogenesis in the avian forebrain and some notes on the amniote strioamygdaloid complex. Acta Anat 17:72–84

Källén B (1962) Embryogenesis of brain nuclei in the chick telencephalon. Ergebn Anat Entw-gesch 36:62–82

Karten HJ (1963) Ascending pathways from the spinal cord in the pigeon (Columba livia). Proc XVI Int Congress Zool 2:23

Karten HJ (1967) The organization of the ascending auditory pathway in the pigeon (Columba livia). I. Diencephalic projections of the inferior colliculus (nucleus mesencephali lateralis, pars dorsalis). Brain Res 6:409–427

Karten HJ (1968) The ascending auditory pathway in the pigeon (Columba livia). II. Telencephalic projections of the nucleus ovoidalis thalami. Brain Res 11:134–153

Karten HJ (1969) The organization of the avian telencephalon and some speculations on the phylogeny of the amniote telencephalon. Ann NY Acad Sci 167:164–179

Karten HJ (1971) Efferent connection of the Wulst of the owl. Anat Rec 169:353

Karten HJ (1979) Visual lemniscal pathways in birds. In: Granda AM, Maxwell JH (eds) Neural mechanisms of behavior in the pigeon. Plenum Press, New York, pp 409–430

Karten HJ (1991) Homology and evolutionary origins of the "neocortex". Brain Behav Evol 38:264–272

Karten HJ, Dubbeldam JL (1973) The organization and projections of the paleostriatal complex in the pigeon (Columba livia). J Comp Neurol 148:61–90

Karten HJ, Finger TE (1976) A direct thalamo-cerebellar pathway in the pigeon and catfish. Brain Res 102:335–338

Karten HJ, Hodos W (1967) A stereotaxic atlas of the brain of the pigeon (Columba livia). Johns Hopkins University Press, Baltimore

Karten HJ, Hodos W (1970) Telencephalic projections of the nucleus rotundus in the pigeon (Columba livia). J Comp Neurol 140:35–52

Karten HJ, Revzin AM (1966) The afferent connections of the nucleus rotundus in the pigeon. Brain Res 2:268–377

Karten HJ, Shimizu T (1989) The origins of neocortex: connections and laminations: distinct events in evolution. J Cognit Neurosci 1:291–301

Karten HJ, Hodos W, Nauta WJH, Revzin AM (1973) Neural connections of the "visual Wulst" of the avian telencephalon. Experimental studies in the pigeon (Columba livia) and owl (Speotyto cunicularia). J Comp Neurol 150:253–278

Karten HJ, Fite KV, Brecha N (1977) Specific projection of displaced retinal ganglion cells upon the accessory optic system in the pigeon (Columba livia). Proc Natl Acad Sci USA 74:1753-1756

Katz DM, Karten HJ (1979) The discrete anatomical localization of vagal aortic afferents within a catecholamine-containing cell group in the nucleus solitarius. Brain Res 171:187-195

Katz DM, Karten HJ (1983a) Visceral representation within the nucleus of the tractus solitarius in the pigeon (Columba livia). J Comp Neurol 218:42-73

Katz DM, Karten HJ (1983b) Subnuclear organization of the dorsal motor nucleus of the vagus nerve in the pigeon, Columba livia. J Comp Neurol 217:31-46

Kelley DB, Nottebohm F (1979) Projections of a telencephalic auditory nucleus - field L - in the canary. J Comp Neurol 183:455-470

Keynes RJ, Stearns CD (1984) Segmentation in the vertebrate nervous system. Nature 310:786-789

Kim JR, Alvarez-Buylla A, Nottebohm F (1991) Production and survival of projection neurons in a forebrain vocal center of adult male canaries. J Neurosci 11:1756-1762

Kingsbury BF (1920) The extent of the floor-plate of His and its significance. J Comp Neurol 32:113-137

Kirsch M, Coles RB, Leppelsack H-J (1980) Unit recordings from a new auditory area in the frontal neostriatum of the awake starling (Sturnus vulgaris). Exp Brain Res 38:375-380

Kishida R, Dubbeldam JL, Goris RC (1985) Primary sensory ganglion cells projecting to the principal trigeminal nucleus in the mallard, Anas platyrhynchos. J Comp Neurol 240:171-179

Kiss JZ, Voorhuis TAM, van Eekelen JAM, de Kloet ER, de Wied D (1987). Organization of vasotocin-immunoreactive cells and fibers in canary brain. J Comp Neurol 263:347-364

Kitt CA, Brauth SE (1981) Projections of the paleostriatum upon the midbrain tegmentum in the pigeon. Nature 6:1551-1566

Kitt CA, Brauth SE (1982) A paleostriatal-thalamic-telencephalic path in pigeons. Neuroscience 7:2735-2751

Kitt CA, Brauth SE (1986a) Telencephalic projections from midbrain and isthmal cell groups in the pigeon. I. Locus coeruleus and subcoeruleus. J Comp Neurol 247:69-91

Kitt CA, Brauth SE (1986b) Telencephalic projections from midbrain and isthmal cell groups in the pigeon. II. The nigral complex. J Comp Neurol 247:92-110

Knechtl G (1954) Untersuchungen und Vergleich der motorischen Hirnnervekernen bei einige Vögeln. Morphol Jahrb 93:364-399

Knudsen EI (1980) Sound location in birds. In: Popper AN, Fay RR (eds) Comparative studies of hearing in vertebrates. Springer-Verlag, Berlin Heidelberg New York, pp 289-322

Knudsen EI (1982) Auditory and visual maps of space in the optic tectum of the owl. J Neurosci 2:1177-1194

Knudsen EI, Knudsen PF (1983) Space-mapped auditory projections from the inferior colliculus to the optic tectum in the barn owl Tyto alba. J Comp Neurol 218:187-196

Knudsen EL, Knudsen PF, Masino T (1993) Parallel pathways mediating both sound localization and gaze control in the forebrain and midbrain of the barn owl. J Neurosci 13:2837-2852

Konishi M, Gurney ME (1982) Sexual differentiation of brain and behaviour. Trends Neurosci 5:20-23

Korf H-W (1982) Afferent connections of physiologically identified neuronal complexes in the paraventricular nucleus of conscious Peking ducks involved in regulation of salt-and water-balance. Cell Tissue Res 226:275-300

Korf H-W (1984) Neuronal organization of the avian paraventricular nucleus: intrinsic, afferent, and efferent connections. J Exp Zool 232:387-395

Korf H-W, Vigh-Teichmann I (1984) Sensory and central nervous elements in the avian pineal organ. Ophthalm Res 16:96-101

Korf H-W, Zimmerman NH, Oksche A (1982) Intrinsic neurons and neural connections of the pineal organ of the house sparrow, Passer domesticus, as revealed by anterograde transport of horseradish peroxidase. Cell Tissue Res 222:243-260

Korzeniewska E (1987) Multisensory convergence in the thalamus of the pigeon (Columba livia). Neurosci Lett 80:55-60

Korzeniewska E, Güntürkün O (1990) Sensory properties and afferents of the N. dorsolateralis posterior thalami of the pigeon. J Comp Neurol 292:457-479

Krayniak PF, Siegel A (1978a) Efferent connections of the hippocampus and adjacent regions in the pigeon. Brain Behav Evol 15:372-388

Krayniak PF, Siegel A (1978b) Efferent connections of the septal area in the pigeon. Brain Behav Evol 15:389-404

Krebs JR, Sherry DF, Healy SD, Perry VH, Vacarino AL (1989). Hippocampal specialization of food-storing birds. Proc Natl Acad Sci USA 86:1388-1392

Krebs JR, Erichsen JT, Bingman VP (1991) The distribution of neurotransmitters and neurotransmitter-related enzymes in the dorsomedial telencephalon of the pigeon (Columba livia). J Comp Neurol 314:467-477

Krol CPM, Dubbeldam JL (1979) On the innervation of taste buds by the N. facialis in the mallard, Anas platyrhynchos L. Neth J Zool 29:267-274

Kuenzel WJ, Masson M (1988) A stereotaxic atlas of the brain of the chick (Gallus domesticus). Johns Hopkins University Press, Baltimore

Kuenzel WJ, Van Tienhoven A (1982) Nomenclature and location of avian hypothalamic nuclei and associated circumventricular organs. J Comp Neurol 206:293-313

Kuhlenbeck H (1938) The ontogenetic development and phylogenetic significance of the cortex telencephali in chick. J Comp Neurol 69:273-295

Kuhlenbeck H (1973) VI. Morphologic pattern of the vertebrate neuraxis. 6. The telencephalon and its secondary zonal system. The central nervous system of vertebrates, vol 3/II. Karger, Basel, pp 471-668

Kuhlenbeck H (1977) Derivatives of the prosencephalon: diencephalon and telecephalon. In: The central nervous system of vertebrates, vol 5/I. Karger, Basel, pp 650-686

Kühn A, Trendelenburg W (1911) Die exogenen und endogenen Bahnen des Rückenmarks der Taube. Arch Anat Physiol (Anat):35-48

Kusunoki T (1969) The chemoarchitectonics of the avian brain. J Hirnforsch 11:477-497

Labandeira-Garcia MJ, Guerra-Seijas MJ, Segade LAG, Suarez-Nunez JM (1987) Identification of abducens motoneurons, accessory abducens motoneurons, and abducens internuclear neurons in the chick by retrograde transport of horseradish peroxidase. J Comp Neurol 259:140-149

Labandeira-Garcia MJ, Guerra-Seijas JA, Labandeira-Garcia JA, Jorge-Barreiro FJ (1989) Afferent connections of the oculomotor nucleus in the chick. J Comp Neurol 282:523-534

Lance-Jones C, Landmesser L (1981) Pathway selection by chick lumbosacral motoneurons during normal development. Proc Roy Soc London B 214:1-18

Landmesser L (1978) The distribution of motoneurons supplying chick hind limb muscles. J Physiol 284:371-389

Larsell O (1948) The development and subdivisions of the cerebellum of birds. J Comp Neurol 89:123-190

Larsell O, Whitlock DG (1952) Further observations on the cerebellum of birds. J Comp Neurol 97:545-566

LaVail JH, Cowan WM (1971) The development of the chicken optic tectum. I. Normal morphology and cytoarchitectonic development. II. Autoradiographic studies. Brain Res 28:391-441

Lavalley AL, Ho RH (1983) Substance P, somatostatin, and methionine enkephalin immunoreactive elements in the spinal cord of the domestic fowl, Gallus domesticus. J Comp Neurol 213:406-413

Lednor AJ, Walcott C (1988) Orientation of homing pigeons at magnetic anomalies. Behav Ecol Sociobiol 22:3-8
Leonard RB, Cohen DH (1975a) A cytoarchitectonic analysis of the spinal cord of the pigeon Columba livia. J Comp Neurol 163:159-180
Leonard RB, Cohen DH (1975b) Spinal terminal fields of dorsal root fibers in the pigeon (Columba livia). J Comp Neurol 163:181-192
Leppelsack H-J (1974) Funktionelle Eigenschaften der Hörbahn im Feld L des Neostriatum caudale des Staren (Sturnus vulgaris L., Aves). J Comp Physiol 88:271-320
Leresche N, Hardy O, Jassik-Gerschenfeld D (1983) Receptive field properties of single cells in the pigeon's optic tectum during cooling of the 'visual Wulst'. Brain Res 267:225-236
Leresche N, Hardy O, Audinat E, Jassik-Gerschenfeld D (1986) Synaptic transmission of excitation from the retina to cells in the pigeon's optic tectum. Brain Res 365:138-144
Lewis JW, Ryan SM, Arnold AP, Butcher LL (1981) Evidence for a catecholaminergic projection to area X in the zebra finch. J Comp Neurol 196:347-354
Lumsden A (1990) The cellular basis of segmentation in the developing hindbrain. Trends Neurosci 13:329-335
Lyser KM (1973) The fine structure of the glycogen body of the chicken. Acta Anat 85:533-549
Macadar AW, Rausch LJ, Wenzel BM, Hutchison LV (1980) Electrophysiology of the olfactory pathway in the pigeon. J Comp Physiol 137:39-46
MacPhail EM, Reilly S, Good M (1993) Lateral hyperstriatal lesions disrupt simultaneous, but not successive conditional discrimination learning of pigeons (Columba livia). Behav Neurosci 107:289-298
Manley GA, Köppl C, Konishi M (1988) A neural map of interaural intensity differences in the brain stem of the barn owl. J Neurosci 8:2665-2676
Manni E, Bortolami R, Azzena GB (1965) Jaw muscle proprioception and mesencephalic trigeminal cells in birds. Exp Neurol 12:320-328
Manogue KR, Nottebohm F (1982) Relation of medullary motor nuclei to nerves supplying the vocal tract of the budgerigar (Melopsittacus undulatus). J Comp Neurol 204:384-291
Margoliash D, Konishi M (1985) Auditory representation of autogenous song in the song system of white-crowned sparrows. Proc Natl Acad Sci USA 82:5997-6000
Martin GR (1979) A brief introduction to the taxonomy and ecology of the Columbiformes (pigeons and doves). In: Granda AM, Maxwell JH (eds) Neural mechanisms of behavior in the pigeon. Plenum Press, New York, pp 1-4
Martin JT, De Lanerolle N, Phillips RE (1979) Avian archistriatal control of fear-motivated behavior and adrenocortical function. Behav Processes 4:283-293
Martínez S, Puelles L (1989) Avian nucleus isthmi ventralis projects to the contralateral optic tectum. Brain Res 481:181-184
Martínez S, Geijo E, Sanchez-Vives M, Puelles L, Gallego R (1993) Diencephalic intersegmental boundaries in the chick: bounding cells and junctional properties. Eur J Neurosci 6:112
Martínez-de-la-Torre M, Martínez S, Puelles L (1990) Acetylcholinesterase-histochemical differential staining of subdivisions within the nucleus rotundus in the chick. Anat Embryol 181:129-135
Matochik JA, Reems CN, Wenzel BM (1991) A brain atlas of the Northern Fulmar Fulmarus glacialis in stereotaxic coordinates. Brain Behav Evol 37:215-244
Matsushita M (1968) Zur Zytoarchitectonik des Hühnerrückenmarkes nach Silberimpregnation. Acta Anat 79:238-259
Maturana HR, Varela FJ (1982) Color-opponent responses in the avian lateral geniculate: a study in the quail (Coturnix coturnix japonica). Brain Res 247:227-242
McCasland JS (1987) Neuronal control of bird song production. J Neurosci 7:23-39
McConnell JA, Sechrist JW (1980) Identification of early neurons in the brainstem and spinal cord I. An autoradiographic study in the chick. J Comp Neurol 192:769-783
McGill JI, Powell TPS, Cowan WM (1966a) The retinal representation upon the optic tectum and isthmo-optic nucleus in the pigeon. J Anat 100:5-33
McGill JI, Powell TPS, Cowan WM (1966b) The organization of the projection of the centrifugal fibres to the retina in the pigeon. J Anat 100:35-49
McKenna OC, Wallman J (1985) Accessory optic system and pretectum of birds: comparisons with those of other vertebrates. Brain Behav Evol 26:91-116
Medina L, Reiner A (1994) Distribution of choline acetyltransferase immunoreactivity in the pigeon brain. J Comp Neurol 342:497-537
Meier RE (1971) Interhemisphärischer Transfer visueller Zweifachwahlen bei kommissurotomierten Tauben. Psychol Forsch 34:220-245
Meier RE (1973) Autoradiographic evidence for a direct retinohypothalamic projection in the avian brain. Brain Res 53:417-421
Meier RE, J. Mihailovic J, Cuénod M (1974) Thalamic organization of the retino-thalamo-hyperstriatal pathway in the pigeon (Columba livia). Exp Brain Res 19:351-364
Melian AP, Fonolla JP, Loyzaga PG (1986) The ontogeny of the cerebellar fissures in the chick embryo. Anat Embryol 175:119-128
Mello NK (1968) The effect of unilateral lesions of the optic tectum on interhemispheric transfer of monocularly trained color and pattern discrimination in pigeon. Physiol Behav 3:725-734
Menaker M, Oksche A (1974) The avian pineal organ. In: Farner DS, King JR (eds) Avian biology, vol 4. Academic Press, New York
Meschini E (1983) Pigeon navigation: some experiments on the importance of olfactory cues at short distances from the loft. J Comp Physiol A 150:493-498
Mestres P, Delius JD (1982) A contribution to the study of afferents to the pigeon optic tectum. Anat Embryol 165:415-423
Miceli D, Repérant J (1982) Thalamo-hyperstriatal projections in the pigeon (Columba livia) as demonstrated by retrograde double-labeling with fluorescent tracers. Brain Res 245:365-371
Miceli D, Repérant J (1983) Hyperstriatal-tectal projections in the pigeon (Columba livia) as demonstrated by the retrograde double-label fluorescence technique. Brain Res 276:147-153
Miceli D, Repérant J, Ptito M, Weidner C (1983) Les structures à projection télencéphalique chez le pigeon. Identification par marquage à la peroxydase du Raifort. J Hirnforsch 24:437-446
Miceli D, Repérant J, Villalobos J, Dionne L (1987) Extratelencephalic projections of the avian visual Wulst. A quantitative autoradiographic study in the pigeon, Columba livia. J Hirnforsch 28:45-58
Miceli D, Marchand L, Repérant J, Rio J-P (1990) Projections of the dorsolateral anterior complex and adjacent thalamic nuclei upon visual Wulst in the pigeon. Brain Res 518:317-323
Miceli D, Repérant J, Rio JP, Medina M (1995) GABA immunoreactivity in the nucleus isthmo-opticus of he centrifugal visual system in the pigeon: a light and electron microscopic study. Vis Neurosci 12:425-441
Mihailovic J, Perisic M, Bergonzi R, Meier ER (1974) The dorsolateral thalamus as a relay in the retino-Wulst pathway in pigeon (Columba livia). An electrophysiological study. Exp Brain Res 21:229-240
Mikami S, Yamada S, Hasegawa Y, Miyamoto K (1988) Localization of avian LHRH-immunoreactive neurons in the hypothalamus of the domestic fowl, Gallus domesticus, and the Japanese quail, Coturnix coturnix. Cell Tissue Res 251:51-58
Mogensen J, Divac I (1982) The prefrontal "cortex" in the pigeon. Brain Behav Evol 21:60-66

Mollà R, Rodriguez J, Calvet S, Garcia-Verdugo JM (1986) Neuronal types of the cerebral cortex of the adult chicken (Gallus gallus). A Golgi study. J Hirnforsch 27:381–390

Molnar M, Casini G, Davis BM, Bagnoli P, Brecha NC (1994) Distribution of preproenkephalin mRNA in the chicken and pigeon telencephalon. J Comp Neurol 348:419–432

Moons L, Gils J van, Ghysels E, Vandesande F (1994) Immunocytochemical localization of L-Dopa and dopamine in the brain of the chicken (Gallus domesticus). J Comp Neurol 346:97–118

Mugnaini E, Dahl AL (1975) Mode of distribution of aminergic fibers in the cerebellar cortex of the chicken. J Comp Neurol 162:417–432

Müller H, Drenckhahn D, Haase E (1979) Vergleichend quantitative und ultrastrukturelle Untersuchungen am Geruchsorgan von vier Haustaubenrassen. Z Mikrosk-Anat Forsch 93:888–900

Narayanan CH, Narayanan Y (1976) An experimental inquiry into the central source of preganglionic fibers to the chick ciliary ganglion. J Comp Neurol 166:101–110

Nau F, Delius JD (1981) Discrepant effects of unilateral and bilateral forebrain lesions on the visual performance of pigeons. Behav Brain Res 2:119–124

Nauta WJH, Karten HJ(1970) A general profile of the vertebrate brain, with sidelights in the ancestry of cerebral cortex. In: Schmitt FO (ed) The neurosciences: second study program. Rockefeller University Press, New York, pp 7–26

Necker R (1985) Projection of a cutaneous nerve to the spinal cord of the pigeon. I Evoked field potentials; II Responses of the dorsal horn neurons. Exp Brain Res 59:388–343;344–352

Necker R (1989) Cells of origin of spinothalamic, spinotectal, spinoreticular and spinocerebellar pathways in the pigeon as studied by the retrograde transport of horseradish peroxidase. J Hirnforsch 30:33–43

Necker R (1990) Sensory representation of the wing in the spinal dorsal horn of the pigeon. Exp Brain Res 81:403–412

Necker R (1991a) A novel spinal pathway and other connections to the spinocerebellum in the pigeon. Brain Res Bull 27:581–586

Necker R (1991b) Cells of origin of avian postsynaptic dorsal column pathways. Neurosci Lett 126:91–93

Necker R (1992) Spinal neurons projecting to anterior or posterior cerebellum in the pigeon. Anat Embryol 185:325–334

Necker R (1994) Sensorimotor aspects of flight control in birds: specializations in the spinal cord. Eur J Morphol 32:207–211

Necker R, Schermuly C (1985) Central projections of the radial nerve and one of its cutaneous branches in the pigeon. Neurosci Lett 58:271–276

Neuhaus W (1963) On the olfactory sense of birds. In: Zotterman Y (ed) Olfaction and taste. Pergamon Press, Oxford, pp 111–124

Newman JD, (1970) Midbrain regions relevant to auditory communication in songbirds. Brain Res 22:259–261

Ngo TD, Német A, Tömböl T (1992) Some data on GABAergic innervation of nucleus rotundus in chicks. J Hirnforsch 33:335–355

Ngo TD, Davies DC, Egedi GY, Tömböl T (1994) A phaseolus-lectin anterograde tracing study of the tectorotundal projections in the domestic chick. J Anat 184:129–136

Nieuwenhuys R (1964) Comparative anatomy of the spinal cord. Progr Brain Res 11:1–55

Nieuwenhuys R (1967a) Comparative anatomy of olfactory centres and tracts. Progr Brain Res 23:1–64

Nieuwenhuys R (1967b) Comparative anatomy of the cerebellum. Progr Brain Res 25:1–93

Nieuwkoop PD (1952) Activation and organization of the central nervous system in amphibians, P.III. Synthesis of a new working hypothesis. J Exp Zool 120:83–108

Nixdorf BE, Bischof H-J (1982) Afferent connections of the ectostriatum and visual Wulst in the zebra finch (Taeniopygia guttata castanotis Gould) – an HRP study. Brain Res 248:9–17

Nordeen KW, Nordeen EJ (1988) Projection neurons within a vocal motor pathway are born during song learning in zebra finches. Nature 334:149–151

Norgren RB, Silver R (1989) Retinal projections in quail (Coturnix coturnix). Vis Neurosci 3:377–387

Norgren RB, Silver R (1990) Distribution of vasoactive intestinal peptide-like and neurophysin-like immunoreactive neurons and acetylcholinesterase staining in the ring dove hypothalamus with emphasis on the question of an avian suprachiasmatic nucleus. Cell Tissue Res 259:331–339

Norgren RB, Lippert J, Lehman N (1992) Luteinizing hormone-releasing hormone in the pigeon terminal nerve and olfactory bulb. Neurosci Lett 135:201–204

Nottebohm F (1981) A brain for all seasons: cyclical anatomical changes in song control nuclei of the canary brain. Science 214:1368–1370

Nottebohm F (1991) Reassessing the mechanisms and origins of vocal learning in birds. Trends Neurosci 14:206–211

Nottebohm F (1993) The search for neural mechanisms that define the sensitive period for song learning in birds. Neth J Zool 431:193–234

Nottebohm F, Kelley DB, Paton JA (1982) Connections of the vocal control nuclei in the canary telencephalon. J Comp Neurol 207:344–357

Oehmke H-J (1971) Vergleichende neurohistologische Studien am nucleus infundibularis einiger australischer Vögel. Z Zellforsch 122:122–138

O'Flaherty JJ (1970) A Golgi analysis of the optic tectum of the mallard duck. J Hirnforsch 12:389–404

Ohmori Y, Watanabe T, Fujioka T (1984) Localization of parasympathetic preganglionic neurons in the sacral spinal cord of the domestic fowl. Jpn J Zootech Sci 55:792–794

Ohmori Y, Watanabe T, Fujioka T (1987) Projections of the visceral and somatic primary afferents to the sacral spinal cord of the domestic fowl revealed by transganglionic transport of horseradish peroxidase. Neurosci Lett 74:175–179

Ohmori Y, Wakita T, Watanabe T (1992) Sympathetic and sensory neurons projecting into the cervical sympathetic trunk in the chicken. J Auton Nerv Syst 40:207–214

Okado N, Oppenheim RW (1985) The onset and development of descending pathways to the spinal cord in the chick embryo. J Comp Neurol 232:143–161

Okado N, Yoshimoto M, Furber SE (1987a). Pathway formation and the terminal distribution pattern of the spinocerebellar projection in the chick embryo. Anat Embryol 176:165–174

Okado N, Ito R, Homma S (1987b.) The terminal distribution pattern of spinocerebellar fibers. An anterograde labelling study in the posthatching chick. Anat Embryol 176:175–182

Okado N, Sako H, Homma S, Ishakawa K (1992) Development of serotonergic system in the brain and spinal cord of the chick. Progr Neurobiol 38:93–123

Oksche A, Wilson WO, Farner DS (1964) The hypothalamic neurosecretory system of Coturnix coturnix japonica. Z Zellforsch 61:688–709

Okuhata S, Nottebohm F (1992) Single units in nucleus UVA respond to sounds and respiration and are part of a loop involved in song production. Proc 3rd Int Congress Neuroethology, Montreal, p 60

Oscarsson O, Rosen I, Uddenburg N (1963) Organization of ascending tracts in the spinal cord of the duck. Acta Physiol Scand 59:143–153

Ostrom JH (1974) Archaeopteryx and the origin of flight. Q Rev Biol 49:27–47

Panzica GC (1985) Vasotocin-immunoreactive elements and neuronal typology in the suprachiasmatic nucleus of the chicken and Japanese quail. Cell Tissue Res 242:371–376

Panzica GC, Viglietti-Panzica C (1983) A Golgi-study of the parvocellular neurons in the paraventricular nucleus of the domestic fowl. Cell Tissue Res 231:603–613

Panzica GC, Aste N, Coscia A, De Bernardi W, Viglietti-Panzica C, Balthazart J (1991a) A sex-dependent influence of testosterone on the dorso-medial neuronal population of the Japanese quail intercollicular nucleus. J Hirnforsch 32:469–475

Panzica GC, Viglietti-Panzica C, Sanchez F, Sante P, Balthazart J (1991b) Effects of testosterone on a selected neuronal population within the preoptic sexually dimorphic nucleus of the Japanese quail. J Comp Neurol 303:443–456

Papi F, Casini G (1990) Pigeons with ablated pyriform cortex home from familiar, but not from unfamiliar sites. Proc Natl Acad Sci USA 87:3783–3787

Papi F, Ioalé P, Fiaschi V, Benvenuti S, Balduccini NE (1974) Olfactory navigation of pigeons: the effect of treatment with odorous air currents. J Comp Physiol 94:187–193

Papi F, Keeton WT, Brown AI, Benvenuti S (1978) Do American and Italian pigeons rely on different homing mechanisms. J Comp Physiol 128A:303–317

Parent A (1986) Comparative neurobiology of the basal ganglia. Wiley, New York

Parent A, Olivier A (1970) Comparative histochemical study of the corpus striatum. J Hirnforsch 12: 73–81

Parks TN, Rubel EW (1975) Organization and development of the brain stem auditory nuclei of the chicken: organization of projections from n. magnocellularis to n. laminaris. J Comp Neurol 164:435–448

Parsons CH, Rogers LJ (1993) Role of the tectal and posterior commissures in lateralization of the avian brain. Behav Brain Res 54:153–164

Paton JA, Manogue KR (1982) Bilateral interactions within the vocal control pathway of birds: two evolutionary alternatives. J Comp Neurol 212:329–335

Paton JA, Manoque KR, Nottebohm F (1981). Bilateral organization of the vocal control pathway in the budgerigar, Melopsittacus undulatus. J Neurosci 1:1279–1281

Patten BM (1957) Early embryology of the chicken, 4th edn. McGraw-Hill, New York

Pearlman AL, Hughes CP (1976) Functional role of efferents to the avian retina. II Effects of reversible cooling of the isthmo-optic nucleus. J Comp Neurol 166:123–132

Pearson R (1972) The avian brain. Academic Press, London

Péczely P, Antoni FA (1984) Comparative localization of neurons containing ovine corticotropin releasing factor (CRF)-like and neurophysine-like immunoreactivity in the diencephalon of the pigeon (Columba livia domestica). J Comp Neurol 228:69–80

Perisic J, Mihailovic J, Cuénod M (1971) Electrophysiology of contralateral and ipsilateral visual projections to the Wulst in pigeon (Columba livia). Intern J Neurosci 2:7–14

Petrovicky P (1966) Reticular formation of the pigeon. Folia Morphol 14:334–346

Petrovicky P (1973) The pattern of avian reticular formation. Acta Univ Carol Med 19:393–398

Petrovicky P, Kolesárová D (1989) Parabrachial nuclear complex. A comparative study of its cytoarchitectonics in birds and some mammals including man. J Hirnforsch 30:539–550

Pettigrew JD, Konishi M (1976) Neurons selective for orientation and binocular disparity in the visual Wulst of the barn owl (Tyto alba). Science 193:675–678

Peusner KD, Morest DK (1977) The neuronal architecture and topography of the nucleus vestibularis tangentialis in the late chick embryo. Neuroscience 2:189–207

Phillips RE (1964) "Wildness" in the mallard duck: effects of brain lesions and stimulation on "escape behavior" and reproduction. J Comp Neurol 122:139–156

Portmann A (1946) Etudes sur la cérébralisation chez les oiseaux I. Alauda 14:2–20

Portmann A (1947) Etudes sur la cérébralisation chez les oiseaux, II, III. Alauda 15:1–15, 161–171

Powers AS (1989) Wulst lesion in pigeons disrupt go/no-go reversal. Physiol Behav 46:337–339

Pritz MB, Mead WR, Northcutt RG (1970) The effect of Wulst ablations on color, brightness and pattern discrimination in pigeons (Columba livia). J Comp Neurol 140:81–100

Puelles L, Medina L (1994) Development of neurons expressing tyrosine hydroxylase and dopamine in the chicken brain: a comparative segmental analysis. In: Smeets WJAJ, Reiner A (eds) Phylogeny and development of catecholamine systems in the CNS of vertebrates. Cambridge University Press, Cambridge, pp 302–310

Puelles L, Rubenstein JLR (1993) Expression patterns of homeobox and other putative regulatory genes in the embryonic mouse forebrain suggest a neuromeric organization. Trends Neurosci 16:472–477

Puelles L, Martínez S, Martínez-de la Torre M (1987) The locus of optic nerve head representation in the chick retinotectal map lacks a retinal projection. Neurosci Lett 79:23–28

Puelles L, Robles C, Martínez-de-la-Torre M, Martínez S (1994) New subdivision schema for the avian torus semicircularis: neurochemical maps in the chick. J Comp Neurol 340:98–125

Rabin A (1975) Labyrinthine and vestibulospinal effects on spinal motoneurons in the pigeon. Exp Brain Res 22:431–448

Rager G (1980) Development of the retinotectal projection in the chicken. Adv Anat Embryol Cell Biol 63:1–92

Rager G, von Oeynhausen B (1979) Ingrowth and ramification of retinal fibers in the developing optic tectum of the chick embryo. Exp Brain Res 35:213–227

Ramirez JM, Delius JD (1979a) Aggressive behavior in pigeons: suppression by archistriatal lesions. Aggress Behav 5:3–17

Ramirez JM, Delius JD (1979b) Nucleus striae terminalis lesions affect agonistic behavior of pigeons. Physiol Behav 22:871–875

Raymond A (1958) Responses to electrical stimulation in the cerebellum of unanesthetised birds. J Comp Neurol 110:299–320

Rehák P, Kostová D, Boda K (1985) The morphology of the medial hypothalamic region in the goose. Z Mikrosk-Anat Forsch 99:425–438

Rehkämper G, Zilles K (1991) Parallel evolution in mammalian and avian brains: comparative cytoarchitectonic and cytochemical analysis. Cell Tissue Res 263:3–28

Rehkämper G, Zilles K, Schleicher A (1984) A quantitative approach to cytoarchitectonics. IX. The areal pattern of the hyperstriatum ventrale in the domestic pigeon, Columba livia f.d. Anat Embryol 169:319–327

Rehkämper G, Zilles K, Schleicher A (1985) A quantitative approach to cytoarchitectonics. X. The areal pattern of the neostriatum in the domestic pigeon, Columba livia f.d. A cyto- and myeloarchitectonic study. Anat Embryol 171:345–355

Rehkämper G, Haase E, Frahm HD (1988) Allometric comparison of brain weight and brain structure volumes in different breeds of the domestic pigeon, Columba livia f.d. (fantails, homing pigeons, strassers). Brain Behav Evol 31:141–149

Rehkämper G, Frahm HD, Ziller K (1991) Quantitative development of brain and brain structures in birds (Galliformes and Passeriformes) compared to that in mammals (insectivores and primates). Brain Behav Evol 37:125–143

Reiner A (1986a) The co-occurrence of substance P-like immunoreactivity and dynorphin-like immunoreactivity in striatopallidal and striatonigral projection neurons in birds and reptiles. Brain Res 371:155–161

Reiner A (1986b) Is prefrontal cortex found only in mammals? Trends Neurosci 9:298–300

Reiner A, Carraway RE (1987) Immunohistochemical and biochemical studies on lys [8]-Asn [9]-neurotensin [8-13] (LANT)-related peptides in the basal ganglia of pigeons, turtles, and hamsters. J Comp Neurol 257:453–476

Reiner A, Karten HJ (1982) Laminar distribution of the cells of origin of the descending tectofugal pathways in the pigeon (Columba livia). J Comp Neurol 204:165–187

Reiner A, Karten HJ (1983) The laminar source of efferent projections from the avian Wulst. Brain Res 275:349–354

Reiner A, Karten HJ (1985) Comparison of olfactory bulb projections in pigeons and turtles. Brain Behav Evol 27:11–27

Reiner A, Brecha NC, Karten HJ (1979) A specific projection of retinal displaced ganglion cells to the nucleus of the basal optic root in the chicken. Neuroscience 4:1679–1688

Reiner A, Brecha NC, Karten HJ (1982a) Basal ganglia pathways to the tectum: the afferent and efferent connections of the lateral spiriform nucleus of pigeon. J Comp Neurol 208:16–36

Reiner A, Karten HJ, Brecha NC (1982b) Enkephalin-mediated basal ganglia influences over the optic tectum: immunohistochemistry of the tectum and the lateral spiriform nucleus in pigeon. J Comp Neurol 208:37–53

Reiner A, Karten HJ, Gamlin PDR, Erichsen JT (1983a) Parasympathetic ocular control: functional subdivisions and circuitry of the avian nucleus of Edinger-Westphal. Trends Neurosci 6:141–145

Reiner A, Karten HJ, Solina AR (1983b) Substance P: localization within paleostriatal-tegmental pathways in the pigeon. Neuroscience 9:61–85

Reiner A, Brauth SE, Karten HJ (1984a) Evolution of the amniote basal ganglia. Trends Neurosci 7:320–325

Reiner A, Davis BM, Brecha NC, Karten HJ (1984b) The distribution of enkephalinlike immunoreactivity in the telencephalon of the adult and developing domestic chicken. J Comp Neurol 228:245–262

Reiner A, Karle EJ, Anderson KJ, Medina L (1994). Catecholaminergic perikarya and fibers in the avian nervous system. In: Smeets WJAJ, Reiner A (eds) Phylogeny and development of catecholamine systems in the CNS of vertebrates. Cambridge University Press, Cambridge, pp 135–181

Reinke H, Necker R (1992) Spinal dorsal column afferent fiber composition in the pigeon: an electrophysiological investigation. J Comp Physiol A171:397–403

Remy N, Güntürkün O (1991) Retinal afferents to the tectum opticum and the nucleus opticus principalis thalami in the pigeon. J Comp Neurol 305:57–70

Rendahl H (1924) Embryologische und morphologische Studien über das Zwischenhirn beim Huhn. Acta Zool 5:241–344

Renggli F (1967) Vergleichend anatomische Untersuchungen über die Kleinhirn- und Vestibulariskerne der Vögel. Rev Suisse Zool 4:701–778

Repérant J (1973) Nouvelles données sur les projections visuelles chez le Pigeon (Columba livia). J Hirnforsch 14:151–187

Revzin AM (1969) A specific visual projection area in the hyperstriatum of the pigeon. Brain Res 15:246–249

Revzin AM (1970) Some characteristics of wide-field units in the brain of the pigeon. Brain Res 3:195–204

Revzin AM, Karten H (1966) Rostral projections of the optic tectum and the nucleus rotundus in the pigeon. Brain Res 3:264–267

Rexed B (1952) The architectonic organization of the spinal cord in the cat. J Comp Neurol 96:415–497

Rexed B (1954) A cytoarchitectonical atlas of the spinal cord in the cat. J Comp Neurol 100:297–380

Rieke GK (1980) Kainic acid lesions of pigeon paleostriatum: a model for study of movement disorders. Physiol Behav 24:683–687

Rieke GK, Wenzel BM (1978) Forebrain projections of the pigeon olfactory bulb. J Morphol 158:41–56

Rio JP, Villalobos J, Miceli D, Repérant J (1983) Efferent projections of the visual Wulst upon the nucleus of the basal optic root in the pigeon. Brain Res 271:145–151

Ritchie TLC (1979) Intratelencephalic visual connections and their relationship to the archistriatum in the pigeon (Columba livia). PhD thesis, University of Virginia, Charlottesville

Robert F, Cuénod M (1969) Electrophysiology of the intertectal commissures in the pigeon. I. Analysis of the pathways. Exp Brain Res 9:116–122

Roeling TAP, Feirabend HKP (1988) Glial fiber pattern in the developing chicken cerebellum: vimentin and glial fibrillary acidic protein GFAP immunostaining. Glia 1:398–402

Rogers KT (1960) Studies on chick brain of biochemical differentiation related to morphological differentiation and onset of function. I. Morphological development. J Exp Zool 144:77–87

Rogers LJ (1991) Development of lateralization. In: Andrew RJ (ed) Neural and behavioural plasticity. The use of the domestic chick as a model. Oxford University Press, Oxford, pp 507–533

Rogers LJ, Sink HS (1988) Transient asymmetry in the projections of the rostral thalamus to the visual hyperstriatum of the chicken, and reversal of its direction by light exposure. Exp Brain Res 70:378–384

Rogers LJ, Robinson T, Ehrlich D (1986) Role of the supraoptic decussation on the development of asymmetry of brain function in the chicken. Dev Brain Res 28:33–39

Romanoff AL (1960). The avian embryo. MacMillan, New York

Rose M (1914) Ueber die cytoarchitektonische Gliederung des Vorderhirns der Vögel. J Psychol Neurol 21:278–352

Rubel EW, Parks TN (1975) Organization and development of the brain stem nuclei of the chicken: tonotopic organization of n. magnocellularis and n. laminaris. J Comp Neurol 164:411–434

Rübsamen R, Dorscheidt GJ (1986) Tonotopic organization of the auditory forebrain in a song bird, the European starling. J Comp Physiol A 158:639–646

Rüdeberg SI (1961) Morphogenetic studies on the cerebellar nuclei and their homologization in different vertebrates including man. Publ Tornblad & Inst Comp Embryol Inst Zool, Lund

Rüdeberg SI (1964) The anlage of the cerebellar nuclei after early extirpation of the vestibular region. J Morphol Exp Embryol 12:56–64

Ryan SM, Arnold AP (1981) Evidence for cholinergic participation in the control of birds song: acetylcholinesterase distribution and muscarinic receptor autoradiography in the zebra finch brain. J Comp Neurol 202:211–219

Saetersdal TAS (1959) On the ontogenesis of the avian cerebellum. Univ Bergen Arb Nat 3:1–44

Saleh CN, Ehrlich D (1984) Composition of the supraoptic decussation of the chick (Gallus gallus). A possible factor limiting interhemispheric transfer of visual information. Cell Tissue Res 236:601–609

Salzen EA (1993) The neural locus for imprinting – IMHV or LN? Eur J Neurosci [Suppl] 6:146

Salzen EA, Parker DM, Williamson AJ (1975) A forebrain lesion preventing imprinting in domestic chicks. Exp Brain Res 24:145–157

Salzen EA, Williamson AJ, Parker DM (1979) The effect of forebrain lesions on innate and imprinted colour, brightness and shape preferences in domestic chicken. Behav Processes 4:295–313

Sanders EB (1929) A consideration of certain bulbar, midbrain and cerebellar centers and fiber tracts in birds. J Comp Neurol 49:155–221

Sansone FM (1977) The craniocaudal extent of the glycogen body in the domestic chicken. J Morphol 153:87–106

Sato T, Wake K (1983) Innervation of the avian pineal organ. Cell Tissue Res 233:237–264

Schall U, Delius JD (1986) Sensory inputs to the nucleus basalis prosencephali, a feeding-pecking centre in the pigeon. J Comp Physiol A 159:33–41

Schall U, Güntürkün O, Delius JD (1986) Sensory projections to the nucleus basalis prosencephali of the pigeon. Cell Tissue Res 245:539–546

Scheich H (1983) Two columnar systems in the auditory neostriatum of the chick: evidence from 2-deoxyglucose. Exp Brain Res 51:199–205

Schneider A (1991) Der Somatosensorische Thalamus der Taube (Columba livia). Thesis, University of Bochum

Schneider A, Necker R (1989) Spinothalamic projections in the pigeon. Brain Res 484:139–149

Schoenwolf GC, Desmond ME (1984) Descriptive studies of occlusion and reopening of the spinal canal of the early chick embryo. Anat Rec 209:251-263

Schriber H (1978) Electrophysiological investigation of the archistriato-hypothalamic pathway in the pigeon (Columba livia). Pflügers Arch 375:31-37

Schroeder DM, Murray RG (1987) Specializations within the lumbosacral spinal cord op the pigeon. J Morphol 194:41-53

Schulte M, Necker R (1994) Projection of wing nerves to spinal cord and brain stem of the pigeon as studied by transganglionic transport of Fast Blue. J Brain Res 35:313-325

Schwarz DWF, Schwarz IE, Tomlinson RD (1978) Avian efferent vestibular neurons identified by axonal transport of [^{3}H]adenosine and horseradish peroxidase. Brain Res 155:103-107

Schwarz IE, Schwarz DW (1983) The primary vestibular projection to the cerebellar cortex in the pigeon (Columba livia). J Comp Neurol 216:438-444

Schwarz IE, Schwarz DW, Fredrickson JM, Landolt JP (1981) Efferent vestibular neurons: a study employing retrograde tracer methods in the pigeon (Columba livia). J Comp Neurol 196:1-12

Scott SA, Dinowitz S, Terhaar K, Sherlock D, Campbell MA, Levine D (1994) Cytochemical characteristics of neurons in the trigeminal mesencephalic nucleus of hatchling chicks. J Comp Neurol 350:302-310

Semm P, Demaine C (1984) Electrophysiology of the pigeon's habenular nuclei: evidence for pineal connections and input from the visual system. Brain Res Bull 12:115-121

Sherry DF, Vaccarino AL, Buckenham K, Herz RS (1989) The hippocampal complex of food-storing birds. Brain Behav Evol 34:308-317

Sherry DF, Jacobs LF, Gaulin SJC (1992) Spatial memory and adaptive specialization of the hippocampus. Trends Neurosci 15:298-303

Shimizu T, Hodos W (1989) Reversal learning in pigeons: effects of selective lesions of the Wulst. Behav Neurosci 103:262-272

Shimizu T, Karten HJ (1990) Immunohistochemical analysis of the visual Wulst of the pigeon (Columba livia). J Comp Neurol 300:346-369

Shiosaka S, Takatsuki K, Inagaki S, Sakanaka M, Takagi H, Senba E, Matsuzaki T, Tohyama M (1981) Topographic atlas of somatostatin-containing neuron system in the avian brain in relation to catecholamine-containing neuron system. II. Mesencephalon, rhombencephalon, and spinal cord. J Comp Neurol 202:115-124

Showers MJC, Lyons P (1968) Avian nucleus isthmi and its relation to hippus. J Comp Neurol 132:589-616

Simpson HB, Vicario DS (1990) Brain pathways for learned and unlearned vocalizations differ in zebra finches. J Neurosci 10:1541-1556

Smeets WJAJ (1991) Comparative aspects of the distribution of substance P and dopamine immunoreactivity in the substantia nigra of amniotes. Brain Behav Evol 37:179-188

Smolen AJ, Glazer EJ, Ross LL (1979) Horseradish peroxidase histochemistry combined with glyoxylic acid-induced fluorescence used to identify brainstem catecholaminergic neurons which project to the chick thoracic spinal cord. Brain Res 160:353-357

Sorenson EM, Parkinson D, Dahl JL, Chiappinelli VA (1989) Immunohistochemical localization of choline acetyltransferase in the chicken mesencephalon. J Comp Neurol 281:641-657

Sterling RJ, Sharp PJ (1982) The localisation of LH-RH neurones in the diencephalon of the domestic hen. Cell Tissue Res 222:283-298

Sterling RJ, Gasc JM, Sharp PJ, Renoir JM, Tuohimaa P, Baulieu EE (1987) The distribution of nuclear progesterone receptor in the hypothalamus and forebrain of the domestic hen. Cell Tissue Res 248:201-205

Stewart MG, Csillag A, Rose SPR (1987) Alterations in synaptic structure in the paleostriatal complex of the domestic chick, Gallus domesticus, following passive avoidance training. Brain Res 426:69-81

Stingelin W (1958) Vergleichend Morphologische Untersuchungen am Vorderhirn der Vögel auf Cytologischer und Cytoarchitektonischer Grundlage. Helbing & Lichtenhahn, Basel

Stingelin W (1965) Qualitative und quantitative Untersuchungen an Kerngebieten der Medulla oblongata bei Vögeln. Bibl Anat, vol 6, Karger, Basel

Stokes TM, Leonard CM, Nottebohm F (1974) The telencephalon, diencephalon and mesencephalon of the canary, Serinus canaria, in stereotaxic coordinates. J Comp Neurol 156:337-374

Stone J, Freeman JA (1971) Synaptic organisation of the pigeon's optic tectum: a Golgi and current source-density analysis. Brain Res 27:203-221

Strasznicky C, Tay D (1983) The localization of motoneuron pools innervating wing muscles in the chick. Anat Embryol 166:209-218

Streeter GL (1903) The structure of the spinal cord of the ostrich. Am J Anat 3:1-27

Streit P, Stella M, Cuénod M (1980) Transneuronal labeling in the pigeon visual system. Neuroscience 5:763-775

Striedter GF (1994) The vocal control pathways in budgerigars differ from those in songbirds. J Comp Neurol 343:35-56

Ströhmann B, Schwarz DWF, Puil E (1994) Mode of firing and rectifying properties of nucleus ovoidalis neurons in the avian auditory thalamus. J Neurophysiol 71:1351-1360

Strutz J, Schmidt CL (1982) Acoustic and vestibular efferent neurons in the chicken Gallus domesticus. Acta Otolaryngol 94:45-51

Székely AD, Boxer MI, Stewart MG, Csillag A (1994) Connectivity of the lobus parolfactorius of the domestic chicken (Gallus domesticus): an anterograde and retrograde pathway tracing study. J Comp Neurol 348:374-393

Takahashi JS, Menaker M (1982) Role of the suprachiasmatic nuclei in the circadian system of the house sparrow, Passer domesticus. J Neurosci 2:815-822

Takahashi TT, Konishi M (1988) Projections of the cochlear nuclei and nucleus laminaris to the inferior colliculus of the barn owl. J Comp Neurol 274:190-211

Takahashi TT, Wagner H, Konishi M (1989) Role of commissural projections in the representation of bilateral auditory space in the barn owl's inferior colliculus. J Comp Neurol 281:545-554

Takatsuki K, Shiosaka S, Inagaki S, Sakanaka M, Takagi H, Senba E, Matsuzaki T, Tohyama M (1981) Topographic atlas of somatostatin-containing neuron system in the avian brain in relation to catecholamine-containing neuron system. I. Telecephalon and diencephalon. J Comp Neurol 202:103-113

Takei Y, Kobayashi H, Yanagisawa M, Bando T (1979) Involvement of catecholaminergic nerve fibers in angiotensin II-induced drinking in the Japanese quail, Coturnix coturnix japonica. Brain Res 174:229-244

Tellegen AJ, Dubbeldam JL (1994) Location of premotor neurons of the motor nuclei innervating craniocervical muscles in the mallard (Anas platyrhynchos L.). Eur J Morphol 32:138-141

Teuchert G, Reissmann T, Vockel A (1986) Olfaction in Peking ducks (Anas platyrhynchos): a comparative study of centrifugal and centripetal olfactory connections in young ducks and in embryos and ducklings (Aves). Zoomorphology 106:185-198

Thanos S, Bonhoeffer F (1984) Development of the transient ipsilateral retinotectal projection in the chick embryo: a numerical fluorescens-microscopic analysis. J Comp Neurol 224:407-414

Thanos S, Bonhoeffer F, Rutishauser U (1984) Fiber-fiber interaction and tectal cues influence of the development of the chicken retinotectal projection. Proc Natl Acad Sci USA 81:1906-1910

Thornton SN (1986) Osmoreceptor localization in the brain of the pigeon (Columba livia). Brain Res 377:96–104

Tohyama M, Maeda T, Hashimoto J, Shrestha GR, Tamura O, Shimizu N (1974) Comparative anatomy of the locus coeruleus. I. Organization and ascending projections of the catecholamine containing neurons in the pontine region of the bird, Melopsittacus undulatus. J Hirnforsch 15:319–330

Tömböl T (1991) Arborization of afferent fibers in ectostriatum centrale. Golgi study. J Hirnforsch 32:563–575

Tömböl T, Davies DC (1994) The Golgi architecture of the domestic chick archistriatum. J Hirnforsch 34:17–29

Tömböl T, Csillag A, Stewart MG (1988a) Cell types of the hyperstriatum ventrale of the domestic chicken (Gallus domesticus): a Golgi study. J Hirnforsch 29:319–334

Tömböl T, Maglóczky Z, Stewart MG, Csillag A (1988b) The structure of chicken ectostriatum. I. Golgi study. J Hirnforsch 29:525–546

Tsai HM, Garber BB, Larramendi LMH (1981) ^{3}H-thymidine autoradiographic analysis of telencephalic histogenesis in the chick embryo. I. Neuronal birthdays. II. Dynamics of neuronal migration, displacement and aggregation. J Comp Neurol 198:275–306

Ueck M (1982) Morphologie und Physiologie des Pinealorgans in der Evolution der Wirbeltiere. Verh Dtsch Zool Ges 1982:61–80

Ueck M, Umar H (1983) Environmental, neural, and endocrine influences on the parenchyma of the avian pineal organ and its various responses. In: Mikami S (ed) Avian endocrinology: environmental and ecological perspective. Jpn Sci Soc Press, Tokyo, pp 201–215

Ulinski PS (1983) Dorsal ventricular ridge. Wiley, New York

Ulinski PS (1990) Nodal events in forebrain evolution. Neth J Zool 40:215–240

Ulinski PS, Margoliash D (1990) Neurobiology of the reptile-bird transition. In: Jones EG, Peters A (eds) Cerebral cortex, vol 8A. Plenum, New York, pp 217–265

Vaage S (1969) The segmentation of the primitive neural tube in chick embryos (Gallus domesticus). Adv Anat Embryol Cell Biol 41:1–88

Van den Akker LM (1970) An anatomical outline of the spinal cord of the pigeon. Thesis, Leiden University

Van Tienhoven A, Juhász LP (1962) The chicken telencephalon, diencephalon and mesencephalon in stereotaxic coordinates. J Comp Neurol 118:185–197

Veenman CL, Gottschaldt K-M (1986) The nucleus basalis-neostriatum complex in the goose (Anser anser L.). Adv Anat Embryol Cell Biol 96:1–85

Veenman CL, Reiner A (1990) Telencephalic and thalamic inputs to the striatal portion of the pigeon basal ganglia. Soc Neurosci Abstr 16:246

Veenman CL, Reiner A (1994) The distribution of GABA-containing perikarya, fibers, and terminals in the forebrain and midbrain of pigeons, with particular reference to the basal ganglia and its projection targets. J Comp Neurol 339:209–250

Veenman CL, Albin RL, Richfield EK, Reiner A (1994) Distributions of GABA$_A$, GABA$_B$, and benzodiazepine receptors in the forebrain and midbrain of pigeons. J Comp Neurol 344:161–189

Veenman CL, Wild JM, Reiner A (1995) Organization of the avian "corticostriatal" projection system: a retrograde and anterograde pathway tracing study in pigeons. J Comp Neurol 354:87–126

Verhaart WJC (1971) Forebrain bundles and fibre systems in the avian brainstem. J Hirnforsch 13:39–64

Verhaart WJC (1974) Identification of fibre systems of the avian midbrain. J Hirnforsch 15:379–386

Vicario DS (1993) A new brainstem pathway for vocal control in the zebrafinch song system. Neuroreport 4:983–986

Vicario DS, Nottebohm F (1988) Organization of the zebra-finch song control system: I Representation of syringeal muscles in the hypoglossal nucleus. J Comp Neurol 271:346–354

Vielvoye GJ (1977) Spinocerebellar tracts in the white leghorn Gallus domesticus. Thesis, Leiden University

Vogt-Nilsen L (1954) The inferior olive in birds. A comparative morphological study. J Comp Neurol 101:447–481

Völker H (1971a) Zytoarchitektonik, Myeloarchitektonik und Zytologie der di-mesencephalen Übergangsregion und des okzipitalen diencephalon vom Haushuhn (Gallus domesticus L.). Anat Anz 128:113–135

Völker H (1971b) Hypo- und Epithalamus des Haushuhnes (Gallus domesticus L.) – eine zyto- und myeloarchitektonische sowie zytologische Studie. Anat Anz 129:159–179

Völker H (1972a) Architektur und Zytologie des Diencephalon unter besonderer Berücksichtigung des Thalamus paraventricularis beim Haushuhn (Gallus domesticus L.). Arch Exp Vet Med 26:153–171

Völker H (1972b) Die aborale zyto- und myeloarchitektonische sowie zytologische ventrale Thalamusstruktur vom Haushuhn (Gallus domesticus L.). Anat Anz 130:312–331

Von Bartheld CS, Lindörfer HW, Meijer DL (1987) The nervus terminalis also exists in cyclostomes and birds. Cell Tissue Res 250:431–434

Von Bartheld CS, Code RA, Rubel EW (1989) GABA-ergic neurons in brainstem auditory nuclei of the chick: distribution, morphology, and connectivity. J Comp Neurol 287:470–483

Von Kupffer C (1906) Die Morphogenesis des Zentralnervensystems. In: Hertwig O (ed) Handbuch der Vergleichenden und Experimentellen Entwicklungslehre der Wirbeltiere, vol II/3, Fischer, Jena, pp 1–273

Voneida TJ, Mello NK (1975) Interhemispheric projections of the optic nerve in pigeon. Brain Behav Evol 11:91–108

Vowles DM, Beazley L, Harwood DH (1975) A stereotaxic atlas of the brain of the barbary dove (Streptopelia risoria). In: Wright P, Caryl PG, Vowles DM (eds) Neural and endocrine aspects of behaviour in birds. Elsevier, Amsterdam, pp 351–394

Vu ET, Mazurek ME, Yu-Chien Kuo (1994) Identification of a forebrain motor programming network for the learned song of zebra finches. J Neurosci 14:6924–6934

Wächtler K (1985) Regional distribution of muscarinic acetylcholine receptors in the telencephalon of the pigeon (Columba livia f. domestica). J Hirnforsch 26:85–89

Wächtler K, Ebinger P (1989) The pattern of muscarinic acetylcholine receptor binding in the avian forebrain. J Hirnforsch 30:409–414

Waldman C, Güntürkün O (1993). The dopaminergic innervation of the pigeon caudolateral forebrain: immunocytochemical evidence for a 'prefrontal cortex' in birds? Brain Res 600:225–234

Waldvogel JA, Phillips JB (1991) Olfactory cues perceived at the home loft are not essential for the formation of a navigational map in pigeons. J Exp Biol 155:643–660

Wallenberg A (1964) Beiträge zur vergleichenden Anatomie des Zentralnervensystems. J Hirnforsch 7:281–284

Wallman J, Saldanha CJ, Silver R (1994) A putative suprachiasmatic nucleus of birds responds to visual motion. J Comp Physiol A174:297–304

Wallraff HG (1988) Navigation mit Duftkarte und Sonnenkompass. Das Heimfindevermögen der Brieftauben. Naturwissenschaften 75:380–392

Wallraff HG, Kiepenheure J, Neumann MF, Sinsch U (1992) Microclimatic origin of inhaled air affects olfactory navigation of homing pigeons. Experientia 48:1153–1158

Wang Y-C, Frost BJ (1991) Visual response characteristics of neurons in the nucleus isthmi magnocellularis and nucleus isthmi parvocellularis of pigeons. Exp Brain Res 87:624–633

Watanabe M, Ito H, Masai H (1983) Cytoarchitecture and visual receptive neurons in the Wulst of the Japanese quail (Coturnix coturnix japonica). J Comp Neurol 213:188–198

Watanabe M, Ito H, Ikushima M (1985) Cytoarchitecture and ultrastructure of the avian ectostriatum: afferent terminals from the dorsal telencephalon and some nuclei in the thalamus. J Comp Neurol 236:241–257

Watanabe S, Hodos W, Bessette BB, Shimizu T (1986) Interocular transfer in parallel visual pathways in pigeons. Brain Behav Evol 29:184-195

Watanabe T, Ohmori Y (1988) Location of motoneurons supplying upper neck muscles in the chicken studied by means of horseradish peroxidase. J Comp Neurol 270:271-278

Webster DMS, Steeves JD (1988) Origins of brainstem-spinal projections in the duck and goose. J Comp Neurol 273:573-583

Webster DMS, Steeves JD (1991) Funicular organization of avian brainstem-spinal projections. J Comp Neurol 312:467-476

Webster DMS, Rogers LJ, Pettigrew JD, Steeves JD (1990) Origins of descending spinal pathways in prehensile birds: do parrots have a homologue to the corticospinal tract of mammals? Brain Behav Evol 36:216-226

Webster KE (1979) Some aspects of the comparative study of the corpus striatum. In: Divac I, Oberg RGE (eds) The neostriatum. Pergamon Press, New York, pp 107-126

Welsch U, Wächtler K (1969) Zum Feinbau des Glykogenkörpers im Rückenmark der Taube. Z Zellforsch Mikrosk Anat 97: 160-168

Wenzel BM (1971) Olfaction in birds. In: Beidler LM (ed) Handbook of sensory physiology, vol IV/I. Springer-Verlag, Berlin Heidelberg New York, pp 432-448

Wenzel BM, Rausch LJ (1977) Does the olfactory system modulate affective behavior in the pigeon? Ann NY Acad Sci 290:314-329

Wenzel BM, Sieck MH (1972) Olfactory perception and bulbar electrical activity in several avian species. Physiol Behav 9:287-293

Whitehead MC, Morest DK (1981) Dual populations of efferent and afferent axons in the chicken. Neuroscience 6:2351-2365

Whitlock DG (1952) A neurohistological and neurophysiological study of afferent fiber tracts and receptive areas of the avian cerebellum. J Comp Neurol 97:567-636

Wild JM (1981) Identification and localization of the motor nuclei and sensory projections of the glossopharyngeal, vagus, and hypoglossal nerves of the cockatoo (Cacatua roseicapilla, Cacatuidae). J Comp Neurol 203:351-377

Wild JM (1985) The avian somatosensory system. I Primary spinal afferent input to the spinal cord and brainstem in the pigeon (Columba livia). J Comp Neurol 240:377-395

Wild JM (1987a) The avian somatosensory system: connections of regions of body representation in the forebrain of the pigeon. Brain Res 412:205-223

Wild JM (1987b) Thalamic projections to the paleostriatum and neostriatum in the pigeon (Columba livia). Neuroscience 20:305-327

Wild JM (1987c) Nuclei of the lateral lemniscus project directly to the thalamic auditory nuclei in the pigeon. Brain Res 408:303-307

Wild JM (1988) Vestibular projections to the thalamus of the pigeon: an anatomical study. J Comp Neurol 271:451-460

Wild JM (1989) Avian somatosensory system. II. Ascending projections of the dorsal column and external cuneate nuclei in the pigeon. J Comp Neurol 287:1-18

Wild JM (1990) Peripheral and central terminations of hypoglossal afferents innervating lingual tactile mechanoreceptor complexes in Fringillidae. J Comp Neurol 298:157-171

Wild JM (1993a) The avian nucleus retroambigualis: a nucleus for breathing, singing and calling. Brain Res 606:319-324

Wild JM (1993b) Descending projections of the songbird nucleus robustus archistriatalis. J Comp Neurol 338:225-241

Wild JM, Arends JJA (1987) A respiratory-vocal pathway in the brainstem of the pigeon. Brain Res 407:191-194

Wild JM, Zeigler HP (1980) Central representation and somatotopic organization of the jaw muscle within the facial and trigeminal nuclei of the pigeon (Columba livia). J Comp Neurol 192:175-201

Wild JM, Zeigler HP (1996) Central projections and somatotopic organisation of trigeminal primary afferents in pigeon (Columba livia). J Comp Neurol 368:136-152

Wild JM, Cabot JB, Cohen DH, Karten HJ (1979) Origin, course and terminations of the rubrospinal tract in the pigeon (Columba livia). J Comp Neurol 187:639-654

Wild JM, Arends JJA, Zeigler HP (1985) Telencephalic connections of the trigeminal system in the pigeon (Columba livia): a trigeminal sensorimotor circuit. J Comp Neurol 234:441-464

Wild JM, Arends JJA, Zeigler HP (1990) Projections of the parabrachial nucleus in the pigeon (Columba livia). J Comp Neurol 293:499-523

Wild JM, Karten HJ, Frost BJ (1993) Connections of the auditory forebrain in the pigeon (Columba livia). J Comp Neurol 337:32-62

Williamson LC, Eagles DC, Brady MJ, Moffett JR, Namboodiri MA, Neale JH (1991) Localization and synaptic release of N-acetylaspartyl-glutamate in the chick retina and optic tectum. Eur J Neurosci 3:441-451

Wilson P (1980a) The organization of the visual hyperstriatum in the domestic chick. I. Topology and topography of the visual projection. Brain Res 188:319-332

Wilson P (1980b) The organization of the visual hyperstriatum in the domestic chick. II. Receptive field properties of single units. Brain Res 188:333-345

Wilson VJ, Anderson JA, Felix D (1972) Semicircular canal input to pigeon vestibulocerebellum. Brain Res 45:230-235

Wilson VJ, Anderson JA, Felix D (1974) Unit and field potential activity evoked in the pigeon vestibulo-cerebellum by stimulation of individual semicircular canals. Exp Brain Res 19:142-157

Wiltschko R, Wiltschko W (1987) Pigeon homing. Olfactory experience with inexperienced birds. Naturwissenschaften 74:94-96

Wiltschko W, Munro U, Ford H, Wiltschko R (1993) Red light disrupts magnetic orientation of migrating birds. Nature 364:525-527

Winterson BJ, Brauth SE (1985) Direction-selective single units in the nucleus lentiformis mesencephali of the pigeon (Columba livia). Exp Brain Res 60:215-226

Wirsig-Wiechmann CR (1990) The nervus terminalis in the chick: a FMRF-amide-immunoreactive and AChE-positive nerve. Brain Res 523:175-179

Witkovski P, Zeigler HP, Silver R (1973) The nucleus basalis of the pigeon: a single-unit analysis. J Comp Neurol 147:119-128

Wold JE (1975) The vestibular nuclei in the domestic hen (Gallus domesticus). II. Primary afferents. Brain Res 95:531-543

Wold JE (1976) The vestibular nuclei in the domestic hen (Gallus domesticus). I. Normal anatomy. Anat Embryol 149:29-46

Wold JE (1978a) The vestibular nuclei of the domestic hen (Gallus domesticus). III. Ascending projections to the mesencephalic eye motor nuclei. J Comp Neurol 179:393-405

Wold JE (1978b) The vestibular nuclei in the domestic hen (Gallus domesticus). IV. The projection to the spinal cord. Brain Behav Evol 15:41-62

Wold JE (1979) The vestibular nuclei in the domestic hen (Gallus domesticus). V Commissural connections. Exp Brain Res 34:217-232

Wold JE, Hall JG (1975) The distribution of primary afferents to the cochlear nuclei in the domestic hen (Gallus domesticus). Anat Embryol 147:75-89

Woodburne RT (1936) A phylogenetic consideration of the primary and secondary centers and connections of the trigeminal complex in a series of vertebrates. J Comp Neurol 65:403-501

Woodbury CJ (1992) Physiological studies of cutaneous inputs to dorsal horn laminae I-IV of adult chickens. J Neurophysiol 67:241-254

Wright P, Spence AM (1976) Changes in emotionality following section of the tractus occipito-mesencephalicus in the barbary dove (Streptopelia risoria). Behav Processes 1:29–40

Würdinger I (1979) Olfaction and feeding behaviour in juvenile geese (Anser a. anser and Anser domesticus). Z Tierpsychol 49:132–135

Wylie DR, Frost BJ (1990a) The visual response properties of neurons in the nucleus of the basal optic root of the pigeon: a quantitative analysis. Exp Brain Res 82:327–336

Wylie DR, Frost BJ (1990b) Binocular neurons in the nucleus of the basal optic root (nBOR) of the pigeon are selective for either translational or rotational visual flow. Vis Neurosci 5:489–495

Wylie DR, Kripalani T, Frost BJ (1993) Responses of pigeon vestibulocerebellar neurons to optokinetic stimulation. I Functional organization of neurons discriminating between translational and rotational visual flow. J Neurophysiol 70:2632–2646

Yamada K (1966) An analytical study of myelinated nerve fibers in the white matter of the cervical cord of a pigeon (Columba livia). Acta Med Fukuoka 36:450–465

Yamamoto K, Tohyama M, Shimizu N (1977) Comparative anatomy of the topography of catecholamine containing neuron system in the brain stem from birds to teleosts. J Hirnforsch 18:229–240

Yamauchi K, Yasuda M (1981) Several cell masses around the commissura anterior in the chicken. J Hirnforsch 22:189–194

Yamauchi K, Yasuda M (1985) Cyto-, dendro- and fibroarchitectonic studies on the chicken hypothalamus. J Hirnforsch 26:509–519

Yamauchi K, Yasuda M (1988). Classification of the neurons in fowl hypothalamic nuclei based on their dendritic patterns. Br Poultry Sci 29:581–588

Yan K, Wang S-R (1986) Visual responses of neurons in the avian nucleus isthmi. Neurosci Lett 64:340–344

Youngren OM, Phillips RE (1978) A stereotaxic atlas of the brain of the three-day-old domestic chick. J Comp Neurol 181:567–600

Youngren OM, Phillips RE (1983) Location and distribution of tracheosyringeal motoneuron somata in fowl. J Comp Neurol 213:86–93

Zappia JV, Rogers LJ (1987) Sex differences and reversal of brain asymmetry by testosterone in chickens. Behav Brain Res 23:261–267

Zaretsky MD, Konishi M (1976) Tonotopic organization in the avian telencephalon. Brain Res 111:167–171

Zecha A (1962) The "pyramidal tract" and other telencephalic efferents in birds. Acta Morphol Neerl-Scand 5:194–195

Zecha A (1966) The medial lemniscus in the pigeon. Acta Morphol Neerl-Scand 6: 322

Zeier HJ, Karten HJ (1971) The archistriatum of the pigeon: organization of afferent and efferent connections. Brain Res 31:313–326

Zeier HJ, Karten HJ (1973) Connections of the anterior commissure in the pigeon (Columba livia). J Comp Neurol 150:201–216

Zeigler HP (1986) Feeding behavior of the pigeon. Adv Study Behav 7:285–389

Zeigler HP, Karten HJ (1973) Brain mechanisms and feeding behavior in the pigeon (Columba livia). I. Quinto-frontal structures. J Comp Neurol 152:59–82

Zeigler HP, Bischof H-J (eds) (1993) Vision, brain and behavior in birds. MIT Press, Cambridge, Mass.

Zeigler HP, Levitt PW, Levine RR (1980) Eating in the pigeon (Columba livia): movement patterns, stereotypy and stimulus control. J Comp Physiol Psych 94:783–794

Zeigler HP, Witkovsky P (1968) The main sensory trigeminal nucleus in the pigeon. a single-unit analysis. J Comp Neurol 134:255–264

Zigmond RE, Detrick A, Pfaff DW (1980) An autoradiographic study of the localization of androgen concentrating cells in the chaffinch. Brain Res 182:369–381

Zijlstra C, Dubbeldam JL (1994) Organization of motor innervation of craniocervical muscles in the mallard, Anas platyrhynchos L. J Brain Res 35:425–440

Zimmerman NH, Menaker M (1975) Neural connections of sparrow pineal: role in circadian control of activity. Science 190:477–479

Zweers GA (1971) A stereotactic atlas of the brainstem of the mallard (Anas platyrhynchos L.). Thesis, Leiden University. Van Gorcum, Assen

Zweers GA (1982a) Drinking of the pigeon (Columba livia L.). Behaviour 80:274–317

Zweers GA (1982b) Pecking of the pigeon (Columba livia L.). Behaviour 81:173–230

CHAPTER 22

Mammals

J. Voogd, R. Nieuwenhuys, P.A.M. van Dongen and H.J. ten Donkelaar

22.1	Introduction (Voogd) 1637	22.9.1	Corticobulbar and Corticospinal Tracts 1820
22.2	Gross Morphology of the Mammalian Brain 1642	22.9.2	Vestibular Nuclei 1832
	Frame 1 Lateral Views of the Brains of Selected Mammals 1643	22.9.3	Red Nucleus and Associated Cell Groups . . . 1837
		22.9.4	Long Descending Tracts and Motor Behaviour 1842
22.3	Spinal Cord (Voogd) 1654	22.10	Diencephalon (Van Dongen and Nieuwenhuys) 1844
22.3.1	Gross Anatomy, Ascensus Medullae 1654	22.10.1	General Plan and Shape 1844
22.3.2	Grey matter 1661	22.10.2	Epithalamus 1845
22.3.3	Primary Afferent Fibres and the Dorsal Funiculus 1666	22.10.3	(Dorsal) Thalamus 1848
22.3.4	White Matter 1667	22.10.4	Ventral Thalamus 1860
22.4	Rhombencephalon (Voogd) 1712	22.10.5	Hypothalamus (Nieuwenhuys) 1862
22.4.1	General relations and topography 1712	22.11	Telencephalon (Nieuwenhuys) 1871
22.4.2	Parasympathetic Visceromotor Nuclei of the Rhombencephalon 1716	22.11.1	Configuration and Subdivision 1871
		22.11.2	Rhinencephalon 1884
22.4.3	Nucleus Ambiguus 1717	22.11.3	Septal Region 1898
22.4.4	Motor Nucleus of the Facial Nerve 1717	22.11.4	Basal Ganglia 1901
22.4.5	Motor Nucleus of the Trigeminal Nerve 1719	22.11.5	Amygdala (ten Donkelaar) 1925
22.4.6	Motor Nucleus of the Hypoglossal Nerve . . . 1719	22.11.6	Cerebral Cortex (Nieuwenhuys) 1932
22.5	Mesencephalon (Voogd) 1720	22.12	Overall Functional Subdivision of the Mammalian Brain (Nieuwenhuys) . . . 2023
22.5.1	General Relations and Topography 1720	22.12.1	Introduction 2023
22.5.2	Eye Muscle Nuclei 1723	22.12.2	Lateral Domain 2024
22.5.3	Parasympathetic Visceromotor Neurons of the Mesencephalon 1724	22.12.3	Medial Domain 2026
		22.12.4	Overview 2039
22.6	Cerebellum and Precerebellar Nuclei 1724		References 2041
22.6.1	Gross Anatomy of the Mammalian Cerebellum 1724		
22.6.2	Histology of the Cerebellar Cortex 1733		
22.6.3	Cerebellar Nuclei 1735		
22.6.4	Corticonuclear Projection: Evidence for Longitudinal Projection Zones 1739		
22.6.5	Olivocerebellar Projection and the Modular Organisation of the Output Systems of the Cerebellum 1741		
22.6.6	Mossy Fibre Afferents of the Cerebellum . . . 1748		
22.6.7	Efferent Connections of the Cerebellum 1752		
22.7	Reticular Formation and the Serotonergic, Catecholaminergic and Cholinergic Cell Groups of the Brain Stem (Voogd) 1753		
22.7.1	Structure and Subdivision of the Reticular Formation 1753		
22.7.2	Monoaminergic Cell Groups and Pathways . . 1757		
22.7.3	Cholinergic Innervation of the Brain Stem . . 1761		
22.7.4	Connections of the Reticular Formation 1761		
22.8	Sensory Systems (Voogd) 1763		
22.8.1	Somatosensory and Viscerosensory Systems 1763		
22.8.2	Acoustic System 1777		
22.8.3	Visual System 1791		
22.9	Motor Systems (Voogd) 1820		

22.1
Introduction

J. Voogd

"Intelligent activity may reasonably be regarded as the key note of mammalian progress" (Romer 1962). This progress became possible with the acquisition of a neocortex, with its great analytic, associative and synthetic potential. Other mammalian characteristics, such as improvements in the circulation and in temperature regulation and, in most mammals, the long gestation period, giving birth to live young, and the development of nursing, with concomitant care and training of the young (Romer 1962), are conditional for the development, imprinting and functioning of a complicated brain. The main characteristics of the mammalian brain all are dependent on the presence of a neocortex. This is true for the prominence of the thalamus and

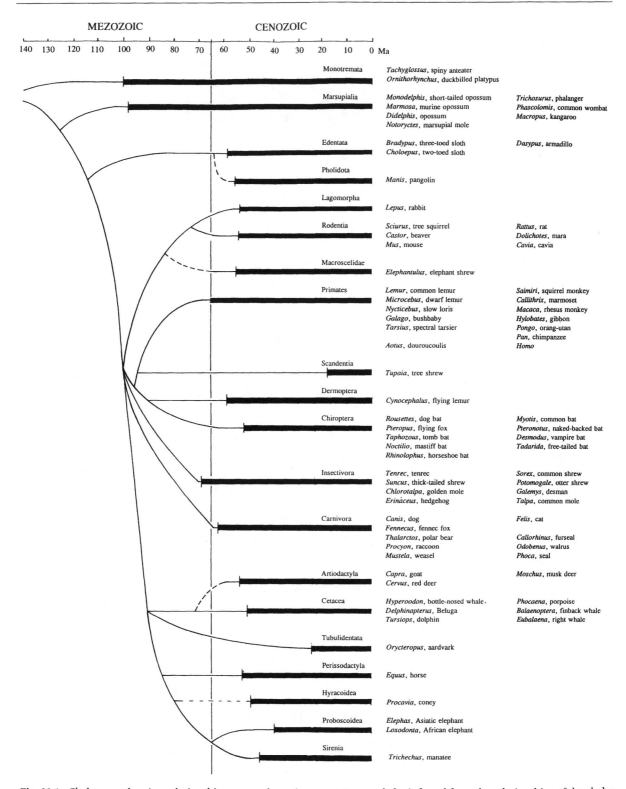

Fig. 22.1. Cladogram showing relationships among the major mammalian orders with extant representatives. The *solid horizontal bars* indicate the age range of the clade of dated first appearance in the fossil record. *Solid lines* indicate the branching sequence, although the date of the actual splitting event can only be inferred from the relationships of the clades and their known ages. *Dashed lines* indicate relatively more ambiguous relationships. The genera mentioned in the text or illustrated in this chapter are listed on the *right*. (Cladogram redrawn from Novacek 1992)

Fig. 22.2. a Opossum. **b** Cat. **c** Rhesus monkey

the striatum and the presence of pontine nuclei, cerebellar hemispheres and the pyramidal tract and for the corpus callosum in eutherian mammals. Sensory systems derive their great analytical power from the neocortex, and the efficiency of the mammalian locomotor apparatus also depends on it.

Many systems of the mammalian brain are not essentially different from those in non-mammalian vertebrates, but for some of their functions, such as digital dexterity in the motor system, they are dependent on the connections with the neocortex, in this case the presence of direct cortico-motoneuronal connections. In other systems, new applications, such as pursuit and target-directed eye movements in the visuomotor system, have developed, using the old wiring for a neocortically dependent task. Some sort of communication with the neocortex is present for almost every structure in the mammalian brain that is permeated by axons from the pyramidal cells of the neocortex, from the striatum to the spinal cord.

With the exception of the echidna and the platypus (the Monotremata or Prototheria), all living mammals can be included in a single monophyletic assemblage, the Theria (Carroll 1988). The modern therian groups – Marsupialia (Metatheria) and placentals (Eutheria) – diverged from a common ancestor early in the Cretaceous period. Information on the brain of the brain of the monotremes is scarce.

The North American opossum *Didelphis virginiana*, the cat *Felis domestica* and the rhesus monkey *Macaca mulatta* are the three species primarily used to illustrate the anatomy of the mammalian central nervous system in this chapter (Figs. 22.1, 22.2). The opossum is a nocturnal, ground-living omnivore, with a strong sense of smell and moderately developed visual and acoustic systems. The cat is a nocturnal predator, a carnivore with strongly developed visual and acoustic senses. The rhesus monkey is one of the Old World monkeys. It is a diurnal omnivore, with an arboreal habitat, a strongly developed visual system, a high degree of manual dexterity and advanced social habits.

Most of our knowledge about the marsupial brain is based on experimental studies in *Didelphis virginiana*. It would be presumptious to state that the cat and the rhesus monkey are truly representative of eutherian mammals. They were chosen for pragmatic reasons. The anatomy of the brain of the cat is probably the best known among eutherian mammals, followed by the brain of the rat and the rhesus monkey. Information on the central nervous system of the rat is readily accessible in a recent monograph (Paxinos 1995) and, therefore, will only be considered when it differs in essential points from the cat and the rhesus monkey. The gross anatomy of the brains of many mammals was illustrated and reviewed by Brauer and Schober (1970, 1976) and Schober and Brauer (1974). The myeloarchitecture of the brain stem and the spinal cord was described in representatives of most mammalian orders by Verhaart (1970b). A fairly extensive literature exists on insectivores (Stephan et al. 1991), Chiroptera (Baron et al. 1996) and lagomorphs. A modern neuroanatomy of ungulates is practically non-existent. Information on the brains of other mammals is either absent or incomplete. Certain mammalian orders or species have received special attention in the study of particular functional systems: Visual systems of the brain stem have been extensively studied in the rabbit and in the tree shrew *Tupaia glis*, studies of the acoustic system have focused on echolocating bats and Cetacea, while the innervation of spinal motoneurons has received special attention in species with a high degree of manual dexterity, such as the raccoon *Procyon lotor*.

The amount of detailed information on the structure and the connections of the brain of rat, cat and monkey is overwhelming. This poses problems regarding selection, but it has the great advantage that most structures in these species can be assigned a place in a meaningful network. With some notable exceptions, these networks are essentially similar in the three species. Knowledge of these networks makes it possible to recognise that entire circuits or certain of their components are hypertrophied or missing in other orders or species. A good example is the relationship between the large nucleus ellipticus (Darkschewitsch's nucleus) in the mesencephalon, the hypertrophied medial accessory olive, the immense posterior interposed cerebellar nucleus and the wide C2 Purkinje cell zone in Cetacea, which was only recognised after this circuit had been studied in rat and cat (Jansen 1969). Detailed knowledge of connections can also be used to extract the essential features and to mitigate differences between species. This is exemplified by the observation that corticospinal fibres always terminate in the contralateral spinal grey matter, irrespective of the great variations in the level and the structure of the pyramidal decussation in different species. For most functional systems, detailed knowledge about neurotransmitters, receptors and molecular aspects of neuromorphology is now available, but data on the chemoarchitecture of the brain have only been collected in a few species, mostly in the rat. Comparative chemoarchitectural studies can be useful in the identification and interpretation of structures, but they are frought with technical problems and require expertise to be adequately reviewed. With some exceptions, neurochemical data are not included in this chapter.

Several authors have taken a systematic approach and have determined allometric relationships for different parts of the brain for certain mammalian groups and have correlated the relative size of these parts with ecological parameters. In their studies of the brain of insectivores, Stephan and Spatz (1962) and Stephan (1967) distinguished a group of 'basal insectivores', characterised by the smallest size of all non-olfactory brain structures, from more progressive insectivores, with a more differentiated morphology of the brain, related to a specialized life-style. The basal insectivores are nocturnal, microptic animals and include the Tenrecidae, with the tenrec and the hedgehog tenrec, and the Erinaceidae, with the European hedgehog and the true shrews (Soricidae). Their brains are characterised by wide ventricles, reminiscent of the ventricles of reptiles, a small and smooth neocortex, a high position of the paleo-neocortical fissure with a complete lack of a basal neocortex, a small and vertically placed corpus callosum, subdivision of the internal capsule into small bundles, a small pes pontis and small cerebellar hemispheres with an underdeveloped flocculus-paraflocculus complex. The broad and unstalked olfactory bulb and the rhinencephalon are large, and the visual system is small, with a tectum uncovered by the cerebral hemispheres and the cerebellum.

The progressive insectivores include the moles (Talpidae; Stephan 1968) and semi-aquatic species from different families, such as *Galemys pyrenaicus* (Pyrenean desman, Talpidae), *Chlorotalpa stuhlmanni* (golden mole, Chrysochloridae) and *Potomogale velox* (otter shrew, Potomogalidae). The visual system of these progressive insectivores is poorly developed (or absent, as in all Talpidae), a feature they share with the basal insectivores, but they show an increase in the size of other regions of the brain, with the neocortex showing the greatest increase in size In the otter shrew, the paradigm of the semi-aquatic insectivores, the paleo-neocortical fissure, is located at the basal surface of the hemisphere, and a basal neocortex can be seen. The anterior commissure is smaller, and the corpus callosum has increased in size. The position of the caudal border of the hemisphere has changed as an expression of progressive temporal rotation of the hemisphere. Adaptations to foraging in the otter shrew lead to a strong development of the vibrissae and a concomitant development of the sensory root and the sensory nuclei of the trigeminal nerve and their crossed ascending connections in this species. In *Chlorotalpa*, the skull is shortened and adapted to grubbing up insects from the bottom of a stream. Secondary adaptations are visible in the brain, which shows a strong anterior-posterior compression.

Stephan (Stephan et al. 1991) also compared the position of the basal, microptic insectivores with the elephant shrews (Macroscelidae), which are macroptic, active, diurnal animals with a strong development of the visual and olfactory systems, with primates (Stephan and Andy 1964, 1969) and chiropteres (Stephan and Pirlot 1970; Baron et al. 1996), all of which are thought to have developed from a basal insectivore-like stock. In elephant shrews, the increase in the neocortex and associated structures is correlated with a more advanced optic system. However, the striate cortex and the lateral geniculate body of the elephant shrews are small; the increased size of the optic system mainly concerns the superior colliculus. The striate cortex, together with the entire neocortex, is increased in size in primates (Stephan 1969).

The Tupaiidae (tree shrews) were formerly included in the Macroscelidae, but Stephan and Spatz (1962) argued against such a position. The progressive development of the lateral geniculate body and the striate cortex, the small rhinencephalon, the strong rotation of the hemisphere and the arboreal habitat of tree shrews argue against this classification and favour inclusion with the primates, as a separate suborder below the prosimians (Stephan and Spatz 1962). Others noticed important differences in brain structure between the tree shrew (*Tupaia glis*) and the primates concerning the localisation of the pyramidal tract in the spinal cord and the extension of the visual map of the superior colliculus in the ipsilateral hemifield (see Sects. 22.8.3.3.2 and 22.9.1). In the cladogram of Fig. 22.1 the Tupaidae are positioned as a separate order (Scandentia, Wagner 1855) most closely related to the Primates, the Dermoptera and the Chiroptera (Gregory, 1910; McKenna, 1975; Novacek 1992).

In Chiroptera, like other mammals, the neocortex is the most progressive structure (Stephan and Pirlot 1970; Baron et al. 1996). Next highest in average progression is the cerebellum, but the progression indices for the neocortex and the cerebellum are not correlated for the individual species, and the rankings for neocortical and cerebellar development are different. Neocortical development in chiropteres is related to their food preferences. The neocortex in Microchiroptera is 1.5–2.5 times greater than in basal insectivores. In frugivorous and nectar-feeding Microchiroptera ,the size of the neocortex is further increased. In the frugivorous macrochiropterean pteropids, the neocortex is 5.2 times the size of that in basal insectivores, and in the fish-eating microchiropteres (*Noctilio leporinus*) and the blood-sucking *Desmodus rotundus*, it is even larger.

More recently, Heffner and Masterton (1975, 1983) and Nudo and Masterton (1988, 1989, 1990a,b) have studied the allometric relations of parameters from brain stem-spinal and corticospinal systems, derived from experimental studies in a great number of species from different mammalian orders, and have correlated them with careful rankings of motor dexterity, coordination and 'visualness' and with certain aspects of their ecological habitat. They attempted to relate these parameters to the evolutionary history of the species. Some of their results are reviewed in Sect. 22.9.1.

Comparative anatomical studies require a taxonomic framework, such as the one provided by Simpson (1945) and by more recent studies of mammalian phylogeny using the cladistic approach (Marshall 1979; Novacek 1986). Cladograms defining the phylogenetic relations between mammalian orders are illustrated in Fig. 22.1 and Poster 1 and used in the sections on the size and the allometric relationships of the mammalian brain and on the patterns of sulci and gyri of the cerebral hemisphere (Chap. 23). Recently, Novacek (1992) reviewed some of the issues in the application of recent palaeontological, morphological and molecular data to mammalian phylogeny. Novacek (1992) no longer includes the Macroscelidae with the microptic insectivores and the tree shrews in a single suborder, but places these orders at different positions in the eutherian tree. The affinity of the tree shrews with primates is confirmed, and new evidence is discussed on the relationship of tree shrews and primates with megachiropteran bats. Such a close relationship between macrochiropteran bats and primates is based on the similarity between the visual map of the superior colliculus in *Pteropus* (Pettigrew 1986) and in primates and is discussed in Sect. 22.8.3.2.2. The question of the diphyletic origin of the Micro- and Megachiroptera was extensively discussed by Baron et al. (1996). Such a diphyletic origin is not supported by other morphological or molecular evidence (Novacek 1992). The close relationship between the Hyracoidea, the Proboscoidea and the Sirenia, proposed by Simpson (1945), is maintained. The combination of the Rodentia and the Lagomorpha into the cohort of the Glires (Simpson 1945) is not supported (Novacek 1992).

This chapter consists of topographical sections on the spinal cord, the rhombencephalon, the mesencephalon, the cerebellum, the diencephalon and the forebrain and of sections devoted to a number of functional systems. For an introduction to the general subdivision of the mammalian brain, the reader is referred to Chap. 4.

22.2
Gross Morphology of the Mammalian Brain

J. Voogd

The gross morphology of the mammalian brain has been reviewed by Haller and Hallerstein (1934) and Schober and Brauer (1974). For a complete bibliography on the subject and excellent illustrations of the brains of mammals, the reader is referred to these publications and to the atlas of mammalian brains by Brauer and Schober (1970, 1976). More recent reviews include the monographs on the brain of Insectivora (Stephan et al. 1991) and Chiroptera (Baron et al. 1996). The literature on the meninges, the blood vessels and the spinal cord of mammals was summarised by Tigges (1964). The ventricular system was reviewed by the latter author and by McFarland et al. (1969). The great variability in size and shape of the mammalian brain is illustrated in poster P1 and in the illustrations of the core species of this chapter (see Figs. 22.2, 22.3 and Frame 1): the North American opossum (see Fig. 22.4), the cat (see Fig. 22.5) and the rhesus monkey (see Fig. 22.6).

In most mammals, as exemplified by the North American opossum and the cat (Fig. 22.3a,b), the brain case is narrow, located between the wide zygomatic arches, and short, located caudal to the facial skeleton of the nasal capsules and the orbitae (Fig. 22.3). The olfactory bulbs extend furthest rostrally and occupy the olfactory fossae. The olfactory fossae are covered by the frontal bones, and their floor is formed by the mesosphenoid. Rostrally, the olfactory fossae communicate with the nasal cavities by perforations into the lamina cribrosa of the ethmoid. The lamina cribrosa occupies an oblique-vertical position. The olfactory fossae are separated from the cerebral fossae by a transverse ridge. The floor of the cerebral cavity is formed by the basosphenoid, with the hypophysial fossa, the alisphenoid (the ala major of the sphenoid), the occipital bone and the pars petrosa of the temporal bone. Its roof is formed by the frontal and parietal bones and the squamous portions of the temporal and occipital bones. The posterior fossa, containing the cerebellum, is separated by the dural fold of the tentorium cerebelli and is distinct in all mammals. The tentorium cerebelli is ossified in carnivores. In higher primates, the facial part of the skull is shortened and withdraws beneath the expanded brain vault (Fig. 22.2c). Frontal (anterior) and temporal (middle) cerebral fossae are formed, containing the corresponding lobes of the hemispheres. The lamina cribrosa becomes located in the floor of the frontal fossa. The frontal bones no longer cover the olfactory bulbs, but form part of the floor of the frontal fossa as the roof of the orbitae.

The subdivision of the mammalian brain into the telencephalon, diencephalon, mesencephalon, metencephalon and myelencephalon is useful for topographical and descriptive purposes and will be followed in this chapter. The morphogenesis and the neuromerism of the mammalian brain are discussed in Chap. 4. The flexures of the embryological neural tube are preserved in the brain of adult mammals, where they are most distinct in the rostrocaudally compressed brains of some burrying species such as the marsupial mole *Notoryctes* (see Fig. 22.7), the golden mole *Chlorotalpa* and in Cetacea (see Figs. 22.8–22.10, 22.12).

The metencephalon and myelencephalon will be considered together as the rhombencephalon, because the border between these two subdivisions, formed by the caudal border of the pons, varies with the rostrocaudal extent of the latter structure (compare Figs. 22.4, 22.5, 22.6c,d and Fig. 22.25). The pons shows a transverse relief, and laterally it

Frame 1.

Lateral Views of the Brains of Selected Mammals
Poster P1 presents lateral views of the brains of selected mammals. The lengths of the brains are logarithmically transformed such that the length of the smallest brain is 2.5 cm and that of the largest brain 9 cm. Consequently, if brain A is larger than brain B in nature, it is also larger in this figure. The phylogenetic relationships of the orders involved are indicated. The full names of the species and their taxonomical relationships are given below. The brains of the various species are numbered in order to facilitate reference in the text.

Class: MAMMALIA
Subclass: Prototheria
Order: Monotremata

1. Duck-billed platypus, *Ornithorhynchus anatinus*
2. Spiny ant-eater, *Tachyglossus aculeatus*
   Subclass: Theria
   Supercohort: Marsupialia
   Order: Marsupicarnivora
3. Opossum, *Didelphis virginiana*
   Order: Paucituberculata (not represented)
   Order: Peramelina (not represented)
   Order: Diprotodonta
4. Red kangaroo, *Macropus rufus*
   Supercohort: Eutheria
   Grandorder: Insectivora
   Order: Tenrecomorpha
5. Tenrec, *Tenrec ecaudatus*
   Order: Erinaceomorpha
6. Hedgehog, *Erinaceus europeus*
   Order: Soricomorpha
7. Lesser shrew, *Sorex minutus*
   Grandorder: Edentata
   Order: Cingulata
8. Armadillo, *Dasypus sexcinctus*
   Order: Pilosa
9. Three-toed sloth, *Bradypus tridactylus*
10. Great ant-eater, *Myrmecophaga tridactyla*
    Order: Pholidota (not represented)
    Grandorder: Glires (taxonomical position uncertain)
    Order: Macroscelidea
11. Elephant shrew, *Elephantulus rozeti*
    Order: Lagomorpha
12. Hare, *Lepus europeus*
    Order: Rodentia (Myomorpha)
13. Rat, *Rattus norvegicus*
14. Beaver, *Castor fiber*
    Order: Hystricomorpha (uncertain position)
15. Cavia, *Cavia procellus*
16. Mara, *Dolichotes patagonica*
    Grandorder: Archonta
    Order: Scandentia
17. Tree shrew, *Tupaia glis*
    Order: Dermoptera
18. Flying lemur, *Cynocephalus volans*
    Order: Chiroptera
19. Little brown bat, *Myotis myotis*
20. Red-necked fruit bat, *Pteropus vampyrus*
    Order: Primates
21. Common marmoset, *Callithrix jacchus*
22. Rhesus monkey, *Macaca mulatta*
23. Man, *Homo sapiens*
    Grandorder: Ferungulata
    Order: Carnivora
24. Fennec fox, *Fennecus zerda*
25. Cat, *Felis catus*
26. Polar bear, *Thalarctos maritimus*
    Order: Tubulidentata
27. Aardvark, *Orycteropus afer*
    Order: Artiodactyla
28. Musk deer, *Moschus moschiferus*
29. Red deer, *Cervus elaphus*
    Order: Cetacea
30. Bottle-nosed dolphin, *Tursiops truncatus*
31. Fin whale, *Balaenoptera physalus*
    Order: Perissodactyla
32. Donkey, *Equus asinus*
    Order: Hyracoidea
33. Hyrax, *Procavia capensis*
    Order: Proboscoidea
34. African elephant, *Loxodonta africana*
    Oder: Sirenia
35. Manatee, *Trichechus manatus*

Sources: These figures are based upon the photographs and outlines by Braüer and Schober (1970), and further upon own material (3, 6, 7, 12, 15, 17, 22, 23, 25). Additional sources are: 1: Rowe (1990); 2: Welker and Lende (1980); 5: Stephan et al. (1990); 8, 9, 27: Elliot Smith (1898); 30: Morgane et al. (1980); 31: Pilleri and Gihr (1969); 34: Flatau and Jacobsohn (1899); and 35: Reep et al. (1990).

continues into the brachium pontis. The crura cerebri (cerebral peduncles) enter the pons rostrally. They emerge on the ventral surface of the brain, caudal to the optic tracts. The depression between the peduncles is known as the fossa interpeduncularis. Rostrally, the fossa interpeduncularis increases in width and depth and contains the corpus mammillare and the tuber cinereum with the hypophysial stalk. Caudally, the nervus oculomotorius emerges from the fossa interpeduncularis. The distance between the optic chiasm with the optic tracts, and the rostral border of the pons, is short in monotremes (see Fig. 22.7a) and in rostrocaudally compressed brains such as those of Cetacea (see Figs. 22.8b, 22.12) and Proboscoidea (see Fig. 22.8c). This distance is long, and the crura cerebri are elongated, in the giant anteater (*Myrmecophaga*), the aardvark (*Orycteropus afer*) and in Artiodactyla and Perissodactyla.

Caudal to the pons, the corpus trapezoideum can be traced from the tuberculum acusticum across the midline. The pyramid is located next to the fissura mediana ventralis, superficial to the corpus trapezoideum (Fig. 22.4, 22.5, 22.6c). The decussa-

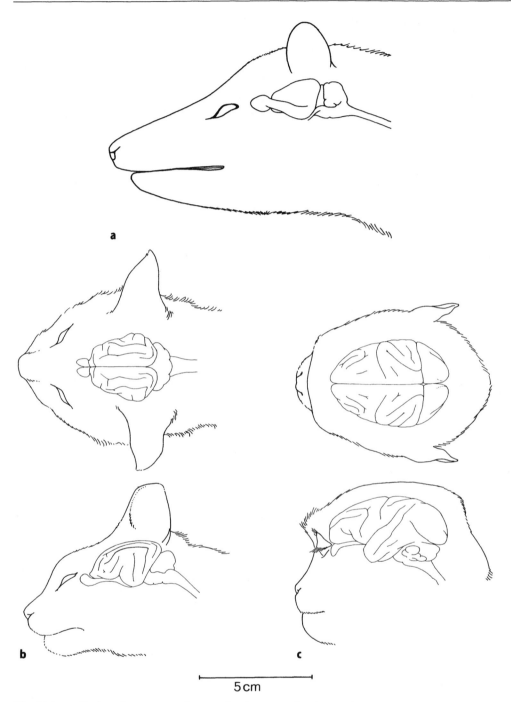

Fig. 22.3a–c. Brains of **a** opossum, **b** cat and **c** rhesus monkey in position

tion of the pyramids may be visible as an interruption of the fissura mediana ventralis (Fig. 22.5c). The inferior olive sometimes forms a prominence lateral to the pyramid. In some species with a large trigeminal nerve and a small olive (monotremes, some insectivores), the spinal tract of the trigeminal nerve and its nucleus form a prominent tuberculum quinti (tuberculum cinereum) on the ventrolateral side of the medulla oblongata (Fig. 22.7a).

The size of the pontine prominence is related to the size of neopallium and the width of the crura cerebri. The shape of the pons in monotremes is triangular, with the apex extending between the short crura cerebri of these species (Fig. 22.7a). The large trigeminal nerve enters the brain stem rostral to the pons, which is therefore located post-trigeminally. In most mammals, the trigeminal nerve enters the brain stem passing through the brachium pontis (Figs. 22.4, 22.5, 22.6b,c). Abbie (1934) thought that the pretrigeminal portion of the pons located rostral to the entrance of the nerve received the frontopontine projection and that the size of this part of the pons was a sign of progressive development. The pons is small in Marsupialia (Fig. 22.4c,d, 22.7b), Insectivora, Rodentia, Edentata and Chiroptera. It is large in some carnivores (Pinnipedia and Ursidae), Cetacea (Fig. 22.8b), Proboscoidea and especially in primates (Fig. 22.4a, 22.8a). A large pons covers part or all of the trapezoid body. A large pons often coincides with a large and prominent olive. A high ratio of the rostrocaudal width of the pons to the width of the portion of the trapezoid body exposed on the ventral surface (in the polar bear *Ursus arctos* 3.8:1; the walrus *Odobenus rosmarus* 7.8:1; the dolphin 7:1; *Homo* 7:1; *Elephas maximus* 7:1; Schober and Brauer 1974) and a small distance between the caudal border of the pons and the rostral pole of the olive (which touch each other in the elephant and the great whales) are characteristic of these species. The size of the pyramid is not related to the size of the neopallium, the crura cerebri or the pons. The pyramids are inconspicuous in monotremes and small in Insectivora, Rodentia, Microchiroptera, Cetacea, Proboscoidea and Ungulates, irrespective of the medium to large size of the crura cerebri and the pons in the latter three groups. The pyramids are prominent in carnivores (Fig. 22.5c) and primates (Fig. 22.6c).

The dorsal aspect of the rhombencephalon contains the fourth ventricle (Fig. 22.13). Caudally, the ventricle tapers into the calamus scriptorius and closes into the central canal at the obex (see Fig. 22.10). At its greatest width, it extends laterally as the canalis communicans ((Fig. 22.13e) Tigges 1964) into the lateral recess of the fourth ventricle. The roof of the fourth ventricle and the caudal wall of the canalis communicans are formed by the tela choroidea. The roof and the rostral wall of the canalis communicans are formed by the white matter of the flocculus. Openings in the tela are present as the apertura lateralis of Luschka in the lateral recess. A midline aperture of Magendie has only been observed in humans (McFarland et al. 1969). Dorsally, the fourth ventricle extends into the base of the cerebellum as the fastigium (Figs. 22.4, 22.5, 22.6d). A prolongation of the fastigium, between the fastigial nuclei of both sides as present in birds, has disappeared in mammals.

The caudal wall of the fastigium consists of the medullary surface of the nodulus and, more laterally, of the velum medullare posterius. This velum is extensive in Pongidae and humans, with a strong development and a paravermal position of the dorsal paraflocculus (tonsilla), but it is inconspicuous in most other mammals. The rostral wall of the fastigium and the roof of the rostral part of the fourth ventricle are formed by the white matter of the lingula and the rostral extension of the cerebellar commissure; together these constitute the velum medullare anterius (see Fig. 22.25), which extends between the brachia conjunctiva and rostrally contains the decussation of the trochlear nerve. The median sulcus and the sulcus limitans divide the floor of the fourth ventricle into an eminentia medialis and lateral vestibular and cochlear areas. The eminentia medialis can sometimes be subdivided into an area hypoglossi, an area vagi (ala cinerea), which extends caudally into the area postrema, and a facial eminence located over the genu internum of the facial nerve (see Fig. 22.10).

The cranial nerves, apart from the trochlear nerve (IV), enter the brain stem in ventral and lateral root entry zones. The oculomotor (III), abducens (VI) and hypoglossal (XII) nerves leave the brain stem at its ventral side. The trigeminal (V), facial-intermediate (VII), statoacoustic (VIII), glossopharyngeal (IX), vagus (X) and bulbar accessory (XI) nerves are situated at the lateral side of the brain stem. The trochlear nerve leaves the brain stem dorsally. Of the nerves entering or leaving the brain, the trigeminal nerve is usually the largest. It is extraordinarily large in monotremes (Fig. 22.7a), some insectivores and rodents, Proboscoidea (Fig. 22.8c) and in Perissodactyla and Artiodactyla. The cochlear division of the statoacoustic nerve is large in echolocating bats and in Cetacea (see Fig. 22.10c). It ends in a large tuberculum acusticum. The tuberculum acusticum should be distinguished from the corpus pontobulbare located at its ventral and caudal side (Essick 1907). The corpus pontobulbare is large in Cetacea. The size of the cranial motor nerves is related to the mass of the muscles they innervate.

The cerebellum is located in the rostral roof of the fourth ventricle. It is covered by a cortex, and cerebellar nuclei are located in its central white matter. Two paramedian sulci subdivide the mammalian cerebellum into the vermis and bilateral hemispheres, and transverse fissures of various depth demarcate transversely oriented lobes, lobules and folia (Figs. 22.4, 22.5, 22.6b–d). As a consequence, the rostrocaudal length of the cortex of vermis and hemispheres is many times larger than its width (Braitenberg and Atwood 1958; Sultan and Braitenberg 1993; see Fig. 22.33). The size and the width of the cerebellar hemispheres and the depth of the fissures in vermis and hemispheres are related to the size of the pons and the development of the premotor and association areas of the neopallium. Other factors, such as muscle mass, oculomotor requirements and specializations related to an aquatic habitat or other ecological factors, are equally important in determining the size of the cerebellum or parts thereof. The great variability in size and shape of the mammalian cerebellum is illustrated in Poster 1. The gross anatomy of the mammalian cerebellum is discussed in more detail in Sect. 22.6.

The mesencephalon is usually subdivided into the pes mesencephali (the crura cerebri and the substantia nigra), the tegmentum and the tectum mesencephali. It contains the aquaeductus cerebri. The aqueduct narrows rostrally. At the transition of the aqueduct into the third ventricle, the posterior commissure with the subcommissural organ protrudes into its roof (Figs. 22.4, 22.5, 22.6d, 22.13). A dorsolateral enlargement of the caudal part of the aqueduct is present in some Cetacea and is related to the massive development of the inferior colliculi to these species (McFarland et al. 1969). A dorsal recess of the caudalmost part of the aqueduct, located between the inferior colliculus and the cerebellum, is present in the dog and rat (Fig. 22.13a,b, McFarland et al. 1969).

The tectum mesencephali can be subdivided into the inferior and superior colliculi. The inferior colliculi form a link in the auditory pathway. They are enlarged in species using echolocation, such as the Microchiroptera and the Cetacea (Fig. 22.10a,c). The superior colliculi are engaged in gaze control. They generally surpass the inferior colliculi in size. In some nocturnal mammals and in other species with highly developed vision, the superior colliculus may reach an extraordinary size (*Capra, Tarsius spectrum*, Fig. 22.8b,d). The tectum may be partially or entirely covered by the cerebral hemispheres and/or the cerebellum. In the platypus *Ornithorhynchus* and in some marsupials (Fig. 22.4d), Insectivora, Chiroptera (Fig. 22.10a) and Rodentia, the tectum is visible in a dorsal view of the brain. However, in larger representatives of these orders and in other mammals, the tectum is hidden from view by the telencephalon and the cerebellum (Figs. 22.5, 22.6d).

On the lateral aspect of the mesencephalon, the lateral lemniscus, the brachium of the inferior colliculus and the cerebral peduncle border a triangular area containing the parabigeminal nucleus (Fig. 22.11). The direction of the lateral lemniscus corresponds to the direction of the mesencephalic portion of the neural tube during early stages of development (Kappel 1981). The angle between the rhombencephalon and the lateral lemniscus represents the pontine flexure; the angle between the lemniscus lateralis and the long axis of the diencephalon represents the cephalic flexure. Ventrally, the cephalic flexure is 'filled in' by the crura cerebri (Figs. 22.11, 22.12).

The diencephalon contains the flattened third ventricle. The floor and the ventral part of the wall of this ventricle are formed by the hypothalamus. Supramammillary and infundibular recesses can be distinguished in most species (Fig. 22.13). The optic chiasm protrudes rostrally into the floor of the third ventricle and gives attachment to the lamina terminalis (Figs. 22.4, 22.5, 22.6d). Shallow, ventricular sulci delimit three longitudinal zones in the wall of the diencephalon: (1) the hypothalamus, (2) the dorsal thalamus and (3) the epithalamus or corpus habenulae. The significance of this zonal pattern is discussed in Chap. 4.

In most mammals, the dorsal thalami are interconnected through the third ventricle by the massa intermedia (Figs. 22.4, 22.5, 22.6d, 22.13). A massa intermedia is lacking in the finwhale *Balaenoptera borealis* (Pilleri 1966a), but is present in other Mysticoceti (Pilleri 1966a,b). The ventricular surface of the dorsal thalamus is relatively small, while the meningeal surface of the thalamus is generally more extensive. The shape and the position of the dorsal thalamus in the different mammalian orders are reviewed in Sect. 12.10.3.

Caudally, the corpora habenulae are interconnected by the habenular commissure. The corpus pineale is attached to the habenular and posterior commissures, which remain separated by the pineal recess. This recess extends for some distance into the gland (Figs. 22.4, 22.5, 22.6d, 22.13). The tela choroidea of the third ventricle is attached to the ganglia habenulae and the habenular commissure. The corpus pineale and the pineal recess are absent in the porpoise (McFarland et al. 1969). The tela choroidea consists of two rows of choroid tufts. Each row continues as part of the wall of the interventricular foramen into the plexus choroideus of the lateral ventricle. Rostrally, the tela choroidea of the third ventricle is attached to the columna of the fornix and to the subfornical organ located in between.

The telencephalon differs greatly in size, shape and complexity among mammalian species (Poster 1). It contains the lateral ventricles, sometimes with an olfactory recess extending into the olfactory bulb (see Fig. 22.147a). The telencephalon can be subdivided into the telencephalon impar, comprising the lamina terminalis and the preoptic region, and the cerebral hemispheres. The lamina terminalis extends as a thin layer between the columnae of the fornix and forms the rostral wall of the third ventricle. The cerebral commissures develop within or in continuity with the lamina terminalis. The anterior commissure retains its position in the lamina terminalis in adult mammals (Figs. 22.4, 22.5, 22.6d, 22.9). In eutherian mammals, it contains commissural fibres interconnecting the olfactory and piriform lobes in its rostral half and fibres interconnecting inferior temporal and parahippocampal neocortical fields caudally (Nieuwenhuys et al. 1988). The commissure of the neopallium, i.e. the corpus callosum, extends dorsally and caudally from the lamina terminalis (Figs. 22.5, 22.6d, 22.9). It is lacking in monotremes and marsupials, where all commissural fibres of the neopallium cross in the anterior commissure (Figs. 22.4d, 22.146). The shape and extent of the corpus callosum are related to the shape and size of the neopallium covering the hemisphere. The third cerebral commissure is the commissure of the hippocampus. It interconnects the formatio hippocampi of both sides and is located between the crura of the fornix, ventral to the corpus callosum (Fig. 22.9).

The exterior of the cerebral hemispheres can be subdivided into the following: (a) olfactory and piriform lobes, which occupy the ventral part of the hemisphere, (b) the formatio hippocampi, located along the taenia fornices of the plexus choroideus of the lateral ventricle and (c) the convexity of the hemisphere. The fissura rhinalis separates the piriform lobe from the convexity of the hemisphere (Figs. 22.4b,c, 22.5, 22.6c). A cortical mantle, or pallium, covers the hemisphere. Deep nuclei (the corpus striatum and amygdala) are found in its interior. The pallium can be subdivided into the three-layered paleopallium, which covers large parts of the piriform lobe, the three-layered archipallium of the formatio hippocampi and the more complex, six-layered neopallium, which extends over the convexity of the hemisphere from the rhinal sulcus and the formatio hippocampi to the corpus callosum.

The lateral ventricles are curved, half moon-shaped cavities (Fig. 22.13). The anterior horn of the lateral ventricle extends ventral to the interventricular foramen into the frontal pole of the hemisphere. Rostrally, dorsally and ventrally, it is bordered by the corpus callosum. Its medial wall is known as the septum (Figs. 22.5, 22.6d). In mammals with a large corpus callosum, the dorsal part of the septum is stretched into a thin sheet, known as the septum pellucidum (Fig. 22.4a). Caudal to the foramen interventriculare, the pars centralis of the lateral ventricle shifts laterally and ventrally to enter the temporal lobe as the inferior horn. In small mammals, a temporal lobe and the inferior horn of the lateral ventricle are absent or inconspicuous (Fig. 22.13a). Similarly, a posterior horn of the lateral ventricle, which extends into the occipital lobe, is only present in the hemispheres of some Cetacea (Pilleri 1964; 1966a,b; McFarland et al. 1969) and most primates (Tigges 1964; McFarland et al. 1969), which have expanded around a central insula (Fig. 22.13d,e, 22.147b,c). The tela choroidea of the lateral ventricle extends into the pars centralis and the inferior horn. It is attached to the fornix (taenia fornicis) medially and to the stria terminalis (taenia choroidea) laterally. The stria terminalis are located in the lamina affixa, a membranous part of the embryonic medial hemisphere wall, which later fuses with the meningeal surface of the thalamus. As a consequence, the taenia choroidea of the plexus becomes attached to the thalamus, and the thalamus appears to become located in the wall of the lateral ventricle.

The olfactory lobe (Figs. 22.4c,d 22.5 22.6c,d 22.9) consists of the bulbus olfactorius, which receives the fila olfactoria from the olfactory epithelium in the roof of the nasal cavity, and, when present, the accessory bulb, which receives the nervus vomeronasalis from the organ of Jacobson, the olfactory tract, which connects the olfactory bulb with the olfactory cortex, and the olfactory tubercle, located between the medial and lateral divisions of the olfactory tract. The piriform lobe caudally adjoins the olfactory tubercle. The amygdala is located within the medial part of the piriform lobe. The lateral olfactory tract extends for some distance over the paleocortex of the piriform lobe, where it terminates in the superficial layer. The medial olfactory tract cannot be traced beyond the olfactory tubercle. A ridge of grey matter, known as the diagonal band of Broca, extends from the tubercle to the septum.

The olfactory lobe is large in macrosmatic mammals with a well-developed sense of smell. In the brains of Marsupialia, many Insectivora, Rodentia and Microchiroptera, the fissura rhinalis is located on the ventrolateral aspect of the hemisphere (Poster 1). In species of other orders, the olfactory lobe is located on the ventral aspect of the hemisphere and is partially covered by the convolutions of the neopallium. The olfactory bulbs and peduncles are absent in anosmatic mammals such as the tooth whales (Odontoceti: Filimonoff 1965). The olfactory tubercle and the piriform lobe are small, but present in these anosmatic and in other microsmatic species (Fig. 22.8b).

The formatio hippocampi consists of the gyrus dentatus, the cornu ammonis and the subiculum (gyrus parahippocampi). Only the gyrus dentatus,

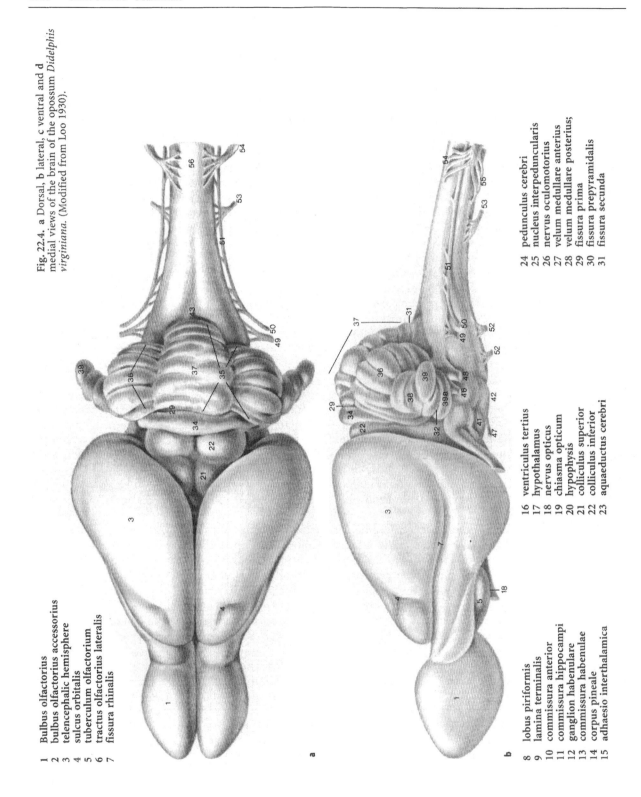

Fig. 22.4. a Dorsal, b lateral, c ventral and d medial views of the brain of the opossum *Didelphis virginiana*. (Modified from Loo 1930).

1 Bulbus olfactorius
2 bulbus olfactorius accessorius
3 telencephalic hemisphere
4 sulcus orbitalis
5 tuberculum olfactorium
6 tractus olfactorius lateralis
7 fissura rhinalis

8 lobus piriformis
9 lamina terminalis
10 commissura anterior
11 commissura hippocampi
12 ganglion habenulare
13 commissura habenulae
14 corpus pineale
15 adhaesio interthalamica

16 ventriculus tertius
17 hypothalamus
18 nervus opticus
19 chiasma opticum
20 hypophysis
21 colliculus superior
22 colliculus inferior
23 aquaeductus cerebri

24 pedunculus cerebri
25 nucleus interpeduncularis
26 nervus oculomotorius
27 velum medullare anterius
28 velum medullare posterius;
29 fissura prima
30 fissura prepyramidalis
31 fissura secunda

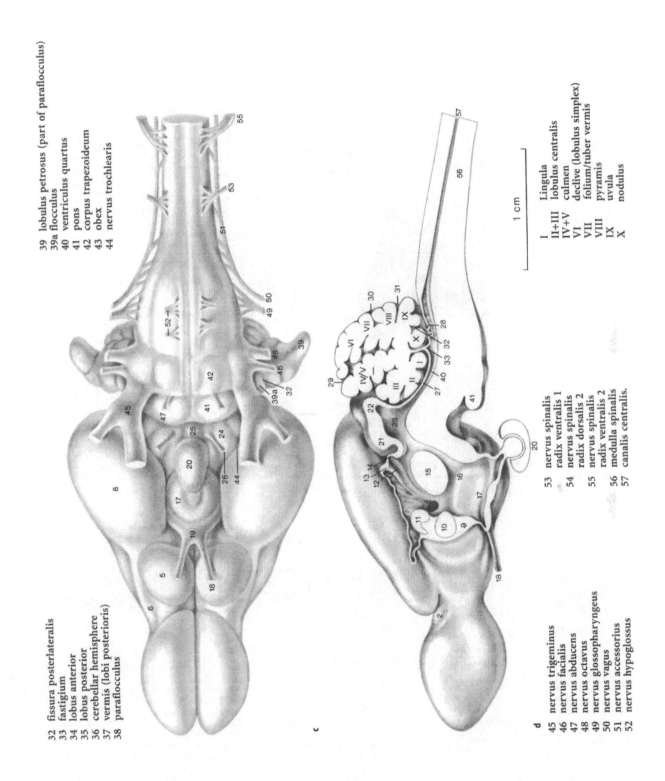

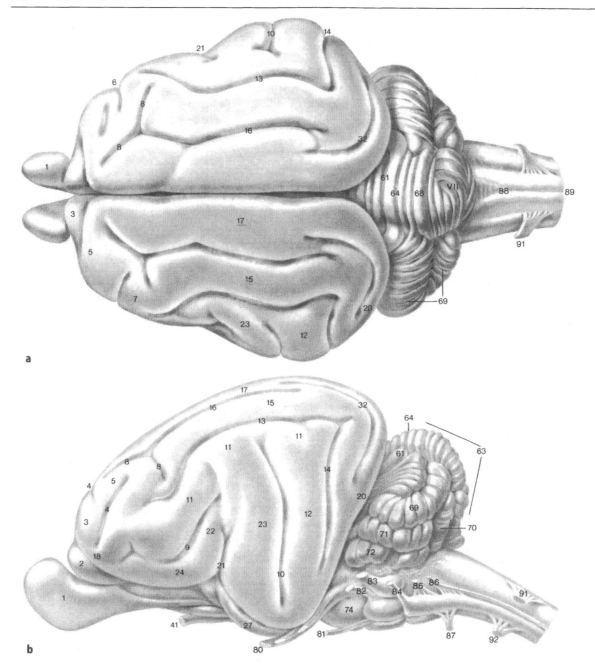

Fig. 22.5. **a** Dorsal, **b** lateral, **c** ventral and **d** medial views of the brain of the cat.

1 Bulbus olfactorius	16 sulcus marginalis	31 gyrus cinguli
2 gyrus proreus	17 gyrus marginalis	32 sulcus marginalis posterior
3 gyrus sigmoideus anterior	18 gyrus compositus anterior	33 corpus callosum
4 sulcus cruciatus	19 gyrus orbitalis	34 septum pellucidum
5 gyrus sigmoideus posterior	20 gyrus compositus posterior	35 fornix
6 sulcus coronalis	21 fissura sylvia (lateralis)	36 foramen interventriculare
7 gyrus coronalis	22 gyrus sylvius anterior	37 commissura anterior
8 sulcus ansatus	23 gyrus sylvius posterior	38 lamina terminalis; tuberculum olfactorium
9 sulcus ectosylvius anterior	24 gyrus diagonalis	40 tractus olfactorius lateralis
10 sulcus ectosylvius posterior	25 fissura rhinalis anterior	41 nervus opticus
11 gyrus ectosylvius anterior	26 fissura rhinalis posterior	42 chiasma opticum
12 gyrus ectosylvius posterior	27 lobus piriformis	43 hypothalamus
13 sulcus suprasylvius	28 sulcus suprasplenialis	44 hypophysis
14 sulcus suprasylvius posterior	29 gyrus splenialis	45 ganglion habenulae
15 gyrus ectomarginalis	30 sulcus splenialis	46 corpus pineale

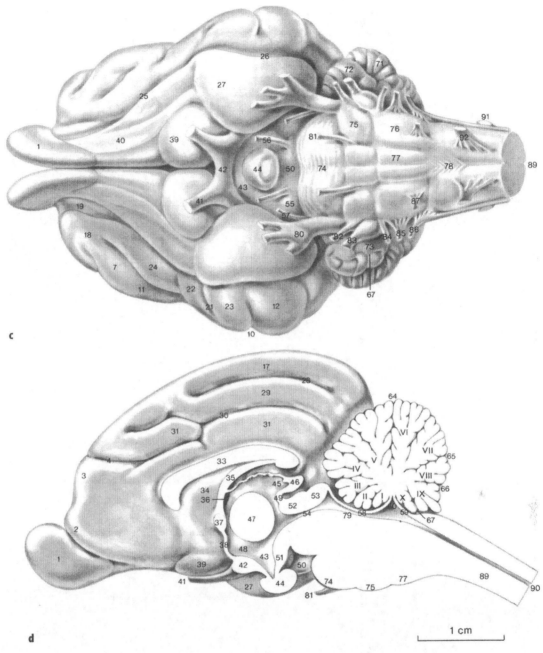

47 adhaesio interthalamica	63 vermis (lobi posterioris)	79 ventriculus quartus
48 ventriculus tertius	64 fissura prima	80 nervus trigeminus
49 commissura posterior	65 fissura prepyramidalis	81 nervus abducens
50 fossa interpeduncularis	66 fissura secunda	82 nervus facialis
51 corpus mammillare	67 fissura posterolateralis	83 nervus octavus
52 colliculus superior	68 lobulus simplex	84 nervus glossopharyngeus
53 colliculus inferior	69 lobulus ansiformis	85 nervus vagus
54 aqueductus cerebri	70 lobulus paramedianus	86 nervus accessorius
55 pedunculus cerebri	71 paraflocculus dorsalis	87 nervus hypoglossus
56 nervus oculomotorius	72 paraflocculus ventralis	88 obex
57 nervus trochlearis	73 flocculus	89 medulla spinalis
58 velum medullare anterius	74 pons	90 canalis centralis
59 velum medullare posterius	75 corpus trapezoideum	91 radix dorsalis nervi spinalis
60 fastigium	76 oliva	92 radix ventralis nervi spinalis
61 lobus anterior	77 pyramis	
62 lobus posterior	78 decussatio pyramidum	

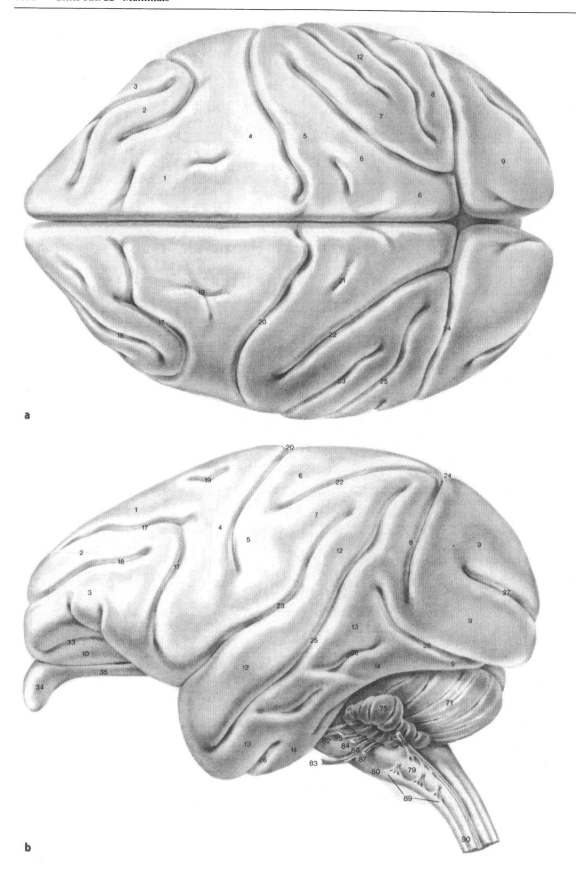

22.2 Gross Morphology of the Mammalian Brain 1653

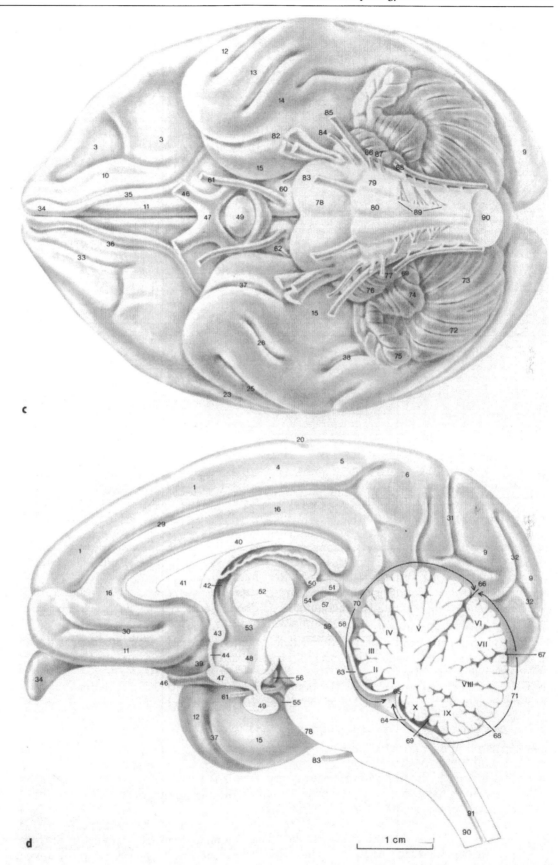

1 cm

◀ **Fig. 22.6. a** Dorsal, **b** lateral, **c** ventral and **d** medial views of the brain of the rhesus monkey.

1	Gyrus frontalis superior	35	tractus olfactorius	69	fissura posterolateralis
2	gyrus frontalis medius	36	sulcus olfactorius	70	lobus anterior
3	gyrus frontalis inferior	37	fissura rhinalis	71	lobus posterior
4	gyrus precentralis	38	sulcus occipitotemporalis	72	lobulus ansiformis
5	gyrus postcentralis	39	gyrus paraterminalis	73	lobulus paramedianus
6	gyrus parietalis superior	40	corpus callosum	74	paraflocculus dorsalis
7	gyrus parietalis inferior	41	septum pellucidum	75	lobulus petrosus
8	gyrus preoccipitalis	42	foramen interventriculare		(=part of paraflocculus dorsalis)
9	gyri occipitalis	43	commissura anterior	76	paraflocculus ventralis
10	gyri orbitales	44	lamina terminalis	77	flocculus
11	gyrus rectus	45	tractus olfactorius lateralis	78	pons
12	gyrus temporalis superior	46	nervus opticus	79	oliva
13	gyrus temporalis medius	47	chiasma opticum	80	pyramis
14	gyrus temporalis inferior	48	hypothalamus	81	ventriculus quartus
15	gyrus parahippocampalis	49	hypophysis	82	nervus trigeminus
16	gyrus cinguli	50	commissura habenulare	83	nervus abducens
17	sulcus arcuates	51	corpus pineale	84	nervus facialis
18	sulcus principalis	52	adhaesio interthalamica	85	nervus octavus
19	sulcus precentralis superior	53	ventriculus tertius	86	nervus glossopharyngeus
20	sulcus centralis	54	commissura superior	87	nervus vagus
21	sulcus postcentralis	55	fissa interpeduncularis	88	nervus accessorius
22	sulcus intraparietalis	56	corpus mammillare	89	nervus hypoglossus
23	sulcus lateralis	57	colliculus superior	90	medulla spinalis
24	sulcus lunatus	58	colliculus inferior	91	canalis centralis.
25	sulcus temporalis superior	59	aquaeductus cerebri	I	Lingula
26	sulcus temporalis inferior	60	pedunculus cerebri	II+III	lobulus centralis
27	sulcus ectocalcarinus	61	nervus oculomotorius	IV+V	culmen
28	sulcus occipitalis inferior	62	nervus trochlearis	VI	declive (lobulus simplex)
29	sulcus cinguli	63	velum medullare anterius	VII	folium/tuber vermis
30	sulcus rostralis	64	velum medullare posterius	VIII	pyramis
31	sulcus parietooccipitalis	65	fastigium	IX	uvula
32	sulcus calcarinus	66	fissura prima	X	nodulus
33	sulcus orbitalis	67	fissura prepyramidalis		
34	bulbus olfactorius	68	fissura secunda		

the subiculum and the fornix (one of the main afferent and efferent systems of the hippocampus) are visible at the meningeal surface of the hemisphere; the cornu ammonis invaginates into the lateral ventricle along the hippocampal fissure. The extent of the hippocampus differs among mammals. In monotremes and marsupials, it extends along the entire taenia fornicis in the wall of the half moon-shaped lateral ventricle, from the amygdala to the base of the septum. In eutherian mammals, the corpus callosum displaces the formatio hippocampi from the central part of the ventricle. In primates, the entire formatio hippocampi occupies a postcommissural position in the wall of the inferior horn of the lateral ventricle. The size of the formatio hippocampi is not related to olfaction. It is relatively large in small brains.

The neopallial surface of the hemisphere is smooth (lissencephalic) in *Ornithorhynchus*, Insectivora, Lagomorpha, Microchiroptera, Sirenia and in small marsupials, rodents and primates. A convoluted (gyrencephalic) brain is present in larger representatives of these orders and in Edentata, Megachiroptera, Proboscoidea, Cetacea, Hyracoidea, Carnivora, Perissodactyla and Artiodactyla (Poster 1, 1–31; Fig. 22.9). A central depression, corresponding to the insula of human anatomy, covered by opercula arising from surrounding convolutions, and accessible through a lateral cerebral (sylvian) fissure, is found in Carnivora, Perissodactyla, Artiodactyla, Proboscoidea, Cetacea and primates (see Fig. 22.149). The opercularisation of the insula and the arching of the hemisphere with the formation of a temporal lobe are most prominent in Cetacea, Proboscoidea and primates (Fig. 22.147). The brain of Cetacea and Proboscoidea is rostrocaudally compressed (Figs. 22.8, 22.9, 22.12), and the temporal lobes protrude far laterally. The sulcal pattern of the hemisphere is discussed in Sect. 22.11.6.6

22.3
Spinal Cord

J. Voogd

22.3.1
Gross Anatomy, Ascensus Medullae

The border of the spinal cord with the medulla oblongata is usually set rostral to the origin of the first cervical nerve, caudal to the appearance of the

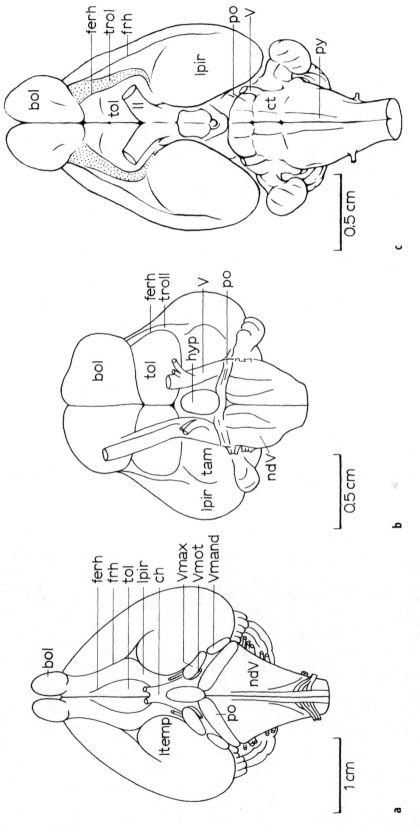

Fig. 22.7a–c. Ventral views of the brains of **a** the duckbilled platypus *Ornithorhynchus anatinus* (after Hines 1929), **b** the greater marsupial mole *Notoryctes typhlops* (after Burkitt 1938) and **c** the common tree-shrew *Tupaia glis*. *bol*, bulbus olfactorius; *ch*, chiasma opticum; *ct*, corpus trapezoideum; *ferh*, fissura endorhinalis; *frh*, fissura rhinalis; *hyp*, hypophysis; *lpir*, lobus piriformis; *ltemp*, lobus temporalis; *ndV*, nucleus tractus descendens nervi trigemini; *po*, pons; *tam*, tuberculum amygdaloideum; *tol*, tuberculum olfactorium; *troll*, tractus olfactorius lateralis; *I* etc., cranial nerve I etc.; *Vm*, radix motorius of V; *Vmand*, radix mandibularis of V; *Vmax*, radix maxillaris of V

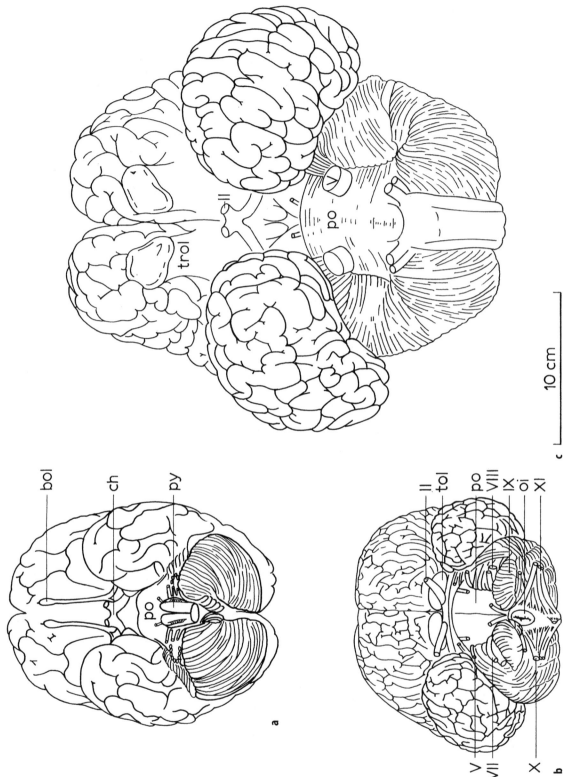

Fig. 22.8a–c. Ventral views of the brains of a man, **b** the porpoise *Tursiops truncatus* (after Langworthy 1932) and **c** the elephant *Loxodonta africana* (after Haug 1970). In C, the olfactory bulbs are lacking. *oi*, oliva inferior. For other abbreviations, see legend to Fig. 22.6

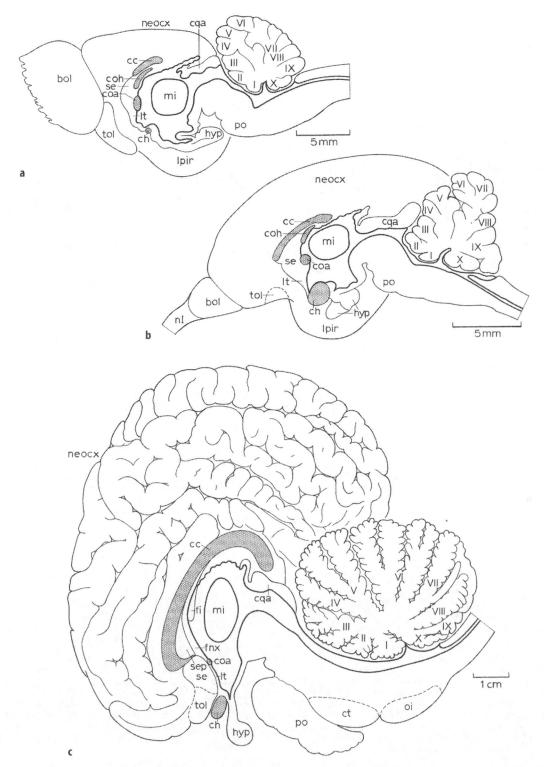

Fig. 22.9a–c. Semi-diagrammatic median sections through the brains of **a** the hedgehog *Erinaceus europaeus* (after Stephan 1975 and Stephan et al. (1991), **b** the Philippine tarsier *Tarsius syrichta* (after Stephan 1984) and **c** the porpoise *Tursiops truncatus* based on figures of Pilleri and Gihr 1969 and Morgane et al. 1980). *bol*, bulbus olfactorius; *cc*, corpus callosum; *ch*, chiasma opticum; *coa*, commissura anterior; *coh*, commissura hippocampi; *ct*, corpus trapezoideum; *fi*, foramen interventriculare; *fnx*, fornix; *hyp*, hypophysis; *lpir*, lobus piriformis; *lt*, lamina terminalis; *mi*, massa intermedia thalami; *neocx*, neocortex; *nI*, nervus olfactorius; *oi*, oliva inferior; *po*, pons; *se*, septum precommissurale; *sep*, septum pellucidum; *tol*, tuberculum olfactorium; *I–X*, cerebellar lobules

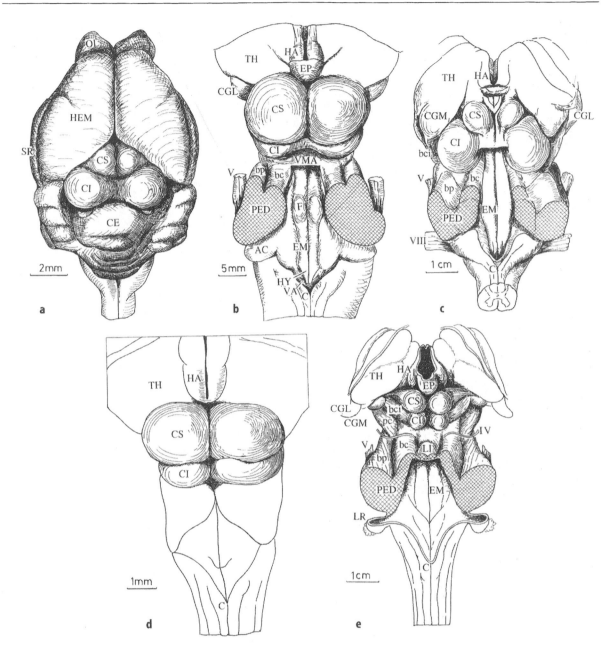

Fig. 22.10a–e. Dorsal views of the brain stem of **a** the tomb bat *Taphozous mauritii* (redrawn from Baron et al. 1996), **b** the ibex *Capra* (redrawn from Schober and Brauer 1974), **c** the porpoise *Tursiops truncatus* (redrawn from Langworthy 1967), **d** the spectral tarsius *Tarsius* (redrawn from Tilney 1927) and **e** man (redrawn from Nieuwenhuys et al. 1988). In **b–e**, the cerebellum was removed. *AC*, tuberculum acusticum; *bc*, brachium conjunctivum; *bci*, brachium of the inferior colliculus; *bp*, brachium pontis; *C*, calamus scriptorius; *CE*, cerebellum; *CGL*, lateral geniculate body; *CGM*, medial geniculate body; *CI*, inferior colliculus; *CS*, superior colliculus; *EM*, eminentia mediana; *EP*, corpus pineale; *F*, colliculus facialis; *HA*, ganglion habenulae; *HEM*, cerebral hemisphere; *HY*, hypoglossal eminence; *IV*, trochlear nerve; *LI*, lingula in anterior medullary velum; *LR*, lateral recess; *OL*, olfactory bulb; *pc*, cerebral peduncle; *PED*, cut surface of the cerebellar peduncles; *SR*, sulcus rhinalis; *TH*, thalamus; *V*, trigeminal nerve; *VA*, vagal eminence; *VIII*, statoacoustic nerve; *VMA*, anterior medullary velum

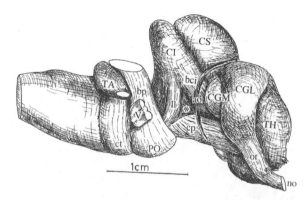

Fig. 22.11. Lateral aspect of the brain stem of the rabbit after removal of the cerebellum. *Asterisk* indicates the triangular area containing the parabigeminal nucleus. *ao*, accessory optc tract; *bp*, brachium pontis; *CGL*, lateral geniculate body; *CGM*, medial geniculate body; *CI*, inferior colliculus; *cp*, cerebral peduncle; *CS*, superior colliculus; *ct*, corpus trapezoideum; *ll*, lateral lemniscus; *no*, optic nerve; *ot*, optic tract; *PO*, pons; *TA*, tuberculum acusticum; *TH*, thalamus; *V*, trigeminal nerve

dorsal column nuclei, at the level of the pyramidal decussation. This border is not well defined. The transition from the motor and sensory nuclei of the spinal cord into the corresponding structures of the caudal medulla oblongata (involved in the innervation of structures which have evolved from occipital somites) is gradual.

The cord occupies the entire length of the vertebral canal during early stages of its development. In adult mammals, it is shorter, due to the continued growth of the vertebral column (Sakla 1969). Caudally, the cord tapers into the conus medullaris and the filum terminale. It extends into the sacral region in the rabbit and in ungulates, and in terrestrial carnivores it reaches almost to the end of the lumbar column; in primates, the conus terminalis is located at the level of the lumbar vertebrae, whereas in the seal *Phoca vitulina* and in some insectivores (*Erinaceus europeus*) the conus is situated as high as the thoracic level (Ariëns Kappers et al. 1936; Tigges 1964; Nieuwenhuys 1964). In the porpoise *Phocaena* the cord reaches the ninth lumbar vertebra. The spinal cord of the right whale (*Eubalaena glacialis*) is short (174 cm) for the body length (11.65 m). The cauda equina is located in the epidural space, where the lower thoracic, lumbar and coccygeal roots have to run for several metres to reach the corresponding intervertebral foramina (Seki 1958). A condition, mentioned by Ariëns Kappers et al. (1936), that may influence the length of the cord with respect to the vertebral column is the atrophy of certain myotomes, producing tail muscles. The difference in extent in the platypus *Ornithorhynchus*, in which the cord extends into the sacral region, and the echidna *Tachyglossus*, with a spinal cord which ends midway in its course through the canal, may be related to the presence of a strong, muscular tail in the former and a thinner, little-used organ in the latter. However, the presence of a long spinal cord extending into the sacral region in almost tail-less mammals, such as the rabbit and the chiropteres, contradicts the general validity of this rule.

It is customary to subdivide the cord into segments. Each segment gives off dorsal and ventral rootlets, which in turn give rise to a pair of dorsal

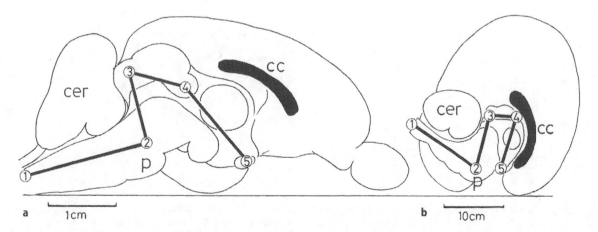

Fig. 22.12a,b. Median sections of the brain of **a** the rabbit and **b** the beluga *Delphinapterus* with the long axis of the brain stem drawn in . The axis interconnects the medulla oblongata (*1*), the pons (*2*), the inferior colliculus (*3*), the commissura posterior (*4*) and the chiasma opticum (*5*). *Axis 1-2* is the axis of the rhombencephalon; *2-3* is based on the direction of the lateral lemniscus and represents the original axis of the caudal mesencephalic neural tube (compare Fig. 22.11); *3-4* is the axis of the rostral mesencephalon; *4-5* is the axis of the diencephalon. The brain of the beluga is rostrocaudally compressed. This compression is also expressed by the direction of the corpus callosum. *cc*, corpus callosum; *cer*, cerebellum; *p*, pons. (Part B is adapted from an original figure by Schober and Brauer 1974)

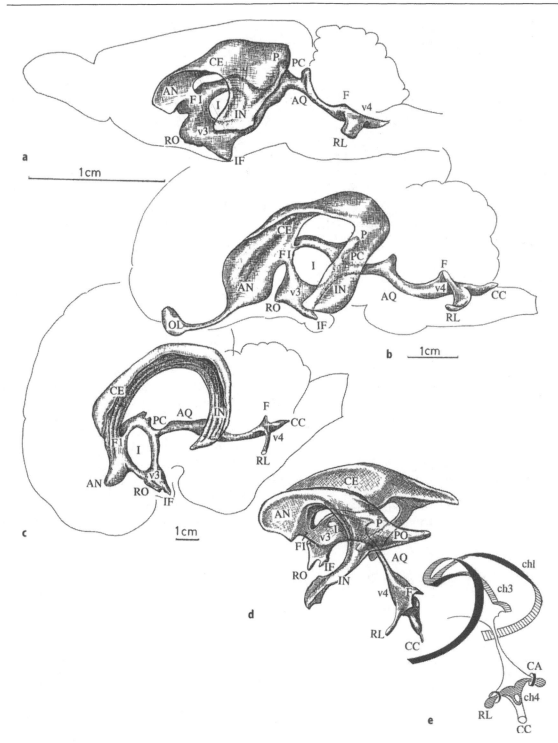

Fig. 22.13a–e. Outlines of the brain with the lateral aspect of the ventricular system of rat (**a**), dog (**b**), porpoise (**d**) and oblique view of the ventricular system of man (**d**). **e** Diagram of an oblique view of the choroid walls of the lateral, third and fourth ventricles. *Black areas*, ventricular surface of choroid; *hatched areas*, meningeal surface. *AN*, lateral ventricle, cornu anterius; *AQ*, aqueductus cerebri; *CA*, canalis communicans; *CC*, canalis centralis; *CE*, lateral ventricle, pars centralis; *ch3*, choroid wall of third ventricle; *ch4*, choroid wall off fourth entricle with apertura mediana of Magendie; *chl*, choroid wall of lateral ventricle; *PC*, commissura posterior; *F*, fastigium; *FI*, foramen interventriculare; *I*, massa intermedia; *IF*, recessus infundibuli; *IN*, lateral ventricle, cornu inferius; *OL*, recessus olfactorius; *P*, recessus pinealis; *PO*, lateral ventricle, cornu posterius; *RL*, recessus lateralis of the fourth ventricle; *RO*, recessus opticus; *v3*, third ventricle; *v4*, fourth ventricle. (**a–c** Redrawn from McFarland et al. 1969. **d,e** Redrawn from Nieuwenhuys et al. 1988)

and ventral roots that unite into a pair of spinal nerves. The number of spinal nerves is fairly constant among mammals. Cervical and lumbar enlargements can be recognised in the spinal cord of most mammalian species, but the lumbar intumescence cannot be seen in species which lack or have poorly developed posterior extremities (e.g. the dugong and the common whale). It appears to be present in *Phocaena* and in dolphins, where it is situated far caudally and may subserve the tail. In the kangaroo, it is large and surpasses the cervical enlargement in size (Kuhlenbeck 1975).

A central canal can be recognised in the cord of most species, although it is not always patent. Caudally, the central canal enlarges into the ventriculus terminalis in the conus medullaris. A deep fissure and scattered ventral rootlets characterise the ventral surface of the cord. Shallow midline and paramedian sulci are present on the dorsal side of the cord; dorsal rootlets enter the cord in the paramedian sulcus on top of the dorsal horn. The butterfly shape of the grey matter can usually be seen. The grey matter reaches the meningeal surface with a bundle of thin and/or unmyelinated fibres (dorsolateral tract of Lissauer) located on top of the dorsal horn. The dorsal roots enter the dorsal funiculus, which is located between the dorsal horn and the glial (dorsal) septum in the midline. The lateral and ventral funiculus are arbitrarily demarcated by the exit of the ventral roots. The ventral funiculus continues accross the midline as the ventral white commissure. The relative proportions of the white and grey matter of the cord generally depend on the size of the animal, and the amount of grey matter is relatively small in large mammals.

22.3.2
Grey Matter

The basic structure of the grey matter of the cord is the same in all mammalian species (see Figs. 22.22–22.24a–e). A laminated dorsal horn, a ventral horn containing the motoneurons innervating the striated muscles and an intermediate zone consisting of cells of different sizes and harbouring the nuclei of the autonomic nervous system can be distinguished. The dorsal horn and the intermediate grey, which constitute the main afferent centres, are derived from the alar plate of the neural tube. The motoneurons of the ventral horn are a derivate of the basal plate. A nomenclature based upon the dorsoventral lamination of the grey matter of the cord of the cat was introduced by Rexed (1952, 1954) and has been applied to other mammalian species (Fig. 22.14; primates: Kuypers and Brinkman 1970; Ralston 1965; Shriver et al. 1968; Miller and Stro-

minger 1973; Schoenen and Faull 1990; marsupials: Martin and Fisher 1968; Rees and Hore 1970; Pindzola et al. 1988; rat: McClung and Castro 1978; Molander and Grant 1995; Molander et al. 1984, 1989). Laminae I–VI belong to the dorsal horn, laminae VII and VIII to the intermediate zone, the motoneuronal cell groups constitute laminae IX and lamina X corresponds to the periaqueductal grey matter surrounding the central canal. The small unmyelinated and myelinated fibres of Lissauer's dorsolateral tract are located on top of the dorsal horn in the root entry zone (Lissauer 1886). Grey commissures, consisting of unmyelinated or thinly myelinated fibres, cross the midline dorsal and ventral to the central canal.

Small, medium-sized and large marginal cells of Waldeyer (1888) are positioned tangentially along the surface of the dorsal horn (see Figs. 22.22–22.24a–e). They form lamina I and are located in a marginal plexus at the border of the dorsolateral tract of Lissauer and the substantia gelatinosa. The substantia gelatinosa corresponds to lamina II of Rexed (Ralston 1965; Brown 1981), although some authors describing the dorsal horn of the cat have included lamina III in this layer (Szentágothai 1964). Lamina II lacks myelin and contains small, densely packed neurons. These largely correspond to the central and the limiting cells of Cajal (1909) or the islet cells and the stalked cells of Gobel (1975, 1978). The dendrites of the cells of lamina II are largely confined to narrow, rostrocaudally extending sagittal slabs or cylinders (Fig. 22.15B). The axons of the substantia gelatinosa either terminate in lamina II or in neighbouring layers of the dorsal horn. Some axons cross in the dorsal grey commissure and terminate in the superficial layers of the contralateral dorsal horn (Szentágothai 1964; for reviews, see Brown 1981; Schoenen and Faull 1990; Willis and Coggeshall 1991). Lamina III cells are slightly larger and less densely packed. Moreover, this layer contains some myelinated fibres. Flattening and rostrocaudal orientation of the dendritic tree are less conspicuous in this layer, which contains some small antenna-like neurons (Brown 1981; Schoenen and Faull 1990). Lamina IV corresponds to the nucleus proprius of the dorsal horn. It contains cells of all sizes, embedded in a longitudinally oriented plexus of myelinated fibres. Conspicuous among the neurons of lamina III and IV are the antenna-like cells with dendrites extending into lamina II (Fig. 22.15C; Cajal 1909; Scheibel and Scheibel 1968; Réthelyi and Szentágothai 1973; Brown 1981; Schoenen 1982). Most neurons in lamina III and IV give rise to projections to the dorsal column nuclei or to the lateral cervical nucleus. Most spinothalamic tract fibres take

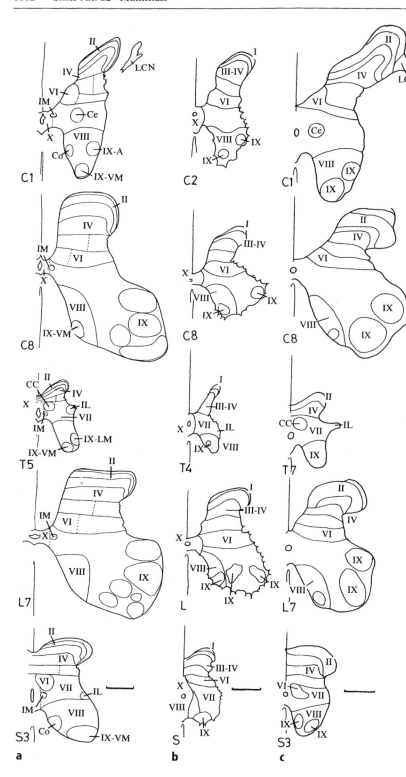

Fig. 22.14a–c. Rexed's laminae in the cat (from Rexed 1954), b the North American opossum *Didelphis virginianna* (from Pindzola et al. 1988) and c the rhesus monkey *Macaca mulatta* (from Shriver et al. 1968). *A*, nucleus spinalis n. accessorii; *C*, cervical cord; *CC*, columna dorsalis, Clark ;*Ce*, central cervical nucleus; *Co*, commissural nucleus; *IL*, intermediolateral nucleus; *IM*, intermediomedial nucleus; *LCN*, lateral cervical nucleus; *LM*, nucleus lateromedialis; *VM*, nucleus ventromedialis; *I–X*, laminae I–X of Rexed. *Bar*, 500 µm

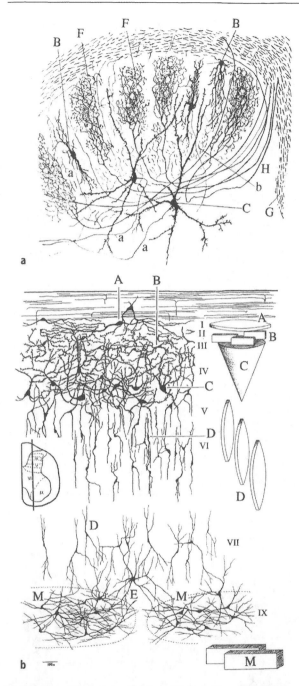

Fig. 22.15. a Transverse Golgi-stained section through the dorsal horn of the cat. (From Cajal 1909–1911). b Sagittal Golgi-stained section through the dorsal horn, the intermediate grey and the ventral horn of the cat. The orientation of the dendrites of the neurons is indicated on the *right*. (Modified from Scheibel and Scheibel 1968). *a*, axon; *A*, marginal cell; *b*, longitudinal axonal plexus of dorsal horn; *B*, small cells of the substantia gelatinosa; *C*, antenna-like cells of lamina IV; *D*, neurons of the intermediate grey (laminae VI and VII); *E*, interneuron with radiating dendrites; *F*, flame-shaped arborisations; *G*, dorsal funiculus; *H*, coarse collaterals of primary afferents; *M*, motoneurons; *I–IX*, laminae I–X of Rexed

their origin from neurons of lamina I and from neurons in the deep laminae V and VI of the dorsal horn (Fig. 22.21).

The demarcation of laminae V–VIII is not sharp and largely serves descriptive purposes in the localisation of neurons identified by specific neurochemical or electrophysiological properties and the delineation of the terminal fields of certain fibre connections. Laminae V and VI constitute the neck and the base of the dorsal horn. They contain cells of all sizes. Lamina VI exists only in the cervical and lumbar enlargements in the cat (Rexed 1954) and rat (McClung and Castro 1978). Lamina VI is narrow in cervical segments in the phalanger *Trichosurus* (Rees and Hore 1970).

The central cervical nucleus was first delimited by Rexed (1954) in lamina VII in the upper four or five cervical segments of the spinal cord of the cat. It consists of large cells, receiving afferents from the upper four cervical dorsal roots and projects to the cerebellum (Wiksten 1985, 1979a,-c). It occupies the same position in the rhesus monkey (Shriver et al. 1968).

A distinct group of smooth-contoured neurons with a characteristic, perinuclear disposition of their Nissl substance is located medially at the base of the dorsal horn at thoracic segments in the medial part of lamina VII. This nucleus is known as the dorsal column of Clarke (1851, 1859) and gives rise to ipsilaterally ascending fibres of the dorsal spinocerebellar tract (Fig. 22.14). Laminae VII and VIII correspond to the intermediate zone. Lamina VIII contains fairly large neurons in the ventromedial part of the intermediate zone (commissural neurons of Cajal 1911). Many of the axons of these cells cross in the ventral white commissure of the cord. One of the features which distinguishes neurons of the intermediate zone from the motoneurons and the cells of lamina IV of the dorsal horn is the disposition of their dendrites in more or less transverse planes (Fig. 22.15D).

Most neurons of laminae V–VIII are interneurons that give rise to propriospinal fibres, terminating on motoneurons or in other parts of the spinal grey matter. Several groups of interneurons have been characterised using electrophysiological methods and localised with intracellular injection of horseradish peroxidase (HRP), lucifer yellow or other substances (for a review, see Jankowska 1992). Some of these interneurons have a restricted, laminar distribution. Ia-inhibitory interneurons, which mediate Ia inhibition of antagonists, are located in the ventral part of lamina VII, just outside the motor nuclei. Interneurons with a dominant input from Golgi tendon organs (Ib interneurons) are located in lamina VI and dorsal lamina VII. The interneurons of lamina VIII, subserving crossed reflexes, have already been mentioned. For many types of interneurons, the localisation is not yet known. This is true of the interneurons mediating presynaptic inhibition of primary afferent fibres. Interneurons mediating presynaptic inhibition of Ia and Ib fibres terminating on motoneurons may correspond to the small GABAergic neurons, located by Ljungdahl and Hökfelt (1973) at the border of laminae V and VI in the cat. Interneurons mediating presynaptic depolarisation of other afferents may correspond to a population of GABAergic neurons in laminae I–III described by the same authors. Renshaw cells are small, inhibitory (GABAergic) interneurons that receive collaterals from ventral root fibres. Renshaw cells distribute their axons to motoneurons and are located medial to or in lamina IX (Matsushita 1969; Jankowska and Lindström 1971).

The intermediomedial and intermediolateral nuclei of the autonomic nervous system are located within the intermediate zone. The intermediolateral nucleus consists of small, spindle-shaped neurons which give origin to preganglionic fibres, leaving the cord with the ventral roots. Preganglionic sympathetic fibres take their origin from the intermediolateral nucleus of the thoracic and high lumbar lateral horn (see Figs. 22.22–22.24c). Neurons giving rise to the preganglionic fibres in a particular ventral root are confined to the corresponding segment in the guinea pig, hamster and cat. Some preganglionic neurons are located in the white matter (lateral to the intermediolateral nucleus), in the dorsal commissural nucleus (dorsolateral to the central canal) or in the intercalated nucleus (located between the commissural nucleus and the intermediolateral column (Fig. 22.16; Rabin and Purves 1980). The dendrites of the neurons of the intermediolateral column are arranged in longitudinal bundles within the confines of this column and in medially directed dendritic bundles. The medially directed dendritic bundles occur at certain intervals. More than one bundle may be present at each thoracic segment (Vera et al. 1987). A similar arrangement of longitudinal and transversely arranged bundles has been visualised with acetylcholinesterase (AChE) and choline acetyltransferase (ChAT) histochemistry in different mammalian species (Fig. 22.16; Galabov and Davidoff 1976; Barber et al. 1984). Preganglionic parasympathetic fibres arise from a cell group located at the lateral border of the intermediate grey with the lateral funiculus and from cells of the intermediomedial nucleus in the second and third sacral segment of the cat spinal cord (Rexed 1954; Oliver et al. 1969).

The localisation of motoneurons in the ventral horn has recently been reviewed by Van der Horst (1996). Motoneurons may be divided into a medial motoneuronal cell group, which innervates the axial muscles by way of the dorsal ramus of the spinal nerve, and a lateral motoneuronal cell group, which innervates the remainder of the musculature of the body and that of the limbs by way of the ventral ramus (Bok 1928; Rexed 1952, 1954). At thoracic levels, the medial and lateral group are fused into one. This observation is in keeping with the find-

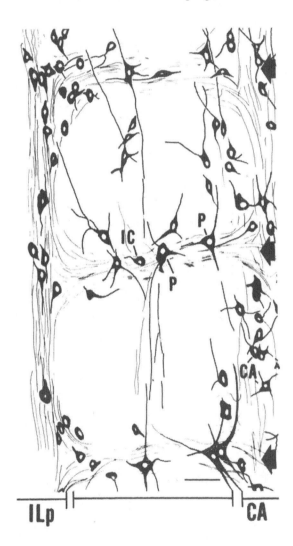

Fig. 22.16. A transverse section through the intermediate part of the thoracic cord showing the spatial relations of the neurons of the principal intermediolateral nucleus (*ILP*), embedded in a longitudinal dendritic bundle The 'ladder rungs' formed by transverse dendritic bundles (*arrows*), interconnecting the intermediolateral nucleus and the central autonomic cell column (*CA*), located dorsal to the central canal are shown. Composite camera lucida drawing made from three serial, horizontal sections of the T6 spinal cord of the rat, immunoreacted with an antibody against choline acetyltransferase. *CA*, central autonomic cell column; *IC*, intercalated nucleus; *P*, partition neurons in the intercalated nucleus. (From Barber et al. 1984)

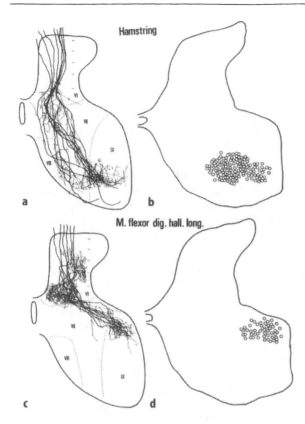

Fig. 22.17. a,c Main collaterals of intra-axonally injected horseradish peroxidase (HRP)-stained Ia fibres in the L7 segment of the cat spinal cord, innervating a the hamstring muscles and c the m. flexor digitorium hallucis longis. (From Ishizuka et al. 1979). b,d Motor neurons in the cat L7 segment, innervating b the hamstring muscles and d the m. flexor digitorum and hallucis longis (Van der Horst and Holstege 1995). Note the precise correspondence between the Ia afferent and efferent innervation of these muscles

ings of Sprague (1951), who demonstrated that, in the thoracic cord of the monkey, the neurons supplying the dorsal and ventral rami of the spinal nerve are not spatially separated. In the lumbar cord, however, a clear separation appeared to exist, such that the dorsal rami were found to be derived exclusively from motoneurons along the ventromedial border of the grey matter, a cell group that corresponds to the medial motoneuronal cell group. At high cervical and low bulbar levels, the motoneurons occupy the medial part of the ventral horn (supraspinal nucleus), and the intermediate zone extends ventrally in the medial part of the ventral horn. Motoneurons which give rise to the spinal root of the accessory nerve are located in the dorsal part of the supraspinal nucleus and in a column in the lateral part of the ventral horn in the first and second cervical segments (Marinesco 1904). A similar organisation exists in rodents, lagomorphs, carnivores and primates.

The somatotopic organisation of spinal motoneurons was studied by Sterling and Kuypers (1967) in the brachial cord of the cat, by Reed (1940) in the brachial cord of the monkey and by Romanes (1951, 1964) and Van der Horst (1996) in the lumbosacral cord of the cat. In the cervical (C4-T1) and lumbosacral (L3-S1) enlargements, the lateral motoneuronal cell group can be subdivided into several nuclei, innervating different groups of muscles of the fore- and hindlimbs, respectively (Fig. 22.17). The motoneurons which innervate the girdle and proximal extremity muscles were found to be located ventromedial to those innervating the distal extremity muscles and the small hand and foot muscles. In the brachial cord, the motoneurons innervating physiological flexors were found dorsal to those innervating the corresponding physiological extensors. The cutaneous trunci muscle which overlies the superficial abdominal, axial and lateral hindlimb muscles in the cat and rat is innervated by motoneurons located in the C8-T1 spinal segments (Haase and Hrycyshyn 1985; Holstege et al. 1987; Haase 1990). In the lumbosacral cord of the cat, Van der Horst (1996) distinguished ten discrete groups of motoneurons innervating sets of muscles. Romanes (1951, 1964) noticed a group of smaller neurons, present in the central part of the ventral horn in sacral segments only, which gives rise to the pudendal nerve. It corresponds to Onuf's nucleus (Onuf 1902). According to Van der Horst (1996), Onuf's nucleus of the cat forms the rostral two thirds of a continuous cell column that innervates the external anal and urethral sphincters, the ischiocavernosus muscle and the bladder. A few parasympathetic preganglionic neurons are present in Onuf's nucleus. In the rhesus monkey, it is located at S1 and L7 (Roppola et al. 1985), and in the rat at L6 (Schröder 1980). A group of neurons in segments L4 and L5 resemble motoneurons, but cannot be labelled from any peripheral nerve in the lumbosacral plexus. These neurons correspond to the 'spinal border cells' described by Cooper and Sherrington (1940), which give rise to fibres of the ventral spinocerebellar tract (Fig. 22.21h,i). In the lumbosacral cord, the flexor motoneurons are located lateral to the motoneurons innervating adductor and abductor muscles and medial to the extensor motoneurons (Van der Horst 1996).

An important question is whether both the dendrites and the somata of cells are restricted to Rexed's laminae. This appears to be the case for the marginal cells of lamina I, for the dorsal column of Clarke and for the cells of the substantia gelatinosa (lamina II). Dendrites of lamina IV cells typically invade the substantia gelatinosa (Fig. 22.15). In the intermediate zone, dendrites generally surpass the borders of the laminae and may even penetrate into the dorsal horn, the motoneuronal cell groups and the white matter. The restriction of dendrites of the intermediate zone to a more or less transverse plane has already been mentioned. Dendrites of motoneurons may extend far beyond the motoneuronal columns, but some dendrites usually remain confined to these columns. Longitudinal bundles of dendrites are typical for the phrenic motor column. Dendritic bundles are prominent in the ventral and ventrolateral motoneuronal groups in the cervical and lumbosacral enlargements, but do not occur in the dorsolateral motoneuronal groups. Dendritic bundles have been observed in the spinal cord of the cat, the monkey and the rat. In the cat, they are absent at birth, but form in the early postnatal period and gain their adult appearance at 4 months (Scheibel and Scheibel 1970; for reviews, see Roney et al. 1979 and Schoenen and Faull 1990).

Some nuclei of the spinal cord are located in the white matter. The lateral cervical nucleus was originally identified in the cat (Rexed and Brodal 1951; Westman 1968), where it occupies a region of the dorsolateral funiculus at C1 and C2 immediately ventrolateral to the dorsal horn (see Fig. 22.23). This nucleus forms a link in an ascending spinothalamic pathway (see Sect. 22.3.4.3), The lateral cervical nucleus and its spinal afferent fibres has been identified in the cat (Rexed and Brodal 1951; Brodal and Rexed 1953) and other carnivores (Kitai and Martin 1965; Craig 1978; Pubols and Haring 1995), insectivores (Waldron 1969; Künzle 1993), rodents and lagomorphs (Giesler et al. 1987, 1988), artiodactyles (Verhaart 1970b) and primates (Ho and Morin 1964; Mizuno et al. 1967; Shriver et al. 1968), including humans (Truex et al. 1970). The lateral cervical nucleus has not been described in other mammalian orders. Notably, it appears to be absent in marsupials and monotremes (see however Fig. 22.22c). The lateral cervical nucleus in the rhesus monkey is small and extends from C1 into the C3 segment (see Fig. 22.24e). In humans, it was present in some subjects, but absent in others. Small neurons of the lateral spinal nucleus are found at the surface of the dorsolateral funiculus in at all levels of the rat spinal cord (Gwyn and Waldron 1968; Menétrey et al. 1980,1982)

22.3.3
Primary Afferent Fibres and the Dorsal Funiculus

Small and darkly staining ganglion cells (B cells) and light ganglion cells of different sizes, with a paucity of cytoplasmatic organelles (A cells), have been distinguished in sensory ganglia (Andres 1961; Jacobs et al. 1975, for a review, see Willis and Coggeshall 1991). The dark cells presumably give rise to small unmyelinated fibres and the light cells to the myelinated fibres of the dorsal root. Most primary afferents enter the spinal cord through the dorsal roots. Unmyelinated primary afferents have been identified in the ventral roots in the cat (Coggeshall et al. 1973, 1974b). Unmyelinated ventral root afferents are less numerous in the rat (Grant 1995). In the monkey, small-diameter, myelinated and unmyelinated fibres located along the periphery of the dorsal root separate from larger-diameter fibres and enter the medial part of the dorsolateral tract of Lissauer (Ranson 1913; Szentágothai 1964; Snyder 1977). Larger-diameter fibres proceed to the dorsal funiculus. Primary root fibres bifurcate in ascending and descending branches that course in Lissauer's tract and in the dorsal funiculus (Cajal 1909; Réthelyi and Szentágothai 1973).

Small fibres terminate in the superficial layers of the dorsal horn. Small myelinated (Aδ) fibres predominate in the marginal plexus of lamina I, and the unmyelinated (C) fibres terminate in lamina II (Ralston 1968, 1979; Heimer and Wall 1968; Light and Perl 1979a,b; Réthelyi et al. 1982, 1979; Snyder 1982; for a review, see Willis and Coggeshall 1991). In accordance with the restricted termination of the unmyelinated fibres, which contain fluoride-resistant acid phosphatase, lamina II displays a strong activity for this enzyme (Nagy and Hunt 1983). Laminae III–IV receive collaterals from dorsal root fibres coursing in the dorsal funiculus. Among them are the hair follicle afferents. These primary root fibres do not bifurcate upon entering the dorsal columns. They enter the dorsal horn passing through the substantia gelatinosa and recurve in the nucleus proprius to terminate as flame-shaped arborisations in lamina III (Fig. 22.15F), sometimes penetrating into lamina II (Brown 1981).

Abundant root fibres supply the base of the dorsal horn and the intermediate zone, where they terminate in a region near the central canal, referred to variously as the dorsal commissural nucleus or the intermediomedial nucleus (Shriver et al. 1968; Imai and Kusama 1969) and in the centre of laminae V and VI. Root fibres innervating Golgi tendon organs (Ib fibres) terminate in lamina VI and the dorsal part of lamina VII (Brown 1981). Dorsal root fibres innervating primary muscles spindle endings (Ia fibres) and secondary endings (group II fibres) proceed to the motoneuronal cell groups. Ia fibres terminate in the medial half of the base of the dorsal horn (lamina VI), in lamina VII and in the homonymous motor cell columns of lamina IX (Fig. 22.17; Ishizuka et al. 1979).

Both in the cat (Morgan et al. 1981) and the

monkey *Macaca mulatta* (Nadelhaft et al. 1983), pelvic nerve afferents terminate at the level of the sacral parasympathetic nucleus and at more rostral and caudal levels. Within lamina I, they form a thin shell around the dorsal horn. Medial and lateral collateral bundles pass ventrally on both sides of the dorsal horn to terminate in laminae V and VI and more medially next to the dorsal grey commissure (see also Matsushita and Tanami 1983). The terminations of these root fibres in lamina I display a rostrocaudal periodicity, similar to the neurons projecting to the mesencephalic periaqueductal grey located in the same layer (Van der Horst and Holstege 1995). Pudendal nerve primary afferents innervating the perineum and pelvic viscera show a similar distribution. Visceral primary afferent fibres do not contact the neurons of the parasympathetic visceral nucleus.

The dorsal funiculus has a laminar organisation. In the cervical cord, the fasciculi gracilis and cuneatus can be recognised by virtue of the discontinuity of this lamination in the dorsal part of the dorsal funiculus, where lumbar and cervical fibre layers are contiguous, thoracic fibres being located in the ventral part of the funiculus (Lietaert Peerbolte 1932). The composition of the dorsal funiculus was studied in the rhesus monkey using Nauta silver impregnation of degenerated axons after transection of the dorsal roots by Shriver et al. (1968) and Carpenter et al. (1968), and in the cat by Imai and Kuzuma (1969).

The dorsal root fibres enter the lateral part of the dorsal funiculus, where they divide into long ascending and shorter descending branches that vary in length between three and eight segments. Ascending fibres from upper thoracic (T1-T7) and cervical roots form narrow laminae that shift to more medial parts of the fasciculus cuneatus, with the largest contribution from C8. Ascending fibres from lower thoracic and lumbar roots form oblique narrow bands in the fasciculus gracilis, with fibres from thoracic segments most lateral. The thoracic roots 5-8 contribute to the fasciculus gracilis and cuneatus. The largest contribution in the monkey is from L4-L6. Ascending fibres from the caudalmost roots initially occupy the medial region of the fasciculus gracilis and shift to a dorsal position. Descending branches mimic the position of the ascending fibres and form laminae in the fasciculus cuneatus and gracilis, with a greater concentration in the ventromedial part of the fasciculus cuneatus close to the spinal grey. Concentrations of descending fibres next to the posterior intermediate septum (comma tract), the posterior median septum (oval bundle) and the extreme dorsomedial part of the fasciculus gracilis (bundle of Gombault and Philippe 1894) were not observed in the rhesus monkey (Shriver et al. 1968; Carpenter et al. 1968).

Ascending and descending branches of dorsal root fibres emit collaterals terminating in the dorsal horn over two to three segments rostral or caudal to the segment of entry. Terminations in the motoneuronal cell groups are found over a more limited rostrocaudal trajectory. Collateralisation of ascending branches to the medial and central lamina VI is more extensive. Sacral, lumbar, thoracic and lower cervical roots project to the columna dorsalis of Clarke (Shriver et al. 1968; Carpenter et al. 1968; Imai and Kuzuma 1969).

A somatotopical localisation is present in the dorsal horn, which roughly corresponds to the topography of the root fibres in the dorsal columns. In the cervical cord of the cat, the terminations of the dorsal root collaterals form oblique slabs that exend from caudolaterally to rostromedially over four to five segments (Imai and Kuzuma 1969). Proximal parts of the body are located medially and the distal extremity laterally in laminae I-IV of the dorsal horn (Brown et al. 1975; Molander and Grant 1985). The somatotopical localisation in the motoneuronal cell groups of the ventral horn is reversed with respect to the dorsal horn; axial and proximal muscles are innervated by medially situated motoneurons, and distal muscles of the limbs by dorsolaterally located cells. The transition of the mediolateral/distoproximal pattern in the dorsal columns and the dorsal horn to the pattern of the ventral horn apparently occurs at the base of the dorsal horn, where primary afferents concentrate in the centre of laminae V and VI (Fig. 22.17a,c).

22.3.4
White Matter

22.3.4.1
Propriospinal Fibres

The topography of ascending and descending tracts in the lateral and ventral funiculi of the cord has been studied using axonal tracing techniques for many individual tracts in different mammalian species. Fibre analysis of the white matter of the cord in Häggqvist-stained transverse sections is one of the most productive methods used to study the myeloarchitecture of the spinal cord. It is based on the different proportions of large, medium-sized and small myelinated fibres in different spinal tracts and/or areas of the spinal white matter (see Fig. 22.20). Tracts can be recognised and traced in such preparations by their texture, and quantification of the fibre pattern is possible. Such a quantitative analysis was first performed by Häggqvist

(1936) for a section through the third thoracic segment of the human spinal cord. The white matter of the lateral and ventral funiculi of the spinal cord was analysed by Van Beusekom (1955) and Busch (1961) for the cat and by Sie (1956) for the human spinal cord. For the cat, these studies on the localisation of myelinated fibres can be compared with more recent topographical reviews, based on axonal transport of tritiated amino acids in different descending spinal tracts, including the unmyelinated raphe-spinal and coeruleospinal projections (Holstege and Kuypers 1982, 1990).

The localisation of propriospinal fibres in the ground bundles of the white matter of the ascending component of the ventral funiculus, and of the ascending fibres in the lateral funiculus, is probably very similar in all mammalian species. Differences in the composition of the white matter mainly concern the position of the pyramidal and the rubrospinal tracts. Propriospinal fibres form the ground bundles or fasciculi proprii of the ventral, lateral and dorsal funiculus. Their localisation is illustrated in Fig. 22.18, representing the relative density of the distribution of fibres surviving in the completely isolated L6 segment of the spinal cord of the cat (Tower et al. 1941; cited in Nathan and Smith 1959). Small (2 μm) fibres predominate in the ventral part of the dorsal funiculus and in the ground bundle of the lateral funiculus. In the ground bundle of the ventral funiculus, larger propriospinal fibres occur (Häggqvist 1936).

Propriospinal fibres arising from lamina VIII and the adjoining part of lamina VII distribute bilaterally to motoneurons in the medial part of the ventral horn. These propriospinal fibres course in the ventral funiculus, cross in the ventral white commissure and have a wide rostrocaudal distribution. Propriospinal fibres from the lateral parts of laminae V–VII are shorter, pass through the lateral funiculus and distribute ipsilaterally to dorsolateral motoneuronal groups (Kuypers 1973, 1982; Molenaar and Kuypers 1978; Rustioni et al. 1971).

Long ascending and descending propriospinal fibres mainly terminate in medial parts of the intermediate grey and avoid the motoneuronal cell groups (Giovanelli and Kuypers 1969; Matsushita and Uyama 1973). In the cat, long descending propriospinal fibres to motoneurons of the cervical enlargement take their origin from a group of neurons in the lateral intermediate grey at C3-C4 (Grant et al. 1980; Alstermark et al. 1987). Long ipsilaterally descending pathways to the lumbar cord arise from the same region of the cervical and the thoracic cord. Crossed pathways take their origin mainly from lamina VIII and the adjoining part of lamina VII both in the cat and rat (Matsushita et

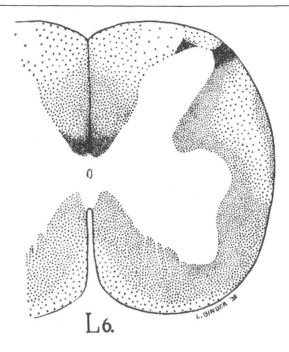

Fig. 22.18. Relative density of distribution of fibres surviving in the white matter of the completely isolated L6 segment of spinal cord of the cat. (From Tower et al. 1941)

al. 1979; Menétrey et al. 1985). Ascending propriospinal pathways from the lumbar to the cervical cord in the cat arise mainly from laminae IV and V of the dorsal horn, while crossed projections take their origin from laminae VIII and VII (English et al. 1985). The collateral origin of propriospinal projections from long ascending tracts to the brain stem and the cerebellum was studied and reviewed by Verburgh and Kuypers (1989) and Verburgh et al. (1989, 1990).

22.3.4.2
Long Descending Fibre Systems

The origin of descending systems to the spinal cord from the brain stem was verified by Nudo and Masterton (1988, 1989) in a great number of mammalian species belonging to many different orders, with retrograde transport of wheat-germ agglutinin (WGA)-coupled HRP from injections at the C1/C2 spinal cord. They identified 27 groups in each of the species, confirming the origin of descending brain stem systems in the cat (e.g., Kuypers and Maisky 1975; Holstege and Kuypers 1982; Holstege 1990) and the grey short-tailed Brazilian opossum *Monodelphis domestica* (e.g., Crutcher et al. 1978; Holst et al. 1991; Fig. 22.19).

The localisation of the main descending fibre systems at or just below the level of the pyramidal decussation was described for the cat by Busch

(1961) and Holstege (1991). The coarse fibres of the medial longitudinal fascicle and of the lateral vestibulospinal tract, which is located lateral to the exit of the first cervical ventral root, constitute distinct bundles (Fig. 22.20). The main components of the medial longitudinal fascicle at this level are the medial reticulospinal tracts, with a medial component from the pontine medial tegmental field and a lateral component of the bulbar medial reticular formation. The crossed bulbospinal tract from the dorsal reticular formation at the level of the inferior olive, the interstitiospinal tract, the tectospinal tract and the crossed and uncrossed medial vestibulospinal tracts occupy the dorsomedial portion of the medial longitudinal fascicle at this level. Fibres from field H of Forel, bordering on the rostral interstitial nucleus of the medial longitudinal fascicle, and from the lateral periaqueductal grey and the adjacent tegmentum are located ventrally along the ventral border of the ventral funiculus, among the fibres of the lateral vestibulospinal tract (Holstege and Cowie 1989). A small number of uncrossed fibres of the pyramidal tract descend below the level of the decussation, next to the ventral fissure to mid-thoracic levels.

The crossed fibres of the pyramidal tract and the rubrospinal tract are the main descending fibre systems of the dorsolateral funiculus. In the cat, the pyramidal tract is located dorsal to the rubrospinal tract. In Häggqvist-stained sections, the border between the large fibres of the rubrospinal tract and the small fibres that predominate in the pyramidal tract is easy to establish. The lateral border of the rubrospinal tract with the large fibres of the dorsal spinocerebellar tract is less distinct (Fig. 22.20). Smaller fibres of the crossed pontospinal tract from the ventral part of the lateral pontine tegmentum and from the lateral medullary tegmental field, including the region surrounding the retroambiguus nucleus, also descend in the dorsolateral funiculus. Coeruleospinal and raphe-spinal systems occupy the periphery of the lateral and ventral funiculus. They are mostly small and unmyelinated (Holstege and Kuypers 1982: cat; Martin et al. 1982b: North American opossum). A mostly crossed, descending pathway from the nucleus retroambiguus consists of scattered fibres in the lateral and ventral funiculus (Van der Horst and Holstege 1995).

The medial and crossed vestibulospinal and interstitiospinal tracts terminate in the cervical cord medially in lamina VIII and the adjacent lamina VII and on motoneurons of prevertebral muscles. According to Holstege (1991), they constitute a system controlling head position. The tectospinal and lateral bulbospinal tracts and the descending fibres from field H and the periaqueductal grey terminate more laterally in the cervical cord on interneurons and motoneurons of the spinal nucleus of the accessory nerve. They control associated gaze movements. The lateral vestibulospinal tract with reticulospinal fibres from the pontine medial tegmental field terminates in lamina VIII and the adjacent lamina VII and descends to thoracic and lumbar levels of the cord. The pyramidal and rubrospinal tracts descend to lumbosacral levels. Their terminations are specified in Sect. 22.9. The crossed pontospinal tract terminates bilaterally in the lateral cervical nucleus and laminae I, II, V and VI over the entire length of the spinal cord.

Projections from the medullary medial reticular formation, including the raphe nuclei, terminate bilaterally over the entire length of the spinal cord. The region rostral to the facial nucleus, including the nucleus raphes magnus, projects to laminae I, III and IV of the dorsal horn and the caudal portion of the spinal trigeminal nucleus, the intermediate zone and the autonomic cell groups. The region caudal to the facial nucleus, including the nucleus raphes pallidus, projects to the intermediate zone and to the somatic motoneuronal cell groups. Coeruleospinal fibres terminate in all parts of the spinal grey matter, with strong terminations in the sympathetic and parasympathetic cell groups and bilaterally in the phrenic nucleus and in Onuf's nucleus (G. Holstege and Kuypers 1982; J.C. Holstege and Kuypers 1982). Direct projections to the spinal motoneurons also take their origin from the nucleus retroambiguus. They terminate on motoneurons innervating respiratory and cutaneous muscles. In the lumbosacral cord, they innervate Onuf's nucleus and motoneurons responsible for lordosis (Van der Horst and Holstege 1995).

The topography and the termination of the reticulospinal, vestibulospinal and interstitiospinal tracts and the raphe-spinal and coeruleospinal projections in marsupials (Beran and Martin 1971; Crutcher et al. 1978; Martin et al. 1979, 1982a,b; Reddy et al. 1990) and other mammalian species, including humans (Nathan and Smith 1955; Haartsen 1961; Schoen 1964; Verhaart 1970b), are very similar to the cat. The variable descent of the tectospinal tract is discussed in Sect. 22.8.3.2. Most variations in the topography of the spinal tracts concern the pyramidal tract and the consequent shift in position of the rubrospinal tract (see Sect. 22.9). Corticospinal fibres descend bilaterally in the dorsal, dorsolateral and ventral funiculi, but in carnivores and primates most of these fibres occupy the contralateral dorsolateral and the ipsilateral ventral funiculus (see Sect. 22.9.1). In carnivores and primates, the rubrospinal tract is located ventral to the

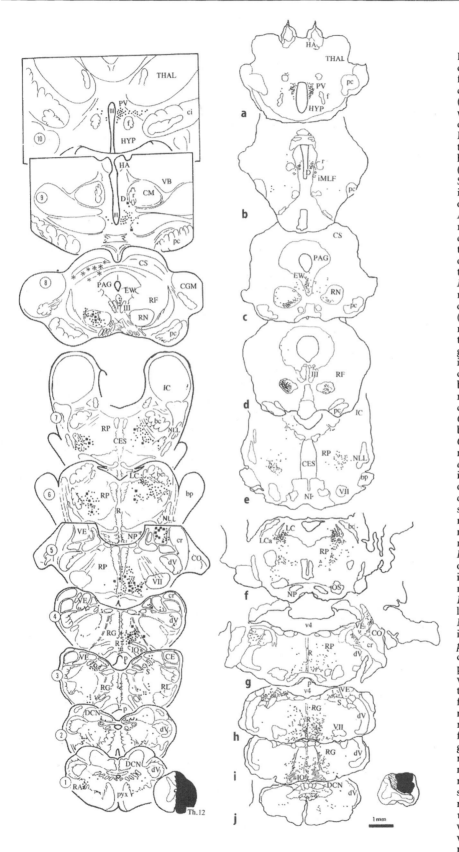

Fig. 22.19. Origin of descending spinal pathways from the brain stem in the cat (*left*) and the opossum (*right*). Diagrams of the cat were relabelled from original drawings by Kuypers and Maisky(1975) from an injection of the retrograde tracer horseradish peroxidase (HRP) in the thoracic cord. Small, labelled neurons are indicated with *dots*, large ones with *filled squares*. *Asterisks* indicate labelled neurons in the superior colliculus after an injection in the cervical cord. Diagrams of the opossum (grey, short-tailed Brazilian opossum *Monodelphis domestica*) were relabelled from original drawings by Holst et al. (1991) of an injection of the retrograde tracer fast blue in the lumbar cord. Retrogradely labelled neurons are indicated as *dots*. *bc*, brachium conjunctivum; *bp*, brachium pontis; *CE*, external cuneate nucleus; *CES*, central superior nucleus; *CGM*, medial geniculate body; *ci*, internal capsule; *CM*, nucleus of the centre médian; *CO*, cochlear nuclei; *cr*, restiform body; *CS*, superior colliculus; *D*, the nucleus of Darkschewitsch; *DCN*, dorsal column nuclei; *dV*, spinal tract of the trigeminal nerve; *EW*, Edinger-Westphal nucleus; *f*, fornix; *HA*, habenulum; *HYP*, hypothalamus; *IC*, inferior colliculus; *III*, oculomotor nucleus; *iMLF*, interstitial nucleus of the medial longitudinal fascicle; *LC*, locus coeruleus; *LCa*, locus coeruleus, pars alpha; *NP*, nuclei pontis; *OI*, oliva inferior; *OS*, superior olive; *p*, pyramid; *PAG*, periaqueductal grey; *pc*, cerebral peduncle; *PV*, nucleus paraventricularis; *pyx*, decussation of the pyramidal tract; *r*, fasciculus retroflexus; *R*, raphe nuclei; *RA*, nucleus retroambiguus; *RF*, reticular formation; *RG*, nucleus gigantocellularis; *RL*, lateral reticular nucleus; *RN*, red nucleus; *RP*, nucleus reticularis pontis; *S*, nucleus of the solitary tract; *t*, fasciculus retroflexus; *THAL*, thalamus; *v4*, fourth ventricle; *VB*, ventrobasal complex; *VE*, vestibular nuclei; *VII*, nucleus of the facial nerve

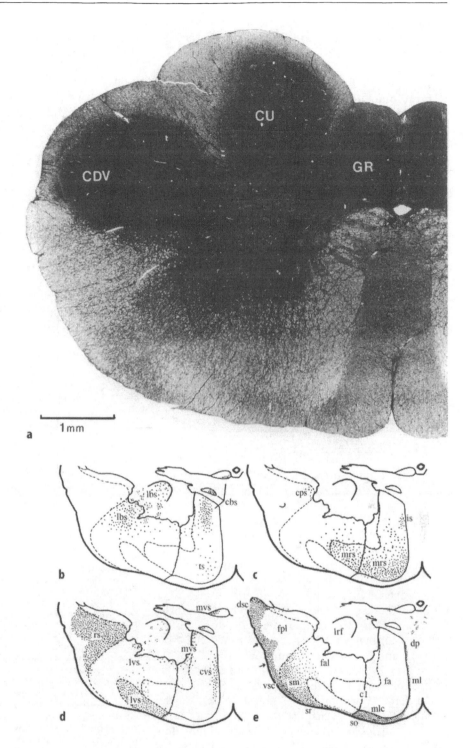

Fig. 22.20. a Transverse Häggqvist-stained section of the brain stem of the cat, taken at the level of the decussation of the pyramidal tract. **b–e** Localisation of ascending and descending tracts at this level. *c1*, first cervical ventral root; *cbs*, crossed bulbospinal tract; *CDV*, nucleus of the ramus spinalis of the trigeminal nerve, pars caudalis; *cps*, crossed pontospinal tract; *CU*, internal cuneate nucleus; *cvs*, crossed vestibulospinal tract; *dp*, decussation of the pyramidal tract; *dsc*, dorsal spinocerebellar tract; *fa*, ventral funiculus; *fal*, anterolateral fasciculus; *fpl*, dorsolateral funiculus; *GR*, gracile nucleus; *is*, interstiospinal tract; *lbs*, lateral bulbospinal tract; *lrf*, lateral reticular fasciculus; *lvs*, lateral vestibulospinal tract; *ml*, medial lemniscus; *mlc*, cervicothalamic tract; *mrs*, medial reticulospinal tract; *mvs*, medial vestibulospinal tract; *rs*, rubrospinal tract; *sm*, spinomedullary fibres; *so*, spino-olivary fibres; *sr*, spinoreticular fibres; *ts*, tectospinal tract; *vsc*, ventral spinocerebellar tract. (From Busch 1961)

pyramidal tract in the dorsolateral funiculus. The rubrospinal tract occupies a dorsal position in the dorsolateral funiculus in species with a corticospinal tract descending in the dorsal funiculus (marsupials, rodents, the tree shrew *Tupaia glis*) or in the ventral funiculus (some insectivores: *Erinaceus and Talpa*, Hyracoidea, Artiodactyla and Perissodactyla) or where it is absent (Cetacea, Lagomorpha). The position in the dorsolateral funiculus of the rubrospinal and the pyramidal tracts is reversed, with the rubrospinal tract located dorsal to the pyramidal tract in Chiroptera, with a high bulbar pyramidal decussation and a crossed pyramidal tract which descends lateral to the inferior olive and ventral to the rubrospinal tract. A similar reversal exists in the living edentates (*Myrmecophaga, Choloepus and manis*), where the fibres of the pyramidal tract enter the lateral funiculus ventrally (Verhaart 1970; see also Sect. 22.9.1). The rubrospinal tract is large in Artiodactyla and Perissodactyla, which lack a spinal pyramidal tract, and small in *Hylobates, Pan, Pongo* and humans, where it is combined with a large spinal pyramid. Such a inverse relationship is not a general rule. In Insectivora, Monotremata and Cetacea, both the rubrospinal and corticospinal tracts are small (Verhaart 1970).

22.3.4.3
Long Ascending Fibre Systems

Long ascending spinal tracts include the primary afferent and the postsynaptic spinal components of the dorsal funiculus, the spinocervical tract, the spinocerebellar tracts, the spinovestibular, spinotectal, spinothalamic and other components of the anterolateral fasciculus and the spino-olivary tract.

The laminar distribution of the primary afferents in the dorsal funiculus was discussed in Sect. 22.3.3. The postsynaptic spinal component of the dorsal funiculus takes its origin mostly from dorsal horn neurons that are concentrated in and around medial laminae IV-VI at cervical levels and from lateral laminae IV-VI at lumbosacral levels of the cord (Fig. 22.21). The primary afferent and postsynaptic components of the dorsal funiculus terminate in different portions of the dorsal column nuclei. Some of the postsynaptic afferents of the dorsal column nuclei do not ascend in the dorsal funiculus, but in the dorsal part of the lateral funiculus. The origin and the course of the postsynaptic dorsal funiculus appear to be very similar in in the monkey (Rustioni 1975,1977; Rustioni and Kaufman 1977; Rustioni et al. 1979; Hayes and Rustioni 1980; Cliffer and Willis 1994), the cat (Uddenberg 1966; Brown and Fyffe 1981; Bennett et al. 1983; Enevoldson and Gordon 1989a) and the rat (Giesler et al. 1984).

The spinocervical tract is the first link in a pathway that connects the dorsal horn with the tectum and the thalamus. In the cat, it takes its origin from the same laminae IV and V of the dorsal horn at all levels of the cord as the postsynaptic dorsal funiculus pathway (Bryan et al. 1973; Craig et al. 1992; Enevoldson and Gordon 1989b). A minor projection to the lateral cervical nucleus arises from lamina I in the cat (Craig et al. 1992) and the raccoon (Pubols and Haring 1995). The spinocervical tract ascends in the dorsal part of the lateral funiculus and terminates on the lateral cervical nucleus. Branching axons of some of the neurons in laminae III–V pass via the dorsolateral and dorsal funiculi to the dorsal column nuclei and the lateral cervical nucleus, respectively (Lu et al. 1988; Lu 1989; Lu and Yang 1989). The lateral cervical nucleus is located lateral to the dorsal horn in the upper cervical segments and gives rise to a crossed ascending spinotectal and spinothalamic pathway (see Sect. 22.8.1.1.2).

Spinocerebellar tracts take their origin from discrete cell groups in laminae IV–IX. Dorsal and ventral spinocerebellar tracts are defined by their position in the lateral funiculus (see Figs. 22.22–22.24) and their course in the brain stem, i.e. dorsal and caudal, or ventral and rostral to the entrance of the trigeminal nerve, respectively. They have been distinguished in most mammalian species (Verhaart 1970b). The definition of the different spinocerebellar tracts was refined by the electrophysiological studies carried out by Oscarsson (1973) in the cat and by the detailed anatomical analysis of the origin, course and termination of the

Fig. 22.21. Transverse sections through the cervical (*top rows*), thoracic (*middle rows*) and lumbar cord (*bottom rows*)-showing the localisation of the neurons giving orgin to the spinothalamic tract, the postsynaptic dorsal column pathway (PSDC), the spinocervical tract and the spinocerebellar tracts. Based on plots of neurons that were retrogradely labelled from their target (**a–d, g–i**) or antidromically activated (**e,f**). **a1–3** Origin of the spinothalamic tract (*filled circles*) and the PSDC (*open circles*) in *Macaca*. (Redrawn from Hayes and Rustioni 1980). **b1–3** Origin of the spinothalamic tract in the cat. (Redrawn from Carstens and Trevino 1978). **c1,2** Origin of the dorsal spinothalamic pathway from lumbar cord in the cat. Neurons were retrogradely labelled from an injection in the contralateral thalamus. At a thoracic level, the contralateral ventrolateral and ventral funiculi were lesioned (**c1**). (Redrawn from Jones et al. 1987). **d1,2** Origin of the ventrolateral spinothalamic pathway in the cat. At a thoracic level, the contralateral dorsolateral funiculus was lesioned (**d1**). (Redrawn from Jones et al. 1987). **e** Origin of the spinocervical tract in the cat. (Redrawn from Bryan et al. 1973). **f** Origin of the spinocervical tract in the *Macaca*. (Redrawn from Bryan et al. 1974). **g1–3** Origin of the PSDC in the cat. (Redrawn from Enevoldson and Gordon 1989). **h1–3** Localisation of chromatolytic neurons after a large lesion of the cerebellum in *Macaca*. (Redrawn from Petras 1977). **i1–3** Origin of the spinocerebellar tracts in the cat. (Redrawn from Matsushita et al. 1979). *CD*, columna dorsalis of Clarke; *SBC*, spinal border cells; *I–IX*, laminae I–IX of Rexed

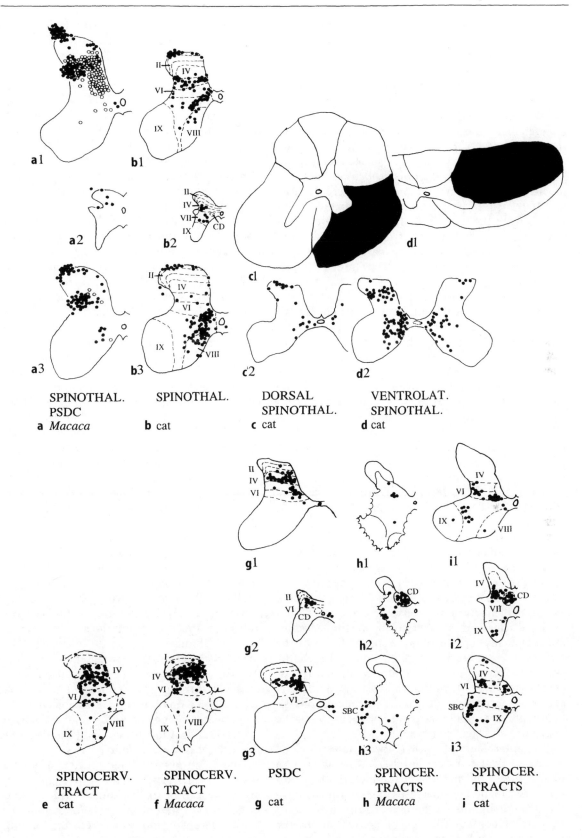

spinocerebellar tracts in the rat and cat by Matsushita (1983; for a review of the literature, see Matsushita et al. 1996) and in primates by Petras (1977), Petras and Cummings (1977), Cummings and Petras (1977) and Snyder et al. (1978). The dorsal spinocerebellar and cuneocerebellar tracts are uncrossed, but the ventral and spinocerebellar and rostral spinocerebellar tracts cross within the spinal cord (Oscarsson 1973). The dorsal spinocerebellar tract arises from Clarke's thoracic nucleus and from neighbouring cells in the dorsal horn and ascends in the superficial part of the posterolateral funiculus. It transmits proprioceptive and exteroceptive signals from the lower limb. Its forelimb equivalent is the cuneocerebellar tract, from the lateral cuneate nucleus and the adjoining part of the medial cuneate nucleus (Gerrits et al. 1985; Grant 1962; Jasmin et al. 1985).

The ventral spinocerebellar tract arises at lumbar levels of the cord, from different cell groups in the substantia intermedia and from the spinal border cells (Fig. 22.21h,i; Matsushita et al. 1979; Petras 1977; Snyder et al. 1978). Its fibres decussate in the ventral commissure and ascend superficially in the ventrolateral funiculus. Certain components of the ventral spinocerebellar tract originate as collaterals from spinal interneurons (Lundberg 1971; Ekerot et al. 1979). This property distinguishes the ventral tract from the dorsal spinocerebellar and cuneocerebellar tracts, which mainly transmit information about events in the periphery. An equivalent, partially crossed pathway arises from the cervical enlargement (Petras and Cummings 1977; Wiksten 1985; Wiksten and Grant 1986). This rostral spinocerebellar tract should be distinguished from the spinocerebellar fibres which arise from the central cervical nucleus located in the substantia intermedia of the upper cervical cord (Cummings and Petras 1977; Matsushita and Okado 1981a,b; Wiksten 1979a–c). This nucleus receives primary input from neck muscles and is under strong vestibular influence. Dorsally located spinocerebellar fibres of the posterior tract reach the cerebellum through the restiform body. The more ventrally located spinocerebellar fibres of the ventral, rostral and central cervical tracts pass rostral to the entrance of the trigeminal nerve and reach the cerebellum via the superior cerebellar peduncle.

The anterolateral fascicle (Gowers 1886), located in the ventral part of the lateral funiculus, is a mixed ascending pathway. Its spinoreticular components can be considered as the continuation of the ground bundle of the lateral funiculus. Additionally, it contains the spinal projections to the lateral reticular nucleus (reviewed by Ruigrok and Cella 1995), spinovestibular connections (reviewed by Rubertone et al. 1995), spino(pre)tectal fibres (see Sects. 22.8.3.2.3 and 22.8.3.3.1) and the dorsal and ventrolateral spinothalamic tracts. The majority of the long ascending fibres of the anterolateral fascicle originate in the contralateral half of the spinal cord and cross in the ventral white commissure. They include an important component that takes its origin from the marginal cells in lamina I (Fig. 22.21c). These fibres are thin and ascend in the dorsal part of the anterolateral fascicle, at its border with the dorsolateral funiculus. They contribute both to the dorsal spinothalamic tract (cat: Fig. 22.21C; Jones et al. 1985, 1987; Craig 1991; monkey: Apkarian and Hodge 1989b; Ralston and Ralston 1992) and to the ascending viscerosensory pathway from lamina I cells at L7-S2, which conveys pelvic and pudendal nerve afferent information from the pelvic viscera and the skin of the perineum and the sex organs to the mesencephalic periaqueductal grey (Van der Horst and Holstege 1995). The coarser fibres of the ventrolateral spinothalamic tract take their origin from laminae IV and V of the dorsal horn and from laminae VII and VIII of the intermediate grey (Fig. 22.21d; Jones et al. 1987).

Spino-olivary fibres arise from all levels of the cord, but mainly from the lumbosacral enlargement (Armstrong and Schild 1979, 1980; Armstrong et al. 1982; Whitworth and Haines 1983; Molinari 1984, 1985). The great majority of the spino-olivary fibres cross in the spinal cord and ascend in the ventral funiculus. They terminate in the accessory olive (Brodal et al. 1950; Blackstad et al. 1951). Most studies have been done in the cat. The cells of origin were located in lamina V of the dorsal horn at segments L5/L6, in the central and medial portion of lamina VII and as a compact cell cluster located in lamina VIII along the ventromedial border of the ventral horn mainly at L7-S1 (Armstrong and Schild 1979, 1980; Armstrong et al. 1982; Molinari 1984). Spino-olivary neurons in the cervical cord are fewer and are mainly located in ventromedial lamina VII and in the central intermediate zone in the caudal cervical enlargement. They include a slightly higher percentage of ipsilaterally located cells (Armstrong and Schild 1980). High cervical segments C1-C4 contribute a sizable, mainly crossed projection (Boesten and.Voogd 1975; Richmond et al. 1982), including a contribution from the lateral cervical nucleus (Molinari 1984).

The spino-olivary tract consists of several subsystems, which differ with respect to their origin and decussation in the spinal cord and their termination in different subnuclei of the inferior olive. The different spino-olivocerebellar climing fibre paths (SOCP) were reviewed by Oscarsson (1973) and Voogd et al. (1996).

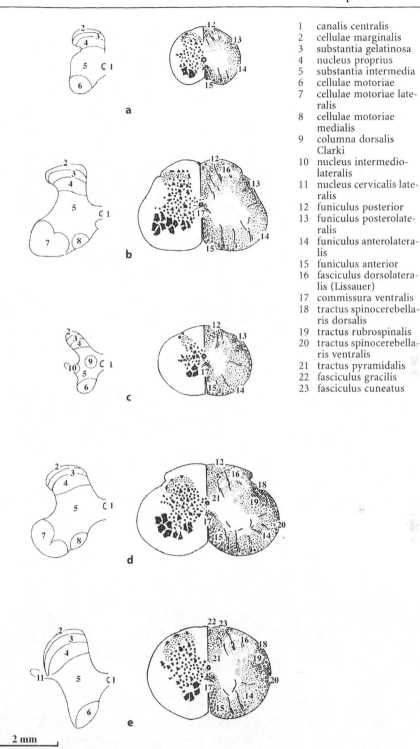

Fig. 22.22. Series of transverse sections through the spinal cord and brain of the North American opossum, *Didelphis virginiana*. The *left half* of each figure shows the cell picture, based on Nissl-stained sections; *the right half* the fibre systems, based on Klüver-Barrera and Häggqvist-stained sections. Nomenclature according to Oswaldo Cruz and Rocha-Miranda (1968).

1 canalis centralis
2 cellulae marginalis
3 substantia gelatinosa
4 nucleus proprius
5 substantia intermedia
6 cellulae motoriae
7 cellulae motoriae lateralis
8 cellulae motoriae medialis
9 columna dorsalis Clarki
10 nucleus intermediolateralis
11 nucleus cervicalis lateralis
12 funiculus posterior
13 funiculus posterolateralis
14 funiculus anterolateralis
15 funiculus anterior
16 fasciculus dorsolateralis (Lissauer)
17 commissura ventralis
18 tractus spinocerebellaris dorsalis
19 tractus rubrospinalis
20 tractus spinocerebellaris ventralis
21 tractus pyramidalis
22 fasciculus gracilis
23 fasciculus cuneatus

1676 CHAPTER 22 Mammals

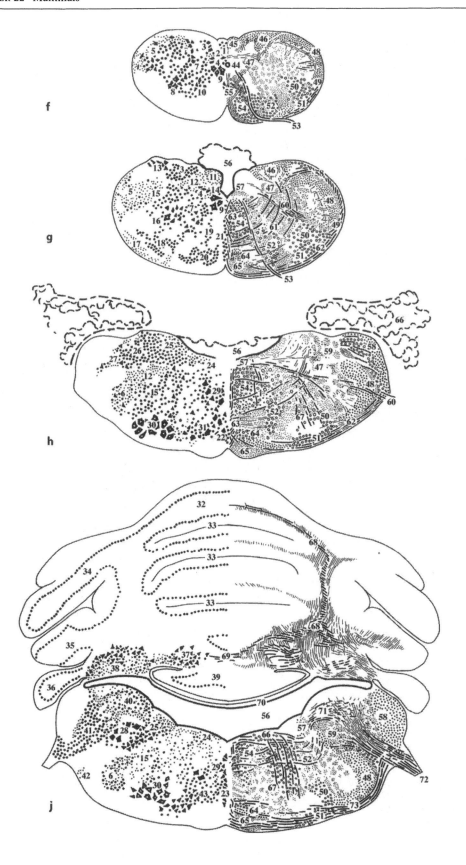

1 nucleus gracilis
2 nucleus commissuralis (of Cajal)
3 nucleus cuneatus medialis
4 nucleus motorius dorsalis vagi
5 nucleus tractus spinalis nervi trigemini, pars caudalis
6 nucleus tractus spinalis nervi trigemini, pars interpolaris
7 nucleus medullae oblongatae centralis
8 nucleus retroambiguus
9 nucleus nervi hypoglossi
10 nucleus supraspinalis
11 area postrema
12 nucleus fasciculi solitarii
13 nucleus cuneatus lateralis
14 nucleus intercalatus
15 nucleus reticularis parvocellularis
16 nucleus ambiguus
17 nucleus reticularis lateralis, pars parvocellularis
18 nucleus reticularis lateralis, pars magnocellularis
19 nucleus reticularis paramedianus
20 oliva inferior
21 nucleus raphes obscurus
22 nucleus raphes pallidus
23 nucleus raphes magnus
24 nucleus prepositus hypoglossi
25 nucleus vestibularis medialis
26 nucleus vestibularis descendens
27 subnucleus × of Brodal and Pompeiano (1957)
28 nucleus vestibularis lateralis
29 nucleus reticularis gigantocellularis
30 nucleus motorius nervi facialis
31 nucleus reticularis gigantocellularis, pars ventralis
32 lobulus simplex
33 fissura prima
34 lobulus ansiformis
35 lobulus paramedianus
36 flocculus
37 nucleus fastigii
38 nucleus interpositus posterior
39 nodulus
40 nucleus cochlearis dorsalis
41 nucleus cochlearis ventralis
42 corpus pontobulare
43 nucleus reticularis gigantocellularis, pars alpha
44 canalis centralis
45 fasciculus gracilis
46 fasciculus cuneatus
47 fasciculus solitarius
48 tractus spinalis nervi trigemini
49 tractus spinocerebellaris dorsalis
50 tractus rubrospinalis
51 tractus spinocerebellaris ventralis
52 tractus vestibulospinalis lateralis
53 nervus hypoglossus
54 fasciculus longitudinalis medialis
55 decussatio pyramidum
56 ventriculus quartus
57 fasciculus longitudinalis dorsalis (Schütz)
58 corpus restiforme
59 corpus juxtarestiforme
60 radix nervi vagus
61 fibrae arcuatae internae
62 fibrae arcuatae externae
63 decussatio fibrae arcuatae internae
64 lemniscus medialis
65 tractus pyramidalis
66 genu nervi facialis
67 radix nervi facialis
68 fibrae semicirculares
69 commissura cerebelli
70 velum medullare posterius
71 striae acusticae dorsales
72 nervus statoacusticus
73 corpus trapezoidum

Fig. 22.22 f-j. Transverse sections through the brain of the North American opossum *Didelphis virginiana*

1678 Chapter 22 Mammals

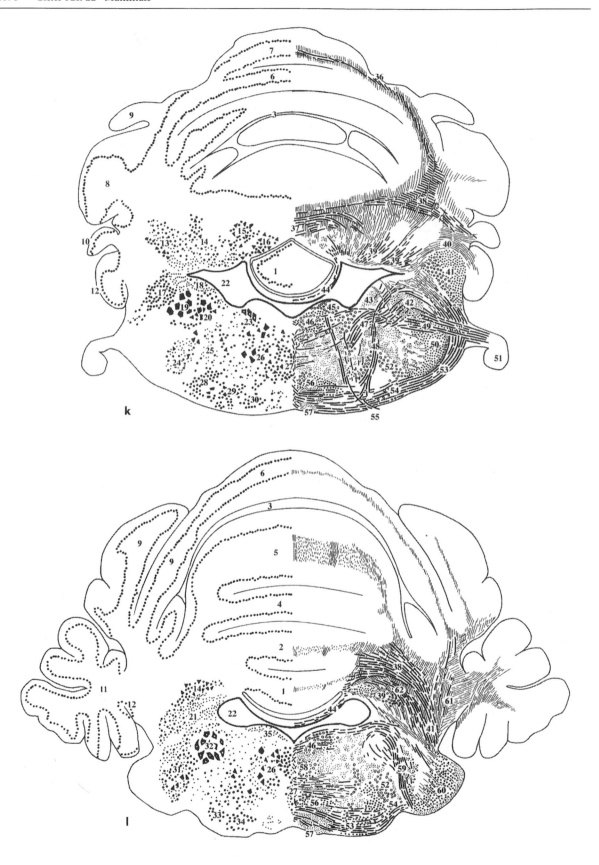

k

l

1 lingula (I)
2 lobulus centralis (II + III)
3 fissura prima
4 culmen, anterior folium (IV)
5 culmen, posterior folium (V)
6 declive (VI)
7 folium/tuber vermis (VII)
8 lobulus ansiformis
9 lobulus simplex
10 paraflocculus
11 paraflocculus (lobulus petrosus)
12 flocculus
13 nucleus cerebellaris lateralis
14 nucleus interpositus anterior
15 nucleus fastigii: protuberantia dorsolateralis
16 nucleus fastigii: pars intermedia
17 nucleus cochlearis ventralis
18 nucleus cochlearis dorsalis
19 nucleus vestibularis lateralis
20 nucleus vestibularis medialis
21 nucleus vestibularis superior
22 ventriculus quartus
23 nucleus nervi abducentis
24 nucleus tractus spinalis nervi trigemini, pars oralis
25 nucleus reticularis parvocellularis
26 nucleus reticularis pontis
27 nucleus raphes magnus
28 nucleus olivaris superior lateralis
29 nucleus olivaris superior medialis
30 nucleus corpus trapezoides
31 nucleus sensibilis principalis nervi trigemini
32 nucleus motorius nervi trigemini
33 nucleus lemnisci lateralis ventralis
34 nuclei periolivaris
35 griseum centrale metencephali
36 area medullaris (interruption of cerebellar cortex)
37 commissura cerebelli
38 fibrae semicirculares
39 brachium conjunctivum
40 pedunculus flocculi
41 corpus restiformis
42 corpus juxtarestiformis
43 fasciculus longitudinalis dorsalis Schütz
44 velum medullare anterius
45 genu nervi facialis
46 fasciculus longitudinalis medialis
47 tractus vestibulospinalis lateralis
48 striae acusticae dorsales
49 nervus vestibularis
50 tractus spinalis nervi trigemini
51 nervus statoacusticus
52 tractus rubrospinalis
53 corpus trapezoideum
54 tractus spinocerebellaris ventralis
55 nervus abducens
56 lemniscus medialis
57 tractus pyramidalis
58 fasciculus predorsalis
59 radix motoria nervi trigemini
60 radix sensibilis nervi trigemini
61 brachium pontis
62 tractus uncinatus

Fig. 22.22 k-l. Transverse sections through the brain of the North American opossum *Didelphis virginiana*

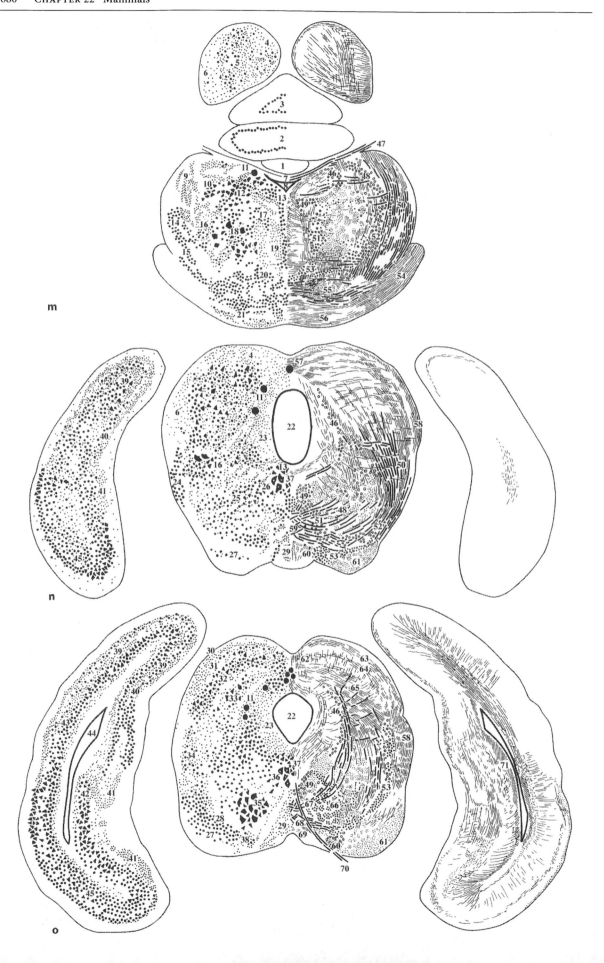

1 lingula (lobulus I)
2 lobulus centralis (lobulus II)
3 lobulus centralis (lobulus III)
4 colliculus inferior, cortex dorsalis
5 colliculus inferior, nucleus centralis
6 colliculus inferior, cortex lateralis (nucleus externus)
7 velum medullare anterior
8 ventriculus quartus
9 nucleus parabrachialis lateralis
10 nucleus parabrachialis medialis
11 nucleus mesencephalicus nervi trigemini
12 locus coeruleus
13 nucleus raphes dorsalis
14 nucleus lemnisci lateralis, subnucleus dorsalis
15 nucleus lemnisci lateralis, subnucleus ventralis
16 nucleus cuneiformis
17 nucleus ventralis tegmenti
18 nucleus reticularis pontis
19 nucleus centralis superior
20 nucleus reticularis tegmenti pontis
21 griseum pontis
22 aqueductus cerebri
23 griseum centralis mesencephali
24 nucleus sagulum
25 nucleus tegmentalis pedunculopontinus
26 nucleus nervi trochlearis
27 substantia nigra, pars reticulata
28 substantia nigra, pars compacta
29 nucleus interpeduncularis
30 colliculus superior, stratum zonale
31 colliculus superior, stratum griseum superficiale
32 colliculus superior, stratum griseum intermedium
33 colliculus superior, stratum griseum profundum
34 nucleus parabigeminalis
35 nucleus ruber, pars magnocellularis
36 nucleus nervi oculomotorii
37 nucleus Edinger Westphall
38 area ventralis tegmenti
39 cortex striatus
40 cortex cingularis
41 subiculum
42 presubiculum
43 cortex prestriatus
44 ventriculus lateralis
45 area piriformis posterior
46 radix mesencephalica nervi trigemini
47 radix nervi trochlearis
48 brachium conjunctivum
49 fasciculus longitudinalis medialis
50 lemniscus lateralis
51 fasciculus predorsalis
52 tractus rubrospinalis
53 lemniscus medialis
54 brachium pontis
55 tractus pyramidalis
56 fibrae transversae pontis
57 commissura colliculi inferioris
58 brachium colliculi inferioris
59 decussatio pedunculi cerebellaris superioris
60 pedunculus mammillaris
61 pedunculus cerebri
62 commissura colliculi superiores
63 colliculus superior, stratum opticum
64 colliculus superior, stratum album intermedium
65 colliculus superior, stratum album profundum
66 brachium conjunctivum, ramus ascendens
67 decussatio tegmenti dorsalis
68 decussatio tegmenti ventralis
69 tractus habenulo-interpeduncularis
70 radix nervi oculomotorii
71 alveus

Fig. 22.22 m-o. Transverse sections through the brain of the North American opossum *Didelphis virginiana*

1682 Chapter 22 Mammals

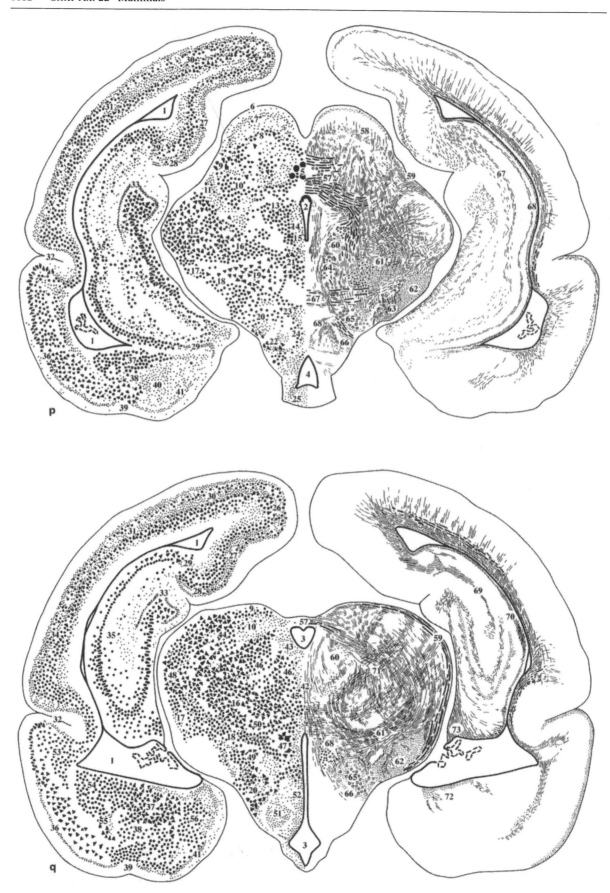

1	ventriculus lateralis	38	nucleus basalis amygdalae (pars parvocellularis)
2	aqueductus cerebri	39	fissura endorhinalis
3	ventriculus tertius	40	area posterior amygdalae
4	recessus infundibularis	41	nucleus corticalis amygdalae
5	nucleus mesencephalicus nervi trigemini	42	massa intermedia
6	colliculus superior, stratum zonale	43	nucleus paraventricularis thalami posterior
7	colliculus superior, stratum griseum superficiale	44	nucleus lateralis dorsalis thalami
8	colliculus superior, stratum griseum intermedium	45	nucleus intermedialis dorsalis thalami
9	nucleus tractus opticus	46	nucleus parafascicularis
10	area pretectalis	47	nucleus paraventricularis hypothalami
11	nucleus suprageniculatus	48	corpus geniculatum laterale (pars dorsalis)
12	corpus geniculatum mediale	49	nucleus subthalamicus
13	nucleus commissurae posterioris	50	nucleus ventrobasalis thalami, subnucleus gelatinosus
14	griseum centralis mesencephali	51	nucleus ventromedialis hypothalami
15	nucleus Darkschewitsch	52	nucleus periventricularis hypothalami
16	nucleus ventrobasalis thalami	53	nucleus endopiriformis
17	corpus geniculatum laterale, pars ventralis	54	nucleus lateralis amygdalae
18	zona incerta	55	nucleus intercallatus amygdalae
19	nucleus campi Foreli	56	nucleus basalis accessorius amygdalae
20	area hypothalamicus lateralis	57	commissura posterior
21	nucleus interstitialis fasciculi longitudinalis medialis	58	colliculus superior, stratum opticum
22	nucleus premammillaris dorsalis	59	tractus opticus
23	nucleus supramammalillaris	60	fasciculus retroflexus
24	nucleus mammillaris, pars anterior	61	lamina medullaris externa thalami
25	nucleus infundibularis (arcuata)	62	pedunculus cerebri
26	cortex cingularis	63	fibrae striatonigralis
27	cortex retrosplenialis	64	fasciculus longitudinalis medialis
28	presubiculum	65	fasciculus medialis telencephali
29	subiculum	66	columna fornicis
30	cortex striatus	67	decussatio supramammillaris
31	cortex parastriatus	68	tractus mammillothalamicus
32	fissura rhinalis	69	cingulum ammoniale
33	fascia dentata	70	alveus
34	cortex temporalis	71	lamina medullaris interna thalami
35	cornu ammonis	72	stria terminalis
36	cortex piriformis	73	fimbriae hippocampi
37	nucleus basalis amygdalae (pars magnocellularis)		

Fig. 22.22 p–q. Transverse sections through the brain of the North American opossum *Didelphis virginiana*

1684   Chapter 22   Mammals

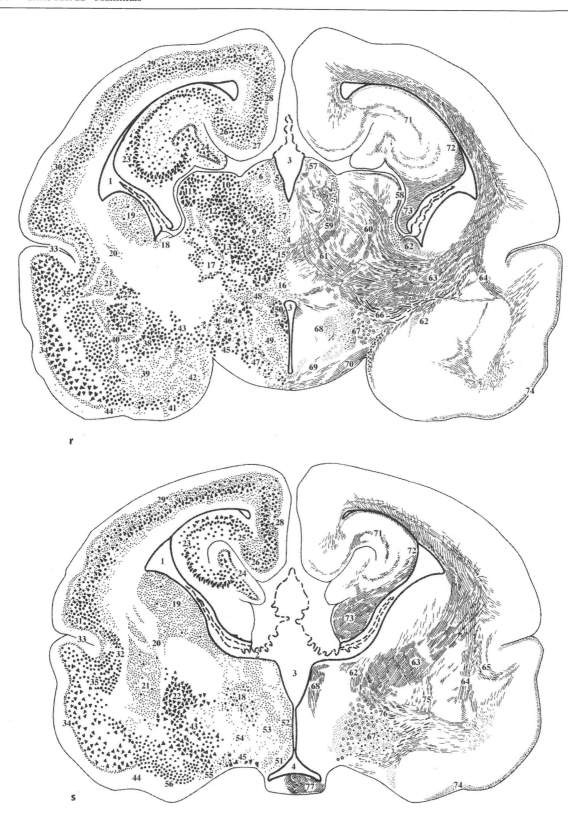

1 ventriculus lateralis
2 foramen interventricularis
3 ventriculus tertius
4 massa intermedia
5 nucleus medialis habenulae
6 nucleus lateralis habenulae
7 nucleus paraventricularis thalami anterior
8 nucleus medialis dorsalis thalami
9 nucleus centralis lateralis thalami
10 nucleus anterodorsalis rhalami
11 corpus geniculatum laterale, pars dorsalis
12 nucleus lateralis thalami
13 nucleus ventralis anterior thalami
14 nucleus anteromedialis thalami
15 nucleus centralis medialis thalami
16 nucleus reuniens
17 nucleus reticularis thalami
18 nucleus interstitialis stria terminalis
19 nucleus caudatus
20 striatum, ponticuli grisei
21 putamen
22 globus pallidus
23 cornu ammonis
24 fascia dentata
25 subiculum
26 presubiculum
27 cortex retrosplenialis
28 cortex cingularis
29 cortex parietalis
30 cortex temporalis
31 cortex insularis
32 claustrum
33 fissura rhinalis
34 cortex piriformis
35 nucleus endopiriformis
36 nucleus lateralis amygdalae
37 nucleus centralis amygdalae
38 nucleus basalis amygdalae, pars magnocellularis
39 nucleus basalis amygdalae, pars parvocellularis
40 nucleus intercalatus amygdalae
41 nucleus corticalis amygdalae, pars anterior
42 nucleus medialis amygdalae
43 substantia innominata
44 fissura amygdalae
45 nucleus supraopticus
46 area hypothalamica lateralis
47 area hypothalamica anterior
48 nucleus hypothalamicus dorsalis
49 nucleus hypothalamicus anterior, pars dorsalis
50 nucleus paraventricularis hypothalami
51 nucleus suprachiasmaticus
52 nucleus preopticus periventricularis
53 area preoptica medialis
54 area preoptica lateralis
55 nucleus fasciculi diagonalis Brocae
56 nucleus tractus olfactorii lateralis
57 stria medullaris
58 radiatio optica
59 lamina medullaris interna thalami
60 lamina medullaris externa thalami
61 radiatio thalami inferior
62 stria terminalis
63 capsula interna
64 capsula externa
65 capsula externa
66 ansa lenticularis
67 fasciculus medialis telencephali
68 columna fornicis
69 decussatio supraoptica
70 tractus opticus
71 cingulum ammonale
72 alveus
73 fimbria hippocampi
74 tractus olfactorius lateralis
75 globus pallidus, lamina medullaris externa
76 ansa peduncularis
77 chiasma opticum

Fig. 22.22 r-s. Transverse sections through the brain of the North American opossum *Didelphis virginiana*

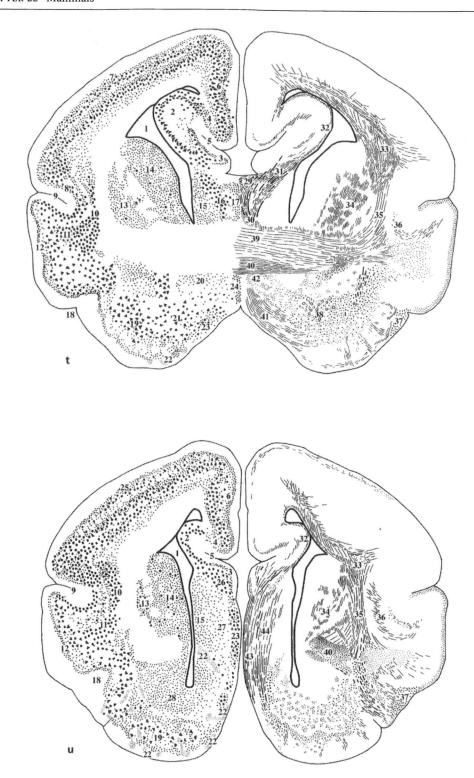

1 ventriculus lateralis	23 nucleus fasciculi diagonalis Brocae, pars ventralis
2 cornu ammonis	24 nucleus preopticus medianus
3 fascia dentata	25 cortex preorbitalis
4 subiculum	26 nucleus dorsalis septi
5 fissura hippocampi	27 nucleus medialis septi
6 cortex cingularis anterior	28 nucleus accumbens septi
7 cortex parietalis	29 commissura hippocampi (psalterium)
8 cortex insularis	30 columna fornicis
9 fissura rhinalis	31 fimmbria hippocampi
10 claustrum	32 alveus
11 nucleus endopiriformis	33 centrum semiovale
12 cortex piriformis	34 capsula interna
13 putamen	35 capsula externa
14 nucleus caudatus	36 capsula extrema
15 nucleus lateralis septi	37 tractus olfactorius lateralis
16 nucleus fimbrialis septi	38 fasciculus medialis telencephali
17 nucleus fasciculi diagonalis Brocae, pars dorsalis	39 commissura anterior, pars neocorticalis
18 fissura endorhinalis	40 commissura anterior, pars olfactoria
19 tuberculum olfactorium	41 fasciculus diagonalis Brocae
20 nucleus interstitialis striae terminalis	42 striae terminalis
21 area preoptica lateralis, pars magnocellularis	43 tractus olfactorius septi
22 insulae Callejae	

Fig. 22.22 t–u. Transverse sections through the brain of the North American opossum *Didelphis virginiana*

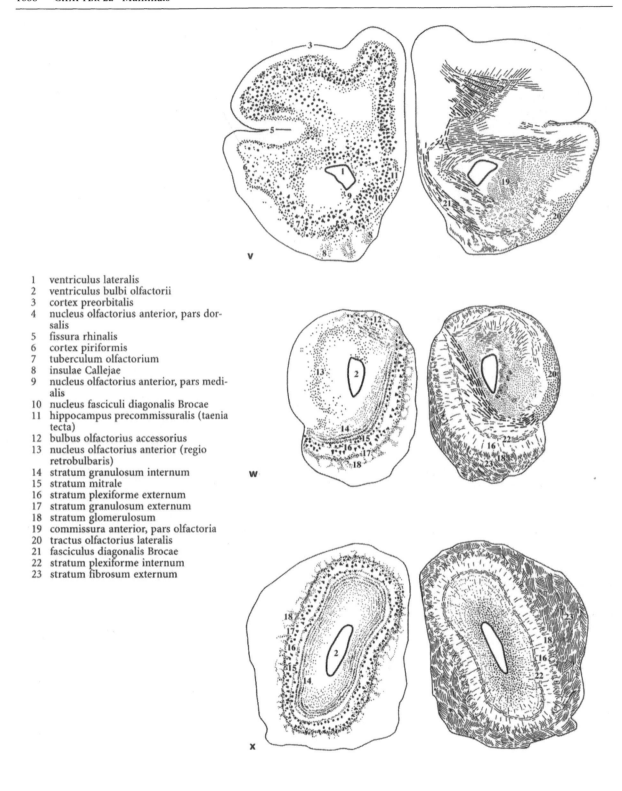

1 ventriculus lateralis
2 ventriculus bulbi olfactorii
3 cortex preorbitalis
4 nucleus olfactorius anterior, pars dorsalis
5 fissura rhinalis
6 cortex piriformis
7 tuberculum olfactorium
8 insulae Callejae
9 nucleus olfactorius anterior, pars medialis
10 nucleus fasciculi diagonalis Brocae
11 hippocampus precommissuralis (taenia tecta)
12 bulbus olfactorius accessorius
13 nucleus olfactorius anterior (regio retrobulbaris)
14 stratum granulosum internum
15 stratum mitrale
16 stratum plexiforme externum
17 stratum granulosum externum
18 stratum glomerulosum
19 commissura anterior, pars olfactoria
20 tractus olfactorius lateralis
21 fasciculus diagonalis Brocae
22 stratum plexiforme internum
23 stratum fibrosum externum

Fig. 22.22 v-x. Transverse sections through the brain of the North American opossum *Didelphis virginiana*

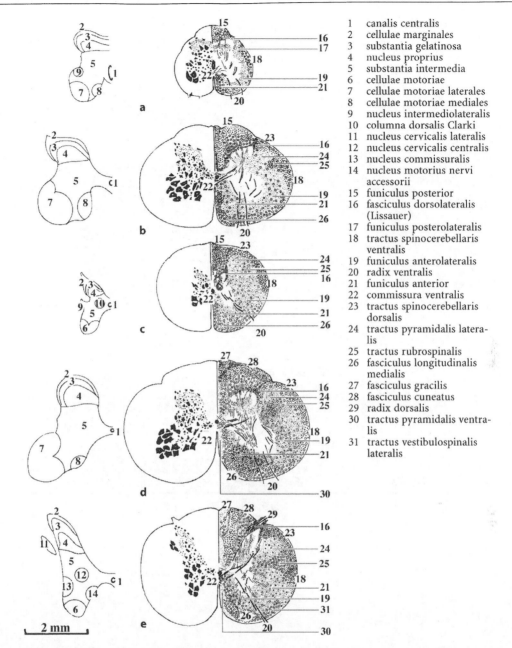

Fig. 22.23 Series of transverse sections through the spinal cord and brain of the cat. The *left half* of each figure shows the cell picture, based on Nissl-stained sections; the *right half* the fibre systems, based on Klüver-Barrera and Häggqvist-stained sections. Nomenclature of the amygdaloid complex according to Krettek and Price, (1975) and of the thalamus according to Jones (1985).

1 canalis centralis
2 cellulae marginales
3 substantia gelatinosa
4 nucleus proprius
5 substantia intermedia
6 cellulae motoriae
7 cellulae motoriae laterales
8 cellulae motoriae mediales
9 nucleus intermediolateralis
10 columna dorsalis Clarki
11 nucleus cervicalis lateralis
12 nucleus cervicalis centralis
13 nucleus commissuralis
14 nucleus motorius nervi accessorii
15 funiculus posterior
16 fasciculus dorsolateralis (Lissauer)
17 funiculus posterolateralis
18 tractus spinocerebellaris ventralis
19 funiculus anterolateralis
20 radix ventralis
21 funiculus anterior
22 commissura ventralis
23 tractus spinocerebellaris dorsalis
24 tractus pyramidalis lateralis
25 tractus rubrospinalis
26 fasciculus longitudinalis medialis
27 fasciculus gracilis
28 fasciculus cuneatus
29 radix dorsalis
30 tractus pyramidalis ventralis
31 tractus vestibulospinalis lateralis

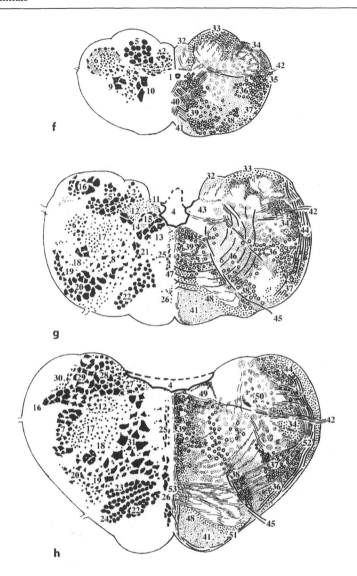

1 canalis centralis
2 nucleus gracilis
3 nucleus commissuralis
4 ventriculus quartus
5 nucleus cuneatus medialis
6 nucleus tractus spinalis nervi trigemini, pars caudalis
7 nucleus tractus spinalis nervi trigemini, pars interpolaris
8 nucleus medullae oblongatae centralis
9 nucleus retroambiguus
10 nucleus supraspinalis
11 area postrema
12 nucleus fasciculi solitarii
13 nucleus nervi hypoglossi
14 nucleus intercalatus
15 nucleus motorius dorsalis vagi
16 nucleus cuneatus lateralis
17 nucleus reticularis parvocellularis
18 nucleus ambiguus
19 nucleus reticularis lateralis, pars parvocellularis
20 nucleus reticularis lateralis, pars magnocellularis
21 nucleus reticularis paramedianus
22 oliva inferior, nucleus accessorius medialis
23 oliva inferior, nucleus accessorius dorsalis
24 oliva inferior, nucleus principalis
25 nucleus raphes obscurus
26 nucleus raphes pallidus
27 nucleus prepositus hypoglossi
28 nucleus vestibularis medialis
29 nucleus vestibularis descendens
30 subnucleus × of Brodal and Pompeiano (1957)
31 nucleus reticularis gigantocellularis
32 fasciculus gracilis
33 fasciculus cuneatus
34 tractus spinalis nervi trigemini
35 tractus spinocerebellaris dorsalis
36 tractus rubrospinalis
37 tractus spinocerebellaris ventralis
38 tractus vestibulospinalis lateralis
39 fasciculus longitudinalis medialis
40 decussatio pyramidum
41 tractus pyramidalis
42 radix nervi vagus
43 fasciculus solitarius
44 corpus restiforme
45 radix nervi hypoglossi
46 fibrae arcuatae internae
47 decussatio fibrae arcuatae internae
48 lemniscus medialis
49 fasciculus longitudinalis dorsalis (Schütz)
50 corpus juxtarestiforme
51 fibrae arcuatae externae
52 fibrae olivocerebellares
53 decussatio olivocerebellaris

Fig. 22.23 f–h. Transverse sections through the brain of the cat

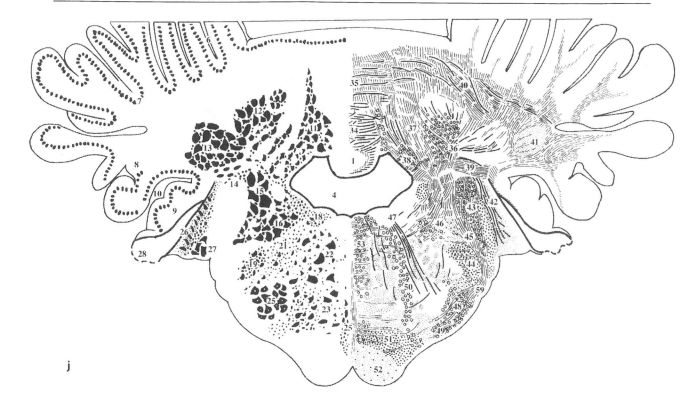

j

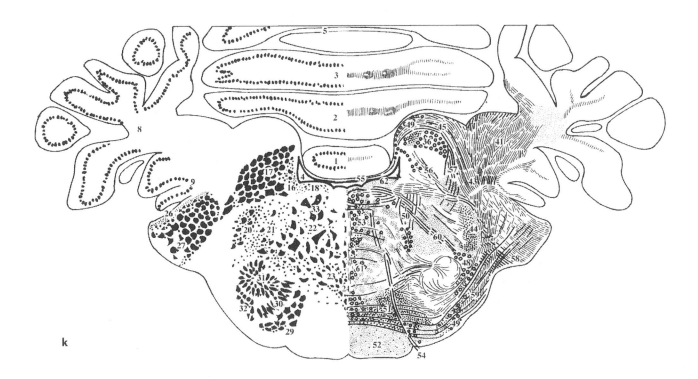

k

1 lingula (lobulus I)
2 lobulus centralis, ventral folium (lobulus II)
3 lobulus centralis, dorsal folium (lobulus III)
4 ventriculus quartus
5 culmen (lobulus IV–V)
6 lobulus simplex
7 lobulus ansiformis
8 paraflocculus
9 flocculus
10 fissura posterolateralis
11 nucleus fastigii
12 nucleus interpositus anterior
13 nucleus lateralis cerebelli
14 group Y
15 nucleus vestibularis lateralis
16 nucleus vestibularis medialis
17 nucleus vestibularis superior
18 nucleus prepositus hypoglossi
19 nucleus spinalis nervi trigemini, pars interpolaris
20 nucleus spinalis nervi trigemini, pars oralis
21 nucleus reticularis parvocellularis
22 nucleus reticularis gigantocellularis
23 nucleus gigantocellularis, pars ventralis
24 nucleus raphes magnus
25 nucleus motorius nervi facialis
26 nucleus cochlearis dorsalis
27 nucleus cochlearis ventralis
28 recessus lateralis ventriculi quarti
29 nucleus corpus trapezoideus
30 oliva superior medialis
31 oliva superior lateralis
32 nucleus periolivaris
33 nucleus abducentis
34 decussatio tractus uncinati
35 commissura cerebelli
36 brachium conjunctivum
37 fibrae perforantes
38 tractus fastigiobulbaris directus
39 pedunculus flocculi
40 fibrae semicirculares
41 brachium pontis
42 striae acusticae dorsales
43 corpus restiformis
44 ramus spinalis nervi trigemini
45 fibrae olivocerebellaris
46 corpus juxta restiformis
47 genu nervi facialis
48 tractus rubrospinalis
49 tractus spinocerebellaris ventralis
50 tractus vestibulospinalis lateralis
51 lemniscus medialis
52 tractus pyramidalis
53 fasciculus longitudinalis medialis
54 radix motorius nervi abducens
55 velum medullare anterius
56 nervus vestibularis, ramus ascendens
57 tractus uncinatus
58 nervus statoacusticus
59 corpus trapezoides
60 radix nervi facialis
61 fasciculus predorsales
62 fasciculus dorsalis Schütz

Fig. 22.23 j–k. Transverse sections through the brain of the cat

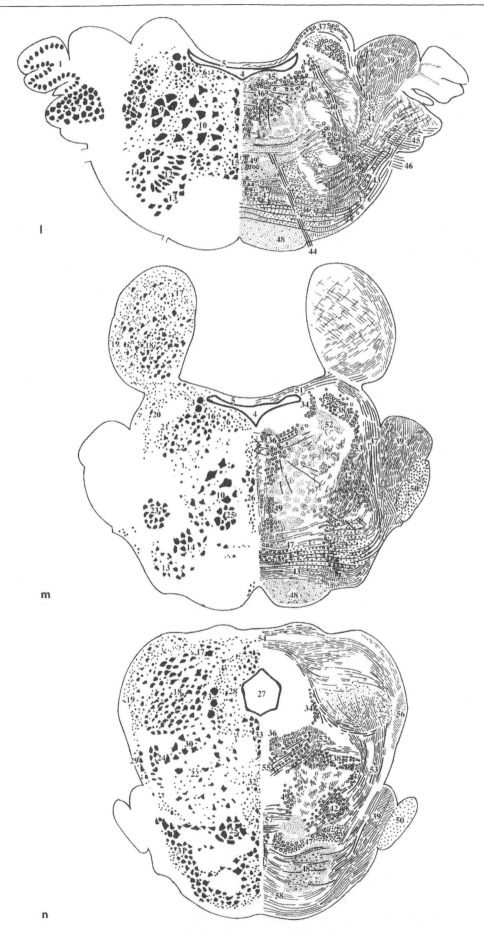

1	flocculus	30	nucleus cuneiformis
2	nuclei parabrachialis	31	nuclei pontis
3	nucleus tractus mesencephalicus nervi trigemini	32	nucleus centralis superior
4	ventriculus quartus	33	nucleus raphes dorsalis
5	velum medullare anterius	34	tractus mesencephalicus nervi trigemini
6	griseum centrale metencephali	35	fasciculus longitudinalis dorsalis Schütz
7	nucleus cochlearis ventralis	36	fasciculus longitudinalis medialis
8	nucleus sensibilis principalis nervi trigemini	37	tractus spinocerebellaris ventralis
9	nucleus motorius nervi trigemini	38	brachium conjunctivum
10	nucleus reticularis pontis	39	brachium pontis
11	nucleus olivaris superior lateralis	40	radix motoria nervi trigemini
12	nucleus olivaris superior medialis	41	radix sensibilis nervi trigemini
13	nucleus corpus trapezoides	42	tractus rubrospinalis
14	nuclei periolivaris	43	corpus trapezoideum
15	nucleus raphes magnus	44	nervus abducens
16	locus coeruleus	45	nervus statoacusticus
17	colliculus inferior: cortex dorsalis	46	radix nervi facialis
18	colliculus inferior: nucleus centralis	47	lemniscus medialis
19	colliculus inferior: cortex lateralis (nucleus externus)	48	tractus pyramidalis
20	nucleus parabrachialis lateralis	49	fasciculus predorsalis
21	nucleus parabrachialis medialis	50	radix nervi trigemini
22	nucleus tegmentalis pedunculopontinus	51	decussatio nervus trochlearis
23	nucleus lemnisci lateralis ventralis	52	tractus trigeminothalamicus dorsalis Wallenberg
24	nucleus lemnnisci lateralis dorsalis	53	lemniscus lateralis
25	nucleus reticularis tegmenti pontis	54	commissura colliculi inferioris
26	nucleus raphes pontis	55	decussatio brachium conjunctivum
27	aquaducus cerebri	56	brachium colliculi inferioris
28	griseum centrale mesencephali	57	decussatio lemnisci trigeminalus
29	nucleus sagulum	58	fibrae transversae pontis

Fig. 22.23 l-n. Transverse sections through the brain of the cat

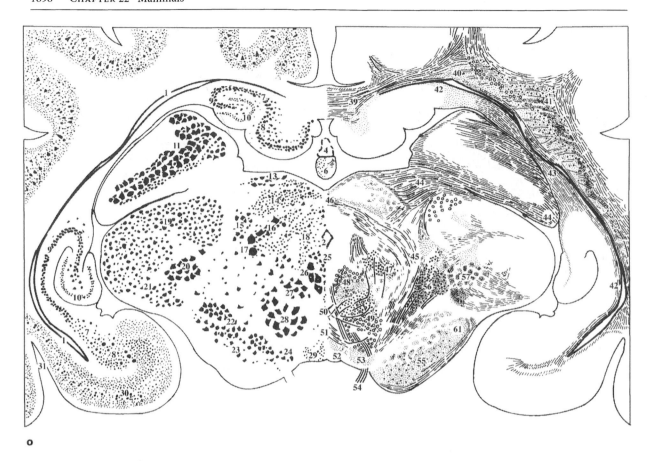

o

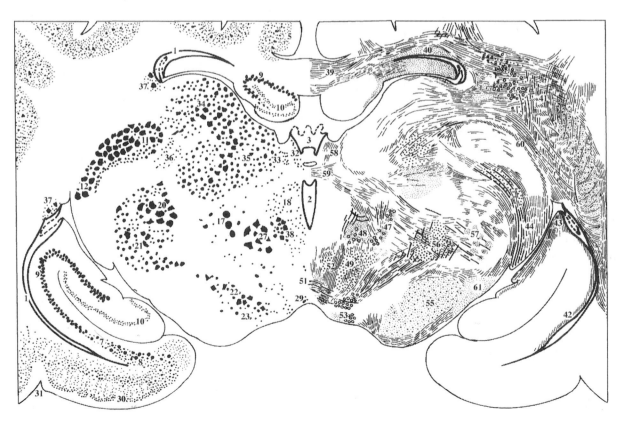

p

1 ventriculus lateralis
2 aqueductus cerebri
3 ventriculus tertius
4 recessus pinealis
5 cortex cingularis
6 corpus pinealis
7 subiculum
8 presubiculum
9 cornu ammonis
10 fascia dentata
11 corpus geniculatum laterale, pars dorsalis
12 corpus geniculatum laterale, pars ventralis
13 nucleus tractus opticus
14 nucleus pretectalis posterior
15 nucleus pretectalis anterior
16 colliculus superior, stratum griseum profundum
17 nucleus mesencephalicus nervi trigemini
18 griseum centralis mesencephali
19 corpus geniculatum mediale, nucleus dorsalis
20 corpus geniculatum mediale, nucleus magnocellularis
21 corpus geniculatum mediale, nucleus ventralis
22 substantia nigra, pars compacta
23 substantia nigra, pars reticulata
24 area ventralis tegmenti
25 nucleus Edinger Westphall
26 nucleus nervi oculomotorii
27 nucleus interstitialis fasciculi longitudinalis medialis
28 nucleus ruber, pars magnocellularis
29 nucleus interpeduncularis
30 cortex entorhinalis
31 fissura rhinalis
32 nucleus habenularis medialis
33 nucleus habenularis lateralis
34 pulvinar
35 nucleus lateralis posterior thalami
36 nucleus posterior thalami
37 cauda nuclei caudati
38 nucleus Darkschewitsch
39 corpus callosum
40 fasciculus subcallosus
41 centrum semiovale
42 alveus
43 fimbria hippocampi
44 tractus opticus
45 colliculus superior, stratum album profundum
46 commissura colliculi superiores
47 fasciculi Foreli
48 fasciculus longitudinalis medialis
49 brachium conjunctivum, ramus ascendens
50 decussatio tegmenti dorsalis
51 decussatio tegmenti ventralis
52 tractus habenulo-interpeduncularis
53 pedunculus mammillaris
54 radix nervi oculomotorii
55 pedunculus cerebri: tractus pyramidalis
56 lemniscus medialis
57 brachium colliculi inferioris
58 stria medullaris
59 commissura posterior
60 radiato optica
61 pedunculus cerebri: tractus corticopontinus parietotemporalis

Fig. 22.23 o-p. Transverse sections through the brain of the cat

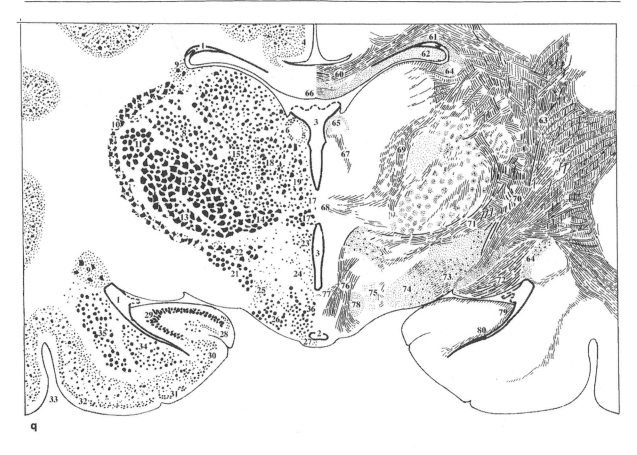
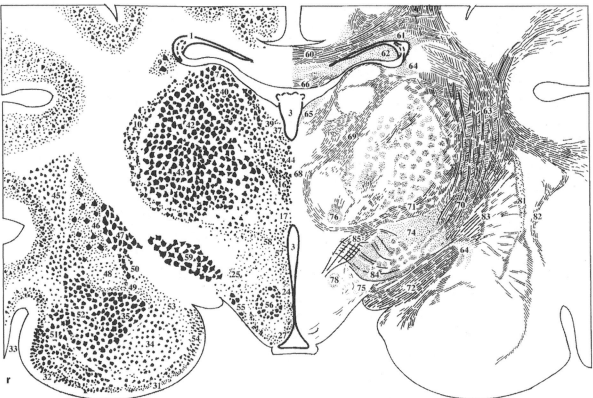

1 ventriculus lateralis
2 recessus infundibularis
3 ventriculus tertius
4 cortex cingularis
5 nucleus habenulae medialis
6 nucleus habenulae lateralis
7 nucleus lateralis dorsalis thalami
8 nucleus lateralis posterior thalami
9 cauda nuclei caudati
10 nucleus reticularis thalami
11 corpus geniculatum laterale (pars dorsalis)
12 nucleus ventralis posterior medialis
13 nucleus ventralis posterior lateralis
14 nucleus ventralis posterior medialis thalami, pars gelatinosa
15 nucleus centralis lateralis thalami
16 nucleus centromedianus thalami
17 nucleus reuniens
18 nucleus medialis dorsalis thalami
19 nucleus parafascicularis
21 nucleus subthalamicus
22 zona incerta
23 area hypothalamica dorsocaudalis
24 area/nucleus hypothalamica periventricularis
25 area hypothalamicus lateralis
26 nucleus mammillaris lateralis
27 nucleus arcuatus
28 fascia dentata
29 cornu ammonis
30 subiculum
31 nucleus amygdalae corticalis
32 cortex piriformis
33 fissura rhinalis
34 nucleus amygdalae basomedialis
35 nucleus amygdalae basolateralis
36 nucleus mammillaris medialis
37 nucleus paraventricularis anterior
38 nucleus anterodorsalis
39 nucleus parataenialis
40 nucleus anteroventralis
41 nucleus anteromedialis
42 nucleus ventralis anterior
43 nucleus ventralis lateralis
44 nucleus rhomboideus
45 claustrum
46 putamen
47 globus pallidus
48 nucleus centralis amygdalae
49 nucleus amygdalae intercallatus
50 nucleus stria terminalis
51 nucleus endopiriformis
52 nucleus amygdalae lateralis
53 nucleus amygdalae medialis
54 area amygdalo-hippocampalis
55 nucleus supraopticus
56 nucleus hypothalamicus ventromedialis
57 nucleus hypothalamicus paraventricularis
58 area hypothalamica dorsalis
59 nucleus entopeduncularis
60 corpus callosum
61 fasciculus subcallosus
62 corpus fornicis
63 centrum semiovale
64 stria terminalis
65 stria medullaris
66 commissura hippocampi
67 fasciculus habenulointerpeduncularis
68 massa intermedia
69 lamina medullaris interna thalami
70 capsula interna
71 lamina medullaris externa thalami
72 tractus opticus
73 fibrae striatonigralis
74 pedunculus cerebri
75 fasciculus medialis telencephali
76 tractus mammillothalamicus
77 decussatio supramammillaris
78 columna fornicis
79 fimbriae fornicis
80 alveus
81 capsula externa
82 capsula extrema
83 lamina medullaris pallidi
84 ansa lenticularis
85 fasciculus lenticularis

Fig. 22.23 q-r. Transverse sections through the brain of the cat

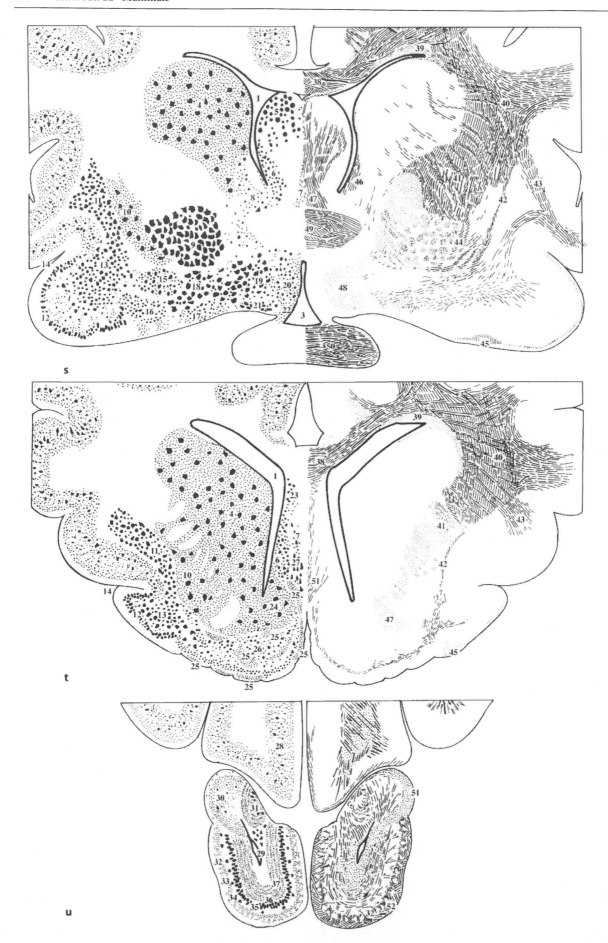

1	ventriculus lateralis, cornu anterius	27	nucleus fasciculi diagonalis Brocae, pars dorsalis
2	cortex cingularis anterior	28	cortex frontalis
3	ventriculus tertius	29	ventriculus bulbi olfactorii
4	caput nuclei caudati	30	nucleus olfactorius anterior
5	nucleus fimbrialis septi	31	bulbus olfactorius accessorius
6	nucleus lateralis septi	32	stratum glomerulosum
7	nucleus medialis septi	33	stratum granulosum externum
8	nucleus interstitialis striae terminalis	34	stratum plexiforme externum
9	globus pallidus	35	stratum mitrale
10	putamen	36	stratum plexiforme internum
11	claustrum	37	stratum granulosum internum
12	cortex piriformis	38	corpus callosum
13	nucleus endopiriformis	39	fasciculus subcallosus
14	fissura rhinalis	40	centrum semiovale
15	nucleus amygdalae intercallatus	41	capsula interna
16	area amygdalae anterior	42	capsula externa
17	nucleus tractus olfactorius lateralis	43	capsula extrema
18	substantia innominata	44	laminamedullaris externa pallidi
19	area preoptica lateralis, pars magnocellularis	45	tractus olfactorius lateralis
20	nucleus preopticus medianus	46	stria terminalis
21	nucleus supraopticus	47	columna fornicis
22	nucleus suprachiasmaticus	48	fasciculus medialis telencephali
23	nucleus dorsalis septi	49	commissura anterior, pars olfactoria
24	nucleus accumbens septi	50	chiasma opticum
25	insulae Callejae	51	fasciculus diagonalis Brocae
26	tuberculum olfactorium	52	stratum fibrosum externum

Fig. 22.23 s-u. Transverse sections through the brain ot the cat

1 canalis centralis
2 cellulae marginales
3 substantia gelatinosa
4 nucleus proprius
5 substantia intermedia
6 cellulae motoriae
7 cellulae motoriae laterales
8 cellulae motoriae mediales
9 nucleus intermediolateralis
10 columna dorsalis Clarki
11 nucleus cervicalis lateralis
12 nucleus cervicalis centralis
13 funiculus posterior
14 radix dorsalis
15 fasciculus dorsolateralis (Lissauer)
16 funiculus posterolateralis
17 funiculus anterolateralis
18 funiculus anterior
19 radix ventralis
20 commissura ventralis
21 tractus spinocerebellaris dorsalis
22 tractus pyramidalis lateralis
23 tractus rubrospinalis
24 tractus spinocerebellaris ventralis
25 fasciculus gracilis
26 fasciculus cuneatus
27 fasciculus longitudinalis medialis
28 tractus pyramidalis ventralis

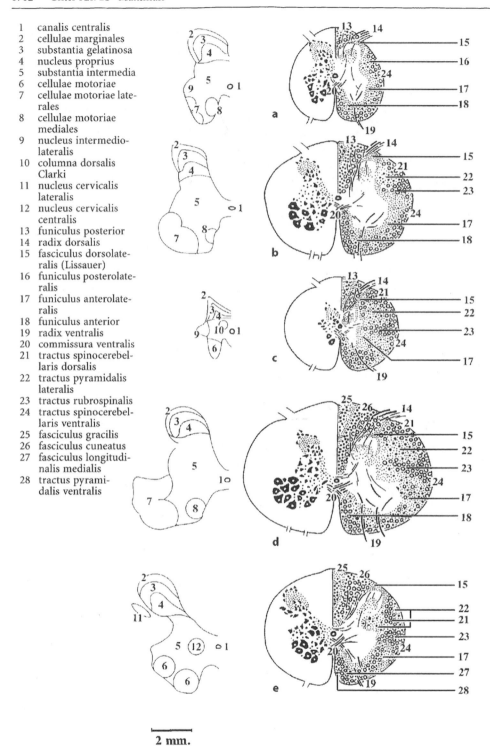

Fig. 22.24 Series of transverse sections through the spinal cord and brain of the rhesus monkey. The *left half* of each figure shows the cell picture, based on Nissl-stained sections; the *right half* the fibre systems, based on Klüver-Barrera and Häggqvist-stained sections. Nomenclature of the amygdaloid complex according to Price et al. (1987) and of the thalamus according to Jones (1985).

1 canalis centralis
2 nucleus gracilis
3 nucleus cuneatus medialis
4 ventriculus quartus
5 nucleus commissuralis
6 nucleus tractus spinalis nervi trigemini, pars caudalis
7 nucleus tractus spinalis nervi trigemini, pars interpolaris
8 nucleus medullae oblongatae centralis
9 nucleus supraspinalis
10 nucleus reticularis lateralis
11 nucleus ambiguus
12 nucleus fasciculi solitarii
13 nucleus cuneatus lateralis
14 nucleus motorius dorsalis vagi
15 nucleus intercalatus
16 nucleus nervi hypoglossi
17 nucleus reticularis parvocellularis
18 nucleus reticularis paramedianus
19 oliva inferior, nucleus accessorius medialis

30 nucleus vestibularis descendens
31 nucleus vestibularis medialis
32 nucleus prepositus hypoglossi
33 nucleus raphes pallidus
34 nucleus reticularis gigantocellularis
35 nucleus reticularis lateralis, pars subtrigeminalis
36 fasciculus gracilis
37 fasciculus cuneatus
38 tractus spinalis nervi trigemini
39 tractus spinocerebellaris dorsalis
40 tractus rubrospinalis
41 tractus spinocerebellaris ventralis
42 tractus vestibulospinalis lateralis
43 fasciculus longitudinalis medialis
44 decussatio pyramidum
45 fibrae arcuatae internae
46 radix nervi vagus
47 fasciculus solitarius
48 corpus restiforme

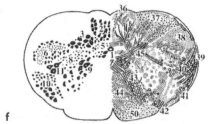
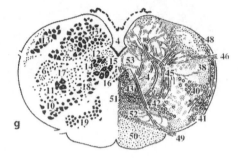
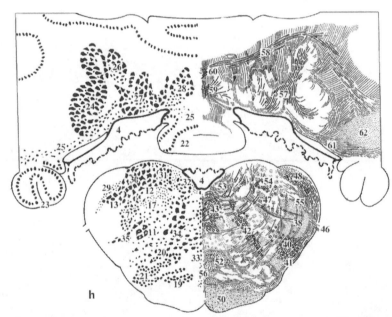

20 oliva inferior, nucleus accessorius dorsalis
21 oliva inferior, nucleus principalis
22 nodulus
23 flocculus
24 nucleus dentatus
25 nucleus basalis interstitialis (Langer)
26 nucleus interpositus anterior
27 nucleus interpositus posterior
28 nucleus fastigii
29 subnucleus × of Brodal and Pompeiano (1957)

49 radix nervi hypoglossi
50 tractus pyramidalis
51 decussatio fibrae arcuatae internae
52 lemniscus medialis
53 fasciculus longitudinalis dorsalis (Schütz)
54 corpus juxtarestiforme
55 fibrae olivocerebellares
56 decussatio olivocerebellaris
57 brachium conjunctivum
58 fibrae semicirculares
59 decussatio tractus uncinati
60 commissura cerebelli
61 pedunculus flocculi
62 brachium pontis

Fig. 22.24 f-h. Transverse sections through the brain of the rhesus monkey

1704   Chapter 22   Mammals

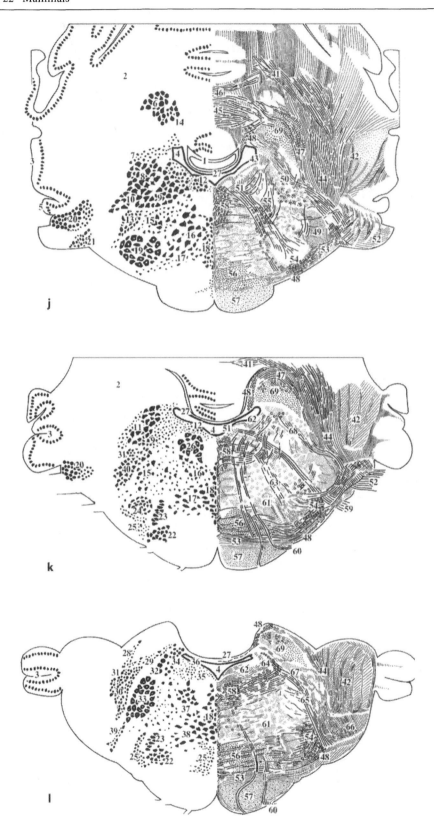

1 lingula (lobulus I)	36 nucleus tegmenti dorsalis (Gudden)
2 lobus anterior cerebelli	37 nucleus reticularis pontis caudalis
3 paraflocculus ventralis	38 nucleus gigantocellularis, pars alpha
4 ventriculus quartus	39 group K
5 flocculus	40 nucleus spinalis nervi trigemini, pars oralis
6 nucleus interpositus anterior	41 fibrae semicirculares
7 group Y	42 brachium pontis
8 nucleus vestibularis lateralis	43 striae acusticae dorsales
9 nucleus vestibularis medialis	44 corpus restiforme
10 nucleus vestibularis descendens	45 decussatio tractus uncinati
11 nucleus vestibularis superior	46 commissura cerebelli
12 nucleus prepositus hypoglossi	47 tractus uncinatus
13 nucleus spinalis nervi trigemini, pars interpolaris	48 tractus spinocerebellaris ventralis
14 nucleus fastigii	49 ramus spinalis nervi trigemini
15 nucleus reticularis parvocellularis	50 corpus juxtarestiforme
16 nucleus reticularis gigantocellularis	51 genu nervi facialis
17 nucleus gigantocellularis, pars ventralis	52 nervus statoacusticus
18 nucleus raphes magnus	53 corpus trapezoideum
19 nucleus nervi facialis	54 tractus rubrospinalis
20 nucleus cochlearis ventralis anterior	55 tractus vestibulospinalis lateralis
21 nucleus cochlearis ventralis posterior	56 lemniscus medialis
22 nucleus corporis trapezoidei	57 tractus pyramidalis
23 oliva superior medialis	58 fasciculus longitudinalis medialis
24 oliva superior lateralis	59 radix nervi facialis
25 nucleus periolivaris	60 radix nervi abducentis
26 nucleus nervi abducentis	61 tractus centralis tegmenti
27 velum medullare anterius	62 fasciculus longitudinalis dorsalis Schütz
28 nucleus parabrachialis lateralis	63 fibrae olivocochleares
29 nucleus parabrachialis medialis	64 tractus mesencephalicus nervi trigemini
30 nucleus sensibilis principalis nervi trigemini	65 radix motoria nervi trigemini
31 nucleus ovalis	66 radix sensibilis nervi trigemini
32 nucleus tractus mesencephalis nervi trigemini	67 tractus trigeminothalamicus dorsalis (Wallenberg)
33 nucleus motorius nervi trigemini	68 nervus vestibularis, ramus ascendens
34 locus coeruleus	69 brachium conjunctivum
35 griseum centrale metencephali	

Fig. 22.24 j-l. Transverse sections through the brain of the rhesus monkey

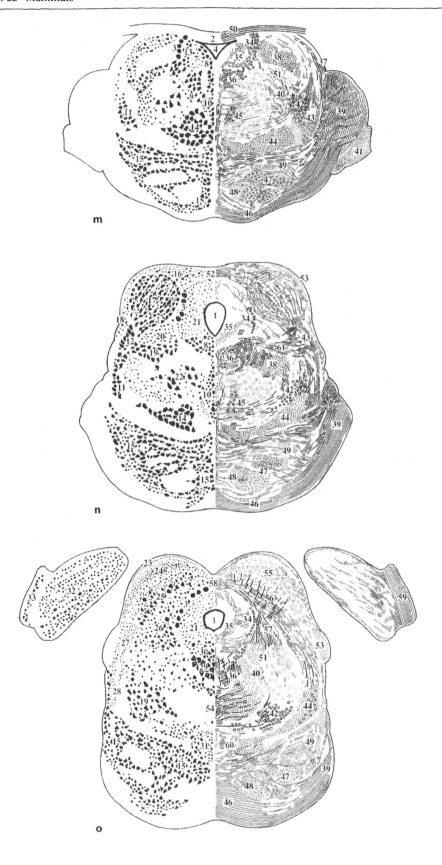

1 aqueductus cerebri
2 velum medullare anterius
3 nucleus tractus mesencephalicus nervi trigemini
4 ventriculus quartus
5 nucleus parabrachialis medialis
6 nucleus parabrachialis lateralis
7 griseum centrale metencephali
8 nucleus raphes dorsalis
9 nucleus reticularis pontis oralis
10 nucleus centralis superior
11 nucleus lemnisci lateralis ventralis
12 nucleus lemnisci lateralis dorsalis
13 locus coeruleus
14 nucleus reticularis tegmenti pontis
15 nuclei pontis
16 colliculus inferior: cortex dorsalis
17 colliculus inferior: nucleus centralis
18 colliculus inferior: cortex lateralis (nucleus externus)
19 nucleus tegmentalis pedunculopontinus
20 nucleus cuneiformis
21 griseum centrale mesencephali
22 substantia reticularis mesencephali
23 colliculus superior: stratum zonale
24 colliculus superior: stratum griseum superficiale
25 colliculus superior: stratum griseum intermedium
26 colliculus superior: stratum griseum profundum
27 nucleus brachii colliculi inferioris
28 nucleus parabigeminus
29 nucleus oculomotorius
30 nucleus Edinger Westphall
31 nucleus interpeduncularis
32 pulvinar
33 nucleus reticularis thalami
34 tractus mesencephalicus nervi trigemini
35 fasciculus longitudinalis dorsalis (Schütz)
36 fasciculus longitudinalis medialis
37 tractus spinocerebellaris ventralis
38 brachium conjunctivum
39 brachium pontis
40 tractus centralis tegmenti
41 radix nervi trigemini
42 tractus rubrospinalis
43 lemniscus lateralis
44 lemniscus medialis
45 fasciculus predorsalis
46 fibrae transversae pontis
47 tractus pyramidalis
48 tractus corticopontinus frontalis
49 tractus corticopontinus parietotemporalis
50 decussatio trochlearis
51 tractus trigeminothalamicus dorsalis (Wallenberg)
52 commissura colliculi inferioris
53 brachium colliculi inferioris
54 decussatio brachii conjunctivi
55 colliculus superior: stratum opticum
56 colliculus superior: stratum album intermedium
57 colliculus superior: stratum album profundum
58 commissura colliculi superioris
59 lamina medullaris externa thalami
60 tractus habenulointerpeduncularis
61 commissura nucleus lemniscus lateralis

Fig. 22.24 m-o. Transverse sections through the brain of the rhesus monkey

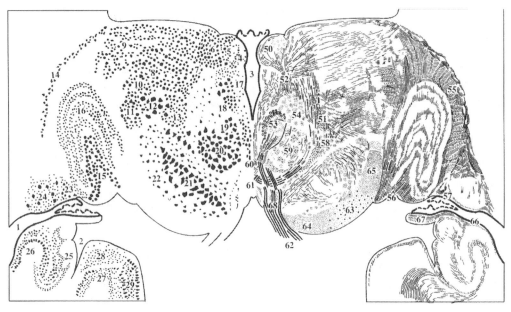

p

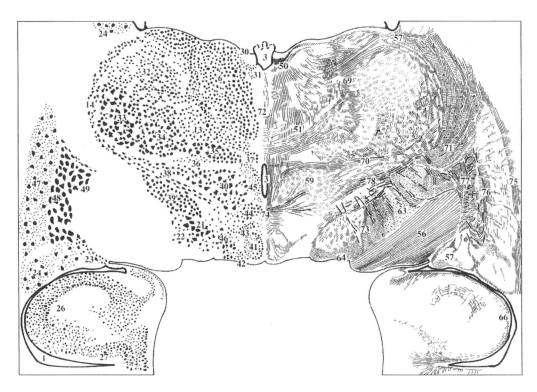

q

1 ventriculus lateralis
2 fissura hippocampi
3 ventriculus tertius
4 nucleus habenulae medialis
5 nucleus habenulae lateralis
6 nucleus medialis dorsalis thalami
7 nucleus centralis lateralis thalami
8 nucleus lateralis posterior thalami
9 pulvinar thalami
10 nucleus posterior thalami
11 corpus geniculatum mediale
12 corpus geniculatum mediale, pars magnocellularis
13 nucleus centromedianus
14 nucleus reticularis thalami
15 corpus geniculatum laterale, laminae magnocellulares
16 corpus geniculatum laterale, laminae parvocellulares
17 substantia grisea centralis mesencephali
18 nucleus Darkschewitsch
19 nucleus interstitialis fasciculi longitudinalis medialis Cajal
20 nucleus ruber
21 substantia nigra, pars compacta
22 substantia nigra, pars reticulata
23 nucleus interstitialis striae terminalis
24 cauda nuclei caudati
25 fascia dentata
26 cornu ammonis
27 subiculum
28 presubiculum
29 parasubiculum
30 nucleus parataenialis
31 nucleus paraventricularis posterior thalami
32 nucleus ventralis lateralis thalami
33 nucleus ventralis posterior lateralis thalami
34 nucleus ventralis posterior medialis thalami
35 nucleus ventralis posterior medialis thalami, pars gelatinosa
36 nucleus ventralis medialis basalis thalami
37 nucleus centralis medialis thalami
38 zona incerta
39 nucleus subthalamicus
40 nucleus campi Foreli
41 nucleus mammillaris medialis
42 nucleus mammillaris lateralis
43 nucleus supramammillaris
44 nucleus paraventricularis
45 area hypothalamica dorsalis
46 area hypothalamica lateralis
47 putamen
48 globus pallidus, pars externa
49 globus pallidus, pars interna
50 stria medullaris thalami
51 centrum medianum thalami
52 fasciculus habenulointerpeduncularis (f. retroflexus)
53 fasciculus longitudinalis medialis
54 tractus centralis tegmenti
55 radiatio optica
56 tractus opticus
57 stria terminalis
58 lemniscus medialis
59 brachium conjunctivum, ramus ascendens
60 decussatio tegmenti dorsalis
61 decussatio tegmenti ventralis
62 radix nervi ocolumotorii
63 pedunculus cerebri: tractus pyramidalis
64 pedunculus cerebri: tractus corticopontinus frontalis
65 pedunculus cerebri: tractus corticopontinus parietotemporalis
66 alveus
67 fimbriae hippocampi
68 pedunculus mammillaris
69 lamina medullaris interna thalami
70 lamina medullaris externa thalami
71 capsula interna
72 massa intermedia
73 fibrae striatonigrales
74 decussatio supramammillaris
75 capsula externa
76 globus pallidus: lamina medullaris externa
77 globus pallidus: lamina medullaris interna
78 fasciculus lenticularis

Fig. 22.24 p-q. Transverse sections through the brain of the rhesus monkey

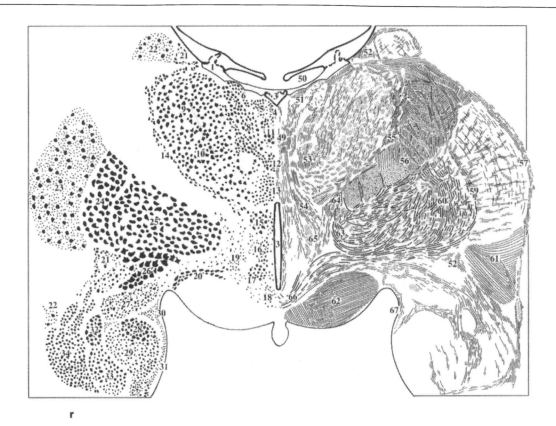

r

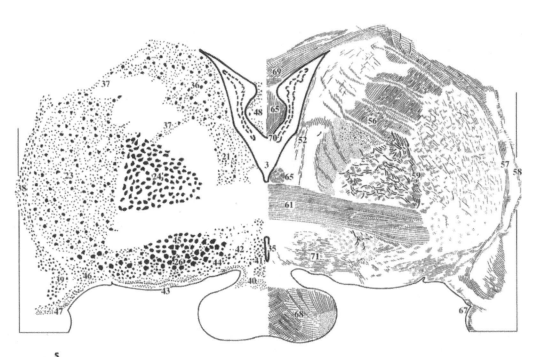

s

1	ventriculus lateralis, pars centralis	36	caput nucleus caudati
2	ventriculus lateralis, cornu anterius	37	pontes grisei caudatolenticulares
3	ventriculus tertius	38	claustrum
4	nucleus paraventricularis anterior thalami	39	nucleus endopiriformis
5	nucleus parataenialis	40	nucleus suprachiasmaticus
6	nucleus anterodorsalis thalami	41	area preoptica medialis
7	nucleus anteroventralis thalmi	42	area preoptica lateralis
8	nucleus anteromedialis thalami	43	tuberculum olfactorium
9	nucleus ventralis anterior thalami	44	nucleus fasciculi diagonalis Brocae
10	nucleus ventralis lateralis thalami	45	globus pallidus, pars ventralis
11	nucleus rhomboideus thalami	46	nucleus tractus olfactorius lateralis
12	nucleus centralis medialis thalami	47	cortex piriformis
13	nucleus medioventralis thalami	48	nuclei septi
14	nucleus reticularis thalami	49	massa intermedia
15	nucleus hypothalamicus paraventricularis	50	fornix
16	nucleus hypothalamicus dorsomedialis	51	stria medullaris thalami
17	nucleus hypothalamicus ventromedialis	52	stria terminalis
18	nucleus arcuatus	53	tractus mammillothalamicus
19	area hypothalamica lateralis	54	pedunculus thalami inferior
20	nucleus supraopticus	55	lamina medullaris externa thalami
21	nucleus interstitialis striae terminalis	56	capsula interna
22	cauda nucleus caudati	57	capsula externa
23	putamen	58	capsula extrema
24	globus pallidus: pars externa	59	globus pallidus: lamina medullaris externa
25	globus pallidus: pars interna	60	globus pallidus: lamina medullaris interna
26	substantia innominata	61	commissura anterior
27	nucleus amygdalae centralis	62	tractus opticus
28	nucleus amygdalae basalis accessorius, pars magnocellularis	63	ansa lenticularis
29	nucleus amygdalae basalis accessorius, pars parvocellularis	64	fasciculus lenticularis
		65	columna fornicis
30	nucleus amygdalae corticalis	66	decussatio supraoptica
31	cortex periamygdaloidea	67	tractus olfactorius lateralis
32	nucleus amygdalae basalis magnocellularis	68	chiasma opticum
33	nucleus amygdalae basalis parvocellularis	69	corpus callosum
34	nucleus amygdalae lateralis	70	commissura fornicis
35	ventriculus tertius, recessus supraopticus	71	fasciculus diagonalis Brocae

Fig. 22.24 r-s. Transverse sections through the brain of the rhesus monkey

22.4
Rhombencephalon
J. Voogd

22.4.1
General relations and topography

The rhombencephalon extends from the decussation of the pyramidal tract and the inferior pole of the dorsal column nuclei caudally to the decussation of the brachium conjunctivum and the trochlear nerve in the anterior medullary velum rostrally (Fig. 22.25). It is subdivided into the *medulla oblongata* and the *metencephalon*. The metencephalon is delimited by and receives its colloquial name "the pons" from the presence of the pons. The pons contains the pontine nuclei, the main relay between the cerebral cortex and the cerebellum. Fibres from the cerebral cortex contained in the cerebral peduncle enter the pons rostrally. Within the pons, the peduncle splits up into bundles and its corticopontine components terminate on the pontine nuclei. Caudal to the pons, the remaining corticobulbar and corticospinal fibres reassemble as the pyramidal tract, which descends ventrally along the midline of the medulla oblongata to its decussation.

In the metencephalon, the region bordered by the pontine nuclei ventrally, the brachium pontis laterally and the fourth ventricle dorsally is referred to as the tegmentum pontis. The term 'tegmentum' is also applied to the region of the medulla oblongata located dorsal to the pyramidal tract.

The pontine nuclei give rise to the bilaterally ascending brachium pontis. Most of the fibres of the brachium pontis decussate within the pons. The size and the caudal extent of the pons are directly related to the size of the neocortex. It is small in Monotremata, Marsupialia (Fig. 22.25a), Edentata, Insectivora and Chiroptera, well developed in Glires (including Rodentia and Lagomorpha), Macroschelidae, Artiodactyla, Perissodactyla and Carnivora (Fig. 22.25a) and large in Cetacea, Proboscoidea and Primates (Fig. 22.25c). As a consequence, the trapezoid body (the decussation of the efferent fibres of the cochlear nuclei) is not covered by the pontine nuclei in the North American opossum (Fig. 22.25a), is partially covered in the cat (Fig. 22.25b) and is almost completely covered by the caudal pons in primates (Fig. 22.25c). The border between the medulla oblongata and the tegmentum pontis is thus not fixed, but is located at a different level in different mammalian species.

The rhombencephalon can be subdivided into a *ventromedial region*, containing the motor nuclei and the reticular formation, and a *dorsolateral region*, containing the sensory nuclei (Figs. 22.22–22.24f–l). The region containing the sensory nuclei is caudally continuous with the dorsal funiculus and the dorsal horn of the spinal cord; the motor nuclei and the reticular formation continue into the motoneuronal groups of the ventral horn and the intermediate grey, respectively. In the caudal part of the medulla oblongata, the sensory nuclei of both sides are fused dorsal to the central canal. Rostral to the obex, the fourth ventricle opens. It reaches its greatest width at the level of the cochlear nuclei, where the fourth ventricle extends to the ventrolateral border of the tuberculum acusticum (Fig. 22.23j). More rostrally, where the fourth ventricle narrows, its roof is formed by the anterior medullary velum, and its floor is formed by the periaqueductal grey, which contains the nucleus tegmentalis dorsalis (dorsal nucleus of Gudden; Fig. 22.24l).

Rostral to the greatest width of the fourth ventricle, the cerebellum is broadly fused with the metencephalon. This connection includes the three cerebellar peduncles (i.e. the brachium pontis, the restiform body and the brachium conjunctivum) and embodies the transition of the vestibular and the cerebellar nuclei and the fibre bundles of the juxtarestiform body contained therein (Fig. 22.22–22.24j,k). The topography of this region is determined by the proportions of the peduncles and the transitional area of the vestibular and cerebellar nuclei. The transitional area becomes more conspicuous in the case of a small cerebellum. The size of the brachium pontis and the brachium conjunctivum is related to the size of the neocortex, and the size of the restiform body is related to the size of the cerebellum as a whole.

Precerebellar nuclei are a conspicuous element of the medulla oblongata and the pons. The lateral and paramedian reticular nuclei are precerebellar

Fig. 22.25a–c. Sagittal sections through the brain stem of **a** the opossum and **b** the cat and **c** a parasagittal section through the brain stem of the rhesus monkey. Commissures containing mainly coarse fibres are indicated with *open circles*, medium-sized fibres with *dots* and small fibres in *black*. *A*, coarse fibre decussation A; *ASK*, dorsal column nuclei; *B*, coarse fibre decussation B; *bcx*, decussation of the brachium conjunctivum; *cp*, posterior commissure; *ct*, trapezoid body; *cxaf*, cerebellar commissure, afferent part; *dtx*, dorsal tegmental decussation; *dX*, dorsal vagal nucleus; *fai*, internal arcuate fibres; *flm*, medial longitudinal fascicle; *gVII*, genu of the facial nerve; *icx*, commissure of the inferior colliculus; *III*, oculomotor nerve; *IP*, interpeduncular nucleus; *IV*, trochlear nucleus; *IVx*, decussation of the trochlear nerve; *mlx*, decussation of the medial lemniscus; *NRTP*, nucleus reticulatris tegmenti pontis; *oix*, decussation of the olivocerebellar fibres; *PO*, pes pontis; *prx*, decussation of the fibres from the principal sensory nucleus of the trigeminal nerve; *py*, pyramidal tract; *pyx*, pyramidal decussation; *r*, fasciculus retroflexus; *ra*, raphe; *scx*, commissure of the superior colliculus; *tx*, tectal decussation; *ux*, decussation of the uncinate tract; *vma*, anterior medullary velum; *vtx*, ventral tegmental decussation; *xHAB*, habenular commissure; *XII*, hypoglossal nucleus; *xP*, Probst's commissure

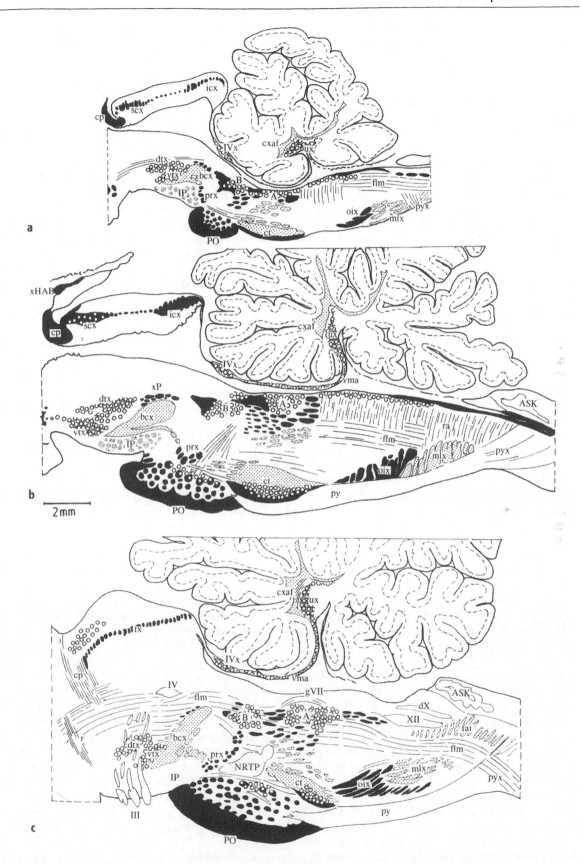

nuclei, located in the caudal bulbar reticular formation. The nucleus reticularis tegmenti pontis is a precerebellar nucleus located in the tegmentum pontis, dorsal to the pontine nuclei (Figs. 22.22–22.24 m,n). The inferior olive is located ventrally in the medulla, dorsolateral to the pyramid. Both the size and the shape of the medial and dorsal accessory olives and the principal olive differ among mammalian species (see Figs. 22.55, 22.56). The medial accessory olive is extremely large in Cetacea and Proboscoidea (see Fig. 22.57). The size of the principal olive is related to the size of the neocortex, and convolutions of its cell sheet are only observed in primates. With the increase in size of the inferior olive, the distance between the decussation of the olivocerebellar fibres and the trapezoid body decreases (Fig. 22.25). In primates, Cetacea and Proboscoidea, the decussations of the pons, the trapezoid body and the olivocerebellar fibres partially overlap (Fig. 22.25).

The rhombencephalon harbours the centres of origin and termination of eight (V–XII) of the 12 cranial nerves (for a review, see Székely and Matesz 1993). At first sight, the arrangement of these cranial nerve nuclei does not show a definite pattern; however, the classical investigations performed by Gaskell (1886, 1889); Herrick (1913) and many others have revealed that these centres form part of seven functional zones, each of which is specifically related to one of the fibre categories of which the cranial nerves are composed. As in the spinal cord, the afferent centres of the cranial nerves are situated in the alar plate, whereas the efferent centres are located in the basal lamina. In the adult, the sulcus limitans, which in the embryonic neuraxis marks the boundary between these two fundamental subdivisions, is only recognisable over a short extent of the wall of the fourth ventricle. It will be noted that most of the zones are only partly occupied by cranial nerve nuclei. This may be related to the reduction or transition of some components of some nerves during fetal development (see Chap. 4). The various zones and their constituent primary afferent or efferent centres will now be briefly reviewed, passing from lateral to medial.

The lateralmost, special somatic afferent zone contains the nuclei of termination of the cochlear and vestibular divisions of nerve VIII. The cochlear nuclei are situated laterally in the floor of the fourth ventricle (Figs. 22.23–22.24 j,k). In marsupials and monotremes, the cochlear nuclei are located medial to the restiform body (see Fig. 22.45). In eutherian mammals, they extend dorsal and lateral to the restiform body, where they form the conspicuous acoustic tubercle in the floor of the lateral recess of the fourth ventricle. Differences in size and structure of the cochlear nuclei and the central acoustic pathways are discussed in Sect. 22.8.2.2. In Microchiroptera and Cetacea, the large size of the acoustic system is obviously related to its role in echolocation (see Figs. 22.81, 22.83).

The vestibular nuclei occupy a more medial position in the floor of the fourth ventricle (see Figs. 22.22–22.24 h–k, 22.122). Rostrally, they extend into the transitional region of the rhombencephalon with the cerebellum. The appearance of the vestibular nuclei, with their efferent pathways contained within the medial longitudinal fascicle, the brachium conjunctivum and the lateral vestibulospinal tract, is fairly constant among mammalian species. The basic similarities in the connections of the vestibular nuclei are exemplified by the presence of two bundles of large decussating fibres (A and B) from the vestibular nuclei in the dorsal tegmentum of the three species illustrated in Figs. 22.25. The vestibular nuclei and their connections are reviewed in Sect. 22.9.2.

The dorsal column nuclei receive ascending branches of dorsal root fibres and afferents from the dorsal horn of the spinal cord, mainly through the dorsal funiculus, and are therefore not included in the functional zones of the cranial nerve nuclei. They occupy a dorsal position in the caudal medulla oblongata, caudal to the fourth ventricle (see Figs. 22.22–22.24 f,h, 22.71, 22.72). They give rise to internal arcuate fibres, which decussate caudally (Fig. 22.25, mlx) and ascend as the medial lemniscus, dorsal to the pyramidal tract. The size, composition and somatotopical organisation of the dorsal column nuclei and the medial lemniscus are related to the representation of somatic sensation at the level of the ventral tier of thalamic nuclei and the sensory cortex (see Fig. 22.69). The dorsal column nuclei attain their largest size in primates (see Sect. 22.8.1.1.3).

The next medial general somatic afferent zone includes the three sensory nuclei of the trigeminal nerve, i.e. the mesencephalic, the principal and the spinal nuclei. The latter nucleus, which also receives some fibres from nerves VII, IX and X, is caudally continuous with the apical part of the dorsal horn of the spinal cord. Several factors appear to determine the size and structure of the trigeminal nuclei. The size of the sensory root of the trigeminal nerve is the primary factor. The sensory root is large in Monotremata (huge in the platypus *Ornithorhynchus*; Hines 1929), Marsupialia, Insectivora (where it is largest in the shrew; Verhaart 1970b), Rodentia, Carnivora, Proboscoidea, Perissodactyla and Artiodactyla. Among the carnivores, it is generally larger in the Pinnipedia (sea lions, seals and walruses) than in the Fissipeda (mainly terrestrial animals). Specializations in the form of sensory fibrissae are an important factor determining the structure of the sensory nuclei of the trigeminal nerve in rodents and aquatic carnivores. In the Fissipeda, this is related to the huge size of the ventral part of the nucleus princeps and the decussation of the trigeminothalamic fibres, which is located caudal and dorsal to the interpeduncular nucleus (Fig. 22.25, prx). The dorsal part of the princeps nucleus and the uncrossed ascending trigeminothalamic tract are well developed in Perissodactyla and Artiodactyla. This feature is probably related to the elaboration of intra- and peri-oral sensibility in these animals. The dorsal column nuclei and the sensory nuclei of the trigeminal nerve are discussed in Sect. 22.8.1.1.4.

The special visceral afferent and general visceral afferent zones are represented in the adult brain by the nucleus of the tractus solitarius, which receives the corresponding components of nerves VII, IX and X. The latter unite in a well-defined fibre system, the tractus solitarius (see Figs. 22.22–22.24f–h, 22.75). The nucleus of the solitary tract is reviewed in Sect. 22.8.1.2.

The general visceral efferent zone in the rhombencephalon contains three nuclei, the nucleus dorsalis of nerve X and the nuclei salivatorii inferior and superior of nerves IX and VII, respectively (Fig. 22.26). These nuclei belong to the cranial division of the parasympathetic system. They give rise to preganglionic fibres that terminate in various autonomic ganglia (see Sect. 22.4.2). The mesencephalic general visceral efferent Edinger-Westphal nucleus is discussed in Sect. 22.5 (see Sect. 22.5.3).

The special visceral efferent or branchiomotor zone contains the motor nuclei of nerves V and VII (see Figs. 22.28, 22.29) as well as the nucleus ambiguus (Fig. 22.27), which gives rise to fibres that pass peripherally as components of nerves IX, X and XI (see Sects. 22.4.3-5). The spinal nucleus of nerve XI, which is situated in the lateral part of the base of the ventral horn of the upper four cervical segments, also belongs to this zone. Like the general somatic afferent nuclei, the cell masses of the branchiomotor zone have migrated away from their original periventricular position.

The course of the axons of the motoneurons of some of these nuclei still reflects this migration pathway. Axons of the nucleus ambiguus and the ventromedial subdivision of the motor nucleus of the trigeminal nerve make dorsomedially directed hairpin bands, similar to the genu of the facial nerve, before they enter their respective nerve roots (Kalia and Mesulam 1980a,b; Székely and Matesz 1982; Bieger and Hopkins 1987).

Finally, the general somatic efferent zone may be considered a rostral continuation of the anterior horn of the spinal cord. It comprises the nucleus of origin of nerve XII (Sect. 22.4.6) and the abducens nucleus (see Sect. 22.5.2). Both nuclei are located ventrolaterally in the brain stem. The abducens nucleus includes a ventrolaterally displaced cell group, known as the accessory nucleus of the abducens nerve, which innervates the retractor bulbi muscle. The axons of the motoneurons of this nucleus pass dorsally before they join the abducens nerve (see Sect. 22.4.2).

The reticular formation occupies a central zone in the rhombencephalon. It can be subdivided into four zones (see Fig. 22.64). The medialmost zone is represented by the raphe nuclei, which contain most of the serotonergic neurons of the lower brain stem. Long ascending and descending pathways arise from the medial, magnocellular reticular formation. The intermediate zone contains the adrenergic, noradrenergic and cholinergic cell groups of the rhombencephalon and rostrally continues into the locus coeruleus, the nucleus subcoeruleus and the cholinergic cell group of the pedunculopontine nucleus. The lateral, parvocellular reticular formation and its rostral continuation in the parabrachial nuclei give rise to propriobulbar connections to the cranial motor nuclei and serve as a relay in viscerosensory and visceromotor pathways. Rostrally, immediately caudal to the decussation of the brachium conjunctivum, the ventral nucleus tegmentalis of Gudden is located dorsally within the nucleus centralis pontis oralis of the medial reticular formation (Fig. 22.24l). The dorsal nucleus of Gudden is situated more dorsally, within the periaqueductal grey. The reticular formation and the monoaminergic and cholinergic nuclei of the brain stem are reviewed in Sect. 22.7.

The comparative anatomy of long fibre systems in the mammalian brain stem was reviewed by Verhaart (1970). Ascending and descending fibre systems, contained in the medial longitudinal fascicle, generally display only little diversity among mammals. The prominence of the lemniscal and extralemniscal sensory projection systems is related to specializations in sensory perception and to the representation of sensation at the level of the thalamus and the sensory cortex (see Sect. 22.8.1). Major differences among mammalian species are present in the relative size, topography and connectivity of the corticobulbar, rubrobulbar and rubrospinal tracts (see Sect. 22.9). Among the precerebellar systems, the spino-, cuneo- and reticulocerebellar tracts are a fairly constant component of the restiform body. The afferent and efferent connections of the principal nucleus of the inferior olive and the pontocerebellar pathway are subject to large variations (see Sect. 22.6). Most of these variations are related to the size of the neocortex. In the case of

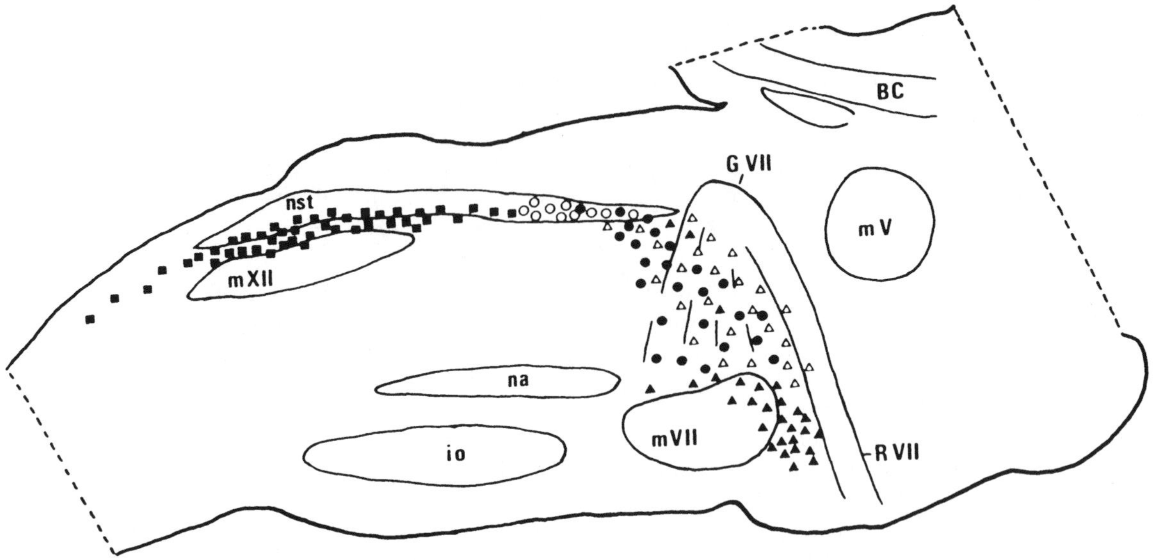

Fig. 22.26. Schematic reconstruction on a parasagittal section through the brain stem of the rat, showing the location of preganglionic parasympathetic neurons of the dorsomedial column and the lateral tegmentum that distribute with the facial, glossopharyngeal, and vagus nerves. *Filled triangles* represent neurons with axons in the greater superficial petrosal component of the VIIth nerve; *open triangles*, neurons distributing with the chorda tympani branch of VII; *filled circles*, neurons whose axons reach the otic ganglion; *open circles*, neurons contributing to the lingual-tonsillar branch of IX; *filled squares*, neurons exiting in the cervical vagus. In reality the neurons are lateral to the motor nucleus and the radiations of the facial nerve, but medial in, or adjacent to, the nucleus of the solitary tract. *BC*, bracium conjunctivum; *GVII*, genu of the facial nerve; *io*, inferior olive; *mV*, motor nucleus of the trigeminal nerve; *mVII*, motor nucleus of facial nerve; *mXII*, motor nucleus of the hypoglossal nerve; *na*, nucleus ambiguus; *nst*, nucleus of the tractus solitarius; *RVII*, root of facial nerve. From Contreras et al. (1980)

the inferior olive, the hypertrophy of other circuits, including the medial accessory olive, may lead to important, if unexplained, changes in the topography of the rhombencephalon of Cetacea and Proboscoidea. Most of the structures in this section are treated in more detail in section 22.7 on the reticular formation, section 22.8 on sensory systems and section 22.9 on motor systems. The motor nuclei of the rhombencephalon are considered in the following sections 22.4.2–22.4.6.

22.4.2
Parasympathetic Visceromotor Nuclei of the Rhombencephalon

Preganglionic neurons of the facial, glossopharyngeal and vagus nerves are located in a dorsomedial column, comprising the dorsal motor nucleus of the vagus nerve and a ventrolateral column, corresponding to the external formation of the nucleus ambiguus (Fig. 22.27). Scattered preganglionic neurons are present in between the two columns. Rostrally, the dorsomedial column turns ventrally into an area of scattered cells in the lateral tegmentum between the rostral pole of the superior olivary complex and the facial nucleus (Contreras et al. 1980; Bieger and Hopkins 1987; Fig. 22.26). Preganglionic fibres from the vagus nerve in the dog, cat, rat and hamster take their origin from the dorsal motor nucleus and the external division of the nucleus ambiguus, without a distinct viscerotopic localisation (Sugimoto et al. 1979; Kalia and Mesulam 1980a,b; Geis and Wurster 1980; Contreras et al. 1980; Kalia and Sullivan 1982; Hopkins and Armour 1982; Miceli and Malsbury 1985; Plecha et al. 1988). The lingual and tympanic branches of the glossopharyngeal nerve, the latter innervating the parotid gland, and the chorda tympani both originate from the scattered neurons in the rostral lateral tegmental field. Some of the preganglionic fibres from this region exit through the motor root of the trigeminal nerve (Jacquin et al. 1983). Inferior and superior salivatory nuclei cannot be distinguished in the rat and cat (Fig. 22.26; Contreras et al. 1980; Nomura and Mizuno 1982; Tramonte and Bauer 1986). The greater petrosal nerve, which supplies the preganglionic parasympathetic innervation of the lacrimal gland, takes its origin from the rostralmost part of the external division of the nucleus ambiguus, located rostrolateral to the facial nucleus (Contreras et al. 1980; Nomura and Mizuno 1982; Bieger and Hopkins 1987). Descending pathways from limbic structures, the hypothalamus and the periaqueductal grey to the parasympathetic motor

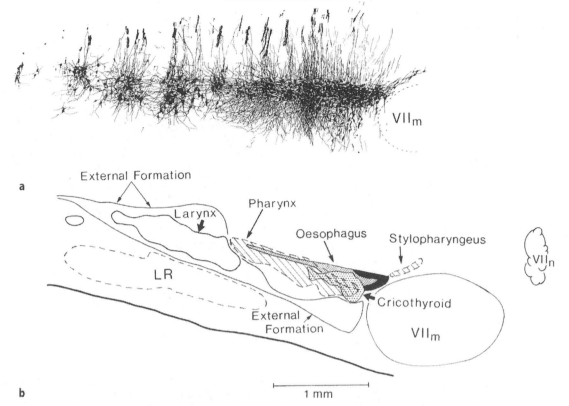

Fig. 22.27. a Distribution of retrograde labelling as reconstructed from serial sagittal sections of the nucleus ambiguus of the rat after injection of the retrograde tracer horseradish peroxidase (HRP) into the supranodosal vagus and glossopharyngeal nerves. **b** Diagrammatic representation of the viscerotopic representation of the upper alimentary and respiratory tract in the nucleus ambiguus of the rat as projected in the sagittal plane. The *dark shading* in the rostral part of the oesophageal representation contains the majority of neurons projecting to the upper cervical oesophagus. *LR*, lateral reticular nucleus; *VIIm*, nucleus of facial nerve; *VIIn*, facial nerve. From Bieger and Hopkins (1987)

nuclei of the medulla were reported and reviewed by Ter Horst et al. (1984) and Holstege (1991).

22.4.3
Nucleus Ambiguus

The nucleus ambiguus can be subdivided into a dorsal, branchiomotor division, corresponding to the nucleus ambiguus in a narrow sense, and a external formation containing preganglionic parasympathetic neurons (for a review, see Bieger and Hopkins 1987; Fig. 22.27). The dorsal division consists of a rostral segment with a compact arrangement of the motoneurons and of intermediate and caudal diffuse segments, where the packing of the cells displays a rostrocaudal periodicity. The structure of the nucleus ambiguus is more diffuse in the cat than in the rabbit and rat (Husten 1924; Lawn 1966; Davis and Nail 1984; Bieger and Hopkins 1987). The stylopharyngeus motor pool of the glossopharyngeal nerve is located dorsal to the facial nucleus in the so-called retrofacial nucleus, together with neurons innervating the cricothyroid muscle by the superior laryngeal nerve. The upper oesophagus and the lower oesophagus with the soft palate and the pharynx are innervated by motoneurons of the rostral, compact and intermediate part of the nucleus ambiguus, respectively. The caudal part of the nucleus contains a second pool of laryngeal motoneurons, which gives rise to the recurrent laryngeal nerve (Yoshida et al. 1981; Holstege et al. 1983; Davis and Nail 1984; Bieger and Hopkins 1987; Fig. 22.27).

Swallowing centres, controlling movements of the soft palate, the pharynx and the oesophagus, and the supra- and infrahyoid and tongue muscles innervated by the hypoglossal and trigeminal nerves were located in the caudal pontine tegmentum, dorsal to the superior olive (Holstege et al. 1983) and in the nucleus of the solitary tract (Barrett et al. 1994).

22.4.4
Motor Nucleus of the Facial Nerve

The subdivision of the motor nucleus of the facial nerve into subnuclei differs for different mamma-

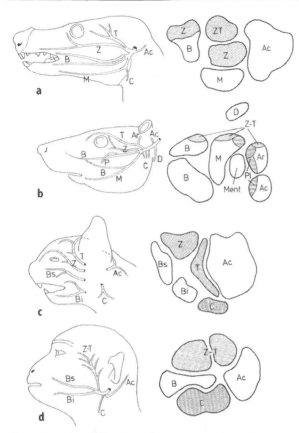

Fig. 22.28a–d. Subdivisions of the motor nucleus of the facial nerve supplying different branches of the nerve, indicated in the figurines on the *left*. **a** The brush-tailed possum (*Trichosuris vulpecula*), redrawn from Provis (1977). **b** The mouse, redrawn from Komiyama et al. (1984). **c** The cat, redrawn from Kume et al. (1978). **d** The monkey (*Macaca fuscata*), redrawn from Satoda et al. (1987). Subnuclei innervating the zygomatic and temporal branches are *lightly shaded*, subnuclei innervating the platysma and the cervical branch are *darkly shaded*. *Ac*, caudal auricular branch; *Ar*, rostral auricular branch; *B*, buccal branch.; *Bi*, inferior buccal (labial) branch; *Bs*, superior buccal (labial) branch; *c*, cervical branch; *D*, branch for caudal belly of digastric muscle; *M*, marginal branch; *Pl*, platysma; *T*, temporal branch; *Z*, zygomatic branch

lian species (Papez 1927). The somatotopical arrangement of the motoneurons in the nucleus, however, is basically similar in marsupials (*Didelphis virginiana*: Dom et al. 1973; the phalanger *Trichosurus*: Provis 1977), rodents (mouse: Komiyama et al. 1984; rat: Hinrichsen and Watson 1984; Friauf and Herbert 1985; Klein and Rhoades 1985), rabbit (Holstege and Collewijn 1982), cat (Szentágothai 1958; Kume et al. 1978; Holstege 1991), Megachiroptera (*Rouettus aegypticus*: Friauf and Herbert 1985) and primates (*Macaca fuscata*: Satado et al. 1987). Pinna muscles innervated by the caudal and rostral auricular branches are located in the medial part of the nucleus (Fig. 22.28). The auricular division of the nucleus is large in the phalanger *Trichosurus*,

the cat and the bat. Friauf and Herbert (1985), who compared the localisation of motoneurons innervating different pinna muscles in the rat and bat, concluded that the superior mobility of the ear in bats is related to the larger size and the discreteness of the localisation of the different muscles in this species. Frontal, periorbital and zygomatic muscles are innervated by motoneurons in the dorsal part of the nucleus. This region is largest in primates. Motoneurons of the perioral muscles and of the muscles of the vibrissae are located laterally. This region of the nucleus is largest in rodents and in the opossum. The musculus mentalis and the platysma are innervated by ventrally located cells in the phalanger *Trichosuris*, the cat and the monkey. In the opossum (Dom et al. 1973) and mouse (Komiyara et al. 1984), these cells occupy a dorsomedial or mediocentral position (Fig. 22.28b). The topical relations of the facial nucleus in the elephant have not been studied. The nucleus is large and subdivided into massive lateral and smaller medial and dorsal subnuclei (Precechtel 1925; Verhaart 1962).

The accessory nucleus of the facial nerve in rodents, primates and carnivores, which innervates the caudal belly of the digastric muscle, is located dorsal to the main nucleus of the facial nerve, medial to the descending facial root. Rostrally, it is continuous with the motor nucleus of the trigeminal nerve (Fig. 22.29AD; Matsuda et al. 1978; Székely and Matesz 1982; Komiyana et al. 1984). Motoneurons of the stapedius muscle lie scattered at the interface between the facial nerve nucleus and the superior olivary complex in the cat (Lyon 1978; Joseph et al. 1985) and in primates (Thompson et al. 1985).

The premotor projections to the facial nucleus were reviewed by Holstege (1991). The medial, ear and platysma motoneuronal groups receive inputs from the cervical cord and the nucleus retroambiguus. The dorsally located periorbital motoneuronal group receives a projection from the ventral spinal trigeminal nucleus. This projection is the second link in the blink reflex and also includes the motoneurons of the accessory motor nucleus of the abducens nerve, which innervate the retracto bulbi muscle. In the cat, these neurons are located medial to the root of the facial nerve, dorsal to the superior olive. Here, they lie interposed between two groups of neurons belonging to the accessory nucleus of the facial nerve innervating the posterior belly of the digastric muscle (Grant et al. 1979, 1981). According to Spencer et al. (1980), the retracter bulbi muscle is also innervated from the medial rectus division of the oculomotor nucleus. The retractor bulbi muscle and the corresponding nucleus are absent in apes and humans (Bolk et al.

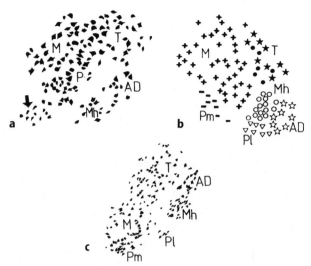

Fig. 22.29a–c. Localization of motoneurons innervating different muscles in the motor nucleus of the trigeminal nerve of a the cat (slightly modified from Mizuno et al. 1975), b the rabbit (slightly modified from Matsuda et al. 1978) and c the macaque monkey (slightly modified from Mizuno et al. 1981). *Arrow* in a indicates clusters of small motoneurons innervating the tensor veli palatini and the tensor tympani muscles. *AD*, anterior belly digastric muscle (*open circles*); *M*, masseter muscle (*crosses*: superficial layer; *filled circles*: deep layer); *Mh*, mylohyoid muscle (*open circles*); *P*, pterygoid muscle; *Pl*, lateral pterygoid muscle (*open triangles*); *Pm*, medial pterygoid muscle (*bars*); *T*, temporal muscle (*filled stars*)

1938). The motoneurons innervating the periorbital and retractor bulbi muscles also receive input from two blink areas in the reticular formation, one in the ventrolateral tegmental field and the other in the caudal dorsomedial reticular formation.

The lateral group of motoneurons innervating the perioral muscles receives projections from the lateral tegmental field and the medial parabrachial nucleus. Mesencephalic projections from the periaqueductal grey, the red nucleus and the tectum terminate in the medial intermediate and lateral subdivisions of the motor nucleus of the facial nerve in the opossum (Panneton and Martin 1983), cat (Courville 1966; Holstege et al. 1984), rat (Isokawa-Akesson and Komisarut 1987), and monkey (Kuypers 1981).

22.4.5
Motor Nucleus of the Trigeminal Nerve

The motor nucleus of the trigeminal nerve can be subdivided in most mammals into a large dorsolateral and a smaller ventromedial subnucleus. The dorsolateral subnucleus contains motoneurons innervating jaw closers, with temporal muscle neurons located dorsomedially and masseter neurons laterally (Fig. 22.29). Neurons innervating the medial and lateral pterygoid muscles are located ventrolateral and ventromedial to the masseter neuronal pool, respectively. The jaw openers are located in a ventromedial group in the caudal part of the nucleus. In the cat (Mizuno et al. 1975; Batini et al. 1976), rat (Mizuno et al. 1975; Jacquin et al. 1983), guinea pig (Uemura-Sami et al. 1982) and in primates (*Macaca fuscata, ira* and *mulatta*: Mizuno et al. 1981), the motoneurons innervating the anterior belly of the digastric muscle are located dorsomedial to the motor pool of the mylohyoid muscle. In the rabbit, which lacks a posterior belly of the digastric muscle, these motoneurons are located ventromedial to the mylohyoid motor pool. Moreover, the motoneurons innervating the anterior belly of the digastric muscle extend caudally, medial to the root of the facial nerve, in the position of the accessory nucleus of the facial nerve, which innervates the posterior belly of the digastric muscle in other mammals (Willems 1910; Matsuda et al. 1978). The small, fusiform motoneurons innervating the tensor tympani muscle constitute the intertrigeminal nucleus ventral to the main motor nucleus of the trigeminal nerve (Jacquin et al. 1983; Rouiller et al. 1986).

Interneuronal projections to the motor trigeminal nucleus were reviewed by Holstege (1991). Mouth-opening motoneurons receive their strongest projections from the lateral tegmental field at levels caudal to the obex. Mouth-closing motoneurons receive their input from the lateral tegmental field, rostral to the hypoglossal nucleus and the supratrigeminal and parabrachial nuclei. The jaw-closing muscles receive collaterals from muscle spindle afferents from the mesencephalic trigeminal nucleus. Collaterals from mesencephalic root axons also terminate in the hypoglossal nucleus and the supratrigeminal nucleus (Lingenhohl and Friauf 1991; Raappana and Arvidsson 1993). The motoneurons innervating the tensor tympani muscle receive projections from the cochlear nuclei and the superior olivary complex (Rouiller et al. 1986; Itoh et al. 1986).

22.4.6
Motor Nucleus of the Hypoglossal Nerve

The nucleus of the hypoglossal nerve of the cat can be subdivided into caudal (ventromedial) and rostral (dorsolateral) parts, which give rise to the medial branch, supplying the protrusor muscles of the tongue, and the lateral branch, innervating the retractor muscles, respectively (Uemura et al. 1979). The geniohyoid and thyrohyoid muscles are innervated by motoneurons located in the ventral part of the nucleus (Miyazaki et al. 1981). A similar subdivision of the hypoglossal nucleus was proposed for the mouse (Stuurman 1906) and for primates

(*Macaca fuscata*: Uemura-Sumi et al. 1981). The differential innervation of the protrusor and retractor muscles of the tongue from the medial and lateral tegmental field, the parabrachial nuclei, the nucleus of the solitary tract and the spinal trigeminal nucleus were discussed by Dobbins and Fieldman (1995) and Travers et al. (1995). The midline medullary nuclei may coordinate complex behaviour, whereas the lateral reticular formation may control the contribution of the tongue to distinct orofacial behaviours.

22.5
Mesencephalon

J. Voogd

22.5.1
General Relations and Topography

The caudal border of the mesencephalon is a curved plane, stretching from the caudal margin of the tectum dorsally, along the decussation of the brachium conjunctivum to the rostral border of the pons ventrally, and along the rostral margin of the brachium pontis laterally (Fig. 22.25). The border of the mesencephalon with the diencephalon is located rostral to the posterior commissure and the rostral pole of the interpeduncular nucleus. Laterally, it passes caudal to the thalamus and the lateral and medial geniculate bodies and rostral to the substantia nigra. The mesencephalon can be subdivided into the pes, tegmentum and tectum mesencephali, the periaqueductal grey and the pretectum.

The *pes mesencephali* comprises the cerebral peduncle and the substantia nigra. The medial lemniscus marks the dorsal border of the pes mesencephali (Figs. 22.22–22.24n–p). The size of the cerebral peduncle is related to the size of the neocortex. In the platypus *Ornithorhynchus* and in Insectivora and Microchiroptera, the peduncle is small. In the pigmy shrew and in Microchiroptera, the fossa interpeduncularis is replaced by a large, prominent interpeduncular nucleus. The dorsomedial part of the peduncle splits up into bundles located in the substantia nigra in many species. The peduncle consists of a narrow, medial portion and a larger, compact lateral part in most Insectivora, Xenarthra, Cetacea, Perissodactyla and Artiodactyla. In the pigmy shrew, the entire peduncle divides into bundles scattered within the rostral substantia nigra. The cerebral peduncle is large in Proboscoidea, Cetacea, carnivores and primates. In primates, the peduncle is partitioned into a medial frontopontine tract, a central area which continues into the medullary pyramid and a lateral parietotemora tract (Fig. 22.24p,q). In species of other mammalian orders, a subdivision into a medial area, which gives rise to the pyramidal tract, and a lateral area is sometimes possible (Fig. 22.23a–q; Verhaart 1970b). The composition of the cerebral peduncle is reviewed in Sect. 22.9.1.

The substantia nigra can be subdivided into a caudal and dorsal pars compacta and a rostral and ventral pars reticulata (Figs. 22.22–22.24o,p). The pars compacta consists of medium to large cells, which are pigmented in humans. In other mammals, they show at most a trace of pigmentation. The pars reticulata consists of a reticulum of long, slender dendrites with an occasional cell body. Rostrally, large numbers of fine striatal fibres occupy the mazes of the reticulum of the pars compacta. The pars reticulata is generally larger than the pars compacta. The pars reticulata is exceptionally large in the great anteater *Myrmecophaga*, Cetacea, Proboscoidea, Artiodactyla and Perissodactyla. In the two-toed sloth *Choloepus*, the pars compacta surpasses the pars reticulata in size. In primates, the size of the two divisions is approximately equal (Verhaart 1970b).

The *tectum* comprises the inferior and superior colliculi. The inferior colliculus extends rostrally, beneath the deep laminae of the superior colliculus (Fig. 22.8). The brachium of the inferior colliculus passes rostrally, ventrolateral to the superior colliculus, to terminate in the medial geniculate body (Figs. 22.22–22.24n–p). The relative size of the colliculi was described and illustrated in Sect. 22.1 (Fig. 22.10). The structure and connections of the inferior and superior colliculu are reviewed in Sects. 22.8.2.5 and 22.8.3.2, respectively. The tectal commissures cross the midline, dorsal to the periaqueductal grey (Fig. 22.25). Laterally, fibres of the commissure of the inferior colliculus fan out along the rostral border of this structure (Figs. 22.22–22.24n). The commissure of the superior colliculus increases in width from caudally to rostrally. Its fibres radiate in the intermediate and deep fibre layers of the superior colliculus. Rostrally, it touches the posterior commissure.

The *periaqueductal grey* is continuous with the grey matter in the floor of the fourth ventricle (Figs. 22.22–22.24m–p). Rostrally, it continues into the periventricular nuclei of the thalamus and the hypothalamus. It contains the fibres of Schütz's fasiculus longitudinalis dorsalis, which continue as a component of the medial forebrain bundle in the lateral hypothalamus. The nucleus raphes dorsalis is located ventral to the aqueduct in the caudal part of the periaqueductal grey. Rostrally, where the posterior commissure borders on the ependyma of the subcommissural organ in the roof of the aqueduct, the circle of the periaqueductal grey is broken (Figs. 22.22, 22.23pq). The nucleus of Darkschewitsch is located at this level in the ventrolateral periaqueductal grey, medial to the fasciculus retroflexus. This nucleus extends ventral to the fasciculus retroflexus, into the subparafascicular nucleus of the diencephalon.

The *pretectum* is located at the level of the posterior commissure (see Figs. 22.103–22.105). Rostrally, it borders on the nucleus limitans of the thalamus; caudally and medially, it abuts on the superior colliculus. A sheet of optic tract fibres covers its meningeal surface. Caudally, this sheet continues into the stratum opticum of the superior colliculus. At the rostral border of the pretectum, fibres of the accessory optic tract detach from the optic tract to pass ventrally, caudal to the medial geniculate body and superficial to the cerebral peduncle (Fig. 22.11).

The interpeduncular nucleus occupies the ventral midline of the mesencephalon. Its caudal pole extends for a short distance dorsal to the pontine nuclei. Fibres of the medial lemniscus from the principal sensory nucleus of the trigeminal nerve decussate immediately caudal to the interpeduncular nucleus (Fig. 22.25). The fibres of the ventral limb of the decussation of the brachium conjunctivum cross along its dorsal border. Rostrally, the interpeduncular nucleus receives the fasciculus retroflexus. The mammilary peduncles are situated lateral to the interpeduncular nucleus.

The reticular formation and the fibres of the medial longitudinal fascicle, the medial lemniscus, the tegmental tracts and the rubrobulbar and rubrospinal tract continue from the tegmentum pontis into the *tegmentum mesencephali*. Laterally, the lateral lemniscus obliquely crosses the metencephalic-mesencephalic border to terminate in the inferior colliculus. Its orientation reflects the original direction of the neural tube, rostral to the pontine flexure (Figs. 22.11, 22.12). The dorsal and ventral nuclei of the lateral lemniscus are contained in this pathway. The nucleus parabigeminus is located at the meningeal surface, rostral to the lateral lemniscus, dorsal to the cerebral peduncle and ventral to the brachium of the inferior colliculus.

The decussation of the brachium conjunctivum extends far rostrally, dorsal to the interpeduncular nucleus (Fig. 22.25). The fibres of the crossed ascending branch of the brachium conjunctivum surround and enter the caudal pole of the red nucleus (Figs. 22.22, 22.23o,p). Rostrally, where the fibres of the brachium deflect to pass into the subthalamus and the internal medullary lamina of the thalamus, this capsule becomes less distinct. The rostral pole of the red nucleus is therefore difficult to delimit in many non-primate species.

In primates, the red nucleus consists of the well-demarcated caudal, magnocellular and rostral parvocellular divisions (Fig. 22.24o,p). The rubrobulbar and rubrospinal tract takes its origin from the magnocellular red nucleus. The central tegmental tract is the efferent pathway from the parvocellular red nucleus. The parvocellular division of the red nucleus is absent or difficult to identify in species of other mammalian orders. Cell groups located at the mesencephalic-diencephalic junction, which give rise to the central and medial tegmental tracts, have been identified using experimental tracing methods in the opossum, rat and cat. The connections of the red nucleus are reviewed in Sect. 22.9.3.

In the caudal mesencephalon, the central tegmental tract is located lateral to the medial longitudinal fascicle, ventrolateral to the periaqueductal grey (Figs. 22.24k–p). The central tegmental tract should be distinguished from the ascending pathways from the medial reticular formation and the raphe nuclei (central reticular fasciculus of Busch 1961) and the dorsal trigeminothalamic tract of Wallenberg. The central reticular fasciculus is located dorsal to the central tegmental tract and is characterised by a central region of fibre bundles containing a proportion of large and medium-sized fibres arising from the nucleus pontis oralis (Fig. 22.23o,p; Forel's fascicles). Wallenberg's tract, which is extremely large in Perissodactyla and Artiodactyla, occupies the same general region as the central reticular fasciculus (see Fig. 22.117; Verhaart 1970b).

The motor nuclei of the trochlear and oculomotor nerves (Sects. 22.5.2 and 3; Fig. 22.30) are located ventral to the periaqueductal grey. The trochlear nucleus is embedded in the medial longitudinal fascicle, and the oculomotor nuclei are laterally bordered by the fibres of this fascicle. The trochlear nerve passes caudally, through the periaqueductal grey, to decussate and to exit from the anterior medullary velum. The rootlets of the oculomotor nerve pass ventrally, through the red nucleus, to exit in the interpeduncular fossa. The large cells of Cajal's interstitial nucleus of the medial longitudinal fascicle are located in the rostral mesencephalon, between the periaqueductal grey and the red nucleus (Fig. 22.22–22.24o,p). This nucleus should be distinguished from the rostral interstitial nucleus of the medial longitudinal fascicle, which is located more rostrally at the junction with the diencephalon.

Four decussations cross the midline of the tegmentum mesencephali (Fig. 22.25). The decussation of the brachium conjunctivum consists of a voluminous dorsal limb and a thin ventral limb, located dorsal to the interpeduncular nucleus. Rostrally, the two limbs fuse and extend dorsal to the interpeduncular nucleus. Probst's commissure of the nucleus

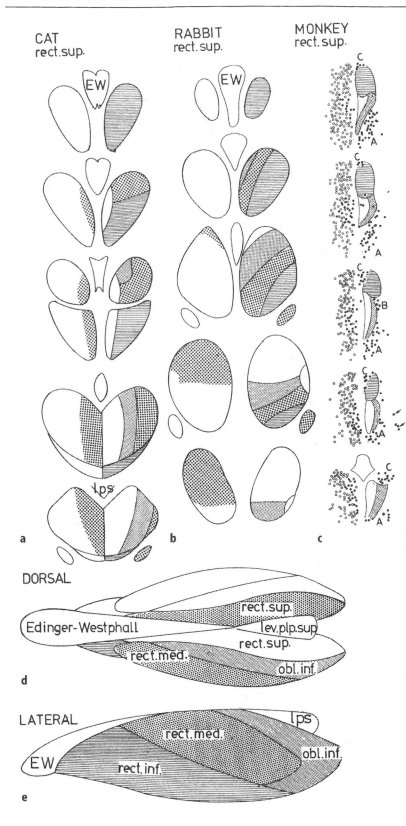

Fig. 22.30a–e. Transverse sections through the oculomotor nuclei of **a** the cat, **b** the rabbit (redrawn from Akagi 1978) and **c** the rhesus monkey (redrawn from Porter et al. 1983). Rostral is *up*. Dorsal and lateral views of a reconstruction of the oculomotor nuclei of the cat are shown in **d** and **e**. The pools of motoneurons innervating different eye muscles are indicated with *different shadings* (for key see **d** and **e**). The superior rectus pool in **a–c** is shown on the *left side*. For the monkey the retrogradely labelled motoneurons innervating the superior rectus muscle (**c**, *left*) and the medial rectus muscle (**c**, *right*) are indicated with *filled circles*. *Open circles* represent unlabelled cells. Three groups of medial rectus motoneurons (A–C) are indicated in the monkey, which were first distinguished in primates by Büttner-Ennever and Akert (1981). *Arrows* indicate medial rectus motoneurons located within the medial longitudinal fascicle. *EW*, Edinger Westphall nucleus; *lev.plp.-sup*, motoneurons innervating the levator palpebrae superior muscle; *lps*, motoneurons innervating the levator palpebrae superior muscle; *obl.inf*, motoneurons innervating the inferior oblique muscle; *rect.inf*, motoneurons innervating the inferior rectus muscle; *rect.med*, motoneurons innervating the medial rectus muscle; *rect.sup*, motoneurons innervating the superior rectus muscle

of the lateral lemniscus is located dorsal to the dorsal limb of the decussation of the brachium conjunctivum. The coarse fibres of the dorsal and ventral tegmental decussations are located more rostrally, at the level of the oculomotor nucleus. They contain fibres of the predorsal fascicle, originating from the superior colliculus, and the decussating rubrobulbar and rubrospinal fibres, respectively. The predorsal fascicle descends ventral to the medial longitudinal fascicle to the level of the rostral tegmentum pontis, where they fuse (Figs. 22.22–22.24). The rubrospinal tract descends in the lateral tegmentum mesencephali, where it traverses the lateral lemniscus (Figs. 22.22–22.24).

Most of the structures in this section are treated in more detail in Sect. 22.7 on the reticular formation, Sect. 22.8 on sensory systems and Sect. 22.9 on motor systems. The motor nuclei of the mesencephalon are considered in Sect. 22.5.2 and 22.5.3.

22.5.2
Eye Muscle Nuclei

Four recti and two oblique muscles move the eye of most vertebrates. The innervation of these muscles by the nuclei of cranial nerves III, IV and VI in mammals follows the same pattern as in all tetrapods (Baker 1986; Evinger et al. 1987; Evinger 1988). The retractor bulbi muscle is innervated by a small group of neurons known as the accessory abducens nucleus, located caudal to the main motor nucleus. The innervation of the levator palpebrae muscle is closely related to the innervation of the superior rectus muscle by the contralateral oculomotor nucleus. Embryologically, the levator palpebrae muscle arises from the superior rectus muscle (Leser 1925; Gilbert 1947; Isomura 1981). γ-Motoneurons are dispersed among the α-motoneurons of the oculomotor nuclei (Akagi 1978).

The motor nucleus of the abducens nerve is located dorsally in the pontine tegmentum at its junction with medulla (Fig. 22.22–22.24k). Fibres of the genu of the facial nerve surround it on its medial, rostral and lateral sides. Apart from motoneurons innervating the lateral rectus muscle, the nucleus also contains internuclear cells, which give rise to a crossed ascending pathway, terminating in the medial rectus subdivision of the oculomotor nucleus (Graybiel and Hartweig 1974; Baker and Highstein 1975; Gacek 1974; Furaya and Markham 1981; Highstein et al. 1982; McCrea et al. 1986). Internuclear cells in the rat are located around and ventral to the genu of the facial nerve and segregated from the motoneurons, which occupy a more dorsomedial position (Cabrera et al. 1988). No segregation was found in the cat or in primates. The internuclear fibres decussate and ascend in the medial longitudinal fascicle.

The motor nucleus of the trochlear nerve is located in the caudal mesencephalon, ventral to the periaqueductal grey, within the medial longitudinal fascicle. The fibres of the nerve pass caudally within the periaqueductal grey and decussate in the anterior medullary velum. The decussation of the trochlear nerve is not complete: 5 % of the motoneurons in the cat and rabbit give rise to uncrossed projections (Miyazaki 1985a; Murphy et al. 1986).

The nucleus of the oculomotor nerve is located in the rostral midbrain, ventral to the periaqueductal grey. The medial longitudinal fascicle borders on its lateral and ventral sides. The oculomotor nerve passes ventrally along the red nucleus to leave the brain stem medial to the cerebral peduncle. The localisation of motoneurons innervating the different eye muscles and the levator palpebrae muscle has been studied in different mammals (Warwick 1953; Porter et al. 1983; Spencer and Porter 1981: monkey; Tarlov and Tarlov 1971; Gacek 1974; Miyazaki 1985b: cat; Agaki 1978: cat and rabbit; Shaw and Alley 1981; Evinger et al. 1987: rabbit; Labandeira-Garcia et al. 1983: rat; Augustine et al. 1981: baboon and opossum). Motoneurons innervating different muscles are separated in rostrocaudally extending columns, with some overlap (Fig. 22.30). Inferior rectus motoneurons occupy the rostral pole and the lateral part of the nucleus. Superior rectus motoneurons are located medially and caudally in the nucleus. The medial rectus and inferior oblique divisions occupy an intermediate position. Dendrites of these motoneurons spread beyond the columns, but most dendrites of the medial rectus motoneurons extend in a longitudinal direction (Durand 1989). One apparent difference in the organisation of the oculomotor nucleus between primates and other mammals is the existence of three anatomically distinct subpopulations of medial rectus motoneurons in the primate (Büttner-Ennever and Akert 1981; Porter et al. 1983; Fig. 22.30c). The ventral portion of the rostral two thirds of the oculomotor nucleus (cell group A) contains the largest number of medial rectus motoneurons. Smaller-diameter medial rectus motoneurons located dorsally (cell group C) also extend throughout the rostral two thirds of the nucleus. These latter motoneurons innervate the small orbital fibres of the medial rectus muscle and are hypothesised to be significant for vergence movements (Büttner-Ennever and Akert 1981). A third population of medial rectus motoneurons (cell group B) occurs dorsolaterally in the caudal two thirds of the oculomotor nucleus. It is probably erroneous to assume that only the primate oculomotor nucleus contains multiple populations of medial rectus motoneurons. For example, re-analysis of the cat oculomotor nucleus shows the existence of a small group of dorsolateral medial rectus motoneurons (possibly homologous to cell group C of the primate) in addition to the principal population of ventrolateral medial rectus motoneurons (Furuya and Markham 1981; Miyazaki 1985b). Whether rabbits have a cell group C homologous to that in the primate is doubtful, since rabbits make relatively little vergence movements.

Axons from motoneurons of the superior rectus and levator palpebrae superius divisions cross within the oculomotor nucleus, innervating the contralateral muscles. The motoneurons innervating the levator palpebrae muscle constitute a separate midline group in the monkey, but lie intermingled with the cells of the superior rectus division in the cat and rabbit (Warwick 1953). Small neurons that cannot be labelled with injections of retrograde tracers in the eye muscles are located in the periphery and the dorsal part of the oculomotor nucleus (Shaw and Alley 1981). Some of these cells project back to the abducens nucleus.

The premotor connections of the oculomotor nuclei were reviewed by Evinger (1988). The internuclear pathways between the abducens and oculomotor nuclei have already been mentioned. Vestibular inputs are reviewed in Sect. 22.9.2. Projections from the pontine paramedian reticular formation and the rostral interstitial nucleus of the medial longitudinal fasciculus are involved in the control of vertical and horizontal saccadic eye movements, respectively. In both cats and primates, the rostral interstitial nucleus of the medial longitudinal fasciculus projects bilaterally to the trochlear nucleus and to all subdivisions of the oculomotor nucleus, except for the medial rectus subdivision (Büttner-Ennever 1977, 1979; Graybiel 1977; Büttner-Ennever and Büttner 1978; Steiger and Büttner-Ennever 1976; Wang and Spencer 1992, 1996).

The interstitial nucleus of Cajal projects bilaterally to motoneurons of the oculomotor nucleus (Fukushima et al. 1978; Graybiel and Hartweig 1974; King et al. 1980; Rutherford and Gwyn 1982; Schwindt et al. 1974; Zuk et al. 1982). A small bilateral projection to the abducens nucleus is present in primates, but not in cats (Langer et al. 1986). The interstitial nucleus also projects upon the vestibular nuclei (King et al. 1980) and the perihypoglossal nuclei (McCrea and Baker 1985). The interstitial nucleus is involved in vertical and torsional eye movements, but its precise role is not clear.

22.5.3
Parasympathetic Visceromotor Neurons of the Mesencephalon

The parasympathetic, preganglionic innervation of the ciliary ganglion is derived from the Edinger-Westphal nuclei. These nuclei are located on both sides of the midline, dorsomedial to the main somatomotor nucleus of the oculomotor nerve. It was shown by Loewy et al. (1978), Loewy and Saper (1978), Toyoshima et al. (1980) and Burde et al. (1982) in the cat that the parasympathetic neurons correspond to medium-sized cells, located in the periphery of the Edinger-Westphal nucleus and in the adjacent periaqueductal grey and medial ventral tegmental area. In primates, the afferent parasympathetic neurons of the ciliary ganglion are restricted to the Edinger-Westphal nucleus, the rostrally adjoining anteromedian nucleus (which was included with the Edinger-Westphal nucleus in the cat) and Perlia's nucleus, located in the midline between the oculomotor nuclei (*Macaca fascicularis*: Burde and Loewy 1980). Spinal projections take their origin from other neurons in the Edinger-Westphal nucleus in the cat and in primates. In addition, a projection of this nucleus to group y, a vestibular subnucleus located in the floccular peduncle, has been documented in the cat (Graybiel and Hartweigk 1974) and monkey (Sugimoto et al. 1978).

22.6
Cerebellum and Precerebellar Nuclei

J. Voogd

22.6.1
Gross Anatomy of the Mammalian Cerebellum

The folial pattern of the mammalian cerebellum has been extensively studied in adult mammals (Bolk 1906; Elliot Smith, 1900, 1902, 1903; Riley 1928; Scholten 1946; Jansen 1954; Schober and Brauer 1974) and in its development (Bradley 1903, 1904; Larsell 1970; Larsell and Jansen 1972). Bolk (1906) studied the gross anatomy of the cerebellum in some 70 mammalian species. On the basis of these observations, exemplified in the cerebellum of *Lemur albifrons* and the study of the development of the human cerebellum, he subdivided the cerebellum into four fundamental regions: (1) the unitary anterior lobe, (2) the lobulus simplex, (3) the posterior vermis (lobus medianus posterior) and (4) the paired hemispheres or lobi lateralis posteriores (Figs. 22.31, 22.32). Bolk described the variability in the appearance of these regions in different species in ontogenetic terms. The anterior lobe "derives its ultimate form from a single growth centre, which exerts its main influence in a sagittal direction; its potency is greater in the midsagittal plane, in a lateral direction its strength gradually diminishes" (Bolk 1906, p. 42). The caudal limit of this unitary growth centre is not formed by the primary fissure, but it extends further caudally, including the unpaired lobule located immediately behind the primary fissure, which Bolk called the lobulus simplex. "Lobulus simplex and the lobus anticus are genetically related" (Bolk 1906, p. 44).

Bolk has been criticised because he failed to recognise a vermis in the anterior lobe and the lobulus simplex. In most mammals, the medial region of the anterior lobe and the lobulus simplex are demarcated from their lateral parts by a faint sulcus that contains the paravermal vein (see Fig. 22.35c). This median region is often referred to as the anterior vermis. The demarcation of the anterior vermis is more distinct in Cetacea (see Fig. 22.39), but even in whales the main characteristic of the anterior lobe and the lobulus simplex remains the continuity of their transverse fissures, across the midline, into the hemispheres.

The remainder of the posterior lobe corresponds to Bolk's lobulus complicatus. The name lobulus simplex has survived, while the name lobulus complicatus has not. A distinct paramedian sulcus is present caudal to the lobulus simplex. It separates the vermis from the hemispheres. Posterior vermis and hemispheres, therefore, constitute mutually independent growth centres which show a tendency to expand in a sagittal direction. The folial chains of vermis and hemispheres are fractionated into segments. Local sagittal expansion of these segments leads to the formation of loops and of secondary fissures located in the centre of these loops. The names used by Bolk for the three segments of the folial chain of the hemisphere (ansiform lobule, paramedian lobule and formatio vermicularis) refer to the presence of loops in the rostral and caudal segments of the chain and the straight course of the chain in the intermediate, paramedian segment. Bolk's names have been retained by later anatomists, with the exception of the formatio vermicularis, which now is known as the paraflocculus (introduced by Stroud 1895) and the flocculus. Hidden folia interconnect the segments of the folial chains (see Figs. 22.34–22.36).

The unitary structure of the anterior lobe and the lobulus simplex, and the presence of distinct folial chains of vermis and hemisphere in the rest of

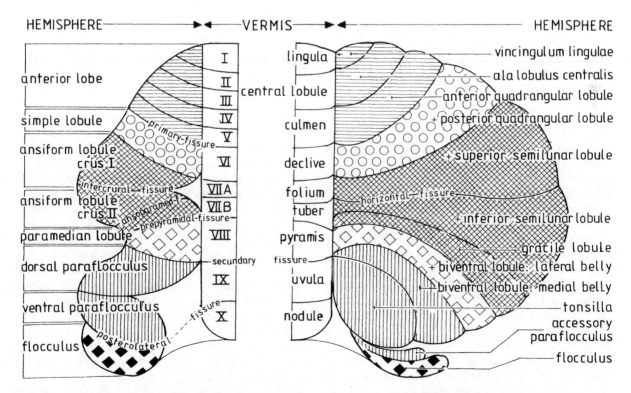

Fig. 22.31. Subdivision and nomenclature of the cerebellum. *Left*, Numbering of cerebellar lobules according to Larsell (1952) and comparative anatomical nomenclature of Bolk (1906) for the hemisphere. *Right*, Nomenclature of the human cerebellum. Homologous subdivisions are indicated with the same symbols. From Voogd et al. (1996)

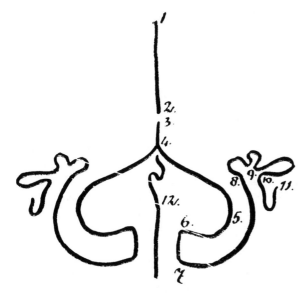

Fig. 22.32. Morphology of the cerebellum according to Bolk (1906). The folia chains of vermis and hemisphere are depicted here as *continuous lines*. Vermis and hemisphere cannot be separated in the anterior lobe (*1–2*) and the simple lobule (*3–4*). They divide caudal to the simple lobule, where the caudal vermis (*12*) and the hemisphere (*4–11*) behave like independent folial chains. The folial chain of the hemisphere can be subdivided into successive segments: ansiform lobule (*5*), paramedian lobule (*6–7*), paraflocculus (crus circumcludens, *7–10*) and the flocculus (uncus terminalis, *11*)

the posterior lobe, is well illustrated in the illustrations by Sultan and Braitenberg (1993) of outlines of the shapes of the cerebellar cortex in different species. These figures also give a good impression of the variations in width and length of the folial chains (Fig. 22.33). The cortex within the folial chains of vermis and hemispheres is continuous. The 'independency' of the folial chains of vermis and hemispheres (Bolk's 'independent growth centers') finds its expression in the constriction of the cortex of the folia at the border of vermis and hemisphere or in the interruption of the molecular layer or the absence of the entire cortex at the bottom of the paramedian fissure. The constriction or interruption of the cortex only affects the parallel fibres, which constitute the only transverse connection within the cortex. At such a site, the number of parallel fibres is diminished or they are completely absent, effectively separating the mossy parallel-fibre systems (see below) terminating on both sides of the gap. This interruption is almost complete between the rostral lobules of the posterior vermis (lobule C2 of Bolk, corresponding to the folium and tuber vermis and the ansiform lobule) and between the caudal lobules a (the nodulus) and b (the uvula) of the caudal vermis and the paraflocculus and the flocculus. It is less complete between lobule C1 (the

pyramis) and the paramedian lobule, where the cortex of the pars copularis (Elliot Smith 1900, 1902, 1903) bridges the paramedian sulcus.

The relative independence of the folial chains of vermis and hemispheres is also expressed in their development. In the 'unitary growth centre' of the anterior lobe and the lobululus simplex, the fissures develop in the midline and subsequently grow laterally. In the middle part of the posterior lobe, the main, interlobular fissures develop in the midline and grow laterally into the hemisphere, but the intralobular fissures develop independently in vermis and hemispheres. This middle region is located between the lobulus simplex rostrally and the uvula and paraflocculus caudally. In the caudal part of the cerebellum corresponding to the uvula, nodulus, paraflocculus and flocculus, both the interlobular and intralobular fissures develop independently in vermis and hemispheres (Bolk 1906). Sagittal continuity of the cortex within the folial chains of vermis and hemispheres, different degrees of transverse (i.e. parallel-fibre) continuity between different segments of vermis and hemispheres, and local expansion of the folial chains into folial rosettes lobules are the key elements of Bolk's generalised plan of the mammalian cerebellum.

Both Bradley (1903, 1904) and Larsell (whose studies are summarised in Larsell 1967, 1970; Larsell and Jansen 1972) used staged fetuses to establish the origin and 'growth' of the transverse fissures. These studies emphasised the importance of the early mediolateral continuities of lobules of vermis and hemispheres contained between these fissures. Larsell's studies culminated in the distinction of ten transverse lobules in the cerebellum of the rat (Larsell 1952), each consisting of a vermal lobule, indicated by a Roman numeral (I–X), and two hemispheral lobules, indicated by a Roman numeral with the prefix H (Figs. 22.31, 22.34). This subdivision could be applied to most cerebella of the mammalian species he studied (Fig. 22.35; see also Figs. 22.36, 22.38–22.40). Apart from these purely descriptive studies, Larsell (1934, 1970) attached particular importance to the posterolateral fissure as the first transverse fissure to develop, separating the caudal flocculonodular lobe from the rostral corpus cerebelli. The flocculonodular lobe, which consists of the flocculus and the nodulus, "is the principal terminus of primary and secondary vestibulocerebellar fibres", and contains "the vestibular commissure corresponding to the vestibulolateral commissure of urodeles" and "is also known as the vestibulocerebellum." "The corpus cerebelli is the terminus of spinal, trigeminal, bulbar, tectal and pontine fibres, related to the proprioceptive, exteroceptive and cerebropontocerebellar systems. An

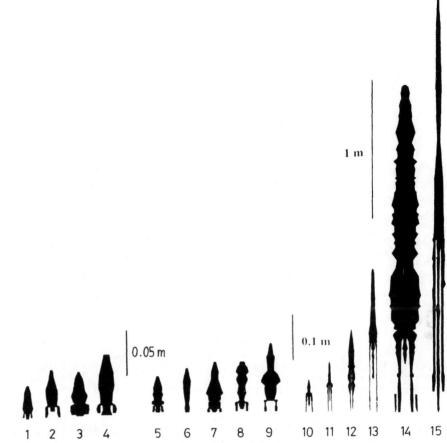

Fig. 22.33. Outlines of the shapes of the cerebellar cortices, obtained by connecting the ends of the most prominent folia. The scale is the same in the laterolateral and anteroposterior direction. 1, Mouse; 2, bat; 3, flying fox; 4, guinea pig; 5, rabbit; 6, pigeon; 7, hare, 8, chinchilla; 9, squirrel; 10, dog; 11, cat: macaque; 13, sheep; 14, human; 15, bovine. Magnifications differ for the diagrams 1-4, 5-9 and 10-15. (From Sultan and Braitenberg 1993)

anterior commissura cerebelli, which is primarily composed of ventral spinocerebellar and trigeminocerebellar fibres corresponds to the cerebellar commissure of Herrick in urodeles" (Larsell 1970, p. 7).

Larsell (1952, 1970) subdivided the anterior lobe into five lobules (lobules I-V and HI-HV). A single, preculminate fissure subdivides the anterior lobe of even the smallest mammals in lobules I-III and IV/V). The depth of the interlobular fissures and the relative size of lobules IV and V, with respect to more rostral lobules, which are greater in Cetacea, Proboscoidea and primates, are signs of progressive development. According to Dillon (1962) and Schober and Brauer (1974), the size of the anterior lobe in Monotremata is disproportionally large with respect to the posterior lobe. Larsell (1970) compared the illustrations of the cerebellum of the platypus *Ornithorhynchus* (Elliot Smith 1899; Hines 1929) and the echidna *Tachyglossus* (Ziehen 1897; Elliot Smith 1899; Hines 1920; Abbie 1934) with Dillon's diagrams and concluded that Dillon (1962) had incorporated lobule VI into the anterior lobe. Larsell's interpretation is followed in Fig. 22.40. The lobulus simplex (lobule VI and HVI of Larsell) varies in size, but is always easy to recognise. The lobulus simplex is large in Perissodactyla and Artiodactyla and reaches huge dimensions in Proboscoidea. The presence of the paramedian sulcus demarcates the caudal border of the lobulus simplex. The posterior superior fissure, which forms the border between vermal lobules VI and VII, and between the lobulus simplex and the folia of the ansiform lobule is therefore discontinuous at the paramedian sulcus.

The cortex between lobule VII and the ansiform lobule is usually interrupted. The extent of this interruption varies between a slight irregularity or gap in the molecular layer of this region in small insectivores, bats and rodents to the large expanse of white matter located in the centre of the ansiform lobule in marsupials (Fig. 22.34) and Lagomorpha. The anterior and posterior limbs of the folial loop of the ansiform lobule (lobule VIIA of Larsell) are known as crus I and crus II. The position of the intercrural fissure separating the folia of crus I and

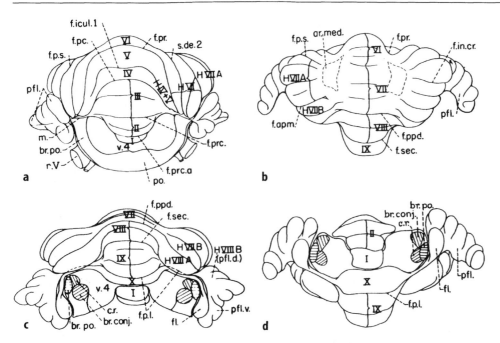

Fig. 22.34a–d. Cerebellum of the North American opossum *Didelphis virginiana*. a Anterior view. b Dorsal view. c Ventroposterior view. d Ventral view. *ar.med*, medullary area; *br.conj*, brachium conjunctivum; *br.po*, brachium pontis; *c.r.*, restiform body; *f.apm*, ansoparamedian fissure; *f.icul1*, intraculminate fissure 1; *f.in.cr*, intercrura; *f.p.l.*, posterolateral fissure; *f.p.p.d.*, prepyramidal fissure; *f.pc*, precentral fissue; *f.pr*, primary fissure; *f.prc a*, precentral fissure a; *f.ps*, posterior superior fissure; *f.sec*, secondary fissure; *fl*, flocculus; *m*, margin of the cerebellar cortex; *Po*, pons; *pfl*, paraflocculus; *pfl.d*, dorsal paraflocculus; *pfl.v*, ventral paraflocculus; *r.V*, root of the trigeminal nerve; *s.de2*, declival sulcus 2; *v.4*, 4th ventricle; *HI–HX*, hemispheral lobules I–X; *I–X*, vermal lobules I–X. From Larsell (1970)

II and the ansoparamedian fissure, between crus II and the paramedian lobule, depends on the changes in direction of the folial chain and therefore often remains arbitrary. In commonly used laboratory animals, the position is based on the conventions published by Larsell (1970). The intercrural fissure is never continuous with a particular fissure in the vermis. The ansiform lobule is relatively small in most mammals. It is fairly large in carnivores and reaches its greatest development in primates (see Figs. 22.34–22.36, 22.38, 22.39).

The ansoparamedian fissure sometimes continues in a fissure that, according to Larsell (1970), subdivides lobule VII into a rostral lobule VIIA and a caudal lobule VIIB. Lobule VII forms a loop in many Felidae, Canidae, Perissodactyla, Artiodactyla and some primates, and this loop sometimes also involves the pyramis (lobule VIII of Larsell; Fig. 22.35). Elliot Smith (1900, 1902, 1903) emphasised the continuity of the pyramis with the caudal portion of the paramedian lobule and the paraflocculus. Laterally, the cortex of the caudal paramedian lobule borders on, and is continuous with, the cortex of the first folium of the dorsal paraflocculus. This connection between the pyramis and the paraflocculus is known as the copula pyramidis. In many mammalian species, including all primates (Fig. 22.36), the copula is subfoliated and the paramedian sulcus between the pyramis and the copula is well developed. In the North American opossum, the copula (HVIIIA) is quite distinct (Fig. 22.34). In the cat, the cortex between the pyramis and copula is even completely interrupted. Two subdivisions, therefore, can be recognised in the paramedian lobule of all mammals. A rostral folial rosette, which is considered to be the hemisphere of lobule VIIA of Larsell, and its caudal, copular portion (lobule HVIIIA of Larsell).

The caudal segments of the folial chain of the hemisphere are represented by the laterally displaced loops of the paraflocculus and the flocculus. The paraflocculus is extremely large in Cetacea (see Fig. 22.39) and is generally larger in aquatic mammals (Jansen 1954). A smaller or larger segment of the paraflocculus is located in the subarcuate fossa of the petrosal bone in many species. This folial loop or rosette is known as the petrosal lobule. The paraflocculus can be arbitrarily divided into dorsal and ventral limbs, which are known as the dorsal and the ventral paraflocculus (Larsell's lobules HVIIIB and HIX). The petrosal lobule may include the entire paraflocculus, as in Rodentia, Lagomorpha and Marsupialia (Fig. 22.34, 22.37A), or the distal part of the dorsal paraflocculus, as in many pri-

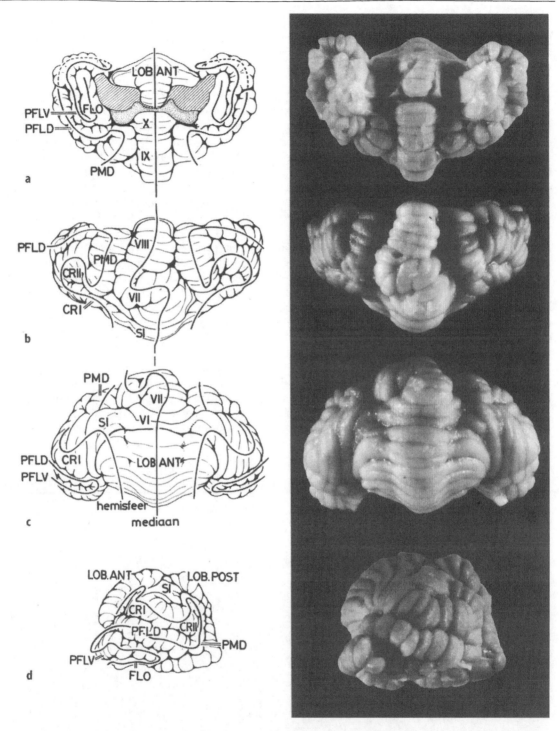

Fig. 22.35a–d. Cerebellum of the cat. **a** Ventral view. **b** Caudal view. **c** Dorsal view. **d** Lateral view. The continuity in the folial chains of vermis and hemisphere is indicated by *lines*. *CRI*, crus I of the ansiform lobule; *CRII*, crus II of the ansiform lobule; *FLO*, flocculus; *LOB ANT*, anterior lobe; *PFLD*, dorsal paraflocculus; *PFLV*, ventral paraflocculus; *PMD*, paramedian lobule; *SI*, simple lobule; *VII–X*, lobules of the caudal vermis. (From Bigaré 1980)

mates (Figs. 22.36, 22.37b). The flocculus (Bolk's uncus terminalis; Larsell's lobule HX) is bent rostrally on the ventral aspect of the ventral paraflocculus.

The flocculus is apposed to the surface of the middle cerebellar peduncle and is located in the roof of the lateral recess of the fourth ventricle (Figs. 22.22–22.24). Because the direction of the distal segments of the folial chain of the hemisphere represented by the paraflocculus and the flocculus is reversed with respect to its proximal part, the outer border of the hemisphere (heavy line in Fig. 22.37) at the level of the dorsal paraflocculus and the flocculus faces medially, and the lateral aspect of these lobules is continuous with the medial border of the folial chain, facing the paramedian sulcus.

The cortex between the dorsal and ventral segments of the paraflocculus and between the paraflocculus and the flocculus at the bottom of the posterolateral fissure is always continuous, with the possible exception of the human cerebellum. The position of the posterolateral fissure, which separates the flocculus from the paraflocculus, is sometimes difficult to establish. This problem was raised for macaque monkeys, when Larsell (1970) changed his original opinion (Larsell 1953) on the demarcation of the flocculus and designated part of his original flocculus as the ventral paraflocculus and included the petrosal lobule with the dorsal paraflocculus (Figs. 22.36, 22.37). The continuity of the cortex of the ventral paraflocculus and the flocculus at the bottom of the posterolateral fissure is usually broad, but the connection between the ventral paraflocculus (Larsell 1970) and the petrosal segment of the dorsal paraflocculus in primates is greatly attenuated. In the human cerebellum, where the dorsal paraflocculus is represented by the medial belly of the biventral lobule and the tonsil (Bolk 1906; Larsell 1947) and the ventral paraflocculus, presumably by the accessory paraflocculus of Henle (for descriptions of this region, see Larsell and Jansen 1972; Tagliavini and Pietrini 1984), this connection is broken. Objective criteria for the subdivision of the distal portion of the folial chain of the hemisphere based on the corticonuclear and olivocerebellar connections of these parts are available (see below), but these borders do not necessarily correspond with the location of the main fissures.

The distal lobules of the caudal vermis are the uvula and the nodulus (lobules IX and X of Larsell). The secondary fissure of Bradley (1903,1904) and the posterolateral fissure are located between VIII/IX and IX/X, respectively. The epithelial roof of the fourth ventricle is attached to the distal margin of the nodulus. The cortex between these lobules and

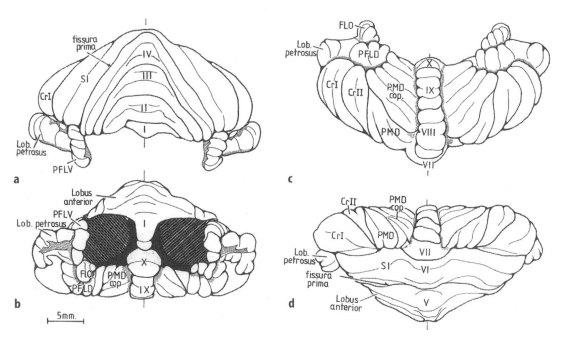

Fig. 22.36a–d. Cerebellum of *Macaca fascicularis*. **a** Anterior aspect. **b** Ventral aspect. **c** Caudal aspect. **d** Dorsal aspect. Regions without cortex, where the white matter comes to the surface, are indicated by *light hatching*. *Heavy hatching* indicates cross-section of the cerebellar peduncles, and *solid black* indicates roof of the fourth ventricle. *CrI*, Crus I of the ansiform lobule; *CrII*, crus II of the ansiform lobule; *FLO*, flocculus; *fpl*, posterolateral fissure; *PFLD*, dorsal paraflocculus; *PFLV*, ventral paraflocculus; *PMD(cop)*, paramedian lobule (copula pyramidis); *SI*, lobulus simplex

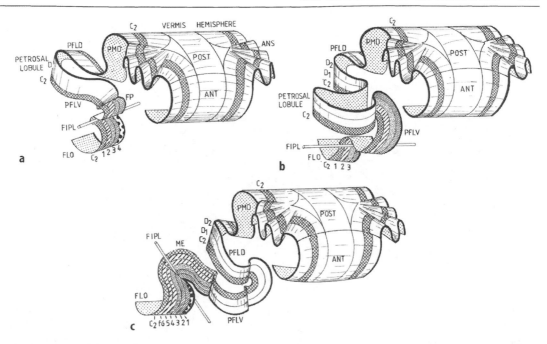

Fig. 22.37a–c. Configuration of the paraflocular and floccular segments of the folial chain of the hemisphere in **a** the rabbit, **b** the rhesus monkey and **c** the cat. The continuity of the folial chain and its subdivision in longitudinal zones is indicated. In the rabbit, the entire paraflocculus participates in the formation of the petrosal lobule; in the monkey, the petrosal lobule belongs to the dorsal paraflocculus. The transitional lobule between the flocculus and the paraflocculus is represented by folium P in the rabbit (Tan et al. 1995a), the ventral paraflocculus in the monkey (Larsell 1970) and the medial extension of the ventral paraflocculus (*ME*) in the cat (Gerrits and Voogd 1982). The zonal configuration of this transitional region is very similar to the flocculus, but different from the paraflocculus. The only zone which continues uninterrupted from the flocculus over the entire hemisphere is the C2 zone. Corresponding cortical zones in different species are indicated by the same symbols. *ANS*, ansiform lobule; *ANT*, anterior lobe; C_2, cortical C2 zone; D_1, D_2, cortical zones D1, D2; *DPFL*, dorsal paraflocculus; *f1–6*, cortical zones of the flocculus of the cat; *FLO*, flocculus; *FIPL*, posterolateral fissure; *FP*, folium P of the rabbit; *HEM*, hemisphere; *ME*, medial; extension of the ventral paraflocculus; *PMD*, paramedian lobule; *VPFL*, ventral paraflocculus, *1–(3)4*, floccular zones 1–(3)4 of rabbit and monkey

the paraflocculus and flocculus is usually completely interrupted. In Microchiroptera, Insectivora, marsupials and some Rodentia, a narrow rim of cortex connects the nodulus with the flocculus along the attachment of the roof of the fourth ventricle (Fig. 22.34). According to Larsell (1970), transient connections between the paraflocculus and lobules VIII and IX justify the designation of the dorsal paraflocculus as lobule HVIIIB and the ventral paraflocculus as lobule HIX, but these connections have not been substantiated in later studies.

Inspection of cerebella in different species has revealed many variations in the width of the different segments of the folial chains. The hemispheres of the cerebellum of monotremes are small (see Fig. 22.40), and these cerebella therefore resemble the avian cerebellum (Hines 1929; Larsell 1970; Schober and Brauer 1974). The width of the hemisphere of the anterior lobe and the folial chain of the hemisphere in Marsupials, Insectivora, Perissodactyla and Artiodactyla is equal to or even smaller than that of the anterior and posterior vermis. The cerebellum of the coney *Procavia* is a fine example of this type (Fig. 22.38). Vermis and hemispheres are of the same width and are oriented as three parallel chains. A folial loop of the ansiform lobule is completely lacking. In rodents and carnivores, the anterior lobe, the lobulus simplex and the lobulus ansiformis are wider, but the folia of the paramedian lobule and the paraflocculus are narrow (Fig. 22.35). The entire hemisphere, with the exception of the paraflocculus, reaches a great width in monkeys (Fig. 22.36), and the width of the dorsal paraflocculus is greatly increased in the great apes, in the human cerebellum and in Cetacea (Fig. 22.39).

The great value of Bolk's basic plan of the mammaliancerebellum and its advantage over Larsell's scheme is that it accounts for two conflicting tendencies in the morphology of the cerebellum, one leading to local elongation of the cortex of the folial chains and the other to maintaining the transverse orientation of the fissures and the continuity of the lobules of vermis and hemispheres. The elongation of the cortex is related to the longitudinal zonal organisation of the projections of the Purkinje cells of the cerebellar cortex to the cerebellar and vesti-

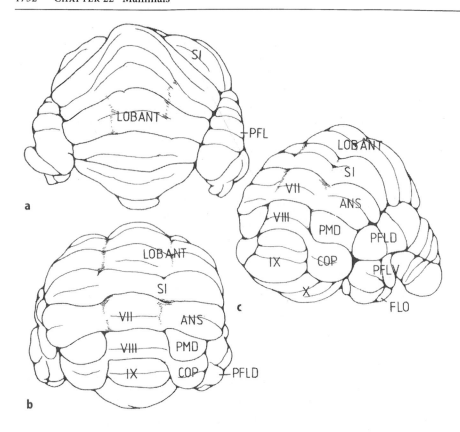

Fig. 22.38a–c. Cerebellum of the coney (*Procavia*). a Anterior view. b Dorsal view. c Caudolateral view. *ANS*, ansiform lobule; *CP*, copula pyramidis; *FL*, flocculus; *lob.ant.*, anterior lobe; *PFL*, paraflocculus; *PMD*, paramedian lobule; *SI*, lobulus simplex; *VII–X*, lobules *VII–X*

bular nuclei. Transverse continuity in the cerebellar cortex is represented by the axons of the granule cells, i.e. the parallel fibres. The longitudinal zonal organisation and the transverse continuity of the cortex find their expression during two successive periods in the morphogenesis of the cerebellum (Korneliussen 1968; Kappel 1981). Purkinje cell zones develop very early, before any of the transverse fissures have appeared. During the histogenesis of the cortex, the granule cells appear much later than the Purkinje cells, and their production coincides with the appearance of the transverse fissures. Maintenance of the ortholinear-perpendicular relations between the parallel fibres and the Purkinje cells is a major condition apparently determining the gross morphology of the mammalian cerebellum.

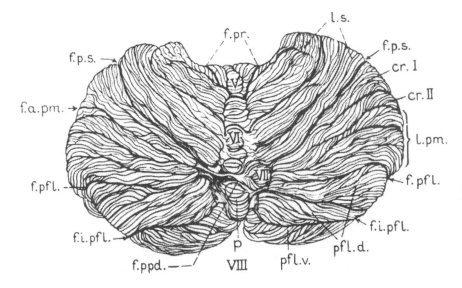

Fig. 22.39. Dorsal view of the cerebellum of a 430-cm C.R. length fin whale embryo (*Balaenoptera physalus*) (C 290)×1. (From Jansen 1954). *cr.I*, crus I of the lobulus ansiformis; *cr.II*, crus II of the lobulus ansiformis; *f.a.pm.*, fissura ansoparamedianus; *f.i.pfl.*, fissura intraparaflocularis; *f.p.s.*, fissura posterior superior; *f.pfl.*, fissura paraflocularis; *f.ppd.*, fissura prepyramidalis; *f.pr.*, fissura prima; *l.pm*, lobulus paramedianus; *l.s.*, lobulus simplex; *p.*, pyramis; *pfl.d.*, dorsal paraflocculus; *pfl.v*, ventral paraflocculus; *V–VIII*, lobulus V–VIII of Larsell

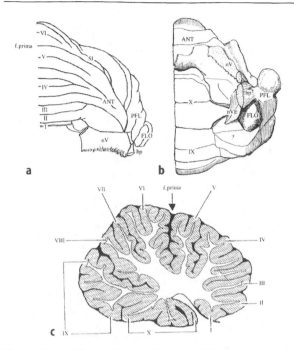

Fig. 22.40a–c. The cerebellum of the duckbill platypus *Ornithorynchus*. **a** Anterior aspect. **b** Ventral aspect. **c** Mid-sagittal section. The *question mark* indicates the possible cortical connection between the uvula (lobule IX) and the paraflocculus. *ANT*, anterior lobe; *bp*, brachium pontis; *FLO*, flocculus; *f.prima*, fissura prima; *nV*, trigeminal nerve; *nVE*, vestibular nerve; *PFL*, paraflocculus; *SI*, lobulus simplex; *I–X*, lobules I–X of Larsell. (Partially redrawn and relabelled from Dillon 1962, according to Larsell 1970)

22.6.2
Histology of the Cerebellar Cortex

The histology of the cortex is uniform over the entire cerebellum and is very similar in all mammals (for reviews, see Cajaliqii; Braitenberg and Atwood 1958; Eccles et al. 1967; Fox et al. 1967; Mugnaini 1972; Palay and Chan-Palay 1974; Figs. 22.41, 22.42). The dendritic and/or axonal arborisations of most cortical neurons are oriented either in the direction of the long axis of the folia, i.e. parallel to the transverse fissures, or in a plane perpendicular to the long axis of the folia, i.e. the parasagittal plane for the cortex near the midline. Purkinje cells constitute the output of the cortex; their axons leave the cortex and terminate with inhibitory GABAergic terminals on cells of the cerebellar or the vestibular nuclei (Fig. 22.42). The somata of the Purkinje cells are localised in a monolayer between the granular layer and the superficial, molecular layer of the cortex. The dendritic tree of the Purkinje cells is flattened in a plane perpendicular to the long axis of the folia (Fig. 22.41a). Purkinje cell axon collaterals terminate on neighbouring Purkinje cells. This collateral plexus, which is emitted shortly after the origin of the myelinated axon from the Purkinje cell soma, is oriented in the same direction as the dendritic trees (Fig. 22.41o).

Granule cells are small, round cells which emit four to five thin dendrites, terminating in claws and a thin, unmyelinated axon (Fig. 22.41g). Granule cells are located in cell nests in the granular layer. Mossy fibres, which are their main afferents, terminate with complex synapses (glomeruli) on the granule cell dendrites. The glomeruli are located in the cell-free spaces (islands of Held) between the granule cells. The axons of the granule cells ascend to the molecular layer, where they divide into two branches that course over considerable distances (up to 5 mm; Pichitpornchai et al. 1994) in the long axis of the folia and are known as the parallel fibres. Small beads along the ascending part of the granule cell axon and the parallel fibres are the sites of synaptic contact of the granule cell with dendrites of Purkinje cells or inhibitory interneurons in the molecular layer.

Most mossy fibres (Figs. 22.41h, 22.42) are of extracerebellar origin, while some are terminals from collaterals of neurons of the cerebellar nuclei and unipolar brush cells (see below). Mossy fibres branch extensively in the cerebellar white matter; their terminal branches, studded with complex boutons (the mossy fibre rosettes), are preferentially oriented in a plane perpendicular to the long axis of the folia. The mossy parallel-fibre afferent system is widely divergent and influences multiple patches of Purkinje cells extending in the long axis of the folia.

Purkinje cells receive two major afferent systems: parallel fibres and climbing fibres that terminate on different compartments of the Purkinje cell dendritic tree. Climbing fibres (Figs. 22.41n, 22.42) take their origin from the contralateral inferior olive, located in the ventral medulla oblongata (Figs. 22.22–22.24j,h). Each Purkinje cell receives a single climbing fibre, which terminates on short, stubby spines on the proximal dendrites. The parallel fibres terminate on long-necked spines of the distal spiny branchlets. In mammals, smooth dendrites of the Purkinje cells and their climbing fibre occupy the entire molecular layer, from the level of the somata to the pial surface. Climbing fibre terminals on the Purkinje cell soma are few or absent.

Climbing fibres and parallel fibres are glutamatergic and excitatory (for reviews of the transmitters in the cerebellar cortex, see Nieuwenhuys 1985; Oertel 1993; Voogd et al. 1996). The double (climbing fibre and parallel fibre) innervation of the Purkinje cell is one of the main features of the cerebellar cortex and can be recognised in most vertebrates. Parallel fibres are the second link in the afferent mossy parallel-fibre input of the Purkinje cell.

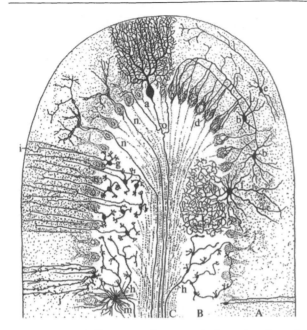

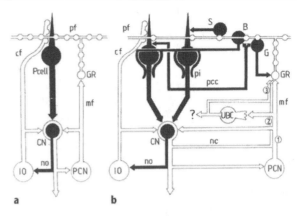

Fig. 22.41. Semi-diagrammatic parasagittal section through a folium of the mammalian cerebellum, based on data from Golgi-stained material. *A*, molecular layer; *B*, granular layer; *C*, white matter; *a*, Purkinje cell; *b*, basket cells of the lower molecular layer; *d*, terminal basket formation of the basket cell axon; *e*, superficial stellate cells; *f*, Golgi cell; *g*, granule cells with their ascending axons; *i*, the bifurcation of the granule cell axons; *h*, mossy fibres; *j*, epithelial glial cell; *m*, astrocyte of the granular layer; *n*, climbing fibre; *o*, branching point of Purkinje cell recurrent axon collaterals. (Redrawn from Cajal 1911)

Fig. 22.42a,b. Cerebellar circuits. Inhibitory neurons are indicated in *black*. **a** Main circuit. **b** Cortical interneurons and recurrent pathways. *Question mark* indicates possible extracortical axonal pathways of the unitpolar brush cell. *B*, basket cell; *cf*, climbing fibre; *G*, Golgi cell; *GR*, granule cell; *IO*, inferior olive; *mf*, mossy fibre; *nc*, nucleocortical axons; *no*, nucleo-olivary axons; *pcc*, recurrent Purkinje cell axon collaterals; *Pcell*, Purkinje cell; *PCN*, precerebellar nuclei; *pf*, parallel fibres; *pi*, pinceau of basket cell axons; *S*, stellate cell; *UBC*, unipolar brush cell. *1*, Extracerebellar mossy fibre; *2*, nucleo-cortical mossy fibre; *3*, mossy fibre collateral of unipolar brush cell

The single ('simple') spikes induced by the parallel fibres in Purkinje cells determine their overall firing rate. Climbing fibres, with their extensive synaptic surface, induce an 'all-or-none' response in the Purkinje cells that consists of a short spike train, known as the complex spike. Climbing fibres fire at very low frequencies (+10 Hz), and the contribution of the complex spikes to the overall firing rate of Purkinje cells is therefore rather small.

The function of the innervation of the Purkinje cells by the climbing fibres is one of the central problems in cerebellar neurobiology. In the early hypotheses of Marr (1969) and Albus (1971), climbing fibres would permanently change the weight of simultaneously activated parallel fibre synapses on the same Purkinje cells. This hypothesis recently received strong support from the observations of long-term depression (LTD) of parallel fibre-Purkinje cell transmission by ionotropic (α-amino-3-hdroxy-5-methyl-4-isoxalone-proprionic acid, AMPA type) glutamate receptors, resulting from simultaneous activation of climbing fibres and a second, metabotropic receptor involved in the parallel fibre-Purkinje cell synapse. AMPA receptors are preferentially located in the postsynaptic density of the spines of the spiny branchlets; metabotropic mGluR1 subunits of the glutamate metabotropic receptor are located immediately outside these active sites in the non-synaptic surface of the spine (Baude et al. 1993; Nusser et al. 1994; for a review, see Voogd et al. 1996). The receptor characteristics of the climbing fibres synapes on the smooth dendrites have not yet been completely elucidated.

Inhibitory interneurons of the molecular layer can be subdivided into the superficially located stellate cells and the basket cells of the deep strata of the molecular layer (Figs. 22.41b-f, 22.42BGS). The dendritic trees of these cells and the axonal arborisation of the basket cells are oriented perpendicular to the long axis of the folia. Stellate cell axons terminate on smooth, proximal dendrites of the Purkinje cells. Basket cell axons emit collaterals that surround Purkinje cell somata as a basket (Figs. 22.41d, 22.42p,i) and form axosomatic synapses. The terminal branches of the axon collaterals of several basket cells surround the non-myelinated initial segment of the Purkinje cell axon as the pinceau. There are no synaptic contacts between the axons of the pinceau and the initial segment.

Golgi cells are the GABAergic neurons of the granular layer (Figs. 22.41f, 22.42G). They extend their smooth dendrites in all directions in the molecular layer. Golgi cell axons course for some distance in the direction of the long axis of the folia (De Zeeuw et al. 1994) before they terminate in a dense plexus in the granular layer. Axon terminals of the Golgi cells synapse with granule cell dendrites at the periphery of the glomeruli. Candelabrum cells located within the Purkinje cell layer resemble Golgi cells, but distribute their axons to the molecular layer (Lainé and Axelrod 1994). Certain neurochemical properties distinguish the Golgi cells from the population of basket and stellate cells (for a review, see Voogd et al. 1996).

Recently, a new cell type, the unipolar brush cell, was identified in the cerebellar cortex of different mammalian species, including humans (Hockfield 1987; Cozzi et al. 1989; Muñoz 1990; for a review, see Mugnaini and Floris 1993). Unipolar brush cells are located within the granular layer, where the brush-like tip of their single dendrite forms a large, crenalated synapse with a mossy fibre terminal (Fig. 22.42UBC). Similar giant synapses were originally thought to contact Golgi cells and are known as the synapse-en-marron of Palay and Chan-Palay (1974). The unipolar brush cell is an excitatory neuron; its axon emits collaterals which terminate as mossy fibres on neighbouring granule cells, but the final destination of the axon is unknown. The distribution of unipolar brush cells is mostly restricted to the vestibulocerebellum. Unipolar brush cells stain strongly with antibodies against calretinin (Resibois and Rogers 1992; Braak and Braak 1993; Floris et al. 1994). The distribution of calretinin immunoreactivity to granule cells and their axons in the molecular layer of the corpus cerebelli and to the unipolar brush cells in the nodulus and flocculus uniquely visualises this fundamental subdivision of the mammalian cerebellum.

A fourth cortical layer was identified by Ogawa (1934) in Pinnipedia (the fur seal *Callorhinus*). It consists of a layer of large neurons located in the white matter under the granular layer which presumably represent displaced neurons of the cerebellar nuclei.

The glial architecture of the cerebellar cortex is remarkable, because the Bergmann glia, with somata located at the level of the Purkinje cell layer and fibrous processes with end-feet at the pial surface (Fig. 22.41j), has retained its embryological, radial disposition. There is a functional interdependence of Bergmann glia and the Purkinje cells, evidenced by the shift of taurine during hypo-osmotic stress from the Purkinje cells to the glial compartment (Nagelhus et al. 1993) and by the dependence of glial enzymes such as 5'-nucleotidase on the integrity of the neighbouring Purkinje cells and their climbing fibre innervation (Hess and Hess 1986; Balaban 1984; Fisher 1983).

22.6.3
Cerebellar Nuclei

The cerebellar nuclei are located in the cerebellar white matter in the roof of the fourth ventricle. Rostrally and ventrally, they border on the vestibular nuclei (Fig. 22.22–22.24h–k). The dominant input of the cerebellar nuclei is derived from the GABAergic axons of the Purkinje cells. Additionally, they receive excitatory collaterals from olivocerebellar and certain mossy fibre afferent systems (Fig. 22.42). Purkinje cell axons also terminate in the vestibular nuclei. The connections of the lateral vestibular nucleus are very similar to the cerebellar nuclei, with a dominant Purkinje cell input and collateral innervation by olivocerebellar fibres (Groenewegen and Voogd 1977) and lateral reticulocerebellar (Ruigrok et al. 1995) and central cervical spinocerebellar (Matsushita and Yaginuma 1990, 1995) mossy fibres. The lateral vestibular nucleus is not innervated by fibres of the vestibular nerve. Purkinje cell axons also terminate in the superior, medial and spinal vestibular nuclei, but here the dominant input of these nuclei is derived from the vestibular nerve. Climbing fibre collaterals to these 'true' vestibular nuclei are a matter of dispute (see Balaban 1984, 1988; Gerrits and Voogd 1982), and the presence of mossy fibre collateral projections to these vestibular nuclei has not been systematically studied.

Different classes of neurons have been distinguished in the cerebellar nuclei on the basis of the size of their perikaryon in Nissl-stained sections, their neurotransmitters and the distribution of their axons (Fig. 22.42). Two populations of small, GABAergic neurons can be distinguished. The first GABAergic population co-localises the inhibitory neurotransmitter glycine and consists of local interneurons (Chen and Hillman 1993: rat). The majority of the small GABAergic neurons give rise to crossed projections to the inferior olive (Fig. 22.42n,o). The population of large and medium-sized neurons of the nuclei consists of relay cells that send their axons to subcortical motor centres and to the thalamus. Axons of the relay cells give rise to collaterals that terminate as mossy fibres in the cerebellar cortex (for a review, see Tolbert 1982).

Four cerebellar nuclei can be identified (Weidenreich 1899; Ogawa 1935; Flood and Jansen 1961; Voogd 1964). Weidenreich distinguished them in some species of insectivores, rodents, lagomorphs and carnivores. The four nuclei are the homologues of the fastigial, globose, emboliform and dentate nucleus of the human cerebellum. Ogawa (1935), who studied the cerebellar nuclei in Cetacea and Pinnipedia, introduced the names 'anterior interpo-

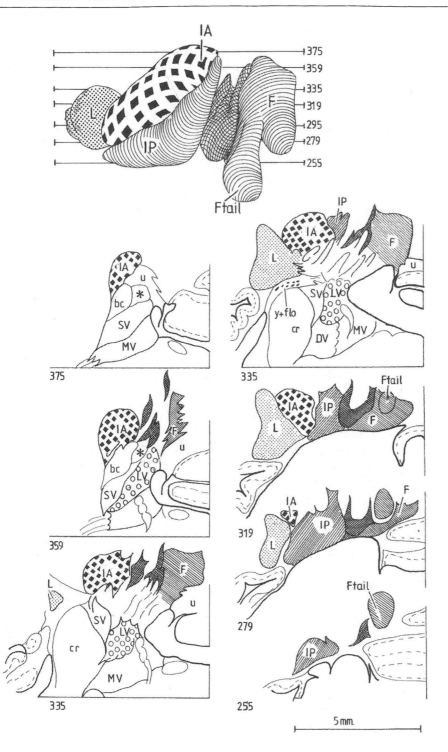

Fig. 22.43. The cerebellar nuclei of the cat. The transitional U-shaped region of the fastigial and posterior interposed nuclei is indicated by *double hatching*. bc, brachium conjunctivum; cr, restiform body; DV, descending vestibular nucleus; F, fastigial nucleus; flo, floccular peduncle; Ftail, tail of the fastigial nucleus; IA, anterior interposed nucleus; IP, posterior interposed nucleus; L, lateral cerebellar nucleus; LV, lateral vestibular nucleus; MV, medial vestibular nucleus; SV, superior vestibular nucleus; u, uncinate tract; Y, group y of Brodal and Pompeiano (1957). (From Voogd et al. 1996)

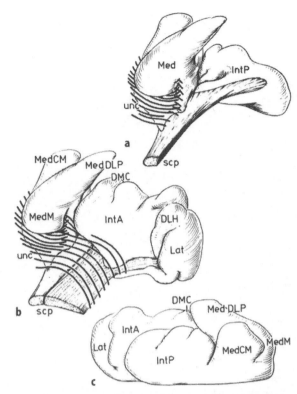

Fig. 22.44a-c. a,b Rostrolateral and c caudomedial views of the central cerebellar nuclei of the rat with the origin of the brachium conjunctivum and the intital course of the uncinate tract. The anterior interposed and lateral nuclei have been removed in a. Drawings from computer reconstructions. *DLH*, dorsolateral hump; *DMC*, dorsomedial crest; *IntA*, anterior interposed nucleus; *IntP*, posterior interposed nucleus; *Lat*, lateral cerebellar nucleus; *Med*, medial cerebellar nucleus; *MedCM*, caudomedial subdivision of the medial cerebellar nucleus; *MedDLP*, dorsolateral protuberance of the medial cerebellar nucleus; *MedM*, middle subdivision of the medial cerebellar nucleus; *scp*, superior cerebellar peduncle; *unc*, uncinate tract. (From Voogd 1995)

sed' and 'posterior interposed' nucleus for the emboliform and globose nuclei, respectively.

In the cat, the nuclei can be divided into two groups of interconnected nuclei (Voogd 1964): a caudal medial group, consisting of the fastigial (or medial) and the posterior interposed nucleus, and a rostrolateral group, consisting of the anterior interposed and the dentate (or lateral) nucleus (Fig. 22.43). The two groups are separated by a fibre lamella. The caudal tail of the fastigial nucleus projects into the white matter of the posterior vermis. The four nuclei of the cat are of roughly the same size. A fifth nucleus can be identified as a small group of mostly large neurons, located at the junction of the fastigial and posterior interposed nuclei (Fig. 22.43).

The four cerebellar nuclei of Weidenreich-Ogawa can also be recognised in the cerebellum of the rat (Korneliussen 1968; Voogd 1995; Fig. 22.44). The medial nucleus bears a dorsolateral protuberance (Goodman et al. 1963), which extends into the white matter of the posterior lobe, and well-developed ridges are present medial (dorsomedial crest) and lateral (dorsolateral hump of Goodman et al. 1963) to the anterior interposed nucleus. The dorsolateral protuberance can be identified in most rodents and in lagomorphs, but not in the cat or monkey.

In marsupials, only three nuclei – the fastigial, interposed and dentate nuclei – were delineated by Brunner (1919). The four cerebellar nuclei of Weidenreich-Ogawa were distinguished by Ohkawa (1957) in several species of marsupials and by Martin et al. (1973) in the North American opossum. The borders between the posterior interposed, anterior interposed and dentate nuclei are less distinct than in the cat. A dorsolateral protuberance of the medial nucleus appears to be present. The lateral nucleus contains a ventrolateral region with large neurons. Ventrally, this subdivision is difficult to distinguish from the large cells of the vestibular nucleus (Fig. 22.45).

In primates, the lateral cerebellar nucleus is larger than any of the other cerebellar nuclei. It consists of lateral and medial limbs, separated by a medially directed hilus. The cytoarchitecture of the dentate nucleus of *Macaca* was extensively studied in Nissl- and Golgi-stained sections by Chan-Palay (1977). A large cell column, ventral to the lateral cerebellar nucleus, represents the group y (Fig. 22.46). Small, scattered neurons located in the white matter of the flocculus and extending into the roof of the fourth ventricle, ventral to the lateral and posterior interposed nuclei, into the white matter of the nodulus, belong to the basal interstitial nucleus of Langer (1985). In higher primates, dentations of the lateral nucleus appear. They are most developed in the human dentate nucleus (Voogd et al. 1990) and are present in the orang-utan *Simia satyrus* and in Hylobatidae. In most of Old and New World monkeys, a few gyrations in the cell band of the lateral nucleus can be distinguished (Ohkawa 1957). A rostromedial region, with narrow dentations and large neurons, can be distinguished in the human dentate nucleus from a caudal and ventrolateral region, with wide gyrations and smaller neurons (Demolé 1927a,b; Gans 1924; Voogd et al. 1990). These two subdivisions have been indicated as the paleodentate and neodentate, respectively. Their homology to subdivisions of the lateral cerebellar nucleus in other mammalian species is uncertain, although a ventral parvocellular subdivision of the dentate nucleus has been recognised in the cat (Flood and Jansen 1961), the rat (Korneliussen 1968) and in several primates (for a review, see Haines 1977).

Detailed information is available on the struc-

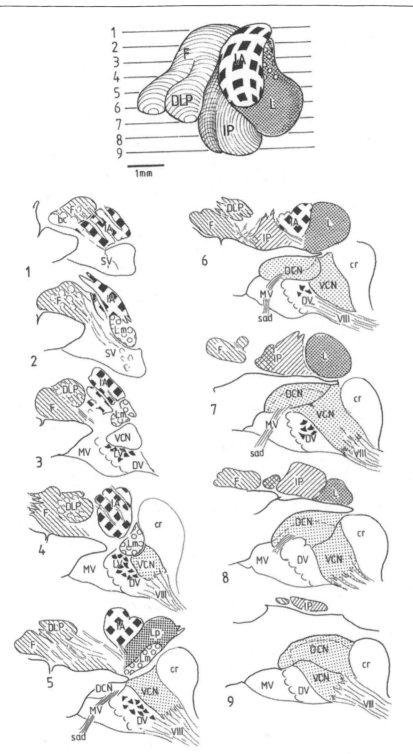

Fig. 22.45. Cerebellar, vestibular and cochlear nuclei of the opossum *Didelphis virginiana*. The magnocellular part of the lateral cerebellar nucleus (Lm) is difficult to distinguish from the lateral vestibular nucleus (LV). Large cells in the lateral and descending vestibular nuclei are indicated by *black triangles*. The level of the sections is indicated in the diagram of the nuclei at the *top*. *bc*, brachium conjunctivum; *F*, fastigial nucleus; *SV*, superior vestibular nucleus; *IA*, anterior interposed nucleus; *Lm*, magnocellular part of lateral cerebellar nucleus; *DLP*, dorsolateral protuberance; *VCN*, ventral cochlear nucleus; *LV*, lateral vestibular nucleus; *DV*, descending vestibular nucleus; *MV*, medial vestibular nucleus; *cr*, restiform body; *VIII*, root of the vestibulocochlear nerve; *Lp*, parvovellular part of lateral cerebellar nucleus; *DCN*, dorsal cochlear nucleus; *sad*, dorsal acoustic stria; *L*, lateral cerebellar nucleus; *IP*, posterior interposed nucleus

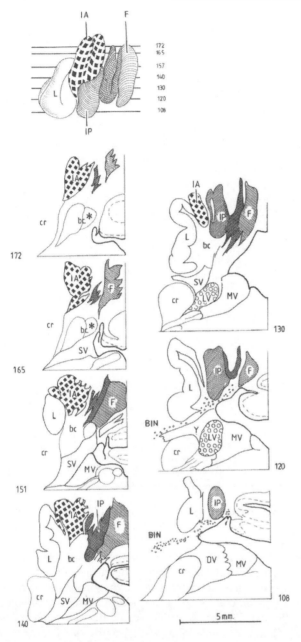

Fig. 22.46. The cerebellar nuclei of *Macaca fascicularis*. The *upper diagram* is a graphical reconstruction of the cerebellar nuclei in a dorsal view. Levels of the transverse sections are indicated. The U-shaped, transitional nucleus located between the fastigial (*F*) and posterior interposed nucleus (*IP*) is indicated by *double hatching*. *Asterisk* indicates medial component of the brachium conjunctivum, issued by the posterior interposed nucleus. *bc*, brachium conjunctivum; *BIN*, basal interstitial nucleus of Langer; *cr*, restiform body; *DV*, descending vestibular nucleus; *IA*, anterior interposed nucleus; *L*, lateral cerebellar nucleus; *LV*, lateral vestibular nucleus (Deiters); *MV*, medial vestibular nuleus; *SV*, superior vestibular nucleus; *Y*, group y; asterisk, medial third of the brachium conjunctivum. (From Voogd et al. 1996)

ture and the development of the cerebellar nuclei of Cetacea (Ogawa 1935; Ohkawa 1957; Korneliussen 1967; Korneliussen and Jansen 1965). The four cerebellar nuclei can be recognised. The posterior interposed nucleus is extremely large in whales and dolphins (Fig. 22.47).

22.6.4
Corticonuclear Projection: Evidence for Longitudinal Projection Zones

Purkinje cell axons terminate in the cerebellar and vestibular nuclei. Purkinje cells projecting to specific target nuclei are arranged in longitudinal zones that may extend over many, and in some cases, over all cerebellar lobules of vermis or hemisphere, crossing the interlobular fissures in their course. The evidence for the zonal arrangement in the corticonuclear projections is based on the myeloarchitecture of the cerebellar white matter, on axonal tracer studies and on the development and the chemoarchitecture of the cerebellar cortex.

Originally, three broad corticonuclear projection zones were distinguished by Jansen and Brodal (1940, 1942) in the anterior lobe of the cat, rabbit and rhesus monkey. Their *medial zone* corresponds to the vermis and projects to the fastigial nucleus, the *intermediate zone* projects to the interposed nucleus, which was not subdivided into anterior and posterior parts, and the *lateral zone* is connected with the dentate nucleus. Voogd (1964, 1969; Voogd and Bigaré 1980) further subdivided these three zones and determined their distribution in the cerebellum of the ferret and the cat. Their conclusions were based on the observation that Purkinje cell axons from a Purkinje cell zone occupy a discrete white matter compartment. These white matter compartments remain separated by small sheets of thin fibres, which continue into the borders of the cerebellar target nuclei of the Purkinje cell zones (Figs. 22.48, 22.49). These borders appear as dark lines when the sections are reacted for AChE (Hess and Voogd 1986, see Fig. 22.51), but correspond to empty slits when the Purkinje cell axons are stained for a Purkinje cell-specific marker (Voogd et al. 1996b).

Three Purkinje cell zones and their compartments can be recognised in the cortex and the white matter of the vermis of the cat (Figs. 22.49, 22.50). Zone A projects to the fastigial nucleus, zone X to a nucleus at the junction of the fastigial and the posterior interposed nucleus and zone B to the lateral vestibular nucleus. The X and B zones are limited to the anterior lobe and the lobulus simplex. Three C zones and compartments are present in the pars intermedia. Two zones, C1 and C3, which project to the anterior interposed nucleus, flank the C2 zone,

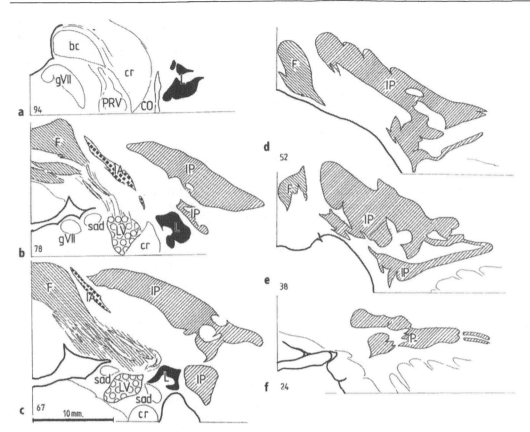

Fig. 22.47a-f. The cerebellar nuclei of the dolphin *Tursio borealis*. **a** Rostralmost and **f** caudalmost section. Note large size of the posterior interposed nucleus. *bc*, brachium conjunctivum; *CO*, cochlear nuclei; *cr*, restiform body; *F*, fastigial nucleus; *gVII*, genu of the facial nerve; *IA*, anterior interposed nucleus; *IP*, posterior interposed nucleus; *L*, lateral cerebellar nucleus; *LV*, lateral vestibular nucleus; *PRV*, principal sensory nucleus of the trigeminal nerve; *sad*, dorsal acoustic stria. (Drawn from photographs from Ogawa 1935)

which projects to the posterior interposed nucleus. C1 and C3 are present in the anterior lobe and the lobulus simplex and in crus II of the ansiform lobule and the paramedian lobule, but they are narrow or interrupted in crus I of the ansiform lobule. C2 continues uninterruptedly from lobule III of the anterior lobe into the flocculus. Two D zones were distinguished which extend over the entire hemisphere. D1 projects to the ventrocaudal part of the lateral cerebellar nucleus, and D2 to its rostromedial portion.

The same compartments have been identified in macaque monkeys, using AChE histochemistry (Fig. 22.51; Hess and Voogd 1986; Voogd et al. 1987, 1996b). The D compartments in the hemisphere are particularly wide. The C2 compartment continues into the flocculus, where three additional compartments that lead into the floccular peduncle – the basal interstitial nucleus of Langer (1985), the dorsal group y nucleus and the vestibular nuclei – can be recognised (Fig. 22.37b). The great similarity in the parasagittal compartmental organisation of the white matter in the cat and monkey is in accordance with the similarity in the zonal organisation of the corticonuclear projection in carnivores and primates (Haines et al. 1982; Fig. 22.52).

White matter compartments and corresponding Purkinje cell zones have been delineated using AChE histochemistry and experimentally verified in the flocculus and the nodulus of the rabbit cerebellum (Tan et al. 1995a-c; Wylie et al. 1994). The C2 zone and four zones with projections to vestibulo-ocular relay cells of the anterior and horizontal canals, one more than in the primate flocculus, are present in the flocculus of the rabbit. In the rabbit, cat and monkey, the floccular zones extend in the adjacent folia of the ventral paraflocculus. The 'floccular' segment of the paraflocculus is small in the rabbit and cat (Gerrits and Voogd 1982), but large in the monkey, where it includes the entire ventral paraflocculus (Fig. 22.37). The relative size of this region may be related to the presence of a fovea and the regulation of smooth pursuit movements in primates and their absence in the rabbit.

The corticonuclear projection in marsupials was studied by Sreesai (1973) and Klinkhachorn et al.

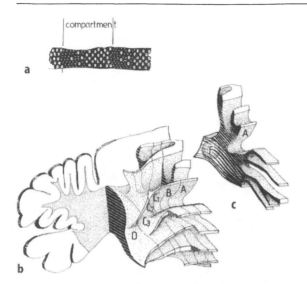

Fig. 22.48a-c. Graphical reconstruction of white matter compartments in the anterior lobe of the cerebellum of the ferret, based on transverse, Häggqvist-stained sections. **a** Part of the white matter, with bundles of large, myelinated Purkinje cell axons, with a sharp medial border, towards a small fibre region, lacking Purkinje cell fibres. A compartment is defined as the region between the medial borders of two adjacent bundles of Purkinje cell fibres. **b** Reconstruction of the A, B, C1, C2, C3 and D compartments. **c** The A compartment with the fastigial nucleus (*F*). (Based on Voogd 1969)

(1984a,b) in the North American opossum. A, B, C1-C3 and D zones were delineated in the anterior lobe (Klinkhachorn et al. 1984a; Fig. 22.52). The uncertainty about the precise borders of the subdivisions of the cerebellar nuclei in marsupials detracts from the value of this diagram. The corticonuclear projection in the rat was reviewed by Buisseret-Delmas and Angaut (1993) and Voogd et al. (1996). Corticonuclear projection zones A, X, B, C1-C3 and three D zones were recognised. Buisseret-Delmas (1988) noticed the presence of a Purkinje cell zone in the medial part of the hemisphere of lobules VI and VII of the posterior lobe, which projects to the dorsolateral pretuberance of the fastigial nucleus (her lateral extension of the A zone). Earlier, a projection from the cerebellar hemisphere to the fastigial nucleus had been noted by Goodman et al. (1963). A similar zone appears to be present in the rabbit (Tan et al. 1995b), but projections from the hemisphere to the fastigial nucleus have never been noticed in the cat or monkey.

During early stages of the development of the cerebellum, prior to the appearance of the transverse fissures, Purkinje cells in the cortical plate are clustered in a number of parasagittal zones, located beneath the external granular layer. Clustering of Purkinje cells was observed in the rat (Korneliussen 1968b; Wassef and Sotelo 1984), Cetacea (Korneliussen 1967), monkey (Kappel 1981) and humans (Korneliussen 1968c). The development of the transverse fissures occurs rather late in the development of the cerebellum and coincides with the inward migration of the granular cells from the external granular layer to their definite position in the internal granular layer. Fissuration and the development of the internal granular layer cause a large increase in the rostrocaudal dimension of the cortical sheet and the transformation of the Purkinje cell clusters into a monolayer. In the process, the borders between the clusters are lost. It appears likely that the Purkinje cell clusters are the primordia of the adult longitudinal Purkinje cell zones, because they display a similar topographical relationship to the cerebellar nuclei. The map of the corticogenetic zones of the fin whale, prepared by Korneliussen (1967), shows that his lateral intermediate zone, which corresponds to the C2 zone of other mammals, is extremely wide in Cetacea (Fig. 22.53). This is in accordance with the huge size of the posterior interposed nucleus (Korneliussen and Jansen 1965) and explains the great width of the paraflocculus in these species (Figs. 22.39, 22.47, 22.53). A similar map of the corticogenetic zones of the human cerebellum shows a similar increase in width of the lateral zone of Korneliussen (corresponding to the D zone of the cat), which projects to the dentate nucleus (Korneliussen 1968c).

Several molecular markers are available which support the longitudinal division of the cerebellar cortex (reviewed by Voogd et al. 1996a). Some of these markers reside in the Bergmann glia (5'-nucleotidase: Scott 1964; Marani 1986; Hess and Hess 1986; AChE: Marani and Voogd 1977), while others are located in the Purkinje cells (the zebrins: Hawkes and Leclerc 1987; Eisenman and Hawkes 1993). 5'-Nucleotidase and the zebrins are distributed in the same longitudinal pattern in the rat and mouse (Fig. 22.54). The distribution of zebrin-positive and -negative Purkinje cells (Fig. 22.54) shows a partial correspondance with the corticonuclear projection zones in the rat (Voogd and Ruigrok 1997). A similar zonal pattern in the distribution of the zebrins is present in marsupials (Dorè et al. 1990) and the squirrel monkey (Leclerc et al. 1990). In the cat and in macaque monkeys, all Purkinje cells express zebrin (Voogd et al. 1996a).

22.6.5
Olivocerebellar Projection and the Modular Organisation of the Output Systems of the Cerebellum

The inferior olive consists of medial and dorsal accessory olives and a curved, principal olive lo-

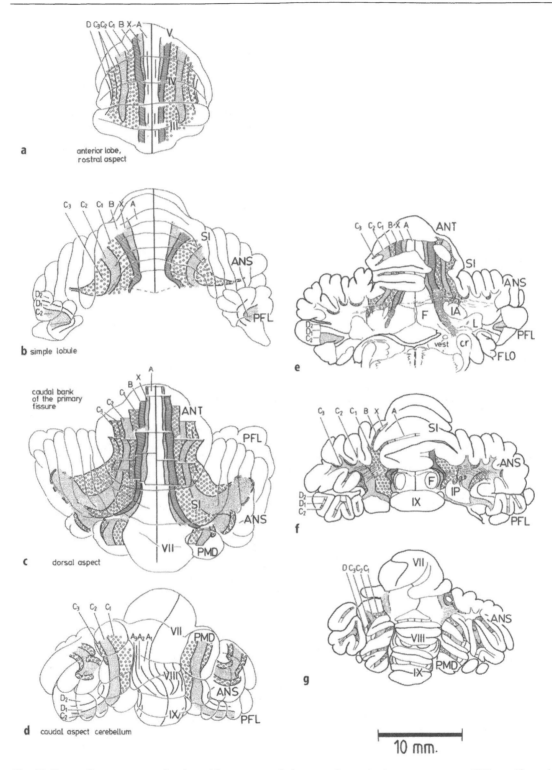

Fig. 22.49a–g. Compartments in the white matter of the cerebellum of the cat. Drawings and reconstructions from Häggqvist- and acetylcholinesterase (AChE)-stained sections. Compartments are indicated by different symbols. **a–d** Graphical reconstructions of the rostral aspect of the anterior lobe (**a**) and the posterior lobe (**b**), the dorsal aspect (**c**) and the caudal aspect (**d**) of the cerebellum. **e–g** Transverse sections. *A*, A compartment; *ANS*, ansiform lobule; *ANT*, anterior lobe; *B*, B compartment; *C1–3*, C1–3 compartments; *cr*, restiform body; *D(1,2)*, D(1,2) compartments; *F*, fastigial nucleus; *IA*, anterior interposed nucleus; *IP*, posterior interposed nucleus; *L*, lateral cerebellar nucleus; *PFL*, paraflocculus; *PMD*, paramedian lobule; *SI*, simple lobule; *vest*, vestibular nuclei; *X*, X compartment; *III–IX*, lobules III–IX

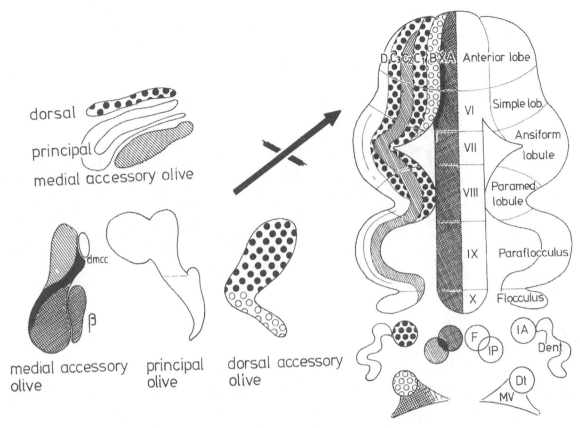

Fig. 22.50. Corticonuclear and olivocerebellar projections, based on experimental studies in the cat reviewed in the text. Left, a transverse section through the inferior olive and a diagram of the unfolded inferior olive. Right, the cerebellar cortex is also unfolded. Regions that are interconnected are indicated by the same symbols. A, A zone; B, B zone; β, group beta; C_1-C_3, C1-C3 zones; D, D zone; Dent, dentate nucleus; dmcc, dorsomedial cell column; Dt, lateral vestibular nucleus of Deiters; F, fastigial nucleus; IA, anterior interposed nucleus; IP, posterior interposed nucleus; MV, medial vestibular nucleys; X, X zone; VI-X, lobules VI-X

cated in between (Figs. 22.22–22.24, 22.55). Comparative anatomical studies of the mammalian inferior olive were published by Kooy (1916) and Whitworth and Haines (1986a). The variations in size and shape of the inferior olive and its subnuclei can best be understood from the organisation of the olivocerebellar projection.

Olivocerebellar fibres from particular subnuclei of the inferior olive terminate contralaterally on the Purkinje cells of the corticonuclear projection zones, described in the previous section, and with collaterals on their cerebellar or vestibular target nucleus. The organisation of the olivocerebellar projection is illustrated for the cat in Fig. 22.50. The caudal halves of the dorsal and medial accessory olives give rise to climbing fibres to the A and B zones of the vermis. The X zones receives climbing fibres from the junction of the caudal and rostral half of the medial accessory olive (Campbell and Armstrong 1985). The C2 and the C1 and C3 zones of the intermediate zone receive projections from the rostral halves of the medial and the dorsal accessory olive, respectively. The ventral and dorsal leaves of the principal olive project to the D1 and D2 zones of the hemisphere, respectively (Groenewegen and Voogd 1977; Groenewegen et al. 1979; Voogd et al. 1996a). The vestibular projection zones of the flocculus and the nodulus receive climbing fibres from small subnuclei (dorsal cap of Kooy; ventrolateral outgrowth, β-group, dorsomedial cell column) located next to the medial accessory olive. These subnuclei relay optokinetic information to the flocculus and nodulus and vestibular information exclusively to the nodulus (Gerrits et al. 1985a; Tan et al. 1995b).

The inferior olive of the rat (Fig. 22.55) was described by Gwyn et al. (1977) and Azizi and Woodward (1984). Distinctive features are the position of the caudal dorsal accessory olive, 'the dorsal fold', which is located dorsal to its rostral portion (the ventral fold of the dorsal accessory olive). The enlarged medial extension of the ventral leaf of the

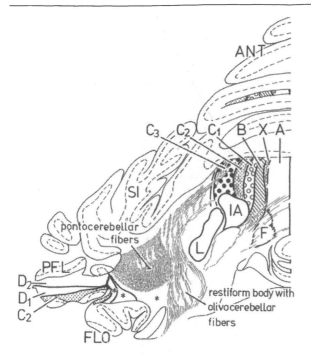

Fig. 22.51. Transverse section through the cerebellum of *Macaca fascicularis*, reacted for acetylcholinesterase, for the staining of the borders of white matter compartments. The compartments in the white matter of the anterior lobe (A, B, C_1, C_2, C_3 and D), the paraflocculus (C_2, D_1, D_2) and the four compartments of the flocculus (C_2; the three medial compartments have been indicated by *asterisks*). *ANT*, anterior lobe; *F*, fastigial nucleus; *FLO*, flocculus; *IA*, anterior interposed nucleus; *L*, lateral cerebellar nucleus; *PFL*, paraflocculus; *SI*, lobulus simplex. (From Hess and Voogd 1986)

principal olive (the dorsomedial subnucleus) may sometimes be confused with the dorsomedial cell column. The dorsomedial cell columns of both sides are fused over the midline (De Zeeuw 1996). The caudal half of the medial accessory olive can be subdivided into four subnuclei: a, b, c and the β-group. Subnucleus c receives a projection from the superior colliculus and projects to the vermal visual area (lobule VII), to the lateral extension of the A zone of Buisseret-Delmas (1988) and sends a collateral projection to the caudal fastigial nucleus and its dorsolateral protuberance.

The primate inferior olive was described for *Galago senegalensis* by Whitworth and Haines (1983), for *Saimiri sciureus* by Whitworth and Haines (1986) and for the rhesus monkey by Bowman and Sladek (1973) and Brodal and Brodal (1981). The principal olive is large in these species, and its cell band is gyrated. The medial extension of the

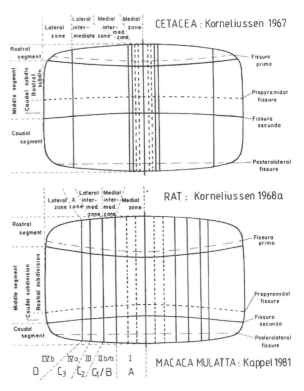

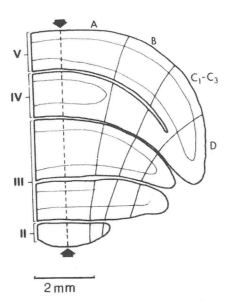

Fig. 22.52. Tracings of anterior lobe of the cerebellum of the North American opossum showing the general zonal configuration of the anterior lobe. For abbreviations see Fig. 22.50. (From Klinkhachorn et al. 1984a)

Fig. 22.53. Diagrams of the corticogenetic zones in the fetal cerebellum of *Cetacea* (Korneliussen 1967), the rat (Korneliussen 1968a) and the rhesus monkey, *Macaca mulatta*. Corticogenitic zones in whale and rat fetus are indicated by the names attributed to them by Korneliussen (1967, 1968a), and in the monkey by *roman numerals* according to Kappel (1981). The "X zone" was distinguished by Korneliussen (1968) in the rat, but not in the whale fetus. The most likely correspondence of the corticogenetic zones in the monkey with the corticonuclear projection zones of rat and whale is indicated at the *bottom* of the diagram. (Modified from Korneliussen 1968b)

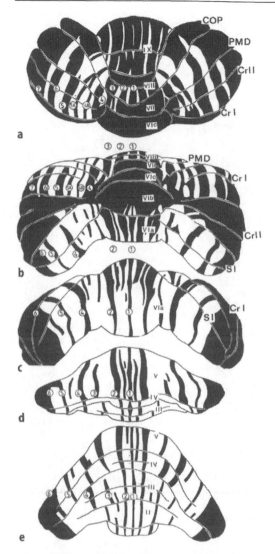

Fig. 22.54a-e. Reconstruction of the localisation of zebrin I-immunoreactive Purkinje cell zones in the cerebellum of the rat. a Caudal aspect, posterior lobe. b Dorsal aspect, posterior lobe. c Rostral aspect, posterior lobe. d Dorsal aspect, anterior lobe. e Rostral aspect, anterior lobe. *Numbers* indicate zebrin-positive Purkinje cell zones P1-P7 of Hawkes and Leclerc (1987). *COP*, copula pyramidis; *CrI, CrII*, crus I and II of the ansiform lobule; *PMD*, paramedian lobule; *SI*, simple lobule; *I–X*, lobules I–X

ventral leaf of the principal olive is enlarged and may be confused with the dorsomedial cell column, as in the rat (Fig. 22.55).

The inferior olive in marsupials shows several distinctive features (Fig. 22.56). The rostral medial accessory olive (subnucleus a) is located lateral or even dorsolateral to the principal olive. Caudally, it can be subdivided into subnuclei a, b and c as in other mammals. Subnucleus c of marsupials presumably corresponds to the β-group of eutherian mammals. The dorsal cap of Kooy occupies its usual position. Cell group a_1 is found in a rostromedial position and may correspond to the dorsomedial cell column. The size and the shape of the principal olive shows variations in different marsupial species (Martin et al. 1975; Watson and Herron 1977). The identification of the principal olive and the rostral cell group a as the rostral pole of the medial accessory olive was supported by experiments in the opossum which showed projections to these subdivisions of the olive from the nuclei at the mesodiencephalic border (Henkel et al. 1975; Martin et al. 1976; Linauts and Martin 1978).

In Cetacea, the rostral pole of the medial accessory olive, which projects to the C2 zone and the posterior interposed nucleus, is enlarged (Fig. 22.57). The inferior olive of the elephant is characterised by a large, broad ventral leaf of the principal olive (Kooy 1916).

The inferior olive receives ascending connections from the spinal cord, the dorsal column nuclei and the spinal nucleus of the trigeminal nerve and descending connections from the parvocellular red nucleus and the nuclei at the mesoencephalic-diencephalic border. These connections have been analysed in great detail using electrophysiological methods and anatomical tracing techniques, mainly in the cat. Spinal and trigeminal afferents terminate diffusely in the caudal half of the medial accessory olive and somatotopically in the entire dorsal accessory olive (Boesten and Voogd 1975). The A and X zones therefore receive diffusely organised spinal projections, whereas projections to the B1, C1 and C3 zones are somatotopically organised in so-called microzones (Anderssen and Oscarsson 1978; Ekerot and Larson 1979a,b; Gellman et al. 1983).

Direct connections from the cerebral cortex to the inferior olive are few. The projections from frontal and parietal areas to the inferior olive are relayed by the ipsilateral parvocellular red nucleus and the nuclei at the mesencephalic-diencephalic border (see Sect. 22.9.3 for further details). Direct cortico-olivary connections are mainly ipsilateral and terminate in the caudal medial accessory olive medial to the β-group (Saint-Cyr and Courville 1979: cat) and at the junction of subnuclei a and b (Martin et al. 1980: opossum). In the cat, they arise from medium-sized pyramidal cells in layer V, mostly from the anterior sigmoid and the presylvian gyrus (areas 4 and 6), including the frontal eye field (Bishop et al. 1976; Saint-Cyr and Courville 1980). In addition, area 4 projects to the junction of the ventral leaf of the principal olive and the dorsal accessory olive (Saint-Cyr 1983). According to some authors, the direct cortico-olivary projection is bilateral (Bishop et al. 1976) or even mainly crossed (Sousa Pinto 1969; Sousa Pinto and Brodal 1979).

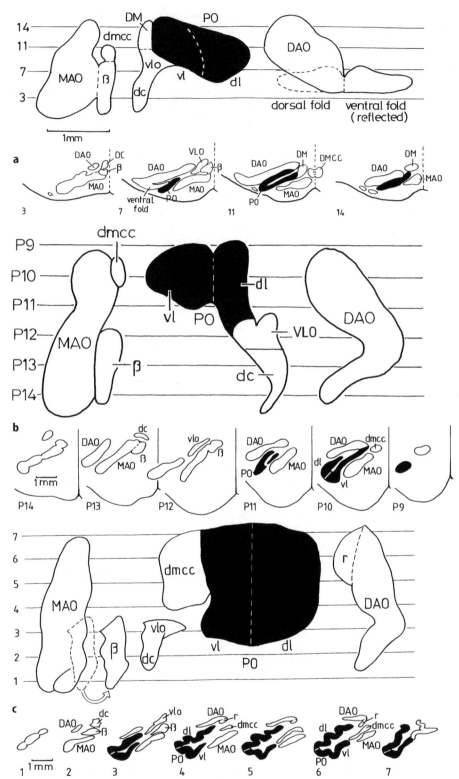

Fig. 22.55a–c. The left inferior olive of **a** the rat, **b** the cat and **c** the rhesus monkey. For each species, a graphical reconstruction of the principal and accessory olives, according to Brodal (1940), and selected transverse sections are illustrated. Obvious species differences concern the large size and the gyration of the principal olive in the monkey, the topographical relations of the ventrolateral outgrowth to the ventral leaf (rat, monkey) or to the dorsal leaf (cat) of the principal olive, the delineation of the dorsomedial cell column as a separate subnucleus in rat and cat, or as part of the ventral leaf in the monkey, the distinction of a dorsomedial subnucleus in the ventral leaf of the principal olive in the rat (Azizi and Woodword 1987) and the reflection of the caudal pole of the dorsal accessory olive as the ventral leaf in the rat (Azizi and Woodword 1987). *DAO*, dorsal accessory olive; *dc*, dorsal cap of Kooy; *dl*, dorsal leaf of the principal olive; *dmcc*, dorsomedial cell column; *DN*, dorsomedial subnucleus; *MAO*, medial accessory olive; *P*, anteroposterior stereotactic plane; *PO*, principal olive; *r*, reflected portion of the dorsal accessory olive; *β*, subnucleus beta; *vl*, ventral leaf of the principal olive; *vlo*, ventrolateral outgrowth

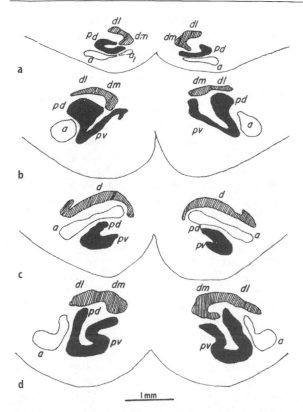

Fig. 22.56a–d. Sections through the inferior olive the three different marsupials. a A coronal section through the inferior olive in the North American opossum *Didelphis virginiana* illustrates the thin sheet the cells connecting subnucleus a and a'. b A coronal section through the inferior olivary complex (IOC) the the wombat *Vombatus ursinus* illustrates the relative increase in the size of the dorsal lamella of the principal nucleus (*pd*). c A coronal section through the IOC in the kangaroo *Macropus rufus* illustrates the relative increase in size of the dorsal nucleus (d). d Another coronal section taken from the kangaroo *Macropus rufus* illustrates the relative increase in the size and complexity of the principal nucleus pd, pv. The medial accessory olive (*a*) is located lateral or dorsolateral to the principal olive. Subnucleus a' may correspond to the dorsomedial cell column of eutherian mammals. *a*, subnucleus a of the medial accessory olive; a_1, subnucleus a_1 of the medal accessory olive; *d*, dorsal accessory olive; *dl*, lateral subnucleus of the dorsal accessory olive; *dm*, medial subnucleus of the dorsal accessory olive; *pd*, principal olive, dorsal leaf; *pv*, ventral leaf of the principal olive

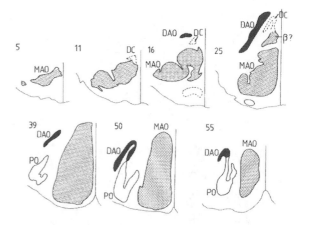

Fig. 22.57. The inferior olive the the porpoise *Phocaena*. 5, Caudalmost section; 55; rostralmost section. β, beta group (tentatively identified); *DAO*, dorsal accessory olive; *DC*, dorsal cap; *MAO*, medial accessory olive; *PO*, principal olive. (Redrawn from Kooy 1916, 1917)

An ipsilateral, cortico-olivary projection in the rat terminates in the medial dorsal and caudomedial medial accessory olive and sparsely in the lateral dorsal and rostral medial accessory olive (Swenson and Castro 1983a,b).

The central tegmental tract is the pathway from the parvocellular red nucleus to the principal olive. The nucleus of Darkschewitsch and other nuclei of the mesoencephalic-diencephalic border project, through the medial tegmental tract, to the rostral half of the medial accessory olive and the ventral leaf of the principal olive (see Sect. 22.9.3). One of these two systems is greatly enlarged in primates, Cetacea and Proboscoidea, i.e. the parvocellular red nucleus with the central tegmental tract and the principal olive in primates, and the nucleus of Darkschewitsch (the nucleus ellipticus of Ogawa 1935a) with the medial tegmental tract and the rostral medial accessory olive in Cetacea (see Figs. 22.57, 22.129) and the ventral leaf of the principal olive in Proboscoidea, respectively.

Neurons of the inferior olive are electrotonically coupled by gap junctions between the spines of their long recurving dendrites (Sotelo et al. 1974; Llinás et al. 1974). Electrotonic coupling synchronises the discharge in the climbing fibre afferents of a module and sometimes may even spread to other modules (Lang et al. 1996). Synchronisation and timing in climbing fibres are important elements in the theory of Llinás/Welsh (Welsh et al. 1996) on the function of climbing fibres in motor control. Other theories on climbing fibre function favour a role in motor learning (see Simpson et al. 1997). In both theories, the close association of the climbing fibres with the output of the cerebellum to different motor centres is an essential element.

The output of the cerebellum is organised in a modular fashion. A module consists of one or more Purkinje cell zones, their cerebellar or vestibular target nucleus and its afferent climbing fibre system (Fig. 22.58). The GABAergic nucleo-olivary system forms a mostly crossed, reciprocal connection between the target nucleus and the olivary subnucleus of the module (Van der Want and Voogd 1987; Van der Want et al. 1989; Ruigrok and Voogd 1990). Relay cells located in the target nucleus of the module give rise to axons that terminate in specific

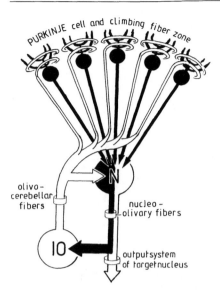

Fig. 22.58. Efferent cerebellar module. A longitudinal strip of Purkinje cells projects to a cerebellar or vestibular target nucleus. A subnucleus of the inferior olive (*IO*) provides the Purkinje cells with climbing fibres and emits collaterals to the cerebellar target nucleus. A GABAergic nucleo-olivary pathway reciprocally connects the target nucleus with the inferior olive

subcortical motor centres or thalamic nuclei. The connections of neighbouring modules do not overlap.

22.6.6
Mossy Fibre Afferents of the Cerebellum

Mossy fibres take their origin from different sources in the spinal cord and the brain stem. Mossy fibres from the spinal cord, the dorsal column nuclei and the precerebellar nuclei of the bulbar reticular formation enter the cerebellum through the restiform body. Most of these fibres pass caudal and dorsal to the entrance of the trigeminal nerve. The ventral spinocerebellar tract passes rostral and ventral to the trigeminal root and enters the cerebellum along the brachium conjunctivum. Mossy fibres from the nucleus reticularis tegmenti pontis and the basal pontine nuclei enter the cerebellum via the middle cerebellar peduncle.

Mossy fibres generally terminate bilaterally in the cerebellar cortex. Spino-, cuneo- and reticulo-cerebellar mossy fibres cross in the rostral part of the cerebellar commissure (Fig. 22.25). Pontocerebellar fibres cross both in the pes pontis and in the cerebellar commissure. The distribution of different mossy fibres systems over the cerebellar cortex is fairly complicated and can be summarised as follows:

1. The main mossy fibre systems roughly terminate in transversely oriented, terminal fields that approximately correspond to a lobule or a set of lobules. On the basis of their distribution, the cerebellum can be subdivided into vestibular-, spinal- and cerebropontine-dominated regions (Fig. 22.59).

2. A more accurate description of the distribution of the mossy fibres is their concentric arrangement. Primary vestibulocerebellar fibres from the vestibular root and secondary vestibulocerebellar fibres from the vestibular nuclei terminate heavily in the nodulus and the adjoining part of the uvula, in the ventral part of the anterior lobe and at the bottom of the deep interlobular fissures (Fig. 22.60). In the hemisphere, the secondary vestibulocerebellar fibres terminate in the flocculus and the adjoining paraflocculus, but primary vestibulocerebellar fibres do not terminate in these lobules (Gerrits et al. 1989; Langer et al. 1985). Spinocerebellar fibres terminate more distally in certain lobules of the anterior and posterior lobes (Fig. 22.61). Pontocerebellar fibres terminate distal to the spinocerebellar fibres in the apex of these lobules, in lobules VII and IX and heavily in more lateral parts of the hemisphere (Fig. 22.61). In rodents, marsupials and carnivores, the spinocerebellar fibres may reach the apex of the lobules of the anterior and posterior lobes. In primates, the spinocerebellar fibres are relegated back to the more proximal parts of the lobules by the increasing numbers of pontocerebellar fibres terminating in the apical and lateral parts of the lobules.

3. The terminal fields of the spino-, reticulo- and pontocerebellar mossy fibres are parcellated in longitudinal aggregates of mossy fibre terminals (Fig. 22.61). Longitudinal aggregates of mossy fibre terminals belonging to different mossy fibre systems may alternate (i.e. the lumbar and thoracic spinocerebellar fibres and the cuneocerebellar fibres in the anterior lobe of the rat; Gravel and Hawkes 1990; Ji and Hawkes 1994) or overlap (i.e. secondary vestibulocerebellar and spinocerebellar fibres from the central cervical nucleus at the bottom of the fissures of the anterior lobe of the cat; Matsushita and Tanami 1987; Matsushita and Wang 1987) in certain parts of the cerebellum. The longitudinal mossy fibre aggregates are topographically related to the longitudinal Purkinje cell zones of the output modules. In some parts of the cerebellum, the mossy fibre terminals are aggregated in patches. The receptive fields of the patches formed by trigeminocerebellar mossy fibres were mapped in electrophysiological recording experiments by

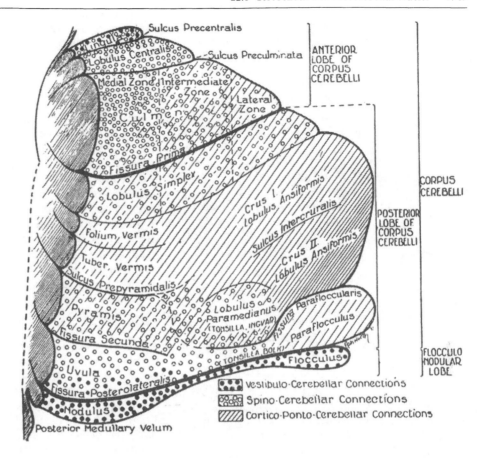

Fig. 22.59. Transverse lobular distribution of evoked potentials on stimulation of spino-, ponto- and vestibulo-cerebellar afferent systems in the cat. The position of the medial, intermediate and lateral corticonuclear projection zones of Jansen and Brodal (1940) is indicated in the anterior lobe and the lobulus simplex. (From Dow 1942)

Welker and co-workers in the rat, cat and slow loris (for a review, see Welker 1987). In the posterior lobe hemisphere and in the uvula, the patches constitute a fractured somatotopical pattern, in which the original topographical relations between the receptive fields have become lost, and the same receptive field is represented in multiple patches (Fig. 22.42).

4. Some mossy fibre systems also project to the cerebellar nuclei (Fig. 22.42). Spinocerebellar fibres from the central cervical nucleus (Matsushita and Yaginuma 1995: rat) and from the lateral reticular nucleus (Künzle 1975; Qvist 1989; Ruigrok and Cella 1995: rat and cat) terminate in the rostral fastigial, the anterior interposed and the lateral vestibular nuclei and in the medial parts of the posterior interposed and lateral cerebellar nuclei. Mossy fibres from the nucleus reticularis tegmenti pontis emit collaterals to complementary regions, i.e. to the caudal fastigial, and the lateral part of the posterior interposed and lateral cerebellar nuclei (Dietrichs and Walberg 1987; Gerrits and Voogd 1987; Qvist 1989; Mihailoff 1993: cat and rat). A projection from the basal pontine nuclei to the cerebellar nuclei is still controversial (Brodal et al. 1986; Shinoda et al. 1992). The distribution of mossy fibre collateral projections to particular subdivisions of several cerebellar and vestibular nuclei differs from the climbing fibre collateral projections, which always distribute to a single cerebellar or vestibular nucleus. This nucleus is always the target nucleus of the Purkinje cells that receive the climbing fibre terminals of the same olivocerebellar fibres (Fig. 22.58). An approximate reciprocity also seems to exist between the nuclei that receive mossy fibre collaterals and the corticonuclear projection of the Purkinje cells in their terminals fields.

Mossy fibres transmit information from different levels in motor control. Exteroceptive and proprioceptive input reaches the cerebellum through the dorsal spinocerebellar and cuneocerebellar tracts. Ventral, rostral and central cervical spinocerebellar tracts monitor the activity in groups of interneurons in the spinal cord (Lundberg 1971). Vestibulo- and rubrocerebellar fibres provide the cerebellum with information from these subcortical motor centres. Precerebellar reticular nuclei are sites of convergence from different systems. The lateral reticular nucleus receives converging information

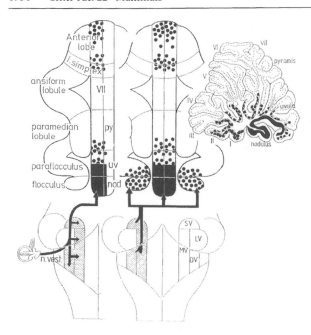

Fig. 22.60. Primary and secondary vestibulocerebellar mossy fibre projections. Vestibular nuclei that receive vestibular root fibres (*left*) and give rise to secondary vestibulocerebellar fibres (*right*) appear as *hatched areas*. Dense projections to the nodulus/uvula are indicated in *black*, and less heavy projections by *filled circles*. Distribution in sagittal section of the cerebellum applies to both primary and secondary vestibulocerebellar projections. *DV*, descending vestibular nucleus; *LV*, lateral vestibular nucleus; *MV*, medial vestibular nucleus; *n.vest*, vestibular nerve; *nod*, nodulus; *py*, pyramis; *SV*, superior vestibular nucleus; *uv*, uvula; *I–X*, lobules I–X. (From Voogd 1996a, based on data from Gerrits et al. 1989 and Thunnissen et al. 1989)

from the medial and lateral descending motor systems of Kuypers (see Sect. 22.9.1) and from medial and lateral interneurons in the spinal cord (Clendelin et al. 1974a–c: Ruigrok and Cella 1995). The nucleus reticularis tegmenti pontis receives converging input from the cerebral cortex, visual centres of the mesencephalon and the cerebellar nuclei.

The pontine nuclei are the main relay stations between the cerebral cortex and the cerebellum. Many corticopontine fibre systems that terminate within the pontine nuclei are collaterals from other systems (Ugolini and Kuypers 1986; Keizer et al. 1987). In the cat, the sensorimotor cortex and adjoining areas, which give origin to the pyramidal tract, also give rise to a cortico-pontocerebellar projection focused on the anterior lobe. The corticopontine fibres from the sensorimotor cortex terminate in the peripeduncular region of the pons (Brodal 1968a,b, 1971, 1982, 1987). Spinopontine fibres and collaterals from the medial lemniscus terminate in the same area (Ramon y Cajal 1911; Björkeland 1983). The collateral origin of the corticopontine fibres from the pyramidal tract was demonstrated by Ugolini and Kuypers (1986) using a retrograde fluorescent fibre labelling technique. The cortico-pontocerebellar projection from the sensorimotor cortex includes the paramedian lobule and extends into the ansiform lobule and the paraflocculus. All these lobules also receive corticopontocerebellar projections from the parietal association cortex and contributions from area 6, the second somatosensory area, the orbital and cingulate gyrus and the visual cortex (for a review, see Brodal 1987).

Corticopontine projections from striate and extrastriate visual areas terminate in different areas in rostral, ventral and dorsolateral parts of the pontine grey of the cat (Brodal 1972, 1987; Bjaali and Brodal 1983; Sanides et al. 1978). The paraflocculus and lobules IX and VII are the main recipients of the visual cortico-pontocerebellar projection (Robinson et al. 1984; Burne et al. 1978; Hoddevik et al. 1977). The basal pontine nuclei do not project to the flocculus. A large proportion of the corticopontine fibres from visual areas are branches from corticotectal fibres. Other projections to the superior and inferior colliculus from parietal, temporal and frontal areas also give rise to branching axons to the tectum and the pons (Fig. 22.97; Baker et al. 1983; Keizer et al. 1987). Auditory corticopontocerebellar connections seem to be less prominent than the visual pathways. The paraflocculus is the main target of the auditory pontocerebellar projection in the rat (Azizi et al. 1985). Direct tectopontine connections from the inferior colliculus terminate in the dorsal lateral pons, which projects to the vermal lobules VI and VII (Azizi et al. 1985; Burne et al. 1981). The tectopontine pathway from the superior colliculus (Münzer and Wiener 1902; Kawamura and Brodal 1974) terminates in the dorsolateral pons, which projects to the same lobules.

The corticopontine projection in primates is more extensive and includes input from the prefrontal cortex in addition to projections from pericentral, parietal, superior temporal, striate and extrastriate areas (Brodal 1978; Schmahmann and Pandya 1995).

The transverse orientation in the terminal fields of different mossy fibre systems, alluded to at the beginning of this section, is enhanced by the transverse orientation of the parallel fibres, the axons of the granule cells on which they terminate. The perpendicular arrangement of the mossy parallel-fibre input with respect to the modular output of the cerebellum optimises the chance that a particular mossy parallel-fibre input may influence Purkinje cells belonging to different output modules. In theories of cerebellar motor learning, the climbing fibres select the appropriate parallel fibre-Purkinje

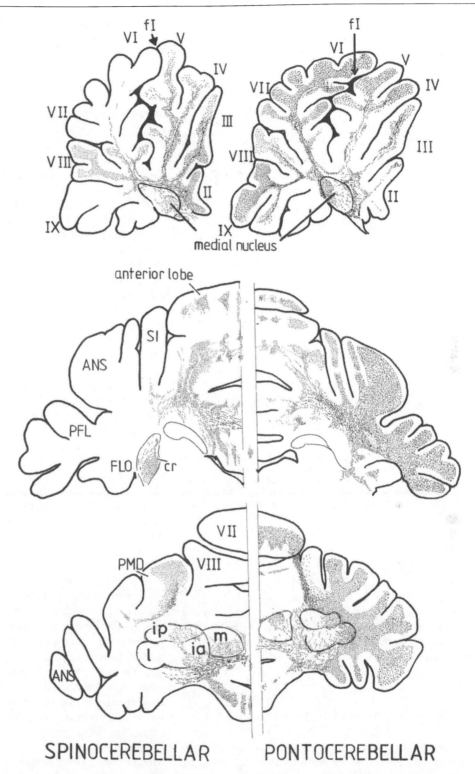

Fig. 22.61. Distribution of degenerated silver-impregnated spinocerebellar and pontocerebellar fibres after lesions of the cervical cord and the pes pontis with the nucleus reticularis tegmenti pontis in sagittal (*upper panels*), transverse (*middle panels*) and horizontal sections (*lower panels*) through the cerebellum of the tree shrew *Tupaia glis*. Note the zonal distribution in the vermis and pars intermedia and complementarity of the two projections to the cortex and to the cerebellar nuclei illustrated in *middle* and *lower panels*. ANS, antiform lobule; *cr*, restiform body; *fl*, primary fissure; *FLO*, flocculus; *ia*, anterior interposed nucleus; *ip*, posterior interposed nucleus; *L*, lateral cerebellar nucleus; *m*, medial cerebellar nucleus; *PFL*, paraflocculus; *SI*, simple lobule; *I–X*, lobules I–X. (From Voogd et al. 1996)

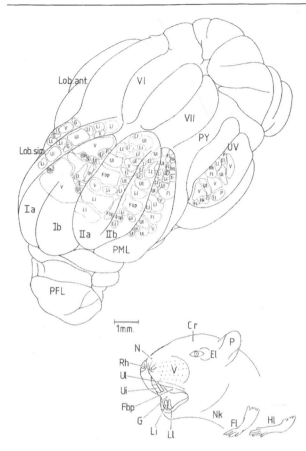

Fig. 22.62. Fractured somatotopy in the somatosensory projections to the granule cell layer in the cerebellum of the rat. Patches with similar receptive fields are indicated by abbreviations for the stimulation sites on the head and the extremities. *Cr*, crown; *El*, eyelid; *Fbp*, furry buccal pad; *FL*, forelimb; *G*, gingiva; *HL*, hindlimb; *I, II*, crus I and II; *Li*, lower incisor; *LL*, lower lip; *Lob.ant.*, anterior lobe; *Lob.sim*, lobulus simplex; *N*, nose; *Nk*, neck; *P*, pinna; *PFL*, paraflocculus; *PMD*, paramedian lobule; *PML*, paramedian lobule; *PY*, pyramis; *Rh*, rhinarium; *Ui*, upper incisor; *UL*, upper lip; *UV*, uvula. (Redrawn from Welker 1987)

cell synapses on the basis of previous experience. In the timing hypothesis, this selection would depend on the Purkinje cells that are synchronously activated by the climbing fibres.

22.6.7
Efferent Connections of the Cerebellum

Efferent connections of the cerebellar nuclei were extensively reviewed by Asanuma et al. (1983a,b), Nieuwenhuys et al. (1988) and Voogd et al. (1990) for primates, by Voogd (1964), Hendry et al. (1979) and Aumann et al. (1994) for the cat, by Voogd (1995) for the rat and by Martin et al. (1974) for the opossum. Efferent connections of the vermis of the cerebellum are relayed by the target nuclei of the A, X and B zones, i.e. by the fastigial nucleus and the vestibular nuclei. The connections of the fastigial nucleus and of the B zone with the vestibular nuclei are considered in Sect. 22.9.2 on the vestibular nuclei. The fastigial nucleus gives rise to crossed and uncrossed pathways. The fibres of the uncinate tract decussate within the caudal part of the cerebellar commissure (Fig. 22.25). The direct fastigiobulbar tract enters the brain stem from the fastigial nucleus immediately lateral to the fourth ventricle. The target nuclei of the X zone are the interstitial cell groups, located between the fastigial and posterior interposed nuclei. In the cat, rat and opossum, they give rise to a major projection to the cervical cord (Matsushita and Hosoya 1978; Bentivoglio and Kuypers 1982).

The anterior and posterior interposed nuclei, which receive Purkinje cell axons from the C1 and C2 zones and the C2 zone of the intermediate part of the cerebellum, respectively, ascend in the superior cerebellar peduncle (Fig. 22.63). They cross in the dorsal part of the decussation of the superior cerebellar peduncle. Ascending branches from the anterior interposed nucleus mainly terminate in the magnocellular red nucleus and in the ventral lateral nucleus of the thalamus. Descending branches from this nucleus terminate in the nucleus reticularis tegmenti pontis and in the bulbar reticular formation. Ascending fibres from the posterior interposed nucleus terminate in the mesencephalic reticular formation, the periaqueductal grey, the deep layers of the superior colliculus (see Sect. 22.8.3.2.3.), the nucleus of Darkschewitsch and other nuclei at the mesoencephalic-diencephalic border which project to the inferior olive. In the thalamus, the projections of the posterior interposed nucleus overlap with those from the anterior interposed and dentate nuclei. Descending branches from the posterior interposed nucleus avoid the nucleus reticularis tegmenti pontis and terminate in the medial reticular formation.

In the rat and mouse, the dorsolateral hump and neighbouring regions of the anterior interposed and lateral cerebellar nuclei give rise to an uncrossed pathway that descends between the motor and principal sensory nuclei of the trigeminal nerve to terminate in the lateral reticular formation or in the deep layers of the principal and spinal sensory nuclei of the trigeminal nerve (Cajal 1911; Mehler 1969, 1969; Achenbach and Goodman 1968; Faull 1978; Woodson and Angaut 1984). The uncrossed descending limb of the superior cerebellar peduncle has only been found in rodents.

Fibres from the dentate nucleus cross in the ventral part of the decussation of the superior cerebellar peduncle (Fig. 22.63). Ascending branches

22.7 Reticular Formation and the Serotonergic, Catecholaminergic and Cholinergic Cell Groups of the Brain Stem

22.7.1 Structure and Subdivision of the Reticular Formation

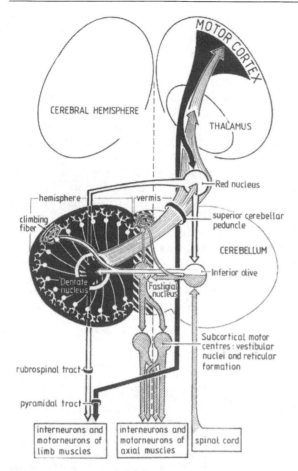

Fig. 22.63. Diagram of the efferent connections of the cerebellum and the climbing fibre projection of the inferior olive. The connections of the interposed nuclei are included with the dentate

terminate in the parvocellular red nucleus and rostrally and medially in the subthalamus and the ventral lateral and ventral anterior nuclei of the thalamus. Descending branches terminate in the nucleus reticularis tegmenti pontis.

Nucleo-olivary pathways from small, GABAergic neurons of the interposed and dentate nuclei remain separated from the main bundle of the brachium conjunctivum. In the caudal pons, they collect in the lateral angle of the fourth ventricle. They ascend ventral to the brachium conjunctivum, decussate somewhat more rostrally and descend lateral to the main descending branch of the superior cerebellar peduncle (Legendre and Courville 1987: cat; Cholley et al. 1989: rat; Tan et al. 1995b: rabbit).

The area which occupies the central portion of the brain stem is known as the reticular formation. Throughout most of its extent, this area is occupied by aggregations of loosely arranged cells of different types and sizes, and the fibre systems that pass through its territory are likewise mostly diffusely organised. The term 'reticular formation' refers to the fact that the dendrites of the cells in this area are arranged in bundles that together form a net-like pattern (Scheibel and Scheibel 1958). The traversing fibre systems pass through the interstices of this network. The reticular formation is surrounded by cranial nerve nuclei and relay centres and also by the long ascending (lemniscal) and descending fibre systems of the brain stem. Caudally, the reticular formation is continuous with the intralaminar nuclei of the thalamus and certain aggregations of subthalamic cells, among which the zona incerta may be mentioned. On cytoarchitectonic, chemoarchitectonic and functional grounds, the reticular formation of the medulla oblongata and the pons was divided into three longitudinal fields or zones (Brodal 1957): (1) a median and paramedian zone, which consists of the raphe nuclei, (2) a medial zone, which contains many large cells, and (3) a lateral, largely parvocellular zone.

Serotonergic neurons of the groups B1-B9 of Dahlström and Fuxe (1964) are found in the raphe nuclei and the adjacent reticular formation. Catecholaminergic and cholinergic neurons occupy a more lateral position. Most of these cells are confined to a fourth, intermediate field of the reticular formation, located between the medial magnocellular and the lateral parvocellular fields of the pontobulbar reticular formation (see Fig. 22.65). This intermediate field contains the ambiguus and the facial nucleus and the preganglionic parasympathetic neurons of cranial nerves VII, IX and X (Paxinos et al. 1990; Jones 1995). Rostrally, the intermediate reticular field extends medial to the trigeminal motor nucleus into the subcoeruleus area and the locus coeruleus, which correspond to the A5, A6 and A7 noradrenergic cell groups of Dahlström and Fuxe (1964). The aggregation of cholinergic neurons in the pedunculopontine nucleus in the dorsal tegmentum mesencephali can be considered as the rostral continuation of the intermediate reticular field (Jones 1995). The medial, magnocellular reticular formation is a somatic motor field connected with the spinal intermediate grey, spinal motoneurons and the oculomotor nuclei. The intermediate reticular field is concerned with branchiomeric and visceromotor functions. The lateral, parvocellular reticular formation is closely allied to the visceral and somatic sensory nuclei of the cranial nerve (Jones 1995).

The long, sparsely branching dendritic trees of the neurons of the medial magnocellular zone of the reticular formation are rostrocaudally compressed. Dendrites preferentially extend into the transverse plane (Scheibel and Scheibel 1958; Valverde 1961; Ramon Molinar 1984). The neurons of the precerebellar reticular nuclei, which are found within the reticular formation, display a glomerular structure, with dendrites confined to these nuclei (Fig. 22.64).

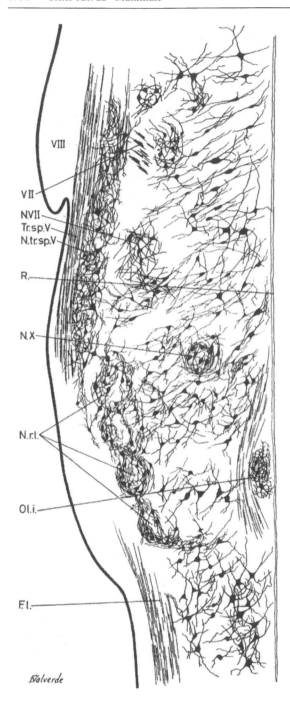

Fig. 22.64. Horizontal, Golgi-stained section through the reticular formation and the facial and ambiguus nuclei of the rat. The glomerular arrangement of the nucleus reticularis lateralis is clearly shown. (Valverde 1961). *F.l.*, lateral funiculus; *N.rl*, lateral reticular nucleus; *N.tr.sp.V*, nucleus of the spinal tract of the trigeminal nerve; *N.X*, dorsal vagus nucleus; *NVII*, motor nucleus of the facial nerve; *Ol.i*, inferior olive; *R*, nuclei raphes; *Tr.sp.V.*, spinal tract of te trigeminal nerve; *VII*, root of the facial nerve; *VIII*, statoacoustic nerve and cochlear nuclei

Fig. 22.65a-f. Subdivision of the reticular formation of the rat. **a** The ventral aspect of the rat brain stem. The subdivision of the reticular formation is shown on the *left*; the position of the adrenergic cell groups (*open circles*) and noradrenergic cell groups (*closed circles*) is indicated on the *right*. The data on the localization of the (nor)adrenergic groups were derived from Hökfelt et al. (1984a, b). **b-f** Transverse sections redrawn from Aston-Jones (1995), showing the subdivision of the reticular formation and the position of cholinergic cell groups (*crosses*) of Mesulam et al. (1983) on the *left*, and the localization of adrenergic and noradrenergic cell groups on the *right*. *A*, nucleus ambiguus; *A1-8*, noradrenergic cell groups A1-8; *bc*, brachium conjunctivum; *bp*, brachium pontis; *C*, nucleus cuneiformis; *C1-3*, adrenergic cell groups C1-3; *CH1-6*, cholinergic cell groups 1-6; *CO*, cochlear nuclei; *cr*, restiform body; *CS*, nucleus centralis superior; *D*, dorsal raphe nucleus; *DC*, dorsal column nuclei; *DM*, dorsomedial tegmental area; *DO*, central nucleus of the medulla oblongata, dorsal part; *DP*, dorsal paragigantocellular nucleus; *DT*, dorsal tegmental nucleus; *dV*, spinal tract of the trigeminal nerve and nucleus; *f*, fasciculus longitudinalis medialis; *G*, gigantocellular nucleus; *GA*, nucleus gigantocellularis, pars alpha; *GV*, ventral gigantocellular nucleus; *IC*, inferior coliculus; *IO*, inferior olive; *IR*, intermediare reticular field; *LC*, locus coerulaeus; *LD*, laterodorsal tegmental nucleus and adjacent reticular formation; *ll*, lateral lemniscus; *LP*, lateral paragigantocellular nucleus; *LP*, lateral parabrachial nucleus; *M*, nucleus raphes magnus; *MP*, medial parabrachial nucleus; *mV*, mesencephalic tract of the trigeminal nerve and nucleus; *MV*, motor nucleus of the trigeminal nerve; *OS*, superior olive; *P*, nucleus raphes pallidus; *PC*, nucleus pontis centralis, pars caudalis; *PH*, nucleus prepositus hypoglossi; *PN*, pontine nuclei; *PO*, nucleus pontis centralis, pars oralis; *PO*, nucleus pontis centralis, pars oralis; *PPL*, nucleus pedunculopontinus, lateral part; *PPM*, nucleus pedunculopontinus, medial part; *PR*, parvocellular reticular nucleus; *PV*, nucleus pontis centralis, pars ventralis; *PV*, principal sensory nucleus of the trigeminal nerve; *py*, pyramidal tract; *R*, nucleus reticularis tegmenti pontis; *RL*, lateral reticular nucleus; *RP*, nucleus raphes pontis; *S*, solitary tract and nucleus; *SA*, nucleus subcoerulaeus, pars alpha; *SS*, nucleus subcoerulaeus; *VEST*, vestibular nuclei; *VII*, facial nucleus; *VL*, ventrolateral intermediate reticular field; *VO*, central nucleus of the medulla oblongata, ventral part; *XII*, hypoglossal nucleus

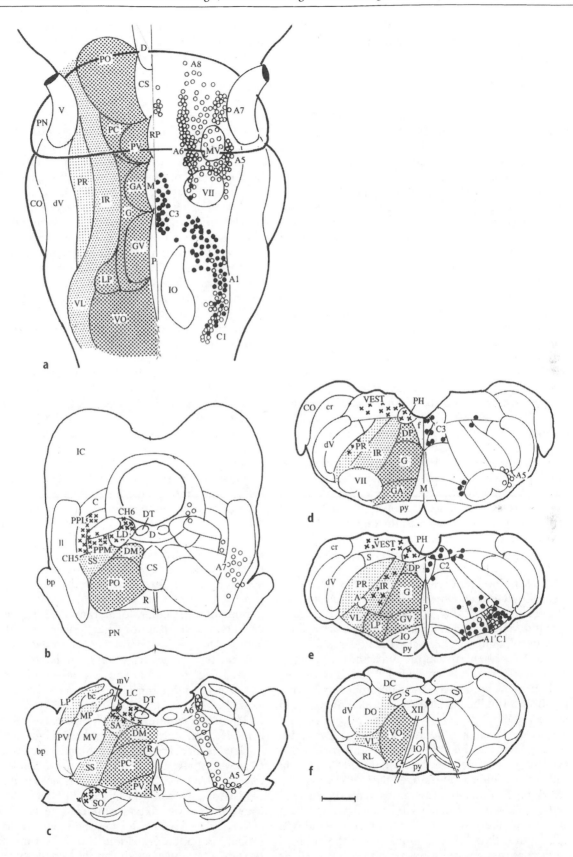

Very similar subdivisions of the reticular formation have been adopted for marsupials (*Didelphis*: Oswaldo-Cruz and Rocha Miranda 1968), Microchiroptera (*Myotis myotis*: Petrovický 1967), insectivores (*Erinaceus europeus*: Petrovicky 1966b), the rabbit (Meessen and Olszewski 1949), the cat (Taber 1961), primates (Olszewski and Baxter 1954; Koikegami 1957; Martin et al. 1990), the rat (Jones 1995) and the guinea pig (Petrovický 1966a). The caudal medullary reticular formation (the central nucleus of the medulla oblongata) is usually subdivided into dorsal and ventral fields. The dorsal field mostly contains small neurons and rostrally continues into the lateral parvocellular reticular formation. At mid-olivary levels, the medium-sized neurons of the ventral field give way to the neurons of the gigantocellular reticular nucleus of the medial reticular formation. The intermediate reticular field of Paxinos et al. (1990) contains the nucleus ambiguus. Ventrolaterally, it extends into the ventrolateral intermediate reticular field, which contains the retroambiguus nucleus (Jones 1995).

Two precerebellar reticular nuclei are contained within the central nucleus (Figs. 22.22–22.24h,g). The lateral reticular nucleus occupies a ventrolateral position. In mammals, it can be subdivided into ventrolateral parvocellular, ventromedial magnocellular and dorsal subtrigeminal portions (Walberg 1952). The subtrigeminal portion of the nucleus is located within the rubrospinal tract, and the magnocellular and parvocellular portions of the nucleus are located among the ascending fibres of the anterolateral funiculus. Medially, the lateral reticular nucleus extends into the region of the lateral vestibulospinal tract. The paramedian reticular nucleus is located next to the midline, within the medial longitudinal fascicle. Its lateral border is formed by the large cells of the nucleus interfascicularis hypoglossi. Dorsally, it borders on the cells of the nucleus of Roller, located lateral to the medial longitudinal fascicle and ventral to the nucleus prepositus hypoglossi. The nucleus of Roller is the caudalmost representative of the nuclei of the paramedian tracts, precerebellar nuclei in the dorsomedial reticular formation with a predominant projection to the vestibulocerebellum (Büttner-Ennever and Horn 1996).

The nucleus gigantocellularis extends from mid-olivary levels to the level of the rostral pole of the nucleus of the facial nerve (Fig. 22.65). The giant cells of the nucleus gigantocellularis proper are located fairly dorsally. The more ventrally located cells of the ventral gigantocellular nucleus (GiV) are slightly smaller and are oriented mediolaterally in cell strands that are continuous across the midline. At a more rostral pontine level, this ventral subdivision of the gigantocellular nucleus is represented by the pars alpha of the gigantocellular nucleus (GiA) of Meessen and Olszewski (1949). The pars alpha is located lateral to the nucleus raphes magnus and contains the lateral wings of the serotoninergic cell group B3 of Dahlström and Fuxe (1964). Both the pars alpha and the ventral subdivision of the gigantocellular nucleus contain a large proportion of GABAergic neurons (Jones 1995). The gigantocellular nucleus is bordered dorsally and ventrolaterally by the dorsal and ventrolateral paragigantocellular nuclei of Olszewski and Baxter (1954). The intermediate reticular field of Paxinos et al. (1990) forms the transition to the parvocellular reticular nucleus, which borders on and, more rostrally, penetrates in between the spinal trigeminal nucleus and the nucleus of the solitary tract. Rostrally, it contains the motor nucleus of the facial nerve.

The nucleus reticularis pontis caudalis (PnC) and oralis (PnO) extend from the motor nucleus of the facial nerve to the decussation of the superior cerebellar peduncle (Fig. 22.65). In the rat, the nucleus reticularis pontis caudalis was further subdivided into central, dorsomedial and ventral fields. At the level of the nucleus reticularis pontis oralis, this ventral field is replaced by the precerebellar nucleus reticularis tegmenti pontis (Jones 1990). The small, medium and scattered giant cells of the nucleus reticularis pontis caudalis are less densely packed. At the level of the motor nucleus of the trigeminal nerve, it merges with the nucleus reticularis pontis oralis. The nucleus reticularis pontis gradually increases in width, and the intermediate reticular field and the parvocellular reticular nucleus are displaced laterally. The dorsal part of the nucleus reticularis pontis caudalis and oralis, located rostral and caudal to the nucleus of the abducens nerve, corresponds to the horizontal gaze centre. This region, also known as the pontine paramedian reticular formation (PPRF), contains excitatory and inhibitory burst neurons with projections to the ipsilateral and contralateral abducens nucleus, respectively, and omnipause neurons, located next to the nucleus raphes pontis, that do not project to this motor nucleus (Büttner-Ennever 1977; Pearson et al. 1990; Graybiel 1977a,b).

The intermediate reticular field is rostrally continuous with the nucleus subcoeruleus and the nucleus subcoeruleus, pars alpha (Meessen and Olszewski 1949; Fig. 22.65). The parvocellular reticular nucleus continues as the small neurons of the field H of Meessen and Olszewski (1949) that surround the motor trigeminal nucleus. A paralemniscal reticular nucleus, located between the nucleus reticularis pontis oralis and the ventral nucleus of the lateral lemniscus, with projections to the spinal cord, was distinguished in the opossum, rat and cat (Martin et al. 1990).

A comprehensive subdivision of the reticular formation of the mesencephalon was proposed by Jones (1990). At the level of the inferior colliculus, the mesencephalic reticular formation is located dorsal and lateral to the superior cerebellar peduncle. It can be subdivided into the laterodorsal tegmental field, adjoining the nucleus in the periaqueductal grey bearing the same name, and the nucleus pedunculopontinus, which surrounds the lateral pole of the superior cerebellar peduncle. Both the laterodorsal tegmental field and the nucleus pedunculopontinus contain a large proportion of cholinergic cells. The small cells of the cuneiform nucleus occupy the oblong region between the inferior colliculus and the nucleus pedunculopontinus. The reticular formation of the rostral mesencephalon can be subdivided into a dorsal deep reticular field and the ventral retrorubral area.

22.7.2
Monoaminergic Cell Groups and Pathways

22.7.2.1
Methods Used in the Study of Monoaminergic Neurons

Monoaminergic systems are diffusely organised and innervate large parts of the spinal cord, the brain stem and the forebrain. Originally, monoaminergic neurons and pathways in the brain stem were visualised using the Falck-Hillarp formaldehyde histofluorescence technique (see Chap. 7). Using this technique, Dahlström and Fuxe (1964, 1965) were the first to describe the distribution of monoaminergic neurons in the brain. In their pioneering studies, they reported the localisation of catecholaminergic and serotonergic cell groups in the brain stem, which they designated A1-A2 and B1-B9, respectively. Since the Falck-Hillarp formaldehyde histofluorescence technique and its subsequent modifications have only limited specificity for the individual catecholamines, specific pharmacological tools such as enzyme inhibitors have to be used to distinguish between them. A great step forward was the introduction of immunocytochemical labelling techniques for the detection of the major synthesising enzymes, such as tryptophan hydroxylase, tyrosine hydroxylase, dopamine β-hydroxylase and phenylethanol-*N*-methyltransferase, which have the distinct advantage of high sensitivity. Since these enzymes play a key role in the catecholamine biosynthetic pathway, the presence of a particular catecholamine within a neuronal structure could be determined indirectly by comparing subsequent sections processed with antibodies against the respective catecholamine enzymes (for reviews, see Hökfelt et al. 1984a,b). When antibodies directed against serotonin (Steinbusch 1981), dopamine and noradrenaline became available (Geffard et al. 1984), they were used extensively for direct immunocytochemical detection of serotonergic, dopaminergic and noradrenergic neuronal structures in the central nervous system of different species. The advantage of direct detection with high specificity and sensitivity makes it the method of choice for investigation of the monoaminergic innervation in the brain stem, especially in areas where this innervation is less abundant or where different monoamines occur.

22.7.2.2
Serotonergic Nuclei

The description by Dahlström and Fuxe (1964) of the serotonergic cell groups B1-B9, which are mainly located in the raphe nuclei of the rat, has been confirmed in other species (rat: Swanson and Hartman 1983; Steinbusch 1984; primates: *Saimiri sciureus*: Felten et al. 1974; Hubbard and DiCarlo 1974b; *Macaca speciosa*: Garver and Sladek 1975; humans: Törk and Hornung 1990; Marsupialia, *Didelphis virginiana*: Crutcher and Humbertson 1978; cat: Wiklund et al. 1981; Jacobs et al. 1984; sheep: Tiller 1987).

The general organisation of the nuclei raphes and the distribution of serotonergic neurons appears to be very similar among mammalian species (Taber et al. 1960; Parent et al. 1981, 1984). The serotonergic neurons can be subdivided into two groups. One is located in the rostral pons and the mesencephalon, and the other in the caudal pons and the medulla oblongata. The two groups are separated by a gap and give rise to ascending and descending projections, respectively (Törk 1990; Törk and Hornung 1990; see Fig. 22.66, 22.67). Serotonergic cells in the medulla are located in the ventral nucleus raphes pallidus (B1) and the more dorsally located nucleus raphes obscurus (B2). In the nucleus raphes obscurus, serotonergic neurons are confined to two sheets on both sides of the midline in all species. Rostrally, these serotonergic groups are continuous with the B3 group in and lateral to the nucleus raphes magnus. The lateral wings of the B3 group are located in the ventral gigantocellular nucleus. Serotonergic neurons in the floor of the fourth ventricle (B4) are absent in some primates, but are more numerous in the cat, where they extend into the vestibular nuclei and the base of the cerebellum (Wiklund et al. 1981). The nucleus raphes pontis (B5) contains only a few neurons. The largest aggregates of serotonergic neurons are found in the nucleus raphes dorsalis (B8), located in the ventromedial part of the periaqueductal grey matter, and in the central superior nucleus. The serotonergic neurons of the B9 group are a lateral extension of B8, located dorsomedial to the lateral lemniscus. The rostralmost serotonergic neurons occur in the nucleus linearis and in the retrorubral field.

A sizable proportion of the serotonergic neurons (22.5 % in the cat; Wiklund et al. 1981) is found outside the raphe nuclei in the reticular formation lateral to the inferior olive, surrounding the lateral reticular nucleus, in the ventral gigantocellular nucleus, scattered in the dorsal tegmentum pontis and among the noradrenaline neurons of the locus coeruleus, the nucleus subcoeruleus and the parabrachial nuclei. The presence of serotonergic neurons outside the raphe nuclei in the nucleus pontis oralis is a particularly strong feature in primates, including humans (Törk and Hornung 1990). Serotonergic neurons outside the raphe nuclei in the mesencephalic tegmentum are more prominent in the cat and rat (Törk 1990). Moreover, all raphe nuclei are populated by serotonergic and non-serotonergic neurons.

The raphe nuclei receive their input mainly from limbic structures, including the interpeduncular nucleus (Groenewegen et al. 1986), the prefrontal cortex (Arnsten and Goldman-Rakic 1984) and the periaqueductal grey. Two efferent serotonergic fibre systems can be distinguished: an ascending system, mostly originating from serotonergic neurons in the mesencephalon and the rostral pons, and a caudally directed system, innervating the spinal cord, that originates from the serotonergic nuclei of the medulla oblongata. The brain stem and the cerebellum are target areas shared by the two divisions (for reviews, see Törk 1990; Törk and Hornung 1990; Halliday et al. 1995).

Thin, varicose serotonergic fibres from the dorsal raphe nucleus ascend in the periaqueductal grey; coarser fibres from the central superior nucleus travel in the ventral tegmentum. The ascending fibres subsequently collect in the medial forebrain bundle (Törk and Hornung 1990). The serotonergic innervation of the hypothalamus, the basal forebrain with strong projections to the septal

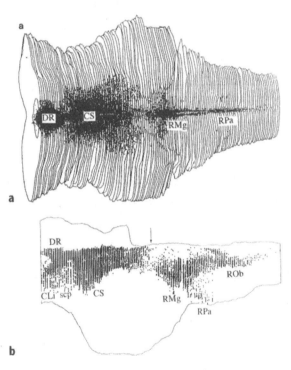

Fig. 22.66a,b. Computer-generated three-dimensional reconstructions of the distribution of serotonergic neurons in the human brain stem. a Dorsal view. b Lateral view. *CLI*, nucleus linearis caudalis; *CS*, nucleus centralis superior; *DR*, nucleus raphes dorsalis; *RMg*, nucleus raphes magnus; *ROb*, nucleus raphes obscurus; *RPa*, nucleus raphes pallidus; *scp*, decussation of the brachium conjunctivum. (Redrawn from Törk and Hornung 1990)

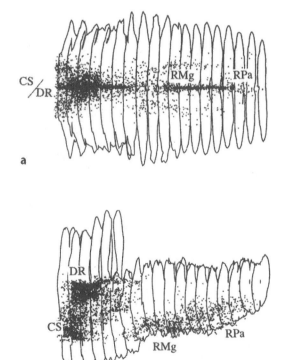

Fig. 22.67a,b. Computer-generated three-dimensional reconstructions of setotonergic cells in the brain stem of the rat. a Dorsal view. b Lateral view. *CS*, nucleus centralis superior; *DR*, nucleus raphes dorsalis; *RMg*, nucleus raphes magnus; *RPa*, nucleus raphes pallidus. (From Halliday et al. 1996)

area, some regions of the thalamus, the striatum and the cerebral cortex was reviewed by Nieuwenhuys et al. (1988). Different types of serotonergic plexus can be distinguished. The most common terminal plexus consists of D fibres, i.e. thin, branching fibres with small varicosities. They originate from the nucleus raphes dorsalis. D fibres make few synaptic contacts with their target cells. M fibres are beaded axons with large, oval or round varicosities that are more frequently engaged in synaptic contacts. M fibres take their origin from the nucleus centralis superior. Törk (1990) also reviewed the literature on species differences in the number of serotonergic synapses in the neocortex between the rat (80 % of the varicosities make synaptic contact) and monkey (3 %).

Descending systems from the nucleus raphes magnus and the adjoining reticular formation, which contain both serotonergic and nonserotonergic components, densely innervate the marginal layer of the spinal trigeminal nucleus and laminae I and II of the dorsal horn in the rat (Marlier et al. 1991), cat (Ruda et al. 1982) and monkey (Kojima and Sano 1983). These nuclei also project to the intermediate grey. These fibres descend in the dorsolateral funiculus (Holstege and Kuypers 1982; Steinbusch 1984). The strong serotonergic innervation of the intermediolateral nucleus is derived from the nucleus raphes magnus and the ventrolateral medulla (Lu et al. 1993). The motoneuronal groups receive serotonergic and nonserotonergic input from the nucleus raphes pallidus and obscurus and the neighbouring reticular formation (see Fig. 22.129), which travel in the lateral and ventral funiculus (Martin et al. 1982a,b; Holstege and Kuypers 1982; Steinbusch 1984; Törk 1990). Some of the serotonergic terminals on motoneurons co-localise GABA (J.C. Holstege 1996) or substance P (Tallaksen-Greene et al. 1993).

In the mesencephalon, the periaqueductal grey (Behbehani 1995), the nucleus of the optic tract, the substantia nigra and the nucleus interpeduncularis receive a particularly strong serotonergic innervation. In the rhombencephalon, the lateral reticular formation, ventral to the brachium conjunctivum, the trigeminal and facial motor nuclei, the locus coeruleus, the spinal mesencephalic and spinal trigeminal nuclei (Pearson and Jennes 1988; Lazaroff and Chouchkof 1995), the dorsolateral and the caudal and commissural subnuclei of the nucleus of the solitary tract and the juxtraventricular portion of the medial vestibular nucleus are the main targets of the serotonergic system (Steinbusch 1981, 1984). The cerebellar cortex and the cerebellar nuclei receive a sizable serotonergic projection (Bishop and Ho 1985). The serotonergic innervation of the inferior olive is mainly directed at the accessory olives (for a review, see Voogd et al. 1996).

22.7.2.3
Adrenergic and Noradrenergic Neurons

The adrenergic and noradrenergic neurons in the medulla oblongata are located in a dorsolateral and a ventrolateral group. The ventrolateral group corresponds to the noradrenergic group A1 of Dahlström and Fuxe (1964) and the adrenergic group C1 of Hökfelt et al. (1973, 1974). The neurons are located in the ventrolateral intermediate reticular field, dorsomedial to the lateral reticular nucleus and scattered among the fibres of the rubrospinal tract (Fig. 22.65). The C1 group extends most rostrally to the level of the motor nucleus of the facial nerve. The A2 and C2 group are located within the nucleus of the solitary tract and along the borders of the dorsal vagal nucleus (Fig. 22.65). Scattered neurons interconnect the two groups. Neurons of the adrenergic C3 group occupy the dorsal midline and the floor of the fourth ventricle at rostral medullary levels. More rostrally, noradrenergic cells are present in the locus coeruleus in the periaqueductal grey (A6), as scattered cells in the nucleus subcoeruleus (A6v), in the subependymal extension of the locus coeruleus (A4), next to the superior olive (A5) and dispersed within and around the ventral nucleus of the lateral lemniscus (A7) (Fig. 22.65). Noradrenergic cell strands interconnect the groups A5, A6 and A7 (Hökfelt et al. 1984a,b). Similar localisations of adrenergic and noradrenergic cells have been described for marsupials (Crutcher and Humbertson 1978), primates (Hubbard and DiCarlo 1974; Garver and Sladek 1975; Pearson et al. 1990) and the cat (Wiklund et al. 1981, Parent et al. 1981). The C3 group is absent in the North American opossum and in primates and could not be identified in the rat by Swanson and Hartman (1983). The A7 group of the North American opossum is located next to, and not within, the ventral nucleus of the lateral lemniscus, as in the rat and in primates. The presence of species differences in the composition of the locus coeruleus was stressed in the review of the groups A4-A7 by Aston-Jones et al. (1995). In the rat and in primates, the locus coeruleus is comprised entirely of noradrenergic neurons, whereas in the cat, rabbit, guinea pig and other species the locus coeruleus consists of a mixture of noradrenergic and non-adrenergic cells. In primates, the noradrenergic cells are pigmented, but in other mammalian orders they are not.

The main afferent connections of the locus coeruleus were summarised by Aston-Jones et al.

(1995) as mainly adrenergic and inhibitory from the C1 group in the lateral paragigantocellular field and as GABAergic and inhibitory from the medial part of the nucleus prepositus hypoglossi and the adjacent reticular formation. Other, less prominent sources of coeruleus input include the paraventricular hypothalamic nucleus, the preoptic area, the ventral tegmental area, the ventrolateral periaqueductal grey, the nucleus of the solitary tract and the intermediate grey of the spinal cord. It receives a serotonergic projection from neurons located ventromedial to the locus coeruleus and the group B9 (Steinbusch 1984). Other areas (central nucleus of the amygdala, prefrontal cortex) project to the region surrounding the locus coeruleus, which contains many of its dendrites.

The efferent connections of the locus coeruleus are both ascending and descending. The noradrenergic groups 5 and 7 mainly give rise to descending connections. Noradrenergic fibres ascend in the central reticular fasciculus, lateral to the medial longitudinal fascicle, in the dorsal longitudinal fascicle in the periaqueductal grey and in a bundle located in the ventral tegmentum. Rostrally, most of these fibres join the medial forebrain bundle. In the forebrain, they are distributed to olfactory structures, the basal forebrain, the cortex cerebri, the thalamus and the hypothalamus. Very few terminate in the striatum (Aston-Jones et al. 1995).

The noradrenergic and adrenergic innervation of the spinal cord is derived from all noradrenergic cell groups. The fibres descend in the dorsolateral and the ventral funiculus and in the superficial layers of the dorsal horn. They terminate strongly in the superficial laminae of the dorsal horn, among the motoneurons of the ventral horn and around the central canal (see Fig. 22.129). The intermediate grey is less densely innervated (Fritschy and Grzanna 1990). The nucleus intermediolateralis receives both noradrenergic input from group A5 and an adrenergic projection (Hökfelt et al. 1984a; Aston-Jones et al. 1995).

Similarly, the brain stem is innervated from all adrenergic and noradrenergic cell groups. (Nor)adrenergic fibres and terminals are distributed ubiquitously throughout the brain stem. The spinal trigeminal nucleus receives a strong projection, mainly from the locus coeruleus (Grzanna and Fritschy 1991). The nucleus of the solitary tract and the dorsal and dorsomedial parabrachial region are densely innervated by adrenergic and noradrenergic fibres.

The periaqueductal grey is mainly innervated by the noradrenergic cell groups A1 and A2 and the adrenergic cell groups C1 and C2 in the medulla. A dense (nor)adrenergic plexus is present in the ventrolateral and dorsolateral parts of the periaqueductal grey (Herbert and Saper 1992). Noradrenergic fibres terminating in the cranial motor nuclei are very thin, with small varicosities (Holstege et al. 1996). The motoneuronal innervation of the spinal and cranial motor nuclei takes its origin from the lateral and dorsal noradrenergic cell groups (Fritschy and Grzanna 1989). Both the cerebellar cortex, the cerebellar nuclei and the inferior olive receive a noradrenergic innervation from the locus coeruleus (for a review, see Voogd et al. 1996).

22.7.2.4
Dopaminergic Neurons

Dopamine-containing neurons are confined to the mesencephalon and the diencephalon. They are located in the olfactory bulb (group A16; Dahlström and Fuxe 1964; Fuxe and Hökfelt 1969; Björklund et al. 1973; Hökfelt et al. 1984b), in the preoptic area and the hypothalamus (A12-A15), in the periventricular region and surrounding the fasciculus retroflexus (A11), in the ventral tegmental area and nucleus linearis (A10), in the substantia nigra (A9) and in the retrorubral field (A8). The dopaminergic neurons of the substantia nigra are mainly found in the pars compacta and lateralis and are scattered in the pars reticulata.

The dopaminergic projections to the spinal cord take their origin from the A11 group (Hökfelt et al. 1979; Björklund and Skagerberg 1979). Strongest dopamine labelling in the cat, rat and monkey was present in the sympathetic intermediolateral cell column. Strong dopamine labelling of varicose fibres was found in all laminae of the dorsal horn, including the central canal area (region X), sparing the substantia gelatinosa, which was only sparsely labelled in the rat and monkey. In the motoneuronal cell groups, dense dopamine labelling with a fine granular appearance was observed. The sexually dimorphic motor nucleus, innervating the cremaster muscle, and Onuf's nucleus showed a much stronger labelling than the surrounding somatic motoneurons. In the parasympathetic area at sacral levels, labelling was moderate. The remaining areas, such as the intermediate zone (laminae VI–VIII), were only sparsely innervated. In all species, the descending fibres were located mostly in the dorsolateral funiculus, but laminae I and III also contained many rostrocaudally oriented fibres (Holstege et al. 1996).

In the brain stem of the rat, the dopaminergic and the (nor)adrenergic innervation generally overlapped. Strongest dopamine labelling was found in the nucleus of the solitary tract, dorsal motor nucleus of the vagus and lateral parabrachial nucleus. Other areas such as the medial parabrachial nucleus, locus coeruleus, periaqueductal grey and all raphe nuclei were strongly labelled. The principal nucleus of the inferior olive was also strongly labelled. Less dense staining was observed in many other areas of the brain stem, such as the inferior and superior colliculi, deep mesencephalic nuclei, pontine reticular nuclei, both sensory and motor trigeminal nuclei, prepositus hypoglossal nucleus, cochlear nuclei, cuneate and gracile nuclei and spinal trigeminal complex. Some fibres were running throughout the brain stem reticular formation. Other brain stem regions, such as the area postrema, cerebellum and external cuneate nucleus, were virtually devoid of dopaminergic fibres, with only an occasional traversing fibre (Van Dijken et al. 1997). The strong dopaminergic innervation of the striatum, the basal telencephalon and the frontal cortex is derived from the substantia nigra and the adjoining groups A9 and A10 (see Sect. 22.11.4).

22.7.3
Cholinergic Innervation of the Brain Stem

Cholinergic neurons and pathways have been identified by AChE histochemistry (Shute and Lewis 1967) and by the more sensitive immunocytochemical methods, using antibodies against ChAT (Levey et al. 1983). In the forebrain, cholinergic neurons are found as large interneurons in the striatum and as smaller cells in the nucleus accumbens and around the islands of Calleja of the olfactory tubercle. A collection of small cholinergic neurons in the medial habenular nucleus gives rise to a cholinergic component of the fasciculus habenulointerpeduncularis (Woolf and Butcher 1985). A nomenclature for the non-striatal cholinergic neurons was introduced by Mesulam et al. (1983) for the rat and by Mesulam et al. (1984) for macaque monkeys. Cholinergic neurons in the medial septal nucleus (group Ch1) and the vertical limb of the nucleus of the diagonal band (Ch2) principally innervate the cortex of the hippocampal formation. The horizontal limb of the nucleus of the diagonal band (Ch3) provides cholinergic fibres to the olfactory bulb. The nucleus basalis (Ch4), including the cholinergic cells of the magnocellular preoptic nucleus and scattered cells within the globus pallidus, innervates the cerebral cortex and the amygdaloid complex. The cholinergic groups Ch5 and Ch6 are located at the junction of the pons with the mesencephalon and correspond to the pedunculopontine nucleus and the laterodorsal tegmental nucleus, situated in the periaqueductal grey at the level of the rostral fourth ventricle, respectively (Fig. 22.65). The connections of these cholinergic groups were reviewed by Nieuwenhuys et al. (1988), Saper (1990) and Butcher (1995).

In the rat, the group Ch5 neurons extend dorsolaterally to the caudal substantia nigra (Armstrong et al. 1983). In the cat, cholinergic cells are also present in the parabrachial nuclei and the locus coeruleus, where they are co-extensive with catecholaminergic neurons (Jones and Beaudet 1987). Axons from the group Ch5 neurons ascend in the central reticular fasciculus, dorsolateral to the medial longitudinal fascicle, and provide a cholinergic innervation to the thalamus, the tectum, the substantia nigra, the lateral hypothalamus and the cerebral cortex (Mesulam et al. 1983; Woolf and Butcher 1986). Projections from the Ch5 group and from local cholinergic neurons of the intermediate reticular field terminate in the raphe nuclei, the pontine nuclei and in the pontine and medial medullary reticular formation (Woolf and Butcher 1989; Jones 1990). The diffuse cholinergic projections to the cerebellar cortex and nuclei are similarly derived from the Ch5 group (for a review, see Voogd et al. 1996). Other cholinergic projection systems not included in the classification of Mesulam et al. (1983, 1984) include the tectal projections from the parabigeminal nucleus (see Sect. 22.8.3.2.3), the mossy fibre projections to the cerebellum from the medial and descending vestibular nuclei (see Sect. 22.6.6), the cholinergic projection from the nucleus prepositus hypoglossi to the inferior olive, which co-localises GABA (De Zeeuw et al. 1993; Caffé et al. 1996) and the olivocochlear bundle from cholinergic neurons in and/or around the lateral superior olive (see Sect. 22.8.2.6)

22.7.4
Connections of the Reticular Formation

22.7.4.1
Medial Reticular Formation

The long efferent connections of the reticular formation are often difficult to distinguish from the connections of the monoaminergic and cholinergic cell groups of the brain stem. The connections of the reticular formation were extensively reviewed by Nieuwenhuys et al. (1988). For references on this subject, the reader is referred to this monograph. Pathways from the contralateral nucleus gigantocellularis and pontis caudalis, from the ipsilateral nucleus pontis oralis and from the deep tegmentum of the mesencephalon ascend in the central reticular fasciculus, dorsolateral to the medial longitudinal fascicle. They innervate the intralaminar and midline nuclei of the thalamus, the subthalamus and the dorsolateral hypothalamus. Ascending fibres in the medial forebrain bundle terminate in the preoptic area, the basal forebrain and the medial frontal cortex.

The ascending connections from the medial reticular formation are a crucial part of the non-specific afferent or extralemniscal system that conveys stimuli, gathered by various kinds of receptors, to the cerebral cortex. The structural components of this system are the spinoreticular projection, reticular afferents from various cranial nerves, direct and indirect reticular thalamic fibres and the 'aspecific' thalamocortical projection.

Spinoreticular fibres from the intermediate grey matter of the spinal cord ascend in the contralateral anterolateral fasciculus and terminate in the medial reticular formation, with areas of maximal termination situated in the rostral central nucleus of the medulla oblongata with the adjoining caudal gigantocellular nucleus and in the nucleus reticularis pontis caudalis. In the pons, the spinoreticular projection shifts laterally and extends along the subcoeruleus nucleus into the parabrachial area (Holstege and Kuypers 1977). Some of the fibres of the anterolateral fasciculus bypass the reticular formation and project directly to the intralaminar nuclei of the thalamus. They are reinforced by the projection from the reticular formation to the intralaminar nuclei, which preferentially arises from the regions of the reticular formation receiving major spinoreticular projections (Brodal 1957; Jones and Yang 1985). The reticulothalamic fibres ascend in the central reticular fasciculus, located lateral to the periaqueductal grey. The central bundles of this fasciculus contain larger fibres, originating from the nucleus pontis oralis (Forel's fascicles, Fig. 22.23o,p; Busch 1961). The widespread, diffusely organised projections of the intralaminar nuclei of the thalamus to the cortex constitute the final link in the non-specific afferent system (Macchi and Bentivoglio 1982).

Descending reticulospinal pathways take their origin from the contralateral nucleus pontis oralis and nucleus paragigantocellularis dorsalis and the ipsilateral nucleus pontis caudalis and gigantocellularis. Both in the cat and the North American opossum, reticulospinal fibres to the dorsal horn and the intermediate grey arise from the paralemniscal reticular nucleus (Martin et al. 1990). Serotonergic and non-serotonergic fibres from the nucleus raphes magnus and the adjoining nucleus gigantocellularis, pars alpha join the long ascending reticular systems and give rise to a descending projection, terminating in the marginal layer of the caudal spinal trigeminal nucleus and the superficial and deep laminae of the dorsal horn. The more caudal nucleus raphes pallidus and obscurus and the adjoining ventral gigantocellular nucleus give rise to a diffuse projection to the intermediate grey and the motoneuronal cell groups of the spinal cord. This projection includes serotonergic, non-serotonergic and GABAergic components (Holstege and Kuypers 1982; J.C. Holstege 1996).

22.7.4.2
Lateral Reticular Formation

The lateral tegmental field, which includes the intermediate reticular field of Paxinos et al. (1990) and the lateral parvocellular reticular formation, gives rise to ascending, descending and propriobulbar connections. The lateral pontine tegmentum and the parabrachial nuclei can be considered as the rostral extensions of the intermediate reticular field and the lateral reticular formation, respectively. These regions receive ascending projections from the nucleus of the solitary tract, the spinal trigeminal nucleus and the intermediate ventrolateral reticular field (see Fig. 22.76; Norgren 1990; see Sect. 22.8.1.2) and contain respiratory, cardiovascular and gastrointestinal control centres. They give origin to ascending pathways to the nucleus ventralis posterior medialis, pars parvocellularis, the intralaminar nuclei, the hypothalamus, the nucleus centralis amygdalae and the bed nucleus of the stria terminalis, with the lateral pontine tegmentum projecting directly to the insular cortex (Saper and Loewy 1980; Takeuchi et al. 1982; Fulwiler and Saper 1984; reviewed by Norgren 1990). Descending connections of the parabrachial nuclei and the lateral pontine tegmentum terminate in the nucleus of the tractus solitarius, the ventrolateral intermediate reticular field and the respiratory and autonomic preganglionic neurons of the spinal cord. The ventrolateral intermediate reticular field forms the caudal extension of the intermediate reticular field of Paxinos et al. (1990) and contains the adrenergic and noradrenergic cell groups C1 and A1 and the retroambiguus nucleus. It similarly projects to autonomic cell groups and to motoneurons engaged in respiration and sexual behaviour (Holstege 1991; Van der Horst and Holstege 1995).

Short premotor connections of the pontomedullary lateral reticular field to the cranial motor nuclei were studied in the cat (Holstege and Kuypers 1977; Holstege et al. 1977) and the rat (Ter Horst et al. 1991). Holstege et al. (1977) distinguished a *medial* propriobulbar system with bilateral projections to the trigeminal, facial and hypoglossal motor nuclei from a *lateral* propriobulbar system with more restricted, ipsilateral connections to those motor nuclei. The lateral propriobulbar system originates from a lateral zone of the lateral tegmental field, i.e. from the parvocellular reticular nucleus. The origin cof the medial propriobulbar system was located in a medial zone of the lateral tegmental field, i.e. in an area that may correspond to the intermediate reticular zone of Paxinos et al. (1990). Premotor neurons of the facial nerve nucleus also are located in the mesencephalic reticular formation (Holstege and Kuypers 1977).

22.8
Sensory Systems

J. Voogd

The emphasis in this section is on the subthalamic centres and connections of the sensory systems. Their relays in the thalamus and their cortical projections are reviewed in Sect. 22.10.3.

22.8.1
Somatosensory and Viscerosensory Systems

Sensory systems convey information from receptors in the skin, the mucous membranes, the musculoskeletal system and the viscera that may lead to a conscious experience. Because conscious experience, at least, involves a relay in the thalamus and/or a projection to the cerebral cortex, most sensory systems consist of three neurons: (1) a primary afferent neuron, (2) a relay neuron in the spinal cord or the brain stem and (3) a thalamocortical neuron.

Of the somatosensory systems that convey exteroceptive and proprioceptive information, the spinothalamic tracts and the dorsal column-medial lemniscus pathway and their trigeminal equivalents present this three-neuronal organisation. These pathways are mainly crossed, with the decussation in the second neuron. The spino-cervicothalamic system is a somatosensory pathway consisting of four neurons. It conveys exteroceptive information from the spinal dorsal horn, with relays in the lateral cervical nucleus and the thalamus. The cervicothalamic tract decussates in the upper cervical cord. A trigeminal equivalent of the spino-cervicothalamic system does not exist.

The three- or four-neuronal somatosensory pathways have been distinguished as 'lemniscal' and/or 'neo-spinothalamic' pathways from the multisynaptic palaeo-spinothalamic connections that possess multiple relays in the spinal cord and brain stem. Bishop (1959) suggested that afferent paths send their output to many levels of the nervous system. Lower levels of termination represent the more primitive condition. During the evolution of higher forms, additions of two types have occurred: First, additions of fibres similar to those already present, but bypassing the primitive levels of termination to reach the newly developed higher centers, and second invasion of these higher levels by successively larger fibres as these centers become more highly differentiated in function (Bishop 1959, p. 92; Fig. 22.68; Kevetter and Willis 1984).

Figure 22.68 illustrates four hypothetical stages in the evolution of monosynaptic spinothalamic pathways from a multisynaptic chain. The stages C and D of spinothalamic fibres with collaterals to and relays in the reticular formation correspond to the palaeo-spinothalamic system proposed by Mehler (1966). Their distinction from the neo-spinothalamic (direct) pathways also rests on a different mode of termination in the thalamus. Palaeo-spinothalamic fibres terminate diffusely in the intralaminar nuclei, whereas neo-spinothalamic fibres deviate laterally and terminate somatotopically in the ventrobasal complex of the thalamus (Mehler et al. 1960; Mehler 1962, 1969). The evolution of direct spinothalamic projections from multisynaptic or strongly collateralised systems has not received support from observations in agnathams, teleosts or amphibians, which confirmed the presence of direct spinothalamic projections and suggested a common pattern throughout all vertebrate classes (Muñoz et al. 1997).

Since the role of multineuronal chains in sensation is difficult to substantiate experimentally, most comparative studies of the somatosensory systems have focused on the amount and the loci of collateralisation in pathways to the intralaminar nuclei and the ventrobasal complex. Additionally, other possible brain stem relays for somatosensory pathways have gained prominence, such as the periaqueductal grey of the mesencephalon, the tectum, the subthalamus and the hypothalamus. The following review of the spino- and trigeminothalamic tracts, the medial lemniscus and the cervicothalamic tract will focus on these aspects.

Viscerosensory pathways differ from somatosensory paths in the presence of multiple brain stem relays, in the avoidance of the thalamus in some of the viscerosensory projections to the cerebral cortex and in the bilaterality of these connections. Viscerosensory pathways from the spinal cord, presumably, follow the same course as the spinothalamic tracts. The best example of such a spinal viscerosensory system is the projection of the sacral lamina I neurons, which convey information from pelvic viscera to the periaqueductal grey via a pathway that coincides with the dorsolateral spinothalamic tract (Van der Horst and Holstege 1995; see Sect. 22.3). The review of the viscerosensory pathways will focus on the connections of the nucleus of the solitary tract.

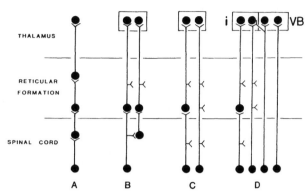

Fig. 22.68a–d. Hypothetical evolution of the spinothalamic tract. **a** A multisynaptic chain of neurons connects the spinal cord with the thalamus. **b,c** Both the direct and the indirect pathways are highly collateralised. **d** In mammals, the multisynaptic and collateralised direct spinothalamic (paleospinothalamic) fibres terminate in the intralaminar nuclei (*i*); direct spinothalamic (neospinothalamic) fibres terminate in the ventrobasal complex (*VB*), with or without collaterals to the intralaminar nuclei. (Modified from Kevetter and Willis 1984)

22.8.1.1
Somatosensory Pathways

The relative importance of the extremities and the trigeminal nerve-innervated structures, such as the lips, the mystacial vibrissae and the oral cavity, in sensory perception was assessed in different mammals by Rose and Mountcastle (1959) by reconstructing the sensory representation in the ventrobasal complex of the opossum, rabbit, cat and rhesus monkey (Fig. 22.69). The general pattern of representation in the ventrobasal complex is rather similar in these species and is characterised by a trigeminal projection from the head located medial to the projections from limbs and trunk, with the dorsum of the body pointing dorsally. Except for a small, ipsilateral face and mouth representation in the medialmost aspect of the ventrobasal complex, the representation is essentially contralateral. The principal differences between species concern the relative size of representations of different body regions. In the rabbit and opossum, the trigeminal representation predominates. The forelimb representation prevails in the North American opossum, and the hindlimb representation gains prominence in the cat and monkey (Pubols and Pubols 1966).

The contralateral projections of the dorsal column nuclei and the sensory nuclei of the trigeminal nerve were quantified in the rat by Kemplay and Webster (1989). The projection of the nucleus princeps of the trigeminal nerve accounts for 50% of the total contralaterally labelled neuron population (Fig. 22.70), followed by approximately equal contributions of the gracile nucleus, the cuneate nucleus and the pars interpolaris of the spinal trigeminal nucleus.

Verhaart (1970) determined the relative contribution of the spinal and trigeminal representations by comparing the size of the components of the medial lemniscus, taking their origin from the dorsal column nuclei and the nucleus princeps of the trigeminal nerve in different mammalian species. In Monotremata, Marsupialia, Insectivora, Xenarthra and Microchiroptera, the contribution of the caudal dorsal column nuclei (mainly from the gracile nucleus, i.e. from the trunk and lower extremity) to the lateral part of the medial lemniscus is smaller than the contribution of the rostral dorsal column nuclei (from the internal cuneate, i.e. the upper extremity) to the dorsomedial medial lemniscus. In some species, these two components also differ in their fibre calibre; in the rat and rabbit, the caudal component consists of thin fibres, and the rostral component of coarse fibres, which presumably take their origin from the cell clusters or columns of the cuneate nucleus (see below). In primates, coarse fibres take their origin preferentially from the gracile nucleus. The projection of the nucleus princeps dominates over that of the dorsal column nuclei in Monotremata, Insectivora, Xenarthra, Pinnepedia, Perissodactyla and Artiodactyla. In Perissodactyla and Artiodactyla, the nucleus princeps gives rise to the ipsilateral dorsal trigeminothalamic tract of Wallenberg (Karamanlidis and Voogd 1970; see Sect. 22.8.1.1.4). In species

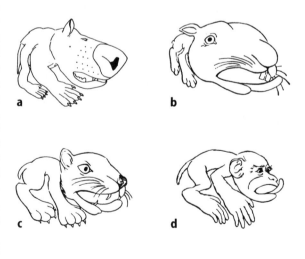

Fig. 22.69a–d. Somatic sensory representation in the left ventrobasal complex of **a** opossum, **b** rabbit, **c** cat and **d** rhesus monkey. Note the overall similarity in patterns of representation and that the opossum appears to fall between the rbbit and cat in the relative degree of trigeminal versus spinal representation. (From Pubols and Pubols 1966)

of most other mammalian orders, crossed trigeminothalamic fibres joining the medial lemniscus at its medial side predominate. In most of these species, the trigeminal component consists of thin fibres; in Pinnepedia, this contingent consists of coarse fibres.

22.8.1.1.1
Spinothalamic Tracts

The spinothalamic tracts arise from the dorsal horn and the intermediate grey at all levels of the cord, decussate in the ventral white commissure and ascend in the anterolateral fascicle of Gowers (1886). The anatomy of the spinothalamic tracts and their role in pain conduction was reviewed by Willis and Coggeshall (1991) and Willis et al. (1995). The distribution of the anterolateral fascicle in the brain stem and the thalamus was described in the opossum, cat and monkey by Mehler (1969).

With respect to their origin and termination, two systems can be distinguished. The *dorsal spinothalamic tract* arises mainly from lamina I cells of the spinal dorsal horn (Fig. 22.21c), is located in the dorsal part of the ventrolateral funiculus of the cord and terminates predominantly in the ventrolateral nuclei of the thalamus. The *ventrolateral spinothalamic tract* is located more ventrally, takes it origin from laminae IV, V, VII and VIII (Fig. 22.21d) and mainly terminates in the medial thalamus, including the intralaminar nuclei (laminae VII and VIII) and the nuclei of the posterior group of the thalamus (IV and V; cat: Jones et al. 1985, 1987; Stevens et al. 1989, 1991; Craig 1991; monkey *Macaca fascicularis*: Willis et al. 1978, 1979; Hayes and Rustioni 1980; Apkarian and Hodge 1989a,b; rat: Giesler et al. 1979, 1981; Burstein et al. 1990).

The largest group of spinothalamic tract cells is located in the dorsal horn of the cervical cord (Fig. 22.21a,b). Some cervical neurons project to the ipsilateral thalamus. In primates, the contribution of the lumbar dorsal horn is relatively increased with respect to the rat and cat (Carstens and Trevino 1978a,b; Craig et al. 1989; Smith et al. 1991; Burstein et al. 1990; Apkarian and Hodge 1989a). Laminae I and V spinothalamic tract cells also give rise to propriospinal connections and to projections to the tectum, the pretectum, the periaqueductal grey, the red nucleus, the nucleus of Darkschewitsch, the hypothalamus and other structures (Kerr 1975; Gordon and Grant 1982; Robertson et al 1983; Wiberg and Blomqvist 1984; Wiberg et al. 1987; Verburgh et al. 1990; Zhang et al. 1990; Cliffer et al. 1991; Burstein et al. 1996). Lamina IV of the dorsal horn also gives rise to ascending postsynaptic dorsal column fibres to the dorsal column nuclei and to a projection terminating in the lateral cervical nucleus. Its contribution to the spinothalamic tract appears to predominate in primates, whereas spinocervical tract neurons prevail in the cat (see Sect. 22.3; Fig. 22.21).

The termination of the spinothalamic tract in the thalamus and the cortical projections of its target nuclei were reviewed by Jones (1985). It terminates in the lateral part of the ventral posterior nucleus of the thalamus, but this termination is much heavier in monkeys than in other species. In primates, the spinothalamic fibres in the ventral posterior nucleus terminate as bursts or rod-like aggregates. In the cat, rat, and opossum, the spinothalamic terminals mainly occupy the periphery of the nucleus. In all species, the termination of the spinothalamic fibres extends into the border region of the ventral posterior and ventral lateral nuclei (Boivie 1971, 1979; Hazlett et al. 1972; Mehler 1969, 1974; Pearson and Haines 1980; Berkley 1980, 1983; Asanuma et al. 1983b; Burton and Craig 1983; Mantyh 1983; Ma et al. 1986). In the rat, cat, and monkey, the spino-

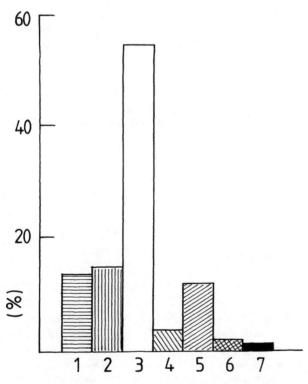

Fig. 22.70. Relative contribution of each nucleus and subnucleus (expressed as percentages) to the total contralateral neuron population of the somatosensory relay nuclei of the brain steim of the rat after injections of the retrograde tracer horseradish peroxidase in the thalamus of the rat. Survival time is 72 h. *1*, Gracile nucleus; *2*, cuneate nucleus; *3*, principal trigeminal nucleus; *4*, nucleus of spinal tract of the trigeminal nerve; *5*, pars interpolaris; *6*, pars caudalis; *7*, Bisschoff's nucleus. (From Kemplay and Webster 1989)

thalamic projection includes the submedius nucleus (Craig and Burton 1981; Mantyh 1983; Peschanski et al. 1983). The main projection to the intralaminar nuclei passes through the subparafascicular and central median nuclei and is located in the central lateral nucleus in all species (Molinari et al. 1987). Spinothalamic fibres terminating in the posterior group of thalamic nuclei arise from laminae I and V, travel in the dorsolateral spinothalamic tract and include a distinct nucleus in macaque monkeys with a particularly dense input from lamina I (Mehler 1966; Ralston and Ralston 1992; Craig et al. 1995).

Mehler (1966) suggested that the neo-spinothalamic system terminates exclusively in the ventrobasal complex of the thalamus and has evolved from a multisynaptic and highly collateralised palaeo-spinoreticulothalamic system, which is mainly aimed at the intralaminar nuclei of the thalamus. It should be noted that, although the study of the spinothalamic connections derives its main importance from their role in nocicepsis, the collateralised, multisynaptic ascending pathways to the thalamus may subserve quite different functions in (viscero)sensory transmission and motor control. The possible stages in this evolution of the spinothalamic connections are diagrammatically illustrated in Fig. 22.68 (taken from Kevetter and Willis 1984). Kevetter and Willis (1984) concluded from their double-labelling experiments in the rat and their review of the available literature that a high proportion of spinoreticular cells in laminae VII and VIII in the rat and cat send descending collaterals in the cord. Spinothalamic cells with collaterals to the medial reticular formation and projecting to either the ventrolateral complex and/or the intralaminar nuclei have been identified in the rat and in primates. Moreover, spinothalamic neurons with an exclusive projection to the lateral thalamus appear to be present in the rat, cat, and monkey. The evidence concerning mammalian evolution of a direct lateral neothalamic system from a more primitive medial, multisynaptic palaeo-spinothalamic system is still meagre. Spino-reticulothalamic, intralaminar and ventrolateral connections appear to coexist as parallel channels, each subserving specific functions in sensory transmission and motor control.

22.8.1.1.2
Cervicothalamic Tract

The cervicothalamic tract was first described by Morin and Catalano (1955) and Busch (1957, 1961) in the cat as a component of the medial lemniscus, originating from the contralateral lateral cervical nucleus. It decussates in the ventral white commissure at upper cervical segments. At the C1 level, its medium-sized fibres are located at the ventral surface of the medulla, lateral to the pyramidal tract (Fig. 22.20). It ascends in the lateral part of the medial lemniscus, and terminates in the ventrobasal complex. Edinger (1889) observed the origin of the cervicothalamic tract in the cat, but mistook it for the ventral spinothalamic tract. In experimental studies of the spinothalamic projection using lesions or transections of the upper cervical segments containing the lateral cervical nucleus, the cervicothalamic tract has often been included with the spinothalamic tracts originating from more caudal levels of the cord.

The projection of laminae IV and V of the dorsal horn to the lateral cervical nucleus forms the first link in the spino-cervicothalamic pathway (Fig. 22.21). Both the spinocervical projection and the lateral cervical nucleus were identified in Insectivora, Rodentia, Carnivora, Perissodactyla, Artiodactyla and primates, but not in marsupials (see Sect. 22.3).

Fibres from the lateral cervical nucleus terminate in the tectum, the rostral part of the medial geniculate body, the posterior nuclear group and the lateral part of the nucleus ventralis posterior, where they overlap with projections from the dorsal column nuclei and more caudal levels of the cord (hedgehog tenrec: Künzle 1993; cat and dog: Hagg and Ha 1970; Boivie 1980; Berkley 1980; Berkley et al. 1980; Flink and Westman 1985; *Macaca fascicularis*: Boivie 1980; Berkley 1980; rat: Giesler et al. 1987, 1988; Granum 1986). The lateral cervical nucleus is relatively small in primates, and its major projection in the monkey is to the mesencephalon rather than to the thalamus (Smith and Apkarian 1991).

22.8.1.1.3
Dorsal Column Nuclei

Gracile, internal cuneate and external cuneate nuclei can be distinguished in all mammals (Zeehandelaar 1920). An unpaired nucleus, located in the midline at the level of the caudal nucleus gracilis, is present in mammals with a prehensive tail (Bischof 1899, cited by Zeehandelaar 1920). The external cuneate nucleus consists of fairly uniform, medium-sized and large neurons. The dorsal parts of the internal cuneate and gracile nuclei of the cat are lobulated and consist of clusters of round cells with short, dense, ramifying dendrites separated by medullary sheets (Fig. 22.71). The ventral parts and the rostral poles of the nuclei consist of a reticulum of fusiform and multipolar cells with long, ramifying dendrites (Kuypers et al. 1961; Kuypers and Tuerk 1964). In the dorsal roots and dorsal columns, afferent fibres from different receptive fields are intermingled to some degree. By the time they reach the dorsal column nuclei, they have become rearranged to terminate in a detailed somatotopical pattern on the dorsal cell clusters of the dorsal column nuclei and, more diffusely, in the ventral and rostral reticular parts of the nuclei (Kuypers and Tuerk 1964; Johnson et al. 1968; Ueyama et al. 1994). The lobulation of the dorsal part of the cuneate nucleus is more extensive in the raccoon *Procyon lotor* than it is in the cat, with a segregation of the representations of the ventral, glabrous surface of the individual digits of the forepaw (Fig. 22.72c; Johnson et al. 1968; Rasmussen 1989).

Both the postsynaptic afferents from the dorsal horn, which reach the dorsal column nuclei by way of the dorsal columns and the dorsolateral funiculus, and the cortical projections to the dorsal column nuclei terminate preferentially in the ventral and rostral reticular subdivisions of the dorsal column nuclei (Walberg 1957; Kuypers et al. 1961; Rustioni 1973; Rustioni and Hayes 1981; Martinez et al. 1995).

The efferent pathway from the dorsal column nuclei decussates as internal arcuate fibres in the caudal medulla (Figs. 22.22–22.24g, 22.25). The fibres ascend, dorsal to the pyramidal tract, in the medial lemniscus. The medial lemniscus shifts laterally along the border of the pes and tegmentum of the pons and the mesencephalon and terminates in the thalamus. Fibres from the caudal half of the dorsal column nuclei are located medially, and those from more rostral levels laterally in the medial lemniscus (Busch 1961; Kuypers and Tuerk 1964).

Graybiel (1972) distinguished the thalamic projections from the dorsal column nuclei, with their ability to preserve the exquisitely specialized spatial and temporal patterns of stimulation of the receptor sheet, as so-called 'lemniscal line systems' from the heterogeneous pathways that accompany the dorsal column-thalamic projection, but that show a much higher degree of extrinsic convergence and divergence along their course and are less closely linked to the detailed response characteristics of the periphery. The lemniscal line systems of the cat take their origin from the cell cluster region of the dorsal column nuclei and terminate in the lateral part of the ventral posterior nucleus of the thalamus, with the nucleus gracilis projecting to the lateral part of this nucleus, and the nucleus cuneatus to its medial part (Boivie 1971; Groenewegen et al. 1975; Berkley 1975, 1986; Hand and Van Winkle 1977; Berkley and Hand 1978; Blomqvist 1980; Bull and Berkley 1984; Tamai et al. 1984; Cooper and Dostrovsky 1985; Waters et al. 1985). In the raccoon, the ventral posterior nucleus displays an intricate sublobulation with an extremely large representation of the glabrous forepaw area, accounting for 50% of the volume of the nucleus (Welker and Johnson 1965). The projection of the ventral posterior nucleus to the primary sensory cortex in the raccoon is characterised by a pattern of sulci, dimples and spurs, demarcating the representations

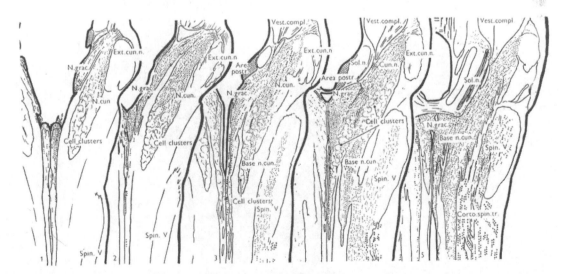

Fig. 22.71. Dorsal column nuclei of the cat, shown in Nissl-stained, horizontal sections. *Left*: most dorsal, *right*, most ventral section. Clustering of the cells is indicated by circular profiles. Redrawn from Kuypers and Tuerk (1964). *Area postr.*, area postrema; *Cortopin.tr*, corticospinal tract; *Ext.cu.nu.*, external cuneate nucleus; *N.cun*, internal cuneate nucleus; *N.grac*, nucleus gracilis; *Sol.n*, nucleus of the solitary tract; *Spin.V*, spinal tact of the trigeminal nerve and nucleus; *vest.compl.*, vestibular nuclei.

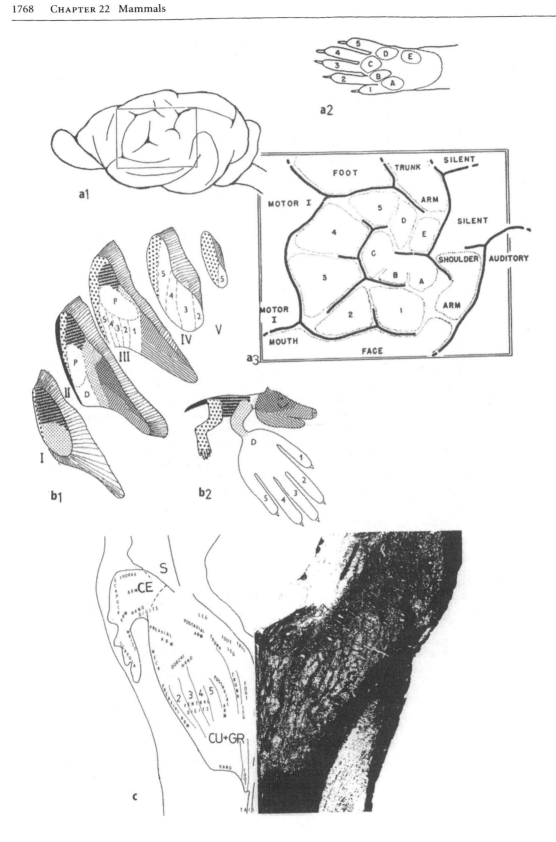

of the digits in the hand area (Welker and Seidenstein 1959; Fig. 22.72a,b).

The ventral and rostral parts of the dorsal column nuclei give rise to ascending projections to the contralateral accessory olives, the external nucleus of the inferior colliculus, the intercollicular area, the deep layers of the superior colliculus, the cuneiform nucleus, the pretectum, the magnocellular part of the medial geniculate body, the posterior nuclear group of the thalamus, the zona incerta, the periaqueductal grey, the red nucleus and the nucleus of Darkschewitsch (Boivie 1971; Dart and Gordon 1973; Groenewegen et al. 1975; Berkley 1975; Hand and Van Winkle 1977; Berkley and Mash 1978; Itoh et al. 1984; Björkeland and Boivie 1984; Bull and Berkley 1984; Wiberg and Blomqvist 1984). Projections of the dorsal column nuclei to the spinal cord descend in the lateral funiculus (Enevoldson and Gordon 1984).

Few specific data are available on the dorsal column nuclei of marsupials (Fig. 22.22f,h). The cytoarchitecture and the dendritic patterns of the dorsal column nuclei, including the midline nucleus of Bischoff, were studied by Penny (1982), and the somatotopical organisation of the nuclei was reported by Hamilton and Johnson (1973), both in *Didelphis marsupialis*. The differentiation of the nuclei into clustered and reticular regions is less distinct than in the cat and raccoon, and a detailed somatotopy with clear rostrocaudal differences in segregation and size of the receptive fields was not observed. The distribution of the medial lemniscal connections of the dorsal column nuclei was demonstrated by Hazlett et al. (1972) and Gray et al. (1981) in the North American opossum and by Clezy et al. (1961) in the phalanger *Trichosuris*.

◀ **Fig. 22.72a–c.** The finger print of the raccoon. Somatotopical localization in the dorsal column nuclei, the ventrobasal complex and the sensory cortex of the raccoon, *Procyon lotor*. **a** Localization in the sensory cortex. **a1**, left cerebral hemisphere of the raccoon: the area outlined in **a1** is enlarged in **a3**. **a2** The forepaw with a key for the nomenclature of the digital and palmar pads. **a3** The intimate topographical association between physiological subdivisions in the SI sensory cortex. Every sulcus in this region delimits palmar or digital pads from another or from more proximal regions of the upper extremity. From Welker and Seidenstein (1959). **b** Correlation between nuclear morphology and somatotopic organization in the ventrobasal complex of the thalamus of the raccoon. **b1** Transverse slabs through the left ventrobasal complex of the thalamus. *I* is the most rostral slab. The key for the somatotopical localization in the nucleus is provided by the figurine in **b2**. Slightly modified from Welker and Johnson (1965). Compare Figure 22.69. **c** Somatotopic organization of the dorsal comumn nuclei in the raccoon indicated in a diagram of a horizontal section. A corresponding, myelin-stained section is illustrated on the *right*. The ventral hand digit are represented as distinct rostro-caudal columns labelled *2–5*, separated from each other by fibre lamellae. From Johnson et al. (1968)

The gracile and internal cuneate nuclei of the rat are very similar to those of the cat with respect to the presence of a region equivalent to the cell clusters of the cuneate nucleus of the cat and raccoon, containing discrete columnar representations of the digits of the forepaw (Belford and Killackey 1978; Gross et al. 1979; Maslany et al. 1991). A similar organisation of the gracile nucleus, if present, is less distinct (Maslany et al. 1991). The nuclei project both to thalamic and extrathalamic targets, but a differential origin of thalamic and extrathalamic projections has not been documented for the rat (Feldman and Kruger 1980; Massopust et al. 1985; Ma et al. 1986). Postsynaptic dorsal column fibres in the rat terminate in all parts of the dorsal column nuclei and synapse with neurons projecting to the thalamus (Cliffer and Giesler 1989). In this respect, the dorsal column nuclei of the rat differ from those of the cat, but resemble those in primates (see below).

Regions corresponding to the cell clusters and the reticular areas of the feline dorsal column nuclei have also been identified in primates. Cell clusters or columns occupy the medial part of the caudal gracile nucleus and the central pars rotunda of the internal cuneate in prosimian primates and macaque monkeys (Fig. 22.24f,h). The structure of the lateral part of the gracile nucleus and the ventral and rostral parts of the cuneate nucleus, including the pars triangularis of Ferrara and Barrera (1935) that surrounds the pars rotunda, is more reticular (Ferrara and Barrera 1935 cited in Boivie 1978; Albright and Haines 1978; Rustioni et al. 1979; Noriega and Wall 1991). The projections of the primary afferent fibres to the dorsal column nuclei have been studied in macaque monkeys (Shriver et al. 1968) and prosimian primates (Albright and Haines 1978). In *Macaca mulatta*, the digits of the forepaw are represented in the central part of the internal cuneate nucleus, including the pars rotunda (Culbertson and Brushart 1989). This pattern is essentially similar to that of the cell cluster region of the cuneate nucleus of the raccoon (Johnson et al. 1968).

Cortical projections (Kuypers 1958a,b; Kuypers et al. 1960, 1961) and postsynaptic dorsal column afferent fibres are more widely distributed in the primate dorsal column nuclei than they are in the cat and include both the columnar pars rotunda and the pars triangularis of the cuneate nucleus (Cliffer and Willis 1994).

The course and the distribution of the components of the medial lemniscus arising from the dorsal column nuclei are very similar in the cat and primates. The fibres terminate in the lateral part of the ventral posterior nucleus, in the posterior nuclear group and in the zona incerta (Boivie 1978; Berkley 1980, 1983, 1986; Berkley et al. 1980). Fibres terminating in the thalamus take their origin from all parts of the primate dorsal column nuclei, including the external cuneate nucleus (Boivie et al. 1975; Rustioni et al. 1979). Projections to the deep layers of the tectum have been described by Wiberg et al. (1987) in *Macaca fascicularis*.

22.8.1.1.4
Trigeminal Sensory System

The three branches of the trigeminal nerve innervate the skin over the dorsum of the head and the nose, the supraorbital vibrissae, the cornea, the conjunctiva, the nasal mucosa (V1, ophthalmic division), the infraorbital skin, the mystacial vibrissae, the upper lip and the mucosa and the tooth pulp of the upper jaw (V2, maxillary division), the skin over the lower lip and mandible, lower jaw mucosa and tooth pulp, the anterior tongue and the temporomandibular joint (V3, mandibular division). The sensory root (the portio major) of the trigeminal nerve arises as the central branches of the neurons of the trigeminal ganglion of Gasser and enters the brain stem along the motor root (the portio minor) by passing through the brachium pontis. Myelinated and unmyelinated fibres of the portio major descend in the spinal tract of the trigeminal nerve, located in the dorsolateral medulla, covered by the restiform body over most of their extent. A proportion of the myelinated fibres of the portio major bifurcate, with one branch terminating in the nucleus princeps in the tegmentum pontis and one branch descending in the spinal tract (Figs. 22.22–22.24f–m, 22.73; Cajal 1911; Li et al. 1992; Kobayashi and Matsumora 1996). In marsupials, the descending branches of these bifurcating fibres descend in the medial part of the spinal tract, segregated from the thin fibres of the portio major, which are located in the lateral part of the tract (Fig. 22.22g–k; Verhaart 1970). Most of the fibres of the mesencephalic root of the trigeminal nerve, which innervate the proprioceptors of jaw and facial muscles, enter the brain stem via the motor root of the trigeminal nerve (portio minor) and can be traced to their origin in the mesencephalic nucleus.

Three sensory nuclei of the trigeminal nerve are distinguished (Fig. 22.73): (1) the principal nucleus, located lateral to the motor nucleus in the tegmentum pontis, (2) the spinal trigeminal nucleus, located along the spinal tract in the medulla oblongata and (3) the nucleus of the mesencephalic root of the trigeminal nerve, which extends along the periaqueductal grey, from the tegmentum pontis into the mesencephalon (Figs. 22.22–22.24l–o). The spinal nucleus can be further subdivided into the pars oralis, the pars interpolaris and the pars caudalis. The pars caudalis shares the same laminated structure and is continuous with the dorsal horn of the spinal cord. The spinal tract continues into Lissauer's dorsolateral bundle at the transition of the medulla oblongata into the spinal cord.

As pointed out in the introduction to Sect. 22.8, the trigeminal nerve differs greatly in size in different mammalian orders and species. The huge size of the portio major in *Ornithorhynchus* (Hines 1929) is probably related to the highly developed sensation in its bill and the presence of electrosensation in the same organ (Schleich et al. 1986; Gregory et al. 1987, 1988; Iggo et al. 1992). In insectivores, rodents and carnivores (especially in Pinnepedia), the large size of the nerve is related to the innervation of the mystacial vibrissae. In Perissodactyla and Artiodactyla, the size of the sensory root is probably related to the importance of peri- and intra-oral sensibility.

The principal sensory nucleus can be subdivided into dorsal and ventral parts. The dorsal part harbours a small, oval subnucleus (nucleus ovalis) in primates and humans (Fig. 22.24j,k), which serves as a relay nucleus in the gustatory system of nerves VII, IX and X (see Sect. 22.8.1.2). In the cat, the dorsal division consists of medium-sized cells and gives rise to an uncrossed ascending pathway to the thalamus, i.e. the dorsal trigeminothalamic tract of Wallenberg (1896, 1900, 1905; Torvik 1957). The ventral part of the nucleus contains closely packed, small and medium-sized cells and gives rise to fibres that decussate caudal and dorsal to the interpeduncular nucleus (Fig. 22.25) and ascend in the medial part of the medial lemniscus, terminating in the medial part of the ventral posterior nucleus of the thalamus and in the posterior group. In the posterior group it overlaps with the termination of the projection of the pars caudalis of the spinal nucleus. The uncrossed dorsal trigeminothalamic tract ascends in the dorsal tegmentum, lateral to the periaqueductal grey (Figs. 22.23, 22.24m). In the cat, its termination differs from that of the crossed fibres and occupies the lateral part of the medial parvocellular portion of the nucleus ventralis posterior of the thalamus, lateral to the gustatory subdivision of this nucleus (Torvik 1957; Burton and Craig 1979; Matsushita et al. 1982; Yasui et al. 1983).

The ipsilateral ascending dorsal trigeminothalamic tract was first described by Wallenberg (1896, 1900, 1905) in the rabbit. Wallenberg found both ipsilaterally and contralaterally ascending fibres arising from the nucleus princeps, but was unable to trace the ipsilaterally ascending fibres beyond the nucleus of the centre médian. Karamanlidis et al. (1982) could not confirm the ipsilateral projection in the rabbit and concluded that the trigemino-

22.8 Sensory Systems 1771

medial part of the nucleus ventralis posterior and in the ventral zona incerta. A small ipsilateral dorsal trigeminothalamic tract from the dorsal, mandibular portion of the nucleus princeps terminates in a discrete dorsomedial paralaminar portion of the medial part of the nucleus ventralis posterior that does not receive a crossed projection. The question of whether part or all of the ipsilateral dorsal trigeminothalamic tract represents a gustatory system arising from the nucleus ovalis was considered by Smith (1975) and others, but was never decided (see Sect. 22.8.1.2).

Experimental studies in Artiodactyla (*Capra*: Karamanlidis and Voogd 1970) and Perissodactyla (donkey: Karamanlidis, unpublished) showed the presence of an extensive ipsilateral projection arising from the nucleus princeps and the absence of a crossed trigeminothalamic projection from that nucleus (Karamanlidis and Voogd 1970). The dorsal trigeminothalamic tract in the goat and other artiodactyls and perissodactyls is large and located lateral to the periaqueductal grey, where it overlaps with the central reticular fasciculus (see Fig. 22.116; Verhaart 1970). It terminates in the medial part of the ventral posterior nucleus. A contribution to this tract of the nucleus ovalis appears likely. In the rat, the ipsilateral projection from the dorsal nucleus princeps is either absent (Smith 1973) or much smaller than in cats and primates (Fukushima and Kerr 1979; Feldman and Kruger 1980).

The nucleus of the spinal tract of the trigeminal nerve has been subdivided into the caudal, laminated and rostral, unlaminated portions (Figs. 22.22–22.24f–k). The rostral, unlaminated part of the spinal nucleus can be further subdivided into a caudal subnucleus interpolaris and a rostral subnucleus oralis. This subdivision was proposed by Meessen and Olszewski (1949) for the rabbit, by Olszewski (1950) for the monkey and by Olszewski and Baxter (1954) for humans. Brodal et al. (1956) and Taber (1961) applied it to the cat and Fukushima and Kerr (1979) to the rat spinal nucleus (for a review, see Waite and Tracey 1995). In Artiodactyla (*Capra*: Karamanlidis 1968; *Bos*: Barone and Doucet 1964), the nucleus interpolaris and oralis could not be distinguished from one another, and a group of large neurons was distinguished, located ventral to the spinal and principal nuclei, which projects to the cerebellum. The nucleus ovalis is present in the dorsal part of the nucleus princeps and the rostral spinal nucleus.

In the caudal, laminated part of the nucleus spinalis, a marginal cell layer, a substantia gelatinosa and a deep layer of loosely packed, small and medium-sized cells can be distinguished. The structure and connections of the pars caudalis are

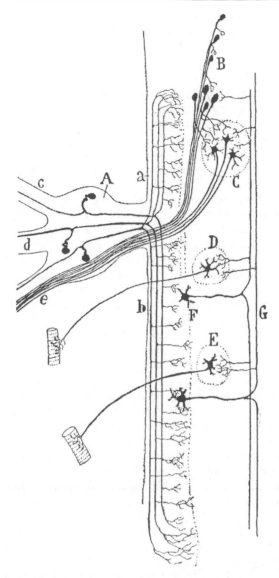

Fig. 22.73. Sensory nuclei of the trigeminal nerve. *A*, trigeminal ganglion; *a*, ascending branches of the portio major; *B*, nucleus of the radix mesencephalicus; *b*, descending branches of the portio major; *C*, motor nucleus of the trigeminal nerve; *c*, nervus ophthalmicus; *D*, motor nucleus of the facial nerve; *d*, nervus maxillaris; *E*, motor nucleus of the hypoglossal nerve; *e*, nervus mandibularis; *F*, neuron of the substantia gelatinosa of the nucleus of spinal tract of the trigeminal nerve *G*, secondary trigeminal connections. (From Cajal 1909–1911)

thalamic connections of the nucleus princeps of the rabbit are entirely crossed. Eponymes apparently have a longer life than the experimental results on which they are based. The extensive literature on the crossed and uncrossed trigeminothalamic projections in primates was reviewed by Smith (1975) and Burton and Craig (1979). They concluded that crossed trigeminothalamic fibres from the nucleus princeps in the rhesus monkey terminate in the

essentially similar to the dorsal horn, with which it is continuous. The marginal cells and the deep layers of the pars caudalis give rise to a crossed ascending pathway that joins the anterolateral fascicle of the spinal cord and terminates in the nucleus ventralis posterior medialis, in the posterior group and in some of the intralaminar nuclei of the thalamus (cat: Burton et al. 1979; Berkley 1980; squirrel monkey: Ganchrow 1978; rat: Craig and Burton 1981; Yoshida et al. 1991) and to an intranuclear ascending system (Ikeda et al. 1982; Panneton and Burton 1983; Nasution and Shigenaga 1987).

The subnucleus interpolaris consist of small and medium-sized neurons, with scattered large cells. The border of the subnucleus interpolaris towards the lateral tegmental field is not sharp. The subnucleus oralis consists of uniform, medium-sized neurons. The pars oralis and interpolaris share some projections with the nucleus princeps. Additionally, the nucleus interpolaris projects to the cerebellum, the inferior olive and the superior colliculus (Wiberg et al. 1986; Van Ham and Yeo 1992).

The somatotopical localisation in the sensory nuclei of the trigeminal nerve is based on anatomical studies of the distribution of the three branches of the trigeminal nerve, on electrophysiological studies (reviewed by Darian-Smith 1973) and on the multiple representations of mystacial vibrissae in the trigeminal nuclear complex of the rat and other mammals. Ascending and descending fibres of the mandibular branch occupy a dorsal position in the nucleus princeps and in the spinal tract, respectively. Fibres from the ophthalmic branch are located most ventrally, while the maxillary branch takes an intermediate position and descends least far caudally (Kruger and Michel 1962; Wall and Taub 1962; Kerr 1963; Rustioni et al. 1971; Arvidsson 1982). The representation of the face in the trigeminal nuclei is thus inverted. The representation of the periphery is repeated in each successive subdivision of the trigeminal nuclei. This is well illustrated for the representation of the mystacial vibrissae, which can be visualised with succinic dehydrogenase histochemistry as discrete patches of reactive neuropil in the nuclei of young rats. Complete patterns of patches or 'barrelettes', each patch representing a single vibrissa, are present in the nucleus princeps, the nucleus interpolaris and the nucleus caudalis in the rat (Belford and Killackey 1979; Hayashi 1980) and the cat (Nomura et al. 1986; Fig. 22.74). A similar localisation has been recognised in the ventrobasal complex of the thalamus and in the somatosensory cortex of the rat (Belford and Killackey 1978; Akers and Killackey 1979; Killackey 1983).

The bifurcating axons of the nucleus of the mesencephalic tract project peripherally into the motor root of the trigeminal nerve and to the trigeminal and hypoglossal motor nuclei (see Frame 5) and to the lateral parvocellular reticular formation (Ruggiero et al. 1982). The fibre system descending from the mesencephalic nucleus is known as Probst's tract (Probst 1899).

22.8.1.2
Viscerosensory Systems

Taste buds occur in five distinct populations in the oral cavity of mammals, i.e. on fungiform papillae on the anterior tongue, on foliate and circumvallate papillae on the posterior tongue and on the palate and the larynx (Norgren 1984). The taste buds are innervated by the chorda tympani, which joins the intermediate central nerve to the geniculate ganglion (anterior tongue), the greater superficial petrosal branch of the facial nerve (palate), at least in non-human animals (Norgren 1990), the lingual branch of the glossopharyngeal nerve (posterior tongue) and the superior laryngeal branch of the vagus nerve (larynx). Taste is difficult to separate from intra-oral sensibility, and in the gustatory nerves the majority of the nerve fibres responds to tactile or thermal sensibility (Norgren 1990).

In the medulla oblongata, the rootlets of the nerves traverse the restiform body, the trapezoid body, where they emerge from the cochlear nuclei, and the spinal tract of the trigeminal nerve and its nucleus, assembling in the solitary tract. Somatosensory components of these nerves join the spinal tract of the trigeminal nerve, descend in its dorsal portion and terminate in the nucleus of the spinal tract and the cervical dorsal horn. Ascending somatosensory fibres from the facial nerve terminate in the principal sensory nucleus (Arvidsson and Thomander 1984).

The gustatory viscerosensory afferent fibres terminate in the nucleus of the solitary tract. In rodents, this nucleus extends from the level of entrance of the facial root to a level below the obex. In cats and primates, the nucleus of the solitary tract extends rostral to the level of entrance of the facial nerve as a group of neurons embedded in a fibrous capsule within the dorsal part of the pars oralis of the spinal trigeminal nucleus and the principal sensory nucleus of the trigeminal nerve (Fig. 22.24). In humans, this rostral extension of the nucleus of the solitary tract was referred to as the nucleus ovalis by Olszewski and Baxter (1954). In the cat and in primates, some facial and glossopharyngeal gustatory visceral afferent fibres ascend to terminate in the nucleus ovalis (Rhoton et al. 1966;

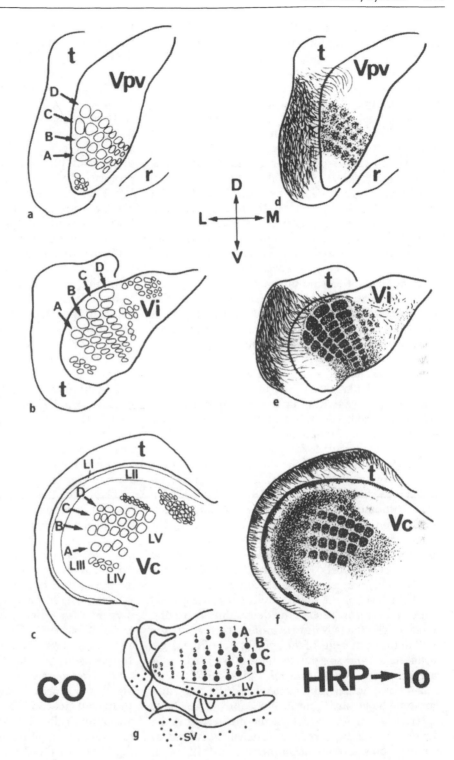

Fig. 22.74a-g. Projections of mystacial vibrissae within the trigeminal sensory nuclei of the cat. **a-c** Distribution patterns of cytochrome oxidase. **d-f** Terminal labelling after application of horseradish peroxidase to the infra-orbital nerve. **g** Position of the mystacial vibrisae on the snout of the cat. The rows of large vibrissae (A-D) implanted on the vibrissal pad are indicated in the diagrams on the *left*. *A-D*, rows mysracial vibrissae A-D; *CO*, cytochrome oxidase; *HRP*, horseradish peroxidase; *Io*, infraorbital nerve; *LI*, lamina I of the pars caudalis; *LII*, lamina II of the pars caudalis; *LIV*, lamina IV of the pars caudalis; *LV* (in part C), lamina V of the pars caudalis; *LV* (in part G), labial vibrissae; *SV*, submental vibrissae; *t*, apinal tract of the trigeminal nerve; *Vc*, nucleus of the spinal tract of the trigeminal nerve, pars caudalis; *Vi*, nucleus of the spinal tract, pars interpolaris; *Vpv*, principal sensory nucleus. (From Nomura et al.1986)

Nomura and Mizuno 1982: cat; Beckstead et al. 1979, 1980: *Macaca*; for a review of the literature on the human solitary tract and nucleus, see Norgran 1990). A rostral extension of the nucleus of the solitary tract and ascending components of the facial, glossopharyngeal and vagus nerves appear to be absent in rodents (see, however, Brining and Smith 1996). The nucleus ovalis is prominent in ungulates (see Sect. 22.8.1.1.).

Gustatory fibres of the facial, glossopharyngeal and vagus nerves descend in the solitary tract in all species and terminate in the rostral third of the nucleus of the solitary tract to the level of the obex. Fibres from the facial nerve terminate most rostrally, followed by gustatory fibres from the glossopharyngeal and vagus nerve. Facial and glossopharyngeal afferents terminate mainly in the rostrolateral subdivision of the nucleus. The vagus nerve monopolises the medial subdivision of the nucleus. The gustatory fibres are joined by the general viscerosensory components of the glossopharyngeal and vagus nerves that innervate the baroreceptors of the carotid sinus, the heart and the great vessels, the airways and the lungs and other viscera and terminat in the caudal two thirds of the nucleus (Kerr 1962; Cottle 1964; Rhoton et al. 1966; Kalia and Mesulam 1980a; Nomura and Mizuno 1981, 1982; Arvidsson and Thomander 1984: cat; Allen 1923a; Kalia and Sullivan 1982; Contreras et al. 1982; Whitehead and Frank 1983; Hamilton and Norgren 1984; Norgren 1995; Brining and Smith 1996: rat and hamster; Beckstead et al. 1979, 1980; Norgren 1990; Satoda et al. 1995: primates). Somatosensory fibres from the trigeminal nerve terminate in the lateral part of the nucleus, overlapping with viscerosensory fibres from the glossopharyngeal and vagus nerves (Torvik 1957; Kerr 1962; Contreras et al. 1982; for a review, see Norgren 1995).

At the level of the facial nerve, the solitary tract is located ventromedial to the descending vestibular nucleus. The nucleus of the solitary tract is situated medial to the nucleus of the spinal tract of the trigeminal nerve and surrounds the solitary tract on its lateral, medial and ventral sides (Figs. 22.22–22.24f–h). Caudally, the nucleus and the tract shift dorsomedially, surfacing in the floor of the fourth ventricle at the level at which the medial vestibular nucleus has disappeared. The nucleus is capped by the area postrema, which protrudes into the ventricle. Caudal to the obex, the nuclei of the solitary tract of both sides fuse dorsal to the central canal as the commissural nucleus. The solitary tract bifurcates; the medial branch enters the commissural nucleus and decussates to innervate the contralateral nucleus, and the lateral branch descends laterally within the nucleus. The dorsal vagus nucleus is located ventromedial to the nucleus of the solitary tract. The caudal extension of the descending vestibular nucleus (parasolitary nucleus of Walberg et al. 1962) is located lateral to the nucleus of the solitary tract and medial to the internal cuneate nucleus.

The nucleus of the solitary tract is usually subdivided into medial and lateral parts (Cajal 1911; Torvik 1957; Fig. 22.75). Different schemes for the subdivision of the nucleus of the solitary tract were proposed by Torvik (1956) and Kalia and Sullivan (1982) for the rat, by Torvik (1957), Taber (1961) and Loewy and Burton (1978) for the cat and by Beckstead and Norgren (1979) for *Macaca*. These different modes of subdivision of the nucleus were reviewed for the rat by Kalia and Mesulam (1980a) and for the hamster by Whitehead (1988).

In the medial nucleus, a parvocellular subnucleus gelatinosus and a medial subnucleus are usually distinguished (Fig. 22.75). The central portion of the nucleus has been referred to as the intermediate subnucleus or has been incorporated into the lateral subdivision. The lateral, or more appropriately the ventrolateral and ventral subnuclei, contain the largest cells. The dorsolateral subnucleus, distinguished by some authors, corresponds to the nucleus parasolitarius of Walberg et al. (1962) and, therefore, does not belong to the nucleus of the solitary tract, but to the vestibular complex. The rostral extension of the nucleus of the solitary tract in the cat and in primates is thought to belong to the central part of the nucleus. The nucleus commissuralis is considered as an extension of the medial subnucleus. Gustatory fibres terminating in the rostral third of the nucleus extend farthest caudally in the central part of the nucleus. The medial, ventral and ventrolateral subnuclei, including the commissural nucleus, mainly receive general viscerosensory afferent fibres (Kerr 1962; Berger 1979; Kalia and Mesulam 1980a: cat; Kalia and Sullivan 1982; Contreras et al. 1982; Whitehead and Frank 1983; Hamilton and Norgren 1984: rat and hamster; Satoda et al. 1995: *Macaca*).

The area postrema (Figs. 22.22, 22.23g, 22.75) is a highly vascularised structure, containing glial cells and small neurons. In the area postrema, the permeability of the blood-brain barrier is increased. It receives fibres from the glossopharyngeal and the vagus nerves (Gwyn and Leslie 1979; Contreras et al. 1982). It projects to the nucleus of the solitary tract, possibly to the dorsal vagus nucleus, the ventrolateral intermediate reticular area and the parabrachial nuclei (Morest 1967; Shapiro and Miselis 1985; Cunningham et al. 1994).

The nucleus of the solitary tract establishes efferent connections with local motor nuclei and give

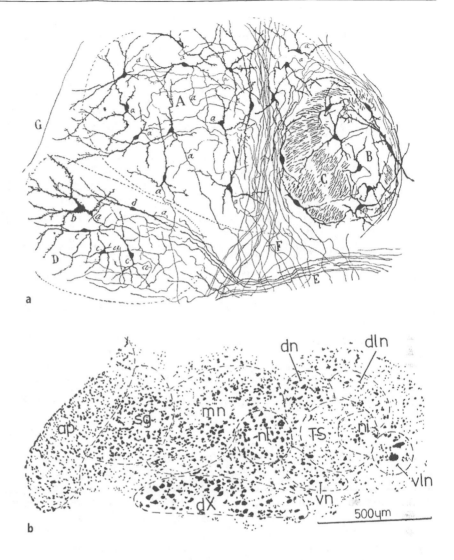

Fig. 22.75a,b. Nucleus of the tractus solitarius of the cat. **a** Golgi-stained nucleus of the kitten From Cajal (1909–1911). **b** Nissl-stained section, showing the subdivision of the nucleus of the solitary tract in the cat, redrawn from Kalia and Mesulam (1980a). *a*, axons; *A*, medial nucleus; *ap*, area postrema; *b*, motoneuron of the vagus nerve; *B*, lateral or interstitial nucleus; *c*, small cells of the dorsal vagus nucleus; *C*, solitary tract; *d*, small cells of the dorsal vagus nucleus; *D*, dorsal vagal nucleus; *dln*, dorsolateral subnucleus; *dn*, dorsal subnucleus; *dX*, dorsal vagus nucleus; *E*, rootfibres of the vagus nerve; *F*, secondary connections of the vagus and glossopharyngeal nerves; *G*, fourth ventricle; *mn*, medial subnucleus; *ni*, interstitial subnucleus; *nI*, intermediate subnucleus; *sg*, subnucleus gelatinosus; *TS*, solitary tract; *vln*, ventrolateral subnucleus; *vn*, ventral subnucleus

rise to descending connections to the spinal cord and ascending connections with the parabrachial nuclei, the thalamus and the basal telencephalon. The thalamic connections are difficult to differentiate from the ascending connections of the principal sensory nucleus of the trigeminal nerve.

Local projections target the dorsal vagus, ambiguus and retrofacial nuclei, the intercalatus and prepositus hypoglossi nuclei, the parvocellular reticular formation and the ventrolateral intermediate reticular field (Morest 1967; Loewy and Burton 1978; Beckstead et al. 1980; Otake et al. 1992; Yu and Gorden 1996). Projections of the nucleus of the solitary tract to the inferior olive (Loewy and Burton 1978) probably do not originate from this nucleus, but from the neighbouring nucleus parasolitarius of Walberg et al. (1962). Descending connections from the lateral division of the nucleus of the solitary tract terminate in the dorsal horn, bilaterally in the intermediolateral nucleus and on contralateral motoneurons of the phrenic nerve (Torvik 1957; Loewy and Burton 1978: cat).

There are important differences with respect to the ascending connections of the nucleus of the solitary tract between the rat and primates. In the rat and the hamster, both the general visceral and gustatory subdivisions of the nucleus of the solitary tract project to the parabrachial nuclei (Fig. 22.76). The general visceral commissural and caudal parts of the nucleus project to the lateral portions of the lateral parabrachial nucleus, the ventrolateral respiratory subdivision to lateral portions of the lateral and medial parabrachial nucleus, and the rostral gustatory subdivision to caudal and medial parts of the parabrachial nuclei. The lateral parvocellular reticular formation and the ventrolateral intermediate reticular field, both of which receive projections from the nucleus of the solitary tract, are reciproc-

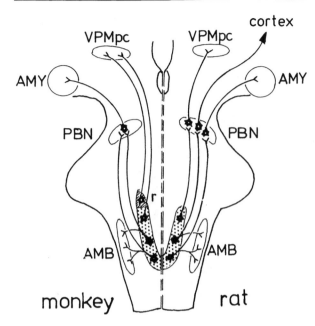

Fig. 22.76. Central connections of the nucleus of the tractus solitarius (*stippled*) in primates (*left*) and rodents (*right*). The rostral extension of te nucleus of the solitary tract in primates is *hatched*. AMB, nucleus ambiguus; AMY, amygdaloid nuclei; PBN, parabrachial nuclei; r, rostral extension of the nucleus of the tractus solitarius; VPMpc, parvocellular subdivision of the nucleus ventralis posterior medialis. (Based on Beckstead et al. 1980)

ally connected with the parabrachial nuclei (Herbert et al. 1990; Whitehead 1990; Jia et al. 1994; Halsell et al. 1996). Additionally, the caudal general viscerosensory subdivisions of the nucleus of the solitary tract of the rat project to the caudal amygdala, the bed nucleus of the stria terminalis and the periventricular hypothalamus (Ricardo and Koh 1978).

Direct connections of the nucleus of the solitary tract with the thalamus were observed by Allen (1923b) in the rat, but have not been confirmed in this species. According to Norgren (1976), the parabrachial nuclei (the 'pontine taste area') serve as the main relay for taste and general visceral sensibility in rodents. The parabrachial nuclei project to the dorsal raphe nucleus and the Edinger-Westphal nucleus in the mesencephalon, the intralaminar and midline nuclei and the ventrobasal complex of the thalamus and the bed nucleus of the stria terminalis. Additionally, the lateral parabrachial nucleus projects to the medial hypothalamus and the amygdala, including its central nucleus. Insular, frontal, retro-olfactory and infralimbic cortical areas receive projections from the medial parabrachial nucleus (Fig. 22.76; Norgren 1976, 1995; Saper and Loewy 1980). Some of these cortical areas, the central nucleus of the amydala and the hypothalamus, which receive fibres from the nucleus of the solitary tract and/or the parabrachial nuclei, send a reciprocal connection back to these structures (Takeuchi et al. 1983; Van der Kooy et al. 1984; Yasui et al. 1991).

Similar projections from the nucleus of the tractus solitarius and the lateral parvocellular reticular formation with the parabrachial nuclei (Loewy and Burton 1978; King 1980; Otake et al. 1992) and from the parabrachial nuclei with the medial parvo cellular ventral posteromedial thalamic nucleus (Nomura et al. 1979) were found in the cat. The parabrachial projections to the amygdala take their origin from the lateral general viscerosensory portion of these nuclei. The projection to the ventral posterior nucleus arises from their medial and caudal gustatory region and is enforced by a similar projection from the dorsal part of the principal sensory nucleus of the trigeminal nerve (Nomura et al. 1979).

In macaque monkeys, the connections of the caudal, general viscerosensory portion of the nucleus of the tractus solitarius with the lateral parabrachial nuclei are similar to those in the rat and cat (Beckstead et al. 1980). The parabrachial nuclei of the monkey are reciprocally connected with the amygdaloid nucleus (Mehler 1980; Price and Amaral 1981). The rostral, gustatory portion of the nucleus of the solitary tract, however, does not project to the parabrachial nuclei, but to the medial parvocellular part of the ventral posteromedial thalamic nucleus (Fig. 22.76). The intermediate part of the nucleus projects both to the dorsal parabrachial nuclei and to the parvocellular ventral posteromedial nucleus (Beckstead et al. 1980). Crossed ascending projections through the medial lemniscus terminating in the magnocellular ventral posterior nucleus have been observed in the cat and monkey (Loewy and Burton 1978; Beckstead et al. 1980) but probably do not originate from the nucleus of the solitary tract.

The presence of (a) a differential projection of the caudal, general viscerosensory division of the nucleus of the solitary tract, via the lateral parabrachial nuclei, to the amygdala, (b) its ventral, gustatory division, through the caudomedial parabrachial nuclei to the medial parvocellular portion of the ventral posterior nucleus of the thalamus and (c) a direct, parallel pathway from the rostral nucleus of the solitary tract to the gustatory portion of the ventral posterior nucleus in primates seems to be well established (Fig. 22.76). This direct, parallel gustatory pathway, arising from the rostral nucleus ovalis portion of the nucleus of the tractus solitarius, and the direct trigeminothalamic projection from the dorsal principal sensory nucleus of the trigeminal nerve (see Sect. 22.8.1.14) apparently share the same ascending system, i.e. the dorsal trigeminothalamic tract of Wallenberg. Whether the two systems have, and can be distinguished remains uncertain. Functionally, the two systems, transmitting intra- and peri-oral sensibility and gustatory information, respectively, may be closely related. This problem was extensively discussed by Norgren (1990).

22.8.2
Acoustic System

22.8.2.1
Cochlear Nerve

The centres in the brain stem of the acoustic system include the cochlear nuclei, the superior olivary complex, the nuclei of the lateral lemniscus and the inferior colliculus (Fig. 22.77). The cochlear nerve arises as the central processes of the cells of the spiral ganglion and terminates in the cochlear nuclei. The cochlear nuclei project through the acoustic stria and the trapezoid body to the superior olivary complex and along the lateral lemniscus to the inferior colliculus. The brachium of the inferior colliculus terminates in the medial geniculate body. Multisynaptic descending systems link the auditory cortex with lower auditory centres. The olivocochlear bundle contains efferent fibres terminating on hair cells of the origin of Corti. The auditory brain stem was reviewed by Irvine (1986), Webster and Garey (1990) and Webster (1995).

The typical mammalian pattern of hair cells of the organ of Corti consists of one row of inner hair cells and three rows of outer hair cells with their typical W-shaped hairs (Ross 1977; Spoendlin 1985). Hair cells are contacted by two systems: *afferent* processes from spinal ganglion cells and *efferent* fibres of the olivocochlear bundle. Outer hair cells are innervated by several large efferent and small afferent terminals. Inner hair cells are only innervated by afferent fibres; efferent fibres

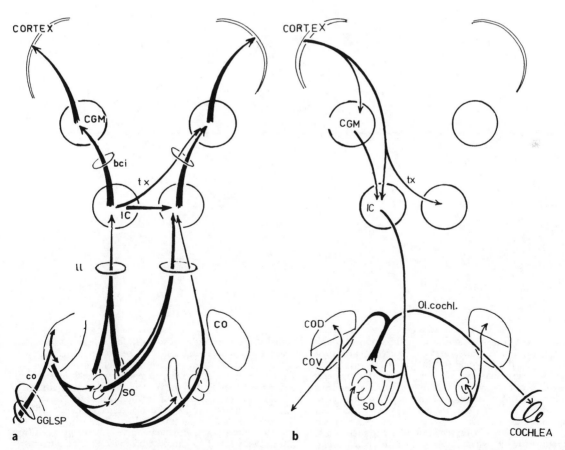

Fig. 22.77a,b. a Ascending and b descending connections of the auditory system. *bci*, brachium of the inferior colliculus; *CGM*, medial geniculate body; *CI*, inferior colliculus; *co*, cochlear nerve; *CO*, cochlear nuclei; *COS*, dorsal cochlear nucleus; *COV*, ventral cochlear nucleus; *GGLSP*, ganglion spirale; *ll*, lateral lemniscus; *Ol.cochl*, olivocochlear bundle; *SO*, superior olivary complex; *tx*, commissure of the inferior colliculus

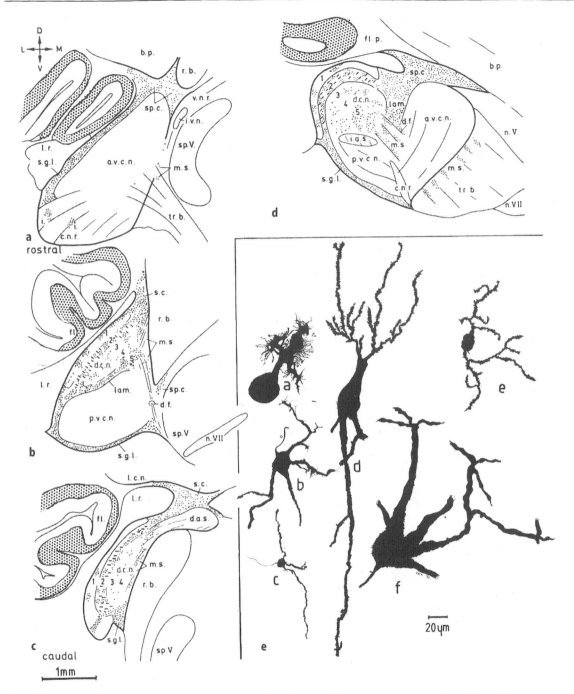

Fig. 22.78a–e. a–c Transverse and **d** sagittal sections through the cochlear nuclei of the cat. Granule cells are *stippled*; fusiform cells of the dorsal cochlear nucleus are indicated as *short bars*. The granular layer of the cerebellar cortex is *shaded*. *1–5*, Layers 1–5 of the dorsal cochlear nucleus; *a.v.c.n.*, anteroventral cochlear nucleus; *ap.V*, spinal tract of the trigeminal nerve; *b.p.*, brachium pontis; *c.n.r.*, cochlear nerve root; *d.a.s.*, dorsal acoustic stria; *d.c.n.*, dorsal cochlear nucleus; *d.f.*, descending fibres; *fl.p.*, floccular peduncle; *i.a.s.*, intermediate acoustic stria; *i.v.n.*, interstitial vestibular nucleus; *l.r.*, lateral recess; *lam.*, lamina of granule cells; *m.s.*, medial sheet of granule cells; *n.VII*, facial nerve; *p.v.c.n.*, posteroventral cochlear nucleus; *r.b.*, restiform body; *s.c.*, strial corner of granule cells; *s.g.l.*, superficial layer of granule cells; *sp.c.*, subpeduncular corner of granule cells; *tr.b.*, trapezoid body; *v.n.r.*, vestibular nerve root. (From Mugnaini et al. 1980). **e** Cell types of the cochlear nucleus of the cat. *a*, bushy cell of the anteroventral cochlear nucleus; *b*, stellate cell from the same nucleus; *c*, granule cell; *d*, fusiform cell; *e*, cart wheel from the dorsal cochlear nucleus; *f*, octopus cell from the posteroventral cochlear nucleus. (Redrawn from the Golgi study by Brawer et al. 1974)

make synaptic contact with the terminals of the afferent fibres (Nadol 1983a,b; Pujol et al. 1980). The efferent innervation of the outer cells is lacking in the horseshoe bat *Rhinolophus rouxi* (Bruns and Schmiessek 1980; Aschoff and Ostwald 1987).

Most ganglion cells of the spiral ganglion are bipolar, and only a minority are pseudo-unipolar (Kiang et al. 1984). Two populations of spiral ganglion cells can be distinguished. Radial fibre neurons innervate the inner hair cells on a one to two basis, and spiral fibre neurons each innervate ten to 60 outer hair cells. Each inner hair cell is innervated by up to 20 afferent neurons, whereas each outer hair cell is only contacted by two spiral fibres (Perkins and Morest 1975). The main transmission of auditory information from the cochlea to the brain, therefore, is subserved by the inner hair cells. Outer hair cells have electromotile properties and act as mechanical effectors, influencing the tuning of the inner hair cells by altering the mechanics of the organ of Corti. The efferent, olivocochlear system influences the motility of the outer hair cells (Brownell 1982).

22.8.2.2
Cochlear Nuclei

The subdivision of the cochlear nuclei is based on detailed studies in carnivores, mainly in the cat (Lorente de Nó 1933; van Noort 1969; Osen 1969; Brawer et al. 1974). The morphology of the cochlear nuclei in the rat and in primates was reviewed by Webster and Garey (1990) and Webster (1995), respectively.

The cochlear nuclei of the cat can be subdivided into the dorsal and the ventral cochlear nucleus (Fig. 22.78). The cochlear nerve enters the cochlear nuclei ventrally and divides the ventral nucleus into anteroventral and posteroventral nuclei. All fibres of the cochlear nerve bifurcate into ascending and descending branches (Fig. 22.79b). The ascending branches terminate in the anteroventral nucleus; the descending branches innervate the posteroventral nucleus and the deep layers of dorsal cochlear nucleus. Root fibres do not terminate in the molecular layer of the dorsal cochlear nucleus. Low-frequency transmission from the apex of the cochlea is subserved by coarse root fibres that bifurcate

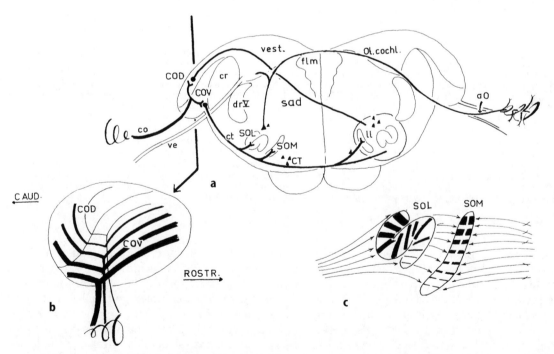

Fig. 22.79a-c. Tonotopical organisation of the cochlear nuclei and the medial and lateral superior olive. **a** Transverse section through the cochlear nuclei and the superior olivary complex of the cat. The projection of the cochlear nerve is indicated on the *left*, the termination of the olivocochlear bundle on the right. **b** Tonotopical organisation of the cochlear nuclei indicated in a parasagittal section through the nuclei. **c** Tonotopical organisation of the medial and lateral superior olive. *aO*, anatomosis of Oort; *caud*, caudal; *co*, cochlear nerve; *COD*, dorsal cochlear nucleus; *COV*, ventral cochlear nucleus; *cr*, restiform body; *ct*, corpus trapezoideum; *CT*, nucleus of the corpus trapezoideum; *drV*, spinal tract of the trigeminal nerve; *flm*, medial longitudinal fascicle; *ll*, lateral lemniscus; *Ol.cochl*, olivocochlear bundle; *rostr*, rostral; *sad*, dorsal acoustic stria; *SOL*, lateral superior olive; *SOM*, medial superior olive; *ve*, vestibular nerve; *vest*, vestibular nuclei. (Based on Van Noort 1969)

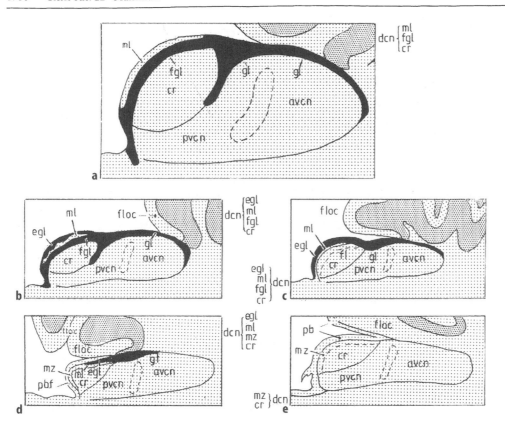

Fig. 22.80a–e. Horizontal sections through the cochlear nuclei in **a** the cat, **b** cat and galago, **c** the marmoset, **d** the macaque and **e** the gibbon. All figures are drawn to the same scale, showing the relative size of the cochlear nuclei in these species as well as the area of distribution of cochlear granule cells within the nuclei. The localisation of cochlear granule cells (*black area*) in the galago is very similar to that of the cat. This cellular layer is less extensive in the marmoset and macaque, retracting from both the anterior and posterior ends of the nuclear complex and from the line of division between the dorsal and ventral nuclei. In the gibbon, no cochlear granule cell area is present. In those primates exhibiting cochlear granule cells, the granular layer of the flocculus (*densely shaded area*) remains in contact with the cochlear granular layer, although the point of contiguity of the two layers is displaced posteriorly in the monkey. Within the dorsal cochlear nucleus (*dcn*) of cat and galago, cochlear granule cells are divided into an external granular layer (*egl*) and a deep or internal granular layer which is coincident with the peripheral layer of fusiform cells (*fgl*). The two granular layers are separated by a molecular layer (*ml*). In the marmoset and macaque, the entire granular layer is external or superficial. Fusiform cells form an irregular peripheral layer in the marmoset, but are also located within the central region of the nucleus (*cr*). In the macaque and gibbon, the portion of the dorsal cochlear nucleus not covered by a granular and a molecular layer is rimmed by a cell-free marginal zone (*mz*). In the macaque and gibbon, much of the free surface of the cochlear nuclei and adjacent areas of the medulla and flocculus are covered by a layer of pontobulbar fibres (*pb, pbf*). This sheet of fibres is thickest and most extensive in the gibbon. (From Moore 1980)

in the ventral part of the nucleus; low frequencies are transmitted from the base of the cochlea by thin fibres that bifurcate in the dorsal part of the nucleus (cat: Rose 1960; Van Noort 1996a,b; Leake and Snyder 1989; Leake et al. 1993; rat: Feldman and Harrison 1969). A component from the vestibular nerve, innervating the sacculus, was traced to the border region of the dorsal and posteroventral cochlear nuclei in the guinea pig (Burian et al. 1989, 1991) and to more extensive granular areas of the cochlear nuclei in the mongolian gerbil (Kevetter and Perachio 1989).

The main cell types of the anteroventral cochlear nucleus are the bushy cell and the stellate cell (Brawer et al. 1974). Bushy cells have a spherical soma and a single dendrite, terminating in a dendritic tangle (Fig. 22.78a, inset). The cochlear nerve terminates with large, complex end-bulbs of Held on the somata of the bushy cells, together with terminals of other origin. The somata of the cells increase in size in the rostral anteroventral nucleus (Osen 1969). Stellate cells (Fig. 22.78b, inset) of the ventral cochlear nucleus receive smaller terminals of the cochlear nerve on their dendrites. Some of the stellate cells are large and have been described as the giant cells of the ventral cochlear nucleus (Brawer et al. 1974). Different cell types have been identified in the posteroventral cochlear nucleus. A group of neurons in the caudal posteroventral cochlear nucleus with multiple long, straight dendrites are known as octopus cells (Fig. 22.78 f, inset). Most of these dendrites are directed dorsally. The central region of the ventral cochlear nucleus includes the bifurcation of the fibres of the cochlear nerve with the cells of the interstitial nucleus of Lorente de Nó (1933). This region was included in the posteroventral cochlear nucleus by Brawer et al. (1974).

The dorsal cochlear nucleus is laminated (Fig. 22.78); five laminae can be distinguished (Lorente de Nó 1933; Mugnaini et al. 1980). The superficial molecular layer contains granule cells and the apical dendrites of the fusiform (pyramidal or bipolar) cells that are located in the second layer (Fig. 22.78d, inset). The third layer contains the basal dendrites of these cells. The fourth and fifth layers consist of a plexus of afferent and efferent fibres and contain the giant cells of the central nucleus of Lorente de Nó (1933). Cartwheel cells are characterised by radiating and recurving dendrites (Fig. 22.78e, inset, Brawer et al. 1974). They occur in the superficial layers of the dorsal cochlear nucleus and share many of their properties with the Purkinje cells of the cerebellum (Wouterlood and Mugnaini 1984; Mugnaini and Morgan 1987). Small cells are ubiquitous in all subdivisions of the cochlear nuclei. The free surface of the ventral cochlear nucleus is covered by a layer of granule cells (Figs. 22.78, 22.80) that is continuous with the granule cells of the superficial layers of the dorsal cochlear nucleus and with the granular layer that separates the dorsal from the ventral cochlear nucleus (Mugnaini et al. 1980). Granule cells extend along the dorsal acoustic stria in its course over the restiform body. Medially, they are continuous with the granular layer of the flocculus (subpeduncular corner). Axons of granular cells converge upon the molecular layer of the dorsal cochlear nucleus, where they form parallel fibres (Mugnaini et al. 1980).

The efferent pathways of the cochlear nuclei are the dorsal and intermediate acoustic striae and the trapezoid body (Fig. 22.77g). The dorsal acoustic striae originate from the dorsal cochlear nucleus, pass dorsally over the restiform body and through the vestibular nuclei to arch through the tegmentum to the base of the lateral lemniscus of the opposite side. The intermediate striae (not illustrated) stem from the octopus cells of the posteroventral nucleus. They pass through the restiform body and either dorsal to or through the spinal tract of the trigeminal nerve to join the trapezoid body before its decussation. The trapezoid body takes its origin from the ventral cochlear nucleus and leaves the cochlear nuclei ventromedially and decussates in the ventral tegmentum, dorsal to the pyramidal tract.

In Monotremata and marsupials, the cochlear nuclei are located medial to the restiform body and lateral to the vestibular nuclei (Fig. 22.45). In the North American opossum, the cochlear nerve passes ventral to the restiform body and caudal to the vestibular nerve, entering the cochlear nuclei ventrolaterally (Willard and Martin 1983). The dorsal cochlear nucleus protrudes into the fourth ventricle. Ventrally, it borders on the vestibular nuclei and the ventral cochlear nucleus. The dorsal cochlear nucleus is composed of four layers. Somata of the fusiform cells (the principal cells of Willard and Martin 1983) are aligned in layer 2, and giant neurons occur in layers 3 and 4. The dorsal acoustic stria originates from the principal cells. The ventral cochlear nucleus is wedged between the restiform body laterally and the spinal tract of the trigeminal nerve ventrally, with the vestibular nuclei ventromedially and the dorsal cochlear nucleus dorsomedially. An extension of the granular layer that separates the dorsal from the ventral cochlear nucleus in the cat is absent in the opossum. The ventral cochlear nucleus cannot be subdivided into anterior and posterior parts. The main cell types of the ventral cochlear nucleus are the principal and stellate cells. Principal cells presumably correspond to the bushy cells of Brawer et al. (1974) in the cat. Giant neurons are found in the posteromedial part of the ventral nucleus. Octopus cells have been identified in the posteromedial aspect of the ventral cochlear nucleus; however, their axons do not constitute intermediate acoustic stria, but join the dorsal acoustic stria. The nucleus of the auditory nerve of the opossum is separated from the ventral cochlear nucleus. Its cells are scattered between the fibres of the nerve. The trapezoid body emerges from the ventral cochlear nucleus and occupies its usual position, lateral to the spinal tract of the trigeminal nerve (Fig. 22.22j,k).

The cochlear nuclei have a dorsal origin from the rhombic lip and the ventricular matrix (Willard and Martin 1986). In the North American opossum, it was shown that the large principal cells of the dorsal and ventral cochlear nuclei originate medially, close to the sulcus limitans, and migrate laterally, over the vestibular nuclei to become located medial to the restiform body. These and other neurons of the cochlear nuclei in eutherian mammals presumably migrate still more laterally, dorsal to the restiform body to settle lateral to it, next to the entrance of nerve VIII.

In the cochlear nuclei of primates, from prosimians, ceboid and cercopethoid monkeys to the gibbon, the granule cell layers are progressively

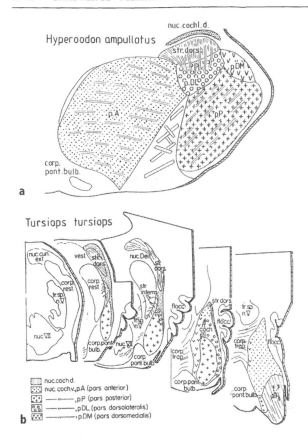

Fig. 22.81a,b. The cochlear nuclei of Cetacea. **a** Sagittal section through the cochlear nuclei of the bottle-nosed whale *Hyperoodon ampullatus*. (Redrawn from De Graaf 1967). **b** A series of transverse sections through the cochlear nucleus of the porpoise *Tursiops tursiops*. (Reproduced from Osen and Jansen 1965). In addition to the anteroventral and the posteroventral cochlear nuclei, the authors distinguished dorsolateral and dorsomedial subnuclei. The dorsal cochlear nucleus is small. *corp.pont.bulb*, corpus pontobulbare; *corp.rest*, restiform body; *corp.trap*, trapezoid body; *flocc*, flocculus; *n.cochl*, cochlear nerve; *nu VII*, facial nerve nucleus; *nu.cun.ex.*, external cuneate nucleus; *nuc.cochl.d.*, dorsal cochlear nucleus; *nuc.Dei*, lateral vestibular nucleus of Deiters; *p.i*, interstitial nucleus; *str.dors*, stria acustica dorsalis; *str.interm*, stria acustica intermedia; *tr.sp.n.V*, spinal tract of the trigeminal nerve; *vest*, vestibular nuclei

reduced and the dorsal cochlear nucleus becomes smaller (Fig. 22.80; Moore 1980). In prosimians, granule cells are present as a superficial stratum in the molecular layer and among the fusiform cells of layer 2 of the dorsal cochlear nucleus. Granule cells extend over the ventricular surface of the anteroventral nucleus. In ceboid and cercopethoid monkeys, granule cells are absent from layer 2, and there is a reduction in their extent over the anteroventral nucleus. In the gibbon, the granule cell layers and the fusiform cells are absent and the dorsal cochlear nucleus is represented by the central nucleus only. A marginal layer containing fibres of the corpus pontobulbare covers the external surface of the cochlear nuclei in the gibbon.

Important modifications of the auditory system are present in Microchiroptera and Cetacea, species that use echolocation. The cochlear nuclei in echolocating bats, like other auditory centers, are hypertrophied. Dorsal, anteroventral and posteroventral nuclei can be recognised and show the same cytoarchitectural features as in other mammals (see Fig. 22.85). Zook and Casseday (1982) found small instead of large spherical cells in the rostral anteroventral nucleus of the echolocating mustache bat *Pteronotus parnelli*. Caudally, this nucleus is demarcated from the posteroventral nucleus by a marginal band of large multipolar neurons that are absent in other mammals (m in Fig. 22.83b). The marginal band is located in the region receiving cochlear afferents from the basal turn of the cochlea, i.e. in the high-frequency domain. These authors remark that similar specializations have not been found in other echolocating bats.

The cochlear nerve in Cetacea is large, i.e. about ten times larger than the vestibular nerve. The cochlear nuclei of Cetacea were described in different species belonging to different families by Jelgersma (1934) and Ogawa and Arifuku (1948). Osen and Jansen (1965) studied the cochlear nuclei in the common porpoise *Phocaena phocaena*. De Graaf (1967) studied the cochlear nuclei in whalebone and toothed whales. The cochlear nuclei in Cetacea mainly consist of large antero- and posteroventral nuclei, and the dorsal cochlear nucleus is inconspicuous and lacks a lamination (Fig. 22.81). A small interstitial nucleus of the cochlear nerve is present. There is no trace of a granule cell layer covering the ventral cochlear nucleus. Ventrally, the ventral cochlear nucleus borders on a conspicuous pontobulbar body. Dorsal and intermediate acoustic striae emerge from the dorsolateral and dorsomedial portion of the posterior ventral cochlear nucleus. The trapezoid body is large (Osen and Jansen 1965; de Graaf 1967).

22.8.2.3
Superior Olivary Complex

The superior olivary complex in the cat (Morest 1968; Van Noort 1969) lies embedded in the trapezoid body. It consists of the medial and lateral nuclei of the superior olive, the medial nucleus of the trapezoid body and the peri-olivary cell groups (Fig. 22.23k,l). The medial superior olive consists of a simple sheet of fusiform neurons with their dendrites pointing medially and laterally. The lateral dendrites receive an ipsilateral projection from the anterior ventral cochlear nucleus, while the medial

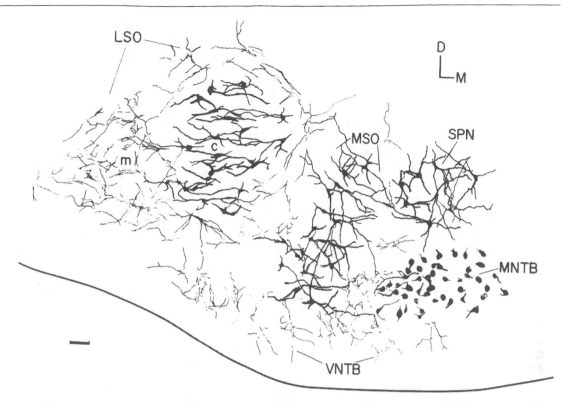

Fig. 22.82. Dendroarchitecture in the nuclei of the superior olivary complex derived from Golgi-Cox-impregnated, coronal sections through the superior olivary complex of the North American opossum. For contrast, the neurons in the core of the lateral superior olive (*LSO*), the medial superior olive (*MSO*), the medial nucleus of the trapezoid body (*MNTB*) and the superior parolivary nucleus (*SPN*) have been shaded *black*, while the remaining cells are outlined. *D*, dorsal; *m*, marginal area of the lateral superior olive; *M*, medial; *VNTB*, ventral nucleus of the trapezoid body. Bar, 100 μm. (From Willard and Martin 1983)

dendrites are contacted by fibres from the contralateral nucleus. The afferent fibres from the ventral cochlear nuclei are ordered tonotopically, with projections from the dorsal, high-frequency region of the cochlear nucleus ventrally and the ventral, low-frequency projection dorsally (Carr and Casseday 1986; Van Noort 1969a,b; Fig. 22.79c). The medial superior olive projects to the ipsilateral central nucleus of the inferior colliculus (Irvine 1986).

The lateral superior olive is S-shaped in the cat (Fig. 22.23k). It contains a longitudinally oriented fibre plexus, originating in the ipsilateral ventral cochlear nucleus. The tonotopic organisation of the lateral superior olive of the cat is illustrated in Fig. 22.79. The dendrites of the principal neurons radiate from the hili in rostrocaudal planes, perpendicular to the curvatures of the nucleus. Marginal neurons extend perpendicular to the principal cells. The principal cells and the marginal neurons project to the nuclei of the lateral lemniscus and bilaterally to the central nucleus of the inferior colliculus. Neurons projecting to the ipsilateral and contralateral inferior colliculus constitute separate populations (Henkel and Brunso-Bechtold 1993). Multipolar cells of the lateral superior olive distribute their axons within the nucleus (Majorossy and Kiss 1990). The medial nucleus of the trapezoid body of the cat consists of large cells that receive fibres of globular bushy cells located in the central part of the ventral cochlear nucleus, near the root of the cochlear nerve. These fibres of the cochlear nucleus terminate with calices of Held (1893) that surround the somata of the neurons with multiple synaptic specializations (Morest 1968; Spirou et al. 1990). The medial nucleus of the trapezoid body projects to the ipsilateral lateral superior olive and the central nucleus of the inferior colliculus (Irvine 1986). The projections of the medial nucleus of the trapezoid body are glycinergic (Adams and Mugnaini 1990; Glendenning et al. 1991: cat; Saint Marie and Baker 1990; Bledsoe et al. 1990: guinea pig). As a result, the lateral superior olive is excited from the ipsilateral ear and inhibited from the contralateral ear. The neurons of the peri-olivary nuclei are heterogeneous, both in their morphology and in their projections. The peri-olivary nuclei surround the

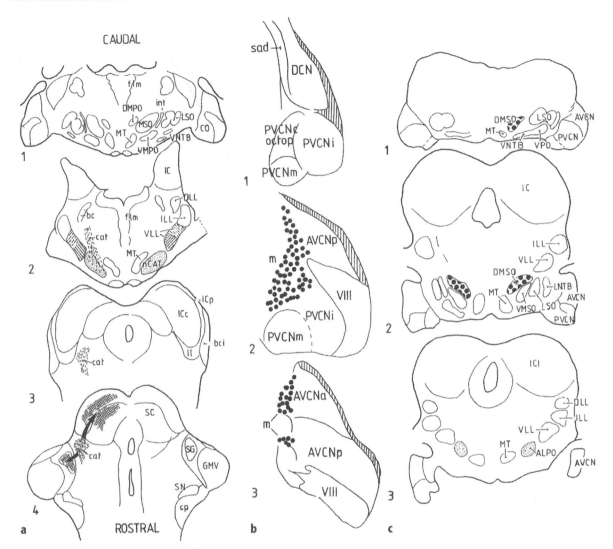

Fig. 22.83. a Acoustic nuclei of the brain stem and b Cochlear nuclei of the mustache bat *Pteronotus parnelli.* c Superior olivary complex of the horseshoe bat *Rhinolophus rouxi.* The anteroventral and posteroventral cochlear nuclei of the mustache bat can be further subdivided. Large, densely staining cells of the marginal division (*m*) are located between the anteroventral and posteroventral cochlear nuclei. The granule cell layer is *hatched.* The two divisions of the ventral nucleus of the lateral lemniscus of the mustache bat, with a different columnar orientation of the neurons, are indicated in **a**. The central acoustic tract (*cat*) is shown in **a**. It takes its origin from a nucleus (*nCAT*) corresponding to the anterolateral preolivary nucleus (*ALPO*) of the horseshoe bat. Of the two nuclei located between the medial nucleus of the trapezoid body (*mt*) and the lateral superior olive (*LSO*) which were distinguished by Vater and Feng (1990) in the horseshoe bat (**c**), the dorsal medial superior olivary nucleus (*DMSO*) gives rise to the olivocochlear bundle, and the ventral medial superior olivary nucleus (*VMSO*) corresponds to the medial superior olive of other mammals. *AVCN(a,p),* anteroventral cochlear nucleus (anterior and posterior subdivisions); *bc,* brachium conjunctivum; *bci,* brachium of the inferior colliculus; *CO,* cochlear nuclei; *cp,* cerebral peduncle; *DCN,* dorsal cochlear nucleus; *DLL,* dorsal nucleus of the lateral lemniscus; *DMP,* dorsomedial preolivary nucleus; *flm,* medial longitudinal fascicle; *GMV,* medial geniculate body; *IC,* inferior colliculus; *ICc,* central nucleus of the inferior colliculus; *ICp,* lateral zone of the inferior colliculus; *ILL,* intermediate nucleus of the lateral lemniscus; *int,* intermediate cell group; *ll,* lateral lemniscus; *LNTP,* lateral nucleus of the trapezoid body; *m,* marginal subdivision of the ventral cochlear nucleus; *MSO,* medial superior olive; *MSO,* medial superior olive; *MT,* medial nucleus of the trapezoid body; *PVCN(c otop,i,m),* posteroventral cochlear nucleus (central octopus, intermediate, medial subdivision); *sad,* dorsal acoustic stria; *SC,* superior colliculus; *SG,* suprageniculate nucleus; *SN,* substantia nigra; *VIII,* cochlear nerve; *VLL,* ventral nucleus of the lateral lemniscus; *VMPO,* ventromedial preolivary nucleus; *VMSO,* ventral medial superior olivary nucleus; *VNTB,* ventral nucleus of the trapezoid body; *VPO,* ventral preolivary nucleus. (Course and termination of the central acoustic tract taken from Casseday et al. 1989; redrawn from Zook and Cassedy 1982, 1980, Cassedy et al. 1989 and Vater and Feng 1990)

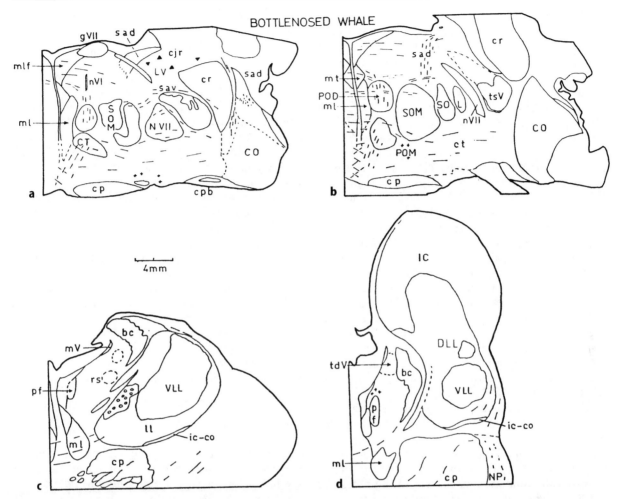

Fig. 22.84a–d. Acoustic nuclei of the bottle-nosed whale *Hyperoodon ampullatus*. Note the large size of the medial and lateral superior olives, the medial nucleus of the trapezoid body and the ventral nucleus of the lateral lemniscus. The dorsal nucleus of the lateral lemniscus is relatively small. *bc*, brachium conjunctivum; *cjr*, corpus juxtarestiforme; *CO*, cochlear nuclei; *cp*, cerebral peduncle; *cpb*, corpus pontobulbare; *cr*, restiform body; *CT*, nucleus of the trapezoid body; *ct*, trapezoid body; *DLL*, dorsal nucleus of the lateral lemniscus; *gVII*, genu of the facial nerve; *IC*, inferior colliculus; *icco*, colliculo-cochlear fibres; *ll*, lateral lemniscus; *LV*, lateral vestibular nucleus; *ml*, medial lemniscus; *mlf*, medial longitudinal fascicle; *mt*, medial tegmental tract; *mV*, mesencephalic root of the trigeminal nerve; *N.VII*, facial nerve nucleus; *NP*, pontine nuclei; *nVI*, abducens nerve; *pf*, predorsal fascicle; *PMO*, medial paraolivary nucleus; *POD*, dorsal paraolivary nucleus; *rs*, rubrospinal tract; *sad*, stria acustica dorsalis; *sav*, stria acustica ventralis, fibres of the trapezoid body; *SOM*, medial superior olive; *SOL*, lateral superior olive; *tdV*, dorsal trigeminothalamic tract of Wallenberg; *tsV*, spinal tract of the trigeminal nerve; *VLL*, ventral nucleus of the lateral lemniscus. (From De Graaf 1967)

principal nuclei like a hollow sphere. In the cat, they can be subdivided into several subnuclei by extension of the principal nuclei in the ring of peri-olivary cells.

Peri-olivary cell groups give rise to projections (Fig. 22.77) to the inferior colliculus (Van Noort 1969; Adams 1979), descending projections to the cochlear nuclei (Van Noort 1969; Adams and Warr 1976; Elverland 1977; Kane and Finn 1977) and crossed and uncrossed olivocochlear pathways passing through the vestibular nerve to the cochlea (Warr 1975; White and Warr 1983; Vetter and Mugnaini 1992). These systems arise from different groups of peri-olivary cells (Adams 1983). Neurons forming the descending olivocochlear projection can be identified by their staining for acetylcholinesterase (AChE; Warr 1975).

The medial superior olive in primates exhibits the same characteristic morphology seen in other mammals (Fig. 22.24k,l). The size of the medial superior olive steadily increases across prosimians, New and Old World monkeys and apes (Moore 1987). The lateral superior olive of primates does not show the convolutions seen in the cat and in prosimians, but is shaped as a single mass. There is no evident systematic orientation of the cells transversing to the convolutions of the nucleus as is the case in lower mammals. The medial nucleus of the trapezoid body is small and dispersed, and the ring of peri-olivary cells surrounding the olivary complex is fairly complete, with distinct dorsal and ventral subnuclei (Moore and Moore 1971; Moore 1987).

The superior olivary complex of the North American opossum consists of the same nuclei as in eutherian mammals (Figs. 22.22k,l, 22.82, Willard and Martin 1983). The medial superior olive consists of stacked bipolar cells. The bilateral projection of these cells to the central nucleus of the inferior colliculus differs from the ipsilateral projection described for the medial superior olive in placental mammals. The lateral superior olive lacks convolutions and consists of a core of planar principal neurons and of marginal cells wrapped around it. Its bilateral projection to the inferior colliculus is similar to the cat. The medial nucleus of the trapezoid body is distinct, and peri-olivary cell groups are well developed. The medial nucleus of the trapezoid body presumably projects to the lateral superior olive. The peri-olivary neurons of the ventral nucleus of the trapezoid body are reciprocally connected with the inferior colliculus (Willard and Martin 1983).

The superior olivary complex of echolocating bats is large (Fig. 22.83). A well-developed medial superior olive, receiving tonotopical projections from the ipsilateral and contralateral anteroventral cochlear nuclei, with an expanded portion of the 61-kHz call component is present in the mustache bat *Pteronotus parnelli*. The lateral superior olive is S-shaped, and the medial nucleus of the trapezoid body is distinct. Several peri-olivary cell groups can be recognised. The rostralmost of these groups, the anterolateral peri-olivary nucleus (Fig. 22.83C; Zook and Casseday 1982), was later identified as the nucleus of origin of the central acoustic tract (Fig. 22.83a, cat; Casseday et al. 1989).

In the horseshoe bat *Rhinolophus rouxi*, two nuclei are present between the lateral superior olive and the medial nucleus of the trapezoid body: the dorsomedial and the ventromedial nuclei (Fig. 22.83c). Identification of one of these nuclei as the medial superior olive was not possible: both receive projections from both anteroventral cochlear nuclei (Vater and Feng 1990). Grothe et al. (1994) similarly localised two nuclei between the lateral superior olive and the medial nucleus of the trapezoid body in the free-tailed bat *Tadarida brasiliensis mexicana* and identified the dorsomedial cell group, which stains for AChE, as the origin of the olivocochlear system and the ventromedial nucleus as the medial superior olive. It receives a bilateral projection from the cochlear nuclei. Its projection to the inferior colliculus is bilateral, differing from the traditional ipsilateral connection. The afferent connections of the lateral superior olive of the horseshoe bat differ from the cat in receiving afferent fibres from the ipsilateral posteroventral nucleus and the dorsal cochlear nucleus, in addition to the anteroventral cochlear nucleus (Casseday et al. 1988; Vater and Feng 1990). Zook and Casseday (1985) raised the question of why the medial superior olive is large (Fig. 22.83a) in an animal with a small head, such as the mustache bat, that would allow for only very small phase shifts in the localisation of low-frequency sounds. They also discussed the literature on other species of small bats that lack a medial superior olive.

In Cetacea, the superior olivary complex is large (Fig. 22.84). It is better developed in toothed whales than in whalebone whales. The lateral superior olive is gyrated, and the medial superior olive differs in shape in different species. The distinction between the medial and lateral superior olives is not always obvious in whales. The medial nucleus of the trapezoid body and the pre- and peri-olivary nuclei are well developed; calices of Held were not observed (Ogawa and Arifuku 1948; De Graaf 1967).

22.8.2.4
Nuclei of the Lateral Lemniscus

Efferent fibres from the ipsilateral and contralateral cochlear nuclei and the superior olives ascend in the lateral lemniscus, terminating on the nuclei of the lateral lemniscus and the inferior colliculus. In the cat, the ventral and dorsal nuclei of the lateral lemniscus lie embedded in this bundle (Fig. 22.23m,n). The neurons of these nuclei are oriented at right angles to the fibres of the lateral lemniscus (Geniec and Morest 1971; Kane and Barone 1980). The ventral nucleus of the lateral lemniscus receives afferent fibres from the contralateral cochlear nuclei, some of which terminate with calices of Held and project to the ipsilateral central nucleus of the inferior colliculus. The dorsal nucleus of the lateral lemniscus receives projections from the superior olive, bilateral from the lateral superior olive and an ipsilateral one from the medial superior olive (Van Noort 1969). The projection of the dorsal nucleus of the lateral lemniscus to the inferior colliculus is bilateral and tonotopically organised (Merchan et al. 1994). Additionally, it projects to the contralateral dorsal nucleus of the lateral lemniscus (Kudo 1981). Efferent fibres of the dorsal nucleus cross dorsal and caudal to the decussation of the brachium conjunctivum in the commissure of Probst (Fig. 22.25). The projections of the dorsal nucleus of the lateral lemniscus probably are GABAergic

(Schneiderman et al. 1988, 1993; Oliver and Schneiderman 1989).

The ventral nucleus of the lateral lemniscus of Old World monkeys is very similar to the cat (Verhaart 1970). It is indistinct in New World monkeys and difficult to define in apes. It is the most poorly developed part of the brain stem auditory system in the human brain (Moore 1987). A distinct dorsal nucleus of the lateral lemniscus can be identified in all primates (Fig. 22.24n).

The structure of the nuclei of the lateral lemniscus of the North American opossum was studied by Willard and Martin (1983). The nuclei consist of a column of neurons with interlacing dendrites, surrounded by a marginal band of neurons with dendrites that extend perpendicular to the fibres of the lateral lemniscus. The distinction of dorsal and ventral nuclei rests on similar differences in the laterality of their projection to the inferior colliculus as have been reported for the cat.

Three large nuclei were distinguished in the lateral lemniscus of echolocating bats (Zook and Casseday 1982; Vater and Feng 1990). The ventral and intermediate nuclei have an unusual appearance, while the dorsal nucleus and the commissure of Probst are very similar to those of other mammals. The cells in the ventral part of the ventral nucleus are arranged in columns, extending parallel to the fibres of the lateral lemniscus; in its dorsal multipolar cell area, there is no columnar arrangement (Fig. 22.83a). The columns are precisely related to the tonotopic organisation of the ventral nucleus, with low frequencies located ventrally, high frequencies dorsally and an over-representation of the frequencies in the range of the echolocating call (Covey and Casseday 1986). The intermediate nucleus forms a lateral prominence on the brain stem. Its cells are heterogeneous; dendrites of elongated cells often are oriented mediolaterally. Tonotopic representations have also been found in the intermediate and dorsal nuclei of the lateral lemniscus in echolocating bats (Matzner and Rodtke-Schushler 1987; Covey and Casseday 1991).

The central acoustic tract is a short-circuit system that connects acoustic centres in the lower brain stem with the thalamus and avoids the inferior colliculus. The central acoustic tract has been described in cats and primates (Papez 1929; Verhaart 1970). It is large in echolocating bats. In the mustache bat, it takes its origin from a nucleus located ventral to the ventral nucleus of the lateral lemniscus, formerly known as the lateral periolivary nucleus (Fig. 22.83a). The nucleus of the central acoustic tract receives a projection from the contralateral anteroventral cochlear nucleus. The central tract ascends in the lateral tegmentum mesencephali and terminates in the deep layers of the superior colliculus and the suprageniculate nucleus (Fig. 22.83a). The latter projects to the auditory and frontal cortex (Casseday et al. 1989).

Neurons located in the same paralemniscal tegmental area as the central acoustic tract provide an interface between the pathways for auditory sensory processing and those for the motor control of vocalisation. In the horseshoe bat, these neurons project to the superior colliculus, the facial nerve nucleus and the vocalisation center in the reticular formation between the facial and ambiguus nuclei (Metzner 1996).

The lateral lemniscus is extremely large in whales, occupying one half to one third of the tegmentum (Fig. 22.84). Ventral and dorsal nuclei of the lateral lemniscus were distinguished; the ventral nucleus is large, while the dorsal nucleus and Probst's commissure are poorly developed.

22.8.2.5
Inferior Colliculus

The inferior colliculus can be subdivided into a central nucleus, a dorsal cortex and a lateral cortex (or external nucleus, Morest and Oliver 1984; Van Noort 1969; Geniec and Morest 1971; Irvine 1986; for a tabulated summary of subdivisions of the inferior colliculus, see Huffman and Henson 1990). Dorsal to the aqueduct, the commissure of the inferior colliculus crosses the midline (Figs. 22.22–22.24). This is a true commissure, interconnecting the inferior colliculi. Additionally, it contains the fibres of the crossed ascending pathway from the colliculus. The dorsal cortex and the external nucleus have an indistinct laminated structure. The terminal plexus of the fibres of the lateral lemniscus and the neurons of the central nucleus are arranged in laminae that extend from ventromedially to dorsolaterally. These laminae roughly correspond to the isofrequency planes of the central nucleus. Low frequencies are represented dorsally, and high frequencies ventrally in the colliculus. Certain types of neurons of the central nucleus are confined to one lamina (disc-shaped cells of Oliver and Morest 1984; Oliver et al. 1991; Paloff et al. 1982; Fig. 22.85). Stellate cells cut across the laminae. The laminae of the dorsal cortex of the inferior colliculus are arranged perpendicular to the laminae of the central nucleus.

Two types of flattened neurons were distinguished by Malmierca et al. (1993) in the rat: flat (F) and less flat (LF) cells; these were confined to the one-cell-thick laminae and the interlaminar compartments of the central nucleus, respectively. Saldaña and Merchán (1992) analysed the intrinsic,

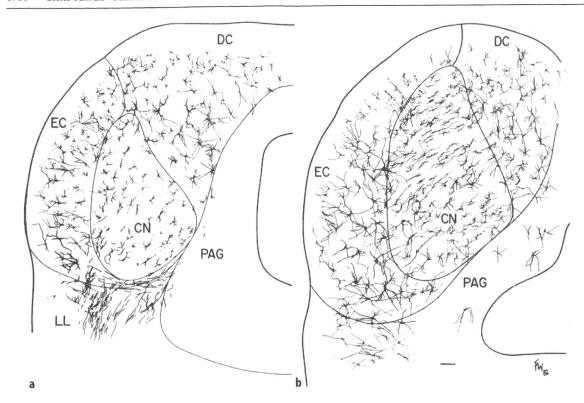

Fig. 22.85a,b. Camera lucida drawings of neurons in a rostral and b caudal, Golgi-Cox-stained, coronal sections taken through the inferior colliculus of the North American opossum. The central nucleus can be recognised by the predominence of disc-shaped principal cells. The external cortex has many large stellate cells oriented perpendicular to the pial surface (more obvious at rostral levels), and the dorsal cortex has large stellate neurons with radiate dendritic distributions. Note that the dorsomedial boundary of the central nucleus is clearly illustrated in Golgi-impregnated material. *CN*, central nucleus; *DC*, dorsal cortex; *EC*, external cortex; *LL*, lateral lemniscus; *PAG*, periaqueductal grey. *Bar*, 100 μm. (From Willard and Martin 1983)

tonotopic organisation of the inferior colliculus of the rat. They injected the anterograde tracer *Phaseolus vulgaris*-leucoagglutinin at different sites along the tonotopic axis of the central nucleus. Each of these injections labelled four bands. One main band extended ipsilaterally from the central nucleus into the dorsal cortex, and a second ipsilateral band extended into the external nucleus. Caudally in the colliculus, the two bands merge. A similar pattern of two fused bands was present on the contralateral side (Fig. 22.86d-f).

The central nucleus receives tonotopically ordered projections from the contralateral cochlear nuclei, both lateral superior olives, the ipsilateral medial superior olive and the ventral nucleus of the lateral lemniscus, bilateral projections from the dorsal nucleus of the lateral lemniscus and commissural afferent fibres from the contralateral central nucleus of the inferior colliculus. The external nucleus receives inputs from the contralateral dorsal cochlear nucleus, the ipsilateral and contralateral central nucleus and the dorsal nucleus of the lateral lemniscus (Van Noort 1969; Osen 1972; Glendenning and Masterton 1983; Oliver 1984; Coleman and Clerici 1987; Moore 1988). Additionally, the external nucleus receives somatosensory input from the spinal cord, the dorsal column nuclei, the spinal trigeminal nucleus, the parabrachial area and the deep layers of the superior colliculus (Van Noort 1969; Aitkin et al. 1981; Coleman and Clerici 1987). The dorsal cortex receives its fibres from the nucleus sagulum, a group of small cells lateral to the dorsal nucleus of the lateral lemniscus (Henkel and Shneiderman 1988), and from the auditory cortex (Van Noort 1969; Anderssen et al. 1980; Herbert et al. 1991).

Efferent fibres of the inferior colliculus collect in the brachium of the inferior colliculus, which terminates in the medial geniculate body (Figs. 22.22-22.24n-p). The medial geniculate projects to the auditory cortex in the superior temporal region. The central nucleus and the cortex of the inferior colliculus differ in their efferent projections. The central nucleus projects bilaterally to the external

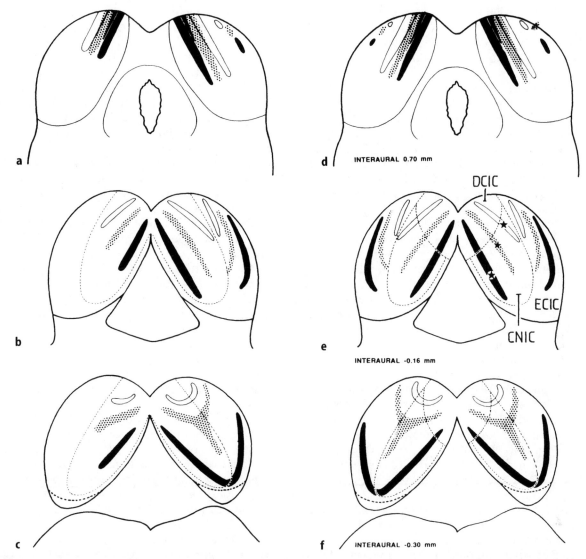

Fig. 22.86a–f. Tonotopic organisation of corticocollicular and intrinsic connections of the inferior colliculus of the rat. **a–c** Distribution of auditory corticocollicular terminal fields from three different regions of the primary auditory cortex: a high-frequency region (*black*), medium-frequency region (*stippled*) and a low-frequency region (*white*). Regions of the ipsi- and contralateral inferior colliculus innervated by each of these regions are shown. **d–f** Distribution of intracollicular terminal fields from Saldaña and Merchán (1992). *Asterisks* indicate the position of three *Phaseolus vulgaris*-leucoagglutinin (PHA-L) injection sites into different tonotopic regions of the central nucleus of the inferior colliculus: high-frequency (*black*), medium-frequency (*stippled*) and low-frequency (*white*) regions. The territories innervated by neurons in these regions are shown with the corresponding shading. Note that, with the exception of the contralateral external cortex of the inferior colliculus, which is innervated by the central nucleus but not by the auditory cortex, the distribution of the corticocollicular projections mimics the topography of the intracollicular projections. *CNIC*, central nucleus of the inferior colliculus; *DCIC*, dorsal cortex of the inferior colliculus; *ECIC*, external cortex of the inferior colliculus. (From Saldaña et al. 1996)

nucleus, the parabrachial region of the lateral tegmentum, the interstitial nucleus of the brachium of the inferior colliculus, the ventral and magnocellular divisions of the medial geniculate body and the posterior group of the thalamus. The external nucleus projects to the dorsal division, and the dorsal cortex to the medial division of the medial geniculate body (Morest 1965; Van Noort 1969; Moore and Goldberg 1963; Huffman and Henson 1990). The pathway from the central nucleus, through the ventral division of the medial geniculate body to the primary auditory AI cortex, is tonotopically organised. The external nucleus and the dorsal cortex are mainly connected with the auditory area

AII, the insular cortex and the anterior auditory field by way of the dorsal and medial divisions of the medial geniculate body, respectively (Andersen et al. 1980; Calford and Aitkin 1983).

The inferior colliculus of marsupials consists of a central nucleus and an overlying cortex, separated by a thin fibre capsule (Willard and Martin 1983; Aitkin et al. 1994). The dendrites of the principal cells of the central nucleus are oriented in a ventrolateral to dorsomedial direction (Figs. 22.24m,n, 22.85). Dorsolaterally, the ventral nucleus reaches the meningeal surface, where it separates the dorsal cortex from the lateral cortex (external nucleus). The cortex can be subdivided into at least two layers populated by stellate cells (Willard and Martin 1983). The pattern of afferent connections is the same as in the cat: crossed projections from the ventral and dorsal cochlear nuclei and bilateral projections from the superior olive to the central nucleus; a crossed projection from the dorsal cochlear nucleus and bilateral projections from the superior olivary complex to the external nucleus, which also receives afferent connections from the reticular formation and from somatosensory relay nuclei; and a projection from the neocortex to the dorsal cortex and the external nucleus of the inferior colliculus (Willard and Martin 1983; Aitkin et al 1994).

The primate inferior colliculus is similar in most respects to the inferior colliculus of the cat (Geniec and Morest 1971; Moore 1987; Garey and Webster 1989). In Cetacea (De Graaf 1967) and echolocating bats (Zook and Casseday 1982, 1987; Frisina et al. 1989; Wenstrup et al. 1994), the inferior colliculus is large. The subdivision and connections of the inferior colliculus conform to the general mammalian pattern exemplified by the cat and opossum. One isofrequency contour, containing neurons tuned to the 60-kHz isofrequency call, is greatly over-represented in the dorsoposterior division of the central nuclei of the mustache bat (Ross et al. 1988).

22.8.2.6
Descending Auditory Pathways

Descending auditory pathways were reviewed by Huffman and Henson (1990). These pathways connect (a) the auditory cortex with the medial geniculate body and the inferior colliculus, (b) the inferior colliculus and, to some degree, the nuclei of the lateral lemniscus with the superior olivary complex and (c) the cochlear nuclei and the superior olivary complex with the cochlear nuclei and the hair cells of the orgen of Corti (Fig. 22.77b).

The descending and ascending connections of the auditory cortex with the subdivisions of the medial geniculate body and the inferior colliculus are roughly reciprocal. The main projection of the primary auditory cortex (AI) of the cat, which receives fibres from the central nucleus of the medial geniculate body, is to the same nucleus and, bilaterally, to the dorsal cortex of the inferior colliculus (Andersen et al. 1980; Huffman and Henson 1990). In primates, the projection from the primary auditory cortex to the inferior colliculus includes the central nucleus (FitzPatrick and Imig 1978, 1982). Area AII is reciprocally connected with the dorsal nucleus of the medial geniculate body and the external nucleus of the inferior colliculus, and the anterior auditory field with the medial nucleus and the dorsal cortex. The relations of the auditory cortical fields to the auditory centres are slightly different in the rat (Huffman and Henson 1990; Fig. 22.86a,c).

The central nucleus of the inferior colliculus projects bilaterally to the deep layers of the dorsal cochlear nucleus. Fibres from the external nucleus and the dorsal cortex terminate more superficially. The descending connections to the superior olivary complex are mainly ipsilateral. These fibres terminate diffusely in the peri-olivary nuclei, but mainly target the cells of the ventral nucleus of the trapezoid body, i.e. the cells of origin of the medial crossed olivocochlear pathway (Thompson and Thompson 1993; Caicedo and Herbert 1993; Vetter et al. 1993; Huffman and Henson 1990).

Separate populations of peri-olivary neurons project to the cochlear nuclei and to the inferior colliculus and give rise to the olivocochlear pathways. Neurons that give rise to the olivocochlear bundles can be stained for AChE and with antibodies against choline acetyltransferase (ChAT; Warr 1975; Thompson and Thompson 1986; Vetter et al. 1991). A subpopulation of olivocochlear neurons in the lateral superior olive of the rat is GABAergic. Medial and lateral groups of olivocochlear neurons can be distinguished. The medial group, which includes the ventral nucleus of the trapezoid body in the cat and consists of fairly large cells, gives rise to fibres that terminate bilaterally, but mainly contralaterally on outer hair cells. The neurons of the lateral group are smaller and, in the cat, are located in the margins of the lateral superior olive. The lateral group gives rise to the uncrossed olivocochlear bundle, which terminates on inner hair cells (Warr and Guinan 1979; Guinan et al. 1983). The same two groups can be recognised in primates (Thompson and Thompson 1986) and in rodents (White and Warr 1983; Aschoff and Oswald 1987). In the rat, guinea pig and echolocating bats, the small neurons of the lateral group are located within the lateral superior olive. In bats, the medial group forms a horseshoe-like nucleus that rostrally surrounds the medial superior olive. The medial group and the projection to the outer hair cells are absent in the horseshoe bat (Aschoff and Oswald 1987). All neurons of the olivocochlear bundle in this species are located in a nucleus (the dorsomedial cell group of Vater and Feng 1990) between the medial superior olive (the ventromedial group of Vater and Feng 1990; Fig. 22.83) and the lateral superior olive.

Olivonuclear fibres originate bilaterally from all peri-olivary nuclei, with a strong contribution of the ventral nucleus of the trapezoid body. Most fibres terminate in the anteroventral and dorsal cochlear nuclei. The granular layers only receive a contralateral projection. Olivocochlear fibres emit collaterals to the granular layers (Spangler et al. 1987). Some olivonuclear projections may be GABAergic (Ostapoff et al. 1990) or glycinergic (Benson and Potashner 1990).

22.8.3
Visual System

22.8.3.1
General Plan

The eyes of all vertebrates share the same general optical and cellular plan. Light from objects in front of the eye is refracted by the cornea and/or the lens, and the object is imaged on an array of photoreceptors, which are the first waystation in visual processing. The retina of all vertebrates have three cellular layers. The receptors and their nuclei are the first of these three layers. The receptors communicate with the ganglion cells, the final layer in the retina, through an array of horizontal, bipolar and amacrine cells whose cell bodies are in the inner nuclear layer between the receptors and ganglion cells. However, although this general organisation is present in the retina of all mammals, and indeed of all vertebrates, there are important differences among species in the details.

There are two fundamentally different types of photoreceptors in the eyes of mammals: the rods and the cones. These two receptor types are distinguished on the basis of the morphology of their outer segments and their sensitivity to light. Rods are exquisitely sensitive to the smallest amount of light and have their peak sensitivity to wavelengths of about 500 nm. Cones are relatively less sensitive to light, and there are usually two or more classes of cones with peak sensitivities at different wavelengths. In humans and some Old World monkeys, there are three classes of cone, whose peak sensitivities are at 420, 531 and 558 nm. The presence of two or more receptor types with differential wavelength sensitivity is an essential prerequisite for colour vision.

Rods and cones differ in their absolute sensitivity and in their response to different wavelengths of light. There are also important differences in the way in which the two receptor types are connected to the ganglion cells. These differences can be seen most clearly in those animals that have retinas with virtually only rods or those with only cones. Rod-dominated animals are strongly nocturnal, while cone-dominated animals are strongly diurnal in their habits. The grey squirrel and the tree shrew *Tupaia glis* are diurnal mammals, and the opossum *Didelphis virginiana* is a nocturnal mammal. The tree shrew *Tupaia glis* has a single array of photoreceptors, nearly all cones, on which the image is formed. The nuclei of the cones form a single row just internal to the outer limiting membrane of the retina. *Didelphis* has a dense array of narrow rods, with a large number of rod nuclei extending in many layers beneath the outer limiting membrane. In *Tupaia*, there are roughly equal numbers of ganglion cells and cones. In *Didelphis*, rods outnumber ganglion cells by more than 100 to one. The retinas of these two species, therefore, reflect two completely different types of retinal organisation. Convergence of many rod signals onto a single ganglion cell is a way of maximising the sensitivity of the eye to light and is typical of the retina of nocturnal animals. Activation of any one of 100 rods will influence the response of the ganglion cell to which they are connected. However, this high sensitivity is associated with a limitation in visual acuity, i.e. the ability to resolve fine detail. The ganglion cell would not be able to distinguish which of the many rods over the region of the retina had been active. In the retina of *Tupaia*, there is an approximately equal number of cones and ganglion cells. Thus sensitivity to light would be lower, but visual acuity would be correspondingly greater.

Many mammals have combined the two plans. Cats, for example, have a single row of cone nuclei lining the outer limiting membrane and a much greater number of rod nuclei lying below them. The cat retina combines the sensitivity of an all-rod retina with some of the advantages in acuity of the all-cone retina. Cats have a region of retina, the area centralis, in which the density of cones is increased, and visual acuity is consequently higher. Nevertheless, the acuity of vision in cats is far below that of humans and the higher primates, and their colour vision is, at best, very poor. In humans and the higher primates, there is a region of retina, the fovea, with extremely high cone density. Rods are excluded from the fovea, and the two other retinal layers and the retinal blood vessels are displaced laterally. The great density of foveal cones allows for acuity in the foveal region that is close to the limits that are imposed by the physics of image formation.

Certain animals, such as the rabbit and deer, have eyes placed far laterally and panoramic vision. Such animals are often at risk from predators appearing over the horizon from any direction. In these animals, the area of most acute vision is not concentrated in a single central retinal region, but is spread out along the horizontal meridian of the retina. The sharper acuity for objects on the horizon serves as part of a distant early-warning system.

Mammals differ in their degree of binocular vision. Some, such as rats and horses, have eyes that are placed far laterally on the head. Some, such as humans and the higher primates, have frontally placed eyes in which nearly all of the visual field is binocular. The extent to which the eyes share the same visual field is paralleled closely by the degree to which the optic nerves cross or remain uncrossed at the optic chiasma. In rats and horses, the great majority of the optic nerve fibres decussate at the chiasm. In humans and monkeys, the optic nerve fibres that arise from ganglion cells in the nasal retina cross at the chiasm. Fibres that arise from ganglion cells in the temporal retina project ipsilaterally. The net result of these decussation patterns is that in all mammals the right side of the visual world is projected onto the left side of the brain, and the left side of the visual world onto the right side of the brain. This simple principle, whereby the visual fields and not the eyes are represented in the central pathways, was first suggested by Isaac Newton and was largely ignored until its gradual re-emergence in the nineteenth century. As late as 1876, David Ferrier, in his book on the *Functions of the Brain*, was still describing the optic nerve as decussating completely. It was a combination of the self-observation by Wollaston (1824) describing his own hemianopia, the anatomical studies carried out by von Gudden (1870a,b) and the experimental evidence of hemianopia in an occipitally lesioned monkey obtained by Munk (1881) that led to the gradual recognition and acceptance of the principle of partial decussation.

The main targets of the optic tract in the mesencephalon are the superior colliculus, the pretectum and the accessory optic system. The pretectum is located between the superior colliculus and the thalamus. Left and right pretectum are connected through the posterior commissure. It contains relays for the pupillary light reflex (olivary pretectal nucleus) and for the horizontal optokinetic reflex (nucleus of the optic tract). The latter nucleus functions in close association with the nuclei of the accessory optic system. Other functions and connections of the pretectum are poorly understood.

The nuclei of the accessory optic system and the nucleus of the optic tract are centres for compensatory reflexes that stabilise the image on the retina using visual optokinetic information.

In all vertebrates, the superior colliculus or optic tectum is located at the roof of the midbrain and receives one of its major inputs from the optic tract. In non-mammalian vertebrates, the tectum is the major visual structure of the brain. It functions in form vision and visual learning and also in the visual guidance of movement. In some mammals, it retains both of these function, but in many mammals, and particularly in humans and the higher primates, the colliculus plays a less direct role in conscious vision. However, it retains an important function in the guidance of certain classes of movement, particularly of the eyes and the head. The anatomical connections of the colliculus can best be understood in terms of these two functions, i.e. as a central visual structure and in the visual control of certain classes of movement. These two functions are clearly segregated in the colliculus. The superficial laminae retain their input from the optic tract and also receive inputs from visual areas of the cerebral cortex. Cells in these superficial laminae are briskly activated by appropriate visual targets. Their major efferent targets are thalamic structures that are involved in visual processing. The intermediate and deeper layers of the colliculus receive auditory and somatosensory as well as visual input, and they target brain stem and spinal structures that control movement of the eyes and the head.

The presence of a fovea in primates would be of little advantage if it could not be freely aimed to inspect objects of interests. Humans and the higher primates have several mechanisms that serve to keep the visual scene steady during movement of the head and allow them to follow a moving target or inspect a novel object. Saccadic movements are one of these mechanisms. Saccades are the rapid shifts of the eyes' position which are used to inspect a new target in the visual field and as a reset mechanism to centre the eyes when they have been displaced far laterally. The superior colliculus plays a major role in the generation of saccades. Cells in the deep layers of the colliculus are briskly active prior to saccades, and collicular lesions impair the generation of saccadic movement. Since the frontal eye fields, area 8 of the frontal cortex, can also initiate saccadic eye movements, the initial deficit which follows collicular damage can recover. If the colliculus is damaged, animals and humans suffer only a transient impairment. A similar transient disability results from damage to the frontal eye fields. If both frontal eye fields and colliculus are destroyed, saccadic movements are virtually abolished.

22.8.3.2
Superior Colliculus

22.8.3.2.1
Structure

The superior colliculus is a laminated structure (Fig. 22.87). Most recent authors have followed the subdivision of the superior colliculus of the cat of Kaneseki and Sprague (1974) into seven laminae. The *stratum zonale* (lamina I) can be subdivided into sublamina I_1, containing thin, myelinated fibres and few small cells, and sublamina I_2, which consists of a narrow, somewhat discontinuous row of small cells. The neurons of the broad *stratum griseum superficiale* (lamina II) increase in size from dorsally to ventrally. Lamina II can be further subdivided into superficial, small-cell and deeper, large-cell layers (Graham and Casagrande 1980: the tree shrew *Tupaia glis*). The *stratum opticum* (lamina III) is composed predominantly of rostrocaudally oriented fibres with small cells scattered among them. The thick *stratum griseum intermedium* (lamina IV) contains small, medium and some large multipolar neurons. The *stratum album intermedium* (stratum lemnisci or lamina V) is dominated by transversely oriented fibres; many of its fibres can be traced ventrally into the tegmentum or medially into the tectal commissure. Small and medium-sized cells and a few large neurons are

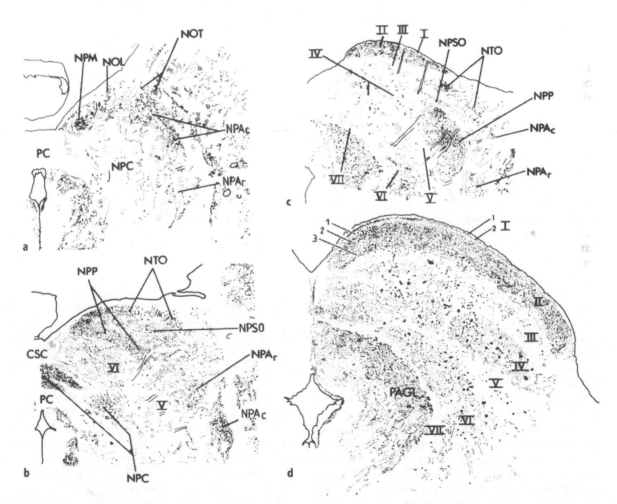

Fig. 22.87a–d. Transverse, Nissl-stained sections through **a,b** the pretectum and **c,d** the superior colliculus of the cat. *NPM*, medial pretectal nucleus; *NOL*, olivary pretectal nucleus; *PC*, posterior commissure; *NPC*, nucleus of the posterior commissure; *NOT*, nucleus of the optic tract; *NPAc*, anterior pretectal nucleus, compact part; *NPAr*, anterior pretectal nucleus, reticular part; *NPP*, posterior pretectal nucleus; *NPSO*, suboptic pretectal nucleus; *CSC*, commissure of the superior colliculus; *II(1–3)*, (sublayers 1–3 of the) stratum griseum superficiale; *I(1,2)*, (sublayers 1,2 of the) stratum zonale; *III*, stratum opticum; *IV*, stratum griseum intermedium; *V*, stratum album intermedium; *VI*, stratum griseum profundum; *VII*, stratum album profundum; *PAGL*, central grey. (From Kaneseki and Sprague 1974)

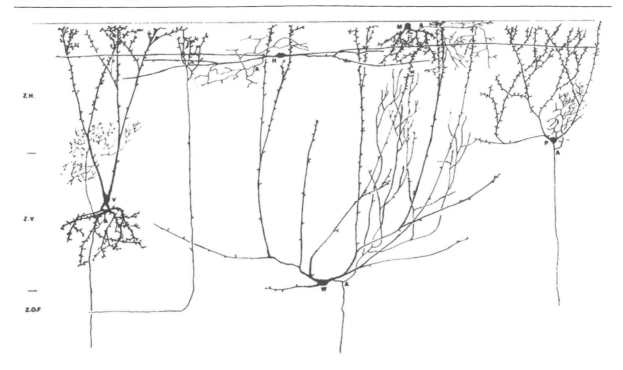

Fig. 22.88. Principal neurons of the upper layers of the superior colliculus of the rat. Golgi impregnation. *A*, axon; *H*, horizontal cell; *M*, marginal cell; *P*, piriform cell; *V*, narrow field vertical cell; *W*, wide field vertical cell; *Z.H.*, zone of horizontal cells; *Z.V.*, zone of vertical cells; *Z.O.F.*, zone of optic fibres. (From Langer and Lund 1974)

situated between the fibre bundles. The *stratum griseum profundum* (lamina VI) is a densely cellular layer that extends from a rostral position, dorsal to the posterior commissure and its nucleus, to the caudal and lateral border of the superior colliculus with the cuneiform nucleus of the reticular formation. It contains small and medium-sized and a scattering of large neurons. The border between the nucleus of the posterior commissure and lamina VI is not well defined. The *stratum album profundum* (lamina VII) dorsally covers the periaqueductal grey. Medially, it is continuous with the tectal commissure. It contains the coarse fibres issued by the large neurons of laminae IV–VI on their way to the dorsal (*Fontainen-artige Haubenkreuzung*; Obersteiner 1912; Bechterew 1899) tegmental decussation and the predorsal bundle.

The light and electron microscope morphometric analysis of Golgi-impregnated neurons of the rat superior colliculus by Alberts (1990) and Alberts and Meek (1991) showed that the laminar organisation of the superior colliculus is much less distinct than in the tectum of lower vertebrates. The only elements that show a fairly distinct lamination pattern are the nerve fibres in laminae III and V, but cell bodies, dendrites and synaptic contacts are not laminarly organised in this structure. Wiener (1986) correctly identified lamina V in the rodent superior colliculus on the basis of the transverse orientation of its axons, which distinguishes this layer from the rostrocaudally oriented fibre bundles in deep lamina IV (IVb of Wiener).

Golgi studies of the superior colliculus (Cajal 1911: rabbit; Sterling 1971; Norita 1980: cat; Tokunaga and Otani 1976; Langer and Lund 1974; Laemle 1981, 1983: *Saimiri sciureus*, *Macaca mulatta* and humans) indicate that few of its neurons extend their dendrites beyond the confines of the main layers. Dendrites of the marginal cells of layer I and of small neurons in layer II participate in a dense neuropil in the superficial half of layer II. These cells belong to the ovoid or piriform cells of Cajal (1911) with their dense dendritic bouquets (*P* in Fig. 22.88). Neurons in the deep part of layer II are generally larger. Their dendrites are oriented vertically or obliquely with respect to the surface of the colliculus and may arise from one pole or from opposite poles of the cell body. Thus narrow- and wide-field vertical cells (Langer and Lund 1974; Laemle 1981), cylindrical-type and reversed conical-type neurons (Tokunaga and Otani 1976) and stellate or multipolar neurons have been distinguished in the superficial grey layers (Figs. 22.88, 22.89). Large, vertically oriented neurons in the deep part of the stratum griseum superficiale and the stratum opticum (fusiform cells of Cajal 1911)

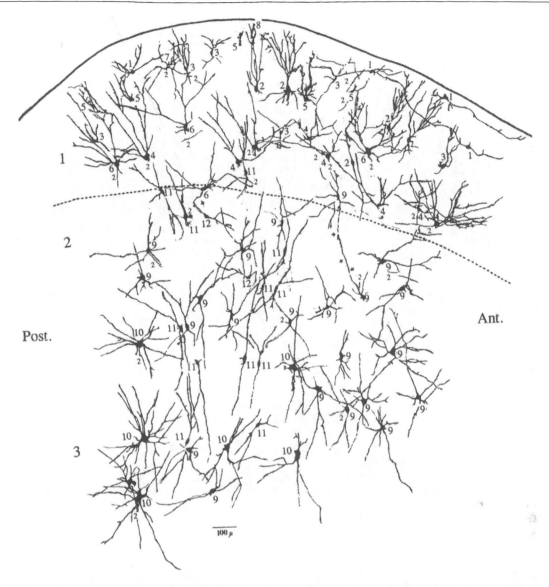

Fig. 22.89. Cell types in a sagittal section of the superior colliculus of the rat. Golgi impregnation. Three layers of the superior colliculus were distinguished: *Layer 1* corresponds to the stratum zonale, the stratum griseum superficiale and the stratum opticum; *layer 2* corresponds to the stratum griseum and album intermedium and the stratum griseum profundum; *layer 3* represents the stratum album profundum. The following cell types can be distinguished: *1,12*, horizontal type; *2,3,5*, cylindrical type; *4*, reversed conical type; *6,7,9,10*, multipolar type; *8*, marginal cells of Cajal (1909–1911); *11*, vertical type. Axons are indicated with a small number *2*. (From Tokunaga and Otani 1976)

may extend their dendrites into the superficial strata of the colliculus. Narrow-field vertical neurons of the superficial layer II and larger, presumably wide-field vertical neurons of deep layer II are neurons with ascending projections to the different parts of the thalamus (Graham and Casagrande 1980: the tree shrew *Tupaia glis*; Mooney and Rhoades 1993: hamster). The horizontal cells of the superficial laminae are probably intrinsic neurons (*H* in Fig. 22.88). Projection neurons of the deep layers of the colliculus with commissural and/or descending axons have been characterised in the cat and squirrel monkey using electrophysiological and intracellular injection techniques (Moschovakis and Karabelas 1985; Moschovakis et al. 1988a,b). Larger, multipolar cells with radiating dendrites, resembling those of the reticular formation (Edwards 1980), prevail in the deeper layers of the colliculus (Fig. 22.89).

The superficial layers I–III are distinguished from the deep collicular layers IV–VII by their afferent and efferent connections (Edwards 1980).

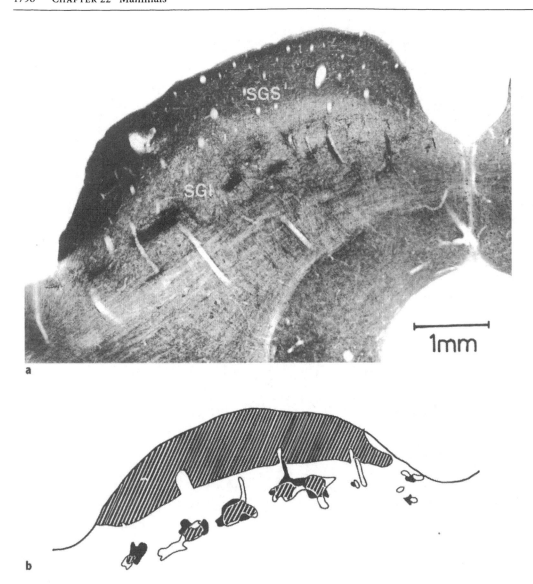

Fig. 22.90. a Photograph of an acetylcholinesterase-reacted section through the superior colliculus of the cat, showing strong reactivity in the stratum griseum superficiale (*SGS*) and clustering of reaction product in the stratum griseum intermedium (*SGI*). Notice double tier of labelled patches in medial SGI. Compare Fig. 22.102. **b** Distribution of choline acetyltransferase-like immunoreactivity (*black*) and acetylcholinesterase staining (*open contours*) in a transverse section through the superior colliculus of the cat. Overlap (*hatched regions*) occurs in the superficial layers of colliculus and in clusters in the stratum griseum intermedium. Compare Fig. 22.94. (Redrawn from Illing 1990)

The superficial layers contain a map of the visual field and are connected with cortical and subcortical visual centres. Visual projections to the deep layers are less prominent, but they receive a strong somatosensory and auditory input in addition to afferents from the cerebellum, the substantia nigra and extrastriate areas of the cerebral cortex.

Terminations of certain visual projections in the superficial layers and of the majority of the non-visual afferent fibres to the deep layers display a mediolateral periodicity. The superficial layers I and II display a strong, but evenly distributed AChE activity. In the stratum griseum intermedium, AChE is distributed in patches. This patchy distribution has been used to compare the lamination and the mediolateral periodicity in the termination of the different afferent systems of this layer. In reconstructions, this patchy distribution appears as an array of rostrocaudally directed, partially interconnected columns that is most pronounced in the caudal part of the superior colliculus (Fig. 22.90). It was first discovered in the superior colliculus of the cat and rhesus monkey (Graybiel 1978a; Graybiel and Ragsdale 1978; Illing 1988). Similar distributions were found for AChE in other species (Graybiel 1979: humans; Wiener 1986; Beninato and Spencer 1986; Wallace 1986a,b: rat and hamster). ChAT activity is high, and ChAT-immunoreactive fibres are concentrated in the superficial layers of the colliculus and in the AChE-positive patches of the intermediate grey stratum in the rat, cat and guinea pig (Jeon et al. 1993; Illing 1990; Schnurr et al. 1992; Ross and Godfrey 1985; Hall et al. 1989). Projections of putative cholinergic (ChAT-positive) axons to the patches were traced from the parabigeminal nucleus and the lateral dorsal and pedunculopontine tegmental nuclei of the pontomesencephalic reticular formation (see Fig. 22.94) in the cat, ferret, rat and guinea pig (Hall et al. 1989; McHaffie et al. 1991; Mufson et al. 1986). Generally, the distribution of AChE and ChAT immunoreactivity in the superficial layers of the superior colliculus is uniform, but its intensity varies in different species. A population of small ChAT-positive nerve cells has been found in the superficial tectal layers in the cat and guinea pig, but not in other species (Hall et al. 1989; Illing 1990; Schnurr et al. 1992). Similar, but certainly not identical, lattice-like distributions of oxidative enzymes (cytochrome oxidase, succinate dehydrogenase), reduced nicotinamide adenine dinucleotide phosphate (NADPH)-diaphorase and enkephalin occur in the intermediate grey stratum of different species (Wallace 1986a,b, 1988; Wallace and Fredens 1989; Graybiel et al. 1984; Wiener 1986).

Afferent connections of the superior colliculus can be classified according to their functional systems (subcortical or cortical; visual, somatosensory or auditory systems; hypothalamic, striatonigral or cerebellar), their laminar patterns of termination (visual systems, including the retinotectal projection and the projections from the striate and most extrastriate visual areas, in the superficial layers of the colliculus; somatosensory, auditory, hypothalamic and cerebellar afferents in the intermediate and deep strata; some extrastriate visual areas and the nigrotectal projection in both the superficial and deeper layers) and their neurotransmitters (GABAergic projections from substantia nigra, zona incerta, pretectum and perihypoglossal nuclei; cholinergic projections from the parabigeminal nucleus, the pedunculopontine nucleus and the dorsolateral nucleus in the ventral periaqueductal grey; noradrenergic projections from the locus coeruleus; serotonergic projections from the raphe nuclei; glutamatergic and/or aspartatergic projections from the retina, the cerebral cortex, the somatosensory and auditory relay nuclei and the cerebellar nuclei).

22.8.3.2.2
Retinotectal Projection

The projection from the retina to the superior colliculus has been extensively studied using anatomical and electrophysiological techniques (Fig. 22.91). In most mammals, the retinotectal projection is bilateral. Retinotectal fibres arise from all classes of ganglion cells. Ganglion cells giving rise to the ipsilateral pathway are located in the temporal retina, and in the rat, cat and opossum they include a higher proportion of larger ganglion cells than the contralateral pathway (Dräger and Olsen 1980; Wässle and Illing 1980; Illing and Wässle 1981; Rapaport and Wilson 1983; Hofbauer and Dräger 1985; Dreher et al. 1985; Dong and Rahman 1992; Moriya and Yamadori 1993). Within the superior colliculus, the fibres of the optic tract course in the superficial layer of the stratum opticum to terminate in the stratum griseum superficiale, where the individual fibres form nests that include groups of neurons (Cajal 1911; Sachs and Schneider 1984). These terminations are found in the superficial part of the stratum griseum superficiale (Sterling 1973; Ortega et al. 1993). Only a few optic nerve fibres were found to terminate below the stratum opticum in the intermediate layers (Beckstead and Frankfurter 1983).

The retinotectal projection in the cat is bilateral. Crossed fibres of the optic nerve terminate over the entire extent of the superior colliculus, with an interruption at the representation of the optic disc and less dense terminations and the appearance of gaps near the location of the area centralis (Fig. 22.92a3). In the caudal half of the colliculus, holes appear in the ventral tier of the band of terminals, and these sometimes correspond with patches in the ipsilateral retinotectal projection. The ipsilateral projection is much weaker than the contralateral one and spares the rostral and caudal poles of the colliculus. The area centralis only receives a weak ipsilateral projection, but a column of ipsilateral terminals marks the representation of the optic disc (Graybiel 1975; Hoffman et al. 1984; Harting and Guillery 1976; Behan 1982). Zhang and Hoffman (1993) studied the retinotectal projection in the ferret.

The map of the contralateral visual field in the superficial layers of the mammalian superior colliculus is oriented slightly obliquely, with the upper half of the visual field represented medially, the inferior half laterally, the temporal visual field caudally and the nasal field in the rostral part of the colliculus. In the cat, the representation in the rostral colliculus of the central 10° temporal to the vertical meridian is not expanded, compared to the repre-

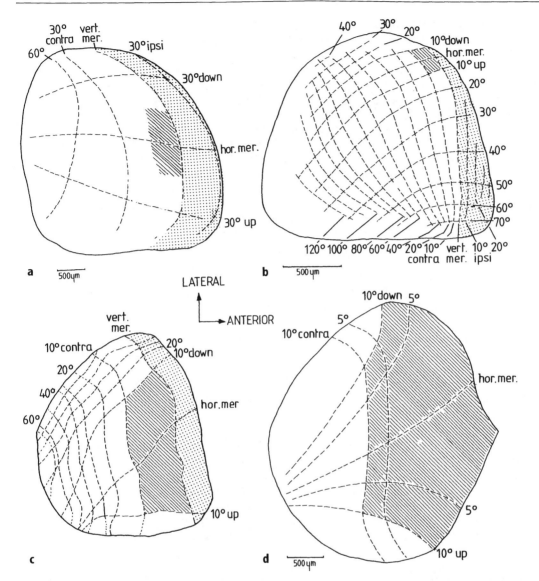

Fig. 22.91a–d. Visual field maps of the contralateral superior colliculus of **a** the South American opossum, *Didelphis marsupialis aurita* (redrawn from Rocha-Miranda et al. 1978), **b** the cat (redrawn from Feldon et al. 1970), **c** the mouse (redrawn from Dräger and Hubel 1976) and **d** the rhesus monkey, *Macaca mulatta* (redrawn from Cynader and Berman 1972). The *shaded area* rostral to the vertical 0° meridian in opossum, mouse and cat corresponds to the representation of the ipsilateral hemifield. The central 10° of the contralateral hemifield is *hatched*. For the opossum no precise data to delineate the central 10 0° were available. For the mouse and the monkey the original digrams of the right uperior colliculus were reversed

sentation of more peripheral parts of the visual field in the caudal colliculus. The representation is binocular, with the exception of the ipsilateral nasal and the contralateral monocular temporal fields, which are projected in the rostralmost and caudalmost sections of the colliculus, respectively (Fig. 22. a,b; Feldon et al. 1970; Stein 1981; Lane et al. 1974).

In the monkey, the central visual field (within 10° of the fovea) is expanded and occupies over one third of the surface of the superior colliculus, while the peripheral parts of the field are crammed into the remaining portion (Cynader and Berman 1972). The topology of the visual field maps in monkeys and cats is very similar (Fig. 22.91d), with the exception of the projection of the ipsilateral nasal field in the rostral colliculus, which was not observed in the monkey colliculus (Cynader and Berman 1972: *Macaca mulatta*; Kadoya et al. 1971: *Saimiri sciureus*; Lane et al. 1973: *Aotes trivirgarus* and *Galago senegalensis*). Optic nerve fibres in *Macaca mulatta* (Wilson and Toyne 1970) were found to terminate in pat-

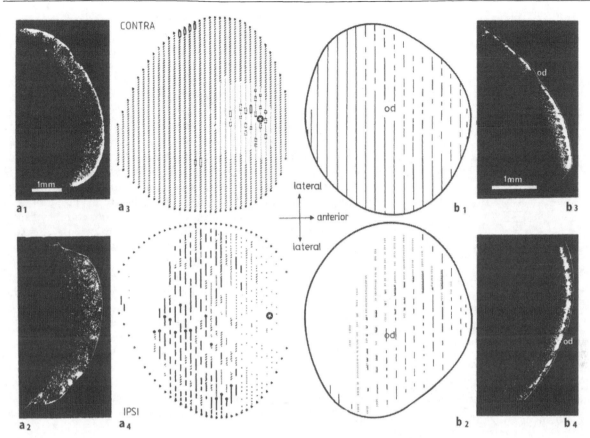

Fig. 22.92a,b. Dorsal views of reconstructions showing patterns of labelling in superficial grey layers of the ipsilateral and contralateral superior colliculus after injections of [³H]proline [³H]fucose in one eye in **a** the cat (Graybiel 1975) and **b** the rhesus monkey (Hubel et al. 1975). Labelling on the contralateral side is located superficially and is diffuse, as shown in the autoradiograms of **a**1 and **b**3. Gaps in the contralateral labelling in the cat superior colliculus are indicated by hollow rectangles (**a**3), and by interruptions of the lines in the colliculus of the monkey (**b**1). The ipsilateral side shows a puff-like pattern of labelling found mainly in the deep layer of the stratum griseum superficiale in the cat (**a**2) and extending into superficial layers in the monkey (**b**4). *Dense lines* indicate location of densely labelled puffs, *hatched lines* less densely labelled grain clusters. Diffuse labelling is shown by *dotted lines*. Area centralis is indicated by a *star*; optic disk by the abbreviation *od*

ches of terminals from one or the other eye in the superficial grey strata of the rostrolateral third of the colliculus that contains the representation of the fovea (Fig. 22.92b₁,b₂). The projections of both eyes to the middle third of the colliculus, which contains the representation of the periphery of the visual field, are complementary, with a uniform contralateral projection to the superficial layers. Tongues of terminals extend more deeply, where they enclose clumps of ipsilateral terminals. The representation of the optic disc is spared on the contralateral side, but is innervated ipsilaterally. The caudal pole of the colliculus, representing the temporal crescent of the visual field, only contains a superficial band of terminals from the contralateral retina (Hubel et al. 1975; Wilson and Toyne 1970). In *Galago crassicaudatus*, the retinotectal projections from both eyes are restricted to the superficial part of the superficial grey stratum, with the contralateral projection terminating superficial to the ipsilateral one (Tigges and Tigges 1970). A bilateral projection of the retina to the superficial layers of the colliculus superior was reported for the tree shrew *Tupaia glis*, with a clustering of the ipsilateral projection in the deep part of the superficial grey stratum, bordering on the stratum opticum (Hubel 1975). The visuotopic map in the contralateral retina is of the non-primate mammalian type, with a projection of the contralateral temporal retina, representing the ipsilateral nasal hemifield in the rostrolateral superior colliculus (Lane et al. 1971; Kaas et al. 1974).

The visuotopic projection in the rat and mouse is uniform and approximately linear (Fig. 22.91d). The orientation and the extent of the map of the visual field on the superior colliculus of the rat and mouse is rather similar to that of the cat and tree

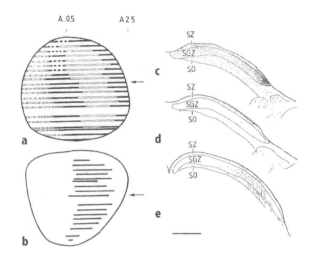

Fig. 22.93a-e. Superior colliculi of the opossum *Didelphis marsupialis aurita* a contralateral and b ipsilateral to the proline-injected eye, from Nissl-stained radioautograms. Anterior is to the *right*. *Open bars* indicate labelling in the stratum griseum superficale (SGZ); *black bars* indicate additional labeling of the stratum zonale (SZ). *Broken lines* denote uncertain labelling. c,d Parasagittal sections of c the contralateral and d the ipsilateral superior colliculus. Levels are indicated with *arrows* in a and b. (From Rocha-Miranda 1978). e Transverse section of the superior colliculus showing the distribution of degenerated fibres from a lesion restricted to the ipsilateral striate cortex (From Linden and Rocha-Miranda et al. 1978). *A*, anterior; *SO*, stratum opticum; *SZ*, stratum zonale. *Bar*, 1 mm

shrew. It includes the entire retina, with 40° of the ipsilateral nasal hemifield being represented in the rostral colliculus (Siminoff et al. 1966; Dräger and Hubel 1974, 1976; Diao et al. 1983). There is a mild anisotropic magnification of the central visual field in the coronal plane of the colliculus, i.e. for the vertical direction in the retina. A similar anisotropy has been observed in the squirrel and tree shrew, and the highest values have been observed in the rabbit. This is probably related to the degree of development of the horizontal streak of high ganglion cell density in the retina, which is most pronounced in the rabbit (Dräger and Hubel 1976). The retinotectal projection in the rat is mainly crossed; a small uncrossed projection from the ipsilateral retina terminates rostrolaterally in the projection area of the ipsilateral hemifield, where binocularly driven cells were found in the rat (Hayhow et al. 1962; Lund and Lund 1976; Lund et al. 1980; Diao et al. 1983). The retinotectal projection in the guinea pig (Giolli and Greel 1973) and rabbit (Giolli and Guthrie 1969) is very similar.

The visuotopic map of the superior colliculus in marsupials is similar to that of the rat (Fig. 22.91a). The vertical dimension of the representation of the retinal visual streak is also magnified with respect to the periphery. The map includes a representation of part of the ipsilateral hemifield (25° in the wallaby) in the rostral colliculus (Volchan et al. 1978: *Didelphis marsupialis autita*; Mark et al. 1993: the wallaby *Macropus eugenii*). The patterns produced by the retinotectal projections in the superficial collicular layers in *Didelphis* (Fig. 22.93) differ from those observed in cats and primates (Rocha-Miranda et al. 1978). In a rostral and a middle sector of the colliculus, the contralateral projection is located in the stratum zonale and the stratum griseum superficiale. The ipsilateral projection is restricted to a middle sector, where clumps of ipsilateral terminals in the stratum zonale are surrounded and bordered by a subjacent continuous band of contralateral terminals. The rostral sector of the colliculus, with its monocular contralateral projection, corresponds to the representation of the ipsilateral nasal 35° of the visual field (Volchan et al. 1978). This rostral sector receives a bilateral projection in *Didelphis virginiana* and *Marmosa mitis*: an ipsilateral projection to the stratum zonale of the rostral third of the colliculus and a contralateral projection to the subjacent superficial grey layer that extends over the entire colliculus and includes the stratum zonale in its caudal two thirds (Royce et al. 1976; Fig. 22.105).

Retinofugal projections to the superior colliculus in Megachiroptera (*Pteropus giganteus*) and Microchiroptera (*Myotis lucifugeus*) are bilateral, with the crossed projection terminating in two layers in the superficial grey stratum: a continuous, superficial layer and a deeper layer of clustered terminals. The ipsilateral projection terminates as clusters intermediate between the two ipsilateral layers (Pierson-Pentney and Cotter 1976; Cotter and Pierson-Pentney 1979). This sandwich structure in the retinal projection to the superficial grey stratum was not observed by Thiele et al. (1991) in another megachiropteran bat (*Rouettes aegyptiacus*). The bilateral projections to the superior colliculus in this species overlap in the superficial grey and the ipsilateral projection spares the rostral and caudal poles of the colliculus. The pattern of projection of the visual field onto the superior colliculus in *Rouettes* resembles the pattern found in most non-primates, with a representation of the contralateral hemifield and 25° of the nasal region of the ipsilateral field in the rostral, contralaterally innervated pole of the colliculus (Thiele et al. 1991). In *Pteropus*, the representation of the ipsilateral hemifield in the rostral superior colliculus is smaller (10–13°; Rosa and Schmid 1974).

Some information is available on the retinofugal projections in Monotremata. Both in the echidna

Tachyglossus (Campbell and Hayhow 1971) and the duckbill platypus *Ojrnithorhynchus* (Campbell and Hayhow 1972), optic fibres are distributed to the surface of the superior colliculus, terminating in its superficial grey, and do not enter the tectum through the stratum opticum. Although a small contingent of optic nerve fibres does not decussate in the optic chiasm, the retinocollicular projection in these species is completely crossed.

The general pattern of the innervation of the superior colliculus by the optic nerve is the same in insectivores. The superficial layer of the superior colliculus is densely innervated by the contralateral retina, and an ipsilateral projection is restricted to the rostromedial aspect of the colliculus (Mizuno et al. 1991: the musk shrew *Suncus murinus*). In the European hedgehog *Erinaceus europaeus*, the ipsilateral projection to the superior colliculus is more extensive (Dinopoulos et al. 1987). The retinal projections in the mole *Talpa europaea* (Lund and Lund 1965) and the fossorial moles of the Mogera group (Sato 1977) are much reduced. The superficial laminae of the superior colliculus are narrow, and the retinotectal projections are absent in *Talpa*.

Bilateral projections of the retina to the superficial layers of the superior colliculus occur in most mammals, with the possible exception of the monotremes and certain *Microchiroptera* (Thiele et al. 1991). The ipsilateral projection is small in rodents, insectivores and in the rabbit and is more extensive in primates, carnivores, chiropteres and marsupials. Clustering and/or interruption of the retinotectal projections to the superficial layers of the superior colliculus occurs in some species. A periodicity in the crossed and uncrossed projections in the projection area of the pericentral visual field is distinct in the macaque monkey, present in the cat and of a variable morphology in marsupials. A patchy distribution of ipsilateral retinal projection has been documented in primates, carnivores, Marsupialia and Chiroptera. The retinotectal projections in rodents, insectivores, *Galago* and the tree shrew *Tupaia glis* appear to be more uniformly distributed. The rostral pole of the superior colliculus of most mammals contains a representation of the nasal part of the ipsilateral hemifield; only in primates is the representation in the superior colliculus restricted to the contralateral hemifield.

22.8.3.2.3
Subcortical and Cortical Projections to the Superior Colliculus

Subcortical Projections to the Superficial Layers I–III. Subcortical projections to superficial layers of the superior colliculus were traced from the anterior pretectal nucleus, pars compacta, in the rat (Cadusseau and Roger 1991). Other pretectal nuclei innervating the superior colliculus are the nucleus of the optic tract, which gives rise to GABAergic and non-GABAergic projections (Nunes Cardozo and Wortel 1993; Nunes Cardozo et al. 1994), and the suprageniculate pretectal nucleus (Lagares et al. 1994), both in the rabbit. The laminar terminations of these projections have not yet been determined. The perihypoglossal nuclei project to superficial and deep strata of the contralateral superior colliculus. In the cat, the projection to the stratum griseum superficiale is derived from the nucleus intercallatus, and the projection to the stratum griseum intermedium is derived from the nucleus prepositus hypoglossi (Higo et al. 1992).

GABAergic projections from the substantia nigra to the superior colliculus (see Fig. 22.96) include the superficial layers (Hopkins and Niessen 1976: rat; May and Hall 1984: the grey squirrel; Huerta et al. 1991: *Galago*). In the cat, the pars lateralis of the substantia nigra projects bilaterally to the stratum griseum superficiale and the stratum opticum of the rostral superior colliculus (Harting et al. 1988).

An extensive bilateral cholinergic projection to the superficial layers of the superior colliculus that overlaps with the retinotectal and visual cortical projections to these layers has been traced from the parabigeminal nucleus in the cat (Graybiel 1978a; Hall et al. 1989), rat (Watanabe and Kawana 1979; Beninato and Spencer 1986), mouse (Mufson et al. 1986) and hamster (Jen et al. 1984). The ipsilateral projection extends over the entire superior colliculus and is derived from large cells of the middle subdivision of the nucleus. The crossed projection arises from smaller neurons of the dorsal and ventral subdivisions of the parabigeminal nucleus and decussates in the supraoptic commissure (Tokunaga and Otani 1978; Watanabe and Kawana 1979; Linden and Perry 1983; Schümann 1987; Jen et al. 1984). The contralateral projection is limited to the rostral part of the colliculus (Graybiel 1978a). A small number of neurons of the parabigeminal nucleus of the guinea pig are GABAergic (Hardy and Corvisier 1991). A reciprocal connection from the superior colliculus to the ipsilateral parabigeminal nucleus was first described in the cat by Van Noort (1969; see Sect. 22.8.3.2.4).

Projections from the cerebellar nuclei and the hypothalamus to the superficial layers of the superior colliculus are considered below together with the afferent projections to the intermediate and deep layers.

Subcortical Projections to the Intermediate and Deep Layers IV–VII. Direct connections from the retina to the intermediate layers of the colliculus are few. Indirect visual connections to the ventral tier of the intermediate grey stratum and deeper layers have been traced from the lateral subdivision of the ipsilateral ventrolateral geniculate in the cat (Swanson et al. 1974; Edwards et al. 1974), rat (Brauer and Schober 1982) and tree shrew (Conley and Friedrich-Escy 1993). The nucleus of the posterior commissure is the only pretectal nucleus that projects to the stratum griseum intermedium and profundum. Its terminations in the dorsal tier of the stratum griseum intermedium are clustered in patches, while the projections to the stratum griseum profundum are more diffuse.

Connections from the spinal cord, the dorsal column nuclei and the sensory nuclei of the trigeminal nerve to the external nucleus of the inferior colliculus, the intercollicular region and the stratum griseum intermedium and profundum of the superior colliculus have been noted in many studies of these pathways in different mammalian species. Crossed spinotectal fibres from all levels of the cord, including the lateral cervical nucleus (Van Noort 1969), terminate in patches in the lower half of the stratum griseum intermedium and the adjoining stratum album intermedium and, diffusely, in the deep layers of the caudal superior colliculus in the cat, rat, monkey and hedgehog tenrec (Wiberg and Blomqvist 1984; Wiberg et al. 1987; Antonetty and Webster 1975; Illing and Graybiel 1986; Yezierski 1988; Künzle 1993). A similar, mainly crossed projection has been traced from the dorsal column nuclei in the cat, rat and opossum and in primates (Berkley and Hand 1978; Hazlett et al. 1972; Lund and Webster 1967; Wiberg and Blomqvist 1984; Wiberg et al. 1987). Trigeminotectal fibres terminate in the same layers, with patches in the lower stratum griseum and album intermedium and more diffusely in the deep layers of the colliculus (see Fig. 22.102). In the rat, the trigeminal projection occupies the lateral part of the superior colliculus (Killackey and Erzurumlu 1981), and in the cat and monkey, it is restricted to its rostral half (Huerta et al. 1981; Wiberg et al. 1986, 1987). The projections are mainly crossed and derived from the non-laminar pars interpolaris of the spinal trigeminal nucleus. A bilateral projection from the principal sensory nucleus has been traced in rat and hamster (Jacquin et al. 1986; Rhoades et al. 1989). Subcortical and cortical somatosensory pathways probably project to the same clusters in the intermediate layers of the colliculus (see

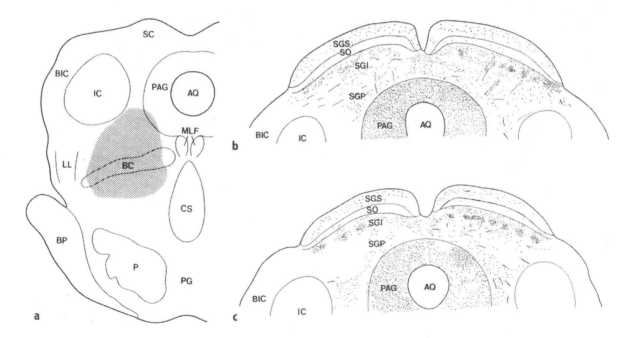

Fig. 22.94a–c. Distribution of presumed cholinergic fibres in two transverse sections through the superior colliculus of the cat (**b,c**) after an injection of wheat germ agglutinin-coupled horseradish peroxidase in the pedunculopontine nucleus (**a**). Compare Fig. 22.90b for the corresponding distribution of choline acetyltransferase in the superior colliculus of the cat. From Hall et al. (1989). *AQ*, aquaductus cerebri; *BC*, brachium conjunctivum; *BIC*, brachium of the inferior colliculus; *BP*, brachium pontis; *CS*, nucleus centralis superior; *LL*, lateral lemniscus; *MLF*, medial longitudinal fascicle; *P*, cerebral peduncle; *PAG*, periaquaductal grey; *PAG*, periaquaductal grey; *PG*, pontine nuclei; *SC*, superior colliculus; *SGI*, stratum griseum intermedium; *SGP*, stratum griseum profundum; *SGS*, stratum griseum superficiale; *SO*, stratum opticum

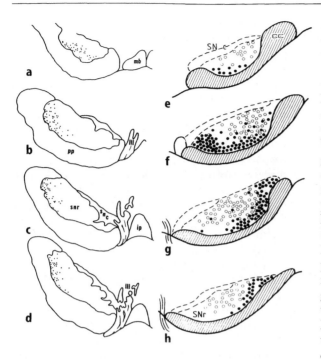

Fig. 22.95a–h. Distribution of neurons with projections to the superior colliculus in the substantia nigra pars reticulata in a–d *Macaca fascicularis* (from Beckstead and Frankfurter 1982) and e–h the rat (*filled circles*; from Faull and Mehler 1978). Neurons with projections to the thalamus in the substantia nigra of the rat are indicated by *open circles*. a and e are the rostralmost sections. *CC*, cerebral peduncle; *III*, oculomotor nerve; *ip*, interpeduncular nucleus; *mb*, mamillary body; *pp*, cerebral peduncle; *SNc*, substantia nigra, pars compacta; *SNr*, substantia nigra, pars reticulata

Fig. 22.102), i.e. in a complementary pattern with respect to the AChE-positive patches and columns in these layers (Illing and Graybiel 1986). Somatotopic maps in the deep layers of the colliculus will be discussed in relation to the corticotectal projection of the somatosensory cortex.

Auditory projections from the dorsal nucleus of the lateral lemniscus, the external nucleus of the inferior colliculus and the nucleus of the brachium of the inferior colliculus have been traced to the stratum griseum intermedium of the rat (Druga and Syka 1984; Tanaka et al. 1985; Wallace and Fredens 1989; Yasui et al. 1993; see also Fig. 22.83a).

Projections to deep layers of the colliculus in cat and rat that are presumed to be cholinergic (ChAT-immunoreactive) arise from the neurons in the cuneiform nucleus, the laterodorsal tegmental nucleus in the periaqueductal grey and the pars compacta of the pedunculopontine nucleus (Hoover and Jacobowitz 1979; Satoh and Fibiger 1986; Beninato and Spencer 1986; Jeon et al. 1993). The distribution of the patches of ChAT-immunoreactive fibres from the pedunculopontine nucleus of the cat (Figs. 22.90, 22.94) corresponds to the distribution of AChE in this layer (Hall et al. 1989).

The substantia nigra is the main source of a GABAergic projection to the deep layers of the colliculus. The inhibitory, GABAergic nature of this connection was established by Vincent et al. (1978), Chevalier et al. (1981), Araki et al. (1984) and Karabelas and Moschovakis (1985) using pharmacological and immunocytochemical methods in the rat and cat. A proportion of the nigrotectal cells in the rat co-localizes ChAT (Moriizumi et al. 1991). The projection arises from neurons of the pars reticulata and the pars lateralis of the substantia nigra (Fig. 22.95; Hopkins and Niessen 1976; Faull and Mehler 1978; Beckstead 1979). Originally, the populations of nigrotectal, nigrothalamic and nigrotegmental neurons were considered to be segregated in different regions of the pars reticulata of the rat and hamster substantia nigra (Faull and Mehler 1978), with the nigrotectal cells concentrated in the ventral pars reticulata, next to the cerebral peduncle, in a region receiving input from visually dominated portions of the striatum (Rhoades et al. 1982; Gerfen et al. 1982). In later studies in the rat, cat and primates, the origin of the nigrotectal projection was found to be more extensive, with a systematic overlap of the nigrothalamic and nigrotegmental projections (Beckstead and Frankfurter 1982; Harting et al. 1988; Harting and Van Lieshout 1991; Huerta et al. 1991; Deniau and Chevalier 1992; Redgrave et al. 1992; Tokuno et al. 1993). Some neurons in the pars lateralis of the rat that project both to the striatum and the superior colliculus were considered as displaced pars compacta cells (Takada et al. 1988).

The main projection of the pars reticulata is to the caudal two thirds of the ipsilateral stratum griseum intermedium, where it distributes in longitudinal bands (Fig. 22.96). In the cat, this distribution of the nigrotectal fibres is in register with the patches of AChE activity in this layer (Graybiel 1978b; Illing and Graybiel 1985). The projection of the pars reticulata to the stratum intermedium is laminated. The projection to the middle lamina is derived from the ventral pars reticulata, overlaps with the cholinergic pedunculopontine input and is complementary to the trigeminotectal projection to this sublamina. The dorsal pars reticulata projects to dorsal and ventral sublamina of the stratum griseum intermedium that do not receive pedunculopontine or trigeminal afferents (Figs. 22.90b, 22.102; Harting et al. 1988; Harting and Van Lieshout 1991). Similar, discontinuous projections to the ipsilateral stratum griseum intermedium are present in primates (Jayaraman et al. 1977), the rat (Beckstead et al.

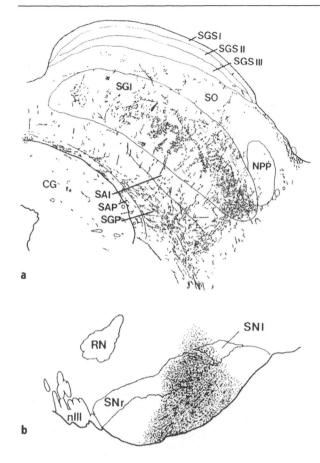

Fig. 22.96a,b. Distribution of labelled fibres in a a transverse section through the superior colliculus of the cat after b an injection of tritiated amino acids into the substantia nigra, pars reticulata. *CG*, central grey; *nIII*, oculomotor nerve; *NPP*, posterior pretectal nucleus; *RN*, red nucleus; *SAI*, stratum album intermedium; *SAP*, stratum griseum profundum; *SGI*, stratum griseum intermedium; *SGP*, stratum album produndum; *SGSI-III*, stratum griseum superficiale, sublaminae I-III; *SNl*, substantia nigra, pars lateralis; *Snr*, substantia nigra pars reticulata; *SO*, stratum opticum. (From Harting et al. 1988)

fibres terminate diffusely in the intermediate layers of the colliculus in the rat (Watanabe and Kawana 1982; Nicolelis et al. 1992; Kim et al. 1992) and cat (Rieck et al. 1986). Afferents from the nucleus prepositus hypoglossi have been documented in the rat and guinea pig (Hardy and Corvisier 1991; Ohtsuki et al. 1992). In the cat, they terminate in clusters in the stratum griseum intermedium (Higo et al. 1992; Corvisier and Hardy 1993).

Extensive projections from the cerebellar nuclei terminate in different sublayers of the colliculus. Projections to the stratum griseum superficiale were traced bilaterally from the fastigial nucleus (Kawamura et al. 1982; Gonzalo-Ruiz and Leichnetz 1987) and the contralateral lateral nucleus (Kurimoto et al. 1995) in the rat. Crossed projections from the posterior interposed nucleus terminate in two layers in the stratum griseum profundum and lower part of the stratum griseum intermedium in the cat (Kawamura et al. 1982; Hirai et al. 1982) and rat (Gonzalo-Ruiz and Leichnetz 1987; Gayer and Faull 1988; Kurimoto et al. 1995), with a contribution of the caudal dentate nucleus in the grey squirrel (May and Hall 1986) and in *Macaca fascicularis* (May et al. 1990). Bilateral projections from the fastigial nucleus are more prominent in species with frontal eyes. In *Macaca fascicularis*, this bilateral projection occupies the rostral pole of the intermediate grey (May et al. 1990). In the grey squirrel (May and Hall 1986) and the rabbit (Uchida et al. 1983), with laterally positioned eyes, an uncrossed projection from the fastigial nucleus is absent, and the contralateral projection is sparse.

Subsets of neurons from the dorsal raphe nucleus and the locus coeruleus project to the superior colliculus (Loughlin et al. 1986; Waterhouse et al. 1993). In the rat, coerulotectal projections preferentially terminate in the deep layers of the superior colliculus (Takemoto et al. 1978). The projection of the hypothalamus to the superior colliculus in the cat is organised in two layers of clusters in superficial and deep sublamina of the stratum griseum intermedium. A diffuse termination in the deep layers of the colliculus is continuous with a heavy projection to the periaqueductal grey (Rieck et al. 1986).

Cortical Projections. In the rat, cortical projections to the superior colliculus involve all of its laminae. The visual cortex projects in a topical manner to the stratum opticum and griseum superficiale and sparsely to the stratum zonale. More rostral cortical areas are connected with the stratum opticum and deeper layers of the colliculus (Lund 1964). Similar observations were made in the hamster (Lent 1982; So and Jen 1982), cat (summarised by Harting et al.

1979; Williams and Faull 1988) and the hamster (Rhoades et al. 1982). In some species, the nigrotectal projection to the stratum griseum intermedium is continuous (the grey squirrel *Sciurus carolinensis*: May and Hall 1984). A projection from the entopeduncular nucleus to the rostral part of the ipsilateral superior colliculus, which may be complementary to the projection of the nigrotectal projections, was found in the rat (Takada et al. 1994). This connection is probably GABAergic and terminates in patches in the stratum griseum intermedium and more diffusely in the stratum griseum profundum.

The zona incerta and the nucleus prepositus hypoglossi are two other sources of GABAergic projections to the intermediate grey. Incertotectal

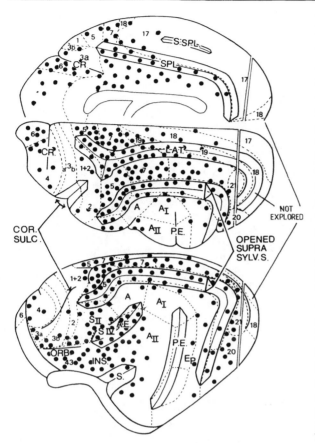

Fig. 22.97. Diagram of the medial, dorsal and lateral aspect of the cerebral hemisphere of the cat, showing the distribution of cortical layer V neurons with projections to the tectum and the pontine nuclei. Their distribution rouphly corresponds with the distribution of all corticotectal neurons. The approximate position of the cytoarchitectonic areas is indicated. A (I,II), acoustic area (I,II); AE, anterior ectosylvian sulcus; CR, cruciate sulcus; LAT, lateral sulcus; ORB, orbital sulcus; PE, posterior ectosylvian sulcus; SPL, splenial sulcus; SS, suprasylvian sulcus; SSPL, supraspleneal sulcus; SYLV.S, sylvian sulcus. (Redrawn from Keizer et al. 1987)

1992), rabbit (Giolli et al. 1978), various Marsupialia (Martin 1968; Martin et al. 1970), Edentata (*Dasypus novemcinclus*: Harting and Martin 1970) and in primates. Kuypers and Lawrence (1967) and Wilson and Toyne (1970) found in the rhesus monkey that occipital (striatal) cortical areas project to the superficial layers and temporal, pericentral and frontal cortical areas project to the layers below the stratum opticum.

Retrograde tracer injections in the superior colliculus labelled layer V pyramidal cells in most areas of the neocortex in all species (Holländer 1974; Kawamura and Konno 1979; Tortelly et al. 1980; Catsman-Berrevoets et al. 1979; Schofield et al. 1987). These cells are located superficial to pyramidal cells with projections to the spinal cord and the lateral geniculate body (Killackey et al. 1989), but extensive collateral branching of pyramidal cells to the superior colliculus and to the basal pontine nuclei and/or the pyramidal tract has been documented in the cat (Fig. 22.97; Keizer et al. 1987).

Projections of areas 17–19 in the cat are focused and restricted to the stratum zonale, the stratum griseum superficiale and the superficial stratum opticum, with small laminar differences between the areas (Fig. 22.98). Posterior suprasylvian visual areas 20 and 21 project slightly deeper in the superficial grey, the stratum opticum and the intermediate grey. Dorsal lateral and anterolateral suprasylvian visual areas project both to the superficial and deep layers of the colliculus. The projection of the parietal cortex is restricted to deep strata (Kawamura et al. 1974; Updyke 1977; Norita et al. 1991; Harting et al. 1992; Maekawa and Ohtsuka 1993). The distribution of the terminals from suprasylvian visual areas in the stratum griseum intermedium does not match the distribution of AChE in this layer (Illing and Graybiel 1986). In primates, the middle temporal visual area in the caudal superior temporal sulcus, which codes for parameters of moving visual stimuli and connects with parietal cortical areas, projects to deep and superficial strata of the colliculus (Maioli et al. 1992). The projection of the parietal cortex in *Macaca mulatta* to the stratum griseum intermedium is very similar to that in the cat (Baizer et al. 1993). Inferior temporal visual areas, involved in pattern recognition, project to the stratum griseum intermedium and profundum (Maioli et al. 1992; Webster 1995; Steele and Weller 1993). Striate and extrastriate visual cortical areas in the rat project in a topical manner to the superior colliculus (Sefton et al. 1981; Olavarria and Sluijters 1982), but their precise laminar projections have not yet been determined.

Frontal and prefrontal corticotectal projections in the cat take their origin from area 4 in the ventral bank of the cruciate sulcus, area 6, the gyrus proreus, the presylvian eye field, the medial frontal cortex and the rostral gyrus cinguli (Fig. 22.97; Keizer et al. 1987). Projections from caudal area 4 and somatosensory areas SI, SII and SIII are few, but a strong projection is derived from SIV in the anterior ectosylvian sulcus (Stein et al. 1983; McHaffie et al. 1988). All these areas project to the stratum griseum intermedium and to deep layers of the colliculus (Fig. 22.98). In the stratum griseum intermedium, these fibres terminate diffusely in its rostral part. More caudally, these fibres distribute in a laminar manner to the superficial and deep tiers and in patches or columns to the middle tier of the stratum griseum intermedium (Segal et al. 1983; Hartwich-Young and Weber 1986; Harting et al.

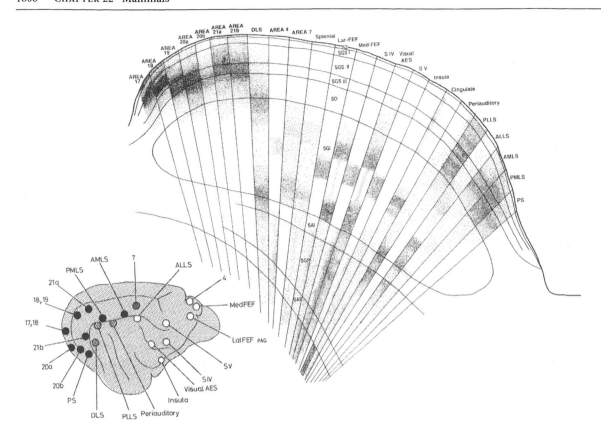

Fig. 22.98.) Laminar organization of corticotectal projections in the cat. The laminar distribution of the labelling from 24 small injections of tritiated aminoacids in different areas of the cerebral cortex is indicated in a transverse section through the right superior colliculus. From Harting et al.(1992). Visual areas located in the occipital part of te hemisphere project to superficial layers (*filled circles* in diagram of the right hemisphere); areas in the posterior ectosylvian and periauditory belt region (*hatched circles*) project to superficial, intermediate and deep layers; area 7 (*double hatched*) projects to intermediate layer and more rostral areas (*open circles*) project in different laminar patterns to the intermediate and deep layers of the colliculus. ALLS, anterolateral lateral suprasylvian area; AMLS, anterolateral medial suprasylvian area; DLS, dorsal lateral suprasylvian area; *Lat-FEF*, lateral frontal eye field area; *Med-FEF*, medial frontal eye field area; PAG, periaquaductal grey; PLLS, posterolateral lateral suprasylvian area; PMLS, posteromedial lateral suprasylvian area; PS, posterior suprasylvian area PTN, pretectal nuclei; S IV, V, 4th, 5th somatosensory cortical area; SAI, stratum album profundum; SAP, stratum album prfundum; SGI, stratum griseum intermedium; SGP, stratum griseum profundum; SGSI–III, stratum griseum superficiale, sublaminae I–III; SNpr, substantia nigra pars compacta SO, stratum opticum; *Visual AES*, visual anterior ectosylvian area

1992). With respect to the distribution of AChE in the caudal stratum griseum intermedium, the patches and the linear arrays of terminals from SIV are distributed in a non-overlapping, complementary manner (see Figs. 22.90, 22.102; Illing and Graybiel 1986). However, the projections from the presylvian eye field and the prefrontal cortex of the gyrus proreus closely match the distribution of AChE in this layer (Illing and Graybiel 1985). Clustering of terminals in the stratum griseum intermedium and profundum is also a prominent feature of the projection of the prearcuate frontal eyefield (area 8) and the lateral and medial prefrontal cortex in primates, and of the prefrontal (dorsomedial thalamic nucleus-receiving) and rostral cingulate cortex of the rat (Beckstead 1979; Wyss and Stipanidkulchai 1984; Neafsey et al. 1986; Zeng and Stuesse 1993) and rabbit (Buchanan et al. 1994). The insular cortex in rat and cat projects to deep layers of the superior colliculus, adjoining the projections of this area to the periaqueductal grey (Shimizu and Norita 1991; Neafsey et al. 1986).

Somatosensory areas in the cat (SIV; Stein et al. 1983) and the rat (SI: Wise and Jones 1977; Kassel 1982; Killackey and Erzurumlu 1981) project in a somatotopical pattern to the stratum griseum intermedium (album intermedium in the rat) and the stratum griseum profundum. The projection to the stratum album intermedium in the rat is distinctly clustered. In the cat, the representation of the face in the rostrolateral part of the colliculus and of the forelimbs caudolaterally is large. The hindlimb occupies a small area in the caudal colliculus (Stein et al. 1983). In rodents, the representation of the vibrissae in the rostrolateral colliculus is magnified (Dräger and Hubel 1974). Neurons activated by the stimulation of the somatosensory cortex generally could also be activated by peripheral somatosensory stimuli of topographically corresponding body regions (Kassel 1982). The somatic representation in the deep layers of the colliculus is in rough topographical register with the representation of the visual field in the superficial layers. Nasal receptive fields are found rostrally, overlying somatic cells with their receptive fields on the face, whereas the temporal visual field and somatic cells with receptive fields on the posterior part of the body are represented in the caudal parts of the colliculus. Superior regions of the visual fields are represented medially in the colliculus, and inferior portions of the visual field and the body are located laterally (Stein 1981). An exquisite topographical register between visual fields and the representations of the individual vibrissae has been shown to exist in the mouse (Dräger and Hubel 1974, 1976).

22.8.3.2.4
Efferent Connections of the Superior Colliculus

The main efferent connections of the superior colliculus comprise the crossed descending predorsal bundle, the ipsilateral tectopontine tract, the reciprocal connection with the parabigeminal nucleus and the ascending connections with the pretectum, the subthalamus and the thalamus. The differential origin of some of these systems from the superficial visual and the deep multimodal layers and the possible correspondence between afferent compartmentalisation in the intermediate layers and the origin of specific efferents are of particular interest. Intrinsic systems include the connections between superficial and deep layers and the commissure of the colliculus.

The predorsal bundle connects the deep layers of the superior colliculus with contralateral gaze centres and the spinal cord. The large and small fibres of the predorsal bundle course from their origin in the stratum griseum intermedium and profundum, in the direction of the periaqueductal grey, where they turn ventrally and medially to decussate in the dorsal tegmental decussation (Figs. 22.22–22.24). Before their decussation, they give off thin collaterals that ascend to the (sub)thalamus (cat: Bucher and Bürgi 1950; Grantyn and Grantyn 1982; Moschovakis and Karabelas 1985; Figs. 22.99, 22.100). The crossed fibres descend along the midline, ventral to the medial longitudinal fascicle. Fibres of the predorsal bundle terminate in the medial reticular formation of the pons and the medulla oblongata, including the rostral nucleus gigantocellularis, the

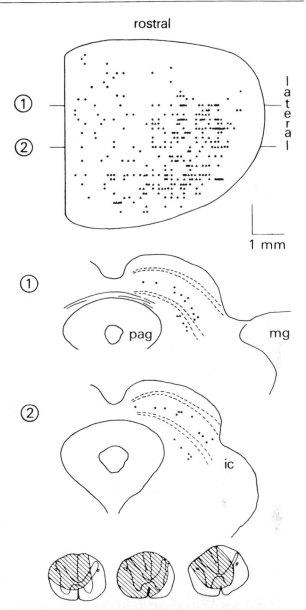

Fig. 22.99. Reconstruction of the right superior colliculus (as seen from a dorsal view) in the cat, demonstrating the location of neurons retrogradely labelled from the rostral cervical spinal cord. Sections *1* and *2* are transverse sections through the superior colliculus at the levels indicated. Labelled neurons occur in both the stratum griseum intermedium (*i*) and stratum griseum profundum (*p*). Note the high relative retrograde labelling density in the caudolateral quadrant of the colliculus. *a*, stratum album intermediate; *ic*, inferior colliculus; *mg*, medial geniculate nucleus of the thalamus; *p*, stratum griseum profundum; *pagm*, periaqueductal grey; *s*, stratum griseum superficiale. (From Murray and Coulter 1982)

caudal nucleus reticularis pontis oralis, the rostral nucleus reticularis pontis caudalis and the region surrounding the abducens nucleus and the nucleus reticularis tegmenti pontis. In the medullary reticular formation, their terminations include the lateral

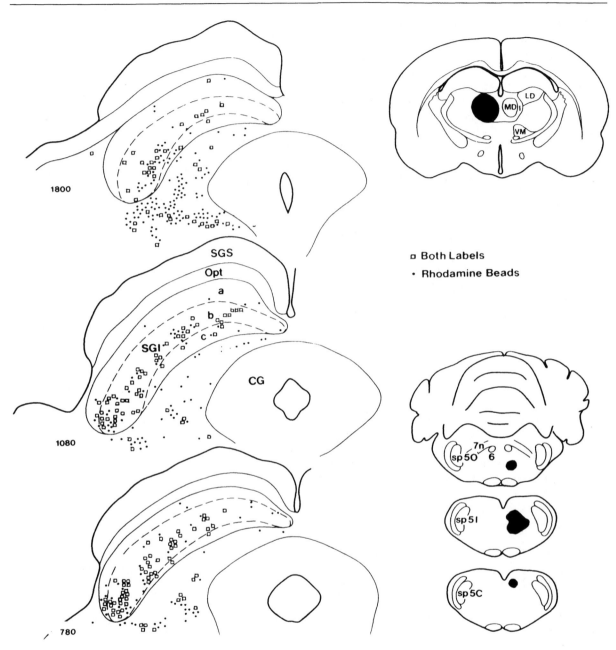

Fig. 22.100. Localization of neurons with descending axons in the contralateral predorsal bundle and collaterals to the ipsilateral intralaminar nuclei in three transverse sections through the superior colliculus of the rat; *1800* is most *rostral* section. Double-labelled neurons (*open squares*) are located in clusters in the stratum griseum intermedium and were filled from an injection of Fluoro-Gold in the thalamus (*upper right*) and an injection of rhodamine-labeled microspheres in the medial pontine reticular formation (*lower right*). *6*, abducens nucleus; *7n*, facial nerve; *CG*, central grey; *I*, intralaminar nuclei; *LD*, lateral dorsal nucleus; *MD*, mediodorsal nucleus; *Opt*, stratum opticum; *SGI (a,b,c)*, sublaminae a–c of the stratum griseum intermedium; *SGS*, stratum griseum superficiale; *Sp5 (0,I,C)*, spinal nucleus of the trigeminal nerve (pars oralis, interpolaris or caudalis). (From Bickford and Hall 1989)

and paramedian reticular nucleus and the inferior olive. Terminations were also found on motoneurons of the facial nerve nucleus and the nucleus prepositus hypoglossi.

In the spinal cord of the cat, the tectospinal tract is located in the ventral funiculus, ventral to the medial vestibulospinal tract and the reticulospinal components of the medial longitudinal fascicle (Fig. 22.20). It terminates on cells of the spinal nucleus of the accessory nerve and in the intermediate grey (laminae V–VIII, mainly VII and VIII). Connections of the superior colliculus with spinal motoneurons innervating neck muscles are mainly indirect, through reticulospinal fibres from the nucleus gigantocellularis and reticularis pontis caudalis. The tectospinal tract in the cat descends to C6 (Bucher and Bürgi 1950; Bürgi 1957; Nyberg-Hansen 1964; Kawamura et al. 1974; Rose et al. 1991; Holstege and Cowie 1992), in the North American opossum to C4 (Martin 1969) and in the rat (Waldron and Gwyn 1969; Redgrave et al. 1987a,b) and hedgehog tenrec (Künzle 1992) to the cervical cord. In the rabbit (Holstege and Collewijn 1982), the monkey Macaca mulatta (Harting 1977) and the tree shrew (Harting et al. 1973), few if any fibres of the predorsal fascicle reach the cord.

Tectopontine fibres descend ipsilaterally in the lateral tegmentum of the mesencephalon to terminate caudally in the dorsolateral pontine nuclei (Münzer and Wiener 1902). These classic observations in the rabbit were confirmed and extended in many other species (cat: Bucher and Bürgi 1950; Kawamura and Brodal 1973; Graham 1977; Mower et al. 1979; Cowie and Holstege 1993; opossum: Martin 1969; rat: Waldron and Gwyn 1969; Burne et al. 1981; Redgrave et al. 1987a,b; rabbit: Holstege and Collewijn 1982; Wells et al. 1989; tree shrew: Harting et al. 1973; Weber and Harting 1980; monkey: Frankfurter et al. 1976; Harting 1977). Apart from the pontine nuclei, the ipsilateral projections of the superior colliculus include parts of the periaqueductal grey, the external nucleus of the inferior colliculus, the reticular formation of the mesencephalon, with the cuneiform and pedunculopontine nuclei, the ventrolateral reticular formation of the pons and the medulla oblongata. Terminations in the medial part of the facial nerve nucleus were noticed in the opossum (Martin 1969). Terminations in the ipsilateral inferior olive occur in the opossum and rabbit (Holstege and Collewijn 1982).

Tectopontine fibres terminate in the ipsilateral dorsolateral pontine nuclei at middle and caudal levels of the pons. Caudally projections from the inferior colliculus overlap with those from the superior colliculus (cat: Kawamura and Brodal 1973; Mower et al. 1979; rat: Burne et al. 1981).

Visual corticopontine fibres terminate more rostrally in the dorsolateral pons in the rat, rabbit and monkey (Burne et al. 1981; Glickstein et al. 1982; Wells et al. 1989), but not in the cat (Kawamura and Brodal 1973; Kawamura et al. 1974). At middle pontine levels, the tectopontine projection overlaps with the projection from the pretectum (Burne et al. 1981). The tectopontine projection in primates is less extensive than in other mammals (Glickstein 1982; Redgrave et al. 1987a). Tecto-olivary fibres terminate in the caudal half of the contralateral medial accessory olive, medial to the β-group (rat, subnucleus C: Hess 1982; Huerta et al. 1983; Akaike 1992; monkey: Frankfurter et al. 1976; Harting 1977; cat: Weber et al. 1978; Kyuhoe and Matzuzaki 1991; rabbit: Holstege and Collewijn 1982).

Ascending projections of the superior colliculus were subdivided by Casagrande et al. (1972), Harting et al. (1973) and Diamond et al. (1991) in the tree shrew into the projections from superficial layers which receive projections from the retina and the striate cortex (stratum zonale, stratum griseum superficiale and stratum opticum) to visual centres (the pulvinar-lateral posterior complex, the pretectum and the dorsal and ventral geniculate nuclei) and from the deeper multimodal layers to the intralaminar nuclei, the subthalamus and the posterior nuclear group of the thalamus. The comparative anatomy of the projections to the lateral geniculate was reviewed by Harting et al. (1991). The possible collateral origin of the tecto-subthalamic projection from the predorsal bundle has been mentioned in connection with the predorsal bundle. Because the pulvinar is the main target of the ascending projection of the superficial layers of the superior colliculus, the tectocortical projection bypasses the lateral geniculate body and focuses on extrastriate visual areas. Similar observations were made in the cat (Bucher and Bürgi 1950; Bürgi 1957; Graham 1977; Edwards 1980), opossum (Benevento and Ebner 1970; Martin 1969), rhesus monkey (Harting et al. 1980; Graybiel 1972), rabbit (Holstege and Collewijn 1982), rat (Perry 1980; Mason and Groos 1981; Pasquier and Villar 1982; Sugita et al. 1983; Yamasaki et al. 1986; Groenewegen 1988; Kim et al. 1992), hedgehog and Galago (Harting et al. 1972).

The laminar origin of the different efferent systems and the modular organisation of the intermediate grey stratum deserves special attention. In the tree shrew, projections from the superficial layers of the superior colliculus to the pulvinar-lateralis posterior complex and the dorsal lateral geniculate body are segregated in deep and superficial laminae of the stratum griseum superficiale, respectively (Huerta and Harting 1984). Similar observations were made for many other mammalian species

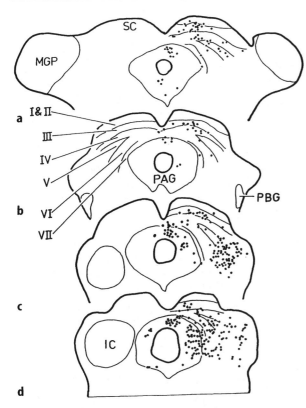

Fig. 22.101a–d. The positions of tectopontine cells at various levels of the superior colliculus after an injection of horseradish peroxidase in the dorsolateral pontine nucleus. Each of these *diagrams* represents a composite of three successive mounted sections, and each filled cells is represented by a *dot*. The demarcation of collicular layers and nuclei is approximate since drawings are composites. *BP*, brachium pontis; *I–VII*, lamina I–VII of the superior colliculus; *P*, pons; *p*, pyramid; *PGR*, pontine nuclei; parabigeminal nucleus; *SC*, superior colliculus; *TB*, trapezoid body. Redrawn from Mower et al. (1979)

(Sugita et al. 1983; Harting et al. 1991). Superficially located tectopontine neurons occupy an intermediate position in the stratum griseum superficiale (Fig. 22.101; Huerta and Harting 1984; Wells et al. 1989; Simp and Donate-Oliver 1991). The parabigeminal nucleus, which provides the superficial layers of the superior colliculus with a bilateral, cholinergic projection, receives a reciprocal connection mainly from tectal neurons of the ipsilateral stratum griseum superficiale and opticum (cat: Graybiel 1978b; Sherk 1979; tree shrew: Harting et al. 1973; monkey: Harting 1977; Tokunaga et al. 1981; rat: Linden and Perry 1983; hamster: Jen et al. 1984).

Predorsal bundle cells occupy the stratum griseum and album intermedium. Tectoreticular cells are small and medium-sized, while tectospinal cells are larger. In the cat, they are located laterally in clusters of three to five tectospinal neurons (Fig. 22.99; Kawamura and Hashikawa 1978; Huerta and Harting 1982; Murray and Coulter 1982; Redgrave et al. 1986). Neurons projecting to the contralateral periabducens reticular formation are situated more medially (Redgrave et al. 1992). In macaque monkeys the neurons projecting to the contralateral pontine reticular formation are located superficially to the tectospinal neurons (May and Porter 1992). Neurons with projections to the intralaminar nuclei are found among the predorsal bundle cells (Fig. 22.100; Bickford and Hall 1989). Generally, the predorsal bundle cells are more or less confined to the stratum griseum intermedium in the rat, squirrel and opossum, but are more widely scattered over the stratum griseum intermedium and profundum in the cat and monkey (Redgrave et al. 1986, Bickford and Hall 1989). In macaque monkeys they occupy the ventral tier of the stratum griseum intermedium. Both neurons with crossed projections to the inferior olive and neurons with projections to the ipsilateral tegmentum, the pontine nuclei and the cuneiform nucleus are located in the superficial stratum griseum intermedium and in the deep strata of the colliculus (Henkel et al. 1975; Huarta and Hall 1982; Redgrave et al. 1986, 1987a,b; Bickford and Hall 1989; Jeon and Mize 1993).

Almost complete overlap exists between the predorsal bundle cells and the neurons projecting to contralateral periabducens area and the GABAergic projection from the substantia nigra to the intermediate grey stratum of rodents (May and Hall 1984; Bickford and Hall 1992; Redgrave et al. 1992). These neurons receive excitatory connections from the cerebellar nuclei, namely from the posterior interposed nucleus (see Sect. 22.8.3.2.3).

The patchy distribution of AChE, ChAT and other markers in the intermediate layers of the superior colliculus in the cat and rat has been used to compare the clustering of afferent and efferent connections of these layers. Illing (1992, 1993) suggested that the patches of AChE-rich neuropil correspond with the sites of termination of the GABAergic nigral fibres and pedunculopontine, ChAT-positive and prefrontal projections (Figs. 22.90, 22.102). These compartments, which are associated with motor systems, alternate with AChE-negative and parvalbumin-positive domains, which receive sensory afferent projections from the nuclei of the trigeminal nerve, and the somatosensory and visual association cortex (Fig. 22.102). The clustering of groups of tectospinal neurons in the cat has been found to coincide with the AChE-positive patches (Jeon and Mize 1993), and neurons receiving input from and projecting to the spinal trigeminal nucleus were reported to occupy the AChE-negative compartments (Huerta and Harting

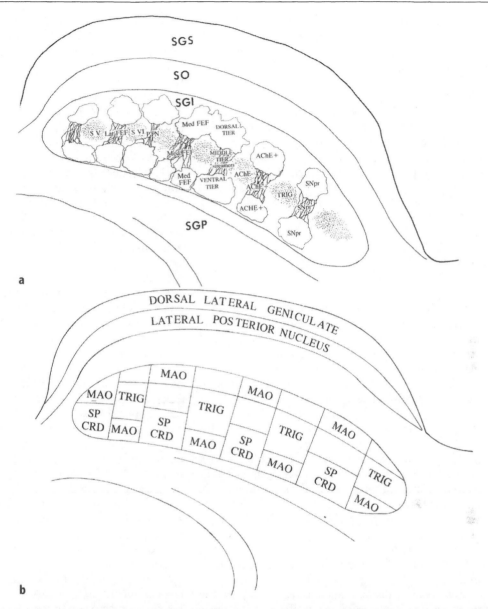

Fig. 22.102. a The spatial relationships of several afferent systems to the two tiers of acetylcholinesterase (AChE)-positive and AChE-negative patches and the interconnecting AChE-positive "streamers" of the stratum griseum intermedium of the cat superior colliculus. Compare Fig. 22.90. While a particular input or cholinergic-rich or -poor zone is labelled only once in its specific region within a tier or tiers, it occurs across the entire tier. For instance, trigeminal (TRIG) inputs and afferents from fourth and fifth somatosensory areas (SIV and SV) innervate the same AChE-negative patches across the middle tier. For abbreviations see Fig. 22.98. From Harting et al. (1992). b Superior colliculus showing the compartmentation of neurons with different projections in the stratum griseum superficiale and intermedium. MAO, tecto-olivary neurons; SP CRD, predorsal bundle and tectospinal neurons; TRIG, tecto-trigeminal neurons. Redrawn from Huerta and Harting (1984)

1984). However, such a precise correlation between efferent neurons and neurochemical compartmentalisation of the intermediate grey was not confirmed by Illing (1992), who found predorsal bundle cells and neurons projecting to the intralaminar nuclei and the pons to be located at the borders rather than being confined to the AChE-positive and -negative compartments of the rat. Similarly, Ma et al. (1991) were unable to find a correlation between the distributions of AChE and saccade-related neurons in the superior colliculus of the rhesus monkey.

Commissural connections are mainly confined to the intermediate and deep layers and are more numerous at rostral levels of the superior colliculus. Some of the commissural fibres continue into the dorsal mesencephalic reticular formation (opossum: Martin 1969; cat: Edwards 1977, 1980; rat: Waldron and Gwyn 1969; Yamasaki et al. 1984; Bickford and Hall 1989). Intrinsic connections between superficial and deep layers have mostly been described as collaterals of projection neurons (Moschovakis and Karabelas 1985; Moschovakis et al. 1988a,b).

22.8.3.2
Pretectum

22.8.3.3.1
Structure and Afferent Connections

The pretectum is located at the level of the posterior commissure, between the superior colliculus caudally and the thalamus rostrolaterally. Medially and ventrally, it borders on the periaqueductal grey and the nucleus of the posterior commissure; laterally, it is bordered by the fibres of the brachium of the superior colliculus. The cytoarchitecture of the pretectum and its nomenclature were reviewed by Fuse (1936), Scalia (1972) in the rat, mouse, rabbit and tree shrew), Kanaseki and Sprague (1974, Fig. 22.87) and Avendaño and Junetschke (1980), both in the cat, Weber (1985), Hutchins and Weber (1985) and Weber (1985) in *Saimiri sciureus* and the tree shrew (Figs. 22.103) and Benevento and Standage (1983) in *Macaca fascicularis*. The five main nuclei of the pretectum are the anterior, posterior and medial pretectal nuclei, the olivary pretectal nucleus of Fuse (1936) and the nucleus of the optic tract.

The olivary pretectal nucleus is a relay in the pupillary light reflex, and the nucleus of the optic tract constitutes a centre for horizontal optokinetic reactions. The anterior and posterior pretectal nuclei receive visual and somatosensory inputs and project to the pulvinar, the lateral posterior nucleus of the thalamus and to several precerebellar nuclei. The borders between the nuclei are often difficult to define, and many discrepancies in the descriptions of the connections in different species may be due to differences in the interpretation of the nuclear borders rather than to differences between the species. The posterior commissure contains decussating and commissural fibres from the ventral nucleus of the lateral geniculate body, the pretectum and the accessory optic system and fibres from the cord, the somatosensory relay nuclei of the brain stem and the cerebellum, on their way to the contralateral pretectum and the thalamus. The nucleus of the posterior commissure is sometimes included in the pretectum. It differs from the other nuclei of the pretectum in that it does not receive retinal, visual cortical or tectal projections (Weber and Harting 1980).

The nucleus of the optic tract contains large cells located between the fibres of the brachium of the superior colliculus, where it covers the pretectum and, more caudally, where it enters the stratum opticum laterally. The olivary pretectal nucleus is a compact nucleus, located medial to the nucleus of the optic tract, within the brachium of the superior colliculus. The neurons of the pretectal olivary nucleus are located at its periphery, surrounding a dense neuropil. In primates, the contour of the olivary pretectal nucleus is indented. A narrow tail of the olivary pretectal nucleus extends caudally between the nucleus of the optic tract and the posterior pretectal nucleus (Fig. 22.103). The rostral pole of the olivary pretectal nucleus can be distinguished in most species, although it is more distinct in primates (Fig. 22.106). The posterior pretectal nucleus is located ventral to the olivary pretectal nucleus. The lateral border region of the pretectum is occupied by the anterior pretectal nucleus. The medial pretectal nucleus is located medial to the olivary pretectal nucleus and dorsal to the fibres of the posterior commissure. Caudally, the tail of the olivary pretectal nucleus and the posterior pretectal nucleus shift laterally, and the anterior pretectal nucleus is displaced ventrally. The complex of the nucleus of the optic tract and the olivary, posterior and anterior pretectal nuclei becomes located lateral to the superior colliculus, where the posterior pretectal nucleus fuses with the intermediate grey stratum of the colliculus (Figs. 22.103–22.106).

Retinopretectal fibres from ganglion cells with large dendritic fields (Rodieck and Watanabe 1993) terminate bilaterally in the olivary pretectal nucleus in rodents (Fig. 22.104; Hayhow et al. 1962; Scalia 1972; Giolli and Greel 1973; Campbell and Lieberman 1985; Young and Lund 1994), including microphthalmic species (Cooper et al. 1993; Herbin et al. 1994), shrews (Mizuno et al. 1991), carnivores (Berman 1977; Berkley and Mash 1978; Hoffmann et al. 1984; Hutchins and Weber 1985; Zhang and Hoffmann 1993), marsupials (Figs. 22.103, 22.105; Benevento and Ebner 1970; Royce et al. 1976) and primates (Fig. 22.106; Hendrickson et al. 1970; Pierson and Carpenter 1974; Tigges et al. 1977; Tokunaga et al. 1981; Weber et al. 1981; Hutchins and Weber 1985; Weber 1985; Benevento and Standage 1983). In pigmented and albino rabbits (Klooster et al. 1983), in the mole *Talpa europea* (Lund and Lund 1965) and in monotremes (Campbell and Hayhow

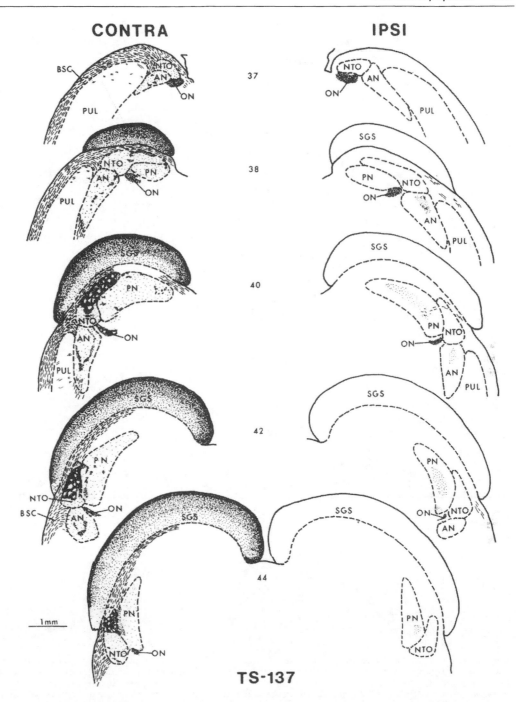

Fig. 22.103. Transverse sections illustrating the locations of transported label after an injection of [^{3}H]proline in one eye, over the pretectal complex and the superior colliculus of the tree shrew *Tupaia glis*. Terminal label is indicated by *black stippling* and labelled fibres are shown as *dashed lines*. Notice bilateral labelling in the olivary pretectal nucleus. *AN*, anterior pretectal nucleus; *BSC*, brachium of the superior colliculus; *NTO*, nucleus of the optic tract; *ON*, olivary pretectal nucleus; *PN*, posterior pretectal nucleus; *PUL*, pulvinar; *SGS*, stratum griseum superficiale. From Weber (1985)

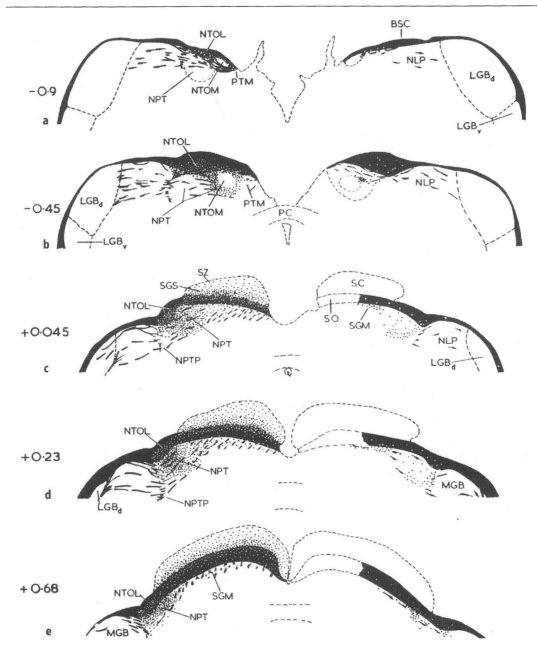

Fig. 22.104a–e. Semidiagrammatic projection drawings of Nauta-stained coronal sections through the pretectal and tectal regions in the rat 7 days following a right unilateral optic nerve section. Contralateral degeneration is represented on the *left* side of the figures. **a–e** form an anteroposterior series, the accompanying *number* indicating the distance in millimetres of the section anterior (-) or posterior to (+) the rostral pole of the superior colliculus. Note bilateral, complementary labelling in the olivary pretectal nucleus (*NTOM*). *BSC*, brachium of the superior colliculus; *LGBd*, dorsal nucleus of the lateral geniculate body; *LGBv*, ventral nucleus of the lateral geniculate body; *MGB*, medial geniculate body; *NLP*, lateral posterior nucleus; *NPT*, pretectal nucleus; *NPTP*, posterior pretectal nucleus; *NTOL*, lateral nucleus of the tractus opticus; *NTOM*, medial nucleus of the optic tract (olivary pretectal nucleus); *PC*, posterior commissure; *PTM*, medial pretectal nucleus; *SC*, superior colliculus; *SGM*, stratum griseum intermedium; *SGS*, stratum griseum superficiale; *SO*, stratum opticum; *SZ*, stratum zonale. From Hayhow et al. (1962)

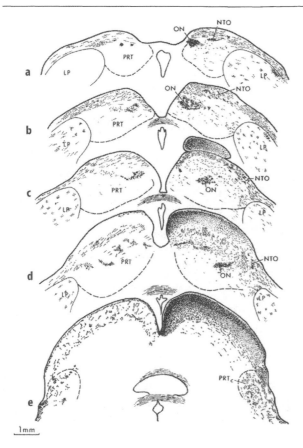

Fig. 22.105a-e. Series of line drawings of transverse sections through the pretectum and the superior colliculus of the South American opossum, *Marmosa mitis*, showing the termination of the left optic nerve. **a** is the rostralmost section. Note bilateral, complementary labelling in the olivary pretectal nucleus (*ON*). *CS*, superior colliculus; *LP*, lateral posterior nucleus; *NTO*, nucleus of the optic tract; *ON*, olivary pretectal nucleus; *PRT*, pretectum; *PRTc*, caudal part of the pretectum. From Royce et al. (1976)

1971, 1972), the projection to the olivary pretectal nucleus is completely crossed. The ultrastructure of the olivary pretectal nucleus of the rat and its retinal afferents were analysed by Campbell and Lieberman (1985). Crossed projections in *Saimiri sciureus* terminate in the central, cell-poor regions of the olivary nucleus (Fig. 22.106) and the ipsilateral projection in the rostral cellular part of the nucleus (Weber et al. 1981). A similar, partial segregation of crossed and uncrossed retinal projections has been noted in the olivary pretectal nucleus of other species (Young and Lund 1994).

Retinal projections to the nucleus of the optic tract are completely crossed in the rat and mouse (Scalia 1972; see, however, Hayhow et al. 1962 and Giolli and Greel 1973), rabbit, shrew and mole, marsupials and monotremes. In the cat and the pigmented ferret *Mustela putorius furo*, but not in the albino of this species (Zhang and Hoffmann 1993), and in primates, the nucleus of the optic tract receives a bilateral projection from the retina. The ultrastructure of the nucleus of the optic tract and the presence of interneurons in this nucleus was discussed by Nunes Cardozo and Van der Want (1987) for the rabbit.

Terminations of optic nerve fibres in other pretectal nuclei are sparser and more variable among species (Figs. 22.103-22.106). Terminations have been observed most frequently in the posterior and anterior pretectal nuclei in several rodents (Hayhow et al. 1962; Scalia 1972), in the musk shrew (Mizuno et al. 1991), the cat (Hutchins 1991), the pigmented ferret (Zhang and Hoffmann 1993) and the rabbit (Klooster et al. 1983). In the guinea pig, rabbit and tree shrew, they occupied the dorsal part of these nuclei (Scalia 1972; Giolli and Greel 1973; Weber 1985). In primates, sparse terminations have only been observed in the posterior and medial prectal nuclei (Hendrickson et al. 1970; Hutchins and Weber 1985; Benevento and Standage 1983). The nucleus of the posterior commissure does not receive a projection from the retina.

Other afferent connections of the pretectum take their origin from the superior colliculus (Benevento and Ebner 1970; Cadusseau and Roger 1991; Lagares et al. 1994), the nucleus prepositus hypoglossi (Ohtsuki et al. 1992) and from the cerebral cortex, the ventral geniculate, the zona incerta, the cholinergic groups Ch5 and Ch6 of Mesulam et al. (1983, 1984), the somatosensory nuclei of the brain stem, the spinal cord and the cerebellum. For the olivary pretectal nucleus, the only non-retinal projection that has been reported stems from the ventral geniculate in the cat (Edwards et al. 1974; Swanson et al. 1974). The nucleus of the optic tract receives a bilateral projection from the ventral geniculate (Edwards et al. 1974) and strong projections from the nuclei of the accessory optic system (see below). The visual cortex in primates (Kuypers and Lawrence 1967), cat (areas 17-19, 21; Kawamura et al. 1974; Updyke 1977), rat (Schmidt et al. 1993) and guinea pig (Lui et al. 1994) projects to the nucleus of the optic tract and the adjacent dorsal terminal nucleus of the accessory optic system (see Sect. 22.8.3.4), but this connection was disputed in the tree shrew (Harting and Noback 1971; Weber 1985) and is absent in the rabbit (Giolli et al. 1978). The primate nucleus of the optic tract also receives afferent projections from the cortex in the superior temporal sulcus (Hoffmann et al. 1991, 1992; Ilg and Hoffmann 1993). Prefrontal projections to the pretectum have been reported in primates (Kuypers and Lawrence 1967) and rodents (Beckstead 1979). A projection of the frontal eye field to the nucleus of

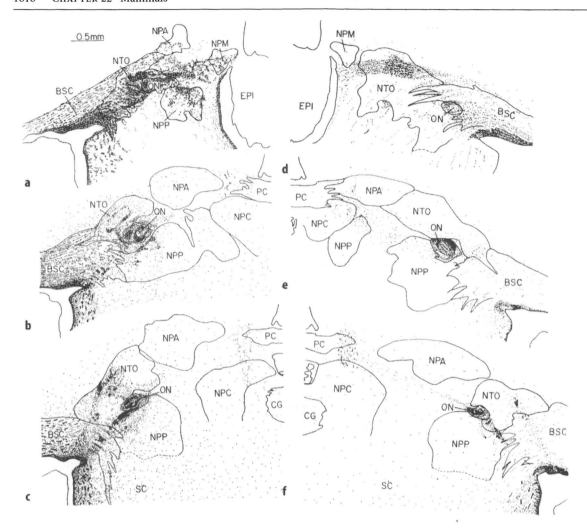

Fig. 22.106a-f. A series of line drawings of horizontal sections illustrating the distribution of transported label over the contralateral (**a-c**) and ipsilateral (**d-f**) pretectal complex of the squirrel monkey *Saimiri sciureus* following an intraocular injection of [³H]proline in one eye. The pattern of silver grains is indicated by the *black stippling*. Rostral is at the *top*. Note bilateral and complementary labelling of the olivary pretectal nucleus. *BSC*, brachium of the superior colliculus; *CG*, peri aquaductal grey; *EPI*, pineal body; *NPA*, anterior pretectal nucleus; *NPC*, nucleus of the posterior commissure; *NPM*, medial pretectal nucleus; *NPP*, posterior pretectal nucleus; *NTO*, nucleus of the optic tract; *ON*, olivary pretectal nucleus; *PC*, posterior commissure; *SC*, superior colliculus. From Hutchins and Weber (1985)

the optic tract has been reported in macaque monkeys (Künzle and Akert 1977; Leichnetz 1982).

The anterior and posterior pretectal nuclei receive visual, somatosensory and motor projections. The spinal cord and the spinal nucleus of the trigeminal nerve project contralaterally to the ventrolateral part of the posterior pretectal nucleus and, sparsely, to the anterior pretectal nucleus in the cat and monkey (Wiberg and Blomqvist 1984; Wiberg et al. 1986, 1987; Björkeland and Boivie 1984; Mehler 1969; Yezierski 1988). Similar projections were traced from the dorsal column nuclei to the anterior and posterior pretectal nuclei in the opossum, rat, cat and monkey (Hazlett et al. 1972; Lund and Webster 1967; Berkley and Hand 1978; Wiberg and Blomqvist 1984; Wiberg et al. 1987). Berkley and Mash (1978) noted that the terminations from the SI and SII somatosensory cortex in the anterior and posterior pretectal nuclei of the cat overlap with the somatosensory projections from the cord and the brain stem, but remain segregated from the retinal input and the visual cortical projections (Kawamura et al. 1974; Updyke 1977; Giolli et al. 1978; Lent 1982; Cadusceau and Roger 1991) to these nuclei.

Additionally, projections from the zona incerta

(Watanabe and Kawana 1982; Cadusceau and Roger 1991: rat), the ventral geniculate (Edwards et al. 1974; Cadusceau and Roger 1991), the lateral and anterior interposed cerebellar nuclei (Kawamura et al. 1982; Kitao and Nakamura 1987), the motor and parietal cortex in the cat (Kitao and Nakamura 1987; Flindt-Egebak and Møller 1984; Kawamura et al. 1974) and the frontal, postcentral and inferior parietal (but not the precentral cortex) in macaque monkeys (Kuypers and Lawrence 1967; Leichnetz 1982) reach the anterior and posterior pretectal nuclei.

22.8.3.3.2
Efferent Connections of the Pretectum

Efferent connections of the pretectum are organised in different functional systems. The olivary pretectal nucleus and the nucleus of the posterior commissure are involved in the pupillary light reflex and project to the Edinger-Westphal nucleus. The nucleus of the optic tract is an optokinetic centre, closely allied to the accessory optic system (see below), and strongly projects to the vestibular and vestibulocerebellar relay nuclei. The anterior and posterior pretectal nuclei give rise to ascending connections with the thalamus and descending connections with somatosensory-dominated subnuclei of the inferior olive and the pontine nuclei.

The olivary pretectal nucleus, the ventral lateral geniculate nucleus and the nucleus of the posterior commissure have been considered as links in the pupillary light reflex. Both the olivary pretectal nucleus and the ventral lateral geniculate nucleus receive a bilateral projection from the retina. Moreover, the ventral lateral geniculate nucleus projects bilaterally in the rat and unilaterally or bilaterally in the cat to the olivary pretectal nucleus (Swanson et al. 1974; Edwards et al. 1974); it gives rise to a reciprocal connection to the contralateral ventral lateral geniculate nucleus in the cat (Berman 1977). A bilateral projection from the olivary pretectal nucleus to the Edinger-Westphal nucleus or its immediate vicinity was traced in macaque monkeys by Benevento et al. (1977), Steiger and Büttner-Ennever (1976) and Itaya and Van Hoesen (1982), in the rat by Trejo and Cicerone (1984) and Young and Lund (1994) and in the cat by Graybiel and Hartweig (1974). No connections between the olivary pretectal nucleus and the visceromotor nuclei of the oculomotor nerve could be established in the tree shrew *Tupaia glis* (Weber and Harting 1980). These authors suggested that the pathway for the pupillary reflex in this species may pass through the olivary pretectal nucleus and the nucleus of the posterior commissure. In primates, no projections of the nucleus of the commissure to the Edinger-Westphal nucleus have been observed (Benevento et al. 1977). In the rat, both a direct and an indirect pathway through the nucleus of the posterior commissure to the Edinger-Westphal nucleus seem to exist (Young and Lund 1994). The olivary pretectal nucleus, together with the posterior nucleus, also gives rise to a substantial, bilateral innervation of the suprachiasmatic nucleus (Mikkelsen and Vrang 1994).

All the nuclei of the pretectum have been reported to give rise to ascending connections to the thalamus and the subthalamus. Descending connections to the superficial and intermediate layers of the superior colliculus, the accessory oculomotor nuclei (nucleus of Darkschewitsch and the interstitial nucleus of Cajal), the reticular formation, the pontine nuclei and the inferior olive arise from all nuclei, with the exception of the olivary pretectal nucleus. Ascending connections to the reticular, intralaminar and midline nuclei of the thalamus and the zona incerta have been reported for the macaque monkey (Benevento et al. 1977; Benevento and Standage 1983), tree shrew (Weber and Harting 1980), rabbit (Holstege and Collewijn 1982) and cat (Berman 1977). The primate pretectum does not project to the visual relay nuclei of the thalamus (Benevento et al. 1977; Benevento and Standage 1983; Mustari et al. 1994). In the rat (Mackay-Sim et al. 1983) and rabbit (Holstege and Collewijn 1982), however, efferents of the pretectum also terminate in the dorsal lateral geniculate nucleus. Some of the connections of the nucleus of the optic tract and the olivary and posterior pretectal nuclei to the dorsal lateral geniculate of the rat are GABAergic (Cucchiaro et al. 1991). A GABAergic component was also described by Nunes Cardozo and Wortel (1993) for the projection of the nucleus of the optic tract to the superior colliculus in the rat. The transmitters in the projections of the anterior pretectal nucleus to the superficial layers of the colliculus in the rat (Cadusseau and Roger 1991) and the suprageniculate pretectal nucleus of the rabbit (Lagares et al. 1994) are not known.

All pretectal nuclei give rise to descending projections to the ipsilateral pontine nuclei and the inferior olive. Pretecto-olivary and pretectothalamic (lateral dorsal nucleus) projections in the rat take their origin from different populations of neurons (Robertson 1983). Pretectopontine projections terminate mainly, but not exclusively, in dorsolateral regions of the pontine nuclei (Benevento et al. 1977; Weber and Harting 1980; Aas 1989; Mihailoff et al. 1989; Mustari 1994; Holstege and Collewijn 1982), where they may overlap with projections from the visual cortical areas and the superior colliculus. A projection to the nucleus reticularis tegmenti pontis has been found in the rat (Terasawa et al. 1979) and in primates (Mustari et al. 1994), but has been denied in the rabbit (Holstege and Collewijn 1982).

Different parts of the inferior olive are innervated by the nucleus of the optic tract and the anterior and posterior pretectal nuclei. Fibres from the nucleus of the optic tract mainly terminate in the dorsal cap (Sect. 22.8.3.4). The anterior and posterior pretectal nuclei are links in a crossed, indirect pathway from the dorsal column nuclei to the dorsal accessory olive. Projections from the anterior and posterior pretectal nuclei (for reviews, see Kitao et al. 1989; Bull et al. 1990) to the inferior olive have been substantiated in the rat (Brown et al. 1977; Weber and Harting 1980; Robertson 1983; Swenson and Castro 1983), opossum (Linauts and Martin 1978) and cat (Itoh et al. 1983; Saint-Cyr and Courville 1982; Kawamura and Onodera 1984; Bull et al. 1990). The anterior and posterior pretectal nuclei project to the rostral part of the dorsal accessory olive (Itoh et al. 1983) and the margin of the ventral leaf of the principal olive (Kawamura and Onodera 1984), where they overlap with the projections from the dorsal column nuclei (Bull et al. 1990). There is no, or very little, overlap of the uncrossed pretecto-olivary projection with the crossed tecto-olivary projection (see Sect. 22.8.3.2.4).

22.8.3.4
Accessory Optic System

The accessory optic system consists of several nuclei that receive projections from the contralateral retina through offshoots of the optic tract. The nuclei of the accessory optic system, like the nucleus of the optic tract, project directly or indirectly to precerebellar nuclei with projections to the flocculus, the nodulus and other visual areas of the cerebellum. Hayhow (1959, 1966) and Hayhow et al. (1962) first described the accessory optic system in the cat, the marsupial phalanger *Trichosurus* and the rat (Fig. 22.107). It consists of an inferior fascicle that detaches from the optic tract shortly after its decussation and courses caudally, medial to the cerebral peduncle. It contains the neurons of the medial terminal nucleus of the accessory optic system. Other fibres detach more distally from the optic tract in its course towards and lateral to the superior colliculus as the superior fascicle of the accessory optic system. These fibres sweep over or around the medial geniculate, the lateral surface of the mesencephalon and the cerebral peduncle, reaching the medial terminal nucleus. In some species, the caudalmost of these fibres constitute a compact bundle, which was recognised by von Gudden (1870) in the rabbit as the transpeduncular tract (Fig. 22.11). The dorsal and lateral terminal nuclei lie embedded among these fibres. The dorsal terminal nucleus is located immediately ventral to the lateral margin of the superior colliculus, dorsal to the medial geniculate. The lateral terminal nucleus is located ventral to the medial geniculate. The cells of the interstitial nucleus of the superior fascicle are located among its fibres, in between the dorsal and lateral terminal nuclei (Giolli et al. 1984). The medial terminal nucleus was subdivided into a ventral and a dorsal part by Cooper (1986) in a prosimian primate. The ventral division of the medial terminal nucleus is located superficially, medial to the cerebral peduncle, while the dorsal division is found ventrolateral to the red nucleus and dorsal to the substantia nigra and receives retinal fibres from the superior fascicle that penetrate the peduncle and the substantia nigra. The medial terminal nucleus should not be confused with the visual tegmental relay zone, a diffuse cell group located in the same region that serves as a relay between the contralateral medial terminal nucleus and the inferior olive in the rabbit (Giolli et al. 1984, 1988).

The nuclei of the accessory optic system and the nucleus of the optic tract supply optokinetic information to brain stem centers, including the pontine nuclei and the inferior olive. The dorsal terminal nucleus is topographically and functionally closely related to the nucleus of the optic tract. Both nuclei respond to movements of the whole visual field around a vertical axis and are involved in the processing of eye movements around this same axis. The medial and lateral terminal nuclei respond to movements around horizontal axes and are engaged in the processing of vertical eye movements. The anatomy and the physiology of the accessory optic system were reviewed by Simpson et al. (1984).

A complete accessory optic system, including a medial terminal nucleus, subdivided into dorsal and ventral parts, is present in rodents (Giolli and Greel 1973; Terubayashi and Fujisawa 1984; Cooper et al. 1990), insectivores (Tigges and Tigges 1969; Cooper et al. 1990; however, for the absence of lateral and medial terminal nuclei in the musk shrew *Suncus murinus*, see Tokunaga et al. 1992), megachiropteran bats (Cooper et al. 1990; but absent in microchiropteran species, Cotter and Pierson Pertney 1979), lagomorphs (Giolli and Guthrie 1969) and marsupials (Royce et al. 1976). Perissodactyla and Artiodactyla possess dorsal and lateral terminal nuclei that receive a contralateral projection

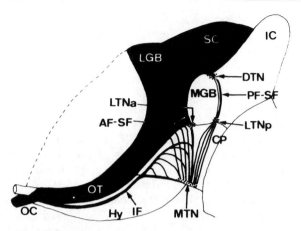

Fig. 22.107. Contralateral retinal pathway of the optic nerve fibres in the rat, based on antegrade filling of the axons with horseradish peroxidase. The main pathway consists of the optic tract and courses towards the lateral geniculate body and the superior colliculus; the accessory optic system consists of multiple fascicles detaching from the optic tract and passing medial to, and over, the cerebral peduncle and rostral and caudal to the medial geniculate body. The medial, lateral and dorsal terminal nuclei are shown by *black dots*. *AF-SF*, anterior branch of superior accessory fascicle; *CP*, cerebral peduncle; *DTN*, dorsal terminal nucleus; *HY*, hypothalamus; *IC*, inferior colliculus; *IF*, inferior fascicle; *LGB*, lateral geniculate body; *LTNa*, lateral terminal nucleus, pars anterior; *LTNp*, lateral terminal nucleus, pars posterior; *MGB*, medial geniculate body; *MTN*, medial terminal nucleus; *OC*, optic chiasm; *OT*, optic tract; *PF-SF*, posterior branch of superior accessory fascicle; *SC*, superior colliculus. (From Terubayashi and Fujisawa 1984)

from the retina through the superior fascicle. A ventral division of the medial terminal nucleus that receives a retinal projection from the superior fascicle has only been definitely identified in the horse, but was absent in sheep, pig and ox (Karamanlidis and Magras 1972, 1974). No data are available on the presence of a dorsal subdivision of the medial terminal nucleus in these species. Carnivores lack an inferior fascicle, but the two divisions of the medial terminal nucleus receiving afferents from the superior fascicle were detected (Terubayashi and Fujisawa 1984; Cooper et al. 1990). Dorsal and lateral terminal nuclei and the two divisions of the medial terminal nucleus were present in Pholidota (*manis tricuspus*: Cooper et al. 1990; Lee et al. 1991). In non-eutherian mammals, a medial terminal nucleus that receives a crossed retinal projection was identified. In the echidna *Tachyglossus*, only the inferior fascicle of the accessory optic system was present, while the superior fascicle appears to be lacking. The dorsal terminal nucleus, if present, cannot be distinguished from the nucleus of the optic tract (Campbell and Hayhow 1971). In the duckbill platypus, the accessory optic system is diffusely organised and no discrete fascicles are present. A dorsal terminal nucleus could not be identified (Campbell and Hayhow 1972).

In primates and prosimians, the termination of retinal fibres in the medial terminal nucleus has been doubted (Fig. 22.108). Giolli (1963) was able to trace fibres of the accessory optic tract from the brachium of the superior colliculus in *Macaca* to a cell group (lateral terminal nucleus) located lateral to and covering the lateral aspect of the cerebral peduncle, but these fibres did not continue accross the peduncle to terminate in the medial terminal nucleus. The absence of a medial terminal nucleus with the presence of lateral and dorsal tract nuclei was also noted in other primates (*Saimiri sciureus*: Tigges and Tigges 1969; *Macaca mulatta*: Hendrickson et al. 1970; the chimpanzee: Tigges et al. 1977) and in the prosimian species *Galago crassicaudatus* (Fig. 22.108; Tigges and Tigges 1969).

A medial terminal nucleus receiving sparse retinal projections was noted in primates by Weber and Giolli (1986). Cooper (1986) and Cooper et al. (1990) pointed out that the medial terminal nucleus consists of two divisions. The ventral division, located superficially, medial to the cerebral peduncle is present in the prosimian primate *Microcebus murinus* (Cooper 1986), but absent in the macaque monkey (Nakagawa et al. 1988; Cooper et al. 1990) and in *Hylobates concolor* (Cooper and Magnin 1987). According to these authors, the dorsal division of the medial tract nucleus is present in all primates.

The terminal nuclei of the accessory optic system and the nucleus of the optic tract establish direct and indirect connections with the dorsal cap and the ventrolateral outgrowth of the inferior olive, other precerebellar and vestibular nuclei and parts of the reticular formation. Moreover, these nuclei are interconnected by a set of partially GABAergic pathways.

In the rabbit, the caudal part of the dorsal cap receives afferent projections from the dorsal and interstitial terminal nucleus and the nucleus of the optic tract (Mizuno et al. 1973, 1975; Takeda and Maekawa 1976; Holstege and Collewijn 1982). This projection is mainly uncrossed and also involves the β-group (Holstege and Collewijn 1982). The rostral part of the dorsal cap and the ventrolateral outgrowth receive afferent projections from the lateral and medial terminal nuclei of the accessory optic system and the visual tegmental relay zone (Maekawa and Takeda 1977, 1979; Giolli et al. 1984, 1985; Simpson 1988). The major projection from these nuclei is ipsilateral and is derived from the visual tegmental relay zone. This zone does not receive direct afferents from the optic tract, but receives a crossed projection from the medial tract thin fibres (Fig. 22.109). The distribution of these

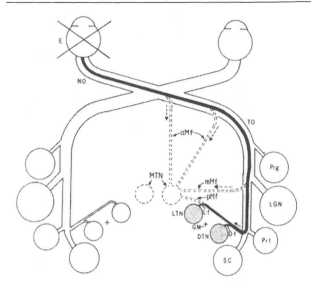

Fig. 22.108. Accessory optic system in the bush-baby *Galago*. Optic tract fibres decussate in the chiasm and detach from the optic tract caudal to the medial geniculate body, to terminate in the dorsal and lateral terminal nuclei. Other components of the accessory optic system (*stippled*; compare Fig. 22.107) are absent in primates. *aMf*, anterior (inferior) fascicle of accessory optic system; *Df*, fibres to the dorsal terminal nucleus; *DTN*, dorsal terminal nucleus; *E*, eye; *lf*, fibres to the lateral terminal nucleus; *LGn*, lateral geniculate body; *LTN*, lateral terminal nucleus; *mMf*, intermediate fascicles of accessory optic system; *MTN*, medial terminal nucleus; *NO*, optic nerve; *pMf*, posterior (superior) fascicle of accessory optic system; *Prg*, ventral nucleus of the geniculate body; *Prt*, pretectum; *SC*, superior colliculus; *TO*, optic tract. (Redrawn from Tigges and Tigges 1969)

large fibres corresponds to the distribution of the nucleus, passing through the posterior commissure (Giolli et al. 1984).

Data on the efferent projections of the nuclei of the optic tract to the inferior olive in other species are less complete. In primates, the lateral terminal nucleus projects to the rostral two thirds of the dorsal cap and the ventrolateral outgrowth, i.e. to the same subdivisions of the inferior olive as in the rabbit (Blanks et al. 1995: *Callithrix jachus*, marmoset monkey). The projection of the dorsal terminal nucleus and/or the nucleus of the optic tract to the caudal dorsal cap was found in the rat (Brown et al. 1977; Terasawa et al. 1979; Swenson and Castro 1983), cat (Walberg et al. 1981; Kawamura and Onodera 1984) and tree shrew (Weber and Harting 1980). The primate nucleus of the optic tract and the dorsal terminal nucleus project to the ipsilateral caudal dorsal cap and the β-group (Mustari et al. 1994). Connections of the pretectum, the superior colliculus and the nuclei of the accessory optic system to precerebellar relay nuclei, such as the pontine nuclei, the nucleus reticularis tegmenti pontis and the inferior olive, are found in all mammals. In a few species, a direct mossy fibre projection of the medial terminal nucleus to the ipsilateral cerebellum has been reported (flocculus in the chinchilla: Winfield et al. 1978; ventral uvula in the tree shrew: Haines and Sowa 1985).

Reciprocal connections are present between the nuclei of the accessory optic system and the nucleus of the optic tract. In the rabbit, rat and tree shrew, the connections of the nucleus of the optic tract with the medial terminal nucleus, the interstitial nucleus of the superior fascicle, the lateral terminals nucleus and the dorsal terminal nucleus are uncrossed (Giolli et al. 1984, 1988; Blanks et al. 1982; Holstege and Collewijn 1982; Weber and Harting 1980). Fibres of the posterior commissure interconnect the nuclei of the optic tract and the dorsal terminal nuclei of both sides. In primates, additional crossed projections of the olivary pretectal nucleus to the lateral terminal nucleus (Baleydier et al. 1990) and of the nucleus of the optic tract to the medial terminal nucleus (Mustari et al. 1994) are present. Some of the reciprocal connections between the nucleus of the optic tract and the dorsal terminal nucleus and between the medial and lateral terminal nuclei and the visual tegmental relay zone are GABAergic (Ottersen and Storm-Mathisen 1984; Van der Want et al. 1992; Giolli et al. 1992).

Other projections from the nucleus of the optic tract and the nuclei of the accessory optic system target the reticular formation, the superior colliculus, the parabigeminal nucleus, the vestibular nuclei and the nucleus prepositus hypoglossi (Holstege and Collewijn 1982; Mustari et al. 1994; Giolli et al. 1984, 1988; Weber and Harting 1980; Blanks et al. 1995). Two separate neuronal populations of the nucleus of the optic tract and the dorsal terminal nucleus give rise to the commissural projection and to a branching pathway to the ipsilateral dorsal cap and nucleus prepositus hypoglossi in the rat and cat (Schmidt et al. 1995). The nucleus prepositus hypoglossi is bilaterally connected with the dorsal cap via a GABAergic pathway (De Zeeuw et al. 1993).

22.9
Motor Systems

J. Voogd

22.9.1
Corticobulbar and Corticospinal Tracts

The corticobulbar and corticospinal tracts enter the brain stem via the cerebral peduncle. Inspection of the cerebral peduncle in Häggqvist-stained sections

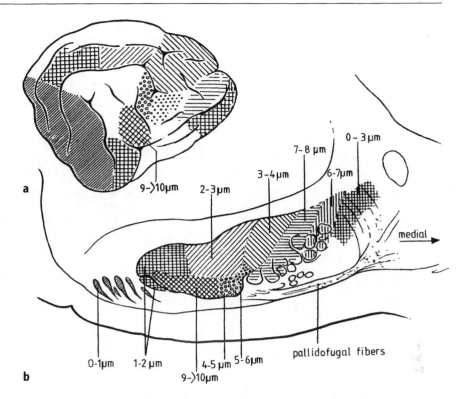

Fig. 22.109a,b. Localisation of fibres of different calibre in the cerebral peduncle of the cat (b). The main origin of these fibres is shown in a diagram of the cerebral hemisphere (a). Small-calibre myelinated fibres of the lateral portion and the extreme medial part of the peduncle take their origin from occipital and frontal portions of the hemisphere. Larger-calibre fibres of the medial half of the peduncle originate from the pericrucial, coronary and ectosylvian areas. The nature of the large-calibre component from the sylvian region has not been explained. (From Mechelse 1957)

(Häggqvist 1936) of carnivores and primates (Figs. 22.23o, 22.24p) shows an area in its medial half or two thirds that contains a relatively high proportion of large myelinated fibres among a majority of thin fibres (Fig. 22.109). The distribution of these large fibres corresponds to the distribution of the corticobulbar and corticospinal fibres from the precentral cortex and makes it possible to establish the size of the contributions of precentral, frontal and parietotemporal cortical areas in the projection of the cerebral cortex to the brain stem.

In the cat and ferret, large (4–8 μm) fibres occupy the medial half of the peduncle (see Fig. 22.112). In primates, the large fibres are restricted to a central portion of the peduncle. On their medial and lateral sides, they are flanked by regions consisting solely of small fibres that are known as the medial frontopontine tract and the lateral parieto-temporopontine tract (Fig. 22.24p). The border of the central area with the parieto-temporopontine tract is sharp. In a medial direction, the large fibres of the central area gradually disappear until they are completely absent from the area of the frontopontine tract. The fibre pattern of the medial half of the cerebral peduncle in the cat and ferret and of the central portion of the peduncle in primates is very similar to that of the medullary pyramid, which is also characterised by the presence of large (4–8 μm) myelinated fibres (see Fig. 22.112; Mechelse 1957; Verhaart and Noorduyn 1961; Verhaart 1970).

In the cat, the area corresponding to the primate frontopontine tract can be identified as a small contingent of thin (1–3 μm) fibres in the extreme medial pole of the peduncle that arise from the gyrus proreus and the medial prefrontal cortex (Mechelse 1957; Fig. 22.109). The lateral half of the peduncle consists exclusively of small (1–3 μm) fibres that belong to a system arising from the sylvian, ectosylvian, suprasylvian, lateral and occipital cortex (Fig. 22.23o). Many fibres of this lateral system do not join the cerebral peduncle, but pass through or along the thalamus on their way to the pretectum and the superior and inferior colliculus. A large proportion of the corticopontine fibres located in the lateral third of the cerebral peduncle are collaterals of the cortico-pretectal and tectal fibres (Keizer et al. 1987; see Sects. 22.8.2.6 and 22.8.3.2.3). The corticopontine projection is considered in more detail in Sect. 22.6.6.

In the cat, the corticobulbar and corticospinal fibres of the pyramidal tract arise from the sigmoid, coronal and the rostral parts of the cingulate, lateral and suprasylvian gyri (Fig. 22.110; Van Crevel 1958; Biedenbach and De Vito 1980; Keizer and Kuypers 1984; Keizer et al. 1987) corresponding to areas 6, 4, 3, 1, 2 and 5 of Hassler and Muhs-Clement (1964). In the bulbar pyramid, the large fibres account for

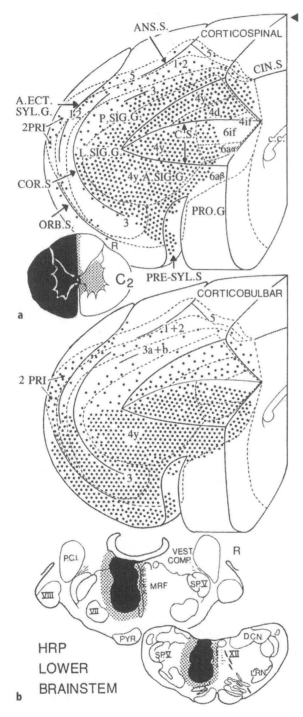

Fig. 22.110a,b. Distributions of the retrogradely horseradish peroxidase (HRP)-labelled neurons in the pericruciate cortex of the cat after contralateral HRP injections at C2 and after contralateral HRP injections in the bulbar medial reticular formation. *Dotted areas* represent HRP diffusion around the injection areas. The cerebral hemisphere is viewed frontomedially. Borders between cortical areas are indicated according to Hassler and Muhs-Clement (1964). After injections of the bulbar retiruclar formation the labelling is displaced rostrally and extends further into area 6 and the region rostral to the orbital sulcus. *A.ECT.SYL.G*, anterior ectosylvian gyrus; *A.SIG.G*, anterior sigmoid gyrus; *ANS.S*, ansate sulcus; *c.c.*, corpus callosum; *C.S*, cruciate sulcus; *CIN.S*, cingulate sulcus; *COR.S*, coronal sulcus; *D.C.N*, dorsal column nuclei; *G.F.c.*, granular frontal cortex; *L.R.N*, lateral reticular nucleus; *MRF*, medial reticular formation; *ORB.S.*, orbital sulcus; *P.C.I.*, inferior cerebellar peduncle; *P.Sig.G*, posterior sigmoid gyrus; *PRESYL.S*, presylvian sulcus; *PRO.G.*, prorean gyrus; *PYR*, pyramid; *SP.V*, spinal trigeminal nucleus; *VEST.COMP*, vestibular complex; *VII*, facial nucleus; *VIII*, cochlear nucleus; *XII*, hypoglossal nucleus. (From Keizer and Kuypers 1984)

7% of the fibres (5%, 4–6 μm; 2%, >6 μm; Van Crevel and Verhaart 1963; see Fig. 22.112). Most of these large fibres take their origin from Betz cells in area 4 of the gyrus sigmoideus. The majority of the corticobulbar and corticospinal fibres are small and originate in pyramidal cells located over a much wider area. Fibres leaving the contingent of corticobulbar and corticospinal fibres in the cerebral peduncle course between the subthalamic nucleus and the substantia nigra (Ramon-Molinar 1979). In the ferret, the number of fibres leaving this contingent in the mesencephalon is estimated at 17%; 35% leave the peduncle or terminate at the level of the pontine nuclei, 11% leave the medullary pyramid before its decussation and 30% enter the spinal cord (Noorduyn and Verhaart 1961).

Fibres from the pericruciate cortex terminate ipsilaterally in the reticular formation of the mesencephalon, the red nucleus, the periaqueductal grey, with the nucleus of Darkschewitsch and the interstitial nucleus. There are no terminations in the motor nuclei of the oculomotor and trochlear nerves. At and below the level of the motor nucleus of the trigeminal nerve, motor fibres leaving the peduncle cross and terminate in the sensory nuclei of the trigeminal nerve, the adjoining lateral reticular formation, parts of the nucleus of the solitary tract, the medial vestibular nucleus and the hilus of the dorsal column nuclei. The medial reticular formation of the pons and the medulla oblongata receive a bilateral projection. Precerebellar nuclei receive uncrossed (basal pontine nuclei, nucleus reticularis tegmenti pontis), bilateral (paramedian reticular nucleus, nucleus of Roller) or crossed projections (medial, magnocellular part of the lateral reticular nucleus). The motor nuclei of the trigemi-

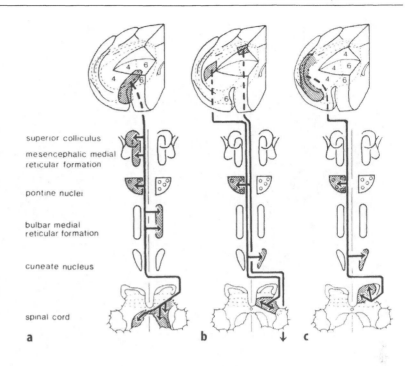

Fig. 22.111a–c. Distribution of corticospinal neurons that project to different parts of the spinal cord and in addition distribute collaterals to different brain stem nuclei. Compare Fig. 22.110. **a** The rostromedial part of the pericruciate cortex in cat contains highly branching neurons which project bilaterally to the ventromedial part of the spinal intermediate zone and also distribute collaterals to the superior colliculus and the mesencephalic and bulbar medial reticular formation. **b** Corticospinal fibres from the fore- and hindlimb regions of area 4 project preferentially to the dorsolateral parts of the intermediate zone. Note absence of collaterals to the brain stem nuclei which give rise to the medially descending brain stem pathways. Corticospinal fibres from the forelimb region distribute collaterals to the cuneate nucleus. **c** Corticospinal fibres from somatosensory area 3 project to the dorsal part of the intermediate zone and the dorsal horn and, additionally, distribute collaterals to the cuneate nucleus. Note that the various cortical regions also give rise to corticospinal collaterals to the pontine nuclei. (From Keizer 1989)

nal, facial and hypoglossal nerves, the dorsal vagal nucleus and the nucleus ambiguus do not receive fibres from the pericruciate cortex (Brodal et al. Torvik 1956; Walberg 1957; Rossi and Brodal 1956; Szentágothai 1958; Kuypers 1958b).

The pyramidal tract of the cat decussates at the junction of the medulla oblongata and the spinal cord (Fig. 22.23f). Postcommissural ascending fibres contribute to the projection to the dorsal column nuclei. Most crossed fibres descend in the dorsolateral funiculus of the cord to sacral levels. Uncrossed fibres descend in the ventral funiculus to mid-thoracic levels. Smaller fibre contingents descend ipsilaterally in the dorsolateral funiculus, contralaterally in the ventral funiculus and bilaterally in the dorsal funiculus (Satomi et al. 1989). Fibres terminate mainly contralaterally in the base of the dorsal horn and, less intensely, in the superficial layers of the dorsal horn, the intermediate grey and the commissural nucleus, but not on the motor neurons of the ventral horn, the intermediolateral nucleus of the lateral horn or on sacral parasympathetic cell groups (Szentagothai-Schimert 1941; Kuypers 1958a).

The projections from the pericruciate cortex to the brain stem and the cord are topically organised. The hindlimb area of the primary sensory cortex in the medial posterior sigmoid gyrus projects to the contralateral lumbar dorsal horn (laminae I, superficial II, IV, V and dorsal VI, especially medially) and to the nucleus gracilis. Similar projections to the contralateral cervical dorsal horn and the cuneate nucleus arise from the sensory forelimb area in the lateral posterior sigmoid gyrus (Fig. 22.111c; Cheema et al. 1984; Armand et al. 1985). The sensory mouth area in the anterior part of the coronal gyrus projects to the sensory nuclei of the trigeminal nerve (nucleus princeps and pars oralis of the spinal nucleus) and to the rostral pole of the nucleus of the solitary tract (Kawana and Kusama 1968). Projections from the forelimb area of the motor cortex (area 4) in the lateral part of the anterior sigmoid gyrus overlap with those from the sensory cortex in the cuneate nucleus. In the cervical cord, they are limited to the lateral part of the intermediate zone (laminae V, VI and dorsal VII; Fig. 22.111b). The intermediate part of area 4 projects to the intermediate zone in both enlargements and bilaterally to ventromedial laminae VII and VIII (Fig. 22.111a; Cheema et al. 1984; Armand et al. 1985). Neurons in rostral area 4 and area 6 project bilaterally to the medial reticular formation (Keizer and Kuypers 1989). Many of these neurons give rise to collaterals terminating in the red nucleus, the dorsal column nuclei and the intermediate zone of the spinal cord (Rustioni and Hayes 1981; Martinez et al. 1995).

In most mammals, with the exception of primates and some carnivores, large fibres are absent from the cerebral peduncle and the pyramidal tract (Figs. 22.220, 22.112e,f), and the region containing corticobulbar and corticospinal fibres arising from

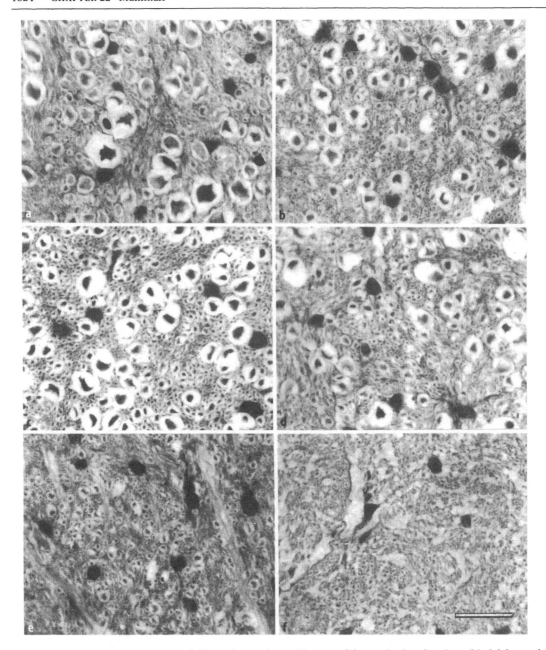

Fig. 22.112a-f. Pattern of myelinated fibres of **a,c,e** the middle part of the cerebral peduncle and **b,d,f** the medullary pyramid in **a,b** the cat, **c,d** the rhesus monkey and **e,f** the North American opossum. Häggqvist stain. Bar = 20 μm

the primary motor cortex cannot be demarcated in normal, Häggqvist-stained material (Verhaart 1970). Experimental axonal tracing techniques should be used to demarcate the subdivisions of the cerebral peduncle. Large numbers of unmyelinated fibres were noted in the bulbar pyramid of the rat (Leenen et al. 1982; Harding and Towe 1985) and cat (Thomas et al. 1984; Biedenbach and De Vito 1986). In the pyramidal tract of the rhesus monkey, unmyelinated fibres are apparently rare (Ralston et al. 1987). Most profiles in electron micrographs that resemble transversely sectioned, unmyelinated fibres turned out to represent astroglial processes when sectioned longitudinally. Joosten and Gribnau (1988) showed that some of the unmyelinated axons of the rat pyramidal tract, at least, take their origin from the cerebral cortex.

In experimental, axonal tracing studies in marsupials, fibres from the frontal cortex, rostral to the orbital sulcus, which contains the motor area for

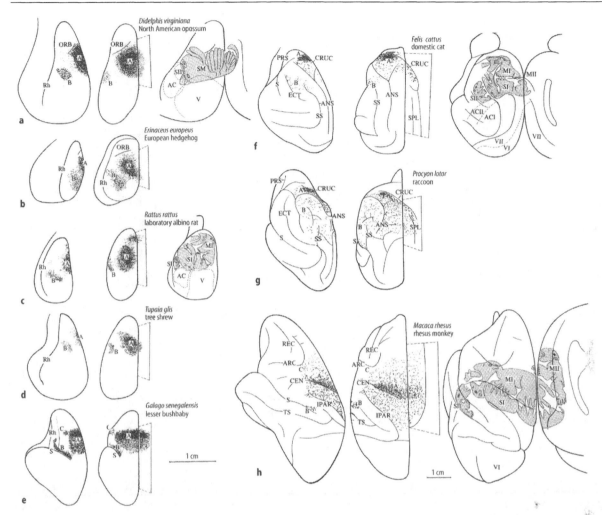

Fig. 22.113a–h. Origin of the corticospinal tract in different species, based on retrograde labelling after injections of the retrograde tracer horseradish peroxidase at the junction of the C1/C2 cervical segments. Note different magnifications for **a–e** and **f–h**. Redrawn and relabelled from Nudo and Masterton (1990a). Two concentrations (**a,b**) of retrogradely labelled neurons are present in all species. They correspond to the primary motor and somatosensory cortex and to the second somatosensory area, respectively. Area C was distinguished as a separate concentration of retrogradely labelled neurons in primates. The figurines of somatotopical localization in the motor and sensory areas of the opossum were copied from Lende (1963a,b) and of the rat, the cat and the monkey from Woolsey (1958). *A*, concentration A of corticospinal neurons; *AC(I,II)*, (first, second) acoustic area; *ACR*, arcuate sulcus; *ANS*, ansate sulcus; *B*, concentration B of corticospinal neurons; *C*, concentration C of corticospinal neurons; *CEN*, central sulcus; *CRUC*, cruciate sulcus; *ECT*, anterior ectosylvian sulcus; *IPAR*, intraparietal sulcus; *MI*, primary motor area; *MII*, supplementary motor area; *ORB*, sulcus orbitalis; *PRS*, presylvian sulcus; *REC*, sulcus rectus; *RH*, rhinal fissure; *S*, sylvian fissure; *SI*, primary sensory area; *SII*, secondary sensory area; *SM*, combined sensory-motor area of the opossum; *SPL*, splenial sulcus; *SS*, suprasylvian sulcus; *TS*, superior temporal sulcus; *V(I.II)*, (first, second) visual area

head and neck in the phalanger *Trichosurus* (for a discussion of the localisation of the motor area in marsupials, see Martin et al. 1975), but which is located rostral to the overlapping motor and sensory areas in the North American opossum (*Didelphis*: Lende 1963a,b; Fig. 22.113), are located in the medial part of the cerebral peduncle in both species. In the opossum, few of these fibres descend caudal to the pons. Fibres from the motor and sensory cortices caudal to the orbital sulcus occupy medial and middle portions of the peduncle. These regions provide the bulk of the fibres of the medullary pyramid. Fibres from the caudal (striate and peristriate) cortex in the North American opossum occupy the lateral peduncle and do not descend beyond the level of the pons (Martin 1968; Martin et al. 1975). Similar observations on the composition of the cerebral peduncle have been made in

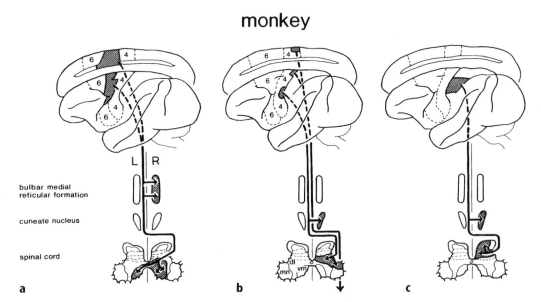

Fig. 22.114a–c. Differences in spinal projections from the anterior and posterior parts of the precentral corticospinal area and the postcentral somatosensory area in the monkey. **a** Corticospinal fibres from the rostral part of of area 4 and the caudal part of area 6 project bilaterally to the ventromedial (*vm*) parts of the intermediate zone. Additionally, several of the corticospinal neurons from the anterior part of the motor cortex and from the supplementary motor cortex give off collaterals to the bulbar medial reticular formation. **b** Corticospinal fibres from the foot and hand regions of the posterior parts of the precentral region project contralaterally to the dorsolateral (*dl*) parts of the intermediate zone and to the motoneuronal (*mn*) pool. Some of the corticospinal fibres from the hand region distribute collaterals to the cuneate nucleus. **c** Corticospinal fibres from the postcentral somatosensory regions typically project to the dorsal part of the intermediate zone and to the dorsal horn. Many fibres from the hand representation area give off collaterals to the cuneate nucleus. (From Keizer 1989)

experimental studies in rodents (rat: Glickstein et al. 1992) and Artiodactyla (goat: Haartsen and Verhaart 1967), with tracing of corticofugal fibres from the different regions of the cerebral cortex.

In primates, the frontal cortex, rostral to the precentral gyrus (area 4), gives rise to the thin fibres of the medial third of the cerebral peduncle. In the macaque monkey, the majority of these fibres take their origin from areas 6 and 8, with a smaller contribution of the prefrontal cortex (Kuypers and Lawrence 1967; Künzle and Akert 1977; Künzle 1978; Wiesendanger et al. 1979). These fibres, which are traditionally known as the frontopontine tract, terminate in the tectum and the tegmentum of the mesencephalon and in the pontine tegmentum and the basal pontine nuclei. Fibres from area 6 join the pyramid and reach the spinal cord (Fig. 22,114). The frontopontine tract is small in prosimians, larger in Cercopithecidae, Cebidae, Callitrichidae and in the chimpanzee and reaches its largest size in Hylobatidae (*Pan* and *Pongo*), where it equals the surface area of the central part of the peduncle containing the large fibres of the motor cortex (Verhaart 1970a,b). The frontopontine tract in humans is even larger, comprising almost the entire medial half of the peduncle (Lankamp 1967). The increase in size of the frontopontine tract in monkeys and apes is probably due to an increased contribution of the prefrontal cortex. In the rhesus monkey, the pre-frontopontine fibres mainly arise from the dorsal prefrontal cortex (areas 8–10, 32, 45 and 46), whereas the corticopontine projections from the orbitofrontal and ventral prefrontal areas are relatively minor (Schmahmann and Pandya 1995).

The transition of the area of the frontopontine tract into the central area of the peduncle containing the large myelinated fibres arising from Betz cells in the precentral cortex is gradual. This is in line with the observation that the giant pyramidal cells in the primate motor cortex gradually diminish in number and size when moving frontally from area 4 towards area 6 (Wise 1985). The largest fibres in the central region of the peduncle and in the medullary pyramid in prosimians measure 4.5 µm. In the macaque monkey, fibres over 8 µm in calibre account for 1 % of the fibres of the central region (Fig. 22.112), whereas in hylobates over 3 % of the fibres of the central region of the peduncle and the medullary pyramid measure 8 µm or more (Verhaart 1948).

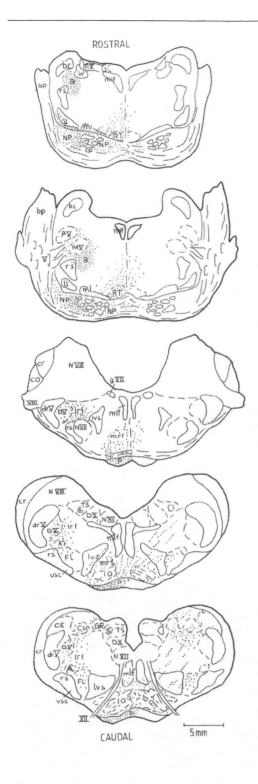

◀ Fig. 22.115. Transverse sections through the pons and medulla oblongata in a goat with a pericentral cortical ablation; degenerated fibres of the pyramidal tracts (p) and Bagley's bundle (B) are indicated by stipple. Nauta stain. A, nucleus ambiguus; bc, brachium conjunctivum; bp, brachium pontis; CE, external cuneate nucleus; CO, cochlear nuclei; cp, cerebral peduncle; cr, restiform body; CU, internal cunenate nucleus; drV, spinal tract of the trigeminal nerve; DV, nucleus of the spinal tract of the trigeminal nerve; DX, dorsal vagus nucleus; FL, lateral reticular nucleus; GR, nucleus gracilis; gVII, genu of the facial nerve; IO, inferior olive; ll, lateral lemniscus; lrf, lateral reticular formation; lvs, lateral vestibulospinal tract; ml, medial lemniscus; mlf, medial longitudinal fascicle; mrf, medial reticular formation; mV, mesencephalic root of the trigeminal nerve; MV, motor nucleus of the trigeminal nerve; NP, nuclei pontis; nVIII, vestibular nuclei; nXII, nucleus of the hypoglossal nerve; p, pyramid; PV, principal sensory nucleus of the trigeminal nerve; rs, rubrospinal tract; RT, nucleus reticularis tegmenti pontis; S, tractus solitarius; TS, nucleus of the tractus solitarius; V, trigeminal nerve; VII, nucleus of the facial nerve; VIII, statoacoustic nerve; W, dorsal trigeminothalamic tract of Wallenberg. (From Haartsen and Verhaart 1967)

The lateral region of the cerebral peduncle, which contains small fibres originating caudal to the central sulcus, shows less variations in size than the frontopontine tract. In the macaque monkey, its corticopontine fibres originate mainly from the postcentral, parietal and prestriate cortex. The contribution of the more ventral, temporal association cortex is much smaller (Keizer et al. 1987; Schmahmann and Pandya 1989, 1991, 1993). Fibres from the somatosensory and rostral parietal cortex join the pyramidal tract and reach the spinal cord.

Standard plots of the origin of the corticospinal tract were prepared and compared in 22 mammalian species by Nudo and Masterton (1990a,b). They confirmed the observations on the origin of this tract in Carnivora (Van Crevel 1958 and others), marsupials (Martin 1968; Martin et al. 1975, 1981) and primates (Kuypers 1958b, 1964, 1973; Kuypers and Brinkman 1970; Jones and Wise 1977; Coulter and Jones 1977; Murray and Coulter 1981; Hutchins and Strick 1988; Keizer and Kuypers 1989; Dum and Strick 1992). They distinguished two segregated regions A and B of layer V pyramidal cells that give rise to the corticospinal tract in all species of their samples, which included Marsupialia, Insectivora, Rodentia, carnivores, Hyracoidiae, Prosimia and higher primates (Fig. 22.113). Region A gives rise to most corticospinal fibres. It is located rostrally in the hemisphere, extending into the cingulate sulcus. It overlaps with the electrophysiologically established somatosensory and motor areas of the cortex. A different localisation of the largely overlapping somatosensory and motor representations in the caudal part of the hemisphere was found in the echidna *Tachyglossus* (Lende 1964). In marsupials

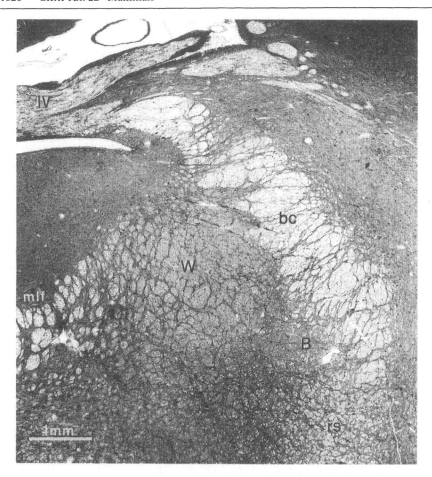

Fig. 22.116. Photograph of the pontine tegmentum in a sheep. Bagley's bundle is found in the triangular region bordered dorsolaterally by the brachium conjunctivum, dorsomedially by the direct trigeminothalamic tract of Wallenberg and ventrolaterally by the rubrospinal tract. It is relatively darkly stained because it consists chiefly of fine fibres. Staining according to Häggqvist's technique. *B*, Bagley's tract; *bc*, brachium conjunctivum; *IV*, decussation of the trochlear nerve; *mlf*, medial longitudinal fascicle; *rs*, rubrospinal tract; *W*, Wallenberg's tract

and insectivores, region A consists of agranular cortex, without Betz cells. In carnivores, it includes areas 6, 4, 3, 1, 2 and 5 and, in primates, also parts of areas 23 and 24. Region B is located lateral and posterior to region A. It roughly corresponds to the second somatosensory area. In marsupials and carnivores, it is small and is contained in the insular area. It is large in the rhesus monkey, where it is largely buried in the sylvian sulcus; its exposed portion corresponds to part of area 7. In primates, a third region C can be distinguished, located in the lateral part of area 6. There is a predominant allometric relationship between the logarithm of the absolute size of the cortex giving rise to the corticospinal tract and the size of the total neocortex. The size of region A and (in primates) region C is directly related to digital dexterity, as measured on the five-point scale proposed by Napier and Napier (1967); for region B, size and dexterity are inversely related (Nudo and Masterton 1990b).

With two exceptions, the pattern of termination of the corticobulbar fibres in the brain stem in most mammals is similar to that in the cat. One exception is the occurrence of a large bundle of corticobulbar fibres that separates from the peduncle at high levels of the mesencephalon and descends in the ipsilateral tegmentum in Perissodactyla and Artiodactyla (the bundle of Bagley 1922; Verhaart and Sopers-Jurgens 1957; Noorduyn and Verhaart 1961; Verhaart 1970b). The second exception concerns the termination of fibres from area 4 in certain motor nuclei of the cranial nerves and on spinal motoneurons in primates (see Fig. 22.117).

In the goat, the fibres of Bagley's bundle take their origin from the same region as the corticospinal tract (gyrus sigmoideus, frontal and occipital quarter of the gyrus coronalis), corresponding to the electrically excitable motor areas for head and face, forelimb and hindlimb. It descends in the ipsilateral dorsolateral tegmentum. Bagley's bundle passes through the decussating fibres of the brachium conjunctivum and takes up a position ventral to the brachium and medial to and surrounding the motor nucleus of the trigeminal nerve (Figs. 22.115, 22.116). It terminates in the spinal trigeminal nucleus and the adjoining lateral reticular

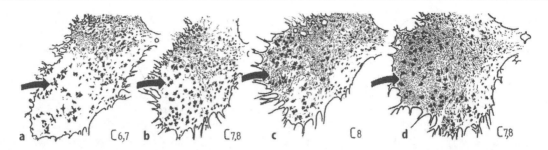

Fig. 22.117. Semidiagrammatic representation of the fiber projections to the spinal grey matter of the cervical spinal cord (*C6-8*) in the adult cat (**a**), the newborn rhesus monkey (**b**), the adult rhesus monkey (**c**) and the adult chimpanzee (**d**). *Arrow* points to the motoneuronal cell groups. Terminations of the corticospinal tract in motoneuronal groups of the ventral horn are absent or scarce in non-primates. In primates they develop postnatally. From Kuypers (1964)

formation (Haartsen and Verhaart 1967). Corticobulbar projections to the spinal trigeminal nucleus in the North American opossum (Martin and West 1967), cat (Kuypers 1958a) and monkey (Kuypers 1958b) are mainly crossed. The presence of a large Bagley's bundle, terminating ipsilaterally in the spinal trigeminal nucleus, may be related to the presence of extensive uncrossed trigeminothalamic projections in the goat and other Perissodactyla and Artiodactyla (see Sect. 22.8.1.1.4). Fibres following the same course as Bagley's bundle have been observed in the North American opossum (Martin et al. 1975) and may be present in small numbers in most other mammals.

Terminations of corticobulbar fibres in the cranial motor nuclei of nerves V, VII and XII and in the ambiguus nucleus have only been observed in primates. Monosynaptic connections of the corticospinal tract with spinal motoneurons are present in primates and in some carnivores, but are absent or extremely rare in other mammalian species that could be studied with axonal tracing methods. Terminations on distal dendrites, which are technically difficult to identify as belonging to motoneurons, generally cannot be excluded.

In the rhesus monkey and chimpanzee, corticobulbar fibres terminate bilaterally, but mainly contralaterally in the motor nucleus of the trigeminal nerve, contralaterally in the facial nucleus and bilaterally in the hypoglossal and ambiguus nuclei (Kuypers 1958b). Corticospinal fibres terminate bilaterally on ventromedial motoneuronal groups and contralaterally in the dorsolateral motoneuronal group (Kuypers et al. 1960). Direct corticomotoneuronal connections are present in increasing numbers in prosimians (slow loris, *Nycticebus coucang*), the rhesus monkey, the chimpanzee and humans (Schoen 1964; for a review, see Kuypers 1973). Direct cortico-motoneuronal connections in the rhesus monkey develop postnatally (Fig. 22,117). The adult pattern is present at the age of 8 months (Kuypers 1962).

In primates, the projections to different groups of motoneurons and their interneurons and to the sensory relay nuclei arise from different regions of the cerebral cortex. Their origin roughly corresponds to the situation in the cat (Fig. 22.111). Ventral and dorsal regions in the caudal precentral cortex (area 4) project to the dorsolateral intermediate grey and the dorsolateral motoneuronal groups in the contralateral cervical and lumbar cord, respectively (Fig. 22.114b). An intermediate and more rostrally extending zone projects bilaterally to the ventromedial parts of the intermediate grey and to motoneurons of the ventromedial group (Coulter and Jones 1977). This intermediate region also gives rise to bilateral projections to the medial reticular formation and an ipsilateral projection to the red nucleus (Fig. 22.114a). The origin of the projection to the medial reticular formation extends most rostrally into the premotor area (area 6) and the supplementary motor cortex (Kuypers and Brinkman 1970; Keizer and Kuypers 1989). A caudal strip in the lower third of the precentral gyrus projects to the cranial motor nuclei, the intertrigeminal area and the contralateral lateral parvocellular reticular formation (Kuypers 1958b).

The dorsal column nuclei receive projections both from pre- and postcentral areas. Most fibres terminating in the cuneate nucleus originate in the postcentral gyrus. The postcentral gyrus (areas 3, 1 and 2) and parietal area 5 also gives rise to somatotopically arranged projections to the nucleus princeps and the contralateral spinal trigeminal nucleus of the trigeminal nerve. These fibres also terminate in the marginal layer (lamina I), the substantia gelatinosa (lamina II), the nucleus proprius (laminae III and IV), the internal basal nucleus of the dorsal horn and the dorsal part of the intermediate grey (laminae V and VI) (Fig. 22.114c). The extragelatin-

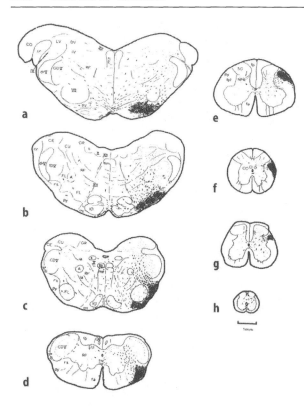

Fig. 22.118a-h. Nauta-stained sections, showing the decussation and distribution of the degenerated fibres of the pyramidal tract in the bat *Myotis myotis* after a cortical ablation. **a-c** Rostral, middle and caudal sections through the medulla oblongata. **d** High cervical cord. **e** Cervical enlargement. **f** Thoracic cord. **g** Lumbar enlargement. **h** Low sacral cord. *A*, nucleus ambiguus; *CC*, central canal; *CDV*, spinal trigeminal nucleus, pars caudalis; *CE*, external cuneate nucleus; *CO*, cochlear nuclei; *cr*, restiform body; *CU*, nucleus cuneatus; *drV*, spinal tract of the trigeminal nerve; *fa*, ventral funiculus; *FL*, lateral reticular nucleus; *fp*, dorsal funiculus; *fpl*, dorsolateral funiculus; *GR*, nucleus gracilis; *ia*, internal arcuate fibres; *IDV*, spinal trigeminal nucleus, pars interpolaris; *IO*, inferior olive; *IV*, inferior vestibular nucleus; *IX*, root of the glossopharyngeal nerve; *LV*, lateral vestibular nucleus; *lvs*, lateral vestibulospinal tract; *ml*, medial lemniscus; *mlf*, medial longitudinal fascicle; *NPR*, nucleus proprius of the dorsal horn; *ODV*, spinal trigeminal nucleus, pars oralis; *py*, pyramidal tract; *pylat*, lateral corticospinal tract in the lateral funiculus; *RF*, reticular formation; *rs*, rubrospinal tract; *S*, nucleus of the solitary tract; *SG*, substantia gelatinosa; *VII*, (motor nucleus of the) facial nerve; *X*, dorsal vagal nucleus; *XII*, (nucleus of the) hypoglossal nerve. (From Broere 1971)

ous part of the nucleus of the solitary tract receives corticobulbar fibres from the ventralmost part of the postcentral gyrus (Kuypers 1958a,b; Kuypers et al. 1960; Coulter and Jones 1977; Ralston and Ralston 1985).

In the North American opossum (Broere 1971), the cat (Satomi et al. 1989) and, presumably, in most other mammals, crossed and uncrossed corticospinal fibres descend in all funiculi, but one funiculus usually contains the majority of the crossed or the uncrossed fibres. The main termination of the corticospinal fibres is generally located in the contralateral grey matter of the cord. There are conspicuous species differences, however, in the magnitude and the level of the decussation of the pyramidal tract, in the funiculi of the cord containing the majority of the corticospinal fibres and in the extent of their descent in the spinal cord (Verhaart 1970b; Armand 1982; see Fig. 22.120). Moreover, direct cortico-motoneuronal projections only occur in primates and in a few non-primate species.

In monotremes (*Tachyglossus*: Addens and Kurotsu 1936; Verhaart 1970b; the platypus *Ornithorhynchus*: Verhaart 1970b), the pyramidal tract decussates in the pes pontis and descends ventral to the spinal tract of the trigeminal nerve into the periphery of the dorsolateral funiculus. In Chiroptera (*Myotis myotis*, *Pteropus giganteus*: Broere 1971), the decussation of the pyramidal tract is found at the level of the exit of the abducens nerve, immediately rostral to the inferior olives (Fig. 22.118). The crossed tract descends lateral to the inferior olive into the lateral funiculus, where it shifts to a more dorsolateral position. Of the Xenarthra, the giant anteater *Myrmecophaga jubata*, the two-toed sloth *Choloepus didactylus* and the armadillo *Dasypus novemcinctus*, which belong to different superfamilies, were studied by Verhaart (1970b). Both in the giant anteater and in the sloth, the pyramidal tract in the lower medulla is located ventrolateral to the inferior olive. A decussation is present in the sloth, but absent in the giant anteater. In both species, the fibres of the pyramidal tract shift along the surface of the medulla to a lateral position in the lateral funiculus. The pyramidal decussation of the armadillo is located at the bulbospinal junction. It is small, and the fibres cannot be traced into the cord.

The decussation of the pyramidal tract of the cat at the bulbospinal junction, and the descent of the majority of its fibres in the contralateral dorsolateral funiculus and the ipsilateral ventral funiculus, is typical for carnivores and primates. In the cat, the corticospinal fibers do not terminate on motoneurons. Direct cortico-motoneuronal projections have been documented in the raccoon *Procyon lotor*. They are limited to the dorsolateral ventral horn at C7/C8 (Petras and Lehman 1966; Buxton and Goodman 1967; Wirth et al. 1974). In primates, corticospinal fibers terminate on contralateral dorsolateral motoneuronal groups and bilaterally on ventromedial motoneurons.

The pyramidal tract in Artiodactyla and Perissodactyla is small. The decussation is found at its usual position, but the crossed fibres collect in a commissural bundle in the dorsal part of the ventral funiculus, dorsal to the uncrossed fibres of the pyramidal tract (Verhaart and Sopers-Jurgens 1957; Seki et al. 1963; Haartsen and Verhaart 1967; Verhaart 1970a,b). The same situation is present in the elephant (Verhaart 1963). The pyramidal decussation appears to be lacking in insectivores (Verhaart 1970b; Broere 1971: the European hedgehog *Erinaceus europeus*; the mole *Talpa europea*: Nudo and Masterton 1990a), where most fibres of the pyramidal tract descend, without crossing, into the ventral funiculus. A similar position is taken by the Hyracoideae (the klipdassie *Procavia capensis*), with an uncrossed pyramidal tract that descends in the ventral funiculus to sacral levels (Verhaart 1967; Nudo and Masterton 1990a). Its fibres cross within the cord ventral to the central canal and terminate, mainly contralaterally, in the dorsal horn and the intermediate grey (Fig. 22.119).

In marsupials (Martin and Fisher 1968; Martin et al. 1970; Broere 1971), rodents (Verhaart 1970b; Broere 1971; for reviews of the corticospinal system in the rat and other rodents, see Paxinos et al. 1990;

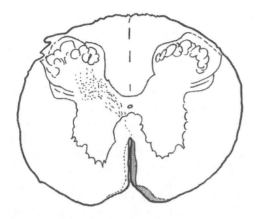

Fig. 22.119. Section through the spinal cord of *Procavia capensis* showing the localisation of the degenerated uncrossed fibres of the descending pyramidal tracts in the ventral funiculus and their termination in the contralateral spinal grey after an ablation of the cerebral cortex. Nauta stain. (From Verhaart 1967)

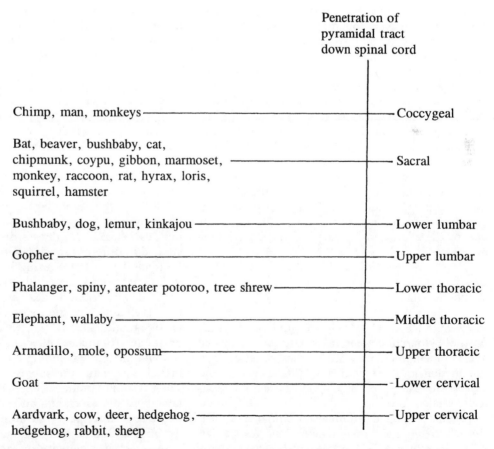

Fig. 22.120. Distribution of penetration or extension of pyramidal tract down spinal cord in 44 mammals. Note concentration of dextrous animals near top of scale. This parameter shows a close correspondence with digital dexterity. (From Heffner and Masterton 1975)

Paxinos 1995) and in the tree shrew *Tupaia glis* (Verhaart 1966), the pyramidal tract decussates at the transition of the medulla oblongata into the cord, and most of its fibres descend in the ventral part of the dorsal funiculus (Fig. 22.22e). Some crossed fibres of the pyramidal tract descend in the lateral and ventral funiculi of the North American opossum, and a few uncrossed fibres are present in all funiculi (Broere 1971). The termination of the corticospinal fibres in the North American opossum and the phalanger *Trichosuris* is restricted to the contralateral dorsal horn and the dorsal region of the intermediate grey (lamina VII). In the dorsal horn, terminations predominate in medial laminae IV–VI; they are fewer in laminae I–III and VII (Martin and Fisher 1968; Rees and Hore 1970; Martin et al. 1975; Martin and Cabana 1985).

Various parameters of the corticospinal tract (size, proportion of large fibres, descent and laminar termination in the cord, size of the cortical areas giving rise to the corticospinal tract) were compared to ratings of digital dexterity in a large number of mammals from different genera (Heffner and Masterton 1975, 1983; Nudo and Masterton 1990a,b). The strongest, positive correlations with digital dexterity were found for the total size of the cerebral cortex, the total size of the area of origin of the corticospinal tract, the descent of the tract into the cord and the ventralmost lamina receiving corticospinal terminals, because the lowest laminae of the spinal cord include the motoneurons of lamina IX. The extent of the descent of the corticospinal fibres was represented for 44 mammals by Heffner and Masterton (1975; Fig. 22.120). They noticed a concentration of dextrous mammals near the top of the scale. Their conclusions on the lower lamina receiving corticospinal fibres confirm the observations made by Lawrence and Kuypers (1968a,b) that direct cortico-motoneuronal connections are essential for the execution of independent finger movements.

22.9.2
Vestibular Nuclei

The vestibular nuclei usually are subdivided on the basis of cytoarchitectonic criteria. In the cat, Brodal and Pompeiano (1957) distinguished superior, lateral, medial and descending vestibular nuclei and several small cell groups (x, y, l) that are topographically related to the vestibular nuclei. This subdivision can be applied without substantial modifications to other mammalian species (Figs. 22.22–22.24h–k, 22.121; opossum: Henkel and Martin 1977a,b; monkey: Brodal 1984; rat: for references, see Rubertone et al. 1995). When the question was raised whether all vestibular nuclei are vestibular in the sense that they receive root fibres from the vestibular nerve, it turned out that one of the principal nuclei or, to be more precise, the dorsal half of the lateral vestibular nucleus of Brodal and Pompeiano (1957) does not receive vestibular afferents (Voogd 1964). The giant neurons of this nucleus receive projections from Purkinje cells of the anterior vermis and, also in other respects, behave like a cerebellar nucleus. This nucleus will be referred to as the lateral vestibular nucleus (Deiters' nucleus; Fig. 22.121b,c). The ventral part of the lateral nucleus of Brodal and Pompeiano (1957) contains neurons of different sizes and merges imperceptibly with the medial vestibular nucleus of these authors. It will be referred to as the magnocellular part of the medial vestibular nucleus (for a review of the early literature, see Voogd 1964).

The neurons of the lateral vestibular nucleus and the descending vestibular nucleus are located among the fibre bundles of the juxtarestiform body (Figs. 22.22–22.24h,j; Fuse 1912). The juxtarestiform body consists of Purkinje cell axons from the anterior vermis on their way to the lateral vestibular and the adjoining part of the descending vestibular nucleus and efferent tracts from the fastigial nucleus. They are flanked by efferent fibres from the fastigial nuclei. Descending branches of the vestibular nerve join the juxtarestiform body in the descending vestibular nucleus. The borders between the nuclei located within (the nucleus vestibularis lateralis and descendens) and outside (the nucleus vestibularis superior and medialis) the juxtarestiform body are generally easy to define. The borders between the lateral and descending vestibular nuclei and between the medial and superior vestibular nuclei are much harder to establish.

Fibres of the vestibular nerve bifurcate lateral to the magnocellular part of the medial vestibular nucleus (Fig. 22.122). Their distribution has been studied in different species using the Golgi method (Lorente de Nó 1933), using intra-axonal injection techniques (Mannen et al. 1982; Ishizuka et al. 1982), using antibodies against parvalbumin during the development of the nerve in the rat (Morris et al. 1988) and using axonal tracing techniques. The descending branches collect in bundles in the ventral and medial descending nucleus. The ascending branches traverse the superior vestibular nucleus. Their terminal branches enter the cerebellum. Collaterals of the ascending and descending branches of the root fibres penetrate all regions of the superior, medial and descending nuclei. Overlap rather than regional specificity in the projections of the individual semicircular canals and maculae appears to be the rule (Carpenter et al. 1972; Carleton and

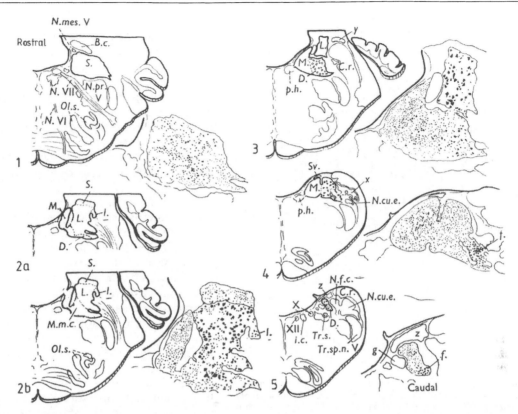

Fig. 22.121. Sections through the vestibular complex of the cat showing borders between the vestibular nuclei, based on cytoarchitectonic criteria. The diagram of section 2b was redrawn to show the alternative border between the lateral vestibular nucleus (*L*) and the magnocellular portion of the medial vestibular nucleus (*M.m.c.*) according to Voogd (1964). *B.c.*, brachium conjunctivum; *S.*, superior vestibular nucleus; *N.VII*, facial nerve; *N.PrV*, principal sensory nucleus of the trigeminal nerve; *Ol.s*, superior olive; *N.VI*, abducens nerve; *n.mes.v*, mesencephalic root of the trigeminal nerve; *M*, medial vestibular nucleus; *l*, subnucleus l; *D*, descending vestibular nucleus; *M.m.c*, medial vestibular nucleus, magnocellular part; *p.h*, nucleus prepositus hypoglossi; *y*, subnucleus y; *Sv*, supravestibular nucleus; *x*, subnucleus x; *N.cu.e*, external cuneate nucleus; *N.f.c*, cuneate nycleus; *z*, subnucleus z; *Tr.s*, soliary tract; *X*, dorsal vagal nucleus; *C.r*, restiform body; *Tr.sp.n*, nucleus of the spinal tract of the trigeminal nerve; *i.c.*, nucleus intercallatus; *f*, subnucleus f; *g*, gracile nucleus. (From Brodal and Pompeiano 1957)

Carpenter 1984; Carpenter and Cowie 1985: monkey; Siegborn and Grant 1983; Siegborn et al. 1991: cat; Kevetter and Perachio 1986: gerbil). Deiters' lateral vestibular nucleus and the small cells of group x, located at the border of the descending vestibular nucleus and the restiform body, do not receive a projection from the vestibular nerve (Korte 1979; Epema 1990). The fusiform cells of the dorsal group y, located dorsal to the restiform body, between the fibres of the floccular peduncle (Fig. 22.23), receive a projection from the sacculus (Frederickson and Thrune 1986). Ventral group y consists of smaller, closely packed neurons along the dorsal border of the restiform body (for a discussion of the subdivision of the group y, see Tan et al. 1995a; De Zeeuw et al. 1994). The nucleus prepositus hypoglossi, which is located medial to the medial vestibular nucleus in the floor of the fourth ventricle, next to the midline, does not receive fibres from the vestibular nerve.

The vestibular nuclei and their connections are very similar among mammals and among vertebrates. The vestibular nuclei of monotremes and marsupials are separated from the restiform body by the cochlear nuclei (Fig. 22.45). The connections of the vestibular nuclei were reviewed by Brodal (1974), by Henkel and Martin (1977a,b) for the opossum, by Rubertone et al. (1995) for the rat and by Brodal (1984) and Gerrits (1990) for primates and humans. There are some variations in size among the nuclei in different species, especially with respect to the lateral vestibular nucleus, which is stated to be smaller in primates, especially in humans (Verhaart 1970; Gerrits 1990). The dorsal group y is much larger in primates than in other mammals (Fig. 22.46). Its size may be related to the

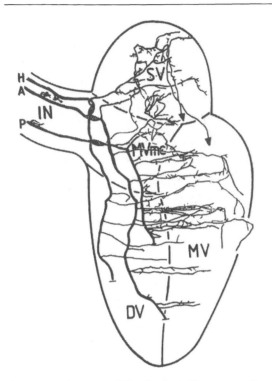

Fig. 22.122. Course and distribution of intra-axonally stained fibres of the vestibular nerve in the rat. The course of the horizontal (*H*), anterior (*A*) and posterior (*P*) canal fibres and their distribution in the four main vestibular nuclei is very similar. Cerebellar collaterals arise from the ascending branches to the superior vestibular nucleus (*arrows*). *DV*, descending vestibular nucleus; *IN*, interstitial nucleus of the vestibular nerve; *MV*, medial vestibular nucleus, caudal part; *MVmc*, medial vestibular nucleus, magnocellular part; *SV*, superior vestibular nucleus. (Redrawn from Mannen et al. 1982)

large ventral paraflocculus in primates, which shares its outflow through the floccular peduncle with the flocculus and may be involved in regulation of smooth pursuit movements of the eyes.

Apart from the vestibular nerve, the vestibular nuclei receive their main afferents from the spinal cord, the cerebellum and the nucleus prepositus hypoglossi. Spinovestibular fibres in cat, opossum and monkey mainly terminate in the lateral and descending vestibular nuclei and in group × (Pompeiano and Brodal 1957; Hazlett et al. 1972; Rubertone and Haines 1982). Some of these terminals arise as collaterals of spinocerebellar fibres, including those from the central cervical nucleus, which receives a strong reciprocal projection from the vestibular nuclei and primary afferents from the high cervical roots.

Cerebellovestibular connections arise from the Purkinje cells in the anterior vermis, the flocculus and the nodulus and the adjoining cortex in the banks of the posterolateral fissure and from the fastigial nucleus. The projections from the anterior vermis terminate in the non-vestibular lateral vestibular nucleus and in adjacent parts of the descending vestibular nucleus in all species examined (Voogd et al. 1991; see Sect. 22.6.4). Purkinje axons from the flocculus and the caudal vermis terminate in complementary regions of the superior, medial and descending nuclei. The projection from the flocculus includes the vestibulo-ocular relay cells of the anterior canal in the superior vestibular nucleus and those of the horizontal canal in the medial vestibular nucleus (for reviews, see Ito 1984; Voogd et al. 1996).

Efferent fibres from the fastigial nucleus constitute the uncrossed, direct fastigiobulbar tract and the crossed, uncinate tract, which decussates within the cerebellum. These fibres enter the brain stem by passing around the lateral vestibular nucleus: the direct fastigiobulbar fibres pass on its medial side in the wall of the fourth ventricle, while the uncinate tract passes lateral to the lateral nucleus, along its border with the restiform body. Fastigial nuclear projections terminate heavily in the magnocellular medial vestibular nucleus and, more caudally, in the medial vestibular nucleus and in the descending nucleus, including its caudal pole, located next to the nucleus of the solitary tract (parasolitary nucleus). After passing through the vestibular nuclei, the uncinate and direct fastigiobulbar tract fibres proceed to the medial reticular formation, where they terminate (Voogd 1964; Batton et al. 1977). Descending fibres from the interstitial nucleus of Cajal (Spence and Saint-Cyr 1988) and the parietal cortex (Faugier-Grimaud and Ventre 1989) terminate in the vestibular nucleus, with the exception of the lateral vestibular nucleus. The nucleus prepositus hypoglossi projects to widespread regions within the vestibular nuclei. The interstitial nucleus serves as a link in a disynaptic pathway from the nucleus of the optic tract and the nuclei of the accessory optic system in the mesencephalon to the vestibular nuclei (Cazin et al. 1982; McCrea and Baker 1985).

The great majority of the neurons of the vestibular nuclei give rise to intrinsic and commissural connections. Extrinsic projections can be subdivided into descending, ascending and cerebellar tracts. Intrinsic systems interconnect the vestibular nuclei and focus on the magnocellular medial vestibular nucleus (Fig. 22.123). The lateral vestibular nucleus does not participate in this intrinsic system or in the vestibular commissure. The commissural fibres cross in the dorsal tegmentum and in the cerebellar commissure. Their terminations diverge somewhat and include other vestibular nuclei, apart from the true commissural, symmetrical connections between the individual nuclei in rat, cat

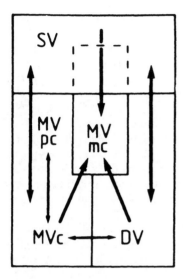

Fig. 22.123. Summary of the intrinsic connections of the vestibular nuclei of the rabbit. *DV*, descending vestibular nucleus; *MVC*, medial vestibular nucleus, caudal port; *MVmc*, medial vestibular nucleus, magnocellular part; *MVpc*, medial vestibular nucleus, parvocellular part; *SV*, superior vestibular nucleus. (From Epema et al. 1988)

rabbit and primates (Rubertone and Mehler 1980; Rubertone et al. 1983; Carleton and Carpenter 1983; Ito et al. 1983; Epema et al. 1988).

Descending vestibulospinal tracts arise from the lateral, medial and descending vestibular nuclei. Accurate descriptions of the vestibulospinal tracts are few. The older literature, as summarised by Busch (1961), and Busch's own observations on the vestibulospinal system in the cat are the most reliable. Busch (1961) distinguished lateral, medial (uncrossed) and crossed vestibulospinal tracts. The coarse fibres of the lateral vestibulospinal tract take their origin from the giant cells of the lateral vestibular nucleus and descend through the reticular formation to a position in the lateral ventral funiculus (Figs. 22.20, 22.23e–k). The fibres descend to sacral levels and terminate on commissural neurons of lamina VIII and in the ventromedial lamina VII (Nyberg-Hansen 1964, 1975). A contribution from the magnocellular medial vestibular nucleus to the lateral vestibulospinal tract cannot be excluded, but appears unlikely. The medial (uncrossed) vestibulospinal tract consists of thin and medium-sized fibres from the magnocellular and adjoining regions of the medial vestibular nucleus. It descends in the dorsolateral part of the bulbar medial longitudinal fascicle into the ventral funiculus. The crossed vestibulospinal tract is much larger than the medial vestibulospinal tract and contains a large proportion of coarse fibres. It descends in the medial longitudinal fascicle next to the raphe and enters the ventral funiculus. The termination of the medial and crossed vestibulospinal tracts is mainly restricted to the cervical cord. With respect to the course and the calibre of the vestibulospinal tracts, this account is very similar to the description given by Glover and Petursdottir (1988) of the vestibulospinal system in 11-day-old chick embryos (Fig. 22.124). This organisation may, therefore, be characteristic of most, if not all vertebrates. For more recent observations on vestibulospinal systems originating from the caudal vestibular nuclei, their termination in more dorsal laminae of the spinal grey, the collateralisation of the vestibulospinal tracts to the brain stem (reticular formation, including precerebellar nuclei, inferior olive) and the distribution of excitatory and inhibitory transmitters in the vestibulospinal system, the reader is referred to the review by Rubertone et al. (1995).

The main, ascending connections of the vestibular nuclei terminate in the oculomotor nuclei. The pattern of the vestibulo-oculomotor connections, like the vestibulospinal tracts, is highly conserved in the vertebrate central nervous system. The medial, descending and superior vestibular nuclei and group y give rise to crossed excitatory and uncrossed ascending inhibitory pathways to the oculomotor nuclei. The organisation of the vestibulo-oculomotor system is such that each of the three semicircular canals is connected with two pairs of eye muscles that move the ipsilateral and contralateral eye in the plane of the semicircular canal. Displacement of the endolymph towards the ampulla (in the horizontal canal) or away from the ampulla (in the vertical canals) gives rise to conjugate compensatory movements of the eyes in the direction of the endolymph stream, through the contraction of the muscles whose motoneurons are innervated by the excitatory vestibulo-ocular pathway and the inhibition of their antagonists (Fig. 22.125). The organisation and the spatial coordinate system of the vestibulo-ocular system were discussed by Highstein and Reisine (1979), Ito (1984), Simpson et al. (1957, 1984), Simpson (1988) and McCrea et al. (1987).

The excitatory vestibulo-ocular relay cells for the posterior and the horizontal canal and the inhibitory horizontal canal cells are located in the magnocellular and adjoining regions of the medial vestibular nucleus (Fig. 22.125a,b). Excitatory neurons for the anterior canal are located in the dorsal part of the superior vestibular nuclei and the group y (Fig. 22.125c). Inhibitory, anterior and posterior canal cells occupy the centre of the superior vestibular nucleus.

Excitatory and inhibitory connections from the horizontal canal terminate in the contralateral and

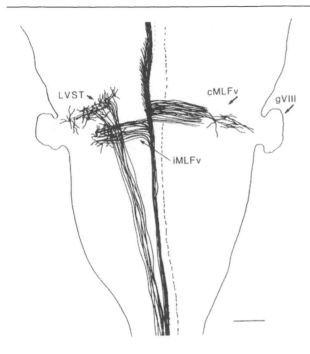

Fig. 22.124. Spatial relationships of retrogradely labelled vestibulospinal groups and their initial axon trajectories. This configuration closely corresponds to the situation in the mammalian brain stem. Composite camera lucida drawing of two consecutive 100-μm horizontal sections near the floor of the fourth ventricle of an 11-day-old chick embryo. The rostralmost area is at the top. Note the different rostrocaudal extents of the initial axon trajectories of the ipsilateral (*iMLF*) and the contralateral (*cMLFv*) groups of the vestibulospinal axons. The lateral vestibulospinal tract runs in a more lateral position (*LVST*). The auditory vestibular ganglion complex (*gVIII*) serves as a landmark. *Bar*, 400 μm. (From Glover and Petursdottir 1988)

ipsilateral abducens nucleus, respectively (Fig. 22.125a). The inhibitory projection is glycinergic (Spencer et al. 1989). Excitation and disexcitation of the motoneurons of the medial rectus division of the oculomotor nucleus, ipsilateral to the horizontal canal, is coupled to the abducens nucleus through an internuclear pathway in the medial longitudinal fascicle, which originates from neurons in the abducens nucleus and terminates in the medial rectus division (see Frame 7).

The axons of excitatory posterior canal cells ascend in the contralateral medial longitudinal fascicle. They terminate on motoneurons of the oculomotor nucleus innervating the superior rectus and the inferior oblique muscles (Fig. 22.125b). The axons of superior rectus motoneurons cross between the oculomotor nuclei to innervate the muscle of the eye ipsilateral to the posterior canal. The axons from the inhibitory (GABAergic) anterior and posterior canal cells ascend from the superior vestibular nucleus into the lateral wing of the medial longitudinal fascicle.

Axons from excitatory anterior canal cells in the superior vestibular nucleus and group y do not travel in the medial longitudinal fascicle, but join the superior cerebellar peduncle and cross in the ventral part of its decussation (Yamamoto et al. 1978). They terminate on motoneurons of the trochlear nerve and the inferior rectus subdivision of the oculomotor nucleus (Fig. 22.125c). The axons from the trochlear nucleus cross in the anterior medullary velum to innervate the superior oblique muscle of the eye ipsilateral to the anterior canal (Thunnissen 1990).

Exactly the same pattern of vestibulo-oculomotor connections can be recognised in 11-day-old chick embryo (Petursdottir 1990). Abberations from this pattern have been described in many studies using tract-tracing techniques (Gerrits 1994). They were discussed and partially explained on the basis of differences in angulation of the planes of the semicircular canals and the planes of action of the three pairs of extraocular muscles by Ezure and Graf (1984). An ipsilaterally ascending system, carrying excitatory fibres for motoneurons of the medial rectus division of the oculomotor nucleus, is known as the 'ascending tract of Deiters'. It does not originate from Deiters' nucleus, but from the magnocellular medial vestibular nucleus, nor is it a separate tract. It is part of the lateral wing of the medial longitudinal fascicle. It is present in cat and monkey (Lang et al. 1974), but absent in the rabbit (Thunnissen 1990).

Ascending fibres from the vestibular nuclei also terminate in the periaqueductal grey, dorsal to the oculomotor nucleus, the interstitial nucleus of Cajal, the deep layers of the superior colliculus and the thalamus (Rubertone et al. 1995). Some of the ascending fibres to the interstitial nucleus are ascending branches from vestibulospinal fibres (Epema 1990). Secondary vestibulocerebellar mossy fibre connections are reviewed in Sect. 22.6.

Centrifugal fibres in the vestibular nerve that terminate in the sensory epithelium of the labyrinth originate from small neurons localised dorsolateral to the genu of the facial nerve and ventromedial to the medial vestibular nucleus (White and Warr 1983; Rubertone et al. 1995). They stain strongly for AChE in the cat, primate species and rodents (Blanks and Palay 1978; Goldberg and Fernandez 1980; Perachio and Kevetter 1989).

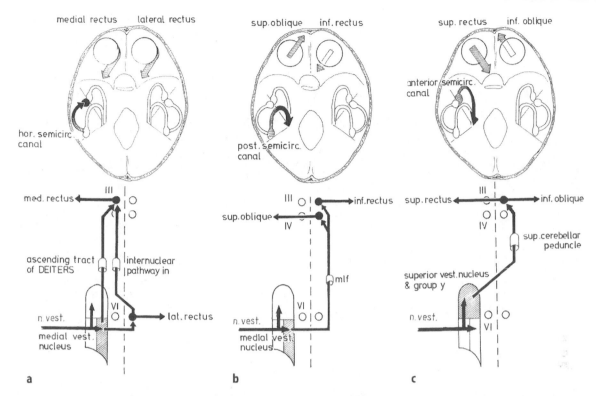

Fig. 22.125a-c. Main excitatory vestibulo-ocular connections. a Excitation of the horizontal canal results from movement of the endolymph in the direction of the ampulla. The motoneurons of the lateral rectus muscle are innervated by a crossed pathway from the medial vestibular nucleus. Motoneurons innervating the ipsilateral rectus muscle are innervated by the internuclear pathway from the abducens nucleus and the ascending tract of Deiters. b,c Excitation of the anterior and posterior semicircular canals results from displacement of the endolymph away from the ampulla. In the case of the posterior canal, motoneurons are innervated by a crossed ascending pathway in the medial longitudinal fascicle, and in the case of the anterior canal by a pathway ascending in the brachium conjunctivum. Ipsilateral muscles are innervated by axons that cross within the brain stem

22.9.3
Red Nucleus and Associated Cell Groups

The red nucleus is a premotor centre located in the tegmentum of the mesencephalon in limb-using vertebrates (ten Donkelaar 1988). It receives cortical, cerebellar and somatosensory afferent projections. It gives rise to the crossed rubrobulbar and rubrospinal tract, which terminates on interneurons and some motoneuronal groups in the brain stem and the cord. In primates, it can be subdivided into a caudal magnocellular and a rostral parvocellular part. This subdivision is much less distinct in Carnivora, Perissodactyla, Artiodactyla and other mammals (Hatschek 1907). The rostral parvocellular red nucleus gives rise to the central tegmental tract (Fig. 22.24k-p), which terminates in the ipsilateral inferior olive.

The parvocellular red nucleus and the central tegmental tract are extremely large in the human brain (for a review, see Voogd et al. 1990). The human magnocellular red nucleus is small, and only a few fibres can be detected in the position of the rubrospinal tract at the level of the pons and the medulla oblongata. A rubrospinal tract is absent in humans (Schoen 1964).

The central tegmental tract is only one of the descending systems from the mesencephalon that terminate in the inferior olive. In the cat, the medial tegmental tract arises from the nucleus of Darkschewitsch, descends along the midline close to the medial longitudinal fascicle and terminates in the medial accessory olive (Ogawa 1939a,b; Busch 1961). At their origin, the central and medial tegmental tracts constitute a single bundle; they do not separate until they reach a more caudal position. The origin of the central and medial tegmental tract is not limited to the red nucleus and the nucleus of Darkschewitsch. In the cat, neurons projecting to the inferior olive are found in the prerubral field, the subparafascicular nucleus, the medial accessory nucleus of Bechterew, the interstitial nucleus of Cajal and as scattered neurons located within the rostral half of the red nucleus and along its dorso-

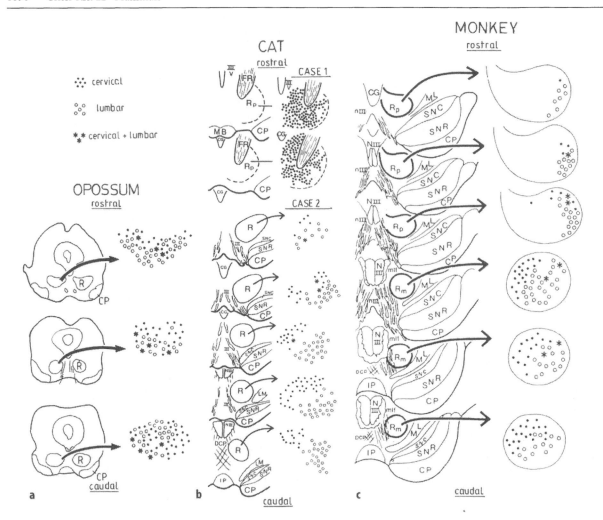

Fig. 22.126. Somatotopical arrangement in the rubrospinal projection in the opossum (Martin 1982), the cat (Huisman 1981) and the monkey (Huisman 1983). The segregation of neurons with projections to the cervical (*dots*) and lumbar cord (*open circles*) is poor in the opossum and almost complete in the cat (*case 2*) and the monkey. Neurons with double projections to the cervical and the lumbar cord are indicated by *asterisks*. The localisation of small neurons projecting to the ipsilateral inferior olive in the rostral pole of the red nucleus of the cat is indicated in *case 1*. The magnocellular part of the red nucleus in the monkey gives rise to rubrospinal fibres and extends rostrally for some distance, lateral to the parvocellular red nucleus. *CG*, central grey; *CP*, cerebral peduncle; *DCP*, dorsal tegmental decussation; *FR*, fasciculus retroflexus; *III*, oculomotor nucleus; *IP*, interpeduncular nucleus; *LM*, medial lemniscus; *MB*, mammillary body; *ML*, medial lemniscus; *mlf*, medial longitudinal fascicle; *nIII*, oculomotor nerve; *R*, red nucleus; *Rm*, magnocellular red nucleus; *Rp*, parvocellular red nucleus; *SNC*, substantia nigra, pars compacta; *SNR*, substantia nigra, pars reticulata

medial and dorsal borders (Figs. 22.126, 22.127; Saint-Cyr and Courville 1980, 1981; Onodera 1984; Holstege and Tan 1988). The central tegmental tract of the cat, which contains the axons from the rostral red nucleus, is rather small (Busch 1961; Edwards 1972; Walberg 1982; Walberg and Nordby 1981; Condé and Condé 1982; Saint-Cyr and Courville 1982; Robinson et al. 1987).

Cytoarchitectonic borders between the neurons that give rise to the rubrospinal spinal tract and neurons in and around the rostral pole of the red nucleus that project to the inferior olive are hard to establish (Davenport and Ranson 1930). In the cat, neurons of all sizes, distributed over the entire red nucleus, give rise to rubrospinal fibres (Pompeiano and Brodal 1957b; Huisman et al. 1982).

The rubrospinal projection in the cat is somatotopically organised, with neurons projecting to the cervical cord located dorsally and neurons projecting to the lumbar cord ventrolaterally (Fig. 22.126b). Few (less than 8%) of the neurons collateralise to both enlargements (Hayes and Rustioni 1981; Huisman et al. 1981, 1982). Rubrobulbar neurons are located dorsal to the magnocellular red nucleus in the cat (Holstege and Tan 1988). At the level of the motor nucleus of the trigeminal nerve,

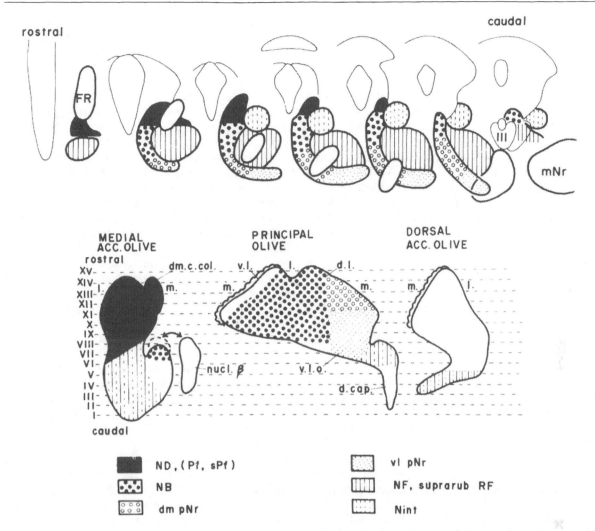

Fig. 22.127. Summary of the topographic relations between the mesodiencephalic nuclei (*top*) and the inferior olive (*bottom*) in the cat. Different symbols indicate the origin and the termination of the mesodiencephalo-olivary projections. *d.cap*, dorsal cap; *d.l.*, dorsal leaf; *dm pNr*, dorsomedial part of parvicellular red nucleus; *dm.c.col.*, dorsomedial cell column; *FR*, fasciculus retroflexus; *l.*, lateral; *m.*, medial; *mNr*, red nucleus; *NB*, Bechterew's nucleus; *ND*, nucleus of Darkschewitsch; *NF*, nucleus of the field of Forel, suprarubral reticular formation; *Nint*, interstitial nucleus; *v.l.*, ventral leaf; *v.l.o.*, ventrolateral outgrowth; *vl pNr*, ventrolateral part of parvicellular red nucleus. (From Onodera 1984)

collaterals from rubrospinal fibres detach from the tract and enter the cerebellum along the brachium conjunctivum. Most of these fibres terminate in the cerebellar nuclei (Brodal and Gogstad 1954; Courville and Brodal 1966), but some proceed to the cerebellar cortex (Dietrichs and Walberg 1983). At least 25% of the rubrospinal tract neurons collateralise to the interposed nucleus of the rat (Huisman et al. 1983). Rubrobulbar fibres terminate in the lateral part of the nucleus of the facial nerve, in the abducens nucleus, in the lateral bulbar reticular formation, in the hilus of the dorsal column nuclei and in the rostrodorsal part of the lateral reticular nucleus (Kuypers 1964, 1981; Holstege and Kuypers 1982; Qvist et al. 1984). Crossed rubrobulbar fibres do not terminate in the inferior olive. The rubrobulbospinal and rubro-olivary neurons of the red nucleus constitute separate populations. Neurons with branching axons to the inferior olive and the spinal cord are not present in the cat (Huisman et al. 1982). Similar observations on the structure of the red nucleus and the origin of the descending pathways were made in the rat (for a review, see Ruigrok and Cella 1995). In some studies employing retrograde double-labelling techniques in the rat, it was suggested that the rubro-olivary pathway, at least in part, may be composed of collaterals of the rubrospinal tract (Kennedy 1990; Tucker and Kennedy 1990; Tucker et al. 1989).

In the spinal cord of the cat, the rubrospinal

fibres are located in the lateral funiculus, ventral to the lateral pyramidal tract (Figs. 22.20, 22.23a–e). They terminate in the dorsolateral intermediate grey (laminae V–VII) over the entire length of the cord (Nyberg-Hansen and Brodal 1964) and on motoneurons in the dorsolateral ventral horn at C8/T1 (Holstege 1987; Fujito et al. 1991).

The organisation of the projections of the red nucleus and the nuclei at the mesoencephalic-diencephalic border to the contralateral brain stem and spinal cord and the ipsilateral olive in the North American opossum are very similar to the cat. The rubrospinal tract arises from the caudal and rostroventral part of the red nucleus (Fig. 22.126a). Cells with projections to the facial nucleus are found rostrodorsally. Small rubro-olivary neurons overlap with rubrospinal cells in the rostromedial part of the nucleus (Martin et al. 1983). The segregation of neurons with projections to the cervical and the lumbar cord is much less distinct than in the cat (Martin et al. 1981). Rubrobulbar fibres terminate in and around the principal sensory nucleus of the trigeminal nerve, the spinal nucleus of the trigeminal nerve, pars oralis and interpolaris, ventrolateral, lateral and dorsal parts of the facial nucleus, the lateral bulbar reticular formation and the lateral reticular nucleus. Fibres reach the cerebellum along the brachium conjunctivum (Martin and Dom 1970; Martin et al. 1974). The rubrospinal tract is of moderate size and descends in the contralateral dorsolateral funiculus to sacral levels (Fig. 22.22a–e). A few fibres are present in the ipsilateral lateral funiculus. The rubrospinal tract terminates dorsolaterally in laminae III–VII, while fewer fibres terminate at thoracic, sacral and coccygeal levels (Cabana and Martin 1986). The degree of collateralisation between the cervical and lumbar enlargements is higher in the opossum than in the cat (Martin et al. 1981; Huisman et al. 1982; Fig. 22.126a). The topography in the projection of the red nucleus and the nuclei of the mesoencephalic-diencephalic border to the inferior olive in marsupials was studied and reviewed by Linauts and Martin (1978) and Martin et al. (1980, 1983).

The red nucleus of Old and New World monkeys consists of well-demarcated caudal magnocellular and rostral parvocellular portions. Caudally, in the magnocellular part, large cells, dispersed between the fibre bundles of the ascending branch of the brachium conjunctivum, prevail. Its rostral portion contains mainly medium-sized neurons. A small group of medium-sized neurons extends over a short distance along the lateral aspect of the parvocellular part (Fig. 22.126c). Neurons of the magnocellular red nucleus give rise to crossed rubrobulbar and rubrospinal fibres. In their distribution, they display a clear somatotopical pattern (Fig. 22.126c; Poirier and Bouvier 1966; Kuypers and Lawrence 1967; Kneisley et al. 1978; Miller and Strominger 1973; Huisman et al. 1982).

The rostral part of the red nucleus in the monkey is better delineated from the surrounding reticular formation than in the cat and opossum. The rostral part of the primate red nucleus consists of relatively small neurons. Rostrally, it is located lateral to the fasciculus retroflexus. In the ventral part of the parvocellular red nucleus, the cells are densely packed, whereas dorsally they form a loose network, merging into the medial accessory nucleus of Bechterew, the subparafascicular nucleus and the nucleus of Darkschewitsch (Fuse 1937; Fukuyama 1940; Huisman 1983). The parvocellular red nucleus in monkeys gives rise to the central tegmental tract (Poirier and Bouvier 1966; Miller and Strominger 1973; Robertson and Stotler 1974; Strominger et al. 1979, 1985; Kennedy et al. 1986). The projection of the nucleus of Darkschewitsch through the medial tegmental tract is also present in primates and humans (Courville and Otabe 1974; Voogd et al. 1990).

The course and the termination of the rubrobulbar and rubrospinal tract in monkeys is very similar to that in the cat (Fig. 22.24a–p; Kuypers et al. 1960, 1962; Kuypers 1964, 1981). In the rhesus monkey, collateralisation of rubrospinal tract fibres between the cervical and lumbar enlargements is limited to less than 10% of the rubrospinal neurons (Huisman et al. 1982). It is not known whether the direct rubromotoneuronal projections that have been reported for the cat (Holstege 1987a) are also present in primates. A rubrocerebellar projection has never been reported in primates.

The parvocellular red nucleus and the nuclei of the mesoencephalic-diencephalic junction serve as the main link in the projection of the cerebral cortex onto the inferior olive. Direct cortico-olivary connections are few. The corticorubral pathways to the parvocellular red nucleus in the monkey arise from the entire motor cortex and from patches in the premotor cortex, including the frontal eye field, the supplementary motor cortex and the cingulate gyrus (areas 4, 6, 8, 23 and 24) and to a limited degree from the superior parietal lobule. The projection is mainly ipsilateral and somatotopically organised (Kuypers and Lawrence 1967; Hartman-von Monakov et al. 1979; Humphrey et al. 1984; Miyara and Sasaki 1984; Leichnetz et al. 1984; Huerta et al. 1986; Stanton et al. 1988). The projection from areas 8, 6 and 4 also reaches the medial accessory nucleus of Bechterew and the nucleus of Darkschewitsch (Leichnetz 1982; Leichnetz et al. 1984; Huerta et al. 1986; Stanton et al. 1988). The nucleus of Darkschewitsch also receives a projection from posterior area 7 in the monkey (Faugier-Grimaud and Ventre 1989). The pyramidal cells of areas 4 and 6 which give rise to these projections are located in layer Va, superficial to the cells projecting to the magnocellular red nucleus (Catsman-Berrevoets et al. 1979; Humphrey et al. 1984). The projection of the motor cortex to the magnocellular red nucleus is much smaller than the projection to the parvocellular red nucleus and consists of both direct fibres and collaterals from larger corticospinal fibres (Humphrey et al. 1984; Canedo and Towe 1986). Similar projections from areas 4 and 6 to the parvocellular red nucleus and other mesencephalic cell groups known to project to the inferior olive, namely the nucleus of Darkschewitsch, are present in the cat (Mabuchi and Kusama 1966; Mizuno et al. 1973; Saint-Cyr 1987; Miyashita and Tamai 1989). A minor projection to the red nucleus of the cat from sensory areas 3a and 3b was found in the cat (Saint-Cyr 1987), but was absent in the monkey (Humphrey et al. 1984). In the studies carried out by Oka et al. (1979) and Oka (1988) on the corticorubral projection to the parvocellular red nucleus in the cat, these connections were traced exclusively from the parietal association cortex and not from the motor cortex.

In the rat, the region lateral to the fasciculus retroflexus, which is thought to correspond to the parvocellular red nucleus, receives a bilateral projection from the sensorimotor cortex (Brown 1974; Rutherford et al. 1989) and the medial agranular cortex, which is thought to correspond to the primate frontal eye field (Stuesse and Newman 1990). An important projection from a subpretectal area, corresponding to the posterior thalamic nucleus, to the parvocellular red nucleus was described by Roger and Cadesseau (1985, 1987) in the rat and cat.

The red nucleus and the nucleus of Darkschewitsch and adjacent structures receive a major projection from the cerebellar nuclei. Those terminating in the parvocellular red nucleus are derived from the dentate nucleus. The interposed nuclei project to the magnocellular division of the red nucleus. This projection is mainly derived from the anterior interposed nucleus and is somatopically organised (Flumerfelt et al. 1973; Chan-Palay 1977; Asanuma et al. 1983a,b; Gonzalo-Ruiz and Leichnetz 1990: monkey; Chan-Palay 1977; Caughall and Flumerfelt 1977; Flumerfelt and Caughall 1978; Daniel et al. 1987: rat; Voogd 1964; Angaut and Bowsher 1965; Condé 1966; Courville 1966; Angaut 1970; Robinson et al. 1987: cat; King et al. 1973: opossum).

The terminations from the posterior interposed nucleus are arranged as a half-shell along the ventral and medial border of the red nucleus. They continue along the oculomotor nuclei in the supra-oculomotor region and the lateral periaqueductal grey to include the nucleus of Darkschewitsch and the subparafascicular nucleus (Voogd 1964; Courville 1966; Angaut 1970; Robinson et al. 1987: cat: Kievit 1979; May et al. 1992: monkey; Daniel et al. 1987: rat). Fastigial projections do not terminate in the red nucleus, but cover the supra-oculomotor region including the Edinger-Westphal nucleus and extend into the lateral periaqueductal grey, the nucleus of Darkschewitsch, Bechterew's medial accessory nucleus and the prerubral field (May and Hall 1986; Gonzalo-Ruiz et al. 1990; May et al. 1992).

Cerebellar projections terminate on the somata and the proximal dendrites of the magnocellular red nucleus, whereas the distal dendrites are contacted by the relatively minor corticorubral projection (Dekker 1981; King et al. 1973; Nakamura 1971; Naus et al. 1985). The lateral cerebellar nucleus (Naus et al. 1985: rat) and the sensorimotor cortex (Naus et al. 1985: rat; Jenny et al. 1991: monkey) both contact only dendrites in the parvocellular red nucleus.

Connections of the spinal cord and the dorsal column nuclei with the red nucleus and the nucleus of Darkschewitsch have been established for different species (Ghez 1975; Berkley and Hand 1978; Jeneskog and Padel 1984; Björkelund and Boivie 1984; Wiberg and Blomqvist 1984: cat; Padel et al. 1986; Wiberg et al. 1987; Yezierski 1988: rat, cat and monkey). These connections, which bypass the cerebellum and the thalamus, may constitute links in the long-latency spino-olivocerebellar climbing fibre paths of Larson et al. (1969a,b).

The parvocellular red nucleus and the nuclei of the mesoencephalic-diencephalic junction, therefore, are links in closed cerebello-mesencephalic-olivary circuits which feed back into the cerebellum (Fig. 22.63). These multisynaptic loops are presumably excitatory and are superimposed on the direct GABAergic nucleo-olivary projections from the same cerebellar nuclei. These multisynaptic loops are organised in the same reciprocal fashion as the corresponding nucleo-olivary pathways. This, at least, seems to be true for the connections of the posterior interposed nucleus, via the nucleus of Darkschewitsch to the rostral medial accessory olive, and for the dentato-rubro-olivary loop which converges upon the principal olive. These loops are under the influence of the motor, premotor and the parietal cortex through the corticorubral and corticomesencephalic pathways. The main output of these loops is through the cerebellothalamic projections to the ventral lateral nucleus and the motor and premotor cortex.

In primates, the size of the pyramidal and the rubrospinal tracts, which both terminate in the dorsolateral intermediate grey of the spinal cord and on spinal motoneurons, are inversely related. The rubrospinal tract is small in the baboon (less than 1300 rubrospinal neurons) and even smaller in Hylobatidae and other Homonoidea (Verhaart 1970a,b; Padel et al. 1981). The rubrospinal tract is extremely large in Artiodactyla and Perissodactyla, which lack a well-developed spinal pyramidal tract. In Insectivora, however, a small or absent pyramidal tract is combined with an inconspicuous rubrospinal tract (Verhaart 1970b).

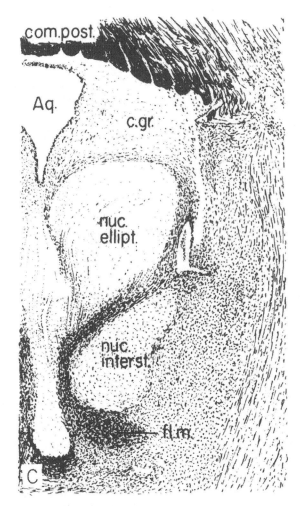

Fig. 22.128. Frontal section through the right half of the mesencephalic periaqueductal grey with the nucleus ellipticus of the porpoise *Phocaena phocaena*. The nucleus ellipticus corresponds to the nucleus of Darkschewitsch in other mammals and gives rise to the medial regmental tract, which terminates in the rostral medial accessory olive, which is greatly enlarged in Cetacea (Fig. 22.57). *Aq,* cerebral aqueduct; *c.gr,* periaqueductal grey; *com.post,* posterior commissure; *flm,* medial longitudinal fascicle; *nuc.ellipt,* nucleus ellipticus; *nuc.interst,* interstitial nucleus of the medial longitudinal fascicle of Cajal. (From Jansen 1969)

The loop consisting of the dentate nucleus-parvocellular red nucleus-principal olive-dentate nucleus and the Purkinje cells of the D zones of the cerebellar hemisphere is well developed in primates. Both the dentate-rubral loop and the recurrent circuit consisting of the posterior interposed nucleus-the nucleus of Darkschewitsch-rostral medial accessory olive-posterior interposed nucleus and Purkinje cells of the C2 zone are present in marsupials and carnivores. The recurrent circuit through the posterior interposed nucleus and the nucleus of Darkschewitsch is greatly enlarged in Cetacea. The large nucleus of Darkschewitsch, which almost replaces the red nucleus in Cetacea is known as the nucleus ellepticus (Fig. 22.128). The rostral half of the medial accessory olive (Fig. 22.57) and the posterior interposed nucleus (Fig. 22.47) are also greatly magnified in these species (Ogawa 1935; De Graaf 1967; Jansen 1969). The wide C2 zone in whales and dolphins (Fig. 22.53) may account for the large paraflocculus in these species; Fig. 22.39).

The question of the functional relation between the magnocellular and parvocellular subdivisions of the red nucleus, i.e. whether these are separate, unrelated structures that have been accidentally joined into a single griseum or whether they are two centres which function in unison, has not received a satisfactory answer. The anatomical evidence favours a strict separation between the two subdivisions. However, at the level of the cerebellar nuclear input to the red nucleus, there may be overlap, in the sense that the dentate and the (posterior) interposed nucleus may have access to both rubrospinal neurons in the magnocellular red nucleus and to rubro-olivary neurons in the parvocellular part and associated nuclei. According to the hypothesis proposed by Kennedy (1990), the rubro-olivo-cerebellar projection is involved in the switch in activity from the corticospinal tract to the rubrospinal tract, which would occur when learned movements become executed in a more or less automatic manner. The access of both the dentate and the interposed nuclei to rubro-olivary, rubrospinal and, through the thalamus, to corticospinal neurons is an essential element of this hypothesis.

22.9.4
Long Descending Tracts and Motor Behaviour

Several attempts have been made to place the corticobulbar and corticospinal connections and the descending brain stem systems in a wider concept of motor regulation. Kuypers (1964, 1981; Kuypers et al. 1960, 1962; Holstege and Kuypers 1982) distinguished the ventromedial brain stem systems

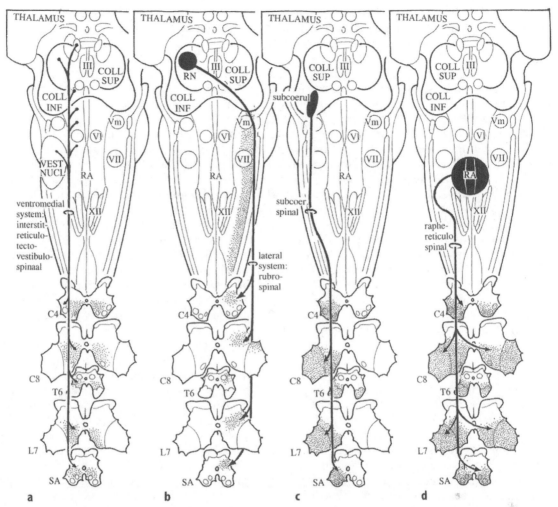

Fig. 22.129a–d. The organization of the descending motor systems from the brain stem in the cat. **a** Ventromedial system, **b** lateral system, **c** subcoeruleospinal system, and **d** raphespinal system. Note that the ventromedially descending pathways distribute fibres bilaterally to the ventromedial part of the intermediate zone. In contrast, the laterally descending pathway distribute fibres contralaterally to the lateral and dorsal parts of the spinal intermediate zone and to the dorsolateral cervical motoneuronal cell groups (Holstege, 1987a). The two diagrams on the right show the diffuse projecions of the subcoeruleospinal and raphe-reticulospinal fibres to the intermediate zone and the spinal motoneuronal cell groups (Halstege (1987a). *C*, cervical cord; *III*, oculomotor nucleus; *INF COLL*, inferior colliculus; *L*, lumbar cord; *RA*, raphe nuclei; *RN*, red nucleus; *SA*, sacral cord; *subcoerul*, subcoerulaean nucleus; *SUP COLL*, superior colliculus; *T*, thoracic cord; *VEST NUCL*, vestibular nuclei; *VI*, abducens nucleus; *VII*, facial nucleus; *Vm*, motor nucleus of the trigeminal nerve; *XII*, hypoglossal nucleus. Modified from Kuypers (1981)

(vestibulospinal, interstitiospinal, tectospinal and medial reticulospinal tracts), which descend in the ventral and the adjacent ventrolateral funiculus and which terminate bilaterally in the ventromedial intermediate grey, from the lateral brain stem systems (rubrospinal tract; crossed pontine reticulospinal tract), which descend in the dorsolateral funiculus and terminate unilaterally in the dorsolateral intermediate grey (Fig. 22.129). Ventromedial systems terminate on long propriospinal neurons, which distribute collaterals bilaterally over large regions of the cord. These systems steer body and integrated limb and body movements and movement synergies of the individual limbs. Lateral systems focus on functionally related discrete groups of interneurons and motoneurons. They steer independent movements of the limb, especially of their distal parts. The spinal projections of the corticospinal system and the ventromedial and lateral subcortical motor systems overlap, but the two main subdivisions in the corticospinal system share characteristics with the ventromedial and lateral subcortical motor system, respectively (Figs. 22.111, 22.114a,b). Kuypers' distinction of ventromedial and lateral motor systems can also be applied to the innervation of the cranial motor nuclei of the brain stem.

Holstege's studies of the premotor connections of the bulbar reticular formation with the cranial motor nuclei in the cat (Holstege and Kuypers 1977; Holstege et al. 1977) and, more importantly, his observation that some raphe-spinal, medial reticulospinal and coerulospinal systems terminate diffusely, but directly on spinal motoneurons (Fig. 22.129c,d; Holstege 1991; see also J.C. Holstege and Kuypers 1987a,b), led him to distinguish the 'emotional motor system' from the volitionally controlled corticospinal, medial and lateral motor systems of Kuypers. Holstege's emotional motor system (Holstege 1992, 1995) includes coerulospinal, raphe-spinal and reticulospinal systems that cited controlled by the limbic system through the hypothalamus and the periaqueductal grey. These concepts cited discussed at great length in a recent publication edited by G. Holstege et al. (1996).

The origin of descending pathways to the spinal cord in 22 species, belonging to 19 different genera, was studied by Nudo and Masterton (1988) using a standard protocol, with large injections of a retrograde tracer in the high cervical spinal cord (Fig. 22.113); cell groups with projections to the spinal cord could be identified in each of the species examined, including reticular and raphe nuclei, the tectum, the red nucleus and associated cell groups, the vestibular complex, several motor and somatosensory relay nuclei, the periaqueductal grey, the hypothalamus and the cerebral cortex. The basic similarity in this catalogue of descending connections to the mammalian spinal cord hides large differences in the absolute or relative size or distribution of individual or certain combinations of descending systems.

Nudo and Masterton (1989) were among the first to study the allometric relations of anatomical parameters of some of these descending systems to the size of the body and the brain or its relevant subdivisions. They designed methods to establish possible relations of these parameters with behavioural characteristics, such as digital dexterity, visual prowess and nocturnal or diurnal habits, and ecological features, such as foraging location, trophic level, nesting habits and food preference. They used cladistic (and hence phylogenetic) relationships to provide the basis for kinship orderings of the parameters of the descending supraspinal pathways. Some of their results on the correlation of anatomical parameters of the corticospinal system with digital dexterity (Nudo and Masterton 1990a) cited cited in Sect. 22.9.1 (Fig. 22.120). Theirs is a necessary and potentially fruitful approach to increase our understanding of the role of the different cortical and subcortical systems in behaviour.

22.10
Diencephalon

P.A.M. van Dongen and R. Nieuwenhuys

22.10.1
General Plan and Shape

Traditionally, the diencephalon is subdivided into four dorsoventrally arranged zones: the epithalamus, dorsal thalamus, ventral thalamus and hypothalamus. This subdivision was first deduced from studies on amphibians and reptiles (Herrick 1910) and soon thereafter confirmed in mammals (Droogleever Fortuyn 1912). It is now clear, however, that some of these zones cited derived from neuromeres and, hence, during early development cited not dorsoventrally arranged, but rather transversely oriented (see Chap. 4). Moreover, discussion continues whether the diencephalon basically consists of four or five zones. In ontogenetic studies of the human diencephalon, Kahle (1956) distinguished five longitudinal zones. The extra zone he distinguished was the subthalamus, consisting of the anlage of the entopeduncular nucleus (i.e. the internal part of the globus pallidus in primates), the subthalamic nucleus and the zona incerta. According to others (e.g. Kuhlenbeck 1954), the entopeduncular nucleus and the subthalamic nucleus cited derived from the hypothalamus, and the zone incerta from the ventral thalamus. Nevertheless, the traditional subdivision into four zones offers a convenient startingpoint for a discussion of the various diencephalic cell masses. The outgrowth of each of these parts determines the eventual shape of the diencephalon.

Early in development, the diencephalon is thin-walled and tube-shaped throughout, but later the anlage of the thalamus expands enormously, a process leading to a remarkable deformation. This deformation process is illustrated by Fig. 22.130, which shows three stages of the development of the human diencephalon. One consequence of the expansion of the dorsal thalamus, and of the related increase in the connections between the thalamus and the cerebral hemispheres, is that the telencephalic-diencephalic boundary plane enlarges and changes its orientation. In early embryonic stages, this plane is more or less transversely oriented, but later assumes an almost rostrocaudal orientation, particularly in 'higher' mammals (Figs. 22.150). The thalamic expansion and the widening and change in position of the telencephalic-diencephalic boundary plane further lead to the early embryonic free lateral surface of the thalamus ultimately becoming its caudal surface and one thalamic centre, the lateral geniculate

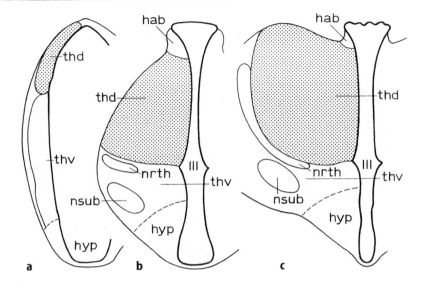

Fig. 22.130a–c. Transverse sections through the diencephalon of human embryos of **a** about 10 mm, **b** about 30 mm and **c** about 100 mm in length, showing the expansion of the dorsal thalamus. *hab*, ganglion habenulae; *hyp*, hypothalamus; *nrth*, nucleus reticularis thalami; *n sub*, nucleus subthalamicus; *thd*, thalamus dorsalis; *thv*, thalamus ventralis; *III*, third ventricle

body, becoming displaced laterocaudally (Figs. 22.130, 22.131, 22.150).

Figure 22.132 shows graphical reconstructions of some major diencephalic cell masses and fibre tracts in the rat, cat and human in medial view. It will be seen that some structures, e.g. the mammillary body and the anterior thalamic nuclei and their main connections, show a remarkable constancy in their position, whereas other structures, including the habenular ganglion and the lateral geniculate body, show considerable positional differences. The habenular ganglion receives afferents via the stria medullaris thalami and projects by way of the habenulo-interpeduncular tract or fasciculus retroflexus to the interpeduncular nucleus, situated in the base of the midbrain. Due to the enormous expansion of the thalamus, the habenular ganglion is displaced caudally in humans. Consequently, the stria medullaris thalami is extraordinarily long, and the habenulo-interpeduncular tract shows a sharp flexure in this species. In fact, the latter tract is truly 'retroflexed' only in humans and other primates.

As regards the lateral geniculate body, it has already been mentioned that the expansion of the thalamus leads to a laterocaudal displacement of this structure (Fig. 22.131). It should be added that, in primates, a particularly strong development and outgrowth of the dorsocaudal portion of the dorsal thalamus, i.e. the pulvinar thalami, effects, in addition, a ventral rotation of the lateral geniculate body. Hence, this cell mass is situated dorsocaudally in the rat (Fig. 22.132a) and cat (Fig. 22.132b), but rather ventrocaudally in humans (Fig. 22.132c).

22.10.2
Epithalamus

The epithalamus encompasses the ganglia habenulae, the epiphysis (pineal gland) and a narrow strip of tissue passing over the dorsomedial surface of the dorsal thalamus, directly adjacent to the line of attachment of the membranous roof of the diencephalon, i.e. the taenia thalami. The fibres of the stria medullaris thalami pass along this taenia to the habenular ganglia. Some authors (e.g. Jones 1985) consider the paraventricular nucleus as part of the epithalamus, but according to others (Groenewegen and Berendse 1994) it is one of the midline nuclei of the dorsal thalamus. It will be discussed in Sect. 22.10.3.

22.10.2.1
Habenula

The medial and lateral habenular nuclei are situated just under the ependyma of the third ventricle. They lie on the posterodorsal part of the thalamus, although they may extend over the whole length of the thalamus in small mammals.

The habenula receives its afferents from the basal forebrain via the stria medullaris. Its efferent path is the fasciculus retroflexus (or habenulo-interpeduncular tract or bundle of Meynert), which terminates in the interpeduncular nucleus. Phylogenetically, this is a very old pathway connecting the basal forebrain via the diencephalon with the mesencephalon. The habenular nuclei receive bilateral afferents; fibres in the stria meduallaris pass through the habenular commissure to the contralateral habenular nuclei. The afferent and efferent connections of

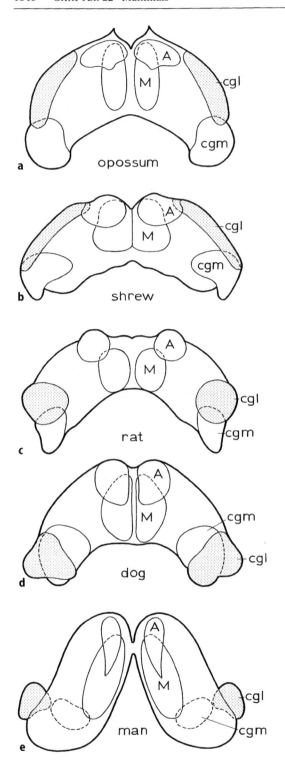

Fig. 22.131a–e. The dorsal thalamus and some of its principal nuclei in **a** the opossum, **b** the shrew, **c** the rat, **d** the dog and **e** the human, projected upon a horizontal plane. *A*, nucleus anterior; *cgl*, corpus geniculatum laterale; *cgm*, corpus geniculatum mediale; *M*, nucleus medialis (or mediodorsalis)

the medial and lateral habenular nucleus are different (Table 22.1; Herkenham and Nauta 1977, 1979).

The *medial habenular nucleus* projects to the interpeduncular nucleus; it receives afferents from the medial septum, the nucleus of the diagonal band and the ventral periaqueductal gray.

The *lateral habenular nucleus* receives afferents from the prepiriform cortex, the lateral hypothalamus, lateral preoptic area, the nucleus of the diagonal band, the medial pallidum (entopeduncular nucleus) and the ventral periaqueductral gray. The lateral habenular nucleus does not project to the interpeducular nucleus, but its efferent fibres pass through this nucleus to end in the ventral periaqueductal grey, raphe nuclei and mesencephalic reticular formation.

Remarkably, few functional studies have been devoted to this very old system. The habenular nuclei have been suggested to be involved (among others) in the generation of sleep patterns, the secretion of hormones (noradrenaline, adrenaline, corticosterone) in response to stress and in avoidance learning (Haun et al. 1992; Murray et al. 1994; Thornton et al. 1994). This incoherent collection of presumed functions indicates that the function of the habenular nuclei is unclear.

22.10.2.2
Epiphysis

Early in development, two anlagen of the epiphysis (pineal gland) are present, but they fuse to a single midline structure. In some fishes and reptiles, the pineal gland is associated with another epithalamic structure, the parietal eye (or third eye) with photoreceptive cells, but in mammals, no photoreceptive cells (or parietal eye) are present. The pineal gland consists mainly of characteristic pinealocytes, while neurons are rare in the pineal gland of most mammals. The pineal gland derives from a finger-shaped evagination of the roof of the diencephalon between the habenular and posterior commissures. A large variation in the eventual shape of the epiphysis is present between mammalian species, even within the same order (Kenny and Scheelings 1979; Reiter 1981; Bhatnagar et al. 1990).

In some species (such as primates), the pineal gland remains a compact structure close to the habenula, between the habenular and posterior commissures. In other species (such as the guinea pig), it is strongly elongated. In still others (such as the Syrian hamster), it seems as if two pineals are present: a deep pineal gland at its original position close to the diencephalon and a remote, superficial pineal gland connected to the deep pineal by a long, thin stalk.

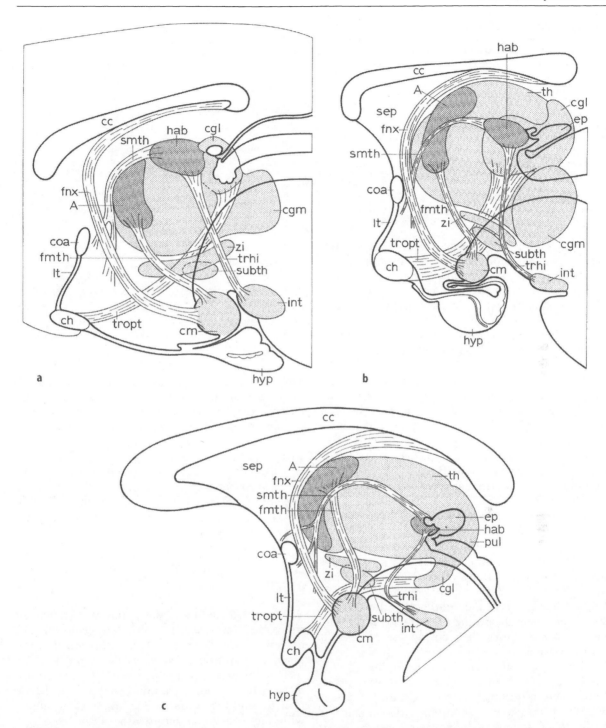

Fig. 22.132a–c. Reconstructions of diencephalic and adjacent structures in medial view. **a** Rat. **b** Cat. **c** Human. *cc*, corpus callosum; *cgl*, corpus geniculatum laterale; *cgm*, corpus geniculatum mediale; *ch*, chiasma opticum; *cm*, corpus mammillare; *coa*, commissura anterior; *cqa*, corpora quadrigemina; *ep*, epiphysis; *fmth*, fasciculus mammillo-thalamicus; *fnx*, fornix; *hab*, ganglion habenulae; *hyp*, hypophysis; *int*, nucleus interpeduncularis; *lt*, lamina terminalis; *pulv*, pulvinar thalami; *sep*, septum pellucidum; *smth*, stria medullaris thalami; *subth*, nucleus subthalamicus; *th*, thalamus; *trhi*, tractus habenulo-interpeduncularis; *tropt*, tractus opticus; *zi*, zona incerta

Table 22-1. Survey of the epithalamus

Epithalamus	Main function	Main afferents	Main efferents
Medial habenula	(Unknown)	Prepiriform cortex Medial septum N. diagonal band Ventral periaqueductal gray	Interpeduncular n.
Lateral habenula	(Unknown)	N. diagonal band Lateral preoptic n. Lateral hypothalamus Substantia innominata Medial pallidal segment Ventral periaqueductal gray	Ventral periaqueductal gray Substantia nigra Dorsal raphe N. centralis superior Mesencephalic reticular formation
Pineal gland	Melatonin secretion	Ganglion cervicale superius	

In the rabbit, only a remote superficial pineal gland is present connected to the diencephalon by a long, thin stalk, and in edentates and sea cows, no pineal gland is visible macroscopically; it is unknown whether scattered pinealocytes are present in these animals.

The pineal gland is a strongly vascularised gland outside the blood-brain barrier. Its main product is the hormone melatonin. The pineal gland is the most important, but not the only source of melatonin. Melatonin is excreted only during the night; this applies to both diurnal and nocturnal mammals. If animals are kept in constant darkness, melatonin is secreted only during the 'circadian night' and not during the 'circadian day'. If the night is illuminated by artificial light, melatonin secretion stops. Information about the illumination and the circadian time reaches the pineal via the suprachiasmatic nucleus, the spinal cord and the ganglion cervicale superius (Moore and Card 1986).

Phylogenetically, melatonin is a very old 'substance of darkness'; in unicellular organisms it has a similar rhythm as in mammals (Hardeland 1993). During winter time, the night is much longer than the day in northern and southern regions of the earth; consequently, during the winter, much more melatonin is excreted than during the summer. It has been argued that melatonin not only codes for the time of the day (a clock function), but also for the season (a calendar function; Reiter 1993). Remarkably, large pineal glands are found in walruses, Weddell seals and lemmings, animals that live far north of the equator. Among megachiropteran bats, some species have a much larger pineal gland than others, but the functional implications of this finding are not understood (Bhatnagar et al. 1990). Melatonin is (among other things) involved in entrainment of the circadian rhythm and in reproduction (which in most species is associated with one season).

22.10.3
(Dorsal) Thalamus

An extensive discussion of the mammalian thalamus is beyond the scope of this work; for a comprehensive survey of the structure and functions of the thalamus, the reader is referred to the monograph by Jones (1985), which was an important source of information for us. In this chapter, especially those aspects of the mammalian thalamus will be mentioned that are interesting from a comparative point of view; the differences between the various mammalian orders will be highlighted. Few comparative studies of the thalamus have been published; a comparison of the thalamus in monotremes, marsupials, cingulates, rodents, carnivores and primates can be found in Ariëns Kappers et al. (1936). An interesting attempt to integrate data on thalamic nuclei in various groups of vertebrates into an evolutionary model has recently been made by Butler (1994).

22.10.3.1
Introduction

The dorsal thalamus is the main relay station for subcortical information to the telencephalon. The thalamus projects to almost all parts of the telencephalon, i.e. not only the whole neocortex, but also the archicortex (hippocampus), palaeocortex, mesocortex, the striatum, amygdala (lateral nucleus) and tuberculum olfactorium. Concomitant with the enormous outgrowth of the cortex in mammals, the thalamus is also considerably enlarged. Compared with reptiles, mammals have a large thalamus. Traditionally, the thalamic nuclei were subdivided into *sensory relay* or *extrinic nuclei* and *association* or *intrinsic nuclei*. Thalamic nuclei projecting to the 'association cortex' were thought to receive their inputs mainly from other thalamic nuclei (Rose and Woolsey 1949). It is now accepted, however, that connections between thalamic nuclei are virtually absent and that all thalamic nuclei transfer subcortical information to the telencephalon.

22.10.3.1.1
Overall Shape

The (dorsal) thalamus develops from a matrix zone situated dorsally in the wall of the diencephalon. During development of the dorsal thalamus, the following changes take place:

1. The external surface of the thalamus, which is initially parallel to the ventricular surface, turns outward and eventually becomes its posterior surface.
2. The boundary plane between thalamus and telencephalon, which is originally rather narrow, expands strongly.
3. The anlagen of centres situated in the rostrolateral part of the thalamus (e.g. the lateral geniculate nucleus) attain a caudolateral position in ferungulates and in primates.

In 'primitive' mammals, such as marsupials and insectivores, the developmental events sketched above are less pronounced than in ferungulates and in primates; hence, in these groups, the lateral geniculate nucleus occupies a rostrolateral rather than a caudolateral position (Fig. 22.131). Given these differences in the caudolateral expansion of the thalamus among various mammalian orders, the names of the various nuclei, which are mainly derived from their topographical position in the adult primate brain, can be misleading, especially in a comparative study.

22.10.3.1.2
Subdivisions

Figure 22.133 presents a survey of the positional relations of the principal cell groups in the human thalamus. From caudal to rostral the thalamus is largely intersected by a sagittal layer of fibres, the lamina medullaris interna. It separates the medial from the lateral cell group. Some small cell masses, the intralaminar nuclei, are embedded in the fibre mass of the lamina medullaris interna. Rostrally, this fibre mass bifurcates into two sheets which embrace the anterior nuclear group. The midline nuclear group consists of bilateral arrays of small nuclei which are united by a median coalescence, the adhaesio interthalamica. In mammals in which the adhaesio interthalamica is large (marsupialis, insectivores, rodents, ferungulates), most midline nuclei are represented by a single cell mass. The human lateral cell group encompasses three nuclei:

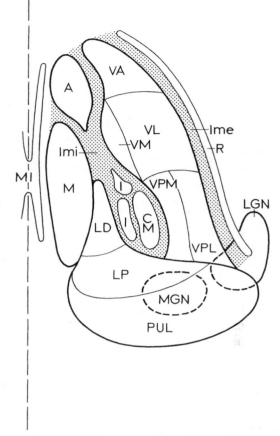

Fig. 22.133. Principal cell masses in the human thalamus, flattened out into a horizontal plane. *A*, anterior nuclear group; *CM*, centromedian nucleus; *I*, intralaminar nuclei; *LD*, nucleus lateralis dorsalis; *LGN*, lateral geniculate nucleus; *LP*, nucleus lateralis posterior; *lme*, lamina medullaris externa; *lmi*, lamina medullaris interna; *M*, medial nucleus; *MI*, midline nuclei; *MGN*, medial geniculate nucleus; *R*, reticular nucleus; *PUL*, pulvinar nuclei; *VA*, nucleus ventralis anterior; *VL*, nucleus ventralis lateralis; *VM*, nucleus ventralis medialis; *VPL*, nucleus ventralis posterior lateralis; *VPM*, nucleus ventralis posterior medialis

the nucleus lateralis dorsalis, the nucleus lateralis posterior and the particularly strongly developed pulvinar. Two prominent relay nuclei in the human thalamus are designated on the basis of their mutual topographical position as the (acoustic) medial geniculate nucleus and the (visual) lateral geniculate nucleus. In other primates and in ferungulates, these nuclei occupy similar positions. In marsupialis, insectivores and rodents, these two cell masses do not occupy medial and lateral, but rather rostral and caudal positions with respect to each other, but yet the 'primate' terminology is maintained in these groups (Fig. 22.131). Table 22.2 presents a survey of the principal fibre connections and functions of the various thalamic nuclei.

Table 22-2. Survey of the groups of thalamic nuclei

	Main function	Main afferents	Main efferents
1. **Ventral group**	Somatosensory Taste Vestibular	Spinothalamic tract Medial lemniscus Trigeminal nuclei Vestibular nuclei Cerebellar nuclei Internal globus pallidus	SI (1, 2, 3a, 3b) SII Area 4 Area 6
2. **Medial geniculate complex**	Auditory	Inferior colliculus	AI, AII Temporal auditory field
3. **Lateral geniculate complex**	Visual	Eye (optic tract)	Striate cortex
4. **Lateral group**	"Second visual system"	Superior colliculus pretectum	Prestriate Superior temporal gyrus Pariotemporal
5. **Posterior group**	Visual Somatosensory Auditory	Superior colliculus Spinothalamic tract	Granular insular Retroinsular Postauditory
6. **Midline/intralaminar complex**	(Unclear)	Cerebellar nuclei Reticular formation Spinothalamic tract Substantia nigra Colliculus superior	Striatum Parietal cortex Frontal cortex
7. **Medial group**	Olfactory	Amygdala	Frontal eye field Orbitofrontal cortex Lateral frontal cortex Amygdala
8. **Anterior complex**	Part of "limbic system"	Mammillary nuclei	Anterior limbic cingulate cortex Retrosplenial (pre)subiculum

Thalamic nuclei have reciprocal connections with the cortical target regions, but these are not mentioned in the column of the afferent connections. Thalamic nuclei receive (apart from the afferents mentioned) GABAergic fibres from the reticular thalamic nucleus, noradrenergic fibres from the locus coeruleus, serotonergic fibres from the nucleus dorsalis raphe and centralis linearis, and cholinergic fibres from the dorsolateral tegmentum.

22.10.3.1.3
Nomenclature

The nomenclature of the thalamic nuclei can be confusing. Various authors use different names for the same nucleus or even the same name for different nuclei. As in other parts of the central nervous system, the relative position of the cell masses and their fibre connections form important clues to their homologisation (see Chap. 6, Sect. 6.2.6). As regards nomenclature, we will follow the policy of Rose (1942) and Jones (1985): the prefix (or the first Latin name) is derived from the nuclear group, and the suffix specifies the position of the nucleus in that group. For instance, the medioventral nucleus (or N. medialis ventralis) is the ventral nucleus of the medial group, while the ventromedial nucleus (or N. ventralis medialis) is the medial nucleus in the ventral group.

22.10.3.2
Ventral Group

The ventral thalamic nuclei are mainly somatosensory. In most mammalian orders, the following four ventral nuclei can be distinguished (Table 22.3):

– *N. ventralis posterior.* This is the main somatosensory thalamic nucleus (proprioceptive, touch, pain). It receives afferents via the medial lemniscus and the spinothalamic and trigeminothalamic tracts. These projections are somatotopically organised; the spinothalamic and trigeminothalamic projections overlap with those of the medial lemniscus. The N. ventralis posterior has a main projection to the somatosensory cortices SI and SII. In many mammals, two main parts of the N. ventralis posterior can be distinguished: the N. ventralis posterior medialis with a somatosensory representation of the head, and the N. ventralis posterior lateralis with a somatosensory representation of the rest of the body. Part of the N. ventralis posterior medialis is the thalamic taste area. The N. ventralis posterior late-

Table 22-3. Survey of the ventral group of nuclei of the dorsal thalamus

Ventral group	Main function	Main afferents	Main efferents
N. ventralis posterior	Somatosensory	Spinal cord	SI (1,2,3a,3b)
– Lateral (VPL): body	(Proprioceptive, touch, pain)	(Spinothalamic tract) N. cuneatus, gracilis (medial lemniscus)	SII
– Medial (VPM): head	(Part of VPM: taste)	Trigeminal nuclei (trigeminothalamic tract)	
N. ventralis medialis	Taste Vagal?	Parabrachial nuclei Spinal cord	SI, frontal and cingulate cortex
N. ventralis lateralis	Vestibular	Vestibular nuclei Cerebellar nuclei Spinal cord Entopeduncular nucleus (=internal globus pallidus)	Primary motor cortex (Areas 4 and 6)
N. ventralis anterior	(Unknown)		Frontal eye field Prefrontal cortex

ralis and medialis are the most important thalamic 'pain nuclei'.
- *N. ventralis medialis.* This has afferents from the parabrachial nuclei and the spinal cord and rather diffuse projections to the frontal, cingulate and somatosensory cortex. The part receiving fibres from the parabrachial nuclei is involved in taste.
- *N. ventralis lateralis.* This has afferents from vestibular nuclei, cerebellar nuclei and the internal globus pallidus and possibly also from a part of the spinothalamic tract. The N. ventralis lateralis projects to the primary motor cortex (areas 4 and 6) and other nuclei.
- *N. ventralis anterior.* This is quite distinct in primates; it might have a common origin with the N. ventralis lateralis.

Monotremes. In contrast to other mammals, the ventral thalamic nucleus of monotremes cannot easily be subdivided (Jones 1985). In *Echidna*, the undifferentiated ventral nucleus could not be demarcated from the medial geniculate nucleus. Nevertheless, in this species, a rostral part of the ventral nucleus projects to the motor cortex (Welker and Lende 1980); in this respect, it resembles the N. ventralis lateralis of other mammals. A posterior part of the ventral nucleus (the ventrobasal part) projects to the somatosensory cortex, like the N. ventralis posterior in other mammals (Welker and Lende 1980).

Marsupials. The opossum is the marsupial whose brain has been most intensively studied. It has a 'primitive' thalamus; the N. ventralis posterior is merged with the medial geniculate nucleus. The medial and lateral ventral nuclei can be distinguished. In the brush-tailed possum (but not in the opossum), a separate anterior ventral nucleus is also found (Haight and Neylon 1978). Some marsupials (such as the brush-tailed possum) have a specialized vibrissae system; in the N. ventralis posterior medialis of this animal, the representation of the individual vibrissae ('barreloids') can be recognised even in thionin-stained sections (Jones 1985, p. 329).

Insectivores. In insectivores, the N. ventralis medialis is fairly large. As in marsupials, the N. ventralis posterior in many insectivores is merged with the medial geniculate nucleus. In shrews, the snout is important in exploring their environment; in these animals, the N. ventralis posterior medialis is relatively large compared to the N. ventralis posterior lateralis (Campbell and Ryzen 1953).

Rodents. Rodents have a specialized somatosensory system, the vibrissae system. Cortical representations of the separate vibrissae have been identified as the 'barrels' in the cortical barrel field (Woolsey and Van der Loos 1970). Representations of the individual vibrissae have been found also in the rodent N. ventralis posterior medialis: the 'barreloids' (Van der Loos 1976). In rodents, the N. ventralis posterior medialis is considerably larger than the N. ventralis posterior lateralis (Cabral and Johnson 1971).

Carnivores. Most carnivores cannot make fine finger movements with their forepaws, but the racoon can. Accordingly, its N. ventralis posterior lateralis is large and shows an extensive representation of the forepaws (Cabral and Johnson 1971).

Primates. Primates can make fine, independent finger movements with their hands. In this order, the representations of hands (and sometimes also the feet) are large in the N. ventralis posterior lateralis (Cabral and Johnson 1971). New World monkeys have a prehensile tail; the tail representation in their N. ventralis posterior lateralis is large. In primates, a N. ventralis posterior inferior has been identified, which receives vestibular inputs.

In primates, but not in other mammals, the N. ventralis lateralis is subdivided into (a) an anterior part with projections from the internal globus pallidus and projections to cortical area 6 and (b) a posterior part with projections from vestibular and cerebellar nuclei and projections to cortical area 4.

Whales. In whales, the total N. ventralis posterior is relatively small. Compared to the N. ventralis posterior lateralis, the N. ventralis posterior medialis is relatively large (Kruger 1959). In these animals, the somatosensory representation of the body is small, as is also reflected in the spinal cord.

Other Groups. In two mammalian orders, the trigeminal system is extremely specialized: elephants (with their proboscis) and anteaters (with their remarkably long, flexible tongue). Unfortunately, no publications have been found on the ventral thalamic nuclei of these groups.

22.10.3.3
Medial Geniculate Complex

In primates, the medial and lateral geniculate complexes are protrusions of the ventral posterior part of the thalamus; the designations 'medial' and 'lateral' geniculate complex are derived from their location in this group (Figs. 22.131, 22.133). However, in marsupials, insectivores and rodents, these nuclei are situated laterally, where the visual nucleus is located anterodorsally and the auditory nucleus posterioventrally (Bodian 1939; Fig. 22.131). In carnivores, the medial geniculate nucleus is located as laterally as the lateral geniculate nucleus, but it has a more ventral position (Fig. 22.132b).

The medial geniculate complex is the main thalamic auditory relay group. In most mammalian groups, three nuclei are distinguished: ventral, dorsal and medial (Table 22.4).

The *ventral medial geniculate complex* (which is the principal nucleus) receives the most direct ascending auditory information from the inferior colliculus; it has the most clear cochleotopic (tonotopic) organisation, and it projects to the primary auditory cortex (AI).

The *dorsal medial geniculate complex* also receives auditory input from the colliculus inferior; it projects to the secondary auditory cortex (AII) and the temporal auditory field.

The *medial medial geniculate complex* (magnocellular or inferior) receives inputs from the inferior colliculus and deep layers of superior colliculus and possibly some somatosensory and vestibular inputs. This nucleus projects diffusely to the insular temporal cortex.

Auditory/Somatosensory Differentiation. On the basis of cytoarchitectonics, no clear medial geniculate complex could be distinguished in monotremes. Even after destruction of the auditory cortex, the retrograde degeneration was not clearly associated with a single nuclear group (Welker and Lende 1980). It is not clear whether this is due to technical problems or whether the thalamic auditory neurons in the monotremes are not grouped together in a nuclear complex. Also at the cellular level, the somatosensory and auditory system are not always fully separated. In the opossum and hedgehog, some auditory neurons are situated in the N. ventralis posterior and some somatosensory neurons in the medial geniculate complex, and some neurons even have both an auditory and a somatosensory receptive field (Erickson et al. 1964).

On the basis of cytoarchitectonics, the medial and dorsal medial geniculate complex cannot be

Table 22-4. Survey of the medial geniculate complex

Medial geniculate complex	Main function	Main afferents	Main efferents
N. gen. med. ventralis (principal nucleus)	Auditory	Central nucleus of inferior colliculus	AI
N. gen. med. dorsalis	Auditory (Sound localisation?)	Pericentral and external nuclei (inf.coll.)	AII Temporal auditory field
N. gen. med. medialis	(Unknown)	Inferior colliculus, deep layers superior colliculus	Insular temporal cortex

distinguished in marsupials and rodents, but in carnivores, rabbits, tree shrews, primates, bats and whales, they can. It has been suggested that the dorsal medial geniculate complex (with projections to AII) is especially large in bats and whales, animals with a specialized sonar system for echolocation (Jones 1985, p. 427). In whales, and especially the toothed whales, the medial geniculate complex is the largest sensory nuclear group of the thalamus (Kruger 1959), which reflects the importance of hearing for whales.

22.10.3.4
Lateral Geniculate Complex

In all mammals investigated, the main visual relay nucleus, the lateral geniculate nucleus, can be easily identified. It is usually called the dorsal lateral geniculate nucleus to differentiate it from the ventral lateral geniculate nucleus, which is derived from the ventral thalamus. Due to transformation and rotation of the primate thalamus during development, the ventral lateral geniculate nucleus is situated dorsally to the dorsal lateral geniculate nucleus in adult primates; in these animals, the ventral lateral geniculate nucleus is often called 'pregeniculate nucleus'. The ventral lateral geniculate nucleus will be described in Sect. 22.10.4.

The dorsal lateral geniculate nucleus receives a massive direct projection from the retina by way of the optic nerves and tracts. The amount of decussation of the optic fibres depends on the position of the eyes (Fig. 22.134). Some mammals have *laterally placed eyes* (e.g. opossums, rats, rabbits, ungulates, whales); this is probably the primitive location of the mammalian eye. In these animals, each eye mainly represents the ipsilateral visual field and only a small part of the contralateral visual field. Most optic fibres cross in these animals; only few optic fibres project to the ipsilateral dorsal lateral geniculate nucleus. Therefore, the dorsal lateral geniculate nuclei of these animals have a large monocular segment and a small binocular one (Fig. 22.135); no clear lamination is seen in their lateral geniculate nucleus. Mammals with laterally placed eyes do not always have poor vision and a poorly developed lateral geniculate nucleus. The ungulates, for instance, have laterally placed eyes, good vision and a large, unlaminated lateral geniculate nucleus (Solnitzky 1938).

Some mammals have *frontally placed eyes*. In these animals, about half of the optic fibres do not cross, but project to the ipsilateral dorsal lateral geniculate nucleus. Their dorsal lateral geniculate nuclei have a large binocular segment. These binocular segments show a distinct laminar pattern; layers receiving fibres from the ipsi- and contralateral eye alternate, separated by the incoming optic tract fibres. In monkeys, the projections of both eyes remain separated via these laminae as far as the primary visual cortex (area 17), i.e. the ocular dominance strips in layer IV.

Retinal ganglion cells with different receptive field properties (X, Y and Wcells) likewise project to different parts of the dorsal lateral geniculate nucleus (Wilson et al. 1976).

The importance of the visual system for mammals varies considerably. At one extreme, we find mammals with a degenerated visual system. Such mammals are virtually blind and belong to different orders: in marsupials, the marsupial moles (*Notoryctes*); in insectivores, the moles; and in rodents,

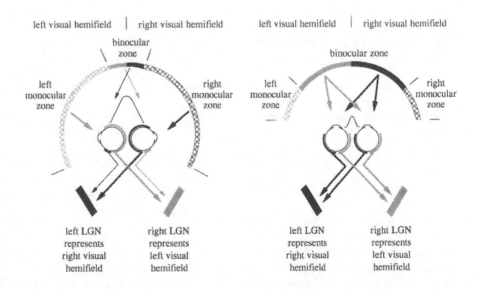

Fig. 22.134. Lateral (*left*) and frontal eyes (*right*). Decussation of the optic fibres. *LGN*, lateral geniculate nucleus

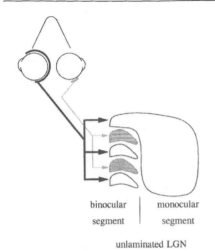
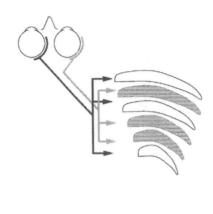

Fig. 22.135. Lateral (*left*) and frontal eyes (*right*). Lamination of the lateral geniculate nucleus (*LGN*). Due to the alternation of the incoming fibres into the LGN in mammals with frontal eyes, a laminated pattern emerges in this structure

the naked mole rat. These animals have a very small lateral geniculate nucleus (Bauchot 1959; Niimi et al. 1962; Rehkämper et al. 1994). For several other mammals, the visual system is not very important. For instance, monotremes have a very small lateral geniculate nucleus (Jones 1985, p. 476). The lateral geniculate nucleus is reported to be small and non-laminated in most whales (Breathnach 1960), but it is large in the bottle-nosed dolphin *Tursiops* (Kruger 1966).

At the other extreme, there are typical visual animals, for which the visual system is the most important sensory system. These animals have a highly developed visual system, often with frontally placed eyes. Visual mammals are present in various orders: within marsupials, the group of phalangers; within bats, the *Macrochiroptera* and fruit-eating *Microchiroptera*; within rodents, the squirrels; within carnivores, the felids; and the tree shrews and primates. In the course of evolution, these groups separated before their visual systems expanded. It is thus assumed that these visual systems expanded independently during evolution. This has resulted in different solutions for the problem of improving vision:

1. In all visual animals, a large and extensively laminated dorsal lateral geniculate nucleus is found, but the patterns of lamination strongly differ.
2. The superior colliculus projects to some circumscribed layers of the lateral geniculate nucleus, but in various orders to different layers in a pattern too complex to be described in a few words (see Jones 1985, p. 506).
3. The 'second visual system' is also different; in primates, for instance, the pulvinar is expanded, while in carnivores the lateral posterior nucleus is greatly enlarged (see below).
4. In some highly visual mammals (tree shrews, squirrels), the ventral lateral geniculate nucleus is large, but in others (primates) it is not.

In the primate lineage, the enlargement of the lateral geniculate nucleus is already present in the prosimians (Simmons 1980). The tarsier is a highly visual prosimian; its lateral geniculate nucleus occupies about one third of the thalamus (Simmons 1982).

22.10.3.5
Lateral Group

The lateral group of thalamic nuclei consists of the lateral posterior complex, the pulvinar complex and the lateral dorsal nucleus (Table 22.5). It is involved in visual orientation, eye movements and accommodation. Especially in higher primates, the pulvinar complex is enlarged (Harting et al. 1972). In humans, the pulvinar nuclei form the largest thalamic nuclear group.

In *marsupials*, *insectivores* and *rodents*, an undifferentiated lateral complex has been found. This implies that no pulvinar can be distinguished in these groups on the basis of cytoarchitectonics. In these groups, projections from the superficial layer of colliculus superior and the primary visual cortex to parts of the lateral complex have been found. In the armadillo, a small lateral posterior nucleus and a small but distinctive pulvinar is described (Papez 1932).

In *tree shrews and primates*, the pulvinar is a separate entity (Simmons 1981). In prosimians, it is larger than in tree shrews (Simmons 1980, 1982), while in monkeys, apes and humans, the pulvinar is huge (Le Gros Clark 1932; Harting et al. 1972). In the primate pulvinar, four distinct nuclei are identified: medial, lateral, inferior and anterior. Lateral to the pulvinar, a small lateral posterior complex is found in primates.

Table 22-5. Survey of the lateral thalamic nuclei

Lateral group	Main function	Main afferents	Main efferents
N. pulvinaris	Second visual system	Superficial superior colliculus Pretectum	Occipital cortex Posterior parietal Posterior temporal *(Species differences)*
N. lateralis posterior	Second visual system	Superficial superior colliculus	Occipital cortex Posterior parietal Posterior temporal *(Species differences)*
N. lateralis dorsalis	(Unknown)	Hippocampus (fornix)	Cingulate cortex Retrosplenial cortex

The anterior pulvinar and the lateral posterior complex project to areas 5 and 7. The lateral and inferior pulvinar nuclei receive fibres from the (visual) superficial layers of colliculus superior; they project to areas 18 and 19, to a lesser extent to area 17 and also to the visual association cortex.

The medial pulvinar nucleus receives fibres from the (non-visual) deep layers of colliculus superior; it sends fibres to the superior temporal gyrus, while some neurons project to the frontal eye fields.

In the *cat*, a relatively small pulvinar and a large lateral posterior complex consisting of several nuclei (Rioch 1929) have been described; this is just the opposite of the situation found in primates. As in primates, however, the cat lateral posterior complex receives its main afferent projection from the superficial layers of colliculus superior and from the pretectum. With respect to afferent and efferent connections of the nuclei of the lateral posterior complex, no single organisational plan has been found that applies to both primates and the cat. The visual systems of primates and felids have almost certainly reached their high degree of development independently. Therefore, there is no a priori reason for expecting homologies between their lateral posterior complex and pulvinar nuclei.

The lateral dorsal nucleus receives its afferents from the pretectum and also from the hippocampus via the fornix, and it projects to the cingulate and retrosplenial cortex. This nucleus is of approximately the same size in most mammals, but in some insectivores, some bats and *Tarsius*, it is relatively large (Jones 1985, p. 685).

22.10.3.6
Posterior Group

A group of nuclei are distinguished in the posterior thalamus that are usually described as a single nuclear complex, the posterior group. Three nuclei are often distinguished: suprageniculatus, posterior medialis and posterior lateralis (Table 22.6).

– The *N. suprageniculatus* receives its input from the deep layers of the colliculus superior; it is mainly a visual nucleus. It projects to the granular insular cortex. In the cat, this nucleus is differentiated into two parts: the suprageniculate and the limitans part.

– The *N. posterior medialis* receives its main input from the tractus spinothalamicus. It is mainly a somatosensory nucleus and projects to the retroinsular cortex.

– The *N. posterior lateralis* receives fibres from the colliculus inferior and is mainly auditory. It projects to the postauditory cortex (Jones and Burton 1976).

The posterior complex is not a single entity with respect to afferents and function. The nuclei of this complex merely share their location in the thalamus and their projections to and around the insular cortex. In whales, the lateral group (lateral posterior and pulvinar) and the posterior group were described as a single complex (Kruger 1959). This complex is very large, comprising almost half of the whale thalamus. The question arises of whether homologues of the lateral or of the posterior group became strongly enlarged in whales. Vision is not very important for whales, so it is unlikely that a homologue of the lateral group (a part of the second visual system) is extremely large in whales; we would guess that in whales the auditory part of the posterior group is strongly enlarged to enable echolocation.

22.10.3.7
Midline/Intralaminar Group

The intralaminar and the dorsal midline nuclei probably form part of a single entity with a well-organised projection to the striatum and the cortex, except the ventral midline nuclei (N. rhomboideus and reuniens; Groenewegen and Berendse 1994; Table 22.7). For the sake of simplicity, the intralami-

Table 22-6. Survey of the posterior thalamic group

Posterior group	Main function	Main afferents	Main efferents
N. suprageniculatus (limitans)	Visual	Deep layers of superior colliculus	Granular insular
N. posterior medialis	Somatosensory (including pain)	Spinal cord (spinothalamic tract)	Retroinsular cortex
N. posterior lateralis	Auditory	Inferior colliculus	Auditory cortex

nar nuclei and the dorsal nuclei of the midline group are referred to here as the intralaminar complex. A topical projection is found from the intralaminar complex to the striatum/accumbens; lateral intralaminar nuclei project to dorsolateral parts of the striatum, and the medial nuclei project to ventromedial parts. The corticopetal efferents of the intralaminar complex project mainly to frontal parts of the cortex; the temporal and occipital cortex receive only a sparse projection. Each nucleus projects to a restricted part of the cortex, with little overlap with the projection of the other nuclei. Until recently, the projection of the intralaminar complex was regarded as diffuse and non-specific. With modern anatomical tracing techniques, however, a well-organised ('specific') projection of the intralaminar complex is found. In general, these nuclei project to a part of the striatum and to that part of the cortex that projects to the same part of the striatum (Groenewegen and Berendse 1994).

The *centromédian nucleus* is difficult to distinguish in marsupials, insectivores, rodents, rabbits and bats. It has been identified in ungulates, carnivores, tree shrews and lower primates and is rather large in the higher primates (including humans), whales and elephants (Simmons 1980, 1982). The centromédian nucleus is remarkably large in humans (Fig. 22.133). This nucleus receives a massive projection from the internal part of the globus pallidus (N. entopeduncularis) and the (pre)frontal cortex.

The *parataenial nucleus* is large in some 'primitive' orders (marsupials and insectivores) and also in rodents and bats, whereas it is small in lagomorphs and primates.

The *rhomboid and reuniens nuclei* do not have a main projection to the striatum, but to the limbic temporal cortex. It is unclear whether these nuclei form an entity with the other midline/intralaminar nuclei. The N. reuniens (medioventral nucleus) is small in some orders (marsupials, insectivores and primates), relatively small in carnivores and well developed in rodents, lagomorphs and ungulates.

The *paraventricular nuclei* consist of a midline part and a thin sheet covering the dorsal, medial and anterior part of the thalamus. The paraventricular nuclei receive afferents from the preoptic area and the anterior and lateral hypothalamus. Little is known about their efferents.

Hypothetical functions of the intralaminar complex include gaze control, nociception, cardiac reflexes and sleep/wakefulness control (Groenewegen and Berendse 1994). These many possibilities demonstrate that the function of the intralaminal complex is unclear.

Table 22-7. Survey of the midline/intralaminar nuclei

Midline/intralaminar nuclei	Main function	Main afferents	Main efferents
Intralaminar group: rostral nuclei (centralis, paracentralis)	(Unknown)	Spinal cord Cerebellar nuclei Trigeminal nuclei Colliculus superior Pretectum	Striatum Prefrontal cortex Posterior parietal cortex
caudal nuclei (centromédian, parafascicularis)	(Unknown)	Internal globus pallidus Central gray Substantia nigra	Striatum Sensorimotor cortex Premotor cortex
Midline group: dorsal nuclei (paraventricularis, parataenialis, intermediodorsalis)	(Unknown)		Striatum, accumbens "Limbic" parts of frontal Cortex amygdala
ventral nuclei (rhomboideus, reuniens)	(Unknown)		Hippocampus Amygdala

Table 22-8. Survey of the medial thalamic nuclei

Mediodorsal nucleus	Main function	Main afferents	Main efferents
Medial part (primates: magnocellular)	Olfactory (memory?)	Basolateral amygdaloid nucleus	Basolateral amygdaloid nucleus Dorsal agranular insular Prelimbic area
Central part (primates: magnocellular)	Olfactory	Piriform cortex	Lateral orbital cortex Ventral agranulr insular
Lateral part (primates: parvocellular)	Unknown	Superior colliculus Vestibular nuclei Midbrain tegmental fields Nucleus accumbens	Medial precentral cortex

22.10.3.8
Medial Nuclei

The mediodorsal nucleus (also called the dorsomedial or medial nucleus) is a large nucleus in most species. It is surrounded by the internal medullary lamina with its nuclei. In several orders, three parts are distinguished, i.e. medial, central and lateral (Table 22.8), while two parts are evident cytoarchitectonically in primates, i.e. magnocellular and parvocellular.

The medial part has reciprocal connections with the basolateral amygdaloid nucleus, the dorsal agranular insular cortex and the prelimbic area. This part also receives a projection from the ventral pallidum; through this pathway, the ventral striatum (nucleus accumbens) is connected with the prefrontal cortex.

The central part receives fibres from the piriform cortex and projects to the lateral orbital and the ventral agranular insular cortex. The medial and central parts form part of the olfactory system. Given their connections, it was concluded that they are equivalents of the magnocellular part in primates.

The lateral part receives afferents from the superior colliculus, substantia nigra, vestibular nuclei and midbrain tegmental fields. It is connected with the dorsal and lateral parts of the prefrontal cortex.

The relationship between the mediodorsal nucleus and the prefrontal cortex is so characteristic that the latter is often defined as the projection area of the former. The sizes of these two structures are strongly correlated. Both are rather small in insectivores, larger in tree shrews (Simmons 1981) and still larger in primates and whales. In the bottle-nosed dolphin *Tursiops*, the mediodorsal nucleus comprizes about 10% of the whole thalamus (Kruger 1966). In the monotreme *Echidna* (the spiny anteater), the prefrontal cortex is surprisingly extensive, and the mediodorsal nucleus is likewise very large (Welker and Lende 1980; Regidor and Divac 1987; Fig. 22.191f).

22.10.3.9
Anterior Nuclei

The anterior group is regarded as a primitive group; it is large in the opossum and insectivores. In the opossum, the anterior nuclei comprise a considerable fraction of the dorsal thalamus (Bodian 1939). In most mammals, this group consists of three nuclei: ventral, medial and dorsal (Table 22.9). They receive afferents from the mammillary nuclei and the (pre)subiculum and project to the cingulate, the anterior limbic cortex and the (pre)subiculum.

The *anteroventral nucleus* is about the same size as the anteromedial nucleus in marsupials, rodents, ungulates, whales and tree shrews. In the rabbit and cat, a dorsal and ventral part of the anteroventral nucleus can be clearly distinguished.

The *anteromedial nucleus* is relatively small in primates, tree shrews and some insectivores, while it is almost absent in some other insectivores, small bats and the armadillo.

The *anterodorsal nucleus* is most distinctive, since it contains large, deeply staining cells. This nucleus is relatively small in marsupials, rodents, ungulates, carnivores, primates and cetaceans. In the hedgehog, it appears absent (Le Gros Clark 1929). However, in lagomorphs, some rodents and some insectivores, the anterodorsal nucleus is approximately the same size as the anteroventral and anteromedial nuclei.

22.10.3.10
Basic Uniformity of the Thalamic Nuclei

The various thalamic nuclei share a number of salient features with regard to their extrinsic and their intrinsic organisation:

1. The thalamic nuclei primarily represent the contralateral side of the body or the external space.
2. The thalamocortical relay neurons receive a direct, massive subcortical input. The sensory relay nuclei (lateral and medial geniculate

Table 22-9. Survey of the anterior thalamic nuclei

Anterior group	main function	Main afferents	Main efferents
N. anterior ventralis	Part of "limbic" system	N. mamm. medialis (mammillothalamic tr.)	Anterior limbic cortex
N. anterior medialis	Part of "limbic" system	N. mamm. medialis (mammillothalamic tr.)	Cingulate cortex (Pre)subiculum Retrosplenial cortex
N. anterior dorsalis	Part of "limbic" system	N. mamm. lateralis (mammillothalamic tr.)	Cingulate cortex (Pre)subiculum Retrosplenial cortex

nucleus, N. ventralis posterior) receive a topically organised input (retinotopic, cochleo/tonotopic and somatotopic).

3. All thalamic nuclei contain relay cells that massively project to the telencephalon: mainly to the cortex, or otherwise to the striatum. The sensory relay nuclei have a strictly topically ordered projection to their cortical target regions, with endings predominantly in layer IV. Non-specific thalamic nuclei have a more diffuse projection to the cortex, with endings not just in layer IV, but also in other layers, especially I and VI.

4. The thalamic nuclei receive a massive input from their cortical target region. For several nuclei, the cortical input is larger than the main subcortical input. Strictly reciprocal connections are present between cortical regions and their thalamic counterparts (the 'principle of reciprocity'; Diamond and Hall 1969). The function of this massive cortical input has been elusive, but for the lateral geniculate nucleus a start has been made in understanding the function of these connections (see below).

5. The main subcortical afferents, the thalamocortical and the corticothalamic neurons, use excitatory amino acids as their main neurotransmitter.

6. The thalamic nuclei contain inhibitory GABAergic interneurons.

7. The thalamic nuclei have reciprocal relationships with the reticular thalamic nucleus (or its equivalents). The axons of the relay cells give off collaterals to the reticular nucleus, where they

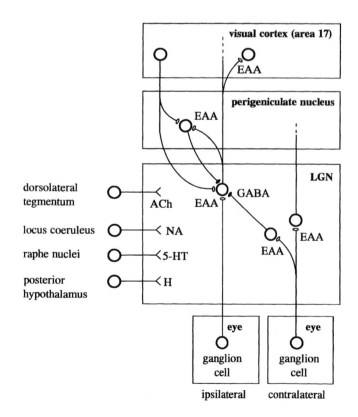

Fig. 22.136. Survey of the overall wiring of the mammalian thalamic nuclei with their connections with the cortex and the thalamic reticular nucleus. *EAA*, excitatory amino acids; *Ach*, acetylcholine; *GABA*, γ-aminobutyric acid; *NA*, noradrenaline; *5-HT*, serotonin; *LGN*, lateral geniculate nucleus; *H*, histamine

excite inhibitory GABAergic neurons that project back to the region of the same relay cells.
8. The thalamic nuclei receive a scattered noradrenergic input from the locus coeruleus.
9. The thalamic nuclei receive a scattered serotonergic input from the nucleus raphes dorsalis and the nucleus centralis superior.
10. The thalamic nuclei receive a cholinergic input from the dorsolateral tegmentum.
11. A few thalamic nuclei receive a histaminergic input from the posterior hypothalamus.

Figure 22.136 gives a schematic survey of the overall wiring of thalamic nuclei. Another connection is conspicious by its absence; in contrast to earlier ideas, interthalamic connections are lacking.

22.10.3.11
Example of Thalamic Functioning: Lateral Geniculate Nucleus

The (dorsal) thalamus is often described as a relay station to the cortex. However, the signals relayed to the cortex are strongly modified by the other afferents of the thalamus. In order to give an impression of the way in which information is processed, the modification of visual messages in the lateral geniculate nucleus of the cat will be described. We have chosen to discuss the lateral geniculate nucleus of the cat because a great deal of information is available on signal processing in this nucleus.

In the lateral geniculate nucleus, the *contralateral visual field* is represented. Fibres from the ipsi- and contralateral retina terminate in alternating layers in the lateral geniculate nucleus (Fig. 22.135).

Retinal ganglion cells can be classified as *X, Y* and *W cells*. The cells are situated intermingled in the retina. All of these cells project to the lateral geniculate nucleus, but the Y and W cells also send branches to the superior colliculus. X and Y cells project to dorsal laminae in the lateral geniculate nucleus, while the W cells project to other (ventral) laminae and to the ventral lateral geniculate nucleus.

The *receptive field properties* of neurons in the lateral geniculate nucleus are similar to those of retinal ganglion cells. This would be expected, since individual lateral geniculate nucleus relay cells receive terminals from only a few optic tract axons. In the lateral geniculate nucleus of the cat, the X, Y and W properties are conserved. The X and Y cells have concentric receptive fields, either on-centre/off-surround or off-centre/on-surround. The sizes of the receptive fields are similar to those of retinal ganglion cells. In the lateral geniculate nucleus, the antagonism between centre and surround might be somewhat stronger than in the retina.

Binocular Integration. The contralateral visual field is represented in each lateral geniculate nucleus by afferents from both eyes. There has been ongoing discussion for a long time now about whether binocular integration takes place in the lateral geniculate nucleus. Single lateral geniculate nucleus relay cells are excited by fibres from one eye and are indirectly inhibited by signals from ganglion cells in the other eye via interneurons (Lindström 1982).

For the lateral geniculate nucleus, the equivalent of the *thalamic reticular nucleus* is the perigeniculate nucleus, a nucleus just dorsal to the lateral geniculate nucleus. The lateral geniculate nucleus has reciprocal connections with the perigeniculate nucleus. Lateral geniculate nucleus relay cells projecting to the neocortex give off collaterals to the perigeniculate nucleus. A single perigeniculate neuron is activated indirectly from both eyes by on-centre and off-centre lateral geniculate nucleus cells. In the perigeniculate nucleus, separate X and Y cells are present. Perigeniculate neurons project back to the parts of lamina A and A1 from which they receive terminals. Their effect is recurrent inhibition rather than lateral inhibition (Lo and Sherman 1994).

Like the other thalamic nuclei, the lateral geniculate nucleus has reciprocal connections with the *cortex*. It receives even more fibres from the cortex than from the eyes. Indications of a function for these cortical afferents come from a study by Sillito et al. (1994). Depending on the stimulus present, the cortical neurons synchronise the activity of a precisely selected subpopulation of lateral geniculate nucleus relay cells. The stimulus characteristics determine which neurons are selected to become synchronised. This synchronised firing of selected lateral geniculate nucleus relay cells increases the probability of some special stimulus features to be detected. This is the first indication of the involvement of reciprocal thalamocortical connections in complex pattern recognition.

Other Inputs. Like other thalamic nuclei, the lateral geniculate nucleus receives afferents from various subcortical sources, i.e. noradrenergic afferents from the locus coeruleus, serotonergic afferents from the nucleus raphes dorsalis and centralis linearis and cholinergic afferents from the dorsolateral tegmentum.

The signal transmission and neuronal activity in the lateral geniculate nucleus (as in other thalamic nuclei) strongly depends on the state of vigilance and the stages of sleep. It is now generally accepted

Table 22-10. Survey of the ventral thalamus

Ventral thalamus	Main function	Main afferents	Main efferents
N. reticularis thalami	Modifies thalamic signal transmission	Thalamic nuclei Cortical areas N. cuneiformis	Thalamic nuclei
N. zona incerta (dorsal)	(Unknown)	Cingulate cortex Hypothalamic ventromedial nucleus	Pontomesencephalic tegmentum, Thalamic parafascicular nucleus
N. zona incerta (ventral)	(Unknown)	Somatosensory cortex Ventral LGN Cerebellar nuclei Colliculus superior Trigeminus nuclei Dorsal column nuclei	Pretectum, Superior colliculus (deep layers), Spinal cord
Fields of Forel	(Unknown)	Internal globus pallidus Spinal cord	Spinal cord
Ventral LGN	Eye movements Pupil reflexes Entrainment circadian clock	Retina Superior colliculus Olivary nucleus N. of optic tract Cortex (17, 18, 19) Cerebellum Subthalamus	Contralateral vLGN Superior colliculus Olivary nucleus N. of optic tract Prerubral field Pontine nuclei N. term. lat. AOT N. suprachiasmaticus

that thalamic nuclei mainly determine the cortical electroencephalogram (EEG) (Steriade and Llinás 1988). Brain stem neurons influence the thalamic nuclei either directly or via the thalamic reticular nucleus; they control signal transmission to the cortex and the degree of synchonisation in the thalamus. Whether and how noradrenergic, serotonergic, histaminergic and cholinergic afferent play a role in this process is unknown.

22.10.4
Ventral Thalamus

Ventral to the anlage of the dorsal thalamus, the anlage of the ventral thalamus is found (Fig. 22.130). The thalamic reticular nucleus, the zona incerta and the ventral lateral geniculate nucleus have a common origin (Rose 1942); these groups comprise the ventral thalamus. A survey of the connections of the ventral thalamus is given in Table 22.10. The entopeduncular and subthalamic nuclei, which according to most authors also form part of the ventral thalamus, will be dealt with in Sect. 22.11.4.2.

22.10.4.1
Reticular Nucleus

Due to the huge expansion of the dorsal thalamus in mammals, the reticular nucleus has been stretched out into a thin sheet of neurons, covering the anterior, lateral and lateroventral surfaces of the dorsal thalamus (Fig. 22.130b,c). In all mammals investigated, including monotremes, a reticular nucleus is found. The part covering the lateral geniculate nucleus in carnivores is usually called the perigeniculate nucleus, but this nucleus is most probably part of the reticular nucleus. In some carnivores (raccoon, mink, seal), the reticular nucleus is remarkably large (Sanderson 1974), while it is only a thin sheet around the thalamus in several larger mammals.

The reticular nucleus receives its main afferents from the thalamic nuclei; these are collaterals of thalamocortical axons (Ahlsén et al. 1978). From a comparison of the receptive field properties of lateral geniculate nucleus and perigeniculate neurons, it was concluded that the axons of several geniculate neurons converge to a single perigeniculate neuron (Cleland et al. 1971); this might be the general pattern for the reticular nucleus. The reticular nucleus projects back to the thalamic nuclei from which it receives afferents; this relationship is strictly reciprocal (Jones 1975). The neurons of the reticular nucleus are inhibitory GABAergic neurons (Houser et al. 1980) with axons that extensively ramify in the dorsal thalamus (Scheibel and Scheibel 1966). The neurons of the the perigeniculate nucleus exert recurrent inhibition on the neurons of the lateral geniculate nucleus, rather than lateral inhibition (Lo and Sherman 1994). Another important source of afferents to the reticular nucleus is the cerebral cortex; corticothalamic axons give off collaterals when passing through the reticular nucleus

(Fig. 22.136). An intricate thalamo-corticoreticular microcircuit is present. The strictly reciprocal thalamocortical and thalamoreticular connections have already been mentioned. A certain part of the cortex projects to the part of the reticular nucleus that has reciprocal connections to the thalamic nucleus with which this cortical part also has reciprocal connections (Rose 1952; Jones 1975).

Apart from the thalamic and cortical afferents, the reticular nucleus also receives fibres from the mesencephalic cuneiform nucleus, the noradrenergic locus coeruleus and serotonergic raphe nuclei.

The reticular nucleus is clearly involved in modifying thalamic signal transmission. The function of the reticular nucleus has been compared to a searchlight (Crick 1984), which seems to be a useful hypothesis. According to this hypothesis, the reticular nucleus selectively improves signal transmission in a part of the thalamus. The question then arises as to how the part of the thalamus that is favoured is determined. It must be determined by the stimulus, memory and internal set-points (motivation). The reticular nucleus receives the cortical and subcortical inputs necessary to fulfill this task. In line with this hypothesis, it is assumed that the reticular nucleus changes thalamic transmission during sleep.

22.10.4.2
Zona Incerta

At the medial ventral surface of the dorsal thalamus, the zona incerta and the fields of Forel are located. The fields of Forel consist mainly of passing fibres with few neurons, while the region with less myelin is called the zona incerta or the nucleus of the zona incerta (to indicate that many neurons are located in this region). These neurons seem to be a ventromedial continuation of the reticular nucleus, but the zona incerta has other afferent and efferent connections than the reticular nucleus. In most studies, the rostral and caudal pole of the zona incerta have not been investigated; these poles are reported to have different projections than the central part (Romanowski et al. 1985). The ventral part of the zona incerta receives afferents from the somatosensory cortex, ventral lateral geniculate nucleus, cerebellar nuclei, colliculus superior, trigeminus nuclei and dorsal column nuclei. This part projects to the pretectum, superior colliculus (deep layers) and the spinal cord. The dorsal part of the zona incerta receives fibres from the cingulate cortex and the hypothalamic ventromedial nucleus (Roger and Cadusseau 1985); it projects to the thalamic parafascicular nucleus and the pontomesencephalic tegmentum. It also contains magnocellular cells projecting to the posterior pituitary; these are probably neurosecretory cells (Kelly and Swanson 1980). The zona incerta is reported to be involved in sensory processing, visuomotor integration, feeding and drinking, cortical activation, learning and locomotion (for references, see Roger and Cadusseau 1985). To our knowledge, no comparative studies on the zona incerta have been carried out so far.

22.10.4.3
Ventral Lateral Geniculate Nucleus

The ventral lateral geniculate nucleus is a conspicuous nuclear mass in most mammals; only in primates is it less overt, and here it is referred to as the pregeniculate nucleus (Woollard and Beattie 1927). In rodents, it is called the intergeniculate leaflet (Botchkina and Morin 1995). In virtually blind mammals (moles, mole lemmings), a clear ventral lateral geniculate nucleus is still found. The name of the ventral lateral geniculate nucleus is derived from its position in embryos of marsupials, rodents, tree shrews and humans, where it is situated just ventral to, and is apparently associated with the (dorsal) lateral geniculate nucleus. However, its original (primitive) position is more anterior, closer to the anterior nuclei, as it still is in monotremes, some insectivores and bats (Campbell and Hayhow 1971, 1972). In large carnivores, it is found rostroventral to the (dorsal) lateral geniculate nucleus. Due to the expansion and rotation of the (dorsal) lateral geniculate nucleus in primates and ungulates, the ventral lateral geniculate nucleus is found dorsal to the dorsal lateral geniculate nucleus in adult animals of these groups.

From its afferent and efferent connections, it can be deduced that the ventral lateral geniculate nucleus is a visual nucleus. It is traversed by the optic tract and receives optic tract afferents. This projection is retinotopic and mainly contralateral, even in primates and carnivores with frontally placed eyes (Holländer and Sanides 1976). Most afferent axons are of the slowly conducting W type, as demonstrated in the cat; the receptive fields of the ventral lateral geniculate nucleus neurons are mainly monocular and large, and the cell responses are of the 'tonic-on' type (Spear et al. 1977). In addition to these optic tract afferents, the ventral lateral geniculate nucleus also receives afferents from the ipsilateral visual cortex (areas 17–19; Updyke 1977), the superficial and intermediate layers of the ipsilateral superior colliculus, the nucleus of the optic tract and the pretectal olivary nucleus. In some groups (tree shrews, ungulates, some rodents), the ventral lateral geniculate nucleus is laminated; this lamination reflects the distribution of retinal and midbrain afferents (Niimi et al. 1963).

The ventral lateral geniculate nucleus projects to intermediate layers of the superior colliculus, where a retinotopic projection from the ventral lateral geniculate nucleus seems to match with the retinotopic organisation in the superior colliculus. It has reciprocal connections with the contralateral ventral lateral geniculate nucleus via the posterior commissure and projects to the nucleus of the optic tract, the olivary nucleus, the prerubral field and pontine nuclei. Moreover, it projects with fibres containing neuropeptide Y to the hypothalamic suprachiasmatic nucleus.

The size of the ventral lateral geniculate nucleus in various groups of mammals varies considerably. In some highly visual animals such as tree shrews and squirrels, it is about as large as the dorsal lateral geniculate nucleus. However, in other visual animals such as the primates, it is much smaller. The size of the ventral lateral geniculate nucleus often reflects the size of the superior colliculus, with which it has reciprocal connnections (Le Gros Clark 1932).

The ventral lateral geniculate nucleus is involved in various visual functions:

- *Eye movements.* The firing pattern of neurons in the ventral lateral geniculate nucleus depends not only on visual stimuli, but also on eye and head movements (Magnin and Fuchs 1977). The connections with superior colliculus and the olivary nucleus are consistent with its role in oculomotor control.
- *Entrainment of the circadian rhythm.* It is now generally accepted that the hypothalamic nucleus suprachiasmaticus is the main endogenous circadian clock. This clock must obtain information on whether it is day or night. Apart from direct retinal afferents, the N. suprachiasmaticus also receives time/light information via the ventral lateral geniculate nucleus (Moore 1978). A role of the ventral lateral geniculate nucleus in the entrainment of the circadian rhythm is consistent with the direct retinal afferents to the ventral lateral geniculate nucleus, the receptive field properties of its neurons (large receptive fields and mainly on-tonic responses) and with its projection to the suprachiasmatic nucleus.
- *Pupillary light reflex.* It has been suggested that the ventral lateral geniculate nucleus is involved in the pupillary light reflex (Polyak 1957). This is consistent with the receptive field properties of its neurons (Spear et al. 1977) and with its connections with the pretectal region.

22.10.5
Hypothalamus (R. Nieuwenhuys)

22.10.5.1
General Relations

The hypothalamus represents the ventralmost part of the diencephalon, where it forms the floor and contributes to the lateral walls of the third ventricle. Its upper boundary is marked on the ventricular side by a shallow groove, the hypothalamic sulcus. Caudally, the hypothalamus passes gradually over into the periventricular and tegmental grey matter of the mesencephalon. However, it is customary to define the posterior margin of the hypothalamus as a vertical plane passing just caudal to the mammillary bodies. The latter are paired nuclear complexes, which in carnivores and primates are externally visible as small eminences on the basal aspect of the brain. The rostral boundary of the hypothalamus coincides with a vertical plane passing from the interventricular foramen to the middle of the optic chiasm. The preoptic region, which flanks the rostral end of the third ventricle, extends from the lamina terminalis to the rostral boundary of the hypothalamus. Although this region is of telencephalic origin, it is structurally and functionally so closely tied up with the hypothalamus that it will be treated here together with the latter.

During ontogenesis, the floor of the hypothalamus forms a hollow, finger-shaped process, i.e. the infundibulum or hypophysial stalk. Two structures, the median eminence and the neurohypophysis, develop from this process. The median eminence is a neurohemal contact zone, which forms a functional interface between the hypothalamus and the anterior lobe of the pituitary (see below). It is situated in the anterior wall of the infundibulum. The neurohypophysis develops from the distalmost part of the infundibulum. The infundibulum, and with it the median eminence and the neurohypophysis, differs considerably in topography and configuration in the different mammalian species (Haymaker 1969). Without going into details, it may be mentioned that in most mammals the infundibulum is caudally directed and the infundibular recess, i.e. the extension of the third ventricle into the hypophysial stalk, is shallow. The median eminence is situated rostral to the pituitary gland, in which the neurohypophysis lies superiorly and the adenohypophysis inferiorly (Fig. 22.132a). In carnivores the inferior recess is deep and surrounded by the neurohypophysis, and the latter structure is situated dorsocaudal to the adenohypophysis (Fig. 22.132b). In primates and humans, the infundibulum forms a ventrally directed stalk, and the neurohypophysis is situated caudal to the adenohypophysis. The median eminence is situated dorsal to the pituitary gland on the rostral side of the stalk, and the infundibular recess extends only over a short distance into the latter (Fig. 22.132c).

22.10.5.2
Subdivision

The cell masses in the preoptico-hypothalamic continuum have been studied in many different mammalian groups, including marsupials (Bodian 1939: opossum; Oswaldo-Cruz and Rocha-Miranda 1968: opossum), insectivores (Campbell and Ryzen 1953: shrew; Bauchot 1959: mole; Bauchot 1963: several species), chiropteres (Baron et al. 1996: many species), rodents (Gurdjian 1927: rat; Krieg 1932: rat; Swanson 1987: rat; Simerly 1995: rat; Broadwell and Bleier 1976: mouse), lagomorphs (Wahren 1957: rabbit), carnivores (Rioch 1929: cat and dog), ungulates (Diepen 1941: sheep), prosimians (Kanagasuntheram et al. 1968: bush-baby and slow lori; Simmons 1979, 1980, 1981, 1982: lemur, bush-baby, tarsier and several tree shrew species) and simians (Crouch 1934: rhesus monkey; Le Gros Clark 1938: several species; Ingram 1940: several species; Nauta and Haymaker 1969: humans). Christ (1969) presented a very useful atlas of sections through the hypothalamus of laboratory animals (rat, rabbit, cat, dog and monkey).

The mammalian hypothalamus can be subdivided into a number of cell masses, the discreteness of which differs considerably. Some are fairly distinct but others are rather ill-defined, and in some regions the cells are diffusely arranged without recognisable cell groupings. Although some clear interspecies differences occur, it should be emphasised that there is a remarkable structural similarity in the hypothalamic region in all mammals. Golgi studies in the rat (Millhouse 1979) have shown that hypothalamic neurons are generally provided with few, poorly ramifying long dendrites that extend well beyond the cytoarchitectonic boundaries of the cell masses to which their somata belong and which tend to be oriented perpendicularly to the longitudinal axis of the hypothalamus (see Chap. 2, Figs. 2.11–2.13, 2.30, 2.31).

Le Gros Clark (1938) divided the preoptico-hypothalamic continuum into four rostrocaudal levels or regions, i.e. preoptic, anterior, tuberal and mammillary, while Crosby and Woodburne (1940) distinguished three mediolaterally arranged zones, i.e. a periventricular, medial and lateral zone (Table 22.11; Figs. 22.137, 22.138).

The periventricular zone consists of a few layers of small cells, among which larger, neurosecretory

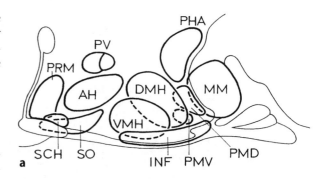

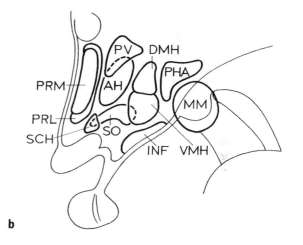

Fig. 22.137a,b. Some major hypothalamic nuclei in medial view in **a** the rat and **b** the human. For abbreviations, see Table 22.11

elements are intermingled. The latter form a single concentration, the medial magnocellular part of the paraventricular nucleus. The cell layers alternate with thin sheets of fine, mostly unmyelinated fibres. In addition to the medial part of the paraventricular nucleus, the periventricular preoptic nucleus, the compact suprachiasmatic nucleus and the infundibular nucleus in the basal hypothalamus represent derivatives of the periventricular zone.

The medial zone, which is relatively cellular, contains a number of variably discrete cell masses, the rather compact medial preoptic nucleus, the more diffuse anterior nucleus, the discrete posterior magnocellular part of the paraventricular nucleus, the dorsomedial and ventromedial nuclei, which are often poorly delimited from each other, the generally discrete dorsal and ventral premammillary nuclei and medial and lateral mammillary nuclei and the diffuse posterior hypothalamic area.

The lateral zone is partly separated from the medial zone by the postcommissural fornix, a large bundle connecting the hippocampal formation with the mammillary body (Fig. 22.138b). The lateral

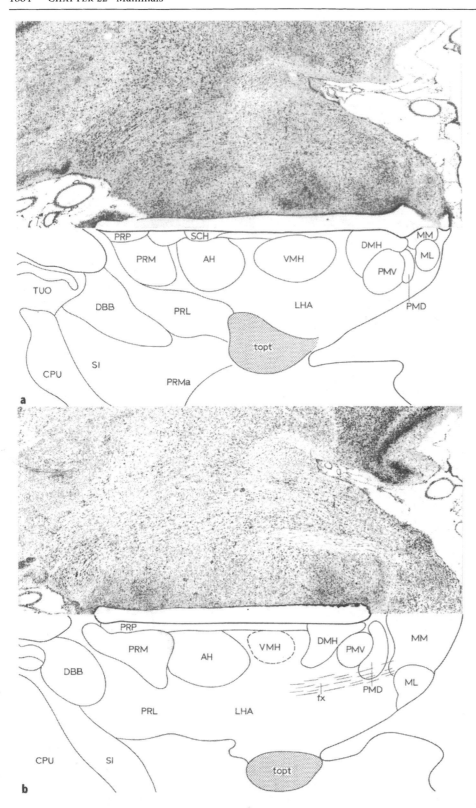

Fig. 22.138a,b. Horizontal, Nissl-stained sections through **a** the ventral and **b** the dorsal hypothalamus of the rat. The interpretation, as presented in the accompanying tracings, is simplified from Geeraedts et al. (1990a,b). *CPU*, caudate-putamen complex; *DBB*, diagonal band of Broca; *fx*, fornix; *SI*, substantia innominata; *topt*, tractus opticus; *TUO*, tuberculum olfactorium. For other abbreviations see Table 22.11

Table 22-11. The preoptico-hypothalamic continuum: principal nuclei (N) and areas (A)

Preoptic region	Anterior region	Tuberal region	Mammillary region
	Periventricular zone		
N. preopticus periventricularis (PRP)	N. suprachiasmaticus (INF) N. paraventricularis (PF)	N. infundibularis (INF)	
	Medial zone		
N. preopticus medialis (PRM)	N. anterior hypothalami (AH) N. paraventricularis (PV)	N. ventromedialis (VMH) N. dorsomedialis (DMH)	N. premammillaris dorsalis N. premammillaris ventralis (PMV) N. mammillaris medialis (MM) N. mammillaris lateralis (ML) A. hypothalamica posterior (PMA)
	Lateral zone		
A. preoptica lateralis (PRL)	A. hypothalamica lateralis (LHA) N. supraopticus (SO)		

zone contains the magnocellular neurosecretory supraoptic nucleus, but most of its territory is occupied by the diffuse lateral preoptic and lateral hypothalamic areas.

22.10.5.3
Circuitry and Major Conduction Channels

The hypothalamus forms part of the greater limbic system, a functional entity which will be discussed below (see Sect. 22.12). Suffice it to mention here that this entity extends throughout the brain and that the hypothalamus is interposed between its rostral, telencephalic and its caudal, mesencephalic and rhombencephalic parts. The septal region, the hippocampus and the amygdaloid complex are major components of the telencephalic part of the greater limbic system, while its caudal part includes the mesencephalic central grey, the parabrachial nuclei, the solitary nucleus, the raphe nuclei and an array of cytoarchitectonically poorly differentiated cell masses known as the lateral paracore (Figs. 22.219, 22.222).

Several hypothalamic fibre systems, including the medial forebrain bundle, the dorsal longitudinal fasciculus of Schütz, the mammillothalamic bundle, the fornix, the fasciculus mammillaris princeps and the mammillary peduncle, form part of the circuitry of the greater limbic system.

The *medial forebrain bundle* or *fasciculus medialis telencephali* may be considered as the central longitudinal pathway of the limbic continuum. It is an assemblage of loosely arranged, mostly thin fibres and extends from the septal region to the tegmentum of the rhombencephalon. It traverses the lateral preoptico-hypothalamic area, the scattered neurons of which are collectively designated as the bed nucleus of the medial forebrain bundle.

The bundle is highly complex, comprising a variety of short and long, ascending and descending links (Nieuwenhuys et al. 1982; Veening et al. 1982; Vertes 1984a,b). In the transitional area of the diencephalon and the mesencephalon, the medial forebrain bundle fibres are rearranged in a smaller medial and a larger lateral stream (Hosoya and Matsushita 1981; Holstege 1987). The medial fibre stream roughly maintains the sagittal orientation of the hypothalamic trajectory of the bundle. It passes through the medial parts of the mesencephalic and rhombencephalic tegmental areas, just next to the raphe nuclei. The medial stream contains descending fibres, by which several hypothalamic centres project to the raphe nuclei and to the adjacent parts of the medial reticular formation; it also comprises fibres which ascend from the raphe nuclei to the lateral hypothalamus, from where they pass to a variety of diencephalic and telencephalic centres, including most preoptic and hypothalamic cell masses, the amygdaloid complex and the hippocampus (Bobillier et al. 1976, 1979; Moore et al. 1980). The lateral stream of fibres extending from the medial forebrain bundle to the brain stem sweeps laterally and caudally over the dorsal border of the substantia nigra and descends through the mesencephalic central tegmental area to the lateral tegmental field of the pons and the medulla oblongata. This fibre stream contains fibres descending from the central nucleus of the amygdala (Price and Amaral 1981; Van der Kooy et al. 1984), the bed nucleus of the stria terminalis (Holstege et al. 1985) and several hypothalamic areas (Holstege 1987; Luiten et al. 1987). These descending fibres terminate in a variety of brain stem centres, among which the pars compacta of the substantia nigra, the parabrachial nuclei, the locus coeruleus, the noradrenergic groups A1, A2 and A5, the superficial ventrolateral reticular area and the dorsal vagal complex deserve special mention. Most of these descending medial forebrain bundle projections are reciprocated by corresponding ascending projections (Vertes 1984a,b). Indeed, the medial forebrain bundle is a major descending and ascending link between the forebrain and the brain stem.

The *dorsal longitudinal fasciculus of Schütz*, much like the medial forebrain bundle, is a composite system consisting of thin ascending and descending fibres. This fascicle extends from the posterior of the hypothalamus to the caudal medulla oblongata and occupies a periventricular position over its entire length. In the older literature (e.g. Crosby et al. 1962), it was reported that most of the ascending and descending projections contained within the dorsal longitudinal fasciculus are synaptically interrupted in either the mesencephalic central grey matter or in the dorsal tegmental nucleus of Gudden; however, more recently it has been established that substantial fibre contingents passing directly from the forebrain to the autonomic centres of the lower medulla oblongata and vice versa are also present (Saper et al. 1976; Ricardo and Koh 1978).

The *fasciculus mammillaris princeps*, containing the efferent of the mammillary nuclei, constitutes a large compact bundle that pass dorsally for a short distance and then splits up into two components, the larger mammillothalamic and the smaller mammillotegmental tract. The mammillothalamic tract, which passes to the anterior thalamic nucleus (Fig. 22.132) forms part of the so-called Papez circuit. The mammillotegmental tract curves caudally into the tegmentum of the midbrain and terminates in the dorsal tegmental nucleus and in the nucleus reticularis tegmenti pontis of Bechterew (Cruce 1977; Ricardo 1983).

The *mammillary peduncle* receives fibres from the dorsal tegmental nucleus. It passes ventrally and then ascends along the ventral surface of the midbrain to the mammillary body, where most of its fibres terminate. Some of its fibres join the medial forebrain bundle and spread to the lateral preoptico-hypothalamic zone and the septum (Nauta and Kuypers 1958; Morest 1961).

22.10.5.4
Survey of Fibre Connections

In the following survey, our current knowledge of the fibre connections of the mammalian hypothalamus will be summarised. Key references from which the pertinent literature can easily be found are mentioned in connection with each of the fibre systems mentioned.

Hypothalamic afferents (Fig. 22.139a) include fibres originating from the following structures:

1. Laminae X, V and I of the spinal cord (Burstein et al. 1987, 1996; Newman et al. 1996)
2. The caudal part of the spinal trigeminal nucleus (Burstein 1996)
3. The nucleus of the solitary tract (Ricardo and Koh 1978; Ciriello and Calaresu 1980a,b)
4. The area reticularis superficialis ventrolateralis, usually referred to as the ventrolateral medulla (Ciriello and Caverson 1984)
5. The parabrachial nuclei (Fulwiler and Saper 1984; Krukoff et al. 1993)
6. The noradrenergic cell groups A1, A2, A5, A6 (locus coeruleus) and A7 (Nieuwenhuys 1985)
7. The adrenergic cell groups C1 and C2 (Nieuwenhuys 1985)
8. The serotonergic dorsal raphe and central superior nuclei (Nieuwenhuys 1985)
9. The periaqueductal grey (Berk and Finkelstein 1981; Mantyh 1983)
10. The deep cerebellar nuclei (Dietrichs and Haines 1989)
11. The hippocampal subicular cortex (Swanson and Cowan 1977)
12. The septum (Swanson and Cowan 1979; Berk and Finkelstein 1981)
13. Various parts of the neocortex, including the prefrontal and cingulate cortex in the monkey (Nauta and Haymaker 1969; Müller-Preuss and Jürgens 1975) and the prefrontal, insular, prelimbic and infralimbic areas in the rat (Kita and Oomura 1982; Saper 1982)
14. The olfactory bulb (Smithson et al. 1989; Hatton and Yang 1989)
15. The subfornical organ and the organum vasculosum laminae terminalis (Oldfield and McKinley 1995)
16. The retina (Kudo et al. 1991)
17. The amygdala, passing via the stria terminalis and the so-called ventral amygdalofugal pathway
18. The bed nucleus of the stria terminalis (Swanson and Cowan 1979)

Hypothalamic efferents (Fig. 22.139b) project to the following structures:

1. Various parts of the neocortex (Kievit and Kuypers 1975; Kita and Oomura 1981; Saper 1985)
2. The septum (Veening et al. 1987; Risold et al. 1994)
3. The hippocampus (Wyss et al. 1979; Haglund et al. 1984)
4. The bed nucleus of the stria terminalis (Krieger et al. 1979)
5. The amygdaloid complex via both the stria terminalis and the ventral amygdalofugal pathway (Amaral et al. 1982; Canteras et al. 1994)

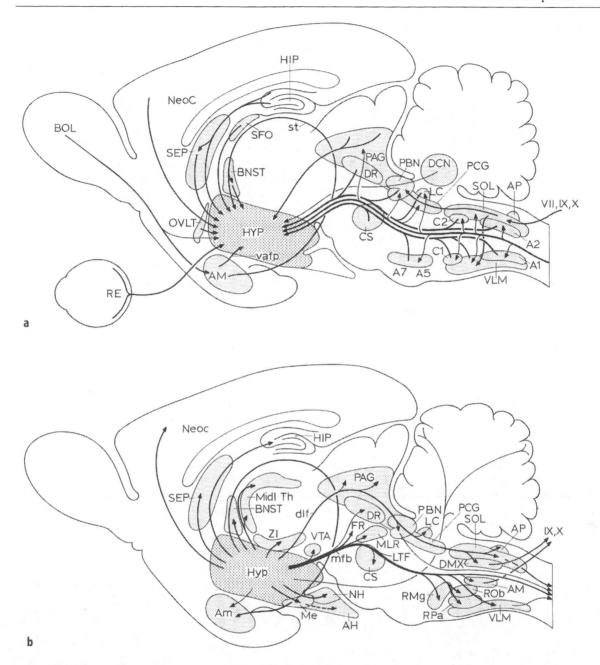

Fig. 22.139a,b. Summary of **a** the afferent and **b** the efferent connections of the hypothalamus in the rat. *A1, A2* etc., noradrenergic cell masses; *AH*, adenohypophysis; *AM*, amygdaloid complex; *AP*, area postrema; *BNST*, bed nucleus of stria terminalis; *BOL*, bulbus olfactorius; *C1, C2*, adrenergic cell groups; *CS*, nucleus centralis superior; *DCN*, deep cerebellar nuclei; *DMx*, dorsal motor vagus nucleus; *DR*, nucleus raphes dorsalis; *FR*, formatio reticularis; *HIP*, hippocampal formation; *HYP*, hypothalamus; *LC*, locus coeruleus; *Me*, median eminence; *Midl Th*, thalamic midline nuclei; *NeoC*, neocortex; *NH*, neurohypophysis; *OVLT*, organum vasculosum laminae terminalis; *PAG*, periaqueductal grey; *PBN*, parabrachial nuclei; *PCG*, pontine central grey; *RE*, retina; *RMg*, nucleus raphes magnus; *ROb*, nucleus raphes obscures; *RPa*, nucleus raphes pallidus; *SEP*, septal nuclei; *SFo*, organum subfornicale; *SOL*, nucleus tractus solitarii; *st*, stria terminalis; *vafp*, ventral amygdalofugal pathway; *VLM*, ventrolateral medulla

6. Thalamic midline nuclei (Risold et al. 1994; Canteras et al. 1994)
7. The zona incerta (Canteras et al. 1994)
8. The periaqueductal grey, the parabrachial nuclei, the locus coeruleus, the pontine central grey, the nucleus of the solitary tract and the area postrema by way of the dorsal longitudinal fasciculus of Schütz (Beitz 1982; Veening et al. 1987; Roeling et al. 1994)
9. The ventral tegmental area, the mesencephalic and rhombencephalic raphe nuclei, the mesencephalic reticular formation, the mesencephalic locomotor region, the rhombencephalic lateral tegmental field, the ambiguus nucleus, the ventrolateral medulla, the marginal zone (lamina I), the central grey and the intermediolateral column of the spinal cord (Swanson et al. 1984, 1987; Luiten et al. 1985; Holstege 1987) (several centres in the brain stem, including the mesencephalic reticular formation and the nucleus of the solitary tract, are innervated by fibres passing via both the dorsal longitudinal fasciculus and the medial forebrain bundle)
10. The cerebellar cortex and nuclei (Dietrichs and Haines 1989; Dietrichs et al. 1992)
11. The pituitary gland

The *hypothalamo-hypophyseal pathways* are shown in Fig. 22.140. The magnocellular nuclei of the anterior hypothalamus, i.e. the supraoptic and (magnocellular) paraventricular nuclei, give rise to axons that descend through the infundibular stalk to the posterior lobe of the pituitary. These axons, which together form the supraoptico-paraventriculo-hypophyseal tract, transport colloid droplets containing the hormones oxytocin and vasopressin to the posterior pituitary lobe or neurohypophysis, where they are released into the blood. The cells of the nucleus arcuatus are involved in controlling the secretion of the anterior pituitary hormones. They exert this control by means of regulating hormones that stimulate or inhibit the liberation of the hormones produced in the pituitary. Each of the latter has a corresponding regulating hormone. The regulating hormones pass from the arcuate nucleus, along the axons of its constituent cells, to the median eminence, where they are released from the axon terminals into the capillaries of the hypophyseal portal system. The latter system forms a vascular link between the infundibulum and the adenohypophysis. It is noteworthy that the median eminence, a conspicuous neurohemal organ situated in the ventrocaudal wall of the hypothalamus, in addition to the axons of arcuate nucleus neurons also receives neurosecretory fibres from several other centres, including the parvocellular part of the paraventricular nucleus and the medial septal nucleus.

22.10.5.5
Functional Considerations

The hypothalamus is critically involved in the regulation of endocrine functions, the control of autonomic reactions and the generation of basic behavioural patterns. The endocrine functions are regulated along the hypothalamo-hypophyseal pathways discussed above. The cells of origin of these pathways may be considered as neuroendocrine motoneurons, representing the final common pathway for central neural influences on hormone secretion from the anterior and posterior lobes of the pituitary gland (Nauta 1963; Swanson 1989).

Several hypothalamic centres exert regulatory influences on preganglionic autonomic neurons in the brain stem and spinal cord. Thus the paraven-

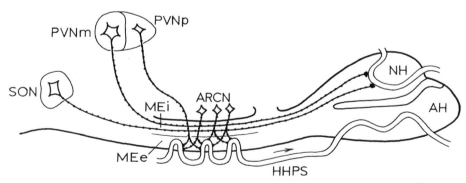

Fig. 22.140. Sagittal representation of hypothalamo-hypophysial relationships in the rat. The width of the median eminence has been exaggerated. *AH*, adenohypophysis; *ARCN*, arcuate nucleus; *HHPS*, hypothalamo-hypophysial portal system; *MEe*, external zone of median eminence; *MEi*, internal zone of median eminence; *NH*, neurohypophysis; *PVNm*, magnocellular part of paraventricular nucleus; *PVNp*, parvicellular part of paraventricular nucleus; *SON*, supraoptic nucleus. (Modified from Buma and Nieuwenhuys 1987)

tricular nucleus, referred to as 'the master controller of the autonomic system' (Loewy 1991), projects directly to the dorsal vagal motor nucleus, the nucleus ambiguus and the thoracic (sympathetic) and sacral (parasympathetic) preganglionic cell columns (Swanson 1987; Saper 1995). This projection shows a distinct functional differentiation in that selective clusters of paraventricular neurons address specific sites of preganglionic neurons (Strack et al. 1989). Indirect projections from the paraventricular nucleus to the sympathetic and parasympathetic efferent centres are synaptically interrupted in autonomic premotor nuclei, such as the A5 noradrenergic cell group and the ventrolateral medulla. Apart from the paraventricular nucleus, the dorsomedial nucleus and the lateral hypothalamic area also provide descending connections to the autonomic nuclei of the lower brain stem and spinal cord (Swanson 1987; Loewy 1991).

The classic experiments performed by Hess and co-workers (e.g. Hess and Brügger 1943), Bard (1928, 1929) and others have shown that by electrical stimulation characteristic behavioural patterns related to feeding, fear, attack, rage and reproduction can be elicited from different hypothalamic loci. During the second half of this century, an enormous body of research has been devoted to unravelling the neural circuitry underlying these basic behavioural patterns, which are all directly related to the maintenance of the internal milieu (homeostasis), the maintenance of the integrity of the individual or the preservation of the species. An exposé of the physiological, ethological and histological techniques used in this research will not be given here. Suffice it to mention that a combination of immunocytochemical mapping of c-Fos (and related immediate-early genes) with other immunocytochemical and tract-tracing techniques is currently being widely and very successfully used to elucidate the neural substrate of different kinds of behaviour.

This section will be concluded with an outline of the neural circuits underlying three specific behaviours: drinking, defensive behaviour and sexual behaviour, but it should not be left unmentioned that all behaviour is superimposed upon a daily rhythm of sleep and wakefulness and that this rhythm is generated in the suprachiasmatic nucleus (Swanson 1987). This small nucleus, which is situated in the rostral part of the hypothalamus (Figs. 22.137, 22.138a), may be considered as the endogenous clock of the brain and is entrained to the light-dark cycle by a direct input from the retina. It imposes circadian organisation on the functional activity in many telencephalic and diencephalic centres, chiefly via a nearby periventricular area designated as the subparaventricular zone (Watts et al. 1987). Interestingly, the retinohypothalamic projection and the suprachiasmatic nucleus are well developed in fossorial mammals, such as moles (Kudo et al. 1991) and mole rats (Cooper et al. 1993), in which the remaining parts of the visual system are greatly reduced in size.

Thirst and Drinking (Swanson 1987, 1989; Fig. 22.141a). Hypovolemic thirst leads to increased circulating levels of the peptide hormone angiotensin II, which acts directly on neurons in the subfornical organ, one of the so-called circumventricular organs. (Circumventricular organs are small receptive loci lying in regions of the brain which lack a blood-brain barrier and are thus accessible to circulating peptides and other compounds.) The subfornical organ, which contains a high density of angiotensin receptors, is directly involved in mediating endocrine, autonomic and behavioural responses to circulating angiotensin II. Its efferents include fibres passing to the supraoptic and magnocellular paraventricular nuclei, the parvocellular paraventricular nucleus, the medial preoptic area and the infralimbic area (area 25) of the medial prefrontal region. The efferents to the magnocellular paraventricular and supraoptic nuclei project directly to large vasopressin neurons in these nuclei. The supraoptic and paraventricular nuclei send neurosecretory fibres to the posterior pituitary gland, where they release anti-diuretic hormone into the general circulation. The parvocellular part of the paraventricular nucleus sends descending projections to the nucleus of the solitary tract, the ventrolateral medulla and the spinal intermediolateral column, all of which are involved in the regulation of blood pressure. The outputs of the medial preoptic area, which include projections to the lateral hypothalamic areas and the mesencephalic locomotor region, are likely to play a role in the procurement and consummatory phases of drinking behaviour. The projection from the subfornical organ to the prefrontal cortex has been suggested to be involved in cognitive aspects of behaviour. It should be emphasised that the circuitry involved in thirst and drinking behaviour is far more intricate than is indicated in Fig. 22.141a. Thus the subfornical organ receives afferents from a variety of sources, including the median preoptic and paraventricular hypothalamic nuclei, the medial preoptic nucleus projects to the paraventricular nucleus and the latter is also in receipt of projections from the nucleus of the solitary tract and the A1 cell group in the ventrolateral medulla, which form routes for ascending baroreceptive information to influence the release of anti-diuretic hormone.

Aggressive Behaviour (Swanson 1987; Roeling et al. 1994; Fig. 22.141b). In rats, typical attack behaviour can be elicited by electrical stimulation from an area including most of the intermediate, perifornical zone of the hypothalamus. The circuitry in which this 'hypothalamic attack area' (HAA) is embedded is shown in Fig. 22.141b. The hypothalamic attack area is strongly and reciprocally connected with the lateral septal nucleus. It is well known that lesioning of the septum results in a type of behaviour that has been referred to as 'septal rage', characterised by unprovoked attacks upon experimenters and an overall high reactivity towards environmental stimuli. The mediodorsal thalamic nucleus, which is also known to be associated with aggressive behaviour, receives a large afferent projection from the hypothalamic attack area and projects to the prefrontal cortex, which in turn has a regulatory influence in neurons within the hypothalamic attack area. The hypothalamic attack area is in receipt of a strong projection from the medial amygdala, which on the basis of stimulation studies in cats has been interpreted as an aggression-facilitating pathway. Attack behaviour can be induced by electrical stimulation of the mesencephalic central grey, and such stimulation has also been shown to facilitate hypothalamically elicited attack. These observations seem to warrant the conclusion that the hypothalamic attack area and parts of the mesencephalic central grey form part of the attack-relevant circuitry. However, lesioning the entire grey does not abolish hypothalamically elicited attack and only temporarily increases the threshold current intensity for eliciting attack behaviour in the hypothalamus. Fibres descending from the hypothalamic attack area to the brain stem have been shown to reach the tegmentum of the midbrain, and it may well be that impulses travelling along these fibres compensate for destruction of the central grey. In cats, other mechanisms have been proposed for the interaction between the hypothalamic attack area and the central grey, in which the latter receives 'attack-associated' afferent input from the anterior hypothalamic area and projects back to the hypothalamic nucleus as a positive feedback. So far, comparable projections have not been found in the rat.

Male Sexual Behaviour (Swanson 1987; Coolen 1995; Fig. 22.141c). The neural substrate of male sexual behaviour in the rat includes the vomeronasal organ, the accessory olfactory bulb, the medial amygdala, parts of the bed nucleus of the stria terminalis, the medial preoptic nucleus and the subparafascicular nucleus, which is situated in the caudal diencephalon.

The vomeronasal organ, or Jacobson's organ, is a chemoreceptor organ that plays a prominent role in chemical (pheromone) communication. Together with the accessory olfactory bulb, it constitutes the medial amydala and parts of the bed nucleus of the stria terminalis, the accessory olfactory system. The medial preoptic nucleus is a major target area of the accessory olfactory system, receiving chemosensory inputs from the medial amygdala and the bed nucleus of the stria terminalis.

Apart from chemosensory inputs, male sexual behaviour is also dependent on somatosensory and genital sensory cues. Information of this type is most probably conveyed to the medial preoptic nucleus by ascending spinal projections that are synaptically interrupted in the caudal diencephalon. Within the latter region, a distinct cluster of Fos-immunoreactive neurons, corresponding to the parvocellular part of the subparafascicular nucleus, appeared to be present following ejaculation.

The medial preoptic nucleus and its outflow via the medial forebrain bundle have been demonstrated to be crucially involved in the consummatory phase of masculine copulatory behaviour. Interestingly, this centre also plays a prominent role in feminine sexual behaviour and in maternal behaviour.

◀ **Fig. 22.141a-c.** Neural circuitry underlying **a** thirst, **b** aggressive behaviour and **c** masculine sexual behaviour. For explanations, see text. *ANGII*, angiotensin II; *AOB*, accessory olfactory bulb; *BNST*, bed nucleus of stria terminalis; *CG*, mesencephalic central grey; *HAA*, hypothalamic attack area; *IML*, nucleus intermediolateralis; *LHA*, lateral hypothalamic area; *LS*, lateral septal nucleus; *MD*, nucleus mediodorsalis thalami; *MeA*, medial part of amygdala; *MFB*, medial forebrain bundle; *MLR*, mesencephalic locomotor region; *MPO*, medial preoptic nucleus; *NTS*, nucleus tractus solitarii; *PFC*, prefrontal cortex; *PP*, posterior pituitary; *PVm*, nucleus paraventricularis, pars magnocellularis; *PVp*, nucleus paraventricularis, pars parvocellularis; *SFO*, subfornical organ; *SO*, nucleus supra opticus; *SPFp*, nucleus subparafascicularis, pars parvocellularis; *TEST*, testosterone; *VLM*, ventrolateral medulla; *VN*, vomeronasal nerve; *25*, area 25 or infralimbic area

22.11
Telencephalon

R. Nieuwenhuys

22.11.1
Configuration and Subdivision

The telencephalon of mammals comprises three main parts: the telencephalon medium, the cerebral hemispheres and the olfactory bulbs. The cerebral hemisperes are, as in all other tetrapods, of the evaginated type. The olfactory bulbs arise as secondary evaginations from the rostrobasal wall of the hemispheres. The telencephalon medium or telencephalon impar represents the caudalmost, non-evaginated part of the telencephalon. It surrounds the rostralmost part of the third ventricle and consists of the lamina terminalis and the preoptic region. The lamina terminalis forms the rostral closure of the third ventricle. All fibres interconnecting the two cerebral hemispheres pass through the lamina terminalis, in which they concentrate in three bundles: the commissura anterior, the commissura hippocampi and the corpus callosum. Due to this passage of fibres, the dorsal part of the lamina terminalis is transformed into a thickened commissural plate. The anterior and hippocampal commisures occur in all mammals and are comparable to the similarly named structures in reptiles. The corpus callosum, which carries neopallial fibres, is confined to eutherians. In monotremes and marsupials, the neopallial fibres decussate through the anterior commissure, which thus attains a considerable size in these forms (see Figs. 22.4d, 22.146a).

The preoptic region flanks the rostralmsot part of the third ventricle. The cell masses which develop in this region structurally and functionally form a continuum with the hypothalamus. They have therefore been dealt with in Sect. 22.10.

The cerebral hemispheres can be divided into a dorsal pallium and a ventral subpallium or basis.

The subpallium comprises a small medial or septal zone and a much larger lateral zone. The latter contains four nuclear formations: the nucleus accumbens, the tuberculum olfactorium, the corpus striatum and the amygdaloid complex. During ontogenesis, the wall of the lateral subpallium thickens and forms on either side two longitudinally oriented intraventricular protrusions, the medial and lateral ganglionic eminences (Fig. 22.142). The nucleus accumbens, the striatum and the amygdaloid complex develop from the rostral, intermediate and caudal parts, respectively, of these ganglionic eminences (see Fig. 22.165). The olfactory tubercle consists of cells which migrate peripherally from the rostral part of the lateral subpallial zone. The expansion of the neopallium (see below) leads to a bending of the cerebral hemispheres. Due to these processes, the caudal part of the striatum is drawn out into a long, curved tail – hence the term nucleus caudatus – and the amygdaloid complex is displaced ventrally (see Fig. 22.147). In primates, the curvature of the hemispheres leads to the formation of a rostrally directed temporal pole, and the amygdaloid complex comes to lie ventral to the rostral portion of the striatum (see Figs. 22.145a, 22.147b, 22.148, 22.165d).

The pallium can be divided into three longitudinally arranged zones: the medial pallium or formatio hippocampi, the dorsal pallium or neopallium and the lateral or piriform pallium. Ariëns Kappers (1909) designated the lateral pallium as the palaeopallium and the medial pallium as the archipallium. In his opinion, the three pallial formations differ markedly with regard to their afferent projections, the palaeopallium, archipallium and neopallium receiving secondary olfactory fibres, tertiary olfactory fibres (via the septum and the olfactory tubercle) and non-olfactory fibres ascending from the dorsal thalamus, respectively. Throughout the three pallial fields, the cellular elements migrate outward during ontogenesis to form distinct cortical formations.

The mammalian hippocampal formation can be subdivided into no less than eight strip-like areas; from (morphologically) ventral to (morphologically) dorsal, these are as follows: (1) the small-celled fascia dentata, (2-5) the large-celled cornu Ammonis, comprising fields CA4-CA1, (6) the subiculum, (7) the presubiculum and (8) the parasubiculum (see Fig. 22.187). The parasubiculum is separated from the neocortex by a transitional zone, the area entorhinalis. The fascia dentata and fields CA1-3 of Ammon are typical trilaminar formations, in which

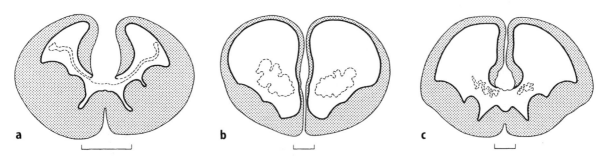

Fig. 22.142a-c. Transverse sections through the telencephalic hemispheres of mammalian embryos. a Hedgehog. b Human. c The blue whale *Balaenoptera musculus*. Bar, 1 mm.

a cell layer is flanked by internal and external plexiform layers. The subiculum, presubiculum and parasubiculum contain three to four layers, whereas in the entorhinal area five layers can be distinguished. The reptilian medial cortex is divisible into medial and dorsomedial fields, but these cannot be homologised in a one-to-one fashion with particular mammalian hippocampal fields (Fig. 22.143). It is known that the axons of the pyramidal cells situated in the medial part of the reptilian dorsal cortex join a periventricular alveus system and leave the pallium via the medial hemisphere wall. Because of this fornix-like course of its efferent, some authors have claimed that the medial sector of the reptilian dorsal or general cortex is homologous to the mammalian subiculum. However, Kuhlenbeck (1977) held that the medial portion of the reptilian dorsal cortex corresponds to the entorhinal cortex.

The reptilian medial cortex is a superficially situated structure throughout its extent (Figs. 22.143b,c), but in mammals, this formation rolls in along itself along a longitudinal groove, the hippocampal sulcus (Figs. 22.143d, 22.144a). Only relatively small parts of the mammalian hippocampus maintain their initial, superficial position (Figs. 22.144a-c, 22.146).

In mammals which lack a corpus callosum, a well-differentiated hippocampal formation extends over the entire length of the medial hemisphere wall (Figs. 22.146a), but in eutherians the development of the hippocampus is strongly influenced by the corpus callosum. This neopallial commissure develops in the dorsalmost part of the lamina terminalis and extends caudally, splitting the rostral part of the hippocampal region into a dorsal and a ventral part. The dorsal part can be divided into a precommissural and a supracommissural hippocampus. The precommissural hippocampus is formed by a small, mostly trilaminar field, extending from the caudal margin of the olfactory bulb into the precommissural part of the medial hemisphere wall (Figs. 22.144b,c, 22.146b,c). The supracommissural hippocampus is represented by a tiny band of grey matter, known as the indusium griseum, which bilaterally follows the dorsal aspect of the corpus callosum (Figs. 22.144d, 22.146b,c). The subcommissural part of the hippocampus is mainly formed by the fornix, a large fibre system that connects the hippocampal formation with the septum and the hypothalamus. During their subcallosal course, the right and left fornices lie side by side and form a commissure, the hippocampal commissure or psalterium. Somewhat more rostrally, the fornix arches around the foramen interventriculare and splits into a smaller precommissural and a larger postcommissural part, names which indicate the position of these parts with respect to the anterior commissure.

The precommissural and supracommissural parts of the hippocampus are small, vestigial structures; hence in callosal mammals, only the (morphologically) caudal, retrocommissural hippocampus is well developed and clearly differentiated into the zones and layers outlined above. Due to the ventral bend of the hemispheres (see below), the retrocommissural hippocampus is displaced ventrally or ventrorostrally, particularly in animals in which this curvature has led to the formation of a rostrally directed temporal lobe (see Figs. 22.146b,c, 22.148).

On the basis of its position and connections, the mammalian lateral pallium, which encompasses the prepiriform and periamygdaloid cortices, is homologous to the reptilian lateral cortex and the amphibian primordium piriforme (Fig. 22.143). All of these structures receive a strong, direct projection from the olfactory bulb. The mammalian prepiriform and periamygdaloid cortices, like the hippocampal cortex, are trilaminar structures, but unlike the situation in the hippocampus, the third layer is not plexiform, but rather cellular, although the neuronal density in this layer is much lower than in the second layer (Fig. 22.143d). It is important to note that the neurons constituting the mammalian lateral cortex stem from the lateral zone of the pallial matrix and reach their definitive position by nonradial, but vectorial migration (Misson et al. 1991; see Chap. 5, Sect. 5.5.5 and Figs. 5.13, 6.11).

The boundary between the olfactory cortex and the non-olfactory neocortex is often marked by a distinct external groove, the fissura rhinalis, whereas another groove, the fissura endorhinalis, marks the boundary between the prepiriform cortex and the olfactory tubercle (Fig. 22.144). In macrosmatic mammals, the region situated between these two sulci forms a large and prominent piriform lobe (Figs. 22.4c, 22.5c, 22.7b,c). In many marsupials and insectivores, the olfactory system, including the olfactory cortex, is strongly developed, while the neocortex is only small. Where this combination occurs, the rhinal fissure is situated far dorsal on the lateral surface of the telencephalon (Fig. 22.144b; see poster P1, 5-7). Formerly, it was widely believed (see e.g. Ariëns Kappers et al. 1936) that primitive mammals possessed a large olfactory system and a small neopallium and that mammalian phylogeny is characterised by a reduction in the rhinencephalon and an expansion of the neopallium. It is true that in primates and cetaceans the olfactory system is much reduced, whereas the neopallium is strongly developed (see poster P1, 21-23, 30, 31), but the 'scala naturae' view men-

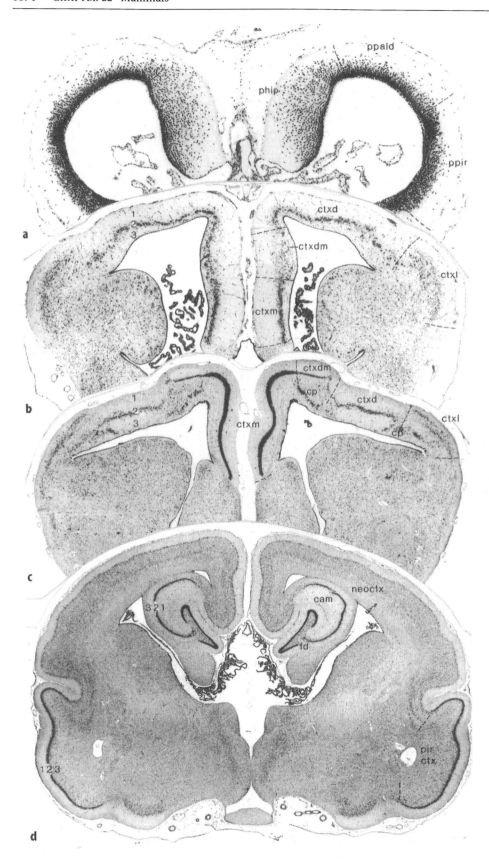

Fig. 22.143a–d. Transverse sections through the telencephalon of tetrapods to show the structure of the pallium and the differentiation of cortical formations. **a** The tiger salamander *Ambystoma tigrinum*. **b** The turtle *Testudo hermanni*. **c** The lizard *Tupinambis nigropunctatus*. **d** The opossum *Didelphis virginiana*. Nissl technique. *cam*, cornu ammonis; *cp*, cell plate; *ctx*, cortex dorsalis; *ctxdm*, cortex dorsomedialis; *ctxl*, cortex lateralis; *ctxm*, cortex medialis; *fd*, fascia dentata; *neoctx*, neocortex; *phip*, primordium hippocampi; *pir ctx*, (pre)piriform cortex; *ppald*, primordium pallii dorsalis; *ppir*, primordium piriforme; *1,2,3*, cortical layers. (Reproduced from Nieuwenhuys 1994)

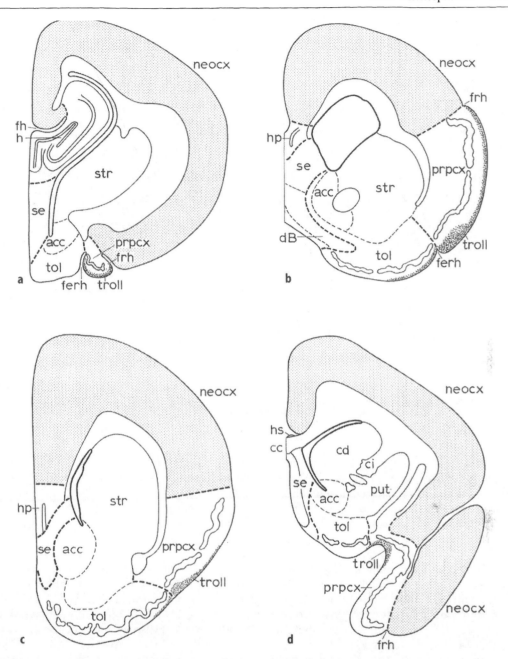

Fig. 22.144a–d. Transverse sections through the rostral part of the telencephalic hemispheres of **a** the duck-billed platypus *Ornithorhynchus anatinus* (after Hines 1929), **b** the hedgehog *Erinaceus europaeus* (after Stephan et al. 1991), **c** the roussette bat *Rousettus amplexicaudatus* (after Baron et al. 1996) and **d** the common marmoset *Callithrix jacchus* (after Stephan et al. 1980). *acc*, nucleus accumbens; *cc*, corpus callosum; *cd*, nucleus caudatus; *ci*, capsula interna; *dB*, nucleus of the diagonal band of Broca; *ferh*, fissura endorhinalis; *fh*, fissura hippocampi; *frh*, fissura rhinalis; *h*, hippocampus; *hp*, hippocampus precommissuralis; *hs*, hippocampus supracommissuralis; *neocx*, neocortex; *prpcx*, prepiriform cortex; *put*, putamen; *se*, septum; *str*, striatum; *tol*, tuberculum olfactorium; *troll*, tractus olfactorius lateralis

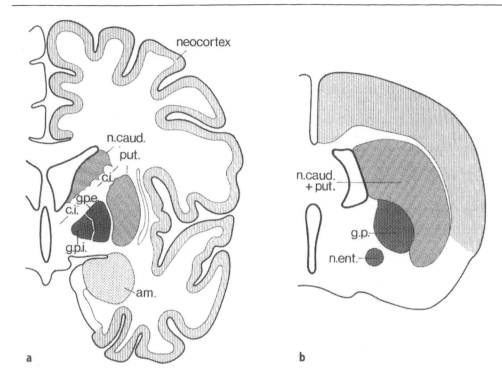

Fig. 22.145a,b. Transverse sections through the telencephalic hemispheres of **a** man and **b** rat. *am*, amygdaloid complex; *c.i.*, capsula interna; *g.p.*, globus pallidus; *g.p.e.*, globus pallidus, pars externa; *g.p.i.*, globus pallidus, pars interna; *n.caud.*, nucleus caudatus; *n.ent.*, nucleus entopeduncularis; *put.*, putamen

tioned above appears much too simple to fit the facts if the full spectrum of variations in the structure of the mammalian telencephalon is taken into consideration. For instance, the monotremes, which in the recent fauna include only the duck-billed platypus and the spiny anteaters of the Australian region, which are generally considered to be direct descendents of the mammal-like therapsid reptiles, possess a relatively small rhinencephalon and an astonishingly large neopallium. Accordingly, the rhinal sulcus is situated far ventral in these animals and cannot be seen from the lateral side (Fig. 22.144a; see poster P1,1, 2).

The neopallium, though rather small in many marsupials and insectivores, is by far the largest telencephalic centre in most mammals and attains amazing proportions in primates, proboscoideans and cetaceans (see poster P1, 21–23, 30, 31, 34). Indeed, in most mammals the expansion of the neopallium is the dominant feature in the ontogenesis of the telencephalon, affecting the development of all other telencephalic components. The dorsal expansion brings the medial walls of the hemispheres, beyond the level of the commissural plate, into apposition. The backward expansion of the neopallium brings the caudal portion of the hemispheres dorsal to other parts of the brain. In animals with a small neopallium, only the diencephalon is covered (Fig. 22.4a), but in most mammals the telencephalic hemispheres reach the cerebellum, covering the midbrain completely (Fig. 22.5a), and in primates the cerebellum is also hidden from view by the large hemispheres (Fig. 22.6a; see poster P2, 29). In the latter group, the caudal expansion of the hemispheres results in the formation of occipital lobes surrounding extensions of the lateral ventricles (Fig. 22.147b). The downward outgrowth of the neopallium leads to a ventral displacement of the rhinal fissure, whereas the rostral expansion leads to the formation of frontal lobes covering the olfactory bulbs.

The outgrowth of the neopallium not only leads to a telencephalic expansion, but also involves a rotation around a transverse axis of the caudal portions of the cerebral hemisperes. This rotation is only slight in animals with small hemispheres, such as the opossum (Fig. 22.147a), is more pronounced in carnivores (Fig. 22.5b; see poster P1, 24–26), ungulates (see poster P1, 28, 29, 32), and cetaceans (Fig. 22.147c; see poster P1, 30, 31) and most advanced in primates (Fig. 22.147b; see poster P1, 22, 23). It leads to the formation of ventrally or ventrorostrally directed temporal lobes and to the concomitant formation of an external depression or

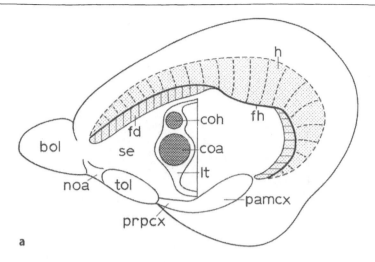

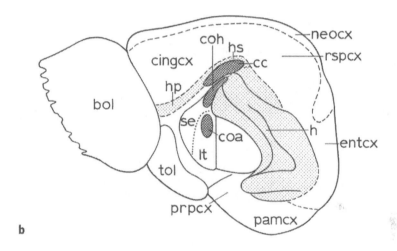

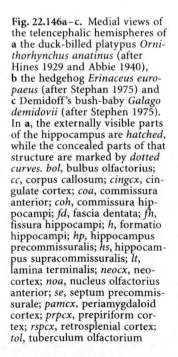

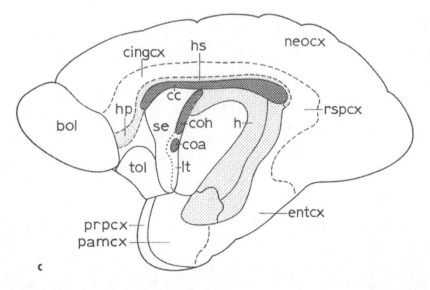

Fig. 22.146a–c. Medial views of the telencephalic hemispheres of **a** the duck-billed platypus *Ornithorhynchus anatinus* (after Hines 1929 and Abbie 1940), **b** the hedgehog *Erinaceus europaeus* (after Stephan 1975) and **c** Demidoff's bush-baby *Galago demidovii* (after Stephen 1975). In **a**, the externally visible parts of the hippocampus are *hatched*, while the concealed parts of that structure are marked by *dotted curves*. *bol*, bulbus olfactorius; *cc*, corpus callosum; *cingcx*, cingulate cortex; *coa*, commissura anterior; *coh*, commissura hippocampi; *fd*, fascia dentata; *fh*, fissura hippocampi; *h*, formatio hippocampi; *hp*, hippocampus precommissuralis; *hs*, hippocampus supracommissuralis; *lt*, lamina terminalis; *neocx*, neocortex; *noa*, nucleus olfactorius anterior; *se*, septum precommissurale; *pamcx*, periamygdaloid cortex; *prpcx*, prepiriform cortex; *rspcx*, retrosplenial cortex; *tol*, tuberculum olfactorium

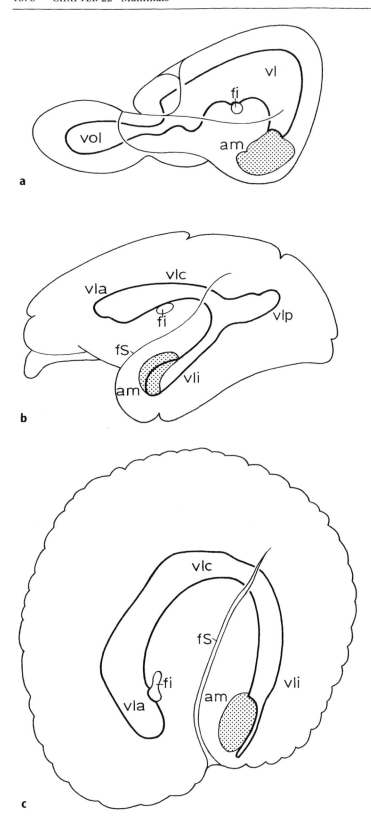

Fig. 22.147a–c. Projections of the lateral ventricle on the lateral surface of the telencephalon. **a** The opossum *Didelphis virginiana* (based on Loo 1930). **b** The rhesus monkey *Macacus rhesus* (based on McFarland et al. 1969). **c** The dolphin *Tursiops trunctatus* (based on McFarland et al. 1969; Jacobs et al. 1971; Morgane et al. 1980). The position of the amygdaloid complex (*am*) is indicated. *fi*, foramen interventriculare; *fl*, fissura lateralis; *vl*, ventriculus lateralis; *vla, vl,* cornu anterius; *vlc, vl,* pars centralis; *vli, vl,* cornu inferius; *vlp, vl,* cornu posterius; *vol,* ventriculus bulbi olfactorii

groove (sulcus lateralis; fissura Sylvii; fossa Sylvii) and of an inferior horn (cornu inferius) of the lateral ventricle (Fig. 22.147b,c). In cetaceans, the curvature of the caudal parts of the hemispheres is coupled with a pronounced downward rotation of its rostral parts. Due to these processes, the cerebral hemispheres in this group have a peculiar, doubled-up appearance, with the frontal pole approaching the temporal pole (McFarland et al. 1969; Figs. 22.147c; see poster P1, 30, 31).

Figure 22.148 illustrates how the neopallial expansion influences the configuration of the deep telencephalic structures and even of the basal olfactory area. Due to the expansion of the neopallium and the concentration of the higher olfactory centres around the 'hilus' of the hemispheres, in primates, osmic cetaceans and sirenians, the connections between the olfactory bulbs and the cerebral hemispheres are transformed into long olfactory stalks or peduncles (Figs. 22.6, 22.8a, 22.157a,b; P1:35).

In many mammalian species, particularly in large ones, the expansion of the neopallium is coupled with convolution of its external surface. This phenomenon, and the resultant patterns of gyri and sulci, will be dealt with in Sect. 22.11.9.

In primates and cetaceans, the part of the pallium which covers the anlage of the corpus striatum lags behind during ontogenesis. Consequently, a depression, the fossa Sylvii, appears on the lateral surface of the cerebral hemispheres. The bottom of this depression is known as the insula of Reil. During further development, the adjacent parts of the frontal, parietal and temporal regions overgrow the depressed area as opercula. Finally, the insula becomes completely covered, and the superficial part of the fossa Sylvii is transformed into a cleft, the lateral fissure of fissura Sylvii (Fig. 22.149). Although a morphologically differentiated insula is confined to cetaceans and primates, the cortex which lines this island has a clear homologue in all mammals,

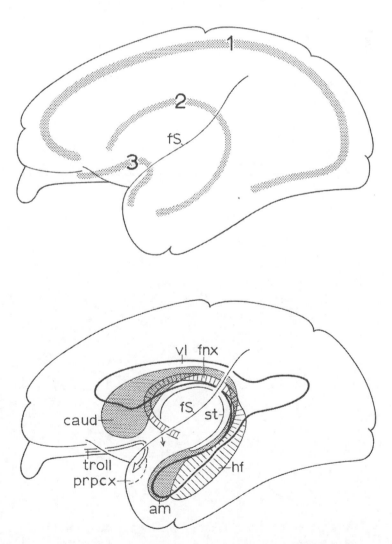

Fig. 22.148. Indication of how, in the telencephalon of the rhesus monkey, the expansion of the neocortex (1) has influenced the shape of the lateral ventricle, strio-amygdaloid complex and stria terminalis, of the hippocampal formation and fornix (2) and finally of the central olfactory system (3). *am*, amygdaloid complex; *bol*, bulbus olfactorius; *caud*, nucleus caudatus; *fnx* fornix; *hf*, hippocampal formation; *prpcx*, prepiriform cortex; *st*, stria terminalis; *troll*, tractus olfactorius lateralis; *vl*, ventriculus lateralis

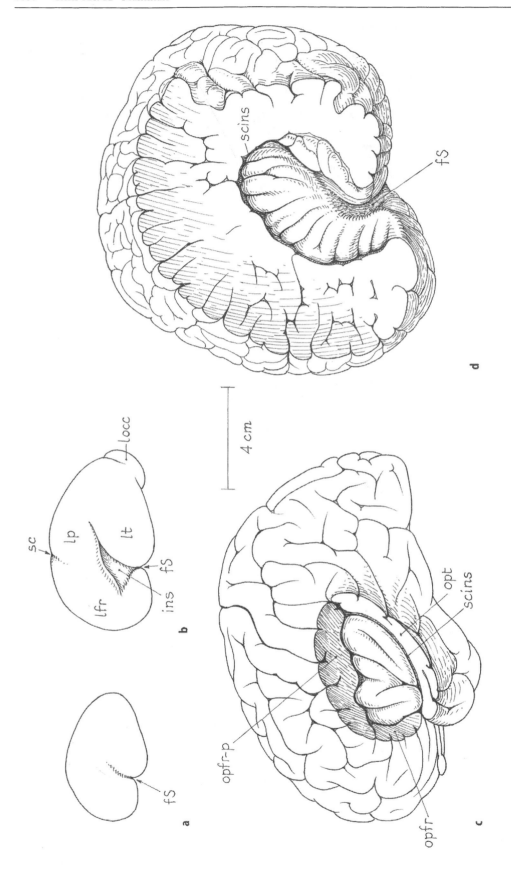

Fig. 22.149a–d. Insula of Reil and its development. Lateral views of the cerebral hemispheres of **a** a 5-month-old human embryo, **b** a 7-month-old human embryo, **c** a human adult and **d** the adult bottle-nosed dolphin *Tursiops truncatus*. In **c** and **d** the opercula have been removed to reveal the insula. For comparison with intact hemispheres, see Poster 1, 23, 30. *fS*, fissura lateralis Sylvii; *ins*, insula of Reil; *lfr*, lobus frontalis; *locc*, lobus occipitalis; *lp*, lobus parietalis; *lt*, lobus temporalis; *opfr*, operculum frontale; *opfr-p*, operculum fronto-parietale; *opt*, operculum temporale; *sc*, sulcus centralis; *fscins*, fissura circularis insulae. (**d** is based on a photograph from Morgane et al. 1980)

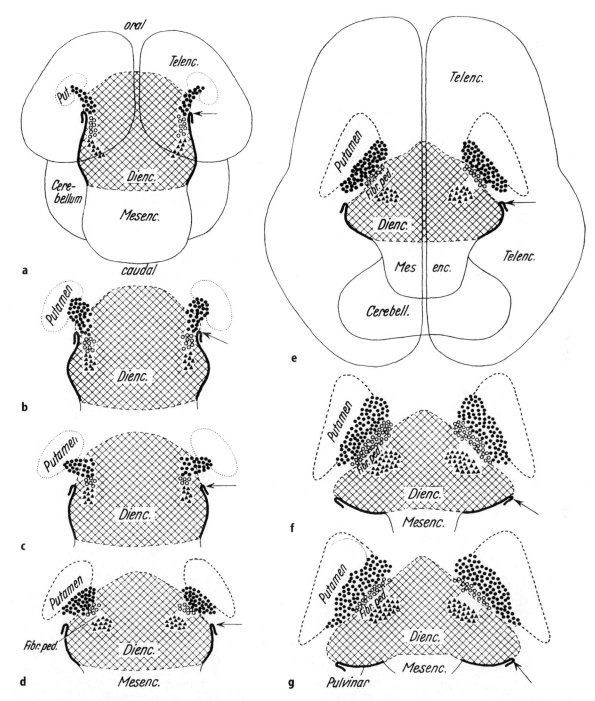

Fig. 22.150a–g. Development of the diencephalon and adjacent parts of the telencephalon, as shown by a series of human embryos ranging from **a** the second to **g** the sixth month. The figures represent graphical reconstructions in which the structures depicted are projected upon a horizontal plane. The anlage of the diencephalon is *cross-hatched*, and its external surface is indicated by *heavy lines*. Arrows indicate transition of the diencephalic surface to the medial hemispheric wall. *Black triangles* indicate the nucleus subthalamicus, *open circles*, the nucleus entopeduncularis or globus pallidus, pars interna and *filled circles*, the globus pallidus, pars externa. *Fibr. ped.*, fibrae pedunculares. (Reproduced from Richter 1966)

which is either overt or covert and is referred to as the insular cortex.

The spatial relations in the mammalian telencephalic-diencephalic border zone are complex and can only be understood in light of the ontogenetic events which take place in this region. The graphical reconstructions of the diencephalon and the adjacent parts of the telencephalon in a series of human embryos, shown in Fig. 22.150, together present a pictorial survey of these events, of which the following are of particular importance:

1. Due to the expansion of the dorsal thalamus, the lateral surface of the diencephalon rotates and ultimately becomes part of the caudal surface of that region.
2. Concomitantly, the boundary plane between the diencephalon and the telencephalon enlarges and gradually assumes an oblique orientation, passing from rostromedial to caudolateral.
3. This change in orientation of the telencephalic-diencephalic boundary plane results in the dorsal thalamus becoming rostrolaterally directly continuous with the telencephalic territory (Fig. 5.14).
4. The lateral part of the subthalamus contains the anlagen of three cell masses, i.e. the nucleus subthalamicus, the nucleus entopeduncularis and the globus pallidus, the latter of which encroaches upon the medial part of the subpallium. Initially, these three anlagen together constitute a caudorostrally oriented cell cord, but in the later course of development this longitudinal orientation changes into an almost transverse arrangement (Richter 1966). The anlage of the globus pallidus unites with the large cell mass that develops from the telencephalic ganglionic eminence to become part of the corpus striatum. In non-primates, the anlage of the entopeduncular nucleus develops into a separate cell mass (Fig. 22.145b), but in primates it takes up a position directly medial to the globus pallidus (Fig. 22.150). Hence, in this group, the globus pallidus is considered as being composed of two parts, the pars externa and the pars interna, or for short, the pallidum externum and the palli-

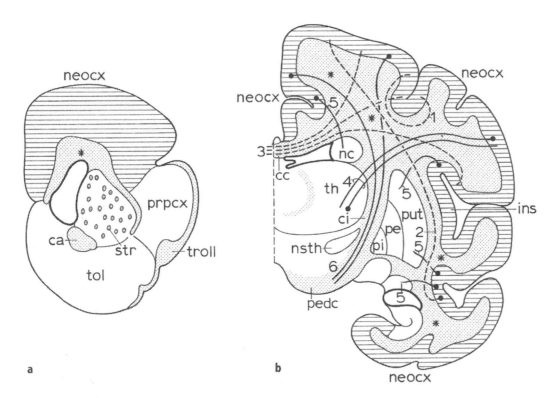

Fig. 22.151a,b. Diagrammatic transverse sections through **a** the telencephalon of the hedgehog and **b** the telencephalon and diencephalon of the macaque monkey, showing the extent and the composition of the subcortical white matter or centrum semiovale (*). In **b**, cortico-cortical connections are indicated by *dashed lines*, and cortico-subcortical and subcortico-cortical projections by *continuous lines*. *ca*, commissura anterior; *cc*, corpus callosum; *ci*, capsula interna; *ins*, insula; *neocx*, neocortex; *nsth*, nucleus subthalamicus; *pe*, globus pallidus, pars externa; *pedc*, pedunculus cerebri; *pi*, globus pallidus, pars interna; *prpcx*, prepiriform cortex; *put*, putamen; *str*, striatum; *th*, thalamus; *tol*, tuberculum olfactorium; *troll*, tractus olfactorius lateralis. For an explanation of the numbers, see text

dum internum (Fig. 22.145a). This implies that the pallidum externum of primates corresponds to the entire (unsegmented) globus pallidus of non-primates and that the primate pallidum internum corresponds to the entopeduncular nucleus of non-primates (Fig. 22.145).

In all mammals, the striatum is pierced by a large number of fibres, many of which pass from the cortex to lower parts of the neuraxis. In most non-primates, these fibres form a number of separate fascicles (Fig. 22.151a), but in primates they unite in a massive fibre plate. This fibre plate, the capsula interna, divides the primate striatum into a medial (and periventricular) nucleus caudatus and a lateral putamen (Fig. 22.145). Putamen, pallidum externum and pallidum internum together form a nuclear complex, designated as the nucleus lentiformis. The caudal part of the capsula interna separates the lentiform nucleus from the thalamus and, more ventrally, from the nucleus subthalamicus (Figs. 22.150, 22.151b).

During early development, the periventricular zone of the pallium is occupied by a wide and very active matrix layer. This matrix produces the cells which, after having migrated outward, go on to form the cortex cerebri. During further development, the anlage of the cortex waxes and the matrix wanes, and ultimately, after having produced the full complement of cortical cells, the latter disappears completely. *Pari passu* with the diminuition of the matrix, the periventricular zone is invaded by fibres, and the pallium eventually shows a condition opposite to that of the spinal cord (and of the entire neuraxis in many lower forms), i.e. grey matter on the outside and white matter on the inside. The pallial central white matter is known as the centrum semiovale. Still small in primitive mammals, such as the hedgehog (Fig. 22.151a), it attains very large proportions in groups with a strongly developed and extensive neocortex, such as cetaceans and primates (Fig. 22.151b).

The centrum semiovale, or subcortical white matter, is first and foremost composed of associational systems, which interconnect different parts of the cortex, but also harbours ascending (subcortico-cortical) and descending (cortico-subcortical) projections. Its various components may be categorised as follows (Fig. 22.151b; N.B. the numbers in this figure correspond to those used in the text):

1. Short ipsilateral association systems
2. Long ipsilateral association systems
3. Long, decussating association systems, most of which pass by way of the corpus callosum, but some via the anterior commissure
4. Thalamocortical and their reciprocating cortico-thalamic projections
5. Corticostriatal projections
6. Long corticofugal systems

The latter comprise the cortico-bulbospinal tract and the frontopontine and parieto-temperopontine projections, all of which attain the rhombencephalon by way of the internal capsule and the cerebral peduncle. In mammals with large cerebral hemispheres, the long association fibres form bundles, some of which are shown in Fig. 22.152. For particulars on these association bundles, the reader is referred to Nieuwenhuys et al. (1988).

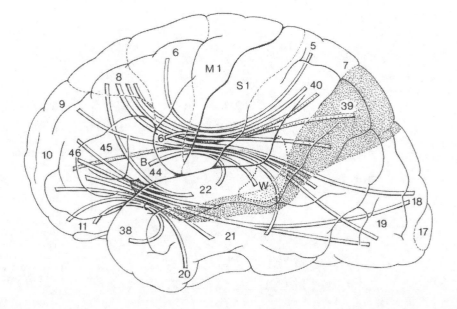

Fig. 22.152. Long cortical association connections in the human left hemisphere. The upper system consists of the fasciculus occipitofrontalis superior and the fasciculus longitudinalis superior; the lower system is composed of the fasciculus occipitofrontalis inferior and the fasciculus uncinatus. The multimodal association cortex is stippled. B, Broca's speech area; *M1*, primary motor cortex; *S1*, primary sensory cortex; W, Wernicke's speech area. The *numbers* indicate cortical areas according to Brodmann (1989). (Reproduced from Nieuwenhuys et al. 1988)

22.11.2
Rhinencephalon

22.11.2.1
Introductory Note

The mammalian rhinencephalon, i.e. the parts of the telencephalon involved in the processing of chemosensory information, encompasses the main olfactory system, the accessory olfactory or vomeronasal system and presumably the terminal nerve system. These three systems will be discussed below.

22.11.2.2
Main Olfactory System

The olfactory bulb and the targets of its efferent projections, i.e. the retrobulbar region, the olfactory tubercle, the prepiriform, periamygdaloid and entorhinal cortices and certain nuclei of the amygdaloid complex together constitute the central part of the main olfactory system (Fig. 22.153). In macrosmatic mammals, all of these structures are well developed and together occupy most of the basal surface of the telencephalon (Figs. 22.5c, 22.7c, 22.154a). In macrosmatic forms with a small neopallium, even the entire basal surface (Figs. 22.4c, 22.7b) and a considerable part of the lateral telencephalic surface (see poster P1, 5–9) belong to the rhinencephalon.

The main olfactory system is extremely well developed in marsupials, like the opossum (Fig. 22.4) and the marsupial mole (Burkitt 1938, Fig. 22.7b), and in most insectivores (see poster P1, 5–7), but it is relatively small in the duck-billed platypus (Fig. 22.7a; see poster P1, 1). Ferungulates generally possess a well-developed main olfactory system, but it is poorly developed in aquatic groups, such as the sirenians (see poster P1, 35) and the cetaceans. Olfactory nerves, bulbs and peduncles are entirely absent in odontocete cetaceans, like

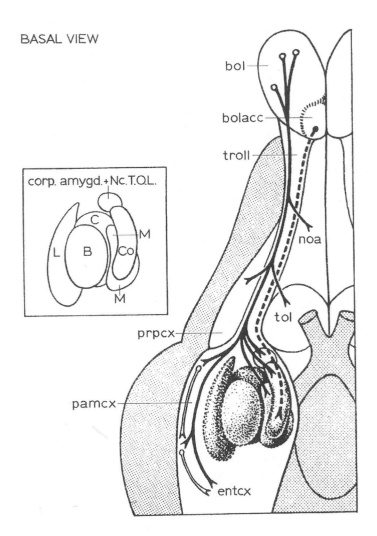

Fig. 22.153. Secondary olfactory projections and their targets. The projections of the main olfactory bulb are in *black*, those of the accessory olfactory bulb are *dashed* and the projections of the periamygdaloid cortex are indicated by *open contours*. *B*, nucleus amygdalae basalis; *bol*, bulbus olfactorius; *bolacc*, bulbus olfactorius accessorius; *C*, nucleus amygdalae centralis; *Co*, nucleus amygdalae corticalis; *entcx*, entorhinal cortex; *L*, nucleus amygdalae lateralis; *tol*, tuberculum olfactorium; *troll*, tractus olfactorius lateralis; *M*, nucleus amygdalae medialis; *Nc.T.O.L.*, nucleus of the tractus olfactorius lateralis; *pamcx*, periamygdaloid cortex; *prpcx*, prepiriform cortex. (Reproduced form Lammers 1972)

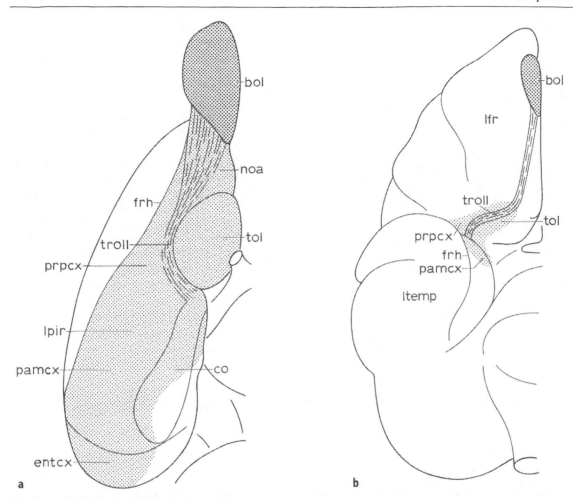

Fig. 22.154a,b. Ventral view of the brain of **a** the rat and **b** the macaque monkey, showing the size of the olfactory bulb and its projection field (*stippled*) in relation to the whole brain. *bol*, bulbus olfactorius; *co*, nucleus amygdalae corticalis; *entcx*, entorhinal cortex; *frh*, fissura rhinalis; *lfr*, lobus frontalis; *lpir*, lobus piriformis; *ltemp*, lobus temporalis; *noa*, nucleus olfactorius anterior; *pamcx*, periamygdaloid cortex; *prpcx*, prepiriform cortex. (Modified from Heimer 1969)

the bottle-nosed dolphin (Breathnach 1960; see Figs. 22.8b, 22.9c, 22.157c). Primates are generally microsmatic (see Fig. 22.6b–d; see poster P1, 21, 22), and the human olfactory system is to be considered rudimentary (see Fig. 22.8a, 22.157a; poster P1, 23). Considerable differences in the development of the main olfactory systems not only exist between the various mammalian orders, but can also be observed within these groups. Thus Stephan et al. (1991) reported that, among insectivores, this system is much better developed in terrestrial species than in semi-aquatic forms.

The *main olfactory bulb* is the primary brain centre in the main olfactory system and serves as a relay station for all olfactory impulses between the olfactory mucosa and the higher olfactory centres. Consequently, the size of the main olfactory bulb may be regarded as an important and direct indicator of the significance of the olfactory system in a given species. The main olfactory bulbs are separate forward extensions of the telencephalon which develop from evaginations of the hemisphere walls. In marsupials (Figs. 22.147a) and insectivores, they contain a large olfactory ventricle, but in most other mammals these cavities are either small or entirely obliterated in the adult stage (Fig. 22.155). Short olfactory peduncles connect the olfactory bulbs with the cerebral hemispheres in most mammals, but, due to a strong rostrally directed expansion of the frontal lobes, these peduncles are drawn out into long stalks in primates (Fig. 22.8a; see poster P1, 21–23), elephants (Fig. 22.8c), sirenians (see poster P1, 35) and osmatic cetaceans (see Fig. 22.157b).

The main olfactory bulb is organised in seven layers, which are concentrically arranged around the bulbar ventricle or its remnant (Fig. 22.155). Passing from superficial to deep, these layers are as follows (Fig. 22.156):

1. A layer of olfactory nerve fibres, consisting of densely interwoven, extremely fine axons originating from the olfactory epithelium.
2. The glomerular layer, which contains the conspicuous glomeruli, special regions of neuropil in which the terminal arborisations of the olfactory nerve fibres synapse with the dendrites of three types of secondary olfactory neurons, i.e. the mitral, tufted and periglomerular neurons. The two cell types first mentioned lie in deeper zones of the olfactory bulb; the latter elements are small and granular and, as their name implies, surround the glomeruli.
3. The external plexiform layer constitutes a complex meshwork of interlacing axonal and dendritic branches. However, it also contains the perikarya of superficial interneurons and of the relatively large tufted cells.
4. The mitral cell layer, consisting of densely packed granule cells, in which the very large somata of the mitral cells are embedded.
5. The internal plexiform layer, which in some places cannot be clearly distinguished as a separate zone; it contains ascending dendrites of deep granule cells, axons of mitral and tufted cells and axons of centrifugal fibres from other regions of the brain. Moreover, in this layer the perikarya of some granule cells and of somewhat larger, intermediate short-axon cells (see below) are found.

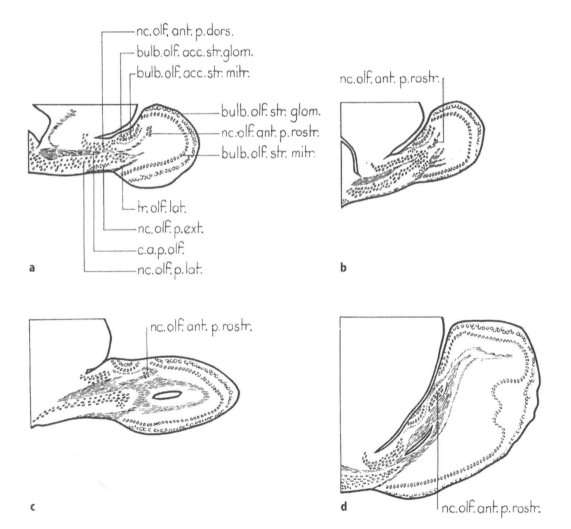

Fig. 22.155a–d. Sagittal sections through the rostral part of the telencephalon of **a** the guinea pig, **b** the rat, **c** the rabbit and **d** the cat, based on Klüver-Barrera preparations. *bulb.olf.*, bulbus olfactorius; *bulb.olf.acc.*, bulbus olfactorius accessorius; *c.a.p.olf.*, commissura anterior, pars olfactoria; *nc. olf.ant.p.dors., ext., lat., rostr.*, nucleus olfactorius anterior, pars dorsalis, externa, lateralis, rostralis; *str.glom., mitr.*, stratum glomerulosum, mitrale; *tr.olf.lat.*, tractus olfactorius lateralis. (Reproduced from Lohman and Lammers 1967)

6. The granular layer, consisting of several concentric zones of densely packed granular cells, separated from each other by bundles of nerve fibres; larger neurons are scattered in this layer.
7. The periventricular zone, which is formed by a layer of ependymal cells or, in animals with obliterated olfactory ventricles, by the remnants of this layer.

The microcircuit of the mammalian main olfactory bulb has been the subject of numerous studies, the older ones of which were summarised by Stephan (1975). Within the frame of the present work, only a brief survey can be presented of the structural and functional relations of the various bulbar elements and of their chemical signatures. In the preparation of this survey, I have relied heavily on the review articles by Macrides and Davis (1983), Halász and Shepherd (1983), Shepherd and Greer (1990) and Shipley et al. (1995). The following aspects will be discussed consecutively: the olfactory projections, the roles of the periglomerular and granule cells, the centrifugal fibres and their targets and the relationships of the deep and superficial interneurons:

1. The *olfactory projection* is constituted by the primary olfactory elements and the second-order olfactory projection neurons, i.e. the mitral and tufted cells. The receptive element of the olfactory apparatus is represented by slender bipolar cells situated in the mucosa of the nasal cavity. These elements give rise to extremely fine (0.2–0.4 µm), unmyelinated axonal processes, which, together forming the first cranial nerve, carry the olfactory impulses directly to the brain. Upon entering the olfactory bulb, the axons of the peripheral olfactory elements interlace in a highly complex fashion and terminate by free arborisations in the glomeruli. In these spherical

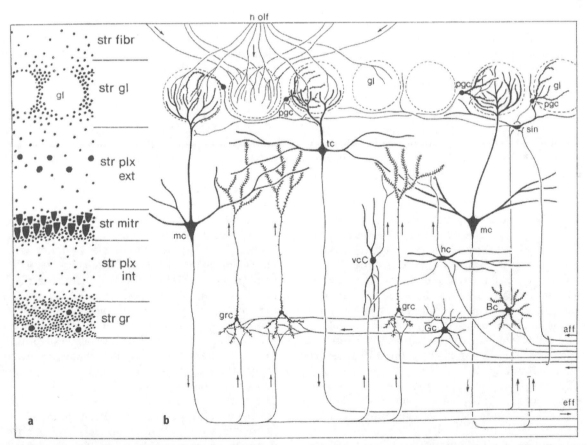

Fig. 22.156a,b. The olfactory bulb. **a** Cytoarchitecture. **b** Neuronal elements and fibres as observed in Golgi preparations (somewhat simplified). *aff*, centrifugal afferents; *Bc*, Blanes cell; *eff*, bulbar efferents; *Gc*, Golgi cell; *gl*, glomeruli; *grc*, granule cell; *hc*, horizontal cell; *mc*, mitral cell; *n olf*, nervus olfactorius; *pgc*, periglomerular cells; *sin*, superficial interneuron; *str fibr*, stratum fibrosum; *str gl*, stratum glomerulosum; *str gr*, stratum granulare; *str mitr*, stratum mitrale; *str plx ext*, stratum plexiforme externum; *str plx int*, stratum plexiforme internum; *tc*, tufted cell; *vcC*, vertical cell of Cajal. (Modified from Nieuwenhuys 1985)

neuropil configurations, the synaptic contacts between the terminals of the incoming fibres and the dendrites of the mitral and tufted cells are established. Electrophysiological studies have shown that the peripheral inputs to the olfactory bulb are excitatory. The primary olfactory elements and their processes have been found to contain high concentrations of the dipeptide carnosine (β-alanyl-L-histidine), as well as a specific (20-kDa) olfactory marker protein. However, the exact roles played by these substances remain to be elucidated.

The large mitral cells and the somewhat smaller tufted cells have one main or primary dendrite entering a glomerulus and several secondary or accessory dendrites branching in the external plexiform layer. The main dendrites enter into synaptic contact with olfactory nerve fibres and with axons and dendrites of periglomerular cells. The synaptic relations of the secondary dendrites will be dealt with below.

The axons of the mitral and tufted cells pass radially through the deeper layers of the olfactory bulb. During their course through the bulb, they emit numerous collaterals which contact granule cells and deep interneurons. The main axons of the mitral and the middle and deep tufted cells gain myelin sheaths and become grouped together in bundles, which, turning caudally, constitute the secondary olfactory projection. The fibres of this projection convey olfactory information to higher-order olfactory structures. The axons of superficially situated tufted cells project mainly to other sites in the same olfactory bulb, thus forming an intrabulbar association system. Several lines of evidence strongly suggest that glutamate and/or aspartate function as excitatory neurotransmitters in the efferents of the olfactory bulb to its principal target, i.e. the prepiriform cortex. As regards the tufted cells, glutamate/aspartate may well be the transmitter of only the more deeply situated elements of that category. Immunohistochemical studies have shown that the great majority of the external tufted cells contain cholecystokinin and that many middle tufted cells contain vasoactive intestinal polypeptide.

2. The *periglomerular cells* and the *granule cells* have several features in common. Both maintain reciprocal dendrodendritic synaptic contacts with mitral and tufted cells, and both have these two cell types as their main targets. However, the periglomerular and granule cells also show marked differences, the most salient of which is that the former are regular short-axon cells, whereas the latter are amacrine, i.e. axon-less elements.

Because of their position and small size, the periglomerular cells are often designated as superficial granule cells. Their dendrites enter glomeruli, where they receive impulses from olfactory nerve terminals and also enter into synaptic contact with the intraglomerular ramifications of the main dendrites of mitral and tufted cells. The ultrastructural features of these dendrodendritic contacts suggest that mitral and tufted cell dendrites are excitatory to the periglomerular dendrites and that these latter dendrites are inhibitory to mitral and tufted cell dendrites. The axons of the periglomerular cells course along the periphery of two to four glomeruli, making inhibitory synapses on the somata and dendrites of other periglomerular cells and on the initial parts of the primary dendrites of mitral and tufted cells. Thus the periglomerular cells exert an inhibitory influence on the mitral and tufted cells in two different places and in two different ways, namely via interglomerular dendrodendritic synapses and via subglomerular axodendritic synapses. The population of periglomerular cells is neurochemically heterogeneous. Immunohistochemical evidence indicates that many of these elements are dopaminergic or GABAergic and that many of these elements contain both of these neurotransmitters. Moreover, with regard to their content of neuroactive principles, the periglomerular cells show considerable interspecies differences. Thus, in the hamster, many of the periglomerular cells contain substance P, and most of these elements also co-localise GABA and dopamine, but in the rat, periglomerular cells containing substance P have not been found.

The amacrine (deep) granule cells have several short, basal dendrites and a long, peripherally coursing dendrite which ramifies in the external plexiform layer among the secondary dendrites of mitral and tufted cells. The branching distal portions of these long granule cell dendrites are densely studded with conspicuous gemmules (Fig. 22.156). The granule cells receive axodendritic synapses from recurrent collaterals of mitral and tufted cells, and there is also a heavy input to the granule cells from the telencephalon proper (see below).

The mitral/tufted cell secondary dendrites and the gemmules of the peripheral granule cell dendrites are richly interconnected by dendrodendritic synapses, organised in reciprocally oriented pairs. There is both ultrastructural and electrophysiological evidence indicating that the mitral/tufted dendrite-to-gemmule synapses are excitatory, whereas the adjacent gemmule-to-mitral/tufted dendrite synapses are inhibitory. These peculiar synapse pairs constitute extremely short inhibitory pathways from mitral cell to mitral cell or from tufted cell to tufted cell. The loops formed by the collaterals of mitral and tufted cells and by granule cells provide for surround inhibition of the former elements. The inhibition delivered by the granule-to-mitral synapses is very powerful and is the principal means of mediating control of the output of the main olfactory bulb. Extensive electrophysiological, immunohistochemical and biochemical evidence strongly suggests that GABA is an important neurotransmitter of the amacrine granule cells. Enkephalin immunoreactivity has been found in a small percentage of these elements. Co-existence of GABA and enkephalin in granule cells has been reported to be rare.

3. The main olfactory bulb receives *centrifugal afferent inputs* from many different sources, including afferents from olfactory-related structures, that reciprocate main olfactory bulb efferents and non-olfactory general modulatory afferents. The former originate principally from the retrobulbar area and from the prepiriform cortex, whereas the latter arise form the horizontal limb of the nucleus of the diagonal band of Broca, the mesencephalic raphe nuclei and the locus coeruleus.

The retrobulbar region projects massively and bilaterally to the main olfactory bulb. These centrifugal afferents terminate mainly in the granular and internal plexiform layers, but some ascend to the subglomerular zone of the external plexiform layer. Immunoreactivity for several neuropeptides, including enkephalin, neuropeptide Y and substance P, has been reported for cells in the retrobulbar area.

The main olfactory bulb afferents arising from the prepiriform cortex originate in pyramidal, presumably glutamatergic neurons and terminate in the granular layer, where they are believed to excite the GABAergic amacrine main olfactory bulb cells, which in turn inhibit the firing of mitral and tufted cells.

Cholinergic fibres from the nucleus of the diagonal band of Broca, serotoninergic fibres from the mesencephalic raphe nuclei and noradrenergic fibres from the locus coeruleus ascend to the main olfactory bulb. The serotoninergic and noradrenergic fibres terminate mainly in the internal plexiform and granular layers, but some serotonergic fibres ascend to the superficial zone of the main olfactory bulb to terminate within the glomeruli. The cholinergic fibres distribute evenly through the main olfactory bulb laminae.

4. It has already been discussed that the activity of the main olfactory bulb output neurons (i.e. the mitral and tufted cells) is regulated by two types of small inhibitory interneurons, the interglomerular or *superficial granule cells* and the *deep amacrine granule cells* (Fig. 22.156b). In addition to these small interneurons, the olfactory bulb contains several types of larger interneurons, all of which together form a link between the centrifugal bulbar afferents and the small interneurons. As indicated by Macrides and Davis (1983), these larger interneurons can be categorised into deep granule cells and periglomerular cells.

The larger, deep interneurons include Blanes cells, Golgi cells, vertical cells of Cajal and horizontal cells. Blanes and Golgi cells are multipolar neurons, the dendrites and axons of which are confined mainly to the granular layer. Their main difference is that the dendritic trees of the former, unlike those of the latter, are densely covered with spines. The vertical cells of Cajal and the horizontal cells are usually located in the internal plexiform layer, their names referring to the orientation of their dendritic trees (Fig. 22.156). The axons of these elements are thought to form axodendritic synapses with the distal part of the long, peripherally directed dendrites of the granule cells.

The larger, deep interneurons as a group are thought to receive excitatory impulses from centrifugal fibres and from mitral and tufted cell axons collaterals and to inhibit granule cell activities, thus ultimately disinhibiting the projection neurons, i.e. the mitral and tufted cells. According to Mugnaini et al. (1984a,b), most, if not all of the deep interneurons are GABAergic. According to the description given by Macrides and Davis (1983), the larger, superficial interneurons are located in the subglomerular zone and in the adjacent superficial zone of the external granular layer. Their dendrites branch among and around glomeruli, while their axons ramify predominantly in the periglomerular regions of the glomerular layer. Electron microscopy studies indicate that they receive asymmetrical (i.e. 'excitatory-type') axodendritic and axosomatic synapses from tufted cell collaterals and centrifugal fibres as well as symmetrical (i.e. 'inhibitory-type') axodendritic and axosomatic synapses from periglomerular cells and that, in turn, they form symmetrical axosomatic and axodendritic synapses with periglomerular cells. On the basis of these ultrastructural features, it is reasonable to assume that the superficial interneurons, by inhibiting inhibitory interglomerular elements, exert a disinhibitory influence on mitral and tufted cells. Mugnaini et al. (1984a,b) suggested that the superficial interneurons may contain GABA. For a detailed description of the interneurons in the main olfactory bulb of the hedgehog, the reader is referred to the Golgi studies performed by López-Mascaraque et al. (1986, 1990).

If we survey the microcircuitry of the main olfactory bulb, it appears that the olfactory projection is constituted by two consecutive sets of excitatory projection neurons, namely the peculiar neurosensory cells, the somata of which are situated in the nasal mucosa, and the mitral and tufted cells, which represent the bulbar output neurons. The activity of these bulbar output elements is regulated by two categories of small, inhibitory interneurons: the periglomerular cells and the amacrine (deep) granule cells. These elements provide for both local and surround inhibition. The output of the olfactory bulb is also strongly influenced by large numbers of excitatory centrifugal afferent fibres. These fibres terminate on various types of medium-sized inhibitory, short-axon cells, which can be categorised as deep and superficial groups. The deep, short-axon cells inhibit the inhibitory granule cells, whereas the superficial, short-axon cells inhibit the inhibitory periglomerular cells. Through the mediation of these sets of inhibitory interneurons, the excitatory bulbopetal fibres exert a disinhibitory influence on the bulbar output neurons.

The *efferent projections* of the main olfactory bulb are formed by the axons of the mitral cells and the middle and deep tufted cells. These axons assemble in the olfactory peduncle, from where they course caudally as a compact bundle along the endorhinal fissure (Figs. 22.4c, 22.5c, 22.7b,c). This bundle – the lateral olfactory tract – can be macroscopically followed to the rostral pole of the amygdaloid complex. During its caudal course, it distributes fibres to the various secondary olfactory areas in the cerebral hemisphere (Fig. 22.153). On the basis of normal material, it has been suggested that in mammals, in addition to a lateral olfactory tract, a medial olfactory tract is also present, and in primates a superficially situated, macroscopically visible bundle has been designated as such. This fibre system was believed to originate from the olfactory bulb and to distribute to the septum and, via the anterior commissure, to the contralateral olfactory bulb (Ariëns Kappers et al. 1936). However, experimental neuroanatomical studies have convincingly shown that the efferent projection from the main olfactory bulb is strictly ipsilateral and that the septum is not supplied by secondary olfactory fibres (Scalia 1966; Heimer 1969; Turner et al. 1978; Shipley and Adamek 1984).

The *regio retrobulbaris* (Rose 1935; Stephan 1975) corresponds to the nucleus olfactorius anterior of Herrick (1924) and his numerous followers, including Obenchain (1925), Young (1936), Crosby and Humphrey (1939) and Lohman (1963). This region intervenes between the bulbar formation rostrally and the various centres of the cerebral hemispheres caudally. It is largely situated in the olfactory peduncle, surrounding the ventricle or its vestige like a ring. Stephan (1975) considered the retrobulbar region as a cortical structure, composed of a superficial molecular layer, an intermedial densocellular layer and a deep multiform layer. He noted that the third layer is not identifiable throughout and that the first layer locally contains a superficially situated sheet of small, densely packed cells. The authors who interpreted the retrobulbar region as a nucleus subdivided it into a pars dorsalis, a pars medialis, a pars ventralis, a pars lateralis and a pars externa, the latter corresponding to the superficially situated sheet of cells already mentioned. Lohman (1963) and Lohman and Lammers (1967) remarked that a more or less separate pars rostralis of the anterior olfactory nucleus extends forward into the dorsal part of the olfactory bulb (Fig. 22.155).

The fibre connections of the retrobulbar region have recently been reviewed by Shipley et al. (1995). In addition to a massive input from the olfactory bulb, it also receives afferent projections from the contralateral retrobulbar region, from the olfactory tubercle, the piriform and entorhinal cortices and from the olfactory parts of the amygdaloid complex. Its efferents project to the ipsilateral and contralateral main olfactory bulbs and to the olfactory tubercle, the piriform cortex, the nucleus accumbens and the lateral hypothalamic area.

The *tuberculum olfactorium* is situated directly caudal to the retrobulbar region and medial to the endorhinal fissure and the lateral olfactory tract. In macrosmatic mammals, such as marsupials (Figs. 22.4c, 22.7b), insectivores (Fig. 22.9a) and rodents (Fig. 22.154a), it forms a prominent mass on the base of the hemispheres. In the armadillo, its surface is thrown into longitudinal folds (see poster Pl, 8). The olfactory tubercle is also well developed in most ferungulates (Fig. 22.5c), but is relatively small in prosimians (Figs. 22.7c, 22.9b).

Microscopically, in most forms the olfactory tubercle has an external plexiform layer, a compact middle layer of pyramidal and ellipsoid cells and a deep stratum of multiform cells. Moreover, this structure is characterised by islands of Calleja (1893), i.e. dense aggregates of granule cells located in the middle layer, but often encroaching on the superficial and deep layers.

The olfactory tubercle has much in common with the adjacent regions of the striatum. Both structures show an intense straining for AChE, and the deeper layers of the tubercle contain numerous medium-sized, densely spiny neurons, which correspond to the most common striatal cell type. The somata of these elements form numerous cell bridges that directly link the olfactory tubercle with the nucleus accumbens and the caudate-putamen (Millhouse and Heimer 1984). Finally, the ventral striatal regions and the olfactory tubercle both receive a projection from the hippocampal formation and numerous dopaminergic afferents from the area tegmentalis ventralis.

How olfactory is the olfactory tubercle? This question is not only relevant for mammals, but also for several groups of non-mammalian vertebrates, e.g. the dipnoans (see Chap. 16). Several previous authors (e.g. Beccari 1943) stated, on the basis of normal material, that the secondary olfactory projection to the mammalian olfactory tubercle is confined to the rostral and lateral parts of that structure, and numerous experimental neuroanatomical studies in different mammalian species (summarised by Allison 1953 and Scalia 1968) have led to the same conclusion. However, Heimer (1968, 1969) claimed that, in the rat, olfactory bulb efferents terminate profusely in the superficial fibre layer throughout the extent of the tubercle.

The disposition of the olfactory tubercle in primates and cetaceans deserves special mention. In primates, the long olfactory peduncle, on attaching to the base of the hemispheres, splits into two diverging, band-like structures: the small medial olfactory stria and the larger lateral olfactory stria (Fig. 22.157a). The lateral olfactory stria contains numerous secondary olfactory fibres, which together constitute the lateral olfactory tract, but, as already mentioned, contrary to the assumptions of numerous previous authors, such fibres are lacking in the medial stria. The region immediately behind the medial and lateral striae is known as the anterior perforated area. It is characterised by the presence of numerous, penetrating, small blood vessels. When these vessels are pulled out, their sites of entrance manifest themselves as a number of perforations, hence the name anterior perforated area (or substance). The rostral part of this area is the homologue of the olfactory tubercle of macrosmatic mammals; its caudal part, which is traversed by numerous fibres connecting the septum with the periamygdaloid cortex, is known as the diagonal band (Stephan 1975; Fig. 22.157a).

As regards cetaceans, in osmatic mysticocetes such as the finback whale, *Balaenoptera physalus*, distinct lateral and medial olfactory striae are present, and these structures form the rostral boundary of the conspicuous anterior perforated area or olfactory lobe (Fig. 22.157b). Remarkably, a large and prominent 'olfactory' lobe, divisible into a rostral 'olfactory' tubercle and a caudal diagonal area, is also present in entirely anosmatic odontocele cetaceans such as the common porpoise *Phocaena phocaena* (Breathnach 1953), the common dolphin *Delphinus delphis* (Filimonoff 1965) and the bottlenosed dolphin *Tursiops truncatus* (Jacobs et al. 1971; Morgane et al. 1980; Fig. 22.157c). Broca (1878), who first remarked upon this formation in a dolphin, designated it as *le désert olfactif* or the 'lobule désert'. It is important to note that the 'olfactory' tubercle of odontocetes, although extensive, is histologically poorly differentiated, largely consisting of a thin rim of cells, which merges imperceptibly with the striatum (Filimonoff 1965). In fact, the tubercle's prominence is entirely extrinsic in this group and reflects the internal encroachment of the head of the caudate nucleus on the basal forebrain surface (Jacobs et al. 1971).

Like the retrobulbar region and the olfactory tubercle, the *piriform cortex*, which is also designated as the primary olfactory cortex or the palaeocortex, is a trilaminar structure, comprising a superficial molecular layer, an intermediate densocellular layer and a deep multiform layer (Fig. 22.143d). It extends from the retrobulbar region rostrally to the entorhinal cortex caudally. Laterally, throughout its extent, it borders on the neocortex. The boundary between the piriform cortex and the neocortex in most mammals is marked by a distinct external groove, the fissura rhinalis (Figs. 22.4b, 22.5c, 22.144, 22.154a; see poster P1). The piriform cortex can be subdivided into a prepiriform cortex and periamygdaloid cortex. In macrosmatic mammals, the prepiriform cortex covers the tapered rostral part of the piriform lobe, whereas the periamygdaloid cortex covers most of the widened, caudal portion of that lobe (Figs. 22.153, 22.154a). In the caudomedial part of the piriform lobe, the amygdaloid complex becomes a superficial structure. Here, the periamygdaloid cortex borders on the cortical amygdaloid nucleus. Like the piriform cortex, the anterior and posteromedial parts of this nucleus receive a direct projection from the main olfactory bulb (Scalia and Winans 1975; see Fig. 22.159a,c).

In the caudal part of the piriform lobe, the periamygdaloid cortex borders on the five-layered entorhinal cortex, usually without any externally visible landmark (Figs. 22.146b,c, 22.153). Secondary olfactory fibres originating from the main olfactory bulb extend over the rostral and lateral parts of this cortex (Fig. 22.154a).

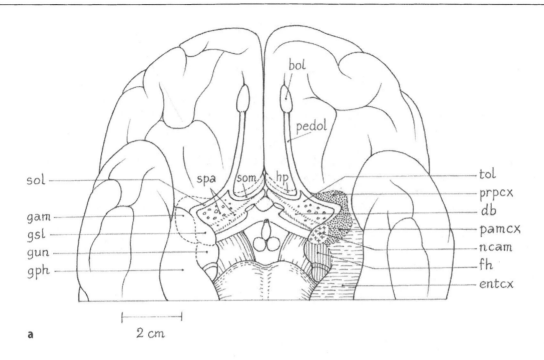

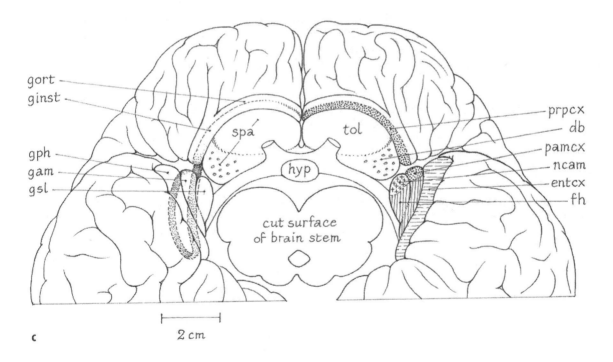

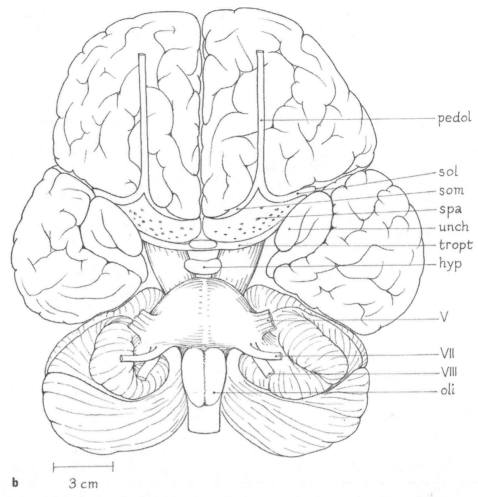

Fig. 22.157a–c. Basal views of the brains of **a** the human, **b** the finback whale *Balaenoptera physalus* and **c** the bottle-nosed dolphin *Tursiops truncatus*. In **a**, the temporal lobes have been displaced laterally in order to expose underlying structures, and the semilunar and uncinate gyri have been rotated towards the plane of page. On the *left side* of **c**, the long, circuitous path by which the piriform cortex reaches the rostral part of the temporal lobe has been indicated by *dotted curves*. *bol*, bulbus olfactorius; *db*, diagonal band; *entcx*, entorhinal cortex; *fh*, formatio hippocampi; *gam*, gyrus ambiens; *ginst*, gyrus insularis transversus; *gort*, gyrus orbitalis transversus; *gph*, gyrus parahippocampalis; *gsl*, gyrus semilunaris; *gun*, gyrus uncinatus; *hp*, hippocampus precommissuralis; *hyp*, hypophysis; *ncam*, nucleus corticalis amygdalae; *unch*, uncus hippocampi; *pamcx*, periamygdaloid cortex; *pedol*, pedunculus olfactorius; *prpcx*, prepiriform cortex; *sol*, stria olfactoria lateralis; *som*, stria olfactoria medialis; *spa*, substantia perforata anterior; *tol*, tuberculum olfactorium; *tropt*, tractus opticus. (**a** Modified from Nauta and Haymaker 1969. **b,c** Based on photographs and drawings from Jacobs et al. 1971 and Morgane et al. 1980)

The relative size of the olfactory area in the forebrain varies greatly in different mammals. In macrosmatic mammals with a poorly or moderately developed neocortex, this area occupies most or even all of the ventral surface of the telencephalon (Figs. 22.4c, 22.7b, 22.154a); however, in microsmatic forms with a large neocortex, it is confined to a small area along the inner curvature of the hemispheres (Figs. 22.148, 22.154b). In primates, the prepiriform cortex is represented by a small area situated on the basal surface of the frontal lobe, just lateral to the anterior perforated substance. It accompanies the lateral olfactory stria to the dorsomedial aspect of the temporal lobe, where it passes over into the periamygdaloid cortex. In humans, most of the prepiriform and periamygdaloid cortical fields are represented in a small convolution, the gyrus semilunaris, and in a narrow strip of the adjacent gyrus ambiens. The surface of the remainder of the latter gyrus is occupied by the cortical amygdaloid nucleus (Fig. 22.157a). Comparable spatial relations are found in cetaceans. In odontocetes, olfactory bulbs, peduncles and tracts are lacking, but the trajectory of the lateral olfactory tract or stria is, nevertheless, marked by a small lateral 'olfactory' gyrus (Jacobs et al. 1971; Morgane et al. 1980). This gyrus represents the prepiriform cortex of osmatic mammals (Filimonoff 1965;

Jacobs et al. 1971). Due to the strong folding of the cetacean hemispheres (compare Fig. 22.147c and poster P1, 30, 31), the prepiriform cortex describes a long loop before reaching the rostroventral part of the temporal lobe, where it passes over into the periamygdaloid cortex (Jacobs et al. 1971; Morgane et al. 1980; Fig. 22.157c), as in other mammals.

The structural relations of the piriform and entorhinal cortices will be discussed in Sect. 22.11.6.

Experimental neuroanatomical studies, ably summarised by Shipley et al. (1995), have shown the following: (a) the retrobulbar region, the piriform cortex and the entorhinal cortex project heavily back to the olfactory bulb, (b) most targets of the secondary olfactory projection are linked by associative connections and (c) several of these target areas project to other parts of the brain, particularly the hypothalamus and the neocortex. Hypothalamic projections arise from the retrobulbar region, the olfactory tubercle, the piriform cortex and the anterior part of the cortical amygdaloid nucleus and via the medial forebrain bundle pass into the lateral hypothalamic area (Price et al. 1991). As regards the rhinencephalic projections to the neocortex, it has been demonstrated that, in both rat and macaque monkey, the piriform cortex projects to the orbitofrontal and insular cortical regions and that these direct cortico-cortical projections are supplemented by indirect, transthalamic inputs (Price et al. 1991; Ray and Price 1992, 1993; Carmichael et al. 1994). Shipley et al. (1995) pointed out that the orbitofrontal and insular cortical regions are also targets of (inter alia) gustatory projections, ascending from the nucleus of the tractus solitarius, and that these neocortical regions may thus be sites that correlate olfactory and gustatory signals to generate an integrated perception of flavour.

22.11.2.3
Accessory Olfactory System

The accessory olfactory system (Halpern 1987; Meredith 1987; Segovia and Guillamón 1993; Dulce Madeira and Lieberman 1995; Shipley et al. 1995), also known as the vomeronasal system, comprises the vomeronasal organ or Jacobson's organ, the vomeronasal nerve, a special part of the olfactory bulb, known as the accessory olfactory bulb, some nuclei of the amygdaloid complex and some parts of the bed nucleus of the stria terminalis. Although the main olfactory system and the accessory olfactory system show many parallels in their structural organisation, the two are morphologically separate and subserve different functions. The accessory olfactory system is primarily involved in the regulation of reproductive behaviour, elicited by pheromones, i.e. chemical messengers from other members of the same species.

The *vomeronasal organ* is an elongated tube contained in a bony capsule, located bilaterally along the base of the nasal septum. The tube opens anteriorly via a narrow duct into the floor of the nasal cavity (as in rodents, lagomorphs and some primates) or into the nasopalatine duct (as in insectivores, ungulates, carnivores and most primates). The nasopalatine duct (Stensen's canal) is a passageway through the palate between the nasal and oral cavities. Thus, in the second group of species, stimuli can reach the organ via either the nose or the mouth.

The vomeronasal chemosensory receptor cells, which are confined to the medial side of the organ, are similar to those in the olfactory epithelium, except that they bear microvilli rather than cilia. The fine axons of these receptor cells assemble in the vomeronasal nerve, which, after having passed through the cribriform plate, terminates in the accessory olfactory bulb (Fig. 22.158).

The *accessory olfactory bulb* commonly appears as a small protrusion on the mediocaudal surface of the olfactory bulb. Its size varies directly with that of Jacobson's organ; thus it is proportionally well developed in insectivores and rodents, of moderate size in carnivores, ungulates and many primates and poorly developed in cetaceans, some bats and humans (Allison 1953). Its internal structure replicates most of the features of the main olfactory bulb, and the same seven layers can be recognised. However, there are also some notable differences between these two structures. Thus the glomeruli in the accessory olfactory bulb are smaller and less

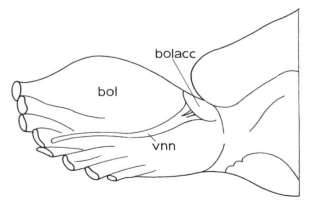

Fig. 22.158. Medial view of the rostral part of the brain of the opossum, showing the course and termination of the vomeronasal nerve (*vnn*). *bol*, bulbus olfactorius; *bolacc*, bulbus olfactorius accessorius. (Redrawn from Loo 1930)

distinct than in the main olfactory bulb, the periglomerular cells are much less numerous, the mitral cells are scattered through the entire external plexiform layer and tufted cells appear to be lacking entirely.

The accessory olfactory bulb projects to the amygdaloid complex and to the bed nucleus of the stria terminalis by way of the accessory olfactory tract. The latter initially forms part of the lateral olfactory tract, but becomes increasingly individuated at more caudal levels (see Fig. 22.160).

The *vomeronasal amygdala* comprises the nucleus of the accessory olfactory tract and the medial and posteromedial cortical amygdaloid nuclei (see Fig. 22.160). The nucleus of the accessory olfactory tract is a superficially situated forward

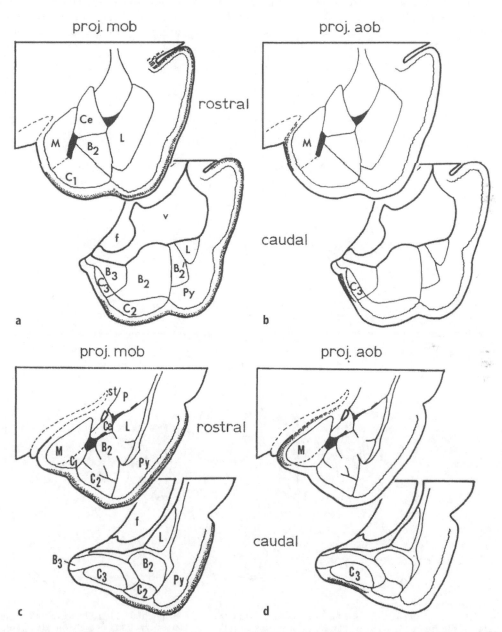

Fig. 22.159a–d. Transverse sections through the rostral and caudal parts of the amygdaloid complex of **a,b** the opossum and **c,d** the rat, showing the pattern of terminal degeneration, as observed in Fink-Heimer preparations, following lesions in **a,c** the main and **b,d** accessory olfactory bulb. *aob*, projections from main olfactory bulb; *B2, B2', B3*, basal amygdaloid nuclei; *C1, C2, C3*, subdivisions of cortical amygdaloid nucleus; *Ce*, central amygdaloid nucleus; *f*, fimbria fornicis; *L*, lateral amygdaloid nucleus; *M*, medial amygdaloid nucleus; *mob*, projections from main olfactory bulb; *P*, putamen; *Py*, periamygdaloid cortex; *st*, stria terminalis; *V*, lateral ventricle. (Reproduced from Scalia and Winans 1975)

extension of the amygdala which, as its name indicates, adjoins the accessory olfactory tract. The superficial zone of the corticomedial amygdala can be subdivided into four quadrants (see Fig. 22.160). The two lateral quadrants represent parts of the cortical nucleus and are designated as the anterior (C1) and posterolateral (C2) cortical nuclei; the anteromedial quadrant comprises the medial amygdaloid nucleus (M), whereas the posteromedial quadrant contains a third part of the cortical nucleus, designated as the posteromedial cortical nucleus (C3). Scalia and Winans (1975), who studied the efferent projections of the main and accessory olfactory bulbs in the opossum, rat and rabbit using the Fink-Heimer technique, demonstrated that lesions in these two bulbar formations cause terminal degeneration in two different and non-overlapping zones of the amygdaloid complex; the main olfactory bulb appeared to project to C1 and C2, whereas the accessory olfactory bulb efferents terminated exclusively in the nucleus of the accessory olfactory tract, M and C3 (Figs. 22.159, 22.160). Because of these salient connectional differences, C1 and C2 are together designated as the olfactory

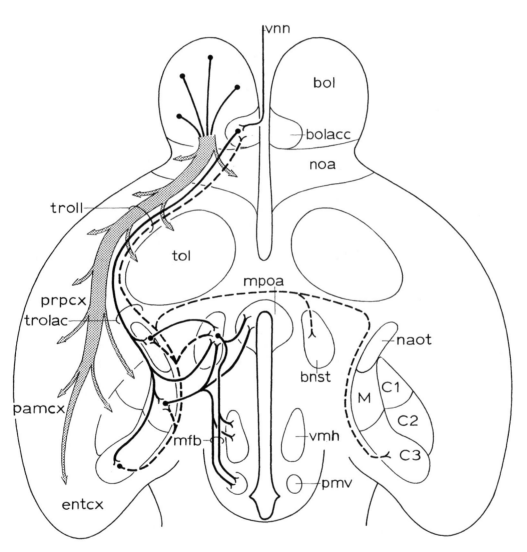

Fig. 22.160. Ventral surface of the hamster brain, showing the efferents of the main olfactory bulb (*dotted arrows*), the accessory olfactory bulb (*dashed lines*), the vomeronasal amygdala and the vomeronasal part of the bed nucleus of the stria terminalis (*solid lines*). *AOT*, accessory olfactory bulb; *BNOAT*, bed nucleus of accessory olfactory tract; *BNST*, bed nucleus of stria terminalis; *C1*, anterior cortical amygdaloid nucleus; *C2*, posterolateral cortical amygdaloid nucleus; *C3*, posteromedial cortical amygdaloid nucleus; *LOT*, lateral olfactory tract; *M*, medial amygdaloid nucleus; *MPOA*, medial preoptic area; *MOB*, main olfactory bulb; *NOA*, nucleus olfactorius anterior; *PAMCX*, periamygdaloid cortex; *PMV*, nucleus premammillaris ventralis; *PRPCX*, prepiriform cortex; *TOL*, tuberculum olfactorium; *VMH*, nucleus ventromedialis hypothalami; *VMN*, vomeronasal nerve. (Largely based on Kevetter and Winans 1981)

amygdala, while the nucleus of the accessory olfactory tract, M and C3 are referred to as the vomeronasal amygdala.

In addition to distributing to the corticomedial amygdaloid region, the accessory olfactory tract enters the stria terminalis to terminate in a small-celled medial subdivision of the bed nucleus of the stria terminalis.

The efferent projections of the various components of the vomeronasal amygdala were studied in the golden hamster by Kevetter and Winans (1981) using tritiated amino acids as tracers. It appeared that C3 projects ipsilaterally to M, the nucleus of the accessory olfactory tract and the accessory olfactory bulb and that additional fibres from C3 terminate in the medial subdivision of the bed nucleus of the stria terminalis bilaterally and in contralateral C3. M appeared to project to the ipsilateral nucleus of the accessory olfactory tract and C3 and bilaterally to the medial subdivision of the bed nucleus of the stria terminalis. 'Extrinsic' efferent projections from M were demonstrated to terminate in an area including parts of the medial preoptic area and the anterior hypothalamic nucleus and in the ventromedial and ventral premammillary hypothalamic nuclei. Experimental studies in the rat (De Olmos and Ingram 1972; Simerly and Swanson 1986) and cat (Krettek and Price 1978) have shown that the intermediate and medial divisions of the bed nucleus of the stria terminalis give rise to projections to the same preoptico-hypothalamic targets as those arising in from M. The connections discussed above are summarised in Fig. 22.160.

It is important to note that the structures comprising the accessory olfactory system, i.e. the vomeronasal organ, the accessory olfactory bulb, M, the nucleus of the accessory olfactory tract and the medial subdivision of the bed nucleus of the stria terminalis are all sexually dimorphic and that the same holds true for the preoptico-hypothalamic targets of the system, i.e. the medial preoptic area and the ventromedial and ventral premammillary hypothalamic nuclei. This sexual dimorphism manifests itself in the fact that males show greater values for volume and number of neurons (or neurosensory cells) in these structures than females (Segovia and Guillamn 1993; Dulce Madeira and Lieberman 1995). Gonadal steroid-concentrating cells are widespread in the accessory olfactory system (Dulce Madeira and Lieberman 1995), and recent studies of the expression patterns of immediate-early genes have shown that several structures belonging to this system become activated by sexual behaviour in male (e.g. Coolen 1995: M, medial subdivision of the bed nucleus of the stria terminalis, medial preoptic area) and female rodents (e.g. Flanagan-Cato and McEwen 1995: M, bed nucleus of the stria terminalis, medial preoptic area, ventromedial hypothalamic nucleus).

22.11.2.4
Terminal Nerve System

The terminal nerve (Demski and Schwanzel-Fukuda 1987; Demski 1993), also referred to as nerve zero, is a ganglionated cranial nerve that extends from the nose to the basal forebrain regions. Intracranially, its fibres form a loose plexus along the medial and ventral sides of the olfactory bulb and olfactory peduncle. Its peripheral branches are associated with olfactory and vomeronasal fila and project to the nasal mucosa, providing free nerve endings in both the olfactory and non-olfactory epithelia and dense plexuses surrounding Bowman's glands. Branches of the terminal nerve have also been observed to form terminations in the walls of blood vessels. Centrally, the terminal nerve establishes connections with various forebrain regions, including the septum, the olfactory tubercle, the preoptic area and the median eminence. A component of this nerve enters the brain in the immediate vicinity of the lamina terminalis (hence its name).

The neurons which contribute fibres to the terminal nerve are either scattered along its course or aggregated in some ganglia. The largest of these, known as the 'ganglion terminale', is located ventral to the caudal border of the olfactory bulb. Bipolar or fusiform and multipolar neurons have been observed in the terminal nerve. The multipolar neurons are believed to represent postganglionic autonomic elements providing for innervation of glands and blood vessels in the nose. The bipolar cells originate in the olfactory placode and migrate inward during ontogenesis (Schwanzel-Fukuda and Pfaff 1994). They form part of a centripetal conduction system conveying neural (perhaps chemosensory) messages from the nasal epithelium directly to the brain. Immunocytochemical studies have shown that some of these cells and their processes contain the reproductive peptide hormone luteinising hormone-releasing hormone. It has been postulated that the terminal nerve, particularly its luteinising hormone-releasing hormone-immunoreactive component, is involved in the (pheromonal) regulation of reproductive behaviour. This postulate has received some experimental support from studies by Wirsig and Leonard (1987) and Wirsig (1987). These authors examined the effects of bilateral terminal nerve transections on the sexual behaviour of male hamsters and found that these lesions produced a decrease in mating frequency and/or an increase in the number of intromissions required to reach ejaculation.

The terminal nerve has been observed in a variety of mammals, including the opossum *Monodelphis domestica* (Schwanzel-Fukuda et al. 1987; Zheng et al. 1990), mouse (Jennes 1987), rat (Bojsen-Moller 1975), hamster (Wirsig and Leonard 1986), rabbit (Huber and Guild 1913), cat and dog (McCotter 1913), pig and sheep (Johnston 1913), horse and the porpoise *Phocaena* (Johnston 1914), several odontocete cetaceans (Ridgway et al. 1987; Demski et al. 1990), several non-human primates (Witkin 1987) and humans (Brookover 1914; Fuller and Burger 1990). It has been reported to be absent in adult insectivorous bats (Brown 1987). Remarkably, in odontocete cetaceans, in which the olfactory nerves, bulbs and tracts degenerate during ontogenesis and are completely absent from the mature brain, the terminal nerve is larger and contains more ganglion cells than in any other mammal studied to date (Ridgway et al. 1987; Fig. 22.161).

22.11.3
Septal Region

22.11.3.1
Organisation and Subdivision

The septal region forms the medial part of the subpallium. Ontogenetically, it arises from the medial wall of the cerebral hemisperes and from the adjacent parts of the commissural plate. It is situated directly in front of the lamina terminalis. Dorsally and rostrally, it is bordered by the rostral part of the corpus callosum and/or the precommissural hippocampal rudiment (Figs. 22.8, 22.144, 22.146) and caudally by the preoptic region. Ventrolaterally, it borders on the nucleus accumbens and on the olfactory tubercle (Fig. 22.144). In primates (Fig. 22.6d), carnivores (Fig. 22.5d), large ungulates and cetaceans (Fig. 22.9c), the dorsal part of the septum is drawn out into a thin plate, composed largely of glial cells, the septum pellucidum. This structure is bounded by the basal surface of the cor-

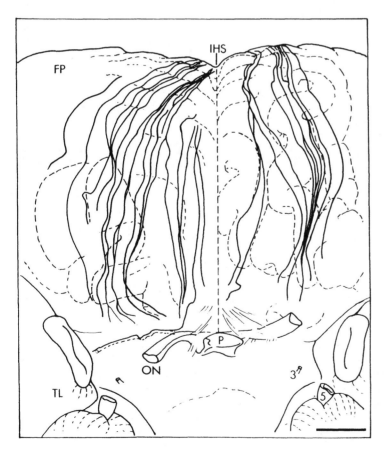

Fig. 22.161. Terminal nerve strands on the ventral aspect of the cerebral hemispheres of a bottle-nosed dolphin, traced from a photograph of a specimen in which the nerves were stained with osmium. *FP*, frontal pole of hemisphere; *IHS*, interhemispheric sulcus; *ON*, optic nerve; *P*, pituitary; *TL*, temporal lobe; *3*, oculomotor nerve; *5*, trigeminal nerve. *Bar*, 1 cm. (Reproduced from Ridgway et al. 1987)

pus callosum and by the rostral convexity of the column of the fornix. In animals possessing a septum pellucidum, the thickened, neuronally differentiated part of the septum is designated as the precommissural septum or septum verum.

In general, the septal grey matter is parcelled into medial and lateral subdivisions, the relations of which closely resemble those of the similarly named cell masses in reptiles (Ariëns Kappers et al. 1936). The lateral septal division was subdivided by Swanson and Cowan (1979) into dorsal, intermediate and ventral parts, but there are no cytoarchitectonically distinct borders between these subdivisions. The medial septal division comprises the dorsally situated medial septal nucleus and the ventrally located nucleus of the diagonal band of Broca. The latter comprises a dorsal or septal limb and a ventral or tubercular limb. As its name implies, this nucleus is embedded in a fibre bundle, the diagonal band of Broca.

The septal region contains a variety of neurotransmitters and other neuroactive substances. Jakab and Leranth (1995) listed 35 different neuropeptides occurring in septal neurons and/or fibres. These authors also pointed out that the septum can be divided into six curved, mediolaterally arranged laminae or zones on the basis of the distribution of immunohistochemically identified neuron populations (Fig. 22.162) and that this organisation is a consequence of the chronotopic inside-out sequence of septal neurogenesis. Without going into details, it may be mentioned that cholinergic neurons are concentrated in lamina II, that laminae I and VI contain numerous GABAergic neurons and that GABAergic and peptidergic neurons are intermingled in laminae III and V. The distribution of many neurochemically specified septal afferents also appeared to adhere to the zonal system.

22.11.3.2
Fibre Connections

The fibre connections of the septum show that this region serves as a waystation between the hippocampal formation on one hand and the hypothalamus and more caudally situated parts of the greater limbic system (see Sect. 22.12) on the other. The following synopsis of these connections is based on three reviews, i.e. Swanson et al. (1987), Nieuwenhuys et al. (1988) and particularly Jakab and Leranth (1995), to which the reader is referred for details and primary sources. The numbers in this synopsis correspond to those in the accompanying diagram (Fig. 22.163):

1. The hippocampal formation, including the subiculum, projects via the precommissural part of the fornix massively to the lateral septum. This projection is topographically organised and uses excitatory amino acids (glutamate, aspartate or both) as neurotransmitter.

2. The principal neurons in the lateral septum are GABAergic, and these elements issue strong inhibitory projections which, mainly via the medial forebrain bundle, reach the lateral and medial preoptic areas, the lateral hypothalamic area, the supramammillary region, the ventral tegmental area and the mesencephalic central grey.

3. Many of the preoptic and hypothalamic targets of the lateral septum project back to the medial septum.

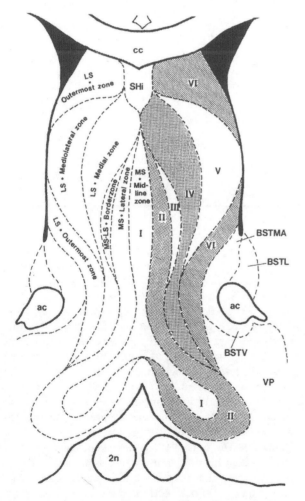

Fig. 22.162. Transverse section through the septal region of the rat, showing its laminar chemoarchitectural organisation. *2n*, optic nerve; *ac*, anterior commissura; *BSTL, BSTMA, BSTV*, divisions of the bed nucleus of the stria terminalis; *cc*, corpus callosum; *SHi*, septohippocampal nucleus; *VP*, ventral pallidum. (After Jakab and Leranth 1995)

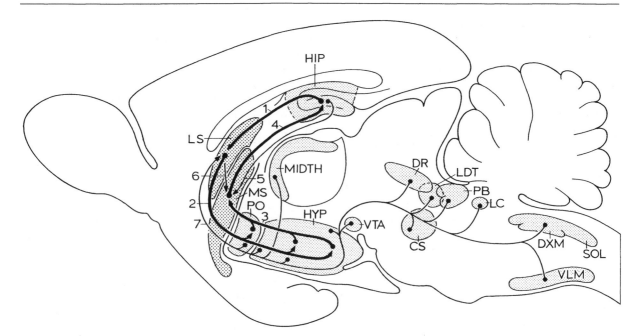

Fig. 22.163. Principal connections of the septal region. For an explanation of the numbers, see text. *CS*, central superior nucleus; *DR*, dorsal raphe nucleus; *DXM*, dorsal motor vagus nucleus; *HIP*, hippocampus; *HYP*, hypothalamus; *LC*, locus coeruleus; *LDT*, laterodorsal tegmental nucleus; *LS*, lateral septum; *MIDTH*, thalamic midline nuclei; *MS*, medial septum; *PB*, parabrachial nuclei; *PO*, preoptic region; *SOL*, nucleus of the solitary tract; *VTA*, ventral tegmental area

4. Cholinergic and GABAergic cells represent the two major populations of projection neurons in the medial septum. Numerous elements of both populations project, mainly via the fornix, to the hippocampal formation and the adjacent entorhinal cortex in a topographically organised manner. These fibres close what appears to be the principal hippocampo-septo-hypothalamo-septo-hippocampal circuit.
5. The cholinergic and GABAergic neurons in the medial septum, which project to the hippocampal formation, are both in receipt of a reciprocal GABAergic projection arising from non-pyramidal hippocampal neurons.
6. The previously described 'massive' projection from the lateral septum to the medial septum appears to be rather sparse in the rat.
7. The lateral septal nucleus receives ascending projections from numerous diencephalic, mesencephalic and rhombencephalic areas and cell groups, including the preoptic region, some thalamic midline nuclei, dopaminergic and non-dopaminergic cells in the supramammillary and ventral tegmental areas, serotonergic cells in the mesencephalic raphe nuclei and cholinergic and non-cholinergic cells in the laterodorsal tegmental nucleus, the mesencephalic central grey, the parabrachial nuclei, the dorsal vagal complex and the ventrolateral medulla.

Numerous other septal connections are not shown in Fig. 22.163. These include efferent projections from the lateral septum to some thalamic midline nuclei and efferents from the medial septum to the medial preoptic, lateral hypothalamic and supramammillary areas.

The cholinergic neurons in the medial septum (i.e. the medial septal nucleus and the nucleus of the diagonal band of Broca) form part of a continuum which also encompasses the substantia innominata. The connections of this cholinergic continuum will be dealt with in Sect. 12.11.4.6.

As regards the function of the septum, it was formerly believed that this region receives a substantial direct projection from the olfactory bulb; thus its massive efferent projections to the hippocampus were thought to represent a tertiary olfactory connection, and septum and hippocampus were both included in the rhinencephalon. However, experimental neuroanatomical studies in the rat, cat and monkey (see Nauta and Haymaker 1969) have conclusively shown that the septum does not receive direct afferent projections from the olfactory bulb. The septum, via its strong reciprocal connections with the hippocampus, is most probably critically involved in higher cognitive functions, such as memory and learning, and it also forms part of the circuitry underlying goal-oriented behaviours, such as agression, feeding and sexual and reproductive behaviour.

22.11.4
Basal Ganglia

22.11.4.1
Introduction

The term basal ganglia refers to a group of closely connected cell masses forming a continuum extending from the basis of the telencephalon, via the central part of the diencephalon, into the tegmentum of the mesencephalon (Fig. 22.164a). This complex classically encompasses the striatum (the nucleus caudatus and the putamen), the globus pallidus, the entopeduncular nucleus, the subthalamic nucleus and the substantia nigra.

During the first decades of this century, the concept was developed that two independent systems, the pyramidal and the extrapyramidal system, converge on the bulbar and spinal motor apparatuses. In contrast to the direct corticospinal, pyramidal system, the extrapyramidal system was thought to be an array of centres which, together with their emergent fibres, constitute a multisynaptic descending system. Striopallidal, pallidoreticular and reticulospinal pathways were considered to be the principal links in this system. Its highest centre, the striatum, was believed to receive its main input from the thalamus. Experimental hodological studies have shown that, although the pathways mentioned above do indeed exist, the striatum and the other 'extrapyramidal' centres are not interconnected in a unidirectional chain-like fashion. Rather, they and their emerging fibre systems constitute a number of interrelated loops or circuits, from which output systems emerge at several levels. However, it appears that the idea of independently operating cortical pyramidal and subcortical extrapyramidal motor systems has to be abandoned, as the cortex, including its sensory and motor fields, projects strongly to the caudate-putamen complex and the principal output channel of the latter, via the globus pallidus and the thalamus, converges on the motor, premotor and prefrontal areas of the cortex. That the term extrapyramidal motor system is still frequently used has to do with the fact that it is a convenient term with which a highly interconnected array of subcortical centres can be denoted and that, in humans, these centres are implicated in the pathogenesis of a number of distinct motor disturbances, including involuntary movements, muscular rigidity, tremor and akinesia.

During the last few decades, the concept of 'basal ganglia' has been considerably widened by Heimer and his associates (Heimer 1976; Heimer and Wilson 1975; Heimer et al. 1982). These authors advanced a view, the essence of which may be summarised as follows: The caudate nucleus, putamen and globus pallidus, as usually delineated, represent only the dorsal part of the striatal complex. The nucleus accumbens, which both cytoarchitectonically and histochemically closely resembles the caudate nucleus and the putamen, and the greater part of the olfactory tubercle should be considered together as a ventral portion of the striatum. The rostral part of the substantia innominata, another basal forebrain structure, represents a ventral extension of the globus pallidus. The ventral striatum and the ventral pallidum are nodal points in a loop system which forms a striking pendant to the principal striatal circuit denoted above. This loop includes the allocortex, the nucleus accumbens (i.e. the principal part of the ventral striatum), the ventral pallidum, the mediodorsal thalamic nucleus, the prefrontal, prelimbic and cingulate cortical regions and premotor cortical area 6. The well-known fact that the nucleus accumbens receives a strong dopaminergic input from the ventral tegmental area, whereas the caudate-putamen complex receives a similar projection from the substantia nigra, further substantiates the interpretation of the nucleus accumbens as a part of the ventral striatum.

Given the fact that the allocortex forms part of the limbic system, Heimer et al. (1982) conjectured that, whereas the dorsal striatopallidal system plays a pre-eminent role in initiating motor activities stemming from cognitive activities, the ventral striatopallidum has a role in initiating movements in response to emotionally or motivationally powerful stimuli.

A concept related to that of Heimer and his associates was put forward by Kelley et al. (1982). These authors found that, in the rat, a voluminous amygdalostriatal projection is present, which is distributed to all parts of the caudate-putamen complex except its antero-dorsolateral quadrant. They pointed out that this projection widely overlaps the striatal projections from the hippocampus, the cingulate cortex, the ventral tegmental area and the mesencephalic raphe nuclei. Like the amygdalostriatal system, all of these striatal afferents avoid the antero-dorsolateral striatal sector. Kelley et al. (1982) further established that the striatal sector mentioned above is the main region to which the corticostriatal projection is distributed from the sensorimotor cortex. In view of these findings, they interpreted the large striatal region receiving a direct projection from the amygdala as the 'limbic' and the remainder as the 'non-limbic' striatal compartment. Kelley et al. (1982) suggested that the amygdala, which is known to play a role in the neural mechanism underlying motivation and adaptive behaviour, has access via its massive projection to

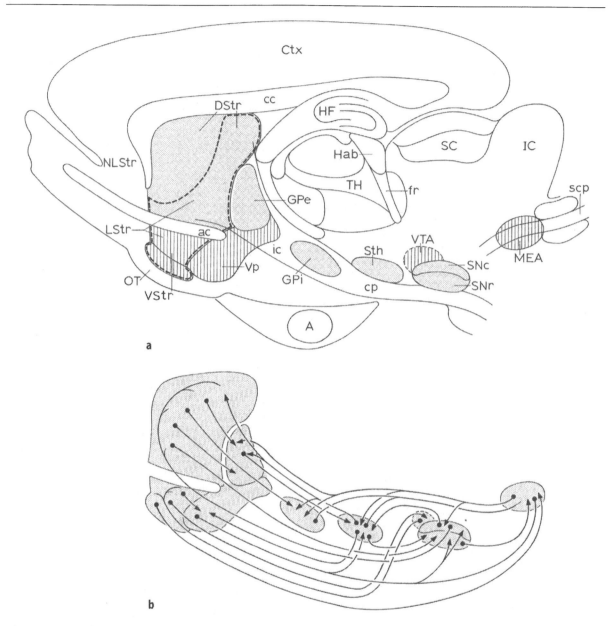

Fig. 22.164. Basal ganglia of the rat. **a** Position of the various cell masses; *darkly hatched areas* are classical 'extrapyramidal' centres, and *lightly hatched areas* more recent additions. The limbic striatum is surrounded by a *heavy, dashed line*. **b** Intrinsic connections of the basal ganglia. *A*, amygdala; *ac*, anterior commissure; *cc*, corpus callosum; *Ctx*, neocortex; *fr*, fasciculus retroflexus; *GPe*, globus pallidus, external segment; *GPi*, globus pallidus, internal segment (or entopeduncular nucleus); *Hab*, habenular ganglion; *IC*, inferior colliculus; *ic*, internal capsule; *LStr*, limbic striatum; *MEA*, mesencephalic extrapyramidal area (forms a complex with the nucleus tegmentalis pedunculopontinus and the mesencephalic locomotor region); *NLStr*, non-limbic striatum; *OT*, olfactory tubercle; *SC*, superior colliculus; *scp*, superior cerebellar peduncle; *SNr*, substantia nigra, pars compacta; *SNr*, substantia nigra, pars reticulata; *Sth*, subthalamic nucleus; *TH*, thalamus; *VP*, ventral pallidum; *VStr*, ventral striatum; *VTA*, ventral tegmental area

the striatum to the initiation and patterning of somatomotor behaviour, in which the latter structure is involved. It is important to note that the nucleus accumbens is included in both the 'ventral' striatum as and the 'limbic' striatum, but that the latter entity extends further dorsally than the former.

22.11.4.2
Cell Masses

Caudate-Putamen Complex or Striatum. In early embryonic stages, the walls of the mammalian hemispheric vesicles are of a uniform thickness. However, the basolateral parts of the wall of these structures soon increase in thickness more than the other parts, bilaterally forming an intraventricular protrusion, which is known as the ganglionic eminence (Fig. 22.165). This ganglionic eminence gives rise to a number of different cell masses: the olfactory tubercle and the nucleus accumbens develop from its rostral portion, its large intermediate portion represents the anlage of the caudate-putamen complex and the corpus amygdaloideum arises from its most caudal part (Fig. 22.165a). The situation outlined above is essentially maintained throughout development in 'simple' mammalian groups such as marsupials, insectivores and rodents. In these forms, the derivatives of the ganglionic eminence together constitute an arch-like formation, the rostral and caudal poles of which are formed by the olfactory tubercle plus the caput nuclei caudati and the amygdaloid complex, respectively (Fig. 22.166). However, in prosimians and more strongly in primates, the general expansion of the hemispheric primordia is followed by a curvature of these structures by which their original posterior pole moves ventrally and then rostrally to become the definitive temporal pole of the hemisphere. The lateral ventricle and the strio-amygdaloid primordium follow this movement. The caudal part of the striatal anlage is drawn out into a long, highly arched tail, and the amygdaloid complex is displaced rostroventrally into the temporal lobe (Fig. 22.165d).

In view of the developmental events outlined above, the general relations of the mammalian striatum may be summarised as follows:

1. Its massive rostral part, which is known as the caput nuclei caudati, projects into the anterior horn of the lateral ventricle.
2. The nucleus accumbens is a medial expansion of the head of the caudate nucleus, which extends around the ventral aspect of the lateral ventricle into the medial or septal wall of the hemisphere.
3. Ventrally, the head of the caudate nucleus and the accumbens nucleus are in direct contact with the olfactory tubercle.
4. The head of the caudate nucleus passes without a definite boundary into the tail of the same cell mass; the latter is short and thick in animals with small or moderately developed hemispheres, but long and strongly curved in animals with large, 'temporalized' hemispheres (Figs. 22.165d, 22.166).
5. The tail of the caudate nucleus remains in direct contact with the lateral ventricle throughout its extent.
6. The morphologically caudalmsot portion of the striatum blends with the amygdaloid complex.
7. During development, increasing numbers of fibres passing to and from the dorsal, pallial parts of the hemispheres penetrate the striatum.

In most mammalian groups, these fibres unite to form a more or less continuous capsula interna, dividing the striatum into a dorsomedial nucleus caudatus and a ventrolateral putamen (Figs. 22.165b,d, 22.166). It is important to note that the separation of these structures is always incomplete. Rostroventrally and caudally, the caudate nucleus and the putamen remain connected and, at intermediate levels, strands of cells joining these two cell masses pass across between the bundles of the internal capsule. In some mammalian groups, including the insectivores and rodents, the corticofugal and corticopetal fibres passing through the striatum assemble in a large number of small bundles which do not unite to form an internal capsule. In these groups, the undivided striatum is usually designated as the caudate-putamen complex.

The caudate-putamen complex, divided or undivided, is by far the longest subcortical cell mass in the mammalian brain. Cytoarchitectonically, the caudate nucleus and the putamen are identical. They have a homogeneous structure throughout and contain numerous medium-sized neurons, among which conspicuous large cells are sparsely scattered. Quantitative analysis (Schröder et al. 1975) has revealed that, in the human striatum, about 100 million medium-sized and some 600 000 large cells are present. A recent stereological study (Oorschot 1996) revealed that the right basal ganglia of the rat consist, on average, of 2.79 million caudate-putamen neurons, 46 000 globus pallidus neurons, 3200 entopeduncular neurons, 13 600 subthalamic neurons, 7200 substantia nigra pars compacta neurons and 26 300 substantia nigra, pars reticulata neurons.

Although the caudate-putamen complex is cytoarchitectonically a relatively homogeneous

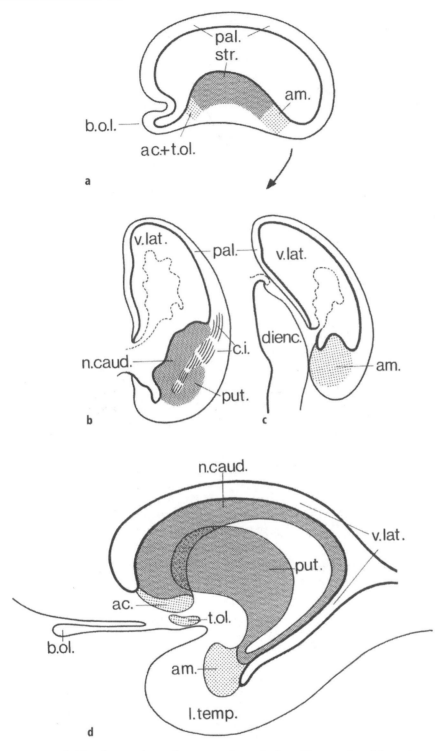

Fig. 22.165a–d. Morphogenesis and topography of the human striatum and some of its adjacent structures. **a** Sagittal section through the cerebral hemisphere of a 40-mm embryo, showing the position of the anlagen of some basal telencephalic structures. **b** Frontal section through the telencephalon of a 40-mm embryo, at the level of the primordial striatum. **c** Section at a more caudal level of the same series, passing through the primordial corpus amygdaloideum. **d** Topography of the striatum and some other components of the basal telencephalon in the adult. *ac.*, nucleus accumbens; *am.*, corpus amygdaloideum; *b.ol.*, bulbus olfactorius; *c.i.*, capsula interna; *dienc.*, diencephalon; *l.temp.*, lobus temporalis; *n.caud.*, nucleus caudatus; *pal.*, pallium; *put.*, putamen; *str.*, striatum; *t.ol.*, tuberculum olfactorium; *v. lat.*, ventriculus lateralis. (From Nieuwenhuys 1977)

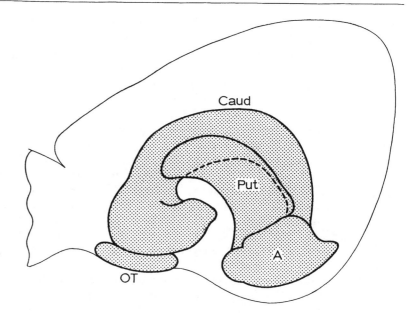

Fig. 22.166. Reconstruction of the strio-amygdaloid complex of the opossum *Didelphis virginiana* in lateral view. *A*, amygdala; *Caud*, caudate nucleus; *OT*, olfactory tubercle; *Put*, putamen

structure, histochemical and particularly immuno-histochemical studies have revealed a remarkable heterogeneity within the complex. The first evidence for this chemoarchitectural heterogeneity came from studies in which a staining technique for the enzyme AChE was applied. These studies showed that, in the caudate-putamen complex, 300- to 600-μm-wide zones of low AChE activity stand out against an otherwise AChE-rich background. Graybiel and Ragsdale (1978, 1979), who first identified these zones, designated them as striatal bodies or striosomes. During recent years, it has gradually become clear that throughout most of the caudate-putamen complex the striosomes and the 'matrix' in which they are embedded represent chemoarchitectonically distinct tissue compartments, which are related to the intrinsic structure of the complex and to the organisation of its afferent and efferent connections. Some aspects of the microcircuitry and the chemodifferentiation of the striatum will be discussed in Sect. 22.11.4.4.

Nucleus Accumbens. Most previous authors (e.g. Obenchain 1925: the marsupials *Caenolestes* and *Orolestes*; Loo 1931: opossum; Smith 1930: anteater, *Tamandua*; Young 1936: rabbit; Crosby and Humphrey 1941: humans) were of the opinion that a cytoarchitectonic boundary between the striatum and the nucleus accumbens cannot be drawn and thus considered the latter as a rostral, ventromedial extension of the former. As has already been pointed out, Heimer and colleagues (Heimer 1976; Heimer and Wilson 1975; Heimer et al. 1982), on structural, histochemical and connectional grounds, proposed that the nucleus accumbens and medium-sized cell territories of the olfactory tubercle together represent the ventral sector of the striatum. This concept is corroborated by the fact that the dorsal and ventral sectors of the striatum both receive a robust dopaminergic projection from the ventral midbrain (Moore and Bloom 1979; Lindvall and Björklund 1983) and by the results of a number of recent experimental hodological studies. Thus it appeared that each of the major cortical regions projects to a defined, but partially overlapping longitudinal zone of the striatum: the sensorimotor area projects onto the dorsolateral part of the caudate-putamen; the associational areas of the neocortex project to a more medial sector of the caudate-putamen; the mesocortex projects mainly to the medial and ventral regions of the caudate-putamen, but also to the ventral striatum; finally, the allocortex projects mainly to the ventral striatum, but also to the medial and ventral parts of the caudate-putamen (Percheron et al. 1984a,b; McGeorge and Faull 1989). Moreover, Haber et al. (1990a) concluded that, with regard to its efferent connections, the nucleus accumbens is indistinguishable from other parts of the striatum.

The structural and connectional continuity of the dorsal and ventral striatal sectors does not necessarily mean that all parts of the striatum are equivalent. Groenewegen and Russchen (1984) demonstrated that in the cat the efferents of the lateral and medial parts of the accumbens, though very similar, are not identical. Using the patterns of

Timm's sulphide-silver technique for zinc, AChE activity and cholecystokinin immunoreactivity, Záborszky et al. (1985) showed that the accumbens nucleus is divisible into a peripheral and a central subterritory, which they referred to as the 'shell' and 'core', respectively. Subsequent studies have revealed that the distinction between shell and core can also be demonstrated with antibodies against substance P and a calcium-binding protein (for a review, see Zahm and Brog 1992). The core region, which merges dorsally with the caudate-putamen, shows a complex striosome/matrix-like compartmentalisation. The shell region also shows chemoarchitectural inhomogeneities, but these bear little resemblance to those in the core (for details, see Groenewegen et al. 1989; Jongen-Rêlo et al. 1993, 1994). Recent experimental hodological studies (summarised in Jongen-Rêlo et al. 1994) indicate that the core and shell regions also show differences in their afferent and efferent connections.

Olfactory Tubercle. The *tuberculum olfactorium* forms part of the basal olfactory area, i.e. the telencephalic territory which is in receipt of direct projections from the olfactory bulb (see Sect. 22.11.2.2). It is situated directly behind the anterior olfactory nucleus and superficially covers the head of the caudate nucleus and the nucleus accumbens. Its size is directly proportional to that of the olfactory bulb (Stephan 1975). In macrosmatic mammals, it forms a prominent mass on the ventral surface of the hemisphere (hence its name).

In most mammals, three layers can be distinguished in the olfactory tubercle: an external plexiform layer, an intermediate densocellular layer of pyramidal and ellipsoid cells and a deeper layer of loosely arranged polymorphic cells. At many places, bridges of cells extend between the polymorphic cell layer and the ventral striatum-accumbens region. A characteristic feature of the olfactory tubercle is the presence of numerous islets of closely crowded, mostly small cells. These so-called islands of Calleja may occur in all three layers.

The olfactory tubercle receives a direct projection from the olfactory bulb which, however, is mainly confined to its rostral and lateral parts. Heimer and Wilson (1975) incorporated the polymorphic layer of the olfactory tubercle into the striatal domain, because (a) this layer is structurally continuous with the head of the caudate nucleus and the nucleus accumbens, (b) the ventral parts of the striatum, the nucleus accumbens and the polymorphic cell layer form a continuous projection field for allocortical fibres originating from the hippocampus and the prepiriform cortex and for dopaminergic fibres from the area tegmentalis ventralis and (c) the projection areas of the accumbens and the tubercle are contiguous and outline a discrete, unitary region underneath the anterior commissure.

Globus Pallidus or Dorsal Pallidum and the Entopeduncular Nucleus. The globus pallidus is a triangular mass of cells lying along the medial aspect of the putamen. It is separated from the latter by a fibre sheet, the lamina medullaris externa. The globus pallidus, which is of diencephalic origin (see Sect. 22.11.1 and Fig. 22.150), differs considerably in its structure from the telencephalically derived striatal nuclei. It is chiefly composed of large, widely spaced, fusiform cells. The total number of pallidal cells in humans is about the same as that of the large striatal neurons, i.e. 600 000 (Thörner et al. 1975). Many bundles of myelinated fibres traverse the globus pallidus and in fresh preparations give it a paler colour than the putamen or the caudate nucleus. In primates, a lamina medullaris interna divides the globus pallidus into medial and lateral segments. In non-primates, the globus pallidus consists of a single nuclear mass, which appears to correspond with the lateral pallidal segment of primates. The medial pallidal segment in non-primates is represented by the nucleus entopeduncularis, a small cellular area lying directly against (cat) or at some distance caudomedially from the globus pallidus (rat, dog) (Fig. 22.145). The cells of the entopeduncular nucleus are intercalated in the course of the ansa lenticularis, the principal stream of efferent fibres from the basal ganglia. In mammals in which the internal capsule divides the striatum into caudate nucleus and putamen, the putamen and the directly adjacent globus pallidus are collectively designated as the lentiform nucleus.

Substantia Innominata, Basal Nucleus of Meynert, Cholinergic Basal Forebrain and Ventral Pallidum. The substantia innominata is a rather ill-defined, flattened cell mass, which is situated directly ventral to the putamen and the globus pallidus; it is partly separated from the latter by the anterior commissure. Its rostral part covers the olfactory tubercle; medially, it borders on the lateral preoptic and lateral hypothalamic area, and its caudal part is situated between the globus pallidus and the nucleus centralis amygdalae. The substantia innominata is mainly composed of rather loosely arranged, medium-sized cells, but also contains large, deeply stained, fusiform and multipolar neurons. The population of large cells is collectively designated as the nucleus basalis of Meynert. It becomes progressively larger and more conspicuous with increasing cerebralisation, attaining its greatest development in cetaceans and primates. The nucleus basalis is known to project mainly to the neocortex (Divac 1975; Kievit and Kuijpers 1975; Pearson et al. 1982).

The basal forebrain contains a population of large cholinergic neurons. This population, which, like many other chemically identified groups, does not respect classically defined anatomical boundaries, extends from the septal region rostromedially to the substantia innominata caudolaterally. Mesulam et al. (1984) subdivided the basal cholinergic neurons into four groups, Ch1-Ch4, a parcellation which is also applicable to the human brain (Hedreen et al. 1984; Saper and Chelimski 1984) and, at least according to some investigators, also to rodents and carnivores (see Butcher and Semba 1989). The Ch1 group is found in the medial septal nucleus, the Ch2 group corresponds to the nucleus of the vertical limb of the diagonal band, the small Ch3 group is embedded in the nucleus of the horizontal limb of the diagonal band and the Ch4 group, which is very extensive in the human brain, corresponds to the above-mentioned nucleus basalis of Meynert. According to Mesulam et al. (1984), the cholinergic neurons which constitute Ch4 can be separated into five subgroups: anteroventral (Ch4av), anterolateral (Ch4al), intermedioventral (Ch4iv), intermediodorsal (Ch4id) and posterior (Chp).

Heimer and Wilson (1975) found that, in the rat, fibres originating from the nucleus accumbens and the olfactory tubercle converge on the rostral, subcommisural part of the substantia innominata. Similar connections have been found in the cat (Groenewegen and Russchen 1984) and monkey (Haber et al. 1990a). Since this ventral part of the substantia innominata appeared to be related to the ventral striatum in the same way as the main or dorsal part of the globus pallidus is related to the caudate-putamen complex, it has been referred to as the ventral pallidum. Within the ventral pallidum, separate subcommissural and ventral parts have been distinguished on the basis of the differential distribution of enkephalin and substance P immunoreactivity and of differences with respect to input and output characteristics (Haber et al. 1990a,b; Spooren et al. 1991; Groenewegen et al. 1993).

The relatively poorly characterised part of the substantia innominata caudal to the ventral pallidum, according Alheid and Heimer (1988), shares many characteristics with the medial and central parts of the amygdala; for this reason, they included this caudal part of the substantia innominata with the bed nucleus of the stria terminalis in a new entity, the extended amygdala. The interpretation by Alheid and Heimer (1988) was corroborated by studies performed by Grove (1988a,b), which showed that the afferent and efferent connections of the caudal part of the substantia innominata of the rat strongly resemble those of neighbouring components of the amygdala. In the rat, cholinergic neurons are scattered throughout the substantia innominata, but in primates the equivalent cell population is mainly concentrated in the nucleus basalis of Meynert, which lies embedded in the caudal part of the substantia innominata.

Nucleus Subthalamicus. This nucleus is situated in the caudal part of the diencephalon, ventral to the zona incerta and dorsal to the internal capsule and its transition into the cerebral peduncle. It is composed of fairly large, triangular and polygonal cells. It is rather small in 'lower' mammals, but well developed in primates. In the latter, the subthalamic nucleus is a conspicuous biconvex mass of cells, the mediocaudal part of which is in direct contact with the substantia nigra.

Substantia Nigra. This forms a major component of the basal ganglia. It overlies the cerebral peduncle and extends throughout the length of the mesencephalon. On the basis of cytoarchitectonic criteria, it can be subdivided in all mammals examined into a dorsal, cell-rich pars compacta and a ventral, less cellular pars reticulata. This subdivision into two components is supported by the chemoarchitecture of this nuclear complex. The pars compacta is mainly composed of large, melanin-containing cells which synthesise dopamine. The cells in the pars reticulata are somewhat smaller than those in the pars compacta; most of these elements are GABAergic. It is noteworthy that the pars reticulata of the substantia nigra has many features in common with the entopeduncular nucleus.

The dopaminergic cells in the pars compacta of the substantia nigra, designated as the A9 group in the terminology of Dahlström and Fuxe (1964), form a continuum with two other dopaminergic cells groups, A10 and A8. The former is embedded in the area tegmentalis ventralis, and the latter in the nucleus parabrachialis pigmentosus.

Area Tegmentalis Ventralis. This area was first described by Tsai (1925) in the opossum and lies in the ventromedial part of the tegmentum of the midbrain. Rostrally, it is continuous with the lateral hypothalamic area; caudally, it extends to the level of the caudal pole of the interpeduncular nucleus.

Nucleus Parabrachialis Pigmentosus. This nucleus is situated in the dorsolateral part of the mesencephalic tegmentum and forms part of the reticular formation.

Nucleus Tegmentalis Pedunculopontinus and Related Cell Groups. The caudalmost griseum, which is clearly included in the circuitry of the basal ganglia, is the nucleus tegmentalis pedunculopontinus. This nucleus is situated in the caudal part of the tegmentum of the midbrain and extends into the rostralmost part of the pontine tegmentum. This nucleus contains numerous cholinergic neurons, which are collectively designated as cell group Ch5 (Mesulam et al. 1983a,b, 1984) and roughly corresponds to the so-called mesencephalic locomotor region. The latter has been given this name because its electrical stimulation in postmammillary decerebrate mammals induces coordinated locomotion on a treadmill (Grillner and Shik 1973; Skinner and Garcia-Rill 1984). It has recently been claimed that, at least in the rat, not the pedunculopontine tegmental nucleus itself, but rather a medially adjacent 'midbrain extrapyramidal area' is specifically included in basal ganglia circuitry (Rye et al. 1987; Lee et al. 1988; Steininger et al. 1992).

The *claustrum* is a sheet of grey matter situated underneath the cerebral cortex in the lateral wall of the hemispheres. It is present in all mammals examined, from marsupials and insectivores to primates. According to Källén (1951a,b; Fig. 4.51b,c), the claustrum develops from a separate pallial cell zone, which he designated as d_v and which gives rise to neither neocortical nor striatal formations. The claustrum consists of two parts: a dorsal part underlying the insular cortex and a ventral part adjoining the prepiriform cortex. In primates, the dorsal part of the claustrum is separated laterally from the insular cortex by a thin medullary layer, the capsula extrema, and medially from the putamen by another white lamina, the capsula externa. The ventral part of the claustrum in macrosmatic mammals forms a distinct cell mass, known as the endopiriform nucleus.

The claustrum is usually considered to form part of the basal ganglia, but this allocation is neither supported by its ontogenesis nor by its fibre connections. Its dorsal part entertains widespread and reciprocal connections with the neocortex, whereas its ventral part is connected with the hippocampal formation, the entorhinal cortex and the amygdala. The neocortical projections to the dorsal claustrum are topographically organised, and discrete visual and somatosensory subdivisions have been delineated with the latter. These hodological data were derived from a review article by Sherk (1986), to which the reader is referred for details. Dinopoulos et al. (1992) have recently established that, in the hedgehog *Erinaceus*, the dorsal claustrum is well developed and is connected with all neocortical areas and with the thalamus.

22.11.4.3
Fibre Connections

22.11.4.3.1
Introductory Note

During the last two decades, an extraordinarily large number of experimental studies on the connections of the striatum and related 'extrapyramidal' centres have appeared. Within the limits set for the present work, only a brief and simplified overview of these connections can be presented. For details, the reader is referred to the papers cited and to the recent reviews by Parent and Hazrati (1995a,b), in which much of the pertinent literature is compiled. Figure 22.164b presents a survey of the intrinsic connections of the basal ganglia.

22.11.4.3.2
Striatal Circuits

Principal Striatal Circuit: Cerebral Cortex-Striatum-Globus Pallidus-Thalamus-Cerebral Cortex. Contrary to the views of previous workers, it has been established that the whole of the neocortex sends fibres to both the caudate nucleus and the putamen and that all parts of these two cell masses receive fibres from the cortex. It was originally held that this corticostriate projection is arranged on a simple topgraphical basis in that particular cortical areas project to proximal portions of the striatum (Carman et al. 1963: rabbit; Martin and Hamel 1967: opossum; Kemp and Powell 1970: monkey). However, more recent studies (Goldman and Nauta 1977: monkey; Selemon and Goldman-Rakic 1985: monkey; McGeorge and Faull 1989: rat) have shown that the terminal fields of the corticostriate fibres originating from circumscribed cortical regions are organised in narrow, longitudinally arranged strips or bands, which often span the entire length of the nucleus. It was formerly held that the sensorimotor cortex projects bilaterally to the striatum, but that the corticostriate projections from the remaining cortical areas are strictly ipsilateral (Carman et al. 1965: rat, rabbit and cat; Kemp and Powell 1971b: monkey). However, it has recently been established by McGeorge and Faull (1989) that, in the rat, all major regions of the cerebral cortex project to the striatum on both sides of the brain, albeit with strong ipsilateral preponderance. The fibres arising from the motor cortex project in a somatotopic fashion on the putamen (dorsal to ventral: leg, arm and face; Künzle 1975). The corticostriate fibres, which are thought to have the excitatory amino acid glutamate as their neurotransmitter (Fonnum et al. 1981), establish direct synaptic contacts with the striatal efferent neurons, i.e. the medium spiny cells (Somogyi et al. 1981; see Sect. 22.11.4.4).

Efferent fibres from the striatum converge towards the globus pallidus, where they constitute a massive fibre system, which passes radially through the two segments of the pallidum, the external segment (GPe or globus pallidus in non-primates) and the internal segment (GPi or entopeduncular nucleus in non-primate). During its transit through the globus pallidus, this system emits numerous collateral and terminal arborisations at right angles to its stem fibres. These disc-like terminal fields synapse with pallidal neurons, which are likewise oriented perpendicular to the main stream of the striatofugal bundle. The latter then leaves the GPe and descends to substantia nigra, where its fibres terminate mainly in the pars reticulata. The striopallidonigral bundle is topographically organised (Percheron et al. 1984 b; Hazrati and Parent 1992). It was previously believed that the striatal projections terminating in the external and internal pallidal segments are collateral branches of striatonigral fibres (Fox and Rafols 1975; Fox et al. 1975), but studies using double retrograde tracing (Beckstead and Cruz 1986; Oertel and Mugnaini 1984) have shown that these three projections come mainly from separate cell populations. The vast majority of the striatopallidal and striatonigral fibres are known to contain GABA and to exert an inhibitory influence. In many of these fibres, GABA co-exists with one or several neuropeptides. Thus neurons projecting to the GPe contain enkephalin, those arborising in GPi are enriched with substance P and dynorphin, and striatonigral neurons contain either enkephalin, substance P and/or dynorphin (Parent and Hazrati 1995a).

GPi and the substantia nigra, pars reticulata constitute the principal output structures of the (dorsal) striatum. Both structures project massively to the thalamus. In primates, the fibres emanating from GPi initially constitute two separate bundles, the fasciculus lenticularis and the ansa lenticularis. These two bundles merge in Forel's field H, after which they ascend as a single bundle, the fasciculus thalamicus, to the rostral part of the thalamus (Nauta and Mehler 1966). The fibres of this pallidothalamic projections, which is topographically organised (DeVito and Andersen 1982: monkey), terminate in the rostral part of the ventral lateral nucleus of the thalamus. The nigrothalamic fibres, which, as already mentioned, arise from cells in the pars reticulata, project rostrally to certain parts of the ventral anterior, ventral lateral and dorsomedial thalamic nuclei (Carpenter et al. 1976; Ilinski et al. 1985). Both the pallidothalamic and nigrothalamic projections are known to be GABAergic (Francois et al. 1984; Penney and Young 1986).

The final link in the principal striatal circuit is formed by fibres passing from the target nuclei of the pallidothalamic and nigrothalamic projections to the cerebral cortex. The ventral anterior and ventral lateral thalamic nuclei send their efferent fibres to the motor, premotor and supplementary motor areas, whereas the dorsomedial thalamic nucleus is known to project to the prefrontal cortex, including the frontal eye field.

Initially, the principal striatal circuit described above was looked upon as the morphological substrate of the motor functions of the basal ganglia (see e.g. Evarts and Thach 1969; Kemp and Powell 1971b). Information derived from the entire neocortex is processed in the striatum, GPe and the thalamus, respectively, and then fed back into the motor and premotor cortices, i.e. the principal sites of origin of the motor part of the pyramidal tract. Subsequent anatomical and physiological findings led to the concept that influences from the sensorimotor and association cortices remain segregated in the basal ganglia thalamocortical pathways. Thus DeLong and Georgopoulos (1981) suggested the presence of two separate loops through the basal ganglia: (1) a 'motor' loop passing largely through the putamen, which receives input from the sensorimotor cortex converging upon certain premotor cortical areas, and (2) a 'complex' loop passing through the caudate nucleus, which receives input from the association areas and whose influences are ultimately returned to certain portions of the prefrontal cortex. Somewhat later, Alexander et al. (1986) suggested the presence of five parallel circuits: motor, oculomotor, orbitofrontal, dorsolateral prefrontal and anterior cingulate. Each of these five basal ganglia-thalamocortical circuits was thought to receive inputs from several separate, but functionally related cortical areas and to be centred on a separate part of the frontal lobe. The question as to whether these various circuits really exist and operate as separate processing channels has been much debated in the literature (see e.g. Alexander and Crutcher 1990; Percheron and Fillion 1991). Without going into details, it may be mentioned that evidence has recently been presented indicating that the different basal ganglia-thalamocortical circuits are interconnected at several levels. Thus Lynd-Balta and Haber (1994) reported that fibres originating from the ventral striatum synapse with dopaminergic neurons in the substantia nigra that innervate the dorsal striatum and suggested that, via this striato-nigrostriatal loop, the limbic-related striatum communicates with the somatomotor-related striatum. Joel and Weiner (1994) emphasised that the thalamocortical connections which close the various basal ganglia-thalamocortical circuits are always supplemented by thalamocortical projections terminating in one or several cortical regions that are not the source of the corticostriate input of that circuit. They held that the various circuits may well interact via these 'asymmetrical' or 'open' thalamocortical pathways. Groenewegen and Berendse (1994) suggested that cortico-cortical and intrastriatal connections may provide a linkage between structures belonging to different circuits. Finally, it may be mentioned that Parent and Hazrati (1995a) recently extensively reviewed the literature on the anatomical substrate of information processing along the corticobasal ganglia-thalamocortical circuit. Their main conclusions may be summarised as follows:

1. On the basis of the spatial arrangement of its cortical afferents, the striatum may be subdivided into separate somatomotor, associative and limbic territories.
2. This segregation is maintained at the level of the target structures of the striatal efferents; GPi and the substantia nigra, pars reticulata are involved in the processing of sensorimotor and associative information, respectively, whereas the ventral pallidum is concerned with the processing of limbic information.
3. Whether this segregation is maintained at the levels of the thalamus and the thalamocortical projections remains to be established.

First Subsidiary Striatal Circuit: Striatum-Globus Pallidus-Thalamus-Striatum. The striopallidal and pallidal efferent fibres, which form the initial part of this circuit, have already been dealt with. It has been shown that a considerable number of fine, presumably collateral fibres leave the fasciculus thalamicus before it reaches the ventral lateral nucleus. These fibres enter the internal medullary lamina and terminate in the centromedian and parafascicular nuclei (Nauta and Mehler 1966; DeVito and Anderson 1982; Fénelon et al. 1990). Fibres arising from these intralaminar nuclei project massively to the striatum (Mehler 1966). There is a clear differential distribution of the striatal input from the centromedian/parafascicular complex, the centromedian nucleus projecting mainly to the putamen and the parafascicular nucleus only to the caudate nucleus (Smith and Parent 1986; Sadikot et al. 1992a,b). The thalamostriate fibres, which most likely use glutamate as a neurotransmitter, termi-

nate, like the corticostriate axons, upon medium spiny neurons (Kemp and Powell 1971b), on which they exert an excitatory influence. According to Sadikot et al. (1992a,b), the centromedian nucleus projects specifically to the sensorimotor territory of the striatum, whereas the parafascicular nucleus projects principally to the associative striatal territory and less abundantly to the limbic striatal circuitry.

In addition to the pallido-thalamostriatal connections described above, direct pallidostriatal projections have also been demonstrated (e.g. Spooren et al. 1996).

Second Subsidiary Striatal Circuit: Globus Pallidus-Subthalamic Nucleus-Globus Pallidus. The external and internal segments of the primate pallidum (i.e. the homologues of GPe and GPi in non-primates) are engaged in different circuits in the basal ganglia. Whereas the efferents of the internal pallidum end in the thalamus and lower part of the principal striatal circuit, those of the external pallidum project in a topically organised fashion onto the subthalamic nucleus (McBride and Larsen 1980: cat; Carpenter et al. 1981a: monkey), and the latter has been shown to project massively back to both parts of the globus pallidus (Nauta and Cole 1978: cat and monkey; Carpenter et al. 1981a,b: monkey). The subthalamic efferents have an excitatory effect on pallidal neurons and presumably use glutamate as a neurotransmitter (Albin et al. 1989). The existence of the pallido-subthalamopallidal circuit just described has recently been questioned by Parent and Hazrati (1995b). These authors reported that neurons of GPe do not project to the subthalamic territory that contains most of the neurons projecting to GPi. In their opinion, GPe is an additional integrative centre intercalated in the principal striatal circuit. With regard to the subthalamic nucleus, Parent and Hazrati (1995b) emphasised that this cell mass receives projections from a wide variety of sources, including the cerebral cortex, the pars compacta of the substantia nigra, the pedunculopontine tegmental nucleus and the parafascicular/centromedian complex, and exerts its driving effect not only on GPi, but also on GPe, the pars reticulata of the substantia nigra and the striatum itself. The results obtained by Parent and Hazrati (1995a,b) are hard to reconcile with those of Shink et al. (1996). The latter authors, who studied the pallido-pallidal, pallido-subthalamic and subthalamo-pallidal connections in the squirrel monkey, found that the connections between the two segments of the globus pallidus and the subthalamic nucleus show a high degree of specificity, such that small groups of neurons in the external segments of the globus pallidus and subthalamic nucleus innervate common regions in the internal segment of the globus pallidus and are reciprocally connected at the synaptic level.

Third Subsidiary Striatal Circuit: Striatum-Substantia Nigra-Striatum. It has already been mentioned that fibres originating from the caudate nucleus and the putamen traverse the globus pallidus and subsequently descend to the substantia nigra. The fibres of this striatonigral projection terminate mainly, but not exclusively, in the external, reticular part of the substantia nigra. Fibres originating from the sensorimotor parts of the striatum overlap widely with those arising from the associative parts, but the latter are more numerous (Parent and Hazrati 1995a: monkey) . The ventral striatum innervates a large area of the substantia nigra, including the medial pars reticulata and much of the pars compacta. This projection from the ventral striatum shows only a very limited overlap with that from the sensorimotor part of the striatum (Lynd-Balta and Haber 1994: monkey). It has already been mentioned that most striatonigral fibres contain GABA and that in many of these fibres this inhibitory amino acid co-exists with substance P, dynorphin or enkephalin.

The major ascending efferent projection of the substantia nigra is formed by a system of extremely fine axons, which originates from the dopaminergic cells in the pars compacta (Ungerstedt 1971: rat; Tanaka et al. 1982: monkey). This projection is topographically organised (Carpenter and Peter 1972: monkey; Fallon and Moore 1978: rat). The nigrostriatal dopaminergic neurons establish direct synaptic contacts with medium spiny striatal projections neurons (Freund et al. 1984), and it has been observed that striatonigral fibres in turn synapse with the somata (Hedreen and DeLong 1991: monkey) and with the dendrites of pars compacta neurons (Wassef et al. 1981: rat). The axodendritic contacts are mainly located in the pars reticulata, into which many of the pars compacta cells extend their long, ventrally directed dendrites. There is evidence indicating that the nigrostriatal fibres reciprocate the striatonigral projection in an orderly topographic fashion. It is, however, not so that the nigrostriatal fibres exclusively form part of closed striato-nigrostriatal loops. The electron microscopy studies performed by Somogyi et al. (1981) have shown that fibres originating in the nucleus accumbens terminate on the somata and proximal dendrites of dopaminergic neurons projecting to the main body of the striatum.

22.11.4.3.3
Additional Striatal Connections

With the aid of the modern anterograde and retrograde tracer techniques, a considerable number of additional 'extrapyramidal' fibre connections have been established, only a few of which will be mentioned here.

As has already been indicated, the nucleus subthalamicus not only sends numerous fibres to the globus pallidus, but also projects to the caudate nucleus and the putamen (Beckstead 1983: cat), the pars reticulata of the substantia nigra (Ricardo 1980: rat; Smith et al. 1990: monkey) and the pedunculopontine tegmental nucleus (Jackson and Crossman 1981: rat; Edley and Graybiel 1983: cat). The latter cell mass, which receives a projection from the globus pallidus (Nauta and Mehler 1966), also takes up fibres from the pars reticulata of the substantia nigra (Beckstead and Frankfurter 1982: monkey; Edley and Graybiel 1983: cat; Spann and Grofova 1991: rat). The neurons constituting the latter projection utilise GABA as their neurotransmitter (Childs and Gale 1983).

The pedunculopontine tegmental nucleus has been shown to give rise to a small descending and a much larger ascending efferent projection. The former will be considered below; the latter is distributed mainly to the pars compacta of the substantia nigra, the subthalamic nucleus and the pallidum (Jackson and Crossman 1983: rat; Edley and Graybiel 1983; cat; Charara and Parent 1994: monkey; Lavoie and Parent 1994a: monkey). The pedunculopontine tegmental nucleus is known to exert an excitatory action upon neurons in the various target structures (Parent and Hazrati 1995a). This excitatory effect is principally mediated by acetylcholine (Woolf and Butcher 1986; Inglis and Winn 1995). It will be clear that the connections discussed above close a considerable number of additional 'extrapyramidal' loop systems.

22.11.4.3.4
Input Systems

The reticular formation of the brain stem has a quantitatively very important access to the 'extrapyramidal' circuitry. The reticular formation, particularly its mesencephalic portion, is one of the principal sources of afferent fibres to the intralaminar nuclei of the thalamus and, as pointed out earlier, the latter project to both the caudate nucleus and the putamen. A second important input system to be mentioned here is the mesostriatal serotoninergic projection, which originates mainly from the dorsal raphe nucleus and terminates throughout the striatum, but more significantly in its ventral and medial regions (Mori et al. 1985: rat, cat and monkey; Smith and Parent 1986: monkey).

22.11.4.3.5
Output Channels

Because the main outflow from the striatum and the nuclei related to it converges via the globus pallidus and the thalamus upon the motor, premotor and supplementary motor cortical areas, the fibres originating from these cortical areas constitute the principal output channel of the basal ganglia. Prominent among the systems emanating from the cortical areas mentioned is, of course, the motor part of the pyramidal tract, but these areas also give rise to a number of other projections, some of which may well consist of collaterals of pyramidal axons. Such projections terminate in the ventral lateral thalamic nucleus (Mehler 1981), the subthalamic nucleus (Petras 1969: monkey and carnivores; Carpenter et al. 1981b: monkey; Canteras et al. 1990: rat), the compact part of the substantia nigra (Usunoff et al. 1982: cat; Sakai 1988: raccoon; Künzle 1975: monkey) or both parts of that structure (Naito and Kita 1994: rat) and the pedunculopontine tegmental nucleus (Edley and Graybiel 1983: cat). All of these pathways feed back into centres included in one of the striatal circuits, thus closing additional subsidiary loops. The remaining output pathways of the basal ganglia include pallidohabenular, nigrotectal, nigrotegmental and pedunculopontino-reticular projections (Fig. 22.167).

The pallidohabenular fibres originate in primates mainly in a peripheral zone of the medial pallidal segment, which encroaches upon the lateral hypothalamus (Parent 1979; Parent and DeBellefeuille 1983). In non-primates, these fibres arise from the entopeduncular nucleus (Nauta 1974: cat; Herkenham and Nauta 1977: rat). In all mammals studied so far, the pallidohabenular projection terminates in the lateral habenular nucleus. Fibres originating from this nucleus descend in the habenulo-interpeduncular tract, bypass the interpeduncular nucleus and terminate in various mesencephalic centres, including the periaqueductal grey matter, the compact part of the substantia nigra and the dorsal raphe nucleus (Herkenham and Nauta 1979: rat). The projections to the two latter centres may well add further loops to the circuitry of the basal ganglia.

The nigrotectal and nigroreticular projections originate both from the pars reticulata of the substantia nigra (Hopkins and Niessen 1976: rat, cat and monkey; Graybiel 1978: cat; Beckstead and Frankfurter 1982: monkey; Huerta et al. 1991: bushbaby; Spann and Grofova 1991: rat); both are

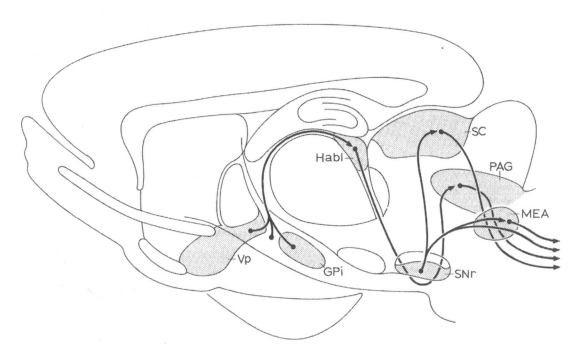

Fig. 22.167. Output channels of the basal ganglia. *GPi*, internal segment of globus pallidus or entopeduncular nucleus; *Habl*, lateral habenular nucleus; *MEA*, mesencephalic extrapyramidal area and nucleus tegmentalis pedunculopontinus; *PAG*, periaqueductal grey; *SC*, superior colliculus; *SNr*, substantia nigra, pars reticulata; *Vp*, ventral pallidum

thought to be inhibitory and to use GABA as their neurotransmitter (Childs and Gale 1983).

The massive nigrotectal projection terminates, mainly ipsilaterally, in the middle grey layer of the superior colliculus, where its fibres enter into synaptic contact with the cells of origin of the predorsal bundle (Bickford and Hall 1992). The latter represents the principal pathway from the superior colliculus to the centres in brain stem and spinal cord involved in initiating orienting movements of the head and eyes. Olazébel and Moore (1989) reported that in rats, bats and cats dopaminergic neurons situated in the lateral part of the substantia nigra and in the adjacent lateral tegmentum project to the ipsilateral inferior colliculus. They conjectured that impulses travelling along these fibres may influence orienting movements of the pinnae, head and eyes to auditory stimuli, mediated by the inferior colliculus.

The nigrotegmental fibres terminate mainly in the pedunculopontine tegmental nucleus, but some descend to the medullary reticular formation. The pedunculopontine tegmental nucleus, in addition to a projection from the substantia nigra, pars reticulata, also receives efferents from the motor cortex (Kuypers and Lawrence 1967: monkey), globus pallidus (De Vito and Anderson 1982: monkey) and the subthalamic nucleus (Nauta and Cole 1978: monkey and cat; Jackson and Crossman 1981: rat; Edley and Graybiel 1983: cat). Steininger et al. (1992) reported that, in the rat, the pedunculopontine tegmental nucleus receives major efferents from the periaqueductal grey, central tegmental field, lateral hypothalamic area, dorsal raphe nucleus, superior colliculus and the pontine and medullary reticular formation. The immediately adjacent midbrain extrapyramidal area shares most of these efferents with the pedunculopontine tegmental nucleus, but receives its principal afferent from the lateral habenula. The ascending efferents from the pedunculopontine tegmental nucleus have already been discussed (see Sect. 22.11.4.3.3); its descending efferents, which, in contrast to its ascending efferents, are principally non-cholinergic, terminate in the pontine and medullary parts of the medial reticular formation (Jackson and Crossman 1983: rat; Edley and Graybiel 1983: cat; Nakamura et al. 1989: rat; Inglis and Winn 1995) and in the cervical spinal cord (Spann and Grofova 1989: rat).

22.11.4.4
Aspects of the Organisation of the Striatum

The so-called medium spiny cells, which are present in enormous numbers in the caudate nucleus and the putamen, occupy a central position in the circuitry of these centres. The name of these neurons refers to the fact that their dendrites, which radiate and ramify in all directions, are densely covered with spines (Kemp and Powell 1971a: cat; DiFiglia et al. 1976; monkey; Graveland et al. 1985: humans). The principal afferent systems to the striatum are glutamatergic fibres from the cerebral cortex (Fonnum et al. 1981: rat) and from the intralaminar thalamic nuclei (Royce 1987), dopaminergic fibres from the substantia nigra (Kubota et al. 1986: rat) and serotoninergic fibres from the dorsal raphe nucleus (Pasik et al. 1981: cat and monkey; Lavoie and Parent 1990: monkey). All of these extrinsic afferents synapse with the medium spiny cells; the glutamatergic and dopaminergic fibres mainly contact the distal portion of the dendritic arbor of these neurons. With regard to the synaptology of these afferents, Parent and Hazrati (1995a; Fig. 22.168) pointed out that the terminals of the corticostriatal fibres make asymmetrical synapses with the head of dendritic spines, whereas the thalamocortical fibres make asymmetrical contacts with dendritic shafts. The nigrostriatal dopaminergic fibres make symmetrical synapses, many of which selectively contact the necks of those dendritic spines that receive cortical input. Dubé et al. (1988) reported that, although in the rat afferents from the somatosensory cortex and from the parafascicular nucleus converge upon the same part of the striatum, they probably do not converge upon the same spiny neurons. They observed that afferents from the neocortex form synapses on the spines of densely spiny, medium-sized neurons, whereas the terminals originating from neurons in the parafascicular nucleus are in synaptic contact with the dendritic shafts of medium-sized spiny neurons with a lower density of spines than the typically very densely spiny neurons. It has already been mentioned that the medium spiny neurons have long axons which terminate in the globus pallidus and in the pars reticulata of the substantia nigra, that they use GABA as their main neurotransmitter, that they also express a number of neuropeptides, such as substance P, enkephalin and dynorphin, and that these neuropeptides are contained in particular subsets of striatal projection neurons.

In addition to the extrinsic afferents discussed above, several groups of intrinsic elements impinge upon the medium spiny cells. Under this heading,

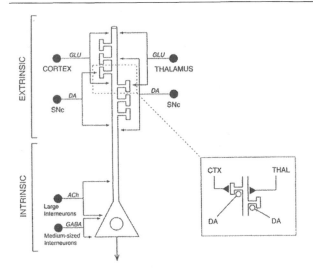
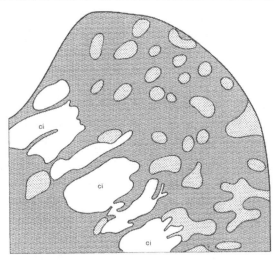

Fig. 22.168. Various extrinsic and intrinsic inputs to the medium spiny projection neurons of the striatum. The *insert* shows the modes of termination of corticostriatal (*CTX*), thalamostriatal (*THAL*) and dopaminergic (*DA*) nigrostriatal afferents. For further explanation, see text. (Reproduced from Parent and Hazrati 1995a)

Fig. 22.169. Representation of a transverse section through the caudate nucleus of the rhesus monkey, stained for acetylcholinesterase (AChE); striosomes poor in AChE are embedded in an AChE-rich matrix. *ci*, capsula interna. (From Nieuwenhuys 1985, based on Graybriel and Ragsdale 1983)

the medium spiny cells themselves should be mentioned first because, apart from their long projection axons, these elements also issue numerous collaterals, both local ones and longer ones, which have been shown to contact other neurons of the same type. Other local-circuit neurons synapsing with medium spiny cells include large, aspiny interneurons, which use acetylcholine as their neurotransmitter (Bolam et al. 1988: rat; Phelps et al. 1985: rat), and (2) medium-sized, aspiny GABAergic interneurons, which also contain the calcium-binding protein parvalbumin (Cowan et al. 1990: rat; Kawaguchi 1993: rat). Whereas the extrinsic inputs terminate on the distal dendritic tree, the inputs from the local circuit neurons favour the proximal somatodendritic domain of the medium spiny cells (Parent and Hazrati 1995a). It is noteworthy that the large cholinergic neurons in the striatum of the rat receive hardly any input from the cortex, but are under prominent synaptic control by the thalamostriatal system (Lapper and Bolam 1992). The group of medium aspiny local circuit neurons encompasses, in addition to GABAergic neurons, elements containing somatostatin and the enzyme nicotinamide adenine dinucleotide phosphate-diaphorase (NADPH-diaphorase) (DiFiglia and Aronin 1982: rat; Vincent et al. 1983: rat). This population of somatostatinergic, NADPH-diaphorase-positive neurons is not homogeneous, because a certain proportion of them also contains neuropeptide Y, but others do not (Rushlow et al. 1995: rat).

In the striatum, the medium spiny projection neurons are much more numerous than the interneurons. In rats, the ratio of projection neurons to interneurons is about 9:1, whereas it is 3:1 in primates (Graveland and DiFiglia 1985). It has already been mentioned that the striatum displays a mosaic-like organisation of interdigitating compartments, known as striosomes and the extrastriosomal matrix (Fig. 22.169), and that these two compartments, which were first recognised on the basis of their marked difference in AChE concentration, are related to the intrinsic structure of the striatum and to the disposition of its efferent connections. An extensive discussion of the organisation of these compartments is beyond the scope of the present work (for a review, see Graybiel 1990; Gerfen 1992); a few features may, however, be mentioned:

1. In addition to a low AChE concentration, the striosomes also show high enkephalin, substance P, GABA and neurotensin immunoreactivity (Gerfen 1984; Graybiel et al. 1981).
2. The complementary matrix compartment, in addition to a high AChE concentration, also shows a dense plexus of somatostatin-containing fibres (Gerfen 1984).
3. The striosomes show a remarkably high concentration of opiate receptors (Pert et al. 1976; Desban et al. 1993).
4. In rats, there are two different sets of nigrostriatal dopaminergic neurons, one projecting to the matrix and the other to the striosomes (Graybiel and Ragsdale 1983; Graybiel 1990; Gerfen 1992).

5. In the caudate nucleus of the rhesus monkey, cell clusters, some with capsules poor in cells, appear to match the striosomes (Goldman-Rakic 1982).
6. In the rat, medium spiny neurons, including their dendritic trees and local axon arborisations, strictly adhere to the striatal compartment boundaries, lying in their entirety either in a striosome or in the matrix (Penny et al. 1988). The same holds true for about three quarters of the medium spiny cells in the squirrel monkey. The remaining medium spiny neurons in this species, however, in both striosomes and matrix appear to have dendrites that cross from one compartment into the other (Walker et al. 1993).
7. The cholinergic, GABAergic and somatostatinergic striatal intrinsic neurons cross the borders between striosomes and matrix and are considered to form association connections between the two compartments (Gerfen 1984; Chesselet and Graybiel 1986; Kita et al. 1990; Kubota and Kawaguchi 1993).
8. Studies with anterograde tracers have revealed that the corticostriatal, thalamostriatal and amygdalostriatal projections terminate in a patchy fashion. With regard to the corticostriatal projection, it has been reported that motor, somatosensory and visual cortical areas preferentially project to the matrix compartment, but that the efferents from the prelimbic cortex are distributed to the striosomes (Donoque and Herkenham 1986: rat). However, it has also been reported that the compartmentalisation of the corticostriatal inputs is related to the laminar rather than to their areal origin. Each cortical area was found to innervate both matrix and striosomes, but corticostriatal neurons in supragranular layers appeared to project principally to the matrix, whereas those from the infragranular layers were observed to send their axons to the striosomes (Arikuni and Kubota 1986: monkey; Gerfen 1990: rat). Recent studies by Cowan and Wilson (1994) and Kincaid and Wilson (1996) suggest that, in the rat, three kinds of corticostriatal neurons are present: (1) cells situated in laminae III–Va, forming extended axonal ramifications that preferentially, but not exclusively innervate the matrix, (2) elements situated in laminae Vb and VI, innervating the patch compartment in a focused fashion and (3) elements forming focused arborisations in the matrix. The latter were found to contribute most strongly to the innervation of the caudal half of the caudate-putamen and appeared to be concentrated in the primary somatosensory cortex. Afferents from the basolateral amygdaloid nucleus selectively innervate striosomes (Ragsdale and Graybiel 1988: cat), but the terminal patches of the thalamostriatal fibres avoid these structures (Herkenham and Pert 1981: rat; Graybiel 1984; Sadikot et al. 1992a,b).
9. The striatal projection neurons are differentially distributed over the two striatal compartments. Following injections of a retrograde tracer in the internal pallidal segment (or in the entopeduncular nucleus of non-primates), in the external pallidal segment (or in the globus pallidus of non-primates) and in the substantia nigra, pars reticulata, labelled neurons appear to be most densely distributed in the matrix compartment, whereas after injections centred in the compact part of the substantia nigra, labelled neurons are located preferentially in the striosomes (Gerfen 1984: rat; Bolam et al. 1988; Giménez-Amaya and Graybiel 1990: monkey). In cats and monkeys, striatal projections terminating in the globus pallidus and entopeducular nucleus (or their homologues) and in the substantia nigra, pars reticulata largely originate from separate cell populations (Beckstead and Cruz 1986; Selemon and Goldman-Rakic 1990). In rats, however, considerable numbers of striatal cells appeared to project to both the globus pallidus and the substantia nigra (Loopuijt and van der Kooy 1985).

In cats and primates, the projection neurons lying in the extrastriosomal matrix are arranged in clusters (Desban et al. 1993; Giménez-Amaya and Graybiel 1990; Selemon and Goldman-Rakic 1990), which have been called matrisomes (Graybiel et al. 1991). Flaherty and Graybiel (1993) tested the hypothesis that these matrisomes consist of neurons projecting to a single extrastriatal target by small paired injections of distinguishable retrograde tracers in the external and internal pallidal segments of squirrel monkeys. It appeared that individual neurons project to only one of these targets, but that, within individual matrisomes, GPe- and GPi-projecting neurons are extensively intermixed. Desban et al. (1995) studied the distribution of stratonigral and striatopallidal neurons in the matrix compartment of the caudate nucleus in the cat. They identified two types of striatonigral neurons: poorly collateralised and well-collateralised cells. The former were distributed in clusters, and the latter outside these clusters. Neurons innervating the entopeducular nucleus were distributed outside the matrix clusters of aggregated striatonigral neurons and appeared to be intermingled with the non-aggregated, well-collateralised striatonigral neurons.

From the foregoing synopsis, it may be concluded that the caudate-putamen complex or dorsal striatum displays an intriguing mosaic-like chemical heterogeneity and that several structural and connectional features fit into this mosaic.

The nucleus accumbens, i.e. the principal component of the ventral striatum, also shows a marked chemoarchitectural heterogeneity. Histochemical and immunohistochemical markers with activity or affinity in the accumbens have revealed two major subterritories: the core and shell (Zahm and Brog 1992; Jongen-Rêlo et al. 1994). These subterritories have also been shown to differ with respect to their afferent (Berendse and Groenewegen 1990: rat) and efferent connections (Heimer et al. 1991: rat; Berendse et al. 1992a,b: rat; Groenewegen and Russchen 1984: cat). Striosome-like and matrix-like compartments have been observed in the nucleus accumbens, but on the whole the chemoarchitecture of the striatal mosaic is different in the dorsal and ventral striatum. Martin et al. (1991) observed that, in the rhesus monkey, the compartmental organisation of some neurotransmitters and neuropeptides in the ventral striatum is variable and not as easily divisible into conventional striosome and matrix areas as in the dorsal striatum. Jongen-Rêlo et al. (1993) studied the organisation of the nucleus accumbens in the rat by comparing the pattern of L-enkephalin immunoreactivity with that of the opioid receptor ligand naloxone, an established marker for the compartmental organisation of the dorsal striatum. They arrived at the conclusion that, rather than a bicompartmental striosome-matrix organisation, the nucleus accumbens has a multicompartmental organisation.

22.11.4.5
Connections of the Ventral Striatum

The nucleus accumbens and adjacent sectors of the caudate nucleus and the putamen receive afferents from the cerebral cortex, particularly from its limbic areas, the midline nuclear complex of the thalamus, the amygdala, the ventral tegmental area and the dorsal raphe nucleus. The cortical afferents, which originate from the prelimbic, infralimbic, medial orbital, agranular insular, perirhinal and entorhinal cortices and from the hippocampal subiculum, project bilaterally with an ipsilateral predominance to the ventral striatum. Each of these cortical areas has a longitudinally oriented striatal terminal field that overlaps slightly with those of adjacent cortical areas (Groenewegen et al. 1982: cat; McGeorge and Faull 1989: rat; Berendse et al. 1992a,b: rat). With regard to the thalamostriate projections, Berendse and Groenewegen (1990: rat) established that the various midline and intralaminar thalamic nuclei project to longitudinally oriented striatal sectors and that the terminal sectors of the midline nuclei (i.e. the paraventricular, parataenial, intermediodorsal and rhomboid nuclei) are mainly situated in the ventral striatum, whereas those of the intralaminar nuclei occupy most of the dorsal striatum. Correlation of these results with those of previous studies on the organisation of the projections of the midline and intralaminar thalamic nuclei on the prefrontal cortex (Berendse and Groenewegen 1991) and those of the prefrontal cortex to the striatum (Berendse et al. 1992a,b; Groenewegen and Berendse 1994: rat) revealed that the projection zones of individual prefrontal cortical areas converge in the striatum with those of 'their own' midline and intralaminar afferent nuclei (Groenewegen and Berendse 1994). The amygdalostriatal projection arises from the basal amygdaloid complex and terminates mainly in the ventral striatum. Its fibres are topographically organised (Russchen and Price 1984: rat; Groenewegen et al. 1980: cat; Russchen et al. 1985: monkey). Apart from the ventral striatum, the prefrontal cortex receives a massive input from the amygdaloid nucleus mentioned (Krettek and Price 1977: rat and cat; Porrino et al. 1981: monkey). Double-labelling experiments have shown that, in rats, numerous neurons in the basolateral amygdaloid nucleus project to both the prefrontal cortex and the nucleus accumbens (Shinonaga et al. 1994). The topographical organisation and the projection patterns of these neurons, and of the amygdaloprefrontal and amygdalostriatal connections in general, correspond to those of the pre-frontostriatal projection.

The ventral tegmental area projects to the entire ventromedial half of the striatum, but most massively to the ventral striatal zone that includes the nucleus accumbens (Beckstead et al, 1979: rat; Szabo 1980a: cat; Szabo 1980b: monkey). The dopaminergic innervation of the ventral striatum, which originates mainly from the ventral tegmental area, is very strong and unevenly distributed. This uneven distribution of dopaminergic terminals is paralleled by the uneven distribution of the neuropeptides enkephalin, substance P and dynorphin and the neurotransmitter GABA (Voorn et al. 1989: rat). It is noteworthy that many neurons in the ventral tegmental area which project to the ventral striatum contain cholecystokinin (Zaborski et al. 1985: rat) and that, in some of these elements, cholecystokinin is co-localized with dopamine (Hökfelt et al. 1980a,b)

The efferents of the ventral striatum pass primarily to extrapyramidal centres, such as the ventral pallidum, the ventral tegmental area and the pars compacta and pars reticulata of the substantia nigra, but a number of limbic-related structures, such as the bed nucleus of the stria terminalis and the lateral hypothalamic area, are also innervated (Nauta et al. 1987: rat; Heimer et al. 1991: rat; Groenewegen and Russchen 1984: cat; Haber et al. 1990a: monkey). The projections from the nucleus accumbens to the ventral pallidum contain a major GABAergic component (Mogenson and Nielsen 1983; rat), and substance P-, enkephalin- and dynorphin-positive fibres have also been shown to be present within this projection (Haber and Nauta 1983: rat; Haber and Watson 1985: humans). The neurons containing these neuropeptides most probably use GABA as their principal neurotransmitter. The enkephalin- and substance P-positive fibres form dense and highly characteristic terminal plexuses within the ventral pallidum (Beach and McGeer 1984: humans; Haber and Nauta 1983: rat; Haber and Watson 1985: humans). It has been demonstrated that the accumbofugal efferents to the ventral pallidum project monosynaptically to output cells of that structure (Yang and Mogenson 1985). The efferents from the ventral striatum, which are topographically organised, also project to limited portions of the external and internal pallidal segments. However, the ventral striatal output remains segregated from the dorsal striatal efferent projections to these pallidal structures (Haber et al. 1990a). The same holds true for the dorsal and ventral striatal projections to the substantia nigra. The ventral striatal fibres passing to this structure terminate mainly in the pars compacta (Lynd-Balta and Haber 1994), where they enter into synaptic contact with dopaminergic neurons that innervate the dorsal striatum (Somogyi et al. 1981). It has already been mentioned that, via this striatonigrastriatal circuit, the ventral striatum, and ultimately the limbic system, may influence the somatomotor-related striatum.

The efferent projections from the core region of the nucleus accumbens differ in some respect from those of the shell region of the same structure. Without going into details, the following may be mentioned: (a) core and shell project to different parts of the ventral pallidum, (b) both regions project to the midbrain (but with a bias for the core projection to innervate the substantia nigra and the shell projection to reach primarily the ventral tegmental area) and (c) the shell, in contrast to the core, projects diffusely throughout the rostrocaudal extents of the lateral hypothalamus (Heimer et al. 1991: rat; Zahm and Brog 1992: rat). Deniau et al. (1994) presented anatomical and physiological evidence indicating that, in the rat, the core of the nucleus accumbens innervates a dorsal region of the substantia nigra, pars reticulata, which, via certain parts of the mediodorsal and ventral medial thalamic nuclei, projects mainly to the prelimbic and to a lesser degree to the orbital areas of the prefrontal cortex.

The efferent projections of the ventral pallidum terminate primarily in the lateral habenular nucleus, the mediodorsal thalamic nucleus, the subthalamic nucleus, the lateral hypothalamus, the substantia nigra and the dorsocaudal part of the mesencephalic tegmentum (Haber et al. 1985: rat; Groenewegen et al. 1993: rat; Haber et al. 1990a,b: monkey). In the rat, efferent fibres from the ventral pallidum have in addition been traced to numerous other areas, including the prefrontal cortex, the ventral striatum, the basolateral, lateral and central amygdaloid nuclei, the lateral septum, the area tegmentalis ventralis, the rostral raphe nuclei, the periaqueductal grey and the locus coeruleus (Groenewegen et al. 1993). The projections from the ventral pallidum to the mediodorsal thalamic nucleus, the subthalamic nucleus and the lateral hypothalamus are distinctly topographically organised (Haber et al. 1985; Groenewegen et al. 1993).

The fibres passing from the ventral pallidum to the nucleus mediodorsalis thalami are a link in a projection from the ventral striatum to the prefrontal cortex. Heimer et al. (1982) held that the impulses travelling along this circuit via the premotor cortex ultimately attain the motor cortex, thus forming a limbic counterpart of the well-known principal or dorsal striatal circuit. Haber et al. (1985) and Nauta (1986) considered it likely that the ventral striatum via this circuit affects the mechanism of the frontal cortex and thus a class of functions more likely to be cognitive than skeletomotor. Groenewegen et al. (1993; see also Pennartz et al. 1994) pointed out that the prefrontal cortex, which is strongly and reciprocally connected with the mediodorsal thalamic nucleus, can be subdivided into several functionally distinct subregions. Thus the dorsal prelimbic area is involved in spatial orientation and visual motor functions, the infralimbic and ventral prelimbic areas probably subserve visceral motor functions and the dorsal and ventral agranular insular areas receive visceral sensory information. On the basis of existing knowledge in the rat about the prefrontal corticostriatal projections, the ventral striatopallidal projections, the ventral pallido-mediodorsal thalamic projections and the reciprocal connections between the mediodorsal thalamic nucleus and the prefrontal cortex, Groenewegen et al. (1993) tentatively identified four parallel circuits that involve the functionally distinct parts of the prefrontal cortex indicated above and particular parts or sectors of the ventral striatum, the ventral pallidum and the mediodorsal thalamic nucleus. They considered it likely that parallel arrangement of the connections between the prefrontal cortex, the ventral striopallidum and the mediodorsal thalamic nucleus is essential for the conjoint functioning of the prefrontal cortex and the basal ganglia in various behavioural processes. In their opinion, parallel organisation is a cerebral feature of corticostriatal processing, and Groenewegen et al. (1996) recently pointed out that the projections from the midline thalamic nuclei, the amygdala and the hippocampal region have specific relationships with the ventral basal ganglia-thalamocortical circuits.

It is noteworthy that, whereas in the rat the ventral pallidothalamic projection is strongly developed, in the cat and in the monkey surprisingly few ventral pallidal terminations were found in the mediodorsal nucleus of the thalamus (Spooren et al. 1993; Haber et al. 1993). This may mean that in these species the loops from the various prefrontal subareas through the ventral striatopallidothalamic system back to the frontal lobe are much less prominent than in the rat. Like the dorsal pallidothalamic fibres, the ventral pallidal efferents presumably carry GABA as their neurotransmitter (Kuroda and Price 1991).

Following this discussion of the ventral striatopallidal fibres and of the circuits of which they form part, brief consideration will be given to some of the remaining efferent projections of the ventral pallidum.

The strongly developed projection from the ventral pallidum to the subthalamic nucleus is reciprocated by an equally strong efferent projection from the latter (Groenewegen and Berendse 1990: rat). The circuit, thus closed, presumably corresponds to the well-known dorsal pallido-subthalamopallidal circuit, fitting in with the general parallel organisation of basal ganglia circuits (Groenewegen et al. 1993; however, see Parent and Hazrati 1995b).

With regard to the ventral pallidonigral projections, Haber et al. (1993) reported that, in the monkey, these efferents from the ventral pallidum overlap widely, suggesting convergence of terminals from different pallidal regions. In the rat, the ventral pallidonigral fibres terminate predominantly in the medial part of the pars reticulata, an area which also receives input from the ventral striatum (Groenewegen et al. 1993). As has already been discussed, the efferents from the substantia nigra pars reticulata terminate in several thalamic nuclei, in the superior colliculus and in the caudolateral mesencephalic tegmentum.

The ventral pallidohabenular projection terminates in the lateral habenula, which sends its efferents to several targets, including the mesencephalic serotoninergic cell groups (Herkenham and Nauta 1977). The latter also receive a direct input from the ventral pallidum.

Input from the nucleus accumbens, an integral part of the ventral striatum, may attain the caudolateral mesencephalic tegmentum along four different routes: (1) accumbo-tegmental, (2) accumbonigro-tegmental, (3) accumbo-ventral pallidonigrotegmental and (4) accumbo-ventral pallidotegmental. The caudolateral mesencephalic tegmentum, which includes the mesencephalic locomotor area, gives rise to a small descending projection, the fibres of which terminate in the pontine and medullary reticular formation and in the spinal cord (Garcia-Rill and Skinner 1987; Spann and Grofova 1989).

Locomotion is an essential component of various adaptive behaviours, such as fight and flight reactions and food and water procurement. Pharmacological (Pijnenburg et al. 1976) and physiological experiments (Mogenson et al. 1980, 1984; Mogenson and Nielsen 1983) have provided evidence suggesting that the nucleus accumbens influences locomotor behaviour and that some of the pathways enumerated above (accumbens → ventral pallidum → caudal mesencephalic tegmentum) may well form part of the extrapyramidal projection system along which these influences attain the spinal motor mechanism. The recent ethopharmacological experiments carried out by Hooks and Kalivas (1995) revealed that a circuit containing the ventral pallidum is critical for the manifestation of noveltyinduced locomotor activity.

The limbic system is known to play a key role in the initiation of the various adaptive or goal-oriented behaviours of which locomotion is a component. Because the nucleus accumbens receives massive projections from the hippocampal formation and the amygdala, both essential components of the limbic system, this cell mass has been hypothesised to constitute the functional interface between the limbic and the motor system (Mogenson et al. 1980; Mogenson 1984; Hooks and Kalivas 1995). Pennartz et al. (1994) recently critically reviewed the morphological, electrophysiological and ethological literature on the nucleus accumbens. They arrived at the conclusion that, although the nucleus accumbens most probably influences locomotion, it is not correct to associate this particular behavioural state with the activity of this centre as a whole. They pointed out that the accumbens nucleus is a collection of neuronal ensembles, or groups, with specific input-output relationships and with different functional and behavioural connotations. This view has been corroborated by the results of tracing experiments carried out by Groenewegen et al. (1996), which showed that the neurons projecting directly or indirectly (i.e. via the ventral pallidum) to the caudolateral mesencephalic tegmentum (including the mesencephalic locomotor region) are located in specific subregions of the nucleus accumbens preferentially innervated by fibres originating from specific limbic territories. For example, the ventral striatal projections to the caudolateral merencephalic tegmentum originate mainly from clusters of cells in the border region between the 'shell' and the 'core' regions of the accumbens nucleus, and these clusters are preferentially innervated by fibres stemming from the prelimbic cortex and from the basolateral amygdaloid nuclei.

22.11.4.6
Connections of the Substantia Innominata

Before presenting a survey of the afferent and efferent connections of the substantia innominata (Fig. 22.170) and related cell groups, it is felt appropriate to recall a few data concerning these grisea (for details, see Sect. 22.11.4.2):

- The substantia innominata is an ill-defined cell mass, situated in the substriatal part of the telencephalon.
- The studies carried out by Heimer and associates (Heimer and Wilson 1975; Heimer et al. 1982) have shown that the rostral third of the substantia innominata is occupied by a ventral extension of the globus pallidus; the term substantia innominata has since been reserved for the area caudal to the ventral pallidum.
- The substantia innominata shares a number of features with the bed nucleus of the stria terminalis and with the amygdala.
- Chemoarchitectural and connectional data indicate that, at least in the rat, the substantia innominata is organized into a dorsal and a ventral division. Each of these divisions is affiliated with a different portion of the amygdala and, together with its amygdalar affiliate, forms part of one of two distinct complexes of interconnected forebrain and brain stem cell groups (Grove 1988a,b).
- The substantia innominata contains numerous large cholinergic projections neurons (Ch4). Similar cells occur in the lateral preoptic area, the horizontal limb of the diagonal band (Ch3), the vertical limb of the diagonal band (Ch2) and the medial septal nucleus (Ch1). Together, these cells form a continuum, which is referred to as the basal forebrain cholinergic projection system (Zaborszki et al. 1991).
- In primates, the Ch4 group is well developed and concentrated in the nucleus basalis of Meynert.

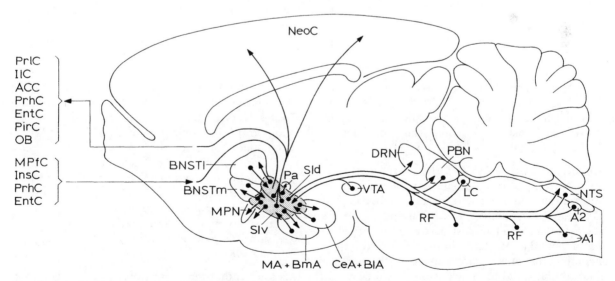

Fig. 22.170. Principal connections of the substantia innominata of the rat. A1, A2, noradrenergic cell groups; ACC, anterior cingulate cortex; BlA, basolateral amygdaloid nucleus; BmA, basomedial amygdaloid nucleus; BNSTl, lateral part of bed nucleus stria terminalis; BNSTm, medial part of BNST; CeA, central amygdaloid nucleus; DRN, dorsal raphe nucleus; EntC, entorhinal cortex; IlC, infralimbic cortex; InsC, insular cortex; LC, locus coeruleus; MA, medial amygdaloid nucleus; MPfC, medial prefrontal cortex; MPN, medial preoptic nucleus; NeoC, neocortex; NTS, nucleus of solitary tract; OB, olfactory bulb; Pa, paraventricular nucleus; PBN, parabrachial nuclei; PirC, piriform cortex; Prhc, perirhinal cortex; RF, reticular formation; SId, substantia innominata, pars dorsalis; SIv, SI, pars ventralis; VTA, ventral tegmental area. (Mainly based on Grove 1988a,b)

The substantia innominata and the large cholinergic cells within it receive afferent fibres from a wide range of structures in both the brain stem and the forebrain. The following survey of these connections is mainly based on the studies by Russchen et al. (1985: monkey) and Grove (1988a: rat) and on the comprehensive review by Zaborszki et al. (1991). These connections include the following:

1. Noradrenergic fibres from the A1 and A2 groups in the caudal medulla oblongata and from the locus coeruleus.
2. A substantial projection from the parabrachial nuclear complex, particularly from its medial part. This projection carries viscerosensory information toward the substantia innominata; it is accompanied by a limited number of fibres originating directly from the nucleus of the solitary tract.
3. A considerable number of serotoninergic fibres originating mainly from the mesencephalic dorsal raphe nucleus.
4. Dopaminergic fibres from the ventral tegmental area.
5. Fibres arising from the full rostrocaudal extent of the brain stem reticular formation.
6. Fibres originating from cells located through most of the hypothalamus.
7. A very large projection from the amygdala. In the monkey, the cells of origin of this projection are concentrated in the parvocellular and magnocellular basal nuclei, the magnocellular accessory basal nucleus and the central nucleus (Russchen et al. 1985).
8. Fibres from the bed nucleus of the stria terminalis.
9. Input from a number of cortical areas, including the orbitofrontal cortex, rostral insula, rostroventral temporal cortex and the piriform and entorhinal cortices in the monkey and the medial prefrontal, insular, perirhinal and entorhinal cortices in the rat.

According to Grove (1988a), the dorsal division of the substantia innominata receives selective innervation from the lateral part of the bed nucleus of the stria terminalis, the central and basolateral nuclei of the amygdala, distinctive zones of the lateral hypothalamus and brain stem structures, including the dorsal raphe nucleus, the parabrachial complex and the nucleus of the solitary tract, whereas projections preferentially directed to the ventral division of the substantia innominata arise from the medial part of bed nucleus of the stria terminalis, the medial and basomedial nuclei of the amygdala and certain hypothalamic areas complementary to those innervating the dorsal substantia innominata. In addition, Grove (1988a) observed that cell groups which preferentially innervate a single substantia innominata division maintain numerous projections to one another and thus form a tightly linked assembly of structures.

The inputs of the cholinergic projection neurons are probably non-specific. Light microscopy observations strongly suggest that the afferents to a given part of the substantia innominata contact both cholinergic and non-cholinergic elements (Grove 1988a). Grove (1988b) considered it likely that the cholinergic neurons within the substantia innominata are integrated into the overall afferent and efferent circuitry of the regions and thus into whatever functional roles the substantia innominata may subserve.

The basal forebrain cholinergic projection system encompasses the following components (Lehman et al. 1980: rat; Amaral and Kurz 1985: rat; Mesulam et al. 1983b, 1986 monkey; Kitt et al. 1987: monkey; Mesulam and Geula 1988: humans):

1. Neurons within the medial septal nucleus (Ch1) and the vertical limb of the nucleus of the diagonal band (Ch2) provide a substantial cholinergic projection to the hippocampus. The fibres pass through the fornix and terminate in the cornu Ammonis and in the fascia dentata.
2. Neurons within the horizontal limb of the nucleus of the diagonal band (Ch3) and the magnocellular preoptic nucleus project to the olfactory bulb and the piriform and entorhinal cortices.
3. Neurons situated within the substantia innominata (Ch4) innervate the basolateral amygdala and provide a major cholinergic projection to the entire neocortex. There is a specific, but overlapping topography in the organisation of this projection. In primates, the corticopetal cholinergic cells are subdivided according to the topography of their projections. Thus Ch4am is the major source of fibres passing to medial cortical areas, including the cingulate cortex, Ch4al projects to the frontoparietal cortex, Ch4i to the prefrontal, lateral peristriate, middle temporal and inferotemporal regions and Ch4p to the superior temporal gyrus and the temporal pole region (Mesulam et al. 1983b, 1986; Mesulam and Geula 1988).

It is known that the projection from the basal forebrain to the amygdala, hippocampus and neocortex in addition to cholinergic fibres also encompass fibres from non-cholinergic neurons (Rye et al. 1984; Amaral and Kurz 1985; Woolf et al. 1986).

The efferent connections of the substantia innominata in the rat were investigated by Grove

(1988b), employing anterograde and retrograde tracer techniques. In order to determine whether the cells of origin of the various projections arising from the substantia innominata are cholinergic or non-cholinergic, appropriate histochemical and immunohistochemical techniques were combined with the retrograde transport of WGA-HRP. The main results may be summarised as follows:

1. The efferent projections of the substantia innominata largely reciprocate its afferent connections.
2. The efferent projections resemble those of the amygdala.
3. The ventral and dorsal divisions can be distinguished in the substantia innominata on the basis of both its afferent and efferent connections.
4. The dorsal substantia innominata projects preferentially to the lateral nuclei of the amygdala, the paraventricular nucleus of the hypothalamus, certain parts of the lateral hypothalamus, the dorsal raphe nucleus, the parabrachial nuclear complex and the nucleus of the solitary tract. The projections to the latter four areas appeared to arise mainly or exclusively from non-cholinergic cells.
5. The efferents of the ventral subdivision are distributed more heavily over the medial part of the bed nucleus of the stria terminalis, a ventromedial amygdaloid region including the medial nucleus, the medial preoptic nucleus and the anterior hypothalamus.
6. Cortical efferents from the mid-rostrocaudal part of the substantia innominata are distributed primarily in prelimbic, infralimbic, anterior cingulate, insular, perirhinal, entorhinal and piriform cortices and in the main and accessory olfactory bulbs.
7. Comparison of the efferent connections of the substantia innominata with those of adjacent subcortical centres revealed that the divisions of the substantia innominata form part of two longitudinally arranged complexes with common efferents, the first consisting of the lateral part of the bed nucleus of the stria terminalis, the dorsal substantia innominata and the central nucleus of the amygdala and the second formed by the medial part of the bed nucleus of the stria terminalis, the ventral substantia innominata and the rostral parts of both the medial nucleus of the amygdala and an amygdaloid region lying ventral to the central nucleus and dorsal to the cortical nuclei.

The basal nucleus has been suggested to act as a cholinergic relay between limbic plus paralimbic centres and the entire neocortex instrumental in influencing complex behaviour according to the prevailing emotional and motivational states (Mesulam et al. 1983b). Russchen et al. (1985) pointed out that, although the structures which project to the substantia innominata are remarkably disparate, virtually all of them are integrative regions of polysensory convergence. This particularly applies to the amygdala, which is itself organised to sample inputs from a variety of cortical, limbic and subcortical structures. The amygdala, which projects massively to the substantia innominata, is believed to play a role in evaluating whether incoming sensory information is relevant to the individual in relation to previous experience or to other inputs from the external or internal environment. Russchen et al. (1985) considered it likely that the integrated sensory input which the amygdala and other areas of the brain provide to the substantia innominata leads to activation of the cholinergic inputs to the cerebral cortex, and that this activation modulates cortical activity such that relevant sensory inputs lead to integrated motor or emotional responses or to learning and memory.

Finally, Grove (1988 a,b) pointed out that the dorsal and ventral parts of the substantia innominata differ considerably with regard to both their afferent and efferent connections and that each of these two parts is strongly affiliated with a particular portion of the amygdaloid complex. Combining the results of her own studies on the connections of the substantia innominata with those of others on the connections and functions of the various part of the amygdala, she conjectured that the functional complex to which the ventral substantia innominata belongs might play a role *inter alia* in the initiation of sexual and investigatory behaviour in response to complex (chemo)sensory stimuli, whereas the dorsal substantia innominata, with its associated forebrain and brain stem structures, forms a functional assembly concerned, at least in part, with monitoring and influencing autonomic and visceral functions, such as blood pressure, heart rate and respiration. Grove (1988a,b) emphasised that the cholinergic cortical projection represents only one of the many output systems of the substantia innominata.

22.11.4.7
General Discussion and Summary

From the foregoing survey, it appears that ideas about the extent, connections and functional significance of the basal ganglia have changed considerably over time. Classically, these ganglia comprise a chain of subcortical centres extending form the basal telencephalon to the midbrain tegmentum which, mainly on the basis of clinicopathological evidence, were thought to influence motor activity quite independently from the control exerted by the motor cortex via the pyramidal tract. The term 'extrapyramidal motor system', still in use, bears witness to this interpretation. Later, it became clear that the basal ganglia do not operate independently from the cortex. It appeared that the caudate-putamen complex, the rostralmost and by far the largest of the basal ganglia, receives a massive direct projection from the entire neocortex and that this complex, via the globus pallidus, another basal ganglion, and the ventral anterior-ventral lateral complex of the thalamus, projects back to the motor and premotor cortices. The realisation that the basal ganglia receive their principal input from the neocortex and that the main outflow of these ganglia is directed towards the motor and premotor cortices represented a significant departure from the concept of separately operating cortical, pyramidal and subcortical, extrapyramidal motor systems.

In addition to the neocortex and the rostral basal ganglia, the telencephalon also contains important components of a third functional domain, the limbic system. As with the neocortex and the basal ganglia, it was long thought that the basal ganglia and the limbic system represent separate and independent functional entities. However, some 20 years ago, Heimer and associates demonstrated that the telencephalic basal ganglia extend much further ventrally than was previously assumed and that the newly discovered ventral striatal territory receives strong projections from two limbic core structures, the hippocampal formation and the amygdala. On the basis of these findings, a ventral or limbic striatum was added to the long-known non-limbic or somatic dorsal striatum. Further analysis of the fibre connections revealed that the ventral striatum forms part of a subcortico-cortical circuit that resembles the classic loop through the main, dorsal parts of the basal ganglia and the ventral anterior-ventral lateral complex of the thalamus. This limbic or limbic system-related circuit is composed of sequences of projections from the ventral striatum to a ventral extension of the globus pallidus, from this ventral pallidum to the mediodorsal thalamic nucleus and from the latter to the prefrontal, prelimbic and anterior cingulate cortices. Fibres passing from these cortical regions back to the ventral striatum close the circuit. The ventral striatum, like the classical, dorsal striatum, was considered to form an integral part of the motor system. It was pointed out that the anterior cingulate gyrus, one of the cortical components of the limbic striatal circuit, gives rise to corticobulbar projections to the medullary reticular formation and projects directly to the premotor cortex (Heimer et al. 1982).

During the last decade, the organisation of the basal ganglia-thalamocortical circuits has received much attention. In the light of new hodological data, the so-called 'information-funnelling theory', according to which information stemming from the entire cortical mantle and from the amygdala via striopallidal, pallidothalamic and thalamocortical connections converges upon the premotor and motor cortices, has been replaced by another hypothesis, designated as the 'parallel processing theory' (see Parent and Hazrati 1995a). This theory infers that cortical information is processed through multiple parallel-arranged, largely separated and at least partly closed corticobasal ganglia-thalamocortical circuits. The cortical components of these circuits are formed by functionally different areas of the frontal cortex, each of which projects to a particular sector of the striatum. Alexander et al. (1986) distinguished five such parallel loops, a motor one, an oculomotor one and three prefrontal ones; Groenewegen et al. (1993) tentatively identified four parallel circuits, each involving a prefrontal cortical field and a distinct sector of the ventral striatum, and Parent and Hazrati (1995a) held that information related to three different functional modalities, sensorimotor, limbic and associative, is processed along well-segregated channels through the basal ganglia. The classical concept, according to which the basal ganglia in their entirety are considered to form an integral part of the motor system, is hard to reconcile with the 'parallel processing theory'. The circuits of which the motor and premotor cortices form part are most probably primarily involved in motor control, but those centred upon association and limbic-prefrontal cortical fields subserve functions that seem more likely to be emotional or affective and cognitive than primarily skeletomotor (Nauta 1986). The parallel arrangement of the various channels from the cortex through the basal ganglia and back to the cortex does not exclude the possibility of interaction between them. Indeed, Groenewegen et al. (1996) argued that such interaction is essential for the production of coherent behaviour. Evidence suggests the presence of linkages between the various circuits at the following four levels: (1) intrastriatal by means of interneurons (see Graybiel et al. 1994), (2) in the substantia nigra (Somogyi et al. 1981; Lynd-Balta and Haber 1994; Bevan et al. 1996), (3) between the thalamus and the cortex by means of 'asymmetrical' or 'open' thalomocortical pathways (Joel and Weiner 1994) and (4) at the cortical level by cortico-cortical connections.

The question arises how the various parallel

trans-striatal circuits are related to the remaining intrinsic and extrinsic connections of the basal ganglia. Because many of the intrinsic connections forming the various subsidiary loops described in Sect. 22.11.4.3.2 are topographically organised, it seems likely that each of these loops is actually composed of a number of subloops, each of which is specifically linked to one of the parallel circuits. With regard to the extrinsic connections, Groenewegen and Berendse (1994) showed that various intralaminar and midline thalamic nuclei are specifically related to one of the parallel trans-striatal circuits. Thus, in the rat, the lateral parafascicular nucleus projects specifically to the cortical and striatal waystations of the circuit involving the sensorimotor cortex, and the same part of the parafascicular nucleus also has a projection to the lateral part of the subthalamic nucleus which, via a projection to the internal segment of the globus pallidus, has access to the same circuit (Fig. 22.171a). Similarly,

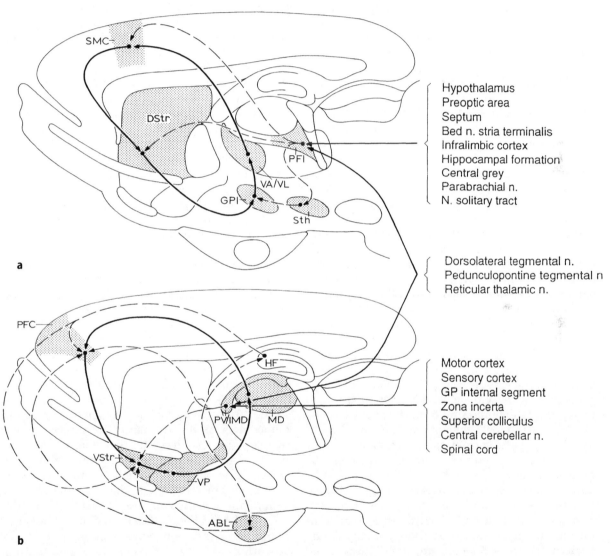

Fig. 22.171a,b. Direct and indirect involvement of the intralaminar parafascicular nucleus (*PFl*), in particular its lateral part, and the midline paraventricular-intermediodorsal thalamic nuclei (*PV/IMD*) in basal ganglia-thalamocortical circuits. The distinctive and common inputs of these thalamic nuclei are also shown. **a** Involvement of the PFl in cortical and striatal way stations of the 'motor circuit'. **b** Influence of the PV/IMD on one of the 'limbic circuits'. For further explanation, see text. *ABL*, basolateral amygdaloid nucleus; *DStr*, dorsal striatum; *GPi*, internal segment of globus pallidus; *HF*, hippocampal formation; *IMD*, intermediodorsal thalamic nucleus; *MD*, mediodorsal thalamic nucleus; *PFC*, prefrontal cortex; *PFl*, lateral part of parafascicular nucleus; *PV*, paraventricular thalamic nucleus; *SMC*, sensorimotor cortex; *Sth*, subthalamic nucleus; *VA*, ventral anterior thalamic nucleus; *VL*, ventral lateral nucleus; *VP*, ventral pallidum; *VStr*, ventral striatum. (Modified from Groenewegen and Berendse 1994)

the midline paraventricular-intermediodorsal thalamic nuclei project to the ventral part of the prelimbic cortical area and the medial part of the nucleus accumbens, which are specifically related to each other through corticostriatal projections, as relay stations in one of the limbic basal ganglia-thalamocortical circuits. Moreover, the paraventricular-intermediodorsal nuclei project to the hippocampal formation and the basolateral amygdaloid nucleus, both of which in turn project to the same cortical and striatal loci that are targeted directly by the thalamic nuclei (Fig. 22.171b). Other parts of the central nervous system are also linked to the trans-striatal circuits. For example, experiments using a trans-synaptic retrograde tracer have shown that, in the monkey, restricted regions of the cerebellar dentate nucleus via the thalamus project to particular prefrontal cortical areas (Middleton and Strick 1994). The general idea emerging from these studies is that fundamental processing units in the central nervous system are composed of interconnected neuronal assemblies belonging to highly different morphological (cortex, striatum, thalamus) or functional (extrapyramidal motor system, limbic system) entities of classical neuroanatomy.

Although much of the output of the basal ganglia is directed to the frontal cortex, there are several connections by which this set of subcortical cell masses have access to the somatic motor mechanisms of the brain stem and the spinal cord. These connections are much smaller than those leading to the cortex, but they are large enough to keep alive the idea of an 'extrapyramidal motor system', operating without the neocortex as intermediary. The largest of these projections is formed by fibres passing from the substantia nigra, pars reticulata to the intermediate layers of the superior colliculus, which synapse directly with collicular efferent neurons, giving rise to the predorsal bundle. Other descending 'extrapyramidal' conduction lines are formed by (a) fibres passing from the substantia nigra, pars reticulata directly to the reticular formation, (b) fibres originating from the dorsal and ventral pallidum terminating in the lateral habenular nucleus and by efferents from the latter which reach the periaqueductal gray and (c) fibres descending from the nucleus tegmentalis pedunculopontinus and the adjacent 'mesencephalic extrapyramidal area' to the medullary reticular formation (Fig. 22.167).

Although the striatum and the neocortex arise from adjacent telencephalic primordia, they differ profoundly in development and structural organisation. In the developing neocortex, the neuroblasts are guided to their destination by radial glial fibres, and in the adult most neurons and their processes are radially arranged. In the striatum, a radial guiding system is absent, the neuroblasts differentiate in situ and the neurons show no prevailing orientation. A chemoarchitechtonic differentiation into striosomes and matrix, so prominent in the striatum, is lacking in the neocortex. Another difference between neocortex and striatum concerns the 'depth' of these two structures. Although in the neocortex thalamocortical fibres synapse directly with efferent neurons, most cortical projection neurons participate either with their main axons or with collateral branches in cortico-cortical connections. Indeed, the pyramidal neurons together constitute a continuous network extending over the entire neocortex, justifying the generalisation that the neocortex communicates first and foremost within itself (Nieuwenhuys 1994). In the striatum, all major input systems (corticostriatal, thalamostriatal and nigrostriatal; see Fig. 22.168) synapse directly with the output neurons, i.e. the medium spiny cells. The axons of these elements are provided with collateral branches that impinge on other medium spiny cells. However, because the striatal output cells use GABA as a neurotransmitter and are inhibitory in nature, their collaterals are not involved in a chain-like propagation of impulses. Interneurons are present, but as far as it is known, these elements fulfil local tasks, e.g. in the communication between the matrix and the striosome compartments. Evidence for the presence of intrastriatal association paths, comparable to those in the neocortex, is lacking. A final difference between the striatum and the neocortex is that, in the former, one of the three principal afferent systems, namely the nigrostriatal dopaminergic projection, influences the activity of striatal neurons not only through synaptic transmission, but most probably also through non-synaptic transmission (Bouyer et al. 1984).

The history concerning our knowledge of the connections of the substantia innominata is less complex and less dramatic than that of the basal ganglia. Up to the mid-1960s, next to nothing was known concerning the afferents and efferents of this amorphous region of the mammalian basal forebrain. It was then discovered that this structure, and particularly the nucleus basalis of Meynert embedded therein, projects massively to the entire neocortex (Kievit and Kuypers 1975) and that this projection is at least partly cholinergic (Mesulam and van Hoesen 1976; Mesulam et al. 1983a,b). Because the basal nucleus appeared to receive afferents from the septal nuclei, the amygdala, the orbitofrontal cortex, the anterior insula and the entorhinal cortex, Mesulam and co-workers (Mesulam et al. 1983a,b; Mesulam and Mufson 1984) considered it likely that this nucleus acts as a cholinergic relay in a major extrathalamic ascending system between limbic plus paralimbic areas and the entire neocortex. Investigations by Grove (1988a,b) showed that the substantia innominata is strongly affiliated with the amygdala and with the bed nucleus of the stria terminalis, that the efferent projections of the substantia innominata largely reciprocate its afferent projections and that the cholinergic neurons within the substantia innominata are integrated into the overall afferent and efferent circuitry of the region (Fig. 22.170).

22.11.5
Amygdala

H.J. Ten Donkelaar

22.11.5.1
Introduction and Subdivision

In most mammals, the amygdala or amygdaloid complex forms the floor of the telencephalon between the olfactory tubercle and the hippocampal region. The amygdaloid complex comprises several irregularly shaped subdivisions, which gives it a thalamic appearance (Amaral 1987). The cytoarchitectonic organisation of the amygdaloid complex has been extensively studied, beginning with the studies by Volsch (1906, 1910) on insectivores and primates. Japanese investigators (Uchida 1950a,b; Koikegami 1963) followed the German tradition (see also Brockhaus 1938) of dividing the amygdaloid complex into numerous small subdivisions. Most recent studies, however, follow the American school of comparative neurology (Johnston 1923; Humphrey 1936; Crosby and Humphrey 1941, 1944), according to which the amygdaloid complex is divided into a few nuclei only. Humphrey (1936) introduced a subdivision of the amygdaloid nuclei into a corticomedial (cortical, central and medial nuclei) and a basolateral (basal and lateral nuclei) group. In contrast, extensive quantitative comparative studies led Stephan and co-workers (Stephan and Andy 1977; Stephan et al. 1987) to separate the cortical amygdaloid nucleus from the corticomedial group and to assign it to the basolateral group. Recently, Alheid et al. (1995) have introduced the term cortical-like compartment for the corticobasolateral group. [^{3}H]Thymidine data on the time of neuron origin in rodents (ten Donkelaar et al. 1979; Bayer 1980; Bayer and Altman 1987; McConnell and Angevine 1983) not only showed a clear rostrocaudal gradient within the amygdaloid complex, but also support a subdivision of the amygdala into two groups comparable with the subdivision suggested by Stephan and Andy (1977): a group of structures that arise early, including the central, medial and anterior cortical nucleus and the bed nucleus of the stria terminalis, and a group of nuclei which originate later during development, including the lateral and basal nuclei and the main part of the cortical nucleus (for data in the rhesus monkey, see Kordower et al. 1992).

There is no consistent nomenclature for the amygdaloid nuclei. For instance, the primate basal nucleus is known as the basolateral nucleus in rodents and carnivores, whereas the accessory basal nucleus of the primate brain is comparable to the basomedial nucleus (see Krettek and Price 1978b; Price et al. 1987). Crosby and Humphrey (1944) showed that the general pattern of nuclear organisation of the amygdaloid complex is more similar between the various mammalian groups than the various nomenclatures would suggest (Fig. 22.172). It appears that, as the neocortex increases, a ventromedial rotation affects the nuclei of the amygdaloid complex: in rodents and insectivores, the lateral amygdaloid nucleus is located dorsal to the basal nucleus; in the cat and rhesus monkey, it is found lateral to the basal nucleus; and in humans, it is placed ventrolateral to the basal nucleus.

Golgi studies on the amygdaloid complex are limited in number and often restricted to the basolateral nuclei (Hall 1972; Kamal and Tömböl 1975; McDonald and Culberson 1981; McDonald 1982; Braak and Braak 1983; Millhouse and de Olmos 1983). The amygdala contains a large number of neurotransmitters and peptides (see Price et al. 1987; Amaral et al. 1992).

Based on its fibre connections, the heterogeneous amygdaloid complex can be divided into a basolateral complex, which is directly connected with many cortical structures, an olfactory amygdala, which is innervated by the (main) olfactory bulb, when present, a vomeronasal amygdala, which is innervated by the accessory olfactory bulb, and a centromedial group, which is characterised by a highly organised system of pathways to many hypothalamic and brain stem areas. Recently, the extended amygdala concept was introduced (Alheid and Heimer 1988). It is based on extensive tract-tracing and immunohistochemical data that support the view of Johnston (1923) on the similarities between the bed nucleus of the stria terminalis and the central and medial amygdaloid nuclei. The extended amygdala can be further subdivided into the central extended amygdala (related to the central amygdaloid nucleus) and the medial extended amygdala (related to the medial amygdaloid nucleus). For extensive discussions of the cytoarchitecture of the amygdaloid complex, the reader is

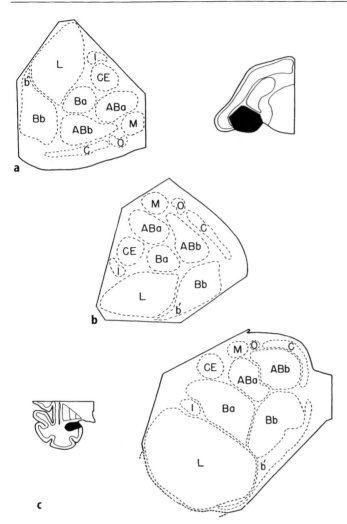

Fig. 22.172. a Amygdala of the shrew in its normal position as seen in transverse sections through the hemisphere (see key figure to show the position of the amygdala). ×30. b Amygdala of the shrew with its position changed to indicate a medial rotation of approximately 140°. ×30. c Human amygdaloid complex as seen in transverse section through the hemisphere (see key figure to illustrate the position of the human amygdala) ×5. *ABa*, accessory basal nucleus – lateral portion in humans, deep portion in the shrew; *ABb*, accessory basal nucleus – medial portion in humans, superficial portion in the shrew; *Ba*, basal nucleus, larger-celled portion; *Bb*, basal nucleus, smaller-celled portion; *b'*, the external portion (so-called superficial part in humans) of the basal nucleus outside the lateral nucleus; *C*, cortical nucleus; *CE*, central nucleus; *I*, intercalated cell mass; *L*, lateral nucleus; *M*, medial nucleus; *O*, nucleus of the olfactory tract. (From Crosby and Humphrey 1944)

directed to the detailed studies by Price et al. (1987) and Alheid et al. (1995).

22.11.5.2
Extended Amygdala

Johnston (1923) argued that the bed nucleus of the stria terminalis forms a continuous structure with the central and medial amygdaloid nuclei. Much later, the similarities between the bed nucleus of the stria terminalis and the centromedial amygdala were noted in Golgi (McDonald 1983), immunohistochemical (Roberts et al. 1982; Woodhams et al. 1983) and tract-tracing studies (de Olmos 1972; de Olmos et al. 1985; Schwaber et al. 1982; Holstege et al. 1985). Cell columns accompany the stria terminalis along its dorsal course (see Fig. 22.173). De Olmos and associates (De Olmos 1972, 1990; de Olmos et al. 1985) appreciated that such cell columns also accompany the ventral amygdalofugal pathway along its course through the sublenticular part of the substantia innominata into the ventral forebrain (see also Grove 1988a,b). The whole continuum, i.e. the centromedial amygdala with its extensions both alongside the stria terminalis and through ventrally located cell columns in the substantia innominata, has been labelled the extended amygdala. Recently, the extended amygdala has been further subdivided into the central extended amygdala and the medial extended amygdala (Alheid et al. 1995; Heimer et al. 1997), which generally form separate corridors both in the sublenticular region and along the supracapsular course of the stria terminalis. The extended amygdala is directly continuous with the caudomedial shell of the nucleus accumbens. Both structures form an extensive forebrain continuum, which establishes specific neuronal circuits with the medial prefrontal-orbitofrontal cortex and the medial temporal lobe. This continuum is particularly characterised by a prominent system of long intrinsic association fibres and a variety of highly differentiated downstream projections to the hypothalamus and the brain stem (Heimer et al. 1997).

The centromedial amygdala and the rest of the extended amygdala contain a rich assortment of peptides and neurotransmitters (see Price et al. 1987; Amaral et al. 1992). The basolateral group lacks the dense innervation by peptidergic fibres that is so characteristic of the centromedial amygdala.

22.11.5.3
Intrinsic Connections

Extensive intrinsic connections exist within the amygdaloid complex. The central nucleus receives the largest intra-amygdaloid projection, mainly from the lateral, basal and accessory basal nuclei (e.g. Krettek and Price 1978b; Ottersen 1982; Pitkänen and Amaral 1991; Amaral et al. 1992; Smith and Paré 1994; Pitkänen et al. 1995; Savander et al. 1995, 1996). Via these intrinsic connections, sensory stimuli from the environment reach the centromedial amygdala, i.e. the main output channel of the complex.

22.11.5.4
Extrinsic Connections

Three large fibre bundles, the lateral olfactory tract, the stria terminalis and the ventral amygdalofugal pathway, connect the amygdala with other parts of the brain. The *lateral olfactory tract* carries secondary olfactory fibres to the cortical and medial amygdaloid nuclei. The *stria terminalis*, i.e. the dorsal

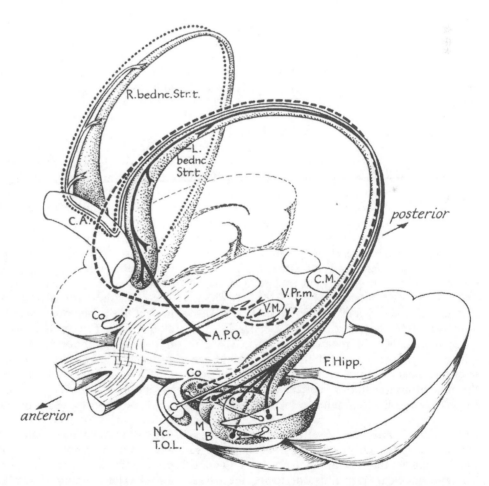

Fig. 22.173. Dorsal amygdalofugal pathway in the cat. Clearly shown is the continuum of the centromedial amygdala with the bed nucleus of the stria terminalis. *A.P.O.*, anterior preoptic area; *B, C, Co, L, M*, basal, central, cortical, lateral and medial amygdaloid nuclei; *C.A.*, commissura anterior; *C.M.*, corpus mamillare; *F.Hipp.*, hippocampal formation; *Nc.T.O.L.*, nucleus of the lateral olfactory tract; *V.M.*, ventromedial hypothalamic nucleus; *V.Pr.M.*, regio premamillaris ventralis. (From Lammers 1972)

amygdalofugal pathway (Fig. 22.173), emerges from the caudomedial aspect of the amygdala, from where it runs a long curved course to the anterior commissure. Immediately dorsocaudal to that commissure, it splits up into precommissural, commissural and postcommissural components (Lammers 1972). The stria terminalis is composed of both amygdalofugal and amygdalopetal fibres. The *ventral amygdalofugal pathway* is a large assembly of rather loosely arranged fibres which extend from the amygdaloid complex to the rostral part of the diencephalon.

Most of the extrinsic connections of the amygdala are reciprocal, with the exception of those to the striatum, the mediodorsal thalamic nucleus and several cortical fields. The extrinsic connections of the amygdala were discussed in several extensive reviews (Amaral 1987; Price et al. 1987; de Olmos 1990; Amaral et al. 1992).

Olfactory Connections. In all mammals studied, the cortical part of the amygdala is tightly linked to the olfactory system. The olfactory bulb projects directly to the nucleus of the lateral olfactory tract, the anterior cortical nucleus and the periamygdaloid cortex (Price 1973; Broadwell 1975; Scalia and Winans 1975; Skeen and Hall 1977; Davis et al. 1978; de Olmos et al. 1978; Turner et al. 1978; Meyer 1981). Apart from the direct olfactory projections, several indirect olfactory routes to the amygdala were also found (Price 1973; Krettek and Price 1978b; Luskin and Price 1983a,b). The prepiriform cortex innervates the basolateral amygdala, whereas the entorhinal cortex projects to the central, basolateral and cortical amygdaloid nuclei. In nonprimates, an accessory olfactory bulb that is innervated by the vomeronasal organ projects to a bed nucleus of the accessory olfactory tract, to the medial and posterior cortical amygdaloid nuclei and to the bed nucleus of the stria terminalis (Broadwell 1975; Scalia and Winans 1975; Skeen and Hall 1977; de Olmos et al. 1978; Davis et al. 1978; Fig. 22.160). In rodents, the nucleus of the lateral olfactory tract and the periamygdaloid cortex project back to the main olfactory bulb, whereas the medial and posterior cortical nuclei innervate the accessory olfactory bulb.

Brain Stem Connections. Brain stem neurons located in the periaqueductal grey, the substantia nigra pars compacta, the ventral tegmental area, the dorsal raphe nucleus, the locus coeruleus, the lateral parabrachial nucleus and the nucleus of the solitary tract project to the central amygdaloid nucleus (Norgren 1976; Veening 1978b; Aggleton et al. 1980; Mehler 1980; Norita and Kawamura 1980; Ottersen 1981; Russchen 1982b; Takeuchi et al. 1982). Most of these projections reach the amygdala via the ventral amygdalofugal pathway. Serotonergic and noradrenergic fibres also pas via the stria terminalis. The lateral parabrachial nucleus is by far the most important brain stem source of amygdalar afferent connections (Mehler 1980; Norita and Kawamura 1980).

The central amygdaloid nucleus gives rise to a large projection to the brain stem via the ventral amygdalofugal pathway. Hopkins (1975) first demonstrated that the amygdaloid complex in the rat, cat and rhesus monkey projects into the brain stem. In the cat, Hopkins and Holstege (1978) found that this projection originates in the central nucleus and terminates heavily on the parabrachial nucleus, the nucleus of the solitary tract and the dorsal motor nucleus of the vagus. These projections were confirmed for the rat (Krettek and Price 1978a; Takeuchi et al. 1982, 1983; Veening et al. 1984; Canteras et al. 1995), rabbit (Schwaber et al. 1982) and monkey (Price and Amaral 1981). The caudalmost fibres of this amygdalotegmental projection reach the cervical cord (Mizuno et al. 1985; Sandrew et al. 1986). The brain stem projections of the bed nucleus of the stria terminalis are virtually identical to the ones derived from the central amygdaloid nucleus (Holstege et al. 1985).

Hypothalamic Connections. Several hypothalamic structures, including the paraventricular, the ventromedial and the infundibular nuclei, as well as the lateral hypothalamic area project to the centromedial amygdala (Conrad and Pfaff 1976; Saper et al. 1976; Swanson 1976; Renaud and Hopkins 1977; Veening 1978b; Krieger et al. 1979; Norita and Kawamura 1980; Russchen 1982b). Most amygdaloid nuclei project to the hypothalamus and to the bed nucleus of the stria terminalis (see Fig. 22.173). The postcommissural fibres of the stria terminalis project to the bed nucleus of the stria terminalis, an elongated cell mass that accompanies the stria terminalis throughout most of its extent, and to the region of the anterior hypothalamic nucleus. These postcommissural fibres originate mainly from the medial, basal and lateral amygdaloid nuclei (Leonard and Scott 1971; Lammers 1972; McDonald and Culberson 1986). The precommissural fibres of the stria terminalis terminate in the medial preoptic and anterior hypothalamic areas as well as in the paraventricular and ventromedial hypothalamic nuclei (Heimer and Nauta 1969; de Olmos 1972; Lammers 1972; McBride and Sutin 1977; Berk and Finkelstein 1981). Fibres originating from the basolateral amygdala pass via the ventral amygdalofugal pathway to the lateral preoptico-hypothalamic zone

(Nauta 1961; Nauta and Haymaker 1969; Lammers 1972; Krettek and Price 1978a).

Thalamic Connections. The amygdala receives input from both dorsal and ventral thalamic regions. The dorsal thalamic nuclei that project to the amygdala include the midline nuclear complex, the parafascicular nucleus and the medial geniculate body (the perigeniculate belt area), all of which project predominantly to the central amygdaloid nucleus (Veening 1978b; Ottersen and Ben-Ari 1979; Aggleton et al. 1980; Mehler 1980; Norita and Kawamura 1980; Russchen 1982b; LeDoux et al. 1985, 1990a). The ventral thalamic peripeduncular nucleus projects to the medial cortical, central and lateral amygdaloid nuclei (Aggleton et al. 1980; Norita and Kawamura 1980; Ottersen 1981; Russchen 1982b).

The amygdala projects to both the mediodorsal thalamic nucleus and its cortical projection area, the prefrontal cortex (Nauta 1961; Krettek and Price 1974a, 1977a,b; Aggleton and Mishkin 1984; McDonald 1987; Russchen et al. 1987). In both the rat and the monkey, the projection to the mediodorsal nucleus (especially its medial, magnocellular part) arises from a population of neurons that are scattered throughout most of the amygdala (except its centromedial division).

Basal Forebrain Connections. The bed nucleus of the stria terminalis receives a heavy, topographically organised innervation from the amygdaloid complex (Krettek and Price 1978a; Price and Amaral 1981; Lehman and Winans 1983). The amygdaloid complex also projects heavily to the substantia innominata and to magnocellular cell groups such as the horizontal limb of the nucleus of the diagonal band of Broca and the basal nucleus of Meynert (Krettek and Price 1978a; Price and Amaral 1981; Záborszky and Heimer 1984; Russchen et al. 1985b; Grove 1988a). The substantia innominata gives rise to a prominent projection to the amygdala, particularly to the basal amygdaloid nucleus and the nucleus of the lateral olfactory tract (Ottersen 1980; Russchen 1982b; Nagai et al. 1982; Woolf and Butcher 1982; Carlsen et al. 1985; Russchen et al. 1985b; Grove 1988b).

Striatal Connections. The amygdaloid complex projects to both the 'limbic' or ventral striatum (the nucleus accumbens and the striatal-like zones of the olfactory tubercle) and the dorsal striatum, which includes the caudate nucleus and putamen (Krettek and Price 1978a; Kelley et al. 1982; Russchen and Price 1984; Russchen et al. 1985a; McDonald and Culberson 1986; McDonald 1991a,b). The amygdalostriatal projection appears to be topo-

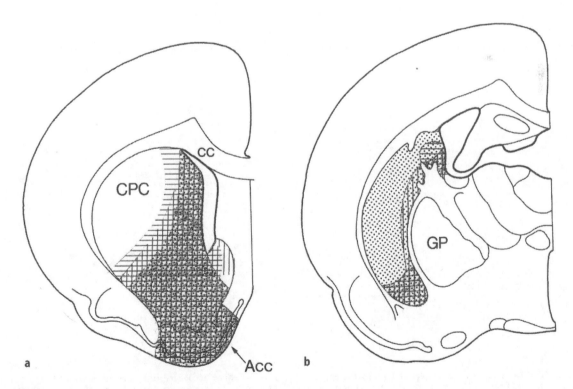

Fig. 22.174a,b. Overlapping striatal projections in the rat from the ventral tegmental area (*vertical lines*), the prefrontal cortex (*horizontal lines*) and the amygdala (*stipple pattern*). (After Kelley et al. 1982)

graphically organised in all mammals. The basolateral amygdala is the main source of amygdaloid fibres to the striatum. Basal amygdaloid complex projections to the rat nucleus accumbens are compartmentally organised (Wright et al. 1996). The amygdalostriatal projection overlaps the striatal projection from the prefrontal cortex, the ventral tegmental area and the mesencephalic raphe nuclei (Fig. 22.174). Like the amygdalostriatal projection, all of these striatal inputs avoid the anterodorsolateral striatal sector that forms the main region to which, in the rat, the corticostriatal projection from the sensorimotor cortex is distributed.

Hippocampal Connections. Prominent, and often bidirectional, projections were found between several fields of the hippocampal formation and the amygdaloid complex (Krettek and Price 1974b, 1977c; Veening 1978a; Amaral and Cowan 1980; Van Hoesen 1981; Ottersen 1982; Russchen 1982a; Aggleton 1986; Amaral 1986; Room and Groenewegen 1986; Witter and Groenewegen 1986; Canteras and Swanson 1992; Petrovich et al. 1996). The basal nucleus projects to the pyramidal cell layer of the subiculum and the periamygdaloid cortex to the plexiform layer of the subiculum, whereas the lateral nucleus and the periamygdaloid cortex innervate the lateral entorhinal cortex. The subiculum projects to the basal nucleus, whereas the entorhinal cortex projects to both the lateral and basal amygdaloid nuclei. The connections between the amygdaloid complex and the hippocampal forma-

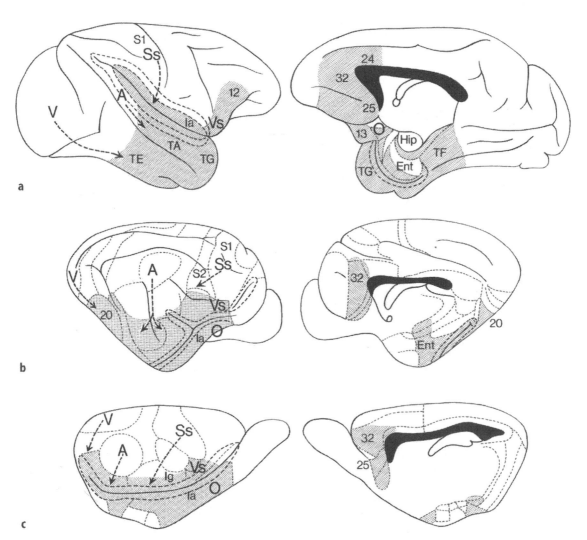

Fig. 22.175a–c. Diagrams of lateral (*left*) and medial (*right*) surfaces of the brains of **a** the monkey, **b** the cat and **c** the rat. Areas that project directly to the amygdaloid complex are indicated by *shading*. Routes by which sensory information reaches these areas are indicated by *arrows*. *A*, auditory; *O*, olfactory; *Ss*, somatosensory; *V*, visual; *Vs*, viscerosensory; *numbers*, cortical areas according to Brodmann. (After Russchen 1986)

tion may be more extensive in primates than in non-primates (Amaral 1986, 1987).

Cortical Connections. The amygdaloid cortex projects to a greater number of cortical regions than those from which it receives projections (Amaral 1987). Moreover, the primate amygdaloid complex has the most extensive connections with the neocortex. Figure 22.175 compares the sources of cortical input to the amygdaloid complex in the rat, cat and monkey. In rats, the major known cortical projections to the amygdala arise in a distinct zone along the rhinal sulcus, the medial frontal lobe and in the piriform cortex (Veening 1978a; Ottersen 1982; Turner and Zimmer 1984; Russchen 1986). Each of these regions receives a reciprocal connection from the amygdala. The amygdalocortical projections arise primarily from the basal amygdaloid complex and to a lesser extent from the lateral and cortical nuclei (Krettek and Price 1974a, 1977b; Saper 1982; Sarter and Markowitsch 1983, 1984). Sripanidkulchai et al. (1984) demonstrated a projection from the basal nucleus to the sensorimotor cortex. In the cat, the amygdaloid complex receives projections from the piriform cortex, the perirhinal cortex, the medial frontal cortex and the anterior and posterior sylvian gyri (Druga 1970; Heath and Jones 1971; Kamal and Tömböl 1976; Russchen 1982a, 1986). Retrograde tracer studies showed that the amygdaloid complex projects to much of the frontal cortex, including the motor and premotor cortices (Llamas et al. 1977), as well as to primary and secondary somatosensory and auditory cortices (Macchi et al. 1978).

In primates, the frontal, temporal and insular cortices project to the amygdaloid complex (Whitlock and Nauta 1956; Pandya et al. 1973; Herzog and Van Hoesen 1976; Leichnetz and Astruc 1977; Aggleton et al. 1980; Turner et al. 1980; Van Hoesen 1981; Mufson et al. 1981; Mufson and Mesulam 1982). The anterior half of the temporal lobe contributes the largest component of the cortical input to the amygdaloid complex. The amygdala receives input from the modality-specific association areas one or more steps removed from the primary sensory areas (Turner et al. 1980). The major amygdalar targets are the following: for vision, the anterodorsal parts of the lateral, basal and basal accessory nuclei; for audition, the posterior parts of the lateral and basal accessory nuclei; for taste, the medial parts of the lateral and basal nuclei; and for olfaction, the cortical and medial nuclei. Thus each part of the amygdala is under the influence of a particular sensory system.

Amygdalocortical projection fields include vast areas of the frontal, temporal, insular and occipital cortices (Jacobson and Trojanowski 1975; Potter and Nauta 1979; Mizuno et al. 1981; Mufson et al. 1981; Porrino et al. 1981; Tigges et al. 1982, 1983; Avendaño et al. 1983; Amaral and Price 1984; Iwai and Yukie 1987). Amaral and Price (1984) showed that all major divisions of the temporal neocortex receive a projection from the amygdaloid complex, with the most prominent projections ending in the cortex of the temporal pole and the rostral superior temporal gyrus. They also found a strong projection to peristriate regions of the occipital lobe.

22.11.5.5
Functional Aspects

Anatomically, the amygdala is a complex and heterogeneous structure. The basolateral amygdaloid complex is directly connected with many cortical structures and receives visual, auditory, somatosensory and gustatory input (Ben-Ari et al. 1974; Sanghera et al. 1979; Turner et al. 1980). Olfactory and vomeronasal input is restricted to the cortical and medial amygdaloid nuclei and the nuclei of the lateral and accessory olfactory tracts. Via the extensive intrinsic connections within the amygdaloid complex, these sensory stimuli from the environment reach the centromedial amygdala. The extended amygdala (the centromedial amygdala, the bed nucleus of the stria terminalis and related structures) gives rise to a highly organised system of pathways to many hypothalamic and brain stem areas and forms the main output channel of the amygdaloid complex. The extended amygdala is ideally suited to generate endocrine, autonomic and somatomotor aspects of emotional and motivational states (Heimer et al. 1997).

Electrical stimulation of a variety of loci in the amygdaloid complex of the cat, the central nucleus in particular, initially leads to an orienting response, which is followed either by flight or defence behaviours (Kaada 1972). The following motor and autonomic concomitants of this stimulation were noted: arrest of spontaneous behaviour followed by searching movements, retraction of the nictitating membrane and pupillary dilation, piloerection, micturition, growling, hissing, posturing for attack, elevation of blood pressure with bradycardia, respiratory alterations, changes of gastric motility and secretion, and masticatory movements with sniffing. The defence reaction is elicited most easily by stimulation of the basomedial amygdaloid nuclei, but the basolateral and central nuclei are also positive sites (Hilton and Zbrozyna 1963). Hilton and co-workers found a column of effective sites for elicitation of the defence reaction or its autonomic correlates which stretches from the

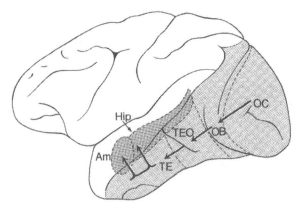

Fig. 22.176. Lateral view of the monkey brain showing one route for the cortical processing of visual information from the striate cortex (*OC*) to the inferotemporal cortex (*TE*). Area TE, like other higher-order cortical sensory areas, projects to the amygdala (*Am*) and to the hippocampus (*Hip*) via the hippocampal gyrus and perirhinal cortex. The fact that medial temporal lobe lesions, which include the hippocampus and the amygdala, cause amnesia but save many premorbid memories suggests that the saved memories are stored upstream from the lesion, i.e. in the neocortex. (After Mishkin 1982)

amygdala through the hypothalamus and into the pontomedullary brain stem as far as the level of the dorsal motor nucleus of the vagus nerve (Abrahams et al. 1962; Hilton and Zbrozyna 1963; Coote et al. 1973).

The amygdala also plays an important role in motivation, memory and visual recognition (Aggleton 1992; LeDoux 1992). Monkeys with lesions of the amygdala are insensitive to stimuli that normally arouse intense fear (Klüver and Bucy 1939; Weiskrantz 1956; for data in rats, see Blanchard and Blanchard 1972) and cannot be conditioned to associate stimuli with fear (LeDoux et al. 1988, 1990b; Davis 1992a,b). Both the monkey (Leonard et al. 1985; Rolls 1992) and human (see Adolphs et al. 1995) amygdala contain neurons that respond selectively to faces. The behavioural impairments of amygdalectomised monkeys are most severe in the wild, resulting in the death of the animal if left unattended in its troop (Kling and Brothers 1992). Neurons within the amygdala are active during social interactions (Kling et al. 1979) and social communication (Jürgens 1982). The integration of complex sensory information, such as the recognition of faces with the motivational valence of the stimulus, is especially important to recognise social intentions and social status and to guide behaviour on the basis of social cues such as threats, warnings and submissive gestures (Kling and Brothers 1992).

The amygdala plays a role in the processing of long-term memory, apparently in concert with the hippocampal formation (Fig. 22.176). In the monkey, Mishkin (1978, 1982) demonstrated that bilateral ablation of the amygdala or the hippocampal formation does not markedly impair the animal's performance on visual recognition memory tests. When the amygdala and hippocampal formation were conjointly ablated, however, the animals showed a very severe memory impairment, which is apparent in both visual and tactile modalities (Murray and Mishkin 1984; Zola-Morgan and Squire 1985). The amygdaloid complex appears to be particularly important for learning associations of stimuli in different modalities (Murray and Mishkin 1985).

22.11.6
Cerebral Cortex

R. Nieuwenhuys

22.11.6.1
Introduction and Subdivision

The mammalian pallium contains a well-developed cortex throughout its extent. This cortex has been divided on structural, connectional and phylogenetic grounds into an olfactory piriform cortex or palaeocortex, a hippocampal cortex or archicortex and a neocortex or isocortex (Ariëns Kappers 1909; Ariëns Kappers et al. 1936). These three types of pallial cortex will be dealt with in the present section. However, in order to be able to place the data and relations to be discussed in a comparative perspective, I will begin with a brief overview of the structure of the pallium in some non-mammalian groups. It should be emphasised that the literature on the cerebral cortex is vast and that what follows has no claim at completeness whatsoever. Readers who are particularly interested in the structure and functions of the cortex are referred to the comprehensive, multi-author work entitled *Cerebral Cortex*, edited by E.G. Jones and A. Peters, of which 12 volumes have appeared so far.

22.11.6.2
Notes on the Pallium of Some Non-mammalian Groups

The *amphibian pallium* (see also Chap. 18, Sect. 18.10.3 and Chap. 19, Sect. 19.10.3.1) can be subdivided into a lateral primordium piriforme, a primordium pallii dorsalis and a medial primordium hippocampi (Herrick 1927, 1933, 1948; Figs. 22.143a, 22.177). In all of these three fields, the great majority of the neuronal perikarya are situated within a zone of periventricular grey. In the lateral and dorsal fields, this grey is confined to the inner half of the pallial wall. Only in the medial field does the cellular zone extend more peripherally.

In all parts of the amphibian pallium, the central grey is mainly composed of spherical or triangular neurons, which extend a tuft of moderately branching, spiny dendrites peripherally. The dendrites originate either directly from the soma or from a short dendritic trunk (Fig. 22.177). The axons of these cells usually emanate from the peripheral or the lateral aspect of the soma or from the proximal-most part of a dendrite. According to Cajal (1911), the great majority of these axons ascend toward the peripheral zone of the pallium, where they pass, either transversely or longitudinally, over considerable distances parallel to the meningeal surface. They issue collaterals in both the central grey and the peripheral fibre zone. Cajal (1911) and Sanides and Sanides (1972) considered the elements described above as the precursors of the pyramidal neurons occurring in the reptilian and the mammalian pallium. The peripherally extending dendrites of the amphibian neurons, according to Sanides and Sanides, are directly comparable to the apical dendritic bouquet of true pyramidal cells. These authors considered the presence of a set of peripherally directed dendrites as a conservative feature and the development of basal dendrites as a progressive feature in pallial neuron evolution. In addition to the cells discussed above, the amphibian pallial grey contains small Golgi type II neurons with very short and highly branched axons (Herrick 1927).

The peripheral zone of the amphibian pallium is occupied by a neuropil and fibre layer. Throughout this layer, scattered fusiform neurons occur, the dendrites and axons of which spread parallel to the external pallial surface (Herrick 1927; Cajal 1911; P. Cajal 1922; Fig. 22.177f–h). There is evidence to suggest that the external layer of the amphibian pallium is divisible into a peripheral zone consisting of extrinsic afferents and a deeper zone containing mainly intrapallial associational fibres. The submarginal zone of the hippocampal primordium contains the fibres of a large afferent system, the fasciculus olfactorius septi (Fig. 22.177). According to Herrick (1927, 1948), this bundle originates *inter alia* from the septal and anterior olfactory nuclei and terminates in the hippocampal and dorsal pallial primordia.

The lateral part of the pallium receives a large, direct projection from the olfactory bulb by way of the dorsolateral olfactory tract (Fig. 22.177). Experimental neuroanatomical studies have shown that the fibres of this projection are confined to the superficial part of the pallial fibre layer (Northcutt and Royce 1975; Northcutt and Kicliter 1980; Neary 1990). The idea that the deeper part of the amphibian pallial fibre layer is mainly occupied by intrapallial associational fibres is derived from the extensive Golgi studies carried out by Herrick (1927, 1933a, 1948), but has received support from the experimental neuroanatomical studies by Scalia (1976) and Neary (1990). These studies revealed that, in the frog, fibres interconnecting the three pallial fields pass mainly through the deeper part of the fibre layer of these structures.

Most of the axons in the amphibian pallium are unmyelinated and may be expected to enter into synaptic contact with all of the dendrites that are

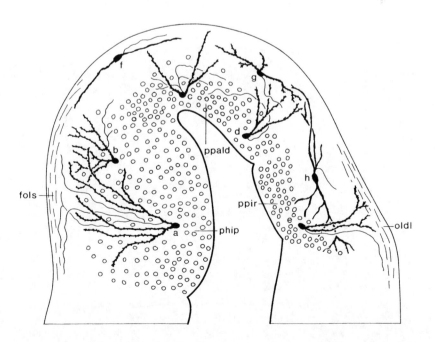

Fig. 22.177. Transverse section through the pallial part of the right cerebral hemisphere of the tiger salamander *Ambystoma tigrinum*. *fols*, fasciculus olfactorius septi; *oldl*, tractus olfactorius dorsolateralis; *phip*, primordium hippocampi; *ppald*, primordium pallii dorsalis; *ppir*, primordium piriforme. Golgi images of periventricular (*a–e*) and horizontal neurons (*f–h*) redrawn from Herrick (1927)

passed at short distance. If this is the case, the periventricular neurons with the longest peripherally extending dendrites presumably receive three more or less sharply separated, proximodistally arranged inputs. The soma and the proximalmost parts of the dendrites will be in contact with local collaterals of other periventricular neurons and with the short axons of Golgi type II cells, whereas the intermediate and distal parts of these dendritic trees will receive their main inputs from intrapallial association fibres and from extrapallial afferents, respectively. In contrast, the tangentially arranged horizontal cells (Fig. 22.177f–h) are presumably monoreceptive, i.e. their input is confined to either the inner, intrinsic or the outer, extrinsic fibre zone. Their axons may be expected to impinge upon other horizontal cells and upon the dendrites of periventricular neurons. It would be interesting to know whether the latter contacts are intermingled with those from other sources or rather occupy a separate segment of the receptive surface of the periventricular cells.

The *pallium of dipnoan fishes* (see also Chap. 16, Sect. 16.9.3) contains two cell layers, an inner zone of periventricular cells and an outer zone which is clearly detached from the periventricular grey (Elliot Smith 1908; Holmgren 1922; Holmgren and van der Horst 1925). The Golgi material studied by Rudebeck (1945) revealed that the periventricular pallial zone is composed of three different kinds of cells, which I designate here as types 1–3, whereas the layer of migrated cells consists of a single type, to be termed type 4 (Fig. 22.178).

The very common type 1 cells have a short stem dendrite which gives rise to a number of peripherally directed, very spiny branches. Most of these branches extend into the superficial fibre layer of the pallium.

The rare type 2 cells are characterised by a rather long principal dendrite, which arborises in a fairly widespread area. The main dendrite and its branches are very slender, and some of the latter attain the superficial fibre zone of the pallium. The axons of these elements could be traced over only a short distance; they are directed toward the ventricular surface.

The type 3 cells are confined to the external parts of the periventricular layer. These elements closely resemble those present in the external cell layer and represent, according to Rudebeck, cortical elements left behind in their matrix layer.

The external cell layer is composed of rather large elements, which in many places are gathered into clusters. Their long, smooth dendrites spread mainly tangentially (t4 in Fig. 22.178). Branches of these dendrites arborise in the superficial pallial fibre zone. According to Rudebeck, the type 4 cells contribute axons to both the external and the internal pallial fibre layer.

The fibre connections of the dipnoan pallium have not been studied using experimental techniques so far. Rudebeck (1945) reported, on the basis of an analysis of Bielschowsky, Bodian and Golgi material, that the internal fibre layer contains secondary olfactory fibres and axons as well as collaterals from periventricular and cortical cells. According to him, the external fibre layer is composed of secondary olfactory fibres and axons of cortical cells and also contains most of the fibres connecting the pallium with other brain parts.

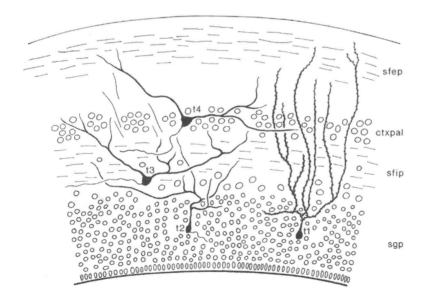

Fig. 22.178. Transverse section through the pallium of the lungfish *Protopterus annectens*. *ctxpal*, cortex pallii; *sfep*, stratum fibrosum externum pallii; *sfip*, stratum fibrosum internum pallii; *sgp*, stratum griseum periventriculare; *t1-t4*, cell types described in the text. Golgi images of neurons redrawn from Rudebeck (1945)

The data available suggest that the structure of the dipnoan pallium is similar to that of amphibians. The dipnoan type 1 cells closely correspond to the amphibian periventricular neurons. The main difference between the pallia of these two groups seems to be that, in the dipnoans, horizontal cells are more numerous than in amphibians and that, in the former group, part of these elements together constitute a distinct separate layer.

The *pallium of reptiles* (see also Chap. 20, Sect. 20.11.3) contains a well-differentiated cerebral cortex throughout its extent (Fig. 22.143b,c). This cortex is composed of three layers: (1) an external plexiform layer, which contains relatively few scattered neurons, (2) an intermediate cellular layer, containing the somata of most cortical neurons and (3) an internal plexiform layer, in which a moderate number of somata are found. A sheet-like aggregation of cells, known as the cell plate (cp in Fig. 22.143c), occurs at one or several places in the internal plexiform layer of lizards and snakes. The periventricular zone of the medial parts of the pallium in all reptiles is occupied by a compact fibre layer called the alveus. Cytoarchitectonic differences allow a subdivision of the reptilian cortex into four longitudinally arranged zones: the medial cortex, dorsomedial cortex, dorsal cortex and lateral cortex (Fig. 22.143b,c).

Golgi studies (Cajal 1911; P. Cajal 1917; Ebbesson and Voneida 1969; Ulinski 1974, 1976, 1977, 1979, 1990; Ulinski and Rainey 1980; Wouterlood 1981; Berbel 1988; Berbel et al. 1987) have shown that the cerebral cortex in reptiles contains several types of neurons. Prominent among these are the pyramidal cells, a type of neurons that occurs only in the reptilian and the mammalian cerebral cortex.

A typical pyramidal cells has a conical cell body from the apex of which a radially oriented stout apical dendrite arises that ascends through the cortex. At a shorter or longer distance from the soma,

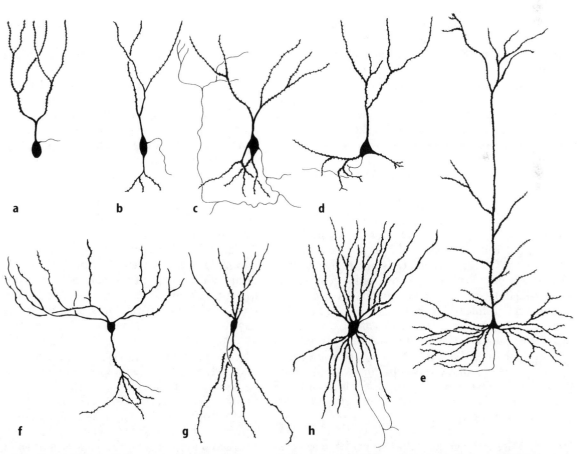

Fig. 22.179a-h. Pyramidal cells. **a** Periventricular cell from the primordium piriforme of a salamander. **b** Fusiform cells from the medial cortex of a lizard. **c** Pyramidal cells from the medial cortex of a chameleon. **d** Pyramidal cells from the lateral cortex of a snake. **e** Typical pyramidal cell from the third layer of the human prefrontal cortex. **f** Sparsely spinous horizontal cell from the medial cortex of a lizard. **g,h** Double pyramidal cells from the dorsomedial cortex of a lizard. Redrawn from illustrations by **a,d** Berbel (1988), **b,f,g,h** Cajal (1911) and **c,e** Ulinski and Rainey (1980)

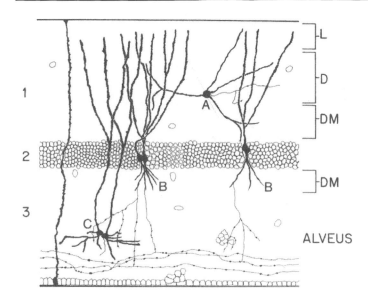

Fig. 22.180. Organisation of the medial cortex of a snake. Semi-diagrammatic depiction of neurons as they appear in Golgi preparations. *Open profiles* indicate the distribution of somata as observed in Nissl preparations. The laminar distribution of afferents from other parts of the cortex is indicated to the *right*. A, stellate cell; B, candelabra cells; C, pyramidal cells; D,DM, L, afferents from the dorsal, dorsomedial and lateral cortex, respectively. *1*, Lamina plexiformis externa; *2*, lamina intermedia; *3*, lamina plexiformis interna. (From Ulinski 1977)

this dendrite breaks up into an apical tuft with branches into the outermost cortical layer. A number of so-called basal dendrites radiate from the broad other side of the soma and arborise in its vicinity. All the dendrites of the pyramidal cells, except for their proximalmost parts, are densely covered with spines. The axon emerges from the base of the soma and descends towards the white matter (Fig. 22.179e). Cells having all these features have been observed in all of the four cortical zones of the reptilian pallium. The perikarya of most of these elements are situated in the intermediate layer (Fig. 22.179c,d), but the somata of some deep pyramids are embedded in the cell plate (C in Fig. 22.180). In the cerebral cortex of reptiles and mammals, cells occur which closely resemble the pyramidal cells described above, but lack one or two of the features characterising these neurons. For example, many of the pyramidal cells in the superficial layers of the mammalian cortex have only a short apical dendrite or even lack this process entirely, so that they are essentially multipolar in form. Nevertheless, these elements are termed pyramidal cells because they are considered to stand at the end of a series of pyramidal cells whose apical dendrites become progressively shorter as their cell bodies lie increasingly closer to the cortical surface (Peters 1987).

It has already been mentioned that Cajal (1911) and Sanides and Sanides (1972) considered the typical pyramidal cells as having developed from the common periventricular neurons which populate the amphibian pallium (Figs. 22.177, 22.179a). The former author pointed out that in the reptilian pallium numerous cells occur which morphologically stand intermediate between periventricular and pyramidal cells. Such 'intermediate' elements are provided with a peripherally directed tuft of spiny dendrites, which closely resembles the dendritic system of periventricular neurons. However, the lower part of the somata of these elements gives rise to a single centrally directed dendrite that is initially smooth, but after a short distance divides into some spiny branches that enter the peripheral zone of the internal plexiform layer. Elements fitting this description are abundant in the medial cortex of lizards and snakes. Ulinski (1977) described these elements as candelabra cells (B in Fig. 22.180), whereas Berbel et al. (1987) termed them spinous pyramidal cells (Fig. 22.179b). The development of the basal dendrites of these elements, and of the pyramidal cells as well, is probably related to the peripheralward migration of their somata and to the consequent appearance of the inner plexiform layer as a new synaptic domain. In this context, the development of basal dendrites can be considered as a progressive feature (Sanides and Sanides 1972). The axons of the reptilian pyramidal and 'intermediate' elements descend through the inner plexiform layer and reach the periventricular fibre layer, i.e. the alveus. Cajal (1911; see also P. Cajal 1917) pointed out that the axons of pyramidal cells emit three types of collaterals: (1) recurrent collaterals that ascend to the external plexiform layer, (2) horizontal collaterals that ramify in the internal plexiform layer and (3) long associational collaterals, which course in a direction opposite to that of the main axon. Ulinski (1977) observed that the axons

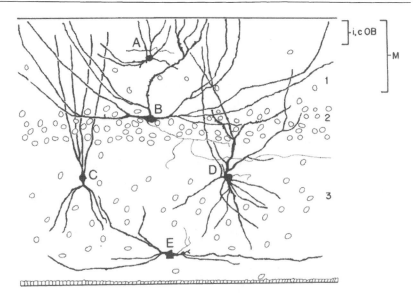

Fig. 22.181. Organisation of the lateral cortex of a snake. Semi-diagrammatic depiction of neurons as they appear in Golgi preparations. *Open profiles* indicate the distribution of somata as observed in Nissl preparations. The laminar distribution of afferents from the ipsilateral and contralateral olfactory bulb (*i,cOB*) and from the medial cortex (*M*) is indicated to the *right*. A, stellate cell; B, bowl cell; C, 'displaced' double pyramidal cell; D, large stellate cell; E, horizontal cell. 1, 2, 3, Cortical layers as indicated in Fig. 2.84. (From Ulinski and Rainey 1980)

of the candelabra cells, after having reached the alveus, bifurcate into a medial and a lateral branch (Fig. 22.180). The medially directed branch gives rise to efferent projections to the septum and hypothalamus; the lateral branch (which most probably corresponds to an associational collateral of Cajal) gives rise to intracortical projections to the dorsomedial, dorsal and lateral cortical areas.

Neurons of the types described so far are mainly found in the intermediate layer. Golgi studies have revealed that this layer of the reptilian cortex contains several other types of cells, most of which can be viewed as variants of one of the members of the series periventricular cell-intermediate cell-pyramidal cell (Fig. 22.179a-e). Three of these cell types, i.e. the bowl cell, the sparsely spinous horizontal cell and the double pyramidal cells, will be briefly discussed.

Bowl cells, which were first described by Ulinski and Rainey (1980), are very common in the lateral cortex of snakes. These neurons have fusiform somata that give rise to an exceptionally wide dendritic arbor, extending into the external plexiform layer (B in Fig. 22.181). The proximal shafts of the dendrites of these elements are smooth, but their more distal primary and secondary branches are densely covered with spines. The axons of bowl cells descend into the internal plexiform layer, where they collateralise extensively. One branch of each cell is efferent from the lateral cortex. Axons of elements situated in the dorsal part of the lateral cortex course dorsally and pass along the dorsal surface of the pallium, ending in the medial cortex (Ulinski and Rainey 1980). The dendrites of most bowl cells emanate only from the lateral and dorsal aspects of the soma, but some are provided with a short basal dendrite (Ulinski and Rainey 1980). The elements of the former group can be considered as variants of the periventricular cell type (Fig. 22.179a), whereas those of the latter group belong to the intermediate cell type (Fig. 22.179b).

Sparsely spinous horizontal cells occur in the medial cortex of lizards (Berbel et al. 1987). The cell bodies of these elements are located at the upper border of the intermediate layer. These neurons send most of their dendrites peripherally into the external plexiform layer, either directly or as branches from tangentially oriented initial dendritic segments (Fig. 22.179f). A centrally directed dendrite with two or three secondary branches arises from the basal aspect of these cells. Like the basal dendrite bearing bowl cells, these spinous horizontal cells belong to the intermediate cell type.

Finally, the double pyramidal cells are provided with two conical dendritic arbors, one ascending into the external plexiform layer and the other descending into the internal plexiform layer (Fig. 22.179 g,h). The axons of these elements descend into the internal plexiform layer, where they collateralise. The dorsomedial cortex of lizards and snakes is characterised by the presence of large double pyramidal cells, with numerous very spiny dendrites (Ebbesson and Voneida 1969; Ulinski 1979; Berbel et al. 1987; Fig. 22.179 h). The axons of some of these conspicuous elements pass medially in the alveus and participate in the formation of a commissural connection with the contralateral dorsomedial cortex (Lohman and Mentink 1972; Ulinski 1976). 'Displaced' double pyramidal cells have been found in the internal plexiform layer of vari-

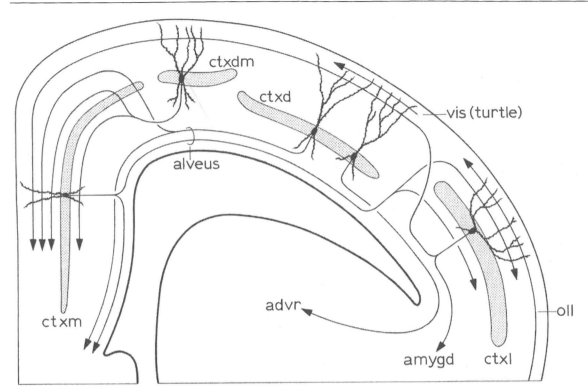

Fig. 22.182. Transverse section through the pallial part of the telencephalon of a reptile, showin the organisation of the principal connections of the various cortical fields. *advr*, anterior dorsal ventricular ridge; *amygd*, amygdala; *ctxd*, cortex dorsalis; *ctxdm*, cortex dorsomedialis; *ctxl*, cortex lateralis; *ctxm*, cortex medialis; *oll*, tractus olfactorius lateralis; *vis*, visual thalamocortical projection

ous parts of the reptilian cortex (Ulinski 1990; C in Fig. 22.181).

In the plexiform layers, two types of neurons prevail, i.e. the stellate cells and the horizontal cells.

Small stellate cells occur in the external plexiform layer of all parts of the reptilian cortex. These elements have spherical somata and bear dendrites that radiate out in all directions. The axon arborises within the vicinity of the cell's dendritic field (A in Figs. 22.180, 22.181). Large stellate cells have been observed in the lateral cortex of snakes (Ulinski and Rainey 1980; D in Fig. 22.181). Some of the dendrites of these large elements pass through the cellular layer and extend into the external plexiform layer.

Horizontal cells have been observed in both the external and the internal plexiform layer, but are most numerous in the latter. These elements have fusiform somata and tangentially spreading dendrites and axons (E in Fig. 22.181).

It is important to note that most of the neurons occurring in the external and internal plexiform layers can be grouped as non-pyramidal cells and that the dendrites of these elements are generally aspiny or sparsely spiny.

Immunohistochemical studies (Schwerdtfeger and López-García 1986; Schwerdtfeger and Lorente 1988a,b; Shen and Kriegstein 1987; Martinez-Guijarro et al. 1991) have revealed that most neurons whose somata are located in the external and internal plexiform layer are GABAergic. It has also been shown that, in the dorsal, dorsomedial and medial parts of the cortex, GABAergic boutons are concentrated in the cellular intermediate layer and that these terminals form baskets around the somata present in that layer. Moreover, it has been demonstrated that these GABAergic boutons form symmetric (and thus presumably inhibitory) contacts with their target somata. Taken together, these data indicate that, in the reptilian cortex, typical basket cells are present. These elements use GABA as a neurotransmitter and probably exert an inhibitory influence on the somata of their target neurons. Physiological evidence suggests that these elements form part of pathways for both feedforward and feedback inhibition of pyramidal and pyramidal-like neurons, the somata of which are situated in the intermediate layer (Kriegstein and Connors 1986).

One other type of local circuit neurons may be briefly mentioned. Golgi studies (Berbel et al. 1987;

López-García et al. 1988) have shown that the inner plexiform layer of the medial cortex of lizards contains a population of large multipolar neurons, the dendrites of which bear distinctive, long spines. These elements use GABA as a neurotransmitter and are hence presumably inhibitory (López-García et al. 1988). Their axons pass peripherally and ramify in the external plexiform layer. Because this layer contains only very few neuronal elements in the medial cortex, it is reasonable to assume that the axons of the long-spined multipolar neurons mainly contact the peripherally extending dendrites of the pyramidal-like principal neurons of the medial cortex. The axons of the latter elements contain zinc, and since it has been established that most of the boutons terminating on the long-spined multipolar elements accumulate this metal, it seems likely that, in the medial cortex of lizards, the long-spined GABAergic neurons establish recurrent inhibitory circuits with the principal neurons (López-García and Martinez-Guijarro 1988; López-García et al. 1988).

With regard to the fibre connections of the various parts of the reptilian cortex, I will confine myself to a number of extrinsic and intrinsic afferents which show a distinct laminar organisation (Fig. 22.182)

The lateral cortex receives afferents from the ipsilateral and contralateral olfactory bulbs, which terminate in the most superficial zone of the external plexiform layers (Ulinski and Rainey 1980; Ulinski and Peterson 1981; Skeen et al. 1984; Lohman et al. 1988; Fig. 22.181). In turtles, the dorsal cortex receives a substantial visual projection from the dorsal lateral geniculate nucleus (Hall and Ebner 1970; Heller and Ulinski 1987; Mulligan and Ulinski 1990). The fibres of this projection terminate in the outer half of the external plexiform layer. Ultrastructural studies have shown that these geniculocortical afferents form asymmetric junctional complexes with the dendritic spines of pyramidal cells and with the smooth dendritic shafts of stellate cells, with each stellate cell receiving about six times more thalamic synapses than each pyramidal cell (Ebner and Colonnier 1975; Smith et al. 1980).

Experimental studies have shown that the four cortical areas are interconnected by several discrete association connections (Lohman and Mentink 1972; Lohman and van Woerden-Verkley 1976; Ulinski 1976; Ulinski and Rainey 1980; Skeen et al. 1984; Hoogland and Vermeulen-van der Zee 1988). Thus the lateral cortex receives a projection from the medial cortex. In snakes, this projection terminates throughout most of the external plexiform layer, but in turtles it involves the inner part of the

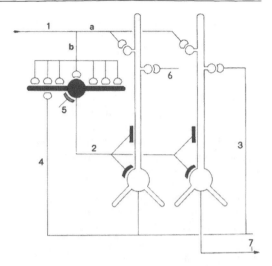

Fig. 22.183. Principal intracortical connections of the dorsal cortex of the turtle. Visual thalamic afferents (*1*) make excitatory contacts with pyramidal cell dendrites (*a*) and with inhibitory stellate cells (*b*). Feedforward inhibition is mediated by contacts between stellate cells and pyramidal cells (*2*). Recurrent collaterals of pyramidal cells mediate reciprocal excitation between pyramidal cells (*3*), as well as feedback inhibition of pyramidal cells via stellate neurons (*4*). Stellate cells are presumably inhibited by contacts of other stellate elements (*5*). Non-thalamic afferents make excitatory contacts with the apical dendrites of pyramidal cells (*6*). Axons of the pyramidal cells leave the cortex via the subependymal fibre zone (*7*). (From Kriegstein and Connors 1986)

external plexiform layer, thus complementing the olfactory projections discussed above, and part of the inner plexiform layer. A remarkably distinct laminar organisation of association afferents has been observed in the medial cortex. In this part of the cortex, the afferents from the other cortical fields appeared to project to restricted subzones of the external plexiform layer, the afferents from the lateral cortex to the outer subzone, those from the dorsal cortex to the intermediate subzone and those from the dorsomedial cortex to the inner subzone and to the internal plexiform layer. Because the external plexiform layer of the medial cortex contains only a small number of neurons, most of the afferents terminating in this layer may be expected to impinge on the dendrites of the candelabra cells (Fig. 22.180).

The extrinsic efferents from the various parts of the reptilian cortex, which like the cortico-cortical connections arise from pyramidal and pyramidal-like neurons, will not be discussed here. Suffice it to mention that the lateral cortex sends fibres to the amygdala and that both the dorsal and the medial cortex project mainly to the septum. The dorsal cortex also projects to the anterior dorsal ventricular ridge. No projections from the dorsomedial cortex to subcortical structures have been reported

(Hoogland and Vermeulen-van der Zee 1988; Ulinski 1990).

Little is known about the intrinsic circuitry of the reptilian cortex. However, some interesting data on the dorsal cortex of the turtle are available. By combining neuroanatomical data provided by Smith et al. (1980) with the results of physiological and pharmacological experiments, Kriegstein and Connors (1986) and Kriegstein (1987) established the following features of the microcircuitry of this cortex (Fig. 22.183):

1. Visual thalamic afferents passing through the superficial zone of the external plexiform layer directly excite both pyramidal neurons and inhibitory stellate cells.
2. Feedforward inhibition is mediated by the axons of stellate cells impinging upon pyramidal elements.
3. Local pathways, presumably provided by recurrent collaterals of pyramidal cells, mediate reciprocal excitation between pyramidal cells and feedback inhibition through contacts of pyramidal cells with stellate cells.
4. The interneurons involved in both feedforward and feedback inhibition use GABA as a neurotransmitter.
5. There is physiological support for inhibition of stellate neurons, presumably arising from contacts between stellate cells.
6. Non-thalamic afferent systems make excitatory contacts with the apical dendrites of pyramidal cells all along their length.
7. The pyramidal cells provide output from the cortex by way of axons passing through the subependymal zone.

If we survey data on the structural and functional organisation of the reptilian pallium presented above, the following conclusions seem to be warranted:

1. The reptilian pallium contains a well-differentiated cortex throughout its extent.
2. This cortex displays a simple, three-layered pattern consisting of a cellular intermediate layer sandwiched between an external and in internal plexiform layer.
3. The cellular layer contains the somata of radially arranged, spiny pyramidal or pyramidal-like neurons, which extend an apical dendritic tree into the external plexiform layer and a basal dendritic tree into the internal plexiform layer. These neurons occupy a central position in the cortical circuitry and may hence be designated as principal cells.
4. The plexiform layers are primarily neuropil zones, but also contain scattered local circuit neurons, the dendrites of which are generally aspiny or sparsely spiny. Many of these elements are inhibitory and use GABA as a neurotransmitter. One group of GABAergic inhibitory interneurons are provided with axons that form baskets around the somata of principal neurons, whereas the axons of another group ascend from the internal plexiform layer, where their somata are situated, to the external plexiform layer.
5. Both extrinsic and intracortical association afferents are tangentially distributed and terminate in distinct sublaminae of the plexiform layers, making excitatory synapses onto principal cells and local circuit neurons.
6. In parts of the cortex that are in receipt of extrinsic sensory projections, the fibres constituting these projections are confined to the superficial zone of the external plexiform layer.
7. Consecutive segments of the radially oriented dendritic trees of the principal cells are in synaptic contact with different afferent systems.
8. Local interneurons that are activated by external afferents mediate feedforward inhibition of principal cells.
9. The axons of the principal neurons descend to the deep fibre layer, where they bifurcate into two branches; one branch leaves the cortex, whereas the other participates in the formation of cortico-cortical association systems. The latter are re-excitatory to principal neurons at long distances.
10. The initial part of the axons of principal neurons gives rise to local collaterals. Interneurons activated by these local collaterals mediate lateral and feedback inhibition of principal cells.

We will now turn to the principal types of mammalian cortex, i.e. the olfactory piriform cortex, the hippocampal cortex and the neocortex. It will be seen that all of these structures have many features in common with the reptilian cortex.

22.11.6.3
Piriform Cortex

The piriform cortex, i.e. that part of the cortex that is in receipt of direct projections from the olfactory bulb, is very large in macrosmatic mammals, such as the opossum (Fig. 22.143d). In primates and in osmatic cetaceans, it is small and inconspicuous but, remarkably, in anosmatic cetaceans it is not entirely lacking (see Chap. 22, Sect. 22.11.2.2 and Fig. 22.157c). In the preparation of the ensuing survey of the structure and circuitry of this part of the mammalian cortex, we have relied heavily upon two reviews by Haberly (1990a,b), but some previous studies (Cajal 1911; O'Leary 1937; Valverde 1965; Stevens 1969) have also been consulted.

The piriform cortex comprises three layers, which are usually designated as laminae I, II and III (Fig. 22.184).

Lamina I is a superficial plexiform layer, which contains only sparsely scattered neurons. It is mainly composed of tangentially passing axons and of the ascending dendritic expansions of pyramidal and semilunar cells. The plexiform layer can be subdivided into layer Ia, a superficial sublamina dominated by afferent fibres from the olfactory bulb, and layer Ib, a deep sublamina dominated by association fibres from other parts of the piriform cortex and other cortical areas.

Lamina II consists of densely packed neuronal somata. It can be subdivided into a superficial lamina IIa, in which the somata of semilunar cells are concentrated, and a deep sublamina IIb, in which pyramidal cell somata prevail.

Lamina III displays a moderate cell density. It contains pyramidal neurons, which tend to be concentrated in the superficial part, and multipolar neurons, which are most common in the deeper parts.

The neurons in the piriform cortex can be classified as belonging to one of two basic types: pyramidal cells and non-pyramidal cells. Two populations of typical pyramidal cells can be distinguished: superficial pyramidal cells, which have their somata in lamina IIb, and deep pyramidal cells, which have their somata in lamina III. All pyramidal neurons are provided with apical and basal dendritic tufts, but the lengths of their apical dendritic trunks are determined by the depth of their somata. The axons of the pyramidal cells descend radially through lamina III and issue numerous collaterals. Short collaterals course in all directions, establishing synaptic contacts with both pyramidal and non-pyramidal neurons. Long collaterals ascend to the superficial plexiform layer, where they take a tangential course. It has been experimentally established that long collaterals from pyramidal neurons situated in the anterior part of the piriform cortex pass to the posterior piriform cortex, where they are concentrated in a superficial sublayer of lamina Ib, whereas long collaterals from the posterior piriform cortex project predominantly to the deeper parts of lamina Ib in the anterior and posterior parts of the piriform cortex (see Fig. 22.186). The distribution of synaptic contacts over the receptive surface of the pyramidal neurons in the piriform cortex closely resembles that observed on pyramidal neurons in other parts of the cortex: asymmetrical synapses with associated round vesicles are

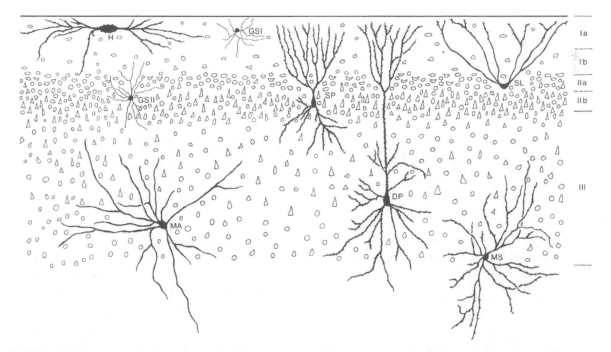

Fig. 22.184. Cytoarchitecture and major cell types of the piriform cortex. *DP*, deep pyramidal cell; *GSI*, globular soma cell in layer I; *GSII*, globular soma cell in layer II; *H*, horizontal cell; *M*, multipolar cell; *MA*, multipolar aspiny neuron; *MS*, multipolar spiny neuron; *SG2*, small Golgi type 2 cell; *SP*, superficial pyramidal cell; *SL*, semilunar cell. (Modified from Haberly 1990a and Haberly 1990b)

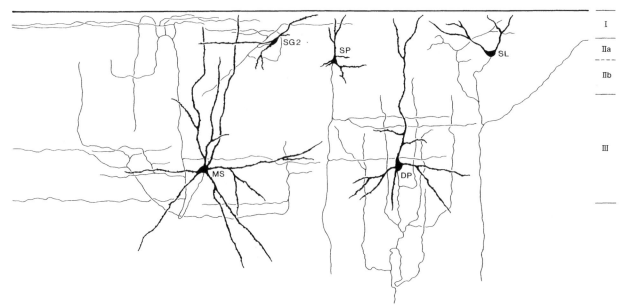

Fig. 22.185. Neurons in the piriform cortex of a 3-day-old mouse. Golgi technique. *MS*, multipolar spiny neuron; *SG2*, small Golgi type 2 cell; *SP*, superficial pyramidal cell; *DP*, deep pyramidal cell; *SL*, semilunar cell. (Redrawn from Valverde 1965)

concentrated on dendritic spines, but are lacking on the somata; a symmetrical synapses with associated pleiomorphic vesicles are found at a very high density on the initial axonal segments and at a moderate density over all other parts, including perikarya and both apical and basal dendrites. The axons of the pyramidal cells in the piriform cortex, like those in other parts of the cortex, make typical asymmetric synapses with their target neurons.

It has already been mentioned that in the superficial part of lamina II the somata of semilunar cells are concentrated. These elements emit dendrites or groups of dendrites from the poles of their crescent-shaped cell bodies. These dendrites extend peripherally and ramify in the plexiform layer. Their axons descend radially through the cortex to enter the underlying white matter. Although the semilunar elements differ at first sight considerably from typical pyramidal cells, they should still be considered as members of the pyramidal cell group. The reasons for this are threefold: (1) like pyramidal cells, they have spiny dendrites and radially oriented centrally directed axons, (2) ultrastructurally, their somata closely resemble those of pyramidal cells (Haberly and Feig 1983) and (3) they resemble what are presumed to be the phylogenetic precursors of pyramidal cells occurring in amphibians and reptiles (Fig. 22.179).

As in other parts of the cortex, the non-pyramidal cells form a heterogeneous group. Most, though not all of these neurons are provided with smooth dendrites, and Golgi studies suggest that their axons generally do not extend beyond the cortex. The following types of non-pyramidal neurons may be briefly mentioned.

Large *horizontal cells* are found exclusively in lamina Ia. The somata of these cells bear spines that presumably receive synaptic input from secondary olfactory fibres. All of these elements contain GABA and its synthesising enzyme glutamic acid decarboxylase.

Small *stellate cells* with globular somata have been observed throughout the piriform cortex, but in layers I and II they are by far the most frequently occurring type of non-pyramidal cell. A high percentage of these elements have been found to contain GABA and glutamic acid decarboxylase (Haberly et al. 1987).

Medium-sized and large *multipolar cells* are abundant in the deeper parts of lamina III. These elements have long, radiating dendrites some, of which may reach the superficial layers. Their axons give rise to many local collaterals that can extend to the surface of the cortex (Valverde 1965; Tseng and Haberly 1989; MS in Fig. 22.185). Golgi studies (Haberly 1983) and experiments with intracellular dye injections (Tseng and Haberly 1989) have revealed the presence of different types of multipolar cells, including elements with smooth or beaded dendrites and elements with spiny dendrites (Fig. 22.184). A large population of the multipolar cells use GABA as their neurotransmitter.

A detailed discussion of the interneuronal relationships and the functional organisation of the

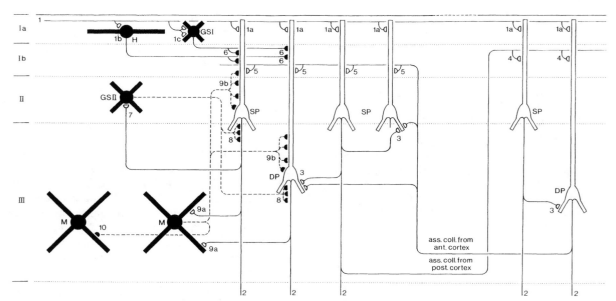

Fig. 22.186. Circuitry of the piriform cortex. Secondary olfactory fibres (*1*) make excitatory synaptic contacts with the apical dendrites of pyramidal cells (*1a*) and with horizontal (*1b*) and superficial stellate cells (*1c*). Main axons of pyramidal cells leave the cortex (*2*). Local collaterals of pyramidal cells make excitatory synaptic contacts with the basal dendrites of other pyramidal cells (*3*). Long associational collaterals of pyramidal cells ascend to lamina Ib, where they make excitatory synaptic contacts with the apical dendrites of other pyramidal cells (*4, 5*). Feedforward inhibition is mediated by horizontal and superficial stellate cells (*6*). Stellate cells in layer II receive an excitatory input via collaterals of pyramidal cells (*7*). A certain proportion of the stellate cells in layer II may give rise to chandelier axons, which terminate in inhibitory axo-axonic contacts on the initial segments of pyramidal cells (*8*). Axon collaterals of pyramidal cells make excitatory synaptic contacts with multipolar cells (*9a*), which in turn are thought to make inhibitory synaptic contacts with pyramidal cells (*9b*). Thus stellate cells and multipolar cells are both assumed to be involved in feedback inhibition. Multipolar cells are inhibited by other multipolar cells (*10*). *H*, horizontal cell; *GSI*, globular soma cell in layer I; *GSII*, globular soma cell in layer II; *SP*, superficial pyramidal cell; *M*, multipolar cell; *DP*, deep pyramidal cell. (Based on Haberly 1990b and Tseng and Haberly 1989)

piriform cortex is beyond the scope of the present survey. However, it is felt appropriate to present a brief synopsis of these aspects. The numbers in this synopsis, which is entirely based on the reviews by Haberly (1990a,b) already mentioned, correspond to those in the summary diagram presented in Fig. 22.186:

1. The mitral cells in the olfactory bulb are the principal source of afferent input to the piriform cortex. These afferents pass via the lateral olfactory tract to lamina Ia of that cortex, making excitatory synapses with the apical dendrites of pyramidal cells (1a), with GABAergic horizontal cells (1b) and with superficial small stellate cells (1c).
2. Main axons of pyramidal cells leave the cortex.
3. Local axon collaterals of superficial pyramidal cells make excitatory synaptic contacts with the basal dendrites of other superficial and deep pyramidal cells.
4. Long associational collaterals of superficial pyramidal cells in the anterior part of the piriform cortex descend to the deepest part of the cortex and then ascend in the posterior part of the piriform cortex to the superficial zone of lamina Ib, where they make excitatory synapses with the apical dendrites of superficial and deep pyramidal cells.
5. Long associational collaterals of deep pyramidal cells in the posterior part of the piriform cortex descend to the deepest part of the cortex, ascending in the anterior piriform cortex to the deep zone of lamina Ib, where they make excitatory synaptic contacts with the apical dendrites of superficial and deep pyramidal cells.
6. It seems likely that the axons of the horizontal cells and the superficial stellate cells impinge upon the apical dendrites of pyramidal cells and mediate feedforward inhibition.
7. Small stellate cells situated in layer II presumably receive axosomatic contacts from superficial pyramidal cells.
8. The initial segments of the pyramidal cells are densely covered with symmetrical synapses. It seems likely that these contacts are made by the axons of one variant of small stellate cells present in layer II. If this assumption is confirmed,

these elements would occupy a position in the circuitry of the piriform cortex comparable to that of the chandelier cells in the hippocampus and the neocortex (Somogyi et al. 1982, 1983, 1985).
9. Physiological studies have shown that interneurons in the deeper parts of the cortex form part of feedback loops. It seems likely that multipolar cells receiving an excitatory synaptic input from superficial and deep pyramidal cells (9a) and making inhibitory synapses with elements of the same types (9b) subserve this function.
10. Physiological evidence indicates that feedback inhibitory interneurons inhibit each other.

22.11.6.4
Hippocampus

22.11.6.4.1
Subdivision

The hippocampus or hippocampal formation is a large C-shaped structure that forms part of the medial wall of the cerebral hemispheres (Figs. 22.143d, 22.187). During ontogenesis, this structure rolls in on itself along a longitudinal groove, the fissura hippocampi. This infolding causes the hippocampus to protrude into the lateral ventricle. The hippocampal formation constitutes the archipallial part of the cerebral hemisphere; it contains a relatively simple, three-layered allocortex throughout its extent. Within the hippocampus, three longitudinally arranged structures, the fascia dentata, the cornu Ammonis (Ammon's horn) and the subiculum, can be distinguished.

The *fascia dentata* forms the morphologically medialmost part of the cerebral cortex. Its three layers are from external to internal: the stratum moleculare, stratum granulare and stratum multiforme. Due to the strong curvature of the fascia dentata, the stratum multiforme occupies a central position within that structure (Fig. 22.187). The stratum granulare, as its name indicates, is made up of the densely packed somata of granule cells. The spiny dendrites of these elements enter the molecular layer, where they branch extensively (Fig. 22.188). In the molecular layer, the dendrites of the granule cells enter into synaptic contact with afferents from the entorhinal cortex and with fibres ascending from the multiform layer. The afferents and terminals from the entorhinal cortex are concentrated in the outer two thirds of the molecular layer, whereas those from the multiform layer occupy the inner third of the molecular layer. The axons of the granule cells pass through the polymorphic layer and enter the cornu Ammonis as mossy fibres. In the cornu Ammonis, these mossy fibres enter into synaptic contact with the dendrites of pyramidal cells.

The *cornu Ammonis* forms the largest part of the hippocampal formation. In addition to a limited number of small elements, its intermediate cellular layer contains the somata of pyramidal cells, which form by far the most numerous class of neurons in the cornu Ammonis. These elements closely resemble their counterparts in other cortices. Their morphology has already been dealt with (see Chap. 2, Sect. 2.7.2 and Fig. 2.39a,b). Each hippocampal pyramidal cell has a stout apical dendrite, which splits up at a shorter or longer distance from the soma into a number of terminal branches. The apical stem dendrites make up the stratum radiatum, whereas the layer in which the terminal branches of the apical dendrites are situated is known as the stratum lacunosum-moleculare. The basal dendrites of the pyramidal cells ramify in the stratum oriens (Fig. 22.188). The axons of the hippocampal pyramidal cells enter a compact subependymal fibre layer known as the alveus. A certain proportion of these axons descend via the fornix to the lateral septal nucleus, others pass to the entorhinal cortex, but most remain within the hippocampal formation.

In addition to a dense neuropil, the internal and external plexiform layers of the cornu Ammonis, i.e. the stratum oriens and the stratum radiatum plus the stratum lacunosum-moleculare, contain a variety of scattered local circuit neurons.

Differences in size and packing density of the pyramidal cells allow a subdivision of the cornu Ammonis into four fields, which have been traditionally designated CA1-CA4. Area CA1 is adjacent to the subiculum, whereas area CA4 is situated in the so-called hilar region of the fascia dentata, where it coincides largely or entirely with the multiform layer of that structure.

The *subiculum* forms the third part of the hippocampal formation. Its three layers are designated as the stratum moleculare, stratum pyramidale and stratum multiforme (Stephan 1975). In many species, including the rat, the pyramidal cell layer is in the subiculum much wider than in the adjacent Ammon's horn (Fig. 22.187). Three transitional or mesocortical areas, i.e. the presubiculum, the parasubiculum and the area entorhinalis, separate the allocortical hippocampal formation from the neocortex. In the area entorhinalis, two principal cell layers (known as the lamina principalis interna and the lamina principalis externa), which are separated by a cell-free lamina dissecans, can be distinguished. Each of these two principal layers can be subdivided into a number of separate laminae (for details, see Lorente de Nó 1934a; Stephan 1975; Braak 1980). The two principal layers, present in the entorhinal cortex, can also be distinguished in the parasubiculum and the presubiculum. It is the lamina principalis interna which continues into the cell layer of the subiculum and the cornu Ammonis (Blackstad 1967; Fig. 22.187).

There is a vast literature on the cytology, ultrastructure, connections and functional significance of the hippocampal formation. No attempt will be made here to summarise all of the data collected. However, within the frame of the present survey, it is felt appropriate to discuss two related topics, i.e. the classical trisynaptic hippocampal circuit and the microcircuitry of the cornu Ammonis, particularly area CA1. A review by Brown and Zador (1990) has been very helpful in the preparation of this discussion.

22.11.6.4.2
Classical Trisynaptic Circuit

The circuit considered to be the morphological substrate of the main flow of information through the hippocampus begins with a projection from the area entorhinalis to the fascia dentata, is relayed to field CA3 of the cornu Ammonis and then proceeds to field CA1. Projections from field CA1 to the subiculum and from the subiculum to the entorhinal area close this circuit (Swanson 1982; Fig. 22.188).

The fibres destined for the fascia dentata arise from the superficial layers of the entorhinal area

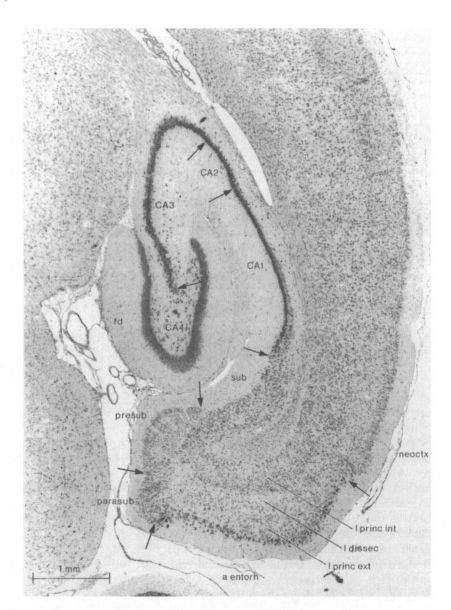

Fig. 22.187. Horizontal section through the caudal part of the right cerebral hemisphere of the rat, showing the hippocampal formation and adjacent cortical areas. Nissl stain. *a entorh*, area entorhinalis; *CA1-CA4*, fields of the cornu ammonis; *fd*, fascia dentata; *l dissec*, lamina dissecans; *l princ ext*, lamina principalis externa; *l princ int*, lamina principalis interna; *neoctx*, neocortex; *parasub*, parasubiculum; *presub*, presubiculum; *sub*, subiculum

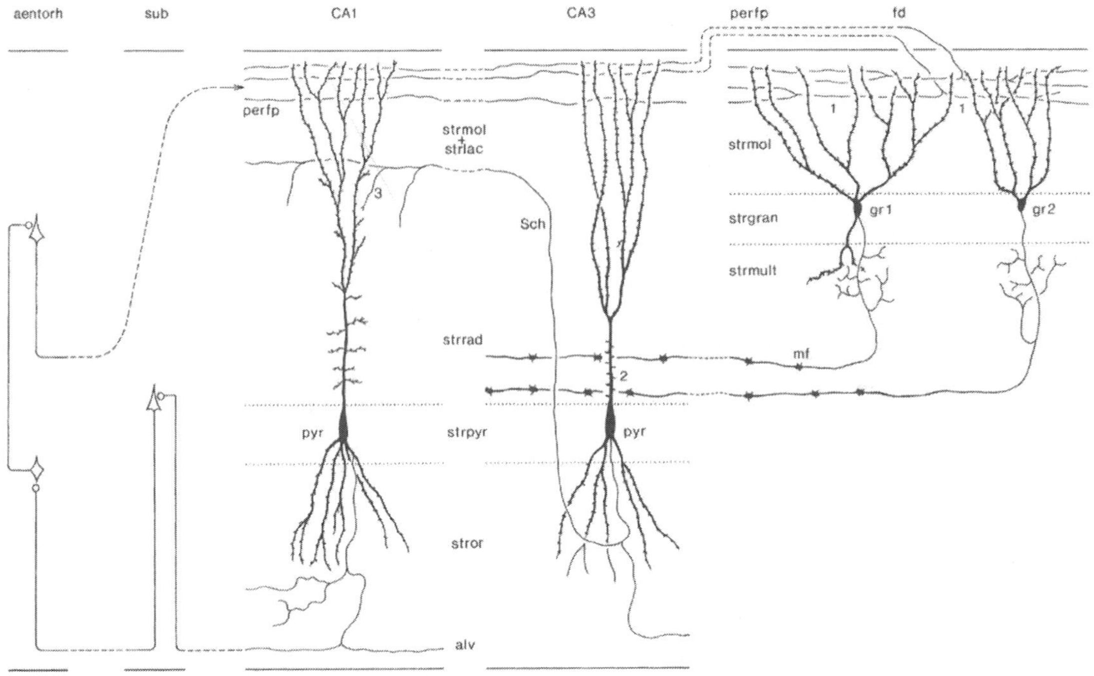

Fig. 22.188. Elements participating in the formation of the principal hippocampal circuit. *Numbers* indicate the location of the three sequential excitatory synapses which define the circuit. *aentorh*, area entorhinalis; *alv*, alveus; *CA1, CA3*, fields of Ammon's horn; *fd*, fascia dentata; *gr1, gr2*, granule cells; *mf*, mossy fibres; *perfp*, perforant path; *pyr*, pyramidal cell; *strgran*, stratum granulare; *strlac*, stratum lacunosum; *strmol*, stratum moleculare; *strmult*, stratum multiforme; *stror*, stratum oriens; *strpyr*, stratum pyramidale; *strrad*, stratum radiatum; *Sch*, Schaffer collateral; *sub*, subiculum. The neuronal elements depicted are modified from Cajal (1911)

and constitute the so-called temporo-ammonic or perforant path. They terminate in the outer two thirds of the molecular layer of the fascia dentata, where they form excitatory synapses onto the dendritic spines of the granule cells. The axons of the granule cells issue numerous collaterals in the hilus region of the fascia dentata and then pass as mossy fibres to the stratum radiatum of field CA3, in which they form a separate suprapyramidal sublayer known as the stratum lucidum. The mossy fibres synapse with the large pyramidal cells in CA3. The proximal part of the apical dendrites of the latter are provided with large, thorny excrescences, and these are embraced by the large, irregular expansions in the mossy fibres to which their name refers. The dendritic excrescences and the axonal expansions together form huge synapse complexes in which several active zones are present. These active zones are of the type I variety, i.e. they are characterised by the presence of spherical synaptic vesicles and asymmetric membrane specializations (Blackstad and Kjaerheim 1961; Hamlyn 1962). The main axons of the large CA3 pyramidal cells pass to the alveus. However, during their descent through the stratum oriens, these axons issue coarse collateral branches, which penetrate the stratum pyramidale and pass to area CA1, where they form a compact sheet of fibres in the stratum lacunosum. Within area CA1, these so-called Schaffer collaterals issue short branches which enter into synaptic contact with the dendrites of pyramidal and probably also non-pyramidal neurons. Physiological experiments have shown that the Schaffer collaterals exert an excitatory action on CA1 pyramidal cells; consistent therewith, the Schaffer collateral synapses, like those of the mossy fibres, are of the type I variety. As the last neuronal link in Ammon's horn, the CA1 pyramidal cells project heavily to the subicular complex. There is a strong projection from the subicular complex to the deep layers of the entorhinal cortex, and these in turn project to the superficial layers of the same cortical area. As has already been mentioned, it is these superficial layers of the entorhinal areas which give rise to the perforant path. The locations of the three sequential excitatory synapses which define the classical trisynaptic hippocampal circuit are indicated in Fig. 22.188.

22.11.6.4.3
Microcircuitry of Ammon's Horn

As regards the microcircuitry of Ammon's horn, it should be emphasised that the highly characteristic pyramidal cells with their large somatodendritic surface should be placed in a central position in any consideration of the interneuronal relationships within this centre. Experimental neuroanatomical studies (for reviews, see Stephan 1975; Swanson 1982; Swanson et al. 1987) have shown that, in field CA1, afferents from the entorhinal cortex terminate on the terminal branches of the apical dendrites, whereas the proximal and intermediate parts of the apical dendritic trees are contacted by septal and commissural afferents, respectively. The basal dendrites of the pyramidal cells are also in receipt of afferents from the septum and from the contralateral hippocampus. The axons of the pyramidal cells issue numerous collaterals in the stratum oriens (r in Fig. 22.189), where they contact *inter alia* the basal dendrites of other pyramidal neurons (Lorente de Nó 1934b). Moreover, we have just seen that two sets of intrahippocampal association fibres, i.e. the mossy fibres and the Schaffer collaterals, are arranged in distinct sublayers and impinge upon particular segments of the apical dendritic systems of the pyramidal neurons (Figs. 2.39a,b, 22.188). In studying the hippocampal afferents in Golgi material, Cajal (1911) observed fibres ascending from the alveus to the stratum moleculare and others which branch mainly in the stratum lacunosum (a,b in Fig. 22.189). If these fibres, the origin of which is unknown, were proven to contact the apical dendrites of the pyramidal cells, they would constitute additional laminary segregated afferent systems to these elements.

Two types of hippocampal local circuit neurons, i.e. the basket cells and the recently discovered axo-axonic or chandelier cells (Somogyi et al. 1983, 1985), are also known to impinge upon particular segments of the receptive surface of the pyramidal cells. The basket cells have long been known to invest with their terminal branches the somata of the hippocampal pyramids (Cajal 1911; Lorente de Nó 1934b), whereas the axo-axonic cells, as their name indicates, make remarkable serial synapses with the initial axonal segments of the same elements. Before further discussing these elements and their integration in hippocampal circuitry, some general remarks on the hippocampal local circuit neurons will be made.

Although, in the hippocampus, local circuit Golgi type 2 neurons are much less numerous than pyramidal cells (according to Dietz and Frotscher, they form only 12% of the total hippocampal neuronal population; unpublished observations quoted in Misgeld and Frotscher 1986), these elements show a remarkable structural diversity. An inventory based on the classical Golgi studies by Cajal (1911) and Lorente de Nó (1934) showed that, in the hippocampus, some 18 different types of interneurons are present. A sample of these, encompassing 11 types, supplemented with an axo-axonal cell as described by Somogyi et al. (1985), is shown in Fig. 22.189. As regards this inventory the following cautionary remarks should be made: (a) it includes cell types observed in different species (Cajal: rabbit; Lorente de Nó: mouse; Somogyi et al.: cat), (b) both Cajal and Lorente de Nó studied exclusively material from very young animals, (c) there is evidence that non-pyramidal cells in the hippocampus may give rise to long axons (Chronister and DeFrance 1979; Alonso and Kohler 1982; Hayes and Totterdell 1985; Ino et al. 1988; Buhl et al. 1989) and (d) in the recent literature, numerous hippocampal local circuit neurons are shown, based on intracellular staining with HRP or nuclear yellow or on treatment with histochemical techniques (e.g. Misgeld and Frotscher 1986; Lacaille et al. 1987; Lacaille and Schwartzkroin 1988a,b; Kawaguchi and Hama 1987; Matthews et al. 1987). It is remarkable that some of these cells are not similar to any of those previously described. Notwithstanding these caveats, we feel that our inventory warrants the following general conclusions. (The letters included refer to elements depicted in Fig. 22.189):

1. Interneurons occur in all of the six hippocampal layers.
2. Many hippocampal interneurons are with regard to the spread of their dendrites (e, g, h, j–n), axons (d, e, g, h, j–q), or both (h, k, n, m, l) mainly confined to a single layer. Thus, in Ammon's horn, lamination is not merely a cytoarchitectonic or myeloarchitectonic feature, but is also manifest in the disposition of the dendritic and axonal systems of many of its constituent neurons. As regards the pyramidal cells, the fact that the axons of several types of interneurons distribute their axons to one single layer or sublayer could imply that not only the extrinsic, commissural and associational fibres and the axon of the basket cell (see Fig. 2.39a,b), but also several other types of hippocampal local circuit neurons terminate at particular, restricted levels of the receptive surface of these elements.
3. Not only pyramidal cells (Figs. 2.39a,b, 22.188, 22.189), but also several types of interneurons extend their dendritic systems more or less radially through most or all of the hippocampal layers (d, f, o–q), thus potentially contributing to

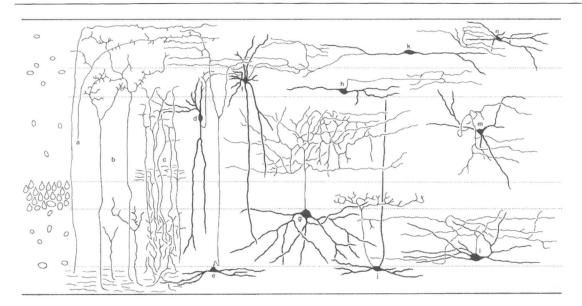

Fig. 22.189.

interlaminar radial coupling. It is remarkable that the axonal ramifications of all of these interneurons are confined to a single layer.

4. Some elements apparently fulfil a specific role in radial interlaminar coupling in that their dendritic and axonal systems spread in radially matched zones of two different layers (e, g, j).
5. Intralaminar radial coupling also occurs. Thus element l interconnects the deep and the superficial zone of the stratum oriens.
6. Basket cells occur in all hippocampal layers except for the stratum moleculare and with regard to both their dendritic and axonal systems show a remarkable multiformity. Two typical basket cells are shown in Fig. 22.189 (j, p), and in Fig. 22.190 all hippocampal neuron types whose axons participate in the formation of pericellular baskets in the stratum pyramidale, described and depicted by Cajal (1911) and Lorente de Nó (1934b), are represented semidiagrammatically. It will be seen that the dendritic trees of some of these elements are mainly confined to the stratum oriens (a, b), that those of others are vertically oriented spanning several layers (e, f), whereas those of still others comprise a system of basal dendrites and large apical dendrites (c, d), which make them resemble pyramidal neurons. As regards the distribution of their terminal axonal ramification, it seems likely that some elements participate mainly in the formation of baskets situated in the superficial zone of the stratum pyramidale (e), whereas others focus on the deep zone of the same layer (c, d).

If we survey the elements depicted in Figs. 22.188–22.190, it will be clear that, in Ammon's horn, the possibilities for interneuronal communication are virtually infinite. What is actually known from the local circuitry of Ammon's horn (particularly area CA1) is shown diagrammatically in Fig. 22.191. With regard to the synaptic relations of interneurons, this diagram is almost entirely based on the combined physiological and morphological studies performed by Schwartzkroin and Mathers (1978), Schwartzkroin and Kunkel (1985), Lacaille and Schwartzkroin (1988a,b) and Lacaille et al. (1987, 1989). In these studies, which were carried out on hippocampal slice preparations, intracellular recordings from neurons were made and these elements were subsequently labelled with intracellular markers and studied at the light and electron microscopy level. The data summarised in Fig. 22.191 may be commented upon as follows. (The numbers and letters in the text correspond to those in the figure.)

1. Perforant path fibres originating from the entorhinal cortex pass through the molecular layer and establish excitatory synaptic contacts with pyramidal cells (1a) and so-called L-M interneurons (1b). Ultrastructurally, fibres passing through the molecular layer have also been found to make synaptic contact with the dendrites of other non-pyramidal cells, possibly basket cells (1c) and so-called O/A interneurons (1d).
2. Afferent fibres passing through the alveus and the stratum oriens make excitatory synaptic contacts with pyramidal cells (2a) and with various

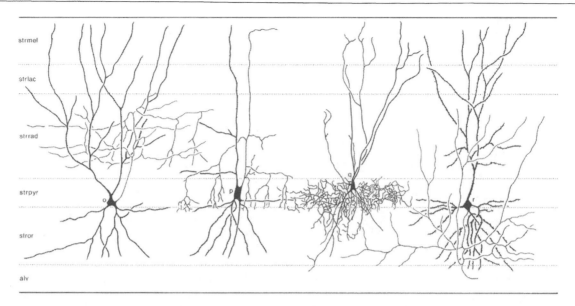

Fig. 22.189. Afferent fibres (*a–c*), local circuit neurons (*d–q*) and a pyramidal cell (*r*), as observed in Golgi material of Ammon's horn in different animals. *Far left*, cytoarchitecture as observed in Nissl preparations. *strmol*, stratum moleculare; *strlac*, stratum lacunosum; *strrad*, stratum radiatum; *strpyr*, stratum pyramidale; *stror*, stratum oriens; *alv*, alveus. Compiled from Cajal (1911: 8- to 30-day-old rabbits), Lorente de Nó (1934b: 12-day-old mice) and Somogyi et al. (1985: adult cat)

classes of interneurons (2b–d). It is known that fibres originating from the medial septal nucleus and from the contralateral hippocampus pass through the alveus and terminate mainly in the stratum oriens and the stratum radiatum (see Fig. 2.39a,b). However, the nature and the precise course of these fibres after they have left the alveus is unknown. Cajal (1911) depicted several kinds of fibres terminating in one or both of the layers mentioned (c in Fig. 22.189).

3. Afferents passing through the stratum radiatum make excitatory synaptic connections with the apical dendrites of pyramidal cells (3a) and with different types of interneurons (3b–d). In field CA1, this group of fibres probably consists of Schaffer collaterals (Fig. 22.188), but may also include commissural fibres.

4. The main axons of the CA1 pyramidal cells (Fig. 22.191: pyr) pass to the alveus, where they bifurcate into a coarse branch passing to the subiculum and a smaller branch projecting toward the fimbria (Knowles and Schwartzkroin 1981).

5. The initial part of the axons of the pyramidal cells issue numerous local collaterals, which ramify in the stratum oriens and to a lesser extent in the stratum radiatum (q in Fig. 22.189). These collaterals make excitatory synapses on basket cells (5b) and on so-called O/A interneurons (5c). Lorente de Nó (1934b; see Fig. 2.39a) concluded from his Golgi studies that the collaterals of pyramidal cells also contact the basal dendrites of other pyramids (5a). MacVicar and Dudek (1980) presented electrophysiological evidence for direct excitatory interaction between CA3 pyramidal cells.

6. With regard to the shape of their dendritic trees, basket cells (Fig. 22.191: b) situated near the border of the strata pyramidale and oriens resemble the hippocampal pyramidal neurons and were hence designated by Cajal (1911) and Lorente de Nó (1934b) as pyramidal basket cells (p in Fig. 22.189, c,d in Fig. 22.190). The dendrites of these elements are aspinous and varicose; their axonal processes, which also display periodic enlargements, form dense, basket-like plexuses around the somata of pyramidal cells. Ultrastructural studies have shown that the axonal terminals of the pyramidal basket cells make symmetrical contacts with the somata and proximal apical and basal dendrites of the pyramidal elements (6a). Contacts onto other interneurons were also found (6b,c; Schwartzkroin and Kunkel 1985). The pyramidal basket cells are under the synaptic control of many sources (Schwartzkroin and Mathers 1978; Knowles and Schwartzkroin 1981; Schwartzkroin and Kunkel 1985). They receive direct excitatory inputs from pyramidal cells (5b) and from afferents in the alveus (2b) and in the stratum radiatum (3b). The latter include commissural fibres (Frotscher and Zim-

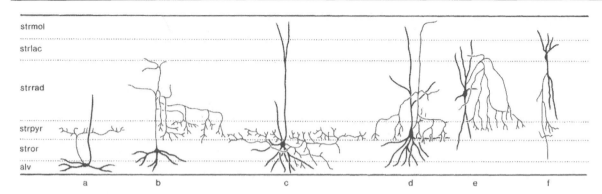

Fig. 22.190. Hippocampal neuron types, the axons of which participate in the formation of pericellular baskets in the stratum pyramidale, as depicted by Cajal (1911) and Lorente de Nó (1934b). *strmol*, stratum moleculare; *strlac*, stratum lacunosum; *strrad*, stratum radiatum; *strpyr*, stratum pyramidale; *stror*, stratum oriens; *alv*, alveus

mer 1983; Frotscher et al. 1984) and most probably also Schaffer collaterals. Other interneurons to be discussed below, i.e. the O/A cells and possibly also the L/M cells, make inhibitory synapses with the pyramidal basket cells (7b, 9c). The pyramidal basket cells inhibit pyramidal cells (6a). Ultrastructurally, it has been observed that synaptic contacts (possibly from basket cell collaterals) are made on non-pyramidal dendrites (synapse 6b on O/A interneuron; synapse 6c on L/M interneuron). Neurons closely resembling pyramidal basket cells have been found to be immunoreactive for GABA (Gamrani et al. 1986) or glutamic acid decarboxylase (Ribak 1978; Kunkel et al. 1986). Thus the pyramidal basket cells are inhibitory interneurons which most likely use GABA as a neurotransmitter. The axon collaterals of pyramidal cells terminating on the pyramidal basket cells (5b) and the latter elements with their axonal ramifications impinging upon the pyramidal cells (6a) together form a pathway for recurrent or feedback inhibition, whereas the direct inputs from hippocampal afferents to the pyramidal basket cells (2b, 3b) have been shown to form part of a feedforward inhibitory circuit (Knowles and Schwartzkroin 1981).

7. Lacaille et al. (1987, 1989) and Lacaille and Williams (1990) obtained intracellular recordings and subsequent intracellular labelling of local circuit neurons, the somata of which are situated in area CA1 at the border of the stratum oriens and the alveus. The smooth or beaded dendrites of these so-called O/A interneurons course parallel to the alveus, but one or two dendrites turn abruptly and ascend through the strata oriens, pyramidale and radiatum. Their axons branch profusely in the stratum pyramidale. These elements closely resemble the cells with horizontal axons described by Cajal (1911) and depicted here in Figs. 22.189 (j) and 22.190 (a). Their morphology and the data concerning their ultrastructure, synaptic connections and electrophysiological properties provided by Lacaille et al. (1987, 1989) all indicate that these O/A interneurons represent a particular type of basket cell.

As indicated in Fig. 22.191, the O/A interneurons receive excitatory synapses from pyramidal cells collaterals (5c) and from afferent fibres in the stratum radiatum (3c) and in the alveus (2c). There is ultrastructural evidence suggesting that pyramidal basket cells and other elements designated as L/M interneurons (see below) make synaptic contacts with O/A interneurons (6b, 9d). The effect produced by these synapses is unknown. The axons of the O/A interneurons make inhibitory synapses with pyramidal cells (7a) and presumably also with pyramidal basket cells (7b) and L/M interneurons (7c). The O/A interneurons mediate, like the pyramidal basket cells, feedforward and feedback inhibition of CA1 pyramidal cells.

8. Chandelier or axo-axonic cells (Fig. 22.191: a–a) are specialized interneurons that make multiple synaptic contacts exclusively with the axon initial segments of pyramidal neurons in the cerebral cortex (Fairén and Valverde 1980; Somogyi et al. 1982). It has been shown that elements of this type are not only present in the neocortex, but also occur in the hippocampus (Somogyi et al. 1983, 1985; Soriano and Frotscher 1989). A drawing of an apparently completely impregnated specimen of this cell type, found in the CA1 region of the cat, has been presented by Somogyi et al. (1985) and is reproduced here in a somewhat simplified form (q in Fig. 22.189). The soma of this element was situated in the stratum pyramidale. Its dendritic tree spanned all hippocampal layers. The axon ramified in the stratum pyrami-

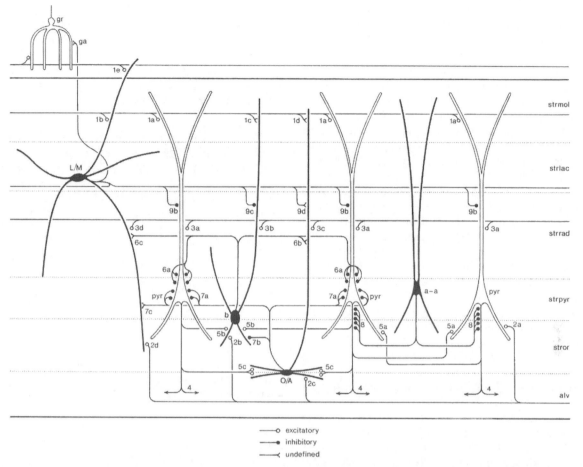

Fig. 22.191. Known and probable interneuronal relationships in Ammon's horn (particularly field CA1) of the mammalian hippocampal formation. For further explanation, see text. *strmol*, stratum moleculare; *strlac*, stratum lacunosum; *strrad*, stratum radiatum; *strpyr*, stratum pyramidale; *stror*, stratum oriens; *alv*, alveus. (Based on Schwartzkroin and Kunkel 1985, Lacaille and Schwartzkroin 1988a,b and Lacaille et al. 1987, 1989)

dale and emitted more than 300 characteristic, specialized terminal segments that climbed along the axonal initial segments of pyramidal neurons. Electron microscopy examination revealed that a single terminal segment provides a series of about eight to 30 symmetrical synapses to a single axonal initial segment of a pyramidal neurons and that terminal segments of several axo-axonic cell converge upon the initial segment of an individual pyramidal cell (Somogyi et al. 1983, 1985). Axo-axonic cells in the neocortex and in various parts of the hippocampus have been shown to be immunoreactive to GABA (Somogyi et al. 1985; Soriano and Frotscher 1989). The ultrastructure and the strategic location of their synapses, and the presence of GABA in these axo-axonal cells, strongly suggest that they play an important role in GABA-mediated inhibition of pyramidal cells (8). Data concerning the input to the axo-axonic cells are lacking; however, the fact that hippocampal axo-axonic cells have roughly the same dendritic distribution as the pyramidal cell may well indicate that they have access to all input available to the latter (Somogyi et al. 1985).

9. Interneurons situated in the CA1 region at the border between the stratum lacunosum-moleculare and the stratum radiatum were penetrated intracellularly, characterised physiologically, subsequently labelled and studied at the light and electron microscopy level (Lacaille and Schwartzkroin 1988a,b; Kunkel et al. 1988; Lacaille et al. 1989). These elements, which were designated as L/M interneurons, have fusiform or multipolar somata and aspinous, beaded dendritic processes ramifying in the strata lacunosum, moleculare and radiatum. The varicose axon originates from a primary dendrite and often projects for millimetres along the stratum lacunosum. It ramifies profusely in the stratum

radiatum, some branches reaching as far as the stratum pyramidale and the stratum oriens. Remarkably, the processes of the L/M interneurons are not restricted to the CA1 region, since dendritic and axonal processes of some of these elements were seen ascending in the stratum moleculare, crossing the hippocampal fissure and entering the adjacent stratum moleculare of the fascia dentata. In my opinion, these L/M interneurons do not resemble any of the hippocampal cell types described by Cajal (1911) and Lorente de Nó (1934b).

The synaptic relations of the L/M interneurons as indicated by Lacaille and Schwartzkroin (1988a,b) and Lacaille et al. (1989) may be summarised as follows. Afferent fibres in the alveus (2d), the stratum radiatum (3d) and the stratum moleculare of the CA1 region (1b) and the fascia dentata (1e) make excitatory synapses with them. Other interneurons, probably including the pyramidal basket cells and the O/A interneurons, make functionally as yet undetermined synapses with L/M cells (6c, 7c). The axonal ramifications of the L/M interneurons make inhibitory contacts with pyramidal cells (9b), pyramidal basket cells (9c) and probably also with O/A interneurons (9d) in the CA1 region. Some of the axonal branches of the L/M interneurons cross the hippocampal fissure and synapse with granule cells in the fascia dentata (9a). From these data, it appears that the O/A elements, like the pyramidal basket cells and L/M cells, are inhibitory interneurons. However, whereas the pyramidal basket cells and the O/A cells mediate both feedforward and feedback inhibition onto CA1 pyramidal cells, L/M interneurons appear to mediate only feedforward inhibition. Lacaille and Schwartzkroin (1988b) did not find any evidence of excitatory synaptic connections from CA1 pyramidal cells onto L/M interneurons.

Although the data discussed above and Fig. 22.191 reveal that the local circuitry of the CA1 region is quite complex, it should be emphasised that this circuitry is in reality doubtless even more complex by far. This appears immediately from the fact that only three of the 18 or more types of interneurons present in this hippocampal region have been included (see Cajal 1911; Lorente de Nó 1934b; Figs. 22.189, 22.190). It should also be noted that the diagram is only based on qualitative data, and that quantitative data, which are available to some extent (see e.g. Amaral et al. 1990), have been omitted. Notwithstanding these limitations, the data presented are considered useful for a comparative consideration of the microcircuitry of the cerebral cortex (see below).

Before leaving the hippocampal formation, attention should be given to the principal neurons of the fascia dentata, i.e the granule cells. Although these elements differ at first sight considerably from the typical hippocampal pyramids (Fig. 22.188), it should be emphasised that both belong to the category of pyramid-like cells (Fig. 22.179). The evidence for this is as follows:

1. Both have typical spiny dendritic trees.
2. Although most hippocampal granule cells have only peripherally directed dendrites, thus resembling the amphibian pallial cells (compare gr2 in Fig. 22.188 with Fig. 22.179a), some of them are provided with a basal dendritic system as observed in many pyramid-like reptilian pallial neurons (compare gr1 in Fig. 22.188 with Fig. 22.179b,d).
3. The granule cells and the hippocampal pyramids both serve as primary output cells for their respective regions.
4. The somata of both cell types are invested by axonal branches of typical basket cells (Lorente de Nó 1934b).
5. The axon initial segment of both cell types is the specific target of GABAergic axo-axonal or chandelier cells (Somogyi et al. 1985; Soriano and Frotscher 1989).

22.11.6.5
Neocortex

22.11.6.5.1
Size, Lamination and Subdivision

Although numerous authors (for a recent review, see Reiner 1991) have claimed that the dorsal or general cortex of reptiles represents, or at least contains, a primordial neocortex, it is generally agreed that a fully developed neocortex is only present in mammals. The size of the neocortex varies greatly among the various mammalian groups (see poster P1). In insectivores, e.g. the hedgehog, its size does not exceed that of the 'older' parts of the cortex, but in primates and cetaceans it attains remarkable dimensions, becoming by far the largest centre in the brain. Stephan and Andy (1964) determined the average index of progression for the neocortex (and for many other brain structures) in a number of insectivores, prosimians and simians, including humans. These indices express how many times larger the neocortex in a particular group or species is than that of a typical basal insectivore of the same size. It was found that in prosimians the neocortex is on the average 14.5 times, and in the simians it is 45.5 times as large as in basal insectivores. In humans, the neocortex appeared to be 156 times as large as that of basal insectivores.

The neocortex shows a laminated structure throughout its extent. Whereas in the piriform and hippocampal parts of the mammalian cortex three layers can be distinguished, in the neocortex six layers are usually recognised. The characterisation and determination of these layers is generally based on Nissl-stained material. Although the study of such material, in which the dendritic and axonal processes of the neurons remain unstained, yields in itself very little insight into the structural and functional organisation of the cortex, the results of such cytoarchitectonic analyses are important, because they provide a very useful general framework for studies with other, more critical techniques.

Beginning at the surface the six neocortical layers are as follows: (I) lamina molecularis, (II) lamina granularis externa, (III) lamina pyramidalis externa, (IV) lamina granularis interna, (V) lamina pyramidalis interna, (VI) lamina multiformis (VI).

I. The *lamina molecularis* or *lamina zonalis* contains only very few cell bodies.
II. The *lamina granularis externa* is composed of small, densely packed, cell bodies. The name of this layer is misleading, because most of its constituent somata belong to small pyramidal neurons which, like all typical cortical pyramids, direct their apices toward the surface.
III. The *lamina pyramidalis externa* is a thick layer in which pyramidal somata prevail. These somata increase progressively in size from superficial to deep.
IV. The *lamina granularis interna* consists of small, densely packed, pyramidal and non-pyramidal somata.
V. The *lamina pyramidalis interna* consists mainly of medium-sized and large, loosely arranged pyramidal somata.
VI. The *lamina multiformis* is composed of relatively tightly packed, spindle-shaped somata. Golgi material reveals that most of these somata belong to modified pyramidal cells.

Apart from the neuronal perikarya, several other structural elements in the cortex show a more or less distinct laminar arrangement. Myelin-stained and reduced silver preparations reveal the presence of tangentially oriented fibre concentrations at several levels, and Golgi material shows that these fibres and their terminal ramifications may contribute to plexiform zones, which are likewise tangentially arranged. Thus lamina I contains a dense plexus of horizontally running extrinsic and intrinsic fibres which contact the apical dendritic bouquets of pyramidal neurons situated in the deeper layers. Tangential plexuses of fibres in layer IV and deep in layer V are known as the outer and inner striae of Baillarger. These striae are probably formed primarily by intrinsic cortical axons, i.e. axons of local circuit neurons and collateral branches of pyramidal cell axons (Jones 1987). The outer stria of Baillarger is particularly well developed in the visual cortex, where it is referred to as Gennari's line.

Although tangential lamination is a prominent feature of the neocortex, many of its constituent elements show an evident radial orientation. Prominent among these are the pyramidal cells. The apical dendritic shafts of these ubiquitous elements extend peripherally, and many of them reach the most superficial layers before forming their terminal tufts. The axons of the pyramidal cells are also radially oriented. They emerge from the base of the cell bodies and descend toward the white matter (see Fig. 2.39c, o,r in Fig. 22.193, s,v,w,x in Fig. 22.194). The pyramidal cell axons, which at a short distance from the soma acquire a myelin sheath, assemble in bundles, which increase in size as they descend and as more axons are added. These bundles are known as radial fasciculi. The apical dendrites of the pyramidal cells are also arranged in bundles (Peters and Walsh 1972; Peters and Sethares 1991; Fleischhauer et al. 1972; Fleischhauer 1974). These axonal and dendritic bundles impose upon the neuronal cell bodies an arrangement in slender radially oriented columns which extend the thickness of the cortex. Not only the main axons of the pyramidal cells, but also their collateral branches often show a radial orientation, and the same holds true for the axonal systems of many intrinsic cortical neurons (see Figs. 22.193, 22.194). Many extrinsic afferents also take a radial course after having entered the cortex from the deep white matter. These structural features suggest that the main pattern of connections between cortical neurons is in the vertical direction. Lorente de Nó (1938) advanced the idea that the cerebral cortex essentially consists of small, radially arranged sets of neurons having a thalamic afferent fibre as their axis. According to Lorente de Nó, these column-like elementary units contain all types of cortical cells and within their confines the whole process of the transmission of impulses from the afferent fibre to the efferent axon is accomplished. Electrophysiological studies of the somesthetic (Mountcastle 1957) and the visual cortex (Hubel and Wiesel 1962) have provided powerful evidence for functionally discrete radial columns or modules. Later morphological studies (Goldman-Rakic and Nauta 1977) showed that the cortical modules are organised around cortico-cortical afferents rather than thalamic inputs. The cortical columns will be further discussed in Chap. 22, Sect. 22.11.6.5.7.

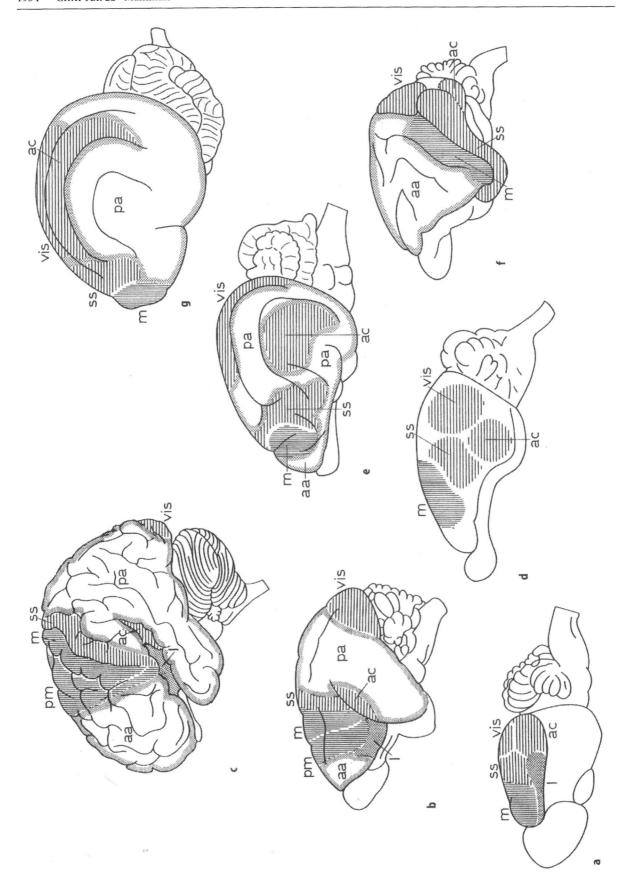

◀ **Fig. 22.192a–g.** Lateral views of the brains of **a** the hedgehog, **b** the prosimian *Galago*, **c** the human, **d** the guinea pig, **e** the cat, **f** the spiny anteater and **g** the dolphin, showing the position of major motor (*vertically hatched*), sensory (*horizontally hatched*) and association areas (surrounded by a rim of *fine dots*). In **g**, only the first-order sulci are shown. *aa*, anterior association area; *ac*, acoustic area; *i*, insular cortex; *m*, motor cortex; *pa*, posterior association cortex; *pm*, premotor cortex; *ss*, somatosensory cortex; *vis*, visual cortex. (**a–c** Reproduced from Nieuwenhuys 1994. **d,e** Based on Johnson 1990. **f** Modified from Rowe 1990. **g** Modified from Morgane et al. 1990)

Although the six-layered basic pattern is recognisable throughout the neocortex, this structure is not homogeneous. Differences in the relative thickness and cell density of the various layers, and in the size, shape and arrangement of the neuronal perikarya, are present and have been used to divide the neocortex into cytoarchitectural areas. Similarly, differences in the pattern of myelinated fibres (local development and distinctness of the striae of Baillarger, length of the radial fasciculi) have been used in myeloarchitectonic parcellations of the cortex. Comparative studies have shown that the neocortex of different mammals may vary not only in size, but also in areal differentiation. This may be exemplified by some of the results obtained by Brodmann (1909), who subjected the cortex of a considerable number of mammals, including humans, to a thorough cytoarchitectonic analysis. In the hedgehog, a 'basal' insectivore, he was able to delineate six neocortical areas, whereas in the prosimian *Lemur*, the simian *Cereopithecus* and humans, 14, 23 and 44 areas were distinguished, respectively. The structural differentiation just touched upon is paralleled by a functional differentiation. Without going into details, it appears that in marsupials, insectivores and rodents much of the neocortex is occupied by areas which either receive, via the thalamus, impulses directly related to the various special senses or are concerned with the steering of motor activity (Fig. 22.192a,d). The sensory fibres comprise the somatosensory cortex receiving impulses from sense organs situated in the skin, the muscles and the joints, then the primary visual cortex and finally the primary auditory cortex. However, in prosimians and ferungulates, the primary sensory cortices are separated from each other by a strip of non-projection or association cortex, and another area of association cortex is present in front of the motor cortex (Fig. 22.192b,d). In primates, the posterior or parietotemporal association area and the anterior or prefrontal association area are considerably expanded (Fig. 22.192c). Highly remarkable spatial arrangements of the primary and association cortical areas have been observed in the spiny anteater and in cetaceans. In the spiny anteater, the motor, somatosensory, visual and auditory areas are tightly grouped in the posterior region of the hemisphere, while the entire anterior region is occupied by a large association area (Lende 1964; Rowe 1990; Fig. 22.192f). In cetaceans, the primary areas also lie close together and surround a vast frontoparietotemporal association cortex (Morgane et al. 1990; Fig. 22.192g). All cortical areas maintain afferent and efferent connections with subcortical centres. However, it should be emphasised that the various association cortical areas are primarily connected with other cortical fields.

Following these introductory notes, the question as to how and to what extent the various elements in the cortex contribute to the laminar pattern of the structure will be addressed. The following elements will be considered: the cortical afferents, the typical pyramidal cells and their recurrent collaterals, primitive pyramid-like and modified pyramidal cells and local circuit neurons. The entorhinal cortex will be included in this analysis. Although this cortex shows a six-layered laminar pattern (Stephan 1975; Amaral et al. 1987; Fig. 22.193), it is considered as part of the so-called mesocortex, i.e. a cortex which both positionally and structurally stands intermediate between the allocortical hippocampal formation and the neocortex (Fig. 22.187).

22.11.6.5.2
Afferents

The extrinsic afferents of the cortex may be classified according to their mode of entrance into two groups, i.e. tangential and radial. The tangential fibres enter a particular area via lamina I, whereas the radial fibres ascend from the deep white matter.

Superficially passing extrinsic afferent fibres are prominent in the entorhinal cortex. Experimental neuroanatomical studies have shown that, in the rat, the lateral part of this cortex receives a direct projection from the olfactory bulb and is also supplied by fibres originating from the piriform cortex (Heimer 1968; Price 1973). Comparable projections have been observed in the cat (Boeijinga and Van Groen 1984; Valverde 1965) and monkey (Turner et al. 1978; Van Hoesen and Pandya 1975; Van Hoesen et al. 1975). In the rat and cat, these two projections terminate at different levels, the fibres from the olfactory bulb terminating in a superficial sublayer IA and the fibres from the piriform cortex terminating in a deeper sublayer IB. The boundary between these two sublayers is sharp, with very little overlap. It may be expected that the two projections discussed above make synaptic contacts with

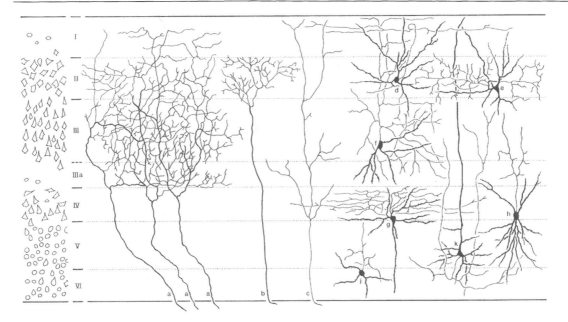

Fig. 22.193.

different segments of the apical dendritic systems of pyramidal and pyramid-like elements in the entorhinal cortex (k,m,o,r,v in Fig. 22.193).

A segregation of tangentially passing fibre systems containing restricted portions of the dendritic systems of pyramidal or pyramid-like cells has been observed in all cortices discussed so far and may hence be considered a general feature of this type of laminar structures.

Apart from direct and indirect olfactory projections, the entorhinal cortex receives a substantial superficially passing projection from the presubicular and parasubicular portions of the hippocampal formation (Lorente de Nó 1934a; Shipley 1974; see also Köhler 1985). The exact spatial relations of these hippocampal afferents to those originating from the olfactory bulb and the piriform cortex are not known at present.

The most superficial layer of the neocortex contains tangentially running extrinsic afferents from many sources. However, all of these fibres represent the end segment of axons that reach this layer after having traversed the cortex from the deep white matter. Fibres of this type are collectively designated here as radial, although many of them take a more or less oblique course through the cortex.

Radially running extrinsic afferents are abundant in the entorhinal cortex. In his classic Golgi study of this area, Lorente de Nó (1934a) reported the presence of the following three types of such afferents (a–c in Fig. 22.193): (a) coarse fibres which traverse the deeper layers of the cortex to form richly branching terminal arborisations in the superficial cortical layers; These arborisations are particularly dense in lamina III, less dense in lamina II and extend with only a few terminal branches into lamina I; (b) coarse fibres ending with dense ramifications exclusively in lamina II; and (c) thin, poorly ramifying fibres, which issue a small number of collateral branches in the upper four layers of the cortex.

Köhler (1985) investigated the projections of the various parts of the hippocampal subicular complex (i.e. the subiculum, the presubiculum and the parasubiculum; see Fig. 22.187) to the entorhinal cortex with the aid of the anterogradely transported *Phaseolus vulgaris*-leucoagglutinin (PHA-L). It was found that each subicular field has projections restricted to separate layers of the entorhinal cortex. Thus fibres originating from the subiculum proper form a dense terminal plexus in lamina IV, whereas fibres from the parasubiculum densely innervate lamina II. Finally, fibres originating from the presubiculum were traced to the outer three layers, lamina III receiving a dense innervation, while in lamina II and deep lamina I restricted zones of innervation were found. If we compare the results of the Golgi study by Lorente de Nó (1934a) with those of the anterograde tracer study by Köhler (1985), it seems likely that fibres originating from the presubiculum are of type a and fibre from parasubiculum of type b. A Golgi equivalent of the fibres arising from the subiculum proper has not been described by Lorente de Nó, and none of the affer-

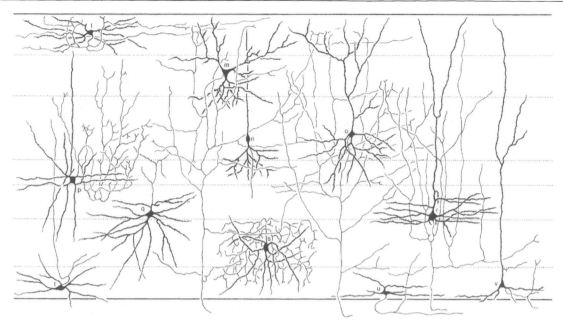

Fig. 22.193. Afferent fibres (*a–c*), local circuit neurons (*d–l, n,p,q,s,t*) and efferent neurons (*m,o,r,u,v*) as observed in Golgi material of the entorhinal cortex of the mouse. *Left*, cytoarchitecture of the same cortical region, as observed in Nissl preparations. (Based on Lorente de Nó 1934a)

ents observed by Köhler matches the thin type c fibres. In addition to hippocampal projections, the entorhinal cortex receives afferents from many other sources, but the laminar organisation of these afferents has not yet received much attention. Amaral and Insausti (1998) and Amaral et al. (1998) recently studied the efferent projections from various cortical regions to the entorhinal cortex in the rhesus monkey. They found that all of these projections have laminar preferences for their termination. As in the rat, layer III appeared to be distinguished by receiving a prominent projection from the presubiculum, whereas layer II receives a projection from the parasubiculum. Projections from the perirhinal cortex terminate predominantly in layer I, whereas projections from the orbitofrontal cortex terminate mainly in layer VI. It has also been shown that, in the rat, the midline nuclei of the thalamus project to the entorhinal cortex and that the fibres constituting this projection terminate mainly in laminae V and VI (Berendse and Groenewegen 1991).

In the neocortex, the fibres of all extrinsic afferent systems follow a radial course. Golgi studies and experimental investigations using anterograde degeneration and tracer techniques have shown that most of these afferent systems, after having entered the cortex from the deep white matter, distribute their terminal branches preferably in one or more layers.

Lorente de Nó (1922, 1938) provided a detailed description of thalamic afferent fibres in the rodent neocortex. His Golgi material displayed two different laminar distributions of terminal arborisations: the 'specific' distribution, which is densely aggregated in lamina IV, with or without extension to lamina III (a,b in Fig. 22.194), and the 'unspecific' distribution, which is sparsely distributed throughout all cortical layers, but appears predominantly in lamina VI (c in Fig. 22.194). The 'specific' afferents were considered to originate from 'specific' thalamic sensory relay nuclei, such as the medial and lateral geniculate bodies and the ventral posterior nucleus, and to terminate in specific sensory cortical fields. The 'unspecific' afferents, which were observed to extend over multiple cortical fields, were believed to originate from other, as yet undetermined thalamic sources. Frost and Caviness (1980) studied the intracortical distributions of the projections of a number of different thalamic loci to the neocortex in the mouse using an axon degeneration technique. They found that in this animal virtually the entire tangential extent of the neocortex receives a projection from the thalamus. The areas of termination of thalamofugal axons appeared to be segregated into three tiers: an outer tier in layer I, a middle tier in layer IV and/or III and an inner tier in layer VI. In most fields, terminating axons belonging to the middle or the inner tier or both were found to extend over some dis-

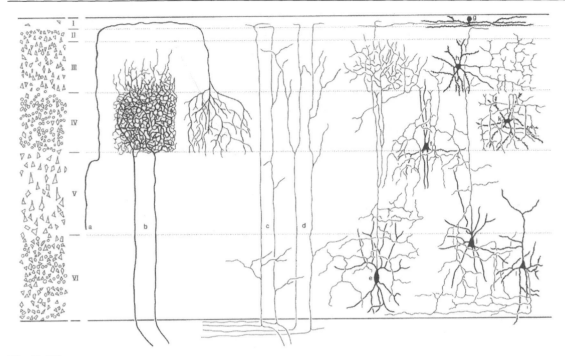

Fig. 22.194.

tance into layer V. Herkenham (1980a,b, 1986) examined the cortical projections of individual thalamic nuclei in the rat using tritiated amino acids as anterograde tracers. He found that the thalamic nuclei according to the laminar patterns of their cortical projections can be grouped into three classes. The first class includes the thalamic relay nuclei for vision, audition and somatic sensibility. Their cortical projections terminate mainly in lamina IV, lamina III or both. The fibres of this class clearly correspond to the 'specific' afferents of Lorente de Nó (1938). The second class includes the intralaminar thalamic nuclei, which issue sparse but widespread projections to deep cortical layers (laminae V, VI or both). The third class encompasses a number of nuclei that share a pattern of dense, widespread projections to lamina I, though terminations on other laminae may or may not be present. Because these nuclei projecting to lamina I generally occupy a position adjacent to the intralaminar nuclei, they were collectively designated by Herkenham (1986) as the paralaminar nuclei. The finding that different (groups of) thalamic nuclei project in a particular laminar fashion to smaller or larger parts of the neocortex has been confirmed by a large number of studies on different animals, although with regard to the projections of certain nuclei different results were obtained by different authors. Thus Berendse and Groenewegen (1991), who studied the cortical projections of the midline and intralaminar thalamic nuclei in the rat using PHA-L as an anterograde tracer, reported that the fibres originating from the intralaminar nuclei do not terminate exclusively in the deep cortical layers (as reported by Herkenham), but also project to lamina I. Corresponding findings have been reported for the cat and the monkey (Kaufman and Rosenquist 1985; Royce and Mourey 1985; Royce et al. 1989; Friedman et al. 1987).

The cortical projections of some thalamic centres have been analysed in detail by placing multiple small focal injections of anterograde tracers within their confines or by intracortical filling with HRP by micropipette after recording from individual thalamocortical fibres. Such an analysis of the ventroanterior-ventrolateral complex of the thalamus of the cat revealed, for instance, that fibres originating from the ventrolateral or the caudal part of this complex are distributed in laminae I, III and IV of the parietal cortex, whereas fibres arising from rostral or dorsomedial portions of the complex are almost confined to lamina I (Kakei and Shinoda 1990). However, the most detailed information of this type available at present concerns the projection from the lateral geniculate body to the primary visual cortex in primates. In this group, the lateral geniculate nucleus is a laminar structure in which separate magnocellular, parvocellular and intercalated zones can be distinguished. In the primate primary visual cortex, the laminar pattern is very dis-

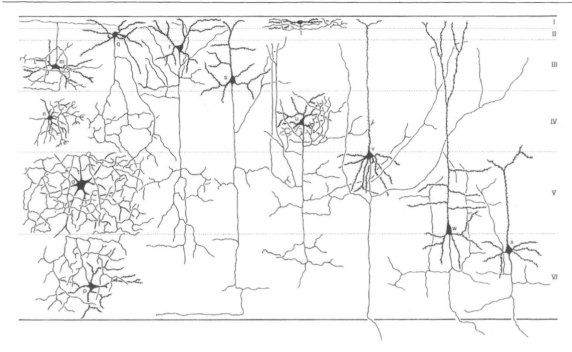

Fig. 22.194. Afferent fibres (*a-d*), local circuit neurons (*e-u*) and efferent neurons (*v-x*), as observed in Golgi material of the somatosensory cortex of the mouse. *Left*, same cortical area as observed in Nissl preparations. (Based on Lorente de Nó 1922)

tinct, and lamina IV can be divided into four subzones, designated as IVA, IVB, IVCα and IVCβ (Fig. 22.195). Investigations of many different authors, summarised by Fitzpatrick et al. (1983, 1985) and Lund (1988), revealed that there are at least six discrete populations of geniculocortical axons, differing markedly from one another in laminar distribution and tangential spread (Fig. 22.195):

1. Coarse fibres originating from the magnocellular layers of the lateral geniculate nucleus projecting to lamina IVCα with wide-spreading terminal fields that span the entire depth of the lamina
2. Fibres resembling those mentioned above, the terminal fields of which are, however, confined to the upper half of lamina IVCα
3. Axons originating presumably from both the parvocellular and the intercalated layers of the lateral geniculate nucleus forming small, dense clusters of terminal branches in lamina IVA and contributing some rising collaterals to the adjacent zone of lamina III
4. Fibres originating from the intercalated layers of the lateral geniculate nucleus, the terminal arborisations of which participate in the formation of small, dome-shaped formations in lamina III
5. Fibres from the parvocellular layers forming small, dense terminal fields in lamina IVCβ
6. Fine fibres likewise originating from the parvocellular layers of the lateral geniculate nucleus, extending over large distances horizontally, that contribute terminals to upper lamina VI and to lamina I

In the prosimian *Tupaia*, the lateral geniculate nucleus is divided into six well-defined cellular laminae separated by cell-sparse interlaminar zones. It has been established that each of these cellular laminae projects mainly to a particular subzone within layers III or IV of the primary visual cortex (Conley et al. 1984; Lund et al. 1985).

Each particular neocortical area, in addition to thalamic afferents, also receives a strong input from other neocortical areas. These association fibres come from either the same or the opposite hemisphere; in the latter case, they are called callosal fibres. According to Lorente de Nó (1922, 1938), the association fibres give off collaterals in the deep laminae, especially VI, but their main territory of distribution is in laminae I-IV, and especially II and III (d in Fig. 22.194). Goldman-Rakic and Nauta (1977) made small focal injections into various areas of the association cortex of monkeys. They found that the anterogradely transported label accumulated in narrow 200- to 300-μm-wide columns in relatively distant regions. Szentágothai (1978, 1983) concluded on the basis of a study of

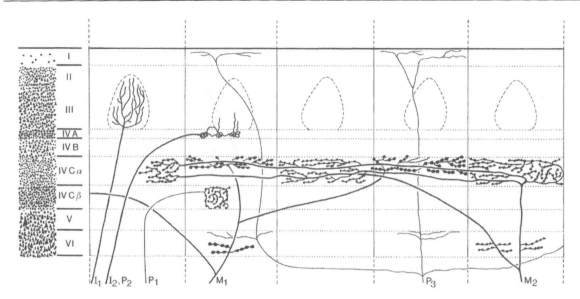

Fig. 22.195. Distribution of thalamic inputs from the lateral geniculate nucleus (LGN) to the primate primary visual cortex. *M1, M2,* input from magnocellular layers; *P1-P3,* input from parvocellular layers; *I1, I2,* input from intercalated layers. The ocular dominance bands are indicated by *dashed lines,* 400–500 µm apart. *Left,* same cortical area as observed in Nissl preparations. (Redrawn from Lund 1988)

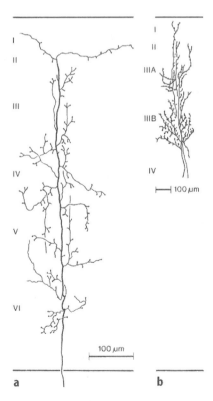

Fig. 22.196a,b. Cortico-cortical fibres in the cerebral cortex of the macaque monkey. **a** Cortico-cortical afferent fibre arborisation in the auditory cortex. (Redrawn from Szentágothai 1978). **b** Callosal axon in the somatosensory cortex, anterogradely labelled by horseradish peroxidase (HRP) injected in the corpus callosum. (Redrawn from Jones 1986)

Golgi material that the arborisation space and pattern of individual cortico-cortical axons correspond in size with the columns revealed by isotopic labelling. He observed that these fibres pass radially through the cortex and issue relatively short branches at all levels (Fig. 22.196a). Cortico-cortical fibres defining a narrow radial zone of termination have also been observed in the primate sensorimotor cortex (Hendry and Jones 1983; Jones 1986). Such a fibre, labelled by HRP injected in the corpus callosum, is depicted in Fig. 22.196b. The major layers of termination of callosal fibres in the primate somatosensory areas are laminae I–IV. In the motor cortex, they terminate in a comparable pattern in laminae I–III. Studies on cortico-cortical projections which, like the one cited above, are based on the tracing of individual fibres are scant. However, an overall picture of the mode of termination of these fibres can be deduced from axon degeneration studies of cortico-cortical projections and from tracer studies in which such fibres are labelled en masse. A detailed survey of this literature, presented by Innocenti (1986), revealed the following features:

1. In rodents, the densest cortico-cortical termination is found in lower layer I – layer III, with an additional, often less conspicuous zone of termination in layers V and VI.
2. In the cat and in primates, most cortico-cortical fibres terminate in layers III and IV.

In one axon degeneration study, in which the distribution of thalamocortical afferents in the cingulate cortex of the rat was compared with that of callosal afferents (Vogt et al. 1981), it was found that these two projections terminate in a largely complementary pattern. Quantitative ultrastructural analysis revealed that there may be as many as seven times more callosal than mediodorsal thalamic terminals in this cortex. Callosal axon terminals contain round vesicles and form asymmetrical synapses on spines (Lund and Lund 1970; Hendry and Jones 1983; Cipolloni and Peters 1983), which in the rat belong almost exclusively to apical or basal dendrites of pyramidal cells.

Investigations with the aid of histofluorescence and immunohistochemistry have shown that the cerebral cortex is innervated throughout its entire extent by extrinsic cholinergic, noradrenergic and serotoninergic fibres. It has been established that the cholinergic fibres stem from the nucleus basalis complex, whereas the noradrenergic and serotoninergic fibres originate, respectively, from the locus coeruleus and the mesencephalic raphe nuclei (Lidov et al. 1980; Levitt et al. 1984; Mesulam et al. 1983, 1984; Morrison and Magistretti 1983; Morrison et al. 1978, 1980, 1982; Takeuchi and Sano 1983). Although these transmitter-specified subcortico-cortical projections, particularly in primates, display marked regional variations in laminar distribution and density, they will not be further considered here.

From the foregoing, it appears that most of the extrinsic cortical afferent systems distribute themselves in layered arrays, and it may be added that in the primary sensory cortices an elaborate laminar segregation of thalamic inputs appears to be reflected in the stratification visible in the cellular architecture. Morphological evidence suggests that all extrinsic neocortical afferents are excitatory, and the question arises how the intrinsic machinery of this cortex is driven by these afferents.

Lamina I contains the apical dendritic bouquets of pyramidal cells of laminae II, III and V. Because in this layer only sparsely scattered intrinsic neurons are present, it is reasonable to assume that the extrinsic thalamic afferents which spread in it synapse principally with pyramidal neurons and thus have a direct access to the long-axonal outflow of the cortex. Superficial pyramids situated in lamina II extend not only their apical dendrites in lamina I; these elements often have basal dendrites which pass laterally from the soma and ascend to layer I (Vogt 1991). These elements are most probably strongly excited by the lamina I extrinsic afferents. It is interesting to note that the afferents from different thalamic nuclei, which, after having traversed the cortex, spread in lamina I, terminate in different subzones of that layer (Herkenham 1979) and the apical dendritic tufts of the pyramids thus receive stratified input from different sources.

Until recently, it was widely held that the specific sensory thalamic afferents which are mainly distributed to lamina IV (or its sublayers) of the primary sensory cortices contact only a single category of neurons, i.e. the spiny stellate cells, and that the input is then processed sequentially by hierarchically organised chains of neurons (see e.g. Hubel and Wiesel 1962, 1968; Eccles 1984). Golgi-electron microscopy studies combined with degeneration and tracer experiments (for a review, see White 1986, 1989) conclusively showed that the specific sensory thalamic afferents to the cortex synapse with pyramidal cells and with non-pyramidal intrinsic cortical elements. According to White (1986, 1989), this finding challenges the concept that thalamic input is processed by hierarchically organised chains of neurons and lends support to the notion that the function of the cerebral cortex depends heavily on parallel processing mechanisms. However, it should be noted that in highly differentiated sensory cortices different groups of local circuit neurons may well be involved in the parallel processing of sensory information. We have seen that, in the primate primary visual cortex, the magnocellular zones of the lateral geniculate nucleus project to sublamina IVCα, whereas the parvocellular zones project to IVCβ (Fig. 22.195). These two sublaminae are characterised by the presence of different classes of spiny stellate cells and by an almost complete lack of pyramidal neurons (Lund 1973, 1984, 1987; Mates and Lund 1983). Moreover, it has been established that the principal projection of the IVCα spiny stellate neurons is to sublamina IVB, whereas the majority of IVCβ spiny stellate neurons project to sublaminae IIIC and IVA (Lund 1973; Fitzpatrick et al. 1985). These findings suggest a continued separation of magnocellularly and parvocellularly derived information within the primate primary visual cortex (Lund 1988). Valverde (1985) noted that, with regard to the targets of specific thalamic afferents in the primary visual cortex, there is an important difference between primates and 'lower' mammalian species. Along with Lund and her associates, he was of the opinion that in primates the input from the lateral geniculate nucleus is relayed to local circuit neurons. However, analyses of Golgi material suggested to him that the thalamic input in the hedgehog is relayed mainly to pyramidal cells projecting to the white matter (Valverde 1983) and that in the visual cortex of the mouse most thalamic afferents synapse with the basal dendrites of pyramidal cells in the deep zone of lamina III, which again have axons projecting to the white matter (Ruiz-Marcos and Valverde 1970).

We have seen that the deeper layers of the cortex (V and VI) also receive direct thalamic inputs. It is known that these layers contain many pyramidal and pyramid-like cells and only a relatively small number of intrinsic elements. The axons of many of these deep pyramidal cells possess recurrent, ascending collaterals (o,r in Fig. 22.193, w,x in Fig. 22.194), and the axons of a certain proportion of the local circuit neurons in laminae V and VI are also directed to more superficial cortical layers (j,h,t in Fig. 22.193, e,j in Fig. 22.194). There is experimental evidence indicating that, in the deeper layers of the primary somatosensory cortex of the mouse, thalamocortical afferents contact both pyramidal and non-pyramidal cells (White 1978; Hersch and White 1981a,b). The deep pyramidal cells may well receive an 'aspecific' thalamic input via collaterals spreading in lamina VI (c in Fig. 22.194) and a 'specific' thalamic input via their apical dendrites in lamina IV.

Finally, the callosal and ipsilateral association fibres form nearly all their synapses with the spiny dendrites of pyramidal cells (Innocenti 1986; White 1989). Because these fibres terminate principally in laminae III and IV, it seems likely that the superficial pyramidal neurons form their main target.

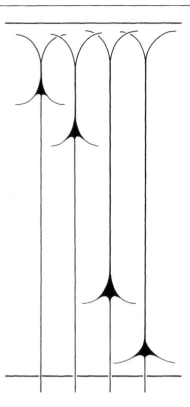

Fig. 22.197. Typical pyramidal neurons in the mammalian cerebral cortex

22.11.6.5.3
Typical Pyramidal Cells

The typical pyramidal cells constitute the largest and the most characteristic category of neocortical neurons. In the visual cortex of rat, cat and monkey, these elements account, according to Winfield et al. (1980), for about 65% of the total neuronal population, and in other cortical areas this percentage may be even higher. As mentioned before, a typical pyramidal neuron is provided with a radially oriented apical dendrite that forms a terminal tuft in laminae I and II, with basal dendrites that radiate out from the base of the soma and with an axon descending downward to leave the cortex. The apical dendrite may give rise to one or more horizontal or obliquely ascending side branches. All of the dendrites of typical pyramidal cells are more or less densely covered with spines (Peters and Jones 1984; o,r,v in Fig. 22.193, s,v,w in Fig. 22.194). Positionally, the apical dendritic tuft with its ramifications in laminae I and II is the only feature shared by all typical pyramidal cells. The somata and their basal dendritic systems may vary in position from lamina II to lamina VI, and the length of the apical dendrites varies accordingly (Fig. 22.197). In the following discussion of the typical pyramidal cells, the afferents impinging upon the various parts of their receptive surface will be considered first; the relation between the position of their somata and the destination of their axons will then be dealt with. Finally, attention will be paid to the patterns of distribution of their collaterals.

The somata of the pyramidal cells are not under the direct influence of any extrinsic afferent system. Rather, they are specifically addressed by one type of local circuit neuron, i.e. the basket cell (Fig. 22.198a). In a recent review article, DeFelipe and Fariñas (1992) point out that several types of aspiny nonpyramidal cells other than basket cells may contribute to some extent to the somatic and proximal dendritic innervation of pyramidal cells. The somata of these elements, which, like those of the pyramidal cells, vary in distance from the pia, are concentrated in laminae III and V (Marin-Padilla 1969, Fairén et al. 1984). Their poorly ramifying dendrites, which bear few or no spines, radiate in all directions, with a tendency toward vertical and horizontal trajectories. The axons are either ascending or descending and provide four or more horizontal collaterals at various levels. The horizontal collaterals, which can extend for 1 mm

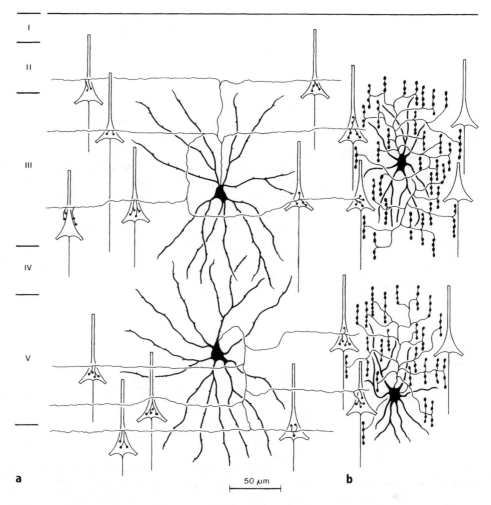

Fig. 22.198a,b. Intrinsic cortical neurons contacting specific parts of the receptive surface of pyramidal cells. **a** Basket cells, the axonal terminals of which participate in the formation of pericellular nests surrounding the somata and proximal dendrites of pyramidal cells. (Partly based on Jones 1986). **b** Chandelier cells; the numerous vertical axonal terminal portions specifically contact the axon initial segments of pyramidal cells. (Based on Peters 1984, Fairén et al. 1984, Somogyi et al. 1985 and Jones 1986)

or more in either direction, issue at intervals short side branches. These side branches terminate in pericellular baskets surrounding the somata and proximal dendrites of pyramidal cells. The basket cell terminals contain flattened vesicles, and the synapses made with the pyramidal cells are of the symmetric variety. They use GABA as a neurotransmitter and most probably exert a powerful inhibitory action on their target elements (Jones and Hendry 1984). There is evidence suggesting that the basket cells are major recipients of thalamic and commissural inputs (White 1978; Sloper and Powell 1979). Moreover, the somata of the basket cells are contacted by numerous GABAergic terminals, which may well arise from the axons of other basket cells (Hendry et al. 1983).

Like the somata, the axon hillocks and initial axonal segments of the pyramidal cells are also the target of a special category of local circuit neurons, i.e. the axo-axonic or chandelier cells (Somogyi 1977; Somogyi et al. 1982, 1983; Fairén and Valverde 1980; Peters 1984a; Fig. 22.198b). These elements occur in laminae II–V, but are most prominent in lamina III. Their dendrites may be grouped in an upper and a lower tuft or spread in all directions. The axon ramifies many times and produces a dense plexus in the vicinity of the parent soma. From this axonal plexus, numerous (up to 300) vertically oriented arrays of terminals arise which contact the initial axonal segments of pyramidal cells. The axo-axonic synapses between the chandelier cells and the pyramidal neurons are symmetrical, and the vesicles in the presynaptic axon terminals are flattened (Jones and Powell 1969; Sloper and Powell 1979; Feldman 1984). It has been shown that the axon terminals of the chandelier cells contain the inhibitory neurotransmitter GABA; hence it may be assumed that these elements inhibit the pyramidal cells (Peters et al. 1982; Somogyi et al. 1983). Because the chandelier cells exert this action at a strategic site, i.e. the trigger zone for the initiation of action potentials, they may be considered to have a decisive influence on the output from the pyramidal neurons. It has been found that the number of axo-axonal synapses along the initial segments of pyramidal cells is much larger in the supragranular layers than in the infragranular layers (Sloper and Powell 1979; Peters et al. 1982). Because the pyramidal neurons which project via the corpus callosum to the contralateral hemisphere are situated mainly in laminae II and III, it has been suggested that the chandelier cells principally influence cortico-cortical circuitry (Somogyi et al. 1979; Peters et al. 1982). It has also been suggested that the chandelier cells are specially activated by thalamocortical afferents via spiny stellate cells (Eccles 1984), but in fact nothing is known about the inputs to these neurons.

In the reptilian cortex and in the mammalian hippocampal cortex, the perikarya of the pyramidal neurons are concentrated in a single layer, and all of these elements extend their basal dendrites in one and the same plexiform zone. In the mammalian mesocortex and neocortex, the pyramidal somata are situated at different levels and, given the laminar organisation of many extrinsic and intrinsic cortical fibre systems, it is to be expected that the basal dendritic systems of different pyramidal neurons are involved in different synaptic relationships, depending on the laminae or sublaminae within which these dendritic systems occur. Thus it has been established that the basal dendrites of pyramidal cells situated in laminae II and III receive callosal afferents (Cipolloni and Peters 1983) and that lamina III border pyramids extend their basal dendrites into lamina IV, where they are contacted by terminals of thalamocortical fibres (Ruiz-Marcos and Valverde 1970; Hornung and Garey 1981). Intrinsic cortical axons, i.e. axons of local circuit neurons and collateral branches of pyramidal cells, are concentrated in lamina IV and deep in lamina V, where they form the external and internal striae of Baillarger (Valverde 1986; Jones 1987). The basal dendrites of certain groups of pyramidal neurons exhibit a specific affinity to the terminal ramifications of the axons within these striae. In the primate primary visual cortex, the external stria of Baillarger (also known there as the line of Gennari) is strongly developed and occupies a discrete sublayer IVB, in which no thalamic afferents are present. According to Valverde (1985), its axonal components include horizontal collaterals of descending axons of pyramidal and stellate cells of sublayer IVA, horizontal branches of axons of neurons located within sublayer IVB and ascending ramifications of spiny stellate cells residing in sublayer IVC. Lund (1973) reported that the basal dendrites of pyramidal neurons situated in sublayer IVA turn sharply downward from the soma to enter sublayer IVB, where they fan out horizontally, markedly avoiding any ramifications within lamina IVA (see Fig. 22.203c). As far as I am aware, the question of which neurons contribute axons or axonal branches to the internal stria of Baillarger has never been specifically addressed; however, the Golgi studies by Cajal (1911) and Valverde (1985, 1986) suggest that the axons of both superficial and deep pyramids issue numerous horizontally oriented collaterals into this stria and that certain local circuit neurons with ascending axons, situated in lamina VI, also contribute collaterals to it (see Valverde 1986 and his Fig. 5C, i,j). In the primate motor and visual cortices, basal dendrites of pyramidal cells form a conspicuous plexus within the domain of the internal stria. In both of these cortices, giant pyramidal neurons occur, known in the motor cortex as Betz cells and in the visual cortex as Meynert cells. Beyond their extraordinary size, these neurons have several other features in common. Both cells are characterised by the presence of long horizontally oriented basal dendrites that extend into the internal stria, and both further contribute to this stria with additional horizontal dendrites emanating from the lateral surface of the soma and even from the proximal stem of the apical dendrite (Cajal 1911; Braak 1976, 1980; Scheibel and Scheibel 1978; Scheibel 1979).

Whereas the horizontal axonal systems discussed so far contact the basal dendrites of pyramidal neurons situated at one particular level, the neocortex also contains vast numbers of vertically oriented axonal elements, which potentially contact the basal dendritic systems of pyramidal elements situated at different levels. These vertical axonal elements, which play a prominent role in the so-called radial coupling of neuronal elements and therewith in the columnar functional organisation of the cortex, may be categorised as follows:

1. *Thalamocortical and cortico-cortical association fibres.* It has already been discussed that thin 'aspecific' thalamic fibres and both ipsilateral and contralateral cortico-cortical association fibre traverse all cortical layers (c,d in Fig. 22.194).

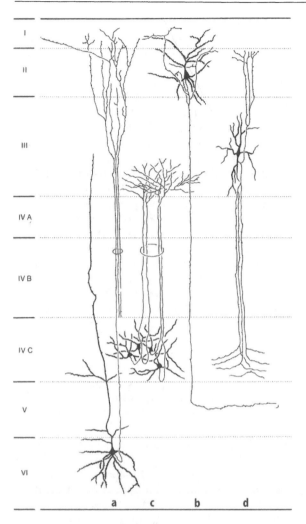

Fig. 22.199. Intrinsic neurons with radially coursing axons in the cerebral cortex of the rhesus monkey. Small pyramidal neuron in lamina VI with an ascending axon participating in the formation of an axonal bundle (*a*). Small pyramidal neuron in lamina II with a thin, unmyelinated, varicose axon terminating in lamina V (*b*). A group of spiny stellate cells, the axons of which form a radially arranged ascending bundle (*c*). Typical double bouquet cell (*d*). (*a–c* Redrawn from Valverde 1985. *d* Based on Jones 1986)

2. *Axons and recurrent collaterals of pyramid neurons.* The axons of both superficial and deep pyramidal cells may emit one or several recurrent collaterals that ascend radially through the cortex (o,r in Fig. 22.193, v in Fig. 22.194). In lamina V of the primate visual cortex, Valverde (1985) observed medium-sized and small pyramidal neurons possessing recurrent ascending axons which could be traced to the superficial cortical layers. Some of these elements had only a single ascending axon, but most commonly the main axon ramified into two or more ascending collaterals. Similar 'modified' pyramidal cells appeared to be present in lamina VI (Fig. 22.199a). The recurrent ascending axons of these elements were observed to assemble in radially oriented small bundles. Valverde (1985) also mentioned the presence of small pyramidal cells in laminae II and III, the axons of which descend but remain within the cortex, terminating in lamina V (Fig. 22.199b). Some of these axons displayed short side appendages and small varicosities all along their path.

Ultrastructural analyses of pyramidal neurons labelled by intracellular injections with HRP have conclusively shown that the axon collaterals of these elements make contact predominantly with the dendrites of other pyramidal cells (see e.g. Kisvárday et al. 1986; Gabbott et al. 1987; McGuire et al. 1991). However, the suggestion made by Scheibel and Scheibel (1970) that the basal dendrite systems of pyramidal cells are specifically contacted by recurrent collaterals of other pyramidal elements and the proposal made here that these systems form the prime targets of all thin, radially directed axons and collateral branches of pyramidal cells requires ultrastructural substantiation.

3. *The vertically elongate axonal systems of some types of cortical local circuit neurons.* The axons of certain populations of the groups of cortical local circuit neurons designated as spiny stellate cells and smooth or sparsely spinous non-pyramidal neurons form highly characteristic bundles of radially oriented translaminar collaterals. Spiny stellate cells are specifically located in lamina IV (or any of its sublayers) of the primary sensory cortical areas. The axon of a certain proportion of these cells first descends but soon forms recurrent branches which ascend in a strictly columnar fashion (Fairén et al. 1984; Lund 1984; Valverde 1985, 1986; Fig. 22.199c). Spiny stellate cells form synapses of the asymmetrical type with their target structures (LeVay 1973). Among the smooth or sparsely spiny non-pyramidal neurons, there are elements whose axons form vertically elongate plexuses of parallel axonal branches. Because the dendrites of these elements arise primarily from the upper and lower poles of the cell body, they were designated by Cajal (1911) as '*cellules à double bouquet dendritique*'. Typical double bouquet cells have so far only been found in laminae II and II of the neocortex of cats and primates, including humans. Their axonal plexuses extend both above and below their parent cell bodies and may span all cortical layers except for the most superficial one (Fairén et al. 1984; Somogyi and Cowey 1984; Peters 1984; Fig. 22.199d). The

individual collaterals are beaded, suggesting that they make synaptic contacts at many different levels throughout their length. The double bouquet cells probably use GABA as a neurotransmitter, and their terminals contain flattened vesicles and make symmetrical contacts with their target structures (Somogyi and Cowey 1984). It has been suggested that the radial axon bundles of the double bouquet cells terminate mainly on the apical dendrites of pyramidal cells (Cajal 1911; Colonnier 1966; Szentágothai 1978), but this proposal could not be substantiated in the studies carried out by Somogyi and Cowey (1981) and DeFelipe et al. (1989, 1990). These studies revealed that double bouquet cells do not form synapses with apical dendrites, but with basal dendrites and oblique branches of the apical dendrites of pyramidal neurons and with postsynaptic structures belonging to non-pyramidal neurons.

The axonal branches of the local circuit neurons discussed above closely resemble the radially oriented recurrent collaterals of pyramidal neurons, and I consider it likely that all of these processes impinge primarily on the basal dendritic system of (other) pyramidal neurons. However, it should be re-emphasised that ultrastructural evidence supporting this notion is wanting.

Turning now to the apical dendritic shafts of the typical pyramidal cells, it is important to note that these conspicuous processes are well placed to receive input from a variety of axonal pathways known to terminate within specific cortical layers. There is considerable variation in apical shaft length, ranging from essentially no shaft at all (pyramidal cells in the superficial zone of lamina II), to apical dendrites of two millimetres or more (pyramidal cells in lamina VI). Moreover, particularly in primates, the apical dendrites of many deep pyramidal cells do not extend into the subpial cortical zone, but rather terminate in a variably developed terminal tuft deeper within the cortex. Given these variations in length and in position of their apical dendrites, it is evident that different pyramidal cells may receive different samples of lamina-specific extracortical and intracortical afferents, and these differences are still further accentuated by the fact that the apical dendrites of different pyramidal neurons may exhibit different specific affinities to particular afferent systems. The evidence for such specific affinities includes (1) the presence of distinct lamina-specific differences in the density of spines along the apical dendrites, (2) the presence of lamina-specific side branches on the apical dendrites and (3) direct proof that apical dendritic segments of different pyramidal cells passing through a particular layer may receive highly different numbers of synapses from the afferents concentrated in that layer. These three aspects will now be briefly discussed.

1. *The presence of distinct lamina-specific differences in the density of spines along the apical dendrites.* Spines are present in abundance on the dendrites of pyramidal neurons, where they function as the primary postsynaptic structures of the cell (Feldman 1984). Quantitative analyses have shown that these spines are not evenly distributed along the dendrites. Characteristically, the most proximal portions of the dendrites emanating from the pyramidal somata are devoid of spines. From these initial segments onward, the concentration of spines increases gradually, attains a maximum some 80 μm away from the soma and then declines again distally (Globus and Scheibel 1967; Valverde 1967; Peters and Jones 1984). Superimposed over this general pattern of spine distribution, the apical dendrites of pyramidal neurons may display distinct lamina-specific changes in spine density. For example, Lund (1973) found that in the primary visual cortex of the monkey, the apical dendrites of pyramidal cells lying in laminae V and VI show a marked reduction in the number of spines in lamina IVCβ, as compared to the densely spiny portion of the same shafts in laminae V and VI. The number of spines on these apical dendrites in general increases again in laminae above IVCβ, but not to the level shown on the proximal portion of the shaft (Fig. 22.200). According to Lund (1973) this reduction in spines on the apical dendritic shaft as it passes through laminae IVCβ is true of the great majority of the pyramidal cells of laminae V and VI, including the giant pyramidal cells of Meynert. It is known that lamina IVCβ receives afferents specifically from the parvocellular layers of the lateral geniculate nucleus. Thus the findings of Lund (1973) might suggest that, in the monkey, the deep pyramids in the visual cortex are not the primary targets of these particular afferents. Marked differences in spine density along the shafts of the apical dendrites have also been observed by Vogt (1991). This author found that, in the posterior cingulate cortex of the rat, the density of spines on the proximal apical shaft of large layer V pyramids reaches a maximum in layer IV. In layer II–III, there is a sharp reduction in spine density, while the branches of the apical dendritic tuft, which extend into layer I, attain an intermediate spine density. Vogt (1991) also

observed that small pyramidal neurons in layers II and IV of the same cortical area in the rat often had essentially spine-free apical shafts, while their apical tufts, which were confined to the superficial zone of layer I, showed a moderate density of spines.

2. *The presence of lamina-specific side branches on the apical dendrites.* The apical dendrites of many pyramidal neurons issue small numbers (some three to six) of oblique or horizontally oriented side branches. At first sight, these side branches seem to be randomly distributed along the apical shafts. However, systematic studies of large numbers of apical dendrites have shown that in several cortical areas these processes are given off at selective levels, i.e. as they pass through particular layers. Thus Lorente de Nó (1934a) observed that in the entorhinal cortex the ascending shafts of the pyramids of the deeper laminae always issue some side branches in lamina III, but give off none at all during their passage through lamina II. Lund and Boothe (1975) observed that the apical dendrites of a particular population of lamina VI pyramidal neurons in the primate visual cortex issue side branches selectively in lamina IVCα, whereas the apical dendrites of another lamina VI pyramidal population in this cortex give off side branches at the border of laminae V and IVCβ and split up into their terminal dendritic tufts in lamina IVA (Fig. 22.200b,c). According to Lund (1988), these various dendritic patterns suggest that deeper pyramidal neurons may receive thalamic input within laminae IVCα and IVA.

It has already been mentioned that, in the prosimian *Tupaia*, the lateral geniculate nucleus is differentiated into six cellular layers and that each of these layers projects mainly to a particular subzone within the layers III or IV of the primary visual cortex (Conley et al. 1984). Lund et al. (1985) reported that this elaborate laminar pattern of lateral geniculate nucleus afferents is reflected in the precise patterns according to which the apical dendrites of lamina V and VI

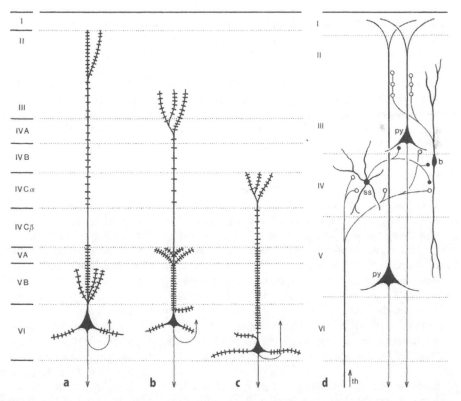

Fig. 22.200a–d. Features of apical dendrites of pyramidal neurons. **a,b** Density of spines along these processes may show distinct lamina-specific differences. **a–c** Side branches and terminal tufts of the apical dendrites may be issued selectively in particular (sub)layers. **d** Apical dendrites of pyramidal neurons receive multiple asymmetric synapses (*open circles*) from the axons of bipolar cells. **d** also shows that smooth or sparsely spiny, non-pyramidal cells (*ss*) form symmetric synapses (*filled circles*) with pyramidal (*py*) and bipolar cells (*b*) and that thalamic afferents (*th*) contact pyramidal, bipolar and smooth or sparsely spiny, non-pyramidal cells forming asymmetric synapses. (**a–c** Redrawn from Lund 1984. **d** Based on Peters 1984b)

pyramidal cells distribute their side branches. Eleven different patterns of distribution were recognised. Ten of these had side branches on their apical dendritic tuft in one, two or even three of the 'visual' sublayers that are in receipt of afferents from the lateral geniculate nucleus. Seven of these ten also emitted side branches in one of the 'nonvisual' sublayers.

3. *Direct proof that apical dendritic segments of different pyramidal cells passing through a particular layer may receive highly different numbers of synapses from the afferents concentrated in that layer.* The thalamocortical connectivity of pyramidal neurons has been the subject of a series of qualitative studies by White and collaborators (White and Hersch 1981, 1982; White et al. 1982; Hersch and White 1981a–c, 1982). Hersch and White (1981a,b) reported that, in layer IV of the primary somatosensory cortex of the mouse, the proportion of thalamocortical synapses received by apical dendrites belonging to various sizes of Golgi-impregnated lamina V and lamina VI pyramidal neurons ranged from 1.3% to 14.6% of all the asymmetrical synapses formed onto these dendrites. The (specific) thalamocortical afferents were labelled by lesion-induced degeneration, and the pyramidal neurons were deimpregnated and gold-toned. In subsequent work in the same cortex, the number of thalamocortical synapses in lamina IV on the apical dendrites belonging to deep pyramids that project from the primary somatosensory cortex to either the ventrobasal nucleus of the thalamus (Hersch and White 1981c; White and Hersch 1982), the primary motor cortex (White et al. 1982) or the ipsilateral striatum (Hersch and White 1982) was determined. These three categories each contained seven apical dendrites of pyramidal neurons retrogradely labelled with HRP. It appeared that these three populations of pyramidal neurons have very different thalamocortical connectivity patterns. Thus only 0.3%–0.9% of the total number of apical dendritic synapses in lamina IV of corticostriatal projection neurons were made with thalamic afferents. For the cortico-cortical and corticothalamic elements examined, these values were 1%–7% and 7%–20%, respectively.

So far I have dealt with tangentially oriented afferent systems establishing synaptic contacts with particular segments of the apical dendrites of pyramidal neurons. However, there is also evidence for the presence of radially oriented afferents that make repeated contacts with these processes. Thus Scheibel and Scheibel (1970b) mentioned that 'nonspecific' cortical afferent fibres originating from the brain stem and the medial thalamus break up into a series of branches that ascend radially through the cortex, establishing sequences of axodendritic contacts with the spines of apical shafts and terminal arches of pyramidal neurons. It is also known that the axons of a certain category of cortical local circuit neurons, known as bipolar cells, typically give rise to vertically oriented branches which parallel the trajectories of clustered pyramidal apical dendrites (Fig. 22.200d). These branches form multiple asymmetrical synapses with the spines of the apical dendrites (Feldman and Peters 1978; Peters 1984b). The following data on the structure and the possible functional significance of bipolar cortical neurons are all derived from Peters (1984b).

Typical bipolar cells have small, spindle-shaped cell bodies from which two or three primary dendrites arise that produce an elongated, radially oriented dendritic field. The axons of bipolar cells frequently arise from one of the primary dendrites and form a plexus that is also radially elongated. Thus far, the bipolar cells have been examined most extensively in the cerebral cortex of the rat, in which they are encountered throughout laminae II–V. However, elements of this type have also been observed in other species, including the cat, monkey and human. As regards the afferents of the bipolar cells, it has been observed that elements of this type situated in the visual cortex of the rat receive geniculocortical axon terminals (Peters and Kimerer 1981), and White (1978) has shown that a bipolar cell in the primary somatosensory cortex of the mouse receives thalamocortical afferents. Thalamocortical fibres make asymmetrical, presumably excitatory contacts with all of their target elements, thalamocortical fibres synapse not only with bipolar neurons, but also directly with pyramidal neurons and bipolar cells make asymmetrical, presumably excitatory synapses with pyramidal neurons; given these facts, it seems likely that bipolar cells reinforce the excitation that pyramidal neurons receive from thalamocortical afferents. Moreover, because the axons of bipolar cells form elongated, radially oriented plexuses, they may activate vertical arrays of pyramidal cells (Peters 1984b). The bipolar cells also bear numerous symmetrical, presumably inhibitory synapses. It is likely that, at least in the visual cortex, most of these synapses are derived from smooth or sparsely spinous stellate and bitufted cells (Peters and Kimerer 1981; Fig. 22.200d). The question as to whether all neocortical bipolar cells fit the description given above will be taken up in a later section of the present chapter.

Apart from 'nonspecific' cortical afferents and the axons of bipolar cells, collateral branches of pyramidal cell axons may also make repeated, climbing fibre-like contacts with the apical dendrites of (other) pyramidal cells. Gabbott et al. (1987) studied the connections between pyramidal neurons in lamina V of the primary visual cortex of the cat. In one case, they observed a fine axon collateral of a pyramidal neuron coming into close contact with the proximal portion of the apical dendrite of another pyramidal neuron. During its ascent along the apical dendrite, the axon collateral showed 14 varicosities. Ultrastructural analysis revealed that most of these varicosities formed asymmetrical synaptic contacts with spines that most probably had arisen from the apical dendrite.

The final part of the dendritic system of typical pyramidal neurons to be considered is the set of dendritic branches which emanates from the tip of the apical dendrite. These so-called apical dendritic tufts or terminal dendritic bouquets extend into lamina I, in which, together with numerous tangentially running axons, they constitute a typical plexiform zone. It has already been mentioned that fibres originating from the intralaminar and midline thalamic nuclei project to lamina I of the cortex and that, according to Herkenham (1979), fibres from different thalamic nuclei terminate in different subzones of that layer (a in Fig. 22.201). However, there is evidence that the plexiform zone in lamina I, in addition to thalamic afferents, also receives fibres from several other sources, including (1) fibres presumably originating from the brain stem, (2) recurrent collaterals of pyramidal neurons, (3) ascending axons from neurons situated in deeper cortical layers, and (4) axons of neurons situated in lamina I itself:

1. *Axons presumably arising from the brain stem* (b in Fig. 22.201). In Golgi material of the developing cerebral cortices of mouse, cat and human embryos, Marin-Padilla (1984) observed numerous early developing fibres which arise from the primordial pallial white matter and reach a superficial position. Here, these fibres split up into two or more long, tangentially oriented collaterals which can be followed for a long distance. These collaterals give off numerous short, ascending terminal branches throughout their entire length. In later developmental stages, these fibres appear to be preferentially distributed through the upper half of lamina I. Marin-Padilla (1984) suggested that these fibres originate from the brain stem and are monoaminergic in nature. He believed that these fibres are still present in the adult stage, although by then they are much 'diluted' by numerous thalamic and other afferents that have arrived in lamina I later in development. According to Marin-Padilla (1984), the main targets of these long, tangentially coursing collaterals are the so-called Cajal-Retzius cells. These elements will be discussed below.

2. *Recurrent collaterals of pyramidal neurons* (c in Fig. 22.201). Recurrent collaterals of the axons of pyramidal neurons represent an important component of tangential fibres in lamina I (Lorente de Nó 1922; Valverde 1985; Valverde and Facal-Valverde 1986; m,o in Fig. 22.193, r in Fig. 22.194). The neurons giving rise to these collaterals are principally situated in laminae II and III. Scheibel and Scheibel (1970b) reported that recurrent collaterals of pyramidal cells terminate not only on the basal dendrites, but also on the apical dendritic tufts of nearby pyramidal elements.

3. *Ascending axons from neurons situated in deeper cortical layers.* Lamina I is also the site of termination of ascending axons of intrinsic cortical neurons (d,f in Fig. 22.193, m in Fig. 22.194). Among these, multipolar or bitufted neurons are found with an axon that follows a straight ascending course toward lamina I, where it ramifies in a terminal arborisation. These elements, which have smooth or sparsely spinous dendrites, are known as Martinotti cells (d in Fig. 22.201). Large elements of this type are found in laminae V and VI, but they are also present in the more superficial cortical layers (Fairén et al. 1984). Marin-Padilla (1984) reported that the morphology of the axonal termination of Martinotti neurons resembles quite closely the arborisation of the apical dendritic tufts of pyramidal cells and considered it likely that Martinotti cells form dual sets with pyramidal neurons of similar cortical depth. With regard to the function of the Martinotti cells, Marin-Padilla (1984) suggested that these elements are inhibitory and that the inhibition takes place specifically between the axonal terminals of a given Martinotti cell and the dendritic tufts of the pyramidal neurons with which it forms a dual set.

4. *Axons of neurons situated in lamina I itself.* Although lamina I is first and foremost a plexiform layer, it is not entirely free of neuronal elements. Prominent among these are the so-called horizontal cells, which have been observed in Golgi material of a variety of mammals, including humans (Cajal 1911). The somata of these elements vary in size. Their dendritic and axonal plexuses are entirely confined to lamina I

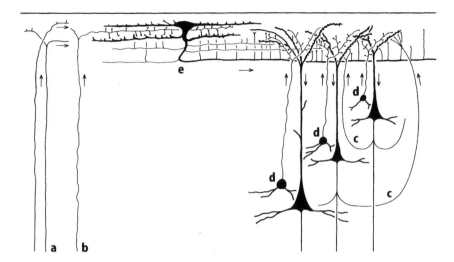

Fig. 22.201. Elements entering into synaptic contact in layer I of the mammalian cerebral cortex, Specific thalamic afferents (*a*). Monoaminergic fibre from the brain stem (*b*). Recurrent collaterals of pyramidal neurons (*c*). Martinotti cells (*d*). Cajal-Retzius cell (*e*). (Largely based on Marin-Padilla 1984)

and extend parallel to the surface of the cortex (l in Fig. 22.193, g,t in Fig. 22.194, E in Fig. 22.201). Some of their axonal branches may attain a considerable length. There is immunohistochemical evidence (summarised by Vogt 1991) indicating that the horizontal neurons are all GABAergic and also contain the neuropeptide cholecystokinin.

It should be mentioned here that the primordial lamina I of the developing neopallium contains numerous neurons, which are provided with thick and irregular tangentially oriented dendrites covered by numerous shorter and longer ascending branchlets. The axons of these conspicuous elements, which are known as the Cajal-Retzius cells, can be followed over long distances. Marin-Padilla (1984) observed that the axons of these elements in the developing cortex constitute a prominent system of tangential fibres in the lower half of lamina I, whereas the terminal branches of the extrinsic fibres, described in the list above under item 1, tend to occupy the upper half of that layer. It has already been mentioned that, according to Marin-Padilla (1984), these extrinsic fibres synapse principally with Cajal-Retzius neurons, with which they establish axodendritic synapses. The axons of the Cajal-Retzius cells, like those of the early-arriving extrinsic fibres, are provided with numerous ascending terminal branches, and Marin-Padilla (1984) observed in Golgi material multiple contacts between these terminal branches and the spines of dendritic branches of pyramidal cells. Thus, in Marin-Padilla's opinion, Cajal-Retzius cells in the developing cerebral cortex constitute a link between certain early-arriving, presumably monoaminergic fibres and the apical dendritic tufts of pyramidal neurons. The extensive literature on the fate of Cajal-Retzius cells will not be discussed here (for reviews, the reader is referred to Marin-Padilla 1984; Valverde and Facal-Valverde 1986). Suffice it to mention here that many authors (e.g. Fox and Inman 1966; Bradford et al. 1977) have suggested that Cajal-Retzius cells should be considered as transient elements which disappear entirely during the maturation of the cortex, but that several others (e.g. Cajal 1911; Marin-Padilla 1984; Valverde and Facal-Valverde 1986) have expressed the opinion that these elements become transformed into the horizontal cells of the adult cortex. The axons of the horizontal cells lack the elaborate system of ascending terminal branches that the typical Cajal-Retzius cells show, but it has been reported that axons of the former elements, like those of the latter, establish synaptic contacts with the apical dendritic ramifications of pyramidal neurons (Lorente de Nó 1938).

In summary, it may be stated that the apical dendritic branches of neocortical pyramidal neurons receive input from various sources. Thalamic afferents and recurrent collaterals of pyramidal cells presumably exert an excitatory influence on these dendrites, whereas the axonal endings of the Martinotti cells and the horizontal cells may well exert an inhibitory influence on them. The horizontal elements have been suggested to receive a specific input from monoaminergic fibres originating from neurons situated in the brain stem, but thalamic afferents may also impinge on these intrinsic lamina I elements and the monoaminergic fibres may also directly contact the pyramidal apical

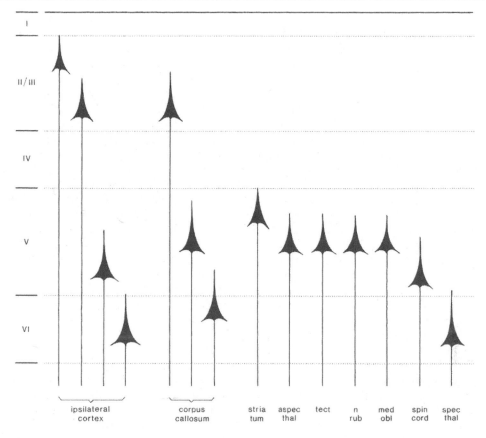

Fig. 22.202. Laminar location of the perikarya of pyramidal cells projecting to other parts of the cerebral cortex and to subcortical centres. *aspec thal*, aspecific thalamic nuclei; *tect*, tectum; *n rub*, nucleus ruber; *med obl*, medulla oblongata; *spec thal*, specific thalamic nuclei; *spin cord*, spinal cord. (Based on White 1989)

dendrites. However, it should be emphasised that so far none of the synaptic contacts suggested by light microscopy material has been verified with the aid of experimental ultrastructural techniques. Finally, it should be noted that the pyramidal apical dendritic tufts constitute the principal, but presumably not the only outlet of lamina I. The Golgi studies performed by Cajal (1911), Lorente de Nó (1922, 1938) and many others showed that dendrites of several types of cortical local circuit neurons, the axons of which project to deeper layers, extend into lamina I (e,k in Fig. 22.193). Among these, the superficially situated basket and chandelier cells observed by Valverde and Facal-Valverde (1986) in the neocortex of the hedgehog may be especially mentioned.

Having discussed the afferents making contact with the various parts of the receptive surface of the typical pyramidal neurons, I will now turn to the axons of these elements. It has already been mentioned that these processes all leave the cortex and pass either to other ipsilateral or contralateral cortical regions or to one or several subcortical centres. The latter may include the striatum, i.e. the nucleus caudatus and the putamen, the various 'specific' and 'non-specific' thalamic nuclei, the nucleus ruber, the colliculus superior or tectum mesencephali, the pontine nuclei, the medulla oblongata and the spinal cord.

Retrograde tracing studies have shown that the cell bodies of pyramidal neurons projecting to particular cortical or subcortical targets are preferably located in particular cortical layers or sublayers (Fig. 22.202). The following summary of the laminar relationships of cortical efferent cells is based on the recent reviews by Jones (1984) and White (1989), to which the reader is referred for the primary sources.

Cortico-cortical and callosally projecting fibres arise predominantly from pyramidal neurons in lamina II and III; however, in rodents and primates, significant numbers of these fibres have been found to originate from elements situated in the infragranular layers. As regards the superficial layers, it has been established that the smaller, more superficially situated pyramids tend to project to ipsilateral cortical areas situated nearby, whereas the larger, more deeply placed cells tend to project to contralateral and to more remote ipsilateral cortical areas.

Pyramidal neurons situated in lamina V have been shown to project subcortically to the intralaminar and other 'aspecific' thalamic nuclei, the striatum, the red nucleus, the tectum, the medulla oblongata and the spinal cord. The smallest and most superficially situated elements in this layer project to the striatum, while the largest and most deeply situated cells project to the spinal cord. The elements projecting to the remaining subcortical sites tend to occupy an intermediate position.

The corticothalamic projections to the 'specific' thalamic relay nuclei arise exclusively from large pyramids in lamina VI.

Although most cortical neuronal populations projecting to a particular cortical or subcortical target show a distinct laminar specificity, it is not uncommon to find some degree of overlap in the boundaries demarcating different populations of projection neurons. This raises the question of the extent to which projections to particular targets of cortical efferent neurons are made up by collaterals of axons projecting to other centres. In this context, it may be recalled that Cajal (1909, 1911) noted in his Golgi studies of the brains of rodents that numerous subcortical centres are supplied by collaterals of corticofugal fibres. Thus he observed that,

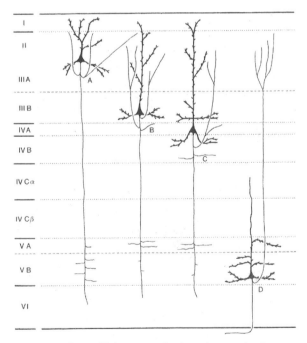

Fig. 22.203. Pyramidal neurons in the primary visual cortex of the rhesus monkey, situated at different levels (A–D), contribute axon collaterals to particular layers. (Based on Lund and Boothe 1975)

during their descent through the internal capsule, such corticofugal fibres issued numerous collaterals to the striatum or thalamus and that pyramidal tract neurons in the brain stem gave off collateral branches to several centres, including the red nucleus, the pontine nuclei and the dorsal column nuclei. Double-labelling experiments, i.e. experiments in which two different and distinctive retrogradely transported labels have been injected into two different known terminal fields, have revealed that double-projecting neurons do occur in the neocortex. Thus Catsman-Berrevoets and Kuypers (1981) reported the presence of double-labelled cells in the motor cortex of the monkey after injections of the magnocellular part of the red nucleus and spinal cord, and Rustioni and Hayes (1981) found double-labelled cells in the sensory cortex of the cat after injections of the dorsal column nuclei and spinal cord. However, in these and other comparable experiments, the number of double-labelled cells appeared to be very small, implying that the degree of subcortical collateralisation of corticofugal fibres is likewise limited (Jones 1984). The abundance of such collaterals observed by Cajal (1909, 1911) might well have to do with the fact that the Golgi material studied by that author was exclusively derived from the brains of very young animals.

The axons of all neocortical pyramidal neurons release a number of collaterals before entering the subgriseal white matter. These collaterals may ramify within close proximity in the parent cell body or may descend, ascend or travel for shorter or longer horizontal distances within the cortex (o,r in Fig. 22.193, s,v,w in Fig. 22.194). Early studies of the intracortical distribution of pyramidal cell axon collaterals were based exclusively on the study of Golgi material, but the real extent of these processes has only recently been revealed by experimental studies in which single pyramidal cells were intracellularly injected with HRP. Given the numerical preponderance of pyramidal neurons, there can be no doubt that the intracortical collateral branches of these neurons together constitute the largest single category of axons in the neocortex. The endings of these collateral branches, like those of the main axons, all make synapses of the asymmetric/round vesicle variety and use the excitatory amino acids glutamate and aspartate as neurotransmitters.

Systematic studies on the intracortical collaterals of pyramidal cell axons have revealed that the distribution of these axonal branches is orderly and obeys certain rules. Lund and Boothe (1975), who studied the organisation of neurons in the primary visual cortex of the macaque in Golgi material,

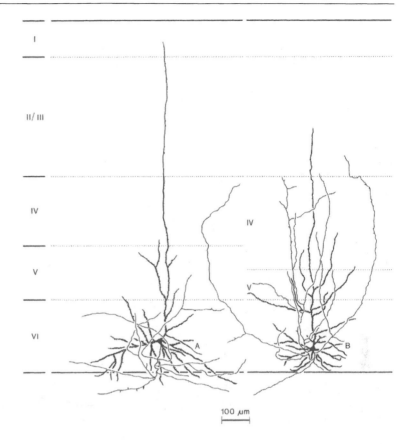

Fig. 22.204. Pyramidal neurons in the sixth layer of the primary visual cortex of the cat. Claustrum-projecting neuron (*A*). Lateral geniculate nucleus-projecting neuron (*B*). (Based on Katz 1987)

found that the axon collaterals of pyramidal neurons situated in various cortical layers often show quite characteristic distribution patterns (Fig. 22.203). Thus pyramidal neurons situated in laminae II and IIIA each have a descending axon trunk that gives off recurrent collaterals in laminae II and IIIA and then descends unbranched to lamina VB, where it issues some relatively short horizontal collaterals. Pyramidal cells situated in lamina IIIB show strong axonal arborisations in laminae IIIA and VA. The axons of neurons with their somata in lamina IVA contribute ascending collaterals to laminae II and III and horizontal collaterals to both IVA and IVB. Their descending part selectively issues some prominent collaterals in lamina VA. Finally, certain pyramidal cells situated in lamina VB, the apical dendrites of which became vestigial above lamina V, characteristically have a strong recurrent axonal branch which arborises in laminae IIIA and II. The data provided by Lund and Boothe (1975) warrant the conclusions that, in the macaque primary visual cortex, many pyramidal neurons are provided with recurrent collaterals that arborise in the superficial laminae II and IIIA, and that the axons of pyramidal neurons situated in laminae II–IV selectively issue horizontal collateral branches in lamina VA or VB.

Another study in which the collateralisation patterns of different populations of pyramidal neurons were compared is that carried out by Katz (1987). Lamina VI of the primary visual cortex of the cat, in addition to elements projecting to the 'specific' visual thalamic nucleus, i.e. the lateral geniculate nucleus, also contains another population of pyramidal neurons projecting to the claustrum (LeVay and Sherk 1981). Katz (1987) labelled neurons giving rise to these two projections retrogradely and found that these two groups of elements have very different patterns of dendrites and local axon collaterals, whereas the patterns within each group appeared to be highly stereotyped (Fig. 22.204). As regards the dendritic patterns, the cells projecting to the claustrum had apical dendrites reaching to lamina I, with branches in lamina V only, while cells projecting to the lateral geniculate nucleus never had an apical dendrite reaching higher than layer III, with side branches in laminae IV and V. The differences between the axonal collaterals appeared to be particularly striking. Cells projecting to the claustrum had fine, horizontally directed collaterals that arborised exclusively in lamina VI and lower lamina V. However, most cells projecting to the lateral geniculate nucleus had virtually no horizontal arborisations in lamina VI. Instead, they

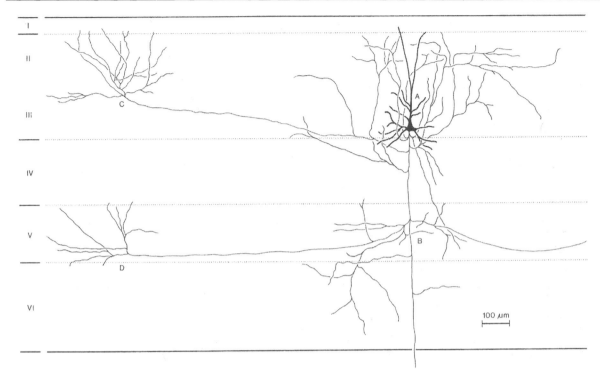

Fig. 22.205. A pyramidal neuron in layer III of the primary visual cortex of the cat. The element was intracellularly filled with horseradish peroxidase (HRP) and reconstructed from 80-μm-thick serial sections. The intracortical system of collaterals forms distinct clusters, one near the cell's dendritic field in layer III (*A*), another just below the cell in layer V (*B*), and two at a distance of some 1000 μm from the soma in layers III and V (*C,D*). (Modified from Kisvárday et al. 1986)

issued widespread ascending collaterals, which arborised extensively in lamina IV.

Apart from local and vertically oriented interlaminar collaterals, the axons of pyramidal neurons may also give rise to long, horizontally disposed branches. Pyramidal neurons emitting such long-range collaterals have been observed in different species, in different cortical areas and in different cortical layers. The ensuing discussion of these remarkable processes is principally based on the following publications (for the sake of brevity, the animals studied, the cortical areas and laminae in which the parent somata were located and an indication of the length of the collaterals observed are included in the references): Gilbert and Wiesel (1979: cat visual cortex, lamina III, 6–8 mm), Landry et al. (1980: cat motor cortex, lamina V, up to 3 mm), Rockland and Lund (1982: tree shrew visual cortex, lamina III, 1 mm), DeFelipe et al. (1986: monkey sensory motor cortex, lamina III, up to 6 mm), Kisvárday et al. (1986: cat visual cortex, lamina III, 1500 μm), Gabbott et al. (1987: cat visual cortex, lamina V, up to 2.64 mm), Ojima et al. (1991: cat auditory cortex, laminae II, III, 0.7–2.5 mm), McGuire et al. (1991: macaque visual cortex, lamina III, 2 mm) and Kisvárday and Eysel (1992: cat visual cortex, lamina III, up to 2.8 mm). In most of these studies, pyramidal neurons were intracellularly injected with tracer substances; however, the studies by DeFelipe et al. (1986) and Kisvárday and Eysel (1992) are based on an analysis of neurons labelled after small extracellular injections of tracers.

It has been repeatedly reported that horizontally oriented axons of pyramidal neurons contribute to the fibre plexuses concentrated in the striae of Baillarger (e.g. Lorente de Nó 1938; Lund 1973; Valverde 1985). The relationship between these striae and the long-range collaterals is not discussed in the literature cited above; however, most of the published illustrations suggest that these collaterals, notwithstanding their overall horizontal orientation, often follow an undulating course, which is difficult to reconcile with adherence to any of the striae (see e.g. Landry et al. 1980; Gabbott et al. 1987; Ojima et al. 1991). As regards the extent of the long-range collaterals, it has been observed that processes of this type do not remain within the cytoarchitectonic area in which their parent soma lies, but may project to adjacent cortical areas (Landry et al. 1980; DeFelipe et al. 1986; Gabbott et al. 1987).

The number of major collaterals with long horizontal trajectories issued by the main axons of pyramidal neurons varies; most of the cells examined by DeFelipe et al. (1986) had one to three long collaterals, whereas the elements studied by Ojima et al. (1991) issued two to five such processes.

The long-range primary collaterals are usually coarse and well myelinated.[1] They give off thin, unmyelinated, bouton-laden secondary branches which are predominantly oriented perpendicular to the cortical surface. Remarkably, these secondary branches are emitted in clusters at regular intervals with a periodicity of about 1 mm (Fig. 22.205). Because of the overall vertical orientation of the individual secondary branches, these clusters often have a column-like appearance. Secondary branches arising from different major collaterals of the same parent axon in different layers often converge upon the same cluster (Kisvárday et al. 1986; DeFelipe et al. 1986; Ojima et al. 1991), and the same holds true for the focused terminal branches of major collaterals of different pyramidal cells (DeFelipe et al. 1986). Kisvárday and Eysel (1992) visualised groups of pyramidal neurons in lamina III of the visual cortex of the cat by making small extracellular injections of the tracer biocytin. The axons of all of these elements emitted long horizontal collaterals which gave rise to clusters of secondary branches. Each pyramidal cell established four to eight clusters of these terminal branches. Three-dimensional reconstruction revealed that these ten neurons and their horizontal axonal processes together constituted a 'patchy' network and that, in each individual 'patch', up to five clusters of terminal branches overlapped. The local axon collaterals of the pyramidal cells, i.e. the collateral branches that ramify within the immediate vicinity of the somata, also participated in the formation of the network. Each axonal 'patch' covered an area of up to 400 µm in diameter, and the centre-to-centre distance between the patches ranged from 0.8 to 1.5 mm, with an average of 1.1 mm for the ten pyramidal neurons. Detailed analyses revealed that many of the pyramidal cells of the network are directly interconnected. Kisvárday and Eysel (1992) hypothesised that the functional role of the network is to establish direct links between remote loci with similar physiological characteristics, such as orientation preference.

In order to gain insight into the way in which pyramidal neurons participate in the intrinsic circuitry of the neocortex, the synaptic connections of axon collaterals belonging to neurons of this type have been examined with the electron microscope in several areas and in different species. In this context, the studies by Winfield et al. (1981), White and Hersch (1981), White and Keller (1987), Elhanany and White (1990), McGuire et al. (1984, 1991), Kisvárday et al. (1986) and Gabbott et al. (1987) should be mentioned. The main results of these studies have been assembled in Table 22.12. In most of these studies, the pyramidal neurons examined were labelled by intracellular injections of HRP, and in several of them it was attempted to identify the postsynaptic targets of the collaterals by serial electron microscopy reconstruction. This type of research is very laborious, which explains why in all of the reports mentioned only a very limited number of neurons were examined and why the study of labelled terminals and their targets is generally confined to one or two layers.

The results of the eight studies summarised in Table 22.12 will now be discussed. For the sake of brevity, the pyramidal cells examined in each of these studies will be designated as groups a–h, as indicated in Table 22.12.

All studies confirmed that pyramidal cell collaterals only form synapses of the asymmetric/round vesicle variety and, hence, may be considered to excite their targets. Axosomatic contacts appeared to be extremely rare, and axo-axonic synapses were not reported. With regard to axodendritic contacts, it was observed that the intracortical collaterals of pyramidal neurons always establish synapses with dendritic spines and with dendritic shafts. Because, in addition to the spiny pyramidal neurons, the cortex also contains spiny non-pyramidal cells and because shaft synapses occur on the dendrites of all types of cortical neurons, this observation as such says little about the identity of the target neurons. However, given the fact that the most common densely spiny, cortical local circuit neurons are concentrated in lamina IV and that in the other layers local circuit neurons are relatively rare and have only smooth or sparsely spiny dendrites, it has been argued that, in laminae I–III and V–VI, most if not all of the spines postsynaptic to local axon collaterals of pyramidal neurons are derived from other pyramidal neurons (Kisvárday et al. 1986; Gabbott et al. 1987; McGuire et al. 1991). On that account, it seems reasonable to assume that the collaterals of the pyramidal neurons of groups b, d and e synapse primarily with other pyramidal neurons. Gabbott et al. (1987) specified the intracortical postsynaptic targets of the pyramidal cells studied by them for layers IV, V and VII. Their data strongly suggest that the collaterals of these elements in layers V and VI also mainly contact the dendrites of other pyra-

---

[1] Myelinated fibres fail to impregnate with the Golgi stain; this may explain why these long-range collaterals remained unnoticed in the classical Golgi studies by Cajal (1911) and Lorente de Nó (1922, 1938).

Table 22-12. Summary of eight studies on the synaptic contacts of collaterals of neocortical pyramidal cells

	Number of cells studied	Localization of soma	Localization collaterals studied	Total number of synapses	% Synapses on spines	% Synapses on shafts (total)	% Synapses on shafts spiny dendrites	% Synapses on shafts smooth dendrites
a. Monkey primary somatosensory cortex (Winfield et al. 1981)	1	III	II,III	13	40	60		
b. Mouse primary somatosensory cortex (Elhanany and White 1990)	?	III	III,IV	215	85	15		
c. Mouse primary somatosensory cortex (White and Hersch 1981)	2	III	IV	12	50	50		
d. Cat primary visual cortex (Kisvárday et al. 1986)	2	III	III,V	191	85	15	10	5
e. Monkey primary visual cortex (McGuire et al. 1991)	2	III	III	117	76	26	4	20
f. Cat primary visual cortex (Gabbott et al. 1987)	2	V	IV–VI	313	80	20		
g. Cat primary visual cortex (McGuire et al. 1984)	2	VI	IV	151	30	70	30	40
h. Mouse primary somatosensory cortex (White and Keller 1987)	?	V–VI	IV–VI	190	8	92	0	92

midal cells. [It may be added here that Czeiger and White (1993) recently analysed the synapses made by callosal projection neurons in the visual cortex of the mouse, both by their main axons in the contralateral hemisphere ('extrinsic callosal axon terminals') and by their local collaterals ('intrinsic callosal axon terminals'). They found that layers II and III contain the highest concentrations of extrinsic and intrinsic callosal axon terminals. Remarkably, 97% of both the extrinsic and intrinsic callosal axon terminals synapsed onto dendritic spines, likely those of pyramidal neurons.]

A considerable number of the endings of the collaterals of the remaining groups of pyramidal neurons (a, c, g, h) contact dendritic shafts, and in two of these groups (g and h) the nature of the dendrites with which these shaft synapses are made has been determined. It appeared that 40% of the synapses made by the collaterals of certain pyramidal cells in lamina VI of the primary visual cortex of the cat are with smooth dendrites (McGuire et al. 1984, group g) and that, in the deep layers of the mouse primary somatosensory cortex, pyramidal cells occur, the local axon collaterals of which form almost all of their synapses with smooth dendrites (White and Keller 1987, group h).

The data discussed indicate that different groups of pyramidal neurons may show striking differences in their local output relationships. In one group (d), over 90% of the intracortical synapses appeared to be made with other pyramidal cells, whereas in another group (h) over 90% of these synapses were found to be made with the smooth dendrites of non-pyramidal neurons. Other groups (e, g) occupy an intermediate position between these two extremes. It is conceivable that a relationship exists between the synaptic output patterns of the local axon collaterals of pyramidal cells and the distant projection site of their main axons (White 1989).

The question of whether the intracortical synaptic contacts made by pyramidal neurons show a similar pattern throughout all parts of their collateral ramification, or whether the synaptic relationships of these collaterals exhibit local differences, has been addressed by several authors. Kisvárday et al. (1986) studied two pyramidal cells in lamina III of the primary visual cortex of the cat, the collaterals of which formed four discrete clusters of terminals, two in lamina III and two in lamina V (Fig. 22.205). One of the clusters in lamina III was located in the immediate vicinity of the parent cell body, and the two clusters in lamina V were in register with those in lamina III. The proportion of synapses made with dendritic spines and with dendritic shafts appeared to be strikingly similar in the four target areas. Closely corresponding findings were reported by McGuire et al. (1991) for similar pyramidal cells with clustered terminal branches in the primary visual cortex of the monkey. According to both groups of authors, these results suggest that the elements studied perform the same operational task in lamina III and lamina V, both in the vicinity and further away from the parent cell body. Evidence suggesting that the output relationships of pyramidal cells may differ in different cortical laminae has also been presented, however. Gabbott et al. (1987) noted that the proportion of synaptic contacts with dendritic shafts made by the collaterals of pyramidal neurons located in lamina V of the primary visual cortex of the cat increases from a value of zero in lamina IV to as high as 25%–30% in lamina VI. White and Hersch (1981) established that the collaterals of pyramidal cells situated in lamina III of the primary somatosensory cortex of the mouse make half of their synapses with dendritic spines and the other half with dendritic shafts. Another study (White and Elhanani, unpublished; see White 1989) revealed that the collaterals of similar elements in the same cortical area of the same species in both lamina III and lamina V form about 85% of their synapses with dendritic spines. White (1989) considered it likely that the cells examined in the two studies actually belong to one and the same population and concluded that the proportions of the different postsynaptic elements contacted by the collaterals of superficial pyramidal neurons may be very different in laminae III and V from those in lamina IV. These results are too limited and too fragmentary to draw any general conclusions. However, it may well be that pyramidal neurons with long-range primary collaterals and clustered terminal branches typically exhibit similar synaptic relationships within all of the clusters in which they participate, whereas the output relationships of other pyramidal neurons are different in different laminae.

From the foregoing, it appears that the collaterals of all neocortical pyramidal neurons most probably contact other pyramidal neurons and that for many of these latter elements other pyramidal neurons represent the principal postsynaptic targets. Because the synaptic contact made by the collaterals of pyramidal neurons are all of the asymmetrical/round vesicle variety, it seems likely that these collaterals and their contacts provide the morphological substrate of feedforward excitation of pyramidal elements. The number of synapses made by the collaterals of a given pyramidal neuron with one other individual pyramidal neuron is presumably generally very limited (Szentágothai 1975, 1979; McGuire et al. 1984, 1991; Kisvárday et al. 1986; Gabbott et al. 1987). However, the collaterals of one pyramidal cell contact numerous other pyramidal cells and, conversely, one pyramidal cells receives the consequent input of numerous other pyramidal cells.

We have seen that the collaterals of pyramidal neurons also contact smooth or sparsely spinous dendrites, which are characteristic of cortical local circuit neurons. In some instances, the type of neuron contacted could be determined. Thus one of these postsynaptic neurons was identified as a non-spiny bipolar cell (McGuire et al. 1984), and another one as a non-spiny multipolar cell (White and Keller 1987), whereas McGuire et al. (1991) adducced evidence strongly suggesting that axon collaterals of pyramidal neurons contact small or medium basket cells. It has also been established that some of the dendrites postsynaptic to pyramidal cells axon collaterals are immunoreactive to GABA (Kisvárday et al. 1986). It is known that most types of non-spiny or sparsely spiny non-pyramidal cells, including basket cells, chandelier cells and double bouquet cells, use GABA as a neurotransmitter and that these elements are the principal source of the GABAergic, symmetrical synapses that impinge upon the somata, proximal dendrites and axon initial segments of pyramidal neurons (Houser et al. 1984). All in all, it seems reasonable to assume that the GABAergic interneurons in the neocortex receive input directly from the axon collaterals of pyramidal neurons and in turn synapse with pyramidal neurons. These circuits probably provide the morphological substrate for both feedforward and feedback inhibition of pyramidal neurons (Houser et al. 1984; White 1989).

If we survey the data concerning the typical pyramidal neurons discussed above, it appears that these elements show a quite remarkable structural diversity. This diversity may concern their size, their laminar position, the branching pattern of their dendrites, the density of spines along their apical dendrites, their affinity to particular afferent systems, the cortical or subcortical target regions to which their main axons project, the distribution of their axonal collaterals and their patterns of intracortical synaptic output. Certain structural properties are clearly correlated. It has been demonstrated that pyramidal neurons projecting to a particular target not only have their somata located in one and the same layer or sublayer, but also show striking similarities with regard to dendritic morphology, thalamocortical connectivity and distribution of their axon collaterals. It seems likely that all pyramidal neurons projecting to a particular target are in receipt of similar extracortical and intracortical inputs and that they participate in a similar way in the intrinsic circuitry of the cerebral cortex.

22.11.6.5.4
Atypical Pyramidal Cells

In the mammalian mesocortex and neocortex, neurons occur which lack one or several of the features characterising typical pyramidal cells, but which are nevertheless considered to belong to the pyramidal cell group. Some of these atypical pyramids are regarded as primitive elements which have not yet attained the status of typical pyramidal cells, but most of them may be designated as aberrant or modified pyramidal neurons.

As regards primitive pyramid-like neurons, in the second layer of the neocortex, cells can be found, most of the dendrites of which extend peripherally, where they form a characteristic, wide, subpial tuft. Basal dendrites are also present, but these are smaller in number and more limited in extent. Cells of this type have been observed in the neocortex of the opossum (Sanides and Sanides 1972), hedgehog (Valverde 1986; Valverde and Facal-Valverde 1986; Fig. 22.206A), rat (Kirsche et al. 1973; Fig. 22.206:B), the bat (Sanides and Sanides 1972) and dolphin (Morgane et al. 1990; Fig. 22.206:C). Sanides and Sanides (1972) designated elements of this type as 'extraverted' neurons. They considered these elements as standing intermediate between pallial neurons provided with only a peripherally extending dendritic tuft (compare a in Fig. 22.179, B in Fig. 22.181, SL in Fig. 22.184 and gr2 in Fig. 22.188) and fully developed pyramidal cells.

The group of aberrant or modified pyramidal cells includes the following seven types: (1) improperly oriented pyramidal cells, (2) elements with a reduced apical dendrite, (3) spiny projection neurons which lack an apical dendrites, (4) sparsely spiny pyramidal cells, (5) 'pure' projection pyramidal cells, (6) intrinsic pyramidal cells, and (7) spiny stellate cells. These seven cells types will now be briefly discussed.

1. *Improperly oriented pyramidal cells.* A radially oriented apical dendrite directed towards the pial surface is a prominent feature of most pyramidal cells. However, in the cortex of the rat, rabbit, cat, monkey and humans, pyramidal neurons have been observed whose long axis is oriented obliquely or parallel to the pial surface or whose main dendritic trunk is directed towards the white matter (van der Loos 1965; Globus and Scheibel 1967; Kirsche et al. 1973; Braak 1980; Fig. 22.206:D). In one sample of pyramidal cells from the cortex of the rabbit, van der Loos found that 18% of the elements were improperly oriented, in the sense that they deviated from the radial by more than 20%. Globus and Scheibel (1967) found that pyramidal neurons make up about 75% of the neuron population of the primary visual cortex of the rabbit and that about 5% are inverted.

2. *Elements with a reduced apical dendrite.* Typical pyramidal cells are provided with a conspicuous apical dendrite which ascends to the subpial zone, where it forms a highly branched terminal tuft. In the deeper layers of the cerebral cortex, numerous pyramidal cells can be observed which should be considered as atypical because their apical dendrite does not fit the description given above. Thus, in lamina VI of the primary visual cortex of the cat (Katz 1987; Fig. 22.204) and monkey (Valverde 1985; Fig. 22.199a), pyramidal neurons occur whose apical dendrites remain unbranched and taper either in lamina I or in lamina II/III. The apical dendrites of a certain proportion of the pyramidal neurons in lamina VB of the primary visual cortex of the monkey issue some side branches in laminae VB and VA, but become vestigial above lamina V (Lund and Boothe 1975; Fig. 22.203:D), and in lamina VA of the same cortex numerous small pyramidal neurons occur whose entire apical dendrite is reduced to a thin, thread-like process (Lund 1973, 1984; Fig. 22.206:E). Finally, in the deeper layers of the cerebral cortex of many different species, including the mouse (Lorente de Nó 1922; x in Fig. 22.194), rat (Koester and O'Leary 1992), cat (Hübener et al. 1990) and monkey (Lund 1984; Valverde 1986; Fig. 22.200b,c), pyramidal cells occur whose apical dendrite does not extend into the subpial zone, but rather ramifies into a terminal bouquet in laminae III, IV or even V, for which reason these elements are characterised as short pyramidal cells. In lamina V of the neocortex, projection

neurons occur that send their axons across the corpus callosum to the contralateral hemisphere. It is known that, in the rat, these callosal neurons display a typical, short pyramidal morphology (Hübener and Bolz 1988), although other lamina V projection cells, e.g. those projecting to the tectum, have a 'tall' appearance with an apical dendrite extending into the first cortical layer. Koester and O'Leary (1992) studied the ontogenetic development of lamina V callosal and tectal projection neurons in the rat. They found that during prenatal development both cell types have a lamina I apical dendrite, but that during the first 3 postnatal weeks the callosal elements gradually acquire the typical, short pyramidal morphology, whereas the tectal projection cells remain 'tall'. Koester and O'Leary (1992) concluded from these observations that the apical dendrites superficial to lamina IV of the lamina V callosal neurons are selectively eliminated during development. They considered it likely that "most, if not all, types of cortical neurons initially develop a radial morphology typified by an apical dendrite extending to the pial surface, and that many of the nonpyramidal morphologies characteristic of the adult cortex are the product of later developmental events" (Koester and O'Leary 1992, p. 1391). Interestingly, Marin-Padilla (1992), studying the prenatal development of the cerebral cortex in Golgi material of a variety of mammals, including the hamster, mouse, rat, cat and humans, arrived at a similar conclusion. He found that, during ontogenesis, all neurons in the mammalian neocortex develop a perpendicular apical dendrite with a terminal bouquet that branches within layer I. During further development, a certain proportion of these neurons elongate their apical dendrite and retain their original connection with layer I, thus becoming real or typical pyramidal neurons, but others lose their initial anchorage to layer I and in Marin-Padilla's opinion become transformed into various types of local circuit neurons. From an evolutionary point of view, it is important to note that in the small and presumably primitive neocortex of the hedgehog the apical dendrites of practically all pyramidal cells form a terminal tuft in layer I (Valverde and Facal-Valverde 1986).

3. *Spiny projection cells which lack an apical dendrite.* In lamina IV of the visual cortex of the cat, monkey and humans, large stellate cells occur whose thorny dendrites either radiate out in all directions or spread mainly horizontally (Cajal 1911, 1922; Braak 1976; Valverde 1986; Fig. 22.206:F,G). The axons of these elements descend towards the white matter and usually emit several horizontal and recurrent collaterals. It has been experimentally established that, in the cat, cells of this type project either ipsilaterally or contralaterally to other parts of the cortex (Einstein and Fitzpatrick 1991; Vercelli et al. 1992). The spiny projection neurons described above are usually considered to be modified pyramidal neurons. Elements provided with one longer, ascending dendrite, clearly intermediary in form between typical large stellate cells and typical pyramidal neurons, have been observed in the monkey (Valverde 1985; Fig. 22.206:H). Moreover, Vercelli et al. (1992) demonstrated that, in the visual cortex of the kitten, callosally projecting pyramidal cells lose their apical dendrite during development and gradually acquire a typical stellate morphology.

4. *Sparsely spiny pyramidal cells.* Physiological studies (Takahashi 1965) have shown that the pyramidal cells in lamina V of the motor cortex can be classified, according to the conduction velocities of their axons, into fast and slow pyramidal tract neurons. Combined physiological and morphological investigations have revealed that in both the cat (Deschênes et al. 1979) and the monkey (Hamada et al. 1981) the pyramidal tract neurons belonging to these two functional categories also show distinct structural differences. The fast pyramidal tract neurons appeared to have larger somata and aspiny apical dendrites, whereas the slow pyramidal tract neurons appeared to have smaller somata and spiny apical dendrites. Pyramidal neurons showing these morphological differences have also been found in lamina V of the somatosensory (Yamamoto et al. 1987a), parietal (Yamamoto et al. 1987b) and visual (Hübener et al. 1990) cortices of the cat. As regards the latter cortex, Hübener et al. (1990) demonstrated that the dendrites of most pyramidal neurons projecting to the tectum are covered with spines, but that, among the corticotectal pyramids with the largest somata, some elements occur whose basal and apical dendrites are almost devoid of spines (Fig. 22.206j,k). It appeared that the spiny and aspiny corticotectal cells also have different intrinsic collaterals and therefore presumably play different roles in the circuitry of the visual cortex. Spiny corticotectal cells generally have axon collaterals that project to lamina VI, but the spine-free corticotectal cells have fewer axon collaterals and these do not ramify in lamina VI (Fig. 22.206:J,K). Interestingly, Hübener et al. (1990) found some cells that, with respect to soma size and spine density, occupied an intermediate position between spiny and spine-free

Fig. 22.206.

corticotectal cell. On the basis of this observation, they suggested that a continuum exists of corticotectal cell morphologies rather than two discrete cells types.
5. *'Pure' projection pyramidal neurons.* It is commonly assumed that the axons of all pyramidal neurons have intracortical collaterals. However, Ghosh et al. (1988) used intracellular injection of HRP to demonstrate that, in lamina VI of the cat motor cortex, pyramidal neurons occur whose axons do not emit any intracortical collaterals.
6. *Intrinsic pyramidal neurons.* Pyramidal cells whose axons do not leave the cortex have been observed in many different species and in all cortical layers except for layer I. In laminae II and III of the visual cortex of the monkey, small pyramidal neurons occur whose varicose, unmyelinated axons descend to the deeper layers of the cortex (Valverde 1985; Fig. 22.199b). In lamina III of the somatosensory cortex of the mouse, Lorente de Nó (1922) observed pyramidal neurons whose richly ramifying, descending axons did not leave the cortex (Fig. 22.206:L). In lamina IV of the sensory cortex, numerous pyramidal neurons with poorly developed apical dendrites and recurving, ascending axons occur. Such elements have been observed, for example, in the temporal cortex of the macaque monkey (Valverde 1986; Fig. 22.206:M), the visual cortex of the cat (Parnavelas 1984) and the somatosensory cortex of the squirrel monkey (Jones 1975). A certain proportion of these neurons are designated as star pyramids because their dendrites radiate out from the soma in all directions (Fig. 22.206:N). The ascending axonal branches of these lamina IV cells are major contributions to the radial fasciculi already discussed. Star pyramids may be considered as intermediary forms between typical pyramidal neurons and the spiny stellate cells to be discussed below. In laminae V and VI of the visual cortex of the monkey, medium-sized pyramidal neurons with recurving axons occur (Valverde 1985; Fig. 22.199a). It has already been mentioned that these axons often split up into two or more branches and participate in the formation of radially coursing bundles which ascend to the superficial cortical layers. Katz (1987) observed that around 20% of the pyramidal neurons in lamina VI of the primary visual cortex of the cat emit numerous intracortical collateral branches, but lack a corticofugal axons.

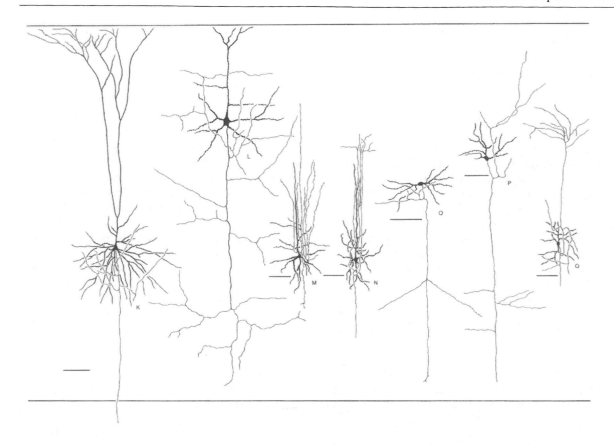

Fig. 22.206. Atypical pyramidal cells in the neocortex of various mammals. 'Extraverted' neurons in lamina II of the neocortex of the hedgehog (Valverde and Facal-Valverde 1986), the rat (Kirsche et al. 1973) and the dolphin (Morgane et al. 1990), respectively (A–C). Inverted pyramidal neuron with poorly developed apical dendrite (*arrowheads*) from lamina VA of the macaque monkey striate cortex (D; Lund 1973). Stellate neuron projecting to area 18 in lamina IV of the primary visual cortex of the cat (F; Einstein and Fitzpatrick 1991). Spinal stellate neurons in lamina IVB of the macaque monkey striate cortex (G,H; Valverde 1985). Typical spiny and spine-free corticotectal pyramidal neurons in lamina V of the primary visual cortex of the cat (J,K; Hübener et al. 1990). Pyramidal neuron with short axon in lamina III of the somatosensory cortex of the mouse (L; Lorente de Nó 1922). Small pyramidal cell with recurving axons in lamina IV of the temporal cortex of the macaque monkey (M; Valverde 1986). A 'star pyramid' in lamina IV of the somatosensory cortex of the squirrel monkey (N; Jones 1975). Spiny stellate cells in laminae IVB, IVA and IVC β (O,P,Q), respectively, of the primary visual cortex of the macaque monkey (Lund 1973; Lund and Boothe 1975). *Bars*, 100 μm

7. *Spiny stellate cells.* Spiny stellate cells occur exclusively in lamina IV of primary sensory areas of the neocortex, where they may be abundant (Lund 1973; Jones 1975; Winer 1982). Their small, spherical or ellipsoid somata in Nissl preparations often form one or several conspicuous granular zones, which is why laminae I–III and V–VI in the sensory cortices are often designated as the supragranular and infragranular layers. The dendrites of these spiny stellate cells are emitted at several points from their soma and are generally confined to the fourth layer or to the sublayer in which their soma is situated. They may be strongly stratified horizontally, or they may be radiate or even vertically elongated in distribution (Lund 1984). The spiny stellate cells are the principal, but not the exclusive targets of the thalamocortical afferents terminating in lamina IV of the sensory cortical areas. Their axons travel for relatively short distances from their cells of origin to the laminae above and/or below lamina IV, but invariably also establish connections within the lamina of origin (Lund 1984). The spiny stellate cells form asymmetric (presumably excitatory) synapses mainly with dendritic spines (LeVay 1973; Mates and Lund 1983, Saint Marie and Peters 1985). Saint Marie and Peters (1985) concluded that, within lamina IV of the visual cortex of the monkey, most dendritic spines contacted by the terminals of spiny stellate cells belong to other stellate cells. However, given the fact that in the supragranular and infragranular layers pyramidal neurons represent the only cell type with densely

spiny dendrites, it seems likely that in these layers dendritic spines arising from pyramidal cell dendrites form important targets of the ascending Boothe descending axonal branches of spinal stellate cells. However, a direct demonstration of such connectivity has not yet been made (De Felipe and Farinas 1992). Taken together, spiny stellate cells in lamina IV of sensory cortical areas are the primary recipients of thalamocortical afferents and most probably serve as the major route by which sensory activity reaches neurons in other layers of these cortical areas.

Spiny stellate cells are absent or poorly represented in the relatively undifferentiated and presumably primitive cerebral cortices of the hedgehog (Valverde 1986) and the dolphin (Morgane et a. 1990). However, the glomérulos (Lorente de Nó 1922) or barrels (Woolsey and van der Loos 1970), which together constitute lamina IV in the whisker area of the somatosensory cortex of rodents (Fig. 2.33) are principally composed of typical spiny stellate cells. The Golgi study by Lorente de Nó (1922) has shown that the dendrites of these elements (his *'células estrelladas de la capa IV'*) ramify within a single barrel. Their axons descend to the deeper cortical layers and issue numerous collateral branches within laminae IV–VI. Some of these branches take an ascending course and reach lamina II (u in Fig. 22.194). Spiny stellate cells have also been found in the somatosensory cortex of the squirrel monkey (Jones 1975), in the primary auditory cortex of the cat (Winer 1982) and in the primary visual cortex of the cat (Lund et al. 1979), the tree shrew (Lund 1984; Lund et al. 1985) and the rhesus monkey (Lund 1973, 1987; Lund and Boothe 1975; Valverde 1985). The investigations carried out by Lund and collaborators have shown that the spiny stellate cells, present in the various sublayers of lamina IV of the monkey primary visual cortex, differ markedly with respect to their projections to other layers. Thus it appeared that the axon of the spiny stellate cells of lamina IVB gives off several collaterals spreading in lamina VI. Collaterals passing obliquely downward into lamina VI are given off from the descending trunk in lamina V (Lund and Boothe 1975; Fig. 22.206:O). The principal axon distribution of the spiny stellate cells in lamina IVA is to laminae IIIB and V (Lund and Boothe 1975; Fig. 22.206:P), whereas the axons of the large population of spiny stellate cells in lamina IVCβ first descend, but soon form recurrent branches, which ascend in a strictly columnar fashion through laminae IVB and IVA, forming fan-shaped terminal arbors in lamina IIIB. The axons also have locally spreading collaterals in lamina IVC (Lund 1973, 1984, 1987; Valverde 1985, 1986; Figs. 22.199c, 22.206:Q).

If the spiny stellate cells are indeed excitatory, as the ultrastructure of their terminals suggests, these elements doubtless play a crucial role in the radial propagation of the activity fed by thalamocortical afferents into lamina IV of primary sensory cortices.

Although spiny stellate cells are typical local circuit neurons and lack an apical dendrite, they nevertheless have to be considered as modified pyramidal neurons (Lund 1984; Valverde 1986). The reasons for this interpretation are as follows:

1. Spiny stellate cells share a number of salient structural features with pyramidal neurons:
   a) The dendrites of both are densely covered with spines.
   b) Both have relatively few synaptic contacts on their somata, and all of these are of the symmetric type (Le Vay 1973).
   c) Their axons leave the soma from its basal side and descend, at least initially, toward the white matter (Figs. 22.199c, 22.206:O-Q).
   d) The axons of both types may participate in the formation of highly characteristic radial bundles (Fig. 22.199a,c).
   e) As already mentioned, the axon terminals of both types form synapses of the asymmetric type, mainly with dendritic spines (Le Vay 1973; Saint Marie and Peters 1985).
2. In lamina IV of the primary sensory areas containing spiny stellate cells, neurons provided with a more or less developed apical dendrite are frequently observed which in all other respects are indistinguishable from typical spiny stellate neurons. The *'pirámides-granos'* observed by Lorente de Nó (1922), i.e. the somatosensory cortex of the mouse (see his Fig. 5a), and the 'star pyramids' seen by Jones (1975; Fig. 22.206:N) in the somatosensory cortex of the squirrel monkey are good examples of such intermediate forms.
3. The fact that pyramidal neurons are relatively rare in lamina IV of primary sensory cortical areas and that these elements are even completely lacking in some sublayers of IV (e.g. lamina IVA of the primary visual cortex of the tree shrew and lamina IVCβ of the same cortex in the macaque monkey; see Lund 1984) suggests that, in the pertinent layer, pyramidal neurons have been replaced by spiny stellate cells.
4. Peinado and Katz (1990) have presented evidence that during ontogenesis lamina IV stellate cells initially extend an apical dendrite to lamina I and only later lose this process and develop their mature stellate morphology.

This completes our long discussion of the neocortical pyramidal neurons.

We started by introducing the 'typical pyramidal neuron' as a cortical element provided with a radially oriented apical dendrite that forms a highly branched terminal tuft in lamina I, with basal dendrites that radiate out from the base of the soma and with an axon descending radially to the white matter. All of the dendrites of typical pyramidal neurons are covered with spines, and their axons emit several intracortical collaterals.

It appeared that cortical neurons that completely fit the description of typical pyramidal neurons show a remarkable structural diversity, a diversity which may concern, inter alia, their laminar position, the branching pattern of their dendrites and the target regions to which their axons project. Evidence was presented that all typical cortical neurons projecting from a given cortical layer to a particular cortical or subcortical target are in receipt of similar extracortical afferents and are embedded similarly in the cortical circuitry.

It has been shown that the typical pyramidal neurons probably developed from 'extraverted cortical neurons' (i.e. elements provided with a well-developed tuft of peripherally extending dendrites, but with only a few relatively short basal dendrites) and that these neurons in turn probably arose from primitive pallial neurons which extend all of their dendrites peripherally.

Whereas in the simple and most probably primitive neocortex of the hedgehog practically all pyramidal neurons may be classified as 'typical', in the more advanced cortices of carnivores and primates numerous 'atypical' or 'aberrant' pyramidal elements are present. In such elements, the apical dendrite may be reduced to a thin, tapering process or may have completely disappeared. Other atypical pyramidal neurons have lost their extracortically projecting main axon and have thus become transformed into local circuit neurons. The spiny stellate cells, which are abundant in lamina IV of the primary sensory cortical areas of carnivores and primates (and in the highly specialized whisker area of rodents), should be considered as strongly modified pyramidal neurons. These elements have lost both their apical dendrites and their main axons. They are the primary recipients of specific thalamocortical afferents and, as excitatory local circuit neurons, provide a major route by which sensory activity is radially propagated to other cortical layers.

22.11.6.5.5
Local Circuit Neurons

In the preceding sections, the neocortical pyramidal neurons have been discussed. It was pointed out that this category not only encompasses true pyramidal cells, but also atypical elements that either have 'not yet' attained the pyramidal morphology or have lost one or several of the typical pyramidal characteristics. Pyramidal neurons account for 60%–85% of the total neuronal population of the neocortex (Globus and Scheibel 1967; Winfield et al. 1980; Powell 1981; Peters et al. 1985). The remaining 15%–40% of neocortical neurons include a variety of morphological types that have the following features in common:

1. They are evidently non-pyramidal, i.e. they have no conical soma and lack a dominant apical dendrite. On that account, the group is often referred to as non-pyramidal, but this designation is not entirely satisfactory, because many neurons belonging to the pyramidal category do not show a pyramidal morphology (Fig. 22.206:E–H, N–Q).
2. Their dendrites bear only few spines or are entirely spine-free. This is a very important distinguishing feature, even though the dendrites of some, otherwise typical pyramidal neurons are also aspinous (Fig. 22.206:K).[2]
3. Their somata have both symmetric and asymmetric axosomatic synapses, whereas pyramidal cell bodies possess only symmetric axosomatic synapses (Peters 1987).
4. Their axons do not leave the cortex, which is why the cells under consideration are often referred to as local circuit neurons. However, it should be recalled that many pyramidal cells, both typical and atypical ones, also possess exclusively intracortical axons (Fig. 22.206:E, L–Q) and, hence, also belong to the category of cortical local circuit neurons.
5. With a single exception, their axon terminals contain flattened vesicles and form symmetric synapses with their postsynaptic targets, both features suggesting an inhibitory function (Le Vay 1973; Parnavelas et al. 1977; Fairén et al. 1984).

---

[2] It is important to note that in previous Golgi studies, including those by Cajal (1911) and Lorente de Nó (1922), the dendrites of several cell types, which according to later workers are aspinous or sparsely spinous, were depicted as bearing a considerable number of spines (see the neurons g,h,k,m,p,t in Fig. 22.194, redrawn from Lorente de Nó 1922). In his studies, Cajal usually employed young specimens, and it is well known that immature local circuit neurons generally bear a greater number of spines than mature ones (Jones 1975). Lorente de Nó (1922) presents no particulars on the age of the animals he employed.

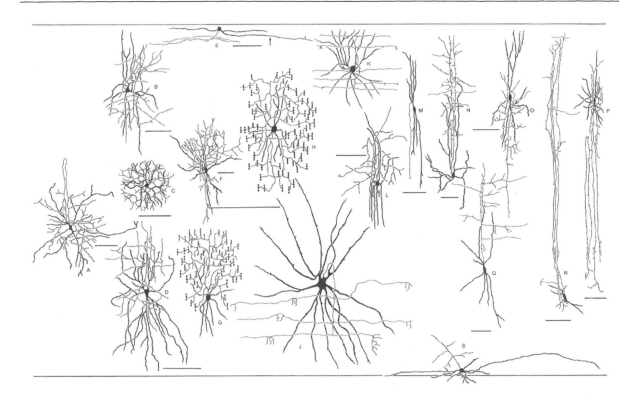

Fig. 22.207. Local circuit neurons with smooth or sparsely spinous dendrites in the neocortex of various mammals. Multipolar local plexus neuron with its cell body in layer V of rat sensorimotor cortex (*A*; Peters and Saint Marie 1984). Multipolar cell with axonal arcades in layer III of the somatosensory cortex of the sqirrel monkey (*B*; Jones 1975). Neurogliaform or spiderweb cell from layer IV of the somatosensory cortex of the squirrel monkey (*C*; Jones 1984). Stellate neuron with sparsely spined dendrites extending from lamina IVC to lamina VI in the primary visual cortex of the macaque (*D*; Lund 1973). Horizontal cell in lamina I of the neocortex of the hedgehog (*E*; Valverde and Facal-Valverde 1986). Stellate neuron from lamina IV in rat visual cortex (*F*; Peters and Saint Marie 1984). Chandelier cells in lamina III and V of the neocortex of the macaque (*G,H*; Jones 1987). Basket cells in laminae II and V of the neocortex of teh macaque (*J,K*; Jones 1987). Neuron with simple local beaded axon in lamina IV of the prefrontal cortex on the macaque (*L*; Lund and Lewis 1993). Bipolar or vertical cascade neuron in lamina III of the prefrontal cortex of the macaque (*M*; Lund and Lewis 1993). Neuron with rising axon in lamina IV of the prefrontal cortex of the macaque (*N*; Lund and Lewis 1993). Neuron with axon connecting lamina II and III in the prefrontal cortex of the macaque (*O*; Lund and Lewis 1993). Double bouquet cell in lamina III of the visual cortex of the macaque (*P*; Werner et al. 1989). Neuron with rising axon in upper tier of lamina V of the primary auditory cortex of the cat (*Q*; Fairén et al. 1984). Martinotti cell in lamina V of the visual cortex of the cat (*R*; Wahle 1993). Horizontally oriented neuron in lamina VI and adjacent white matter of the prefrontal cortex of the macaque (*S*; Lund and Lewis 1993)

6. An inhibitory function is also suggested by the fact that most of the cells under consideration use GABA as their primary neurotransmitter (Peters and Fairén 1978; Ribak 1978; Hendry and Jones 1981; Houser et al. 1984).
7. A certain proportion (25%-30%) of the GABAergic cortical neurons also express one or several neuropeptides. The neuropeptides detected in cortical neurons include substance P, vasoactive intestinal polypeptide, cholecystokinin, neuropeptide Y and somatostatin (Jones et al. 1987).
8. There is evidence suggesting that they are all derived from a set of progenitor cells in the ventricular zone, which differs from that giving rise to the pyramidal neurons (Parnavelas et al. 1991).

In brief, the mammalian neocortex contains a large population of non-pyramidal, inhibitory interneurons with smooth or sparsely spinous dendrites. These elements use GABA as their primary neurotransmitter, and some also produce one or several neuropeptides. The population of neocortical neurons thus outlined is morphologically heterogeneous, and numerous authors, including Lorente de Nó (1938), Jones (1975), Feldman and Peters (1978), Peters and Jones (1984), Fairén et al. (1984), Lund (1987), Lund and Yoshioka (991), Lund and Lewis (1993) and Lund et al. (1988), have attempted to subdivide this population of neurons into different groups or types, using the size and shape of the somata, the shape of the dendritic field and the number, distribution, length and branching patterns of individual dendrites, the preferred direction of axons and axonal branches and the configuration of axonal terminals as criteria. The following discussion of the non-pyramidal, smooth or sparsely spinous local circuit neurons in the mammalian neocortex is based on the publications cited above, on the excellent characterisations of particular cell types in the first volume of the *Cerebral Cortex* by Peters and Jones (1984; Jones and Hendry 1984; Somogyi and Cowey 1984; Peters 1984a,b; Jones 1984; Peters and Saint Marie 1984) and on several other sources to be quoted below. Most of the cell types to be discussed have already been briefly dealt with in a previous section as sources of afferents impinging upon pyramidal neurons.

Stellate neurons are found in all cortical layers (Lorente de Nó 1938; o,p in Fig. 22.194; Fig. 22.198a). The dendrites of these elements radiate out from the soma in all directions and branch infrequently. Their axonal arborisation forms a local plexus occupying approximately the same territory as that covered by the dendrites. Distinctive axonal terminals are lacking (Peters and Saint Marie 1984). Because of the limited spread of their axonal system, the stellate cells are also referred to as local plexus neurons.

In all layers of the neocortex, stellate neurons occur which deviate somewhat from the description given above. In such elements, the dendrites, though initially radiating out in all directions, may show in their further course a tendency to become oriented radially (Fig. 22.207:B,D), or the axonal plexus may be confined essentially to one particular sector of the dendritic field (Fig. 22.107:D,F) or this plexus may extend beyond that field (p in Fig. 22.194). The axons and axonal branches of many stellate cells form arcades, i.e. these processes ascend up to, or even beyond, the upper part of the dendritic field and then break up into a number of descending branches (Peters and Saint Marie 1984; o,p in Fig. 22.194 Fig. 22.207:A).

Neurogliaform or spiderweb cells form a special class of stellate cells (Fig. 22.207:C). These elements have a small, spherical soma and short sinuous dendrites. Their axons arborise profusely around the soma, forming a dense feltwork. Characteristically, small empty spaces appear in this feltwork, representing the positions occupied by the unstained somata of other neurons (Jones 1984). Neurogliaform or spiderweb cells have been observed in all layers of the cortex, but they are particularly concentrated in the primate somatosensory and primary visual cortex. The function of the spiderweb cells is unknown. Jones (1984) considered it likely that the elements concentrated in lamina IV of sensory cortices receive thalamic afferents and synapse mainly with the spiny stellate cells of that layer. Comparing the spiderweb cells with the cerebellar Golgi cells, he speculated that these elements could serve to inhibit the spiny stellate cells and other lamina IV cells less powerfully excited at the periphery of a zone of focal thalamic input. This surround inhibition would effect that the sensory information fed into the cortex by the thalamic input is transferred to discrete radially oriented columns.

There are numerous neurons in the neocortex which, judging from the disposition of their dendritic tree, would fall in the category of stellate cells, but which distinguish themselves from the elements in that category by the course and/or mode of termination of their axons. The chandelier cells and basket cells, which have already been discussed in a previous section, represent two distinct types of such aberrant or specialized stellate cells.

Chandelier cells have been thus named because their profuse axonal plexuses give rise to a large number (up to 300) of highly characteristic, vertically oriented 'candles', each consisting of a series of axonal swellings (Fairén et al. 1984; Peters 1984a; Figs. 22.198b, 22.207:G,H). The axonal candles are arranged along the axon initial segments of pyramidal neurons, with which they form serial synapses. The axon initial segments of spiny stellate cells may also receive synapses from chandelier cells (Lund 1987).

Chandelier cells have been observed in the cortex of a variety of species, including the hedgehog, rat, cat, monkey and humans. They occur in layers II–V, but are most common in layer II. Because the pyramidal neurons projecting to the ipsilateral and contralateral neocortex are mainly situated in layers II and III and because the axon initial segments of these supragranular pyramids receive a much richer synaptic supply from the chandelier cells than the infragranular elements, it has been suggested that chandelier cells principally influence cortico-cortical circuitry (Somogyi et al. 1979; Peters et al. 1982). Although most chandelier cells are of a stellate appearance, spreading their dendrites in all directions, the dendrites of some tend to be concentrated in ascending and descending tufts.

Basket cells are the largest non-pyramidal cells in the neocortex (a in Fig. 22.198; Fig. 22.207:J,K). Their poorly ramifying dendrites radiate in all directions, but in some vertically oriented dendrites prevail and give these cells a bitufted appearance. The axons of the basket cells are either ascending or descending and give rise to four or more horizontal branches at various levels. These collateral branches are myelinated and may reach a length of 1 mm or more. At intervals, they issue short ascending or descending terminal branches which contribute to the formation of paricellular baskets around pyramidal cell bodies. Each terminal branch forms a series of synapses with its target soma. One basket cell contributes to numerous baskets, and terminal axonal branches of several basket cells contribute to a single pericellular plexus. The somata of basket cells are concentrated in laminae II and V.

The cells discussed above with their radiating short dendrites and their long horizontal axonal branches form a distinct group of cortical local circuit neurons, which have been designated as large basket cells (Fairén et al. 1984). Cells of this type have so far only been observed in the cat and monkey (Jones and Hendry 1984). In the neocortex of different mammals, both 'higher' and 'lower', small smooth or sparsely spinous intrinsic neurons have been observed, the axons of which consistently produce multiple synaptic contacts on somata of pyramidal cells. Fairén et al. (1984) named these elements small baskets cells. According to their observations, these small basket cells are multipolar elements with rather wide dendritic fields, sometimes showing a predominant dendritic tuft oriented towards the pial surface. The axon is primary descending and forms a rich local plexus through diverging, recurrent collaterals. Recent studies, summarised by de Félipe and Fariñas (1992, p. 574) have shown that, apart from large and small basket cells, several other intrinsic cortical neurons, including smooth and sparsely spined stellate cells, contribute to a greater or lesser extent to the somatic innervation of pyramidal cells. White (1989, p. 40) emphasised that each pericellular basket is composed of axon collaterals from different sources: "Basket cells contribute to baskets; they do not by themselves form them." In view of this fact, White (1989) proposed including as basket cells all non-pyramidal cell types whose axonal branches contact pyramidal somata, regardless of whether true baskets are formed.

The classical descriptions of basket cells presented by Cajal (1911) created the impression that these elements synapse exclusively with the somata and proximal dendrites of pyramidal neurons. However, in several later studies (Freund et al. 1983; Somogyi et al. 1983; Kisvárday et al. 1985; Somogyi and Soltész 1986), evidence was presented suggesting that axon terminals of basket cells also synapse with the somata of spiny stellate cells, the distal dendrites and axons of pyramidal cells and the somata and dendrites of non-pyramidal neurons.

Kisvárday et al. (1993), using biocytin as a label and taking advantage of the fact that large basket cells are GABAergic and contain parvalbumin, demonstrated that these cells in the visual cortex (area 18) of the cat not only synapse with pyramidal neurons, but also establish an average of four to six perisomatic contacts onto other large basket cells. A large basket cell in lamina II was found to synapse with the somata of 58 other large basket cells, whereas a large lamina V basket cell appeared to contact 33 of its fellow cells. From these observations, Kisvárday et al. (1993) concluded that large basket cells form an interconnected network in area 18 of the visual cortex. Assuming that the GABAergic large basket cells are inhibitory, they proposed that: (a) a large basket cell provided direct perisomatic inhibition onto a number of pyramidal cells (at least 200–300) and in the range of about 50 other basket cells, (b) the target cells of the directly inhibited basket cells become facilitated via a disinhibitory effect and (c) the number of neurons disinhibited through this process may well greatly exceed the number of elements that are directly inhibited by the large basket cell.

Vertically Oriented Neurons. If we consult the extensive inventories of neocortical neurons provided by Cajal (1911), Lorente de Nó (1922), Jones (1975), Feldman and Peters (1978), Fairén et al. (1984), Lund (1973, 1987), Lund and Yoshioka (1991), Lund and Lewis (1993), Lund et al. (1979, 1988) and others, it appears that numerous smooth or sparsely spinous local circuit neurons in this structure show an overall vertical orientation. This orientation may concern their dendritic trees (Fig. 22.207:L), their axonal systems (Fig. 22.207:N–R) or both (Fig. 22.207:M). Among the neurons with vertically oriented axons, elements with rising axons (Fig. 22.207:N,Q,R), descending axons (Fig. 22.207:M,O) and both descending and rising axons (Fig. 22.207:P) may be distinguished. A discussion of all of the types of vertically oriented cortical interneurons described in the literature is not feasible here. However, some, i.e. the bipolar cells, the bitufted cells, the double bouquet cells and the elements with ascending axons known as Martinotti cells, may be briefly commented upon.

Bipolar cells have a small, spindle-shaped, vertically oriented cell body, from which one primary ascending and one primary descending dendrite arise (Peters 1984b; Fig. 22.200d). These two processes and their ramifications, which bear few or no spines, extend for long distances through the depth of the cortex, producing a narrow and very elongated dendritic tree. In the visual cortex of the rat, bipolar cells have been observed extending through all layers of the cortex. The axons of the bipolar cells usually arise from one of the primary dendrites and form a plexus that is likewise narrow in its spread and vertical in its orientation (Peters 1984b). Bipolar cells have been observed throughout laminae II–V in the neocortex of a variety of mammals, including the rat, rabbit, cat, dog, monkey and humans.

Ultrastructural investigations have shown that there are two different populations of bipolar neurons, one forming symmetric synapses and the other forming asymmetric synapses (Connor and Peters 1984; Peters and Harriman 1988).

As regards the bipolar cells with axons forming symmetric synapses, there is evidence suggesting that most of these elements use GABA as a neurotransmitter (Meinecke and Peters 1987). The axons of the bipolar cells forming symmetric synapses preferentially synapse with dendritic shafts. The investigations by Peters and Harriman (1988) and Peters et al. (1987) have shown that the population of bipolar cells that form symmetric synapses include numerous elements that label with antibodies to vasoactive polyptide. The terminals of these peptidergic bipolar cells were also observed to synapse mainly with dendritic shafts, but some appeared to contact cell bodies of pyramidal and non-pyramidal cells.

A certain proportion of the neocortical bipolar neurons which form symmetric synapses can be labelled with antibodies to ChAT, a specific marker for cholinergic neurons. The axon terminals of these cholinergic bipolar cells most commonly synapse with small to medium-sized dendritic shafts and less frequently with apical dendrites and with the somata of neurons (Houser et al. 1985; Parnavelas et al. 1986).

The bipolar cells that form asymmetric, presumably excitatory synapses form the only known exception to the rule that smooth or sparsely spinous local circuit neurons form symmetrical, presumably inhibitory synapses. Such bipolar cells that are presynaptic at asymmetrical synapses were first observed by Peters and Kimerer (1981) in the visual cortex of the rat, and a few years later Fairén et al. (1984) reported the presence of similar cells in the visual cortex of the cat. Peters and Kimerer (1981) found that the axons of the bipolar cells under discussion give rise to vertically oriented branches which parallel the trajectories of clustered apical dendrites of pyramidal neurons, forming multiple asymmetric synapses with spines on these processes (Fig. 22.200d). The axonal branches of these presumably excitatory bipolar cells were also observed to contact the shafts of apical dendrites and the somata and dendrites of non-pyramidal cells. It is not known which neurotransmitter is used by these cells.

Bitufted cells have dendrites arising mainly from the upper and lower poles of the soma, forming, as the name implies, two dendritic tufts (Peters 1987). The dendrites diverge initially, but at some distance from the soma often assume a radial orientation. Many neurons of this type have local axonal plexuses which partly overlap with the field occupied by their dendritic trees (Peters 1987). However, among the bitufted cells elements are found whose axons produce radially oriented long plexuses of thin, parallel axonal branches. These plexuses are either descending or both ascending and descending. Smooth or sparsely spinous cortical neurons provided with these highly characteristic long, fascicular axonal systems have been commonly referred to as double bouquet cells (Peters 1984b; Peters and Jones 1984; Fairén et al. 1984; Somogyi and Cowey 1984). Cajal (1911) used the name '*cellules à double bouquet dendritiques*' to refer to a number of cell types with highly different axonal ramification patterns, including elements with the long radially oriented arrays of axonal branches discussed above. However, it has become customary to designate only smooth or sparsely spinous cells showing this particular axonal pattern as double bouquet cells. In the current literature, even elements with these long vertical axonal branches whose dendritic trees are not of a bitufted appearance are still referred to by this name (Jones 1975; Somogyi and Cowey 1984; Figs. 22.199d, 22.207F:P).

Typical double bouquet cells have only been observed in laminae II and III of the neocortex of cats and primates. Somogyi and Cowey (1984) have shown that the axon terminals of double bouquet cells in the visual cortex of cat and monkey form symmetric synapses, containing flat or pleiomorphic vesicles, with their target structures. The same authors also presented evidence strongly suggesting that double bouquet cells use GABA as a neurotransmitter. It was formerly believed that the vertically oriented axons of double bouquet cells mainly synapse with the apical dendrites of pyramidal neurons (Cajal 1911; Colonnier 1966; Szentágothai 1978). However, the studies by Somogyi and Cowey (1981) and by DeFelipe et al. (1989, 1990) showed that the axon terminals of these cells do not form synapses with apical dendrites, but rather with basal dendrites and with oblique branches of the apical dendrites of pyramidal neurons and with post-synaptic structures belonging to non-pyramidal neurons.

All cortical local circuit neurons with smooth or sparsely spinous dendrites, except for some bipolar cells, use GABA as a neurotransmitter, and it has already been mentioned that 25%–30% of these GABAergic cells additionally contain one or two neuropeptides. The following data concerning the cortical GABA-peptide cells are based on a review article by Jones et al. (1987), who studied these elements in the monkey and compared their findings with those on other species, including the rat and cat, reported in the literature. Jones et al. (1987) concluded that all neocortical GABA-peptide neurons are contained in a limited cell class that can be characterised as having small, rounded somata and a variable number of long dendrites assuming a radial course either directly or at some distance from the soma. Many of these neurons belong to the categories of bipolar and tufted cells discussed above. The somata of the GABA-peptide neurons are found in all cortical layers, but tend to be concentrated in layer II, in the superficial part of layer III and in layer VI. It is difficult to trace the course of individual axons of peptidergic cortical neurons in immunostained material. However, the terminals of all of these neurons appeared to be very similar. They are small, contain flattened or pleiomorphic vesicles and make typical symmetrical synaptic contacts. Points of contact have been demonstrated on dendrites of both pyramidal and non-pyramidal neurons and on dendritic spines. In some of the contacts made by peptide-immunoreactive terminals of cortical local circuit neurons, membrane specialization appeared to be lacking, as was proven by serial thin sectioning of the whole terminal. The neuropeptides released by these terminals are probably involved in diffuse, non-synaptic neurotransmission. As has already ben mentioned, the neuropeptides detected in cortical neurons include substance P, vasoactive intestinal polypeptide, cholecystokinin, neuropeptide Y and somatostatin. Some neurons were found to co-localise cholecystokinin and vasoactive intestinal polypeptide or somatostatin and neuropeptide Y.

Martinotti cells are multipolar or bitufted neurons with smooth or sparsely spinous dendrites. Their distinguishing feature is a long ascending axon which reaches layer I, where it forms a terminal arborisation. The axon emerges either from the upper surface of the soma or from an ascending dendrite. The initial part of the axon gives rise to a number of descending collaterals, which form a local terminal plexus (Martinotti 1890; Cajal 1909, 1911; Valverde 1976; Fairén and Valverde 1979; Ruiz-Marcos and Valverde 1970; Fairén et al. 1984). 'Classical' Martinotti cells have a single ascending axon, but recently Jones et al. (1988) and Wahle (1993) described 'double bouquet type' Martinotti cells, the axon of which branched into a bundle of two to eight long, ascending collaterals (Fig. 22.207:R).

Martinotti cells occur in all cortical layers except layer I, but they have been most frequently found in layers V and VI. They use GABA as a neurotransmitter (Somogyi and Hodgson 1985) and may additionally contain a neuropeptide, e.g. tachykinin (Jones et al. 1988) or somatostatin (Wahle 1993). Little is known with certainty about the afferent and efferent connections of the Martinotti cells. On the basis of a Golgi study of the visual cortex of the mouse, Ruiz-Marcos and Valverde (1970) suggested that Martinotti cells situated in layer V of the cortex receive terminals from both superficial and deep pyramidal cell, that their local axonal plexuses contact deep pyramidal cells and that their terminal plexuses in layer I impinge upon the apical dendritic tufts of pyramidal cells. As has already been mentioned, Marin-Padilla (1984) observed that the axonal terminations of Martinotti cells closely resemble the arborisation of the apical dendritic tufts of pyramidal neurons and considered it likely that Martinotti cells form dual sets with pyramidal neurons of similar cortical depth (d in Fig. 22.201). However, it has also been observed that the axons of certain Martinotti cells, after having reached lamina I, give rise to long, horizontal collaterals (Lorente de Nó 1922). According to Szentágothai (1978; see also Eccles 1984), these branches may run for several millimetres through lamina I, making synapses with the apical dendrites of numerous pyramidal cells.

Horizontal Cells. Local circuit neurons showing an overall horizontal orientation are almost exclusively found in layers I and VI. It is important to note that these layers, although far apart in the adult cortex, are both derivatives of a single embryonic pallial zone, i.e. the primordial plexiform layer. During the formation of the cortex, immature bipolar cells migrate peripherally and together form a compact cortical plate, which splits up the primordial plexiform layer into a superficial and a deep zone. The former becomes layer I, and the latter gives rise to the deep zone of layer VI in the mature cortex. The intervening layers are all derivatives of the cortical plate (Marin-Padilla 1978).

Horizontal cells of Cajal. Bipolar neurons occurring in layer I of the cortex of many different mammals, including the hedgehog, rat, rabbit and humans, are known as the *horizontal cells of Cajal*. These elements are provided with one or a few long, smooth dendrites, which pursue a course parallel to the cortical surface. Their axons, which like the dendrites pass horizontally, may attain a considerable length (Fig. 22.207:E). Curiously enough, horizontal cells in layer I may have more than one axon (Cajal 1911; Valverde and Facal-Valverde 1986). Thus, in the element depicted in Fig. 22.207:E, one axon originates from the soma, whereas one of the dendrites (indicated by an arrow) continues into a second axon. There is immunohistochemical evidence (summarised by Vogt 1991) indicating that the horizontal cells in layer I are GABAergic and additionally contain the neuropeptide cholecystokinin. The axons of the horizontal cells probably enter into synaptic contacts with the apical dendritic branches of pyramidal neurons (Lorente de Nó 1938; Valverde and Facal-Valverde 1986). Several authors (Cajal 1911; Edmunds and Parnavelas 1982; Parnavelas and Edmunds 1983; Marin-Padilla 1984; Valverde and Facal-Valverde 1986) have expressed the opinion that the conspicuous Cajal-Retzius cells found in layer I of the immature cortex (Fig. 22.201:e) undergo morphological changes and transform to horizontal cells.

Horizontal cells in layer VI. The *deep zone of layer VI* contains numerous medium-sized horizontal cells. Such cells have been observed in, among other species, the rat (Feldman and Peters 1978; Peters 1985), cat (Tömböl 1984), macaque (Lund and Lewis 1993; Fig. 22.207:S) and humans (Mrzljak et al. 1988, 1990). One or a few long dendrites arise from the ends of their fusiform cell bodies, but some shorter dendrites may also arise from the upper and lower surface of the soma. Their axons, which like the principal dendrites often pursue a horizontal course, give off varicose side branches. In Golgi material, the axons of the horizontal cells under discussion can rarely be traced to their final destination, and in general it must be stated that very little is known about the afferent and efferent connections of these elements. The fact that the horizontal cells in layer VI most probably synthesise GABA (van Eden et al. 1989) and that their cell bodies have both symmetric and asymmetric axosomatic synapses (Peters 1985) indicates that they represent inhibitory local circuit neurons.

22.11.6.5.6
Microcircuitry

The microcircuitry of the neocortex is summarised in Fig. 22.208. The following brief commentary is intended to supplement this diagram. (The symbols used in the text – Ba,Bi etc. – correspond to those in Fig. 22.208 and are explained in the legend of that figure.)
The pyramidal neurons are doubtless the principal neurons in the neocortex. They are not only by far the most numerous cellular elements in that structure, but also constitute its sole output system and its largest input system.

Separate sets of deep pyramidal neurons project to different subcortical targets, whereas cortico-cortical fibres arise mainly from superficial pyramidal neurons.

The very extensive axon collateral systems of pyramidal neurons primarily contact other pyramidal neurons. There is evidence suggesting that superficial pyramidal neurons contact other superficial and deep pyramidal neurons and that deep pyramidal neurons impinge on other deep and superficial pyramidal neurons. It is known that, in the primary visual cortex, the axon collateral systems of pyramidal neurons constitute reciprocal patchy networks, which can be traced over distances of up to 7 mm (Kisvárday and Eysel 1992). It has been suggested that these networks link sites with similar physiological characteristics, such as orientation preference. This would imply that, within the primary visual cortex, different interwoven networks of pyramidal axon collaterals are present and that the extent of these networks would be confined to that cortical area. However, it is possible (though extremely hard to prove) that in the neocortex much more extensive pyramidal axon collateral networks are present.

A continuous network of excitatory elements involving the entire neocortex and even extending into the hippocampal region is constituted by the ipsilaterally and contralaterally projecting cortico-cortical pyramidal neurons. The continuity of this network is emphasised by the fact that the cortico-cortical fibres terminate throughout the neocortex mainly in the superficial layers, where the cortically projecting pyramidal neurons are concentrated. The presence of this strongly developed, ubiquitous network warrants the conclusion that the neocortex communicates first and foremost with itself. However, the fact that the cortico-cortical fibres most probably impinge not only on superficial pyramidal neurons, but also on the apical dendrites and terminal dendritic bouquets of deep pyramidal neurons indicates that the various subcortical centres to which these elements project are continuously kept informed about the successive transformation of data occurring along the cortico-cortical processing streams. This applies in particular for the caudate-putamen complex and for the pontine nuclei, centres which are known to receive projections from almost the entire neocortex.

Thalamocortical fibres synapse directly with pyramidal neurons, although the number of contacts made by such fibres with particular types of pyramidal neurons is subject to considerable variation (see White 1989, pp. 39–41).

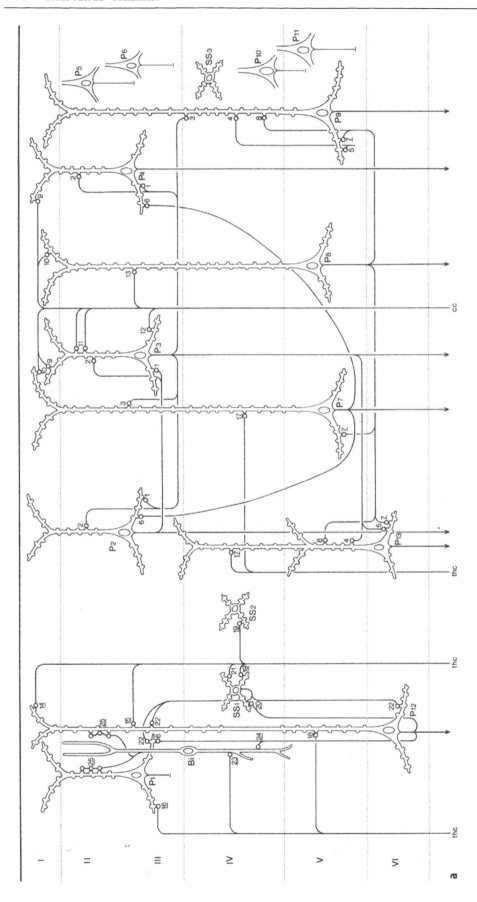

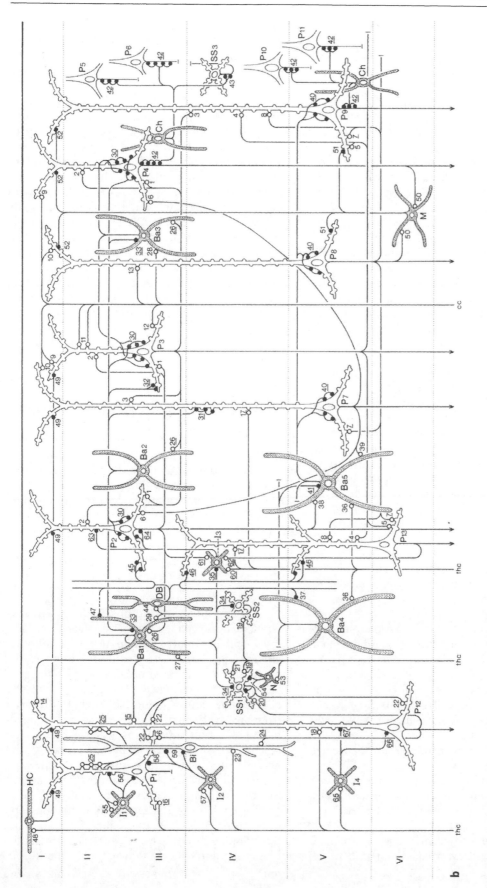

Fig. 22.208a,b. Neocortical circuits showing **a** excitatory elements and **b** excitatory plus inhibitory elements. *Ba1, Ba2* etc., basket cells; *Bi*, bipolar cell; *Ch*, chandelier cell; *cc*, corticocortical fibre; *DB*, double bouquet cell; *HC*, horizontal cell of Cajal; *I1, I2* etc., different types of interneurons; *M*, Martinotti cell; *N*, neurogliaform or spiderweb cell; *P1, P2* etc., pyramidal neurons; *SS1, SS2* etc., spiny stellate cells; *thc*, thalamocortical fibres; *I, II* etc., cortical layers. Excitatory neurons and their synaptic terminals are shown by open profiles, inhibitory neurons and their terminals by filled profiles. For a discussion of the individual synapses, which are indicated by *small numbers*, see Nieuwenhuys (1994). *Underlined numbers* refer to connections that have been conclusively demonstrated by the study of experimental material (labelled or degenerating elements) at the ultrastructural level. (Reproduced from Nieuwenhuys 1994)

All types of neocortical local circuit neurons, except for the neurogliaform cells, have been reported to establish synaptic contacts with pyramidal neurons. Most of these elements are known or have been suggested to receive input from thalamocortical fibres and from axon collaterals of pyramidal neurons (SS, Bi, Ba, I3), while others are intercalated between thalamocortical fibres and pyramidal neurons (H, I2, I4) or between different pyramidal neurons (M). The nature of the input to one type of local circuit neuron, the chandelier cell (Ch), is entirely unknown.

Several types of neocortical interneurons are in receipt of afferent contacts from other interneurons and/or establish efferent contacts with such elements. Prominent among these are the spiny stellate cells and the basket cells. Spiny stellate cells are contacted by basket cells and most probably by neurogliaform cells, chandelier cells and other spiny stellate cells and probably impinge on basket cells and double bouquet cells. Basket cells receive afferents from other basket cells and probably from spiny stellate cells and double bouquet cells and make efferent contacts with spiny stellate cells and probably with smooth stellate elements.

Excitatory local circuit neurons, i.e. spiny stellate cells and some bipolar cells, are, as far as it is known, confined to primary sensory areas. Inhibitory local circuit neurons are of many different types and occur throughout the neocortex. Some types of local circuit neurons are lamina-specific, i.e. their somata are situated mainly in one or in two adjacent layers (SS, DB, HC, M).

There is little direct evidence concerning the specific functions of the various types of local circuit neurons, but their laminar localisation, the geometry of their dendrites and axonal systems and the forms of their axon terminals may yield clues as to their mode of operation, e.g.:

1. The elongated, radially oriented axonal plexuses of bipolar and double bouquet cells indicate that these elements exert their influence on narrow, column-like arrays of postsynaptic elements.
2. The disposition of the axons of horizontal cells of Cajal and Martinotti cells reveals that these elements inhibit pyramidal neurons at the level of their terminal dendritic bouquets.
3. The fact that the axon terminals of chandelier cells impinge selectively on the axon initial segments of pyramidal and spiny stellate cells renders it most likely that chandelier cells exert a strong inhibitory influence on their target elements.
4. In the highly differentiated primary visual cortex of the rhesus monkey, thalamocortical fibres carrying particular kinds of visual information terminate in sharply defined sublayers of lamina IV (Fitzpatrick et al. 1983, 1985; Lund 1988). The detailed Golgi studies carried out by Lund and her associates (Lund 1987; Lund et al. 1988; Lund and Yoshioka 1991) showed that each of these thalamic recipient sublayers, and some other layers as well, contain several types of local circuit neurons. The dendritic trees of all of these elements are strictly confined to the layer or zone in which their soma is situated. Their axons project to more superficial layers, to deeper layers or to both. These interlaminar projections are highly specific, targeting from one to four laminar divisions, depending on the type of neuron. The strictly radial orientation of all of these projections is presumably connected with the preservation of the detailed topographical map existing within one of the thalamic recipient sublayers. Some of these types of local circuit neurons may well be involved in the selective processing and transfer of one particular type of visual information.

22.11.6.5.7
Neocortex in a Comparative Perspective

The preceding sections present a survey of the structural elements that make up the mammalian neocortex. In the present section, the accumulated data will be placed in a comparative perspective, and additional information concerning the circuitry of the neocortex will be presented. Moreover, a number of statements in the literature pertaining to the phylogenetic development of the neocortex will be critically discussed. Three general remarks may serve as a preface to this commentary:

1. Any exposition on the evolution of living structures needs to go beyond observations. The differences observed between the structures analysed will be interpreted in terms of changes, either quantitative, e.g. from less to more, or qualitative, e.g. from simple to complex or from primitive to specialized. The sum total of the quantitative and qualitative changes inferred from observations on a given structure in a number of different animals is often expressed by placing the structures analysed, or rather the animals from which they are taken, in a rank or order, and these linear orders are often referred to as 'phylogenetic series'. Even though there is consensus that the assumed genetic relationships between organisms in the realm of evolutionary morphology can only be adequately expressed in branched patterns (dendrograms), the linear

orders referred to above appear to be tenacious. Without any claim to completeness, I should like to mention the following possible factors why this may be so (see Nieuwenhuys 1977):
a) Linear thinking is simpler than reasoning in patterns.
b) Overall ontogenetic development can be considered as linear, i.e. as passing from immature to mature or from simple to complex. It might well be that the strong influence of Haeckel's biogenetic rule – ontogenesis is an abbreviated recapitulation of phylogenesis – has promoted linear thinking in evolutionary morphology.
c) It cannot be denied that, with respect to certain characteristics, living organisms, or parts of them, can often be arranged in a series-like, hierarchic way. Terms such as 'grades of organisation', 'levels of complexity', 'lower' and 'higher' vertebrates or mammals refer to this phenomenon. There can be little objection to metaphorically ascribing the differences between the structures or organisms thus arranged to 'evolutionary changes'. However, it would be incorrect to state that the elements in the series constructed constitute a 'phylogenetic series'. This would imply a direct genetic relationship between the elements compared and a special driving force behind the progressive development within the series (orthogenesis).
2. The scientific value of any comparative study, including its evolutionary extrapolations, is strongly related to the adequacy of the material selected for that study. A comparative study of the neocortex may be taken as an example. Let us assume for the sake of simplicity that the morphology of the cellular elements in the neocortex can be sufficiently analysed in Golgi material. Given the structural heterogeneity of the neocortex, a comprehensive comparative study of the neocortical neurons would require an equivalent Golgi material (whatever that may mean in light of the well-known vagaries of the Golgi technique) of the entire neocortex of all extant mammalian species or, as such an extensive programme is evidently not workable, at least an equivalent Golgi material of a number of representative cortical areas (e.g. somatosensory, visual, motor and association) in a series of representative mammals, including a monotreme, a marsupial and a considerable number of eutherian species. Even the sum total of all Golgi studies so far devoted to the mammalian neocortex falls hopelessly short of the requirements of this minimum programme. The neocortices of monotremes and marsupials have never been the subject of in-depth Golgi studies, and with regard to the eutherians, these studies are mainly confined to laboratory animals and humans. Nevertheless, many sweeping statements on the phylogenetic development of the neocortex are based on such material. In the light of what is known about the phylogenetic relationships of the rat, cat and monkey, the validity of these statements should be seriously questioned. However, this does not mean that evolutionary extrapolations based on comparative (neuro)anatomical studies of these three laboratory animals have no value whatsoever. At least two justifications, albeit a posteriori, are conceivable: (a) It could appear that comparison of (parts of) mammals constituting a juxtaphylogenetic series, e.g. an insectivore, a prosimian and a simian, would not lead to insights differing substantially from those gained from comparisons of (parts of) the laboratory animals mentioned; (b) evidence could present itself justifying the view that, with regard to the structure of a given part, e.g. the neocortex, the rat, cat and monkey represent three different levels or grades of complexity, which can be hierarchically arranged (see above under 1c).
3. Structures or elements to be compared should be analysed according to a set of distinct criteria. As regards the neocortical neurons, the criteria pyramidal soma, distinct radially oriented apical dendrite, skirt of basal dendrites etc. yield a distinct category of cells, i.e. the typical pyramidal neurons. The remaining neurons form a heterogeneous group, which is commonly negatively defined as those neurons which lack the features of pyramidal cells. We have seen that within this group numerous different neuron types have been distinguished, usually on the basis of the presence of a single dominant dendritic or axonal feature. Several attempts have been made to analyse and classify the non-pyramidal cortical neurons with the aid of a predetermined set of criteria (see e.g. Jones 1975; Feldman and Peters 1978). Unfortunately, these attempts have not yielded a generally accepted classification of these neurons.

Following these general remarks, we will now turn to the comparative anatomy and the evolutionary development of the neocortex.

Classification of Neurons. All neurons in the neocortex can be classified as belonging to one of the following two basic types: pyramidal cells and non-pyramidal cells. This classification, although not perfect (see above), is applicable to the neocortical neurons of all mammalian species studied thus far and hence offers a good point of departure for any consideration of the organisation of the neocortex. With certain qualifications, the pyramidal cells may also be referred to as projection neurons or spiny neurons, and the non-pyramidal cells as local circuit neurons or aspiny neurons.

Pyramidal Neurons. The pyramidal neurons constitute the most conspicuous and the most characteristic elements in the mammalian neocortex. These elements can be considered as members of a typological series ranging from periventricular pallial neurons with peripherally extending dendrites, via typical pyramidal neurons, to spiny stellate cells, whose dendrites are confined to a single cortical sublayer. There is comparative anatomical and ontogenetic evidence justifying an interpretation of this typological series as a phylogenetic series. Within such a phylogenetic or evolutionary context, the periventricular pallial neurons, the typical pyramidal elements and the spiny stellate cells may be designated as primitive, generalised and highly specialized, respectively.

The principal members of the pyramidal neuronal group may be characterized as follows:

1. Neurons with their somata in the periventricular grey, which extend all of their dendrites peripherally into a superficial plexiform layer (Figs. 22.177, 22.179a, 22.209A).
2. Pyramid-like neurons, which extend most of their dendrites into a superficial plexiform layer, but which are also provided with a few centrally directed dendritic branches that enter another synaptic domain (Figs. 22.179b–d,f, 22.180B, 22.209B).
3. Typical pyramidal neurons, with a pyramidal soma, an apical dendrite forming a terminal tuft into a superficial plexiform layer, a skirt of basal dendrites and an axon extending downward into the subcortical white matter and issuing several intracortical collaterals (Figs. 22.179e, 22.209C,D).
4. Pyramidal neurons with a reduced apical dendrite not reaching the superficial plexiform layers (Figs. 22.200B,C, 22.203D, 22.204A,B, 22.206H).
5. Spiny projection neurons which lack both an apical dendrite and a typical basal dendritic spray. Instead, the dendrites radiate out from the soma in all directions (Fig. 22.206F,G).
6. Typical spiny stellate cells with spherical or flattened dendritic trees that are confined to lamina IV, with short ascending or descending axons that do not leave the cortex (Figs. 22.199c, 22.206O-Q).

In the sequence presented, the six cell types outlined above form a typological series that, to a cer-

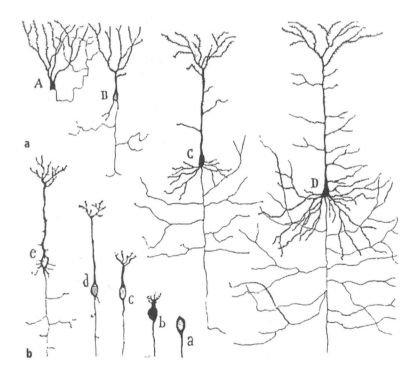

Fig. 22.209a,b. Scheme by Cajal (1893) showing the parallelism between the phylogenetic and the ontogenetic development of the pyramidal cells in adult vertebrates. Frog (*A*). Lizard (*B*). Mouse (*C*). Human (*D*). The *lower row* shows developmental stages of the mammalian pyramidal cell. Neuroblast (*a*). Beginning of development of the apical dendrite and its terminal bouquet (*b*). Lengthening of the apical dendrite (*c*). Beginning of development of a basal dendrite of axon collaterals (*d*). Further development of basal dendrites and axon collaterals (*e*)

tain extent, may be interpreted as a phylogenetic series. The reasons for this are as follows: (a) the periventricular neurons, the pyramid-like neurons and the typical pyramidal neurons constitute the principal elements in the pallium of amphibians, reptiles and mammals, respectively; (b) as shown by Cajal (1893, 1911), the series of amphibian, reptilian and mammalian pallial elements has a remarkable ontogenetic parallel (Fig. 22.209); and (c) all sorts of transitional cells can be interpolated between the elements of the series presented.

Fully developed typical pyramidal cells are abundant in all parts of the neocortex, but also occur in the more primitive piriform (DP in Fig. 22.184), hippocampal (Figs. 2.39, 22.188) and entorhinal cortices (o,r in Fig. 22.193). They are generally considered as mammalian achievements involved in higher functions of the central nervous system, as may appear from the fact that Cajal (1893, 1911) referred to these elements as 'psychic cells'. I designate the typical pyramidal cells as generalised neocortical neurons, because they represent the principal excitatory element in all neocortical circuits and the common precursor cell for diverse types of specialized excitatory neocortical elements. It is appropriate at this juncture to sum up the principal structural properties of the typical pyramidal cells in the neocortex and to comment briefly on the (possible) functional significance of some of these properties:

1. Their dendrites are covered with spines. Spines are small dendritic protrusions or outgrowths, usually 1–3 μm long. Some are finger-like, but most have a widened head, which is connected with the dendritic shaft by a relatively narrow stalk. A cytoplasmatic organelle known as the spine apparatus is found in or near the narrowed neck region of almost all spines. These organelles consist of two to three flattened and interconnected, sac-like cisternae. Spines have been shown to represent postsynaptic elements of axon terminals forming synapses. All spines receive at least one asymmetric (excitatory) synapse, usually on their bulbous terminal part. The vast majority (70%–95%) of the synapses received by a pyramidal cell are on its dendritic spines. The total number of spines on a large neocortical pyramidal neuron may be as high as 10 000. The shape and density of spines on pyramidal neurons may change considerably under normal natural conditions, such as ageing and hormonal fluctuations, in experimentally induced states, such as deafferentation, electrical stimulation and complex as opposed to deprived rearing environments, and in certain pathological states, such as mental retardation and dementia. All of these changes reflect the highly plastic nature of the dendritic spines of the pyramidal neurons. One hypothesis concerning the functional role of the dendritic spines favoured by many authors is that they play an important role in learning and memory processes. In this context, the axospinous synapse is often considered to be of 'Hebbian' nature, which means that, according to a theory first formulated by Hebb (1949), it is strengthened (i.e. its synaptic weight is increased) if the presynaptic activation of the synapse occurs in temporal conjunction with the firing of the cell postsynaptic to the synapse. During synaptic activity, the current injected into the spine head must pass through the narrow stalk of the spine in which the electrical resistance is high. It has been postulated that changing the impedance in the spine neck region could be a mechanism for controlling the efficacy of the synapse and thus the basis of synaptic plasticity. Alterations of the dimensions of the spine neck might effect such changes in electrical resistance. Crick (1982) proposed that alterations in spine length could be brought about by activation of contractile proteins, such as myosin and actin, localised in the spine neck. Calcium ions stored in the spine apparatus and liberated during synaptic activity might play a role in the activation of the contractile proteins. These notes on the structure, ultrastructure and possible functions of the dendritic spines of pyramidal neurons are based on the excellent reviews by Douglas and Martin (1990), DeFelipe and Fariñas (1992) and Horner (1993), to which the reader is referred for primary sources.
2. They are provided with a shorter or longer radially oriented apical dendrite.
3. The distal end of their apical dendrite forms a bouquet or terminal branches in lamina I, where each receives numerous excitatory axospinous synapses from horizontally running fibres. Most of these horizontal fibres arise from the bifurcation of axons that ascend radially through the cortex from the subcortical white matter. These ascending fibres arise partly from pyramidal neurons in other parts of the cortex and partly from the thalamus. The axonal branches of horizontal cells of Cajal and of Martinotti cells form a third source of horizontally running axonal branches in lamina I. Because the apical dendritic ramifications develop early in ontogenesis and are directly comparable to the peripherally extending dendrites of amphibian pallial cells and of reptilian pyramid-like cells, Sanides and Sanides (1972) considered these apical bouquets

as a conservative feature. Marin-Padilla (1992) pointed out that lamina I or the external plexiform layer is an ancient superficial neuropil that has retained its basic structural organisation practically unchanged throughout the phylogeny and ontogeny of terrestrial vertebrates. It should be emphasised that the external plexiform layer constitutes a synaptic domain that is shared by all typical pyramidal neurons.

4. They are provided with a skirt of basal dendrites extending outward from the lower part of the cell body. Because the basal dendrites develop later in ontogeny and phylogeny (Fig. 22.209), Sanides and Sanides (1972) regarded them as a progressive feature in pallial neuron evolution. With the development of these basal dendrites, the pyramidal neurons gained access to a second major synaptic domain, which may be situated at a considerable distance from the first domain in the external plexiform layer. However, whereas the first domain, with regard to both position and fibre content, displays a marked constancy, the second domain varies with the level at which the soma is situated. Afferents impinging on the basal dendrites include thalamocortical fibres, ipsilateral and contralateral cortico-cortical fibres, axon collaterals of other pyramidal neurons, axons of double bouquet cells and most probably axons of spiny stellate cells. Braitenberg (1974) laid great emphasis on what he called the "double dendritic expansion" of typical pyramidal cells. He believed that this particular property offers a clue as to the possible role played by these elements in learning and memory. His views may be summarised as follows: The first cortical layer constitutes the primeval and common input to all typical pyramidal neurons. It contains afferents of both thalamic and cortical origin, i.e. the input to any region of the neocortex may carry both information on events in the sensory system and information on the state of other regions of the cortex. Depending on the particular mixture of external and cortico-cortical afferents, different regions will be more concerned with the elaboration of sensory events or with the elaboration of the output of other cortical regions. Each typical pyramidal neuron is connected in its own special path to a particular set of fibres in the primeval input layer through the branches of its apical dendrite. The basal dendrites of the typical pyramidal neurons receive afferents from various sources. However, by far the most important input comes from neighbouring pyramidal neurons via their axon collaterals. Thus each typical pyramidal cell receives a major input in the first layer through its apical expansion and a second major input, representing a certain constellation of activity in neighbouring pyramidal neurons, through its basal dendritic expansion. Both inputs act on the typical pyramidal neurons in an excitatory fashion. In harmony with Hebb's (1949) conjunction hypothesis already alluded to, Braitenberg (1974) proposed the following mechanism for the fixing of memory traces in the neocortex: These traces are laid down either as a growth of existing spines or the formation of new spines in response to activation of a set of afferent axons impinging in the apical dendrites of a pyramidal neuron, occurring repeatedly in temporal conjunction with activation of the basal dendrites through its afferent pyramidal collaterals. A similar memory theory was developed by Eccles (1984), who surmised, however, that the selective potentiation of the synapses made by horizontal fibres in lamina I on the apical dendritic branches of pyramidal neurons leads to enlargement of the presynaptic terminals rather than of the postsynaptic spines. Moreover, Eccles believed that this selective potentiation occurs when impulses in the horizontal fibres are in approximate temporal conjunction with the activation of the apical dendritic shaft of the same cell by serial synapses made by the axons of spiny stellate cells (rather than activation of basal dendrites by axon collaterals of pyramidal neurons). Given the fact that pyramidal collaterals are abundant throughout the neocortex of all mammals and that spiny stellate cells occur only in highly specialized primary sensory cortices (see below), we believe that the memory hypothesis proposed by Braitenberg (1974) is preferable to that proposed by Eccles (1984).

5. The axon of typical neocortical pyramidal cells descends to the subcortical white matter, projecting either to other parts of the neocortex or to one or more subcortical centres.

6. Their axon gives rise to an extensive and elaborate system of intracortical axon collaterals. Together, these collaterals form a major part of the intrinsic cortical circuitry and by far the largest sources of intracortical axon terminals forming aymmetric (excitatory) synapses.

Typical pyramidal cells, i.e. the conspicuous and ubiquitous neocortical elements possessing all of the six structural properties enumerated above are doubtless the principal neurons in the neocortex. They occupy a central position in all cortical circuits, give rise to all cortical output systems, display a remarkable plasticity, which enables them to play a crucial role in learning and memory processes, and participate by means of their axon collaterals in extensive reciprocal patchy networks. Indeed, the pyramidal axon collaterals and their synapses on dendrites of other pyramidal neurons contrtibute the main internal synaptic system of the neocortex (Braitenberg 1974; Fig. 22.208a). This means that, although the neocortex receives input from various subcortical sources, it communicates first and foremost with itself.

Although all typical pyramidal cells belong to one single morphological type and are all anchored to lamina I by their apical dendritic ramifications, these elements may nevertheless show considerable differences with regard to the nature of their extrinsic and intrinsic afferents, the cortical or subcortical targets to which their principal axon projects and the local output relationship via their axon collaterals. However, all of these differences are not randomly distributed over the population of typical pyramidal cells. As mentioned earlier, it seems likely that all typical pyramidal neurons projecting from a given cortical layer to a particular cortical or subcortical target are in receipt of similar subcortical afferents and are similarly embedded in the intrinsic circuitry of the neocortex. This means that all typical pyramidal neurons projecting to other cortical areas, to the striatum, the thalamus, the superior colliculus etc. form clearly recognisable subtypes.

Whereas the evolutionary development of the pyramidal neurons dealt with so far could be described in terms of progression (see Figs. 22.179, 22.209), the further development of these elements is marked by reduction. Indeed, none of the six morphological acquisitions of typical pyramidal cells summarised above has appeared to be unassailable. Pyramidal neurons with aspinous dendrites, without an apical dendritic bouquet, with a reduced or an entirely vanished apical dendritic shaft, without a clearly recognisable skirt of basal dendrites, without an axon that leaves the cortex or without a set of intracortical axon collaterals have all been observed in the neocortex of one or more mammalian groups (Fig. 22.206), and in the sensory cortices of some species, elements lacking two or more of the structural properties of typical pyramidal cells are common. The typological series presented above, i.e. typical pyramidal neuron – pyramidal neuron with a reduced apical dendrite – spiny projection neuron lacking an apical dendrite and a typical basal dendritic spray – typical spiny stellate cell probably approximates only one of the many routes along which atypical or aberrant pyramidal neurons have come into being. We have placed the spiny stellate cells (Fig. 22.206=–Q) at the end of this series because in these elements only two of the six characteristics of typical pyramidal neurons (i.e. spiny dendrites and intra – cortical axonal ramifications) have been retained.

If we survey the pyramidal neurons observed in the mammalian neocortex, we are struck by their pluriformity. It is remarkable that elements extending all of their dendrites peripherally into one single synaptic domain have become transformed into the conspicuous typical pyramidal neurons with two, often widely separated, principal dendritic domains, and that elements of this type in turn, *inter alia* by giving up their original attachments with lamina I, have ultimately given rise to spiny stellate cells again with a single synaptic domain, confined to lamina IV. This particular development culminates in the primary visual cortex of the prosimian *Tupaia*, in which spiny stellate cells with very flattened dendritic trees are concentrated in a narrow subzone of lamina IVA, a subzone which coincides exactly with a tier of thalamocortical input from a particular component of the lateral geniculate nucleus (Lund 1984; Lund et al. 1985).

The evolutionary interpretation of the structural variety of pyramidal neurons discussed above has often been employed to judge the 'phylogenetic status' of the neocortex of particular mammalian species or groups. Thus the neocortex of insectivores and cetaceans is considered primitive, because (a) its very wide lamina II contains numerous 'extraverted' neurons, extending most of their dendrites into lamina I (Fig. 22.206A,B), (b) the apical dendrites of practically all pyramidal neurons form a terminal tuft in lamina I and (c) spiny stellate cells are lacking in these cortices (Valverde 1986; Glezer et al. 1988; Morgane et al. 1990). I should like to emphasise, however, that the transformation of typical pyramidal neurons into spiny stellate cells in itself cannot be interpreted as progressive. The progression or advancement of specialization is contextual. In response to the arrival of thalamo-cortical fibres carrying specific kinds of sensory innervation in lamina IV of the neocortex, a certain proportion of the pyramidal neurons present in that layer become transformed into spiny stellate cells, i.e. into highly specifically oriented excitatory elements. The integration of these 'new' elements into the circuitry of sensory cortices enables these cortices to carry out a more refined processing of the sensory information fed into them.

Evolutionary Development. With regard to the evolutionary development of the neocortex, the following two notions are frequently encountered in the literature: (1) the pyramidal neurons constitute the relatively constant basic framework of the neocortex and (2) the cortical local circuit neurons, on the other hand, form a dynamic component; they increase during phylogenetic development in both number and diversity. Evidence in support of these notions is lacking, however.

The first notion (see e.g. Lorente de Nó 1938; Poljakow 1979; Morgane et al. 1990) is difficult to reconcile with the considerable structural 'dynamics' of the pyramidal cell group discussed above. The second notion dates back to Cajal (1911) and Lorente de Nó (1938). Thus Cajal stated that, as one descends the phylogenetic scale, the morphology of cortical neurons becomes more uniform. In his opinion, the functional superiority of the human brain is directly related to the abundance and wealth of forms of cells with short axons.

With regard to the cortical local circuit neurons, Lorente de Nó (1938, p. 309) stated succinctly that "..the larger the number, the higher is the brain in the series." The validity of this statement cited cannot be rigorously assessed, because studies carried out on neocortices of mammalian species forming an adequate juxtaphylogenetic series have not been carried out so far. However, as both Cajal and Lorente de Nó studied exclusively laboratory animals and humans, the findings of Rockel et al. (1980) should be quoted here. The electron microscopy studies performed by these authors revealed that the ratio of pyramidal cells to non-pyramidal cells is essentially similar in the cerebral cortex of the rat, cat, monkey and human.

Cajal's idea that the diversity of the cortical local circuit neurons increases in phylogeny has received support from several authors, as is shown by the following quotations:

On the basis of Golgi studies, it seems that while local plexus neurons are probably the dominant form of smooth or sparsely spinous multipolar and bitufted cells with unmyelinated axons in the rodent cortex, in which they have rather stereotyped forms, in cats and primates such neurons become more diversified. At the same time, other non-pyramidal cells with more specialized forms of axonal plexuses make their appearance in these higher mammalian forms. Thus, the rodent cortex appears to have no well-defined counterpart of the basket cells and double bouquet cells (Peters and Saint Marie 1984, p. 434).

In primates, the number of different kinds of non-pyramidal cells in the cortex seems to increase (Peters and Jones 1984, p. 113).

With increasing complexity in advanced mammals, several types of intrinsic neurons with local axons, barely present in the hedgehog, appear intercalated in the local circuits (Morgane et al. 1990, pp. 231–232).

Other authors, including Szentagothai (1978) and Fairén et al. (1984), have emphasised that the various types of non-pyramidal cells described in the literature are present in all mammals:

A comparison of the vast Golgi material from various mammalian species shows surprisingly little essential change in the cell types. All characteristic types so far described can be found from mouse to monkey (even up to man) with only small variations (Szentagothai 1978, p. 225).

The most noticeable fact is that most of the non-pyramidal cell varieties have been found in all species of mammals which so far have been examined by the present authors or by others, and little variation has been found ever for certain of the cell types previously considered as specialized (Fairén et al. 1984, p. 245).

If we survey the vast literature on the cortical local circuit neurons, it appears that the horizontal cells of Cajal and the Martinotti cells in the neocortex represent ancient cell types, occurring in the cortex of all amniotes, and that stellate cells with long, radiating dendrites might well be the precursors of most other neocortical interneurons. The horizontal cells of Cajal occur exclusively in lamina I, and the Martinotti cells are found mainly in lamina VI; it should be remembered that these two laminae are direct derivatives of the early embryonic primordial plexiform layer, which during later development becomes split by the cortical plate (Marin-Padilla 1978). Elements closely resembling the neocortical horizontal cells of Cajal have been observed in the pallium of amphibians (Herrick 1948; P. Cajal 1922; Fig. 22.177), in the superficial plexiform layer of the reptilian cortex (P. Cajal 1917) and in the mammalian piriform (Figs. 22.184, 22.186), hippocampal (h,k,n in Fig. 22.189) and entorhinal cortex (l in Fig. 22.193). The 'phylogenetic' record of the mammalian Martinotti cells is less complete than that of the horizontal cells. However, probable equivalents of these elements have been found in the amphibian pallium (P. Cajal 1922), the reptilian cortex (Cajal 1911), the hippocampal formation (e in Fig. 22.189) and the entorhinal cortex (t in Fig. 22.193).

Stellate cells with smooth or sparsely spinous, radiating dendrites have been found in the reptilian cortex (D in Fig. 22.181) and in the mammalian piriform (MA in Fig. 22.184), hippocampal (m in Fig. 22.189) and entorhinal cortex (d,f in Fig. 22.193). Elements of this type, with an axon branching without particular orientation, making synaptic contacts with dendrites, somata and initial axon segments of pyramidal neurons may well have been the common precursor of all neocortical local circuit neurons (except for the spiny stellate cells). The two main processes of differentiation through which the various types of cortical local circuit neurons conceivably arose are the following: (1) orientation of dendrites (bipolar cells) or axons (double bouquet cells) into one prevailing direction and (2) concentration of their synaptic terminals on the dendrites (bipolar cells, double bouquet cells), the somata (basket cells) or the initial axonal segments (chandelier cells) of the pyramidal neurons. There is some evidence favouring the idea that the evolutionary development of the cortical local circuit neurons proceeded along these lines. Thus, in the neocortex of the hedgehog and the bat, stellate cells with long radiating dendrites are abundant (Sanides and Sanides 1974), and in the neocortex of the hedgehog, neurons with morphological characteristics of both basket and chandelier cells have been observed (Valverde and Facal-Valverde 1986). However, the differentiation process outlined should not be thought of as having taken place entirely within the neocortical domain. Well-differentiated basket and chandelier cells are present in the mammalian hippocampal formation (Figs. 22.189, 22.190), and it should be remembered that Lorente de Nó (1934b) observed more than 18 different types of local circuit neurons in this part of the cortex of the mouse (Fig. 22.189). Finally, it should be emphasised that there is no evidence to support the widely held thesis that throughout the neocortex the differentiation and diversification of local circuit neurons is further advanced in 'higher' mammals than in 'lower' mammals. An investigation of the evolutionary development of the neocortical local circuit neurons should be based on extensive Golgi material of a number of representative cortical areas (e.g. somatosensory, visual, motor and association) of a number of representative mammals, forming together a juxtaphylogenetic series. As mentioned earlier, such an investigation has never been carried out. What we do know is that the sets of local circuit neurons populating different neocortical areas in one and the same species may differ considerably, and that unfamiliarity with this feature has led to curious conclusions. Thus, on the basis of a study of the motor cortex of the mouse Cajal (1911, p. 825), in line with his general thesis discussed above, stated that "la substance grise du cerveau se simplifie considérablement chez les rongeurs et en particulier chez la souris" (the substantia grisea of the brain is considerably simplified in rodents, particularly in the mouse). In the second and third layer of that cortex, he observed no short-axon cells at all, and in the deepest layer only a few of these elements were detected. Some 10 years later, Cajal's pupil Lorente de Nó (1922) published a thorough analysis of a cortical area in the mouse which he believed to be the acoustic cortex, but which is now known to be the barrel field of the somatosensory cortex. He observed more than 30 different types of local circuit neurons in that cortical area (see Fig. 22.194) and concluded that: "La corteza cerebral del ratón presenta una textura complicadísima, que la hace comparable a la de los mamíferos superiores y aun a la humana" (the cerebral cortex of the mouse presents an ultracomplex structure which renders it comparable to the cortex of higher mammals and even to that of humans) (Lorente de Nó 1922, p. 76).

Lund and collaborators have recently published a series of Golgi studies on the local circuit neurons in the visual areas VI (Lund 1987; Lund et al. 1988; Lund and Yoshioka 1991), VII (Lund et al. 1981) and in the prefrontal cortex (Lund and Lewis 1993) of the rhesus monkey. Lund and her associates distinguished in laminae IIIB–VI of area VI (a study on the superficial layers I–IIIA has not yet appeared) 41 different types of local circuit neurons, including four types of chandelier cells and nine types of basket cells. In the prefrontal cortex, 13 types of local circuit neurons appeared to be present. Five of these cell types, i.e. basket, chandelier, neurogliaform, double bouquet and lamina VI beaded dendritic local circuit neurons, resembled cell types identified in the human prefrontal cortex. According to Lund and Lewis (1993), these five cell types are, in addition, probably the most commonly described varieties of non-pyramidal neurons in other cortical regions and in other species. Comparing the results of their study on the prefrontal cortex with those on visual areas VI and VII, Lund and Lewis (1993) concluded that these three areas clearly have some types of interneurons in common, but that the many different cell types having very precisely organised patterns of interlaminar axon projections seen in area V1 do not have clear counterparts in the prefrontal region. They considered it likely that at least certain types of local circuit neurons re-occur across many areas of the cortex and in different species, but that the precise morphology of these cells (in terms of features such as bouton size and density of terminal branches) differs across cortical areas, and that these differences reflect adaptations to the unique functional architecture of each region. This means that the diversity of local circuit neurons in a given cortical area of a given mammal says more about the degree of functional specialization of that particular area than about the 'phylogenetic status' of the entire cortex of that mammal. Analogous 'functional' conclusions may be drawn from the results of comparisons of corresponding cortical areas in different mammals. It is known that some types of local circuit neurons (double bouquet cells, basket cells), commonly encountered in the visual cortex of the cat and the monkey, are rare (double bouquet cells), or even completely lacking (basket cells) in the rat visual cortex. This structural difference most probably mirrors a difference in functional specialization. Cats and monkeys, as Parnavelas (1984:, p. 218) put it, are 'more visual' than rats.

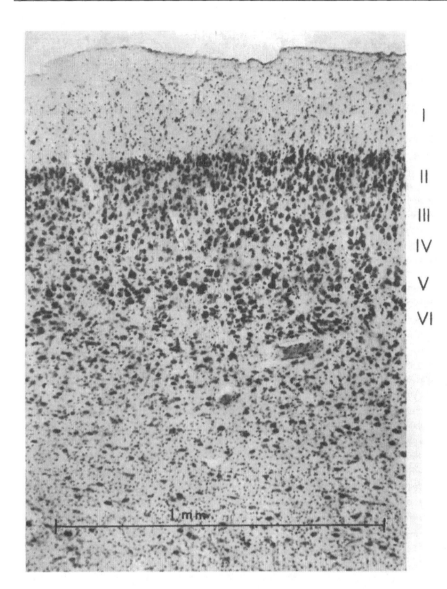

Fig. 22.210. Transverse section through the neocortex of the Northern pilot whale *Globicephala melaena*; thionin staining. (Reproduced from Pilleri and Kraus 1969)

Reduction of the Molecular Layer. During the evolutionary development of the neocortex, the molecular layer (lamina I) tends to become reduced, whereas lamina IV tends to become accentuated.

The molecular layer of the neocortex mainly consists of the apical dendritic tufts of pyramidal neurons and of horizontally running fibres. These fibres represent the end-segments of axons, which are often very long; they ascend from the central white matter and, upon reaching lamina I, bifurcate and turn into the tangential direction. These extrinsic fibres are reinforced by intrinsic fibres, including the ascending axons of Martinotti cells and collateral branches of the axons of pyramidal neurons.

The neocortical molecular layer in 'lower' mammals is much thicker than in 'higher' species. Morgane et al. (1990), quoting data collected by Sanides (1970, 1972), mentions a reduction in thickness from about 32% in the basal insectivores to about 12% in the higher primates. According to Morgane et al. (1990), the molecular layer in the very primitive neocortex of the cetaceans occupies on average 34% of the entire cortical thickness (Fig. 22.210). Preliminary measurements carried out by the present author indicate that this layer accounts for 26%, 14% and 10% of the entire cortical thickness in the opossum, cat and rhesus monkey, respectively.

Two factors are probably involved in the reduction of the molecular layer:

1. The disappearance of the so-called 'extraverted neurons' and the reduction in the number of pyramidal neurons extending their apical dendrites into lamina I. Extraverted neurons, i.e. elements with their soma in lamina II, which extend most of their dendrites peripherally into lamina I, are abundant in the neocortex of the opossum (Sanides and Sanides 1972), hedgehog (Valverde and Facal-Valverde 1986; Fig. 22.206A) and dolphin (Morgane et al. 1990; Fig. 22.206C) and have also been observed in the bat (Sanides and Sanides 1972) and rat (Kirsche et al. 1973; Fig. 22.206B). However, according to Sanides (1970), elements of this type are lacking throughout most of the neocortex of 'higher' mammalian species.

   As regards neocortical pyramidal neurons, Valverde and Facal-Valverde (1986) reported that, in the hedgehog, the apical dendrites of practically all of these elements form a terminal bouquet in lamina I. It is known that; in the monkey; the apical dendrites of numerous lamina V and VI pyramidal neurons terminate in a variably developed terminal tuft deeper within the cortex and that in many other pyramidal neurons the apical dendrite has become vestigial or has disappeared altogether. It is conceivable that, during the evolutionary development of the mammalian neocortex, the number of pyramidal neurons which have lost their anchorage to lamina I gradually increases, although sound quantitative evidence supporting this theory will be hard to obtain.

2. A gradual shift of specific thalamic afferent fibres from lamina I to lamina IV. In this context, the following observations are relevant:
   a) The dorsal cortex of the turtle (Fig. 22.143b), which may be considered as a primordial neopallium, receives a substantial (visual) thalamocortical projection, which terminates exclusively in the superficial zone of the external plexiform layer, i.e. the equivalent of the neocortical molecular layer (Hall and Ebner 1970; Heller and Ulinski 1987).
   b) In all mammals in which the laminar organisation of thalamic projections to the neocortex has been analysed, axons terminating in lamina I have been observed. However, in primitive mammals, this lamina I projection tends to be very strong, whereas in 'higher' mammals projections to deeper laminae prevail, particularly to lamina IV. This difference is particularly pronounced in the sensory parts of the neocortex. In these regions, the development of strong projections to lamina IV leads to the appearance of numerous, densely packed spiny stellate cells in that layer, a phenomenon known as 'granularisation'. Axon degeneration studies following unilateral thalamectomies in the opossum and hedgehog (Ebner 1969; Hall and Ebner 1970) showed that, in these primitive mammals, thalamocortical fibres form a dense, superficial zone in lamina I throughout the neocortex. In their studies on the neocortex of the dolphin, Glezer et al. (1988) and Morgane et al. (1990) were unable to distinguish thalamocortical fibres from other subcortical afferents. However, they observed that, in this cetacean, subcortical efferents in general, although issuing collaterals to all deeper layers, feed mainly into lamina I. This tallies with the observations made by Krasnoshekova and Figurina (1980), who traced a heavy contingent of degenerating fibres to layer I following lesions of the medical geniculate nucleus in the porpoise. In cetaceans, lamina IV is agranular and poorly differentiated.

   With regard to the laminar organisation of the somatosensory projection to the cortex, striking differences appear to be present between the hedgehog on the one hand and the rat and monkey on the other. Autoradiographic studies have shown that, in the rat (Herkenham 1980a) and monkey (Jones 1986), the somatosensory relay nucleus projects mainly to a zone which typically includes all of lamina IV and the deep part of lamina III and that a second, less dense zone of termination is situated in lamina VI and in the deep part of lamina V. Using wheatgerm-agglutinin horseradish peroxidase (WGA-HRP) as an anterograde tracer, Valverde et al. (1986) demonstrated that these two zones of termination are also present in the hedgehog, but that in this species the thalamocortical fibres originating from the somatosensory relay nucleus also form a dense terminal field in the superficial zone of lamina I (Fig. 22.211b). Flores (1911), studying the myeloarchitectonics of the cortex of the hedgehog, observed in many parts of the cortex of this species bundles of coarse, heavily myelinated fibres, which detach from the subcortical white matter and ascend to the superficial part of the cortex, where they bifurcate or bend again to follow a horizontal course in the deepest zone of lamina I (Fig. 22.211a). Valverde and Facal-Valverde (1986) observed that these so-called radii longi are particularly strongly developed in the somatosensory cortex, and Valverde et al. (1986) demonstrated that they are composed of thalamocortical fibres in this region (Fig. 22.211b). Vogt (1985) noticed what he called a progressive shift in the laminar preference of the input of the anterior thalamic nucleus in the cingulate cortex. In the rat, most of these terminals appeared to be located in lamina I, with fewer in lamina IV. In the rabbit and cat, an even distribution between laminae I and IV was found, while in the monkey most of these terminals were observed to end in laminae II–IV, with only a minimal contribution to lamina I.

c) There is evidence suggesting that the ontogenetic development of the neocortical lamina I parallels the evolutionary development discussed above. Thus Molliver and Van der Loos (1970) found that, in the somatosensory cortex of the newborn dog, the density of synapses is higher in the uppermost 100 µm of the cortical depth than in any of the lower layers. Kato et al. (1984, 1986) studied the development of the geniculocortical projection to the cat using an anterograde tracer and found that, in newborn kittens, lamina I receives a strong projection from the dorsal lateral geniculate nucleus, whereas the projections to lamina IV are only sparse. With age, the terminals in layer I appeared to become less dense, while those in layer IV became denser. By 1 month of age, the terminal distribution of the geniculocortical projection appeared to be similar to that found in adult cats, in which the terminals are sparse in lamina I and dense in depth, particularly in lamina IV. A similar developmental change in the terminal distribution of thalamocortical projections was found in the lateral suprasylvian area, which forms part of the visual association cortex (Kato et al. 1986), and in the primary auditory cortex (Miyata et al. 1984). Interestingly, Kato et al. (1986) observed that the terminal distribution of the cortico-cortical projection from areas 17 and 18 to the lateral suprasylvian area also changes during aging. The terminals in kittens younger than 2 weeks appeared to be distributed mainly to lamina I and less densely to laminae V and VI, whereas those in kittens older than 1 month and in the adult cat were found to be distributed almost exclusively in deep layers.

Arrangement of Modules. During the last 25 years, physiological and morphological evidence has been accumulated indicating that several parts of the neocortex of many different mammalian species are composed of radially oriented, column-like units or

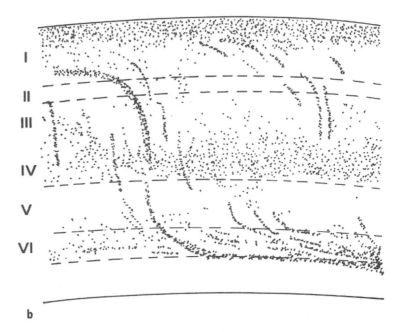

Fig. 22.211a,b. The neocortex of the hedgehog *Erinaceus europaeus*. **a** Myeloarchitecture of the somatosensory cortex; a large radix longus ascends from the subcortical white matter to the lower part of layer I. (Adapted from Flores 1911). **b** Anterograde labelling of fibres and terminals in the same cortical area, after an implantation of wheat-germ agglutinin-horseradish peroxidase (WGA-HRP) in the posterior region of the nucleus ventralis thalami. (Redrawn from Valverde et al. 1986)

modules. However, caution should be exercised in attributing profound general importance to these iterated arrangements.

The concept that column-like modules represent fundamental units of the mammalian neocortex has gained wide acceptance in the literature. In the following, a survey will be presented of the findings and considerations which have led to this concept. This survey will be followed by some critical and cautionary remarks:

1. The first to proposed a modular structure of the neocortex was Lorente de Nó (1938). He claimed that, in small radially oriented cylinders having a specific thalamocortical fibre as the axis, all elements of the cortex are represented and that in these elementary units the whole process of transmission of impulses from the afferent fibre to the efferent axon may theoretically be accomplished. Lorente de Nó's concept was based on the observations that (a) the terminal branches of thalamocortical fibres may form discrete cylindrical patches in lamina IV of the cortex, (b) the efferent system of the cortex is formed by radially oriented axons of pyramidal neurons and (c) the axons of many cortical local circuit neurons are likewise radially oriented.

2. In lamina IV of what he believed to be the auditory cortex of the mouse, Lorente de Nó (1922) described discrete cylindroid aggregations of cells, which he termed glomérulos (Fig. 2.33a). Golgi material revealed that the specific thalamic afferents to this cortex constitute dense patches of terminal ramifications which coincide with the glomérulos (Fig. 2.45b) and that the dendrites of the neurons in lamina IV are largely confined to the glomérulo in which their soma is located (Fig. 2.33b). The glomérulos appeared to contain numerous spinous stellate cells. The axons of these local circuit neurons were found to descend to the deeper cortical layers, where they issue numerous collaterals. Some of these collaterals were observed to ascend to lamina III (U in Fig. 22.194). Some 50 years later, it became clear that Lorente de Nó (1922) had actually analysed the portion of the somatosensory cortex to which the large mystacial vibrissae on the snout project via relays in the main trigeminal nucleus and in the thalamus. It appeared that each 'glomérulo' or 'barrel', as they came to be called, is the primary cortical representation of one vibrissal follicle. The vibrissae are arranged in a stereotyped grid-like pattern, in which each vibrissa has a unique position. Within the somatosensory cortex, each barrel also occupies a unique position and the topographical arrangement of these structures closely correspond to that of the vibrissae. The biggest barrels in mice are elliptical in cross-section, with a greatest diameter of 300 μm (Woolsey and van der Loos 1970; van der Loos and Woolsey 1973; Woolsey 1987). According to van der Loos (1979), barrels are the visible lamina IV counterparts of cortical columns.

3. Mountcastle (e.g. 1957, 1979) analysed the functional organisation of the somatosensory cortex, using microelectrodes to record the activity of single cells. During radial penetrations, he encountered cells with similar receptive fields throughout the depth of the cortex that responded to stimulation of the same type of cutaneous receptors located at a particular site. He concluded from these experiments that the basic functional unit of the neocortex is a vertically oriented column or cylinder of cells extending across all the cellular layers. Mountcastle believed that functionally active columns are able to isolate themselves from their surroundings by exerting an inhibitory action on neurons in their inactive neighbours, a process which he designated as 'pericolumnar inhibition'. Discrete anatomical correlates of Mountcastle's functional columns have not yet been detected in the somatosensory cortex of the cat or monkey.

4. The detailed physiological and morphological studies performed by Hubel and Wiesel and collaborators (Hubel and Wiesel 1968, 1972, 1977; Hubel et al. 1978; Livingstone and Hubel 1984a; LeVay et al. 1975) on the primary visual cortex of the monkey have led to the identification of three different kinds of column-like structure in that area: orientation columns, ocular dominance columns and blobs. Sets of these three column-like structure are conjectured to be united in larger entities, termed hypercolumns.

The *orientation columns* were detected by electrophysiological recordings. During radial penetrations with microelectrodes, cells having identical axes of orientation (i.e. cells which responded strongest to a bar of light in one particular orientation) were encountered throughout the thickness of the cortex. During tangential penetrations, the electrode encountered every 300–100 μm a shift in the axis of orientation of the light bar of about 10°. The morphological substrate of the functional columns thus detected could be visualised with the aid of the 2-deoxyglucose technique of mapping relative changes in metabolic activity following peripheral stimulation.

Microelectrode recordings also revealed the presence of parallel stripes, about 500 μm wide, alternately receiving their input from the ipsilateral and the contralateral eye. The morphological correlate of these *ocular dominance columns* could be strikingly visualised using silver impregnation techniques and autoradiographically by means of transneural transport of radiolabelled amino acids injected into one eye. In the monkey striate cortex, staining for the mitochondrial enzyme cytochrome oxidase revealed an array of densely staining, peg-like structures in laminae II and III, which can be seen lying in register with less densely staining formations in the cortical layers below lamina IV. These *blobs*, as they were termed, are ellipsoid in tangential sections, measuring roughly 150×200 µm. Their most conspicuous superficial (i.e. supragranular) parts receive a separate projection from the intercalated layers of the lateral geniculate body (Fig. 22.195). Physiological studies have shown that the cells within the blobs respond selectively to the colour of a stimulus, without regard to its orientation.

A *hypercolumn* is considered to represent the circuitry necessary for the analysis of a given, discrete region of the visual field. It is conceived to contain a set of nine orientation columns, together encompassing a complete cycle of orientation through 180°, an adjoining pair of right and left ocular dominance columns and several blobs.

In a previous section, it has already been mentioned that the axons of many neocortical pyramidal neurons give rise to long, horizontally running collaterals, which issue clusters of terminal branches at regular intervals (Fig. 22.205). Interestingly, recent experimental studies using different techniques, e.g. retrograde tracing, 2-deoxyglucose autoradiography and cross-correlation analysis, have produced evidence suggesting that, in the primary visual cortex, these horizontal collaterals mediate communication between functionally related columns, e.g. blobs of similar colour opponency and columns of similar orientation preference (for a review, see Gilbert 1992).

5. Evidence of a columnar organisation of the primary auditory cortex has also been presented. During microelectrode penetrations normal to the cortical surface, cells with the same characteristic frequency were encountered throughout the entire depth of the cortex (Abeles and Goldstein 1970). Bineural interaction bands have also been described, exhibiting either bineural summation or inhibition. Experiments in which single-unit mapping was combined with autoradiographic tract tracing revealed that, in the primary auditory cortex of the cat, clusters of summation responses coincide with bands receiving a heavy innervation from the contralateral primary auditory cortex, whereas suppression responses were recorded in regions of sparse contralateral innervation (Brugge and Reale 1985).

6. Within the motor area of the neocortex, there are small loci at which stimulation with weak currents produce movements executed by a single muscle. It has been demonstrated that the loci for producing contraction of a given muscle extend perpendicularly from the surface to the depth of the cortex. These radial arrays of stimulation points have been called *cortical motor columns*. The diameter of these columns is approximately 1 mm in cross-section (Asanuma 1975). Using two penetrating microelectrodes, one for stimulation and the other for recording, it was demonstrated that stimulation of a given column produces a pericolumnar zone of inhibition (Asanuma and Rosen 1973). Whereas the afferent input arriving at the somatosensory columns is modality specific, the cortical motor columns receive polymodal input from sensors in the target muscle itself, from the joint to which the muscle inserts and from the skin overlaying the muscle, i.e. sensors which are stimulated when the target muscle contracts (Asanuma 1987). In the motor cortex, the corticospinal fibres arise from pyramidal neurons with somata in lamina V. Experiments in which HRP was injected at different levels of the spinal cord have shown that these neurons form patches with intervening zones containing few or no corticospinal cells. These patches tend to aggregate in strips 0.5–1 mm wide (Jones and Wise 1977). Whether these patches and stripes somehow correspond to the functional motor columns is unknown.

7. Strong anatomical evidence for the existence of vertical columns or bands in the cerebral cortex has been provided by studies on the origin and termination of cortico-cortical connections. Thus Jones et al. (1975), using an anterograde tracer, demonstrated that both commissural and ipsilateral cortico-cortical fibres arising and terminating in the somatic sensory cortex of monkeys terminate in distinct vertically oriented columns with a variable width of 500–800 µm. Each of these columns appeared to be separated from its neighbours by a gap of approximately 500 µm. Each column was considered to represent the terminals of a bundle of cortico-cortical axons emanating from cells at the centre of each of the loci in which the tracer was injected. Using a retrograde tracer, Jones et al. (1975) also

demonstrated that the cells of origin of the cortico-cortical projections investigated, found predominantly in lamina III, are also aggregated in vertically oriented clusters. Goldman and Nauta (1977) injected an anterograde tracer into three cytoarchitectonically distinct regions (areas 4, 9 and 12) of the frontal lobe of rhesus monkeys. Neurons in these various regions are known to project both contralaterally and ipsilaterally to cytoarchitectonically diverse areas within the frontal, temporal and parietal lobes. They found that all of these cortico-cortical projections terminate in vertically oriented columns, 300–700 μm wide, which alternate in regular sequence with zones of comparable width that are free of such terminals. Interestingly, the pattern and dimensions of cortical columns in the prefrontal cortex, defined by cortical afferent inputs appeared to be strikingly similar in the rat, squirrel monkey and rhesus monkey, although the brains of these animals differ considerably in size (Bugbee and Goldman-Rakic 1983; Isseroff et al. 1984). In order to determine the pattern of termination of two converging cortico-cortical systems in the same animal, Goldman-Rakic and Schwartz (1982) followed a double anterograde labelling strategy. Using rhesus monkeys, they implanted HRP pellets in area 7 of the parietal lobe in one hemisphere and injected a mixture of tritiated amino acids in area 9 of the frontal lobe of the other hemisphere. It appeared that, in the prefrontal cortex, contralateral callosal fibre columns interdigitate with ipsilateral associational fibre columns. Goldman-Rakic and Schwartz (1982) also investigated the spatial organisation of the populations of cells in the prefrontal cortex projecting to the parietal associational cortex and to the contralateral prefrontal cortex using HRP or fluorescent dyes as retrograde tracers. They demonstrated that these two populations are inversely related in their relative densities over portions of the prefrontal cortex examined. Moreover, in the HRP material, the high-density patches of retrogradely labelled neurons appeared to coincide with the afferent fibre columns revealed by anterograde transport. In a later study (Selemon and Goldman-Rakic 1988), terminal labelling originating from prefrontal and parietal injections in the same hemisphere was investigated in a large number of areas of convergence in both the ipsilateral and the contralateral hemisphere. In most of the target areas, prefrontal and parietal terminal fields formed an array of interdigitating columns, but in some other areas prefrontal and parietal projections converged on the same column or cluster of adjacent columns but terminated within different laminae. For instance, in the depths of the superior temporal sulcus, prefrontal terminals occupied laminae I, III and V, with parietal terminals filling complementary laminae IV and VI.

From the experiments of Goldman-Rakic and Schwartz (1982) and Selemon and Goldman-Rakic (1988) discussed above, it appears that projections from two different (heterotopic) cortical regions (in the same or opposite hemispheres) remain segregated within common cortical targets, either by terminating in separate alternating columns or by concentrating in complementary laminae within the same column. The relationship between the termination zones of projections from paired *homotopic* areas was studied by McGuire et al. (1991b) in the cerebral cortex of the rhesus monkey. Injections with HRP and tritiated amino acids were made in two topographically matched regions of the frontal lobe in each hemisphere. The cortical projections from both of these homotopic pairs of areas appeared to converge in common columnar territories.

8. In 1979, Mountcastle introduced what he believed to be 'the basic modular unit of the neocortex' under the name *minicolumn*. He defined this unit as follows:

It is a vertically oriented cord of cells formed by the migration of neurons from the germinal epithelium of the neural tube along the radial glial cells to their destined locations in the cortex, as described by Rakic. If this minicolumn is comparable in size to the cortical units in which Rockel et al. (1974) made neuronal counts, it contains about 110 cells. This figure is almost invariant between different neocortical areas and different species of mammals, except for the striate cortex of primates, where it is 260. Such a cord of cells occupies a gently curving, nearly vertical cylinder of cortical space with a diameter of about 30 μm.

With regard to ontogenesis, Mountcastle (1979) referred to the work of Rakic (1971, 1972, 1978) on the development of the neocortex in the rhesus monkey, some salient aspects of which may be summarised as follows: The ventricular zone of the embryonic pallium is divided by glial septa into columns of precursor or stem cells termed 'proliferative units'. The postmitotic cells produced by these proliferative units find their way to the primordial cortex by following the shafts of elongated, radially oriented glial cells, which stretch across the embryonic pallial wall. Eventually, all postmitotic cells generated in a single proliferative unit form a morphologically identifiable stack of neurons in the cortex. These entities may be designated as 'ontogenetic' or 'embryonic' columns. Each proliferative unit produce multiple neuronal phenotypes. In a later review article, Rakic (1988) advanced the hypothesis that the ventricular zone of the embryonic pallium consists of proliferative units that constitute a proto-map of prospective cytoarchitectonic areas. Each of these areas is composed of a large number of ontogenetic columns, which become the basic processing units in the cerebral cortex. According to Rakic (1988), the relative constant size of these units in different mammalian species suggests that, during evolution, the cortex has expanded by the addition of such radial units rather than by their enlargement. Similar ideas concerning the phylogenetic development of the neocortex have been expressed by Szentágothai (1978) and Sawaguchi and Kubota (1986).

Mountcastle's estimate that a minicolumn contains about 110 cells was, as he indicates, based on the work of Rockel et al. (1974). In this publication and in a later, more extensive study (Rockel et al. 1980), the results of a quantitative comparative analysis of the neocortex were reported. The number of neuronal cell bodies were counted in a narrow strip, 30 µm wide, through the depth of the neocortex in several different functional areas (somatic sensory, visual, motor, frontal, parietal, temporal) in a number of different species (mouse, rat, cat, monkey and human). These counts gave remarkably constant values of 110 ± 10 cells in all areas and in all species studied. The only exception to this similarity appeared to be the binocular part of the visual cortex in a number of primate brains, in which approximately 2.5 times as many neurons were found. Rockel et al. (1980) remarked that they chose a width of 30 µm within which to count the cortical perikarya because, according to Hubel and Wiesel (1972) and others, this is approximately the width of the simplest functional column in the neocortex. However, they pointed out that this is not necessarily the size of the simplest anatomical unit. In fact, their findings imply that, regardless of their constant areal size, modules will contain constant cell numbers (Swindale 1990).

According to Mountcastle (1979), studies on the primary sensory and motor cortices have revealed that within the neocortex it is possible to identify a much larger processing unit than the microcolumn. The diameter of the larger units, which Mountcastle designated as *macrocolumns*, varies according to that author between 500 and 1000 µm for different areas. Each macrocolumn was considered as an aggregation of several hundreds of minicolumns.

9. Two noted neuroscientists, Szentágothai (1978) and Eccles (1984), attempted to give an idea of the microcircuitry of the neocortical columns. Szentágothai (1978) was of the opinion that "the basic unit of cortical architecture – the true cortical column – is anatomically based not on the mode of termination of the specific sensory afferents, as one might logically assume, but on that of cortico-cortical afferents". The size of these columnar units of cortico-cortical afferent terminations, according to Szentágothai, is remarkably constant, measuring around 200–300 µm in diameter. On the basis of light and electron microscopy data, Szentágothai indicated how the various types of presumed excitatory and inhibitory interneurons are connected with the cortical afferent systems and with the cortical output neurons, i.e. the pyramidal cells. He emphasised that the internal connectivity of each columnar unit is predominantly vertical, in a way that would favour a subdivision of each column into narrower microcolumns, and that the space of many of these microcolumns is occupied by the dendritic and/or axonal ramifications of individual interneurons. Eccles' (1984) sketch of the neural structure of the cortical module conforms to that of Szentágothai (1978) and is largely based on the work of the latter.

In the preceding pages, the evidence which has led to the concept that the mammalian neocortex is composed of repetitive radially oriented columnar units has been reviewed. Although this concept has dominated the neurobiological literature on the cortex during that last 25 years, it is by no means generally accepted. Numerous authors, including Towe (1975), Valverde (1986), Swindale (1990) and Purves et al. (1992), have challenged the validity of this concept. Some of the main objections may be summarised as follows:

1. The suggestion made by Rakic (1988) that the periventricular matrix of the embryonic neopallium consists of proliferative units, each of which gives rise to a sharply defined column of cortical neurons, is untenable. Recent ontogenetic studies summarised by Swindale (1990) have shown that clonally related cells in the developing pallium show a considerable tangential dispersion and hence do not adhere to narrow radial columns.

2. The thesis that columnar structures form the basic structural and functional units of the neocortex has so far not been substantiated by direct evidence concerning the specific neuronal composition and the specific mode of operation of these units. It is important to note that the

models presented by Szentágothai (1978) and Eccles (1984) are not based on analyses of the local connectivity within particular morphological units; rather, findings on cortical neurons and their connections derived from different cortical areas in different species were assembled and placed within the spatial framework of a cylinder with a diameter of 300 µm. Hence, these models are to be considered as theoretical constructs, showing at best how the local circuitry within columns of cortical tissue, defined by streams of cortico-cortical afferents, could be organised. Mountcastle's (1979) definition, stating that "a cortical column is a complex processing and distributing unit that links a number of inputs to several outputs", does not contain any information concerning the specific functional design of these columns.

3. The various column-like cortical entities have been distinguished on the basis of highly different morphological and/or physiological properties, e.g. similarity in physiological response properties, similarity in motor responses following stimulation, spatial segregation of sets of thalamocortical afferents, high concentrations of the enzyme cytochrome oxidase, concentrations of spiny stellate cells in lamina IV and the mode of termination of bundles of cortico-cortical fibres. It is unlikely that all of these entities are derivatives of one and the same basic cortical module, especially as they may differ considerably in size. It cannot be excluded that, as Szentágothai (1979) surmised, the units defined by cortico-cortical afferents represent the 'true cortical columns'. With regard to these structures, however, it should be emphasised that (a) their presence has so far not been established in monotremes, marsupials or insectivores, (b) their intrinsic structure is unknown, (c) their relationship to other column-like cortical formations is not clear and (d) their functional significance remains to be elucidated.

4. Although column-like entities of a given type are readily apparent in the neocortex of some species, they are often not detectable in other, sometimes closely related animals. The following examples of such inconsistencies are quoted from Purves et al. (1992), to whom the reader is referred for primary sources and details:
   a) Ocular dominance columns are present in Old World monkeys, but absent in most New World monkeys.
   b) Blobs in the primary visual cortex of the rhesus monkey are reportedly concerned with processing information about colour. However, these structures are also present in nocturnal primates with cone-poor retinas (and, therefore, poor colour vision). Moreover, blobs are absent in some non-primate species with cone-rich retinas and excellent colour vision, such as the tree shrew and the ground squirrel.
   c) Barrels are well developed in the mouse, rat, squirrel, porcupine and walrus, but are absent in other species that have prominent facial whiskers such as the dog, cat, raccoon and tree shrew. Furthermore, barrels occur in the guinea pig, which hardly uses its whiskers, and in the chinchilla, another cavimorph that has no whisking behaviour at all.

If we survey that data discussed above, the following conclusions seem to be warranted:

1. During the last decades, numerous different types or classes of column-like entities have been detected in different parts of the neocortex of many different mammalian species with the aid of a variety of physiological and anatomical techniques. There is little evidence of a systematic relation between these classes of 'columns'.
2. There is very little real evidence in favour of the concepts that (a) the entire mammalian neocortex is composed of column-like entities, (b) all of these entities represent variations on one and the same theme, (c) all of these entities essentially have the same structure and (d) they all essentially subserve the same function.

22.11.6.5.8
Summary and Conclusions

If we compare the neocortex with the reptilian cortex and with the 'older' parts of the mammalian cortex, the following features are found:

1. Pyramidal neurons occupy a central position in the circuitry of all of these cortices. These elements are provided with two, spatially separated, dendritic systems, an apical dendritic system, expanding in the most superficial cortical layer, and a basal dendritic system, the laminar position of which may vary considerably.
2. In all cortices, the pyramidal neurons are excitatory in nature and presumably constitute a continuous network extending throughout the cortex. Nearby connections in this network are provided by axon collaterals of pyramidal neurons, whereas longer intracortical connections consist of the main axons or axonal branches of more remote pyramidal neurons. In the neocortex, each pyramidal neuron receives by far its largest input from other pyramidal neurons.

3. Pyramidal neurons constitute the output system of all cortices. In the reptilian cortex, the piriform cortex and the hippocampal cortex, the pyramidal neurons are provided with bifurcating axons, one branch projecting to another cortical area and the other to one or more subcortical targets. However, in mammals, cortico-cortical fibres and cortico-subcortical efferents arise from separate sets of neurons, situated principally in the supraglanular layers and infraglanular, respectively.
4. In all cortices, the most superficial layer contains numerous tangentially running afferent fibres, making excitatory synaptic contacts with the apical dendritic expansions of pyramidal neurons. The sources of these fibres include the olfactory bulb, the thalamus, and other cortical regions. The fibre composition of the superficial layer varies from region to region. Thus in the lateral part of the reptilian cortex and in the mammalian piriform cortex, secondary olfactory fibres occupy a superficial position, whereas cortico-cortical fibres form a deeper zone. In sensory parts of the neocortex of primitive mammals, numerous thalamocortical fibres attain lamina I, but in the hippocampus and in neocortical association areas, cortico-cortical fibres prevail in this layer. In lamina I of all cortices fibres, from different sources tend to be arranged in different sublaminae.
5. In neocortical sensory regions, thalamic afferents terminate massively in lamina IV. In the primary visual cortex of prosimians and primates, different laminae of the lateral geniculate nucleus project to different sublayers of lamina IV.
6. Inhibitory interneurons, using GABA as their neurotransmitter, occur in all cortices. All inhibitory interneurons in all cortices impinge directly with some or all of their terminals on pyramidal neurons. Some inhibitory interneurons receive their principal input from extrinsic afferents and mediate feedforward inhibition of pyramidal neurons, e.g. the horizontal cells in lamina I of the piriform cortex (Fig. 22.186) and in the neocortex (Fig. 22.208b). Other inhibitory interneurons receive their main input via axon collaterals of pyramidal neurons, thus forming part of feedback loops (e.g. the large multipolar-cells in lamina III of the piriform cortex; Fig. 22.186), and still others, receiving excitatory inputs from extrinsic afferents and from axon collaterals of pyramidal cells, are involved in both feedforward and feedback inhibition of pyramidal neurons, e.g. stellate cells in the reptilian cortex (Fig. 22.183) and basket cells in the hippocampus (Fig. 22.191) and the neocortex (Fig. 22.208b).
7. Inhibitory interneurons, exerting their influence on other inhibitory interneurons, and thus a disinhibition of pyramidal neurons, are present in all cortices. The elements involved in such inter-interneuronal inhibitory circuits may be of the same type, e.g. the large multipolar cells in the piriform cortex and the basket cells in the neocortex, or of different types, e.g. the so-called L/M cells and the basket cells in the hippocampus (Fig. 22.191) and the basket cells and double bouquet cells in the neocortex (Fig. 22.208b).
8. Pericellular baskets, i.e. axonal networks producing numerous inhibitory synapses around the somata of pyramidal neurons, have been observed in all cortices. Inhibitory interneurons participate in the formation of several baskets, and axonal branches of several interneurons are involved in the formation of each basket.
9. Excitatory interneurons have so far only been demonstrated in the neocortex. These elements, the spiny stellate cells, represent transformed pyramidal neurons. They are abundant in primary sensory cortices, where they play a prominent role in the radial propagation of the activity fed by thalamocortical afferents into lamina IV of these cortices.

22.11.6.6
Neocortex: Quantitative Aspects and Folding

P. A. M. van Dongen

22.11.6.6.1
Size

The proportion of the total brain that the neocortex accounts for (neocorticalisation) varies considerably between species. In some insectivores (tenrecs), the neocortex occupies only 8% of the total brain weight, while in the sperm whale this figure is 87%, and in humans 76% (Mangold-Wirz 1966; Stephan et al. 1981). It is generally believed that, when the brain enlarged during evolution in some groups, the neocortex increased disproportionally more, i.e. larger brain – much larger neocortex. As a general principle, this is partly correct, as is demonstrated by Figs. 22.212, 22.213, which show that the pro-

Fig. 22.212. Size of the neocortex as a percentage of the total brain, plotted against brain weight. The fraction of the neocortex depends not only on brain size, but also on the taxonomical position. (Based on data of Mangold-Wirz 1966; Kraus and Pilleri 1969; Stephan and Pirlot 1970; Stephan et al. 1974, 1981; Pirlot and Nelson 1978; Pirlot 1981; Frahm et al. 1982)

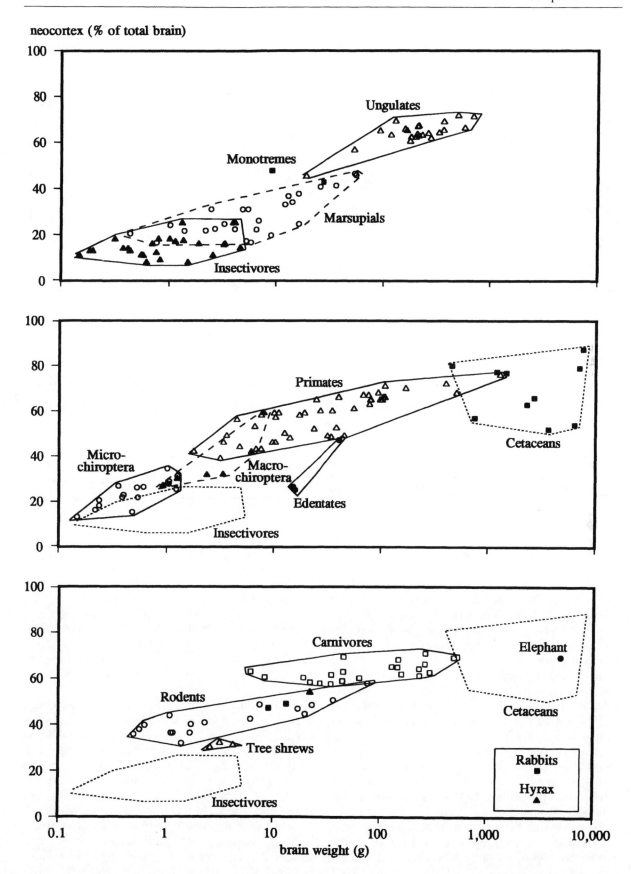

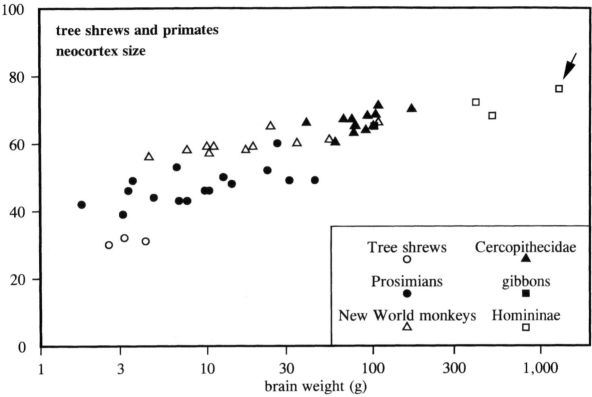

Fig. 22.213. Tree shrews and primates (human marked by an *arrow*). The size of the neocortex as a percentage of the total brain, plotted against brain weight. The fraction of the neocortex depends mainly on the taxonomical position. (Based on data of Mangold-Wirz 1966; Kraus and Pilleri 1969; Stephan and Pirlot 1970; Stephan et al. 1974, 1981; Pirlot and Nelson 1978; Pirlot 1981; Frahm et al. 1982)

portion of the total brain that the neocortex accounts for depends more on the taxonomical position than on brain size in itself.

Compare, for example, the very small brains (about 0.5 g). In insectivores, the neocortex occupies 10 %–15 %, in marsupials 20 %, in bats 20 %–30 % and in rodents approximately 40 % of the total brain.

If we look at small brains (about 3 g), in insectivores, the neocortex occupies 10 %–20 %, in marsupials 20 %–30 %, in the tree shrew 30 %, in rodents and prosimians 40 % and in monkeys 55 % of the total brain.

As far as brains of about 15 g are concerned, in armadillos, the neocortex occupies about 25 %, in monotremes 40 %–50 %, in sloths, rodents, rabbits, hyraxes and prosimians 50 % and in carnivores and monkeys 60 % of the total brain.

In all brains larger than 100 g, the neocortex occupies at least 50 % of the total brain. In groups with large brains, a huge, disproportional enlargement of the neocortex is found, i.e. in primates, ungulates, carnivores, whales and elephants.

The monotremes are usually considered as the most 'primitive' group of mammals, but with respect to neocorticalisation, they are advanced (platypus 48 %, echidna 43 %). In another 'primitive' group, the marsupials, the neocorticalisation ranges from 17 % to 46 %. The lowest figures for neocorticalisation (less than 10 %) are found in insectivores.

In primates, neocorticalisation seems to increase with increasing brain size (Fig. 22.212). However, even within the primates, neocorticalisation depends on the taxonomical position rather than on brain size (Fig. 22.213). Within each subgroup of the primates, neocorticalisation is fairly constant and almost independent of brain size. Prosimians tend to have small brains (for a primate), with a relatively small neocortex (39 %–55 %). New World monkeys have equally small brains, but a larger neocortex. In monkeys (excluding *Hominidae*),

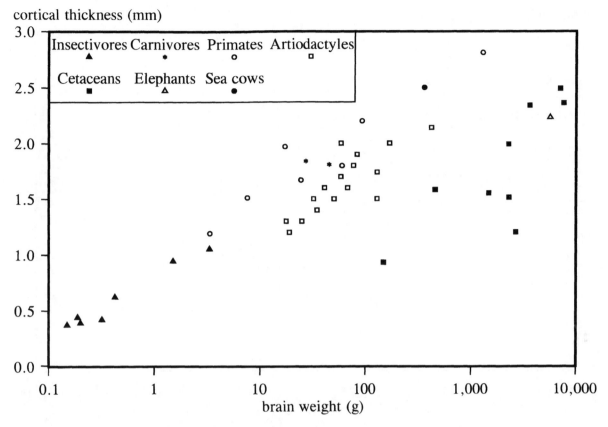

Fig. 22.214. Thickness of the neocortex as a percentage of the total brain, plotted against brain weight. The average cortical thickness increases with brain weight, but a considerable variation is found, expecially in the cetaceans. (Based on data of Mangold-Wirz 1966; Kraus and Pilleri 1969; Stephan and Pirlot 1970; Stephan et al. 1974, 1981; Pirlot and Nelson 1978; Pirlot 1981; Frahm et al. 1982)

necorticalisation varies between 56% and 71%, with brain size having only a small influence. The chimpanzee has a brain of about 440 g, about 72% of which is neocortex. In the gorilla, the brain is somewhat larger (500 g), and the relative proportion of the neocortex somewhat smaller (68%). Compared to the chimpanzee, humans have a much larger brain (about 1300 g, i.e. a factor of 3–4 larger), but the proportion of the neocortex is only slightly larger (76% in humans vs. 72% in the chimpanzee; Stephan et al. 1981). Thus one should not think that, during hominid brain evolution, the neocortex was particularly enlarged; the brain stem also enlarged considerably and to a similar extent.

22.11.6.6.2
Thickness

While mammalian brain size and cortical size varies by more than a factor of 100 000, average cortical thickness varies only by a factor of 7, i.e. from 0.4 mm in the very small shrew brain to 2.8 mm in humans (Fig. 22.214). Figure 22.214 shows that cortical thickness depends on brain size and on taxonomic position. The relationship between brain size and cortical thickness is fairly straightforward, but some striking features are present.

In the cat and fox, the average neocortical thickness is 1.8 mm, which is normal for a brain of that size. However, a miniaturized carnivore, the weasel, has a small brain (2.4 g) with an estimated neocortical thickness of only 0.5 mm (from Fig. 64 in Welker 1990).

Dolphins have large, but thin cortices. It has been discussed whether this can be regarded as true neocortex (Morgane and Jacobs 1972; Kesarev et al. 1977).

The thickest neocortices are found in fin whales (2.3–2.5 mm) and humans (2.5–2.8 mm).

Overall, in mammalian orders (except possibly in dolphins and whales), a striking basic uniformity is found in the structure of the neocortex (Rockel et al. 1980). Apparently, the basic cellular structure of

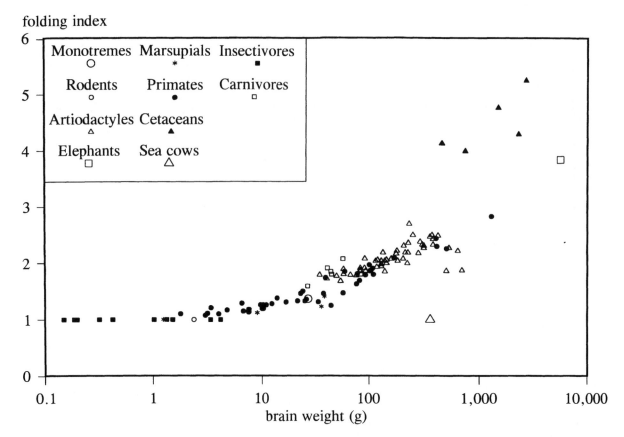

Fig. 22.215. Folding index of the neocortex as a percentage of the total brain, plotted against brain weight. Smooth, lissencephalic brains have a folding index of 1. The largest folding indices are found in cetaceans and elephants. (Based on data of Mangold-Wirz 1966; Kraus and Pilleri 1969; Stephan and Pirlot 1970; Stephan et al. 1974, 1981; Pirlot and Nelson 1978; Pirlot 1981; Frahm et al. 1982)

the neocortex poses constraints on neocortical thickness. If the neocortex cannot become thicker than 2–3 mm, a inevitable consequence of increasing brain and cortex size during mammalian evolution is cortical folding, which is the only solution to pack a cortex with a large surface into a compact box (the skull).

22.11.6.6.3
Folding Index

To express the degree of cortical folding in a single figure, Von Bonin (1941) used a folding index, i.e. the ratio between the total neocortical surface and the external neocortical surface. For a smooth (*lissencephalic*) brain, the folding index is 1, and for a convoluted (*gyrencephalic*) brain, it is larger than 1. Figure 22.215 shows the folding index for various mammalian orders. Cortical folding depends on the brain size and on the thickness of the cerebral cortex (Hofman 1985). Most brains smaller than 5 g are lissencephalic, and most brain

larger than 50 g are gyrencephalic. However, there are some notable exceptions (see below). The highest values of the folding index are found in whales (4.0–5.3) and elephants (3.8). This is considerably larger than the folding index of the human brain (2.8). Some dolphins with a smaller brain than humans (450–750 g), have a larger folding index than humans; this is mainly due to their thin cortex.

22.11.6.6.4
Surface Area

The cortical surface area is often considered to be a measure of cerebral information-processing capacity. Figure 22.216 shows the cortical surface area for various mammals. In this case, the vertical axis is logarithmical; this was necessary, since there is a very large variation (a factor 10 000) in the cortical surface area. The very small lesser shrew (*Suncus etruscus*) has a cortical surface area of 0.8 cm^2 (two hemispheres), the rat (*Rattus norvegicus*) 6 cm^2, the cat (*Felis sylvestris*) 83 cm^2 and humans (*Homo*

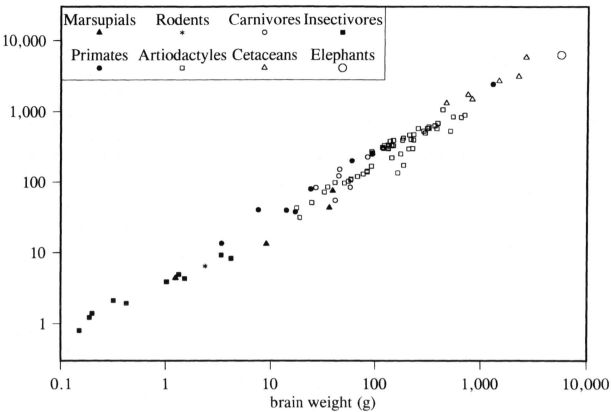

Fig. 22.216. Surface area of the neocortex plotted against brain weight. A regular relationship between brain size and the surface area of the neocortex exists. (Based on data of Mangold-Wirz 1966; Kraus and Pilleri 1969; Stephan and Pirlot 1970; Stephan et al. 1974, 1981; Pirlot and Nelson 1978; Pirlot 1981; Frahm et al. 1982)

sapiens) 2400 cm². The largest cortical surface areas are found in whales and elephants: for the African elephant (*Loxodonta africana*) 6300 cm², the Northern pilot whale (*Globicephala melaena*) 5800 cm² and the false killer whale (*Pseudorca crassidens*) 7400 cm².

22.11.6.6.5
Gyri and Sulci

The most conspicuous difference between the brains of mammals is whether the cerebral cortex is smooth or convoluted. The degree of cortical folding has long been used to judge brain complexity. In this section, we will try to give a survey of the cortical folding patterns and some of the principles behind cortical folding. For an extensive discussion of this subject, the reader is referred to the monograph by Welker (1990).

Deducing the Evolution of Gyrification Patterns. In various mammalian groups, lissencephalic or gyrencephalic brains are found. The closer animals are related, the more similar the patterns of gyrification are (Welker 1990). Three strategies will be followed to deduce the evolution of the various patterns of gyrification:

1. *Classical comparative anatomy.* In the tradition of comparative anatomy, patterns of cortical gyri in presently living species are compared. In this chapter, we wish to show how brain gyri might have evolved. Therefore, we need to distinguish between primitive and derived brain features in a phylogenetic context. For this purpose, late brain features of various mammals need to be related to mammalian phylogeny. The phylogeny of several mammalian orders is still disputed, but a phylogenetic tree is now gradually emerging supported by biochemical evidence (Novacek 1992; see Fig. 2313).
2. *Palaeoneurology.* As far as possible, the impressions of gyri and sulci in fossil skulls are investigated.

3. *Embryology.* The development of cortical gyri is studied during various stages of embryological life, where it is presumed that sulci emerging early in phylogeny also emerged early in ontogeny. However, only for a few species could an ontogenetic series be collected that was sufficient to deduce the morphogenesis of the cortical gyri and sulci.

Basis of Gyrification. Two types of theories exist about the basis of gyrification:

1. *Mechanical theories.* Several authors regard gyrification as a mechanical solution to the problem of how to pack a cortex of a given thickness and surface into a compact box (Bok 1929; Le Gros Clark 1945). Prothero and Sundsten (1984) developed a quantitative model for cortical gyrification. An essential feature of this model is that a gyrus must be broad enough to contain the cortical cell layers and their afferent and efferent fibres. According to these theories, the (genetically determined) pattern of gyri and sulci is mainly determined by random processes during evolution leading to species- and group-specific patterns of gyrification.
2. *Functional theories.* At the beginning of the nineteenth century, Gall and other phrenologists believed that each gyrus represented a specific mental function. This idea has now been abandoned. More recently, Welker (1990) has strongly advocated the theory that sulci separate functionally different gyri:

There is already sufficient evidence to suggest the general working hypothesis that crown-wall-fundic complexes, and gyral-sulcal aggregates may be analogous to the nuclei and nuclear complexes in the rest of the brain, with the exception that these cortical complexes are structural-functional entities of greater hierarchical complexity (Welker 1990).

Some sulci separate cytoarchitectonically different gyri (which are also functionally different), e.g. the rhinal, cingulate and central sulci in primates. Such sulci are called *limiting sulci.* Other sulci lie between cytoarchitectonically similar gyri, e.g. the calcarine sulcus in primates. These sulci are called *axial sulci.* Some gyri consist of cytoarchitectonically different fields without intervening sulci.

Nature and Nurture in Gyrification. The more closely mammals are related, the more similar their patterns of gyrification are (see below). This implies that an heritable component leads to a species-specific pattern of gyri and sulci. However, small differences are present in the pattern of fissuration between normal individuals of a single species, as has been found in the raccoon, orangutan and humans (Connolly 1950; Welker and Seidenstein 1959; Welker 1990). It is completely unclear whether such interindividual differences are due to genetic differences, environmental factors or chance.

Various experimental manipulations during brain development influence the pattern of gyrification. In intact monkeys, the occipital (visual) cortex is rather smooth, but removal of one eye in a fetal monkey causes the emergence of several occipital gyri on the contralateral side (Rakic 1988). During development, cortical fissuration begins at the time that thalamocortical and cortico-cortical connections are made, and these connections apparently influence the emerging pattern of gyri. Cortical ablations in neonatal monkeys, kittens and raccoons caused aberrant patterns of gyrification not only close to the region of the lesion, but also at more remote sites (Welker 1990).

There appears to be a species-specific pattern of gyrification which comes about during normal development due to normal patterns of connections and spatial constraints.

Designations. The cortical neurons are contained in the gyri, while the sulci are merely the gaps between the gyri. It has therefore been advocated that gyral patterns rather than sulcal patterns should be emphasised (Welker 1990). However, for practical reasons, patterns of gyrus formation will be described in this section in terms of sulci, since an emerging sulcus can be described by a single name, whereas two names are needed to describe it in terms of the gyri it separates.

Primitive Mammalian Cortex. From various sources, we can deduce features of the anchestral mammalian cortex. Firstly, the anchestral mammalian neocortex must have been small, since it evolved from the reptile brain. Investigations on the skulls of extinct mammals show that archaic mammals had proportionally small brains (Jerison 1973; see also Chap. 23). During mammalian evolution, brain size has increased independently in various orders. Secondly, the anchestral mammalian cortex must have been smooth, as it evolved from the smooth reptilian telencephalon. Moreover, from the fossil skulls of archaic mammals, it can be deduced directly that the primitive cortex was smooth. Thirdly, the anchestral mammalian cortex had a relatively large piriform cortex and a relatively small neocortex. As a consequence, the border between the piriform cortex and the neocortex (the lateral rhinal fissure) was located dorsally. The ancestral mammalian brain most probably resembled the brain of some insectivores (see below; Stephan et al. 1990).

Three Primitive Fissures. In the brains of all mammals, three primitive fissures are recognised, cytologically if not macroscopically (Ariëns Kappers et al. 1936):

1. The hippocampal fissure between the hippocampus and neocortex
2. The endorhinal fissure between the olfactory tubercle and the piriform cortex
3. The lateral rhinal fissure between the piriform and neocortex

These are often visible at superficial inspection, but sometimes only as an infolding of the cortical laminae. The lateral rhinal fissure will be mentioned frequently in this chapter, as it is an important landmark for comparative neuroanatomy.

Homologies. A central question in comparative neuroanatomy is the question of homology, i.e. are certain features in different species or groups derived from a common ancestral feature? Do we have reason to believe that neocortical gyri and sulci in various mammalian groups are homologous? In all mammalian grandorders (except the *Ferungulata*), small brains are present in which the basic, smooth mammalian pattern still can be recognised (see below). We therefore hypothesise that patterns of neocortical gyri and sulci developed independently in the various grandorders. If this hypothesis is correct, we must conclude that several neocortical gyri and sulci within one grandorder might be homologous, but not between grandorders. An exception are the sulci around the cingulate cortex. All mammals have a cingulate cortex, and in several groups a ventral and dorsal groove limiting the cingulate cortex (cingulate gyrus) are found, i.e. the callosal and cingulate (splenial) sulcus, respectively. Even within an order, it can be difficult to establish homologies between neocortical sulci in advanced brains (Hershkovitz 1970). Attempts have been made to deduce homologies between neocortical gyri from functional studies (Johnson 1980; Welker 1990); however, a discussion of these ideas is beyond the scope of this chapter.

Monotremes. The egg-laying monotremes are usually considered the most primitive mammals. Two families exist: the platypus (duckbill, *Ornythorhynchus*) has a smooth brain (Pl.1, 9 g), and the lateral rhinal fissure is the only landmark on the cerebral surface (Pirlot and Nelson 1978); the echidna (*Tachyglossus*) has a larger, gyrencephalic brain (Pl.2, 28 g). The lateral rhinal fissure has a ventral position in the monotremes. At the medial aspect, the monotreme brain is remarkable for its lack of a corpus callosum; also at the medial aspect of the Echidna cortex, several sulci are present.

A comparison of the platypus and echidna brains illustrates that lissencephalic and gyrencephalic brains can exist within one group. The phylogenetic history of the fissures in echidna is unknown, but is it unlikely that the fissures in echidna are homologous with those of any other mammal. In echidna, all primary cortical projection areas are located caudally, so neither the cortical localisation nor the pattern of gyrification provides an indication of convergent evolution between echidna and other mammalian groups.

Marsupials. Within the group of marsupials, a large variation in body size, brain size, behaviour and ecology is found. Four orders of marsupials are distinguished (see below); within three of these four orders, small-brained and large-brained species are found. For a detailed description of the marsupial cortex, the reader is referred to Johnson (1977) and Haight and Murray (1981). The four orders are as follows:

1. *Paucituberculata.* These include the rat opossums, which are all small animals with small lissencephalic brains. The small brain of *Lestoros inca* (0.8 g) is a characteristic primitive mammalian brain, i.e. small and smooth with a dorsal rhinal fissure.
2. *Peramelina.* These include the bandicoots. Their body sizes vary from that of a rat to that of a cat. The brain of the long-nosed bandicoot (*Isoodon*) is smooth with a dorsally located rhinal fissure. In the long-nosed bandicoot (*Perameles*), a ventrally located lateral rhinal fissure is found, as well as a frontal sulcus that is characteristic for marsupials, called α.
3. *Marsupicarnivora.* These include the opossums, marsupial mice, marsupial mole and the Tasmanian marsupial wolf. Within this order, very small species are found (e.g. the marsupial mole, *Notoryctes*), intermediate species (opossum, *Didelphis*) and rather large species (Tasmanian wolf, *Thylacinus*). The smallest brains (e.g. of the marsupial mole) are lissencephalic with a dorsally situated lateral rhinal fissure. The opossums have larger brains (5–10 g) with relatively more neocortex and a ventrally situated lateral rhinal fissure (Pl.3). This is a general feature in mammals: with neocortical expansion, the lateral rhinal fissure is 'forced' downwards (Edinger 1966). In the opossum brain, too, sulcus α is found (Pl.3). In larger marsupicarnivoral brains (e.g. the Tasmanian wolf, 53 g), additional fissures are present.
4. *Diprotodonta.* This is a heterogeneous group that includes the possums, koala, wombats and kan-

garoos. Some members of the Diprotodonta are as small as mice (e.g. some possums), while the group also includes the large kangaroos. The small possums (*Burramys*) and sugar gliders (*Petaurus*) have lissencephalic brains with a ventrally located lateral rhinal fissure. The sulcus α and the beginning of other sulci are found in the brush-tailed possum (*Trichosurus*). A large, gyrencephalic brain is found in the wombat (*Vombatus*). Among the Diprotodonta, species of the kangaroo family (*Macropodidae*) have a characteristic pattern of sulci (Dillon 1962; Haight and Murray 1981). Small species, such as the rat kangaroo (*Potorous*), have a smooth brain with a ventrally located lateral rhinal fissure. In the larger macropodid brains (*Thylogale*, 25 g; *Macropus*, 50 g), more sulci are present (Pl.4), but the pattern of fissures in the kangaroo is different from that found in wombat. An indication of marsupial brain evolution can be found in the endocast of the Miocean (21 million years ago) diprotodont *Wynyardia bassiana*. This species has a primitive lissencephalic brain (18 g) with dimples indicating the lateral rhinal fissure and the neocortical sulci α, β, δ and μ (Haight and Murray 1981).

Eutheria. The Eutheria are often subdivided into five grandorders: Insectivora, Edentata, Glires (rodents and rabbits), Archonta (including bats and primates) and Ferungulata (carnivores and ungulates):

1. *Insectivores.* Insectivores have a special position in comparative anatomy of the mammalian brain, as the brains of some insectivores are considered to be the most primitive mammalian brains (Stephan et al. 1990). In several respects, the brains of some insectivores are more primitive than those of monotremes and marsupials. All insectivore brains are lissencephalic (folding index, 1; Hofman 1985). Three orders of insectivores are distinguished (tenrecs, hedgehogs and shrews; see below). Some species that used to be included as insectivores are now included in separate orders, e.g. the elephant shrews in the Macroscelidea and the tree shrews in the Scandentia. For an extensive survey of the insectivore brain, the reader is referred to Stephan et al. (1990).
   a) *Tenrecomorpha* (tenrecs, golden moles, solenodons, otter shrews). For the evolution of the mammalian brain, the order Tenrecomorpha is most interesting, since some of its members have the most primitive brains of all mammals. With respect to size and composition, the small, compact brain (0.13 g) of *Geogale aurita* and the larger, elongated brain (2.6 g) of the tenrec (*Tenrec ecaudatus*) come "close to the starting point of mammalian brain evolution" (Stephan et al. 1990, p. 299). These brains have a very small neocortex (less than 10 % of the brain weight). Although the tenrec has a primitive brain, a depression indicating a primitive sulcus is visible; this is a rostral (orbital) sulcus which bends caudally at the lateral sides (Pl.5; Stephan et al. 1990). This primitive insectivore sulcus is also visible in other larger brains in Tenrecomorphs (e.g. *Solenodon* and hedgehogs). In the large, but still primitive brain of *Solenodon*, a second depression is present, i.e. a dorsal sulcus parallel to the interhemispheric fissure. Among the Tenrecomorpha, the brain of the golden moles (*Chrysochloris*) is remarkable; it is a wide, high brain with a lateral rhinal fissure visible at the rostral end. More advanced brains are also encountered among the Tenrecomorpha; the otter shrews (*Potamogale*) have a relatively large neocortex and a ventrally located lateral rhinal fissure.
   b) *Erinaceomorpha* (hedgehogs). These have primitive brains with a dorsally located lateral rhinal fissure. In the hedgehog (*Erinaceus*), a faint orbital sulcus is present (Pl.6). The moon rat (*Echinosorex*) has a larger brain with relatively more neocortex; an orbital sulcus and a dorsal sulcus is present, as in *Solenodon*.
   c) *Soricomorpha* (shrews, moles). The smallest mammals are found among the shrews (*Soricidae*; the smallest species weigh less than 5 g). The shrews have small brains (0.15–0.4 g) which are completely smooth (Pl.7); no lateral rhinal fissure is present (Stephan et al. 1990). Among the moles (*Talpidae*), fossorial and semi-aquatic species are found. A fossorial species such as the European mole (*Talpa europaea*) has a small brain (1 g) with a very faint orbital sulcus and a visible lateral rhinal sulcus in the rostral part of the cortex. In semi-aquatic moles (e.g. *Galemys* and *Desmana*), the brains are larger with a larger contribution of the neocortex.
2. *Edentates.* This group consists of three orders: *Cingulata* (armadillos), *Pilosa* (anteaters and sloths) and *Pholidota* (pangolins). Extensive comparative studies have been made by Elliot Smith (1898) and Pohlenz-Kleffner (1969).
   a) *Cingulata* (armadillos). The armadillos vary considerably in body size and brain size. The smallest brain (2.5 g) is found in the fairy armadillo (*Chlamyphorus*). This is a smooth,

lissencephalic brain without a visible lateral rhinal fissure, but histological investigations reveal that the junction between the piriform cortex and neocortex is high on the lateral wall. In the brains of larger armadillos (*Tatusia peba*), an anterior and posterior lateral rhinal fissure is evident, as well as an anterior neocortical fissure, called β. In even larger brains (*Dasypus*), the posterior rhinal fissure merges with fissure β, and another neocortical fissure becomes evident, called δ (Pl.8). This is the basic pattern of neocortical fissures in *Cingulata* and *Pilosa*. At the medial aspect of the cerebrum of *Dasypus*, a callosal and cingulate (splenial) sulcus is present.
b) *Pilosa* (anteaters, sloths). The smallest anteater (fairy anteater, *Cyclopes*) has a small, lissencephalic brain (4.4 g) without visible lateral rhinal fissure. The lesser anteater (*Tamandua*) and the giant anteater (*Myrmecophaga*) have gyrencephalic brains (Pl.9,10). At the medial aspect of the cerebrum of the giant anteater, a callosal and cingulate (splenial) sulcus is present. Likewise, in the sloths, the larger brain has more fissures (compare the smaller three-toed sloth *Bradypus* with the larger two-toed sloth *Choloepus*). The basic pattern in intermediate-sized anteaters and sloths (*Tamandua* and *Bradypus*) resembles that of the intermediate-sized armadillos (*Cabassous*).
c) *Pholidota* (pangolins). Few studies deal with the morphology of the pangolin brain. Clear pictures of full-grown brains are found in Weber (1892) and Hackethal (1976) and of embryos in Friant (1944). Compared to the *Cingulata* and *Pilosa*, the pangolins have smaller and more delicate cortical gyri. At the medial aspect of the pangolin cerebrum, a callosal and cingulate (splenial) sulcus is present. The pangolin cortex is different from the cortices of *Cingulata* and *Pilosa*.
3. *Glires.* The phylogenetic positions of the Macroscelidea (elephant shrews), Rodentia and Lagomorpha (rabbits) are a matter of dispute. Novacek (1992) placed them together, a suggestion we will follow here, although in some recent studies, the Lagomorpha were regarded as related to the Archonta (Graur et al. 1996).
   a) *Macroscelidea.* For a long time, the Macroscelidea (elephant shrews) were included in the Insectivores. The elephant shrew (*Elephantulus*) was considered as an advanced insectivore, since it has a small (1.3 g), but rather advanced brain. It has a relatively large, smooth neocortex and a ventrally located lateral rhinal fissure (Pl.11).
   b) *Lagomorpha* (rabbits). Lagomorphs (rabbits and hares) are considered as an order that is separate from the rodents. The basic lagomorph pattern of cerebral folding is seen in the smallest species, the pika (*Ochotona*); its brain is small (2.5 g) and lissencephalic with only a lateral rhinal fissure. In larger lagomorph brains (hare, *Lepus*; 13 g), an additional sulcus is present, i.e. a longitudinal sulcus on the dorsal surface of the brain (the marginal sulcus; Pl.12).
   c) *Rodentia.* Three main groups of rodents are distinguished: Myomorphs (mouse-like), Sciuromorphs (squirrel-like) and Hystricomorphs (porcupine-like). It has been argued that the Hystricomorphs are not monophyletic with the Myomorphs and Sciuromorphs (Honeycutt et al. 1995). For an extensive description of the rodent brains, the reader is referred to the series of monographs by Pilleri (1959a–c, 1960a–c).
   – *Myomorphs.* The Myomorphs have lissencephalic brains; the lateral rhinal fissure can be distinguished, but not very clearly (Pl.13; Pilleri 1960). Overall, Myomorphs have rather small brains (but fairly large for their body weights; see Chap. 23).
   – *Sciuromorphs.* The Sciuromorphs (e.g. the squirrel, *Sciureus*) have lissencephalic brains with a clear lateral rhinal fissure. The largest Sciuromorph brains are found in the marmot (*Marmota*, 17 g) and the beaver (*Castor*, 45 g; Pl.14); in these brains, a dorsal, longitudinal dimple can be identified (Pilleri 1959a).
   – *Hysticomorphs.* The porcupines (*Hystrix*) and the South American rodents belong to the Hystricomorphs. They range from the cavia (*Cavia*) to the largest rodent, the capybara (*Hydrochoerus*). Most Hystricomorphs have a more or less gyrencephalic brain. In the small brain of the cavia (5 g), a clear dorsal longitudinal and a pseudosylvian fissure is found (Pl.15). In the larger Hystricomorph brains (e.g. the mara, *Dolichotes*), more fissures become evident (Pl.16). The largest brains (e.g. of the capybara, 76 g) are gyrencephalic. One remarkable exception within the Hystricomorphs is the North American porcupine (*Erethizon*), which has a large (25 g), but lissencephalic brain.
In several studies, the pattern of cerebral cortical convolutions has been used to argue in favour of one specific phylogenetic hypothesis, e.g. in the *Bovidae* and *Mustelidae*

(Oboussier 1972; Radinsky 1973). The gyrencephalic Hystricomorph brain might be an additional argument in favour of rodent polyphyly.

4. *Grandorder Archonta*. The grandorder of Archonta consists of four orders: the tree shrews, the flying lemurs, bats and primates:

   a) *Scandentia* (tree shrews). The tree shrews (*Tupaia*) have small (3 g), lissencephalic brains with a clear, ventrally situated lateral rhinal fissure and some longitudinal dimples (Pl.17).

   b) *Dermoptera* (flying lemurs). The flying lemur (*Cynocephalus volans*) has a considerably larger brain (6 g) than the tree shrews. Apart from the lateral rhinal fissure, a prominent longitudinal sulcus is present at the dorsolateral surface of the cerebral cortex, and there are a few, less prominent dimples (Pl.18). At the ventral part of the palaeocortex, a fissure becomes visible. Since the flying lemur belongs to the same grandorder as the primates, and since this fissure has a similar position as the sylvian fissure in small primates (see below), this fissure might be homologous with the sylvian fissure. At the medial aspect of the cerebrum of the flying lemur, a callosal and cingulate (splenial) sulcus is present.

   c) *Chiroptera* (bats). The order of Chiroptera (bats) consists of two subgroups: the Macrochiroptera and the Microchiroptera. An extensive survey of the brains of bats is presented by Baron et al. (1996).

   – *Microchiroptera*. Most Microchiroptera are small and have small brains. These small brains are smooth (Pl.19). In some species, the lateral rhinal fissure is visible rostrally. In larger brains, some sulci are visible, some of which are fairly deep. In several species, a rostral (orbital) sulcus is present (e.g. in *Noctilio* and *Vampyrum*). In several families, the larger brains have a lateral sulcus, which might be homologous to the sylvian sulcus of primates. In large brains of some species (e.g. *Vampyrum*), a dorsal depression or sulcus is present, parallel to the interhemispheric fissure; this is called the suprasylvian sulcus. In some larger brains, a cingulate sulcus is also present.

   – *Macrochiroptera*. In the Macrochiroptera, several large bats are found (e.g. flying foxes), in addition to rather small species. At the rostral side, the lateral rhinal fissure is macroscopically visible. In the brains of the Macrochiroptera, a sulcus is present that might be homologous to the sylvian sulcus, either as a dimple or as a real, deep sulcus (e.g. in *Pteropus*; Pl.20). In the cortex of *Pteropus*, several additional sulci are present. At the medial aspect of the cerebrum of large Macrochiroptera, a callosal and cingulate (splenial) sulcus is present.

   d) *Primates*. During evolution of the grandorder Archonta, brain size has most strongly increased in the primates. Four main groups of primates are distinguished: Strepsirhini (lemurs and loris), Haplorhini (Tarsiers, often referred to together with the Strepsirhini as 'prosimians'), Platyrrhini (New World monkeys) and Catarrhini (Old World monkeys, including apes and humans).

   – *Strepsirhini*. An extensive comparative study of the cerebral cortex in *Strepsirhini* is given by Stephan et al. (1977). The smallest primate, the mouse lemur (*Microcebus*), has a small (1.8 g), lissencephalic brain, which shows the most primitive new sulcus of primate cerebral cortex, the sylvian fissure. The sylvian sulcus separates the palaeocortex and neocortex into a frontal and an occipitotemporal part (see Fig. 44 in Stephan et al. 1977). In this respect, the primate sylvian sulcus is similar to ventral sulci in Dermoptera and large Chiroptera and differs from the so-called sylvian sulcus in other orders (e.g. Carnivora and Artiodactyla). The sylvian fissure is the only conspicuous fissure that is present in all primates. A sylvian fissure is also present in virtually all endocasts of extinct Strepsirhini skulls (Gurche 1982). The genus *Galago* deserves special attention. Some members (*Galago senegalensis*) are small and have a rather small (3.3 g) brain with only a sylvian and a postsylvian sulcus. Larger species (*Galago crassicaudatus*) have a larger brain (10 g) with an intraparietal, rectal and postsylvian sulcus as well (Radinsky 1977; Carlson and Welt 1981). In the brain of Strepsirhini, the rectal and intraparietal sulci remain apart, but in some endocasts they seem to merge (Radinsky 1977; Carlson and Welt 1981). The rectal sulcus crosses the primary somatosensory and motor cortex. An additional sulcus, the orbital sulcus, emerges in *Lepilemur* and the postsylvian sulcus in *Lemur*. In still larger Strepsirhini brains (*Indri* and the aye-aye *Daubentonia*), still more sulci are added. At the medial aspect of the cortex, a callosal and a cingulate sulcus are present in Strepsirhini such as *Galago crassicaudatus*.

- *Haplorhini*. Tarsiers have small (3–4 g), lissencephalic brains, in which a sylvian fissure and a postsylvian dimple are present. In the endocasts of the skulls of extinct tarsiers, a sylvian fissure is also present (Gurche 1982).
- *Platyrrhini*. An extensive comparative study of the external cerebral cortex in Platyrhini is given by Hershkovitz (1970). Constant features are the anterior and posterior rhinal fissure and the sylvian, orbital and postsylvian fissure. These are found in the smallest New World monkey, the pygmy marmoset (*Cebuella pygmaea*). These are supplemented by the rectal sulcus in the common marmoset (*Callithrix jacchus*; Pl.21) and, in larger species of these genera, by the intraparietal and the central sulcus (e.g. in Goeldi's marmoset *Callimico*). In the larger Platyrhini brains (e.g. the spider monkey *Ateles*), the pattern of sulci becomes more complex. In contrast to the Strepsirhini, the Platyrhini do not have a prominent coronolateral sulcus. At the medial aspect of the cortex in small Platyrhini (*Callitrix, Aotes*), a callosal and calcarine sulcus is present. In somewhat larger Platyrhini (*Saimiri*), moreover, a cingulate and parietooccipital sulcus is present.
- *Catarrhini*. All *Catarrhini* have large brains of at least 40 g. All these brains are gyrencephalic (Pl.22). The common features of the Platyrrhini are also found in the Catarrhini, i.e. anterior and posterior rhinal fissure and sylvian and postsylvian fissure. An additional prominent feature, found in all Catarrhini brains, is the central sulcus, which separates the somatosensory cortex from the motor cortex. A central sulcus is visible in the endocast of *Aegyptopithecus* (Falk 1982), an extinct Catarrhini monkey living 27 million years ago. Other prominent and general features of the Catarrhini cortex are the intraparietal, lunate, ventral occipital, rectal, arcuate and cingulate fissure. In the very large Catarrhini brains (apes and humans), the pattern of convolutions is much more complex (Pl.23).

In several respects, the human brain is asymmetrical. In humans, the right frontal pole is wider than the left one, while the left occipital pole is wider than the right one (Bradshaw and Rogers 1993). In humans, the planum temporale at the left side is broader than at the right side. Moreover, the presylvian (speech-related) cortex is larger at the left side. As a consequence, the sylvian sulcus is asymmetrical in humans; at the right side, it is shorter and bends upward at a more rostral position. Asymmetries are larger in humans than in non-human primates, both anatomically and behaviourally (hand preference, speech), but several asymmetries found in humans are also found in other Old World monkeys (Lemay and Geschwind 1975; Bradshaw and Rogers 1993). The characteristic pattern of human cerebral asymmetries is also found in *Australopithecus* and *Homo erectus* (Holloway and de la Coste-Lareymondie 1982; Bradshaw and Rogers 1993). The gyrogenesis of the Catarrhini cortex can also be deduced from ontogenesis, which has been intensively studied in humans (Chi et al. 1977; Armstrong et al. 1995). The first fissures to emerge are the sylvian and callosal sulci (at week 14), followed at week 16 by the olfactoral sulcus (at the base of the brain), the postsylvian sulcus (parieto-occipital, which separates the parietal and occipital lobe) and the calcarine sulcus (at the medial aspect of the occipital lobe). At about week 18, the insula is defined by the formation of the circular sulcus at the dorsal side. At the same time, the cingulate sulcus demarcates the cingulate gyrus. At about week 20, the central (rolandic) sulcus emerges. Overall, the pattern of sulci emerging during ontogenesis resembles the pattern as deduced from comparative anatomy and palaeoneurology.

5. *Grandorder Ferungulata*. The grandorder *Ferungulata* consists of eight orders, the Carnivora (including the sea carnivores, *Pinnipedia*), the aardvark (*Tubulidentata*), Artiodactyla, whales (*Cetacea*), *Perissodactyla*, hyraxes (*Hyracoidea*), elephants (*Proboscidea*) and sea cows (*Sirenia*):

a) *Carnivora*. In the brains of most adult carnivores, the pattern of convolutions is so complex that the primitive pattern can hardly been deduced. However, the evolution of carnivore brain convolutions can be deduced from the endocasts of fossil skulls.
- *Canidae*. The basic pattern of canid brain convolutions can be seen on the endocast of *Hesperocyon* (30 million years ago; Radinsky 1973). Apart from the lateral rhinal fissure, this brain shows only two well-developed sulci: the coronolateral and suprasylvian sulci. These sulci are fairly straight in *Hesperocyon*. The late Oligocean canids (25 million years ago) developed additional sulci: ectosylvian, ectolateral and presylvian. The suprasylvian sulcus

became curved. Much later in canid brain evolution, a pseudosylvian sulcus had developed, dorsally to the lateral rhinal fissure. The pattern of gyri in recent canids is fairly uniform (Radinski 1973). The primitive pattern can be easily recognised in the brain of the smallest canid, the fennec fox (*Fennecus*, Pl.24), and of the raccoon dog (*Nyctereutes*), which is considered the most primitive carnivore. At the dorsal part of the frontal cortex, a cruciate sulcus is present. At the medial aspect, a callosal and cingulate (splenial) sulcus is present; the cingulate sulcus merges with the cruciate sulcus.

– *Felidae*. The cerebral cortical evolution in the Felidae is in several respects similar to that in the Canidae. In endocasts of an ancient sable-tooth tiger (*Hoplophoneus*, 30 million years ago), three sulci are evident: coronolateral, suprasylvian and ectosylvian (and the lateral rhinal fissure; Radinsky 1975). About 25 million years ago, the rhinal fissure had moved ventrally, the suprasylvian sulcus had became curved and additional sulci had emerged, including the pseudosylvian and presylvian sulci. The cruciate sulcus emerged later (15–20 million years ago). In the present felids, a characteristic pattern of gyri is present with some variations (Pl.25; Radinsky 1975). When the felid brains become larger (e.g. in the tiger *Panthera tigris*), the basic felid organisation can still be recognised, while several additional sulci have emerged.

– *Mustelidae*. A few mustelids have a cortical organisation that is fairly close to the primitive carnivore pattern (Radinsky 1973). In skunks (*Mephitis*) and stink badgers (*Suillotaxus*), the cortical pattern is characterised by a fairly dorsally located lateral rhinal fissure, two prominent, rather straight cortical sulci (coronolateral and suprasylvian) and a pseudosylvian, presylvian and cruciate sulci (Fig. 22.217). This is different from the advanced pattern found in most mustelids (e.g. in the brain of the marten *Martes*, which is approximately the same size). This characteristically advanced mustelid pattern is also found in the smallest mustelid brains; the 2-g brain of the weasel *Mustela nivalis* is gyrencephalic, with an exquisite pattern of gyri. It is presumed that the small Mustelidae evolved from larger carnivores with gyrencephalic brains. The fine gyrencephalic weasel brain is thus an example of miniaturisation. The weasel brain is an unique example of a small, but gyrencephalic mammalian brain. In larger mustelid brains (e.g. the giant otter *Pteronura*), more sulci emerge, in species-specific patterns (Thiede 1966; Radinsky 1968). At the medial aspect, a callosal and cingulate sulcus are present. Early in ontogenetic development of the mustelid (ferret) brain, a cingulate sulcus has already developed (Smart and McSherry 1986).

– *Procyonidae*. The brain of the raccoon (*Procyon*) has been intensively studied by Welker and his group (Welker 1990). This brain is gyrencephalic. It has been described that, in the somatosensory cortex, the representations of separate fingers correspond to separate gyri. The ontogenetic development of the raccoon brain has been shown by Welker (1990, his Fig. 11).

– *Other carnivores*. Several other carnivores have large brains, e.g. the polar bear *Thalarctos maritimus* (500 g; Pl.26) and the walrus *Odobenus rosmarus* (1000 g). In such brains, the carnivore pattern of gyrification is present, although the pattern of gyri has become complex.

b) *Tubulidentata* (aardvarks). The aardvark *Orycteropus* is the only living species of the order *Tubulidentata*. It is a fairly large animal, with a rather large (72 g), gyrencephalic but primitive brain (Pl.27). Its piriform cortex is large, and the lateral rhinal fissure is prominent and dorsally located. The overlying neocortex contains some gyri. The most prominent is a longitudinal dorsal sulcus, referred to as γ by Elliot Smith (1898) as in Edentates, to which the aardvark was considered to be related to. The aardvark is now regarded as related to the ungulates; therefore, it is more probable that the aardvark's dorsal longitudinal sulcus is homologous to it's ungulate counterpart, which is called the marginal (or ectomarginal) sulcus and which might be homologous to the carnivore coronolateral sulcus. More ventrally, some sulci are present, called δ as in Edentates, which might be homologous to the suprasylvian sulcus.

c) *Artiodactyla*. The main Artiodactyla groups are pigs, hippopotamuses, camels, deer and bovids. All Artiodactyla have a gyrencephalic brain. A characteristic feature of the brains in Artiodactyla (and Perissodactyla) is that these brains have a large palaeocortex. From a lateral view, a considerable part of the piriform cortex is visible, and this piriform cortex is

convoluted. In this respect, the ungulates differ from carnivores and primates.
- *Tragulidae* (chevrotains). This is a family of small deer ('mouse deer'). The water chevrotain (*Hyemoschus*) has a brain (25 g) in which the most primitive artiodactyle is present (Fig. 22.217). Three longitudinal main sulci are present: a lateral rhinal, suprasylvian and dorsomedial sulcus, the latter being a continuation of the splenial sulcus. The medial aspect of the cortex of the water chevrotain (and other artiodactyles) differs from that of carnivores. In chevrotains, no clear cingulate gyrus is demarcated by a cingulate sulcus; instead, a splenial sulcus is present, whose caudal part bends dorsally and becomes an externally visible dorsmedial sulcus (Haarmann 1975).
- *Bovidae* (cattle). The convolutions of the bovid brains have been intensively been studied by Oboussier's group (for a survey, see Oboussier 1972). The dwarf antilopes (*Nesotragus*) are among the smallest Bovidae; they have a small (33 g) brain with a simple pattern of sulci, but not as primitive as the pattern of the water chevrotain. The larger brains of larger bovids are considerably more complex (Oboussier 1972).
- *Cervidae* (deer). In the brains of the smallest deer (e.g. the musk deer *Moschus*; Pl.28), the primitive artiodactyle pattern can be recognised. In larger deer, the brain convolutions become much more complex (Pl.29). In this respect, Cervidae are similar to Bovidae.
- *Tylopoda* (camels). Monographs on the evolution of the camel brains have been written by Edinger (1966) and Kruska (1982). The Eocene (40 million years ago) camel *Protylopus* has a brain that shows the primitive artiodactyle features, i.e. a rather dorsally located lateral rhinal fissure and some longitudinal neocortical sulci. The Pliocene (5 million years ago) camel *Procamelus* had a more ventrally situated lateral rhinal fissure and convolutions on its neocortex and palaeocortex. In recent members of this group (e.g. the lama), the neocortex has increased considerably, but the palaeocortex has still remained rather large and convoluted.
- *Other groups*. In the other artiodactyle groups (pigs, hippopotamusses, giraffes), rather large, gyrencephalic brains are found.
- *Ancestral patterns*. The ancestral patterns of convolutions in Artiodactyla, Perissodactyla (see below) and Carnivora are similar; they start with two longitudinal, dorsal fissures. Such a resemblance is expected, since the Ungulata and Ferae are phylogenetically related and placed together in the group of Ferungulata.
- *Cetacea* (dolphins and whales). The huge brains of dolphins and whales are very gyrencephalic (Pl.30,31). The cetacean gyri are finer and smaller than those in primates, which is related to a thinner neocortex in most cetacea (Fig. 22.214). The basic pattern can be observed most clearly in embryos. The development of the cetacean cerebral fissures has been studied in the brains of several blue whale and fin whale fetuses (Hammelbo 1972). In whales, the first neocortical sulci to emerge are splenial (cingulate), ectosylvian, pseudosylvian, confinal (often called marginal or lateral), ectolateral, suprasylvian and coronal (in that order). Overall, in embryological and comparative anatomical studies of cetacean brains, the resemblance with brains of carnivores and ungulates is mentioned (Kojima 1951; Hammelbo 1972; Morgane et al. 1980). This was expected, since the Cetacea are probably closely related to the Bovidae (Graur and Higgins 1994). A prominent difference between the brains of Cetacea and those of other Ferungulata is that, in Cetacea, the diencephalon and frontal telencephalon are rotated ventralwards (Morgane et al. 1980). This is true of both the toothed and baleen whales. As a consequence, the cetacean forebrain is rostrocaudally foreshortened ('telescoped'; Pl.30,31). Possibly as a consequence of this curvature, a prominent, deep lateral groove is present in the brains of all cetaceans. This groove is usually called the sylvian fissure, but it is questionable whether it is homologous to the sylvian fissure of primates or to the pseudosylvian fissure in carnivores and ungulates.

e) *Perissodactyla*. This order consists of the rhinoceroses, tapirs and *Equidae* (horse-like animals).
- *Equidae*. The evolution of the Equidae is well documented by fossil material, as is the evolution of its brain. The evolution of the brain in Equidae has been extensively described by Edinger (1948); her views are partly modified by Radinsky (1976). The most primitive 'horse', *Hyracotherium* (*Eohippus*), was small and had a fairly

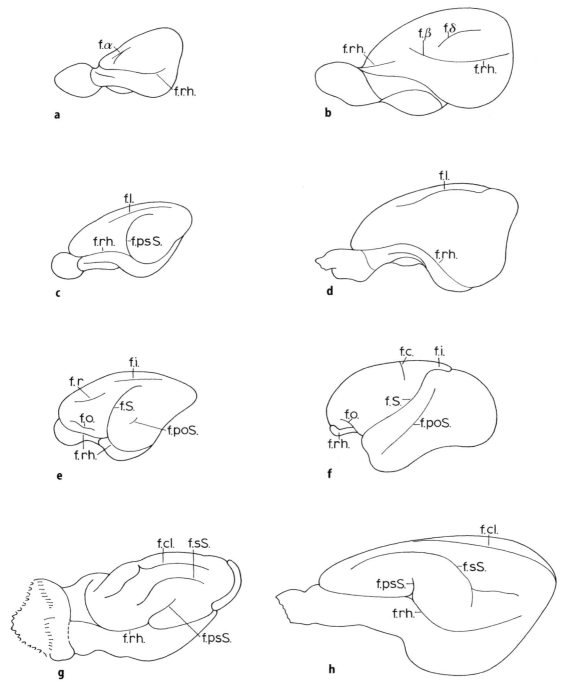

Fig. 22.217a–h. Examples of the basic patterns of gyrification in various groups of mammals at the same magnification. For all groups, a living species is selected whose brain has some sulci, but still in the primitive pattern of this group. **a** Marsupials: opossum (*Didelphis virginiana*); **b** Edentates: six-banded armadillo (*Dasypus sexcinctus*); **c** Hystricomorphs: cavia (*Cavia porcellus*); **d** Lagomorphs: hare (*Lepus europeus*); **e** Primates, *Strepsirhini*: bush-baby (*Galago alleni*); **f** Primates, *Platyrrhini*: squirrel monkey (*Saimiri sciureus*); **g** Carnivores: striped skunk (*Mephitis mephitis*); **h** Artiodactyles: water chevrotain (*Hyemoschus aquaticus*). *f.α*, fissura α; *f.β*, *fissura β*; *f.δ*, *fissura δ*; *f.c.*, fissura centralis; *f.cl.*, fissura coronolateralis (marginalis); *f.i.*, fissura intraparietalis; *f.l.*, fissura longitudinalis (marginalis); *f.o.*, fissura orbitalis; *f.poS.*, fissura postsylvica; *f.psS*, fissura pseudosylvica; *f.r.*, fissura recta; *f.rh.*, fissura rhinalis lateralis; *f.S.*, fissura Sylvii; *f.sS.*, fissura suprasylvica

small brain (15 g) with a clear lateral rhinal fissure and three neocortical sulci (suprasylvian, ectolateral and lateral). In the larger brains (78 g) of a further ancestor of the horse, *Mesohippus* (30 million years ago), more convolutions can be identified. In the horse (*Equus*), the pattern of convolutions has become highly complex (Pl.32).
  – Other groups. In tapirs and rhinoceroses, large and gyrencephalic brains are found (Kruska 1973).

f) *Hyracoidea* (hyraxes). The hyraxes are rather small animals with brain weights between 12 and 25 g. Their brains are gyrencephalic, and the primitive ferungulate pattern can be recognised in the pattern of sulci (Pl.33).

g) *Proboscidea* (elephants). Several skeletons of extinct proboscids have been preserved, but few cranial endocasts have been studied. Endocasts are known from the Eocean (40 million years ago) *Moeritherium* (Jerison 1973). This endocast resembles the sea cow's brain; it is a smooth brain with a lateral dorsoventral groove. Moreover, endocasts are available only from relatively recent mammoths and mastodons; these resemble the brains of recent elephants (Jerison 1973). The gyrencephalic elephant brain has rather broad gyri. A prominent feature is a deep lateral groove, which is usually called the sylvian fissure (Pl.34).

h) *Sirenia* (sea cows). Compared to other mammalian brains, the sirenian brain is a strange, almost kidney-shaped organ. It does not have the characteristic pattern of gyri and sulci, but only a deep groove (Pl.35). This is particularly remarkable, because the sea cow's brain is large (250–370 g). While the folding index of the sirenian brain is 1.06 (Pirlot and Kamiya 1985; Reep and O'Shea 1990), other mammals with a similar brain weight have brains with a folding index of at least 2 (compare Fig. 22.215). In the sea cow brains, the ancestral ferungulate pattern cannot be recognised. Instead, it can be hypothesised that, in the elephant-sirenian clade, the most primitive sulcus is a lateral dorsoventral groove.

22.11.6.6.6
Conclusions

The usual pattern of brain convolutions in mammals is that small brains are lissencephalic and large brains are gyrencephalic. The most remarkable exceptions to this rule are the small, gyrencephalic weasel brain and the large, lissencephalic sea cow brain. Some small species in a group display the ancestral pattern of convolutions (e.g. *Didelphis, Microcebus, Fennecus, Suillotaxus, Hyemoschus, Moschus*).

In the course of evolution, brain size and encephalisation has increased in several groups of mammals independently (see Chap. 22). As a consequence of this increase in brain size, brain sulci have been 'invented' independently in several mammalian groups, i.e. in *Monotremata, Marsupialia, Insectivora, Edentata, Hystricomorpha, Lagomorpha, Primates* and *Ferungulata*. As a consequence of such independent evolution of brain convolutions, the ancestral pattern of cerebral fissure is different between these groups. The following ancestral patterns can be distinguished:

– In *Marsupialia*, the rostral fissure α, dorsal to the rhinal fissure (Fig. 22.217a)
– In *Edentata*, the anterior fissure β and the posterior fissure δ (Fig. 22.217b)
– In *Hystrocomorpha*, a dorsal longitudinal and a pseudosylvian fissure (Fig. 22.217c)
– In *Lagomorpha*, a dorsal longitudinal fissure (Fig. 22.217d)
– In *Primates*, the sylvian, coronolateral (in prosimians), intraparietal and frontal rectal fissures, and in *Catarrhini* a central sulcus (Fig. 22.217e,f)
– In *Carnivora* and *Ungulata*, longitudinal sulci: coronolateral, ectosylvian and suprasylvian (Fig. 22.217g,h)

22.12
Overall Functional Subdivision of the Mammalian Brain

R. Nieuwenhuys

22.12.1
Introduction

The central nervous system of vertebrates can be, and has been, subdivided in many different ways. Most investigators have confined themselves to delineating purely morphological macroscopic or microscopic entities, but a minority have attempted to recognise the fundamental functional units of the neuraxis. The latter approach has been central in the endeavours of the American school of comparative neuroanatomy. As discussed in Chap. 4, Sect. 4.6.2 and Chap. 6, Sect. 6.3.5, the efforts of the American school have led to the subdivision of the brain stem into four longitudinally arranged functional zones: the somatosensory, viscerosensory, visceromotor and somatomotor zone (see Fig. 6.29). This subdivision was based essentially on tracing centrally the various components of the cranial nerves. Attempts at functional subdivisions of the entire central nervous system, which by necessity also encompass those parts which have no direct relations with the periphery, are few in number. Among these more comprehensive subdivisions, the schemes presented by Herrick (1948), Edinger (1908a,b) and MacLean (1970, 1972, 1990, 1992) deserve special mention.

Although Herrick (e.g. 1899) has contributed substantially to the 'four-column doctrine' just touched upon, in his later writings (e.g. Herrick 1948) he distinguished three longitudinal zones that perform three general classes of functions: a dorsal receptive or sensory zone, a ventral emissive or motor zone and an intermediate zone of correlation and integration. He emphasised that the intermediate zone, in contrast to the other two, shows a progressive evolutionary development and in humans comprises more than half of the total weight of the brain.

Edinger held that the brains of amniotes can be subdivided into two fundamental functional units: the palaeencephalon and the neencephalon (see also Chap. 6, Sect. 6.3.3 and Fig. 6.23). He believed that the palaeencephalon is present, with all its characteristic subdivisions, from cyclostomes to humans. It is, as its name implies, the oldest part of the central nervous system and most fishes possess only this part. The activities which depend on this entity are common to all vertebrates and include locomotion, foraging behaviour and reproduction. The palaeencephalon is the seat of all reflex mechanisms and of innate, instinctive behaviour. With the appearance of the neencephalon, of which the neocortex is the most prominent part (see poster P1), the behavioural repertoire gradually changes. This development culminates in mammals, in which the (palaeencephalic) reflexes and instincts become subordinated to (neencephalic) associative and intelligent actions.

The ideas of MacLean, although developed some 60 years later, nevertheless show a striking resemblance to those of Edinger. According to MacLean, the brain of higher primates is composed of three neural formations that reflect an ancestral relationship to reptiles, early mammals and late mammals. In the latter group, these formations constitute 'three brains in one', a triune brain. The reptilian component forms the matrix of the reticular formation, midbrain and basal ganglia. It programmes stereotyped behaviours according to instructions based on ancestral learning and ancestral memories, and it has no neural machinery for learning to cope with new situations. The palaeomammalian brain is represented by the limbic system. This entity derives information in terms of emotional feelings that guide behaviour required for self-preservation and the preservation of the species. The neomammalian brain, which may be defined as the neocortex plus the structures of the brain stem with which it is primarily connected, in contrast to the other two formations, shows a markedly progressive development. Its evolution goes hand in hand with the differentiation of the visual, auditory and somatosensory systems, which carry out refined analysis and discrimination of events in the external environment. In MacLean's opinion, the neomammalian component has afforded a progressive capacity for problem solving, learning and memory of details.

In the following, a new classification of entities within the mammalian central nervous system will be discussed. According to this classification, the brain comprises a sensorimotor and cognitomotor domain, which occupies grossly the lateral part of the brain, and a medial domain, which may be designated as the greater limbic system. The latter differs considerably in structural, chemical and functional terms from the lateral domain. It consists of an array of highly interconnected structures, extending from the medial wall of the telencephalon to the caudal rhombencephalon, which is concerned with specific motivated or goal-oriented behaviour, directly aimed at the maintenance of homeostasis and at the survival of the individual (organism) and of the species. Both domains can be subdivided into smaller functional entities. Thus the lateral domain encompasses the great sensory systems, the association system, the pyramidal and extrapyramidal motor systems and the cerebellum, whereas the medial or greater limbic domain comprises units designated as the core of the neuraxis and the median and lateral paracores (Nieuwenhuys 1985, 1996; Nieuwenhuys et al. 1988, 1989). It is important to note that some of the functional units distinguished correspond to entities distinguished by previous authors. Thus the core region roughly corresponds to Edinger's palaeencephalon and to MacLean's palaeomammalian brain, whereas the association system corresponds to the rostral part of Herrick's intermediate zone.

22.12.2
Lateral Domain

22.12.2.1
Sensory Systems

The general somatosensory projection which relays impulses originating from sensors situated on the body surface, from muscle spindles and joint receptors, passes by way of the gracile and cuneate nuclei and the ventral posterior thalamic nucleus to the primary somatosensory area of the neocortex. The conscious component of this system, which subserves what is called gnostic sensibility, is primarily involved in active tactile exploration. The auditory projection reaches the primary auditory cortex via relays in the cochlear nuclei, the superior olive, the inferior colliculus and the medial geniculate body. The visual projection, which extends from the retina to the primary visual cortex, is synaptically interrupted in the lateral geniculate body. In all mammals, the thalamic relay nuclei of the three sensory systems discussed above lie close together, but the spatial relations of their cortical projection areas show considerable differences among the various mammalian groups (see Sect. 22.11.6.5.1). Thus, in primitive mammals, such as insectivores and rodents, these areas are almost directly adjacent to each other and together occupy a large area of the still relatively small neocortex, but in carnivores and prosimians they are separated by strips of association cortex and in primates they lie far apart (Fig. 22.192).

22.12.2.2
Association Systems

Experimental studies carried out in monkeys (see e.g. Jones and Powell 1970; Pandya 1987; Pandya and Yeterian 1985) revealed that each primary sensory area is flanked by a cortical belt which receives its main input from cascades of short association fibres originating from the same adjoining primary sensory cortical regions. These cortical belts, which are concerned with the further processing of modality-specific inputs, are termed unimodal parasensory association areas. In these areas, both sequential and parallel processing of information takes place. They are concerned with operations such as the discrimination of form, shape, texture or tone, whereas the distalmost zones of these areas serve complex functions such as pattern recognition.

The somatosensory association cortex is situated in the parietal lobe directly behind the postcentral gyrus; the auditory association cortex occupies much of the superior temporal gyrus, whereas the visual association area extends far into the lower parts of the temporal lobe.

Interestingly, at the junction of the three unimodal or modality-specific parasensory association areas, a multimodal association area is found which receives converging afferents from its surrounding cortical fields. This multimodal association area forms a strip, extending forward from the junction of the occipital and parietal lobes, along the superior temporal sulcus into the anterior part of the temporal lobe (Fig. 22.152).

The next step in the processing of sensory information may be characterised as 'the great leap forward'. A vast region of the large frontal lobe of the telencephalon neither receives sensory projections nor sends out fibres concerned with motor functions and should therefore be characterised as an association area. This area, the prefrontal cortex, went through a remarkable expansion very late in the evolution of the mammalian brain. Its most substantial input comes via two long fibre systems, originating from the parietal, temporal and occipital lobes. These fibre systems, the longitudinal superior and uncinate fascicles, arise from the modality-specific parasensory association areas and the multimodal area in the occipital, temporal and parietal lobes (Fig. 22.152). An important output system of the prefrontal cortex is formed by sequences of short association fibres, which successively link the orbital frontal cortex, the various prefrontal areas (9, 10 and 46), the premotor area 6 and the primary motor area 4. This suggests that highly processed visual, somatosensory, auditory and multimodal sensory information is transferred to the frontal lobe and that this lobe, by a recurrent stream of short association fibres, might participate in the planning, timing and sequencing of complex motor tasks (Fuster 1991).

Thus it requires a long, circuitous route to get from the various sensory systems and their central representations to the area where the large highway of voluntary motor control originates, i.e. the motor part of the pyramidal tract.

Comparable association systems are present in the cat, although in this species the prefrontal cortex is much smaller than in primates (Fig. 22.192:E). There is evidence suggesting that even in primitive mammals the various projection areas are separated from each other by narrow strips of non-projection cortex (see Cajal 1911, p. 830 and Fig. 531). However, to my knowledge, nothing is known about the connections of these strips.

22.12.2.3
Motor Systems

The motor systems of the lateral domain comprise a cortical or pyramidal system and a subcortical system (see Sect. 22.9).

The pyramidal tract represents a conspicuous and compact bundle which in humans contains approximately 1 million myelinated fibres. These fibres originate from pyramidal neurons in the fifth layer of the motor and premotor cortical areas and descend to the lower brain stem and the spinal cord. The motor and premotor cortices and their outflow via the pyramidal tract are mainly concerned with the programming and execution of skilled movements that require detailed control of the distal limb musculature (Porter 1985; Kuypers 1987).

The pyramidal tract and its target neurons represents the lateral motor system of Kuypers (1964, 1981). The subcortical or medial motor system of Kuypers (1964, 1981) originates mainly from centres in the brain stem such as the interstitial nucleus of Cajal, the superior colliculus, the vestibular nuclei and the medial reticular formation and is principally concerned with eye and neck movements and axial and proximal body movements. The coarse-fibred fasciculus longitudinalis medialis, which caudally continues into the anterior funiculus of the spinal cord, forms an important communication channel of the medial system. The function of this system is maintenance of erect posture (anti-gravity movements), integration of body and limbs, synergy of the whole limb and orientation of body and head (Kuypers 1981).

22.12.2.4
Motor Control Systems

The motor and premotor cortices are supported during both the programming and the execution phases of their tasks by two large control systems: the extrapyramidal and cerebellar systems. After signals, originating mainly from the association cortices, have been processed through the basal ganglia (caudate nucleus, putamen and globus pallidus) and the cerebellar hemisphere (cerebellar cortex and deep cerebellar nuclei), they converge via the ventral lateral thalamic nucleus on the motor and premotor cortices. Both systems play an important role in motor integration. Whereas the cerebellum is involved in the initiation and proper timing of movements, the basal ganglia play a role in regulating the speed of movements.

It should be emphasised that all of the structures discussed so far are morphologically discrete and well differentiated and that the various pathways – lemniscal, motor and cerebellar – are compact and well myelinated. Some of the systems belonging to the lateral domain, i.e. the general somatosensory, visual, pyramidal and cerebellar efferent systems, are depicted in Fig. 22.218.

22.12.3
Medial Domain

22.12.3.1
Limbic System

In all mammals, the neocortex is not confined to the lateral surface of the cerebral hemispheres, but also expands over the medial surface of these struc-

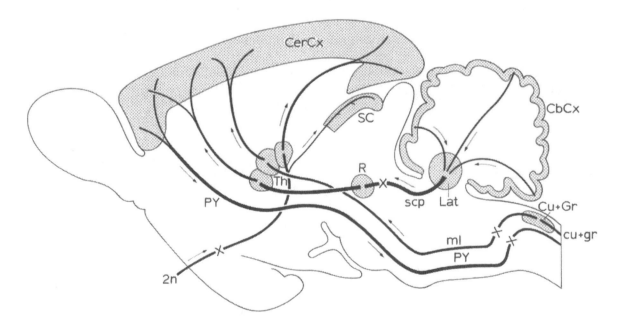

Fig. 22.218. 'Lateral domain systems' in the brain of the rat. *Cbcx*, cerebellar cortex; *CerCx*, cerebral cortex; *Cu*, nucleus cuneatus; *cu*, fasciculus cuneatus; *Gr*, nucleus gracilis; *gr*, fasciculus gracilis; *Lat*, lateral cerebellar nucleus; *ml*, medial lemniscus; *Py*, pyramidal tract; *R*, nucleus ruber; *SC*, superior colliculus; *scp*, superior cerebllar peduncle; *Th*, thalamus; *2n*, optic nerve. (Reproduced from Nieuwenhuys et al. 1989)

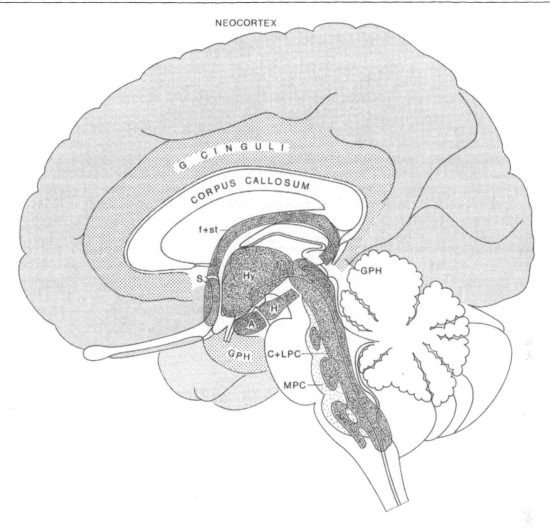

Fig. 22.219. Medial view of the human brain. The position of the gyrus cinguli, the gyrus parahippocampalis (*GPH*), the septum (*S*), the fornix and stria terminalis (*f+st*), the hypothalamus (*HY*), the hippocampus (*H*), the amygdala (*A*), the core and lateral paracore (*C+LP*) and the median paracore (*MPC*) is indicated. (Reproduced from Nieuwenhuys 1996)

tures. However, on this side of the brain, a number of non-neocortical structures also exist. Two of these, the central olfactory system and the septum, occupy a superficial position, whereas two others, the amygdaloid complex and the hippocampal formation, lie largely hidden within the hemispheres (Figs. 22.9, 22.144). In primates, due to the strong development of the neocortex and the related expansion of the temporal lobe, the main efferent systems of the amygdaloid complex and the hippocampal formation have been drawn out into long, arch-shaped bundles, the stria terminalis and the fornix, respectively (Fig. 22.148). The septum, the amygdala, the hippocampal formation and their efferent bundles constitute together the ring-shaped, medialmost zone of the hemisphere (Fig. 22.219). This zone is largely flanked by a second zone, which in primates encompasses two convolutions, the cingulate and the parahippocampal gyri (see Figs. 22.219, 22.227). In 1878, Broca designated these two gyri together as *le grand lobe limbique* ('the great limbic lobe'). He emphasised that he had introduced this term to denote an anatomical structure and not a functional unit. During the last few decades of the nineteenth and the first few decades of the twentieth century, it was generally believed that most, if not all of the structures included in Broca's limbic lobe are dominated by olfactory projections and thus form part of the rhinencephalon.

In 1937, Papez published a notable paper in which he claimed on theoretical grounds that a circuit, of which the hippocampal formation and the cingulate gyrus form important components, constitutes the neural substrate of emotional behaviour. This theory received some substantiation from the work carried out by Klüver and Bucy (1937, 1939), who demonstrated that, in monkeys, resections of the anterior portions of the temporal lobes (which included the hippocampal formation and the amygdaloid complex) have, among other effects, a profound influence on affective responses. Somewhat schematically, it may be said that the impact of the publication by Papez and those by Klüver and Bucy was threefold: (1) the idea that the rhinencephalon encompasses almost the entire limbic lobe fell into the background, (2) a direct linkage between emotion and Broca's limbic lobe became established and (3) the amygdaloid complex, a subcortical structure, became incorporated into the limbic lobe.

MacLean (e.g. 1952, 1970, 1990, 1992) drew attention to the fact that the various components of Broca's great limbic lobe are strongly and reciprocally connected with a number of subcortical structures, particularly the septum, the amygdala, the midline thalamic nuclei, the habenula and the hypothalamus. He suggested that the cortical limbic ring is rostrally closed by two subcortical nodal points, i.e. the more dorsally situated septum and the more ventrolaterally located amygdaloid complex. MacLean cited clinical and experimental evidence suggesting that the lower part of the ring, fed by the amygdaloid complex, is primarily concerned with emotional feelings and with behaviour that ensures self-preservation. As he put it, the circuits of this lower part of the ring are "so to speak, kept busy with the selfish demands of feeding, fighting and self-protection" (MacLean 1970, p. 340). The structures associated with the septum in the upper part of the ring, on the other hand, would be involved in "expressive and feeling states that are conducive to sociability and the procreation and preservation of the species" (MacLean 1970, p. 340). These data and aspects led MacLean to the conclusion that the limbic cortex, together with the subcortical struc-

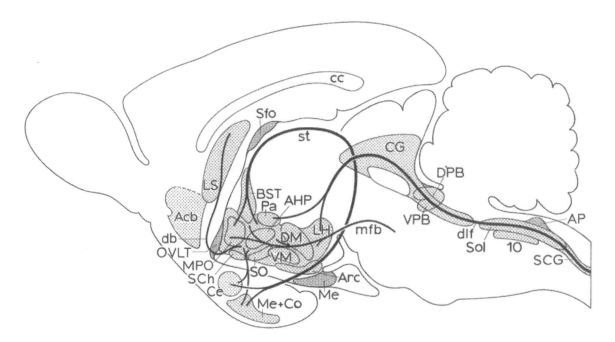

Fig. 22.220. Core of the central nervous system of the rat. *Acb*, accumbens nucleus; *AHP*, anterior hypothalamic area; *AM*, amygdaloid complex; *AP*, area postrema; *Arc*, arcuate hypothalamic nucleus; *A1, A2* etc., noradrenergic cell groups; *BST*, bed nucleus of the stria terminalis; *cc*, corpus callosum; *Ce*, central amygdaloid nucleus; *CG*, central (periaqueductal) grey; *Co*, cortical amygdaloid nucleus; *CSu*, central superior nucleus; *C1, C2* etc. adrenergic cell groups; *DB*, nucleus of the diagonal band of Broca; *db*, diagonal band of Broca; *dlf*, dorsal longitudinal fasciculus; *DM*, dorsomedial hypothalamic nucleus; *DPB*, dorsl parabrachial nucleus; *DR*, dorsal raphe nucleus; *f*, fornix; *fr*, fasciculus retroflexus; *Hi*, hippocampus; *IP*, interpeduncular nucleus; *LH*, lateral hypothalamic area; *lpcb*, lateral paracore bundle; *LS*, lateral septal nucleus; *ME*, median eminence; *Me*, medial amygdaloid nucleus; *mfb*, medial forebrain bundle; *MHb*, medial habenular nucleus; *MnR*, median raphe nucleus; *MPO*, medial preoptic nucleus; *MS*, medial septal nucleus; *OVLT*, organum vasculosum laminae terminalis; *Pa*, paraventricular nucleus; *PHC*, preoptic-hypothalamic continuum; *RLi*, rostral linear nucleus of the raphe; *RMg*, raphe magnus nucleus; *ROb*, raphe obscurus nucleus; *RPa*, raphe pallidus nucleus; *SCG*, spinal central grey; *SCh*, suprachiasmatic nucleus; *SfO*, subfornical organ; *sm*, stria medullaris of the thalamus; *SO*, supraoptic nucleus; *Sol*, nucleus of the solitary tract; *st*, stria terminalis; *VM*, ventromedial hypothalamic nucleus; *VPB*, ventral parabrachial nucleus; *10*, dorsal motor vagal nucleus. (Reproduced from Nieuwenhuys et al. 1989)

tures to which it is directly connected, comprises a functionally integrated system which he designated (in keeping with Broca's terminology) as the limbic system (for the first time in MacLean 1952).

A notable extension to the limbic system concept was made by Nauta (1958, 1973; Nauta and Haymaker 1969). Nauta added to the telencephalic limbic 'arch' (within which he included the hippocampal formation and the amygdaloid complex, but not the gyrus cinguli and the gyrus parahippocampalis) a neural continuum which may be designated as the 'limbic axis'. This continuum includes, from rostral to caudal, the septal and preoptic regions, the hypothalamus and a number of paramedian mesencephalic structures, including the mesencephalic central grey and the dorsal raphe nucleus (Nauta's 'limbic midbrain area'). Nauta pointed out that these various entities are structurally heterogeneous, but that all of them are strongly interconnected by shorter and longer ascending and descending fibres. Taken together, these connections constitute, in Nauta's opinion, one large functional system which he designated as the 'limbic system-midbrain circuit'. He emphasised that the large telencephalic limbic structures, i.e. the hippocampus and the amygdaloid complex, are both reciprocally connected with the rostral pole of the limbic axis and thereby with the limbic system-midbrain circuit.

The fasciculus medialis telencephali or medial forebrain bundle may be considered the central longitudinal pathway of the limbic forebrain-midbrain continuum. It is an assemblage of loosely arranged, mostly thin fibres, which traverses the lateral hypothalamic area. The bundle is highly complex, comprising a variety of short and long ascending and descending links (Nieuwenhuys et al. 1982; Veening et al. 1982; Figs. 3.21, 3.22, 3.29). Moreover, the central limbic axis contains a continuous network of thin, unmyelinated fibres. Although this network harbours numerous quite specific projections interlinking the various septal, preoptic and hypothalamic centres (see e.g. Larsen et al. 1994), only in a few places do these projections manifest themselves as discrete fibre systems.

22.12.3.2
Greater Limbic System

22.12.3.2.1
Introductory Notes

The central limbic continuum and its circuitry does not end at the caudal diencephalic or mesencephalic levels, but rather extends throughout the brain stem. The hypothalamic grey matter is caudally directly continuous with a periventricular mesencephalic and rhombencephalic zone. The mesencephalic part of this zone is constituted by the periaqueductal grey, whereas its rhombencephalic extension comprises the pontine central grey (as defined by Olszewski and Baxter 1954), the parabrachial nuclei and the dorsal vagal complex. In some previous publications (Nieuwenhuys 1985; Nieuwenhuys et al. 1988, 1989), my colleagues and I brought the periventricular brain stem structures mentioned above and the more rostrally situated components of the classical limbic system together under the name 'core of the neuraxis' (Fig. 22.220). In the same publications, it was pointed out that, at the level of the brain stem, this core has two adjuncts, the *median paracore* and the (bilateral) *lateral paracore* (Fig. 22.22.219). The median paracore is constituted by the series of raphe nuclei which extends throughout the brain stem. In most places, the raphe nuclei are directly adjacent to the core region, and in some they even penetrate into it. Moreover, fibres of the core region project heavily towards most raphe nuclei (Fig. 22.221).

The lateral paracore consists of a series of grisea which extends from the core region ventrolaterally into the tegmentum. At the mesencephalic level, this series includes the lateral part of the tegmental grey. In the rhombencephalon, it is the locus coeruleus or A6 group, the nucleus subcoeruleus (A6sc), the Kölliker-Fuse nucleus, the M and L regions of Holstege et al. (1986), the nucleus reticularis parvocellularis, the area reticularis superficialis ventrolateralis and the cytoarchitectonically ill-defined cell groups A1, A2, A5, A7, C1, C2 and Ch5 which constitute this series. All of the rhombencephalic centres mentioned form part of, or are embedded in, the lateral reticular zone or lateral tegmental field (Fig. 22.222).

The median and lateral paracores have the following features in common:

1. As already mentioned, they are both directly continuous with the core region.
2. Both contain large numbers of monoaminergic cells. In the median paracore, numerous serotoninergic cells are found, while in the lateral paracore, catecholaminergic neurons prevail; adrenergic neurons are found in the cell groups C1 and C2, whereas the cell groups A1-A7 contain numerous noradrenergic elements.
3. Both paracores lie clearly beyond the trajectories of the large, compact, well-myelinated ascending and descending pathways, discussed in Sect. 22.12.2 and depicted in Fig. 22.218.
4. Both paracores contain assemblies of thin, longitudinally arranged fibres. Ascending and descending serotoninergic axons contribute sub-

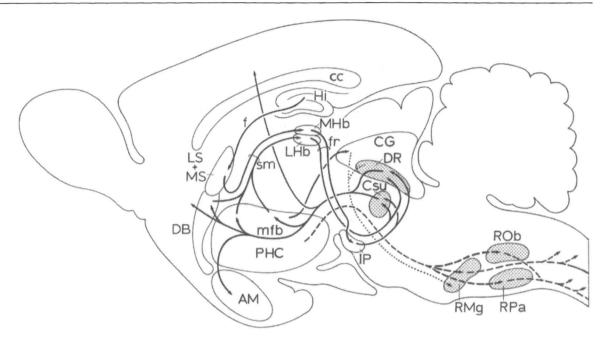

Fig. 22.221. Median paracore of the brain of the rat. For abbreviations, see legend to Fig. 22.220

stantially to the median paracore fibre assembly, whereas the large longitudinal catecholamine bundle described by Jones and Friedman (1983) occupies a central position in the lateral paracore (Fig. 22.223). The grisea discussed above, i.e. the brain stem portion of the core and the median and lateral paracores, not only form a structural continuum with the limbic forebrain, but also share a number of salient functional and connectional features with the latter. For this reason, I proposed in a recent paper (Nieuwenhuys 1996), from which much of the present section is derived, to unite the classical limbic system and its caudal extensions into a new entity, the *greater limbic system*. The functional, structural and chemical features of this system, its input systems, output channels and connections with the neocortex will now be discussed.

22.12.3.2.2
Functional, Structural and Chemical Features

Somatomotor, Visceromotor and Endocrine Responses. The greater limbic system generates integrated somatomotor, visceromotor and endocrine responses directly aimed at the survival of the individual and of the species. The classical experiments performed by Bard (1929), Hess (1954) and others (for reviews, see Jürgens 1974; Swanson 1987) showed that numerous loci are present in the hypothalamus, from which quite characteristic integrated behavioural patterns can be elicited on electrostimulation, including eating, drinking, grooming, fear, attack, rage and reproductive behaviour. Later studies showed that similar responses can be elicited from higher (e.g. septum, amygdala) and lower parts of the brain. In the brain stem, the lateral paracore bundle (see below) is surrounded by a continuous array of loci involved in integrated behavioural and autonomic responses (Klemm and Vertes 1990). This array, which extends from the mesencephalic periaqueductal grey to the obex, includes 'centres' implicated in defence reactions, lordosis, vocalisation, locomotion, swallowing, micturition and cardiovascular and respiratory regulation (for references, see Nieuwenhuys 1996). All of these reactions are directly related to the maintenance of the integrity of the internal milieu (homeostasis) or to the preservation of the species.

Predominance of Thin, Unmyelinated Fibres. A network of thin and ultrathin, mostly varicose fibres extends from the hypothalamus into the limbic brain stem regions. Within this network, longitudinally running axons tend to aggregate into two loose-textured fibre systems, the fasciculus longitudinalis dorsalis of Schütz and the medial forebrain bundle (Figs. 22.220, 22.222).

The dorsal longitudinal fasciculus of Schütz extends from the posterior part of the hypothalamus to the caudal medulla oblongata and occupies a periventricular position over its entire length. It is a composite fibre system which, in addition to long

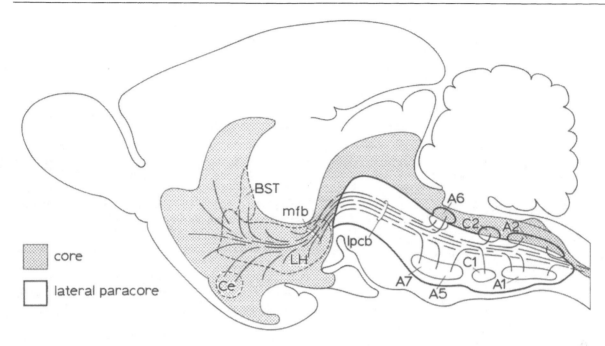

Fig. 22.222. Core and lateral paracore of the brain of the rat. For abbreviations, see legend to Fig. 22.220

ascending and descending fibres directly connecting the hypothalamus with the nucleus solitarius and related autonomic centres, also contains numerous shorter links. Many of its fibres are synaptically interrupted in the periaqueductal grey and in the parabrachial nuclei (Ricardo and Koh 1978; Fulwiler and Saper 1984; Veening et al. 1991).

The medial forebrain bundle, much like the dorsal longitudinal fasciculus, is a composite fibre system. It has already been mentioned that the rostral part of this bundle occupies a central position in the lateral hypothalamic area. In the transitional area of the diencephalon and the mesencephalon, the medial forebrain bundle fibres are rearranged into a smaller medial and a larger lateral stream (Hosoya and Matsushita 1981; Holstege 1987). The medial fibre stream roughly maintains the sagittal orientation of the hypothalamic trajectory of the bundle. It passes through the medial parts of the mesencephalic and rhombencephalic tegmental areas, just next to the raphe nuclei. The medial stream (or medial paracore bundle) is composed of descending fibres, by which several hypothalamic centres project to the raphe nuclei and to the adjacent parts of the medial reticular formation, and also comprises numerous ascending and descending fibres originating from the raphe nuclei themselves.

The lateral stream of fibres extending from the medial forebrain bundle to the brain stem, or lateral paracore bundle, sweeps laterally and caudally over the dorsal border of the substantia nigra into a ventrolateral tegmental position, from where it descends through the lateral tegmented field of the pons and the medulla oblongata (Fig. 22.223). This stream passes through a band-like array of grisea which are all involved in the regulation of specific, motivated behaviours and of related autonomic responses. It contains the innumerable shorter and longer ascending and descending pathways by which these grisea are interconnected.

It will be appreciated that the organisation of the behavioural system touched upon above differs considerably from that observed in the various 'classical' sensory and motor systems discussed in Sect. 22.12.2 and depicted in Fig. 22.218. In these 'classical' systems, generally discrete centres are interconnected by well-myelinated, through-conducting fibres, assembled in discrete bundles. However, in the system concerned with the generation of behaviour directly related to homeostasis and reproduction, 'open-line' polysynaptic connections, as described by Ricardo and Koh (1978; Fig. 3.29) prevail. The consideration of these authors is worth quoting in full; after indicating that such connections are accessible at each synaptic interruption to side inputs of related ('re-entrant' circuits) or of unrelated origins, they continue as follows:

The 'open-line' componentry of such systems appears to reflect the need of many homeostatic functions to be guided by several rather than by a single modality of afferent signals;

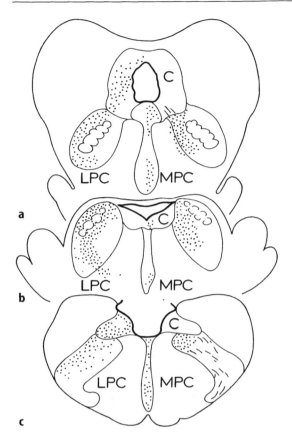

Fig. 22.223a-c. Transverse sections through the brain stem of the cat at **a** caudal mesencephalic, **b** rostral metencephalic and **c** myelencephalic levels to show positions of the core, the median paracore (*MPC*) and the (bilateral) lateral paracores (*LPC*). The position of catecholaminergic fibres, as depicted by Jones and Friedman (1983), is shown to the *right*, whereas the distribution of fibres and terminals originating from the central amygdaloid nucleus, as depicted by Hopkins and Holstege (1978), is shown to the *left*

e.g. grooming, drinking, micturition, aggression and sexual behaviour (see Sect. 22.10.5.5). It is important to note that also at this 'level of resolution' the efferents of behaviourally defined centres do not form well-defined bundles, but rather spread out, forming diffuse fibre streams (Roeling et al. 1994). However, the diffuseness of these fibre streams does not exclude connectional specificity. For instance, injections of the anterograde tracer *Phaseolus* in two behaviourally defined hypothalamic areas, i.e. the 'grooming area' and the 'attack area', revealed that fibres of both of these areas project (*inter alia*) to the periaqueductal grey, but terminate in different sectors of that centre (Roeling et al. 1994).

It has already been mentioned that the median and lateral paracores contain groups of monoaminergic neurons and that these neuron groups contribute substantially to the caudal extensions of the medial forebrain bundle, which occupy a central position in both domains. The axons of the noradrenergic (and adrenergic) cells in the lateral paracore as well as those of the serotoninergic cells in the median paracore are thin and varicose and ramify profusely. Collectively, they constitute extensive noradrenergic and serotoninergic networks, which spread far beyond the limits of the greater limbic system over virtually all parts of the central nervous system. It is intriguing, however, that all monoaminergic cell groups receive their principal afferents from limbic centres and that the initial parts of their open-line axonal systems pass through limbic domains.

Presence of Circumventricular Organs. Circumventricular organs are small, highly specialized brain structures which, as their name implies, are situated in the immediate vicinity of the cerebral ventricular system. Most of these organs contain neurons, and, with regard to both their afferent and efferent connections, these neurons are entirely embedded in the limbic circuitry. Their most important distinguishing feature is the lack of a blood-brain barrier in their vasculature. The circumventricular organs include the subfornical organ and the organum vasculosum of the lamina terminalis, which are both situated in the telencephalic preoptic region, and the area postrema, which is located in the caudalmost part of the brain, in direct contact with the nucleus of the solitary tract (Fig. 22.220). These organs are to be considered to be chemosensitive zones, which monitor the changing levels of circulating hormones and other substances, forming part of the circuitry underlying, inter alia, homeostatic functions of water-electrolytic balance and cardiovascular regulation.

simultaneously, it might serve as a device allowing selective and finely graded modulation of the impulse flow by reentering circuits (Ricardo and Koh 1978, p. 20).

The open-line character of the communication within the system under consideration is further accentuated by the fact that most of its connections are mainly composed of unmyelinated, thin and ultrathin, varicose fibres (Fig. 22.224a,b,d). Assuming that the varicosities are concerned with neural transmission – synaptic, non-synaptic or both (see below) – this means that these fibres are not only involved in interneuronal communication at their end, but rather throughout their extent, or at least at several different levels along their course (Fig. 3.29).

Combined physiological and experimental neuroanatomical studies have led to the elucidation of the circuitry underlying several specific behaviours,

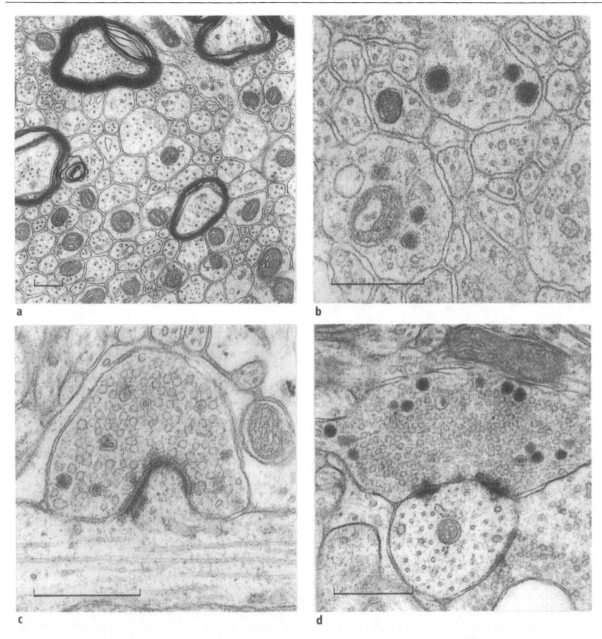

Fig. 22.224a–d. Aspects of the ultrastructure of a typical component of the greater limbic system: the mesencephalic periaqueductal grey (PAG) (rat). a Neuropil in the lateral PAG. Note the presence of numerous thin and ultrathin fibres, intermingled with small-calibred myelinated fibres. b Axonal varicosities with large dense-core vesicles. c Synaptic terminal filled with many small spherical synaptic vesicles and few dense-core vesicles, making synaptic contact with a small dendrite. d Axonal varicosity making synaptic contact with a dendrite. Note the presence of dense-core vesicles at a distance from the active sites. Bars 0.5 µm (Courtesy of Dr. P. Buma)

Presence of Gonadal Steroid-Receptive Neurons. Experimental studies using autoradiographic methods for detecting binding of labelled steroids (e.g. Stumpf 1975; Stumpf and Sar 1978; Wood et al. 1992) or the more recent immunohistochemical techniques for detecting the receptor protein itself (e.g. Koch and Ehret 1989; Blaustein 1992) have revealed that both the forebrain and the brain stem components of the greater limbic system contain aggregations of gonadal steroid-receptive neurons. In the forebrain component, such aggregations occur inter alia in the septal nuclei, the bed nucleus of the stria terminalis, the amygdala, the medial preoptic nucleus, the nucleus ventromedialis hypothalami and the nucleus infundibularis, whereas in the brain stem component the periaqueductal grey, the parabrachial nuclei and the nucleus solitarius contain numerous steroid-receptive neurons. In many of these centres, the neurons concentrating oestrogen and androgen hormones show a distinct sexual dimorphism with regard to number and distribution (for a review, see Dulce Madeira and Lieberman 1995).

Extraordinary Richness and Density of Neuromediators (particularly neuropeptides, which may well suggest extensive non-specific neurotransmission). Some time ago, we gathered data from the literature on the localisation (Nieuwenhuys 1985) and density (Nieuwenhuys et al. 1989) in the central nervous system of 25 different neuromediators, including 16 neuropeptides. Numerous 'classical' limbic forebrain centres appeared to contain an extraordinary diversity of neuropeptides and an extraordinary density of peptidergic neurons and/or fibres and terminals. However, interestingly, the same appeared to hold true for a number of brain stem centres, including the periaqueductal grey, the lateral parabrachial nucleus, the locus coeruleus and the nucleus solitarius.

Neuropeptides may influence behaviour. Characteristic of their action on behavioural processes is their slow onset and long duration (De Wied 1987). It has been frequently suggested that neuropeptides mainly exert a neuromodulatory (facilitating or attenuating) influence on neural transmission by 'classical' non-peptidergic neuromediators (see e.g. Leah et al. 1988). However, there can be no doubt that many neuropeptides play a key role in the regulation of specific behavioural responses. Thus neuropeptide Y and cholecystokinin control food intake (e.g. Baile et al. 1986; Dube et al. 1994), whereas angiotensin II evokes concerted hormonal, vegetative and behavioural responses aimed at the maintenance of fluid homeostasis (see Sect. 22.10.5.5).

The greater limbic system encompasses many different neuronal networks subserving specific, motivated behaviours, such as eating, drinking, aggression and sexual behaviour. Interestingly, high concentrations of the same peptide are often found in different terminal areas of neuronal circuits related to specific, motivated behaviours, which suggests that a single peptide is used at multiple anatomical levels to regulate a particular behavioural process (Herbert 1993). This feature may be interpreted as a 'peptidergic' specification of the open-line components of neuronal systems subserving behavioural functions discussed above.

Some time ago, I expressed the opinion that paracrine or non-synaptic neurotransmission may well play an important role in the greater limbic

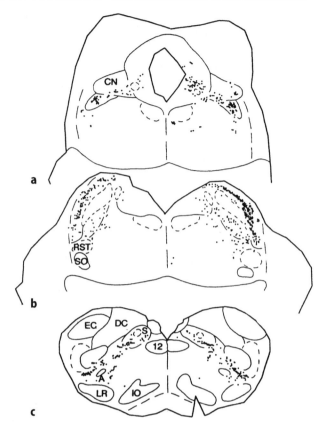

Fig. 22.225a–c. Plots of labelled terminations on transverse sections through **a** the lower mesencephalon, **b** the pontine region and **c** the lower medulla oblongata of a Cynomolgus monkey, following some small, iontophoretic injections of *Phaseolus vulgaris*-leucoagglutinin in lamina I of the cervical spinal cord (C6-C8); the *left side* is ipsilateral to the injection site. *CN*, cuneiform nucleus; *DC*, dorsal column nuclei; *EC*, external cuneate nucleus; *DC*, dorsal column nuclei; *IO*, inferior olive; *LR*, lateral reticular nucleus; *RST*, rubrospinal tract; *S*, solitary nuclear complex; *SO*, superior olive; *12*, hypoglossal nucleus. (Reproduced from Craig 1995)

system (then designated as the core-paracore continuum; Nieuwenhuys 1985). The evidence in favour of this hypothesis has been repeatedly discussed (Nieuwenhuys 1985, 1996; Nieuwenhuys et al. 1989; see also Chap. 2, Sect. 2.7.4) and will not be repeated here. Suffice it to briefly mention some relevant ultrastructural findings in a typical and prominent component of the greater limbic system, the periaqueductal grey. As discussed in Chap. 2, Sect. 2.7.4 (Fig. 2.51), two ultrastructural features are strongly suggestive of non-synaptic chemical neurotransmission: (1) the presence of axonal varicosities containing accumulations of synaptic vesicles which do not show any trace of synaptic membrane specializations and (2) the presence of typical synaptic endings containing, in addition to numerous clear synaptic vesicles in the vicinity of the synaptic contact zone, a number of dense-core vesicles located at some distance from the active synaptic site. Both phenomena have been observed in the periaqueductal grey of the rat (Veening et al. 1991; Buma et al. 1992). In this structure, numerous axonal varicosities containing dense core vesicles occur. Synaptic specializations were not found in this type of varicosity, not even when the serial sectioning approach was used for detection (Fig. 22.224b). Typical synaptic terminals containing dense core vesicles at a distance from the active synaptic site were also frequently observed in the periaqueductal grey (Fig. 22.224c,d).

22.12.3.2.3
Input Systems

Humoral Input. The sodium concentration of the blood and the concentration of glucose, free fatty acids and other nutrients are monitored by specialized hypothalamic and extrahypothalamic (e.g. nucleus of the solitary tract) neurons. Moreover, numerous hormones, such as angiotensin II, neuropeptide Y and the gonadal and adrenal steroids, act directly on endocrine-neural transducer cells, which are abundant in limbic domains. Finally, the circumventricular organs, which topographically and functionally lie entirely embedded in the greater limbic system, represent specialized trigger zones for blood- and cerebrospinal fluid-borne substances.

Interoceptive Input. Cells in lamina I of the spinal dorsal horn give rise to fibres which ascend to the limbic brain stem zone. These fibres, which most probably carry information concerning the physiological status of the various tissues of the body, follow the trajectory of the lateral paracore bundle and terminate *inter alia* in the nucleus of the solitary tract, the ventrolateral medulla, the parabrachial nuclei, the lateral pontine tegmentum, the cuneiform nucleus and the mesencephalic periaqueductal grey and in the catecholaminergic cell groups A1–A7 and C1 (Craig 1995, 1996; Burstein 1996; Fig. 22.225). Other information concerning the internal body state converges via the vagus nerve on the nucleus of the solitary tract. This information is conveyed partly directly and partly indirectly via relays in the parabrachial nuclei or in the periaqueductal grey to prosencephalic limbic regions, including the amygdala (Jia et al. 1994). It is important to note that the nucleus of the solitary tract, the parabrachial complex and the periaqueductal grey all entertain strong, reciprocal connections with the hypothalamus, the central nucleus of the amygdala and the bed nucleus of the stria terminalis, all of which represent sites of origin of substantial descending limbic projections. A third route by which interoceptive (particularly cardiovascular) information is conveyed to the limbic prosencephalon passes via the ventrolateral superficial reticular area (commonly referred to as the ventrolateral medulla) and the catecholaminergic cell groups A1 and C1 embedded therein. The principal afferents of this complex arise from the general viscerosensory zone of the nucleus of the solitary tract (Ross et al. 1985). It projects primarily to autonomic centres in the spinal cord, but a certain proportion of its efferents ascend to the forebrain, where they terminate in the hypothalamic supraoptic and paraventricular nuclei (Sawchenko and Swanson 1982; Ciriello and Caverson 1984) and in a number of telencephalic grisea, including the septum, the nucleus accumbens, the hippocampus and the medial prefrontal cortex (Zagon et al. 1994).

Exteroceptive Input. Protopathic fibres project to two limbic cell masses in the brainstem: the periaqueductal grey and the parabrachial nuclei. The fibres projecting to the periaqueductal grey originate mainly from laminae I and IV/V of the spinal cord and from the most superficial part of the spinal trigeminal nucleus. They carry primarily nociceptive stimuli to the periaqueductal grey and terminate in the lateral zone of that structure (Bandler et al. 1991). The fibres projecting in the parabrachial complex terminate mainly in the lateral portion of that complex. Some of these fibres have been demonstrated to arise from nociceptive and thermoreceptive lamina I neurons (Slugg and Light 1994). It is noteworthy that the parabrachial complex contains numerous neurons which are exclusively activated by noxious stimuli applied to several areas of the body and that these 'nociceptive-specific' neurons have been demonstrated to project to the nucleus centralis amygdalae and the retrochiasmatic area of the hypothalamus (Bernard et al. 1993, 1996; Jhamandas et al. 1996).

It has been recently established that neurons situated in lamina I of the spinal and medullary dorsal horn and in the grey matter surrounding the spinal central canal project directly to many parts of the hypothalamus, including the posterior, ventromedial and paraventricular nuclei and the lateral hypothalamic area and, less densely, to a number of limbic telencephalic regions, including the amygdala, septum, nucleus accumbens, ventral pallidum and orbital cortex (Burstein 1996).

Visual Input. A direct visual input attains the hypothalamus via the retinohypothalamic tract, which terminates in the suprachiasmatic nucleus. This cell group, which projects to various intrahypothalamic and extrahypothalamic limbic centres, may be considered as the endogenous clock of the brain. It is critically involved in the generation and entrainment of circadian rhythms, including patterns of general activity (sleeping/waking), feeding and drinking behaviour and hormonal secretion.

Olfactory Input. The olfactory system, which in most mammals (and non-mammalian vertebrates) exerts a strong influence on feeding, mating and several related goal-oriented behaviours, has 'privileged access' (Nauta and Haymaker 1969) to limbic circuitry. Price et al. (1991) reported that the anterior olfactory nucleus, the olfactory tubercle, the piriform cortex and the anterior cortical nucleus of the amygdala, which are all in receipt of direct afferents from the main olfactory bulb, all project by way of the medial forebrain bundle to the caudal part of the lateral hypothalamic area. The accessory optic system projects via the medial part of the amygdala and the bed nucleus of the stria terminalis to the medial preoptic area and the ventromedial hypothalamic nucleus (Scalia and Winans 1975; Fig. 22.160). Shipley et al. (1996) recently reported that both the main and the accessory olfactory systems project to the medial preoptic area and that this centre has a dense and highly organised projection to the periaqueductal grey and the rostral ventrolateral medulla.

Extrapyramidal Input. The extrapyramidal system has been briefly dealt with in Sect. 22.12.2.4. It was pointed out that this system plays an important role in motor integration. It should be added here that the striatum can be subdivided on connectional grounds into a ventral part or ventral striatum and

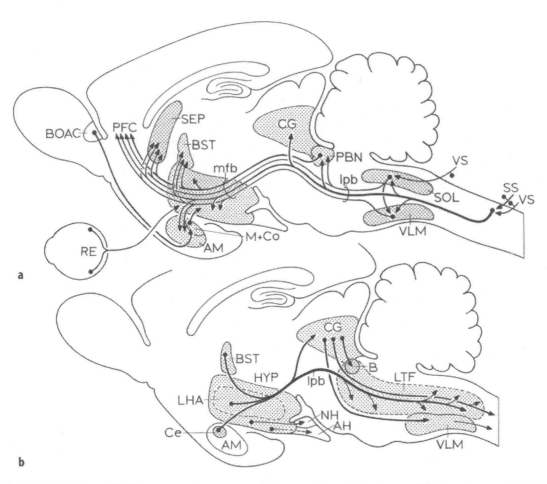

Fig. 22.226a,b. Greater limbic system. **a** Input systems. **b** Output channels. *AH*, adenohypophysis; *AM*, amygdaloid complex; *B*, Barrington's area; *BOAC*, bulbus olfactorius accessorius; *BST*, bed nucleus of stria terminalis; *Ce*, nucleus centralis amygdalae; *Co*, nucleus corticalis amygdalae; *CG*, griseum centrale mesencephali; *HYP*, hypothalamus; *LHA*, lateral hypothalamic area; *LTF*, lateral tegmental field; *lpb*, lateral paracore bundle; *M*, nucleus medialis amygdalae; *NH*, neurohypophysis; *mfb*, medial forebrain bundle; *PBN*, parabrachial nuclei; *PFC*, prefrontal cortex; *RE*, retina; *SEP*, septal nuclei; *SS*, somatic sensory input; *VS*, viscerosensory input

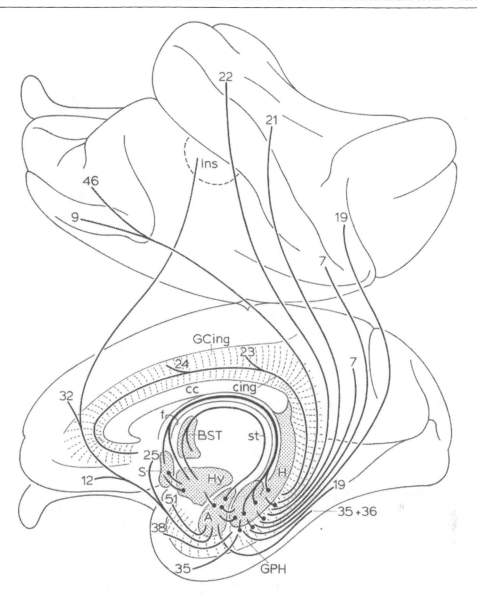

Fig. 22.227. Greater limbic system. Principal connections of its rostral parts. The amygdaloid complex (*A*) receives afferents from many cortical areas and projects to a still larger number of cortical areas. Only areas with which there are reciprocal connections are included here. The hippocampus (*H*) is reciprocally connected with numerous cortical areas via the adjacent parahippocampal cortex. *BST*, bed nucleus of stria terminalis; *cc*, corpus callosum; *cing*, cingulum bundle; *f*, fornix; *G Cing*, gyrus cinguli; *GPH*, gyrus parahippocampalis; *Hy*, hypothalamus; *Ins*, insula; *S*, septum; *st*, stria terminalis. *Numbers* indicate fields of Brodmann. (Partly based on Van Hoesen 1982)

a dorsal part or dorsal striatum (see Sect. 22.11.4). The ventral striatum, which includes the nucleus accumbens and the ventral part of the caudate nucleus and the putamen, receives its main input from typical limbic structures, such as the hippocampal formation and the amygdala, which is why it is often designated as the limbic striatum. The dorsal or non-limbic striatum, on the other hand, receives its principal telencephalic input from the somatosensory, motor and association cortex. The two parts of the striatum are the respective origins of two separate exit lines. The dorsal striatum projects to the globus pallidus proper or dorsal pallidum, whereas the ventral striatum sends it efferents mainly to a ventral extension of the globus pallidus, situated ventrally, beneath the anterior commissure. This ventral pallidum projects to several typical limbic structures, including the lateral hypothalamic area, the ventrolateral periaqueductal grey and the mesencephalic locomotor region. The ventral pallidal projection to these grisea is reinforced by fibres arising from the nucleus accumbens and the prefrontal cortex. Thus it appears that the ventral striatum, within which the nucleus accumbens occupies a prominent position, receives strong projections from the hippocampus and the amygdala and that this ventral striatum projects, either directly or via the ventral pallidum, to certain diencephalic and mesencephalic limbic centres, including the mesencephalic locomotor region. Locomotion is an essential component of the procurement phase of many motivated behaviours, and it seems likely that the ventral striatum, which is situated in a transitional zone between the limbic and the extrapyramidal systems, is critically implicated in the initiation of this activity (Mogenson et al. 1980).

22.12.3.2.4
Output Channels

The limbic motor system comprises a large number of thin-fibred pathways descending from telencephalic, diencephalic and mesencephalic structures to the lower medulla and the spinal cord. Holstege (1991) thoroughly analysed this system and distinguished separate medial and lateral components within it.

The fibres of the medial component originate from cell groups in the medial zone of the hypothalamus and the midbrain. These fibres terminate in the locus coeruleus and in the subcoereleus area and in the most ventromedial part of the pontine and medullary tegmentum. Within the latter region, the caudal raphe nuclei, i.e. the nuclei raphes magnus, pallidus and obscurus, are heavily innervated. The medial component is thought to exert a global influence via the diffuse coerulospinal and raphe-spinal projections on the level of excitability of spinal sensory and motor neurons. It is conceivable that impulses travelling along the descending projections mentioned above also enhance the level of activity of motoneurons and suppress pain transmission in the spinal dorsal horn. These influences would enable an animal in life-threatening situations to combine maximal motor activity with neglect of painful stimuli.

The fibres of the lateral component originate mainly from the lateral periaqueductal grey, the central nucleus of the amygdala, the bed nucleus of the stria terminalis and the lateral hypothalamic area.

Relative large neurons in the lateral periaqueductal grey and laterally adjoining tegmentum project to the lateral tegmental field of the rhombencephalon and to the intermediate zone of the spinal grey matter, where they impinge on numerous, mostly diffuse groups of cells involved in specific functions. This projection includes pathways to Barrington's area (micturition), the ventrolateral medulla (control of blood pressure and respiration), the pontine reticular formation (swallowing), the nucleus retroambiguus (vocalisation) and the intermediolateral cell column in spinal segments T1,T2 (pupil dilatation).

The various lateral tegmental cell groups involved in specific functions and behaviours are also under the direct and indirect control of more rostral limbic areas. Thus the central nucleus of the amygdala, the bed nucleus of the stria terminalis, and the lateral hypothalamic area have been shown to project to the lateral periaqueductal grey and to distribute to these various tegmental cell groups.

In 1992, Holstege replaced the term 'limbic motor system' by 'emotional motor system'. This is understandable, because the general level-setting activities of the medial component are strongly influenced by the emotional state of the individual. Furthermore, most of the specific, motivated behaviours elaborated by the lateral component are always accompanied by signs of emotional arousal under normal circumstances. The 'emotional-motivational motor system' would have been a more comprehensive functional designation. Remarkably, none of the structures included by Papez (1937; see Sect. 22.12.3.1) in his hypothetical emotional circuit forms part of Holstege's (1992) emotional motor system. It is important to note in this context that experimental studies have shown that the amygdaloid complex, which was not included in Papez' circuit, plays a key role in the integration of emotional behaviour (for a review, see Murray 1991; LeDoux 1992).

From the foregoing, it may be concluded that the emotional motor system lies entirely embedded in the greater limbic system (Fig. 22.226) and forms the principal effector apparatus of the latter.

22.12.3.2.5
Connections with the Neocortex (Fig. 22.227)

The hippocampal formation and the amygdaloid complex together constitute, morphologically and functionally, the rostral pole of the limbic domain. Via the septal region, the hippocampus is strongly and reciprocally connected with the preoptico-hypothalamic continuum (see Sect. 22.11.3.2; Fig. 22.163), and comparable reciprocal connections are present between the hypothalamus and the amygdaloid complex, including its dorsomedial extension, the bed nucleus of the stria terminalis (see Sect. 22.11.5.4). It is of paramount importance that the hippocampus and the amygdala, and with it the entire greater limbic system, are under the control of the neocortex. Before discussing this feature further, we should return for a moment to the lateral domain (see Sect. 22.12.2). As already discussed, the primary somatosensory, visual and auditory cortices project via short connections to adjacent unimodal sensory association areas. These unimodal sensory association areas project in turn to a strip of multimodal association cortex, and the unimodal and multimodal sensory association areas project massively to the prefrontal cortex, which should thus be considered as a higher association cortex. Interestingly, fibres originating from all of these association areas (unimodal, e.g. area 19; multimodal, e.g. area 7; prefrontal, e.g. areas 9, 12, 46) enter the medial domain and converge in the medial part of the temporal lobe on the amygdala and the hippocampus. Most of these corticolimbic projections are synaptically interrupted in the cingulate and/or parahippocampal gyri, which may thus be designated as paralimbic cortical areas (Mesulam 1985; Pandya 1987).

It is important to note that the neocortex not only receives and processes information related to the external world, but that the viscera, and hence the internal world of the organism, also have a cortical representation. In primates, the cortical regions which are in receipt of interoceptive information include the anterior cingulate, posterior orbitofrontal, insular and temporal pole areas (for a review, see Reep 1984). All of these visceral cortices project, either directly or indirectly, to the amygdalohippocampal complex in the temporal lobe. However, some also project directly to lower limbic centres, including the bed nucleus of the stria terminalis, the lateral hypothalamic area, the parabrachial complex and the nucleus of the solitary tract (see e.g. Saper 1982; Van der Kooy et al. 1984; Willett et al. 1986). It may be concluded that, via the various conduction routes discussed, highly processed information concerning both the external and internal environment is fed into the circuitry of the greater limbic system.

The amygdaloid complex and the hippocampal formation form an interface between the 'cognitive brain' and the lower domains of the greater limbic system. It is known that hippocampal formation and the adjacent entorhinal cortex contribute significantly to memory function (Zola-Morgan and Squire 1986; Victor and Agamanolis 1990) and that the amygdala plays a key role in emotional behaviour. At first sight, these seem to be quite different functions. However, it should be realised that the emotional significance of newly arriving sensory stimuli depends on past experience and that the evaluation of incoming signals by comparing them with information stored in memory is thus an essential step in the processing sequence leading from the perception of environmental events to emotional behaviour as such, or to the 'attachment' of an emotional component to motivated behaviours (see LeDoux 1987). The amygdala and the hippocampal formation, which entertain direct and indirect reciprocal connections, may well be jointly involved in this evaluation process.

The descending corticolimbic projections discussed above are reciprocated by substantial ascending limbicocortical paths. These ascending and descending conduction routes are both largely funnelled through, and synaptically interrupted in the amygdalo-hippocampal complex. The ascending system terminates mainly in cortical association areas (for references, see Nieuwenhuys et al. 1988). Impulses travelling along the hippocampo- and amygdalocortical projections may lead to conscious emotional experiences and may be instrumental in adapting the activities of the cognitive brain to the prevailing motivational state of the organism.

The preceding discussion of limbico-cortical and corticolimbic connections is entirely based on the results of experimental neuroanatomical studies in the monkey, which, of course, has a large and highly differentiated neocortex. One of the most salient differences between primates and 'primitive' mammals, such as insectivores and rodents, is that in the former the 'higher-order' or association cortex is much more extensive than in the latter (Fig. 22.192). This implies that, in general, the sensory information, which ultimately reaches the somatomotor cortex and the amygdalohippocampal complex, has gone through a much more detailed and refined analysis in primates than in the primitive groups mentioned. What actually takes place in the large association areas of the monotreme Echidna and the cetaceans (Fig. 22.192 f,g) remains an intriguing enigma.

22.12.4
Overview

Information related to occurrences in the external world enter the brain via the somatosensory, auditory and visual systems. This information is subjected to a refined analysis in the sensory association and multimodal cortices and transferred from there to the prefrontal association cortex. The latter is concerned, among other things, with the planning and sequencing of complex and skilled behaviour aimed at manipulation of the external world. The premotor and motor cortices and the large pyramidal tract are instrumental in the planning and execution of the movements related to this behaviour. The motor and premotor cortices are supported by two large control systems, the extrapyramidal and the cerebellar systems.

The sensory and motor systems involved in the cognitive-motor cycle outlined above and the control systems attached to it occupy grossly the lateral part of the brain. All of these systems are mainly composed of discrete and well-differentiated grisea, and their fibre connections are compact and well myelinated.

The medial part of the brain harbours a neural entity which structurally, chemically and functionally differs considerably from the classical systems mentioned above. This entity is designated here as the greater limbic system. It consists of an array of highly interconnected structures, extending from the medial wall of the telencephalon to the caudal rhombencephalon, which is concerned with specific, motivated or goal-oriented behaviours, directly aimed at the maintenance of homeostasis and at the survival of the individual (organism) and of the species. All of these specific behaviours include integrated endocrine, autonomic and skeletomotor responses, and the latter generally pass through three sequential phases: the initiation, procurement, and consummatory phase (Swanson and Mogenson 1981; Swanson 1989). Within the framework of these specific tasks, the greater limbic system, by means of sets of monoaminergic neurons, influences the level of excitability of its own neuronal network, of its effector mechanisms and of virtually all other parts of the brain. By this general activation, the organism is brought to a high level of alertness and is prepared for maximal physical action.

The greater limbic system is characterised by the presence of (a) enormous amounts of thin, unmyelinated varicose fibres, participating in the formation of numberless 'open-line' conduction channels, (b) numerous diffuse and inconspicious grisea, (c) specialized chemosensitive loci, (d) large numbers of neurons receptive for oestrogen and androgen hormones and (e) an extraordinary amount and diversity of neuropeptides. This wealth of neuropeptides indicates that the way in which the greater limbic system operates is distinctly different from other parts of the brain (Nieuwenhuys 1985; Herbert 1993). Many neuropeptides have been demonstrated to play a key role in the regulation of specific behavioural responses. Moreover, there is evidence suggesting that, within the domain of the greater limbic system, these substances are involved in non-synaptic or paracrine interneuronal communication.

The concept of an emotional motor system has recently been presented by Holstege (1992, 1995). This concept emphasises that, within the brain stem, assemblies of thin, descending fibres are involved in the elaboration of emotional behaviours (such as crying and laughter in humans and vocalisation in animals) and that these fibres are completely distinct from those forming the voluntary somatomotor systems. The emotional motor system is entirely embedded in the greater limbic system and forms the principal effector apparatus of the latter. Its fibres originate mainly from the periaqueductal grey, the central nucleus of the amygdala, the bed nucleus of the stria terminalis and the lateral hypothalamic area. These fibres are most likely not only involved in emotional behaviour per se, but also in specific, motivated behaviours and their emotional correlates. According to Holstege (1992, 1995), the emotional motor system also encompasses fibre systems descending from the medial hypothalamus and midbrain which, via monoaminergic cell groups in the brain stem, exert a global influence on the level of activity of spinal sensory and motor neurons.

The greater limbic system interacts with many other parts of the central nervous system. All sensory systems have access to it, and it is in receipt of substantial direct interoceptive, nociceptive and olfactory projections. A large fibre system, carrying interoceptive and nociceptive information, ascends through the limbic brain stem area to higher limbic domains (Figs. 22.222, 22.226a). This fibre system originates mainly in the spinal and medullary dorsal horn and in the nucleus of the solitary tract. Some of its fibres ascend directly to diencephalic or even to telencephalic levels, but most are synaptically interrupted in the parabrachial nuclear complex or in the periaqueductal grey. The nucleus of the solitary tract and the parabrachial nuclear complex and the periaqueductal grey are all strongly and reciprocally connected with the central nucleus of the amygdala, the bed nucleus of the stria terminalis and the hypothalamus. Thus it may be concluded that the ascending projection discussed above, including its waystations in the brain stem, forms the sensory counterpart of the emotional (motivational) motor system.

The ventral part of the striatum is dominated by limbic afferents, and this ventral striatum projects, both directly and indirectly, via a ventral sector of the pallidum back to the limbic domain, in particular to the lateral hypothalamic area and the caudolateral mesencephalic tegmentum. This striatolimbic projection is considered to play an important role in the initiation of locomotor activity.

Finally, the neocortex projects strongly to the greater limbic system. Sensory information originating from the external world, which is processed successively in subcortical centres, primary sensory cortices, unimodal association cortices and polymodal association cortices, is not only transferred to the motor and premotor cortices and thus to the voluntary motor system, but is also transferred in the same measure to the rostral parts of the greater limbic domain and thus to the emotional (motivational) motor system. These descending cortico-limbic projections are paralleled by strong ascending projections connecting rostral limbic regions with (mainly) the various association cortices. It is these extensive reciprocal connections between the cognitive brain and the emotional-motivational domain of the neuraxis which enable the organism to harmonise the reality of the external world with its internal urges (Mesulam 1985).

Acknowledgements. *The authors greatfully acknowledge the assistance of Edith Klink in the preparation of the manuscript and of Eddy Dalm in the photography of the figures. The right halves of the figures 23 and 24 were drawn by Philip Wilson FMAA, AIMI (Orpington, Kent U.K.). The figures 22.22–22.24 were labelled by Karin Voogd (Rotterdam, The Netherlands).*

References

Aas JA (1989) Subcortical projections to the pontine nuclei in the cat. J Comp Neurol 282:331–354
Abbie AA (1934) The brain stem and cerebellum of Echidna. Philos Trans R Soc (Lond) Ser B 224:1–74
Abbie AA (1940) Cortical lamination in Monotremata. J Comp Neurol 72:429–467
Abeles M, Goldstein MH Jr (1970) Functional architecture in cat primary auditory cortex: columnar organization and organization according to depth. J Neurophysiol 33:172–187
Abrahams VC, Hilton SM, Malcolm JL (1962) Sensory connections to the hypothalamus and midbrain, and their role in the reflex activation of the defence reaction. J Physiol (Lond) 164:1–16
Achenbach KE, Goodman DC (1968) Cerebellar projection to pons, medulla and spinal cord in the albino rat. Brain Behav Evol 1:43–57
Adams JC (1979) Ascending projections to the inferior colliculus. J Comp Neurol 183:519–538
Adams JC (1983) Cytology of periolivary cells and the organization of their projections in the cat. J Comp Neurol 215:275–289
Adams JC, Mugnaini E (1990) Immunocytochemical evidence for inhibitory and disinhibitory circuits in the superior olive. Hearing Res 49:281–298
Adams JC, Warr WB (1976) Origins of axons in the cat's acoustic striae determined by injection of horseradish peroxidase into several tracts. J Comp Neurol 170:107–122
Addens JL, Kurotsu T (1936) Die Pyramidenbahnen von Echidna. Proc Kon Acad Wetensch Amsterdam 34:3–12
Adolphs R, Tranel D, Damasio H, Damasio AR (1995) Fear and the human amygdala. J Neurosci 15:5879–5891
Aggleton JP (1986) Description of the amygdalo-hippocampal interconnections in the macaque monkey. Exp Brain Res 64:515–526
Aggleton JP (1992) The functional effects of amygdala lesions in humans: a comparison with findings from monkeys. In: Aggleton JP (ed) The amygdala: neurobiological aspects of emotion, memory and mental dysfunction. Wiley-Liss, New York, pp 485–504
Aggleton JP, Mishkin M (1984) Projections of the amygdala to the thalamus in the cynomolgus monkey. J Comp Neurol 222:56–68
Aggleton JP, Burton MJ, Passingham RE (1980) Cortical and subcortical afferents to the amygdala of the rhesus monkey (Macaca mulatta). Brain Res 190:347–368
Aggleton JP, Petrides M, Iversen SD (1981) Differential effects of amygdaloid lesions on conditioned taste aversion learning by rats. Physiol Behav 27:397–400
Ahlsén G, Lindström S, Sybirska E (1978) Subcortical axon collaterals of principal cells in the lateral geniculate body of the cat. Brain Res 156:106–109
Aitkin L, Nelson J, Farrington M, Swann S (1994) The morphological development of the inferior colliculus in a marsupial, the Northern quoll (Dasyurus hallucatus). J Comp Neurol 343:532–541
Aitkin LM, Kenyon CE, Philpott P (1981) The representation of the auditory and somatosensory systems in the external nucleus of the cat inferior colliculus. J Comp Neurol 196:25–40
Akagi Y (1978) The localization of the motor neurons innervating the extraocular muscles in the oculomotor nuclei of the cat and rabbit, using horseradish peroxidase. J Comp Neurol 181:745–762
Akaike T (1992) The tectorecipient zone in the inferior olivary nucleus in the rat. J Comp Neurol 32:398–414
Akers RM, Killackey HP (1979) Segregation of cortical and trigeminal afferents to the ventrobasal complex of the neonatal rat. Brain Res 161:527–532
Albers FJ (1990) Structure and organization of the superior colliculus of the rat. Thesis, University of Nijmegen
Albers FJ, Meek J (1991) Dendritic and synaptic properties of collicular neurons: a quantitative light and electron microscopical study of Golgi-impregnated cells. Anat Rec 231:524–537
Albin RL, Aldridge JW, Young AB, Gilman S (1989) Feline subthalamic nucleus neurons contain glutamate-like but not GABA-like or glycine-like immunoactivity. Brain Res 491:185–188
Albright BC, Haines DE (1978) Dorsal column nuclei in a prosimian primate (Galago senegalensis). II. Cuneate and lateral cuneate nuclei: morphology and primary afferent fibers from cervical and upper thoracic spinal segments. Brain Behav Evol 15:165–184
Albus JS (1971) A theory of cerebellar function. Math Biosci 10:25–61
Alexander GE, Crutcher MD (1990) Functional architecture of basal ganglia circuits: neural substrates of parallel processing. Trends Neurosci 13:266–271
Alexander GE, DeLong MR, Strick PL (1986) Parallel organization of functionally segregated circuits linking basal ganglia and cortex. Annu Rev Neurosci 9:357–381
Alheid GF, Heimer L (1988) New perspectives in basal forebrain organization of special relevance for neuropsychiatric disorders: the striatopallidal, amygdaloid, and corticopetal components of substantia innominata. Neuroscience 27:1–39
Alheid GF, de Olmos JS, Beltramino CA (1995) Amygdala and extended amygdala. In: Paxinos G (ed) The rat nervous system, 2nd edn. Academic, San Diego, pp 495–578
Allen WF (1923a) Origin and distribution of the tracts solitarius in the guinea pig. J Comp Neurol 35:171–204
Allen WF (1923b) Origin and destination of the secondary visceral fibers in the guinea-pig. J Comp Neurol 35:275–312
Allison AC (1953) The morphology of the olfactory system in the vertebrates. Biol Rev 28:195–244
Alonso A, Kohler C (1982) Evidence for separate projections of hippocampal pyramidal and non-pyramidal neurons to different parts of the septum in the rat brain. Neurosci Lett 31:209–214
Alstermark B, Kummel, Tantisira B (1987) Monosynaptic raphe spinal and reticulospinal projection to forelimb motoneurones in cats. Neurosci Lett 74:286–290
Amaral DG (1986) Amygdalohippocampal and amygdalocortical projections in the primate brain. In: Ben-Ari Y, Schwarcz R (eds) Excitatory amino acids and epilepsy. Plenum, New York, pp 3–17

Amaral DG (1987) Memory: anatomical organization of candidate brain regions. In: Mountcastle VB, Plum F, Geiger SR (eds) The nervous system. American Physiological Society, Bethesda, pp 211–294 (Handbook of physiology, vol 5, part 1)

Amaral DG, Cowan WM (1980) Subcortical afferents to the hippocampal formation in the monkey. J Comp Neurol 189:573–591

Amaral DG, Insausti R (1998) The entorhinal cortex of the monkey. IV. Topographical and laminar organization of cortical afferents. In preparation

Amaral DG, Kurz J (1985) An analysis of the origins of the cholinergic and non-cholinergic septal projections to the hippocampal formation of the rat. J Comp Neurol 240:37–59

Amaral DG, Price JL (1984) Amygdalo-cortical projections in the monkey (Macaca fascicularis). J Comp Neurol 230:465–496

Amaral DG, Veazy RB, Cowan WM (1982) Some observations on hypothalamo-amygdaloid connections in the monkey. Brain Res 252:13–27

Amaral DG, Insausti R, Cowan WM (1988) The entorhinal cortex of the monkey. I. Cytoarchitectonic organization. J Comp Neurol 264:326–355

Amaral DG, Ishizuka N, Claiborne B (1990) Neurons, numbers and the hippocampal network. Prog Brain Res 83:1–11

Amaral DG, Insausti R, Witter MP (1998) The entorhinal cortex of the monkey. VI. Projections from the hippocampus and subicular complex. In preparation

Amaral DG, Price JL, Pitkänen A, Carmichael ST (1992) Anatomical organization of the primate amygdaloid complex. In: Aggleton JP (ed) The amygdala: neurobiological aspects of emotion, memory, and mental dysfunction. Wiley-Liss, New York, pp 1–66

Andersen RA, Knight PL, Merzenich MM (1980) The thalamocortical and corticothalamic connections of AI, AII, and the anterior auditory field (AAF) in the cat: evidence for two largely segregated systems of connections. J Comp Neurol 194:663–701

Andersson G, Oscarsson O (1978) Climbing fiber microzones in the cerebellar vermis and their projection to different groups of cells in the lateral vestibular nucleus. Exp Brain Res 32:565–579

Angaut P (1970) The ascending projections of the nucleus interpositus posterior of the cat cerebellum. An experimental anatomical study using silver impregnation methods. Brain Res 24:377–394

Angaut P, Bowsher D (1965) Cerebello-rubral connexions in the cat. Nature 208:1002

Antonetty CM, Webster KE (1975) The organisation of the spinotectal projection. An experimental study in the rat. J Comp Neurol 163:449–467

Apkarian AV, Hodge CJ (1989a) Primate spinothalamic pathways: I. A quantitative study of the cells of origin of the spinothalamic pathway. J Comp Neurol 288:447–473

Apkarian AV, Hodge CJ (1989b) Primate spinothalamic pathways: III. Thalamic terminations of the dorsolateral and ventral spinothalamic pathways. J Comp Neurol 288:493–511

Apkarian AV, Hodge CJ (1989c) Primate spinothalamic pathways: III. Thalamic terminations of the dorsolateral and ventral spinothalamic pathways. J Comp Neurol 288:493–511

Araki M, McGeer PL, McGeer EG (1984) Presumptive γ-aminobutyric acid pathways from the midbrain to the superior colliculus studied by a combined horseradish peroxidase,-γ-aminobutyric acid transaminase pharmacohistochemical method. Neuroscience 13:433–439

Ariëns Kappers CU (1909) The phylogenesis of the palaeocortex and archi-cortex compared with the evolution of the visual neo-cortex. Arch Neurol Psychiatr (London) 4:161–173

Ariëns-Kappers CU, Huber GC, Crosby E (1936) The comparative anatomy of the nervous system of vertebrates, including man, 2 vols (English edition) New York: Macmillan Co

Arikuni T, Kubota K (1986) The organization of prefrontocaudate projections and their laminar origin in the macaque monkey: a retrograde study using HRP gel. J Comp Neurol 244:429–510

Armand J (1982) The origin, course and terminations of corticospinal fibers in various mammals. Prog Brain Res 57:329–360

Armand J, Holstege G, Kuypers HGJM (1985) Differential corticospinal projections in the cat. An autoradiographic tracing study. Brain Res 343:351–355

Armstrong DM, Schild RF (1979) Spino-olivary neurones in the lumbosacral cord of the cat demonstrated by retrograde transport of horseradish peroxidase. Brain Res 168:176–179

Armstrong DM, Schild RF (1980) Location in the spinal cord of neurones projecting directly to the inferior olive in the cat. In: Courville J (ed) The inferior olive. Anatomy and physiology. Raven, New York, pp 125–144

Armstrong DM, Campbell NC, Edgley SA, Schild RF, Trott JR (1982) Investigations of the olivocerebellar and spino-olivary pathways. In: Palay SL, Chan-Palay V (eds) The cerebellum – new vistas. Exp Brain Res Suppl 6:195–232

Armstrong DM, Saper CB, Levey A, Wainer BH, Terry RD (1983) Distribution of cholinergic neurons in rat brain: demonstrated by the immunocytochemical localization of choline acetyltransferase. J Comp Neurol 216:53–68

Armstrong E, Schleicher A, Omran H, Curtis M, Zilles K (1995) The ontogeny of human gyrification. Cerebral Cortex 5:56–63

Arnsten AFT, Goldman-Rakic PS (1984) Selective prefrontal cortical projections to the region of the locus coeruleus and raphe nuclei in the rhesus monkey. Brain Res 306:9–18

Arvidsson J (1982) Somatotopic organization of vibrissae afferents in the trigeminal sensory nuclei of the rat studied by transganglionic transport of HRP. J Comp Neurol 211:84–92

Arvidsson J, Thomander L (1984) An HRP study of the central course of sensory intermediate and vagal fibers in peripheral facial nerve branches in the cat. J Comp Neurol 223:35–45

Asanuma C, Thach WT, Jones EG (1983a) Distribution of cerebellar terminations and their relation to other afferent terminations in the ventral lateral thalamic region of the monkey. Brain Res Rev 5:237–265

Asanuma C, Thach WT, Jones EG (1983b) Anatomical evidence for segregated focal groupings of efferent cells and their terminal ramifications in the cerebellothalamic pathway of the monkey. Brain Res Rev 5:267–297

Asanuma H (1975) Recent developments in the study of the columnar arrangement of neurons in the motor cortex. Physiol Rev 55:143–156

Asanuma H (1987) Cortical motor columns. In: Adelman G (ed) Encyclopedia of neuroscience, vol I. Birkhäuser, Boston, pp 281–282

Asanuma H, Rosen I (1973) Spread of mono- and polysynaptic connections within cat's motor cortex. Exp Brain Res 16:507–520

Aschoff A, Ostwald J (1987) Different origins of cochlear efferents in some bat species, rats, and guinea pigs. J Comp Neurol 264:56–72

Aschoff A, Oswald J (1988) Distribution of cochlear efferents and olivocollicular neurons in the brainstem of rat and guinea pig. A double labeling study with fluorescent tracers. Exp Brain Res 71:241–251

Aston-Jones G, Shipley MT, Grzanna R (1995) The locus coeruleus, A5 and A7 noradrenergic cell groups. In: Paxinos G (ed) The rat nervous system, 2nd edn. Academic, New York, pp 183–213

Augustine JR, DesChamps EG, Ferguson JG Jr (1981) Functional organization of the oculomotor nucleus in the baboon. Am J Anat 161:393–403

Aumann TD, Rawson JA, Finkelstein DI, Horne MK (1994) Projections from the lateral and interposed cerebellar nuclei to the thalamus of the rat: a light and electron microscopic study using single and double anterograde labelling. J Comp Neurol 349:165–181

Avendaño C, Juretschke MA (1980) The pretectal region of the cat: a structural and topographical study with stereotaxic coordinates. J Comp Neurol 193:69–88

Avendaño C, Price JL, Amaral DG (1983) Evidence for an amygdaloid projection to premotor cortex but not to motor cortex in the monkey. Brain Res 264:111–117

Azizi SA, Woodward DJ (1987) Inferior olivary nuclear complex of the rat: morphology and comments on the principles of organization within the olivocerebellar system. J Comp Neurol 263:467–484

Azizi SA, Burne RA, Woodward DJ (1985) The auditory corticopontocerebellar projection in the rat: inputs to the paraflocculs and mid-vermis. An anatomical and physiological study. Exp Brain Res 59:36–49

Bagley C (1922) Cortical motor mechanism of the sheep brain. Arch Neurol Psychiat 7:417–453

Baile CA, McLaughlin CL, Della-Fera MA (1986) Role of cholecystokinin and opioid peptides in control of food intake. Physiol Rev 66:172–230

Baizer JS, Whitney JF, Bender DB (1991) Bilateral projections from the parabigeminal nucleus to the superior colliculus in monkey. Exp Brain Res 86:467–470

Baizer JS, Desimone R, Ungerleider LG (1993) Comparison of subcortical connections of inferior temporal and posterior parietal cortex in monkeys. Vis Neurosci 10:59–72

Baker J, Gibson A, Mower G, Robinson F, Glickstein M (1983) Cat visual corticopontine cells project to the superior colliculus. Brain Res 265:227–232

Baker R (1986) Brainstem neurons are peculiar for oculomotor organization. Prog Brain Res 64:257–271

Baker R, Highstein SM (1975) Physiological identification of interneurons and motoneurons in the abducens nucleus. Brain Res 83:292–298

Balaban CD (1984) Olivo-vestibular and cerebello-vestibular connections in albino rabbits. Neuroscience 12:129–149

Balaban CD (1988) Distribution of inferior olivary projections to the vestibular nuclei of albino rats. Neuroscience 24:119–134

Baleydier C, Magnin M, Cooper HM (1990) Macaque accessory optic system: II. connections with the pretectum. J Comp Neurol 302:405–416

Bandler R, Carrive P, Zhang SP (1991) Integration of somatic and autonomic reactions within the midbrain periaqueductal grey: Viscerotopic, somatotopic and functional organization. Prog Brain Res 87:269–305

Barber RP, Phelps PE, Houser CR, Crawford GD, Salvaterra PM, Vaughn JE (1984) The morphology and distribution of neurons containing choline acetyltransferase in the adult rat spinal cord: an immunocytochemical study. J Comp Neurol 229:329–346

Bard P (1928) A diencephalic mechanism for the expression of rage with special reference to the synpathetic nervous system. Am J Physiol 84:490–515

Bard P (1929) The central representation of the sympathetic nervous system as indicated by certain physiologic observations. Arch Neurol Psychiatry 22:230–246

Baron G, Stephan H, Frahm HD (1996) Comparative neurobiology in chiroptera, vol 1. Birkhäuser, Basel

Barone R, Doucet J (1964) Recherches sur la morphologie et la topographie de la substance grise dans le bulbe rachidien du boeuf. Ann Biol Anim Bioch Biophys 4:307–343

Barrett RT, Bao X, Miselis RR, Altschuler SM (1994) Brain stem localization of rodent esophageal premotor neurons revealed by transneuronal passage of pseudo-rabies virus. Gastroenterology 107:883–885

Batini C, Buisseret-Delmas C, Corvisier J (1976) Horseradish peroxidase localization of masticatory muscle motoneurons in cat. J Physiol (Paris) 72:301–309

Batton RRI, Jayaraman A, Ruggiero D, Carpenter MB (1977) Fastigial efferent projections in the monkey: an autoradiographic study. J Comp Neurol 174:281–306

Bauchot R (1959) Etude des structures cytoarchitectoniques du diencéphale de Talpa europaea (Insectivora Talpidae). Acta Anat 39:90–140

Bauchot R (1963) L'architectonique comparée, qualitative et quantitative, du diencéphale des insectivores. Mammalia 27 [Suppl 1]:1–400

Baude A, Nusser Z, Roberts JDB, Mulvihil E, McIlinney J, Somogyi (1993) The metabotropic glutamate receptor (mGluRalpha) is concentrated at perisynaptic membrane of neuronal subpopulations as detected by immunogold reaction. Neuron 11:771–787

Bayer SA (1980) Quantitative ^{3}H-thymidine radiographic analyses of neurogenesis in the rat amygdala. J Comp Neurol 194:845–875

Bayer SA, Altman J (1987) Directions in neurogenetic gradients and patterns of anatomical connections in the telencephalon. Prog Neurobiol 29:57–106

Beach TG, McGeer EG (1984) The distribution of substance P in the primate basal ganglia: an immunohistochemical study of baboon and human brain. Neuroscience 13:29–52

Beccari N (1943) Neuroglia comparata. Sansoni Edizione Scientifiche, Firenze

Bechterew W (1899) Die Leitungsbahnen im Gehirn und Ruckenmark. Verlag Arthur Georgi, Leipzig

Beckstead RM (1979) An autoradiographic examination of corticocortical and subcortical projections of the mediodorsal-projection (prefrontal) cortex in the rat. J Comp Neurol 184:43–62

Beckstead RM (1983) A reciprocal axonal connection between the subthalamic nucleus and the neostriatum in the cat. Brain Res 275:137–142

Beckstead RM, Cruz CJ (1986) Striatal axons to the globus pallidus entopeduncular nucleus and substantia nigra come mainly from separate cell populations in the cat. Neuroscience 19:147–158

Beckstead RM, Frankfurter A (1982) The distribution and some morphological features of substantia nigra neurons that project to the thalamus, superior colliculus and pedunculopontine nucleus in the monkey. Neuroscience 7:2377–2388

Beckstead RM, Frankfurter A (1983) A direct projection from the retina to the intermediate gray layer of the superior colliculus demonstrated by anterograde transport of horseradish peroxidase in monkey, cat and rat. Exp Brain Res 52:261–268

Beckstead RM, Norgren R (1979) An autoradiographic examination of the central distribution of the trigeminal, facial, glossopharyngeal, and vagal nerves in the monkey. J Comp Neurol 184:455–472

Beckstead RM, Domesick VB, Nauta WJH (1979) Efferent connections of the substantia nigra and ventral tegmental area in the rat. Brain Res 175:191–217

Beckstead RM, Morse JR, Norgren R (1980) The nucleus of the solitary tract in the monkey: projections to the thalamus and brain stem nuclei. J Comp Neurol 190:259–282

Behan M (1982) A quantitative analysis of the ipsilateral retinocollicular projection in the cat: an EM degeneration and EM autoradiographic study. J Comp Neurol 206:253–258

Behbehani MM (1995) Functional characteristics of the midbrain periaqueductal gray. Prog Neurobiol 46:575–605

Beitz AJ (1982) The organization of afferent projections to the midbrain periaqueductal gray of the rat. Neuroscience 7:133–159

Belford GR, Killackey HP (1978) Anatomical correlates of the forelimb in the ventrobasal complex and the cuneate nucleus of the neonatal rat. Brain Res 158:450–455

Belford GR, Killackey HP (1979) Vibrissae representation in subcortical trigeminal centers of the neonatal rat. J Comp Neurol 183:305–322

Ben-Ari Y, le Gal la Salle G, Champagnat JC (1974) Lateral amygdala unit activity. EEG Clin Neurophysiol 37:449–461

Benevento LA, Ebner FF (1970) Pretectal, tectal, retinal and cortical projections to thalamic nuclei of the opossum in stereotaxic coordinates. Brain Res 18:171–175

Benevento LA, Standage GP (1983) The organization of projections of the retinorecipient and nonretinorecipient nuclei of the pretectal complex and layers of the superior colliculus to the lateral pulvinar and medial pulvinar in the macaque monkey. J Comp Neurol 217:307–336

Benevento LA, Rezak M, Santos-Anderson R (1977) An autoradiographic study of the projections of the pretectum in the rhesus monkey (Macaca mulatta): evidence for sensorimotor links to the thalamus and oculomotor nuclei. Brain Res 127:197–218

Beninato M, Spencer RF (1986) A cholinergic projection to the rat superior colliculus demonstrated by retrograde transport of horseradish peroxidase and choline acetyltransferase immunohistochemistry. J Comp Neurol 253:525–538

Bennett GJ, Seltzer Z, Lu GW, Nishikawa N, Dubner R (1983) The cells of origin of the dorsal column postsynaptic projection in the lumbosacral enlargements of cats and monkeys. Somatosens Res 1:131–149

Benson CG, Potashner SJ (1990) Retrograde transport of [3H]glycine from the cochlear nucleus to the superior olive in the guinea pig. J Comp Neurol 296:415–426

Bentivoglio M, Kuypers HGJM (1982) Divergent axon collaterals from rat cerebellar nuclei to diencephalon, mesencephalon, medulla oblongata and cervical cord. Exp Brain Res 46:339–356

Beran RL, Martin GF (1971) Reticulospinal fibers of the opossum. J Comp Neurol 141:453–466

Berbel PJ (1988) Cytology of medial and dorso-medial cerebral cortices in lizards: a Golgi study. In: Schwerdtfeger WK, Smeets WJAJ (eds) The forebrain of reptiles. Karger, Basel, pp 12–19

Berbel PJ, Martinez-Guijarro FJ, Lopez-Garcia C (1987). Intrinsic organization of the medial cerebral cortex of the lizard Lacerta pityusensis: a Golgi study. J Morphol 194:275–286

Berendse HW, Groenewegen HJ (1990) Organization of the thalamostriatal projections in the rat, with special emphasis on the ventral striatum. J Comp Neurol 299:187–228

Berendse HW, Groenewegen HJ (1991) Restricted cortical termination fields of the midline and intralaminar thalamic nuclei in the rat. Neuroscience 42:73–102

Berendse HW, Groenewegen HJ, Lohman AHM (1992a) Compartmental distribution of ventral striatal neurons projecting to the mesencephalon in the rat. J Neurosci 12:2079–2103

Berendse HW, Galis-de Graaf Y, Groenewegen HJ (1992b) Topographical organization relationship with ventral striatal compartments of prefrontal corticostriatal projections in the rat. J Comp Neurol 316:314–347

Berger AJ (1979) Distribution of carotid sinus nerve afferent fibers to solitary tract nuclei of the cat using transganglionic transport of horseradish peroxidase. Neurosci Lett 14:153–158

Berk ML, Finkelstein JA (1981) Afferent projections to the preoptic area and hypothalamic regions in the rat brain. Neuroscience 6:1601–1624

Berkley KJ (1975) Different targets of different neurons in nucleus gracilis of the cat. J Comp Neurol 163:285–303

Berkley KJ (1980) Spatial relationships between the terminations of somatic sensory and motor pathways in the rostral brain stem of cats and monkeys. I. Ascending somatic sensory inputs to the lateral diencephalon. J Comp Neurol 193:283–317

Berkley KJ (1983) Spatial relationships between the terminations of somatic sensory motor pathways in the rostral brainstem of cats and monkeys. II. Cerebellar projections compared with those of the ascending somatic sensory pathways in lateral diencephalon. J Comp Neurol 220:229–251

Berkley KJ (1986) Specific somatic sensory relays in the mammalian diencephalon. Rev Neurol 142:283–290

Berkley KJ, Hand PJ (1978a) Efferent projections of the gracile nucleus in the cat. Brain Res 153:263–283

Berkley KJ, Hand PJ (1978b) Projections to inferior olive of the cat. II. Comparisons of input from the gracile, cuneate and the spinal trigeminal nuclei. J Comp Neurol 180:253–264

Berkley KJ, Mash DC (1978) Somatic sensory projections to the pretectum in the cat. Brain Res 158:445–449

Berkley KJ, Blomqvist A, Pelt A, Flink R (1980) Differences in the collateralization of neuronal projections from the dorsal column nuclei and lateral cervical nucleus to the thalamus and tectum in the cat: an anatomical study using two different double-labeling techniques. Brain Res 202:273–291

Berman N (1977) Connections of the pretectum in the cat. J Comp Neurol 174:227–254

Bernard J-F, Alden M, Besson J-M (1993) The organization of the efferent projections from the pontine parabrachial area to the amygdaloid complex: a Phaseolus vulgaris leucoagglutinin (PHA-L) study in the rat. J Comp Neurol 329:201–229

Bevan MD, Smith AD, Bolam P (1996) The substantia nigra as a site of synaptic integration of functionally diverse information arising from the ventral pallidum and the globus pallidus in the rat. Neuroscience 75:5–12

Bhatnagar HP, Frahm HD, Stephan H (1990) The megachiropteran pineal organ: a comparative mophological and volumetric investigation with special emphasis on the remarkably large pineal of Dobsonia praedatrix. J Anat 168:143–166

Bickford ME, Hall WC (1989) Collateral projections of predorsal bundle cells of the superior colliculus in the rat. J Comp Neurol 283:86–106

Bickford ME, Hall WC (1992) The nigral projection to predorsal bundle cells in the superior colliculus of the rat. J Comp Neurol 319:11–33

Biedenbach MA, De Vito JL (1980) Origin of the pyramidal tract determined with horseradish peroxidase. Brain Res 193:1–17

Biedenbach MA, De Vito JL, Brown AC (1986) Pyramidal tract of the cat: axon size and morphology. Exp Brain Res 61:303–310

Bieger D, Hopkins DA (1987) Viscerotopic representation of the upper alimentary tract in the medulla oblongata in the rat: The nucleus ambiguus. J Comp Neurol 262:546–562

Bigar F (1980) De efferente verbindingen van de cerebellaire schors in de kat. Een onderzoek naar de corticonucleaire en corticovestibulaire projekties door middel van retrograad axonaal transport van mierikswortel peroxidase (HRP). Thesis

Bishop GA, Ho RH (1985) The distribution and origin of serotonin immunoreactivity in the rat cerebellum. Brain Res 331:195–207

Bishop GA, McCrea RA, Kitai ST (1976) A horseradish peroxidase study of the cortico-olivary projection in the cat. Brain Res 116:306–311

Bishop GH (1959) The relation between nerve fiber size and sensory modality: phylogenetic implications of the afferent innervation of cortex. J Nerv Mental Dis 128:89–114

Bjaalie JG, Brodal P (1983) Distribution in area 17 of neurons projecting to the pontine nuclei: a quantitative study in the cat with retrograde transport of HRP-WGA. J Comp Neurol 221:289–303

Björkeland M (1983) Projections from dorsal column nuclei and spinal cord to pontine nuclei in cat. Neurosci Lett [Suppl] 14:S30

Björkeland M, Boivie J (1984a) An anatomical study of the projections from the dorsal column nuclei to the midbrain in cat. Anat Embryol (Berl) 170:29–43

Björkeland M, Boivie J (1984b) The termination of spinomesencephalic fibers in the cat. An experimental anatomical study. Anat Embryol (Berl) 170:265–277

Björklund A, Skagerberg G (1979) Evidence for a major spinal cord projection from the diencephalic A11 dopamine cell group in the rat using transmitter-specific fluorescent retrograde tracing. Brain Res 177:170–175

Björklund A, Moore RY, Nobin A, Stenevi U (1973) The organization of tubero-hypophyseal and reticulo-infundibular catecholamine neuron systems in the rat brain. Brain Res 51:171–191

Blackstad TW (1967) Cortical gray matter – a correlation of light and electron microscopic data. In: Hydén H (ed) The NEURON. Elsevier, Amsterdam, pp 49–118

Blackstad TW, Kjaerheim A (1961) Special axo-dendritic synapses in the hippocampal cortex. Electron and light microscopic studies on the layer of mossy fibers. J Comp Neurol 117:133–159

Blackstad T, Brodal A, Walberg F (1951) Some observations on normal and degeneration terminal boutons in the inferior olive of the cat. Acta Anat (Basel) 11:461–477

Blanchard CD, Blanchard RJ (1972) Innate and conditioned reactions to threat in rats with amygdaloid lesions. J Comp Physiol Psychol 81:281–290

Blanks RH, Giolli RA, Pham SV (1982) Projections of the medial terminal nucleus of the accessory optic system upon pretectal nuclei in the pigmented rat. Exp Brain Res 48:228–237

Blanks RHI, Palay S (1978) The location and form of efferent vestibular neurons in rat. Anat Rec 190:34

Blanks RHI, Clarke RJ, Lui F, Giolli RA, Van Pham S, Torigoe Y (1995) Projections of the lateral terminal accessory optic nucleus of the common marmoset (Callithrix jacchus). J Comp Neurol 354:511–532

Blaustein JD (1992) Cytoplasmic estrogen receptors in rat brain: immunocytochemical evidence using three antibodies with distinct epitopes. Endocrinology 131:1336–1342

Bledsoe SCJ, Snead CR, Helfert RH, Prasad V, Wenthold RJ, Altschuler RA (1990) Immunocytochemical and lesion studies support the hypothesis that the projection from the medial nucleus of the trapezoid body to the lateral superior olive is glycinergic. Brain Res 517:189–194

Blomqvist A (1980) Gracilo-diencephalic relay cells: a quantitative study in the cat using retrograde transport of horseradish peroxidase. J Comp Neurol 193:1087–1125

Bobillier R, Seguin S, Petitjean F, Salvert D, Touret M, Jouvet M (1976) The raphe nuclei of the cat brain stem: a topographical atlas of their efferent projections as revealed by autoradiography. Brain Res 113:449–486

Bobillier P, Seguin S, Degueurce A, Lewis BD, Pujol J-F (1979) The efferent connections of the nucleus raphe centralis superior in the rat as revealed by autoradiography. Brain Res 166:1–8

Bodian D (1939) Studies on the diencephalon of the Virginia opossum. Part I. The nuclear pattern in the adult. J Comp Neurol 71:259–324

Boeijinga PH, van Groen T (1984) Inputs from the olfactory bulb and olfactory cortex to the entorhinal cortex in the cat. Exp Brain Res 57:40–48

Boesten AJP, Voogd J (1975) Projections of the dorsal column nuclei and the spinal cord on the inferior olive in the cat. J Comp Neurol 161:215–238

Boivie J (1971) The termination in the thalamus and the zona incerta of fibres from the dorsal column nuclei (DCN) in the cat. An experimental study with silver impregnation methods. Brain Res 28:459–490

Boivie J (1978) Anatomical observations on the dorsal column nuclei, their thalamic projection and the cytoarchitecture of some somatosensory thalamic nuclei in the monkey. J Comp Neurol 178:17–48

Boivie J (1979) An anatomical reinvestigation of the termination of the spinothalamic tract in the monkey. J Comp Neurol 186:343–370

Boivie J (1980) Thalamic projections from lateral cervical nucleus in monkey. A degeneration study. Brain Res 198:13–26

Boivie J (1988) Projections from the dorsal column nuclei and the spinal cord to the red nucleus in cat. Behav Brain Res 28:75–79

Boivie J, Grant G, Albe-Fessard D, Levante A (1975) Evidence for a projection to the thalamus from the external cuneate nucleus in the monkey. Neurosci Lett 1:3–8

Bojsen-Moller F (1975) Demonstration of terminalis, olfactory, trigeminal and perivascular nerves in the rat nasal septum. J Comp Neurol 15:245–256

Bok ST (1928) Der Einfluss in den Furchen und Windungen auftretenden Krummungen der Grosshirnrinde auf die Rindenarchitektur. Z Ges Neurol Psychiat 121:682–750

Bolam JP, Izzo PN, Graybiel AM (1988) Cellular substrate of the histochemically defined striosome/matrix system of the caudate nucleus: a combined Golgi and immunocytochemical study in cat and ferret. Neuroscience 24:853–875

Bok ST (1928) Das Rückenmark. In: von Mollendorf (ed) Handbuch der mikroskopischen Anatomie des Menschen, vol 4. Springer, Berlin

Bolk L (1906) Das Cerebellum der Saugetiere. Fisher, Haarlem

Bolk L, Göppert E, Kallius E, Lubosch E (1938) Handbuch der vergleichenden Anatomie den Wirbeltiere. Urban and Schwarzenberg, Berlin

Bowman JP, Sladek JRF (1973) Morphology of the inferior olivary complex of the rhesus monkey (Macaca mulatta). J Comp Neurol 152:299–316

Botchkina GI, Morin LP (1995) Specialized neuronal and glial contributions to development of the hamster lateral geniculate complex and circadian visual system. J Neurosci 15 (1/1):190–201

Bouyer JJ, Park DH, Joh TH, Pickel VM (1984) Chemical and structural analysis of the relation between cortical inputs and tyrosine hydroxylase-containing terminals in rat neostriatum. Brain Res 302:267–275

Braak E, Braak H (1993) The new monodendritic neuronal type within the adult human cerebellar granule cell layer shows calretinin-immunoreactivity. Neurosci Lett 154:199–202

Braak H (1976) On the striate area of the human isocortex. A Golgi- and pigmentarchitectonic study. J Comp Neurol 166:341–364

Braak H (1980) Architectonics of the human telecephalic cortex. Springer, Berlin Heidelberg New York (Studies of brain function, vol 4)

Braak H, Braak E (1983) Neuronal types in the basolateral amygdaloid nuclei of man. Brain Res Bull 11:349–365

Bradford R, Parnavelas JG, Lieberman AR (1977) Neurons in layer I of the developing occipital cortex of the rat. J Comp Neurol 176:121–132

Bradley O (1903) On the development and homology of the mammalian cerebellar fissures. J Anat Physiol 37:112–130

Bradley O (1904) The mammalian cerebellum: its lobes and fissures. J Anat Physiol 38:448–475

Bradshaw J, Rogers L (1993) The evolution of lateral asymmetries, language, tool use and intellect. Academic, New York

Braitenberg V (1974) Thoughts on the cerebral cortex. J Theor Biol 46:421–447

Braitenberg V, Atwood RP (1958) Morphological observations on the cerebellar cortex. J Comp Neurol 109:1–33

Brauer K, Schober W (1970) Katalog der Säugetieregehirne. Fischer, Jena

Brauer K, Schober W (1976) Katalog der Säugetiergehirne. VEB Gustav Fischer, Jena. Supplement 2

Brauer K, Schober W (1982) Identification of geniculo-tectal relay neurons in the rat's ventral lateral geniculate nucleus. Exp Brain Res 45:84–88

Brawer JR, Morest DK, Kane EC (1974) The neuronal architecture of the cochlear nucleus of the cat. J Comp Neurol 155:251–300

Breathnach AS (1953) The olfactory tubercle, prepyriform cortex, and precommissural region of the porpoise (Phocaena phocaena). J Anat (Lond) 87:96–113

Breathnach AS (1960) The cetacean central nervous system. Biol Rev Cambr Philos Soc 35:187–230

Brining SK, Smith DV (1996) Distribution and synaptology of glossopharyngeal afferent nerve terminals in the nucleus of the solitary tract of the hamster. J Comp Neurol 365:556–574

Broadwell RD (1975) Olfactory relationships of the telencephalon and diencephalon in the rabbit. I. An autoradiographic study of the efferent connections of the main and accessory olfactory bulb. J Comp Neurol 163:329–346

Broadwell RD, Bleier R (1976) A cytoarchitectonic atlas of the mouse hypothalamus. J Comp Neurol 167:315–340

Broca P (1878) Anatomie comparJe des circonvolutions cJrJbrales: le grand lobe limbique et la scissure limbique dans la serie des mammif. Pres Rev Anthropol 1:385–498

Brockhaus H (1938) Zur normalen und pathologischen Anatomie des Mandelkerngebietes. J Psychol Neurol 49:1–136

Brodal A (1940) Experimentelle Untersuchungen über die olivo-cerebellare Lokalisation. Z Ges Neurol Psychiat 169:1–153

Brodal A (1957) The reticular formation of the brain stem. Anatomical aspects and functional correlations. The Henderson trust lectures, vol XVIII. Oliver and Boyd, Edinburgh

Brodal A (1974) Anatomy of the vestibular nuclei and their connections. In: Kornhuber HH (ed) Handbook of sensory physiology, vol 6. Springer, Berlin Heidelberg New York, pp 239–352

Brodal A (1984) The vestibular nuclei in the macaque monkey. J Comp Neurol 227:252–266

Brodal A, Gogstad AC (1954) Rubro-cerebellar connection. An experimental study in the cat. Anat Rec 118:455–486

Brodal A, Pompeiano O (1957) The vestibular nuclei in the cat. J Anat 91:438–454

Brodal A, Szabo T, Torvik A (1956) Corticofugal fibers to sensory trigeminal nuclei and nucleus of solitary tract. J Comp Neurol 106:527–552

Brodal P (1968a) The corticopontine projection in the cat. I. Exp Brain Res 5:210–234

Brodal P (1968b) The corticopontine projections in the cat. Arch Ital Biol 106:310–332

Brodal P (1971) The corticopontine projection in the cat. II. Projection from the orbital gyrus. J Comp Neurol 142:141–152

Brodal P (1972) The corticopontine projection from the visual cortex in the cat. II. The projection from areas 18 and 19. Brain Res 39:319–335

Brodal P (1978) The corticopontine projection in the rhesus monkey. Origin and principles of organization. Brain 101:251–283

Brodal P (1982) Further observations on the cerebellar projections from the pontine nuclei and the nucleus reticularis tegmenti pontis in the rhesus monkey. J Comp Neurol 204:44–55

Brodal P (1987) Organization of cerebropontocerebellar connections as studied with anterograde and retrograde transport of HRP-WGA in the cat. In: Liss AR (ed) New concepts in cerebellar neurobiology. Liss, New York, pp 151–182

Brodal P, Brodal A (1981) The olivocerebellar projection in the monkey. Experimental studies with the method of retrograde tracing of horseradish peroxidase. J Comp Neurol 201:375–393

Brodal P, Dietrichs E, Walberg F (1986) Do pontocerebellar mossy fibres give off collaterals to the cerebellar nuclei? An experimental study in the cat with implantation of crystalline HRP-WGA. Neurosci Res 4:12–24

Brodmann K (1909) Vergleichende Lokalisationslehre der Groszhirnrinde. Barth, Leipzig

Broere G (1971) Corticofugal fibers in some mammals. An experimental study in which special attention is paid to the cortico-spinal system. Proefschrift. De Kempenaer, Oegstgeest

Brookover C (1914) The nervus terminalis in adult man. J Comp Neurol 24:131–135

Brown AG, Fyffe REW (1981) Form and function of dorsal horn neurones with axons ascending the dorsal columns in cat. J Physiol (Lond) 321:31–47

Brown AG (1981) Organization in the spinal cord. The anatomy and physiology of identified neurones. Springer, Berlin Heidelberg New York

Brown JT, Chan-Palay V, Palay SL (1977) A study of afferent input to the inferior olivary complex in the rat by retrograde axonal transport of horseradish peroxidase. J Comp Neurol 176:1–22

Brown JW (1987) The nervus terminalis in insectivorous bat embryos and notes on its presence during human ontogeny. Ann NY Acad Sci 519:184–200

Brown LT (1974) Corticorubral projections in the rat. J Comp Neurol 154:149–167

Brown PB, Fuchs JL, Tapper DN (1975) Parametric studies of dorsal horn neurons responding to tactile stimulation. J Neurophysiol 38:19–25

Brown TH. Zador AM (1990) Hippocampus. In: Shepherd GM (ed) The synaptic organization of the brain, 3rd edn. Oxford University Press, New York-Oxford, pp 346–388

Brownell WE (1982) Cochlear transduction: an integrative model and review. Hear Res 6:335–360

Brugge JF, Reale RA (1985) Auditory cortex. In: Peters A, Jones EG (eds) Association and auditory cortices. Plenum, New York, pp 229–271 (Cerebral cortex, vol 4)

Brunner H (1919) Die zentralen Kleinhirnkerne bei den Säugetieren. Arb Neur Inst Wien Univ 22:200–277

Bruns V, Schmiessek E (1980) Cochlear innervation in the greater horseshoe bat. Demonstration of an acoustic fovea. Hear Res 3:27–43

Bryan RN, Trevino DL, Coulter JD, Willis WD (1973) Location and somatotopic organization of the cells of origin of the spino-cervical tract. Exp Brain Res 17:177–189

Bryan RN, Coulter JD, Willis WD (1974) Cells of origin of the spinocervical tract in the monkey. Exp Neurol 42:574–586

Buchanan SL, Thompson RH, Maxwell BL, Powell DA (1994) Efferent connections of the medial prefrontal cortex in the rabbit. Exp Brain Res 100:469–483

Bucher VM, Bürgi SM (1950) Some observations on the fiber connections of the di- and mesencephalon in the rat. I. Fiber connections of the tectum opticum. J Comp Neurol 93:139–172

Bugbee NM, Goldman-Rakic PS (1983) Columnar organization of corticocortical projections in squirrel and rhesus monkeys: similarity of column width in species differing in cortical volume. J Comp Neurol 220:355–364

Buhl EH, Schwerdtfeger WK, Germroth P (1989) New anatomical approaches to reveal afferent and efferent hippocampal circuitry. In: Chan-Palay V, Köhler C (eds) The hippocampus – new vistas. Liss, New York, pp 71–83 (Neurology and neurobiology, vol 52)

Buisseret-Delmas C (1988) Sagittal organization of the olivocerebellonuclear pathway in the rat. I. Connections with the nucleus fastigii and the nucleus vestibularis lateralis. Neurosci Res 5:475–493

Buisseret-Delmas C, Angaut P (1993) The cerebellar olivocorticonuclear connections in the rat. Progr Neurobiol 40:63–87

Bull MS, Berkley KJ (1984) Differences in the neurons that project from the dorsal column nuclei to the diencephalon, pretectum, and tectum in the cat. Somatosens Res 1:281–300

Bull MS, Mitchell SK, Berkley KJ (1990) Convergent inputs to the inferior olive from the dorsal column nuclei and pretectum in the cat. Brain Res 525:1–10

Buma P, Nieuwenhuys R (1987) Ultrastructural demonstration of oxytocin and vasopressin release sites in the neural lobe and median eminence of the rat by tannic acid and immunogold methods. Neurosci Lett 74:151–157

Buma P, Veening J, Hafmans T, Joosten H, Nieuwenhuys R (1992) Ultrastructure of the periaqueductal grey matter of the rat: an electron microscopical and horseradish peroxidase study. J Comp Neurol 319:519–535

Burde RM, Loewy AD (1980) Central origin of oculomotor parasympathetic neurons in the monkey. Brain Res 198:434–439

Burde RM, Parelman JJ, Luskin M (1982) Lack of unity of Edinger-Westphal nucleus projections to the ciliary ganglion and spinal cord: a double-labeling approach. Brain Res 249:379–382

Burgi S (1957) Das tectum opticum. Seine Verbindungen bei der Katze und seine Bedeutung beim Menschen. Dtsch Z Nervenheilkd 176S:701-729

Burian M, Gstoettner W, Zundritsch R (1989) Saccular afferent fibers to the cochlear nucleus in the guinea pig. Arch Otorhinolaryngol 246:238-241

Burian M, Zundritsch R, Mayr R (1991) The origin of the vestibulo-cochlear projection in the guinea pig. Neurosci Lett 122:163-166

Burkitt AN (1938) The external morphology of the brain of Notoryctes typhlops. R Acad Sci Amsterdam, Proc Sect Sci 41:921-933

Burne RA, Mihailoff GA, Woodward DV (1978) Visual corticopontine input to the paraflocculus: a combined autoradiographic and horseradish peroxidase study. Brain Res 143:139-146

Burne RA, Azizi SA, Mihailoff GA, Woodward DJ (1981) The tectopontine projection in the rat with comments on visual pathways to the basilar pons. J Comp Neurol 202:287-307

Burstein H, Dado RJ, Giesler GJ (1990) The cells of origin of the spinothalamic tract of the rat: a quantitative reexamination. Brain Res 511:329-337

Burstein R (1996) Somatosensory and visceral input to the hypothalamus and limbic system. Prog Brain Res 107:257-267

Burstein R, Cliffer KD, Giesler GJ Jr (1987) Direct somatosensory projections from the spinal cord to the hypothalamus and telencephalon. J Neurosci 7:4159-4164

Burstein R, Falkowsky O, Borsook D, Strassman A (1996) Distinct lateral and medial projections of the spinohypothalamic tract of the rat. J Comp Neurol 373:549-574

Burton H, Craig ADJ (1979) Distribution of trigeminothalamic projection cells in cat and monkey. Brain Res 161:515-521

Burton H, Craig AD (1983) Spinothalamic projection in cat, raccoon and monkey. A study based on anterograde transport of horseradish peroxidase. In: Macchi G, Rustioni A, Spreafico R (eds) Somatosensory integration in the Thalamus. Elsevier, Amsterdam, pp 17-42

Burton H, Craig AD, Poulos DA, Molt JT (1979) Efferent projections from temperature sensitive recording loci within the marginal zone of the nucleus caudalis of the spinal trigeminal zone of the nucleus caudalis of the spinal complex in the cat. J Comp Neurol 183:753-778

Busch HFM (1957) Les connexions entre la moelle epiniere et le thalamus chez le chat. Psychiatr Neurol 80:305-307

Busch HFM (1961) White matter in the brain stem of the cat. Thesis, University of Leiden

Butcher LL (1995) Cholinergic neurons and networks. In: Paxinos G (ed) The rat nervous system, 2nd edn. Academic, San Diego, pp 1003-1015

Butcher LL, Semba K (1989) Reassessing the cholinergic basal forebrain: nomenclature schemata and concepts. Trends Neurosci 12:483-485

Butler AB (1994) The evolution of the dorsal thalamus of jawed vertebrates, including mammals – cladistic analysis and a new hypothesis. Brain Res Rev 19:29-65

Büttner-Ennever JA (1977) Pathways from the pontine reticular formation to structures controlling horizontal and vertical eye movements in the monkey. In: Baker R, Berthoz A (eds) Control of gaze by brainstem neurons. Elsevier/North Holland, New York, pp 89-98

Büttner-Ennever JA (1979) Organization of reticular projections onto oculomotor motoneurons. In: Granit R, Pompeiano O (eds) Reflex control of posture and movement. Prog Brain Res 50:619-630

Büttner-Ennever JA, Büttner U (1978) A cell group associated with vertical eye movements in the rostral mesencephalic reticular formation of the monkey. Brain Res 151:31-47

Büttner-Ennever JA, Akert K (1981) Medial rectus subgroups of the oculomotor nucleus and their abducens internuclear input in the monkey. J Comp Neurol 197:17-27

Büttner-Ennever JA, Horn AKE (1996) Pathways from cell groups of the paramedian tracts to the floccular region. Ann NY Acad Sci 1:532-540

Buxton DF, Goodman DC (1967) Motor function and the corticospinal tracts in rat and raccoon. J Comp Neurol 129:341-360

Cabana T, Martin GF (1986) The adult organization and development of the rubrospinal tract. An experimental study using the orthograde transport of WGA-HRP in the North-American Opossum. Dev Brain Res 30:1-11

Cabral RJ, Johnson JI (1971) The organization of mechanoreceptive projections in the ventrobasal thalamus of the sheep. J Comp Neurol 141:17-36

Cabrera B, Portillo F, Pasaro R, Delgad-Garcia JM (1988) Location of motoneurons and internuclear neurons within the rat abducens nucleus by means of horseradish peroxidase and fluorescent double labeling. Neurosci Lett 87:1-6

Cadusceau J, Roger M (1985) Afferent projections to the superior colliculus in the rat, with special attention to the deep layers. J Hirnforsch 26:667-681

Cadusceau J, Roger M (1991) Cortical and subcortical connections of the pars compacta of the anterior pretectal nucleus in the rat. Neurosci Res 12:83-100

Caffé AR, Hawkins RK, De Zeeuw CI (1996) Coexistence of choline acetyltransferase and GABA in axon terminals in the dorsal cap of the rat inferior olive. Brain Res 724:136-140

Caicedo A, Herbert H (1993) Topography of descending projections from the inferior colliculus to auditory brainstem nuclei in the rat. J Comp Neurol 328:377-392

Cajal P (1917) Nuevo estudio del encéfalo de los reptiles. Trab Lab Invest Biol Univ Madrid 15:83-99

Cajal P (1922) El cerebro de los batracios. In: Libro en honor de D. Santiago Ramon y Cajal, Tomo 1. Madrid, pp 13-59

Cajal Ramon y S (1909) Histologie du système nerveux de l'homme et des vertebres. vol. 1. Maloine, Paris

Cajal Ramon y S (1911) Histologie du système nerveux. Maloine, Paris, 39. Reprint 1972: Consejo Sup Invest Cient Inst Ramon y Cajal

Cajal SR (1893) Neue Darstellung vom histologischen Bau des Centralnervensystems. Arch Anat Physiol Anat Abt (Lpz):319-428

Cajal SR (1922) Studien über die Sehrinde der Katze. J Psychol Neurol (Lpz) 29:161-181

Calford MB, Aitken LM (1983) Ascending projections to the medial geniculate body of the cat: evidence for multiple, parallel auditory pathways through thalamus. J Neurosci 3:2365-2380

Calleja C (1893) La Region Olfatoria del Cerebro. N Moya, Madrid

Campbell B, Ryzen M (1953) The nuclear anatomy of the diencephalon of Sorex cinereus. J Comp Neurol 99:1-22

Campbell CBG, Hayhow WR (1971) Primary optic pathways in the echidna, Tachyglossus aculeatus: an experimental degeneration study. J Comp Neurol 143:119-136

Campbell CBG, Hayhow WR (1972) Primary optic pathways in the duckbill platypus, Ornithorhynchus anatinus: an experimental degeneration study. J Comp Neurol 145:195-208

Campbell G, Lieberman AR (1985) The olivary pretectal nucleus: experimental anatomical studies in the rat. Phil T Roy Soc Lond B 310:573-609

Canedo A, Towe AL (1986) Pattern of pyramidal tract collateralization to medial thalamus, lateral hypothalamus and red nucleus in the cat. Exp Brain Res 61:585-596

Canteras NS, Swanson LW (1992) Projections of the ventral subiculum to the amygdala, septum, and hypothalamus: a PHAL anterograde tract-tracging study in the rat. J Comp Neurol 324:180-194

Canteras NS, Shammah-Lagnado SJ, Silva BA, Ricardo JA (1990) Afferent connections of the subthalamic nucleus: a combined retrograde and anterograde horseradish peroxidase study in the rat. Brain Res 513:43-59

Canteras NS, Simerly RB, Swanson LW (1994) Organization of projections from the ventromedial nucleus of the hypothalamus: a Phaseolus vulgaris-leucoagglutinin study in the rat. J Comp Neurol 348:41-79

Canteras NS, Simerly RB, Swanson LW (1995) Organization of projections from the medial nucleus of the amydala: a PHAL study in the rat. J Comp Neurol 360:213-245

Carleton S, Carpenter MB (1984) Distribution of primary vestibular fibers in the brainstem and cerebellum of the monkey. Brain Res 294:281-298

Carleton SC, Carpenter MB (1983) Afferent and efferent connections of the medial, inferior and lateral vestibular nuclei in cat and monkey. Brain Res 278:29-61

Carlsen J, Zaborsky L, Heimer L (1985) Cholinergic projection form the basal forebrain to the basolateral amygdaloid complex: a combined retrograde fluorescent and immunohistochemical study. J Comp Neurol 234:155-167

Carlson M, Welt C (1981) The somatic sensory cortex: SmI in prosimian primates. In: Woolsey CN (ed) Cortical sensory organization, vol 1. Humana, Clifton, pp 1-27

Carman JB, Cowan WM, Powell TPS (1963) The organization of the cortico-striate connexions in the rabbit. Brain 86:525-562

Carman JB, Cowan WM, Powell TPS, Webster KE (1965) A bilateral corticostriate projection. J Neurol Neurosurg Psychiat 28:71-77

Carmichael ST, Clugnet M-C, Price JL (1994) Central olfactory connections in the macaque monkey. J Comp Neurol 346:403-434

Carpenter MB, Cowie RJ (1985) Transneuronal transport in the vestibular and auditory systems of the squirrel monkey and the arctic ground squirrel. I. Vestibular system. Brain Res 358:249-263

Carpenter MB, Peter P (1972) Nigrostriatal and nigrothalamic fibers in the rhesus monkey. J Comp Neurol 144:93-117

Carpenter MB, Stein BM, Shriver JE (1968) Central projections of spinal dorsal roots in the monkey. II. Lower thoracic, lumbosacral and coccygeal dorsal roots. Am J Anat 123:75-118

Carpenter MB, Stein BM, Peter P (1972) Primary vestibulocerebellar fibers in the monkey: distribution of fibers arising from distinctive cell groups of the vestibular ganglia. Am J Anat 135:221-249

Carpenter MB, Nakano K, Kim R (1976) Nigrothalamic projections in the monkey demonstrated by autoradiographic technics. J Comp Neurol 165:401-416

Carpenter MB, Batton RR, Carleton SC, Keller JT (1981a) Interconnections and organization of pallidal and subthalamic nucleus neurons in the monkey. J Comp Neurol 197:579-603

Carpenter MB, Carleton SC, Keller JT, Conte P (1981b) Connections of the sub-thalamic nucleus in the monkey. Brain Res 224:1-29

Carpenter MB, Chang L, Pereira AB, Hersh LB, Bruce G, Wu JY (1987) Vestibular and cochlear efferent neurons in the monkey identified by immunocytochemical methods. Brain Res 408:275-280

Carstens E, Trevino DL (1978a) Laminar origins of spinothalamic projections in the cat as determined by the retrograde transport of horseradish peroxidase. J Comp Neurol 182:161-165

Carstens E, Trevino DL (1978b) Anatomical and physiological properties of ipsilaterally projecting spinothalamic neurons in the second cervical segment of the cat's spinal cord. J Comp Neurol 182:167-184

Casagrande VA, Harting JK, Hall WC, Diamond IT (1972) Superior colliculus of the tree shrew: a structural and functional subdivision into superficial and deep layers. Science 177:444-447

Casseday JH, Covey E, Vater M (1988) Connections of the superior olivary complex in the rufous horseshoe bat Rhinolophus rouxi. J Comp Neurol 278:313-329

Casseday JH, Kobler JB, Isbey SF, Covey E (1989) Central acoustic tract in an echolocating bat: an extralemniscal auditory pathway to the thalamus. J Comp Neurol 287:247-259

Catsman-Berrevoets CE, Kuypers HGJM (1976) Cells of origin of cortical projections to dorsal column nuclei, spinal cord and bulbar medial reticular formation in the rhesus monkey. Neurosci Lett 3:245-252

Catsman-Berrevoets CE, Kuypers HGJM, Lemon RN (1979) Cells of origin of the frontal projections to magnocellular and parvocellular red nucleus and superior colliculus in cynomolgus monkey. An HRP study. Neurosci Lett 12:41-46

Catsman-Berrevoets CE, Kuypers HGJM (1981) A search for corticospinal collaterals to thalamus and mesencephalon by means of multiple retrograde fluorescent tracers in cat and rat, frontal projections to magnocellular and parvollular red nucleus and superior colliculus in cynomolgus monkey: an HRP study. Neurosci Lett 12:41-46

Caughell KA, Flumerfelt BA (1977) The organisation of the cerebellorubral projection: an experiment study in the rat. J Comp Neurol 176:295-306

Cazin L, Magnin M, Lannou J (1982) Non-cerebellar visual afferents to the vestibular nuclei involving the prepositus hypoglossal complex: an autoradiographic study in the rat. Exp Brain Res 48:309-313

Chan-Palay V (1977) Cerebellar dentate nucleus. Organization, cytology and transmitter. Springer, Berlin Heidelberg New York

Charara A, Parent A (1994) Brainstem dopaminergic, cholinergic and serotoninergic afferents to the pallidum in the squirrel monkey. Brain Res 640:155-170

Cheema SS, Rustioni A, Whitsel BL (1984) Light and electron microscopic evidence for a direct corticospinal projection to superficial laminae of the dorsal horn in cats and monkeys. J Comp Neurol 225:276-290

Chen S, Hillman DE (1993) Colocalization of neurotransmitters in the deep cerebellar nuclei. J Neurocytol 22:81-91

Chesselet MF, Graybiel AM (1986) Striatal neurons expressing somatostatin-like immuno-reactivity: evidence for a peptidergic interneuronal system in the cat. Neuroscience 17:547-571

Chevalier G, Thierry AM, Shibazaki T, FJger J (1981) Evidence for a GABAergic inhibitory nigrotectal pathway in the rat. Neurosci Lett 21:67-70

Chi JG, Dooling EC, Gilles FH (1977) Gyral development of the human brain. Ann Neurol 1:86-93

Childs JA, Gale K (1983) Neurochemical evidence for a nigrotegmental GABAergic projection. Brain Res 258:109-114

Cholley B, Wassef M, Arsenio-Nunes L, Brehier A, Sotelo C (1989) Proximal trajectory of the brachium conjunctivum in rat fetuses and their early association with the parabrachial nucleus. A study combining in vitro HRP anterograde axonal tracing and immunocytochemistry. Dev Brain Res 45:185-202

Christ JF (1969) Derivation and boundaries of the hypothalamus, with atlas of hypothalamic grisea. In: Haymaker W, Anderson E, Nauta WJH (eds) The hypothalamus. Thomas, Springfield, pp 13-60

Chronister RB, DeFrance J (1979) Organization of projection neurons of the hippocampus. Exp Neurol 66:509-523

Cipolloni PB, Peters A (1983) The termination of callosal fibres in the auditory cortex of the rat. A combined Golgi-electron microscope and degeneration study. J Neurocytol 12:713-726

Cireillo J, Calaresu FR (1980a) Monosynaptic pathway from cardiovascular neurons in the nucleus tractus solitarii to the paraventricular nucleus in the cat. Brain Res 193:529-533

Ciriello J, Calaresu FR (1980b) Autoradiographuc study of ascending projections from cardiovascular sites in the nucleus tractus solitarii in the cat. Brain Res 180:448-453

Ciriello J, Caverson MM (1984) Direct pathway from neurons in the ventrolateral medulla relaying cardiovascular afferent information to the supraoptic nucleus in the cat. Brain Res 292:221-228

Clarke J (1851) Researches into the structure of the spinal cord. Philos Trans R Soc Lond (Biol) 1:607-621

Clarke J (1859) Further researches on the gray substance of the spinal cord. Philos Trans R Soc Lond (Biol) 149:437-467

Cleland BG, Dubin MW, Levick WR (1971) Sustained and transient neurones in the cat's retina and lateral geniculate nucleus. J Physiol (Lond) 217:473–496

Clendenin M, Ekerot C-F, Oscarsson O, Rosen I (1974a) The lateral reticular nucleus in the cat. I. Mossy fibre distribution in cerebellar cortex. Exp Brain Res 21:473–486

Clendenin M, Ekerot C-F, Oscarsson O, Rosen I (1974b) The lateral reticular nucleus in the cat. II. Organization of component activated from bilateral ventral flexor reflex tract (bVFRT). Exp Brain Res 21:487–500

Clendenin M, Ekerot C-F, Oscarsson O, Rosen I (1974c) The lateral reticular nucleus in the cat. III. Organization of component activated from ipsilateral forelimb tract. Exp Brain Res 21:501–513

Clezy JKA, Dennis BJ, Kerr DIB (1961) A degeneration study of the somaesthetic afferent systems in the marsupial phalanger, Trichosurus vulpecula. Aust J Exp Biol 39:19–28

Cliffer KD, Giesler GJJ (1989) Postsynaptic dorsal column pathway of the rat. III. Distribution of ascending afferent fibers. J Neurosci 9:3146–3168

Cliffer KD, Willis WD (1994) Distribution of the postsynaptic dorsal column projection in the cuneate nucleus of monkeys. J Comp Neurol 345:84–93

Cliffer KD, Burstein R, Giesler GJJ (1991) Distributions of spinothalamic, spinohypothalamic, and spinotelencephalic fibers revealed by anterograde transport of PHA-L in rats. J Neurosci 11:852–868

Coggeshall RE (1973) Unmyelinated fibers in the ventral root. Brain Res 57:229–233

Coleman JR, Clerici WJ (1987) Sources of projections to subdivisions of the inferior colliculus in the rat. J Comp Neurol 262:215–226

Colonnier ML (1966) The structural design of the neocortex. In: Eccles JC (ed) Brain and conscious experience. Springer, Berlin Heidelberg New York, pp 1–23

Condé F, Condé H (1982) The rubro-olivary tract in the cat, as demonstrated with the method of retrograde transport of horseradish peroxidase. Neuroscience 7:715–724

Condé H (1966) Analyse electrophysiologique de la voie dentato-rubrothalamique chez le chat. J Physiol (Paris) 58:218–219

Conley M, Friederich-Ecsy B (1993) Functional organization of the ventral lateral geniculate complex of the tree shrew (Tupaia belangeri): II. connections with the cortex. J Comp Neurol 328:21–42

Conley M, Fitzpatrick D, Diamond IT (1984) The laminar organization of the lateral geniculate body and the striate cortex in the three shrew (Tupaiia glis). J Neurosci 4:171–197

Connolly CJ (1950) External morphology of the primate brain. Thomas, Springfield

Connor JR, Peters A (1984) Vasoactive intestinal polypeptide immunoreactive neurons in rat visual cortex. Neuroscience 4:1027–1044

Conrad LC, Pfaff DW (1976) Autoradiographic tracing of nucleus accumbens efferents in the rat. Brain Res 113:589–596

Contreras RJ, Beckstead RM, Norgren R (1982) The central projections of the trigeminal, facial, glossopharyngeal and vagus nerves: an autoradiographic study in the rat. J Auton Nerv Syst 6:303–322

Contreras RJ, Gomez MM, Norgren R (1980) Central origins of cranial nerve parasympathetic neurons in the rat. J Comp Neurol 190:373–394

Coolen LJMM (1995) The neural organization of sexual behavior in tha male rat; a functional neuroanatomical Fos-study. PhD thesis, University of Nijmegen, the Netherlands

Cooper HM (1986) The accessory optic system in a Prosimian primate (microcebus murinus): evidence for a direct retinal projection to the medial terminal nucleus. J Comp Neurol 249:28–47

Cooper HM, Magnin M (1987) Accessory optic system of an anthropoid primate, the gibbon (Hylobates concolor): evidence of a direct retinal input to the medial terminal nucleus. J Comp Neurol 259:467–482

Cooper HM, Baleydier C, Magnin M (1990) Macaque accessory optic system: I. Definition of the medial terminal nucleus. J Comp Neurol 302:394–404

Cooper HM, Herbin M, Nevo E (1993a) Ocular regression conceals adaptive progression of the visual system in a blind subterranean mammal. Nature 361:156–159

Cooper HM, Herbin M, Nevo E (1993b) Visual system of a naturally microphthalmic mammal: the blind mole rat, Spalax ehrenbergi. J Comp Neurol 328:313–350

Cooper LL, Dostrovsky JO (1985) Projection from dorsal column nuclei to dorsal mesencephalon. J Neurophysiol 53:183–200

Cooper S, Sherrington CS (1940) Gower's tract and spinal border cells. Brain 63:123–134

Coote JH, Hilton SM, Zbrozyna AW (1973) The pontomedullary area integrating the defence reaction in the cat and its influence on muscle blood flow. J Physiol (Lond) 229:257–274

Corvisier J, Hardy O (1993) Distribution of synaptic terminals from prepositus neurones on the collicular maps. Neuroreport 4:511–514

Cotter JR, Pierson Pentney RJ (1979) Retinofugal projections of nonecholocating (Pteropus giganteus) and Echolocating (Myotis lucifugus) bats. J Comp Neurol 184:381–400

Cottle MK (1964) Degeneration studies of primary afferents of IXth and Xth cranial nerves in the cat. J Comp Neurol 122:329–346

Coulter JD, Jones EG (1977) Differential distribution of corticospinal projections from individual cytoarchitectonic fields in the monkey. Brain Res 129:335–340

Courville J (1966a) Rubrobulbar fibres to the facial nucleus and the lateral reticular nucleus (nucleus of the lateral funiculus). An experimental study in the cat with silver impregnation methods. Brain Res 1:317–337

Courville J (1966b) Somatotopical organization of the projection from the nucleus interpositus anterior of the cerebellum to the red nucleus. An experimental study in the cat with silver impregnation methods. Exp Brain Res 2:191–215

Courville J, Brodal A (1966) Rubrocerebellar connections in the cat. An experimental study with silver impregnation methods. J Comp Neurol 126:471–486

Courville J, Otabe S (1974) The rubro-olivary projection in the macaque: an experimental study with silver impregnation methods. J Comp Neurol 158:479–491

Covey E, Casseday JH (1986) Connectional basis for frequency representation in the nuclei of the lateral lemniscus of the bat Eptesicus fuscus. J Neurosci 6:2926–2940

Covey E, Casseday JH (1991) The monaural nuclei of the lateral lemniscus in an echolocating bat: parallel pathways for analyzing temporal features of sound. J Neurosci 11:3456–3470

Cowie RJ, Holstege G (1993) Dorsal mesencephalic projections to pons, medulla, and spinal cord in the cat: limbic and non-limbic components. J Comp Neurol 319:536–559

Cowan RL, Wilson CJ (1994) Spontaneous firing patterns and axonal projections of single corticostriatal neurons in the rat medial agranular cortex. J Neurophysiol 71:17–32

Cowan RL, Wilson CJ, Emson PC, Heizmann CW (1990) Parvalbumin-containing GABAergic interneurons in the rat neostriatum. J Comp Neurol 302:197–205

Cozzi MG, Rosa P, Greco A et al (1989) Immunohistochemical localization of secretogranin II in the rat cerebellum. Neuroscience 28:423–441

Craig AD (1978) Spinal and medullary input to the lateral cervical nucleus. J Comp Neurol 181:729–743

Craig AD (1991) Spinal distribution of ascending lamina I axons anterograde labeled with Phaseolus vulgaris leucoagglutinin (PHA-L) in the cat. J Comp Neurol 313:377–393

Craig AD (1995) Distribution of brainstem projections from spinal lamina I neurons in the cat and the monkey. J Comp Neurol 361:225-248

Craig AD (1996) An ascending general homeostatic afferent pathway originating in lamina I. Prog Brain Res 107:225-242

Craig AD, Broman J, Blomqvist A (1992) Lamina I spinocervical tract terminations in the medial part of the lateral cervical nucleus in the cat. J Comp Neurol 322:99-110

Craig AD, Bushnell MC, Zhang E-T, Blomqvist A (1995) A thalamic nucleus specific for pain and temperature sensation. Nature 372:770-773

Craig ADJ, Burton H (1981) Spinal and medullary lamina I projection to nucleus submedius in medial thalamus: a possible pain center. J Neurophysiol 45:443-466

Craig ADJ, Linington AJ, Kniffki KD (1989) Cells of origin of spinothalamic tract projections to the medial and lateral thalamus in the cat. J Comp Neurol 289:568-585

Crick F (1982) Do dendritic spines twitch? Trends Neurosci 5:44-46

Crick F (1984) Function of the thalamic reticular complex: the searchlight hypothesis. Proc Natl Acad Sci USA 81:4586-4590

Crosby EC, Humphrey T (1939) Studies on the vertebrate telencephalon. I. The nuclear configuration of the olfactory and accessory olfactory formations and of the nucleus olfactorius anterior of certain reptiles, birds and mammals. J Comp Neurol 71:121-213

Crosby EC, Humphrey T (1941) Studies of the vertebrate telencephalon. II. The nuclear pattern of the anterior olfactory nucleus, tuberculum olfactorium and the amygdaloid complex in adult man. J Comp Neurol 74:309-352

Crosby EC, Humphrey T, Lauer EW (1962) Correlative anatomy of the nervous system. MacMillan, New York

Crosby EC, Humprey T (1941) Studies of the vertebrate telencephalon. II. The nuclear pattern of the anterior olfactory nucleus, tuberculum olfactorium and the amygdaloid complex in adult man. J Comp Neurol 74:309-352

Crosby EC, Humphrey T (1944) Studies of the vertebrate telencephalon. III. The amygdaloid complex in the shrew (Blarina brevicauda). J Comp Neurol 81:285-305

Crosby EC, Woodburne RT (1940) The comparative anatomy of the preoptic area and the hypothalamus. Res Publ Assoc Res Nerv Ment Dis 20:52-169

Crouch RL (1934) The nuclear configuration of the hypothalamus and subthalamus of Macacus rhesus. J Comp Neurol 59:431-449

Cruce JAF (1977) An autoradiographic study of the descending connections of the mammillary nuclei of the rat. J Comp Neurol 176:631-644

Crutcher KA, Humbertson AOJ (1978) The organization of monoamine neurons within the brainstem of the North American Opossum (Didelphis virginiana). J Comp Neurol 179:195-222

Crutcher KA, Humbertson AO, Martin GF (1978) The origin of brainstem-spinal pathways in the North American Opossum (Didelphis virginiana). Studies using the horseradish peroxidase method. J Comp Neurol 179:169-194

Cucchiaro JB, Bickford ME, Sherman SM (1991) A GABAergic method. J Comp Neurol 179:169-194

Culberson JL, Brushart TM (1989) Somatotopy of digital nerve projections to the cuneate nucleus in the monkey. Somatosens Mot Res 6:319-330

Cummings JF, Petras JM (1977) The origin of spinocerebellar pathways. I. The nucleus cervicalis centralis of the cranial cervical spinal cord. J Comp Neurol 173:655-692

Cunningham ETJ, Miselis RR, Sawchenko PE (1994) The relationship of efferent projections from the area postrema to vagal motor and brain stem catecholamine-containing cell groups: an axonal transport and immunohistochemical study in the rat. Neuroscience 58:635-648

Cynader M, Berman N (1972) Receptive-field organization of monkey superior colliculus. J Neurophysiol 35:187-201

Czeiger D, White EL (1993) Synapses of extrinsic and intrinsic origin made by callosal projection neurons in mouse visual cortex. J Comp Neurol 330:502-513

Dahlstrom A, Fuxe K (1964) Evidence for the existence of monoamine-containing neurons in the central nervous system. I. Demonstration of monoamines in the cell bodies of brain stem neurons. Acta Physiol Scand [Suppl] 62 (232):1-55

Dahlstrom A, Fuxe K (1965) Evidence for the existence of monoamine neurons in the central nervous system. II. Experimentally induced changes in the intraneuronal amine levels of bulbospinal neuron systems. Acta Physiol Scand Suppl 64 (247):1-36

Daniel H, Billard JM, Angaut P, Batini C (1987) The interposito-rubrospinal system. Anatomical tracing of a motor control pathway in the rat. Neurosci Res 5:87-112

Darian-Smith J (1973) The trigeminal system. In: Iggi A (ed), Handbook of sensory physiology. Somatosensory system. Springer, Berlin Heidelberg New York. pp 271-314

Dart AM, Gordon G (1973) Some properties of spinal connections of the cat's dorsal column nuclei which do not involve the dorsal columns. Brain Res 58:61-68

Davenport HA, Ranson SW (1930) The red nucleus and adjacent cell groups. A topographic study in the cat and in the rabbit. Arch Neurol Psychiat 24:257-266

Davis PJ, Nail BS (1984) On the location and size of laryngeal motoneurons in the cat and rabbit. J Comp Neurol 230:13-32

Davis BJ, Macrides F, Youngs WM, Schneider SP, Rosene DL (1978) Efferents and centrifugal afferents of the main and accessory olfactory bulbs in the hamster. Brain Res Bull 3:59-72

Davis M (1992a) The role of the amygdala in conditioned fear. In: Aggleton JP (ed) The amygdala: neurobiological aspects of emotion, memory, and mental dysfunction. Wiley-Liss, New York, pp 255-306

Davis M (1992b) The role of the amygdala in fear and anxiety. Annu Rev Neurosci 15:353-375

DeFelipe J, Fariñas J (1992) The pyramidal neuron of the cerebral cortex: morphological and chemical characteristics of the synaptic input. Prog Neurobiol 39:563-607

DeFelipe J, Conley M, Jones EG (1986) Long-range focal collateralization of axons arising from corticocortical cells in monkey sensory-motor cortex. J Neurosci 6:3749-3766

DeFelipe J, Hendry SHC, Jones EG (1989) Synapses of double bouquet cells in monkey cerebral cortex visualized by calbindin immunoreactivity. Brain Res 503:49-54

DeFelipe J, Hendry SHC, Hashikawa T, Molinari M, Jones EG (1990) A microcolumnar structure of monkey cerebral cortex revealed by immunocytochemical studies of double bouquet cell axons. Neuroscience 37:655-673

de Graaf AS (1967) Anatomical aspects of the cetacean brain stem. Thesis, University of Leiden, pp 1-169

Dekker JJ (1981) Anatomical evidence for direct fiber projections from the cerebellar nucleus interpositus to rubrospinal neurons. A quantitative EM study in the rat combining anterograde and retrograde intra-axonal tracing methods. Brain Res 205:229-244

DeLong MR, Georgopoulos AP (1981) Motor functions of the basal ganglia. In: Brookhart JM, Mountcastle VB, Brooks VB (eds) Handbook of physiology, sect 1: the nervous system, vol 2: motor control, part 2. American Physiological Society, Bethesda, pp 1017-1061

Demole V (1927A) Structure et connexion des noyeaux denteles du cervelet. I. Schweiz Arch Psychiat u Neurol 20:271-294

Demole V (1927b) Structure et connexion des noyeaux denteles du cervelet. II. Schweiz Arch Psychiat u Neurol 21:73-110

Demski LS (1993) terminal nerve complex. Acta Anat 148:81-95

Demski LS, Schwanzel-Fukuda M (1987) The terminal nerve (nervus terminalis): structure, function and evolution. Ann NY Acad Sci 519:469

Demski LS, Ridgway SH, Schwanzel-Fukuda M (1990) The terminal nerve of dolphins: gross structure, histology and luteinizing-hormone-releasing hormone immunocytochemistry. Brain Behav Evol 36:249-261

Deniau JM, Chevalier G (1992) The lamellar organization of the rat substantia nigra pars reticulata: distribution of projection neurons. Neuroscience 46:361-377

Deniau JM, Menetrey A, Thierry AM (1994) Indirect nucleus accumbens input to the pre-frontal cortex via the substantia nigra pars reticulata: a combined anatomical and electrophysiological study in the rat. Neuroscience 61:533-545

de Olmos JS (1972) The amygdaloid projection field in the rat as studied with the cupric-silver method. In: Eleftheriou BE (ed) The neurobiology of the amygdala. Plenum, New York, pp 145-204

de Olmos JS (1990) Amygdaloid nuclear gray complex. In: Paxinos G (ed) The human nervous system. Academic, San Diego, pp 583-750

de Olmos JS, Ingram WR (1972) The projection field of the stria terminalis in the rat brain. An experimental study. J Comp Neurol 146:303-334

de Olmos JS, Hardy H, Heimer L (1978) The afferent connections of the main and the accessory bulb formations in the rat: an experimental HRP study. J Comp Neurol 181:213-244

de Olmos JS, Alheid GF, Beltramino CA (1985) Amygdala. In: Paxinos G (ed) The rat nervous system, vol 1. Academic, Sydney, pp 223-334

Desban M, Gauchy C, Glowinski J, Kemel M-L (1995) Heterogeneous topographical distribution of the striatonigral and striatopallidal neurons in the matrix compart-ment of the cat caudate nucleus. J Comp Neurol 352:117-133

Desban M, Kemel ML, Glowinski J, Gauchy C (1993) Spatial organization of patch and matrix compartments in the rat striatum. Neuroscience 57:661-671

Deschênes M, Labelle A, Landry P (1979) Morphological characterization of slow and fast pyramidal tract cells in the cat. Brain Res 178:251-274

DeVito JL, Anderson ME (1982) An autoradiographic study of efferent connections of the globus pallidus in macaque mulatta. Exp Brain Res 46:107-117

De Wied D (1987) Neuropeptides and behavior. In: Adelman G (ed) Encyclopedia of neuroscience, vol 2. Birkhäuser, Boston, pp 839-841

De Zeeuw CI, Wentzel P, Mugnaini E (1993) Fine structure of the dorsal cap of the inferior olive and its GABAergic and non-GABAergic input from the nucleus prepositus hypoglossi in rat and rabbit. J Comp Neurol 327:63-82

De Zeeuw CI, Gerrits NM, Voogd J, Leonard CS, Simpson JI (1994) The rostral dorsal cap and ventrolateral outgrowth of the rabbit inferior olive receive a GABAergic input from dorsal group Y and the ventral dentate nucleus. J Comp Neurol 341:420-432

De Zeeuw CI, Lang EJ, Sugihara I, Ruigrok TJH, Eisenman LM, Mugnaini E, Llinas R (1996) Morphological correlates of bilateral synchrony in the rat cerebellar cortex. J Neurosci 16:3412-3426

Diamond IT, Hall WC (1969) Evolution of neocortex. Science 164:251-262

Diamond IT, Conley M, Fitzpatrick D, Raczkowski D (1991) Evidence for separate pathways within the tecto-geniculate projection in the tree shrew. Proc Natl Acad Sci USA 88:1315-1319

Diao YC, Wang YK, Xiao YM (1983) Representation of the binocular visual field in the superior colliculus of the albino rat. Exp Brain Res 52:67-72

Diepen RBWFM (1941) The hypothalamic nuclei and their ontogenetic development in ungulates (Ovis aries). Thesis, University of Amsterdam, the Netherlands

Dietrichs E, Haines DE (1989) Interconnections between hypothalamus and cerebellum. Anat Embryol (Berl) 179:207-220

Dietrichs E, Walberg F (1983) Cerebellar cortical afferents from the red nucleus in the cat. Exp Brain Res 50:353-358

Dietrichs E, Walberg F (1987) Cerebellar nuclear afferents - where do they originate? A re-evaluation of the projections from some lower brain stem nuclei. Anat Embryol 177:165-172

Dietrichs E, Wiklund L, Haines DE (1992) The hypothalamo-cerebellar projection in the rat: origin and transmitter. Arch Ital Biol 130:203-211

DiFiglia M, Aronin N (1982) Ultrastructural features of immunoreactive somatostatin neurons in the rat caudate nucleus. J Neuroscience 2:1267-1274

DiFiglia M, Pasik P, Pasik T (1976) A Golgi study of neuronal types in the neo-striatum of monkeys. Brain Res 114:245-256

Dillon LS (1962) Comparative studies of the brain in the Macropodidae. Contribution to the phylogeny of the mammalian brain. J Comp Neurol 120:43-52

Dinopoulos A, Papadopoulos GC, Michaloudi H, Parnavelas JG, Uylings HBM, Karamanlidis AN (1987) Claustrum in the hedgehog (Erinaceus europaeus) brain: cytoarchitecture and connections with cortical and subcortical structures. J Comp Neurol 316:187-205

Divac I (1975) Magnocellular nuclei of the basal forebrain project to neocortex, brain stem, and olfactory bulb: review of some functional correlates. Brain Res 93:385-398

Dobbins EG, Feldman JL (1995) Differential innervation of protruder and retractor muscles of the tongue in rat. J Comp Neurol 357:376-394

Dom R, Falls W, Martin GF (1973) The motor nucleus of the facial nerve in the opossum (Didelphis marsupialis virginiana). Its organization and connections. J Comp Neurol 152:373-402

Dong K, Rahman HA (1992) The retrograde fluorescence double labeling study of the cat's optic nerve cell which has a bifurcating axon. Kaibogaku Zasshi 67:207-213

Donoghue JP, Herkenham M (1986) Neostriatal projections from individual cortical fields conform to histochemically distinct striatal compartments in the rat. Brain Res 365:397-403

Doré L, Jacobson CD, Hawkes R (1990) Organization and post-natal development of zebrin II antigenic compartmentation in the cerebellar vermis of the grey opossum, Monodelphis domestica. J Comp Neurol 291:431-449

Douglas RJ, Martin KAC (1990) Neocortex. In: Shepherd GM (ed) The synaptic organization of the brain, 3rd edn. Oxford University Press, New York-Oxford, pp 389-438

Dow RS (1942) The evolution and anatomy of the cerebellum. Biol Rev 17:179-220

Dräger UC, Hubel DH (1974) Responses to visual stimulation and relationship between visual, auditory, and somatosensory inputs in mouse superior colliculus. J Neurophysiol 38:690-713

Dräger UC, Hubel DH (1976) Topography of visual and somatosensory projections to mouse superior colliculus. J Neurophysiol 39:91-101

Dräger UC, Olsen JF (1980) Origins of crossed and uncrossed retinal projections in pigmented and albino mice. J Comp Neurol 191:383-412

Dreher B, Sefton AJ, Ni SYK, Nisbett G (1985) The morphology, number, distribution and central projections of class I retinal ganglion cells in albino and hooded rats. Brain Behav Evol 26:10-48

Drooglever Fortuyn AB (1912) Die Ontogenie der Kerne des Zwischenhirns beim Kaninchen. Arch Anat Physiol Anat Abt 36:303-352

Druga R (1970) Neocortical projections on the amygdala (an experimental study with the Nauta method). J Hirnforsch 11:467-476

Druga R, Syka J (1984) Projections from auditory structures to the superior colliculus in the rat. Neurosci Lett 45:247-252

Dubé L, Smith AD, Bolam JP (1988) Identification of synaptic terminals of thalamic or cortical origin in contact with distinct medium-size spiny neurons in the rat neo-striatum. J Comp Neurol 267:455-471

Dubé MG, Xu B, Crowley WR, Kalra PS, Kalra SP (1994) Evidence that neuropeptide Y is a physiological signal for normal food intake. Brain Res 646:341-344

Dulce Madeira M, Lieberman AR (1995) Sexual dimorphism in the mammalian limbic system. Prog Neurobiol 45:275-333

Dum RP, Strick PL (1992) Medial wall motor areas and skeletomotor control. Curr Opin Neurobiol 2:836-839

Durand J (1989) Intracellular study of oculomotor neurons in the rat. Neuroscience 30:639-649

Ebbesson SOE, Voneida TJ (1969) The cytoarchitecture of the pallium in the tegu lizard (Tupinambis nigropunctatus). Brain Behav Evol 2:431-466

Ebner FF (1969) A comparison of primitive forebrain organization in metatherian and eutherian mammals. Ann NY Acad Sci 167:241-257

Ebner FF, Colonnier M (1975) Synaptic patterns in the visual cortex of turtles: an electron microscopic study. J Comp Neurol 160:51-80

Eccles J (1984) The cerebral neocortex: a theory of its operation. In: Jones EG, Peters A (eds) Functional properties of cortical cells. Plenum, New York, pp 1-36 (Cerebral cortex, vol 2)

Eccles JC, Ito M, Szentagothai J (1967) The cerebellum as a neuronal machine. Springer, Berlin Heidelberg New York

Edinger L (1889) Vergleichende Entwicklungsgeschichte und anatomische Studien im Bereiche des Centralnervensystems. 2) Über die Fortsetzung der hinteren Rückenmarkswurzeln zum Gehirn. Anat Anz 4:121-128

Edinger L (1908a) The relations of comparative anatomy to comparative psychology. J Comp Neurol Psychol 18:437-457

Edinger L (1908b) Vorlesungen über den Bau der nervösen Zentralorgane des Menschen und der Tiere. II. Vergleichende Anatomie des Gehirns. Vogel, Leipzig

Edinger T (1948) Evolution of the horse brain. Memoir Series of the Geological Society of America (Memoir 25). Waverly, Baltimore

Edinger T (1966) Brains from 40 million years of camelid history. In: Hassler R, Stephan H (eds) Evolution of the forebrain: phylogenesis and ontogenesis of the forebrain. Thieme, Stuttgart, pp 153-161

Edley SM, Graybiel AM (1983) The afferent and efferent connections of the feline nucleus tegmenti pedunculopontinus, pars compacta. J Comp Neurol 217:187-215

Edmunds SM, Parnavelas JG (1982) Retzius-Cajal cells: an ultrastructural study in the developing visual cortex of the rat. J Neurocytol 11:427-446

Edwards SB (1972) The ascending and descending projections of the red nucleus in the cat: an experimental study using an autoradiographic tracing method. Brain Res 48:45-63

Edwards SB (1977) The commissural projection of the superior colliculus in the cat. J Comp Neurol 173:23-40

Edwards SB (1980) Deep cell layers of superior colliculus, vol 6. In: Hobson JA, Brazier MAB (eds) The reticular formation revisited: specifying function for a nonspecific system. International Brain Research Organization, New York, pp 193-209

Edwards SB, Rosenquist AC, Palmer LA (1974) An autoradiographic study of ventral lateral geniculate projections in the cat. Brain Res 72:282-287

Edwards SB, Ginsburgh CL, Henkel CK, Stein BE (1979) Soures of subcortical projections to the superior colliculus in the cat. J Comp Neurol 184:309-330

Einstein G, Fitzpatrick D (1991) Distribution and morphology of area 17 neurons that project to the cat's extrastriate cortex. J Comp Neurol 303:132-149

Eisenman LM, Hawkes R (1993) Antigenic compartmentation in the mouse cerebellar cortex: Zebrin and HNK-1 reveal a complex, overlapping molecular topography. J Comp Neurol 335:586-605

Ekerot C-F, Larson B (1979a) The dorsal spino-olivocerebellar system in the cat. I. Functional organization and termination in the anterior lobe. Exp Brain Res 36:201-218

Ekerot C-F, Larson B (1979b) The dorsal spino-olivocerebellar system in the cat. II. Somatotopical organization. Exp Brain Res 36:219-232

Ekerot CF, Larson B, Oscarsson O (1979) Information carried by the spinocerebellar paths. In: Granit R, Pompeiano O (eds) Reflex control of posture and movement. North Holland/Elsevier, Amsterdam, pp 79-90

Elhanany E, White EL (1990) Intrinsic circuitry: synapses involving the local axon collaterals of corticocortical projection neurons in the mouse primary somatosensory cortex. J Comp Neurol 291:43-54

Elliot Smith G (1898) The brain in the Edentata. Trans Linn Soc Lond Ser 2 (Zool) 7:277-394

Elliot Smith G (1903) Notes on the morphology of the cerebellum. J Anat Physiol 37:329-332

Elliot-Smith G (1899) The brain in the Edentata. Trans Linn Soc Lond Sec Ser 7:277-394

Elliot-Smith G (1900) On the morphology of the brain in the mammalia, with special reference to that of lemurs, recent and extinct. Trans Linn Soc 8:319-432

Elliot-Smith G (1902) The primary subdivision of the mammalian cerebellum. J Anat (Lond) 36:381-385

Elliot Smith G (1908) The cerebral cortex in Lepidosiren, with comparative notes on the interpretation of certain features of the forebrain in other vertebrates. Anat Anz 33:513-540

Elverland HH (1977) Descending connections between superior olivary and cochlear nuclear complexes in the cat studied by autoradiographic and horseradish peroxidase methods. Exp Brain Res 27:397-412

Enevoldson TP, Gordon G (1984) Spinally projecting neurons in the dorsal column nuclei: distribution, dendritic trees and axonal projections. A retrograde HRP study in the cat. Exp Brain Res 54:538-550

Enevoldson TP, Gordon G (1989a) Postsynaptic dorsal column neurons in the cat: a study with retrograde transport of horseradish peroxidase. Exp Brain Res 75:611-620

Enevoldson TP, Gordon G (1989b) Spinocervical neurons and dorsal horn neurons projecting to the dorsal column nuclei through the dorsolateral fascicle: a retrograde HRP study in the cat. Exp Brain Res 75:621-630

English AW, J. T, Lennard PR (1985) Anatomical organization of long ascending propriospinal neurons in the cat spinal cord. J Comp Neurol 240:349-358

Epema AH (1990) "Connections" of the vestibular nuclei in the rabbit. Doctoral thesis

Epema AH, Gerrits NM, Voogd J (1988) Commissural and intrinsic connections of the vestibular nuclei in the rabbit: A retrograde labeling study. Exp Brain Res 71:129-146

Erickson RP, Jane A, Waite R, Diamond IT (1964) Single neuron investigation of sensory thalamus of the opossum. J Neurophysiol 27:1026-1047

Essick CR (1907) The corpus pontobulbare - a hitherto undescribed nuclear mass in the human. Am J Anat 7:1101

Evarts EV, Thach WT (1969) Motor mechanism of the CNS: cerebrocerebellar inter-relations. Annu Rev Physiol 31:451-498

Evinger C (1988) Extraocular motor nuclei: location, morphology and afferents. In: Buttner-Ennever (ed) Neuroanatomy of the oculomotor system, chap 3. Elsevier Science, Amsterdam, pp 81-117

Evinger C, Graf WM, Baker R (1987) An extra- and intracellular HRP analysis of the organization of extraocular motoneurons and internuclear neurons in the guinea pig and rabbit. J Comp Neurol 262:429-445

Ezure K, Graf W (1984) A quantitative analysis of the spatial organization of the vestibulo-ocular reflexes in lateral- and frontal-eyed animals-II. Neuronal networks underlying vestibulo-oculomotor coordination. Neuroscience 12:95-109

Fairén A, Valverde F (1979) Specific thalamo-cortical afferents and their presumptive targets in the visual cortex. A Golgi study. Prog Brain Res 51:419-438

Fairén A, Valverde F (1980) A specialized type of neuron in the visual cortex of cat: a Golgi and electron microscope study of chandelier cells. J Comp Neurol 194:761-779

Fairén A, DeFelipe J, Regidor J (1984) Nonpyramidal neurons: general account. In: Peters A, Jones EG (eds) Cellular components of the cerebral cortex. Plenum, New York, pp 201-254 (Cerebral cortex, vol 1)

Falk D (1982) Mapping fossil endocasts. In: Armstrong E, Falk D (eds) Primate brain evolution: methods and concepts. Plenum, New York, pp 217-226

Fallon JH, Moore RY (1978) Catecholamine innervation of the basal forebrain. IV. Topography of the dopamine projection to the neostriatum. J Comp Neurol 180:545-580

Faugier-Grimaud S, Ventre J (1989) Anatomic connections of inferior parietal cortex (area 7) with subcortical structures related to vestibulo-ocular function in monkey (Macaca fascicularis). J Comp Neurol 280:1-14

Faull RLM (1978) The cerebellofugal projections in the brachium conjunctivum of the rat. II. The ipsilateral and contralateral descending pathways. J Comp Neurol 178:519-536

Faull RLM, Mehler WR (1978) The cells of origin of nigrotectal, nigrothalamic and nigrostriatal projections in the rat. Neuroscience 3:989-1002

Feldman ML (1984) Morphology of the neocortical pyramidal neuron. In: Peters A, Jones EG (eds) Cellular components of the cerebral cortex. Plenum, New York, pp 123-200 (Cerebral cortex, vol 1)

Feldman ML, Harrison JM (1969) The projection of the acoustic nerve to the ventral cochlear nucleus of the rat. A Golgi Study. J Comp Neurol 137:267-294

Feldman ML, Peters A (1978) The forms of non-pyramidal neurons in the visual cortex of the rat. J Comp Neurol 179:761-794

Feldman SG, Kruger L (1980) An axonal transport study of the ascending projection of medial lemniscal neurons in the rat. J Comp Neurol 192:427-454

Feldon S, Feldon P, Kruger L (1970) Topography of the retinal projection upon the superior colliculus of the cat. Vision 10:135-143

Felten DL, Crutcher KA (1979) Neuronal vascular relationships in the raphe nuclei, locus coeruleus, and substantia nigra in primates. Am J Anat 155:467-482

Felten D, Laties A, Carpenter M (1974) Localization of monoamine-containing cell bodies in the squirrel monkey brain. Am J Anat 139:153-166

Ferraro A, Barrera SE (1935) The nuclei of the posterior funiculi in Macacus rhesus. Arch Neurol Psychiat (Chic) 33:262-275

Ferrier D (1876) The functions of the brain. Smith, London Elder

Filimonoff IN (1965) On the so-called rhinencephalon in the dolphin. J Hirnforsch 8:1-23

Fisher M (1983) Neuron-glia interactions and glial enzyme expression in the mouse cerebellum. Int Soc Dev Neurosci Abstr Salt Lake City, Utah

Fitzpatrick D, Itoh K, Diamond IT (1983) The laminar organization of the lateral geniculate body and the striate cortex in the squirrel monkey (Saimiri sciureus). J Neurosci 3:673-702

Fitzpatrick D, Lund JS, Blasdel GG (1985) Intrinsic conenctions of macaque striate cortex: afferent and efferent connections of lamina 4C. J Neurosci 5:3329-3349

FitzPatrick KA, Imig TJ (1978) Projections of auditory cortex upon the thalamus and midbrain in the owl monkey. J Comp Neurol 177:537-556

FitzPatrick KA, Imig TJ (1982) Organization of auditory connections. The primate auditory cortex. In: Woolsey CN (ed) Cortical sensory organization, vol 3. Humana, Clifton, New Jersey, pp 71-109

Fénelon G, François C, Percheron G, Yelnik J (1990) Topographic distribution of pallidal neurons projecting to the thalamus in macaques. Brain Res 520:27-35

Flaherty AW, Graybiel AM (1993) Output architecture of the primate putamen. J Neurosci 13:3222-3232

Flanagan-Cato LM, McEwen BS (1995) Pattern of Fos and Jun expression in the female rat forebrain after sexual behavior. Brain Res 673:53-60

Flatau E, Jacobsohn L (1899) Handbuch der Anatomie und vergleichende Anatomie des Centralnervensystems der Säugetiere. I. Makroskopischer Teil. Karger, Berlin

Fleischhauer K (1974) On different patterns of dendritic bundling in the cerebral cortex of the cat. Z Anat Entw Gesch 143:115-126

Fleischhauer K, Petche H, Wittkowski W (1972) Vertical bundles of dendrites in the neocortex. Z Anat Entw Gesch 136:213-223

Flindt-Egebak P, Møller HU (1984) Topographical arrangements of feline motor cortical projections onto the pretectum. Neurosci Lett 52:85-89

Flink R, Westman J (1985) Convergence on the same neurons in the feline ventrobasal thalamus of terminals from the dorsal column and the lateral cervical nuclei: an ultrastructural study combining orthograde degeneration and anterograde axonal transport of lectin conjugated horseradish peroxidase. Neurosci Lett 61:243-248

Flood S, Jansen J (1961) On the cerebellar nuclei in the cat. Acta Anat 46:52-72

Flores A (1911) Die Myeloarchitektonik und die Myelogenie des Cortex Cerebri beim Igel (Erinacaeus europaeus). J Psychol Neurol (Lpz) 17:215-247

Floris A, Dino M, Jacobowitz DM, Mugnaini E (1994) The unipolar brush cells of the rat cerebellar cortex and cochlear nucleus are calretinin-positive: a study by light and electron microscopic immunocytochemistry. Anat Embryol (Berl) 189:495-520

Flumerfelt BA, Caughell KA (1978) A horseradish peroxidase study of the cerebellorubral pathway in the rat. Exp Neurol 58:95-101

Flumerfelt BA, Otabe S, Courville J (1973) Distinct projections to the red nucleus from the dentate and interposed nuclei in the monkey. Brain Res 50:408-414

Fonnum F, Storm-Mathisen J, Divac I (1981) Biochemical evidence for glutamate as neurotransmitter in corticostriatal and corticothalamic fibres in rat brain. Neuroscience 6:863-873

Fox CA, Rafols JA (1975) The radial fibers in the globus pallidus. J Comp Neurol 159:177-200

Fox CA, Hillman DE, Siegesmund KA, Dutta CR (1967) The primate cerebellar cortex: a golgi and electron microscopic study. Prog Brain Res 25:174-225

Fox CA, Rafols JA, Cowan WM (1975) Computer measurements of axis cylinder dia- meters of radial fibers and "comb" bundle fibers. J Comp Neurol 159:201-224

Fox MW, Inman O (1966) Persistence of Retzius-Cajal cells in developing dog brain. Brain Res 3:192-194

Frahm HD, Stephan H, Stephan M (1982) Comparison of brain structure volumes in Insectivora and Primates. I. Neocortex. J Hirnforsch 23:375-389

Francois C, Percheron G, Yelnik J (1984) Localization of nigrostriatal, nigro- thalamic and nigrotectal neurons in ventricular coordinates in macaques. Neuroscience 13:61-76

Frankfurter A, Weber JT, Royce GJ, Strominger NL, Harting JK (1976) An autoradiographic analysis of the tecto-olivary projection in primates. Brain Res 118:245-257

Frederickson CJ, Trune DR (1986) Cytoarchitecture and saccular innervation of nucleus Y in the mouse. J Comp Neurol 252:302-322

Freund TF, Martin KAC, Smith AD, Somogyi P (1983) Glutamate decarboxylase-immunoreactive terminals of Golgi-impregnated axoaxonic cells and of presumed basket cells in synaptic contact with pyramidal neurons of the cat's visual cortex. J Comp Neurol 221:263-278

Freund TF, Powell JF, Smith AD (1984) Tyrosine hydroxylase-immunoreactive boutons in synaptic contact with identified striatonigral neurons, with particular reference to dendritic spines. Neuroscience 13:1189-1215

Friant M (1944) Le cerveau des Pangolins arboricoles d'Afrique. Rev Zool Bot Afr 38:104–109

Friauf E, Herbert H (1985) Topographical organization of facial motoneurons to individual pinna muscles in rat (Rattus rattus) and bat (Rousettus aegyptiacus). J Comp Neurol 240:161–170

Friedman DP, Bachevalier J, Ungerleider LG, Mishkin M (1987) Widespread projections to layer I of primate cortex. Soc Neurosci Abstr 13:251

Frisina RD, O'Neill WE, Zettel ML (1989) Functional organization of mustached bat inferior colliculus: II. Connections of the FM2 region. J Comp Neurol 284:85–107

Fritschy J-M, Grzanna R (1989) Immunohistochemical analysis of the neurotoxic effects of DSP-4 identifies two populations of noradrenergic axon terminals. Neuroscience 30:181–197

Fritschy JM, Grzanna R (1990) Demonstration of two separate descending noradrenergic pathways to the rat spinal cord: evidence for an intragriseal trajectory of locus coeruleus axons in the superficial layers of the dorsal horn. J Comp Neurola 291:553–582

Frost DO, Caviness VS (1980) Radial organization of thalamic projections to the neocortex in the mouse. J Comp Neurol 194:369–393

Frotscher M, Zimmer J (1983) Commissural fibers terminate on non-pyramidal neurons in the guinea pig hippocampus – a combined Golgi/EM degeneration study. Brain Res 265:289–293

Frotscher M, Léránth CS, Lübbers K, Oertel WH (1984) Commissural afferents innervate glutamate decarboxylase immunoreactive non-pyramidal neurons in the guinea pig hippocampus. Neurosci Lett 46:137–143

Fujito Y, Imai T, Aoki M (1991) Monosynaptic excitation of mononeurons innervating forelimb muscles following stimulation of the red nucleus in cats. Neurosci Lett 127:137–140

Fukushima K, Pitts NG, Peterson BW (1978) Direct excitation of neck motoneurons by interstitiospinal fibers. Exp Brain Res 33:565–581

Fukushima T, Kerr FWL (1979) Organization of trigeminothalamic tracts and other thalamic afferent systems of the brainstem in the rat: presence of gelatinosa neurons with thalamic connections. J Comp Neurol 183:169–184

Fukuyama U (1940) Über eine substantielle Verschmelzung des roten Kerns mit den Nebenokulomotorius-Kernen (Bechterew and Darkschewitsch) bei Affen. Arb Anat Inst Kaiserl Jpn Univ Sendai 23:1–122

Fuller GN, Burger PC (1990) Nervus terminalis (cranial nerve zeor) in the adult human. Clin Neuropathol 9:279–283

Fulwiler CE, Saper CB (1984) Subnuclear organization of the efferent connections of the parabrachial nucleus in the rat. Brain Res Rev 7:229–259

Furuya N, Markham CH (1981) Arborization of axons in oculomotor nucleus identified by vestibular stimulation and intra-axonal injection of horseradish peroxidase. Exp Brain Res 43:289–303

Fuse G (1912) Die innere Abteilung des Kleinhirnstiels (Meynert, IAK) und der Deiterssche Kern. Hirnanatom Inst Univ Zurich VI:29–267

Fuse G (1936) Das gewundene Grau oder der Olivenkern des vorderen Zweihügels, nucleus olivaris corporis quadrigemini anterioris, bei Mensch und Tier. Arb Anat Inst Kaiserl Jpn 19:49–486

Fuse G (1937) Ein neuer Versuch zur Unterteilung des Nucleus ruber tegmenti bei den Karnivoren und zur phylogenetischen Bewertung seiner Entwicklung unter Berücksichtigung der an Karnivoren und Affen gewonnen Ergebnisse. Arb Anat Inst Kaiserl Jpn Univ Sendai 20:123–188

Fuster JM (1991) The prefrontal cortex and its relation to behavior. Prog Brain Res 87:201–211

Fuxe K, Hokfelt T (1969) Catecholamines in the hypothalamus and the pituitary gland. In: Ganong WF, Martini L (eds) Frontiers in neuroendocrinology. Oxford University Press, Oxford, pp 47–96

Gabbott PLA, Martin KAC, Whitteridge D (1987) Connections between pyramidal neurons in layer 5 of cat visual cortex (area 17). J Comp Neurol 259:364–381

Gacek RR (1974) Localization of neurons supplying the extraocular muscles in the kitten using horseradish peroxidase. Exp Neurol 44:381–403

Galabov P, Davidof M (1976) On the vegatative network of guinea pig thoracic spinal cord. Histochemistry 47:247–255

Gamrani H, Ontoniente B, Seguela P, Geffard M, Calas A (1986) Gamma-aminobutyric acid-immunoreactivity in the rat hippocampus. A light and electron microscopic study with anti-GABA antibodies. Brain Res 364:30–38

Ganchrow D (1978) Intratrigeminal and thalamic projections of nucleus caudalis in the squirrel monkey (Saimiri sciureus): a degeneration and autoradiographic study. J Comp Neurol 178:281–312

Gans A (1924) Beitrag zur Kenntnis des Aufbaus des Nucleus Dentatus aus zwei Teilen namentlich auf Grund von Untersuchungen mit der Eisenreaktion. Z Ges Neurol Psychiat 93:750–755

Garcia-Rill E, Skinner RD (1987) The mesencephalic locomotor region. II. Projections to reticulospinal neurons. Brain Res 411:13–20

Garey LJ, Webster WR (1989) Functional morphology in the inferior colliculus of the marmoset. Hear Res 38:67–79

Garver D, Sladek JRJ (1975) Monoamine distribution in primate brain. I. Catecholamine-containing perikarya in the brain stem of Macaca speciosa. J Comp Neurol 159:289–304

Gaskell WH (1886) On the structure, distribution and function of the nerves which innervate the visceral and vascular systems. J Physiol 7:1–81

Gaskell WH (1889) On the relation between the structure, function, distribution and origin of the cranial nerves: together with a theory of the origin of the nervous system of vertebrata. J Physiol 10:153–211

Gayer NS, Faull RLM (1988) Connections of the paraflocculus of the cerebellum with the superior colliculus in the rat brain. Brain Res 449:253–270

Geeraedts LMG, Nieuwenhuys R, Veening JG (1990a) Medial forebrain bundle of the rat: III. Cytoarchitecture of the rostral (telencephalic) part of the medial forebrain bundle bed nucleus. J Comp Neurol 294:507–536

Geeraedts LMG, Nieuwenhuys R, Veening JG (1990b) Medial forebrain bundle of the rat: VI. Cytoarchitecture of the caudal (lateral hypothalamic) part of the medial forebrain bundle bed nucleus. J Comp Neurol 294:537–568

Geffard M, Buijs RM, Seguela P, Pool CW, Le Moel M (1984) First demonstration of highly specific and sensitive antibodies against dopamine. Brain Res 294:161–165

Geis GS, Wurster RD (1980)) Horseradish peroxidase localization of cardiac vagal preganglionic somata. Brain Res 1820:19–30

Gellman R, Houk JC, Gibson AR (1983) Somatosensory properties of the inferior olive of the cat. J Comp Neurol 215:228–243

Geniec P, Morest DK (1971) The neuronal architecture of the human posterior colliculus. A study with the Golgi method. Acta Otolaryngol [Suppl] 295:1–33

Gerfen CR (1984) The neostriatal mosaic: compartmentalization of cortico-striatal input and striatonigral output systems. Nature 311:461–464

Gerfen CR (1990) The neostriatal mosaic: striatal patch-matrix organization is related to cortical lamination. Science 246:385–388

Gerfen CR (1992) The neostriatal mosaic: multiple levels of compartmental organization in the basal ganglia. Annu Rev Neurosci 15:285–320

Gerfen CR, Staines WA, Arbuthnott GW, Fibiger HC (1982) Crossed connections of the substantia nigra in the rat. J Comp Neurol 207:283–303

Gerrits NM (1990) 26. Vestibular nuclear complex. In: Paxinos G (ed) The human nervous system. Academic, London, pp 863–888

Gerrits NM (1994) Vestibular and cerebellar connections subserving eye movements. In: Delgado-Garcia JM, Godaux E, Vidal P-P (eds) Information processing underlying gaze control. Pergamon, Oxford, pp 341–350

Gerrits NM, Voogd J (1982) The climbing fiber projection to the flocculus and adjacent paraflocculus in the cat. Neuroscience 7:2971–2991

Gerrits NM, Voogd J (1987) The projection of the nucleus reticularis tegmenti pontis and adjacent regions of the pontine nuclei to the central cerebellar nuclei in the cat. J Comp Neurol 258:52–69

Gerrits NM, Voogd J, Magras IN (1985a) Vestibular afferents of the inferior olive and the vestibulo-olivo-cerebellar climbing fiber path-way to the flocculus in the cat. Brain Res 332:325–336

Gerrits NM, Voogd J, Nas WSC (1985b) Cerebellar and olivary projections of the external and rostral internal cuneate nuclei in the cat. Exp Brain Res 57:239–255

Gerrits NM, Epema AH, Van Linge A, Dalm E (1989) The primary vestibulocerebellar projection in the rabbit: absence of primary afferents in the flocculus. Neurosci Lett 290:262–277

Ghez C (1975) Input-output relations of the red nucleus in the cat. Brain Res 98:93–118

Ghosh S, Fyffe REW, Porter R (1988) Morphology of neurons in area 4gamma of the cat's cortex studied with intracellular injection of HRP. J Comp Neurol 269:290–312

Giesler GJ, Spiel HR, Willis WD (1981) Organization of spinothalamic tract axons within the rat spinal cord. J Comp Neurol 195:243–252

Giesler GJJ, Menetrey D, Basbaum AI (1979) Differential origins of spinothalamic tract projections to medial and lateral thalamus in the rat. J Comp Neurol 184:107–126

Giesler GJJ, Nahin RL, Madsen AM (1984) Postsynaptic dorsal column pathway of the rat. I. Anatomical studies. J Neurophysiol 51:260–275

Giesler GJJ, Miller LR, Madsen AM, Katter JT (1987) Evidence for the existence of a lateral cervical nucleus in mice, guinea pigs, and rabbits. J Comp Neurol 263:106–112

Giesler GJJ, Bjorkeland M, Xu Q, Grant G (1988) Organization of the spinocervicothalamic pathway in the rat. J Comp Neurol 268:223–233

Gilbert CD (1992) Horizontal integration and cortical dynamics. Neuron 9:1–13

Gilbert CD, Wiesel TN (1979) Morphology and intracortical projections of functionally identified neurons in cat visual cortex. Nature 280:120–125

Gilbert PW (1947) The origin and development of the extrinsic ocular muscles in the domestic cat. J Morphol 81:151–193

Giménez-Amaya JM, Graybiel AM (1990) Compartmental origins of the striatopallidal projection in the primate. Neuroscience 34:111–126

Giolli RA (1963) An experimental study of the accessory optic system in the cynomolgus monkey. J Comp Neurol 121:89–99

Giolli RA, Greel DJ (1973) The primary optic projections in pigmented and albino guinea pigs: an experimental degeneration study. Brain Res 55:25–39

Giolli RA, Guthrie MD (1969) The primary optic projections in the rabbit. An experimental degeneration study. J Comp Neurol 136:99–126

Giolli RA, Towns LC, Takahashi TT, Karamanlidis AN, Williams DD (1978) An autoradiographic study of the projections of visual cortical area 1 to the thalamus, pretectum and superior colliculus of the rabbit. J Comp Neurol 180:743–752

Giolli RA, Blanks RHI, Torigoe Y (1984) Pretectal and brain stem projections of the medial terminal nucleus of the accessory optic system of the rabbit and rat as studied by anterograde and retrograde neuronal tracing methods. J Comp Neurol 227:228–251

Giolli RA, Blanks RHI, Torigoe Y, Williams DD (1985) Projections of medial terminal accessory optic nucleus, ventral tegmental nuclei, and substantia nigra of rabbit and rat as studied by retrograde axonal transport of horseradish peroxidase. J Comp Neurol 232:99–116

Giolli RA, Torigoe Y, Blanks RHI, McDonald HM (1988) Projections of the dorsal and lateral terminal accessory optic nuclei and of the interstitial nucleus of the superior fasciculus (posterior fibers) in the rabbit and rat. J Comp Neurol 277:608–620

Giolli RA, Torigoe Y, Clarke RJ, Blanks RHI, Fallon JH (1992) GABAergic and non-GABAergic projections of accessory optic nuclei, including the visual tegmental relay zone, to the nucleus of the optic tract and dorsal terminal accessory optic nucleus in rat. J Comp Neurol 319:349–358

Giovanelli Barilari M, Kuypers HG (1969) Propriospinal fibers interconnecting the spinal enlargements in the cat. Brain Res 14:321–330

Glendenning KK, Masterton RB (1983) Acoustic chiasm, efferent projection of the lateral superior olive. J Neurosci 3:1521–1537

Glendenning KK, Masterton RB, Baker BN, Wenthold RJ (1991) Acoustic chiasm. III: Nature, distribution, and sources of afferents to the lateral superior olive in the cat. J Comp Neurol 310:377–400

Glezer II, Jacobs MS, Morgane PJ (1988) Implications of the "initial brain" concept for brain evolution in Cetacea. Behav Brain Sci 11:75–116

Glickstein M, May JG III, Mercier BE (1982) Corticopontine projection in the macaque: the distribution of labelled cortical cells after large injections of horseradish peroxidase in the pontine nuclei. J Comp Neurol 235:343–359

Glickstein M, Kralj-Hans I, Legg C, Mercier B, Ramna-Rayan M, Vaudano E (1992) The organisation of fibres within the rat basis pedunculi. Neurosci Lett 135:75–79

Globus A, Scheibel AB (1967) Pattern and field in cortical structure: the rabbit. J Comp Neurol 131:155–172

Glover JC, Petursdottir G (1988) Pathway specificity of reticulospinal and vestibulospinal projections in the 11-day chicken embryo. J Comp Neurol 270:25–38

Gobel S (1975) Golgi studies of the substantia gelatinosa neurons in the spinal trigeminal nucleus. J Comp Neurol 162:397–416

Gobel S (1978) Golgi studies of the neurons in layer II of the dorsal horn of the medulla (Trigeminal Nucleus Caudalis). J Comp Neurol 180:395–413

Goldberg JM, Fernandez C (1980) Efferent vestibular system in the squirrel monkey: anatomical location and influence on afferent activity. J Neurophysiol 43:986–1025

Goldman PS, Nauta WJH (1977) An intricately patterned prefronto-caudate projection in the rhesus monkey. J Comp Neurol 171:369–384

Goldman-Rakic PS (1982) Cytoarchitectonic heterogeneity of the primate neostriatum: subdivision into island and matrix cellular compartments. J Comp Neurol 205:398–413

Goldman-Rakic PS, Nauta WJH (1977) Columnar distribution of cortico-cortical fibers in the frontal association, limbic and motor cortex of the developing rhesus monkey. Brain Res 122:393–414

Goldman-Rakic PS, Schwartz ML (1982) Interdigitaion of contralateral and ipsilateral columnar projections to frontal association cortex in primates. Science 216:755–757

Gonzalo-Ruiz A, Leichnetz GR (1987) Lateralization of cerebellar efferent projections to the paraoculomotor region, superior colliculus, and medial pontine reticular formation in the rat: fluorescent double-labeling study. Brain Res 68:365–378

Gonzalo-Ruiz A, Leichnetz GR (1990) Connections of the caudal cerebellar interpositus complex in a new world monkey (Cebus apella). Brain Res Bull 25:919–927

Gonzalo-Ruiz A, Leichnetz GR, Hardy SG (1990) Projections of the medial cerebellar nucleus to oculomotor-related midbrain areas in the rat: an anterograde and retrograde HRP study. J Comp Neurol 296:427–436

Goodman DC, Hallett RE, Welch Rb (1963) Patterns of localization in the cerebellar cortico-nuclear projections of the albino rat. J Comp Neurol 121:51–68

Gordon G, Grant G (1982) Dorsolateral spinal afferents to some medullary sensory nuclei. An anatomical study in the cat. Exp Brain Res 46:12–23

Gowers WR (1886) Bemerkungen uber die antero-laterale aufsteigende Degeneration im Rückenmark. Neurol Centralbl 5:97–99

Graham J (1977) An autoradiographic study of the efferent connections of the superior colliculus in the cat. J Comp Neurol 173:629–654

Graham J, Casagrande VA (1980) A light microscopic and electron microscopic study of the superficial layers of the superior colliculus of the tree shrew (Tupaia glis). J Comp Neurol 191:133–151

Grant G (1962) Projection of the external cuneate nucleus onto the cerebellum in the cat. An experimental study using silver methods. Exp Neurol 5:179–195

Grant G (1995) Primary afferent projections to the spinal cord. In: Paxinos G (ed) The rat nervous system, 2nd edn. Academic, London, pp 61–66

Grant G, Gueritaud JP, Horcholle-Bossavit G, Tyc-Dumont S (1979) Anatomical and electrophysiological identification of motoneurones supplying the cat retractor bulbi muscle. Exp Brain Res 34:541–550

Grant G, Illert M, Tanaka R (1980) Integration in descending motor pathways controlling the forelimb in the cat. 6. Anatomical evidence consistent with the existence of C3–4 propriospinal neurones projecting to forelimb motor nuclei. Exp Brain Res 38:87–93

Grant K, Guegan M, Horcholle-Bossavit G (1981) The anatomical relationship of the retractor bulbi and posterior digastric motoneurones to the abducens and facial nuclei in the cat. Arch Ital Biol 119:195–207

Grantyn A, Grantyn R (1982) Axonal patterns and sites of termination of cat superior colliculus neurons projecting in the tecto-bulbo-spinal tract. Exp Brain Res 46:243–256

Granum SL (1986) The spinothalamic system of the rat. I. Locations of cells of origin. J Comp Neurol 247:159–180

Graur D, Higgins DG (1994) Molecular evidence for the inclusion of cetaceans within the order artiodactyla. Mol Biol Evol 11(3):357–364

Graur D, Duret L, Gouy M (1996) Phylogenetic position of the order Lagomorpha (rabbits, hares and allies). Nature 379:333–335

Gravel C, Hawkes R (1990) Parasagittal organization of the rat cerebellar cortex: direct comparison of Purkinje cell compartments and the organization of the spinocerebellar projection. J Comp Neurol 291:79–102

Graveland GA, DiFiglia M (1985) The frequency and distribution of medium-sized neurons with indented nuclei in the primate and rodent neostriatum. Brain Res 327:307–311

Graveland GA, Williams RS, Difiglia M (1985) A Golgi study in the human neostriatum: neurons and afferent fibers. J Comp Neurol 234:317–333

Gray TS, Hazlett JC, Martin GF (1981) Organization of projections from the gracile, medial cuneate and lateral nuclei in the North American opossum. Horseradish peroxidase study of the cells projecting to the cerebellum, thalamus and spinal cord. Brain Behav Evol 18:140–156

Graybiel AM (1972) Some fiber pathways related to the posterior thalamic region in the cat. Brain Behav Evol 6:363–393

Graybiel AM (1975) Anatomical organization of retinotectal afferents in the cat: an autoradiographic study. Brain Res 96:1–23

Graybiel AM (1977a) Direct and indirect preoculomotor pathways of the brainstem: an autoradiographic study of the pontine reticular formation in the cat. J Comp Neurol 175:37–78

Graybiel AM (1977b) Organization of oculomotor pathways in the cat and rhesus monkey. In: Baker R, Berthoz A (eds) Control of gaze by brainstem neurons. Developments in Neuroscience Vol 1, Elsevier, Amsterdam, pp 79–88

Graybiel AM (1978a) A satellite system of the superior colliculus: the parabigeminal nucleus and its projections to the superficial collicular layers. Brain Res 145:365–374

Graybiel AM (1978b) Organization of the nigrotectal connection: an experimental tracer study in the cat. Brain Res 143:339–348

Graybiel AM (1979) Periodic-compartmental distribution of acetylcholinesterase in the superior colliculus of the human brain. Neuroscience 4:643–650

Graybiel AM (1984) Neurochemically specified subsystem in the basal ganglia. In: Evered D, O'Connor M (eds) Functions of the basal ganglia. Pitman, London, pp 114–149 (Ciba foundation symposium 107)

Graybiel AM (1990) Neurotransmitters and neuromodulators in the basal ganglia. Trends Neurosci 13:244–254

Graybiel AM (1975) Anatomical organization of retinotectal afferents in the cat: an autoradiographic study. Brain Res 96:1–23

Graybiel AM, Hartweig EA (1974) Some afferent connections of the oculomotor complex in the rat (1975). An experimental study with tracer techniques. Brain Res 8:543–551

Graybiel AM, Ragsdale CW (1978) Histochemically distinct compartments in the striatum of human, monkey and cat demonstrated by acetylcholinesterase staining. Proc Natl Acad Sci USA 75:5723–5726

Graybiel AM, Ragsdale CW Jr (1979) Fiber connections of the basal ganglia. In: Bloom RE, Kreutzberg GW, Cuénod M (eds) Development and chemical specificity of neurons. Elsevier, Amsterdam, pp 239–283

Graybiel AM, Ragsdale CW Jr (1983) Biochemical anatomy of the striatum. In: Emson PC (ed) Chemical neuroanatomy. Raven, New York, pp 427–504

Graybiel AM, Ragsdale CW, Yoneoka ES, Elde RH (1981) An immunohistochemical study of enkephalins and other neuropeptide in the striatum of the cat with evidence that the opiate peptides are arranged to form mosaic patterns in register with the striosomal compartments visible by acetylcholinesterase staining. Neuroscience 6:377–397

Graybiel AM, Brecha N, Karten HJ (1984) Cluster- and-sheet pattern of enkephalin-like immunoreactivity in the superior colliculus of the cat. Neuroscience 12:191–214

Graybiel AM, Flaherty AW, Giménez-Amaya J-M (1991) Striosomes and matrisomes. In: Bernardi G, Carpenter MB, di Chiara G, Morelli M, Stanzione P (eds) The basal ganglia, vol 3. Plenum, New York, pp 3–12

Graybiel AM, Aosaki T, Flaherty AW, Kimura M (1994) The basal ganglia and adaptive motor control. Science 265:1826–1831

Gregory JE, Iggo A, McIntyre AK, Proske U (1987) Electroreceptors in the platypus. Nature 326:386–388

Gregory JE, Iggo A, McIntyre AK, Proske U (1988) Receptors in the bill of the platypus. J Physiol (Lond) 400:349–366

Gregory WK (1910) The orders of mammals. Bull Am Mus Nat Hist 27:1–534

Grillner S, Shik MI (1973) On the descending control of the lumbosacral spinal cord from the 'mesencephalic locomotor region'. Acta Physiol Scand 87:320–333

Groenewegen HJ (1988) Organization of the afferent connections of the mediodorsal thalamic nucleus in the rat, related to the mediodorsal-prefrontal topography. Neuroscience 24:379–431

Groenewegen HJ, Berendse HW (1990) Connections of the subthalamic nucleus with ventral striatopallidal parts of the basal ganglia in the rat. J Comp Neurol 294:607–622

Groenewegen HJ, Berendse HW (1994) The specificity of the 'nonspecific' midline and intralaminar thalamic nuclei. Trends Neurosci 17:52–57

Groenewegen HJ, Russchen FT (1984) Organization of the efferent projections of the nucleus accumbens to pallidal hypothalamic, and mesencephalic structures: a tracing and immunohistochemical study in the cat. J Comp Neurol 223:347–367

Groenewegen HJ, Voogd J (1977) The parasagittal zonation within the olivocerebellar projection. I. Climbing fiber dis-

tribution in the vermis of cat cerebellum. J Comp Neurol 174:417–488
Groenewegen HJ, Boesten AJP, Voogd J (1975) The dorsal column nuclear projections to the nucleus ventralis posterior lateralis thalami and the inferior olive in the cat: an autoradiographic study. J Comp Neurol 162:505–518
Groenewegen HJ, Voogd J, Freedman SL (1979) The parasagittal zonation within the olivocerebellar projection. II. Climbing fiber distribution in the intermediate and hemispheric parts cat cerebellum. J Comp Neurol 183:551–602
Groenewegen HJ, Becker NEK, Lohman AHM (1980) Subcortical afferents of the nucleus accumbens septi in the cat, studied with retrograde axonal transport of horseradish peroxidase and bisbenzimid. Neuroscience 5:1903–1916
Groenewegen HJ, Room P, Witter MP, Lohman AHM (1982) Cortical afferents of the nucleus accumbens in the cat, studied with anterograde and retrograde transport techniques. Neuroscience 7:977–995
Groenewegen HJ, Ahlenius S, Haber SN, Kowall NW, Nauta WJH (1986) Cytoarchitecture, fiber connections, and some histochemical aspects of the interpeduncular nucleus in the rat. J Comp Neurol 249:65–102
Groenewegen HJ, Meredith GE, Berendse HW, Voorn P, Wolters JG (1989) The compartmental organization of the ventral striatum in the rat. In: Crossman AR, Sambrook AM (eds) Neural mechanisms in disorders of movement. Libbey, London, pp 45–52
Groenewegen HJ, Berendse HW, Haber SN (1993) Organization of the output of the ventral striatopallidal system in the rat: ventral pallidal efferents. Neuroscience 57:113–142
Groenewegen HJ, Beijer AV, Wright CR (1996) The ventral striatum: gateway for limbic structures to reach the motor system? Prog Brain Res 107:485–511
Gross MH, Fox JA, Curtis JD (1979) A horseradish peroxidase study of primary afferent projections to the medullary cuneate nucleus in the rat. Neurosci Lett 14:147–152
Grothe B, Schweizer H, Pollak GD, Schuller G, Rosemann C (1994) Anatomy and projection patterns of the superior olivary complex in the Mexican free-tailed bat, Tadarida brasiliensis mexicana. J Comp Neurol 343:630–646
Grove EA (1988a) Neural associations of the substantia innominata in the rat: afferent connections. J Comp Neurol 277:315–346
Grove EA (1988b) Efferent connections of the substantia innominata in the rat. J Comp Neurol 277:347–364
Grzanna R, Fritschy J-M (1991) Efferent projections of different subpopulations of central noradrenaline neurons. Prog Brain Res 88:89–101
Guinan JJ Jr, Warr WB, Norris BE (1983) Differential olivocochlear projections from lateral versus medial zones of the superior olivary complex. J Comp Neurol 221:358–370
Gurche JA (1982) Early primate brain evolution. In: Armstrong E, Falk D (eds) Primate brain evolution. Methods and concepts. Plenum, New York, pp 227–246
Gurdjian ES (1927) The diencephalon of the albino rat. J Comp Neurol 43:1–114
Gwyn DG, Leslie RA (1979) A projection of vagus nerve to the area subpostrema in the cat. Brain Res 161:335–341
Gwyn DG, Waldron HA (1969) Observations on the morphology of a nucleus in the dorso-lateral funiculus of the spinal cord of the guinea-pig, rabbit, ferret and cat. J Comp Neurol 136:233–236
Gwyn DG, Waldron HA (1968) A nucleus in the dorsolateral funiculus of the spinal cord of the rat. Brain Res 10:342–351
Gwyn DG, Nicholson GP, Flumerfelt BA (1977) The inferior olivary nucleus of the rat: a light and electron microscopic study. J Comp Neurol 174:489–520
Haarmann K (1975) Morphologische und histologische Untersuchungen am Neocortex von Boviden (Antilopinae, Cephalopinae) und Traguliden mit Bemerkungen zur Evolutionshöhe. J Hirnforsch 16:93–116
Haartsen AB (1961) The fibre content of the record in small and large mammals. Acta Morphol Neerl Scand III:331–340

Haartsen AB, Verhaart WJC (1967) Cortical projections to brain stem and spinal cord in the goat by way of the pyramidal tract and the bundle of bagley. J Comp Neurol 129:189–202
Haase P (1990) Explanation for the labeling of cervical motoneurons in young rats following the introduction of horseradish peroxidase into the calf. J Comp Neurol 297:471–478
Haase P, Hrcycyshyn AW (1985) Labeling of motoneurons supplying the cutaneous maximus muscle in the rat, following injection of the triceps brachii muscle with horseradish peroxidase. Neurosci Lett 60:313–318
Haber SN, Nauta WJH (1983) Ramifications of the globus pallidus in the rat as indicated by patterns of immunohistochemistry. Neuroscience 9:245–260
Haber SN, Watson SJ (1985) The comparative distribution of enkephalin dynorphin and substance P in the human globus pallidus and basal forebrain. Neuroscience 14:1011–1024
Haber SN, Groenewegen HJ, Grove EA, Nauta WJH (1985) Efferent connections of the ventral pallidum: evidence for a dual striato-pallidofugal pathway. J Comp Neurol 235:322–335
Haber SN, Lynd E, Klein C, Groenewegen HJ (1990a) Topographic organization of the ventral striatal efferent projections in the rhesus monkey: an anterograde tracing study. J Comp Neurol 293:282–298
Haber SN, Wolfer DP, Groenewegen HJ (1990b) The relationship between ventral striatal efferent fibers and the distribution of peptide-positive woolly fibers in the forebrain of the rhesus monkey. Neuroscience 39:323–338
Haber SN, Lynd-Balta E, Mitchell SJ (1993) The organization of the descending ventral pallidal projections in the monkey. J Comp Neurol 329:111–128
Haberly LB (1983) Structure of the piriform cortex of the opossum. I. Description of neuron types with Golgi methods. J Comp Neurol 213:163–187
Haberly LB (1990a) Comparative aspects of olfactory cortex. In: Jones EG, Peters A (eds) Comparative structure and evolution of cerebral cortex, part II. Plenum, New York, pp 137–166 (Cerebral cortex, vol 8B)
Haberly LB (1990b) Olfactory cortex. In: Shepherd GM (ed) The synaptic organization of the brain, 3rd edn. Oxford University Press, New York, pp 317–345
Haberly LB, Feig SL (1983) Structure of the piriform cortex of the opossum. II. Fine structure of cell bodies and neuropil. J Comp Neurol 216:69–88
Haberly LB, Hansen DJ, Feig SL, Presto S (1987) Distribution and ultrastructure of neurons in opossum displaying immunoreactivity to GABA and GAD and high-affinity tritiated GABA uptake. J Comp Neurol 266:269–290
Hackethal H (1976) Morphologische Untersuchungen am Hirn der Schuppentiere (Mammalia, Pholidota) unter besonderer Berücksichtigung des Kleinhirns. Zool Anz 197:313–331
Hagg S, Ha H (1970) Cervicothalamic tract in the dog. J Comp Neurol 139:357–374
Häggvist G (1936) Analyse der Faserverteilung in einem Rückenmarkquerschnitt (Th3). Z Mikr Anat Forsch 39:1–34
Haglund L, Swanson LW, Köhler C (1984) The projection of the supramammillary nucleus to the hippocampal formation: an immunohistochemical and anterograde transport study with the lectin PHA-L in the rat. J Comp Neurol 229:171–185
Haight JR, Murray PF (1981) The cranial endocast of the early Miocene marsupial Wynyardia bassiana. An assessment of taxonomic relationships based upon comparisons of recent forms. Brain Behav Evol 19:17–36
Haight JR, Neylon L (1978) An atlas of the dorsal thalamus of the marsupial brush-tailed possum, Trichosurus vulpecula. J Anat 126:225–245
Haines DE (1977a) A proposed functional significance of parvicellular regions of the lateral and medial cerebellar nuclei. Brain Behav Evol 14:328–340

Haines DE (1977b) Cerebellar corticonuclear and corticovestibular fibers of the flocculonodular lobe in a Prosimian primate (Galago senegalensis). J Comp Neurol 174:607–630

Haines DE, Sowa TE (1985) Evidence of a direct projection from the medial terminal nucleus of the accessory optic system to lobule IX of the cerebellar cortex in the tree shrew (Tupaia Glis). Neurosci Lett 55:125–130

Haines DE, Patrick GW, Satrulee P (1982) Organization of cerebellar corticonuclear fiber systems. Exp Brain Res [Suppl] 6:320–367

Halász N, Shepherd GM (1983) Neurochemistry of the vertebrate olfactory bulb. Neuroscience 10:579–619

Hall E (1972) Some aspects of the structural organization of the amygdala. In: Eleftheriou BE (ed) The neurobiology of the amygdala. Plenum, New York, pp 95–121

Hall WC, Ebner FF (1970) Thalamotelencephalic projections in the turtle (Pseudemys scripta). J Comp Neurol 140:101–122

Hall WC, Fitzpatrick D, Klatt LL, Raczkowski D (1989) Cholinergic innervation of the superior colliculus in the cat. J Comp Neurol 287:495–514

Haller V, Hallerstein B (1934) Die äussere Gliederung des Zentralnervensystems des Menschen und der Säugetiere. In: Bolk L, Göppert E, Kallins E, Lubosch W (eds) Handbuch der vergleichenden Anatomie der Wirbeltiere, vol 2. Urban, Vienna, pp 1–318

Halliday G, Harding A, Paxinos G (1995) Serotonin and tachylkinin systems. In: Paxinos G (ed) The rat nervous system. Academic, San Diego, pp 929–974

Halpern M (1987) The organization and function os the vomeronasal system. Annu Rev Neurosci 10:325–362

Halsell CB, Travers SP, Travers JB (1996) Ascending and descending projections from the rostral nucleus of the solitary tract originate from separate neuronal populations. Neuroscience 72:185–197

Hamada J, Sakai M, Kubota K (1981) Morphological differences between fast and slow pyramidal tract neurons in the monkey motor cortex. Neurosci Lett 22:233–238

Hamilton RB, Norgren R (1984) Central projections of gustatory nerves in the rat. J Comp Neurol 222:560–577

Hamilton TC, Johnson JI (1973) Somatotopic organization related to nuclear morphology in the cuneate-gracile complex of opossums didelphis marsupialis virginiana. Brain Res 51:125–140

Hamlyn LH (1962) The fine structure of the mossy fibre endings in the hippocampus of the rabbit. J Anat 96:112–120

Hammelbo T (1972) On the development of the cerebral fissures in cetacea. Acta Anat 82:606–618

Hand PJ, Van Winkle T (1977) The efferent connections of the feline nucleus cuneatus. J Comp Neurol 171:83–109

Hardeland R (1993) The presence and function of melatonin and structurally related indoleamines in a dinoflagellate, and a hypothesis on the evolutionary significance of these tryptophan metabolites in unicellulars. Experientia 49:614–623

Harding GW, Towe AL (1985) Fiber analysis of the pyramidal tract of the laboratory rat. Exp Neurol 87:503–518

Hardy O, Corvisier J (1991) GABA and non-GABA immunostained neurones in the nucleus prepositus and the periparabigeminal area projecting to the guinea pig superior colliculus. Neurosci Lett 127:99–104

Harting JK (1977) Descending pathways from the superior colliculus: an autoradiographic analysis in the rhesus monkey (macaca mulatta). J Comp Neurol 173:583–612

Harting JK, Guillery RW (1976) Organization of retinocollicular pathways in the cat. J Comp Neurol 166:133–144

Harting JK, Martin GF (1970) Neocortical projections to the mesencephalon of the armadillo, dasypus novemcinctus. Brain Res 17:447–462

Harting JK, Noback CR (1971) Subcortical projections from the visual cortex in the tree shrew (Tupaia glis). Brain Res 25:21–33

Harting JK, Van Lieshout DP (1991) Spatial relationships of axons arising from the substantia nigra, spinal trigeminal nucleus, and pedunculopontine tegmental nucleus within the intermediate gray of the cat superior colliculus. J Comp Neurol 305:543–558

Harting JK, Hall WC, Diamond IT (1972) Evolution of the pulvinar. Brain Behav Evol 6:424–452

Harting JK, Hall WC, Diamond IT, Martin GF (1973) Anterograde degeneration study of the superior colliculus in Tupaia glis: evidence for a subdivision between superficial and deep layers. J Comp Neurol 148:361–386

Harting JK, Huerta MF, Frankfurter AJ, Strominger NL, Royce GJ (1980) Ascending pathways from the monkey superior colliculus: an autoradiographic analysis. J Comp Neurol 192:853–882

Harting JK, Huerta MF, Hashikawa T, Weber JT, Van Lieshout DP (1988) Neuroanatomical studies of the nigrotectal projection in the cat. J Comp Neurol 278:615–631

Harting JK, Huerta MF, Hashikawa T, Van Lieshout DP (1991a) Projection of the mammalian superior colliculus upon the dorsal lateral geniculate nucleus: organization of tectogeniculate pathways in nineteen species. J Comp Neurol 304:275–306

Harting JK, Van Lieshout DP, Hashikawa T, Weber JT (1991b) The parabigeminogeniculate projection: connectional studies in eight mammals. J Comp Neurol 305:559–581

Harting JK, Updyke BV, Van Lieshout DP (1992) Corticotectal projections in the cat: anterograde transport studies of twenty-five cortical areas. J Comp Neurol 324:379–414

Hartmann-von Monakow K, Akert K, Kunzle H (1979) Projections of precentral and premotor cortex to the red nucleus and other midbrain areas in macaca fascicularis. Exp Brain Res 34:91–105

Hartwich-Young R, Weber JT (1986) The projection of frontal cortical oculomotor areas to the superior colliculus in the domestic cat. J Comp Neurol 253:342–357

Hassler R, Muhs-Clement K (1964) Architektonischer Aufbau des sensomotorischen und parietalen Cortex der Katze. J Hirnforsch 6:377–420

Hatschek R (1907) Zur vergleichenden Anatomie des Nucleus ruber tegmenti. Arb Neurol Inst Wiener Univ 15:89–136

Hatton GI, Yang QZ (1989) Supraoptic nucleus afferents from the main olfactory bulb. II. Intracellularly recorded responses to lateral olfactory tract stimulation in rat brain slices. Neuroscience 31:289–297

Haug H (1970) Der makroskopische Aufbau des Grosshirns. Springer, Berlin Heidelberg New York

Haun F, Eckenrode TC, Murray M (1992) Habenula and thalamus cell transplants restore normal sleep behaviors disrupted by denervation of the interpeduncular nucleus. J Neurosci 12:3282–3290

Hawkes R, Leclerc N (1987) Antigenic map of the rat cerebellar cortex: the distribution of sagittal bands as revealed by monoclonal anti-Purkinje cell antibody mabQ113. J Comp Neurol 256:29–41

Hayashi H (1980) Distributions of vibrissae afferent fiber collaterals in the trigeminal nuclei as revealed by intra-axonal injection of horseradish peroxidase. Brain Res 183:442–446

Hayes L, Totterdell S (1985) A light and electron microscopic study of non-pyramidal hippocampal cells that project to the medial nucleus accumbens. Neurosci Lett [Suppl] 22:S507

Hayes NL, Rustioni A (1980) Spinothalamic and spinomedullary neurons in macaques: a single and double retrograde tracer study. Neuroscience 5:861–874

Hayes NL, Rustioni A (1981) Descending projections from brainstem and sensorimotor cortex to spinal enlargements in the cat. Single and double retrograde tracer studies. Exp Brain Res 41:89–107

Hayhow WR (1959) An experimental study of the accessory optic fiber system in the cat. J Comp Neurol 113:281–313

Hayhow WR (1966) The accessory optic system in the marsupial phalanger. Trichosurus vulpecula. An experimental degeneration. J Comp Neurol 126:653–672

Hayhow WR, Sefton A, Webb C (1962) Primary optic centers of the rat in relation to the terminal distribution of the crossed and uncrossed optic nerve fibers. J Comp Neurol 118:295–321

Haymaker W (1969) Hypothalamo-pituitary neural pathways and the circulatory syste, of the pituitary. In: Haymaker W, Anderson E, Nauta WJH (eds) The hypothalamus. Thomas, Springfield, pp 219–250

Hazlett JC, Dom R, Martin GF (1972) Spinobulbar, spinothalamic and medial lemniscal connections in the American Opossum, Didelphis marsupialis virginiana. J Comp Neurol 146:95–118

Hazrati L-N, Parent A (1992) The striatopallidal projection displays a high degree of anatomical specificity in the primate. Brain Res 592:213–227

Heath CJ, Jones EG (1971) The anatomical organization of the suprasylvian gyrus of the cat. Ergeb Anat Entwicklungsgesch 45:1–64

Hebb DO (1949) The organization of behavior. Wiley, New York, pp 62–66

Hedreen JC, DeLong MR (1991) Organization of striatopallidal, striatonigral, and nigrostriatal projections in the macaque. J Comp Neurol 304:569–595

Hedreen JC, Struble RG, Whitehouse PJ, Price DL (1984) Topography of the magnocellular basal forebrain system in the human brain. J Neuropathol Exp Neurol 43:1–21

Heffner R, Masterton B (1975) Variation in form of the pyramidal tract and its relationship to digital dexterity. Brain Behav Evol 12:161–200

Heffner RS, Masterton RB (1983) The role of the corticospinal tract in the evolution of human digital dexterity. Brain Behav Evol 23:165–183

Heimer L (1968) Synaptic distribution of centripetal and centrifugal nerve fibers in the olfactory system of the rat. An experimental anatomical study. J Anat 103:413–432

Heimer L (1969) The secondary olfactory connections in mammals, reptiles and sharks. Ann NY Acad Sci 167:129–146

Heimer L (1976) The olfactory cortex and the ventral striatum, In: Livingston KE, Hornykiewicz O (eds) Limbic mechanisms: the continuing evolution of the limbic system concept. Plenum, New York, pp 95–187

Heimer L, Nauta WJH (1969) The hypothalamic distribution of the stria terminalis in the rat. Brain Res 13:284–297

Heimer L, Wilson RD (1975) The subcortical projections of the allocortex: similarities in the neural associations of the hippocampus, the piriform cortex, and the neocortex. In: Santini M (ed) Golgi centennial symposium proceedings. Raven, New York, pp 177–193

Heimer L, Harlan RE, Alheid GF, Garcia MM, de Olmos J (1997) Substantia innominata: a notion which impedes clinical-anatomical correlations in neuropsychiatric disorders. Neuroscience 76:957–1006

Heimer L, Switzer RD, Van Hoesen GW (1982) Ventral striatum and ventral pallidum. Components of the motor system? Trends Neurosci 5:83–87

Heimer L, Zahm DS, Churchill L, Kalivas PW, Wohltmann C (1991) Specificity in the projection patterns of accumbal cord and shell in the rat. Neuroscience 41:89–125

Held H (1893) Die centrale Gehörleitung. Arch Anat Physiol Anat Abt 201–248

Helfert RH, Schwartz IR (1986) Morphological evidence for the existence of multiple neuronal classes in the cat lateral superior olivary nucleus. J Comp Neurol 22:533–549

Heller SB, Ulinski PS (1987) Morphology of geniculocortical axons in turtles of the genera Pseudemys and Chrysemys. Anat Embryol (Berl) 175:505–515

Hendrickson A, Wilson ME, Toyne MJ (1970) The distribution of optic nerve fibers in Macaca mulatta. Brain Res 23:425–427

Hendry SHC, Houser CR, Jones EG, Vaughn JE (1983) Synaptic organization of immunocytochemically characterized GABAergic neurons in the monkey sensory-motor cortex. J Neurocytol 12:639–660

Hendry SHC, Jones EG (1981) Sizes and distribution of intrinsic neurons incorporating tritiated GABA in monkey sensory-motor cortex. J. Neurosci 1:390–408

Hendry SHC, Jones EG (1983a) The organization of pyramidal and non-pyramidal cell dendrites in relation to thalamic afferent terminations in the monkey somatic sensory cortex. J Neurocytol 12:277–298

Hendry SHC, Jones EG (1983b) Thalamic inputs to identified commissural neurons in the monkey somatic sensory cortex. J Neurocytol 12:299–316

Hendry SH, Jones EG, Graham J (1979) Thalamic relay nuclei for cerebellar and certain related fiber systems in the cat. J Comp Neurol 185:679–713

Henkel CK, Martin GF (1977a) The vestibular complex of the American opossum (Didelphis virginiana). I. Conformation cytoarchitecture and primary vestibular input. J Comp Neurol 172:299–320

Henkel CK, Martin GF (1977b) The vestibular complex of the American opossum (Didelphis virginiana). II. Afferent and efferent connections. J Comp Neurol 172:321–348

Henkel CK, Shneiderman A (1988) Nucleus sagulum: projections of a lateral tegmental area to the inferior colliculus in the cat. J Comp Neurol 271:577–588

Henkel CK, Brunso-Bechtold JK (1993) Laterality of superior olive projections to the inferior colliculus in adult and developing ferret. J Comp Neurol 331:458–468

Henkel CK, Linauts M, Martin GF (1975) The origin of the annulo-olivary tract with notes on other mesencephalo-olivary pathways. A study by the horseradish peroxidase method. Brain Res 100:145–150

Herbert H, Saper C (1992) Organization of medullary adrenergic and noradrenergic projections to the periaqueductal gray matter in the rat. J Comp Neurol 315:34–52

Herbert H, Moga MM, Saper CB (1990) Connections of the parabrachial nucleus with the nucleus of the solitary tract and the medullary reticular formation in the rat. J Comp Neurol 293:540–580

Herbert H, Aschoff A, Ostwald J (1991) Topography of projections from the auditory cortex to the inferior colliculus in the rat. J Comp Neurol 304:103–122

Herbert J (1993) Peptides in the limbic system: neurochemical codes for co-ordinated adaptive responses to behavioural and physiological demand. Prog Neurobiol 41:723–791

Herbin M, Reperant J, Cooper HM (1994) Visual system of the fossorial mole-lemmings, Ellobius talpinus and Ellobius lutescens. J Comp Neurol 346(2):253–275

Herkenham M (1979) The afferent and efferent connections of the ventromedial thalamic nucleus in the rat. J Comp Neurol 183:487–518

Herkenham M (1980a) Laminar organization of thalamic projections to the rat neocortex. Science 207:532–535

Herkenham M (1980b) The laminar organization of thalamic projections to neocortex. Trends Neurosci 3:17–18

Herkenham M (1986) New perspectives on the organization and evolution of nonspecific thalamocortical projections. In: Jones EG, Peters A (eds) Sensory-motor areas and aspects of cortical connectivity. Plenum, New York, pp 403–446 (Cerebral cortex, vol 5)

Herkenham M, Nauta WJH (1977) Afferent connections of the habenular nuclei in the rat: a horseradish peroxidase study, with a note on the fiber-of-passage problem. J Comp Neurol 173:123–146

Herkenham M, Nauta WJH (1979) Efferent connections of the habenular nuclei in the rat. J Comp Neurol 187:19–48

Herkenham M, Pert CB (1981) Mosaic distribution of opiate receptors, parafascicular projection and acetylcholinesterase in rat striatum. Nature 291:415–418

Herrick CJ (1899) The cranial and first spinal nerves of Menidia: a contribution upon the nerve components of the bony fishes. J Comp Neurol 9:153–455

Herrick CJ (1910) The morphology of the forebrain in Amphibia and Reptilia. J Comp Neurol 20:413–547

Herrick CJ (1913) Anatomy of the brain. In: The reference handbook of the medical sciences, vol 12. Wood, New York, pp 274–342

Herrick CJ (1924) The nucleus olfactorius anterior of the opossum. J Comp Neyrol 37:317–359

Herrick CJ (1927) The amphibian forebrain. IV. The cerebral hemispheres of Amblystoma. J Comp Neurol 43:231–325

Herrick CJ (1933) The amphibian forebrain. VI. Necturus. J Comp Neurol 58:1–288

Herrick CJ (1948) The brain of the tiger salamander. University of Chicago Press, Chicago

Hersch SM, White EL (1981a) Thalamocortical synapses involving identified neurons in mouse primary somatosensory cortex: a terminal degeneration and Golgi/EM study. J Comp Neurol 195:252–263

Hersch SM, White EL (1981b) Quantification of synapses formed with apical dendrites of Golgi impregnated pyramidal cells: variability in thalamocortical inputs and consistency in the ratios of asymmetrical to symmetrical synapses. Neuroscience 6:1043–1051

Hersch SM, White EL (1981c) Thalamocortical synapses with corticothalamic projection neurons in mouse SmI cortex: electron microscopic demonstration of a monosynaptic feedback loop. Neurosci Lett 24:207–210

Hersch SM, White EL (1982) A quantitative study of the thalamocortical and other synapses in layer IV of pyramidal cells projecting from mouse SmI cortex to the caudate-putamen nucleus. J Comp Neurol 211:217–225

Hershkovitz P (1970) Cerebral fissural patterns in platyrrhine monkeys. Folia Primatol 13:213–240

Herzog AG, Van Hoesen GW (1976) Temporal neocortical afferent connections to the amygdala in the rhesus monkey. Brain Res 115:57–69

Hess DT (1982) The tecto-olivo-cerebellar pathway in the rat. Brain Res 250:143–148

Hess DT, Hess A (1986) 5'-Nucleotidase of cerebellar molecular layer: reduction in Purkinje cell-deficient mutant mice. Dev Brain Res 29:93–100

Hess DT, Voogd J (1986) Chemoarchitectonic zonation of the monkey cerebellum. Brain Res 369:383–387

Hess WR (1954) Diencephalon: autonomic and extrapyramidal functions. Grune and Stratton, New York

Hess WR, Brugger M (1943) Das subkortikale Zentrum der affektiven Abwehrreaktion. Helv Physiol Acta 1:33–52

Highstein SM, Reisine H (1979) Synaptic and functional organization of vestibulo-ocular reflex pathways. Prog Brain Res 50:431–442

Highstein SM, Karabelas A, Baker R, McCrea RA (1982) Comparison of the morphology of physiologically identified abducens motor and internuclear neurons in the cat: a light microscopic study employing the intracellular injection of horseradish peroxidase. J Comp Neurol 208:369–381

Higo S, Kawano J, Matsuyama T, Kawamura S (1992) Differential projections to the superior collicular layers from the perihypoglossal nuclei in the cat. Brain Res 599:19–28

Hilton SM, Zbrozyna AW (1963) Amygdaloid region for defence reactions and its efferent pathways to the brainstem. J Physiol (Lond) 165:160–173

Hines M (1929) The brain of Ornithorhynchus anatinus (Monotremata). Philos Trans R Soc (Lond) Ser B 217:155–187

Hinrichsen CFL, Watson CD (1984) The facial nucleus of the rat: representation of facial muscles revealed by retrograde transport of horseradish peroxidase. Anat Rec 209:407–415

Hirai T, Onodera SK (1982) Cerebellotectal projections studied in cats with horseradish peroxidase or tritiated amino acids axonal transport. Exp Brain Res 48:1–12

Hockfield S (1987) A Mab to a unique cerebellar neuron generated by immunosuppression and rapid immunization. Science 237:67–70

Hoddevik GH, Brodal A, Kawamura K, Hashikawa T (1977) The pontine projection to the cerebellar vermal visual area studied by means of retrograde axonal transport of horseradish peroxidase. Brain Res 123:209–227

Hofbauer A, Dräger UC (1985) Depth segregation of retinal ganglion cells projecting to mouse superior colliculus. J Comp Neurol 234:465–474

Hoffmann K-P, Ballas I, Wagner H-J (1984) Double labelling of retinofugal projections in the cat: a study using anterograde transport of 3H-proline and horseradish peroxidase. Exp Brain Res 53:420–430

Hoffmann KP, Distler C, Erickson R (1991) Functional projections from striate cortex and superior temporal sulcus to the nucleus of the optic tract (NOT) and dorsal terminal nucleus of the accessory optic tract (DTN) of macaque monkeys. J Comp Neurol 313(4):707–724

Hoffmann KP, Distler C, Ilg U (1992) Callosal and superior temporal sulcus contributions to receptive field properties in the macaque monkey's nucleus of the optic tract and dorsal terminal nucleus of the accessory optic tract. J Comp Neurol 321:150–162

Hofman MA (1985) Size and shape of the cerebral cortex in mammals. I. The cortical surface. Brain Behav Evol 27:28–40

Hökfelt T, Fuxe K, Goldstein M, Johansson O (1973) Evidence for adrenaline neurons in the rat brain. Acta Physiol Scand 89:286–288

Hökfelt T, Fuxe K, Goldstein M, Johansson O (1974) Immunohistochemical evidence for the existence of adrenaline neurons in the rat brain. Brain Res 66:235–251

Hökfelt T, Philipson G, Goldstein M (1979) Evidence for a dopaminergic pathway in the rat descending from the A11 cell group to the spinal cord. Acta Physiol Scand 107:393–395

Hökfelt T, Johansson O, Goldstein M (1984a) Central catecholamine neurons as revealed by immuno- histochemistry with special reference to adrenaline neurons. In: Bjorklund A, Hökfelt T (eds) Handbook of chemical neuroanatomy, classical transmitters in the CNS, vol 2, part I. Elsevier Science, Amsterdam, pp 157–276

Hökfelt T, Martensson R, Bjorklund A, Kleinau S, Goldstein M (1984b) Distributional maps of tyrosine-hydroxylase-immuno reactive neurons in the rat brain. In: Bjorklund A, Hökfelt T (eds) Handbook of chemical neuroanatomy. Classical transmitters in the CNS, part I. Elsevier Science, Amsterdam, pp 277–379 (Handbook of chemical neuroanatomy, vol 2)

Holländer H (1974) On the origin of the corticotectal projections in the cat. Exp Brain Res 21:433–439

Holländer H, Sanides D (1976) The retinal projection to the ventral part of the lateral geniculate nucleus: an experimental study with silver-impregnation and axoplasmatic protein tracing. Exp Brain Res 26:329–342

Holloway RL, LaCoste-Lareymondie MC (1982) Brain endocast asymmetry in Pongids and Hominids: some preliminary findings on the paleontology of cerebral dominance. Am J Phys Anthropol 58:108–110

Holmgren N (1922) Points of view concerning forebrain morphology in lower vertebrates. J Comp Neurol 34:391–440

Holmgren N, van der Horst CJ (1925) Contribution to the morphology of the brain of Ceratodus. Acta Zool 6:59–161

Holst M-C, Ho RH, Martin GF (1991) The origins of supraspinal projections to lumbosacral and cervical levels of the spinal cord in the gray short-tailed brazilian opossum, Monodelphis domestica. Brain Behav Evol 38:273–289

Holstege G (1987a) Anatomical evidence for an ipsilateral rubrospinal pathway and for direct rubrospinal projections to motoneurons in the cat. Neurosci Lett 74:269–274

Holstege G (1987b) Some anatomical observations on the projections from the hypothalamus to brainstem and spinal cord: an HRP and autoradiographic tracing study in the cat. J Comp Neurol 260:98–126

Holstege G (1990) Subcortical limbic system projections to caudal brainstem and spinal cord. In: Paxinos G (ed) The human nervous system. Academic, San Diego, pp 261–286

Holstege G (1991) Descending motor pathways and the spinal motor system: limbic and non-limbic components. Prog Brain Res 87:307–421

Holstege G (1992) The emotional motor system. Eur J Morphol 30:67–79

Holstege G (1995) The basic, somatic and emotional components of the motor system in mammals. In: Paxinos G (ed) The rat nervous system, 2nd edn. Academic, San Diego, pp 137–154

Holstege G, Collewijn H (1982) The efferent connections of the nucleus of the optic tract and the superior colliculus in the rabbit. J Comp Neurol 209:139–175

Holstege G, Cowie RJ (1989) Projections from the rostral mesencephalic reticular formation to the spinal cord. An HRP and autoradiographical tracing study in the cat. Exp Brain Res 75:265–279

Holstege G, Kuypers HGJM (1977) Propriobulbar fibre connections to the trigeminal facial and hypoglossal motor nuclei. I. An anterograde degeneration study in the cat. Brain 100:239–264

Holstege G, Kuypers HGJM (1982a) The anatomy of brain stem pathways to the spinal cord in cat. In: Kuypers HGJM, Martin GF (eds) Descending pathways to the spinal cord. Elsevier, Amsterdam (Progress in brain research, vol 57)

Holstege JC, Kuypers HG (1982b) Brain stem projections to spinal motoneuronal cell groups in rat studied by means of electron microscopy autoradiography. Progr Brain Res 57:177–183

Holstege G, Tan J (1988) Projections from the red nucleus and surrounding areas to the brainstem and spinal cord in the cat. An HRP and auto- radiographical tracing study. Behav Brain Res 28:33–57

Holstege G, Kuypers HGJM, Dekker JJ (1977) The organization of the bulbar fibre connections to the trigeminal, facial and hypoglossal motor nuclei. II. An autoradiographic tracing study in cat. Brain 100:265–286

Holstege G, Graveland G, Bijker-Biemond C, Schuddeboom I (1983) Location of motoneurons innervating soft palate, pharynx and upper esophagus. Anatomical evidence for a possible swallowing center in the pontine reticular formation. An HRP and autoradiographical tracing study. Brain Behav Evol 23:46–62

Holstege G, Tan J, Van Ham J, Bos A (1984) Mesencephalic projections to the facial nucleus in the cat. An autoradiographical tracing study. Brain Res 311:7–22

Holstege G, Meiners L, Tan K (1985) Projections of the bed nucleus of the stria terminalis to the mesencephalon, pons, and medulla oblongata in the cat. Exp Brain Res 58:379–391

Holstege G, Griffiths D, De Wall H, Dalm E (1986) Anatomical and physiological observations on supraspinal control of bladder and urethral sphincter muscles in the cat. J Comp Neurol 250:449–461

Holstege G, Van Neerven J, Evertse F (1987) Spinal cord location of the motoneurons innervating the abdominal, cutaneous maximus, latissimus dorsi and longissimis dorsi muscles in the cat. Exp Brain Res 67:179–194

Holstege G, Bandler R, Saper CB (1996) The emotional motor system. Prog Brain Res 107:1–627

Holstege JC (1996) The ventro-medial medullary projections to spinal motoneurons: ultrastructure, transmitters and functional aspects. In: Holstege G, Bandler R, Saper CB (eds) The emotional motor system. Prog Brain Res 107:159–181

Holstege JC, Kuypers HGJM (1987a) Brainstem projections to lumbar motoneurons in rat. I. An ultrastructural study using autoradiography and the combination of autoradiography and horseradish peroxidase histochemistry. Neuroscience 21:345–367

Holstege JC, Kuypers HGJM (1987b) Brainstem projections to spinal motoneurons: an update. Neuroscience 23:809–821

Honeycutt RL, Nedbal MA, Adkins RM, Janecek LL (1995) Mammalian mitochondrial DNA evolution: a comparison of the cytochrome b and cytochrome c oxidase II genes. J Mol Evol 40:260–272

Hoogland PV, Vermeulen-van der Zee E (1988) Intrinsic and exttrinsic connections of the cerebral cortex of lizards. In: Schwerdtfeger WK, Smeets WJAJ (eds) The forebrain of reptiles. Karger, Basel, pp 20–29

Hooks MS, Kalivas PW (1995) The role of mesoaccumbens-pallidal circuitry in novelty-induced behavioral activation. Neuroscience 64:587–597

Hoover DB, Jacobowitz DM (1979) Neurochemical and histochemical studies on the effect of a lesion of the nucleus cuneiformis on the cholinergic innervation of discrete areas of the rat brain. Brain Res 170:113–122

Hopkins DA (1975) Amygdalotegmental projections in the rat, cat and rhesus monkey. Neurosci Lett 1:263–270

Hopkins DA, Armour JA (1982) Medullary cells of origin of physiologically identified cardiac nerves in the dog. Brain Res Bull 8:359–365

Hopkins DA, Holstege G (1978) Amygdaloid projections to the mesencephalon, pons and medulla oblongata in the cat. Exp Brain Res 32:529–547

Hopkins DA, Holstege G (1978) Amygdaloid projections to the mesencephalon, pons and medulla oblongata in the cat. Exp Brain Res 32:529–548

Hopkins DA, Niessen LW (1976) Substantia nigra projections to the reticular formation, superior colliculus and central gray in the rat, cat and monkey. Neurosci Lett 2:253–259

Hopkins DA, Niessen LW (1976) Substantia nigra projections to the reticular formation, superior colliculus and central gray in the rat, cat and monkey. Neurosci Lett 2:253–259

Horner CH (1993) Plasticity of the dendritic spine. Prog Neurobiol 41:281–321

Hornung JP, Garey LJ (1981) The thalamic projection to cat visual cortex: ultrastructure of neurons indentified by Golgi impregnation of retrograde horseradish peroxidase transport. Neuroscience 6:1053–1068

Hosoya Y, Matsushita M (1981) Brainstem projections from the lateral hypothalamic area in the rat, as studied with autoradiography. Neurosci Lett 24:111–116

Houser CR, Vaughn JE, Barber RP, Roberts E (1980) GABA neurons are the major cell type of the nucleus reticularis thalami. Brain Res 200:341–354

Houser CR, Vaughn JE, Hendry SHC, Jones EG, Peters A (1984) GABA neurons in the cerebral cortex. In: Jones EG, Peters A (eds) Functional properties of cortical cells. Plenum, New York, pp 63–89 (Cerebral cortex, vol 2)

Houser CR, Crawford GD, Salvaterra PM, Vaughn JE (1985) Immunocytochemical localization of choline acetyltransferase in rat cerebral cortex: a study of cholinergic neurons and synapses. J Comp Neurol 234:17–34

Hubbard JE, DiCarlo V (1974a) Fluorescence histochemistry of monoamine-containing cell bodies in the brain stem of the squirrel monkey (Saimiri sciureus) (1974) II. Catecholamine-containing groups. J Comp Neurol 153:369–384

Hubbard JE, DiCarlo V (1974b) Fluorescence histochemistry of monoamine-containing cell bodies in the brain stem of the squirrel monkey (Saimiri sciureus). III. Serotonin-containing groups. J Comp Neurol 153:385–398

Hubel DH (1975) An autoradiographic study of the retinocortical projections in the tree shrew (Tupaia glis). Brain Res 96:41–50

Hubel DH, LeVay S, Wiesel TN (1975) Mode of termination of retinotectal fibers in macaque monkey: an autoradiographic study. Brain Res 96:25–40

Hubel DH, Wiesel TN (1962) Receptive fields, binocular interaction and functional architecture in the cat's visual cortex. J Physiol (Lond) 160:106–154

Hubel DH, Wiesel TN (1968) Receptive fields and functional architecture of monkey visual cortex. J Physiol (Lond) 195:215–243

Hubel DH, Wiesel TN (1972) Laminar and columnar distribution of geniculo-cortical fibers in the macaque monkey. J Comp Neurol 146:421–450

Hubel DH, Wiesel TN (1977) Functional architecture of macaque monkey cortex. Proc R Soc Lond B 198:1–59

Hubel DH, Wiesel TN, Stryker MP (1978) Anatomical demonstration of orientation columns in macaque monkey. J Comp Neurol 177:361–379

Hübener M, Schwarz C, Bolz J (1990) Morphological types of projection neurons in layer 5 of cat visual cortex. J Comp Neurol 301:655-674

Huber GC, Guild SR (1913) Observations on the peripheral distribution of the nervus terminalis in mammals. Anat Rec 7:253-272

Huerta M, Harting JK (1982a) Projections of the superior colliculus to the supraspinal nucleus and the cervical spinal cord gray of the cat. Brain Res 242:326-331

Huerta M, Harting JK (1982b) The projection from the nucleus of the posterior commissure to the superior colliculus of the cat: patch-like endings within the intermediate and deep grey layers. Brain Res 238:426-432

Huerta MF, Harting JK (1984) Connectional organization of the superior colliculus. Trends Neurosci 7:286-289

Huerta MF, Frankenfurter AJ, Harting JK (1981) The trigeminocollicular projection in the cat: patch-like endings within the intermediate gray. Brain Res 211:1-15

Huerta MF, Frankfurter A, Harting JK (1983) Studies of the principal sensory and spinal trigeminal nuclei of the rat: projections to the superior colliculus, inferior olive, and cerebellum. J Comp Neurol 220:147-167

Huerta MF, Krubitzer LA, Kaas JH (1986) Frontal eye field as defined by intracortical microstimulation in squirrel monkeys, owl monkeys, and macaque monkeys: I. Subcortical connections. J Comp Neurol 253:415-439

Huerta MF, Van Lieshout DP, Harting JK (1991) Nigrotectal projections in the primate Galago crassicaudatus. Exp Brain Res 87:389-401

Huffman RF, Henson JOW (1990) The descending auditory pathway and acousticomotor systems: connections with the inferior colliculus. Brain Res Rev 15:295-323

Huisman AM (1983) Collateralization of descending spinal pathways from red nucleus and other brainstem cell groups in rat, cat and monkey. Thesis, Erasmus University Rotterdam

Huisman AM, Kuypers HGJM, Conde F, Keizer K (1983) Collaterals of rubrospinal neurons to the cerebellum in rat. A retrograde fluorescent double labeling study. Brain Res 264:181-196

Huisman AM, Kuypers HGJM, Verburgh CA (1981) Quantitative differences in collateralization of the descending spinal pathways from red nucleus and other brain stem cell groups in rat as demonstrated with the multiple fluorescent retrograde tracer technique. Brain Res 209:271-286

Huisman AM, Kuypers HGJM, Verburgh CA (1982) Differences in collateralization of the descending spinal pathways from red nucleus and other brainstem cell groups in cat and monkeys. In: Kuypers HMG (ed) Anatomy of descending pathways to the spinal cord. Elsevier, Amsterdam (Progress in brain research, vol 57)

Humphrey DR, Gold R, Reed DJ (1984) Sizes, laminar and topographical origins of cortical projections to the major divisions of the red nucleus in the monkey. J Comp Neurol 225:75-94

Humphrey T (1936) The telencephalon of the bat. I. The noncortical nuclear masses and certain pertinent fiber connections. J Comp Neurol 65:603-711

Husten K (1924) Experimentelle Untersuchungen über die Beziehungen der Vaguskerne zu den Brust- und Bauchorganen. Z Ges Neurol Psychiat 93:763-773

Hutchins B (1991) Evidence for a direct retinal projection to the anterior pretectal nucleus in the cat. Brain Res 561:169-173

Hutchins B, Weber JT (1985) The pretectal complex of the monkey: a reinvestigation of the morphology and retinal terminations. J Comp Neurol 232:425-442

Hutchins KD, Martino AM, Strick PL (1988) Corticospinal projections from the medial wall of the hemisphere. Exp Brain Res 71:667-672

Iggo A, Gregory JE, Proske U (1992) The central projection of electrosensory information in the platypus. J Physiol (Lond) 447:449-465

Ikeda M, Matsoshita M, Tanami T (1982) Termination and cells of origin of the ascending intranuclear fibers in the spinal trigeminal nucleus of the cat. A study with the horseradish peroxidase technique. Neurosci Lett 31:215-220

Ilg UJ, Hoffmann K-P (1993) Functional grouping of the cortico-pretectal projection. J Neurophysiol 70:867-869

Ilinsky IA, Jouandet ML, Goldman-Rakic PS (1985) Organization of the nigro-thalamocortical system in the rhesus-monkey. J Comp Neurol 236:315-330

Illing RB (1988) Spatial relation of the acetylcholinesterase-rich domain to the visual topography in the feline superior colliculus. Exp Brain Res 73:589-594

Illing RB (1990) Choline acetyltransferase-like immunoreactivity in the superior colliculus of the cat and its relation to the pattern of acetylcholinesterase staining. J Comp Neurol 296:32-46

Illing RB (1992) Association of efferent neurons to the compartmental architecture of the superior colliculus. Proc Natl Acad Sci USA 89:10900-10904

Illing RB (1993) More modules. Trends Neurosci 16:179-180

Illing RB, Graybiel AM (1985) Convergence of afferents from frontal cortex and substantia nigra into acetylcholinesterase-rich patches of the cat's superior colliculus. Neuroscience 14:455-482

Illing RB, Graybiel AM (1986) Complementary and nonmatching afferent compartments in the cat's superior colliculus innervation of the acetylcholinesterase-poor domain of the intermediate gray layer. Neuroscience 18:373-384

Illing RB, Wässle H (1981) The retinal projection to the thalamus in the cat: a quantitative investigation and a comparison with the retinotectal pathway. J Comp Neurol 202:265-285

Imai Y, Kusama T (1969) Distribution of the dorsal root fibers in the cat. An experimental study with the Nauta method. Brain Res 13:338-359

Inglis WL, Winn P (1995) The pedunculopontine tegmental nucleus: where the striatum meets the reticular formation. Prog Neurobiol 47:1-29

Ingram WR (1940) Nuclear organization and chief connections of the primate hypothalamus. Res Nerv Ment Dis Proc 20:195-244

Innocenti GM (1986) General organization of callosal connections in the cerebral cortex. In: Jones EG, Peters A (eds) Sensory-motor areas and aspects of cortical connectivity. Plenum, New York, pp 291-354 (Cerebral cortex, vol 5)

Ino T, Itoh K, Kamiya H, Shigemoto R, Akiguchi I, Mizuno N (1988) Direct projections of non-pyramidal neurons of Ammon's horn to the supramamillary region in the cat. Brain Res 460:173-177

Irvine DRF (1986) The auditory brainstem. Prog Sens Physiol 7:1-279

Isa T, Itouji T (1992) Axonal trajectories of single Forel's field H neurones in the mesencephalon, pons and medulla oblongata in the cat. Exp Brain Res 89:484-495

Ishizuka N, Mannen H, Hongo T, Sasaki S (1979) Trajectory of group Ia afferent fibers stained with horseradish peroxidase in the lumbosacral spinal cord of the cat: three dimensional reconstructions from serial sections. J Comp Neurol 186:819-212

Ishizuka N, Sasaki S-I, Mannen H (1982) Central course and terminal arborizations of single primary vestibular afferent fibers from the horizontal canal in the cat. Neurosci Lett 33:135-139

Isokawa-Akesson M, Komisaruk BR (1987) Difference in projections to the lateral and medial facial nucleus: anatomically separate pathways for rhythmical vibrissa movement in rats. Exp Brain Res 65:385-398

Isomura G (1981) Comparative anatomy of the extrinsic ocular muscles in vertebrates. Anat Anz Jena 150:498-515

Isseroff A, Schwartz ML, Dekker JJ, Goldman-Rakic PS (1984) Columnar organization of callosal and associational projections from rat frontal cortex. Brain Res 293:213-223

Itaya SK, Van Hoesen W (1982) WGA-HRP as a transneuronal marker in the visual pathways of monkey and rat. Brain Res 236:199-204

Ito J, Matsuoka I, Sasa M, Takaori S, Morimoto M (1983) Input to lateral vestibular nucleus as revealed by retrograde horseradish peroxidase. Adv Oto Rhino Laryngol 30:64-70

Ito M (1984) The cerebellum and neural control. Raven, New York

Itoh J, Matsuoka I, Sasa M, Takaori S, Morimoto M (1983a) Input to lateral vestibular nucleus as revealed by retrograde horseradish peroxidase. Adv Oto Rhino Laryngol 30:64-70

Itoh K, Takada M, Yasui Y, Kudo M, Mizuno N (1983b) Direct projections from the anterior pretectal nucleus to the dorsal accessory olive in the cat: an anterograde and retrograde WGA-HRP study. Brain Res 272:350-353

Itoh K, Kaneko T, Kudo M, Mizuno N (1984) The intercollicular region in the cat: a possible relay in the parallel somatosensory pathways from the dorsal column nuclei to the posterior complex of the thalamus. Brain Res 308:166-171

Itoh K, Nomura S, Konishi A, Yasui Y, Sugimoto T, Mizuno N (1986) A morphological evidence of direct connections from the ocular nuclei to tensor tympani motoneurons in the cat: a possible afferent limb of the acoustic middle ear reflex pathways. Brain Res 375:214-219

Iwai E, Yukie M (1987) Amygdalofugal and amygdalopetal connections with modality-specific visual cortical areas in macaques (Macaca fuscata, M. mulatta and M. fascicularis). J Comp Neurol 261:362-387

Jackson A, Crossman AR (1981) Subthalamic projections to nucleus tegmenti pedunculopontinus efferent connections with special reference to the basal ganglia, studied in the rat by anterograde and retrograde transport of horseradish peroxidase. Neuroscience 22:17-22

Jackson A, Crossman AR (1983) Nucleus tegmenti pedunculopontinus efferent connections with special reference to the basal ganglia, studied in the rat by anterograde and retrograde transport of horseradish peroxidase. Neuroscience 10:725-765

Jacobs BL, Gannon PJ, Azmitia EC (1984) Atlas of serotonergic cell bodies in the cat brainstem: an immunocytochemical analysis. Brain Res Bull 13:1-31

Jacobs MS, Morgane PJ, McFarland WL (1971) The anatomy of the brain of the bottlenose dolphin (Tursiops truncatus). Rhinic lobe (Rhinencephalon) I. The paleocortex. J Comp Neurol 141:205-272

Jacobson S, Trojanowski JQ (1975) Amygdaloid projections to prefrontal granular cortex in rhesus monkey demonstrated with horseradish peroxidase. Brain Res 100:132-139

Jacquin MF, Rhoades RW, Enfiejian HL, Egger MD (1983) Organization and morphology of masticatory neurons in the rat: a retrograde HRP study. J Comp Neurol 218:239-256

Jacquin MF, Mooney RD, Rhoades RW (1986) Morphology, response properties, and collateral projections of trigeminothalamic neurons in brainstem subnucleus interpolaris of rat. Exp Brain Res 61:457-468

Jakab RL, Leranth C (1995) Septum. In: Paxinos G (ed) The rat nervous system. Academic, San Diego, pp 405-442

Jankowska E (1992) Interneuronal relay in spinal pathways from proprioceptors. Prog Neurobiol 38:335-378

Jankowska E, Lindström S (1971) Morphological identification of Renshaw cells. Brain Res 20:323-326

Jansen J (1954) On the morphogenesis and morphology of the mammalian cerebellum. In: Jansen J, Brodal A (eds) Aspects of cerebellar anatomy, chap 1. Grundt Tanum, Oslo, pp 13-81

Jansen J (1969) Neurobiology of cerebellar evolution and development. In: Llinas R (ed) Proceedings of the first international symposium of the Institute for Medical Research. American Medical Association, Chicago, pp 881-893

Jansen J, Brodal A (1940) Experimental studies on the intrinsic fibers of the cerebellum. II. The cortico-nuclear projection. J Comp Neurol 73:267-321

Jansen J, Brodal A (1942) Experimental studies on the intrinsic fibers of the cerebellum. III Cortico-nuclear projection in the rabbit and the monkey. Avh Norske Vid Akad.Avh I Math Nat Kl 3:1-30

Jasmin L, Courville J, Bakker DA (1985) Afferent projections from forelimb muscles to the external and main cuneate nuclei in the cat. A study with trans-ganglionic transport of horseradish peroxidase. Anat Embryol (Berl) 171:275-284

Jayaraman A, Batton RR III, Carpenter MB (1977) Nigrotectal projections in the monkey: an autoradiographic study. Brain Res 135:147-152

Jelgersma G (1934) Das Gehirn der Wassersäugetiere. Barth, Leipzig

Jen LS, Dai Z-G, So K-F (1984) The connections between the parabigeminal nucleus and the superior colliculus in the golden hamster. Neurosci Lett 51:189-194

Jeneskog T, Padel Y (1984) An excitatory pathway through dorsal columns to rubrospinal cells in the cat. J Physiol (Lond) 353:355-373

Jennes L (1987) The nervus terminalis in the mouse: light and electron microscopic immunocytochemical studies. Ann NY Acad Sci 519:165-173

Jenny AB, Smith JM, Bernardo KL, Woolsey TA (1991) Distribution of motor cortical neuron synaptic terminals on monkey parvocellular red neurons. Somatosens Mot Res 8:23-26

Jeon CJ, Mize RR (1993) Choline acetyltransferase-immunoreactive patches overlap specific efferent cell groups in the cat superior colliculus. J Comp Neurol 337:127-150

Jeon CJ, Spencer RF, Mize RR (1993) Organization and synaptic connections of cholinergic fibers in the cat superior colliculus. J Comp Neurol 333:360-374

Jerison HJ (1973) Evolution of the brain and intelligence. Academic, New York

Jhamandas JH, Petrov T, Harris KH, Vu T, Krukoff TL (1996) Parabrachial nucleus projection to the amygdala in the rat: electrophysiological and anatomical observations. Brain Res Bull 39:115-126

Ji Z, Hawkes R (1994) Topography of Purkinje cell compartments and mossy fiber terminal fields in lobules II and III of the rat cerebellar cortex: spino-cerebellar and cuneo-cerebellar projections. Neuroscience 61:935-954

Jia H-G, Rao Z-R, Shi J-W (1994) An indirect projection from the nucleus of the solitary tract to the central nucleus of the amygdala via the parabrachial nucleus in the rat: a light and electron microscopic study. Brain Res 663:181-190

Joel D, Weiner I (1994) The organization of the basal ganglia-thalamocortical circuits: open interconnected rather than closed segregated. Neuroscience 63:363-379

Johnson JJI, Welker WI, Pubols JBH (1968) Somatotopic organization of raccoon orsal column nuclei. J Comp Neurol 132:1-44

Johnson JI (1980) Morphological correlates of specialized elaborations in somatic sensory cerebral neocortex. In: Ebbesson SOE (ed) Comparative neurology of the telencephalon. Plenum, New York, pp 423-447

Johnson JI (1990) Comparative development of somatic sensory cortex. In: Jones EG, Peters A (eds) Comparative structure and evolution of cerebral cortex, part II. Plenum, New York, pp 335-449 (Cerebral cortex, vol 8B)

Johnson JI Jr (1977) Central nervous system of marsupials. In: Hunsaker D II (ed) The biology of marsupials. Academic, New York, pp 157-278

Johnston JB (1913) Nervus terminalis in reptiles and mammals. J Comp Neurol 23:97-120

Johnston JB (1914) The nervus terminalis in man and mammals. Anat Rec 8:185-198

Johnston JB (1923) Further contributions to the study of the evolution of the forebrain. J Comp Neurol 35:337-481

Jones BE (1990) Immunohistochemical study of choline acetyltransferase-immunoreactive processes and cells innervating the pontomedullary reticular formation in the rat. J Comp Neurol 295:485-514

Jones BE (1995) Reticular formation: cytoarchitecture, transmitters, and projections. In: Paxinos G (ed) The rat nervous system, 2nd edn. Academic, San Diego, pp 155-171

Jones BE, Beaudet A (1987) Distribution of acetylcholine and catecholamine neurons in the cat brainstem: a choline acetyltransferase and tyrosine hydroxylase immunohistochemical study. J Comp Neurol 261:15-32

Jones BE, Friedman L (1983) Atlas of catecholamine perikarya, varicosities and pathways in the brainstem of the cat. J Comp Neurol 215:382-396

Jones BE, Yang TZ (1985) The efferent projections from the reticular formation and the locus coeruleus studied by anterograde and retrograde axonal transport in the rat. J Comp Neurol 242:56-92

Jones EG (1975a) Some aspects of the organization of the thalamic reticular complex. J Comp Neurol 162:285-308

Jones EG (1975b) Varieties and distribution of non-pyramidal cells in the somatic sensory cortex of the squirrel monkey. J Comp Neurol 160:205-268

Jones EG (1984a) Laminar distribution of cortical efferent cells. In: Peters A, Jones EG (eds) Cellular components of the cerebral cortex. Plenum, New York, pp 521-553 (Cerebral cortex, vol 1)

Jones EG (1984b) Neurogliaform or spiderweb cells. In: Peters A, Jones EG (eds) Cellular components of the cerebral cortex. Plenum, New York, pp 409-418 (Cerebral cortex, vol 1)

Jones EG (1985) The thalamus. Plenum, New York

Jones EG (1986) Connectivity of the primate sensory-motor cortex. In: Jones EG, Peters A (eds) Sensory-motor areas and aspects of cortical connectivity. Plenum, New York, pp 113-184 (Cerebral cortex, vol 5)

Jones EG (1987) Cerebral cortex. In: Adelman G (ed) Encyclopedia of neuroscience, vol 1. Birkhäuser, Boston, pp 209-211

Jones EG, Burton H (1976) A projection from the medial pulvinar to the amygdala in primates. Brain Res 104:142-147

Jones EG, Peters A (1984-1992) Cerebral cortex, vol 1-9. Plenum, New York

Jones EG, Powell TPS (1969) Synapses on the axon hillocks and initial segments of pyramidal cell axons in the cerebral cortex. J Cell Sci 5:495-507

Jones EG, Powell TPS (1970) An anatomical study of converging sensory pathways within the cerebral cortex of the monkey. Brain 93:793-824

Jones EG, Wise SP (1977) Size, laminar and columnar distribution of efferent cells in the sensory-motor cortex of monkeys. J Comp Neurol 175:391-438

Jones EG, Burton H, Porter R (1975) Commissural and cortico-cortical "columns" in the somatic sensory cortex of primates. Science 190:572-574

Jones EG, Hendry SHC, DeFelipe J (1987) GABA-peptide neurons of the primate cerebral cortex: a limited cell class. In: Jones EG, Peters A (eds) Further aspects of cortical function, including hippocampus. Plenum, New York, pp 237-266 (Cerebral cortex, vol 6)

Jones EG, DeFelipe J, Hendry SHC, Maggio JE (1988) A study of tachykinin-immunoreactive neurons in monkey cerebral cortex. J Neurosci 8:1206-1224

Jones MW, Hodge CJ, Apkarian AV, Stevens RT (1985) A dorsolateral spinothalamic pathway in cat. Brain Res 335:188-193

Jones MW, Apkarian AV, Stevens RT, Hodge CJJ (1987) The spinothalamic tract: an examination of the cells of origin of the dorsolateral and ventral spinothalamic pathways in cats. J Comp Neurol 260:349-361

Jongen-Rélo AL, Groenewegen HJ, Voorn P (1993) Evidence for a multi-compartmental histochemical organization of the nucleus accumbens in the rat. J Comp Neurol 337:267-276

Jongen-Rélo AL, Voorn P, Groenewegen HJ (1994) Immunohistochemical characterization of the shell and core territories of the nucleus accumbens in the rat. Eur J Neurosci 6:1255-1264

Joosten EA, Gribnau AA (1988) Unmyelinated corticospinal axons in adult rat pyramidal tract. An electron microscopic tracer study. Brain Res 459:173-177

Jordan H, Holländer H (1972) The structure of the ventral part of the lateral geniculate nucleus. A cyto- and myeloarchitectonic study in the cat. J Comp Neurol 145:259-272

Joseph MP, Guinan JJJ, Fullerton BC, Norris BE, Kiang NYS (1985) Number and distribution of stapedius motoneurons in cats. J Comp Neurol 232:43-54

Jürgens U (1974) The hypothalamus and behavioral patterns. Prog Brain Res 41:445-463

Kaada BR (1972) Stimulation and regional ablation of the amygdaloid cortex with reference to functional representations. In: Eleftheriou BE (ed) The neurobiology of the amygdala. Plenum, New York, pp 205-282

Kaas JH, Harting JK, Guillery RW (1974) Representation of the complete retina in the contralateral superior colliculus of some mammals (1971) Brain Res 65:343-346

Kadoya S, Wolin LR, Massopust LC Jr (1971) Photically evoked unit activity in the tectum opticum of the squirrel monkey. J Comp Neurol 142:495-508

Kahle W (1956) Zur entwicklung des menschlichen Zwischenhirnes. Dtsch Z Nervenhkd 175:259-318

Kakei S, Shinoda Y (1990) Parietal projection of thalamocortical fibers from the ventroanterior-ventrolateral complex of the cat thalamus. Neurosci Lett 117:280-284

Kalia M, Mesulam M-M (1980a) Brain stem projections of sensory and motor components of the vagus complex in the cat. I. The cervical vagus and nodose ganglion. J Comp Neurola 193:435-465

Kalia M, Mesulam M-M (1980b) Brain stem projections of sensory and motor components of the vagus complex in the cat. II. Laryngeal, tracheobroncial, pulmonary, cardiac, and gastrointestinal branches. J Comp Neurol 193:467-508

Kalia M, Sullivan M (1982) Brainstem projections of sensory and motor components of the vagus nerve in the rat. J Comp Neurol 211:248-264

Källén B (1951a) The nuclear development in the mammalian forebrain with special regard to the subpallium. Kungl Fysiografiska Sällskapets Handlingar N F vol 61, no 9. Gleerup, Lund

Källén B (1951b) Embryological studies on the nuclei and their homologization in the vertebrate forebrain. Kungl Fysiografiska Sällskapets Handlingar N F vol 62, no 5. Gleerup, Lund

Kamal AM, Tömböl T (1975) Golgi studies on the amygdaloid nuclei of the cat. J Hirnforsch 16:175-201

Kamal AM, Tömböl T (1976) Olfactory and temporal projections to the amygdala. Verh Anat Ges 70:283-288

Kanagasuntheran R, Wong WC, Krishnamurti A (1968) Nuclear configuration of the diencephalon in some Lorisoids. J Comp Neurol 133:241-268

Kanaseki T, Sprague JM (1974) Anatomical organization of pretectal nuclei and tectal laminae in the cat. J Comp Neurol 38:319-338

Kane ES, Barone LM (1980) The dorsal nucleus of the lateral lemniscus in the cat: neuronal types and their distributions. J Comp Neurol 192:797-826

Kane ES, Finn RC (1977) Descending and intrinsic inputs to dorsal cochlear nucleus of cats: a horseradish peroxidase study. Neuroscience 2:897-912

Kappel RM (1981) The development of the cerebellum in macaca mulatta. A study of regional differences during corticogenesis. Thesis, Leiden

Karabelas AB, Moschovakis AK (1985) Nigral inhibitory termination on efferent neurons of the superior colliculus: an intracellular horseradish peroxidase study in the cat. J Comp Neurol 239:309-329

Karamanlidis A (1968) Trigemino-cerebellar fiber connections in the goat studied by means of the retrograde cell degeneration method. J Comp Neurol 133:71-88

Karamanlidis AN, Magras J (1972) Retinal projections in domestic ungulates. I. The retinal projections in the sheep and the pig. Brain Res 44:127-145

Karamanlidis AN, Magras J (1974) Retinal projections in domestic ungulates (1978) II. The retinal projections in the horse and the ox. Brain Res 66:209-225

Karamanlidis AN, Voogd J (1970) Trigemino-thalamic fibre connections in the goat. An experimental anatomical study. Acta Anat 75:596-622

Karamanlidis AN, Michaloudi H, Mangana O, Saigal RP (1978) Trigeminal ascending projections in the rabbit, studied with horseradish peroxidase. Brain Res 156:110-116

Kassel J (1982) Somatotopic organization of SI corticotectal projections in rats. Brain Res 231:247-255

Kato I, Harada K, Hasegawa T, Igarashi T, Koike Y, Kawasaki (1973) T. Role of the nucleus of the optic tract in monkeys in relation to optokinetic nystagmus. Brain Res 364:12-22

Kato N, Kawaguchi S, Miyata H (1984) Geniculocortical projection to layer I of area 17 in kittens: orthograde and retrograde HRP studies. J Comp Neurol 225:441-447

Kato N, Kawaguchi S, Miyata H (1986) Postnatal development of afferent projections to the lateral suprasylvian visual area in the cat: an HRP study. J Comp Neurol 252:543-554

Katz LC (1987) Local circuitry of identified projection neurons in cat visual cortex brain slices. J Neurosci 7:1223-1249

Kaufman EFS, Rosenquist AC (1985) Efferent projections of the thalamic intralaminar nuclei in the cat. Brain Res 335:257-279

Kawaguchi Y (1993) Physiological, morphological and histochemical characterization of three classes of interneurons in rat neostriatum. J Neurosci 13:4908-4923

Kawaguchi Y, Hama K (1987) Two subtypes of non-pyramidal cells in rat hippocampal formation identified by intracellular recording and HRP injection. Brain Res 411:190-195

Kawamura K, Brodal A (1973) The tectopontine projection in the cat: an experimental anatomical study with comments on pathways for teleceptive impulses to the cerebellum. J Comp Neurol 149:371-390

Kawamura K, Hashikawa T (1978) Cell bodies of origin of reticular projections from the superior colliculus in the cat: an experimental study with the use of horseradish peroxidase as a tracer. J Comp Neurol 1978:1-16

Kawamura K, Konno T (1979) Various types of corticotectal neurons of cats as demonstrated by means of retrograde axonal transport of horseradish peroxidase. Exp Brain Res 35:161-175

Kawamura K, Onodera S (1984) Olivary projections from the pretectal region in the cat studied with horseradish peroxidase and tritiated amino acids axonal transport. Arch Ital Biol 122:155-168

Kawamura K, Brodal A, Hoddevik G (1974) The projection of the superior colliculus onto the reticular formation of the brain stem. An experimental anatomical study in the cat. Exp Brain Res 19:1-19

Kawamura S, Sprague JM, Niimi K (1974) Corticofugal projections from the visual cortices to the thalamus, pretectum and superior colliculus in the cat. J Comp Neurol 158:339-362

Kawamura S, Hattori S, Higo S, Matsuyama T (1982) The cerebellar projections to the superior colliculus and pretectum in the cat: an autoradiographic and horseradish peroxidase study. Neuroscience 7:1673-1689

Kawana E, Kusama T (1968) Projections from the anterior part of the coronal gyrus to the thalamus, the spinal trigeminal complex and the nucleus of the solitary tract in cats. Proc Jpn Acad 44:176-181

Keizer K (1989) Collateralization of the pathways descending from the cerebral cortex to brain stem and spinal cord in cat and monkey. Thesis, Erasmu University Rotterdam

Keizer K, Kuypers HGJM (1984) Distribution of corticospinal neurons with collaterals to lower brain stem reticular formation in cat. Exp Brain Res 54:107-120

Keizer K, Kuypers HGJM (1989) Distribution of corticospinal neurons with collaterals to the lower brain stem reticular formation in monkey (Macaca fascicularis). Exp Brain Res 74:311-318

Keizer K, Kuypers HGJM, Ronday HK (1987) Branching cortical neurons in cat which project to the colliculi and to the pons: a retrograde fluorescent double-labeling study. Exp Brain Res 67:1-15

Kelley AE, Domesick VB, Nauta WJH (1982) The amygdalostriatal projection in the rat - an anatomical study by anterograde and retrograde tracing methods. Neuroscience 7:615-630

Kelly J, Swanson LW (1980) Additional forebrain regions projecting to the posterior pituitary: preoptic region, bed nucleus of the stria terminalis and zona incerta. Brain Res 197:1-10

Kemp JM, Powell TPS (1970) The cortico-striate projection in the monkey. Brain 93:525-547

Kemp JM, Powell TPS (1971a) The structure of the caudate nucleus of the cat: light and electron microscopy. Philos Trans R Soc Lond [Biol] 262:383-401

Kemp JM, Powell TPS (1971b) The site of the termination of afferent fibers in the caudate nucleus. Philos Trans R Soc Lond [Biol] 262:413-427

Kemplay S, Webster KE (1989) A quantitative study of the projections of the gracile, cuneate and trigeminal nuclei and of the medullary reticular formation to the thalamus in the rat. Neuroscience 32:153-167

Kennedy PR (1990) Corticospinal, rubrospinal and rubro-olivary projections: a unifying hypothesis. Trends Neurosci 13:474-479

Kennedy PR, Gibson AR, Houk JC (1986) Functional and anatomic differentiation between parvi- cellular and magnocellular regions of red nucleus in the monkey. Brain Res 364:124-136

Kenny GC, Scheelings FT (1979) Observations of the pineal region of non-eutherian mammals. Cell Tissue Res 198:309-324

Kerr FWL (1962) Facial, vagal and glossopharyngeal nerves in the cat. Arch Neurol 6:264-281

Kerr FWL (1963) The divisional organization of afferent fibers of the trigeminal nerve. Brain 86:721-732

Kerr FW (1975) The ventral spinothalamic tract and other ascending systems of the ventral funiculus of the spinal cord. J Comp Neurol 159:335-356

Kesarev VS, Malofeyeva LI, Trykova OV (1977) Ecological specificity of cetacean neocortex. J Hirnforsch 18:447-460

Kevetter GA, Perachio AA (1986) Distribution of vestibular afferents that innervate the sacculus and posterior canal in the gerbil. J Comp Neurol 254:410-424

Kevetter GA, Perachio AA (1989) Projections from the sacculus to the cochlear nuclei in the Mongolian gerbil. Brain Behav Evol 34:193-200

Kevetter GA, Willis WD (1984) Collateralization in the spinothalamic tract: new methodology to support or deny phylogenetic theories. Brain Res Rev 7:1-14

Kevetter GA, Winans SS (1981) Connections to the corticomedial amygdala in the golden hamster. I. Efferents of the 'vomeronasal amygdala'. J Comp Neurol 197:81-98

Kiang NYS, Liberman MC, Gage JS, Northrop CC, Dodds LW, Oliver ME (1984) Afferent innervation of the mammalian cochlea. In: Bolis L, Keynes RD, Madrell SHP (eds) Comparative physiology of sensory systems. Cambridge University Press, Cambridge, pp 143-161

Kievit J (1979) Cerebello-thalamische projecties en de afferente verbindingen naar de frontaalschors in de rhesus aap. Thesis, Erasmus University Rotterdam

Kievit J, Kuypers HGJM (1975) Basal forebrain and hypothalamic connections to frontal and parietal cortex in the rhesus monkey. Science 187:660-662

Killackey HP (1983) The somatosensory cortex of the rodent. Trends Neurosci 6:425-429

Killackey HP, Erzurumlu RS (1981) Trigeminal projections to the superior colliculus of the rat. J Comp Neurol 201:221-242

Killackey HP, Koralek K-A, Chiaia NL, Rhoades RW (1989) Laminar and areal differences in the origin of the subcortical projection neurons of the rat somatosensory cortex. J Comp Neurol 282:428-445

Kim U, Gregory E, Hall WC (1992) Pathway from the zona incerta to the superior colliculus in the rat. J Comp Neurol 321:555–575

Kincaid AE, Wilson CJ (1996) Corticostriatal innervation of the patch and matrix in the rat neostriatum. J Comp Neurol 374:578–592

King GW (1980) Topology of ascending brainstem projections to nucleus parabrachialis in the cat. J Comp Neurol 191:651–638

King JS, Dom RM, Conner JB, Martin GF (1973) An experimental light and electron microscopic study of cerebellorubral projections in the opossum, didelphis marsupialis virginiana. Brain Res 52:61–78

King WM, Precht W, Dieringer N (1980) Synaptic organization of frontal eye field and vestibular afferents to interstitial nucleus of Cajal in the cat. J Neurophysiol 43:912–928

Kirsche W, Kunz G, Wenzel J, Wenzel M, Winkelmann A, Winkelmann E (1973) Neurohistologische Untersuchungen zur Variabilität der Pyramidenzellen des sensorischen Cortex der Ratte. J Hirnforsch 14:117–135

Kisvárday ZF, Eysel UT (1992) Cellular organization of reciprocal patchy networks in layer III of cat visual cortex (area 17). Neuroscience 46:275–286

Kisvárday ZF, Martin KAC, Whitteridge D, Somogyi D (1985) Synaptic connections of intracellularly filled clutch cells. A type of small basket cell in the visual cortex of the cat. J Comp Neurol 241:111–137

Kisvárday ZF, Martin KAC, Freund TF, MaglóczkyZS, Whitteridge D, Somogyi P (1986) Synaptic targets of HRP-filled layer III pyramidal cells in the cat striate cortex. Exp Brain Res 64:541–552

Kisvárday ZF, Beaulieu C, Eysel UT (1993) Network of GABAergic large basket cells in cat visual cortex (area 18): implication for lateral disinhibition. J Comp Neurol 327:398–415

Kita H, Oomura Y (1981) Reciprocal connections between the lateral hypothalamus and the frontal cortex in the rat: electrophysiological and anatomical observations. Brain Res 213:1–16

Kita Y, Oomura Y (1982) An HRP study of the afferent connections to rat lateral hypothalamic region. Brain Res Bull 8:53–71

Kita H, Kosaka T, Heizmann CW (1990) Parvalbumin-immunoreactive neurons in the rat neostriatum: a light and electron microscopic study. Brain Res 536:1–15

Kitao Y, Nakamura Y (1987) An ultrastructural analysis of afferent terminals to the anterior pretectal nucleus in the cat. J Comp Neurol 259:348–363

Kitao Y, Nakamura Y, Kudo M, Morrizumi T, Tokuno H (1989) The cerebral and cerebellar connections of pretecto-thalamic and pretecto-olivary neurons in the anterior pretectal nucleus of the cat. Brain Res 484:304–313

Kitt CA, Mitchell SJ, DeLong MR, Wainer BH, Price DL (1987) Fiber pathways of basal forebrain cholinergic neurons in monkeys. Brain Res 406:192–206

Klein BG, Rhoades RW (1985) Representation of whisker follicle intrinsic musculature in the facial motor nucleus of the rat. J Comp Neurol 232:55–69

Klemm WR, Vertes RP (eds) (1990) Brainstem mechanisms of behavior. Wiley, New York

Klüver H, Bucy PC (1937) Psychic blindness and other symptoms following bilateral temporal lobectomy in rhesus monkeys. Am J Physiol 119:352–353

Klüver H, Bucy PC (1939) Preliminary analysis of functions of the temporal lobes in monkeys. Arch Neurol Psychiatry 42:979–1000

Kling A, Steklis HD, Deutsch S (1979) Radiotelemetered activity from the amygdala during social interactions in the monkey. Exp Neurol 66:688–696

Kling AS, Brothers LA (1992) The amygdala and social behavior. In: Aggleton JP (ed) The amygdala: neurobiological aspects of emotion, memory, and mental dysfunction. Wiley-Liss, New York, pp 353–378

Klinkhachorn PS, Haines DE, Culberson JL (1984a) Cerebellar cortical efferent fibers in the North American Opossum, Didelphis virginiana. I. The anterior lobe. J Comp Neurol 227:424–438

Klinkhachorn PS, Haines DE, Culbertson JL (1984b) Cerebellar cortical efferent fibers in the North American Opossum, Didelphis virginiana. II. The Posterior Vermis. J Comp Neurol 227:439–451

Klooster J, Van der Want JJL, Vrensen G (1983) Retinopretectal projections in albino and pigmented rabbits: an autoradiographic study. Brain Res 288:1–12

Kneisley LW, Biber MP, LaVail JH (1978) A study of the origin of brain stem projections to monkey spinal cord using the retrograde transport method. Exp Neurol 60:116–139

Knowles WD, Schwartzkroin PA (1981) Axonal ramifications of hippocampal CA1 pyramidal cells. J Neurosci 1:1236–1241

Kobayashi Y, Matsumura G (1996) Central projections of primary afferent fibers from the rat trigeminal nerve labeled with isolectin B4-HRP. Neuroscience 217:89–92

Koch M, Ehret G (1989) Immunocytochemical localization and quantitation of estrogen-binding cells in the male and female (virgin, pregnant, lactating) mouse brain. Brain Res 489:101–112

Koester SE, O'Leary DM (1992) Functional classes of corical projection neurons develop dendritic distinctions by class-specific sculpting of an early common pattern. J Neurosci 12:1382–1393

Köhler C (1985) Intrinsic projections of the retrohippocampal region in the rat brain. I. The subicular complex. J Comp Neurol 236:504–522

Koikegami H (1957) On the correlation between cellular and fibrous patterns of the human brain stem reticular formation with some cytoarchitectonic remarks on the other mammals. Acta Med Biol 5:21–72

Koikegami H (1963) Amygdala and other related limbic structures; experimental studies on the anatomy and function. Acta Med Biol 10:161–277

Kojima T (1951) On the brain of the sperm whale (Physeter catadon). Sci Rep Whales Res Inst Tokyo 6:49–72

Kojima M, Sano Y (1983) The organization of serotonin fibers in the anterior column of the mammalian spinal cord. An immunohistochemical study. Anat Embryol (Berl) 167:1–11

Komiyama M, Shibata H, Suzuki T (1984) Somatotopic representation of facial muscles within the facial nucleus of the mouse. Brain Behav Evol 24:144–151

Kooy FH (1916) The inferior olive in vertebrates. Proefschrift. De Erven F Bohn, Haarlem

Kooy FH (1971) The inferior olive in vertebrates. Folia Neurobiol 10:205–369

Kordower JH, Piecinski P, Rakic P (1992) Neurogenesis of the amygdaloid nuclear complex in the rhesus monkey. Dev Brain Res 68:9–15

Korneliussen HK (1967) Cerebellar corticogenesis in Cetacea, with special references to regional variations. J Hirnforsch 9:151–185

Korneliussen HK (1968a) On the morphology and subdivision of the cerebellar nuclei of the rat. J Hirnforsch 10:109–122

Korneliussen HK (1968b) On the ontogenetic development of the cerebellum (nuclei, fissures and cortex) of the rat, with special reference to regional variations in corticogenesis. J Hirnforsch 10:379–412

Korneliussen HK (1968c) Comments on the cerebellum and its division. Brain Res 8:229–236

Korneliussen HK, Jansen J (1965) On the early development and homology of the central cerebellar nuclei in cetacea. J Hirnforsch 8:47–66

Korte G (1979) The brainstem projection of the vestibular nerve in the cat. J Comp Neurol 184:279–292

Krasnoshchekova EI, Figurina II (1980) The cortical projection of the medial geniculate body of the dolphin brain. Arkh Anat Gistol Embriol 78:19–24

Kraus G, Pilleri G (1969) Quantitative Untersuchungen über die Grosshirnrinde der Cetaceen. Inv Cetacea 1:127–150

Krettek JE, Price JL (1974a) A direct input from the amygdala to the thalamus and the cerebral cortex. Brain Res 67:169–174

Krettek JE, Price JL (1974b) Projections from the amygdala to the perirhinal and entorhinal cortices and the subiculum. Brain Res 71:150–154

Krettek JE, Price JL (1977a) The cortical projections of the mediodorsal nucleus and adjacent thalamic nuclei in the rat. J Comp Neurol 171:157–192

Krettek JE, Price JL (1977b) Projections from the amygdaloid complex to the cerebral cortex and thalamus in the rat and cat. J Comp Neurol 172:687–722

Krettek JE, Price JL (1977c) Projections from the amygdaloid complex and adjacent olfactory structures to the entorhinal cortex and to the subiculum in the rat and cat. J Comp Neurol 172:723–752

Krettek JE, Price JL (1978a) Amygdaloid projections to subcortical structures within the basal forebrain and brainstem in the rat and cat. J Comp Neurol 178:225–253

Krettek JE, Price JL (1978b) A description of the amygdaloid complex in the rat and cat with observations on intra-amygdaloid axonal connections. J Comp Neurol 178:255–280

Krieg WJS (1932) The hypothalamus of the albino rat. J Comp Neurol 55:19–89

Krieger MS, Conrad LCA, Pfaff DW (1979) An autoradiographic study of the efferent connections of the ventromedial nucleus of the hypothalamus. J Comp Neurol 183:785–816

Kriegstein AR (1987) Synaptic responses of cortical pyramidal neurons to light stimulation in the isolated turtle visual system. J Neurosci 7:2488–2492

Kriegstein AR, Connors BW (1986) Cellular physiology of the turtle visual cortex: synaptic properties and intrinsic circuitry. J Neurosci 6:178–191

Kruger L (1959) The thalamus of the dolphin (Tursiops truncatus) and comparison with other mammals. J Comp Neurol 111:133–194

Kruger L (1966) Specialized features of the Cetacean brain. In: Norris KS (ed) Whales, dolphins, and porpoises. University of California Press, Berkeley, pp 232–254

Kruger L, Michel F (1962) A morphological and somatotopic analysis of single unit activity in the trigeminal sensory complex of the cat. Exp Neurol 5:139–156

Krukoff TL, Harris KH, Jhamandas JH (1993) Efferent projections from the parabrachial nucleus demonstrated with the anterograde tracer Phaseolus vulgaris leucoagglutinin. Brain Res Bull 30:163–172

Kruska D (1973) Cerebralisation, Hirnevolution und domestikationbedingte Hirngrössenänderungen innerhalb der Ordnung Perissodactyla Owen, 1848 und ein Vergleich mit der Ordnung Artiodactyla Owen, 1848. Z Zool Syst Evol Forsch 11:81–103

Kruska D (1982) Hirngrösseänderungen bei Tylopoden während der Stammesgeschichte und in der Domestikation. Verh Dtsch Zool Ges 75:173–183

Kubota Y, Kawaguchi Y (1993) Spatial distributions of chemically identified intrinsic neurons in relation to patch and matrix compartments of rat neostriatum. J Comp Neurol 332:499–513

Kubota Y, Inagaki S, Kito S (1986) Innervation of substance P neurons by catecholaminergic terminals in the neostriatum. Brain Res 375:163–167

Kudo M (1981) Projections of the nuclei of the lateral lemniscus in the cat: an autoradiographic study. Brain Res 221:57–69

Kudo M, Yamammoto M, Nakamura Y (1991) Suprachiasmatic nucleus and retinohypothalamic projections in moles. Brain Behav Evol 34:332–338

Kuhlenbeck H (1975) The central nervous system of vertebrates. Spinal cord and deuterencephalon, vol 4. Karger, Basel

Kuhlenbeck H (1954) The human diencephalon. Karger, Basel

Kuhlenbeck H (1977) The central nervous system of vertebrates, vol 5, part 1: derivatives of the prosencephalon: diencephalon and telencephalon. Karger, Basel

Kume M, Uemura M, Matsuda K, Ryotaro M, Mizuno N (1978) Topographical representation of peripheral branches of the facial nerve within the facial nucleus: a HRP study in the cat. Neurosci Lett 8:5–8

Kunkel DD, Hendrickson AE, Wu JY, Schwartzkroin PA (1986) Glutamic acid decarboxylase (GAD) immunocytochemistry of developing rabbit hippocampus. J Neurosci 6:541–552

Kunkel DD, Lacaille J-C, Schwartzkroin PA (1988) Ultrastructure of stratum lacunosum-molecular interneurons of hippocampal CA1 region. Synapse 2:382–394

Künzle H (1975a) Autoradiographic tracing of the cerebellar projections from the lateral reticular nucleus in the cat. Exp Brain Res 22:255–266

Künzle H (1975b) Bilateral projections from precentral motor cortex to the putamen and other parts of the basal ganglia an autoradiographic study in the macace fascicularis. Brain Res 88:195–209

Künzle H (1978) An autoradiographic analysis of the efferent connections from premotor and adjacent prefrontal regions (areas 6 and 9) in Macaca fascicularis. Brain Behav Evol 15:185–234

Künzle H (1992) Meso-diencephalic regions projecting to spinal cord and dorsal column nuclear complex in the hedgehog-tenrec. Echinops telfairi. Anat Embryol (Berl) 185:57–68

Künzle H (1993) Tectal and related target areas of spinal and dorsal column nuclear projections in hedgehog tenrecs. Somatosens Mot Res 10:339–353

Künzle H, Akert K (1977) Efferent connections of cortical, area 8 (frontal eye field) in Macaca fascicularis. A reinvestigation using the autoradiographic technique. J Comp Neurol 173:147–164

Kurimoto Y, Kawaguchi S, Murata M (1995) Cerebellotectal projection in the rat: anterograde and retrograde WGA-HRP study of individual cerebellar nuclei. Neurosci Res 22:57–71

Kuroda M, Price JL (1991) Synaptic organization of projections from basal forebrain structures to the mediodorsal thalamic nucleus of the rat. J Comp Neurol 303:513–533

Kuypers HGJM (1958a) An anatomical analysis of corticobulbar connexions to the pons and lower brain stem in the cat. J Anatomya 92:198–218

Kuypers HGJM (1958b) Some projections from the pericentral cortex to the pons and lower brain stem in monkey and chimpanzee. J Comp Neurol 110:221–255

Kuypers HGJM (1962) Corticospinal connections: postnatal development in the Rhesus monkey. Science 138:678–680

Kuypers HGJM (1964) The descending pathways to the spinal cord, their anatomy and function. Prog Brain Res 11:178–200

Kuypers HGJM (1973) The anatomical organization of the descending pathways and their contributions to motor control especially in primates. In: Desmedt JE (ed) New developments in electromyography and clinical neurophysiology, vol 3. Karger, Basel, pp 38–68

Kuypers HGJM (1981) Anatomy of the descending pathways. In: Brookhart JM, Mountcastle VB, Brooks VB, Geiger SL (eds) Handbook of physiology, the nervous system, vol II: motor control, part I. American Physiological Society, Bethesda, pp 597–666

Kuypers HGJM (1982) A new look at the organization of the motor system. Prog Brain Res 57:382–403

Kuypers HGJM (1987) Pyramidal tract. In: Adelman G (ed) Encyclopedia of neuroscience, vol II. Birkhäuser, Boston, pp 1018–1020

Kuypers HGJM, Brinkman J (1970) Precentral projections to different parts of the spinal intermediate zone in the rhesus monkey. Brain Res 24:29–48

Kuypers HGJM, Lawrence DG (1967) Cortical projections to the red nucleus and the brain stem in the rhesus monkey. Brain Res 4:151–188

Kuypers HGJM, Lawrence DG (1967) Cortical projections to the red nucleus and the brain stem in the rhesus monkey. Brain Res 4:151–188

Kuypers HGJM, Maisky V (1975) Retrograde axonal transport of horseradish peroxidase from spinal cord to brain stem cell groups in the cat. Neuroscience Lett 1:9–14

Kuypers HGJM, Tuerk JD (1964) The distribution of the cortical fibres within the nuclei cuneatus and gracilis in the cat. J Anat (Lond) 98:143–162

Kuypers HGJM, Fleming WR, Farinholt JW (1960) Descending projections to spinal motor and sensory cell groups in the monkey: cortex versus subcortex. Science 132:38–40

Kuypers HGJM, Hoffman AL, Beasley RM (1961) Distribution of cortical "feedback" fibers in the nuclei cuneatus and gracilis. Soc Biol Med 108:634–637

Kuypers HGJM, Fleming WR, Farinholt JW (1962) Subcorticospinal projections in the Rhesus monkey. J Comp Neurol 118:107–137

Kyuhou S, Matzuzaki R (1991) Topographical organization of the tecto-olivo-cerebellar projection in the cat. Neuroscience 41:227–241

López-Mascaraque L, de Carlos JA, Valverde F (1986) Structure of the olfactory bulb of the hedgehog (Erinaceus europaeus): description of cell types in the granular layer. J Comp Neurol 253:135–152

López-Mascaraque L, de Carlos JA, Valverde F (1990) Structure of the olfactory bulb of the hedgehog (Erinaceus europaeus): a Golgi study of the intrinsic organization of the superficial layers. J Comp Neurol 301:243–261

Labandeira-Garcia JL, Gomez Segade LA, Suarez-Nunez J (1983) Localisation of motoneurons supplying the extraocular muscles of the rat using horseradish peroxidase and fluorescent double labeling. J Anat 137:247–261

Lacaille J-C, Schwartzkroin PA (1988a) Stratum lacunosum-moleculare interneurons of hippocampal CA1 region. I. Intracellular response characteristics, synaptic responses, and morphology. J Neurosci 8:1400–1410

Lacaille J-C, Schwartzkroin PA (1988b) Stratum lacunosum-moleculare interneurons of hippocampal CA1 region. II. Intrasomatic and intradendritic recordings of local circuit synaptic interactions. J Neurosci 8:1411–1424

Lacaille J-C, Williams S (1990) Membrane properties of interneurons in stratum oriens-alveus of the CA1 region of rat hippocampus in vitro. Neuroscience 36:349–359

Lacaille J-C, Mueller AL, Kunkel DD, Schwartzkroin PA (1987) Local circuit interactions between oriens/alveus interneurons and CA1 pyramidal cells in hippocampal slices: electrophysiology and morphology. J Neurosci 7:1979–1993

Lacaille J-C, Kunkel DD, Schwartzkroin PA (1989) Electrophysiological and morphological characterization of hippocampal interneurons. In: Chan-Palay V, Köhler C (eds) The hippocampus – new vistas. Liss, New York, pp 287–305 (Neurology and neurobiology, vol 52)

Laemle LK (1981) A Golgi study of cellular morphology in the superficial layers of superior colliculus man, Saimiri, and Macaca. J Hirnforsch 22:253–263

Laemle LK (1983) A Golgi study of cell morphology in the deep layers of the human superior colliculus. J Hirnforsch 24:297–306

Lagares C, Caballero-Bleda M, Fernandez B, Puelles L (1994) Reciprocal connections between the rabbit suprageniculate pretectal nucleus and the superior colliculus: tracer study with horseradish peroxidase and fluorogold. Vis Neurosci 11:347–353

Lainé J, Axelrad H (1994) The candelabrum cell: a new interneuron in the cerebellar cortex. J Comp Neurol 339:159–173

Lammers HJ (1972) The neural connections of the amygdaloid complex in mammals. In: Eleftheriou BE (ed) The neurobiology of the amygdala: proceedings of a symposium on the neurobiology of the amygdala, Bar Harbor, Maine, 6–17 June, 1971, pp 123–144

Landry P, Labelle A, DeschLnes M (1980) Intracortical distribution of axonal collaterals of pyramidal tract cells in the cat motor cortex. Brain Res 191:327–336

Lane RH, Allman JM, Kaas JH (1971) Representation of the visual field in the superior colliculus of the grey squirrel (Sciurus Carolinensis) and the tree shrew (Tupaia glis). Brain Res 26:277–292

Lane RH, Allman JM, Kaas JH, Miezin FM (1973) The visuotopic organization of the superior colliculus of the owl monkey (Aotus Trivirgatus) and the bush baby (Galago Senegalensis). Brain Res 60:335–349

Lane RH, Kaas JH, Allman JM (1974) Visuotopic organization of the superior colliculus in normal and Siamese cats (1979). Brain Res 70:413–430

Lang EJ, Sugihara I, Llinas R (1996) GABAergic modulation of complex spike activity by the cerebellar nucleoolivary pathway in rat. J Neurophysiol 76:255–275

Lang W, Buttner-Ennever JA, Buttner U (1974) Vestibular projections to the monkey thalamus: an autoradiographic study. Brain Res 177:3–17

Langer T, Fuchs AF, Scudder CA, Chubb MC (1985) Afferents to the flocculus of the cerebellum in the rhesus macaque as revealed by retrograde transport of horseradish peroxidase. J Comp Neurol 235:1–25

Langer T, Kaneko CRS, Scudder CA, Fuchs AF (1986) Afferents to the abducens nucleus in the monkey and cat. J Comp Neurol 245:379–400

Langer TP (1985) Basal interstitial nucleus of the cerebellum: cerebellar nucleus related to the flocculus. J Comp Neurol 235:38–47

Langer TP, Lund RD (1974) The upper layers of the superior colliculus of the rat: a golgi study. J Comp Neurol 158:405–436

Langworthy OR (1932) A description of the central nervous system of the porpoise (Tursiops truncatus). J Comp Neurol 54:437–486

Langworthy OR (1967) A study of the brain of the porpoise, Tursiops truncatus. Brain 31 (54):225–236

Lankamp DJ (1967) Fiber composition of the pedunculus cerebri (Crus cerebri) in man. Thesis. Luctor and Emergo, Leiden

Lapper SR, Bolam JP (1992) Input from the frontal cortex and the parafascicular nucleus to cholinergic interneurons in the dorsal striatum of the rat. Neuroscience 51:533–545

Larsell O (1934) Morphogenesis and evolution of the cerebellum. Arch Neurol 31:373–395

Larsell O (1952) The development of the cerebellum in man in relation to its comparative anatomy. J Comp Neurol 87:85–129

Larsell O (1953) The morphogenesis and adult pattern of the lobules and tissues of the cerebellum of the white rat. J Comp Neurol 97:281–356

Larsell O (1967) The comparative anatomy and histology of the cerebellum from myxinoids birds. In: Jansen J (ed) University of Minnesota Press, Minneapolis

Larsell O (1970a) Cerebellum of cat and monkey. J Comp Neurol 99:135–200

Larsell O (1970b) The comparative anatomy and histology of the cerebellum from momotremes through apes. University of Minnesota Press, Minneapolis

Larsell O, Jansen J (1972) The comparative anatomy and histology of the cerebellum from monotremes through apes. The human cerebellum, cerebellar connections, and the cerebellar cortex. University of Minnesota Press, Minneapolis

Larsen PJ, Hay-Schmidt A, Mikkelsen JD (1994) Efferent connections from the lateral hypothalamic region and the lateral preoptic area to the hypothalamic paraventricular nucleus of the rat. J Comp Neurol 342:299–319

Larson B, Miller S, Oscarsson O (1969a) A spinocerebellar climbing fibre path activated by the flexor reflex afferents from all four limbs. J Physiol (Lond) 203:641–649

Larson B, Miller S, Oscarsson O (1969b) Termination and functional organization of the dorsolateral spino-olivocerebellar path. J Physiol (Lond) 203:611–640

Lavoie B, Parent A (1990) Immunohistochemical study of the serotoninergic innervation of the basal ganglia in the squirrel monkey. J Comp Neurol 299:1–16

Lavoie B, Parent A (1994) The pedunculopontine nucleus in the squirrel monkey. Projections to the basal ganglia as revealed by anterograde tract-tracing methods. J Comp Neurol 344:210–231

Lawn AM (1966) The nucleus ambiguus of the rabbit. J Comp Neurol 127:307–320

Lawrence DG, Kuypers HGJM (1968a) The functional organization of the motor system in the monkey. I. The effects of bilateral pyramidal lesions. Brain 91:1–14

Lawrence DG, Kuypers HGJM (1968b) The functional organization of the motor system in the monkey. II. The effects of lesions of the descending brain-stem pathways. Brain 91:15–36

Lazarov NE, Chouchkov CN (1995) Serotonin-containing projections to the mesencephalic trigeminal nucleus of the cat. Anat Rec 241:136–142

Le Gros Clark WE (1929) Studies of the optic thalamus of the Insectivora: the anterior nuclei. Brain 52:334–358

Le Gros Clark WE (1932) The structure and connections of the thalamus. Brain 55:406–470

Le Gros Clark WE (1938) Morphological aspects of the hypothalamus. In: Le gros Clark WE, Beattie J, Riddoch G, Dott NM (eds) The hypothalamus. Oliver and Boyd, Edinburgh, pp 1–68

Le Gros Clark WE (1945) Deformation patterns in the cerebral cortex. In: Le Gros Clark WE, Medawar PB (eds) Essays on growth and form presented to D'Arcy Wentworth Thompson. Clarendon, Oxford

Leah J, Menetrey D, De Pommery J (1988) Neuropeptides in long ascending spinal tract cells in the rat: evidence for parallel processing of ascending information. Neuroscience 24:195–207

Leake PA, Snyder RL (1989) Topographic organization of the central projections of the spiral ganglion in cats. J Comp Neurol 281:612–629

Leake PA, Snyder RL, Hradek GT (1993) Spatial organization of inner hair cell synapses and cochlear spiral ganglion neurons. J Comp Neurol 333:257–270

Leclerc N, Dore L, Parent A, Hawkes R (1990) The compartmentalization of the monkey and rat cerebellar cortex: zebrin I and cytochrome oxidase. Brain Res 506:70–78

LeDoux JE (1987) Emotion. In: Plum F (ed) Handbook of physiology, vol 1: the nervous system, vol V, higher functions of the brain. American Physiological Society, Bethesda, pp 419–460

LeDoux JE (1992) Emotion and the amygdala. In: Aggleton JP (ed) The amygdala: neurobiological aspects of emotion, memory, and mental dysfunction. Wiley-Liss, New York, pp 339–351

LeDoux JE, Ruggiero DA, Reis DJ (1985) Projections to the subcortical forebrain from anatomically defined regions of the medial geniculate body of the rat. J Comp Neurol 242:172–213

LeDoux JE, Iwata J, Cicchetti P, Reis DJ (1988) Different projections of the central amygdaloid nucleus mediate autonomic and behavioral correlates of conditioned fear. J Neurosci 8:2517–2529

LeDoux JE, Farb C, Ruggiero DA (1990a) Topographic organization of neurons in the acoustic thalamus that project to the amygdala. J Neurosci 10:1043–1054

LeDoux JE, Cicchetti P, Xagoraris A, Romanski LM (1990b) The lateral amygdaloid nucleus: sensory interface of the amygdala in fear conditioning. J Neurosci 10:1062–1069

Lee GY, Chen ST, Shen CL (1991) Autoradiographic study of the retinal projections in the Chinese pangolin, Manis pentadactyla. Brain Behav Evol 37(2):104–110

Lee HJ, Rye DB, Hallenger AE, Levey AI, Wainer BH (1988) Cholinergic vs non-cholinergic efferents from the mesopontine tegmentum to the extrapyramidal motor system nuclei. J Comp Neurol 275:469–492

Leenen L, Meek J, Nieuwenhuys R (1982) Unmyelinated fibers in the pyramidal tract of the rat: a new view. Brain Res 246:297–301

Legendre A, Courville J (1987) Origin and trajectory of the cerebello-olivary projection: an experimental study with radioactive and fluorescent tracers in the cat. Neuroscience 21:877–891

Lehman J, Nagy IJ, Atmadja S, Fibiger HC (1980) The nucleus basalis magnocellularis: the origin of a cholinergic projection to the neocortex to the rat. Neuroscience 5:1161–1174

Lehman MN, Winans SS (1983) Evidence for a ventral nonstriatal pathway from the amygdala to the bed nucleus of the stria terminalis in the male golden hamster. Brain Res 268:139–146

Leichnetz GR (1982a) Connections between the frontal eye field and pretectum in the monkey: an anterograde/retrograde study using HRP gel and TMB neurohistochemistry. J Comp Neurol 207:394–402

Leichnetz GR (1982b) Comment on the center for vertical eye movements in the medial prerubral subthalamic region of the monkey considering some of its frontal cortical afferents. Neurosci Lett 30:95–101

Leichnetz GR, Astruc J (1977) The course of some prefrontal corticofugals to the pallidum, substantia innominata, and amygdaloid complex in monkeys. Exp Neurol 54:104–109

Leichnetz GR, Spencer RF, Smith DJ (1984) Cortical projections to nuclei adjacent to the oculomotor complex in the medial diencephalic tegmentum in the monkey. J Comp Neurol 228:359–387

LeMay M, Geschwind N (1975) Hemispheric differences in the brain of the great apes. Brain Behav Evol 11:48–52

Lende RA (1963a) Sensory representation in the cerebral cortex of the opossum (Didelphis virginiana). J Comp Neurol 121:395–403

Lende RA (1963b) Motor representation in the cerebral cortex of the opossum (Didelphis virginiana). J Comp Neurol 121:405–415

Lende RA (1964) Representation in the cerebral cortex of a primitive mammal. Sensorimotor, visual and auditory fields in the echidna (Tachyglossus aculeatus). J Neurophysiol 27:37–48

Lent R (1982) The organization of subcortical projections of the hamster's visual cortex. J Comp Neurol 206:227–242

Leonard CM, Scott JW (1971) Origin and distribution of the amygdalofugal pathways in the rat: an experimental neuroanatomical study. J Comp Neurol 141:313–329

Leonard CM, Rolls ET, Wilson FA, Baylis GC (1985) Neurons in the amygdala of the monkey with responses selective for faces. Behav Brain Res 15:159–176

Leser O (1925) On the development of the extra-ocular muscles in some mammals. Br J Ophthalmol 9:154–161

LeVay S (1973) Synaptic patterns in the visual cortex of the cat and monkey: Electron microscopy of Golgi preparations. J Comp Neurol 150:53–86

LeVay S, Sherk H (1981) The visual claustrum of the cat. I. Structure and connections. J Neurosci 1:956–980

LeVay S, Hubel DH, Wiesel TN (1975) The pattern of ocular dominance columns in macaque visual cortex revealed by a reduced silver stain. J Comp Neurol 159:559–576

Levitt P, Rakic P, Goldman-Rakic P (1984) Region-specific distribution of catecholamine afferents in primate cerebral cortex: a fluorescence histochemical analysis. J Comp Neurol 227:23–36

Li YQ, Takada M, Ohishi H, Shinonaga Y, Mizuno N (1992) Trigeminal ganglion neurons which project by way of axon collaterals to both the caudal spinal trigeminal and the principal sensory trigeminal nuclei. Brain Res 594:155–159

Lidov HGW, Grzanna R, Molliver ME (1980) The serotonin innervation of the cerebral cortex in the rat – an immunohistochemical analysis. Neuroscience 5:207–227

Lietaert Peerbolte M (1932) De loop van de achterwortelvezels, die in de achterstrengen van het ruggemerg opstijgen. Thesis University of Leiden, The Netherlands

Light AR, Perl ER (1979a) Reexamination of the dorsal tract projection to the spinal dorsal horn including observations on the differential termination of coarse and fine fibers. J Comp Neurol 186:117-132

Light AR, Perl ER (1979b) Spinal termination of functionally identified primary afferent neurons with slowly conducting myelinated fibers. J Comp Neurol 186:133-150

Linauts M, Martin GF (1978) An autoradiographic study of midbrain-diencephalic projections to the inferior olivary nucleus in the opossum (Didelphis virginiana). J Comp Neurol 179:325-354

Linden R, Perry VH (1983a) Massive retinotectal projection in rats. Brain Res 272:145-149

Linden R, Perry VH (1983b) Retrograde and anterograde-transneuronal degeneration in the parabigeminal nucleus following tectal lesions in developing rats. J Comp Neurol 218:270-281

Linden R, Rocha-Miranda CE (1978) Projections from the striate cortex to the superior colliculus in the opossum (Didelphis marsupialis aurita). In: Rocha-Miranda CE, Lent R (eds) Opossum neurobiology (neurobiologia do Gamba). Academia Brasileira de Ciencias, Rio de Janeiro, pp 137-150

Lindström S (1982) Synaptic organization of inhibitory pathways in the cat's lateral geniculate nucleus. Brain Res 164:304-308

Lindvall O, Björklund A (1983) Dopamine- and norepinephrine-containing neuron systems: their anatomy in the rat brain. In: Emson PC (ed) Clinical neuroanatomy. Raven, New York, pp 229-255

Lingenhohl K, Friauf E (1991) Sensory neurons and motoneurons of the jaw-closing reflex pathway in rats: a combined morphological and physiological study using the intracellular horseradish peroxidase technique. Exp Brain Res 83:385-396

Lissauer H (1886) Beitrag zum Faserverlauf im Hinterhorn des menschlichen Rückenmarks und zum Verhalten desselben bei Tabes doralis. Arch Psych 17:377-438

Livingstone MS, Hubel DH (1984) Anatomy and physiology of a color system in the primate visual cortex. J Neurosci 4:309-356

Ljungdahl A, Hökfelt T (1973) Autoradiographic uptake patterns of [3H]GABA and [3H]glycine in central nervous tissues with special. Brain Res 62:587-595

Llamas A, Avendaño C, Reinoso-Suárez F (1977) Amygdaloid projections to prefrontal and motor cortex. Science 195:794-797

Llinás R, Baker R, Sotelo C (1974) Electrotonic coupling between neurons in the cat inferior olive. J Neurophysiol 37:560-571

Lo FS, Sherman SM (1994) Feedback inhibition in the cat's lateral geniculate nucleus. Exp Brain Res 100(2):365-368

Loewy AD (1991) Forebrain nuclei involved in autonomic control. Prog Brain Res 87:253-268

Loewy AD, Burton H (1978) Nuclei of the solitary tract: efferent projections to the lower brain stem and spinal cord of the cat. J Comp Neurol 181:421-450

Loewy AD, Saper CB (1978) Edinger-Westphal nucleus: projections to the brain stem and spinal cord in the cat. Brain Res 150:1-27

Loewy AD, Saper CB, Yamodis ND (1978) Re-evaluation of the efferent projections of the Edinger-Westphal nucleus in the cat. Brain Res 141:151-159

Lohman AHM (1963) The anterior olfactory lobe of the guinea pig. A descriptive and experimental study. Acta Anat 53 [Suppl 49]:1-109

Lohman AHM, Lammers HJ (1967) On the structure and fibre connections of the olfactory centres in mammals. Prog Brain Res 23:65-82

Lohman AHM, Mentink GM (1972) Some cortical connections of the tegu lizard (Tupinambis teguixin). Brain Res 45:325-344

Lohman AHM, Hoogland PV, Witjes RJGM (1988) Projections from the main and accessory olfactory bulbs to the amygdaloid complex in the lizard gekko gecko. In: Schwerdtfeger, WK, Smeets, WJAJ (eds) The forebrain of reptiles. Karger, Basel, pp 41-49

Lohman AHM, van Woerden-Verkley I (1976) The reptilian cortex and some of its connections in the tegu lizard. In: Creutzfeldt OD (ed) Afferent and intrinsic organization of laminated structures in the brain. Exp Brain Res [Suppl] 1:166-170

Loo YT (1930) The forebrain of the opossum, Didelphis virginiana; Part I. Gross anatomy. J Comp Neurol 51:1-64

Loo YT (1931) The forebrain of the opossum, didelphis virginiana. J Comp Neurol 52:1-48

Loopuijt LD, van der Kooy D (1985) Organization of the striatum: collateralization of its efferent axons. Brain Res 348:86-99

López-Garcia C, Martinez-Guijarro FJ (1988a) Neurons in the medial cortex give rise to Timm-positive boutons in the cerebral cortex of lizards. Brain Res 463:205-217

López-Garcia C, Martinez-Guijarro FJ, Berbel P, Garcia-Verdugo JM (1988b) Long-spined polymorphic neurons of the medial cortex of lizards: a Golgi, Timm, and electron-microscopic study. J Comp Neurol 272:409-423

López-Mascaraque L, de Carlos JA, Valverde F (1986) Structure of the olfactory bulb of the hedgehog (Erinaceus europaeus): description of cell types in the granular layer. J Comp Neurol 253:135-152

López-Mascaraque L, de Carlos JA, Valverde F (1990) Structure of the olfactory bulb of the hedgehog (Erinaceus europaeus): a Golgi study of the intrinsic organization of the superficial layers. J Comp Neurol 301:243-261

Lorente de Nó R (1922) La corteza cerebral del ratón. Trabajos Cajal Madrid 20:41-80 (con 26 grabados)

Lorente de Nó R (1933) Vestibulo-ocular reflex arc. Arch Neurol Psychiat 30:245-291

Lorente de Nó R (1934a) Studies on the structure of the cerebral cortex. I. The area entorhinalis. J Psychol Neurol 45:381-439

Lorente de Nó R (1934b) Studies on the structure of the cerebral cortex. II. Continuation of the study of the ammonic system. J Phys Neurol 46:113-177

Lorente de Nó R (1938) The cerebral cortex: architecture, intracortical connections and motor projections. In: Fulton JF (ed) Physiology of the nervous system. Oxford University Press, London, pp 291-325

Loughlin SE, Foote SL, Bloom FE (1986) Efferent projections of nucleus locus coeruleus: topographic organization of cells of origin demonstrated by three-dimensional reconstruction. Neuroscience 18:291-306

Lu GW (1989) Spinocervical tract-dorsal column postsynaptic neurons: a double-projection neuronal system. Somatosens Mot Res 6:445-454

Lu GW, Yang CT (1989) The morphology of cat spinal neurons projecting to both the lateral cervical nucleus and the dorsal column nuclei. Neurosci Lett 101:29-34

Lu GW, Jiao SS, Zhang GF (1988) Morphological evidence for newly discovered double projection spinal neurons. Neurosci Lett 93:181-185

Lu Y, Dy Y-J, Qin B-Z, Li J-S (1993) The subdivisions of the intermediolateral nucleus in the sacral spinal cord of the cat. Brain Res 632:351-355

Lui F, Giolli RA, Blanks RH, Tom EM (1994) Pattern of striate cortical projections to the pretectal complex in the guinea pig. J Comp Neurol 344:598-609

Luiten PGM, Ter Horst GJ, Karst H, Steffens AB (1985) The course of paraventricular hypothalamic efferents to autonomic structures in medulla and spinal cord. Brain Res 329:374-378

Luiten PGM, ter Horst GJ, Steffens AB (1987) The hypothalamus, intrinsic connections and outflow pathways to the enndocrine system in relation to the control of feeding and metabolism. Prog Neurobiol 28:1-54

Lund JS (1973) Organization of neurons in the visual cortex, area 17, of the monkey (Macaca mulatta). J Comp Neurol 147:455-469

Lund JS (1984) Spiny stellate neurons. In: Peters A, Jones EG (eds) Cellular components of the cerebral cortex. Plenum, New York, pp 255-308 (Cerebral cortex, vol 1)

Lund JS (1987) Local circuit neurons of macaque monkey striate cortex: I. Neurons of laminae 4C and 5A. J Comp Neurol 257:60-92

Lund JS (1988a) Anatomical organization of macaque monkey striate visual cortex. Annu Rev Neurosci 11:253-288

Lund JS (1988b) Local circuit neurons of macaque monkey striate cortex: II. Neurons of laminae 5B and 6. J Comp Neurol 276:1-29

Lund JS, Boothe RG (1975) Interlaminar connections and pyramidal neuron organisation in the visual cortex, area 17, of the macaque monkey. J Comp Neurol 159:305-334

Lund JS, Lewis DA (1993) Local circuit neurons of developing and mature prefrontal cortex: Golgi and immunocytochemical characteristics. J Comp Neurol 328:282-312

Lund JS, Lund RD (1970) The termination of callosal fibers in the paravisual cortex of the rat. Brain Res 17:25-45

Lund JS, Yoshioka T (1991) Local circuit neurons of macaque monkey striate cortex: III. Neurons of laminae 4B, 4A, and 3B. J Comp Neurol 311:234-258

Lund JS, Henry GH, MacQueen CL, Harvey AR (1979) Anatomical organization of the primary visual cortex (Area 17) of the cat. A comparison with Area 17 of the macaque monkey. J Comp Neurol 184:599-618

Lund JS, Hendrickson AE, Ogren MP, Tobin EA (1981) Anatomical organization of primate visual cortex area VII. J Comp Neurol 202:19-45

Lund JS, Fitzpatrick D, Humphrey AL (1985) The striate visual cortex of the tree shrew. In: Peters A, Jones EG (eds) Visual cortex. Plenum, New York, pp 157-205 (Cerebral cortex, vol 3)

Lund JS, Hawken MJ, Parker AJ (1988) Local circuit neurons of macaque monkey striate cortex: II. Neurons of laminae 5B and 6. J Comp Neurol 27:1-29

Lund RD (1964) Terminal distribution in the superior colliculus of fibers originating in the visual cortex. Nature 204:1283-1285

Lund RD, Lund JS (1965) The visual system of the mole, talpa europaea. Exp Neurol 13:302-316

Lund RD, Lund JS (1976) Plasticity in the developing visual system: the effects of retinal lesions made in young rats. J Comp Neurol 169:133-154

Lund RD, Webster KE (1967) Thalamic afferents from the spinal cord and trigeminal nuclei. An experimental anatomical study in the rat. J Comp Neurol 130:313-328

Lund RD, Land PW, Boles J (1980) Normal and abnormal uncrossed retinotectal pathways in rats: an HRP study in adults. J Comp Neurol 189:711-720

Lundberg (1971) Function of the ventral spinocerebellar tract. A new hypothesis. Exp Brain Res 12:317-330

Luskin MB, Price JL (1983a) The topographic organization of associational fibers of the olfactory system in the rat including centrifugal fibers to the olfactory bulb. J Comp Neurol 216:264-291

Luskin MB, Price JL (1983b) The laminar distribution of intracortical fibers originating in the olfactory cortex of the rat. J Comp Neurol 216:292-302

Lynd-Balta E, Haber SN (1994) Primate striatonigral projections: a comparison of the sensorimotor-related striatum and the ventral striatum. J Comp Neurol 345:562-578

Lyon MJ (1978) The central location of the motor neurons to the stapedius muscle in the cat. Brain Res 143:437-444

Ma TP, Graybiel AM, Wurtz RH (1991) Location of saccade-relatedneurons in the macaque superior colliculus. Exp Brain Res 85:21-35

Ma W, Peschanski M, Besson JM (1986) The overlap of spinothalamic and dorsal column nuclei projections in the ventrobasal complex of the rat thalamus: a double anterograde labeling study using light microscopy analysis. J Comp Neurol 245:531-540

Mabuchi M, Kusama T (1966) The cortico-rubral projection in the cat. Brain Res 2:254-273

Macchi G, Bentivoglio M (1982) The organization of the efferent projections of the thalamic intralaminar nuclei: past, present and future of the anatomical approach. Ital J Neurol Sci 2:83-96

Macchi G, Bentivoglio M, Rossini P, Tempesta E (1978) The basolateral amygdaloid projections in the cat. Neurosci Lett 9:347-351

Mackay-Sim A, Sefton FO, Martin PR (1983) Subcortical projections to lateral geniculate and thalamic reticular nuclei in the hooded rat. J Comp Neurol 213:24-35

MacLean P (1952) Some psychiatric implications of physiological studies on frontotemporal portions of limbic system (visceral brain). Electroenceph Clin Neurophysiol 4:407-418

MacLean PD (1970) The triune brain, emotion, and scientific bias. In: Schnitt FO (ed) The neurosciences, 2nd study program. Rockefeller University Press, New York, pp 336-349

MacLean PD (1972) Cerebral evolution and emotional processes: new findings on the striatal complex. Ann NY Acad Sci 193:137-155

MacLean PD (1990) The triune brain in evolution: role in paleocerebral function. Plenum, New York

MacLean PD (1992) The limbic system concept. In: Trimble MR, Bolwig TG (eds) The temporal lobes and the limbic system. Wrightson, Petersfield, pp 1-14

Macrides F, Davis BJ (1983) The olfactory bulb. In: Emson PC (ed) Chemical neuroanatomy. Raven, New York, pp 391-426

MacVicar BA, Dudek FE (1980) Local synaptic circuits in rat hippocampus: interactions between pyramidal cells. Brain Res 184:220-223

Maekawa H, Ohtsuka K (1993) Afferent and efferent connections of the cortical accommodation area in the cat. Neurosci Res 17:315-323

Maekawa K, Takeda T (1977) Afferent pathways from the visual system to the cerebellar flocculus in the rabbit. In: Baker R, Berthoz A (eds) Control by gaze of brain stem neurons. Elsevier, Amsterdam, pp 187-195

Maekawa K, Takeda T (1979) Origin of descending afferents to the rostral part of the dorsal cap of inferior olive which transfers contralateral optic activities to the flocculus. A horseradish peroxidase study. Brain Res 172:393-405

Magnin M, Fuchs AF (1977) Discharge properties of neurons in the monkey thalamus tested with angular acceleration, eye movement and visual stimuli. Exp Brain Res 28:293-299

Maioli MG, Domeniconi R, Squatrito S, Sanseverino R (1992) Projections from cortical visual areas of the superior temporal sulcus to the superior colliculus, in macaque monkeys. Arch Ital Biol 130:157-166

Majorossy K, Kiss A (1990) Types of neurons and synaptic relations in the lateral superior olive of the cat: normal structure and experimental observations. Acta Morphol Hung 38:207-215

Malmierca MS, Blackstad TW, Osen KK, Karaghlle T, Molowny RL (1993) The central nucleus of the inferior colliculus in rat: a Golgi and computer reconstruction study of neuronal and laminar structure. J Comp Neurol 333:1-27

Mangold-Wirz K (1966) Cerebralisation und Ontogenesemodus bei Eutherien. Acta Anat 63:449-508

Mannen H, Sasaki S-I, Ishizuka N (1982) Trajectory of primary vestibular fibers originating from the lateral, anterior, and posterior semicircular canals in the cat. Proc Jpn Acad 58 B:237-242

Mantyh PW (1983a) The spinothalamic tract in the primate: a re-examination using WGA-HRP. Neuroscience 9:847-862

Mantyh PW (1983b) The terminations of the spinothalamic tract in the cat. Neurosci Lett 38:119-124

Mantyh PW (1983c) Connections of midbrain periaqueductal gray in the monkey. I. Ascending efferent projections. J Neurophysiol 49:567-581

Marani E (1986) Topographic histochemistry of the cerebellum. Prog Histol Cytochem 16:1-169

Marani E, Voogd J (1977) An acetylcholinesterase band pattern in the molecular layer of the cat cerebellum. J Anat (Lond) 124:335-345

Marinesco G (1904) Travaux originaux. Semaine Med 29:225-231

Marín-Padilla M (1969) Origin of the pericellular baskets of the pyramidal cells of the human motor cortex: a Golgi study. Brain Res 14:633-646

Marín-Padilla M (1978) Dual origin of the mammalian neocortex and evolution of the cortical plate. Anat Embryol 152:109-126

Marin-Padilla M (1984) Neurons of layer I. A development analysis. In: Peters A, Jones EG (eds) Cellular components of the cerebral cortex. Plenum, New York, pp 447-478 (Cerebral cortex, vol 1)

Marín-Padilla M (1992) Ontogenesis of the pyramidal cell of the mammalian neocortex and developmental cytoarchitectonics: A unifying theory. J Comp Neurol 321:223-240

Mark RF, James AC, Sheng XM (1993) Geometry of the representation of the visual field on the superior colliculus of the wallaby (Macropus eugenii). I. Normal projection. J Comp Neurol 330:303-314

Marlier L, Sandillon F, Poulat P, Rajaofetra N, Geffard M, Privat A (1991) Serotonergic innervation of the dorsal horn of rat spinal cord: light and electron microscopic immunocytochemical study. J Neurocytol 20:310-322

Marr D (1969) A theory of cerebellar cortex. J Physiol (Lond) 202:437-470

Marin-Padilla M (1978) Dual origin of the mammalian neocortex and evolution of the cortical plate. Anat Embryol (Berl) 152:109-126

Marin-Padilla M (1992) Ontogenesis of the pyramidal cell of the mammalian neocortex and developmental cytoarchitectonics: a unifying theory. J Comp Neurol 321:223-240

Marshall LG (1979) Evolution of metatherian and eutherian (mammalian) characters: a review based on cladistic methodology. Zool J Linn Soc 66:369-410

Martin GF (1968) The pattern of neocortical projections to the mesencephalon of the opossum, didelphis virginiana. Brain Res 11:593-610

Martin GF (1969) Efferent tectal pathways of the opossum (Didelphis virginiana). J Comp Neurol 135:209-224

Martin GF, Cabana T (1985) Cortical projections to superficial laminae of the dorsal horn and to the ventral horn of the spinal cord in the North American opossum. Studies using the orthograde transport of WGA-HRP. Brain Res 337:188-192

Martin GF, Dom R (1970) Rubrobulbar projections of the opossum (Didelphis virginiana). J Comp Neurol 139:199-214

Martin GF, Fisher AM (1968) A further evaluation of the origin, the course and the termination of the opossum corticospinal tract. J Neurol Sci 7:177-187

Martin GF Jr, Hamel EG Jr (1967) The striatum of the opossum, Didelphis virginiana. Description and experimental studies. J Comp Neurol 131:491-516

Martin GF, West HJ (1967) Efferent neocortical projections to sensory nuclei in the brain stem of the opossum (Didelphys virginiana). J Neurol Sci 5:287-302

Martin GF, King JS, Dom R (1973) The projections of the deep cerebellar nuclei of the opossum, Didelphis marsupialis virginiana. J Hirnforsch 15:545-573

Martin GF, Dom R, Katz S, King JS (1974) The organization of projection neurons in the opossum red nucleus (1975). Brain Res 278:17-34

Martin GF, Megirian D, Roebuck A (1975) The corticospinal tract of the marsupial phalanger (Trichosurus vulpecula). J Comp Neurol :245-258

Martin GF, Henkel CK, King JS (1976) Cerebello-olivary fibers: their origin, course and distribution in the North American Opossum. Exp Brain Res 24:219-236

Martin GF, Bresnahan JC, Henkel CK, Megirian D (1996) Corticobulbar fibres in the North American opossum (Didelphis marsupialis virginiana) with notes on the Tasmanian brush-tailed possum (Trichosurus vulpecula) and other marsupials. J Anat 120:439-484

Martin GF, Humbertson AO, Laxson C, Panneton WM (1979) Evidence for direct bulbospinal projections to laminae IX, X and the intermediolateral cell column. Studies using axonal transport techniques in the North American opossum. Brain Res 170:165-171

Martin GF, Culbertson J, Laxson C, Linauts M, Panneton M, Tschismadia I (1980) Afferent connections of the inferior olivary nucleus with preliminary notes on their development: studies using the North American Opossum. In: Courville J, deMontigny C, Lamarre Y (eds) The inferior olivary nucleus: anatomy and physiology. Raven, New York, pp 35-72

Martin GF, Cabana T, Humbertson AOJ (1981) Evidence for a lack of distinct rubrospinal somatotopy in the North American Opossum and for collateral innervation of the cervical and lumbar enlargements by single rubral neurons. J Comp Neurol 201:255-263

Martin GF, Cabana T, Ditirro FJ, Ho RH, Humbertson AOJ (1982a) Raphespinal projections in the North American Opossum: evidence for connectional heterogeneity. J Comp Neurol 208:67-84

Martin GF, Cabana T, Ditirro FJ, Ho RH, Humbertson AOJ (1982b) Reticular and raphe projections to the spinal cord of the North Americain opossum. Evidence for connectional heterogeneity. Prog Brain Res 57:109-129

Martin GF, Cabana T, Waltzer R (1983) Anatomical demonstration of the location and collateralization of rubral neurons which project to the spinal cord, lateral brainstem and inferior olive in the North American opossum. Brain Behav Evol 23:93-109

Martin GF, Holstege G, Mehler WR (1990) Reticular formation of the pons and medulla. In: Paxinos G (ed) The human nervous system. Academic, San Diego, pp 203-220

Martin KAC (1984) Neuronal circuits in cat striate cortex. In: Jones, EG, Peters A (eds.) Functional properties of cortical cells. Plenum, New York, pp 241-284 (Cerebral cortex, vol 2)

Martin LJ, Hadfield MG, Dellovade TL, Price DL (1991) The striate mosaic in primates: patterns of neuropeptide immunoreactivity differentiate the ventral striatum from the dorsal striatum. Neuroscience 43:397-417

Martinez L, Lamas JA, Canedo A (1995) Pyramidal tract and corticospinal neurons with branching axons to the dorsal column nuclei of the cat. Neuroscience 68:195-206

Martinez-Guijarro FJ, Soriano E, del Rio JA, Lopez-Garcia C (1991) Parvalbumin-immunoreactive neurons in the cerebral cortex of the lizard Podarcis hispanica. Brain Res 547:339-343

Martinotti C (1890) Beitrag zum Studium der Hirnrinde und dem Centralursprung der Nerven. Int Monatsschr Anat Physiol 7:69-90

Maslany S, Crockett DP, Egger MD (1991) Somatotopic organization of the dorsal column nuclei in the rat: transganglionic labelling with B-HRP and WGA-HRP. Brain Res 564:56-65

Mason R, Groos GA (1981) Cortico-recipient and tecto-recipient visual zones in the rat's lateral posterior (pulvinar) nucleus: An anatomical study. Neurosci Lett 25:107-112

Massopust LC, Hauge DH, Ferneding JC, Doubek WG, Taylor JJ (1985) Projection systems and terminal localization of dorsal column afferents: an autoradiographic and horseradish peroxidase study in the rat. J Comp Neurol 237:533-544

Mates SL, Lund JS (1983) Neuronal composition and development in lamina 4C of monkey striate cortex. J Comp Neurol 221:60-90

Matsuda K, Uemura M, Kume M, Matsushima R, Mizuno N (1978) Topographical representation of masticatory muscles in the motor trigeminal nucleus in the rabbit: a HRP study. Neurosci Lett 8:1-4

Matsushita M (1969) Some aspects of the interneuronal connections in cat's spinal gray matter. J Comp Neurol 136:57-80

Matsushita M (1983) Anatomical organization of the spinocerebellar system, as studied by the HRP method. Acta Morphol Hung 31:73–86

Matsushita M, Gao XH (1995) Y. Spinovestibular projections in the rat, with particular reference to projections from the central cervical nucleus to the lateral vestibular nucleus. J Comp Neurol 361:334–344

Matsushita M, Hosoya Y (1978) The location of spinal projection neurons in the cerebellar nuclei (cerebellospinal tract neurons) of the cat. A study with the HRP technique. Brain Res 142:237–248

Matsushita M, Okado N (1981a) Spinocerebellar projections to lobules I and II of the anterior lobe of the cat, as studied by retrograde transport of horseradish peroxidase. J Comp Neurol 197:411–424

Matsushita M, Okada N (1981b) Cells of origin of brain stem afferents to lobule I and II of the cerebellar anterior lobe in the cat. Neuroscience 6:2392–2405

Matsushita M, Tanami T (1983) Contralateral termination of primary afferent axons in the sacral and caudal segments of the cat, as studied by anterograde transport of horseradish peroxidase. J Comp Neurol 220:206–218

Matsushita M, Tanami T (1987) Spinocerebellar projections from the central cervical nucleus in the cat, as studied by antero-grade transport of wheat germ agglutinin-horseradish peroxidase. J Comp Neurol 266:376–397

Matsushita M, Wang C-L (1987) Projection pattern of vestibulocerebellar fibers in the anterior vermis of the cat: an anterograde wheat germ agglutnin-horseradish peroxidase study. Neurosci Lett 74:25–30

Matsushita M, Ueyama T (1973) Ventral motor nucleus of the cervical enlargement om some mammals; its specific afferents from the lower cord levels and cytoarchitecture. J Comp Neurol 150:33–52

Matsushita M, Yaginuma H (1990) Afferents to the cerebellar nuclei from the cervical enlargement in the rat, as demonstrated with the Phaseolus vulgaris leucoagglutinin method. Neurosci Lett 113:253–259

Matsushita M, Yaginuma H (1995) Projections from the central cervical nucleus to the cerebellar nuclei in the rat, studied by anterograde axonal tracing. J Comp Neurol 353:234–246

Matsushita M, Hosoya Y, Ikeda M (1979a) Anatomical organization of the spinocerebellar system in the cat, as studied by retrograde transport of horseradish peroxidase. J Comp Neurol 184:81–106

Matsushita M, Ikeda M, Hosoya Y (1979b) The location of spinal neurons with long descending axons (long descending propriospinal tract neurons) in the cat: a study with the horseradish peroxidase technique. J Comp Neurol 184:63–80

Matsushita M, Ikeda M, Okado N (1982) The cells of origin of the trigeminothalamic, trigeminospinal and trigeminocerebellar projections in the cat. Neuroscience 7:1439–1454

Matthews DA, Salvaterra PM, Crawford GD, Houser CR, Vaughn JE (1987) An immunocytochemical study of choline acetyltransferase-containing neurons and axon terminals in normal and partially deafferented hippocampal formation. Brain Res 402:30–43

May PJ, Hall WC (1984) Relationships between the nigrotectal pathway and the cells of origin of the predorsal bundle. J Comp Neurol 226:357–376

May PJ, Hall WC (1986) The cerebellotectal pathway in the grey squirrel. Exp Brain Res 65:200–212

May PJ, Hartwich-Young R, Nelson J, Sparks DL, Porter JD (1990) Cerebellotectal pathways in the macaque: implications for collicular generation of saccades. Neuroscience 36:305–324

May PJ, Porter JD, Gamlin PDR (1992) Interconnections between the primate cerebellum and mid-brain near-response regions. J Comp Neurol 315:98–116

McBride RL, Larsen KD (1980) Projections of the feline globus pallidus. Brain Res 189:3–14

McBride RL, Sutin J (1977) Amygdaloid and pontine projections to the ventromedial nucleus of the hypothalamus. J Comp Neurol 174:377–396

McConnell J, Angevine JB (1983) Time of neuron origin in the amygdaloid complex of the mouse. Brain Res 272:150–156

McCotter RE (1913) The nervus terminalis in the adult dog and cat. J Comp neurol 23:145–155

McCrea RA, Baker R (1985) Anatomical connections of the nucleus prepositus of the cat. J Comp Neurol 237:377–407

McCrea RA, Strassman A, Highstein SM (1986) Morphology and physiology of abducens motoneurons and internuclear neurons intracellularly injected with horseradish peroxidase in alert squirrel monkey. J Comp Neurol 243:291–308

McCrea RA, Strassman A, Highstein SM (1987) Anatomical and physiological characteristics of vestibular neurons mediating the vertical vestibulo-ocular reflexes of the squirrel monkey. J Comp Neurol 264:571–594

McDonald AJ (1982) Neurons of the lateral and basolateral amygdaloid nuclei: a Golgi study in the rat. J Comp Neurol 212:293–312

McDonald AJ (1983) Neurons of the bed nucleus of the stria terminalis: a Golgi study in the rat. Brain Res Bull 10:111–120

McDonald AJ (1987) Organization of amygdaloid projections to the mediodorsal thalamus and prefrontal cortex: a fluorescence retrograde transport study in the rat. J Comp Neurol 242:46–58

McDonald AJ (1991a) Organization of amygdaloid projections to the prefrontal cortex and associated striatum in the rat. Neuroscience 44:1–14

McDonald AJ (1991b) Topographical organization of amygdaloid projections to the caudatoputamen, nucleus accumbens, and related striatal-like areas of the rat brain. Neuroscience 44:15–33

McDonald AJ, Culberson JL (1981) Neurons of the basolateral amygdala: a Golgi study in the opossum (Didelphis virgiania). Am J Anat 162:327–342

McDonald AJ, Culberson JL (1986) Efferent projections of the basolateral amygdala in the opossum, Didelphis virgiana. Brain Res Bull 17:335–350

McFarland WL, Morgane PJ, Jacobs MS (1969) Ventricular system of the brain of the dolphin, Tursiops truncatus, with comparative anatomical observations and relations to brain specializations. J Comp Neurol 135:275–368

McGeorge AJ, Faull RLM (1989) The organization of the projection from the cerebral cortex to the striatum in the rat. Neuroscience 29:503–537

McGuire BA, Hornung J-P, Gilbert CD, Wiesel TN (1984) Patterns of synaptic input to layer 4 of cat striate cortex. J Neurosci 4:3021–3033

McGuire BA, Gilbert CD, Rivlin PK, Wiesel TN (1991a) Targets of horizontal connections in macaque primary visual cortex. J Comp Neurol 305:370–392

McGuire PK, Bates JF, Goldman-Rakic PS (1991b) Interhemispheric integration: I. symmetry and convergence of the corticocortical connections of the left and the right principal sulcus (PS) and the left and the right supplementary motor area (SMA) in the Rhesus monkey. Cerebral Cortex 1:390–407

McHaffie JG, Kruger L, Clemo HR, Stein BE (1988) Corticothalamic and corticotectal somatosensory projections from the anterior ectosylvian sulcus (SIV cortex) in neonatal cats: an anatomical demonstration with HRP and 3H-leucine. J Comp Neurol 274:115–126

McHaffie JG, Beninato M, Stein BE, Spencer RF (1991) Postnatal development of acetylcholinesterase in, and cholinergic projections to, the cat superior colliculus. J Comp Neurol 313:113–131

McKenna MC (1975) Towards a phylogenetic classification of the mammalia. In: Luckett WP, Szalay FS (eds) Phylogeny of the primates. A multidisciplinary approach. Plenum Press New York, pp 21–46

Mechelse K (1957) The pedunculus cerebri of the cat. Psychiatr Neurol (Basel) 133:257–275

Meesen H, Olszewski J (1949) A cytoarchitectonic atlas of the rhombencephalon of the rabbit. Karger, Basel

Mehler WR (1962) The anatomy of the so-called 'pain-tract' in man: an analysis of the course and distribution of the ascending fibers of the fasciculus anterolateralis. In: French JD, Porter RW (eds) Basic research in paraplegia. Thomas, Springfield
Mehler WR (1966a) Further notes on the centre médian, nucleus of Luys. In: Purpura DP, Yahr MO (eds) The thalamus. Columbia University Press, New York, pp 109–127
Mehler WR (1966b) Some observations on secondary ascending afferent systems in the central nervous system. In: Knighton RS, Dumke PR (eds) Pain. Little Brown, Boston, pp 11–32
Mehler WR (1966c) The posterior thalamic region in man. Confin Neurol 27:18–29
Mehler WR (1969) Some neurological species differences – a posteriori. Ann NY Acad Sci 167:424–468
Mehler WR (1974) Central pain and the spinothalamic tract (1980). Adv Neurol 4:127–146
Mehler WR (1980) Subcortical afferent connections of the amygdala in the monkey. J Comp Neurol 190:733–762
Mehler WR (1981) The basal ganglia-circa 1982: a review and commentary. Appl Neurophysiol 44:261–290
Mehler WR, Feferman ME, Nauta WJH (1960) Ascending axon degeneration following anterolateral cordotomy. An experimental study in the monkey. Brain 83:718–751
Meinecke DL, Peters A (1987) GABA immunoreactive neurons in rat visual Cortex. J Comp Neurol 261:388–404
Menétrey D, De Pommery J (1985) Propriospinal fibers reaching the lumbar enlargement in the rat. Neurosci Lett 58:257–261
Menétrey D, Chaouch A, Besson JM (1980) Location and properties of dorsal horn neurons at origin of spinoreticular tract in the lumbar enlargement of the rat. J Neurophysiol 44:862–877
Merchan MA, Saldana E, Plaza I (1994) Dorsal nucleus of the lateral lemniscus in the rat: concentric organization and tonotopic projection to the inferior colliculus. J Comp Neurol 342:259–278
Meredith M (1987) Vomeronasal organ and nervus terminalis. In: Adelman G (ed) Encyclopedia of neuroscience. Birkhäuser, Boston, pp 1303–1305
Mesulam M-M (1985) Patterns in behavioral neuroanatomy. In: Mesulam M-M (ed) Principles of behavioral neurology. Davis, Philadelphia, pp 1–70
Mesulam MM, Geula C (1988) Nucleus basalis (Ch4) and cortical cholinergic innervation in the human brain: observations based on the distribution of acetylcholinesterase and choline acetyltransferase. J Comp Neurol 275:216–240
Mesulam MM, Mufson EJ (1984) Neural inputs into the nucleus basalis of the substantia innominata (Ch4) in the rhesus monkey. Brain 107:253–274
Mesulam MM, Van Hoesen GW (1976) Acetylcholinesterase-rich projections from the basal forebrain of the rhesus monkey to neocortex. Brain Res 109:152–157
Mesulam MM, Mufson EJ, Wainer BH, Levey AI (1983a) Central cholinergic pathways in the rat: an overview based on an alternative nomenclature (Ch1-Ch6). Neuroscience 10:1185–1201
Mesulam MM, Mufson EJ, Levey AI, Wainer BH (1983b) Cholinergic innervation of cortex by the basal forebrain: cytochemistry and cortical connections of the septal area, diagonal band nuclei, nucleus basalis (substantia innominata), and hypothalamus in the rhesus monkey. J Comp Neurol 214:170–197
Mesulam MM, Mufson EJ, Levey AI, Wainer BH (1984a) Atlas of cholinergic neurons in the forebrain and upper brainstem of the macaque based on monoclonal choline acetyltransferase immunohistochemistry and acetylcholinesterase histochemistry. Neuroscience 12:669–686
Mesulam MM, Rosen AD, Mufson EJ (1984b) Regional variations in cortical cholinergic innervation: chemoarchitectonics of acetylcholinesterase-containing fibers in the macaque brain. Brain Res 311:245–258
Mesulam MM, Mufson EJ, Wainer BH (1986) Three-dimensional representation and cortical projection topography of the nucleus basalis (Ch4) in the macaque: concurrent demonstration of choline acetyltransferase and retrograde transport with a stabilized tetramethylbenzidine method for horseradish peroxidase. Brain Res 367:301–308
Metzner W (1996) Anatomical basis for audio-vocal integration in echolocating horseshoe bats. J Comp Neurol 368:252–269
Metzner W, Radtke-Schuller S (1987) The nuclei of the lateral lemniscus in the rufous horseshoe bat, Rhinolophus rouxi. A neurophysiological approach. J Comp Physiol 160:395–411
Meyer RP (1981) Central connections of the olfactory bulb in the American opossum (Didelphis virgiana): a light microscopic degeneration study. Anat Rec 201:141–156
Müller-Preuss P, Jürgens U (1975) Projections from the "cingular" vocalization area in the squirrel monkey. Brain Res 103:29–43
Miceli MO, Malsbury CW (1985) Brainstem origins and projections of the cervical and abdominal vagus in the golden hamster: a horseradish peroxidase study. J Comp Neurol 237:65–76
Middleton FA, Strick PL (1994) Anatomical evidence for cerebellar and basal ganglia involvement in higher cognitive function. Science 266:458–461
Mihailoff GA (1993) Cerebellar nuclear projections from the basilar pontine nuclei and nucleus reticularis tegmenti pontis as demonstrated with PHA-L tracing in the rat. J Comp Neurol 330:130–146
Mihailoff GA, Kosinski RJ, Azizi SA, Border BG (1989) Survey of noncortical afferent projections to the basilar pontine nuclei: a retrograde tracing study in the rat. J Comp Neurol 282:617–643
Mikkelsen JD, Vrang N (1994) A direct pretectosuprachiasmatic projection in the rat. Neuroscience 62:497–505
Miller RA, Strominger NL (1973) Efferent connections of the red nucleus in the brainstem and spinal cord of the rhesus monkey. J Comp Neurol 152:327–346
Millhouse OE (1979) A Golgi anatomy of the rodent hypothalamus. In: Morgane PJ, Panksepp J (eds) Handbook of the hypothalamus, vol 1. Dekker, New York, pp 221–265
Millhouse OE, de Olmos J (1983) Neuronal configurations in lateral and basolateral amygdala. Neuroscience 10:1269–1300
Millhouse OE, Heimer L (1984) Cell configurations in the olfactory tubercle of the rat. J Comp Neurol 228:571–597
Misgeld U, Frotscher M (1986) Postsynaptic-GABAergic inhibition of non-pyramidal neurons in the guinea-pig hippocampus. Neuroscience 19:193–206
Mishkin M (1978) Memory in monkeys severely impaired by combined but not separate removal of amygdala and hippocampus. Nature 273:297–298
Mishkin M (1982) A memory system in the monkey. Philos Trans R Soc Lond B Biol Sci 298:85–95
Misson J-P, Austin CP, Takahashi T, Cepko CL, Caviness VS Jr (1991) The alignment of migrating neural cells in relation to the murine neopallial radial glial fiber system. Cerebral Cortex 1:221–229
Miyara M, Sasaki K (1984) Horseradish peroxidase studies on thalamic and striatal connections of the medial part of area 6 in the monkey. Neurosci Lett 49:127–133
Miyashita E, Tamai Y (1989) Subcortical connections of frontal 'oculomotor' areas in the cat. Brain Res 502:75–87
Miyata H, Kawaguchi S, Kato N (1984) Postnatal development of the thalamocortical projection of the primary auditory cortex in the cat. J Physiol Soc Jpn 46:399
Miyazaki S (1985a) Bilateral innervation of the superior oblique muscle by the trochlear nucleus. Brain Resa 348:52–56
Miyazaki S (1985b) Location of motoneurons in the oculomotor nucleus and the course of their axons in the oculomotor nerve. Brain Resb 348:57–63

Miyazaki T, Yoshida Y, Hirano M, Shin T, Kanaseki T (1981) Central location of the motoneurons supplying the thyrohyoid and the geniohyoid muscles as demonstrated by horseradish peroxidase method. Brain Res 219:423-427

Mizuno N, Matsuda K, Iwahori N, Uemura-Sumi M, Kume M, Matsushima R (1981) Representation of the masticatory muscles in the motor trigeminal nucleus of the macaque monkey. Neurosci Lett 21:19-22

Mizuno N, Nakano K, Imaizumi M, Okamoto M (1967) The lateral cervical nucleus of the Japanese monkey (Macaca fuscata). J Comp Neurol 129:375-384

Mizuno N, Mochizuki K, Aikimoto C, Matsushima R (1973) Pretectal projection of the inferior olive in the rabbit. Exp Neurol 39:498-506

Mizuno N, Nakamura Y, Iwahori N (1974) An electron microscope study of the dorsal cap of the inferior olive in the rabbit, with special reference to the pretecto-olivary fibers. Brain Res 77:385-395

Mizuno N, Mochizuki K, Akimoto C, Matsushima R, Sasaki K (1975) Projections from the parietal cortex to the brain stem nuclei in the cat, with special reference to the parietal cerebro-cerebellar system. J Comp Neurol 147:511-522

Mizuno N, Matsuda K, Iwahori N, Uemurasumi M, Kume N, Matsushima R (1981) Representation of the masticatory muscles in the motor trigeminal nucleus of the macaque monkey. Neurosci Lett 21:19-22

Mizuno N, Takahashi O, Satoda T, Matsushima R (1985) Amygdaloid projections in the macaque monkey. Neurosci Lett 53:327-330

Mizuno N, Sumi MU, Tashiro T, Takahashi O, Satoda T (1991) Retinofugal projections in the house musk shrew, Sunctus murinus. Neurosci Lett 125:133-135

Mogenson GJ (1984) Limbic-motor integration - with emphasis on initiation of exploratory and goal-directed locomotion. In: Bandler R (ed) Modulation of sensorimotor activity during alterations in behavioral states. Liss, New York, pp 121-137

Mogenson GJ, Nielsen MA (1983) Evidence that an accumbens to subpallidal GABA-ergic projection contributes to Locomotor activity. Brain Res Bull 11:309-314

Mogenson GJ, Jones DL, Yim CY (1980) From motivation to action: functional interface between the limbic system and the motor system. Prog Neurobiol 14:69-97

Mogenson GJ, Swanson LW, Wu M (1984) Evidence that projections from substantia innominata zona incerta mesencephalic locomotor region contribute to locomotor activity. Brain Res 334:65-76

Molander C, Grant G (1985) Cutaneous projections from the rat hindlimb foot to the substantia gelatinosa of the spinal cord studied by transganglionic transport of WGA-HRP conjugate. J Comp Neurol 237:476-484

Molander C, Grant G (1995) Spinal cord cytoarchitecture. In: Paxinos G (ed) The rat nervous system, 2nd edn. Academic, San Diego, pp 39-45

Molander C, Xu Q, Grant G (1984) The cytoarchitectonic organization of the spinal cord in the rat. I. The lower thoracic and lumbosacral cord. J Comp Neurol 230:133-141

Molander C, Xu Q, Rivero-Melian C, Grant G (1989) Cytoarchitectonic organization of the spinal cord in the rat: II. The cervical and upper thoracic cord. J Comp Neurol 289:375-385

Molenaar I, Kuypers HG (1978) Cells of origin of propriospinal fibers and of fibers ascending to supraspinal levels. A HRP study in cat and rhesus monkey. Brain Res 152:429-450

Molinari HH (1984) Ascending somatosensory projections to the dorsal accessory olive: an anatomical study in cats. J Comp Neurol 223:110-123

Molinari HH (1985) Ascending somatosensory projections to the medial accessory portion of the inferior olive: a retrograde study in cats. J Comp Neurol 232:523-534

Molinari M, Bentivoglio M, Minciacchi D, Granato A, Macchi G (1987) Spinal afferents and cortical efferents of the anterior intralaminar nuclei. An anterograde-retrograde tracing study. Neurosci Lett 72:258-264

Molliver ME, Van der Loos H (1970) The ontogenesis of cortical circuitry: the spatial distribution of synapses in somesthetic cortex of newborn dog. Ergeb Anat Entw Gesch 42:1-53

Mooney RD, Rhoades RW (1993) Determinants of axonal and dendritic structure in the superior colliculus. In: Hicks TP, Molotchnikoff S, Ono (eds) Progress in brain research, vol 95. Elsevier Science, Amsterdam, pp 57-67

Moore JK (1980) The primate cochlear nuclei: loss of lamination as a phylogenetic process. J Comp Neurol 193:609-629

Moore JK (1987) The human auditory brain stem: a comparative view. Hearing Res 29:1-32

Moore JK (1988) Auditory brainstem of the ferret: sources of projections to the inferior colliculus. J Comp Neurol 269:342-354

Moore JK, Moore RY (1971) A comparative study of the superior olivary complex in the primate brain. Folia Primatol 16:35-51

Moore RY (1978) Central neural control of circadian rhythm. Front Neuroendocrinol 5:185-206

Moore RY, Bloom FE (1979) Central catecholamine neuron systems: anatomy and physiology of the norepinephrine and epinephrine systems. Annu Rev Neurosci 2:113-168

Moore RY, Card JP (1986) Visual pathways and the entrainment of circadian rhythms. Ann NY Acad Sci 453:123-133

Moore RY, Goldberg JM (1963) Ascending projections of the inferior colliculus in the cat. J Comp Neurol 121:109-135

Moore RY, Halaris AE, Jones BE (1980) Serotonin neurons of the midbrain raphe: ascending projections. J Comp Neurol 180:417-438

Morest DK (1961) Connexions of the dorsal tegmental nucleus in rat and rabbit. J Anat 95:229-246

Morest DK (1965) The laminar structure of the medial geniculate body in the cat. J Anat 99:143-160

Morest DK (1967a) The collateral system of the medial nucleus of the trapezoid body of the cat, its neuronal architecture and relation to the olivo-cochlear bundle. Brain Res 9:288-311

Morest DK (1967b) Experimental study of the projections of the nucleus of the tractus solitarius and the area postrema in the cat. J Comp Neurol 130:277-300

Morest DK, Oliver DL (1984) The neuronal architecture of the inferior colliculus in the cat: defining the functional anatomy of the auditory midbrain. J Comp Neurol 222:209-236

Morgan C, Nadelhaft I, De Groat WC (1981) The distribution of visceral primary afferents from the pelvic nerve to Lissauer's tract and the spinal gray matter and its relationship to the sacral parasympathetic nucleus. J Comp Neurol 201:415-440

Morgane PJ, Glezer II, Jacobs MS (1990) Comparative and evolutionary anatomy of the visual cortex of the dolphin. In: Jones EG, Peters A (eds) Comparative structure and evolution of cerebral cortex, part II. Plenum, New York, pp 215-262 (Cerebral cortex, vol 8B)

Morgane PJ, Jacobs MS (1972) Comparative anatomy of the cetacean nervous system. In: Harrison RJ (ed) Functional anatomy of marine mammals. Academic, New York, pp 117-244

Morgane PJ, Jacobs MS, Mcfarland WL (1980) The anatomy of the brain of the bottlenose dophin (Tursiops truncatus). Surface configurations of the telencephalon of the bottlenose dolphin with comparative anatomical observations in four other cetacean species. Brain Res Bull 5 [Suppl 3]:1-108

Mori S, Ueda S, Yamada H, Takino T, Sano Y (1985) Immunohistochemical demonstration of serotonin nerve fibers in corpus striatum of the rat, cat and monkey. Anat Embryol (Berl) 173:1-5

Moriizumi T, Leduc-Cross B, Hattori T (1991) Cholinergic nigrotectal projections in the rat. Neurosci Lett 132:69-72

Morin F, Catalano JV (1955) Central connections of a cervical nucleus (nucleus cervicalis lateralis of the cat). J Comp Neurol 103:17-32

Moriizumi T, Leduc-Cross B, Hattori T (1991) Cholinergic nigrotectal projections in the rat. Neurosci Lett 132:69–72

Moriya T, Yamadori T (1993) Correlative study of the morphology and central connections of ipsilaterally projecting retinal ganglion cells in the albino rat. Exp Eye Res 56:79–83

Morris RJ, Beech JN, Heizmann CW (1988) Two distinct phases and mechanisms of axonal growth shown by primary vestibular fibres in the brain, demonstrated by parvalbumin immunohistochemistry. Neuroscience 27:571–596

Morrison JH, Magistretti PJ (1983) Monoamines and peptides in cerebral cortex. Contrasting principles of cortical organization. Trends Neurosci 6:146–151

Morrison JH, Grzanna R, Molliver ME, Coyle JT (1978) The distribution and orientation of noradrenergic fibers in neocortex of the rat: an immunofluorescence study. J Comp Neurol 181:17–40

Morrison JH, Molliver ME, Grzanna R, Coyle JT (1980) Noradrenergic innervation patterns in three regions of medial cortex: an immunofluorescence characterization. Brain Res Bull 4:849–857

Morrison JH, Foote SL, O'Connor D, Bloom FE (1982) Laminar, tangential and regional organization of the noradrenergic innervation of monkey cortex: dopamine-β-hydroxylase immunohistochemistry. Brain Res Bull 9:309–319

Moschovakis AK, Karabelas AB (1985) Observations on the somatodendritic morphology and axonal trajectory of intracellularly HRP-labeled efferent neurons located in the deeper layers of the superior colliculus of the cat. J Comp Neurol 239:276–308

Moschovakis AK, Karabelas AB, Highstein SM (1988a) Structure function relationships in the primate superior colliculus. I. Morphological classification of efferent neurons. J Neurophysiol 60:232–262

Moschovakis AK, Karabelas AB, Highstein SM (1988b) Structure-function relationships in the primate superior colliculus. II. Morphological identity of presaccadic neurons. J Neurophysiol 60:263–302

Mountcastle VB (1957) Modality and topographic properties of single neurones of cat's somatic sensory cortex. J Neurophysiol 20:408–434

Mountcastle VB (1979) An organizing principle for cerebral function: the unit module and distributed system. In: Schmitt FO, Worden FG (eds) The neurosciences fourth study program. MIT Press, Cambridge, pp 21–42

Mower G, Gibson A, Glickstein M (1979) Tectopontine pathway in the cat: laminar distribution of cells of origin and visual properties of target cells in dorsolateral pontine nucleus. J Neurophysiol 42:1–15

Mower G, Gibson A, Robinson F, Stein J, Glickstein M (1980) Visual pontocerebellar projections in the cat. J Neurophysiol 43:355–366

Mrzljak L, Uylings HBM, Kostovic I, Van Eden CG (1988) Prenatal development of neurons in the human prefrontal cortex: I. A qualitative Golgi study. J Comp Neurol 271:355–386

Mrzljak L, Uylings HBM, Van Eden CG, Judás M (1990) Neuronal development in human prefrontal cortex in prenatal and postnatal stages. Progr Brain Rese 85:185–222

Mufson EJ, Mesulam M-M (1982) Insula of the old world monkey. II. Afferent cortical input and comments on the claustrum. J Comp Neurol 212:23–27

Mufson EJ, Mesulam MM, Pandya DN (1981) Insular interconnections with the amygdala in the rhesus monkey. Neuroscience 6:1231–1248

Mufson EJ, Martin TL, Mash DC, Wainer BH, Mesulam M-M (1986) Cholinergic projections from the parabigeminal nucleus (Ch8) to the superior colliculus in the mouse: a combined analysis of horseradish peroxidase transport and choline acetyltransferase immunohistochemistry. Brain Res 370:144–148

Mugnaini E (1972) The histology and cytology of the cerebellar cortex. The comparative anatomy and histology of the cerebellum. The human cerebellum, cerebellar connections, and cerebellar cortex. University of Minnesota, Minneapolis, pp 201–264

Mugnaini E, Floris A (1994) The unipolar brush cell: a neglected neuron of the mammalian cerebellar cortex. J Comp Neurol 339:174–180

Mugnaini E, Morgan JI (1987) The neuropeptide cerebellin is a marker for two similar neuronal circuits in rat brain. Proc Natl Acad Sci USA 84:8692–8696

Mugnaini E, Warr WB, Osen KK (1980) Distribution and light microscopic features of granule cells in the cochlear nuclei of cat, rat, and mouse. J Comp Neurol 191:581–606

Mugnaini E, Oertel WH, Wouterlood FF (1984a) Immunocytochemical localization of GABA neurons and dopamine neurons in the rat main and accessory olfactory bulbs. Neurosci Lett 47:221–226

Mugnaini E, Wouterlood FG, Dahl A-L, Oertel WH (1984b) Immunocytochemical identification of GABAergic neurons in the main olfactory bulb of the rat. Arch Ital Biol 122:83–112

Mulligan K, Ulinski PS (1990) Organization of the geniculocortical projection in turtles: isoatimuth lamellae in the visual cortex. J Comp Neurol 296:531–547

Munk H (1881) Über die Funktionen der Grosshirnrinde. Hirschwald, Berlin

Muñoz A, Muñoz M, Gonzalez A, Ten Donkelaar HJ (1997) Spinal ascending pathways in amphibians: cells of origin and main targets. J Comp Neurol 378:205–228

Muñoz DG (1990) Monodendritic neurons: a cell type in the human cerebellar cortex identified by chromogranin A-like immunoreactivity. Brain Res 528:335–338

Münzer E, Wiener H (1902) Das Zwischen und Mittelhirn des Kaninchens und die Beziehungen dieser Teile zum übrigen Zentralnervensystem, mit besonderer Berucksichtigung der Pyramidenbahn und der Schleife. Monatsschr Psychiatr Neurol 12:241–279

Murphy EH, Garone M, Tashayyod D, Baker R (1986) Innervation of extraocular muscles in the rabbit. J Comp Neurol 254:78–90

Murray EA (1991) Contributions of the amygdalar complex to behavior in macaque monkeys. Prog Brain Res 87:167–180

Murray EA, Coulter JD (1981) Organization of corticospinal neurons in the monkey. J Comp Neurol 195:339–365

Murray EA, Coulter JD (1982) Organization of tectospinal neurons in the cat and rat superior colliculus. Brain Res 243:201–214

Murray EA, Mishkin M (1984) Severe tactual as well as visual memory deficits follow combined removal of the amygdala and hippocampus in monkeys. J Neurosci 4:2563–2580

Murray EA, Mishkin M (1985) Amygdalectomy impairs crossmodal association in monkeys. Science 228:604–606

Murray M, Murphy CA, Ross LL, Haun F (1994) The role of the habenula-interpeduncular pathway in modulating levels of circulating adrenal hormones. Restor Neurol Neurosci 6(4):301–307

Mustari MJ, Fuchs AF, Kaneko CRS, Robinson FR (1994) Anatomical connections of the primate pretectal nucleus of the optic tract. J Comp Neurol 349:111–128

Nadelhaft I, Roppolo J, Morgan C, De Groat WC (1983) Parasympathetic preganglionic neurons and visceral primary afferents in monkey sacral spinal cord revealed following application of horseradish peroxidase to pelvic nerve. J Comp Neurol 216:36–52

Nadol JB (1983a) Serial section reconstruction of the neural poles of hair cells in the human organ of Corti. I. Inner hair cells. Laryngoscopea 93:599–614

Nadol JB (1983b) Serial section reconstruction of the neural poles of hair cells in the human organ of Corti. II. Outer hair cells. Laryngoscopeb 93:780–791

Nagai T, Kimura H, Maeda T, McGeer PL, Peng F, McGeer EG (1982) Cholinergic projections from the basal forebrain of the rat to the amygdala. J Neurosci 2:513–520

Nagelhus FA, Lehmann A, Ottersen OP (1993) Neuronal-glial exchange of taurine during hypo-osmotic stress: a combined immunocytochemical and biochemical analysis in rat cerebellar cortex. Neuroscience 54:615-631

Nagy JI, Hunt SP (1983) The termination of primary afferents within the rat dorsal horn: evidence for rearrangement following capsaicin treatment. J Comp Neurol 218:145-158

Naito A, Kita H (1994) The cortico-nigral projection in the rat: an anterograde tracing study with biotinylated dextran amine. Brain Res 637:317-322

Nakagawa S, Hasegawa Y, Tokushige A, Kubozono T, Nakano K (1988) Retinal projection to the formatio reticularis tegmenti mesencephali in the old world monkeys. Exp Brain Res 69:373-377

Nakamura Y, Mizuno N (1971) An electron microscopic study of the interposito-rubral connections in the cat and rabbit. Brain Res 35:283-286

Nakamura Y, Tokuno H, Moriizumi T, Kitao Y, Kudo M (1989) Monosynaptic nigral inputs to the pedunculopontine tegmental nucleus neurons which send their axons to the medial reticular formation in the medulla oblongata. An electron microscopic study in the cat. Neurosci Lett 103:145-150

Napier JR, Napier PH (1967) A handbook of living primates. Academic, New York

Nasution ID, Shigenaga Y (1987) Ascending and descending internuclear projections within the trigeminal sensory nuclear complex. Brain Res 425:234-247

Nathan PW, Smith MC (1955) Long descending tracts in man. I. Review of present knowledge. Brain 78:248-303

Nathan PW, Smith MC (1959) Fasciculi proprii of the spinal cord in man: review of present knowledge. Brain 82:610-668

Naus CG, Flumerfelt BA, Hrycyshyn AW (1985) An HRP-TMB ultrastructural study of rubral afferents in the rat. J Comp Neurol 239:453-465

Nauta HJW (1974) Evidence of a pallidohabenular pathway in the cat. J Comp Neurol 156:19-27

Nauta HJW, Cole M (1978) Efferent projections of the subthalamic nucleus: an autoradiographic study in monkey and cat. J Comp Neurol 180:1-16

Nauta WJH (1958) Hippocampal projections and related neural pathways to the midbrain in the cat. Brain 81:319-340

Nauta WJH (1961) Fibre degeneration following lesions of the amygdaloid complex in the monkey. J Anat (Lond) 95:515-532

Nauta WJH (1963) Central nervous organization and the endocrine motor system. In: Nalbanov AV (ed) Advances in neuroendocrinology. University of Illinois, Urbana, pp 5-21

Nauta WJH (1973) Connections of the frontal lobe with the limbic system. In: Laitinen LV, Livingston KE (eds) Surgical approaches in psychiatry. Medical and Technical Publishing, Lancester, pp 303-314

Nauta WJH (1986) Circuitous connections linking cerebral cortex, limbic system, and corpus striatum. In: Doane BK, Livingston KE (eds) The limbic system: functional organization and clinical disorders. Raven, New York, pp 43-54

Nauta WJH, Haymaker W (1969) Hypothalamic nuclei and fiber connections. In: Haymaker W, Anderson E, Nauta WJH (eds) The hypothalamus. Thomas, Springfield, pp 136-209

Nauta WJH, Kuypers HGJM (1958) Some ascending pathways in the brain stem reticular formation. In: Jasper HH, Procter LD (eds) Reticular formation of the brain. Little Brown, Toronto, pp 3-31

Nauta WJH, Mehler WR (1966) Projections of the lentiform nucleus in the monkey. Brain Res 1:3-42

Nauta WJH, Smith GP, Faull RLM, Domesick VB (1987) Efferent connection and nigral afferents of the nucleus accumbens septi in the rat. Neuroscience 3:385-401

Neafsey EJ, Hurley-Gius KM, Arvanitis D (1986) The topographical organization of neurons in the rat medial frontal, insular and olfactory cortex projecting to the solitary nucleus, olfactory bulb, periaqueductal gray and superior colliculus. Brain Res 377:261-270

Neary TJ (1990) The pallium of anuran amphibians. In: Jones EG, Peters A (eds) Comparative structure and evolution of cerebral cortex, part I. Plenum, New York, pp 107-138 (Cerebral cortex, vol 8A)

Newman HM, Stevens RT, Apkarian AV (1996) Direct spinal projections to limbic and striatal areas: anterograde transport studies from the upper cervical spinal cord and the cervical enlargement in squirrel monkey and rat. J Comp Neurol 365:640-658

Nicolelis MA, Chapin JK, Lin RC (1992) Somatotopic maps within the zona incerta relay parallel GABAergic somatosensory pathways to the neocortex, superior colliculus, and brainstem. Brain Res 577:134-141

Nieuwenhuys R (1964) Comparative anatomy of the spinal cord. In: Eccles JC, Schade JP (eds) Organization of the spinal cord. Elsevier, Amsterdam, pp 1-57 (Progress in brain research, vol 11)

Nieuwenhuys R (1977) Aspects of the morphology of the striatum. In: Cools AR, Lohman AHM, Van den Bercken JHL (eds) Psychobiology of the striatum. Elsevier/North-Holland, Amsterdam, pp 1-19

Nieuwenhuys R (1985) Chemoarchitecture of the brain. Springer, Berlin Heidelberg New York

Nieuwenhuys R (1994) The neocortex: an overview of its evolutionary development, structural organization and synaptology. Anat Embryol (Berl) 190:307-337

Nieuwenhuys R (1996) The greater limbic system, the emotional motor system and the brain. Progr Brain Res 107:551-580

Nieuwenhuys R, Geeraedts LMG, Veening JG (1982) The medial forebrain bundle of the rat: I General introduction. J Comp Neurol 206:49-81

Nieuwenhuys R, Voogd J, Van Huijzen C (1988) The human central nervous system. A synopsis and atlas, 3rd rev edn. Springer, Berlin Heidelberg New York

Nieuwenhuys R, Veening JG, Van Domburg P (1989) Core and paracores; some new chemoarchitectural entities in the mammalian neuraxis. Acta Morphol Neerl Scand 26:131-163

Niimi K, Takemura A, Suzuki H, Sasaki J (1962) The nuclear configuration of the dorsal thalamus of the mole. Tokushima J Exp Med 9:75-98

Niimi K, Kanaseki T, Takimoto T (1963) The comparative anatomy of the ventral nucleus of the lateral geniculate body in mammals. J Comp Neurol 121:313-323

Nomura S, Mizuno N (1981) Central distribution of afferent and efferent components of the chorda tympani in the cat as revealed by the horseradish peroxidase method. Brain Res 214:229-237

Nomura S, Mizuno N (1982) Central distribution of afferent and efferent components of the glossopharyngeal nerve: an HRP study in the cat. Brain Res 236:1-13

Nomura S, Mizuno N (1986) Histochemical demonstration of vibrissae-representing patchy patterns of cytochrome oxidase activity within the trigeminal sensory nuclei in the cat. Brain Res 380:167-171

Nomura S, Mizuno N, Itoh K, Matsuka K, Sugimoto T, Nakamura Y (1979) Localization of parabrachial nucleus neurons projecting to the thalamus or the amygdala in the cat using horseradish peroxidase. Exp Neurol 64:375-385

Nomura S, Itoh K, Sugimoto T, Yasui Y, Kamiya H, Mizuno N (1986) Mystacial vibrissae representation within the trigeminal sensory nuclei of the cat. J Comp Neurol 253:121-133

Norgren R (1976) Taste pathways to hypothalamus and amygdala. J Comp Neurol 166:17-30

Norgren R (1984) Central neural mechanisms of taste. In: Darien-Smith (ed) Handbook of physiology, Sect 1. American Physiology Society, Washington DC, vol III, pp 1087-1128

Norgren R (1990) Gustatory system. In: Paxinos G (ed) The human nervous system. Academic, San Diego, pp 845-861

Norgren R (1995) Gustatory system. In: Paxinos G (ed) The rat nervous system, 2nd edn. Academic, San Diego, pp 751-771

Noriega AL, Wall JT (1991) Parcellated organization in the trigeminal and dorsal column nuclei of primates. Brain Res 565:188-194

Norita M (1980) Neurons and synaptic patterns in the deep layers of the superior colliculus of the cat. A Golgi and electron microscopic study. J Comp Neurol 190:29-48

Norita M, Kawamura K (1980) Subcortical afferents to the monkey amygdala: an HRP study. Brain Res 190:225-230

Norita M, McHaffie JG, Shimizu H, Stein BE (1991) The corticostriatal and corticotectal projections of the feline lateral suprasylvian cortex demonstrated with anterograde biocytin and retrograde fluorescent techniques. Neurosci Res 10:149-155

Northcutt RG, Kicliter E (1980) Organization of the amphibian telencephalon. In: Ebbesson SOE (ed) Comparative neurology of the telencephalon. Plenum, New York, pp 203-255

Northcutt RG, Royce GJ (1975) Olfactory bulb projections in the bullfrog, R. catesbeina Shaw. J Morphol 145:251-268

Novacek MJ (1986) The skull of leptictid insectivorans and the higher-level classification of eutherian mammals. Bull Am Mus Nat Hist 183:1-112

Novacek MJ (1992) Mammalian phylogeny: shaking the tree. Nature 356:121-125

Nudo RJ, Masterton RB (1988) Descending pathways to the spinal cord: a comparative study of 22 mammals. J Comp Neurol 277:53-79

Nudo RJ, Masterton RB (1989) Descending pathways to the spinal cord: II. Quantitative study of the tectospinal tract in 23 mammals. J Comp Neurol 286:96-119

Nudo RJ, Masterton RB (1990a) Descending pathways to the spinal cord. III. Sites of origin of the corticospinal tract. J Comp Neurol 296:559-583

Nudo RJ, Masterton RB (1990b) Descending pathways to the spinal cord. IV. Some factors related to the amount of cortex devoted to the corticospinal tract. J Comp Neurol 296:584-597

Nunes Cardozo JJ, Van der Want JJL (1987) Synaptic organization of the nucleus of the optic tract in the rabbit: a combined Golgi-electron microscopic study. J Neurocytol 16:389-401

Nunes Cardozo B, Wortel J (1993) Projections from and to the superior colliculus in the nucleus of the optic tract combined with postembedding GABA immunocytochemistry in the rabbit. Eur J Morphol 31:92-96

Nunes Cardozo B, Mize RR, Van der Want JJ (1994) GABAergic and non-GABAergic neurons in the nucleus of the optic tract project to the superior colliculus: an ultrastructural retrograde tracer and immunocytochemical study in the rabbit. J Comp Neurol 350(4):646-656

Nusser Z, Mulvihill E, Streit P, Somogyi P (1994) Subsynaptic segregation of metabotropic and ionotropic glutamate receptors as revealed by immunogold localization. Neuroscience 61:421-427

Nyberg-Hansen (1964) Origin and termination of fibers from the vestibular nuclei descending in the medial longitudinal fasciculus. An experimental study with silver impregnation methods in the cat. J Comp Neurol 122:355-383

Nyberg-Hansen R (1975) Anatomical aspects of the functional organization of the vestibulospinal pathways. In: Naunton R (ed) The vestibular system, pp 71-96

Nyberg-Hansen R, Brodal A (1964) Sites and mode of termination of rubrospinal fibres in the cat. An experimental study with silver impregnation methods. J Anat (Lond) 98:235-253

Obenchain JB (1925) The brains of the South American marsupials Caenolestes and Orolestes. Publ 224 Field Mus Nat Hist Zool Ser 14:175-232

Obersteiner H (1912) Anleitung beim Studium des Baues der Nervösen Zentralorgane im gesunden und kranken Zustande. Deuticke, Leipzig and Vienna

Oboussier H (1972a) Evolution of the mammalian brain: some evidence on the phylogeny of the antelope species. Acta Anat 83:70-80

Oboussier H (1972b) Morphologische und quantitative Neocortexuntersuchungen bei Boviden, ein Beitrag zur Phylogie dieser Familie. III. Formen grossen Körpergewichts (über 75 kg). Mitt Hamburg Zool Mus Inst 68:271-292

Oertel WH (1993) Neurotransmitters in the cerebellum. Scientific aspects and clinical relevance. In: Harding AE, Deufel T (eds) Advances in neurology, vol 61. Raven, New York

Oertel WH, Mugnaini E (1984) Immunocytochemical studies of GABA-ergic neurons in the rat basal ganglia and their relations to other neuronal systems. Neurosci Lett 15:159-164

Ogawa T (1934) Beitrage zur vergleichenden Anatomie des Zentralnervensystems der Wassersäugetiere: über das vierte oder subkortikale graue Lager, Stratum griseum quartum s. subcorticale, im Kleinhirn des Seebären (Callorhinus ursinus Gray). Arb Anat Inst Sendai 16:83-96

Ogawa T (1935a) Beiträge zur vergleichenden Anatomie des Zentralnervensystems der Wassersäugetiere: über die Kleinhirnkerne der Pinnepedien und Cetaceen. Arb Anat Inst Sendai 17:63-136

Ogawa T (1935b) Über den Nucleus ellipticus und den Nucleus ruber beim Delphin. Arb Anat Inst Sendaia 17:55-61

Ogawa T (1939a) The tractus tegmenti medialis and its connections with the inferior olive in the cat. J Comp Neurol 70:181-191

Ogawa T (1939b) Experimentelle Untersuchungen über die mediale und zentrale Haubenbahnen bei der Katze. Arch Psychiatr Nervenkr 110:365-444

Ogawa T, Arifuku S (1948) On the acoustic system in the cetacean brains. Sci Rep Whales Res Inst Tokyo 2:1-20

Ohkawa K (1957) Comparative anatomical studies of cerebellar nuclei of mammals. Arch Hist Jpn 13:21-58

Ohtsuki H, Tokunaga A, Ono K, Hasebe S, Tadokoro Y (1992) Distribution of efferent neurons projecting to the tectum and cerebellum in the rat prepositus hypoglossi nucleus. Invest Ophthalmol Vis Sci 33:2567-2574

Ojima H, Honda CN, Jones EG (1991) Patterns of axon collateralization of identified supragranular pyramidal neurons in the cat auditory cortex. Cerebral Cortex 1:80-94

Oka H (1988) Functional organization of the parvocellular red nucleus in the cat. Behav Brain Res 28:23-240

Oka H, Jinnai K, Yamamoto T (1979) The parieto-rubro-olivary pathway in the cat. Exp Brain Res 37:115-125

O'Keefe J, Nadel L (1978) The hippocampus as a cognitive map. Clarendon, Oxford

Olavarria J, Van Sluyters RC (1982) The projection from striate and extrastriate cortical areas to the superior colliculus in the rat. Brain Res 242:332-336

Olazébal UE, Moore JK (1989) Nigrotectal projection to the inferior colliculus: horseradish peroxidase transport and tyrosine hydroxylase immunohistochemical studies in rats, cats, and bats. J Comp Neurol 282:98-118

Oldfield BJ, McKinley MJ (1995) Circumventricular organs. In: Paxinos G (ed) The rat nervous system. Academic, San Diego, pp 391-403

O'Leary JL (1937) Structure of the primary olfactory cortex of the mouse. J Comp Neurol 67:1-31

Oliver DL (1984) Dorsal cochlear nucleus projections to the inferior colliculus in the cat: a light and electron microscopic study. J Comp Neurol 224:155-172

Oliver DL, Morest DK (1984) The central nucleus of the inferior colliculus in the cat. J Comp Neurol 222:237-264

Oliver DL, Shneiderman A (1989) An EM study of the dorsal nucleus of the lateral lemniscus: inhibitory, commissural, synaptic connections between ascending auditory pathways. J Neurosci 9:967-982

Oliver JE, Bradley WE, Fletcher TF (1969) Identification of preganglionic parasympathetic neurons in the sacral spinal cord of the cat. J Comp Neurol 137:321-328

Oliver DL, Kuwada S, Yin TC, Haberly LB, Henkel CK (1991) Dendritic and axonal morphology of HRP-injected neurons in the inferior colliculus of the cat. J Comp Neurol 303:75–100

Olszewski J (1950) On the anatomical and functional organization of the spinal trigeminal nucleus. J Comp Neurol 92:401–413

Olszewski J, Baxter D (1954) Cytoarchitecture of the human brain stem. Karger, Basel

Onodera S (1984) Olivary projections from the mesodiencephalic structures in the cat studied by means of axonal transport of horseradish peroxidase and tritiated aminoacids. J Comp Neurol 227:37–49

Onuf B (1902) On the arrangement and function of the cell groups of the sacral region of the spinal cord in man. Arch Neurol Psychopathol 3:387–412

Oorschot DE (1996) Total number of neurons in the neostriatal, pallidal, subthalamic, and substantia nigral nuclei of the rat basal ganglia: a stereological study using the Cavalieri and optical disector methods. J Comp Neurol 366:580–599

Ortega F, Donate-Oliver F, Grandes P (1993) Retinal afferents on Golgi-identified vertical neurons in the superior colliculus of the rabbit. A Golgi-EM, degenerative and autoradiographic study. Histol Histopathol 8:105–111

Oscarsson O (1973) Functional organization of spinocerebellar paths. In: Iggo A (ed) Somatosensory systems. Springer, Berlin Heidelberg New York, pp 339–380 (Handbook of sensory physiology, vol 2)

Osen KK (1969) Cytoarchitecture of the cochlear nuclei in the cat. J Comp Neurol 136:453–484

Osen KK (1972) Projection of the cochlear nuclei on the inferior colliculus in the cat. J Comp Neurol 144:355–372

Osen KK, Jansen J (1965) The cochlear nuclei in the common porpoise, Phocaena phocaena. J Comp Neurol 125:223–258

Ostapoff EM, Morest DK, Potashner SJ (1990) Uptake and retrograde transport of [3H]GABA from the cochlear nucleus to the superior olive in the guinea pig. J Chem Neuroanat 3:285–295

Oswaldo-Cruz E, Rocha-Miranda CE (1967) The diencephalon of the opossum in stereotaxic coordinates. II. The ventral thalamus and hypothalamus. J Comp Neurol 129:39–48

Oswaldo-Cruz E, Rocha-Miranda CE (1968) The brain of the opossum (Didelphis Marsupialis). Inst de Biofisica, Rio de Janeiro, Brasil

Otake K, Ezure K, Lipski J, Wong She RB (1992) Projections from the commissural subnucleus of the nucleus of the solitary tract: an anterograde tracing study in the cat. J Comp Neurol 324:365–378

Ottersen OP (1980) Afferent connections to the amygdaloid complex of the rat and cat. II. Afferents from the hypothalamus and the basal telencephalon. J Comp Neurol 194:267–289

Ottersen OP (1981) Afferent connections to the amygdaloid complex of the rat with some observations in the cat. III. Afferents from the lower brainstem. J Comp Neurol 202:335–356

Ottersen OP (1982) Connections of the amygdala of the rat. IV. Corticoamygdaloid and intraamygdaloid connections as studied with axonal transport of horseradish peroxidase. J Comp Neurol 205:30–48

Ottersen OP, Ben-Ari Y (1979) Afferent connections to the amygdaloid complex of the rat and cat. I. Projections from the thalamus. J Comp Neurol 187:401–424

Ottersen OP, Storm-Mathisen J (1984) GABA-containing neurons in the thalamus and pretectum of the rodent. Anat Embryol (Berl) 170:197–207

Padel Y, Angaut P, Massion J, Sedan R (1981) Comparative study of the posterior red nucleus in baboons and gibbons. J Comp Neurol 202:421–438

Padel Y, Bourbonnais D, Sybirska E (1986) A new pathway from primary afferents to the red nucleus. Neurosci Lett 64:75–88

Palay SL, Chan-Palay V (1974) Cerebellar cortex: cytology and organization. Springer, Berlin Heidelberg New York

Paloff AM, Usunoff KG (1992) Projections to the inferior colliculus from the dorsal column nuclei. An experimental electron microscopic study in the cat. J Hirnforsch 33:597–610

Paloff AM, Usunoff KG, Hinova-Palova DV (1982) Ultrastructure of Golgi-impregnated and gold-toned neurons in the central nucleus of the inferior colliculus in the cat. J Hirnforsch 33:361–407

Pandya DN (1987) Association cortex. In: Adelman G (ed) Encyclopedia of neuroscience, vol I. Birkhäuser, Boston, pp 80–83

Pandya DN, Yeterian EH (1985) Architecture and connections of cortical association areas. In: Peters A, Jones EG (eds) Association and auditory cortices. Plenum, New York, pp 3–61 (Cerebral cortex, vol 4)

Pandya DN, Van Hoesen GW, Domesick VB (1973) A cingulo-amygdaloid projection in the rhesus monkey. Brain Res 61:369–373

Panneton WM, Burton H (1983) Origin of intratrigeminal pathways in the cat. Brain Res 236:463–470

Panneton WM, Martin GF (1983) Brainstem projections to the facial nucleus of the opossum. A study using axonal transport techniques. Brain Res 267:19–33

Papez JW (1927) Subdivisions of the facial nucleus. J Comp Neurol 43:159–191

Papez JW (1929) The central acoustic tract in cat and man. Anat Rec 42:60

Papez JW (1932) The thalamic nuclei of the nine-banded armadillo (Tatusia novemcincta). J Comp Neurol 56:49–103

Papez JW (1937) A proposed mechanism of emotion. Arch Neurol Psychiatry 42:725–743

Parent A (1979) Identification of the pallidal and peri-pallidal cells projecting to the habenula in monkey. Neurosci Lett 15:159–164

Parent A, DeBellefeuille L (1983) The pallidointralaminar and pallidonigral projections in primate as studied by retrograde double-labeling method. Brain Res 278:11–28

Parent A, Hazrati L-N (1995a) Functional anatomy of the basal ganglia. I. The cortico-basal ganglia-thalamo-cortical loop. Brain Res Rev 20:91–103

Parent A, Hazrati L-N (1995b) Functional anatomy of the basal ganglia. II. The place of subthalamic nucleus and external pallidum in basal ganglia circuitry. Brain Res Rev 20:128–154

Parent A, Descarries L, Beaudet A (1981) Organization of ascending serotonin systems in adult rat brain. A radioautographic study after intraventricular administration of [3H]5-HT. Neuroscience 6:115–138

Parent A, Poitras D, Dubé L (1984) Comparative anatomy of central monoaminergic systems. In: Bjorklund A, Hökfelt T (eds) Classical transmitters in the CNS, part 1. Elsevier Science, Amsterdam, pp 409–439 (Handbook of chemical neuroanatomy, vol 2)

Parnavelas JG (1984) Physiological properties of identified neurons. In: Jones EG, Peters A (eds) Functional properties of cortical cells. Plenum, New York, pp 205–240 (Cerebral cortex, vol 2)

Parnavelas JG, Sullivan K, Lieberman AR, Webster KE (1977) Neurons and their synaptic organization in the visual cortex of the rat: electron microscopy of Golgi preparations. Cell Tissue Res 183:499–517

Parnavelas JG, Edmunds SM (1983) Further evidence that Cajal-Retzius cells transform to nonpyramidal neurons in the developing rat visual cortex. J Neurocytol 12:863–871

Parnavelas JG, Kelly W, Franke E, Eckenstein F (1986) Cholinergic neurons and fibres in the rat visual cortex. J Neurocytol 15:329–336

Parnavelas JG, Barfield JA, Luskin MB (1991) Separate progenitor cells give rise to pyramidal and nonpyramidal neurons in the rat telencephalon. Cerebral Cortex 1:463–468

Pasik P, Pasik T, Pecci Saavedra J, Holstein GR (1981) Light and electron microscopic immunocytochemical localization of serotonin in the basal ganglia of cats and monkeys. Anat Rec 199:194

Pasquier DA, Villar MJ (1982) Subcortical projections to the lateral geniculate body in the rat. Exp Brain Res 48:409–419

Paxinos G (1995) The rat nervous system. Academic, San Diego

Pearson JC, Haines DE (1980) Somatosensory thalamus of a prosimian primate (Galago senegalensis). I. Configuration of nuclei and termination of spinothalamic fibers. J Comp Neurol 190:533–558

Pearson JC, Jennes L (1988) Localization of serotonin-and substance P-like immunofluorescence in the caudal spinal trigeminal nucleus of the rat. Neurosci Lett 88:151–156

Pearson RCA, Gatter KC, Brodal P, Powell TSP (1982) The projection of the basal nucleus of Meynert upon the neocortex in the monkey. Brain Res 259:132–136

Paxinos G, Tork I, Halliday G, Mehler WR (1990a) Human homologes to brainstem nuclei identified in other animals as revealed by acetylcholinesterase activity. In: Paxinos G (ed) The human nervous system. Academic, San Diego, pp 149–202

Pearson J, Halliday G, Sakamoto N, Michel J-P (1990b) Catecholaminergic neurons. In: Paxinos G (ed) The human nervous system. Academic, San Diego, pp 1023–1049

Peerbolte ML (1932) De loop van de achterwortelvezels, die in de achterstrengen van het ruggemerg opstijgen. Proefschrift. N. Samson N.V. – Alphen aan den Rijn

Peinado A, Katz LC (1990) Development of cortical spiny stellate cells: retraction of a transient apical dendrite. Soc Neurosci Abstr 16:1127

Pennartz CMA, Groenewegen HJ, Lopes Da Silva FH (1994) The nucleus accumbens as a complex of functionally distinct neuronal ensembles: an integration of behavioural electrophysiological and anatomical data. Prog Neurobiol 42:719–761

Penney JB Jr, Young AB (1986) Striatal inhomogeneities and basal ganglia function. Mov Disord 1:3–15

Penny GR, Wilson CJ, Kitai ST (1988) Relationship of the axonal and dendritic geometry of spiny projection neurons to the compartmental organizaiton of the neostriatum. J Comp Neurol 269:275–289

Penny JE (1982) Cytoarchitectural and dendritic patterns of the dorsal column nuclei of the opossum. J Hirnforsch 23:315–330

Perachio AA, Kevetter GA (1989) Identification of vestibular efferent neurons in gerbil: histochemical and retrograde labeling. Exp Brain Res 78:315–326

Percheron G, Filion M (1991) Parallel processing in the basal ganglia: up to a point. Trends Neurosci 14:55–56

Percheron G, Yelnik J, Francois C (1984a) The primate striato-pallido-nigral system: an integrative system for cortical information. In: MacKenzie JS, Kemm RE, Wilcock LN (eds) The basal ganglia: structure and function. Plenum, New York, pp 87–105

Percheron G, Yelnik J, Francois C (1984b) A Golgi analysis of the primate globus pallidus. III. Spatial organization of the striatopallidal complex. J Comp Neurol 227:214–227

Perkins RE, Morest DK (1975) A study of cochlear innervation patterns in cats and rats with the Golgi method and normarski optics. J Comp Neurol 163:129–158

Perry VH (1980) A tectocortical visual pathway in the rat. Neuroscience 5:915–927

Pert CB, Kuhar MJ, Snyder SH (1976) Opiate receptor: autoradiographic localization in rat brain. Proc Natl Acad Sci USA 73:3729–3733

Peschanski M, Mantyh PW, Besson JM (1983) Spinal afferents to the ventrobasal thalamic complex in the rat: an anatomical study using wheat-germ agglutinin conjugated to horseradish peroxidase. Brain Res 278:40–244

Peters A (1984a) Chandelier cells. In: Peters A, Jones EG (eds) Cellular components of the cerebral cortex. Plenum, New York, pp 361–380 (Cerebral cortex, vol 1)

Peters A (1984b) Bipolar cells In: Peters A, Jones EG (eds) Cellular components of the cerebral cortex. Plenum, New York, pp 381–408 (Cerebral cortex, vol 1)

Peters A (1985) The visual cortex of the rat. In: Jones EG, Peters A (eds) Visual cortex. Plenum, New York, pp 19–80 (Cerebral cortex, vol 3)

Peters A (1987a) Cortical neurons. In: Adelman G (ed) Encyclopedia of neuroscience, vol 1. Birkhäuser, Boston, pp 282–284

Peters A (1987b) Number of neurons and synapses in primary visual cortex. In: Jones EG, Peters A (eds) Further aspects of cortical function, including hippocampus. Plenum, New York, pp 267–294 (Cerebral cortex, vol 6)

Peters A, Fairén A (1978) Smooth and sparsely-spined stellate cells in the visual cortex of the rat: a study using a combined Golgi-electron microscope technique. J Comp Neurol 181:129–172

Peters A, Harriman KM (1988) Enigmatic bipolar cell of rat visual cortex. J Comp Neurol 267:409–432

Peters A, Jones EG (1984a) Classification of cortical neurons. In: Peters, A, Jones EG (eds.) Cellular components of the cerebral cortex. Plenum, New York, pp 107–122 (Cerebral cortex, vol 1)

Peters A, Jones EG (eds) (1984b) Cellular components of the cerebral cortex. Plenum, New York (Cerebral cortex, vol 1)

Peters A, Kimerer LM (1981) Bipolar neurons in rat visual cortex: a combined Golgi-electron microscope study. J Neurocytol 10:921–946

Peters A, Saint Marie PL (1984) Smooth and sparsely spinous nonpyramidal cells forming local axonal plexuses. In: Peters A, Jones EG (eds) Cellular components of the cerebral cortex. Plenum, New York, pp 419–445 (Cerebral cortex, vol 1)

Peters A, Sethares C (1991) Organization of pyramidal neurons in area 17 of monkey visual cortex. J Comp Neurol 306:1–23

Peters A, Walsh TM (1972) A study of the organization of apical dendrites in the somatic sensory cortex of the rat. J Comp Neurol 144:253–268

Peters A, Proskauer CC, Ribak CE (1982) Chandelier cells in rat visual cortex. J Comp Neurol 206:397–416

Peters A, Kara DA, Harriman KM (1985) The neuronal composition of area 17 of rat visual cortex III. Numerical considerations. J Comp Neurol 238:263–274

Peters A, Meinecke DL, Karamanlidis AN (1987) Vasoactive intestinal polypeptide immunoreactive neurons in cat primary visual cortex. J Neurocytol 16:23–38

Petras JM (1969) Some efferent connections of the motor and somatosensory cortex of simian primates and felid, canid and procyonid carnivores. Ann NY Acad Sci 167:469–505

Petras JM (1977) Spinocerebellar neurons in the rhesus monkey. Brain Res 130:146–151

Petras JM, Cummings JF (1977) The origin of spinocerebellar pathways. III. The nucleus centrobasalis of the cervical enlargement and the nucleus dorsalis of the thoracolumbar spinal cord. J Comp Neurol 173:693–716

Petras JM, Lehman RAW (1966) Corticospinal fibers in the raccoon. Brain Res 3:195–197

Petrovich GD, Risold PY, Swanson LW (1996) Organization of projections from the basomedial nucleus of the amygdala: a PHAL study in the rat. J Comp Neurol 374:387–420

Petrovicky P (1966a) A comparative study of the reticular formation of the guinea pig. J Comp Neurol 128:85–108

Petrovicky P (1966b) Formatio reticularis of the hedgehog (Erinaceus europaeus). Acta Univ Carol Med 12:293–307

Petrovicky P (1967) The reticular formation in the bat (Myotis myotis Borkh). Folia Morphol (Praha) 146–152

Pettigrew JD (1986) Flying primates? Megabats have the advanced pathway from eye to midbrain. Science 231:1304–1306

Petursdottir G (1990) Vestibulo-ocular projections in the 11-day chicken embryo: pathway specificity. J Comp Neurol 297:283–297

Phelps PE, Houser SR, Vaughn JE (1985) Immunocytochemical localization of choline acetyltransferase within the rat neostriatum: a correlated light and electron microscopic study of cholinergic neurons and synapses. J Comp Neurol 238:286–307

Pierson Pentney R, Cotter JR (1976) Retinofugal projections in an echolocating bat. Brain Res 115:479-484

Pierson RJ, Carpenter MB (1974) Anatomical analysis of pupillary reflex pathways in the rhesus monkey (1976). J Comp Neurol 158:121-144

Pijnenburg AJJ, Honig WMM, Van der Heyden JAM, Van Rossum JM (1976) Effects of chemical stimulation of the mesolimbic dopamine system upon locomotor activity. Eur J Pharmacol 35:45-58

Pilleri G (1959a) Beiträge zur vergleichenden Morphologie des Nagetiergehirnes. 1. Sciuromorpha. Acta Anat 38 [Suppl]:1-42

Pilleri G (1959b) Beiträge zur vergleichenden Morphologie des Nagetiergehirnes. 2. Hystricomorpha. Acta Anat 38 [Suppl]:43-95

Pilleri G (1959c) Beiträge zur vergleichenden Morphologie des Nagetiergehirnes. 3. Das Gehirn der Wassernager (Castor canadensis, Ondatra zibethica, Myocastor coypus). Acta Anat 38 [Suppl]:96-123

Pilleri G (1960a) Beiträge zur vergleichenden Morphologie des Nagetiergehirnes. 4. Zentralnervensystem, Körperorgane und stammesgeschichtliche Verwandtschaft der Aplodontia rufa Refinesque (Rodentia, Aplodontoidea). Acta Anat 40 [Suppl]:5-35

Pilleri G (1960b) Beiträge zur vergleichenden Morphologie des Nagetiergehirnes. 5. Vergleichend-morphologische Untersuchungen über das Zentralnervensystem nearktischer Sciuromorpha und Bemerkungen zum Problem Hirnform und Taxonomie. Acta Anat 40 [Suppl]:36-68

Pilleri G (1960c) Beiträge zur vergleichenden Morphologie des Nagetiergehirnes. 6. Materiale zur vergleichenden Anatomie des Gehirns der Myomorpha. Acta Anat 40 [Suppl]:69-88

Pilleri G (1964) Morphologie des Gehirnes des Southern Right Whale, Eubalaena australis Desmoulins (Cetacea, Mysticeti, Balaenidae). Acta Zool (Stockh) 46:245-272

Pilleri G (1966a) Morphologie des Gehirnes des Seiwals, Balaenoptera borealis Lesson (Cetacea, Mysticeti, Balaenopteridae). J Hirschforscha 8:221-267

Pilleri G (1966b) Morphologie des Gehirnes des Buckelwals, megaptera novaeangliae Borowski (Cetacea, Mysticeti, Balaenopteridae). J Hirnforsch 8:437-491

Pilleri G, Gihr M (1969a) Das Zentralnervensystem der Zahn- und Bartenwale. Rev Suisse Zool 76:995-1037

Pilleri G, Gihr M (1969b) On the anatomy and behavior of Risso's dolphin (Grampus griseus G. Cuvier). Invest Cetacea 1:74-93

Pilleri G, Kraus C (1969) Zum Aufbau des Cortex bei Cetaceen. Rev Suisse Zool 76:760-767

Pinchitpornchai C, Rawson JA, Rees S (1994) Morphology of parallel fibres in the cerebellar cortex of the rat: an experimental light and electron microscopic study with biocytin. J Comp Neurol 342:206-220

Pindzola RR, Ho RH, Martin GF (1988) Catecholaminergic innervation of the spinal cord in the North American opossum, Didelphis virginiana. Brain Behav Evol 32:281-292

Pirlot P (1981) A quantitative approach to the marsupial brain in an eco-ethological perspective. Rev Can Biol 40:229-250

Pirlot P, Kamiya T (1985) Qualitative and quantitative brain morphology in the sirenian Dujong dujon. Z Zool Syst Evol Forsch 23:147-155

Pirlot P, Nelson J (1978) Volumetric analysis of Monotreme brains. Austr Zool 20:171-179

Pitkänen A, Amaral DJ (1991) Demonstration of projections from the lateral nucleus to the basal nucleus: a PHA-L study in the monkey. Exp Brain Res 83:465-470

Pitkänen A, Stefanacci L, Farb C, Go C-G, LeDoux JE, Amaral DG (1995) Intrinsic connections of the rat amygdaloid complex: projections originating in the lateral nucleus. J Comp Neurol 356:288-310

Plecha DM, Randall WC, Geis GS, Wurster RD (1988) Localization of vagal preganglionic somata controlling sinoatrial and atrioventricular nodes. Am J Physiol 255:703-717

Pohlenz-Kleffner W (1969) Vergleichende Untersuchungen zur Evolution der Gehrine von Edentaten. Hirnform und Hirnfurchen. Z Zool Syst Evol Forsch 7:181-208

Poirier LJ, Bouvier G (1966) The red nucleus and its efferent nervous pathways in the monkey. J Comp Neurol 128:223-244

Poljakow GI (1979) Entwicklung der Neuronen der menschlichen Grosshirnrinde. Thieme, Leipzig

Polyak S (1957) The vertebrate visual system. University of Chicago Press, Chicago

Pompeiano O, Brodal A (1957a) Spinovestibular fibers in the cat: an experimental study. J Comp Neurol 108:353-382

Pompeiano O, Brodal A (1957b) Experimental demonstration of a somatotopical origin of rubrospinal fibers in the cat. J Comp Neurol 108:225-252

Porrino JL, Crane AM, Goldman-Rakic PS (1981) Direct and indirect pathways from the amygdala to the frontal lobe in the rhesus monkey. J Comp Neurol 198:121-136

Porter JD, Guthrie BL, Sparks DL (1983) Innervation of monkey extraocular muscles: localization of sensory and motor neurons by retrograde transport of horseradish peroxidase. J Comp Neurol 218:208-219

Porter R (1985) The corticomotoneuronal component of the pyramidal tract: corticomotoneuronal connections and functions in primates. Brain Res Rev 10:1-26

Potter H, Nauta WJH (1979) A note on the problem of olfactory associations of the orbitofrontal cortex in the monkey. Neuroscience 4:361-367

Powell TPS (1981) Certain aspects of the intrinsic organization of the cerebral cortex. In: Pompeiano O, Ajmone Marsan C (eds) Brain mechanisms and perceptual awareness. Raven, New York, pp 1-19

Precechtl A (1925) Some notes upon the finer anatomy of the brain stem and basal ganglia of Elephas indicus. Verh Kon Adak Wetensch Amsterdam Proc 28:81-92

Price JL (1973) An autoradiographic study of complementary laminar patterns of termination of afferent fibers to the olfactory cortex. J Comp Neurol 150:87-108

Price JL, Amaral DG (1981) An autoradiographic study of the projections of the central nucleus of the monkey amygdala. J Neurosci 1:1242-1259

Price JL, Russchen FT, Amaral DG (1987) The limbic region. II. The amygdaloid complex. In: Bjorklund A, Hökfelt T, Swanson LW (eds) Integrated systems of the CNS, part I. Elsevier, Amsterdam, pp 279-388 (Handbook of chemical neuroanatomy, vol 5)

Price JL, Slotnick BM, Revial M-F (1991) Olfactory projections to the hypothalamus. J Comp Neurol 306:447-461

Probst M (1899) Über vom Vierhugel, von der Brucke und vom Kleinhirn absteigende Bahnen. Dtsch Ztschr Nervenheilkd 15:192-221

Prothero JW, Sundsten JW (1984) Folding of the cerebral cortex in mammals. A scaling model. Brain Behav Evol 24:152-167

Provis J (1977) The organization of the facial nucleus of the brush-tailed possum (Trichosurus vulpecula). J Comp Neurol 172:177-188

Pubols BHJ, Haring JH (1995) The raccoon spinocervical and spinothalamic tracts: a horseradish peroxidase study. Brain Res Rev 20:196-208

Pubols BH, Pubols LM (1966) Somatic sensory representation in the thalamic ventrobasal complex of the virginia opossum. J Comp Neurol 127:19-34

Pujol R, Carlier E, Lenoir M (1980) Ontogenetic approach to inner and outer hair cell function. Hearing Res 2:223-230

Purves D, Riddle DR, LaMantia AS (1992) Iterated patterns of brain circuitry (or how the cortex gets its spots). Trends Neurosci 15:362-368

Qvist H (1989a) Demonstration of axonal branching of fibres from certain precerebellar nuclei to the cerebellar cortex and nuclei: a retrograde fluorescent double-labelling study in the cat. Exp Brain Res 75:15-27

Qvist H (1989b) The cerebellar nuclear afferent and efferent connections with the lateral reticular nucleus in the cat as studied with retrograde transport of WGA-HRP. Anat Embryol 179:471-483

Qvist H, Dietrichs E, Walberg F (1984) An ipsilateral projection from the red nucleus to the lateral reticular nucleus in the cat. Anat Embryol (Berl) 170:327-330

Raappana P, Arvidsson J (1993) Location, morphology, and central projections of mesencephalic trigeminal neurons innervating rat masticatory muscles studied by axonal transport of choleragenoid-horseradish peroxidase. J Comp Neurol 328:103-114

Radinsky L (1968) Evolution of somatic sensory specialization in otter brains. J Comp Neurol 134:495-506

Radinsky L (1973a) Are sting badgers skunks? Implications of neuroanatomy for mustelid phylogeny. J Mammal 54:585-594

Radinsky L (1973b) Evolution of the canid brain. Brain Behav Evol 7:169-202

Radinsky L (1975) Evolution of the felid brain. Brain Behav Evol 11:214-254

Radinsky L (1976) Oldest horse brains: more advanced than previously realized. Science 194:626-627

Radinsky L (1977) Early primate brains: facts and fiction. J Hum Evol 6:79-86

Ragsdale CW Jr, Graybiel AM (1988) Fibers from the basolateral nucleus of the amygdala selectively innervate striosomes in the caudate nucleus of the cat. J Comp Neurol 269:506-522

Rakic P (1971) Guidance of neurons migrating to the fetal monkey neocortex. Brain Res 33:471-476

Rakic P (1972) Mode of cell migration to the superficial layers of fetal monkey neocortex. J Comp Neurol 145:61-84

Rakic P (1978) Neuronal migration and contact guidance in the primate telencephalon. Postgrad Med J 54:25-40

Rakic P (1988) Specification of cerebral cortex areas. Science 241:170-176

Ralston DD, Milroy AM, Ralston H Jr (1987) Non-myelinated axons are rare in the medullary pyramids of the macaque monkey. Neurosci Lett 73:215-219

Ralston DD, Ralston H Jr (1985) The terminations of corticospinal tract axons in the macaque monkey. J Comp Neurol 242:325-337

Ralston HJI (1965) The organization of the substantia gelatinosa Rolandi in the cat lumbosacral spinal cord. Z Zellforsch 67:1-23

Ralston HJI (1968) Dorsal root projections to dorsal horn neurons in the cat spinal cord. J Comp Neurol 132:303-330

Ralston HJI, Ralston DD (1992) The primate dorsal spinothalamic tract: evidence for a specific termination in the posterior nuclei (Po/SG) of the thalamus. Pain 48:107-118

Ramon y Cajal S (1909) Histologie du systeme nerveux de l'homme et des vertebres. vol. 1. Maloine Paris

Ramon-Moliner E (1979) A retrothalamic system of collateral fibers from the cerebral peduncle. Brain Res 170:1-21

Ramon-Moliner E (1984) Subcortical projections of the pericruciate cortex of cat. An autoradiography study. J Hirnforsch 25:445-459

Ranson SW (1913) The course within the spinal cord of the non-medullated fibres of the dorsal roots. A study of Lissauer's tract in the cat. J Comp Neurol 23:259-281

Rapaport DH, Wilson PD (1983) Retinal ganglion cell size groups projecting to the superior colliculus and the dorsal lateral geniculate nucleus in the North American opossum. J Comp Neurol 213:74-85

Rasmussen DD (1989) The projection pattern of forepaw nerves to the cuneate nucleus of the raccoon. Neurosci Lett 98:129-134

Ray JP, Price JL (1992) The organization of the thalamocortical connections of the mediodorsal thalamic nucleus in the rat, related to the ventral forebrain-prefrontal cortex topography. J Comp Neurol 323:167-197

Ray JP, Price JL (1993) The organization of the projections from the mediodorsal nucleus of the thalamus to orbital and medial prefrontal cortex in the monkey. J Comp Neurol 337:1-31

Reddy VK, Cassini P, Ho RH, Martin GF (1990) Origins and terminations of bulbospinal axons that contain serotonin and either enkephalin or substance-P in the North American Opossum. J Comp Neurol 294:96-108

Redgrave P, Marrow L, Dean P (1992) Topographical organization of the nigrotectal projection in rat: evidence for segregated channels. Neuroscience 50:571-595

Redgrave P, Mitchell IJ, Dean P (1987) Descending projections from the superior colliculus in rat: a study using orthograde transport of wheatgerm- agglutinin conjugated horseradish peroxidase. Exp Brain Resb 68:147-167

Redgrave P, Mitchell IJ, Dean P (1987) Further evidence for segregated output channels from superior colliculus in rat: ipsilateral tecto-pontine and tecto-cuneiform projections have different cells of origin. Brain Resa 413:170-174

Redgrave P, Odekunle A, Dean P (1986) Tectal cells of origin of predorsal bundle in rat: location and segregation from ipsilateral descending pathway. Exp Brain Res 63:279-293

Reep R (1984) Relationship betwen prefrontal and limbic cortex: a comparative anatomical review. Brain Behav Evol 25:5-80

Reep RL, O'Shea TJ (1990) Regional brain morphometry and lissencephaly in the Sirenia. Brain Behav Evol 35:185-194

Rees D, Hore J (1970) The motor cortex of the brush-tailed possum (Trichosurus vulpecula): motor representation, motor function and the pyramidal tract. Brain Res 20:439-451

Regidor J, Divac I (1987) Architectonics of the thalamus in the echidna. Brain Behav Evol 30:328-41

Rehkämper G, Necker R, Nevo E (1994) Functional anatomy of the thalamus in the blind mole rat Spalax ehrenbergi: an architectonic and electrophysiologicaal controlled tracing study. J Comp Neurol 347:570-584

Reiner A (1991) A comparison of neurotransmitter-specific and neuropeptide-specific neuronal cell types present in the dorsal cortex in turtles with those present in the isocortex in mammals: implications for the evolution of isocortex. Brain Behav Evol 38:53-91

Reiter RJ (1981) The mammalian pineal gland: structure and function. Am J 162:287-313

Reiter RJ (1993) The melatonin rhythm – both a clock and a calendar. Experientia 49:654-664

Renaud LP, Hopkins DA (1977) Amygdala afferents from the mediobasal hypothalamus: An electrophysiological and neuroanatomical study in the rat. Brain Res 121:201-213

Rethelyi M, Szentagothai J (1969) The large synaptic complexes of the substantia gelatinosa. Exp Brain Res 7:258-274

Rethelyi M, Szentagothai J (1973) Distribution and connections of afferent fibres in the spinal cord. In: Iggo A (ed) Handbook of sensory physiology. Vol. II. Somatosensory system. Berlin Heidelberg New York vol. II.: pp. 207-252

Rethelyi M, Light AR, Perl ER (1982) Synaptic complexes formed by functionally defined primary afferent units with fine myelinated fibers. J Comp Neurol 207:381-393

Rexed B (1952) The cytoarchitectonic organization of the spinal cord in the cat. J Comp Neurol 96:415-496

Rexed B (1954) A cytoarchitectonic atlas of the spinal cord in the cat. J Comp Neurol 100:297-379

Rexed B, Brodal A (1951) The nucleus cervicalis lateralis. A spino-cerebellar relay nucleus. J Neurophysiol 14:399-407

Rhoades RW, Fish SE, Chiaia NL, Bennett-Clarke C, Mooney RD (1989) Organization of the projections from the trigeminal brainstem complex to the superior colliculus in the rat and hamster: anterograde tracing with Phaseolus vulgaris Leucoagglutinin and intra-axonal injection. J Comp Neurol 289:641-656

Rhoades RW, Kuo DC, Polcer JD, Fish SE, Voneida TJ (1982) Indirect visual cortical input to the deep layers of the hamster's superior colliculus via the basal ganglia. J Comp Neurol 208:239-254

Rhoton AL, O'Leary JL, Ferguson P (1966) The trigeminal, facial, vagal, and glossopharyngeal nerves in the monkey. Arch Neurol 14:530-540

Ribak CE (1978) Aspinous and sparsely-spinous stellate neurons in the visual cortex of rats contain glutamic acid carboxylase. J. Neurocytol. 7:461-478

Ricardo JA (1980) Efferent connections of the subthalamic region in the rat. I. The subthalamic nucleus of Luys. Brain Res 202:257-271

Ricardo JA (1983) Hypothalamic pathways involved in metabolic regulatory functions, as identified by tracktracing methods. Adv Metab Dis 10:1-30

Ricardo JA, Koh ET (1978) Anatomical evidence of direct projections from the nucleus of the solitary tract to the hypothalamus, amygdala, and other forebrain structures of the rat. Brain Res 153:1-26

Ricardo JA, Koh ET (1978) Anatomical evidence of direct projections from the nucleus of the solitary tract to the hypothalamus, amygdala, and other forebrain structures in the rat. Brain Res 153:1-26

Richmond FJ, Courville J, Saint-Cyr JA (1982) Spino-olivary projections from the upper cervical spinal cord: an experimental study using autoradiography and horseradish peroxidase. Exp Brain Res 47:239-251

Richter E (1966) Über die Entwicklung des Globus pallidus und des Corpus subthalamicum beim Menschen. In: Hassler R, Stephan H (eds) Evolution of the forebrain: phylogenesis and ontogenesis of the forebrain. Thieme, Stuttgart

Ridgway SH, Demski LS, Schwanzel-Fukuda M (1987) The terminal nerve in odontocete cetaceans. Ann NY Acad Sci 519:201-212

Rieck RW, Huerta MF, Harting JK, Weber JT (1986) Hypothalamic and ventral thalamic projections to the superior colliculus in the cat. J Comp Neurol 243:249-265

Riley HA (1928) A comparative study of the Arbor Vitae and the folial pattern of the mammalian cerebellum. Archiv Neurol Psychiatr 20:895-1034

Rioch DM (1929) Studies on the diencephalon of Carnivora, part I. The nuclear configuration of the thalamus, epithalamus, and hypothalamus of the dog and cat. J Comp Neurol 49:1-119

Risold PY, Canteras NS, Swanson LW (1994) Organization of projections from the anterior hypothalamic nucleus: a Phaseolus vulgaris-leucoagglutinin study in the rat. J Comp Neurol 348:1-40

Roberts GW, Woodhams PL, Polak JM, Crow TJ (1982) Distribution of neuropeptides in the limbic system of the rat: The amygdaloid complex. Neuroscience 7:99-131

Robertson B, Grant G, Bjorkeland M (1983) Demonstration of spinocerebellar projections in cat using anterograde transport of WGA-HRP, with some observations on spinomesencephalic and spinothalamic projections. Exp Brain Res 52:99-104

Robertson LT, Stotler WA (1974) The structure and connections of the developing inferior olivary nucleus of the rhesus monkey. J Comp Neurol 158:167-182

Robertson RT (1983) Efferents of the pretectal complex: separate populations of neurons project to lateral thalamus and to inferior olive. Brain Res 258:91-95

Robinson FR, Cohen JL, May J, Sestokas AK, Glickstein M (1987a) Cerebellar targets of visual pontine cells in the cat. J Comp Neurol 223:471-482

Robinson FR, Houk JC, Gibson AR (1987b) Limb specific connections of the cat magnocellular red nucleus. J Comp Neurol 257:553-577

Rocha-Miranda CE, Cavalcante LA, Gawryszewski LG, Linden R, Volchan E (1978) The vertical meridian representation and the pattern of retinotectal projections in the opossum. In: Rocha-Miranda CE, Lent R (eds) Opossum neurobiology (Neurobiologia do Gamba). Academia Brasileira de Ciencias, Rio de Janeiro, pp 113-126

Rockel AJ, Hiorns RW, Powell TPS (1974) Numbers of neurons through the full depth of neocortex. J. Anat 118:371

Rockel AJ, Hiorns RW, Powell TPS (1980) The basic uniformity in structure of the neocortex. Brain 103:221-244

Rockland KS, Lund JS (1982) Widespread periodic intrinsic connections in the tree shrew visual cortex. Science 215:1532-1534

Rodieck RW, Watanabe M (1993) Survey of the morphology of macaque retinal ganglion cells that project to the pretectum, superior colliculus, and parvicellular laminae of the lateral geniculate nucleus. J Comp Neurol 338(2):289-303

Roeling TAP, Veening JG, Kruk MR, Peters JPW, Vermelis MEJ, Nieuwenhuys R (1994) Efferent connections of the hypothalamic "aggression area" in the rat. Neuroscience 59:1001-1024

Roger M, Cadusseau J (1985) Afferents to the zona incerta in the rat: a combined retrograde and anterograde study. J Comp Neurol 241:480-492

Roger M, Cadusseau J (1987) Anatomical evidence of a reciprocal connection between the posterior thalamic nucleus and the parvocellular division of the red nucleus in the rat. A combined retrograde and anterograde study. Neuroscience 21:573-583

Rogers JH, Resibois A (1992) Calretinin and calbindin-D28K in rat brain: patterns of partial co-localization. Neuroscience 51:853-865

Rolls ET (1992) Neurophysiology and functions of the primate amygdala. In: Aggleton JP (ed) The amygdala: neurobiological aspects of emotion, memory and mental dysfunction. Wiley-Liss, New York, pp 143-167

Romanes GJ (1951) The motor cell columns of the lumbosacral cord of the cat. J Comp Neurol 94:313-364

Romanowski CAJ, Mitchell IJ, Crossman AR (1985) The organization of the efferent projections of the zona incerta. J Anat 143:75-95

Romer AS (1962) The vertebrate body, 3rd edn. Saunders, Philadelphia

Roney KJ, Scheibel AB, Shaw GL (1979) Dendritic bundles: survey of anatomical experiments and physiological theories. Brain Res Rev 1:225-271

Room P, Groenewegen HJ (1986) The connections of the parahippocampal cortex in the cat. II. Subcortical afferents. J Comp Neurol 251:451-473

Roppolo JR, Nadelhaft I, De Groat WC (1985) The organization of pudendal motoneurons and primary afferent projections in the spinal cord of the rhesus monkey revealed by horseradish peroxidase. J Comp Neurol 234:475-488

Rosa MG, Schmid LM (1994) Topography and extent of visual-field representation in the superior colliculus of the megachiropteran Pteropus. Vis Neurosci 11:1037-1057

Rose JE (1942) The thalamus of the sheep: cellular and fibrous structure and comparison with pig, rabbit and cat. J Comp Neurol 77:469-523

Rose JE (1952) The cortical connections of the reticular complex of the thalamus. Res Publ Assoc Res Nerv Ment Dis 30:454-479

Rose JE (1960) Organization of frequency sensitive neurons in the cochlear nuclear complex of the cat. In: Grant L, Rasmussen GL, Windle WF (eds) Neural mechanisms of the auditory and vestibular systems, chap 9. Thomas, Springfield, pp 116-136

Rose JE, Mountcastle VB (1959) Touch and kinesthesis. In: Field J (ed) Neurophysiology. American Physiology Society, Washington, DC, Sect 1, vol 1, pp 387-429

Rose JE, Woolsey CN (1949) Organization of the mammalian thalamus and its relationships to the cerebral cortex. Electroencephalogr Clin Neurophysiol 1:391-403

Rose M (1935) Cytoarchitektonik und Myeloarchitektonik der Grosshirnrinde. In: Bumke O, Foerster O (eds) Handbuch der Neurologie, vol I: Allgemeine Neurologie I, Anatomie. Springer, Berlin Heidelberg New York, pp 588-778

Rose PK, MacDonald J, Abrahams VC (1991) Projections of the tectospinal tract to the upper cervical spinal cord of the cat: a study with the anterograde tracer PHA-L. J Comp Neurol 314:91-105

Ross CA, Ruggiero DA, Reis DJ (1985) Projections from the nucleus tractus solitarii to the rostral ventrolateral medulla. J Comp Neurol 242:511-534

Ross CD, Godfrey DA (1985) Distributions of choline acetyltransferase and acetylcholinesterase activities in layers of rat superior colliculus. J Histochem Cytochem 33:631-641

Ross MD (1977) The tectorial membrane of the rat. Am J Anat 139:449-482

Ross LS, Pollak GD, Zook JM (1988) Origin of ascending projections to an isofrequency region of the mustache bat's inferior colliculus. J Comp Neurol 270:488–505

Rossi GF, Brodal A (1956) Spinal afferents to the trigeminal sensory nuclei and the nucleus of the solitary tract. Confin Neurol (Basel) 16:321–332

Rouiller EM, Capt M, Dolivo M, De Ribaupierre F (1986) Tensor tympani reflex pathways studied with retrograde horseradish peroxidase and transneuronal viral tracing techniques. Neurosci Lett 72:247–252

Rowe M (1990) Organization of the cerebral cortex in Monotremes and Marsupials. In: Jones EG, Peters A (eds) Cerebral cortex, vol 8B, part II. Plenum, New York, pp 263–334

Royce G, Ward JP, Harting JK (1976) Retinofugal pathways in two marsupials. J Comp Neurol 170:391–414

Royce GJ (1987) Recent research on the centromedian and parafascicular nuclei. In: Carpenter MB, Jayaraman A (eds) The basal ganglia II: structure and function-current concepts, vol 32: advances in behavioural biology. Plenum, New York, pp 293–319

Royce GJ, Mourey RJ (1985) Efferent connections of the centromedian and parafascicular thalamic nuclei: an autoradiographic investigation in the cat. J Comp Neurol 235:277–300

Royce GJ, Bromley S, Gracco C, Beckstead RM (1989) Thalamocortical connections of the rostral intralaminar nuclei: an autoradiographic analysis in the cat. J Comp Neurol 288:555–582

Rubertone JA, Haines DE (1982) The vestibular complex in a prosimian primate (Galago senegalensis): morphological and spinovestibular connections. Brain Behav Evol 20:129–155

Rubertone JA, Mehler WR (1980) Afferents to the vestibular complex in rat: a horseradish peroxidase study. Neurosci Abstr 6:225

Rubertone JA, Mehler WR, Cox GE (1983) The intrinsic organization of the vestibular complex: evidence for internuclear connectivity. Brain Res 263:137–141

Rubertone JA, Mehler WR, Voogd J (1995) The vestibular nuclear complex. In: Paxinos G (ed) The rat nervous system, 2nd edn. Academic, San Diego, pp 773–796

Rubin E, Purves D (1980) Segmental organization of sympathetic preganglionic neurons in the mammalian spinal cord. J Comp Neurol 192:163–174

Ruda MA, Coffield J, Steinbusch HWM (1982) Immunocytochemical analysis of serotonergic axons in laminae I and II of the lumbar spinal cord of the cat. J Neurosci 2:1660–1671

Rudebeck B (1945) Contributions to forebrain morphology in dipnoi. Acta Zool 26:10–157

Ruggiero DA, Ross CA, Kumada M, Reis DJ (1982) Reevaluation of projections from the mesencephalic trigeminal nucleus to the medulla and spinal cord: new projections. A combined retrograde and anterograde horseradish peroxidase study. J Comp Neurol 206:278–292

Ruigrok TJH, Cella F (1995) Precerebellar nuclei and red nucleus. In: Paxinos G (ed) The rat nervous system, 2nd edn. Academic, San Diego, pp 277–306

Ruigrok TJH, Voogd J (1990) Cerebellar nucleo-olivary projections in the rat: an anterograde tracing study with Phaseolus vulgaris Leucoagglutinin (PHA-L). J Comp Neurol 298:315–333

Ruigrok TJ, Cella F, Voogd J (1995) Connections of the lateral reticular nucleus to the lateral vestibular nucleus in the rat. An anterograde tracing study with Phaseolus vulgaris leucoagglutinin. Eur J Neurosci 7:1410–1413

Ruiz-Marcos A, Valverde F (1970) Dynamic architecture of the visual cortex. Brain Res 19:25–39

Rushlow W, Flumerfelt BA, Naus CCG (1995) Colocalization of somatostatin, neuro- peptide Y, and NADPH-diaphorase in the caudate-putamen of the rat. J Comp Neurol 351:499–508

Russchen FT (1982a) Amygdalopetal projections in the cat. I. Cortical afferent connections. A study with retrograde and anterograde tracing techniques. J Comp Neurol 206:159–179

Russchen FT (1982b) Amygdalopetal projections in the cat. II. Subcortical afferent connections. A study with retrograde tracing techniques. J Comp Neurol 207:157–176

Russchen FT (1986) Cortical and subcortical afferents of the amygdaloid complex. In: Ben-Ari Y, Schwarcz R (eds) Excitatory amino acids and epilepsy. Plenum, New York, pp 35–52

Russchen FT, Price JL (1984) Amygdalostriatal projections in the rat. Topographical organizational and fiber morphology shown using the lectin PHA-L as an anterograde tracer. Neurosci Lett 47:15–22

Russchen FT, Bakst I, Amaral DG, Price JL (1985a) The amygdalostriatal projections in the monkey. An anterograde tracing study. Brain Res 329:241–257

Russchen FT, Amaral DG, Price JL (1985b) The afferent connections of the substantia innominata in the monkey, Macaca fascicularis. J Comp Neurol 242:1–27

Russchen FT, Amaral DG, Price JL (1987) The afferent input to the magnocellular division of the mediodorsal thalamic nucleus in the monkey, Macaca fascicularis. J Comp Neurol 256:175–210

Rustioni A (1973) Non-primary afferents to the nucleus gracilis from the lumbar cord of the cat. Brain Res 51:81–95

Rustioni A (1975) Dorsal column nuclei afferents in the lateral funiculus of the cat: distribution pattern and absence of sprouting after chronic deafferentation. Exp Brain Res 23:1–12

Rustioni A (1977) Spinal neurons project to the dorsal column nuclei of rhesus monkeys. Science 196:656–658

Rustioni A, Hayes NL (1981) Corticospinal tract collaterals to the dorsal column nuclei of cats: an anatomical single and double retrograde tracer study. Exp Brain Res 43:237–245

Rustioni A, Kaufman AB (1977) Identification of cells or origin of non-primary afferents to the dorsal column nuclei of the cat. Exp Brain Res 18:1–14

Rustioni A, Sanyal S, Kuypers HGJM (1971a) A histochemical study of the distribution of the trigeminal divisions in the substantia gelatinosa of the cat. Brain Res 32:45–52

Rustioni A, Kuypers HGJM, Holstege G (1971b) Propriospinal projections from the ventral and lateral funiculi of the motoneurons in the lumbosacral cord of the cat. Brain Res 34:255–276

Rustioni A, Hayes NL, O'Neill S (1979) Dorsal column nuclei and ascending spinal afferents in macaque. Brain 102:95–125

Rutherford JG, Gwyn DG (1982) A light and electron microscopic study of the interstitial nucleus of Cajal in rat. J Comp Neurol(205):327–340

Rutherford JG, Zulk-Harper A, Gwyn DG (1989) A comparison of the distribution of the cerebellar and cortical connections of the nucleus of Darkschewitsch (ND) in the cat: a study using anterograde and retrograde HRP tracing techniques. Anat Embryol (Berl) 180:585–496

Rye DB, Wainer BH, Mesulam M-M, Mufson EJ, Saper CB (1984) Cortical projections arising from the basal forebrain: a study of cholinergic and noncholinergic components combining retrograde tracing and immunohistochemical localization of choline acetyltransferase. Neuroscience 13:627–643

Rye DB, Saper CB, Wainer BH (1987) Pedunculopontine tegmental nucleus of the rat: cytoarchitecture, cytochemistry, and some extrapyramidal connections of the meso-pontine tegmentum. J Comp Neurol 259:483–528

Sachs GM, Schneider GE (1984) The morphology of optic tract axons arborizing in the superior colliculus of the hamster. J Comp Neurol 230:155–167

Sadikot AF, Parent A, Francois C (1992a) Efferent connections of the centromedian and parafascicular thalamic nuclei in the squirrel monkey: a PHA-L study of subcortical projections. J Comp Neurol 315:137–159

Sadikot AF, Parent A, Smith Y, Bolam JP (1992b) Efferent connections of the centromedian and parafascicular thalamic

nuclei in the squirrel monkey: a light and electron microscopic study of the thalamostriatal projection in relation to striatal heterogeneity. J Comp Neurol 320:228–242
Saint-Cyr JA (1983) The projection from the motor cortex to the inferior olive in the cat. Neuroscience 10:667–684
Saint-Cyr JA (1987) Anatomical organization of cortico-mesencephalo-olivary pathways in cat, as demonstrated with axonal transport techniques. J Comp Neurol 257:39–59
Saint-Cyr JA, Courville J (1980) Projections from the motor cortex, mid-brain and vestibular nuclei to the inferior olive in the cat. Anatomical organization and functional correlates. In: Courville J, De Montigny C, Lamarre Y (eds) The Inferior Olivary Nucleus. Raven, New York, pp 97–124
Saint-Cyr JA, Courville J (1981) Sources of descending afferents to the inferior olive from the upper brain stem in the cat as revealed by the retrograde transport of horseradish peroxidase. J Comp Neurol 198:567–581
Saint-Cyr JA, Courville J (1982a) Projection from the vestibular nuclei to the inferior olive in the cat: an autoradiographic and horseradish peroxidase study. Brain Res 165:189–200
Saint-Cyr JA, Courville J (1982b) Descending projections to the inferior olive from the mesencephalon and superior colliculus in the cat. An autoradiographic study. Exp Brain Res 45:333–348
Saint Marie RL, Baker RA (1990) Neurotransmitter-specific uptake and retrograde transport of [3H]glycine from the inferior colliculus by ipsilateral projections of the superior olivary complex and nuclei of the lateral lemniscus. Brain Res 524:244–253
Saint Marie RL, Peters A (1985) The morphology and synaptic connections of spiny stellate neurons in monkey visual cortex (area 17): a Golgi-electron microscopic study. J Comp Neurol 233:213–235
Sakai ST (1988) Corticonigral projections from area 6 in the raccoon. Exp Brain Res 73:498–504
Sakla FB (1969) Quantitative studies on the postnatal growth of the spinal cord and the vertebral column of the albino mouse. J Comp Neurol 136:237–252
Saldaña E, Merchán MA (1992) Intrinsic and commissural connections of the rat inferior colliculus. J Comp Neurol 319:417–437
Saldaña E, Feliciano M, Mugnaini E (1996) Distribution of descending projections from primary auditory neocortex to inferior colliculus mimics the topography of intracollicular projections. J Comp Neurol 371:15–50
Sanderson KJ (1974) Lamination of the dorsal lateral geniculate nucleus in carnivores of the weasel (Mustellidae), Raccoon (Procyonidae) and fox (Canidae) families. J Comp Neurol 153:239–266
Sandrew BB, Edwards DL, Poletti CE, Foote WE (1986) Amygdalospinal projections in the cat. Brain Res 373:235–239
Sanghera MK, Rolls ET, Roper-Hall A (1979) Visual responses of neurons in the dorsolateral amygdala of the alert monkey. Exp Neurol 63:610–626
Sanides D, Sanides F (1974) A comparative Golgi study of the neocortex in insectivores and rodents. Z Mikrosk Anat Forsch 88:957–977
Sanides D, Fries W, Albus K (1978) The corticopontine projection from the visual cortex of the cat. An autoradiographic investigation. J Comp Neurol 179:77–88
Sanides F (1970) Functional architecture of motor and sensory cortices in primates in the light of a new concept of neocortex evolution. In: Noback CR, Montagna W (eds) The primate brain. Meredith, New York, pp 137–208 (Advances in primatology, vol 1)
Sanides F (1972) Representation in the cerebral cortex and its areal lamination patterns. In: Bourne G (ed) The structure and function of nervous tissue, vol V. Academic, New York, pp 329–453
Sanides F, Sanides D (1972) The 'extraverted neurons' of the mammalian cerebral cortex. Z Anat Entw Gesch 136:272–293

Saper CB (1982) Convergence of autonomic and limbic connections in the insular cortex of the rat. J Comp Neurol 210:163–173
Saper CB (1985) Organization of cerebral cortical afferent systems in the rat. II. Hypothalamocortical projections. J Comp Neurol 237:21–46
Saper CB (1990) Cholinergic system. In: Paxinos G (ed) The human nervous system. Academic, San Diego, pp 1095–1113
Saper CB (1995) Central autonomic system. In: Paxinos G (ed) The rat nervous system. Academic, San Diego, pp 107–135
Saper CB, Chelimski TC (1984) A cytoarchitectonic and histochemical study of nucleus basalis and associated cell groups in the normal human brain. Neuroscience 13:1023–1037
Saper CB, Loewy AD (1980) Efferent connections of the parabrachial nucleus in the rat. Brain Res 197:291–317
Saper CB, Loewy AD, Swanson LW, Cowan WM (1976a) Direct hypothalamo-autonomic connections. Brain Res 117:305–312
Saper CB, Swanson LW, Cowan WM (1976b) The efferent connections of the ventromedial nucleus of the hypothalamus of the rat. J Comp Neurol 169:409–442
Sarter M, Markowitsch HJ (1983) Convergence of basolateral amygdaloid and mediodorsal thalamic projections in different areas of the frontal cortex in the rat. Brain Res Bull 10:607–622
Sarter M, Markowitsch HJ (1984) Collateral innervation of the medial and lateral prefrontal cortex by amygdaloid, thalamic, and brainstem neurons. J Comp Neurol 224:445–460
Sato Y (1977) Comparative morphology of the visual system of some Japanese species of Soricoidea (Superfamily) in relation to leif habits. J Hirnforsch 18:531–546
Satoda T, Takahashi O, Tashiro T, Matsushima R, Uemura-Sumi M, Mizuno N (1987) Representation of the main branches of the facial nerve within the facial nucleus of the Japanese monkey (Macaca fuscata). Neurosci Lett 78:283–287
Satoda T, Takahashi O, Uchida T, Mizuno N (1995) An anterograde-retrograde labeling study of the carotid sinus nerve of the Japanese monkey (Macaca fuscata). Neurosci Res 22:381–387
Satoh K, Fibiger HC (1986) Cholinergic neurons of the laterodorsal tegmental nucleus: efferent and afferent connections. J Comp Neurol 253:277–302
Satomi H, Takahashi K, Kosaka I, Aoki M (1989) Reappraisal of projection levels of the corticospinal fibers in the cat, with special reference to the fibers descending through the dorsal funiculus: a WGA-HRP study. Brain Res 492:255–260
Savander V, Go C-G, LeDoux JE, Pitkänen A (1995) Intrinsic connections of the rat amygdaloid complex: projections originating in the basal nucleus. J Comp Neurol 361:345–368
Savander V, Go C-G, LeDoux JE, Pitkänen A (1996) Intrinsic connections of the rat amygdaloid complex: projections originating in the accessory basal nucleus. J Comp Neurol 374:291–313
Sawaguchi T, Kubota K (1986) A hypothesis on the primate neocortex evolution: column-multiplication hypothesis. Int J Neurosci 30:57–64
Sawchenko PE, Swanson LW (1982) The organization of noradrenergic pathways form the brainstem to the paraventricular and supraoptic nuclei in the rat. Brain Res Rev 4:275–325
Scalia F (1966) Some olfactory pathways in the rabbit brain. J Comp Neurol 126:285–310
Scalia F (1968) A review of recent experimental studies on the distribution of the olfactory tracts in mammals. Brain Behav Evol 1:101–123
Scalia F (1972) The termination of retinal axons in the pretectal region of mammals. J Comp Neurol 145:223–258

Scalia F (1976) Structure of the olfactory and accessory olfactory systems. In: Llinás R, Precht W (eds) Frog neurobiology. Springer, Berlin Heidelberg New York, pp 213-233

Scalia F, Winans SS (1975) The differential projections of the olfactory bulb and the accessory olfactory bulb in mammals. J Comp Neurol 161:31-56

Scheibel AB (1979) Development of axonal and dendritic neuropil as a function of evolving behavior. In: Schmidt FO, Worden FG (eds) The neurosciences fourth study program. MIT Press, Cambridge, pp VII+1185

Scheibel ME, Scheibel AB (1958) Structural substrates for integrative patterns in the brain stem reticular core. In: Jasper HH, Procter LD Reticular formation of the brain, chap II. Little Brown, Boston, pp 31-55

Scheibel ME, Scheibel AB (1966a) Spinal motorneurons, interneurons and Renshaw cells. A Golgi study. Arch Ital Biol 104:328-353

Scheibel ME, Scheibel AB (1966b) The organization of the nucleus reticularis thalami: a Golgi study. Brain Res 1:43-62

Scheibel ME, Scheibel AB (1968) Terminal axonal patterns in cat spinal cord. II. The dorsal horn. Brain Res 9:32-58

Scheibel ME, Scheibel AB (1970a) Organization of spinal motoneuron dendrites in bundles. Exp Neurol 28:106-112

Scheibel ME, Scheibel AB (1970b) Elementary processes in selected thalamic and cortical subsystems - the structural substrates. In: Schmitt FO (ed) The neurosciences second study program. Rockefeller University Press, New York, pp 443-457

Scheibel ME, Scheibel AB (1978) The dendritic structure of the human Betz cell. In: Brazier MAB, Petsche H (eds) Architectonics of the cerebral cortex. Raven, New York, pp 43-57

Scheich H, Langner G, Tidemann C, Coles RB, Guppy A (1986) Electroreception and electrolocation in platypus. Nature 319:401-402

Schmahmann JD, Pandya DN (1989) Anatomical investigation of projections to the basis pontis from posterior association cortices in the rhesus monkey. J Comp Neurol 289:53-73

Schmahmann JD, Pandya DN (1991) Projections to the basis pontis from the superior temporal region in the rhesus monkey. J Comp Neurol 308:224-248

Schmahmann JD, Pandya DN (1993) Prelunate, occipitotemporal, and parahippocampal projections to the basis pontis in rhesus monkey. J Comp Neurol 337:94-112

Schmahmann JD, Pandya DN (1995) Prefrontal cortex projections to the basilar pons in rhesus monkey: implications for the cerebellar contribution to higher function. Neurosci Lett 199:175-178

Schmidt M, Zhang H-Y, Hoffmann K-P (1993) OKN-related neurons in the rat nucleus of the optic tract and dorsal terminal nucleus of the accessory optic system receive a direct cortical input. J Comp Neurol 330:147-157

Schmidt M, Schiff D, Bentivoglio M (1995) Independent efferent populations in the nucleus of the optic tract: an anatomical and physiological study in rat and cat. J Comp Neurol 360:271-285

Schnurr B, Spatz WB, Illing R-B (1992) Similarities and differences between cholinergic systems in the superior colliculus of guinea pig and rat. Exp Brain Res 90:291-296

Schober W, Brauer K (1974) Makromorphologie des Zentralnervensystems. II. Teil. Das Gehirn. In: Helmcke JG, Starck D, Wermuth H (eds) Handbuch der Zoologie, vol 8(52). De Gruyter, Berlin, New York, pp 1-26

Schoen JHR (1964) Comparative aspects of the descending fibre systems in the spinal cord. Progress in Brain Resarch 11:203-222

Schoenen J (1982) The dendritic organization of the human spinal cord: the dorsal horn. Neuroscience 7:2057-2087

Schoenen J, Faull RLM (1990) 2. Spinal cord: cytoarchitectural, dendroarchitectural, and myeloarchitectural organization. In: Paxinos G (ed) The human nervous system. Academic, San Diego, p 19-5

Schofield BR, Hallman LE, Lin CS (1987) Morphology of corticotectal cells in the primary visual cortex of hooded rats. J Comp Neurol 261:85-97

Scholten JM (1946) De plaats van den paraflocculus in het geheel der cerebellaire correlaties. Thesis, University of Amsterdam

Schröder KF, Hopf A, Lange H, Thörner G (1975) Morphometrisch-statistische Strukturanalysen des Striatum, Pallidum und Nucleus subthalamicus beim Menschen. I. Striatum. J Hirnforsch 16:333-350

Schümann R (1987) Course and origin of the crossed parabigeminotectal pathway in rat. A retrograde HRP-study. J Histochem 28:585-590

Schwaber JS, Kapp BS, Higgins GA, Rapp PR (1982) Amygdaloid and basal forebrain direct connections with the nucleus of the solitary tract and the dorsal motor nucleus. J Neurosci 2:1424-1438

Schwanzel-Fukuda M, Fadem BH, Soledad Garcia M, Pfaff DW (1987) The immunocytochemical localizatio of luteinizing hormone-releasing hormone in the brain of the gray short-tailed opossum (Monodelphis domestica). Ann NY Acad Sci 519:213-228

Schwanzel-Fukuda M, Pfaff DW (1994) Luteinizing hormone-releasing hormone (LHRH) and neural cell adhesion molecule (NCAM)-immunoreactivity in development of the forebrain and reproductive system. Ann Endocrinol [Paris] 55:235-241

Schwartzkroin PA, Kunkel DD (1985) Morphology of identified interneurons in the CA1 regions of guinea pig hippocampus. J Comp Neurol 232:205-218

Schwartzkroin PA, Mathers LH (1978) Physiological and morphological identification of a nonpyramidal hippocampal cell type. Brain Res 157:1-10

Schwerdtfeger WK, López-García C (1986) GABAergic neurons in the cerebral cortex of the brain of a lizard (Podarcis hispanica). Neurosci Lett 68:117-121

Schwerdtfeger WK, Lorente M-J (1988a) GABA-immunoreactive neurons in the medial and dorsomedial cortices of the lizard telencephalon. Some data on their structure, distribution and synaptic relations. In: Schwerdtfeger WK, Smeets WJAJ (eds) The forebrain of reptiles. Karger, Basel, pp 110-121

Schwerdtfeger WK, Lorente M-J (1988b) Laminar distribution and morphology of gamma-aminobutyric-acid (GABA)-immunoreactive neurons in the medial and dorsomedial areas of the cerebral cortex of the lizard Podarcis hispanica. J Comp Neurol 278:473-485

Schwindt PC, Precht W, Richter A (1974) Monosynaptic excitatory and inhibitory pathways from medial midbrain nuclei to trochlear motoneurons (1964). Exp Brain Res 20:223-238

Scott TG (1981) A unique pattern of localization within the cerebellum of the mouse. J Comp Neurol 122:1-8

Sefton AJ, Mackay-Sim A, Baur LA, Cottee LJ (1981) Cortical projections to visual centres in the rat: an HRP study. Brain Res 215:1-13

Segovia S, Guilamón A (1993) Sexual dimorphism in the vemeronasal pathway and sex differences in reproductive behaviors. Brain Res Rev 18:51-74

Seki Y (1958) Observations on the spinal cord of the right whale. Sci Rep Whales Res Inst 13:231-251

Seki Y, Murakoshi F, Miyoshi M (1963) Comparative anatomical and experimental studies on the anterior pyramidal tract. Nihon Univ J Med 5:1-32

Selemon LD, Goldman-Rakic PS (1985) Longitudinal topography and interdigitation of corticostriatal projections in the rhesus monkey. J Neurosci 5:776-794

Selemon LD, Goldman-Rakic PS (1988) Common cortical and subcortical targets of the dorsolateral prefrontal and posterior parietal cortices in the Rhesus monkey: evidence for a distributed neural network subserving spatially guided behavior. J Neurosci 8:4049-4068

Selemon LD, Goldman-Rakic PS (1990) Topographic intermingling of striatonigral and striatopallidal neurons in the rhesus monkey. J Comp Neurol 297:359-376

Shapiro RE, Miselis RR (1985) The central neural connections of the area postrema of the rat. J Comp Neurol 234:344-364

Shaw MD, Alley KE (1981) Generation of the ocular motor nuclei and their cell types in the rabbit. J Comp Neurol 200:69-82

Shen JM, Kriegstein AR (1987) Turtle hippocampal cortex contains distinct cell types, burstfiring neurons, and an epileptogenic subfield. J Neurophysiol 56:1616-1649

Shepherd GM, Greer CA (1990) Olfactory bulb. In: Shepherd GM (ed) The synaptic organization of the brain, 3rd edn. Oxford University Press, Oxford, pp 133-169

Sherk H (1979) Connections and visual-field mapping in cat's tectoparabigeminal circuit. J Neurophysiol 42:1656-1668

Sherk H (1986) The claustrum and the cerebral cortex. In: Jones EG, Peters A (eds) Sensory-motor areas and aspects of cortical connectivity. Plenum, New York, pp 467-499 (Cerebral cortex, vol 5)

Shimizu H, Norita M (1991) Connections of the insular cortex in kittens: an anatomical demonstration with wheatgerm agglutinin conjugated to horseradish peroxidase technique. Int J Dev Neurosci 9:479-491

Shink E, Bevan MD, Bolam JP, Smith Y (1996) The subthalamic nucleus and the external pallidum: two tightly interconnected structures that control the output of the basal ganglia in the monkey. Neuroscience 73:335-357

Shinoda Y, Sugiuchi Y, Futami T, Izawa R (1992) Axon collaterals of mossy fibers from the pontine nucleus in the cerebellar dentate nucleus. J Neurophysiol 67:547-560

Shinonaga Y, Takada M, Mizuno N (1994) Topographic organization of collateral projections from the basolateral amygdaloid nucleus to both the prefrontal cortex and nucleus accumbens in the rat. Neuroscience 58:389-397

Shipley MT (1974) Presubiculum afferents to the entorhinal area and the Papez circuit. Brain Res 67:162-168

Shipley MT, Adamek GD (1984) The conncetions of the mouse olfactory bulb: a study using orthograde and retrograde transport of wheat germ agglutinin conjugated to horseradish peroxidase. Brain Res Bull 12:669-688

Shipley MT, McLean JH, Ennis M (1995) Olfactory system. In: Paxinos G (ed) The rat nervous system, 2nd edn. Academic, San Diego, pp 899-926

Shipley MT, Murphy AZ, Rizvi TA, Ennis M, Behbehani MM (1996) Olfaction and brainstem circuits of reproductive behavior in the rat. Prog Brain Res 107:355-377

Shneiderman A, Oliver DL, Henkel CK (1988) Connections of the dorsal nucleus of the lateral lemniscus: an inhibitory parallel pathway in the ascending auditory system? J Comp Neurol 276:188-208

Shneiderman A, Chase MB, Rockwood JM, Benson CG, Potashner SJ (1993) Evidence for a GABAergic projection from the dorsal nucleus of the lateral lemniscus to the inferior colliculus. J Neurochem 60:72-82

Shriver JE, Stein BM, Carpenter MB (1968) Central projections of spinal dorsal roots in the monkey. I. Cervical and upper thoracic dorsal roots. Am J Anat 123:27-74

Shute CC, Lewis PR (1967) The ascending cholinergic reticular system: neocortical, olfactory and subcortical projections. Brain 90:497-520

Sie PG (1956) Localization of fibre systems within the white matter of the medulla oblongata and the cervical cord in man. Thesis, University of Leiden

Siegborn J, Grant G (1983) Brainstem projections of different branches of the vestibular nerve: an experimental study by transganglionic transport of horseradish peroxidase in the cat. I. The horizontal ampullar and utricular nerves. Arch Ital Biol 121:237-248

Siegborn J, Yingcharoen K, Grant G (1991) Brainstem projections of different branches of the vestibular nerve: an experimental study by transganglionic transport of horseradish peroxidase in the cat. II. The anterior and posterior ampullar nerves. Anat Embryol 184:291-299

Sillito AM, Jones HE, Gerstein GL, West DC (1994) Feature-linked synchronization of thalamic relay cell firing induced by feedback from the visual cortex. Nature 369:479-482

Simerly RB (1995) Anatomical substrates of hypothalamic integration. In: Paxinos G (ed) The rat nervous system. Academic, San Diego, pp 353-376

Simerly RB, Swanson LW (1986) The organization of neural imputs to the medial preoptic nucleus of the rat. J Comp Neurol 246:312-342

Siminoff R, Schwassmann O, Kruger L (1966) An electrophysiological study of the visual projection to the superior colliculus of the rat. J Comp Neurol 127:435-444

Simmons RMT (1979) The diencephalon of Ptilocercus lowii (pen-tailed tree-shrew). J Hirnforsch 20:69-92

Simmons RMT (1980a) The morphology of the diencephalon in the Prosimii. II. The lemuroidea and lorisoidea, part I. Thalamus and metathalamus. J Hirnforsch 21:449-491

Simmons RMT (1980b) The morphology of the diencephalon in the Prosimii. II. The Lemuroidea and Lorisoidea, part II. Epithalamus, subthalamus and hypothalamus. J Hirnforsch 21:493-514

Simmons RMT (1981) Bearing of the diencephalon on the taxonomic status of the Tupaioidea. J Hirnforsch 22:129-152

Simmons RMT (1982) The morphology of the diencephalon in the Prosimii. III. The Tarsioidea. J Hirnforsch 23:149-173

Simp LS, Donate-Oliver F (1991) Superficial tectal neurons projecting to the dorsolateral pontine nucleus in the rabbit. Exp Brain Res 87:696-699

Simpson GG (1981) The principles of classification and a classification of mammals. Bull Am Mus Nat Hist 85:vii, 349

Simpson JI (1988) The accessory optic system. Annu Rev Neurosci 7:13-41

Simpson JI, Leonard CS, Soodak RE (1957) The accessory optic system of rabbit. II. Spatial organization of direction selectivity. J Neurophysiol 60:2055-2072

Simpson JI, Graf W, Leonard C (1981) The coordinate system of visual climbing fibers to the flocculus. In: Fuchs A, Becker (eds) Progress in oculomotor research. Elsevier/North Holland, Amsterdam, pp 475-484

Simpson JI, Wylie DR, De Zeeuw CI (1996) On climbing fiber signals and their consequence(s). Behav Brain Sci 19:384-398

Skeen LC, Hall WC (1977) Efferent projections of the main and accessory olfactory bulb in the tree shrew (Tupaia glis). J Comp Neurol 172:1-36

Skeen LC, Pindzola RR, Schofield BR (1984) Tangential organization of olfactory association, and commissural projections to olfactory cortex in a species of reptile (Trionyx spiniferus), bird (Aix sponsa), and mammal (Tupaia glis). Brain Behav Evol 25:206-216

Skinner RD, Garcia-Rill E (1984) The mesencephalic locomotor region (MLR) in the rat. Brain Res 323:385-389

Sloper JJ, Powell TPS (1979) An experimental electron microscopic study of afferent connections to the primate motor and somatic sensory cortices. Philos Trans R Soc Lond Ser B 285:199-226

Slugg RM, Light AR (1994) Spinal cord and trigeminal projections to the pontine parabrachial region in the rat as demonstrated with Phaseolus vulgaris leucoagglutinin. J Comp Neurol 339:49-61

Smart IHM, McSherrey GM (1986) Gyrus formation in the cerebral cortex in the ferret. I. Description of the external changes. J Anat 146:141-152

Smith LM, Ebner FF, Colonnier M (1980) The thalamocortical projection in Pseudemys turtles: a quantitative electron microscope study. J Comp Neurol 190:445-462

Smith MV, Apkarian AV (1991) Thalamically projecting cells of the lateral cervical nucleus in monkey. Brain Res 555:10-18

Smith OC (1930) The corpus striatum, amydala, and stria terminalis of tamandua tetradactyla. J Comp Neurol 51:65-127

Smith PH, Joris PX, Carney LH, Yin TC (1991) Projections of physiologically characterized globular bushy cell axons from the cochlear nucleus of the cat. J Comp Neurol 304:387–407

Smith RL (1975) The ascending fiber projections from the principal sensory trigeminal nucleus in the rat. J Comp Neurol 148:423–446

Smith RL (1991) Axonal projections and connections of the principal sensory trigeminal nucleus in the monkey. J Comp Neurol 163:347–376

Smith Y, Parent A (1986) Differential connections of caudate nucleus and putamen in the squirrel monkey (Saimiri sciureus). Neuroscience 18:347–371

Smith Y, Paré D (1994) Intra-amygdaloid projections of the lateral nucleus in the cat: PHA-L anterograde labeling combined with postembedding GABA and glutamate immunocytochemistry. J Comp Neurol 342:232–248

Smith Y, Hazrati L-N, Parent A (1990) Efferent projections of the subthalamic nucleus in the squirrel monkey as studied by the PHA-L anterograde tracing method. J Comp Neurol 294:306–323

Smithson KG, Weiss ML, Hatton GI (1989) Supraoptic nucleus afferents from the main olfactory bulb. I. Anatomical evidence from anterograde and retrograde tracers in rat. Neuroscience 31:277–287

Snyder RL (1977) A comparative study of the neurons of origin of cerebellar afferents in the rat, cat and squirrel monkey based on the horseradish peroxidase (HRP) retrograde tracer technique: the spinal afferents. Anat Rec 187:719

Snyder RL, Faull RLM, Mehler WR (1978) A comparative study of the neurons of origin of the spino-cerebellar afferents in the rat, cat and squirrel monkey based on the retrograde transport of horseradish peroxidase. J Comp Neurol 181:833–852

Solnitzky O (1938) The thalamic nuclei of Sus scrofa. J Comp Neurol 69:121–169

Somogyi P (1977) A specific axo-axonal interneuron in the visual cortex of the rat. Brain Res 136:345–350

Somogyi P, Cowey A (1981) Combined Golgi and electron microscopic study on the synapses formed by double bouquet cells in the cerebral cortex of the cat and monkey. J Comp Neurol 195:547–566

Somogyi P, Cowey A (1984) Double bouquet cells. In: Peters A, Jones EG (eds) Cellular components of the cerebral cortex. Plenum, New York, pp 337–360 (Cerebral cortex, vol 1)

Somogyi P, Hodgson AJ (1985) Antisera to GABA III. Demonstration of GABA in Golgi-impregnated neurons and in conventional electron microscopic sections of cat striate cortex. J Histochem Cytochem 33:249–257

Somogyi P, Soltész I (1986) Immunogold demonstration of GABA in synaptic terminals of intracellularly recorded, horseradish peroxidase-filled basket cells and clutch cells in the cat's visual cortex. Neuroscience 19:1051–1065

Somogyi P, Hodgson AJ, Smith AD (1979) An approach to tracing neuron networks in the cerebral cortex and basal ganglia: combination of Golgi staining, retrograde transport of horseradish peroxidae and anterograde degeneration of synaptic boutons in the same material. Neuroscience 4:1805–1852

Somogyi P, Bolam JP, Tottersdell S, Smith AD (1981) Monosynaptic input from the nucleus accumbens ventral striatum region to retrogradely labelled nigrostriatal neurons. Brain Res 217:245–263

Somogyi P, Freund TF, Cowey A (1982) The axo-axonic interneuron in the cerebral cortex of the rat, cat and monkey. Neuroscience 7:2577–2607

Somogyi P, Kisvarday ZF, Martin KAC, Whitteridge D (1983a) Synaptic connections of morphologically identified and physiologically characterized large basket cells in the striate cortex of cat. Neuroscience 10:261–294

Somogyi P, Nunzi MG, Gorio A, Smith AD (1983b) A new type of specific interneuron in the monkey hippocampus forming synapses exclusively with the axon inital segments of pyramidal cells. Brain Res 259:137–142

Somogyi P, Freund TF, Hodgson AJ, Somogyi J, Beroukas D, Chubb IW (1985) Identified axo-axonic cells are immunoreactive for GABA in the hippocampus and visual cortex of the cat. Brain Res 332:143–149

Soriano E, Frotscher M (1989) A GABAergic axo-axonic cell in the fascia dentata controls the main excitatory hippocampal pathway. Brain Res 503:170–174

Sotelo C, Llinas R, Baker R (1974) Structural study of inferior olivary nucleus of the cat: morphological correlates of electronic coupling (1969). J Neurophysiol 37:541–559

Sousa-Pinto A (1969) Experimental anatomical demonstration of a cortico-olivary projection from area 6 (supplementary motor area?) in the cat. Brain Res 16:73–83

Sousa-Pinto A, Brodal A (1969) Demonstration of a somatotopical pattern in the cortico-olivary projection in the cat: an experimental anatomical study. Exp Brain Res 8:364–386

Spangler KM, Cant NB, Henkel CK, Farley GR, Warr WB (1987) Descending projections from the superior olivary complex to the cochlear nucleus of the cat. J Comp Neurol 259:452–465

Spann BM, Grofova I (1989) Origin of ascending and spinal pathways from the nucleus tegmenti pedunculopontinus in the rat. J Comp Neurol 283:13–27

Spann BM, Grofova I (1991) Nigropedunculopontine projection in the rat: an antero- grade tracing study with phaseolus vulgaris-leucoagglutinin (PHA-L). J Comp Neurol 311:375–388

Spear PD, Smith DC, Williams LL (1977) Visual receptive-field properties of single neurons in the cat's ventral-lateral geniculate nucleus. J Neurophysiol 40:390–409

Spence SJ, Saint-Cyr JA (1988) Mesodiencephalic projections to the vestibular complex in the cat. J Comp Neurol 268:375–388

Spencer RF, Baker R, McCrea RA (1980) Localization and morphology of cat retractor bulbi moto- neurons. J Neurophysiol 43:754–770

Spencer RF, Porter JD (1981) Innervation and structure of extraocular muscles in the monkey in comparison to those of the cat. J Comp Neurol 198:649–665

Spencer RF, Wenthold RJ, Baker R (1989) Evidence for glycine as an inhibitory neurotransmitter of vestibular, reticular and prepositus hypoglossi neurons that project to the cat abducens nucleus. J Neurosci 9:2718–2736

Spirou GA, Brownell WE, Zidanic M (1990) Recordings from cat trapezoid body and HRP labeling of globular bushy cell axons. J Neurophysiol 63:1169–1190

Spoendlin H (1985) Anatomy of cochlear innervation. Am J Otolaryngol 6:453–467

Spooren WP, Veening JG, Cools AR (1993) Descending efferent connections of the subpallidal areas in the cat: projections to the subthalamic nucleus, the hypothalamus, and the midbrain. Synapse 15:104–123

Spooren WPJM, Veening JG, Groenewegen HJ, Cools AR (1991) Efferent connections of the striatopallidal and amygdaloid components of the substantia innominata in the cat: projections to the nucleus accumbens and caudate nucleus. Neuroscience 44:431–447

Spooren WPJM, Lynd-Balta E, Mitchell S, Haber SN (1996) Ventral pallidostriatal pathway in the monkey: evidence for modulation of basal ganglia circuits. J Comp Neurol 370:295–312

Sreesai M (1973) Cerebellar cortical projections of the opossum (Didelphis marsupialis virginiana). J Hirnforsch 15:530–544

Sripanidkulchai K, Sripanidkulchai B, Wyss JM (1984) The cortical projection of the basolateral amygdaloid nucleus in the rat: a retrograde fluorescent dye study. J Comp Neurol 229:419–431

Stanton GB, Goldberg ME, Bruce CJ (1988) Frontal eye field efferents in the macaque monkey: II. Topography of terminal fields in midbrain and pons. J Comp Neurol 271:493–506

Steele GE, Weller RE (1993) Subcortical connections of subdivisions of inferior temporal cortex in squirrel monkeys. Vis Neurosci 10:563–583

Steiger HJ, Büttner-Ennever JA (1976) Oculomotor nucleus afferents in the monkey demonstrated with horseradish peroxidase. Brain Res 160:1–15

Stein BE (1981) Organization of the rodent superior colliculus: some comparisons with other mammals. Behav Brain Res 3:175–188

Stein BE, Spencer RF, Edwards SB (1983) Corticotectal and corticothalamic efferent projections of SIV somatosensory cortex in cat. J Neurophysiol 50:896–909

Steinbusch HWM (1981) Distribution of serotonin-immunoreactivity in the central nervous system of the rat – cell bodies and terminals. Neuroscience 6:557–618

Steinbusch HWM (1984) Serotonin-immunoreactive neurons and their projections in the CNS. In: Bjorklund A, Hokfelt T, Kuhar MJ (eds) Classical transmitters and transmitter receptors in the CNS, part II, vol 3. Elsevier, Amsterdam, pp 68–113

Steininger TL, Rye DB, Wainer BH (1992) Afferent projections to the cholinergic pedunculopontine tegmental nucleus and adjacent midbrain extrapyramidal area in the albino rat. I. Retrograde tracing studies. J Comp Neurol 321:515–543

Stephan H (1967) Zur Entwicklungshohe der Insektivoren nach Merkmalen des Gehirns und die Definition der "Basalen Insektivoren". Zool Anz 179:177–199

Stephan H (1969) Quantitative investigations on visual structures in primate brains. II. Neurology. Proceedings of the 2nd internationl congress on primatology, Atlanta, vol 3, pp 34–42

Stephan H (1975) Handbuch der Mikroskopischen Anatomie des Menschen, vol 4, part 9: Allocortex. Springer, Berlin Heidelberg New York

Stephan H (1984) Morphology of the brain in Tarsius. In: Niemitz C (ed) Biology of tarsiers. Fisher, Stuttgart, pp 319–344

Stephan H, Andy OJ (1964) Quantitative comparisons of brain structures from insectivores to primates. Am Zool 4:59–74

Stephan H, Andy O (1969) Quantitative comparative neuroanatomy of primates: an attempt at a phylogenetic interpretation. Comparative and evolutionary aspects of the vertebrate central nervous system. Ann N Y Acad Sci 167:370–387

Stephan H, Andy OJ (1977) Quantitative comparison of the amygdala in insectivores and primates. Acta Anat 98:130–153

Stephan H, Pirlot P (1970) Volumetric comparisons of brain structures in bats. Z Zool Syst Evol Forsch 8:200–236

Stephan H, Spatz H (1962) Vergleichend-anatomische Untersuchungen an Insektivorengehirnen. IV. Gehirne afrikanischer Insektivoren. Versuch einer Zuordnung von Hirnbau und Lebensweise. Morphol Jahrbuch 103:108–174

Stephan H, Baron G, Schwerdtfeger WK (1980) The brain of the common marmoset (Callithrix jacchus). Springer, Berlin Heidelberg New York

Stephan H, Pirlot P, Schneider R (1974) Volumetric analysis of pteropid brains. Acta Anat 87:161–192

Stephan H, Frahm H, Bauchot R (1977) Vergleichende Untersuchungen an den Gehirnen madagassischer Halbaffen I. Encephalisation und makromorphologie. J Hirnforsch 18:115–147

Stephan H, Frahm H, Baron G (1981) New and revisited data on volumes of brain structures in insectivores and primates. Folia Primatol 35:1–29

Stephan H, Frahm HD, Baron G (1987) Comparison of brain structure volumes in Insectivora and primates. VII. Amygdaloid components. J Hirnforsch 28:571–584

Stephan H, Baron G, Frahm HD (1990) Comparative brain research in mammals, vol 1: insectivora with a stereotaxic atlas of the hedgehog brain. Springer, Berlin Heidelberg New York

Stephan H, Baron G, Frahm HD (1991) Insectivora. Springer, Berlin Heidelberg New York

Steriade M, Llinás RR (1988) The functional states of the thalamus and the associated neuronal interplay. Physiol Rev 68:649–742

Sterling P (1971) Receptive fields and synaptic organization of the superficial gray layer of the cat superior colliculus. Vision Res [Suppl] 3:309–328

Sterling P (1973) Quantitative mapping with the electron microscope: retinal terminals in the superior colliculus. Brain Res 54:347–354

references>Sterling P, Kuypers HGJM (1967) Anatomical organization of the brachial spinal cord of the cat. II. The motoneuron plexus. Brain Res 4:16–32

Stevens CF (1969) Structure of cat frontal olfactory cortex. J Neurophysiol 32:184–192

Stevens RT, Hodge CJJ, Apkarian AV (1989) Medial, intralaminar, and lateral terminations of lumbar spinothalamic tract neurons: a fluorescent double-label study. Somatosens Mot Res 6:285–308

Stevens RT, Apkarian AV, Hodge CJJ (1991) The location of spinothalamic axons within spinal cord white matter in cat and squirrel monkey. Somatosens Mot Res 8:97–102

Strack AM, Sawyer WB, Hughes JH, Platt KB, Loewy AD (1989) A general pattern of CNS innervation of the sympathetic outflow demonstrated by transneuronal pseudorabies viral infections. Brain Res 491:156–162

Strominger NL, Truscott C, Miller RA, Royce GJ (1979) An autoradiographic study of the rubro-olivary tract in the rhesus monkey. J Comp Neurol 83:33–46

Strominger NL, Nelson LR, Strominger RN (1985) Banding of rubro-olivary terminations in the principal inferior olivary nucleus of the chimpanzee. Brain Res 16:185–187

Stroud BB (1895) The mammalian cerebellum. J Comp Neurol 5:71–118

Stuesse SL, Newman DB (1990) Projections from the medial agranular cortex to brain stem visuomotor centers in rats. Exp Brain Res 80:532

Stumpf WE (1975) The brain: an endocrine gland and hormone target. In: Stumpf WE, Grant LD (eds) Anatomical neuroendocrinology. Karger, Basel, pp 2–8

Stumpf WE, Sar M (1978) Anatomical distribution of estrogen, androgen, progestin, corticosteroid and thyroid hormone target sites in the brain of mammals: phylogeny and ontogeny. Am Zool 18:435–445

Stuurman FJ (1906) Die Localisation der Zungenmuskeln im Nucleus Hypoglossi. Anat Anz 48:593–610

Sugimoto T, Itoh K, Mizuno N (1978) Direct projections from the Edinger-Westphal nucleus to the cerebellum and spinal cord in the cat: an HRP study. Neurosci Lett 9:17–22

Sugimoto T, Itoh K, Mizuno N, Nomura S, Konishi A (1979) The site of origin of cardiac preganglionic fibers of the vagus nerve: an HRP study in the cat. Neursci Lett 12:53–58

Sugita S, Otani K, Tokunaga A, Terasawa K (1983) Laminar origin of the tecto-thalamic projections in the albino rat. Neurosci Lett 43:143–147

Sultan F, Braitenberg V (1993) Shapes and sizes of different mammalian cerebella. A study in quantitative comparative neuroanatomy. J Hirnforsch 34:79–92

Swanson LW (1976) An autoradiographic study of the efferent connections of the preoptic region in the rat. J Comp Neurol 167:227–256

Swanson LW (1982) Normal hippocampal circuitry. Neurosci Res Prog Bull 20:624–637

Swanson LW (1987) The hypothalamus. In: Björklund A, Swanson LW (eds) Integrated systems of the CNS, part I. Elsevier, Amsterdam, pp 125–277 (Handbook of chemical neuroanatomy, vol 5)

Swanson LW (1989) The neural basis of motivated behavior. Acta Morphol Neerl-Scand 26:165–176

Swanson LW, Cowan WM (1977) An autoradiographic study of the organization of the efferent connections of the hippocampal formation in the rat. J Comp Neruol 172:49–84

Swanson LW, Cowan WM (1979) The connections of the septal region in the rat. J Comp Neurol 186:621–655

Swanson LW, Hartman BK (1983) The central adrenergic system. An immunofluorescence study of the location of cell bodies and their efferent connections in the rat utilizing dopamine-B-hydroxylase as a marker. J Comp Neurol 163:467–505

Swanson LW, Mogenson GJ (1981) Neural mechanisms for the functional coupling of autonomic, endocrine and somatomotor responses in adaptive behavior. Brain Res Rev 3:1–34

Swanson LW, Cowan WM, Jones EG (1974) An autoradiographic study of the efferent connections of the ventral lateral geniculate nucleus in the albino rat and cat (1975). J Comp Neurol 156:143–164

Swanson LW, Mogenson GJ, Gerfen CR, Robinson P (1984) Evidence for a projection from the ateral preoptic area and substantia innominata to the 'mesencephalic locomotor region' in the rat. Brain Res 295:161–178

Swanson LW, Köhler C, Björklund A (1987a) The limbic region. I: The septohippocampal system. In: Björklund A, Hökfelt T, Swanson LW (eds) Integrated systems of the CNS, part I. Elsevier, Amsterdam, pp 125–277 (Handbook of chemical neuroanatomy, vol 5)

Swanson LW, Mogenson GJ, Simerly RB, Wu M (1987b) Anatomical and electrophysiological evidence for a projection from the medial preoptic area to the 'mesencephalic and subthalamic locomotor regions' in the rat. Brain Res 405:108–122

Swenson RS, Castro AJ (1983a) The afferent connections of the inferior olivary complex in rat: a study using the retrograde transport of horseradish peroxidase. Am J Anat 166:329–341

Swenson RS, Castro AJ (1983b) The afferent connections of the inferior olivary complex in rats. An anterograde study using autoradiographic and axonal degeneration techniques. Neuroscience 8:259–275

Swindale NV (1990) Is the cerebral cortex modular? Trends Neurosci 13:487–492

Szabo J (1980a) Distribution of striatal afferents from the mesencephalon in the cat. Brain Res 188:3–21

Szabo J (1980b) Organization of the ascending striatal afferents in monkeys. J Comp Neurol 189:307–321

Szekely G, Matesz C (1948) The efferent system of cranial nerve nuclei: a comparative neuromorphological study. Adv Anat Embryol Cell Biol 128:5103–5109

Szekely G, Matesz C (1982) The accessory motor nuclei of the trigeminal, facial, and abducens nerves in the rat. J Comp Neurol 210:258–264

Szentagothai J (1958) The representation of facial and scalp muscles in the facial nucleus. J Comp Neurol 88:207–220

Szentágothai J (1964a) Neuronal and synaptic arrangement in the substantia gelatinosa Rolandi. J Comp Neurol 122:40

Szentágothai J (1964b) The anatomical basis of synaptic transmission of excitation and inhibition in motoneurons. Acta Morphol Acad Sci Hung 8:287–309

Szentágothai J (1975) The 'module-concept' in cerebral cortex architecture. Brain Res 95:475–496

Szentágothai J (1978) The neuron network of the cerebral cortex: a functional interpretation. Proc R Soc Lond Ser B 201:219–248

Szentágothai J (1979) Local neuron ciruits of the neocortex. In: Schmitt FO, Worden FG (eds) The neurosciences fourth study program. MIT Press, Cambridge, pp 399–415

Szentágothai J (1983) The modular architectonic principle of neural centers. Rev Physiol Biochem Pharmacol 98:11–61

Szentágothai-Schimert J (1941) Die Endigungsweise der absteigenden Rückenmarksbahnen. Z Anat Entwickl Gesch 111:322–330

Taber E (1961) The cytoarchitecture of the brain stem of the cat. I. Brain stem nuclei of cat. J Comp Neurol 116:27–70

Taber E, Brodal A, Walberg F (1960) The raphe nuclei of the brain stem in the cat. I. Normal topography and cytoarchitecture and general discussion. J Comp Neurol 114:161–187

Tagliavini F, Pietrini V (1984) On the variability of the human flocculus and paraflocculus accessorius. J Hirnforsch 25:163–170

Takada M, Li ZK, Hattori T (1988) Collateral projection from the substantia nigra to the striatum and superior colliculus in the rat. Neuroscience 25:563–568

Takada M, Tokuno H, Ikai Y, Mizuno N (1994) Direct projections from the entopeduncular nucleus to the lower brain stem in the rat. J Comp Neurol 342:409–429

Takahashi K (1965) Slow and fast groups of pyramidal tract cells and their respective membrane properties. J Neurophysiol 28:980–924

Takeda T, Maekawa K (1976) Origin of the pretecto-olivary tract. A study using the horseradish peroxidase method. Brain Res 117:319–325

Takemoto I, Sasa M, Takaori S (1978) Role of the locus coeruleus in transmission onto anterior colliculus neurons. Brain Res 158:269–278

Takeuchi Y, Sano Y (1983) Immunohistochemical demonstration of serotonin nerve fibers in the neocortex of the monkey (Macaca fuscata). Anat Embryol (Berl) 166:155–168

Takeuchi Y, McLean JH, Hopkins DA (1982) Reciprocal connections between the amygdala and parabrachial nuclei: ultrastructural demonstration by degeneration and axonal transport of horseradish peroxidase in the cat. Brain Res 239:583–588

Takeuchi Y, Matsushima S, Matsushima R, Hopkins DA (1983) Direct amygdaloid projections to the dorsal motor nucleus of the vagus nerve: a light and electron microscopic study in the rat. Brain Res 280:143–147

Tallaksen-Greene SJ, Eelde R, Wessendorf MW (1993) Regional distribution of serotonin and substance P coexisting in nerve fibers and terminals in the brainstem of the rat. Neuroscience 53:1127–1142

Tamai Y, Waters RS, Asanuma H (1984) Caudal cuneate nucleus projection to the direct thalamic relay to motor cortex in cat: an electrophysiological and anatomical study. Brain Res 323:360–364

Tan J, Simpson J, Voogd J (1995a) Anatomical compartments in the white matter of the rabbit flocculus. J Comp Neurol 356:1–22

Tan J, Gerrits NM, Nanhoe R, Simpson JI, Voogd J (1995b) Zonal organization of the climbing fiber projection to the flocculus and nodulus of the rabbit. A combined axonal tracing and acetylcholinesterase histochemical study. J Comp Neurol 356:23–50

Tan J, Epema AH, Voogd J (1995c) Zonal organization of the flocculovestibular nucleus projection in the rabbit. A combined axonal tracing and acetylcholinesterase histochemical study. J Comp Neurol 356:51–71

Tanaka C, Ishikawa M, Shimada S (1982) Histochemical mapping of catecholaminergic neurons and their ascending fiber pathways in the rhesus monkey brain. Brain Res Bull 9:255–270

Tanaka K, Otani K, Tokunaga A, Sugita S (1985) The organization of neurons in the nucleus of the lateral lemniscus projecting to the superior and inferior colliculi in the rat. Brain Res 341:252–260

Tarlov E, Tarlov SR (1971) The representation of extraocular muscles in the oculomotor nuclei: experimental studies in the cat. Brain Res 34:36–52

Ten Donkelaar HJ (1988) Evolution of the red nucleus and rubrospinal tract. Behav Brain Res 28:9–20

Ten Donkelaar HJ, Lammers GJ, Gribnau AAM (1979) Neurogenesis in the amygdaloid nuclear complex in a rodent (the Chinese hamster). Brain Res 165:348–353

Ter Horst GJ, Copray JCVM, Liem RSB, Van Willigen JD (1991) Projections from the rostral parvocellular reticular formation to pontine and medullary nuclei in the rat: involvement in autonomic regulation and orofacial motor control. Neuroscience 40:735–758

Ter Horst GJ, Luiten PG, Kuipers F (1984) Descending pathways from hypothalamus to dorsal motor vagus and ambiguus nuclei in the rat. J Auton Nerv Syst 11:59–75

Terasawa K, Otani K, Yamada J (1979) Descending pathways of the nucleus of the optic tract in the rat. Brain Res 173:405–417

Terubayashi H, Fujisawa H (1984) The accessory optic system of rodents: a whole-mount HRP study. J Comp Neurol 227:285–295

Thiede U (1966) Zur Evolution von Hirneigenschaften mitteleuropäischer und südamerikanischer Musteliden. Z Zool Syst Evol Forsch 4:318–377

Thiele A, Vogelsang M, Hoffmann K-P (1991) Pattern of retinotectal projection in the megachiropteran bat Rousettus aegyptiacus. J Comp Neurol 314:671–683

Thomas A, Westrum LE, De Vito JL, Biedenbach MA (1984) Unmyelinated axons in the pyramidal tract of the cat. Brain Res 301:162–165

Thompson AM, Thompson GC (1993) Relationship of descending inferior colliculus projections to olivocochlear neurons. J Comp Neurol 335:402–412

Thompson GC, Thompson AM (1986) Olivocochlear neurons in the squirrel monkey brainstem. J Comp Neurol 254:246–258

Thompson GC, Igarashi M, Stach BA (1985) Identification of stapedius muscle motoneurons in squirrel monkey and bush baby. J Comp Neurol 231:270–279

Thörner G, Lange H, Hopf A (1975) Morphometrisch-statistische Strukturanalysen des Striatum, Pallidum und Nucleus subthalamicus beim Menschen. II. Pallidum. J Hirnforsch 16:404–413

Thornton EW, Murray M, Connorseckenrode T, Haun F (1994) Dissociation of behavioral changes in rats resulting from lesions of the habenula versus fasciculus retroflexus and their possible anatomical substrates. Behav Neurosci 108:1150–1162

Thunnissen I (1990) Vestibulocerebellar and vestibulo-oculomotor relations in the rabbit. Thesis, Erasmus University Rotterdam

Thunnissen IE, Epema AH, Gerrits NM (1989) Secondary vestibulocerebellar mossy fiber projection to the caudal vermis in the rabbit. J Comp Neurol 290:262–277

Tigges J (1964) Morphogenese, Haute, Blutversorgung, Ventrikelsystem und Ruckenmark. In: Hofer H, Tigges J (eds) Makromorphologie des Zentralnervensystems, part I. Helmcke J-G, Starck D, Wermuth H (eds) Handbuch der Zoologie, vol 8(34). DeGruyter, Berlin, pp 1–42

Tigges J, Tigges M (1969) The accessory optic system in Erinaceus (insectivora) and Galago (primates). J Comp Neurol 137:59–70

Tigges M, Tigges J (1970) The retinofugal fibers and their terminal nuclei in Galago crassicaudatus (Primates). J Comp Neurol 138:87–102

Tigges J, Bos J, Tigges M (1977) An autoradiographic investigation of the subcortical visual system in chimpanzee. J Comp Neurol 172:367–380

Tigges J, Tigges M, Cross NA, McBride RL, Letbetter WD, Anschel S (1982) Subcortical structures projecting to visual cortical areas in squirrel monkey. J Comp Neurol 209:29–40

Tigges J, Walker LC, Tigges M (1983) Subcortical projections to the occipital and parietal lobes of the chimpanzee brain. J Comp Neurol 220:106–115

Tiller Y (1987) Immunocytochemical localization of serotonin-containing neurons in the myelencephalon, brainstem and diencephalon of the sheep. Neuroscience 23:501–527

Tilney F (1927) The brain stem of tarsius. A critical comparison with other primates. J Comp Neurol 43:371–432

Tokunaga A, Otani K (1976) Dendritic patterns of neurons in the rat superior colliculus. Exp Neurol 52:189–205

Tokunaga A, Otani K (1978) Neuronal organization of the corpus parabigeminum in the rat. Exp Neurol 58:361–375

Tokunaga A, Akert K, Garey LJ, Otani K (1981) Primary and secondary subcortical projections of the monkey visual system. An autoradiographic study. Brain Res 214:137–143

Tokunaga A, Ono K, Kondo S, Tanaka H, Kurose K, Nagai H (1992) Retinal projections in the house musk shrew, Suncus murinus, as determined by anterograde transport of WGA-HRP. Brain Behav Evol 40:321–329

Tokuno H, Takada M, Kondo Y, Mizuno N (1993) Laminar organization of the substantia nigra pars reticulata in the macaque monkey, with special reference to the caudato-nigro-tectal link. Exp Brain Res 92:545–548

Tolbert DL (1982) The cerebello nucleocortical pathway. Exp Brain Res 6:296–317

Tömböl T (1984) Layer VI cells. In: Jones EG, Peters A (eds) Cellular components of the cerebral cortex. Plenum, New York, pp 479–519 (Cerebral cortex, vol 1)

Tork I (1990) Anatomy of the serotonergic system. Ann NY Acad Sci 600:9–35

Tork I, Hornung JP (1990) Raphe nuclei and the serotonergic system. In: Paxinos G (ed) The human nervous system. Academic, San Diego, pp 1001–1022

Tortelly A, Reinoso-Suarez F, Llamas A (1980) Projections from non-visual cortical areas to the superior colliculus demonstrated by retrograde transport of HRP in the cat. Brain Res 188:543–549

Torvik A (1957) The spinal projection from the nucleus of the solitary tract. An experimental study in the cat. J Anat 91:314–322

Towe AL (1975) Notes on the hypothesis of co;lumnar organization in somatosensory cerebral cortex. Brain Behav Evol 11:16–47

Tower S, Bodian D, Howe H (1941) Isolation of intrinsic and motor mechanism of the monkeys cord. J Neurophysiol 4:388–398

Toyoshima K, Kawana E, Sakai H (1980) On the neuronal origin of the afferents to the ciliary ganglion in cat. Brain Res 185:67–76

Tramonte R, Bauer JA (1986) The location of the preganglionic neurons that innervate the submandibular gland of the cat. A horseradish peroxidase study. Brain Res 375:381–384

Travers JB, Montgomery N, Sheridan J (1995) Transneuronal labeling in hamster brainstem following lingual injections with herpes simples virus-1. Neuroscience 68:1277–1293

Trejo LJ, Cicerone CM (1984) Cells in the pretectal olivary nucleus are in the pathway for the direct light reflex of the pupil in the rat. Brain Res 300:49–62

Tsai C (1925) The optic tracts and centers of the opossum, Didelphis virginiana. J Comp Neurol 39:173–216

Tseng GF, Haberly LB (1989) Deep neurons in piriform cortex. I. Morphology and synaptically evoked responses including a unique high-amplitude paired shock facilitation. J Neurophysiol 62:369–385

Tucker CL, Kennedy PR (1990) Re-defining rat red nucleus: Cytoarchitectural analysis of red nucleus neurones singly and doubly labeled from spinal cord and inferior olivary nucleus. Neurosci Abstr 16:729

Tucker CL, Lee SA, Kennedy PR (1989) Re-defining rat red nucleus: Cytoarchitecture and connectivity. Neurosci Abstr 15:405

Turner BH, Zimmer J (1984) The architecture and some of the interconnections of the rat's amygdala and lateral periallocortex. J Comp Neurol 227:540–557

Turner BH, Gupta KC, Mishkin M (1978) The locus and cytoarchitecture of the projection areas of the olfactory bulb in Macaca mulatta. J Comp Neurol 177:381–396

Turner BH, Mishkin M, Knapp M (1980) Organization of the amygdalopetal projections from modality-specific cortical association areas in the monkey. J Comp Neurol 191:515–543

Uchida K, Mizuno N, Sugimoto T, Itoh K, Kudo M (1983) Direct projections from the cerebellar nuclei to the superior colliculus in the rabbit: an HRP study. J Comp Neurol 216:319–326

Uchida Y (1950a) A contribution to the comparative anatomy of the amygdaloid nuclei in mammals, especially in rodents, part I: rat and mouse. Folia Psychiatr Neurol Jpn 4:25–42

Uchida Y (1950b) A contribution to the comparative anatomy of the amygdaloid nuclei in mammals, especially in rodents, part II: Guinea pig, rabbit and squirrel. Folia Psychiatr Neurol Jpn 4:91–109

Uddenberg N (1966) Studies on modality segregation and second order neurones in the dorsal funiculus. Experientia 22:441–442

Uemura M, Matsuda K, Kume M, Takeuchi Y, Matsushima R, Mizuno N (1979) Topographical arrangement of hypoglossal motoneurons: an HRP study in the cat. Neurosci Lett 13:99–104

Uemura-Sumi M, Mizuno N, Nomura S, Iwahori N, Takeuchi Y, Matsushima R (1981) Topographical representation of the hypoglossal nerve branches and tongue muscles in the hypoglossal nucleus of macaque monkeys. Neurosci Lett 22:31–35

Uemura-Sumi M, Takahashi O, Matsushima R, Takata M, Yasui Y, Mizuno N (1982) Localization of masticatory motoneurons in the trigeminal motor nucleus of the guinea pig. Neurosci Lett 29:219–224

Ueyama T, Houtani T, Ikeda M, Sato K, Sugimoto T, Mizuno N (1994) Distribution of primary afferent fibers projecting from hind-limb cutaneous nerves to the medulla oblongata in the cat and rat. J Comp Neurol 341:145–158

Ugolini G, Kuypers HGJM (1986) Collaterals of corticospinal and pyramidal fibres to the pontine grey demonstrated by a new application of the fluorescent fibre labelling technique. Brain Res 365:211–227

Ulinski PS (1974) Cytoarchitecture of cerebral cortex in snakes. J Comp Neurol 158:243–266

Ulinski PS (1976) Intracortical connections in the snakes Natrix sipedon and Thamnophis sirtalis. J Morphol 150:463–484

Ulinski PS (1977) Intrinsic organization of snake medial cortex: an electron microscopic and Golgi study. J Morphol 152:247–280

Ulinski PS (1979) Intrinsic organization of snake dorsomedial cortex: an electron microscopic and Golgi study. J Morphol 161:185–210

Ulinski PS (1990) The cerebral cortex of reptiles. In: Jones EG, Peters A (eds) Comparative structure and evolution of cerebral cortex, part I. Plenum, New York, pp 139–215 (Cerebral cortex, vol 8A)

Ulinski PS, Peterson EH (1981) Patterns of olfactory projections in the desert iguana, Dipsosaurus dorsalis. J Morphol 168:189–228

Ulinski PS, Rainey WT (1980) Intrinsic organization of snake lateral cortex. J Morphol 165:85–116

Ungerstedt U (1971) Stereotaxic mapping of the monoamine pathways in the brain. Acta Physiol Scand [Suppl] 367:1–49

Updyke BV (1977) Topographic organization of the projections from cortical areas 17, 18 and 19 onto the thalamus, pretectum and superior colliculus in the cat. J Comp Neurol 173:81–122

Usunoff KG, Romansky KV, Malinov GB, Ivanon DP, Blagov ZA, Galabov GP(1982) Electron microscopic evidence for the existence of a cortical tract in the rat. J Hirnforsch 23:23–29

Valverde F (1961) Reticular formation of the pons and medulla oblongata. A Golgi study. J Comp Neurol 116:71–99

Valverde F (1965) Studies on the piriform lobe. Harvard University Press, Cambridge, Massachusetts

Valverde F (1967) Apical dendritic spines of the visual cortex and light deprivation in the mouse. Exp Brain Res 3:337–352

Valverde F (1976) Aspects of cortical organization related to the geometry of neurons with intra-cortical axons. J Neurocytol 5:509–529

Valverde F (1983) A comparative approach to neocortical organization based on the study of the brain of the hedgehog (Erinaceus europaeus). In: Grisolía S, Guerri C, Samson F, Norton S, Reinoso-Suárez F (eds) Ramon y Cajal's contribution to the neurosciences. Elsevier, Amsterdam, pp 149–170

Valverde F (1985) The organizing principles of the primary visual cortex in the monkey. In: Peters A, Jones EG (eds) Visual cortex. Plenum, New York, pp 207–257 (Cerebral cortex, vol 3)

Valverde F (1986) Intrinsic neocortical organization: some comparative aspects. Neuroscience 18:1–23

Valverde F, Facal-Valverde MV (1986) Neocortical layers I and II of the hedgehog (Erinaceus europaeus). I. Intrinsic organization. Anat Embryol (Berl) 173:413–430

Valverde F, De Carlos JA, Lopez-Mascaraque L, DoZate-Oliver F (1986) Neocortical layers I and II of the hedgehog (Erinaceus europaeus). II. Thalamo-cortical connections. Anat Embryol (Berl) 175:167–179

Van Beusekom GT (1929) Fibre analysis of the anterior and lateral funiculi of the cord in the cat. Thesis

Van Crevel H (1958) The rate of secondary degeneration in the central nervous system. Thesis,University of Leiden

Van Crevel H, Verhaart WJC (1963) The rate of secondary degeneration in the central nervous system. II. The optic nerve of the cat. J Anat (Lond) 97:451–464

Van der Horst VGJM (1996) The basic neural circuitry for sexual behavior; pathways and plasticity. Thesis, University of Groningen

Van der Horst VGJM, Holstege G (1990) Caudal medullary pathways to lumbosacral motoneuronal cell groups in the cat: evidence for direct projections possibly representing the final common pathway for lordosis. J Comp Neurol 359:457–475

Van der Kooy D, Koda LY, McGinty JF, Gerfen CR, Bloom FE (1984) The organization of projections from the cortex, amygdala, and hypothalamus to the nucleus of the solitary tract in rat. J Comp Neurol 224:1–24

Van der Loos H (1965) The "improperly" oriented pyramidal cell in the cerebral cortex and its possible bearing on problems of neuronal growth and cell orientation. Bull John Hopkins Hosp 117:228–250

Van der Loos H (1976) Barreloids in mouse somatosensory thalamus. Neuroscience 2:1–6

Van der Loos H (1979) The development of topological equivalencies in the brain. In: Meisami A, Brazier MAB (eds) Neuronal growth and differentiation. Raven, New York. pp 331–336

Van der Loos H, Woolsey TA (1973) Somatosensory cortex: structural alterations following early injury to sense organs. Science 179:395–398

Van der Want JJL, Voogd J (1987) Ultrastructural identification and localization of climbing fiber terminals in the fastigial nucleus of the cat. J Comp Neurol 258:81–90

Van der Want JJL, Wiklund L, Guegan M, Ruigrok T, Voogd J (1989) Anterograde tracing of the rat olivocerebellar system with phaseolus vulgaris leucoagglutinin (PHA-L). Demonstration of climbing fiber collaterals innervation of the cerebellar nuclei. J Comp Neurol 288:1–18

Van der Want JJL, Nunes Cardozo JJ, Van der Togt C (1992) GABAergic neurons and circuits in the pretectal nuclei and the accessory optic system of mammals. Prog Brain Res 90:283–304

Van Dijken H, Holstege JC (1997) The distribution of dopamine immunoreactive fibers and presumptive terminals in the rat brain stem. A comparison with the distribution of dopamine-B-hydroxylase. J Comp Neurol (in press)

Van Eden CG, Mrzljak L, Voorn P, Uylings HBM (1989) Prenatal development of GABA-ergic neurons in the neocortex of the rat. J Comp Neurol 289:213–227

Van Gorcum NV, Assen Busch HFM (1964) Anatomical aspects of the anterior and lateral funiculi at the spinobulbar junction. In: Eccles JC, Schade JP (eds) Progress in brain research, vol 11. Elsevier, Amsterdam, pp 223–237

Van Ham JJ, Yeo C (1992) Somatosensory trigeminal projections to the inferior olive, cerebellum and other precerebellar nuclei in rabbits. Eur J Neurosci 4:302–317

Van Hoesen GW (1981) The differential distribution, diversity and sprouting of cortical projections to the amygdala in the rhesus monkey. In: Ben-Ari Y (ed) The amygdaloid complex. Elsevier/North-Holland, Amsterdam, pp 77-90

Van Hoesen GW (1982) The parahippocampal gyrus. Trends Neurosci 5:345-350

Van Hoesen GW, Pandya DN (1975) Some connections of the entorhinal area (area 28) and perirhinal area (area 35) cortices of the rhesus monkey. I. Temporal lobe afferents. Brain Res 95:1-24

Van Hoesen GW, Pandya DN, Butters M (1975) Some connections of the entorhinal area (area 28) and perirhinal area (area 35) cortices of the rhesus monkey. II. Frontal afferents. Brain Res 95:25-38

Van Noort J (1969) The structure and connections of the inferior colliculus: an investigation of the lower auditory systems. Van Gorcum, Assen

Van Noort J (1996) The anatomical basis for frequency analysis in the cochlear nuclear complex. Psychiatr Neurol Neurochir 72:109-114

Vater M, Feng AS (1990) Functional organization of ascending and descending connections of the cochlear nucleus of horseshoe bats. J Comp Neurol 292:373-395

Veening J, Buma P, Ter Horst GJ, Roeling TAP, Luiten PGM and Nieuwenhuys R (1991) Hypothalamic projections to the PAG in the rat: topographical, immuno-electronmicroscopical and functional aspects. In: Depaulis A, Bandler R (eds) The midbrain periaqueductal gray matter. Plenum, New York, pp 387-415

Veening JG (1978a) Cortical afferents of the amygdaloid complex in the rat: an HRP study. Neurosci Lett 8:191-195

Veening JG (1978b) Subcortical afferents of the amygdaloid complex in the rat: An HRP study. Neurosci Lett 8:197-202

Veening JG, Swanson LW, Cowan WM, Nieuwenhuys R, Geeraedts LMG (1982) The medial forebrain bundle of the rat: II. An autoradiographic study of the topography of the major descending and ascending components. J Comp Neurol 206:82-108

Veening JG, Swanson LW, Sawchenko PE (1984) The organization of projections from the central nucleus of the amygdala to brainstem sites involved in central autonomic regulation: a combined retrograde transport-immunohistochemical study. Brain Res 303:337-357

Veening JG, Te Lie S, Posthumus P, Geeraedts LMG, Nieuwenhuys R (1987) A topographical analysis of the origin of some efferent projections from the lateral hypothalamic area in the rat. Neuroscience 22:537-551

Vera PL, Ellenberger HH, Halselton JR, Haselton CL, Schneiderman N (1987) The intermediolateral nucleus: an open or closed nucleus? Brain Res 386:84-92

Verburgh CA, Kuypers HGJM (1989) Branching neurons in the cervical spinal cord: a retrograde fluorescent double-labeling study in the rat. Exp Brain Res 68:565-578

Verburgh CA, Voogd J, Kuypers HGJM, Stevens HP (1990a) Propriospinal neurons with ascending collaterals to the dorsal medulla, the thalamus and the tectum: a retrograde fluorescent double-labeling study of the cervical cord of the rat. Exp Brain Res 80:577-590

Verburgh CA, Kuypers HGJM, Voogd J, Stevens HP (1990b) Spinocerebellar neurons and propriospinal neurons in the cervical spinal cord: a fluorescent double-labeling study in the rat and the cat. Exp Brain Res 75:73-82

Vercelli F, Assal F, Innocenti GM (1992) Emergence of callosally projecting neurons with stellate morphology in the visual cortex of the kitten. Exp Brain Res 90:346-358

Verhaart WJC (1948) The pes pedunculi and pyramid. J Comp Neurol 88:139-155

Verhaart WJC (1957) The pes pedunculi and pyramid in hylobates. J Comp Neurol 89:71-78

Verhaart WJC (1962) Anatomy of the brain stem of the elephant. J Hirnforsch 5:455-522

Verhaart WJC (1963) Pyramidal tract in the cord of the elephant. J Comp Neurol 121:45-49

Verhaart WJC (1966) The pyramidal tract of Tupaia, compared to that in other primates. J Comp Neurol 126:43-50

Verhaart WJC (1967) The non-crossing of the pyramidal tract in Procavia capensis (Storr) and other instances of absence of the pyramidal crossing. J Comp Neurol 131:387-392

Verhaart WJC (1970a) The pyramidal tract in the primates. The primate brain. Adv Primatol 1:83-108

Verhaart WJC (1970b) Comparative anatomical aspects of the mammalian brain stem and the cord. I. Van Gorcum, Assen

Verhaart WJC, Noorduyn NJA (1961) The cerebral peduncle and the pyramid. Acta Anat 45:315-343

Verhaart WJC, Sopers-Jurgens MR Aspects of the comparative anatomy of the mammalian brain stem. Acta Morphol Neerl Scand I:246-255

Vertes RP (1984a) A lectin horseradish peroxidase study of the origin of ascending fibers in the medial forebrain bundle of the rat. The lower brainstem. Neuroscience 11:651-668

Vertes RP (1984b) A lectin horseradish peroxidase study of the orginin of ascending fibers in the medial forebrain bundle of the rat. The upper brainstem. Neuroscience 11:669-690

Vetter DE, Mugnaini E (1992) Distribution and dendritic features of three groups of rat olivocochlear neurons. A study with two retrograde cholera toxin tracers. Anat Embryol (Berl) 1992:11-16

Vetter DE, Adams JC, Mugnaini E (1991) Chemically distinct rat olivocochlear neurons. Synapse 7:21-43

Vetter DE, Saldana E, Mugnaini E (1993) Input from the inferior colliculus to medial olivocochlear neurons in the rat: a double label study with PHA-L and cholera toxin. Hearing Res 70:173-186

Victor M, Agamanolis J (1990) Amnesia due to lesions confined to the hippocampus: a clinical-pathological study. J Cognit Neurosci 2:246-257

Vincent SR, Hattori T, McGeer EG (1978) The nigrotectal projection: a biochemical and ultrastructural characterization. Brain Res 151:159-164

Vincent SR, Johansson O, Hökfelt T, Skirboll L, Elde RP, Terenius L, Kimmel J, Goldstein M (1983) NADPH-diaphorase: a selective histochemical marker for striatal neurons containing both somatostatin- and avian pancreatic polypeptide (APP)-like immunoreactivities. J Comp Neurol 217:252-263

Vogt BA (1985) Cingulate cortex. In: Peters A, Jones EG (eds) Association and auditory cortices. Plenum, New York, pp 89-149 (Cerebral cortex, vol 4)

Vogt BA (1991) The role of layer I in cortical function. In: Peters A, Jones EG (eds) Normal and altered states of function. Plenum, New York, pp 49-80 (Cerebral cortex, vol 9)

Vogt BA, Rosene DL, Peters A (1981) Synaptic termination of thalamic and callosal afferents in cingulate cortex of the rat. J Comp Neurol 201:265-283

Volchan E, Rocha-Miranda, Lent R, Gawryszewski LG (1978) The retinotopic organization of the superior colliculus in the opossum (Didelphis marsupialis aurita). In: Rocha-Miranda CE, Lent R (eds) Opossum neurobiology (Neurobiologia do Gamba). Academia Brasileira de Ciencias. Rio de Janeiro, pp 107-113

Volsch M (1906) Zur vergleichenden Anatomie des Mandelkerns und seine Nachbargebilde, part I. Arch Mikrosk Anat 68:573-683

Volsch M (1910) Zur vergleichenden Anatomie des Mandelkerns und seine Nachbargebilde, part II. Arch Mikrosk Anat 76:373-523

Von Bonin G (1941) Side lights on cerebral evolution: brain size of lower vertebrates and degree of cortical folding. J Gen Psychol 25:273-282

von Gudden B (1870a) Über einen bisher nicht beschriebenen Nervenfasernstrang im Gehirne der Säugerthiere und des Menschen. Arch Psychiatr Nerv Krankh 2:364-366

von Gudden B (1870b) Experimentaluntersuchungen über das peripherische und centrale Nervensystem. Arch Psychiatr Nerv Krankh 2:693-723

Voogd J (1964) The cerebellum of the cat. Structure and fiber connections. Thesis, Van Gorcum, Assen

Voogd J (1969) The importance of fiber connections in the comparative anatomy of the mammalian cerebellum. In: Llinas R (ed) Neurobiology of cerebellar evolution and development. AMA-ERF Institute for Biomedical Research, Chicago, pp 493–541

Voogd J (1995) The cerebellum of the rat. In: Paxinos G (ed) The rat nervous system, 2nd edn. Academic, San Diego, pp 309–350

Voogd J, Bigaré F (1980) Topographical distribution of olivary and cortico-nuclear fibers in the cerebellum. A review. In: Courville J, De Montigny C, Lamarre Y (eds) The inferior olivary nucleus. Raven, New York, pp 207–234

Voogd J, Ruigrok TJH (1997) Transverse and longitudinal patterns in the mammalian cerebellum. Prog Brain Res 114 (in press)

Voogd J, Hess DT, Marani E (1987) The parasagittal zonation of the cerebellar cortex in cat and monkey. Topography, distribution of acetylcholinesterase and development. In: King JS (ed) New concepts in cerebellar neurobiology. Liss, New York

Voogd J, Feirabend HKP, Schoen JHR (1990) Cerebellum and precerebellar nuclei. In: Paxinos G (ed) The human nervous system. Academic, San Diego, pp 321–386

Voogd J, Epema AH, Rubertone JA (1991) Cerebellovestibular connections of the anterior vermis: a retrograde tracer study in different mammals including primates. Arch Ital Biol 129:3–19

Voogd J, Gerrits NM, Ruigrok TJH (1996a) Organization of the vestibulocerebellum. Ann NY Acad Sci 781:553–579

Voogd J, Jaarsma D, Marani E (1996b) The cerebellum, chemoarchitecture and anatomy. In: Swanson LW, Björklund A, Hökfelt T (eds). Integrated systems of the CNS, part III. Cerebellum, Basal Ganglia, olfactory system. Elsevier, Amsterdam, pp 1–369 (Handbook of chemical neuroanatomy, vol 12)

Voorn P, Jorritsma-Bijham B, Dijk Ch van, Buijs RM (1987) The dopaminergic innervation of the ventral striatum in the rat: a light- and electron-microscopical study using antibodies against dopamine. J Comp Neurol 251:84–99

Wahle P (1993) Differential regulation of substance P and somatostatin in Martinotti cells of the developing cat visual cortex. J Comp Neurol 329:519–538

Wahren W (1957) Das Zwischenhirn des Kaninchens. J Hirnforsch 3:143–242

Waite PME, Tracey DJ (1995) Trigeminal sensory system. In: Paxinos G (ed) The rat nervous system, 2nd edn. Academic, San Diego, pp 705–724

Walberg F (1952) The lateral reticular nucleus of the medulla oblongata in mammals. A comparative-anatomical study. J Comp Neurol 96:283–344

Walberg F (1957) Corticofugal fibres to the nuclei of the dorsal columns. An experimental study in the cat. Brain 80:273–287

Walberg F (1982) The origin of olivary afferents from the central grey and its surroundings in the cat. Anat Embryol (Berl) 164:139–151

Walberg F, Pompeiano O, Brodal A, Jansen J (1962) The fastigiovestibular projection in the cat. An experimental study with silver impregnation methods. J Comp Neurol 118:49–75

Walberg F, Nordby T, Hoffmann KP, Holländer H (1981) Olivary afferents from the pretectal nuclei in the cat. Anat Embryol (Berl) 161:291–304

Waldeyer H (1888) Das Gorilla-Rückenmark. Abh Preuss Akad Wiss Berl 3:1–147

Waldron HA (1969) The morphology of the lateral cervical nucleus in the hedgehog. Brain Res 16:301–306

Waldron HA, Gwyn DG (1969) Descending nerve tracts in the spinal cord of the rat. I. Fibers from the midbrain. J Comp Neurol 137:143–154

Walker RH, Arbuthnott GW, Baughman RW, Graybiel AM (1993) Dendritic domains of medium spiny neurons in the primate striatum: relationships to striosomal borders. J Comp Neurol 337:614–628

Wall PD, Taub A (1962) Four aspects of trigeminal nucleus and a paradox. J Neurophysiol 25:110–126

Wallace MN (1986a) Lattice of high oxidative metabolism in the intermediate grey layer of the rat and hamster superior colliculus. Neurosci Lett 70:320–325

Wallace MN (1986b) Spatial relationship of NADPH-diaphorase and acetylcholinesterase lattices in the rat and mouse superior colliculus. Neuroscience 19:381–391

Wallace MN (1988) Lattices of high histochemical activity occur in the human, monkey, and cat superior colliculus. Neuroscience 25:569–583

Wallace MN, Fredens K (1989) Relationship of afferent inputs to the lattice of high NADPH-diaphorase activity in the mouse superior colliculus. Exp Brain Res 78:435–445

Wallenberg A (1896) Die sekundäre Bahn des sensiblen Trigeminus. Anat Anz 12:95–110

Wallenberg A (1900) Sekundäre sensible Bahnen im Gehirnstamme des Kaninchens, ihre gegenseitige Lage und ihre Bedeutung fur den Aufbau des Thalamus. Anat Anz 18:81–105

Wallenberg A (1905) Sekundäre Bahnen aus dem frontalen sensibeln Trigeminuskerne des Kaninchens. Anat Anz 26:145–155

Wang S-F, Spencer RF (1992) Spatial organization and neurotransmitter utilization of premotor neurones related to vertical saccadic eye movements in the cat. Soc Neurosci Abstr Anaheim 18:19.7

Wang SF, Spencer RF (1996) Spatial organization of premotor neurons related to vertical upward and downward saccadic eye movements in the rostral interstitial nucleus of the medial longitudinal fasciculus (riMLF) in the cat. J Comp Neurol 366:163–180

Warr WB (1975) Olivocochlear and vestibular efferent neurons of the feline brainstem: their location, morphology, and number determined by retrograde axonal transport and acetylcholinesterase histochemistry. J Comp Neurol 161:159–182

Warr WB, Guinan JJJ (1979) Efferent innervation of the organ of corti: two separate systems. Brain Res 173:152–155

Warwick R (1953) Representation of the extraocular muscles in the oculomotor nuclei of the monkey. J Comp Neurol 98:599–611

Wassef M, Sotelo C (1984) Asynchrony in the expression of cyclic GMP dependent protein kinase by clusters of Purkinje cells during the perinatal development of rat cerebellum. Neuroscience 13:1219–1243

Wassef M, Berod A, Sotelo C (1981) Dopaminergic dendrites in the pars reticulata of the rat substantia nigra and their striatal input: combined immunocytochemical localization of tyrosine hydroxylase and anterograde degeneration. Neuroscience 6:2125–2139

Wässle H, Illing R-B (1980) The retinal projection to the superior colliculus in the cat: a quantitative study with HRP. J Comp Neurol 190:333–356

Watanabe K, Kawana E (1979) Efferent projections of the parabigeminal nucleus in rats: a horseradish peroxidase (HRP) study. Brain Res 168:1–11

Watanabe K, Kawana E (1982) The cells of origin of the incertofugal projections to the tectum, thalamus, tegmentum and spinal cord in the rat: a study using the autoradiographic and horseradish peroxidase methods. Neuroscience 7:2389–2406

Waterhouse BD, Border B, Wahl L, Mihailoff GA (1993) Topographic organization of rat locus coeruleus and dorsal raphe nuclei: distribution of cells projecting to visual system structures. J Comp Neurol 336:345–361

Waters RS, Tamai Y, Asanuma H (1985) Caudal cuneate nucleus projection to the direct thalamic relay to the motor cortex: an electrophysiological study. Brain Res 360:361–365

Watson CR, Herron P (1977) The inferior olivary complex of marsupials. J Comp Neurol 176:527–537

Watts AG, Swanson LW, Sanchez-Watts G (1987) Efferent projections of the suprachiasmatic nucleus: I. Studies using anterograde transport of Phaseolus vulgaris leucoagglutinin in the rat. J Comp Neurol 258:204–229

Weber JT (1985) Pretectal complex and accessory optic system of primates. Brain Behav Evol 26:117–140

Weber JT, Giolli RA (1986) The medial terminal nucleus of the monkey: evidence for a 'complete' accessory optic system. Brain Res 365:164–168

Weber JT, Harting JK (1980) The efferent projections of the pretectal complex: an autoradiographic and horseradish peroxidase analysis. Brain Res 194:1–28

Weber JT, Partlow GD, Harting JK (1978) The projection of the superior colliculus upon the inferior olivary complex of the cat: an auto- radiographic and horseradish peroxidase study. Brain Res 144:369–377

Weber JT, Young R, Hutchins B (1981) Morphologic and autoradiographic evidence for a laminated pretectal olivary nucleus in the squirrel monkey. Brain Res 224:153–159

Weber M (1892) Beiträge zur Entwicklung and Anatomie des Genus Manis. Ergebnisse einer Reise nach Niederl. Ostindien. Leiden

Webster WR (1995) Auditory system. In: Paxinos G (ed) The rat nervous system, 2nd edn. Academic, San Diego, pp 797–831

Webster WR, Garey LJ (1990) 27. Auditory system. In: Paxinos G (ed) The human nervous system. Academic, San Diego, pp 889–944

Weidenreich F (1899) Zur anatomie der centralen Kleinhirnkerne der Sauger. Z Morphol Anthropol I:260–312

Weiskrantz L (1956) Behavioral changes associated with ablation of the amygdaloid complex in monkeys. J Comp Physiol Psychol 49:381–391

Welker W (1987a) Comparative study of cerebellar somatosensory representations the importance of micromapping and natural stimulation. In: Glickstein M, Yeo C, Stein J (eds) Cerebellum and neuronal plasticity. Plenum, New York

Welker W (1987b) Spatial organization of somatosensory projections to granule cell cerebellar cortex: functional and connectional implications of fractured somatotopy (Summary of Wisconsin Studies). In: King J (ed) New concepts in cerebellar neurobiology. Liss, New York, pp 239–280

Welker W, Lende RA (1980) Thalamocortical relationships in echidna (Tachyglossus aculeatus). In: Ebbeson SOE (ed) Comparative neurology of the telencephalon. Plenum, New York, pp 449–481

Welker W, Seidenstein S (1959) Somatic sensory representation in the cerebral cortex of the raccoon (procyon lotor). J Comp Neurol 111:469–501

Welker WI (1990) Explaining the morphology of cerebral convolutions: a review of determinants of gyri and sulci. In: Jones EG, Peters A (eds) Cerebral cortex, vol 8. Plenum, New York (in press)

Welker WI, Johnson J (1965) JI. Correlation between nuclear morphology and somatotopic organization in ventro-basal complex of the raccoon's thalamus. J Anat 99:761–790

Welker WI, Seidenstein S (1959) Somatic sensory representation in the cerebral cortex of the raccoon (Procyon lotor). J Comp Neurol 111:469–501

Wells GR, Hardman MJ, Yeo CH (1989) Visual projections to the pontine nuclei in the rabbit: orthograde and retrograde tracing studies with WGA-HRP. J Comp Neurol 279:629–652

Welsh JP, Lang EJ, Sugihara I, Llinas R (1995) Dynamic organization of motor control within the olivocerebellar system. Nature 374:453–457

Wenstrup JJ, Larue DT, Winer JA (1994) Projections of physiologically defined subdivisions of the inferior colliculus in the mustached bat: targets in the medial geniculate body and extrathalamic nuclei. J Comp Neurol 346:207–236

Werner L, Winkelmann E, Koglin A, Neser J, Rodewohl H (1989) A Golgi deimpregnation study of neurons in the rhesus monkey visual cortex (areas 17 and 18). Anat Embryol 180:583–597

Westman J (1968) The lateral cervical nucleus in the cat. I. A Golgi study. Brain Res 10:352–368

White EL (1978) Identified neurons in mouse Sm I cortex, which are postsynaptic to thalamocortical axon terminals: a combined Golgi-electron microscopic and degeneration study. J Comp Neurol 181:627–662

White EL (1986) Termination of thalamic afferents in the cerebral cortex. In: Jones EG, Peters A (eds) Sensory-motor areas and aspects of cortical connectivity. Plenum, New York, pp 271–289 (Cerebral cortex, vol 5)

White EL (1989) Cortical circuits. Synaptic organization of the cerebral cortex. Structure, function and theory. Birkhäuser, Boston

White EL, Hersch SM (1981) Thalamocortical synapses of pyramidal cells which projects from SmI to MsI cortex in the mouse. J Comp Neurol 198:167–181

White EL, Hersch SM (1982) A quantitative study of thalamocortical and other synapses involving the apical dendrites of corticothalamic projection cells in mouse SmI cortex. J Neurocytol 11:137–157

White EL, Keller A (1987) Intrinsic circuitry involving the local axonal collaterals of corticothalamic projection cells in mouse SmI cortex. J Comp Neurol 262:13–26

White JS, Warr WB (1983) The dual origins of the olivocochlear bundle in the albino rat. J Comp Neurol 219:203–214

White EL, Hersch SM, Belford GR (1982) Quantitative studies of thalamocortical synapses with labeled pyramidal cells in mouse SmI cortex. Soc Neurosci Abstr 8:853

Whitehead MC (1988) Neuronal architecture of the nucleus of the solitary tract in the hamster. J Comp Neurol 276:547–572

Whitehead MC (1990) Subdivisions and neuron types of the nucleus of the solitary tract that project to the parabrachial nucleus in the hamster. J Comp Neurol 301:554–574

Whitehead MC, Frank ME (1983) Anatomy of the gustatory system in the hamster: central projections of the chorda tympani and the lingual nerve. J Comp Neurol 220:378–395

Whitlock DG, Nauta WJH (1956) Subcortical projections from the temporal neocortex in Macaca mulatta. J Comp Neurol 106:183–212

Whitworth RH, Haines DE (1983) The inferior olive of a prosimian primate Galago senegalensis. I. Conformation and spino-olivary projections. J Comp Neurol 219:215–227

Whitworth RHJ, Haines DE (1986a) On the question of nomenclature of homologous subdivisions of the inferior olivary complex. Arch Ital Biol 124:271–317

Whitworth RHJ, Haines DE (1986b) The inferior olive of Saimiri sciureus: olivocerebellar projections to the anterior lobe. Brain Res 372:55–71

Wiberg M, Blomqvist A (1984a) The projection to the mesencephalon from the dorsal column nuclei. An anatomical study in the cat. Brain Res 311:225–244

Wiberg M, Blomqvist A (1984b) The spinomesencephalic tract in the cat: its cells of origin and termination pattern as demonstrated by the intra-axonal transport method. Brain Res 291:1–18

Wiberg M, Westman J, Blomqvist A (1987a) The projection to the mesencephalon from the sensory trigeminal nuclei. An anatomical study in the cat. Brain Res 399:51–68

Wiberg M, Westman J, Blomqvist A (1987b) Somatosensory projection to the mesencephalon: an anatomical study in the monkey. J Comp Neurol 264:92–117

Wiener SI (1986) Laminar distribution and patchiness of cytochrome oxidase in mouse superior colliculus. J Comp Neurol 244:137–148

Wiesendanger R, Wiesendanger M, Ruegg DG (1979) An anatomical investigation of the corticopontine projection in the primate (Macaca fascicularis and Saimiri sciureus). II. The projection from frontal and parietal association areas. Neuroscience 4:747–765

Wiklund L, Leger L, Persson M (1981) Monoamine cell distribution in the cat brain stem. A fluorescence histochemical study with quantification of indolaminergic and locus coeruleus cell groups. J Comp Neurol 203:613–647

Wiksten B (1979a) The central cervical nucleus in the cat. I. A Golgi study. Exp Brain Res 36:143–154

Wiksten B (1979b) The central cervical nucleus in the cat. II. The cerebellar connections studied with retrograde transport of horseradish peroxidase. Exp Brain Res 36:155–173

Wiksten B (1979c) The central cervical nucleus in the cat. III. The cerebellar connections studied with anterograde transport of 3H-leucine. Exp Brain Res 36:175–189

Wiksten B (1985) Retrograde HRP study of neurons in the cervical enlargement projecting to the cerebellum in the cat. Exp Brain Res 58:95–101

Wiksten B, Grant G (1986) Cerebellar projections from the cervical enlargement. An experimental study with silver impregnation and autoradiographic techniques in the cat. Exp Brain Res 61:513–531

Willard FH, Martin GF (1983) The auditory brainstem nuclei and some of their projections to the inferior colliculus in the North American opossum. Neuroscience 10:1203–1232

Willard FH, Martin GF (1986) The development and migration of large multipolar neurons into the cochlear nucleus of the North American opossum. J Comp Neurol 248:119–132

Willems E (1910) Les noyaux masticateur et mesencephalique du trijumeau chez le lapin. Le Nevraxe 12:7–221

Willett CJ, Gwyn DG, Rutherford JG, Leslie RA (1986) Cortical projections to the nucleus of the tractus solitarius: an HRP study in the cat. Brain Res Bull 16:497–505

Williams MN, Faull RLM (1988) The nigrotectal projection and tectospinal neurons in the cat. A light and electron microscopic study demonstrating a monosynaptic nigral input to identified tectospinal neurons. Neuroscience 25:533–562

Willis WD, Coggeshall RE (1991) Sensory mechanisms of the spinal cord, 2nd edn. Plenum, New York

Willis WD, Leonard RB, Kenshalo DRJ (1978) Spinothalamic tract neurons in the substantia gelatinosa. Science 202:986–988

Willis WD, Kenshalo DRJ, Leonard RB (1979) The cells of origin of the primate spinothalamic tract. J Comp Neurol 188:543–573

Willis WD, Westlund KN, Carlton SM (1995) Pain. In: Paxinos G (eds) The rat nervous system, 2nd edn. Academic, San Diego, pp 725–750

Wilson ME, Toyne MJ (1970) Retino-tectal and cortico-tectal projections in Macaca mulatta. Brain Res 24:395–406

Wilson PD, Rowe MH, Stone J (1976) Properties of relay cells in the cat's lateral geniculate nucleus: a comparison of W-cells with X- and Y-cells. J Neurophysiol 39:1193–1209

Winer JA (1982) The stellate neurons in layer IV of primary auditory cortex (AI) of the cat: a study of columnar organization. Soc Neurosci Abstr 8:1020

Winfield DA, Gatter KC, Powell TPS (1980) An electron microscopic study of the types and proportions of neurons in the cortex of the motor and visual areas of the cat and rat. Brain 103:245–258

Winfield DA, Brooke RNL, Sloper JJ, Powell TPS (1981) A combined Golgi-electron microscopic study of the synapses made by the proximal axon and recurrent collaterals of a pyramidal cell in the somatic sensory cortex of the monkey. Neuroscience 6:1217–1230

Winfield JA, Hendrickon A, Kimm J (1978) Anatomical evidence that the medial terminal nucleus of the accessory optic tract in mammals provides a visual mossy fiber input to the flocculus. Brain Res 151:175–182

Wirsig CR (1987) Effects of lesions of the terminal nerve on mating behavior in the male hamster. Ann NY Acad Sci 519:241–251

Wirsig CR, Leonard CM (1986) The terminal nerve projects centrally in the hamster. Neuroscience 19:709–717

Wirsig CR, Leonard CM (1987) Terminal nerve damage impairs the mating behavior of the male hamster. Brain Res 417:293–303

Wirth FP, O'Leary JL, Smith JM, Jenny AB (1974) Monosynaptic corticospinal-motoneuron path in the raccoon (1977). Brain Res 77:344–348

Wise SP (1981) The primate premotor cortex: past, present, and preparatory. Annu Rev Neurosci 8:1

Wise SP, Jones EG (1987) Somatotopic and columnar organization in the corticotectal projection of the rat somatic sensory cortex. Brain Behav 133:223–235

Witkin JW (1987) Nervus terminalis, olfactory nerve, and optic nerve representation of luteinizing hormone-releasing hormone in primates. Ann NY Acad Sci 519:174–183

Witter MP, Groenewegen HJ (1986) The connections of the parahippocampal cortex in the cat. IV. Subcortical efferents. J Comp Neurol 252:51–77

Wollaston WH (1824) On semi-decussation of the optic nerves. Phil Trans Roy Soc (London) part 1, pp. 22–231

Wood RI, Brabec RK, Swann JM, Newman SW (1992) Androgen and estrogen concentrating neurons in chemosensory pathways of the male Syrian hamster brain. Brain Res 596:89–98

Woodhams PL, Roberts GW, Polak JM, Crow TJ (1983) Distribution of neuropeptides in the limbic system of the rat: the bed nucleus of the stria terminalis, septum and preoptic area. Neuroscience 8:677–703

Woodson W, Angaut P (1984) The ipsilateral descending limb of the brachium conjunctivum: an autoradiographic and HRP study in rats. Neurosci Lett [Suppl] 18:S58

Woolf NJ, Butcher LL (1982) Cholinergic projections to the basolateral amygdala: a combined Evans Blue and acetylcholinesterase analysis. Brain Res Bull 8:751–763

Woolf NJ, Butcher LL (1985) Cholinergic systems in the rat brain. II. Projections to the interpeduncular nucleus. Brain Res Bull 14:63–83

Woolf NJ, Butcher LL (1986) Cholinergic systems in the rat brain. III. Projections from the pontomesencephalic tegmentum to the thalamus, tectum, basal ganglia, and basal forebrain. Brain Res Bull 16:603–637

Woolf NJ, Butcher LL (1989) Cholinergic systems in the rat brain. IV. Descending projections of the pontomesencephalic tegmentum. Brain Res Bull 23:519–540

Woolf NJ, Hernit MC, Butcher LL (1986) Cholinergic and noncholinergic projections from the rat basal forebrain revealed by combined choline acetyltransferase and Phaseolus vulgaris leucoagglutinin immunohistochemistry. Neurosci Lett 66:281–286

Woollard HH, Beattie J (1927) The comparative anatomy of the lateral geniculate body. J Anat 61:414–423

Woolsey TA (1987) Barrels, vibrissae, and topographic representations. In: Adelman G (ed) Encyclopedia of neuroscience, vol 1. Birkhäuser, Boston, pp 111–11

Woolsey TA, Van der Loos H (1970) The structural organization of layer IV in the somatosensory region (SI) of mouse cerebral cortex. Brain Res 17:205–242

Wouterlood FG (1981) The structure of the mediodorsal cerebral cortex in the lizard Agama agama: a Golgi study. J Comp Neurol 196:443–458

Wouterlood FG, Mugnaini E (1984) Cartwheel neurons of the dorsal cochlear nucleus: a Golgi-electron microscopic study in rat. J Comp Neurol 227:136–157

Wright CI, Beijer AVJ, Groenewegen HJ (1996) Basal amygdaloid complex afferents to the rat nucleus accumbens are compartmentally organized. J Neurosci 16:1877–1893

Wylie DR, Zeeuw DC, DiGiorgi PL, Simpson JI (1994) Projections of individual Purkinje cells of identified zones in the ventral nodulus to the vestibular and cerebellar nuclei in the rabbit. J Comp Neurol 349:448–463

Wyss JM, Sripanidkulchai K (1984) The topography of the mesencephalic and pontine projections from the cingulate cortex of the rat. Brain Res 293:1–15

Wyss JM, Swanson LW, Cowan WM (1979) A study of subcortical afferents to the hippocampal formation in the rat. Neuroscience 4:463-476

Yamamoto M, Shimoyama I, Highstein SM (1978) Vestibular nucleus neurons relaying excitation from the anterior canal to the oculomotor nucleus. Brain Res 148:31-42

Yamamoto T, Samejima A, Oka H (1987a) Morphology of layer V pyramidal neurons in the cat somatosensory cortex: an intracellular HRP study. Brain Res 437:369-374

Yamamoto T, Samejima A, Oka H (1987b) Morphological features of layer V pyramidal neurons in the cat parietal cortex: an intracellular HRP study. J Comp Neurol 265:380-390

Yamasaki DS, Krauthamer G, Rhoades RW (1984) Organization of the intercollicular pathway in rat. Brain Res 300:368-371

Yamasaki DSG, Krauthamer GM, Rhoades RW (1986) Superior collicular projection to intralaminar thalamus in rat. Brain Res 378:223-233

Yang CR, Mogenson GJ (1985) An electrophysiological study of the neural projections from the hippocampus to the ventral pallidum and the subpallidal areas by way of the nucleus accumbens. Neuroscience 15:1015-1024

Yasui Y, Itoh K, Mizuno M et al (1983) The posteromedial ventral nucleus of the thalamus (VPM) of the cat: direct ascending projections to the cytoarchitectonic subdivision. J Comp Neurol 220:219-228

Yasui Y, Itoh K, Kaneko T, Shigemoto R, Mizuno N (1991) Topographical projections from the cerebral cortex to the nucleus of the solitary tract in the cat. Exp Brain Res 85:75-84

Yasui Y, Kayahara T, Shiroyama T, Nakano K (1993) Neurons in the intertrigeminal region of the rat send projection fibers to the superior colliculus. Neurosci Lett 159:39-42

Yezierski RP (1988) Spinomesencephalic tract: projections from the lumbosacral spinal cord of the rat, cat, and monkey. J Comp Neurol 267:131-146

Yoshida A, Dostrovsky JO, Sessle BJ, Chiang CY (1991) Trigeminal projections to the nucleus submedius of the thalamus in the rat. J Comp Neurol 307:609-625

Yoshida Y, Miyazaki T, Hirano M, Shin T, Totoki T, Kanaseki T (1981) Localization of efferent neurons innervating the pharyngeal constrictor muscles and the cervical esophagus muscle in the cat by means of the horseradish peroxidase method. Neurosci Lett 22:91-95

Young MJ, Lund RD (1994) The anatomical substrates subserving the pupillary light reflex in rats: origin of the consensual pupillary response. Neuroscience 62(2):481-496

Young MW (1936) The nuclear pattern and fiber connections of the non-cortical centers of the telencephalon of the rabbit (Lepus cuniculus). J Comp Neurol 65:295-401

Yu F, Gordon FJ (1996) Anatomical evidence for a bi-neuronal pathway connecting the nucleus tractus solitarius to caudal ventrolateral medulla to rostral ventrolateral medulla in the rat. Neurosci Lett 205:21-24

Zaborsky LC, Heimer L (1984) Ultrastructural evidence of amygdalofugal axons terminating on cholinergic cells of the rostral forebrain. Neurosci Lett 52:219-225

Zaborszky L, Alheid GF, Beinfeld MC, Eiden LE, Heimer L, Palkovits M (1985) Cholecystokinin innervation of the ventral striatum: a morphological and radio-immunological study. Neuroscience 14:427-453

Zaborszky L, Cullinan WE, Braun A (1991) Afferents to basal forebrain cholinergic projection neurons: an update. In: Napier TC, Kalivas PW, Henin I (eds) The basal forebrain. Plenum, New York, pp 44-100

Zagon A, Totterdell S, Jones RSG (1994) Direct projections from the ventrolateral medulla oblongata to the limbic forebrain: anterograde and retrograde tract-tracing studies in the rat. J Comp Neurol 340:445-468

Zahm DS, Brog JS (1992) On the significance of subterritories in the "accumbens" part of the rat ventral striatum. Neuroscience 50:751-767

Zeehandelaar I (1920) Ontogenese en phylogenese der achterstrengkernen in verband met de sensibiliteit. Thesis, Bohn, Haarlem

Zeng D, Stuesse SL (1993) Topographic organization of efferent projections of medial frontal cortex. Brain Res Bull 32:195-200

Zhang DX, Carlton SM, Sorkin LS, Willis WD (1990) Collaterals of primate spinothalamic tract neurons to the periaqueductal gray. J Comp Neurol 296:277-290

Zhang HY, Hoffmann KP (1993) Retinal projections to the pretectum, accessory optic system and superior colliculus in pigmented and albino ferrets. Eur J Neurosci 5:486-500

Zheng LM, Pfaff DW, Schwanzel-Fukuda M (1990) Synaptology of luteinizing hormone-releasing hormone (LHRH)-immunoreactive cells in the nervus terminalis of the gray short-tailed opossum (Monodelphis domestica). J Comp Neurol 295:327-337

Ziehen T (1897) Das Centralnervensystem der Monotremen und Marsupialier, part I. Makrosk Anat Jena Denkschr 6:168-187

Zola-Morgan S, Squire L (1985) Amnesia in monkeys after lesions of the mediodorsal nucleus of the thalamus. Ann Neurol 17:558-564

Zola-Morgan S, Squire LR (1986) Memory impairment in monkeys following lesions of the hippocampus. Behav Neurosci 100:165-170

Zook JM, Casseday JH (1982a) Cytoarchitecture of auditory system in lower brainstem of the mustache bat, Pteronotus parnellii. J Comp Neurol 207:1-13

Zook JM, Casseday JH (1982b) Origin of ascending projections to inferior colliculus in the mustache bat, Pteronotus parnellii. J Comp Neurol 207:14-28

Zook JM, Casseday JH (1985) Projections from the cochlear nuclei in the mustache bat, Pteronotus parnellii. J Comp Neurol 237:307-324

Zook JM, Casseday JH (1987) Convergence of ascending pathways at the inferior colliculus of the mustache bat, Pteronotus parnelli. J Comp Neurol 261:347-361

Zuk A, Gwyn DG, Rutherford JG (1982) Cytoarchitecture, neuronal morphology, and some efferent connections of the interstitial nucleus of Cajal (INC) in the cat. J Comp Neurol 212:278-292

III. GENERAL CONCLUDING PART

23 Brain Size in Vertebrates 2099

24 The Meaning of It All 2135

Subject Index 2197

CHAPTER 23

Brain Size in Vertebrates

P.A.M. van Dongen

23.1	Introduction	2099
23.1.1	Absolute and Relative Brain Sizes	2099
23.1.2	Experimental and Theoretical Body-Brain Relationships	2100
23.1.3	Inherent Problems with Allometric Analyses	2102
23.1.4	r- and K-Selection: Body and Brain Size	2103
23.1.5	Selection for Large Bodies or Large Brains?	2103
23.2	Origin of the Chordate Brain	2105
23.3	Agnathans and Jawed Fishes	2105
23.3.1	Cyclostomes	2105
23.3.2	Jawed Fishes	2106
23.4	Amphibians	2108
23.5	Reptiles	2109
23.5.1	Recent Reptiles	2109
23.5.2	Dinosaurs	2109
23.6	Birds	2111
23.6.1	Evolution of Brain Size in Birds	2111
23.6.2	Brain Weights in Various Bird Orders	2111
23.6.3	Brain Size and Ecology in Birds	2112
23.7	Mammals	2114
23.7.1	The Brain in Extinct Mammals	2114
23.7.2	Ecology, Ethology and Brain Size in Mammals	2116
23.7.3	Monotremates	2119
23.7.4	Marsupials	2119
23.7.5	Eutheria	2119
23.8	Concluding Remarks	2131
	Appendix	2131
	References	2131

23.1
Introduction

Humanity and human intelligence are considered to be derived from the large human brain; therefore brain size is regarded as a relevant and interesting parameter. This chapter covers brain size in an evolutionary perspective. Such a starting point of course has an inherent limitation: attention is only paid to overall brain size, and not to the size of brain subsystems. Nevertheless, overall brain size is an interesting parameter.

23.1.1
Absolute and Relative Brain Sizes

Figure 23.1 shows the relationship between body weight and brain weight of 20 mammals in a double-logarithmic graph. The body weights range from 3 g to 150 metric tons: from the smallest shrew (*Suncus etruscus*) to the blue whale (*Balaenopterus musculus*). The smallest brain weight in an adult mammal (74 mg) is found in a bat (*Tylonycteris pachypus*, Stephan et al. 1981b), while brain weights of up to 10 kg have been described in sperm and killer whales (Kojima 1951). The mammalian brain weights differ by a factor of 130 000, while the body weights differ by a factor of 50 million. The human brain is large (about 1.4 kg), but still considerably smaller than the brains of elephants and some large whales (5–10 kg).

Figure 23.1 shows several elements we will often encounter in this chapter: a *regression line* and a *convex polygon*. A convex polygon is formed by straight lines enclosing all data points of a group; this polygon is convex, because all the inner angles are less than 180°.

We know intuitively that brain weight should be related to body weight, but our intuition fails us if we use a simple ratio, a percentage, to analyse this relationship. This is illustrated by Fig. 23.2, which shows the data for the same 20 mammals as relative brain weights (brain weight as a percentage of body weight). The brains of some small rodents comprise about 10% of their body weight (Mace et al. 1981). In man this figure is about 2%, in the pig less than 0.1% and in the blue whale less than 0.01%. Evidently, small animals have relatively large brains (Cuvier 1805; Weber 1896; Dubois 1897). When a selected body parameter does not scale proportionally with another body parameter, this phenomenon is called allometry. Such allometry is essential to evaluate the brain size of a given species. We need to compare its actual brain size with the brain size expected for an animal of this size. This is defined as the encephalisation quotient (*EQ*):

brain weight (g)

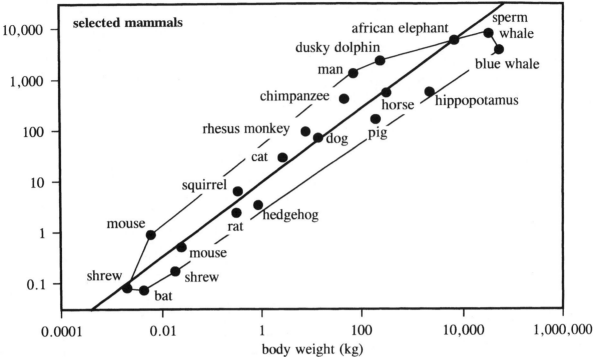

Fig. 23.1. Relationship between body weight and brain weight of 20 mammals, including those with the largest and smallest body and brain weights (double-logarithmic graph). Large mammals have larger brains. In a double-logarithmic graph like this one, the variation in brain weight between mammals of the same body weight seems rather small, but it can actually amount to a factor of 10. (See Table 23.2 for the sources of the data)

$EQ = E_a/E_e$

where E_a = actual brain weight and E_e = expected brain weight.

The key question is: what brain weight do we expect for a given species?

23.1.2
Experimental and Theoretical Body-Brain Relationships

The expected brain weight may be calculated by one of two strategies, either from regression analysis or theoretically (Harvey and Krebs 1990).

23.1.2.1
Regression Analysis

In 1891, Snell noted that the data points of brain versus body weight more or less follow a straight line on a double-logarithmic graph (Fig. 23.1). The equation for such a straight line is:

$\log(E) = \alpha \cdot \log(S) + \log(k)$

where E = brain weight, S = body weight, α = slope of the line and k = intercept.

If this straight line is a correct description for these data points, then the relationship between brain weight and body weight is:

$E = k \cdot S^\alpha$

A few authors (Count 1947; Bauchot et al. 1989b) claim that the data points are better fitted by a second-order function:

$\log(E) = \beta(\log(S))^2 + \alpha \cdot \log(S) + \log(k)$

The great majority of investigators, however, regard a straight line in a double-logarithmic graph as the adequate description of the relationship between brain weight and body weight.

23.1.2.2
Theoretical Relationships

Another approach is fundamentally different: one starts with a theory about the relationship between brain and body size. Snell (1891) assumed that brain size would be proportional to the size of the body surface. Using this theory, the slope of the the-

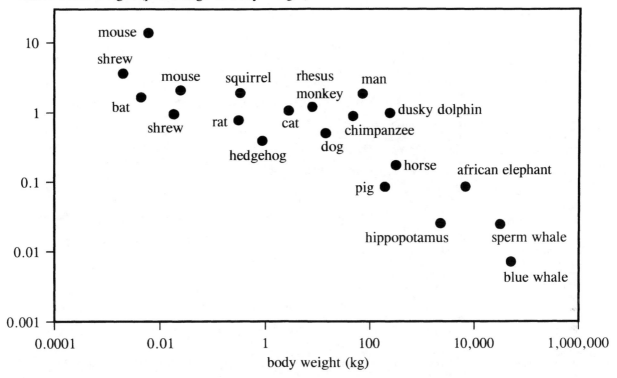

Fig. 23.2. Brain weight as a percentage of body weight for the same 20 mammals as in Fig. 23.1 (double-logarithmic graph). Small animals have large relative brain weights. (See Table 23.2 for the sources of the data)

oretical line between body weight and brain weight would be:

$$\alpha = {}^2/_3$$

For 47 warm-blooded animals (22 mammals and 25 birds), Snell (1891) actually found a slope of 0.68, which he considered supported his theory. Jerison (1973) adopted this theory, implying that the true slope is 2/3. When larger samples of mammals were studied, however, the actual slope of the regression line was closer to 0.75 rather than 0.67. In a modification of the theory, Jerison (1988, 1994) argued that in mammals the body surface is mapped to the neocortex, and that the neocortex is thicker in larger species. Maintaining the theory that body surface is mapped to cortical volume, and, after a correction for cortical thickness, Jerison concluded that the expected slope of a regression line for mammals would be about 0.78.

Table 23.1 shows the values of α for seven main groups of vertebrates. Most values range from 0.53 to 0.64, with a lower extreme of 0.21 (cyclostomes) and an upper extreme of 0.739 (mammals). Given this variation in the values of α, I am not really

Table 23.1. Allometric parameters for brain size of some main vertebrate groups. The values have been calculated with the linear regression function of EXCEL 5.0, based on the logarithmic values of body and brain weight (both in grams)

	Number of species	α (±SEM)	k	r^2	Variation in EQ (SD/average)
Cyclostomes	5	0.21 (±0.17)	0.0127	0.35	39%
Chondrichthyes	61	0.56 (±0.04)	0.0519	0.74	68%
Osteichthyes	878	0.639 (±0.009)	0.0104	0.85	52%
Amphibians	118	0.55 (±0.02)	0.0116	0.81	37%
Reptiles	74	0.53 (±0.02)	0.0179	0.91	43%
Birds	221	0.59 (±0.01)	0.1169	0.89	46%
Mammals	1174	0.739 (±0.005)	0.0626	0.96	55%

Table 23.2. Sources of the data used

Brain and body weight	
Vertebrates	Crile and Quiring 1940
Cyclostomes	Platel and Delfini 1981, 1986; Ebinger et al. 1983; Platel and Vesselkin 1989
Chondrichthyes	Bauchot et al. 1976; Myagkov 1991
Osteichthyes	Platel et al. 1977; Platel, personal communication
Teleosts	Ridet 1982; Bauchot et al. 1989a–c; Ridet and Beauchot 1990a
Amphibians	Thireau 1975; Bauchot et al. 1983; Taylor et al. 1995; Roth et al. 1995
Reptiles	Platel 1976, 1989
Birds	Lapicque and Girard 1905; Portmann 1947; Armstrong and Bergeron 1985; Rehkämper et al. 1991a,b
Mammals	Mangold-Wirz 1966
Monotremes	Pirlot and Nelson 1978
Marsupials	Elias and Schwartz 1969; Möller 1973; Eisenberg and Wilson 1981; Pirlot 1981
Cingulata	Röhrs 1966; Pohlenz-Kleffner 1969; Pirlot 1980; Pirlot and Kamiya 1983
Pilosa	Röhrs 1966; Pohlenz-Kleffner 1969; Pirlot 1980
Pholidota	Weber 1891; Elliot Smith 1898
Macroscelidea	Stephan et al. 1981
Lagomorpha	Mace et al. 1981
Rodents	Brummelkamp 1939; Pilleri 1959a–c; 1960a–c; Kretschmann 1966; Zepelin and Rechtschaffen 1974; Mace et al. 1981; Pirlot and Kamiya 1982; Meddis 1983; Hafner and Hafner 1984
Carnivores	Thiede 1966; Bronson 1979; Sheppey and Bernard 1984; Gittleman 1986; Kruska 1988
Pinnipedia	Worthy and Hickie 1986; Robin 1973
Tenrecomorpha	Bauchot and Stephan 1966; Stephan et al. 1981, 1990
Insectivores	Bauchot and Stephan 1966; Mace et al. 1981; Stephan et al. 1981, 1990
Scandentia	Stephan et al. 1981
Dermoptera	Pirlot and Kamiya 1982
Bats	Pirlot and Stephan 1970; Eisenberg and Wilson 1978; Stephan et al. 1974, 1981; Jürgens and Prothero 1987; Bhatnagar et al. 1990; Eisenberg, personal communication
Primates	Von Bonin 1937; Bauchot and Stephan 1966; Stephan et al. 1977, 1981; Hofman 1983; Armstrong 1985; Harvey and Clutton-Brock 1985
Tubulidentata	Pirlot and Kamiya 1983
Artiodactyles	Oboussier 1966; 1972; Haarmann and Oboussier 1972; Ronnefeld 1970; Kruska 1973; Haarmann 1975
Whales	Worthy and Hickie 1986; Gihr and Pilleri 1969; Kraus and Pilleri 1969; Pilleri and Gihr 1970
Perissodactyles	Kruska 1973
Hyracoidea	Meddis 1983
Elephants	Jerison 1973
Sirens	Worthy and Hickie 1986; Pirlot and Kamiya 1985; Reep et al. 1989; Reep and O'Shea 1990
Fossil vertebrates	Jerison 1969; 1973; Russell 1972; Radinsky 1981; Kruska 1982
Fossil hominids	Holloway 1983; Leigh 1992
Metabolic rates	McNab 1969, 1988; Eisenberg 1981; Bartels 1982; Hayssen and Lacy 1985;
Longevity	Crandall 1964; Jones 1979; Eisenberg 1981; Jürgens and Prothero 1987;
Ecology and diet	
Fishes	Bauchot et al. 1989a–c
Birds	Portmann 1947; Bennett and Harvey 1985a,b
Mammals	Eisenberg and Wilson 1978; Clutton-Brock and Harvey 1980; Mace et al. 1981; Stephan et al. 1981; Smuts et al. 1987

inclined to adhere to (or to develop) a general theory on the relationship between body weight and brain weight.

23.1.3
Inherent Problems with Allometric Analyses

For this chapter, a database of 1174 mammalian species has been constructed, but the various orders have different absolute and relative contributions to the database: for the Proboscidea (elephants) and Tubulidentata (aardvark), 100% of the species are represented, for the primates 65%, and for the rodents only 10%. Even if we had a database for all the mammalian species for regression analysis, some fundamental – and insoluble – problems would remain.

First, from a statistical point of view, data points from species cannot be treated as independent (Harvey and Krebs 1990); neither can genera or families be regarded as independent. In my database of 1174 mammalian species, the rodents are represented by 267 species, the bats by 315 species, but for the orders Pholidota (pangolins), Dermoptera (flying lemur) and Tubulidentata (aardvark) there is only 1 species. Rodents and bats therefore contribute disproportionately to the slope and intercept of the mammalian regression line.

A second, insoluble, problem is the choice of reference group. The steepness of the regression line depends on the taxonomic level (Bennett and Harvey 1985). Suppose we are interested in the brain weights of wolves and races of dogs. Which regression line is the 'correct' reference line: that of mammals, carnivores, Canidae or races of dogs? The slopes of these regression lines differ considerably; so the choice of reference line has great influence on the EQ values.

23.1.3.1
Allometry in This Chapter

We must conclude that allometric analyses do not give straightforward evaluations of brain sizes of species with different body sizes. I do not want to enter into a discussion here on allometry of brain size. Therefore, allometric analyses will receive little mention. For the various groups, as far as possible the original body weight and brain weight data will be shown. Only when the relevance of various ecological factors has been evaluated will a combination of allometric and correlation analyses be used (see 'Appendix'); in these cases, the allometric analyses of brain weight and ecological factors will be performed on the same group of species, all of which contribute to the brain and ecological data.

23.1.4
r- and K-Selection: Body and Brain Size

Two general strategies in evolutionary biology may be relevant for the evolution of brain size. What is the optimum strategy for an animal to increase its inclusive reproductive success (a combination of inclusive fitness and reproductive success): investment primarily in the production of many offspring, or investment primarily in the increased fitness of a smaller number of offspring? The optimum strategy depends on the circumstances (MacArthur and Wilson 1967; Wilson 1975). When a region has a population density far below its carrying capacity, then an animal producing more offspring than its competitors will probably have a higher inclusive reproductive success. For a population in an unpredictable environment, producing large numbers of offspring is usually an effective strategy. For a population in a fairly stable region at a density close to the region's carrying capacity, investing in a small number of descendants to increase their fitness is often a better strategy than producing many descendants, and investing less in them after birth or after hatching.

Selection favouring investment in a large number of offspring is called '*r*-selection' (derived from the parameter *r*, denoting the intrinsic rate of natural increase). Selection where investment in a small number of descendants to increase their fitness is more effective is called '*K*-selection' (*K* denotes the carrying capacity). *r*- and *K*-selection are relevant for brain evolution, because they favour different traits. *K*-selection favours slower development, larger body size, longer life, more reproductive episodes, smaller litter size, lifetime production of fewer descendants and greater investment in individual descendants (Pianka 1970; Eisenberg 1981). Prolonged *K*-selection promotes greater longevity, slower development and a longer period for adult-offspring interactions; this provides the opportunity for a longer learning period. A prolonged history of *r*-selection leads to reduced longevity, rapid development, rapid sexual maturity and fewer reproductive episodes. There is, therefore, less time for learning, and consequently the necessity for genetic programming of several behavioural programs.

23.1.5
Selection for Large Bodies or Large Brains?

Body size is the single most important factor influencing brain size. Large animals have larger brains, but in general these larger brains do not contain larger neurons (for mammals see Haug 1987; for salamanders see Roth et al. 1995). As a consequence, large animals have more neurons, and – theoretically – a larger information-processing capacity, since information capacity is strictly coupled to the number of neurons. A theory on the evolution of brain size should be based on two more independent selection pressures: the selection pressure on body size and that on brain size (or number of neurons).

23.1.5.1
Selection Pressure in Favour of a Large Body

Selection pressure in favour of a large body is present when physical strength has a net positive influence on fitness. Also, for another reason, homeothermic animals with large bodies have an advantage: thanks to a smaller volume to surface ratio they lose proportionally less energy (although larger animals, of course, need a large absolute amount of food.) *K*-selection often promotes the increase in body size.

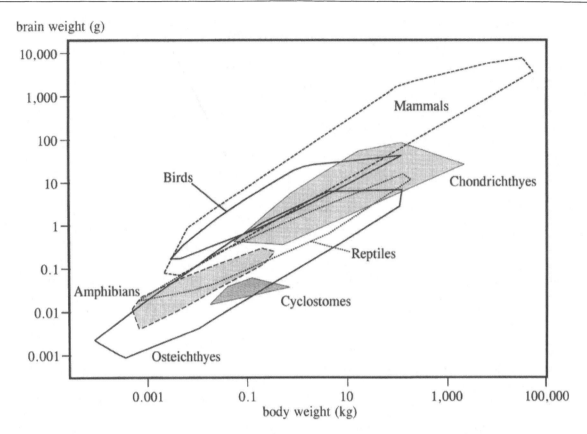

Fig. 23.3. Convex polygons for brain sizes of the main vertebrate groups

23.1.5.2
Selection Pressure in Favour of a Small Body

In general, *r*-selection is consistent with selection in favour of a small body size. Moreover, in flying, burrowing and arboreal vertebrates, selection pressure is active in keeping (or making) body weights low. Homeothermic vertebrates cannot become very small. Small homeothermic animals have a distinct problem. Their small bodies can contain only small energy stores (fat, glycogen), while they have a high metabolic rate (oxygen consumed per gram body weight); small mammals and birds are in constant danger of starvation if they cannot take food every few hours (Lindstedt and Boyce 1985). Small homeothermic animals consume much more energy per gram body weight than large ones. It has been suggested that the maximum attainable cellular metabolic rate for a homeothermic vertebrate is associated with a body weight of about 2 g (Dobson and Headrick 1995). The smallest adult birds and mammals weigh about 2 g. The smallest fishes, amphibians and reptiles are considerably smaller; the smallest fish (*Pandaka*) weighs about 0.1 g. During the process of minaturisation, strategies must be used to preserve a sufficient neural information-processing capacity. In miniaturised salamanders, the strategies actually used are: obtaining larger *EQ*s, obtaining more densely packed neurons and/or obtaining smaller neurons (Roth et al. 1995).

23.1.5.3
Selection Pressure in Favour of a Large Brain

Theoretically, a brain containing more neurons has a larger information-processing capacity. It is often assumed that selection in favour of intelligence promotes an enlargement of the brain (Jerison 1973). Such a relationship between brain size and intelligence has not been demonstrated in a comparative study on animals, since no fair comparison of intelligence or learning capacity has been made between species. It is conceivable that such a comparison cannot be made fairly. In various groups, different parts of the brain have become enlarged (Barton et al. 1995; see also Chap. 22). Brain enlargement can be a consequence of selection pressures in favour of some sensory or motor systems, or in favour of intelligence

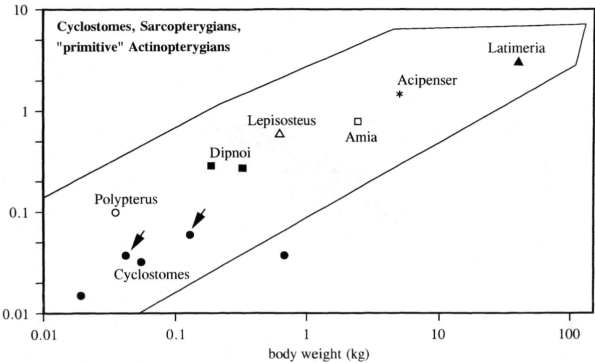

Fig. 23.4. Relationship between body weight and brain weight for cyclostomes (hagfishes marked by *arrows*), sarcopterygians and 'primitive' actinopterigians, in a double-logarithmic graph. Part of the convex polygon of the teleosts is also shown

23.1.5.4
Selection Pressure in Favour of a Small Brain

Few theories have been formulated for selection in favour of a proportionally small brain. (a) A strong selection pressure in favour of a small body mass in general might also reduce EQ (as suggested for insectivore bats, Eisenberg and Wilson 1978). (b) The brain is a rather energy-consuming organ; a strong selection pressure in favour of energy conservation would also keep (or make) brain mass small [as suggested for diving mammals (Robin 1973) and insect-eating bats, see below]. (c) r-Selection is suggested to reduce brain size. As far as I know, there are only two reliable examples of an actual reduction of the EQ: parasitic worms and domestic animals (Kruska 1988).

In many instances, the regression lines between body size and brain size for more related species are less steep than those for less related species. "One interpretation of this phenomenon is that body size responds more readily to selection over evolutionary time, and that changes in brain size lag somewhat behind" (Bennett and Harvey 1985a,b). As a consequence, selection in favour of a small body usually promotes the evolution of larger EQs, and vice versa.

23.2
Origin of the Chordate Brain

A characteristic feature of all chordates is the presence of a single dorsal nerve cord. The most 'primitive' chordates, the cephalochordates, have a rostral differentiation of their nerve cord, but very little or no brain enlargement (Chap. 9). Adult tunicates do not have a real brain, but a small brain is present in their larvae. Therefore it is assumed that the first rostral enlargement of the spinal cord (brain formation) occurred in the ancestors of the tunicates and vertebrates. In the most 'primitive' vertebrates, the cyclostomes (lampreys and hagfishes), a rather small, but well-differentiated brain is present. A survey of the brain weights in vertebrates is presented in Fig. 23.3, where the convex polygons of the main groups are shown in one figure.

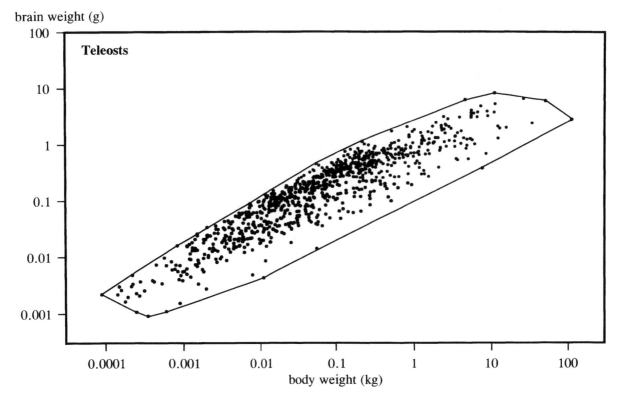

Fig. 23.5. Relationship between body weight and brain weight for teleosts, in a double-logarithmic graph. (This convex polygon for teleosts is preliminary, since data on the largest teleosts are not available.) (Based mainly on data by Ridet 1982 and Bauchot et al. 1989a–c)

23.3
Agnathans and Jawed Fishes

23.3.1
Cyclostomes

The brain of the cyclostomes is the most 'primitive' brain found in the vertebrates (see Chap. 10). Figure 23.4 surveys body and brain weights of the cyclostomes. To facilitate comparison, the convex polygon of the teleosts is also shown. The brains of adult cyclostomes range from 15 to 60 mg (Platel and Delfini 1981, 1986; Ebinger et al. 1983; Platel and Vesselkin 1989). The hagfishes (*Myxine*) have somewhat larger brains than the lampreys. The brains of the cyclostomes are small, but they are by no means the smallest chordate brains; these are only 1 mg in weight (about 1 mm^3 in a teleost fish, *Kraemeria*, Bauchot et al. 1989a,b). When cyclostome brains are compared with the brains of teleosts, the cyclostomes occupy the lower part of the teleosts' convex polygon (Fig. 23.4). This lower part of the convex polygon is occupied by only a few (eel-shaped) teleost species (cf. Fig. 23.5). Therefore the cyclostomes' brain is considerably smaller (often by a factor of 6–10) than that of most bony fishes of their size. Lampreys and hagfishes have elongated bodies, and we will see below that animals with elongated bodies often have proportionally small brains.

23.3.2
Jawed Fishes

The two main groups of jawed fishes are the cartilaginous fishes (Chondrichthyes) with several hundreds of species, and the bony fishes (Osteichthyes) with over 30 000 species. The bony fishes again comprise two main groups: those with ray fins (the Actinopterygii, about 30 000 species), and those with lobe fins (Sarcopterygii, with 7 species). The group of Actinopterygii consists of some 'primitive' groups with a small number of species (see below) and a very large group of 'advanced' species, the Teleostei.

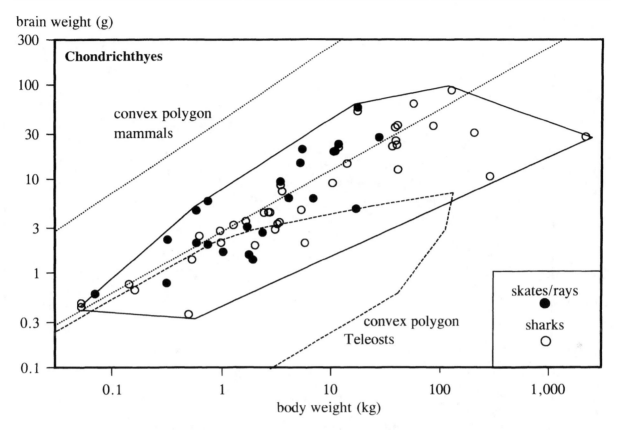

Fig. 23.6. Relationship between body weight and brain weight for Chondrichthyes, in a double-logarithmic graph. Note the large brains of the sharks and rays. (This convex polygon for Chondrichthyes is preliminary, since data on the largest Chondrichthyes not available.) For comparison the convex polygons for teleosts and mammals are also shown

23.3.2.1
Osteichthyes

23.3.2.1.1
'Primitive' Actinopterygii

Among the Actinopterygii, some groups are regarded as primitive: Polypterini (4 species), sturgeons (Acipenseroidei, 25 species) and Holostei (11 species, *Lepisosteus, Amia*). These 'primitive' bony fishes have brain weights within the range of the teleosts and above the values found in cyclostomes (Platel et al. 1977, Fig. 23.4). Therefore, it is assumed that the brain enlargement from cyclostomes to fish occurred early in fish evolution, i.e. around the Ordovician, some 450 million years ago.

23.3.2.1.2
'Advanced' Bony Fishes

Teleosts comprise some 30 000 species. Investigators in Bauchot's group have performed systematic comparative studies on teleost brains (Ridet 1982; Bauchot et al. 1989a,b,c). Figure 23.5 is based mainly on data from Bauchot's group. Some conclusions from the comparative studies on hundreds of teleost species by this group will be mentioned here. Within the group of teleosts, some orders are considered primitive (for instance eels and herrings) and others advanced (mackerels and flatfishes). However, brain size is hardly related to the degree of advancement as established by other criteria.

Elongated fishes (such as eels) have small brains for their body weight. This point will be commented on during the discussion on snakes below. Some fishes have a passive defence against predators, using thick scales or spines, protective colouring, burrowing into the ground, retreating into burrows, or skin toxins. Such fishes often have proportionally small brains. Fishes with more than one highly developed sensory system often have proportionally large brains.

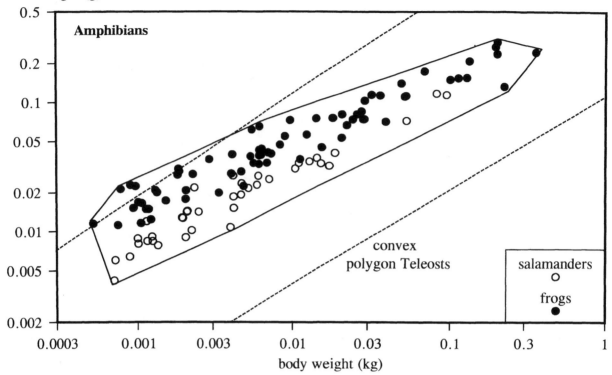

Fig. 23.7. Relationship between body weight and brain weight for amphibians, in a double-logarithmic graph. Salamanders have smaller brains than frogs. (This convex polygon for amphibians is preliminary, since data on the smallest and largest amphibians are not available.) For comparison the convex polygon for teleosts is also shown. (See Table 23.2 for the sources of the data)

23.3.2.1.3
Sarcopterygii

The Sarcopterygii deserve special interest, because they are related to the ancestors of the amphibians (and consequently of the reptiles, birds and mammals). Two orders are present: the Dipnoi (lungfishes with six species) and the Crossopterygii (with one living species, the coelacanth, *Latimeria*). With respect to brain size, the Sarcopterygii fall within the range of the teleosts (Fig. 23.4).

23.3.2.2
Chondrichthyes

Chondrichthyes (sharks and rays) are conspicuous by their large brains (Bauchot et al. 1976; Northcutt 1989; Myagkov 1991). Most species have considerably larger brains than the bony fishes in their weight class: the difference is up to a factor of 10 (Fig. 23.6). The brain weights of several Chondrichthyes lie within the mammalian range. Sharks and rays have about the same brain size, with a relatively large telencephalon (Chap. 23.12). In these respects, their brains are 'advanced'. What can sharks and rays do better thanks to their advanced brains? They have well-developed sensory systems, but so do several bony fishes. Extensive comparative behavioural studies in fishes and sharks have not yet been carried out. Compared to other fishes, the Chondrichthyes have a very small number of offspring; in this respect they are a clear example of *K*-selection.

23.4
Amphibians

Only a small number of studies have been devoted to the comparative study of brain size in amphibians (Thireau 1975; Bauchot et al. 1983; Roth et al. 1995; Taylor et al. 1995). The brain weights of the amphibians fall within the range of the teleost fishes (Fig. 23.7). Within the class of amphibians, frogs (*Anura*) have larger brains than the salamanders (*Urodela*); the average difference is almost a factor of 2. Within the frogs, the arboreal species

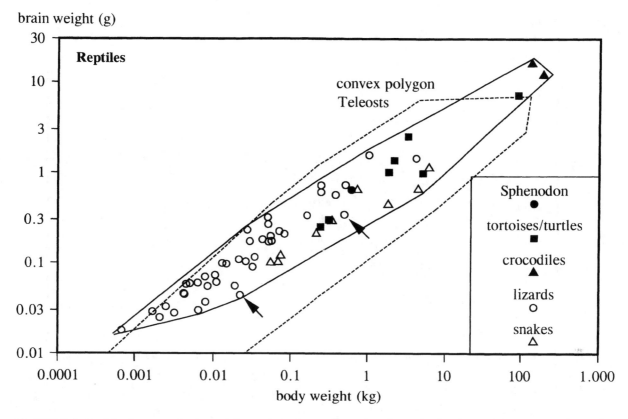

Fig. 23.8. Relationship between body weight and brain weight for living reptiles, in a double-logarithmic graph. Note the small brains of tortoises, turtles, snakes and slowworms (slowworms marked by *arrows*). For comparison the convex polygon for teleosts is also shown. (The convex polygon for reptiles is preliminary, since data on the largest reptiles are not available.) (See Table 23.2 for the sources of the data)

have larger brains than the terrestrial species (Bauchot et al. 1983; Taylor et al. 1995). Arboreal frogs have slightly larger cerebella than frogs from other habitats. In fossorial frogs, olfactory and auditory structures are enlarged, while visual structures are diminished. Special attention has been paid to miniaturisation in a group of salamanders (plethodontid salamanders, Roth et al. 1995). During the process of body and brain miniaturisation, the most rigorous solution for preserving information capacity (and cell numbers) would be reduction of cell size. However, cell size depends strongly on genome size (Roth et al. 1995). Among the vertebrates, salamanders have large genomes. Genome size and cell size has been reduced in only a few species, but this has occurred in miniaturised and non-minaturised species (Roth et al. 1995). Why do salamanders have smaller brains than frogs? Most salamanders in Fig. 23.7 are aquatic species, but the aquatic salamanders have smaller brains than the aquatic frogs. So the small brains of salamanders cannot be explained by their habitat. Salamanders are elongated amphibians; in general, elongated animals tend to have proportionally small brains. This probably applies too for salamanders.

23.5
Reptiles

23.5.1
Recent Reptiles

Reptiles have brain weights in the same range as teleost fishes (Fig. 23.8, Platel 1976, 1989). The lowest values for brain weights corrected for body weights are found in the Chelonia (tortoises and turtles) and the snakes; lizards of the same body weight have brain weights about twice as large. The small brains of the Chelonia are unremarkable, since these reptiles are regarded as 'primitive': the real Chelonia lived 220 million years ago. Chelonia have a simple locomotion, and a passive protection against predators, which often goes together with low brain weight. However, another 'primitive' reptile, *Sphenodon*, has a brain weight close to the average for reptiles. The low brain weights of snakes

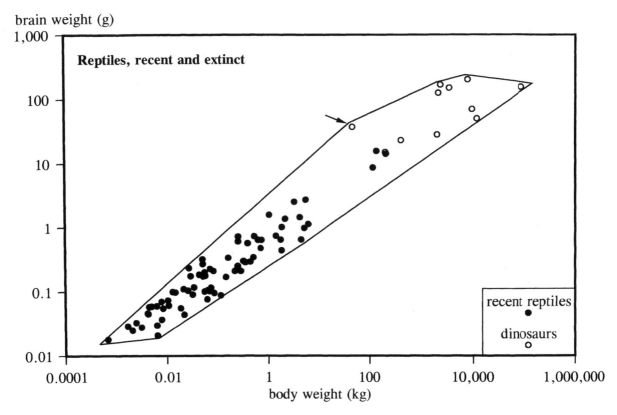

Fig. 23.9. Relationship between body weight and brain weight for living reptiles and dinosaurs, in a double-logarithmic graph, where the estimated brain volume is supposed to be half of the cranial capacity). *Arrow* points to *Stenonychosaurus*, an ostrich-like dinosaur with a large brain. (Data on dinosaurs from Jerison 1973)

are remarkable, because snakes are advanced animals with a sometimes complicated locomotion and specialised sensory systems. Snakes are characterised by an elongated body. The lizards with the smallest brain weights are the slowworms (Anguidae), elongated, legless lizards which the layman sometimes mixes up with snakes.

Apparently, it is a general trait that elongated animals (lampreys, eels, salamanders, snakes, slowworms) have small brain weights for their body weights. Do elongated animals have small brains or large bodies? Given their locomotion, selection pressure in favour of long bodies is plausible. Therefore: snakes, slowworms, salamanders, eels, lampreys and hagfishes have long (and consequently heavier) bodies rather than small brains.

23.5.2
Dinosaurs

Several skulls of dinosaurs have been preserved so well that the cranial cavity is still intact or can be reconstructed. The volume of the cranial cavity (or of the endocast) is the maximum volume the brain could have. *Tyrannosaurus* with a body weight of 7.7 metric tons had a cranial cavity of 400 cm^3, and *Brachiosaurus* weighing 87 metric tons had a cranial cavity of about 300 cm^3. This compares with brain weights between 5 and 10 kg for elephants and large whales. Consequently, dinosaurs have been thought of as extremely small-brained – 'stupid' – animals. The dinosaur brain size could even be smaller, since in several reptiles the brain occupies only a proportion of the cranial cavity (for references see Hopson 1979). For that reason, the brain volume cannot be directly deduced from the endocast. To deal with dinosaurs' brain size despite these limitations, the best Jerison (1969, 1973) could do was to assume that the brain volume was half of the cranial cavity. Another problem with the encephalisation of extinct animals is the estimation of their body size. In some instances, the body weight and brain weight were estimated from remains of different individuals, or even different species (Hopson 1979). Nevertheless, after a careful study of dinosaur endocasts, several conclusions could be drawn about dinosaur brain size and shape (Hopson 1979). Allometric analysis demon-

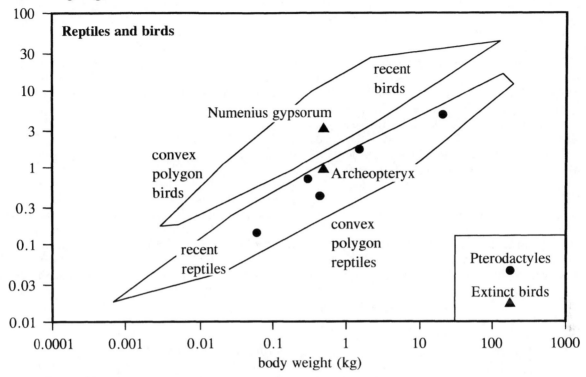

Fig. 23.10. Evolution of brain size in birds. For comparison the convex polygons of living reptiles and birds are shown, and some data points of pterodactyls and extinct birds. Birds have larger brains than reptiles. (See Table 23.2 for the sources of the data)

strated that most dinosaur brains are within the normal range – for reptilian brains (Fig. 23.9, Jerison 1973, Hopson 1979). Some small ostrich-like, carnivorous dinosaurs (*Stenonychosaurus* and *Dromiceiominus*) appeared to have remarkably large brains – for reptiles: their brains fall into the range of birds' brains (Russell 1972). Dinosaurs with body weights similar to those of the crocodiles have cranial capacities similar to those of the crocodiles. It has been suggested that carnivorous dinosaurs would have had larger brains than herbivorous ones, but this has been questioned (Hopson 1979).

23.6 Birds

23.6.1 Evolution of Brain Size in Birds

In general, birds have brain weights 6–10 times larger than reptiles of similar body weight (Fig. 23.10, Portmann 1947; Armstrong 1985; Rehkämper et al. 1991a,b). One might suppose that these larger brains are necessary for a well-coordinated flight. But the extinct flying reptiles, the pterodactyls, have cranial capacities within the reptilian range (Fig. 23.10). Apparently a large brain is not absolutely necessary for flying (as flying insects have already demonstrated). The ancient fossil bird, *Archaeopteryx*, from the late Jurassic (some 150 million years ago), already had a brain weight in the lower range for birds (Fig. 23.10). A fossil bird from the upper Eocene (about 40 million years ago), *Numenius gypsorum*, had a larger brain within the range for birds (Jerison 1973). In the evolution of the birds, a strong degree of brain enlargement probably occurred before the late Eocene.

23.6.2 Brain Weights in Various Bird Orders

The relationship between body weight and brain weight for birds is shown in Fig. 23.11. The smallest hummingbirds have brains of only 0.17 g. The largest living bird, the ostrich, has the largest brain: 42 g. Within the class of birds, considerable variation in brain weight is found. Proportionally the smallest brains are found in the chicken-like birds (Galliformes) (Fig. 23.11b); this is not primarily due to domestication, since small brain weights are also found in non-domesticated Galliformes. Also pigeons (Columbiformes) have small brains (Fig. 23.11a). Corrected for body weight, the largest brain weights in birds are found in the perching birds (Passeriformes), woodpeckers (Piciformes) and parrots (Psittaciformes, Fig. 23.11). The birds of prey (Falconidae) have rather large brains (Fig. 23.11b), but not as large as the perching birds. Owls (Strigiformes) have somewhat larger brains than the birds of prey (Fig. 23.11b). The ostrich has the largest absolute brain weight in the birds (Fig. 23.11a), but its brain weight lies considerably below the regression line for birds in general.

23.6.3
Brain Size and Ecology in Birds

In birds, brain size is associated with various ecological aspects (Bennett and Harvey 1985a). Three aspects were statistically significantly associated with overall brain size in two-way analyses of variance: neonatal development, mating system and mode of prey capture. However, some other aspects that were significant in a one-way analysis of variance will still be mentioned.

23.6.3.1
Neonatal Development

It has long been known that bird brain size is strongly associated with neonatal development (Portmann 1947, Fig. 23.12). Other studies have corroborated this finding (Bennett and Harvey 1985a,b). Immediately after hatching, some birds (for instance newborn chickens) have feathers, leave the nest, walk around and gather all their food themselves (precocial birds). At the other extreme, newborn perching birds are naked and absolutely helpless (altricial birds). Adults of altricial bird species have brain weights 1.5–2 times as large as those of adult precocial birds of the same body weight. This difference is found over various orders of birds. Can we explain this difference? *Precocial birds* are a product of *r*-selection. They invest a lot of energy in the eggs, i.e in their offspring before hatching. *Altricial birds* invest much in their offspring after hatching: in feeding and protecting them. They are a product of *K*-selection.

A clear difference in pre- and post-hatching brain growth is present between precocial and altricial birds. Precocial birds have a larger pre-hatching brain growth and development, while altricial birds have a larger post-hatching brain growth and development (Bennett and Harvey 1985a,b). There is reason to believe that the primitive situation was: precocial offspring with relatively less patental investment after hatching. Why did then altricial birds develop larger brains? It is difficult to identify cause and effect. Did an increase in parental investment promote inclusive reproductive success, and was a larger brain needed for better parental investment? Or did the increased parental investment enable a larger post-hatching brain growth and development? To my mind this is still an open question.

23.6.3.2
Mating System

Monogamous birds have larger brains than polygamous species (Bennett and Harvey 1985a). This association remained significant in the two-way analysis of variance, which implies that it is not due to other confounders (such as neonatal development). Almost all species with altricial offspring have a monogamous mating system, but the mating system still seems to be a relevant factor on its own.

23.6.3.3
Stratification and Habitat

Arboreal birds tend to have larger brains than terrestrial and water birds. Birds living in forest or woodland tend to have larger brains than those living in grassland, marshland or water (Bennett and Harvey 1985a).

23.6.3.4
Diet

Birds use a great variety of food: leaves, fruit, nectar, insects, fish, other birds, mammals or carrion. As regards brain size, two aspects are remarkable (Bennett and Harvey 1985a). Birds that feed on other birds or mammals have rather large brains. On the other hand, birds that feed on plants, without specialising on leaves or fruit, have rather small brains.

23.6.3.5
Migration

One would have expected that migratory birds would need rather large brains for orientation, but, remarkably, migration has not been associated with brain size (Bennett and Harvey 1985a). Pigeons, for instance, have rather small brains.

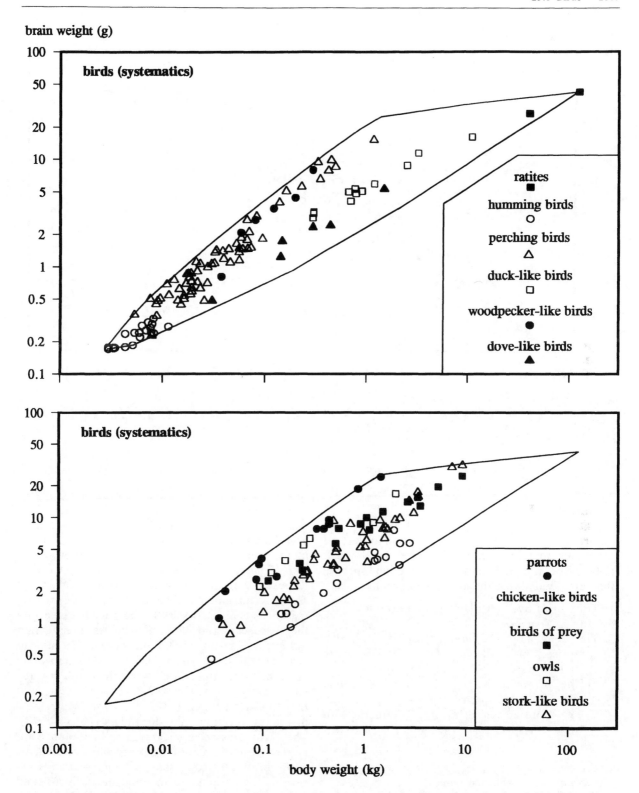

Fig. 23.11a,b. Relationship between body weight and brain weight for various orders of birds, in a double-logarithmic graph. Note the small brains of the gallinaceous birds and pigeons, and the large brains of the perching birds, parrots and woodpeckers. Note also the remarkable position of the ostrich. (See Table 23.2 for the sources of the data; for the classification of birds, the system of Sibley and Ahlquist 1990 is followed)

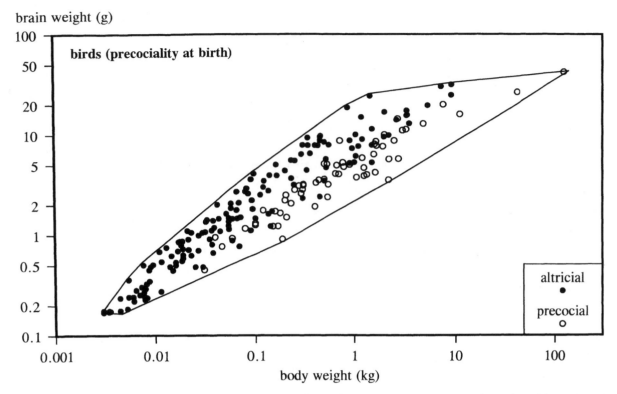

Fig. 23.12. Relationship between body weight and brain weight for birds coded for their neonatal development, in a double-logarithmic graph. Note the small brains of the precocial birds, and the large brains of the altricial birds. (See Table 23.2 for the sources of the data)

23.7 Mammals

Figure 23.13 is a cladogram of several orders of mammals, mainly based on the figure by Novacek (1992). The mammals comprise about 5000 species; they are divided into the following main groups:

1. The 'primitive', egg-laying monotremes with only six species
2. The more advanced, viviparous Theria, which comprise almost all mammals. The Theria are again subdivided into two main groups:
   a) The 'primitive' marsupials without placenta, with about 240 species
   b) The more advanced, placental Eutheria with over 4500 species

23.7.1 The Brain in Extinct Mammals

Overall, mammals have brain weights a factor of 10 higher than reptiles of similar body weights (Fig. 23.14). For many extinct mammals belonging to various orders, data are available on brain size, so we can trace more or less the evolution of mammalian brain size. Most mammal-like reptiles (Permian and Triassic, 290–210 million years ago) had brain weights within the reptilian range, so the increase in brain size in the mammalian evolution started later. The mammals from the Cretaceous, Paleocene and Eocene had brains larger than the reptiles, but smaller than those of most existing mammals (Fig. 23.14). With one notable exception: the Paleocean and Eocean primates already had rather large brains. From the Eocene (38 million years ago) to the Miocene/Pliocene (5 million years ago), the brains in several mammalian groups gradually enlarged (Jerison 1973). It has been argued that the increases in brain size in carnivores and ungulates were mutually dependent (Jerison 1973). The carnivores were suggested to have somewhat larger brains than the ungulates; the larger carnivores' brain was suggested to be a selection pressure for brain increase in ungulates, which urged the carnivores to develop still larger brains. It has been suggested that, in South America, the hoofed mammals and their predators (carnivorous marsupials) had a different evolutionary development. According to Jerison (1973), the brains of preys and

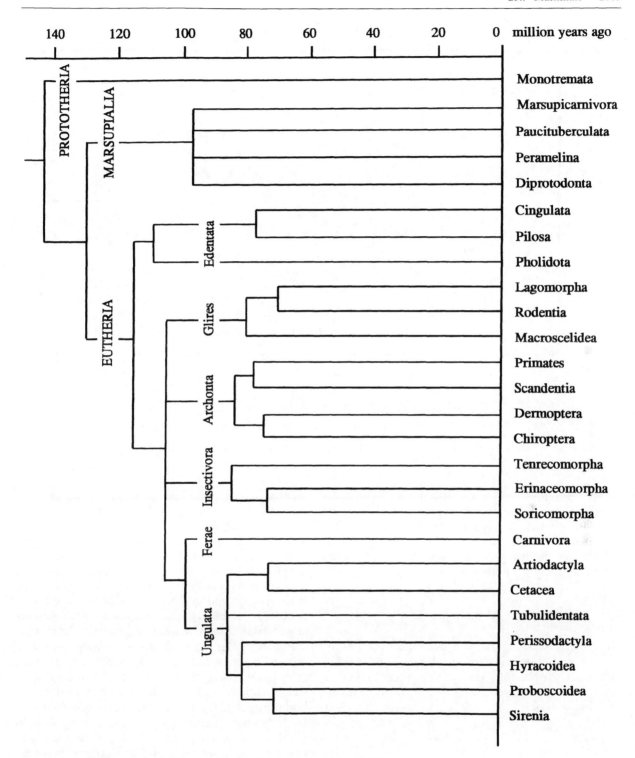

Fig. 23.13. Cladogram of mammalian orders. (Based mainly on Novacek 1992)

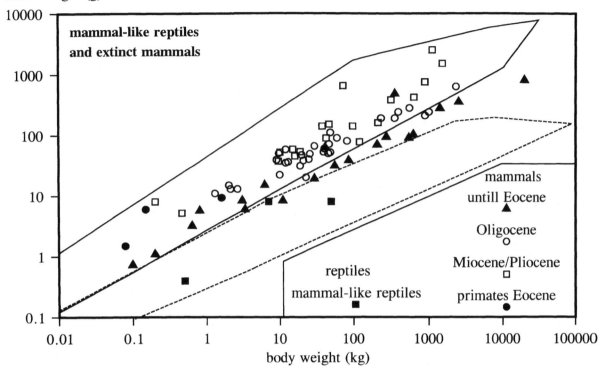

Fig. 23.14. Evolution of brain size in mammals. For comparison the convex polygons of living reptiles and mammals are shown. Most mammal-like reptiles have reptile-sized brains. During mammalian evolution, brain size has gradually increased. By the Eocene, the primates already had large brains for their body size. (Based on data from Jerison 1973)

predators in South America did not grow progressively: these groups seemed not to stimulate one another to develop larger brains (Jerison 1973). According to Radinsky (1981), however, the South American hoofed mammals had larger *EQs* than was originally presumed.

23.7.2
Ecology, Ethology and Brain Size in Mammals

In the literature of the past decades, various ecological and ethological features have been related to brain size. In the following sections these and other features will be discussed for some selected mammalian orders.

23.7.2.1
Metabolic Rate

It has been suggested that brain size is related to the basal metabolic rate (Martin 1981; Armstrong 1983; Hofman 1983a). The basal metabolic rate is the metabolism during a behaviourally inactive state (often sleep). Like brain size, the basal metabolism is an allometrically scaled parameter. From the literature, I have collected data on 264 mammalian species where body weight and brain weight as well as basal metabolic rate are known. The regression lines (based on the same species) have the same slopes: brain weight 0.71 (SEM 0.01) and metabolism 0.72 (SEM 0.01). Since these slopes are similar, it has been suggested that brain weight in some way is causally related to the basal metabolic rate (Martin 1981). The crucial question is: do mammals with large brains (for their body weight) also have large metabolic rates (for their body weight)? By a method described in the 'Appendix', the correlation coefficient between brain weight corrected for body weight (*EQ*) and basal metabolic rate corrected for body weight (*MEQ*) has been calculated for these 264 mammals: the correlation coefficient was 0.08 (Table 23.3). So for mammals in general, the *EQ* is not associated with the *MEQ*. However, in some selected groups (bats, primates), stronger correlations between *MEQ* and *EQ* have been found (see below).

Table 23.3. Correlation coefficients between the *EQ* and various ecological parameters. Longevity, home range, group size and metabolic rate have been corrected for body weight by a method explained in the text. For the feeding pattern such a correction did not apply (primate diet data from Smuts et al. 1987)

	Mammals		Primates	
	Number	r	Number	r
Longevity	389	0.37**	80	0.70**
Home range	114	0.33**	62	0.45**
Group size	Only primate data		89	0.46**
Basal metabolic rate	264	0.08	22	0.40*
Feeding pattern				
% Fruit in diet	Only primate data		25	0.35*
% Herbs in diet	Only primate data		25	−0.55**

* $P < 0.05$
** $P < 0.01$

23.7.2.2
Longevity

Like metabolism, longevity has been related to encephalisation (Hofman 1983a, 1993). Longevity is an interesting feature, because it is part of the life history strategy. Prolonged *K*-selection tends to increase longevity. Moreover, *K*-selection is expected to increase body and brain size. Longevity is related to body size: large animals tend to live longer. Again the question becomes: do mammals with large brains (for their body weight) also have long life expectancies (for their body weight)? For mammals in general, the *EQ* is associated with the 'longevity quotient' (*LOQ*): $r=0.37$ (Table 23.3, $n=389$, $P < 0.01$). In primates, this correlation is stronger (see below).

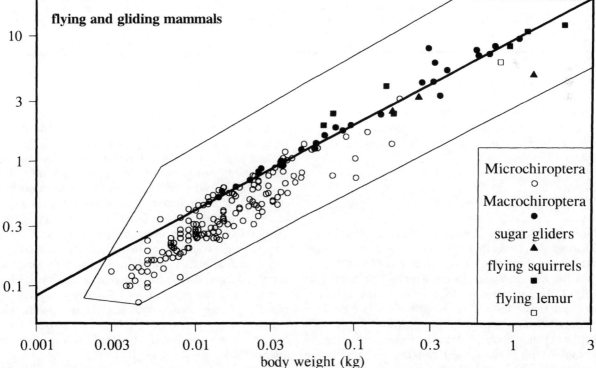

Fig. 23.15. Brain size in flying and gliding mammals. Flying foxes (Macrochiroptera) have brain sizes in the upper range of the other bats (Microchiroptera). For comparison the convex polygon of mammals is also shown: Microchiroptera occupy the lower part of this convex polygon

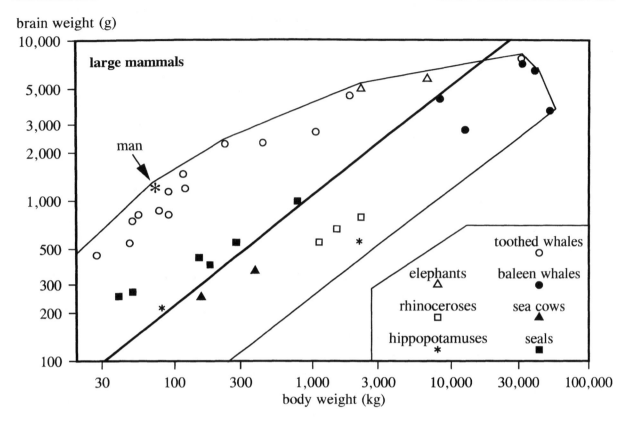

Fig. 23.16. Brain size in whales, sea cows, seals, elephants and a few large ungulates, in a double-logarithmic graph. Most toothed whales (especially dolphins) have very large brains. Sea cows, rhinoceroses and hippopotamusses have rather small brains. For comparison the convex polygon of mammals is also shown. (See Table 23.2 for the source of the data)

23.7.2.3
Neonatal Development

Unlike the situation in birds (see above), brain weight of mammals is not associated with precociality at birth (Eisenberg 1981, p. 325; Bennett and Harvey 1985b).

23.7.2.4
Flying and Gliding Mammals

Bats are the only mammals that can really fly. Among the bats, the Microchiroptera have small brains for their body size, while the Macrochiroptera have brain sizes around average for mammals (Fig. 23.15). In three other orders, species have evolved that can bridge considerable distances by gliding through the air. These are found in the marsupials (sugar glider, *Petaurus*), Dermoptera (flying lemur, *Cynocephalus*) and rodents (several flying squirrels). The brains of these gliding mammals are about average for mammals of their body size (Fig. 23.15, Pirlot and Kamiya 1982).

23.7.2.5
Aquatic Mammals

An aquatic life has other demands and opportunities for mammals than a terrestrial life. The larger an animal is, the larger the ratio of its volume to surface area, which is better for heat conservation (in homeothermic animals). Since mammals are likely to lose more energy under water than on the earth, an additional selection pressure is present to increase the body size of aquatic mammals. Moreover, in terrestrial animals, a rapid increase in body size during evolution is restricted, because bones and muscles must grow disproportionally large to carry a larger body. But in aquatic animals, this constraint is not present, so body size can increase rapidly. As a consequence, all sea mammals (whales, seals, sea cows) are large. It has been suggested that a small brain would be advantageous for aquatic mammals, since it would enable longer diving times because of the smaller oxygen demand (Robin 1973). However, in aquatic mammals, no correlation was found between diving time and *EQ* (Worthy and Hickie 1986). The brain sizes of

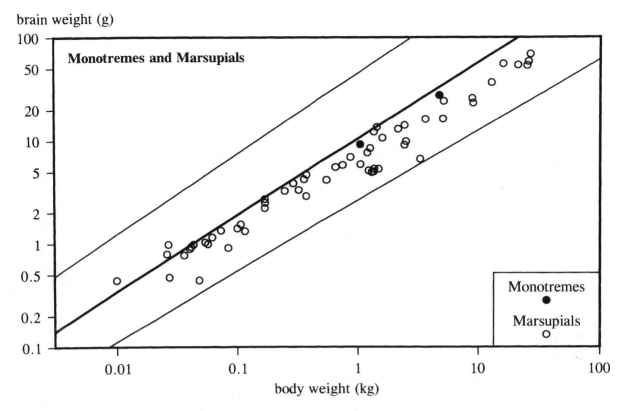

Fig. 23.17. Relationship between body weight and brain weight for monotremes and marsupials, in a double-logarithmic graph. For a comparison the convex polygon of mammals is shown. (See Table 23.2 for the sources of the data)

aquatic mammals differ greatly (Fig. 23.16): whales have large brains, seals slightly above average for mammals and sea cows slightly below average.

23.7.3
Monotremes

Although the monotremes are regarded as the most 'primitive' group of mammals, their brains have about the average size for a mammal of their body size (Fig. 23.17).

23.7.4
Marsupials

The marsupials show a large morphological and ecological variation. In this group we find small bandicoots with a body weight of 10 g, and a brain weight of 0.5 g; but also kangaroos of 30 kg with brain weights of 60 g. Overall, marsupials have small brains, compared to other mammals (Fig. 23.17).

23.7.5
Eutheria

The Eutheria are discussed in this chapter under five groups, which are suggested to be monophyletic (Novacek 1992):

1. The Insectivora, which are often subdivided into three orders: tenrecs, hedgehogs and shrews (including moles)
2. The Edentata
3. The rodents and their presumed relatives, the elephant shrews and the rabbits
4. The Archonta, which comprise the tree shrews, flying lemurs, bats and primates
5. The Ferungulata, which comprise among others the carnivores, ungulates, whales, elephants and sea cows

23.7.5.1
Insectivores

The insectivores show considerable variation in brain weight. Some insectivores have remarkably small brains for their body size: tenrecs, shrews and

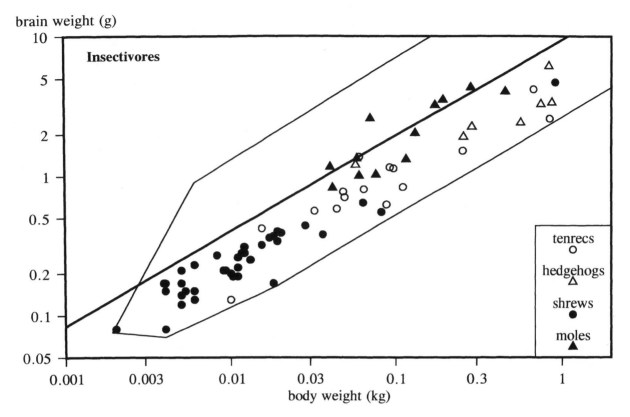

Fig. 23.18. Relationship between body weight and brain weight for various groups of insectivores, in a double-logarithmic graph; note the relatively large brains of moles. For comparison the convex polygon of mammals is shown. (See Table 23.2 for the sources of the data)

hedgehogs (Fig. 23.18); these are sometimes included together as 'basal insectivores' (Bauchot and Stephan 1966). The moles are conspicuous by their large brains – at least for insectivores of their body sizes, but it is not clear why they evolved large brains. At the moment, no clear relationship has been found for insectivores between their brain weight and their ecology or ethology. Semi-aquatic insectivores have advanced brains, but this is expressed in internal brain structure rather than in brain weight (Stephan et al. 1990).

23.7.5.2
Edentates

Edentates have smaller brains than the average mammal of their body size (Fig. 23.19). The Cingulata (armadillos) and Pilosa (sloths and anteaters) have similar brain sizes. Among the edentates, the pangolins (*Manis*) have the smallest brain.

23.7.5.3
Rodents and Their Presumed Relatives

The taxonomic position of rodents, rabbits and elephant shrews is not at all clear, but some authors regard these groups as related. It is even not clear whether the rodents form a monophyletic group. Arguments have been presented that the guinea-pig-like rodents might not be monophyletic with the other rodents (Graur et al. 1992). Macroscelidea (elephant shrews) used to be included in the group of the insectivores, but now they are regarded as a separate order, possibly a sister group of the Lagomorpha. The brain sizes of elephant shrews are only average for a mammal of their body size. So they have considerably larger brains than the real shrews. The brains of the Lagomorpha (hares, rabbits and piping hares) are somewhat smaller than those of the average mammal of their body size.

Three main groups of rodents are distinguished: squirrel-like (sciuromorphs), mouse-like (myomorphs) and guinea-pig-like (caviomorphs). Small myomorphs have remarkably large brains for mammals of their size: up to a factor of 10 larger than

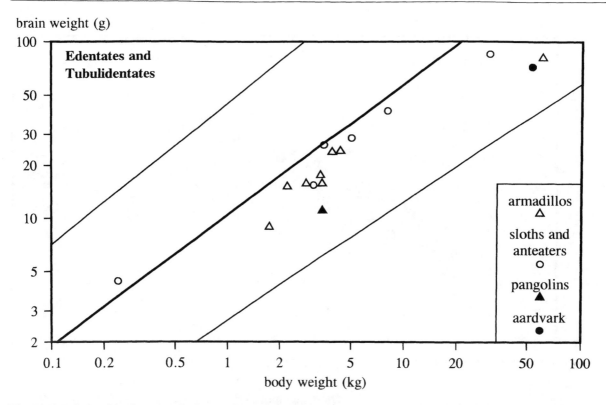

Fig. 23.19. Relationship between body weight and brain weight for edentates and tubulidentates, in a double-logarithmic graph. For comparison the convex polygon of mammals is shown. (See Table 23.2 for the sources of the data)

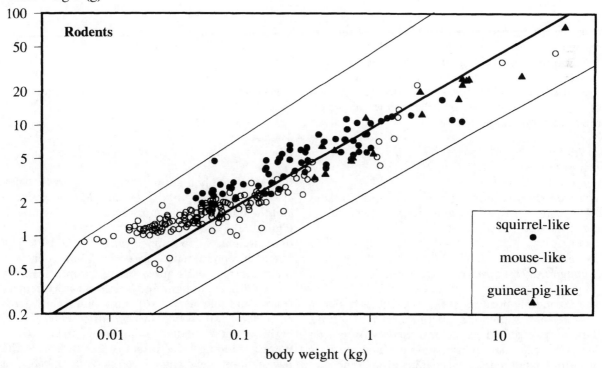

Fig. 23.20. Relationship between body weight and brain weight for rodents, in a double-logarithmic graph. For comparison the convex polygon of mammals is also shown. (See Table 23.1 for the sources of the data)

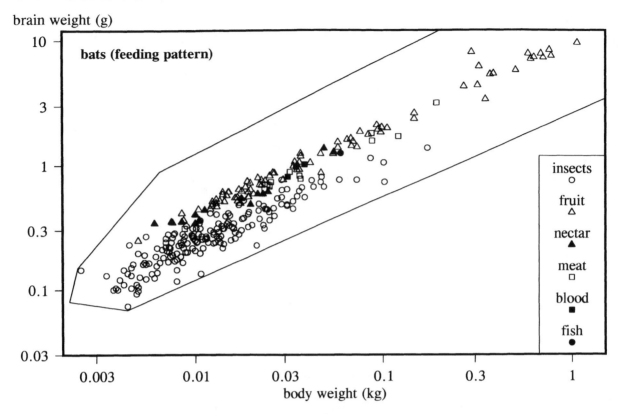

Fig. 23.21. Brain size in bats in a double-logarithmic graph. Insect-eating bats have proportionally small brains. For comparison the convex polygon of mammals is also shown. (See Table 23.2 for the sources of the data)

insectivores or bats (Fig. 23.20). The squirrels are conspicious by their proportionally large brains. This applies for the ground and the tree squirrels; the large squirrel brain cannot therefore simply be explained by their arboreal way of life. The brains of the caviomorphs are somewhat below average for mammals of their body size.

23.7.5.4
Archonta

The tree shrews, flying lemurs, bats and primates are usually regarded as one monophyletic group, the Archonta (Novacek 1992).

23.7.5.4.1
Scandentia (Tree Shrews)

The taxonomic position of the tree shrews has been a matter of dispute for some time. For a long time, they were included in the insectivores; thereafter they were regarded as primates, and now they are usually considered as a separate order. The tree shrews have brain weights just above average for mammals.

23.7.5.4.2
Dermoptera (Flying Lemurs)

The brain weight of the flying lemur is slightly below the average for a mammal of the same body weight (Fig. 23.15).

23.7.5.4.3
Bats

Bats are the only mammals that can really fly. They are distinguished in the (small) Microchiroptera and the (larger) Macrochiroptera (the flying foxes). Figure 23.15 shows that the Macrochiroptera have brain weights in the upper range of the bats. The Macrochiroptera have large visual and olfactory brain centres, while most Microchiroptera have a large auditory system, and small olfactory centres (Baron and Jolicoeur 1980). Also within the Microchiroptera, brain weights vary considerably. In bats, brain weight is related to diet rather than to taxonomic grouping. The insect-eating bats have the lowest brain weights: only 60%–70% of those of the nectar- or fruit-eating bats, which is in the same range as the basal insectivores. Proportionally the

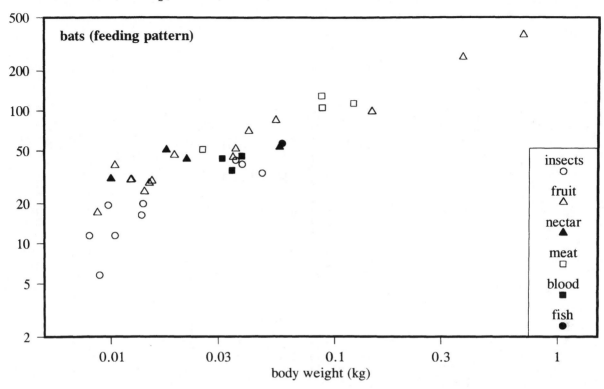

Fig. 23.22. Basal metabolism in bats. Insect-eating bats have the lowest basal metabolism. (Based on data by McNab 1969)

highest brain weights were found in the fruit-, nectar- and blood-eating bats (Fig. 23.21, Pirlot and Stephan 1970; Eisenberg and Wilson 1978; Stephan et al. 1981). The carnivorous bats occupy an intermediate position. Although many bats use echolocation, it seems to be a special quality in insect-eating bats, for they must quickly detect small moving targets by their sonar mechanism, and they must make quick, precisely timed manoeuvres in three dimensions guided by their sonar mechanism to catch their prey (Simmons and Stein 1980). Therefore, it is surprising that insect-eating bats have such small brains.

A similar trend is found with the basal metabolic rate. In bats, the EQ is rather strongly associated with the MEQ ($r = 0.58$, $n = 36$, $P < 0.01$). Figure 23.22 shows the metabolic rate for various bats coded for their diet. Again the insect-eating bats occupy the lower part of the figure (McNab 1969). Among the bats, the insect-eating bats are most economical with energy. It is hypothesised that the small brain size and the small metabolic rate are the product of the same selection pressure. During the night, insect-eating bats can only catch a limited amount of insects, since they must remain light enough for flight and quick manoeuvring. It is impossible for them to store supplies of insects. Therefore, a strong selection pressure to conserve energy is suggested for insect-eating bats. This energy conservation is also evident from another trait: insect-eating bats easily enter torpor, an energy-sparing state of hypothermia (McNab 1988). Since the brain is an energy-consuming organ, a selection pressure to conserve energy would also reduce the size of the brain. Energy conservation might explain both the low metabolic rate and the proportionally small brain weight in insect-eating bats.

23.7.5.4.4
Primates

A simplified phylogenetic tree of primates is shown in Fig. 23.23. For the classification of the Hominoidea (gibbons, apes and man), the terminology of Goodman et al. (1990) is followed. This figure reflects the recent opinion of several authors that the chimpanzee is more closely related to man than to the gorilla or other apes.

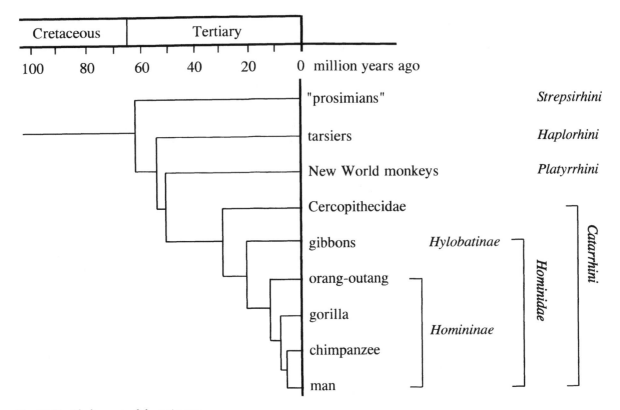

Fig. 23.23. Cladogram of the primates

Brain Size and Taxonomic Relationships. Figure 23.24 gives a double-logarithmic survey of the body and brain weights of 118 living primates. Some differences are more evident in the linear graph of Fig. 23.25a, which shows data on primates with body weights between 1 and 12 kg. The following facts are notable:

1. Primates have large brains for mammals of their body weight.
2. Within the primates, the 'prosimians' (Strepsirhini and Haplorhini) have the smallest brain weight, but most prosimians still have larger brains than the average mammal of their body size. We have seen in Sect. 23.7.1 that the prosimians of the Paleocene and Eocene already had proportionally large brains. These extinct prosimians had overlapping binocular fields and grasping forepaws; they probably had an arboreal (squirrel-like) way of life, and this could explain their large brains.
3. Since man is a member of the Old World monkeys (Catarrhini), we are inclined to regard the Catarrhini as being 'higher' than the New World monkeys (Platyrrhini). Yet some New World monkeys have larger brains than Old World monkeys.
4. Gibbons (Hylobatinae) have large brains for their body size.
5. Within the Homininae, man has an exceptionally large brain, while the brain weights of the other Homininae (the apes) are a continuation of the monkey brain weights.

Brain Size and Ecology. Brain size in primates has been related to various ecological factors: longevity (Sacher 1975; Allman et al. 1993; Hofman 1993), group size (Dunbar 1992), home range (Dunbar 1992), basal metabolic rate (Armstrong 1985) and feeding patterns (Clutton-Brock and Harvey 1980). The problem with primate brain size is that it is associated with all these parameters (Table 23.3).

1. Longevity: Large primates tend to live longer. When the longevity of primates is corrected for body size (by a method described in the 'Appendix'), it is still strongly associated with the *EQ*. In fact, of the parameters investigated, *EQ* is most strongly associated with longevity ($r=0.70$, Table 23.3). With a longer life span, the period for learning increases. A plausible interpretation is that some groups of primates have passed through a period of *K*-selection, promoting a longer life span and a larger brain.

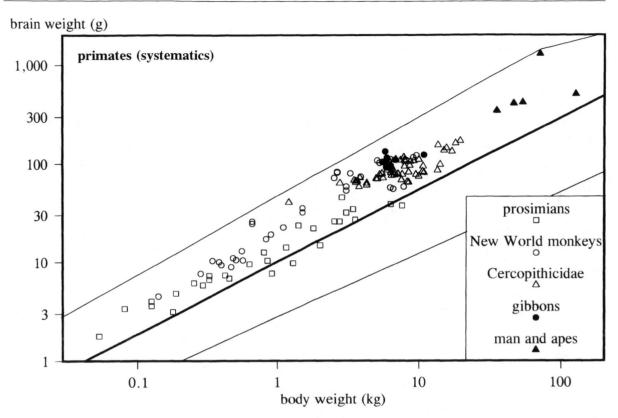

Fig. 23.24. Relationship between body weight and brain weight in primates in a double-logarithmic graph. For comparison the convex polygon and regression line (based on $\alpha = 0.739$ and $k = 0.0626$) for the mammals are shown. Most primates have larger brains than the average mammal of their body size. (See Table 23.2 for the source of the data)

2. Diet: Herbivore primates have proportionally smaller brains than frugivore and insectivore primates (Fig. 23.25b, Table 23.3). In some frugivore species, the brains are twice as large as in herbivore species with the same body weight. In the Old and the New World monkeys, brain size is related to diet rather than to taxonomy. In Fig. 23.25a, puzzling data have been presented on the brain weights of Old and New World monkeys. Herbivore monkeys of both groups have rather small brains. Fruit-eating monkeys in the Old and the New World developed larger brains independently of each other. Living on a fruit diet is more difficult than living on herbs. The various types of fruit are only available during separate seasons, and they are to be found over a much wider region. Fruit-eating primates benefit from their better colour vision for the detection of ripe fruits. These factors might explain the larger brains of fruit-eating primates.

3. Group size: Large primates tend to live in large groups. The group size of primates, corrected for body size, is associated with the EQ ($r = 0.46$, Table 23.3). It is easy to imagine that a monkey living in a larger group needs more social intelligence to know how to deal with its group members. Primates living in large groups benefit from improved vision for the recognition of group members at a large distance. Moreover, a larger group needs a larger home range to collect enough food (see below).

4. Home range: Large primates tend to have large home ranges. A confounding factor is present: often one cannot speak about the home range of an individual monkey, but only of its group. Here an analysis is made for the home range of the social unit of primates, irrespective of the group size. The home range of primates, corrected for body size, is associated with the EQ ($r = 0.45$, Table 23.3). For obvious reasons, home range and group size are correlated: it takes a larger group to defend a large territory, and it takes a larger home range to feed a large group. Primates with a large home range also need greater geographical insight. Fruit-eating primates need a larger home range to find a good supply of fruit throughout the year. Diet, home range, group size and brain size are probably causally related through these factors.

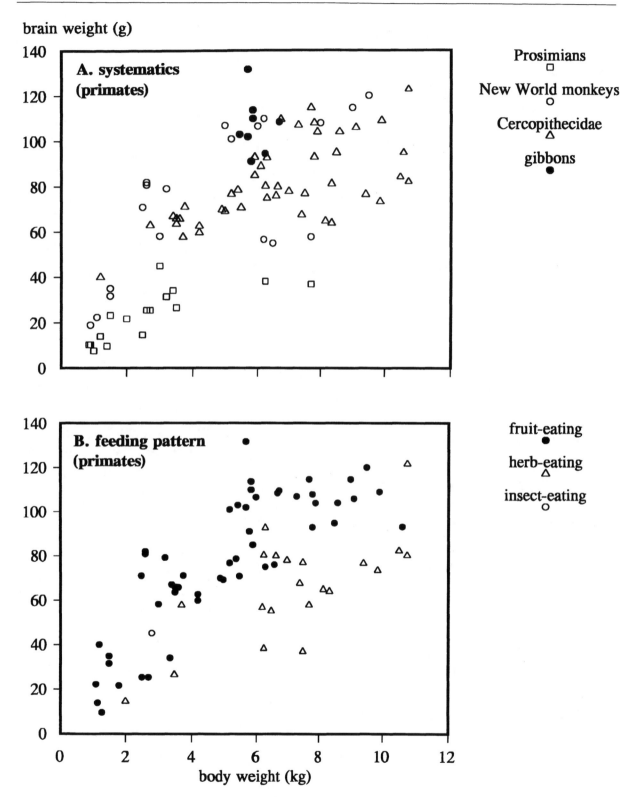

Fig. 23.25. a Linear graph showing the relationship between body weight and brain weight of primates with a body weight between 1 and 12 kg. The actual differences are clearly visible in this linear graph. For primates, prosimians have small brains. Some New World monkeys have rather small brains, others large. The same applies to Old World monkeys. Gibbons have large brains. **b** Linear graph showing the relationship between body weight and brain weight of the same primates. Herb-eating primates have rather small brains, while fruit-eating primates have large brains

5. Basal metabolic rate: For mammals in general, the EQ is not associated with the MEQ, but for primates a positive association between EQ and MEQ is present ($r = 0.40$, Table 23.3). This positive relationship has been noted by Armstrong (1985), but the relevance of this finding is hard to interpret. The basal metabolic rate in primates might be related to feeding pattern. [A positive association between EQ and MEQ has been found in bats (see above), but the hypothetical explanation for that association does not hold for primates.]
6. Conclusion: I hoped to be able to propose a plausible scenario for primate brain evolution after these analyses. However, too many factors seem to be associated with brain size in primates. Independently of each other, in some groups of New and Old World monkeys, K-selection has been active, leading to an increase in longevity, the use of higher quality food (fruit) and an increase in group size and home range. However, which factors prevailed for the increase in brain size, and in which order, is still unclear.

Increase in the Hominid Brain Size. The evolution of the brain of man will be discussed somewhat more extensively. Figure 23.26b shows the brain weights of fossil hominids and man. The brain of *Australopithecus afarensis* is only slightly larger than that of the chimpanzee; therefore it is generally assumed that the common ancestor of man and chimpanzee, who lived some 7 million years ago, probably had a brain of about 350 g. About 3.5 million years ago, *Australopithecus afarensis*, a probable ancestor of man, still had a similar brain weight. However, the pattern of cortical sulci in *Australopithecus afarensis* deviates in some respects from the ape-like pattern (Holloway 1983). In the evolution from *Australopithecus afarensis* to the recent *Homo sapiens*, brain size increased to the present value of about 1400 g, an increase by a factor of 3–4 (Fig. 23.26b). The semi-logarithmic graph of Fig. 23.26b shows that the data points follow a straight line; this implies that the average percentage increase per unit of time was constant over this period of time (but no clear indications are present for or against gradualism or punctuated equilibria, cf. Hofman 1983b; Holloway 1983). Over the last 3.5 million years, the increase in brain weight has been 43% per million years [or 0.43 darwin (d), Haldane 1949]. Over that time, body weight has increased by 26% per million years (0.26 d, Fig. 23.26a). An increase by 0.43 d over a period of 3.5 million years is rather slow (Gingerich 1983). Although the chimpanzee has a much smaller brain than man, the proportion of its neocortex is about the same (72% in the chimpanzee vs. 76% in man). So contrary to popular belief, it was not primarily the neocortex that increased during hominid evolution, but rather a similar increase in brain stem, cerebellum and neocortex took place. Which selection pressures caused this increase or these increases in brain size?

Selection Pressures and the Hominid Brain. Several selection pressures have been suggested as the main cause of the increase (or the increases) in brain size during hominid evolution (e.g. Gibson and Ingold 1993). Most of these are plausible, but at the moment it is not really possible to test these suggestions critically.

1. Motor skills: With respect to motor skills, various differences are present between the chimpanzee and man. (a) Anatomically and behaviourally, man is much better equipped for a *bipedal* life than the chimpanzee. Convincing evidence has been presented that the small-brained *Australopithecus afarensis* walked on two legs (Hay and Leakey 1982; Lovejoy 1988). Apparently, hominids do not need a large brain to walk on two legs. (b) The chimpanzee is far superior to man for a life in the trees, including *arboreal acrobatics*, and consequently in the sensorimotor skills required for it. (c) Kortlandt (1972) suggested that a main selection pressure contributing to hominid brain enlargement was improvement of *throwing accuracy*, which is probably important for a hunter. In the hominid evolution to *Homo sapiens*, some motor skills were lost and others gained. It is unclear why precision throwing (which was gained) would require more brain than arboreal acrobatics (which was lost). Therefore, it is unlikely that changes in motor skills have contributed considerably to the increase in brain size in hominid evolution.
2. Intelligence: Our intelligence is due to our large brain. No one doubts this when man and chimpanzee are compared, but there is no fair way to measure and compare human and chimpanzee intelligence. Human intelligence is usually measured with IQ tests. In older studies on head size and IQ, the correlation coefficients between head size and IQ were small – usually between 0.1 and 0.2 (Van Valen 1974; Passingham 1979). But in recent studies, magnetic resonance imaging has been used to measure brain size directly in vivo: larger correlation coefficients between brain size and IQ have now been found (0.3–0.5, Willerman et al. 1991). Great scientists (such as Einstein or von Helmholtz) stand out by their brain

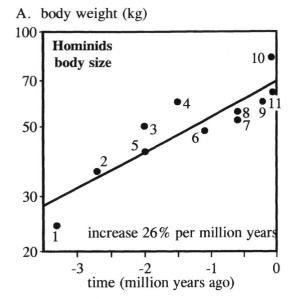
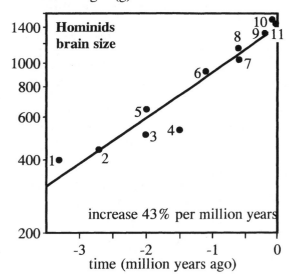

1. Australopithecus afarensis
2. Australopithecus africanus
3. Australopithecus robustus
4. Australopithecus boisei
5. Homo habilis
6. Homo erectus (Java)
7. Homo erectus (Peking)
8. Homo erectus (Solo)
9. Homo sapiens archaicus
10. Homo sapiens neanderthalensis
11. Homo sapiens sapiens

Fig. 23.26a,b. Evolution of body and brain size in the hominids. **a** Over the past 3.5 million years, hominid body size has increased by a factor of 2.5, or 26 % per million years. **b** Over the past 3.5 million years, hominid brain size has increased by a factor of 3.5, or 43 % per million years. (Based on data from Hofman 1983b)

function, but not by their brain size (Hansemann 1899; Cobb 1965; Diamond et al. 1985).

3. Language: The greatest achievement of the human brain is probably speech/language. In the human brain, asymmetries are present between the left and right hemispheres, which are related to speech. Some of these asymmetries have also been found in the brains of apes (Holloway and De la Coste-Lareymondie 1982). However, a vivid discussion is going on about the linguistic capacities of the apes (Terrace et al. 1979; Savage-Rumbaugh et al. 1983; Gardner and Gardner 1989). Nevertheless, a selective pressure in favour of linguistic capacities probably contributed to the emergence of the large human brain.

4. Social learning and culture: Social learning is the basis of culture. Two aspects of social learning are distinguished: learning by imitation and intentional teaching. Learning by imitation has been seen in monkeys, but only anecdotal evidence is offered for intentional teaching by chimpanzees (of offspring by the mother, McGrew 1992). Whether a consequence of social learning or not, regional differences in chimpanzees' behaviour are found, for instance in the use of stone hammers to break open nuts, and ways of grooming (McGrew 1992). When such regional differences are found in human societies, they are invariably called a product of culture. However, culture in human societies is infinitely more elaborate than in chimpanzees. An example of culture which leaves behind archeological traces is tool making. Apes are capable of making simple tools to solve an actual problem (for instance ripping off leafs from a twig to obtain termites), whereas humans sometimes spend much effort to make a tool for future use. Examples of such tools are the durable stone tools. The oldest reliably identified, manufactured stone tools were associated with *Homo habilis*. Probably it is no coincidence that tools are associated with the larger-brained *Homo habilis*. From that time on, brain size increased, as did the quality of the tools. The making of tools should not be discussed in isolation. The production of stone tools in present cultures living 'in the Stone Age' is characterised by (a) the emergence of a group of specialised tool makers (labour specialisation), (b) a tutor system to transmit these skills

and (c) a language enabling the transfer of knowledge (Toth et al. 1992). Remarkably, the brain sizes of the Neanderthalers and of modern man (*Homo sapiens sapiens*) do not differ. Yet, the Neanderthaler artefacts (Mousterièn) are stable and characterise a relatively conservative culture. Modern man, on the other hand, is continuously improving his tools from stones to computers. Modern man is really the Great Innovator.

On the basis of the material available, it is not possible to critically test the various hypotheses which selection pressure mainly contributed to the emergence of the large human brain. I would expect solutions to come from a different field of investigation. When the genes have been identified that distinguish human from chimpanzee brains, the products of these genes can be specified: molecular products, as well as their effects on brain size and structure. Such a scientific program is not simple, but its development is now in progress. It will take many years before we have some idea about how natural selection could eventually produce the brains of Kant, Einstein, Mozart, Michelangelo, Confucius or Buddha.

23.7.5.5
Ferungulata

The carnivores and ungulates are often considered sister groups. The basic pattern of cortical folding in these groups is similar (see also Chap. 22).

23.7.5.5.1
Carnivores

Most carnivores have brain weights above average for mammals of their body size.

1. Taxonomic relationships: Among the carnivores, the bears and canines have the largest brains, while the civets (Viverridae) and hyenas have proportionally the smallest brain weights (Gittleman 1986). However, these differences are rather small: the brain weights of civets and hyenas are about 70% of those of canines of comparable body weight. These differences in brain weight are hard to appreciate, since the behaviour of canines and hyenas does not seem to be too different.
2. Ecological relationships: Contrary to what their name suggests, several carnivores are not primarily meat eating. The bears are mainly fruit and leaf eating, and in various families of carnivores species eating mainly insects are found. A relationship between brain size and feeding pattern in carnivores is less clear than in bats, whales and primates, but it is present. Insect-eating carnivores have somewhat smaller brains than meat-, fruit/leaf-eating or omnivore carnivores (Gittleman 1986). Catching insects is more easy than catching mammalian prey; this could explain the smaller brains in insect-eating carnivores. In bats as well as carnivores, the insect-eating species are small brained, but in bats this is explained by a selection pressure to preserve food energy.
3. Parental care: In some carnivores, the offspring is raised by the mother alone (for instance the domestic cat), in others by both parents (for instance foxes) and in others by a larger social community (for instance wolves). Female carnivores that raise their offspring alone have somewhat larger brains than females that share parental care with their partner or other co-specifics (Gittleman 1994). Remarkably, this small difference is only present in females; carnivore male brain size does not depend on the system of parental care.

23.7.5.5.2
Tubulidentata

For a long time, the aardvark (*Orycteropus*) was considered an edentate. Elliot Smith (1898) noted large differences between the aardvark's brain and those of the edentates. The aardvark is now placed in a separate order (Tubulidentata), related to the ungulates. The aardvark has a small brain (Fig. 23.19).

23.7.5.5.3
Artiodactyla

The artiodactyles have brain weights close to average for mammals of their body weight. Corrected for body weight, small brains are found in pigs (Suidae) and the hippopotamus (but not in the pygmy hippopotamus, Fig. 23.16). Proportionally large brains are found in deer (Cervidae).

23.7.5.5.4
Cetaceans: Dolphins and Whales

The whales are now considered to be closely related to a group of artiodactyles, the Bovidae. Two main groups of whales are distinguished. The *baleen whales* are very large animals with body weights above 5 metric tons. They feed on plankton. Most *toothed whales* are smaller, with body weights between 50 and 2000 kg; only the adult sperm whale is larger than 30 metric tons. Most toothed whales eat fish or sea mammals.

Toothed whales (except the sperm whale) have brain weights a factor of 2–5 above those of average mammals of their body size (Fig. 23.16). Their brains are three to five times heavier than those of seals and sea cows with similar body weights, so the large brains of whales are not simply due to their aquatic habitat. The brain weights of baleen whales on the other hand lie below the regression line for mammals. Must we then regard their brains with weights between 2 and 10 kg as small brains? I am not really inclined to do so.

Dolphins are the only animals with body as well as brain weights in the range of man (Fig. 23.16). The 91-kg white-beaked dolphin (*Lagenorhynchus albirostris*) has an average brain weight of 1.15 kg, and the 234-kg Risso's dolphin (*Grampus griseus*) a brain weight of 2.27 kg (Pilleri and Gihr 1970). As early as 15–20 million years ago (Miocene), toothed whales existed with large brains; the dolphin-like *Argyrocetus* of about 72 kg with a brain of 650 g, and the sperm-whale-like *Aulophyseter* with a presumed body weight of 1100 kg and a brain of about 2.5 kg (Jerison 1973). So for 15 million years, the whales were the mammals with the largest EQs, and they still have the largest brains ever during the history of life. At the moment, we can only guess what selection pressures gave the Miocenic whales such large brains.

Toothed and baleen whales are often regarded as intelligent animals. Both groups use echolocation. But we have already seen in bats that excellent echolocation can be achieved by a very small brain. Whales make large migrations, which requires an elaborate guiding system. However, the homing specialists among birds, the pigeons, have small brains, and also migratory birds can accomplish long journeys with a small brain. The social organisation of groups of whales can be rather complex: conspecific helping and formation of alliances have been described (Caldwell and Caldwell 1966; Connor et al. 1992).

Aquatic mammals tend to develop large bodies (Sect. 23.7.2.5). Most fresh waters are not large enough for really big animals, but the oceans are. In toothed whales several selection pressures work together in favour of a large body, but they cannot become too big, since they must be able to make quick manoeuvres to catch their prey. This constraint does not apply to plankton-eating baleen whales. (This applies generally: also in sharks and rays, the plankton-eating species are huge: the whale shark and the manta ray). Therefore, it is suggested that especially in the evolution of baleen whales a strong net selection pressure is working in favour of a large body. Probably, no selection pressure was then working to further increase their large brains. Baleen whales have large bodies rather than small brains.

23.7.5.5.5
Perissodactyla

Many good fossils are available of the ancestors of horses. During horse evolution, the sizes of body and brain gradually increased to their present size. The brain of the Eocean ancestor of horses, *Hyracotherium (Eohippus)*, had a size in the lowest range for a mammal of its body size. The present horses are much larger and have brains slightly smaller than the average mammal of their body size: not only the body and brain size has increased, but also the EQ of the horses. Other members of the Perissodactyla, the rhinoceroses and tapirs, have small brains: about half of the weight of that of the average mammal (Fig. 23.16).

23.7.5.5.6
Hyracoidea

The hyraxes have brain weights around the average for mammals of their body size.

23.7.5.5.7
Elephants

Elephants have large brains: the Indian elephant about 5 kg and the African elephant about 5.7 kg. The elephants' brains are in the same order of magnitude of those of whales with similar body sizes, but they are six to ten times larger than those of other large terrestrial mammals, the hippopotamus or rhinoceroses (Fig. 23.16).

23.7.5.5.8
Sirenia

Sea cows have rather small brains (Fig. 23.16), somewhat below the mammalian average, and about a tenth of the size of the brains of dolphins with similar body weights.

23.8
Concluding Remarks

A comparison of brain size in various groups of vertebrates does not lead to a general summarising conclusion, except that brain size in various groups is the product of different selection pressures. We are only beginning to understand why some animals have comparatively large brains. Writing and understanding a detailed history of the evolution of brain size and ecology will still take many years. This chapter presents many facts about brain size and ecology that need to be explained in future analyses.

Appendix

In most discussions on brain size, allometric regression lines play a central role. But for instance in the discussion of the EQ of dogs (Sect. 23.1.3), the question was raised whether the dog data should be related to the regression line of dogs or of mammals. In my opinion, this problem is insoluble. In some instances, however, the problems of choosing the 'right' regression line are not relevant. Suppose one wants to investigate the relationship in mammals between brain size and the basal metabolic rate. Brain size and basal metabolic rate are allometrically scaled with body weight. Do then mammals with a large brain for their body weight also have a large metabolic rate for their body size? A solution then is to construct a database of species with data on body weight and brain weight as well as basal metabolic rate. The regression line between body weight and brain weight can be calculated. For each species in this database, the encephalisation quotient (EQ) is then calculated in the conventional way. Similarly, the regression line between body weight and metabolic rate is calculated. A new statistical parameter is calculated, the 'metabolic quotient' (MEQ), by exactly the same arithmetic as for the EQ. MEQ and EQ are calculated based on the regression lines for the same mammals (thereby escaping the problem of the choice of the correct regression line). The correlation coefficient between EQs and MEQs measures the strength of the association between brain weight (corrected for the body weight) and metabolic rate (corrected for the body weight). This applies not only for the basal metabolic rate, but also for other parameters that depend on body weight, such as longevity, home range and group size. The results of such calculations are presented in Table 23.3 and are mentioned in the text.

Acknowledgements. The extensive and constructive remarks by Harry Jerrison and Michel Hofman on a draft of this chapter are greatly appreciated.

References

Allman J, McLaughlin T, Hakeem A (1993) Brain weight and life-span in primate species. Proc Natl Acad Sci [USA] 90:118–122

Armstrong E (1983) Relative brain size and metabolism in mammals. Science 220:1302–1304

Armstrong E (1985) Relative brain size in monkeys and prosimians. Am J Phys Anthropol 66:263–273

Armstrong E, Bergeron R (1985) Relative brain size and metabolism in birds. Brain Behav Evol 26:141–143

Baron G, Jolicoeur P (1980) Brain structure in Chiroptera: some multivariate trends. Evolution 34:386–393

Bartels H (1982) Metabolic rate of mammals equals the 0.75 power of their body weight? Exp Biol Med 7:1–11

Barton RA, Purvis A, Harvey PH (1995) Evolutionary radiation of visual and olfactory brain systems in primates, bats and insectivores. Philos Trans R Soc Lond Biol 348:381–392

Bauchot R, Stephan H (1966) Données nouvelles sur l'encéphalisation des insectivores et des prosimiens. Mammalia 30:160–196

Bauchot R, Platel R, Ridet JM (1976) Brain-body weight relationships in Selachii. Copeia 2:3–75

Bauchot R, Thireau M, Diagne M (1983) Relations pondérales encéphalo-somatiques interspécifiques chez les amphibiens Anoures. Bull Mus Natl Hist Nat Paris 4:383–398

Bauchot ML, Ridet JM, Diagne M, Bauchot R (1989a) Encephalization in Gobioidei (Teleostei). Jpn J Ichthyol 36:63–74

Bauchot R, Randall JE, Ridet JM, Bauchot ML (1989b) Encephalisation in tropical Teleost fishes and comparison with their mode of life. J Hirnforsch 30:645–669

Bauchot R, Ridet JM, Bauchot ML (1989c) The brain organization of butterflyfishes. Environm Biol Fishes 25:205–219

Bennett PM, Harvey PH. (1985a) Relative brain size and ecology in birds. J Zool [London] 207:151–169

Bennett PM, Harvey PH (1985b) Brain size, development and metabolism in birds and mammals. J Zool [London] 207:491–509

Bhatnagar HP, Frahm HD, Stephan H (1990) The megachiropteran pineal organ: a comparative mophological and volumetric investigation with special emphasis on the remarkably large pineal of Dobsonia praedatrix. J Anat 168:143–166

Bronson RT (1979) Brain weight-body weight scaling in breeds of dogs and cats. Brain Behav Evol 16:227–236

Brummelkamp R (1939) Das Wachstum der Gehirnmasse mit kleine Cephalisierungssprünge (sog. V2-Sprüngen) bei den Rodentiern. Acta Neerl Morphol 2:188–194

Caldwell DK, Caldwell M (1966) Epimeletic (care-giving) behavior in Cetacea. In: Norris KS (ed) Whales, dolphins and porpoises. University of California Press, Berkeley, pp 677–717

Clutton-Brock TH, Harvey PH (1980) Primates, brains and ecology. J Zool 190:309–323

Cobb S (1965) Brain size. Arch Neurol 12:555–561

Connor RC, Smolker RA, Richards AF (1992) Two levels of alliance formation among male bottlenose dolphins (Tursiops sp.). Proc Natl Acad Sci [USA] 89:987–990

Count FW (1947) Brain and body weight in man: their antecedents in growth and evolution. Ann NY Acad Sci 46:993–1122

Crandall LS (1964) The management of wild mammals in captivity. University of Chicago Press, Chicago

Crile G, Quiring DP (1940) A record of the body weight and certain organ and gland weights of 3690 animals. Ohio J Sci 40:219–259

Cuvier G (1805) Leçons d'anatomie comparée. Tome III:77–81

Diamond MC, Scheibel AB, Murphy GMJ, Harvey T (1985) On the brain of a scientist: Albert Einstein. Exp Neurol 88:198–204

Dobson GP, Headrick JP (1995) Bioenergetic scaling: metabolic design and body-size constraints in mammals. Proc Natl Acad Sci [USA] 92:7317–7321

Dubois E (1897) Sur le rapport du poids de l'encéphale avec la grandeur du corps chez mammifères. Bull Soc Anthropol Paris 8:337–376

Dunbar R (1992) Neocortex size as a constraint on group size in primates. J Hum Evol 20:469–493

Ebinger P, Wächtler K, Stähler S (1983) Allometrical studies in the brain of Cyclostomes. J Hirnforsch 24:545–550
Eisenberg JF (1981) The mammalian radiations An analysis of trends in evolution, adaptation and behavior. University of Chicago Press, Chicago
Eisenberg JF, Wilson D (1978) Relative brain size and feeding strategies in the Chiroptera. Evolution 32:740–751
Eisenberg JF, Wilson DE (1981) Relative brain size and demographic stategies in Didelphid marsupials. Am Naturalist 118:1–15
Elias H, Schwartz D (1969) Surface areas of the cerebral cortex of mammals determined by stereological methods. Science 166:111–113
Elliot Smith G (1898) The brain in the Edentata. Trans Linnean Soc London Ser 2 (Zoology) 7:277–394
Frahm HD, Stephan H, Stephan M (1982) Comparison of brain structure volumes in Insectivora and Primates. I. Neocortex. J Hirnforsch 23:375–389
Gardner BT, Gardner RA (1989) Chimp-language wars. Science 252:1046
Gibson KR, Ingold T (1993) Tools, language and cognition in human evolution. Cambridge University Press, Cambridge
Gihr M, Pilleri G (1969) Hirn-Körpergewichts-Beziehungen bei Cetaceen. Invest Cetacea 1:109–126
Gingerich PD (1983) Rates of evolution: effect of time and temporal scaling. Science 2222:159–161
Gittleman JL (1986) Carnivore brain size, behavioral ecology and phylogeny. J Mammalogy 67:23–26
Gittleman JL (1994) Female brain size and parental care in carnivores. Proc Natl Acad Sci [USA] 91:5495–5497
Goodman M, Tagle DA, Fitch DHA et al. (1990) Primate evolution at the DNA level and a classification of Hominoids. J Mol Evol 30:260–266
Graur D, Hide WA, Zharkikh A, Li WH (1992) The biochemical phylogeny of guinea-pigs and gundis, and the paraphyly of the order rodentia. Comp Biochem Physiol B Comp Biochem 101:495–498
Haarmann K (1975) Morphologische und histologische Untersuchungen am Neocortex von Boviden (Antilopinae, Cephalopinae) und Traguliden mit bemerkungen zur Evolutionshöhe. J Hirnforsch 16:93–116
Haarmann K, Oboussier H. (1972) Morphologische und quantitative Neocortexuntersuchungen bei Boviden, ein Beitrag zur Phylogenie dieser Familie. II. Formen geringen Körpergewichts (3 kg–25 kg) aus den Subfamilien Cephalophinae und Antilopinae. Mitt Hamburg Zool Mus Inst 68:231–269
Hafner MS, Hafner JC (1984) Brain size, adaptation and heterochrony in Geomyoid rodents. Evolution 38:1088–1098
Haldane JBS (1949) Suggestion as to a quantitative measurement of rates of evolution. Evolution 3:51–56
Hansemann D (1899) Über das Gehirn von Hermann v. Helmholtz. Z Psychol Physiol Sinnesorg 20:1–15
Harvey PH, Bennett PM (1983) Brain size, energetics, ecology and life history patterns. Nature 306:314–315
Harvey PH, Clutton-Brock TH (1985) Life history variation in primates. Evolution 39:559–581
Harvey PH, Krebs JR (1990) Comparing brains. Science 249:140–145
Haug H (1987) Brain sizes, surfaces and neuronal sizes of the cortex cerebri. A stereological investigation of man and his variability and a comparison with some mammals (Primates, whales, Marsupialia, Insectivores and one elephant). Am J Anat 180:126–142
Hay RL, Leakey MD (1982) The fossil footprints of Laetoli. Sci Am 246(2):38–45
Hayssen V, Lacy RC (1985) Basal metabolic rates in mammals: taxonomic differences in the allometry of BMR and body mass. Comp Biochem Physiol 81A:741–754
Hofman MA (1982) Encephalization in mammals in relation to the size of the cerebral cortex. Brain Behav Evol 20:84–96
Hofman MA (1983a) Energy metabolism, brain size and longevity in mammals. Q Rev Biol 58:495–512
Hofman MA (1983b) Encephalization in Hominids: evidence for the model of punctuationalism. Brain Behav Evol 22:102–117
Hofman MA (1989) On the evolution and geometry of the brain in mammals. Progr Neurobiol 32:137–158
Hofman MA (1993) Encephalization and the evolution of longevity in mammals. J Evol Biol 6:209–227
Holloway RL (1983) Human brain evolution: a search for units, models and synthesis. Can J Anthropol 3:215–230
Holloway RL, LaCoste-Lareymondie MC (1982) Brain endocast assymetry in Pongids and Hominids: some preliminary findings on the paleontology of cerebral dominance. Am J Phys Anthropol 58:108–110
Hopson JA (1979) Paleoneurology. In: Gans C, Northcurr RG, Ulinski P (eds) Biology of the reptilia, volume 9. Academic Press, London, pp 39–146
Jerison HJ (1969) Brain evolution and dinosaur brains. Am Naturalist 103:575–588
Jerison HJ (1973) Evolution of the brain and intelligence. Academic Press, New York
Jerison HJ (1988) The evolutionary biology of intelligence: afterthoughts. In: Jerison HJ, Jerison I (eds) Intelligence and evolutionary biology. Springer-Verlag, Berlin Heidelberg New York, pp 447–466
Jerison HJ (1994) Evolution of the brain. In: Zaidel D (ed) Neuropsychology. Academic Press, London, pp 53–82
Jones ML (1979) Longevity of mammals in captivity. Int Zoo News, Apr/May:16–26
Jürgens KD, Prothero JW (1987) Scaling of maximal lifespan in bats. Comp Biochem Physiol 88A:361–367
Kleiber M (1947) Body size and metabolic rate. Physiol Rev 27:511–541
Kojima T (1951) On the brain of the sperm whale (Physeter catadon). Sci Rep Whales Res Inst Tokyo 6:49–72
Kortlandt A (1972) New perspectives on ape and human evolution. Stichting voor Psychobiologie, Amsterdam
Kraus G, Pilleri G (1969) Quantitative Untersuchungen über die Grosshirnrinde der Cetaceen. Inv Cetacea 1:127–150
Kretschmann HJ (1966) Über die Cerebralisation eines Nestflüchters (Acomys cahirinus dimidiatus Cretschmar 1826) im Vergleich mit Nesthockern (Albinomaus, Apodemus sylvaticus Linaeus, 1758 und Albinoratte). Gegenbaurs Morphol Jahrb 109:376–410
Kruska D (1973) Cerebralisation, Hirnevolution und domestikationbedingte Hirngrössenänderungen innerhalb der Ordnung Perissodactyla Owen, 1848 und ein Vergleich mit der Ordnung Artiodactyla Owen, 1848. Z Zool Syst Evol Forsch 11:81–103
Kruska D (1982) Hirngrösseänderungen bei Tylopoden während der Stammesgeschichte und in der Domestikation. Verh Dtsch Zool Ges 75:173–183
Kruska D (1988) Mammalian domestication and its effects on brain structure and behavior. In: Jerison HJ, Jerison I (eds) Intelligence and evolutionary biology. Springer-Verlag, Berlin Heidelberg New York, pp 211–250
Kruska D, Röhrs M (1974) Comparative-quantitative investigations on brains of feral pigs from the Galapagos Islands and of European domestic pigs. Z Anat Entw Gesch 144:61–73
Lapicque L, Girard P (1905) Poids de l'encéphale en fontion du poids du corps chez les oiseaux. C R Soc Biol 57:665–668
Leigh SR (1992) Cranial capacity evolution in Homo erectus and early Homo sapiens. Am J Phys Anthropol 87:1–13
Lemen C (1980) Relationship between relative brain size and climbing ability in Peromyscus. J Mammalogy 61:360–364
Lindstedt SL, Boyce MS. (1985) Seasonality, fasting endurance, and body size in mammals. Am Naturalist 125:873–878
Lovejoy CO (1988) Evolution of human walking. Sci Am 259(5):82–89
MacArthur RH, Wilson EO (1967) The theory of island biogeography. Princeton University Press
Mace GM, Harvey PH, Clutton-Brock TH (1980) Is brain size an ecological variable? Trends Neurosci 3:193–196

Mace GM, Clutton-Brock TH, Harvey PH (1981) Brain size and ecology in small mammals. J Zool 193:333–354

Mangold-Wirz K (1966) Cerebralisation und Ontogenesemodus bei Eutherien. Acta Anat 63:449–508

Martin RD (1981) Relative brain size and basal metabolic rate in terrestrial vertebrates. Nature 293:57–60

McGrew WC (1992) Chimpanzee material culture: implications for human evolution. Cambridge University Press, Cambridge

McNab BK (1969) The economics of temperature regulation in neotropical bats. Comp Biochem Physiol 31:227–268

McNab BK (1988) Complications inherent in scaling the basal rate of metabolism in mammals. Q Rev Biol 63:25–54

Meddis R (1983) The evolution of sleep. In: Mayes A (ed) Sleep mechanisms and functions. Van Nostrand Reinhold, London, pp 57–106

Möller H (1973) Zur Evolutionshöhe des Marsupialiagehirns. Zool Jahrb Anat 91:434–448

Myagkov NA (1991) The brain sizes of living Elasmobranchii as their organization level indicator. J Hirnforsch 32:553–561

Northcutt RG (1989) Brain variation and phylogenetic trends in elasmobranch fishes. J Exp Zool Suppl 2:83–100

Novacek MJ (1992) Mammalian phylogeny: shaking the tree. Nature 356:121–125

Oboussier H (1966) Das Grosshirnfurchenbild als Merkmal der Evolution. Untersuchungen an Boviden II. Mitt Hamburg Zool Mus Inst 63:159–182

Oboussier H (1967) Das Grosshirnfurchenbild als Hinweis auf die Verwandtschaftbeziehungen der heutigen Afrikanischen Bovidae. Acta Anat 68:577–596

Oboussier H (1971) Quantitative und morphologische Studien am Hirn der Bovidae, ein Beitrag zur Kenntnis der Phylogenie. Gegenbaurs Morphol Jahrb 117:162–168

Oboussier H (1972) Morphologische und quantitative Neocortexuntersuchungen bei Boviden, ein Beitrag zur Phylogenie dieser Familie. III. Formen grossen Körpergewicht (über 75 kg). Mitt Hamburg Zool Mus Inst 68:271–292

Oboussier H, Möller G (1971) Zur Kenntnes des Gehirns der Giraffidae (Pecora, Artiodactyla, Mammalia) - ein Vergleich der Neocortex-Oberflächegrösse. Z Säugetierknd 36:291–296

Passingham RE (1979) Brain size and intelligence in man. Brain Behav Evol 16:253–270

Pianka ER (1970) On r- and K-selection. Am Naturalist 104:592–597

Pilleri G (1959a) Beiträge zur vergleichenden Morphologie des Nagetiergehirnes. 1. Sciuromorpha. Acta Anat 38 (Suppl):1–42

Pilleri G (1959b) Beiträge zur vergleichenden Morphologie des Nagetiergehirnes. 2. Hystricomorpha. Acta Anat 38 (Suppl):43–95

Pilleri G (1959c) Beiträge zur vergleichenden Morphologie des Nagetiergehirnes. 3. Das Gehirn der Wassernager (Castor canadensis, Ondatra zibethica, Myocastor coypus). Acta Anat 38 (Suppl):96–123

Pilleri G (1960a) Beiträge zur vergleichenden Morphologie des Nagetiergehirnes. 4. Zentralnervensystem, Körperorgane und stammesgeschichtliche Verwandtschaft der Aplodontia rufa Rafinesque (Rodentia, Aplodontoidea). Acta Anat 40 (Suppl):5–35

Pilleri G (1960b) Beiträge zur vergleichenden Morphologie des Nagetiergehirnes. 5. Vergleichend-morphologische Untersuchungen über das zentralnervensystem nearktischer Sciuromorpha und Bemerkungen zum Problem Hirnform und Taxonomie. Acta Anat 40 (Suppl):36–68

Pilleri G (1960c) Beiträge zur vergleichenden Morphologie des Nagetiergehirnes. 6. Materialien zur vergleichenden Anatomie des Gehirns der Myomorpha. Acta Anat 40 (Suppl):69–88

Pilleri G, Gihr M (1970) The central nervous system of the Mysticete and Odontocete whales. Invest Cetacea 2:890–128

Pirlot P (1980) Quantitative composition and histological features of the brain in two South American edentates. J Hirnforsch 21:1–9

Pirlot P (1981) A quantitative approach to the marsupial brain in an eco-ethological perspective. Rev Can Biol 40:229–250

Pirlot P (1987) Contemporary brain morphology in ecological and ethological perspectives. J Hirnforsch 28:145–211

Pirlot P, Jolicoeur P (1982) Correlations between major brain regions in Chiroptera. Brain Behav Evol 20:172–181

Pirlot P, Kamiya T (1982) Relative size of brain and brain components in three gliding placentals (Dermoptera: Rodentia). Can J Zool 60:565–572

Pirlot P, Kamiya T (1983) Quantitative brain organisation in anteaters (Edentata-Tubilidentata). J Hirnforsch 24:677–689

Pirlot P, Kamiya T (1985) Qualitative and quantitative brain morphology in the sirenian Dujong dujon. Z Zool Syst Evol Forsch 23:147–155

Pirlot P, Nelson J (1978) Volumetric analysis of Monotreme brains. Austr Zool 20:171–179

Pirlot P, Stephan H (1970) Encephalization in Chiroptera. Can J Zool 48:433–444

Platel R (1976) Analyse volumétrique comparée des principales subdivisions encéphaliques ches les Reptiles Sauriens. J Hirnforsch 17:513–537

Platel R (1979) Brain weight-body weight relationships. In: Gans C, Northcurr RG, Ulinski P (eds) Biology of the reptilia, volume 9. Academic Press, London, pp 147–171

Platel R (1989) L'Encéphalisation chez le Tuatara de nouvelle-Zélande Sphenodon punctatus Gray (Lepidosauria, Sphenodonta). Étude quantifiée des principales subdivisions encéphaliques. J Hirnforsch 30:325–337

Platel R, Delfini C (1981) L'encéphalisation ches la myxine (Myxine glutinosa L). Analyse quantifiée des principales subdivisions encéahliques. Cah Biol Marine 22:407–430

Platel R, Delfini C (1986) L'Encéphalisation chez la Lamproie marine, Petromyzon marinus (L.). Analyse quantifiée des principales subdivisions encéphaliques. J Hirnforsch 27:279–293

Platel R, Vesselkin NP (1986) Analyse des allometries encéphalo-somatiques chez l'adulte de Lampetra fluviatilis. Cybium 10:143–153

Platel R, Vesselkin NP (1989) Etude comparé de l'encéphalisation chez 3 espèces de Pétromyzonidae (Agnatha): Petromyzon marinus, Lampetra fluviatilis et Lampetra planeri. J Hirnforsch 30:23–32

Platel R, Ridet JM, Bauchot R, Diagne M (1977) L'organisation encéphalique chez Amia, Lepisosteus et Polypterus: Morphologie et analyse quantitative compareés. J Hirnforsch 18:69–73

Pohlenz-Kleffner W (1969) Vergleichende Untersuchungen zur Evolution der Gehirne von Edentaten. Hirnform und Hirnfurchen. Z Zool Syst Evol Forsch 7:181–208

Portmann A (1947) Étude sur la cérébralisation chez les oiseaux. II. Les indices intracérébraux. Alauda 15:1–15

Prothero JW, Jürgens KD (1987) Scaling of maximal lifespan in mammals. In: Woodhead A, Thompson KH (eds) Evolution of longevity in animals. Plenum, New York, pp 49–74

Radinsky L (1981) Brain evolution in extinct South American ungulates. Brain Behav Evol 18:169–187

Reep RL, O'Shea TJ (1990) Regional brain morphometry and lissencephaly in the Sirenia. Brain Behav Evol 35:185–194

Reep RL, Johnson JI, Switzer RC, Welker WI (1989) Manatee cerebral cortex: cytoarchitecture of the frontal region in Trichechus manatus latirostris. Brain Behav Evol 34:365–386

Rehkämper GK, Frahm HD, Zilles J (1991a) Quantitative development of brain and brain structures in birds (Galliformes and Passeriformes) compared to that in mammals (Insectivores and Primates). Brain Behav Evol 37:125–143

Rehkämper G, Schuchmann KL, Schleicher A, Zilles K (1991b) Encephalization in hummingbirds (Trochilidae). Brain Behav Evol 37:85–91

Ridet JM (1982) Analyse quantitative de l'encéphale des Téléostéens: Charactères évolutifs et adaptifs de l'encéphalisation. Thesis, University of Paris

Ridet JM, Bauchot R (1990a) Analyse quantitative de l'encéphale chez les Téléostéens caractères évolutifs et adaptifs de l'encéphalisation. I. Généralités et analyse globale. J Hirnforsch 31:51-63

Ridet JM, Bauchot R (1990b) Analyse quantitative de l'encéphale des Téléostéens caractères évolutifs et adaptifs de l'encéphalisation. II. Les grandes subdivisions encéphaliques. J Hirnforsch 31:433-458

Ridet JM, Bauchot R, Diagne M, Platel R (1977) Croissance ontogénétique et phylogénétique de l'encéphale des Téléostéens. Cah Biol Mar 18:163-176

Robin ED (1973) The evolutionary advantages of being stupid. Perspect Biol Med 16:369-380

Röhrs M (1966) Vergleichende Untersuchungen zur Evolution der Gehirne von Edentaten. I. Hirngewicht-Körpergewicht. Z Zool Syst Evol Forsch 4:196-207

Röhrs M, Kruska D (1969) Des Einfluss der Domestikation auf das Zentralnervensystem und Verhalten von Schweinen. Dtsch Tierarzt Wochenschr 75:514-518

Ronnefeld U (1970) Morphologische und quantitative Neocortexuntersuchungen bei Boviden, ein Beitrag zur Phylogenie dieser Familie. I. Formen mittlerer Körpergewicht (25 kg-75 kg). Gegenbaurs Morphol Jahrb 115:163-230

Roth G, Blanke J, Ohle M (1995) Brain size and morphology in miniaturized plethodontid salamanders. Brain Behav Evol 45:84-95

Russell DA (1972) Ostrich dinosaurs from the Late Cretaceous of Western Canada. Can J Earth Sci 9:375-402

Sacher GA (1975) Maturation and longevity in relation to cranial capacity in hominid evolution. In: Tuttle RH (ed) Primate functional morphology and evolution. Mouton, The Hague, pp 417-441

Savage-Rumbaugh ES, Pate JL, Lawson J, Smith ST, Rosenbaum S (1983) Can a chimpanzee make a statement? J Exp Psychol Gen 112:457-492

Schultz W (1969) Zur Kenntnis des Hallstromhundes (Canis hallstromi Thoughton, 1957). Zool Anz 183:47-72

Sheppey K, Bernard RTF (1984) Relative brain size in the mammalian carnivores of the Cape province of South Africa. S Afr J Zool 19:305-308

Sibley CG, Ahlquist JE (1990) Phylogeny and classification of birds. A study in molecular evolution. Yale University Press, New Haven

Simmons JA, Stein RA (1980) Acoustic imaging in bat sonar: echolocation signals and the evolution of echolocation. J Comp Physiol 135A:61-84

Smuts BB, Cheney DL, Seyfarth RM, Wrangham RW, Struhsaker TT (eds) (1987) Primate societies. University of Chicago Press, Chicago

Snell O (1891) Die Abhängigkeit des Hirngewichtes von dem Körpergewicht und den geistigen Fähigkeiten. Arch Psychiat Nervenkrankh 23:436-446

Stephan H, Pirlot P (1970) Volumetric comparisons of brain structures in bats. Z Zool Syst Evol Forsch 8:200-236

Stephan H, Pirlot P, Schneider R (1974) Volumetric analysis of pteropid brains. Acta Anat 87:161-192

Stephan H, Frahm H, Bauchot R (1977) Vergleichende Untersuchungen an den Gehirnen madagassischer Halbaffen I. Encephalisation und makromorphologie. J Hirnforsch 18:115-147

Stephan H, Frahm H, Baron G (1981a) New and revisited data on volumes of brain structures in insectivores and primates. Folia Primatol 35:1-29

Stephan H, Nelson J, Frahm HD (1981b) Brain size comparisons in Chiroptera. Z Zool Syst Evol Forsch 19:195-222

Stephan H, Baron G, Frahm HD (1990) Comparative brain research in mammals, vol 1. Insectivora. With a stereotaxic atlas of the hedgehog brain. Springer-Verlag, Berlin Heidelberg New York

Taylor GM, Nol E, Boire D (1995) Brain regions and encephalization in anurans: adaptation or stability? Brain Behav Evol 45:96-109

Terrace HS, Pettito LA, Sanders RJ, Bever TG. (1979) Can an ape create a sentence? Science 206:891-902

Thiede U (1973) Zur Evolution von Hirneigenschaften mitteleuropäischer und südamerikanischer Musteliden. II. Quantitativen Untersuchungen an Gehirne südamerikanischer Musteliden. Z Säugetierk 38:208-215

Thireau M (1975) L'allometrie pondérale encéphalosomatique chez les Urodèles. I. Relations intraspécifiques. Bull Mus Natl Hist Nat 279, Zool 207:483-502

Toth N, Clark D, Ligabue G (1992) The last stone axe makers. Sci Am 266:66-71

Van Valen L (1974) Brain size and intelligence in man. Am J Phys Anthropol 40:417-424

Von Bonin G (1937) Brain weight and body weight of mammals. J Gen Psychol 16:379-389

Weber M (1891) Beiträge zur Entwicklung und Anatomie des Genus Manis. Ergebnisse einer Reise nach Niederländisch Ostindien, vol II, Brill, Leiden

Weber M (1896) Vorstudien über das Hirngewicht der Säugetiere. Festschr C Gegenbaur 3:102-123

Willerman L, Schultz R, Rutledge JN, Bigler ED (1991) In vivo brain size and intelligence. Intelligence 15:223-228

Wilson EO (1975) Sociobiology: the new synthesis. Belknap, Cambridge, Mass

Worthy GAJ, Hickie JP (1986) Relative brain size in marine mammals. Am Naturalist 128:445-459

Zepelin H, Rechtschaffen A (1974) Mammalian sleep, longevity and energy metabolism. Brain Behav Evol 10:425-470

CHAPTER 24

The Meaning of It All

R. Nieuwenhuys, H. J. ten Donkelaar, and C. Nicholson

24.1 Introductory Note 2135
24.2 Macromorphology and General Relationships 2135
24.3 Neuroembryology 2164
24.4 Cell Masses and Fibre Connections 2165
24.5 Macrocircuity 2167
24.6 Microcircuity 2180
24.6.1 Microcircuits at the Interface –
       Retina and Spinal Cord 2182
24.6.2 Central Microcircuits 2184
24.6.3 Microcircuits in a Wider Context 2187
24.7 Conclusion 2189
       References 2191

24.1
Introductory Note

The preceding chapters of this work have presented a survey of our present knowledge of the structural organisation of the CNS of the various groups of vertebrates. This last chapter will highlight some of the major features revealed by our survey and offer some final comments.

24.2
Macromorphology and General Relationships

Figure 24.1 shows the entire CNS, that is to say, brain *and* spinal cord, of nine different vertebrates. It illustrates the remarkable feature that in almost all non-mammalian vertebrates the spinal cord represents the largest part of the CNS.

Of all parts of the CNS, *the spinal cord* preserves the early embryonic tubelike shape most clearly and, although secondary form changes occur in some groups, this organ is generally cylindrical. The length of the spinal cord varies considerably. In most groups – the various classes of fish, the tailed amphibians, the reptiles and birds – the cord extends throughout the entire vertebral canal, but in some – the tailless amphibians, the mammals – it occupies only a part of this space. In most mammals, the end of the spinal cord is found in the lumbar region of the vertebral column, but in a few species of this class such as the echidna *Tachyglossus aculeatus* and the seal *Phoca vitulina*, the conus medullaris is situated as high as the thoracic level. The most shortened spinal cords are not found among the mammals, however, but among the anamniotes. In the South American toad *Pipa pipa*, the spinal cord terminates at the level of the third vertebra (Tensen 1927), and in *Mola mola*, a highly specialised teleost with a very short trunk, the cord does not even extend beyond the limits of the skull (Fig. 24.1b). In this species the vertebral canal is filled only with a bundle of nerve roots, the so-called cauda equina (Haller 1891; Burr 1928). A cauda equina is found in all vertebrates with a foreshortened cord, and in all of these species this bundle surrounds the filum terminale, a stringlike glial continuation of the cord, which extends from the conus medullaris to the caudal end of the vertebral canal. As far as is known, the anlage of the spinal cord extends throughout the length of the spinal canal in early embryonic stages of all vertebrates; thus the ontogenetic process leading to spinal foreshortening is known as the 'ascent' of the spinal cord or ascensus medullae spinalis. It is remarkable that among vertebrates the distance between the sensory and motor centres in the brain on the one hand, and the sensory and motor periphery on the other, is bridged in two different ways: (a) relatively short spino-encephalic and encephalo-spinal connections and long peripheral connections (ascensus medullae), and (b) long spino-encephalic and encephalo-spinal connections and relatively short peripheral connections (no ascensus). Comparison of the relationships of the spinal somatomotor neurons and the striated musculature in vertebrates with those in the cephalochordate Amphioxus reveals a comparable, though even more radical difference. In vertebrates, the axons of the somatomotor neurons leave the cord (and brain) and approach their targets via the peripheral nerves. However, in Amphioxus these axons accumulate at

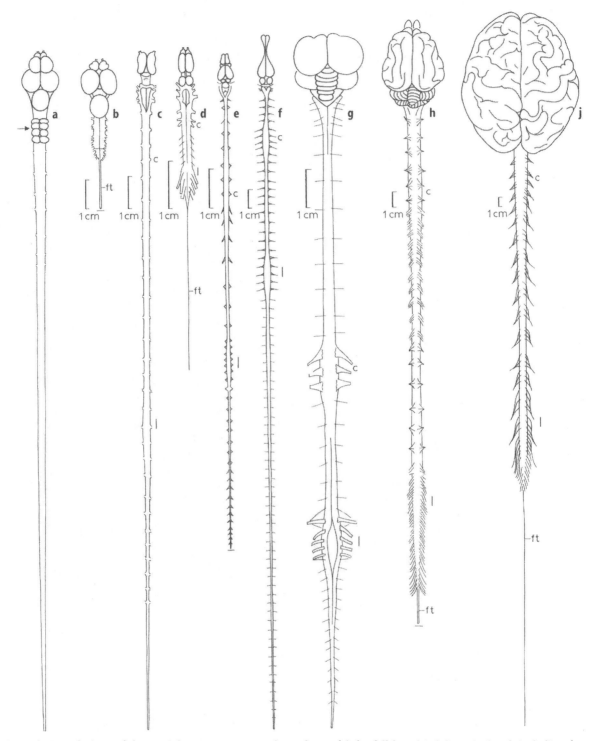

Fig. 24.1a–j. Dorsal views of the central nervous systems of the teleosts *Trigla hirundo* (**a**) and *Mola mola* (**b**), the urodele *Ambystoma tigrinum* (**c**), the anuran *Xenopus laevis* (**d**), the tortoise *Testudo hermanni* (**e**), the tegu lizard *Tupinambis teguixin* (**f**), the pigeon (**g**), the cat (**h**) and man (**j**). In **a, c, d, e, f, g** and **j** the full length of the spinalcord, including the filum terminale (where present) is shown; in **b** and **h** most of the filum terminale is cut (**a** and **g** are modified from Ariëns Kappers et al. 1936, **b** is based on Haller 1891 and Burr 1928, and **h** is simplified from Duvernoy et al. 1970)

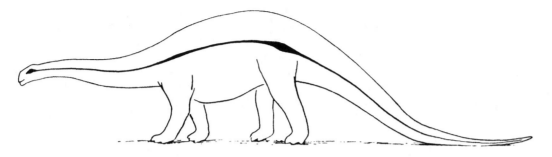

Fig. 24.2. Reconstruction of the CNS of *Brontosaurus*, based on casts of the spinal canal and the endocranial cavity (reproduced from Moodie 1915)

the surface of the cord, where they are contacted by medially extending processes of the myotomal muscle cells (Chap. 9, Sect. 9.6.6).

It is convenient to consider the spinal cord as consisting of a series of segments. Each segment gives off a dorsal root and a ventral root on each side, which unite in all vertebrates, except the petromyzontoids, to form a pair of spinal nerves. The number of these spinal nerves varies widely, ranging from ten pairs in tailless amphibians to more than 400 pairs in certain snakes.

In most tetrapods, the spinal cord shows enlargements or intumescentiae in the cervical and lumbar regions that are associated with the innervation of the paired limbs. Similar spinal enlargements associated with the paired fins have been observed in the crossopterygian fish *Latimeria chalumnae* (Millot and Anthony 1965). In general, the size of the cervical and lumbar enlargements is clearly correlated with the size of the corresponding extremities. Casts made from the endocranial cavity and the spinal canal of certain Mesozoic giant dinosaurs with strongly developed posterior limbs strongly suggest that in these animals the size of the lumbar intumescence exceeded that of the brain by far (Moodie 1915; Fig. 24.2). The anecdotal view that these now extinct dinosaurs possessed a 'rear-brain', allowing rather profound 'afterthoughts' has found no mercy in the eyes of Jerison (1973). In snakes, and in limbless lizards and amphibians as well, cervical and lumbar intumescences are entirely lacking. Planimetric determination of the cross-sectional area of the spinal cord at different levels revealed in *Python* the presence of an intumescentia trunci, i.e. an enlargement related to the well-developed main part of its trunk (Kusuma et al. 1979). As regards mammals, in sea cows, in which hindlimbs are absent, a lumbar enlargement is lacking too (Dexler and Eger 1911). Although in cetaceans hindlimbs are also absent, a lumbar intumescence – believed to be associated with the innervation of the strongly developed tail muscles (Ariëns Kappers et al. 1936; Breathnach 1960) – is nevertheless present.

Before leaving the spinal cord we should mention that gurnards and sea robins, teleosts belonging to the family Triglidae, possess a series of swellings on the dorsal surface of the rostral spinal cord (Fig. 24.1a, Poster 3:14). These swellings, known as accessory spinal lobes, relate to the pectoral fins, particularly to their remarkable free fin rays. The free fin rays are highly mobile and densely covered with chemoreceptors that play an important role in locating food. Finger (1982) showed that the free fin rays are in one-to-one correspondence with the accessory spinal lobes.

Poster 2 shows an array of chordate brains in dorsal view, extending from Amphioxus to the chimpanzee. The first impression is of immense variety without an obvious pattern. The selection of brains in Poster 2 is meant not only to show the variety of brains within the chordate phylum per se, but also to give an impression of the interspecific differences within certain groups. Compare, for example, the brains of the hagfish (2) and the lamprey (3) (both cyclostomes), of the ratfish (4) and the electric ray (9) (both cartilaginous fishes), of the shovelnose sturgeon (11) and of the elephant-nose (17) (both actinopterygians) and, finally, of the hedgehog and the chimpanzee (both mammals). The olfactory bulbs, telencephalic hemispheres, tectum mesencephali and cerebellum are indicated by different colours. Obviously, not only do the brains vary enormously, but so do the constituent parts.

In Poster 3, lateral views of brains, related, but not always identical, to those in Poster 2 are shown. For full names and taxonomic relationships of the various species see Frame 1. This illustration introduces the boundaries of the brain, rhombencephalon, cerebellum, mesencephalon, diencephalon and telencephalon.

We make the assumption that the gross structure of the brain is related to its internal organisation and to the functional capacity of its parts (although, as will be discussed later, it is rather difficult to clarify such a statement). In general, the brains of 'generalised', 'primitive' vertebrates are slender and elongated – e.g. the brains of hexanchiform sharks (5), dipnoans (19, 20) and salamanders (22) – and particular specialisations and advanced features are often reflected in externally visible expansions. Starting with the rhombencephalon and proceeding towards the telencephalon, some of the more noteworthy enlargements will be discussed. It will appear that unambiguous functional

Frame 1 – Posters P2 and P3

The posters P2 and P3 present dorsal and lateral views, respectively, of the brains of a number of representative chordates. The phylogenetic relationships of the species involved are indicated in the plates; their full names and taxonomic relationships are given below. The brains of the various species are numbered consecutively, in order to facilitate reference to them in the text.
Phylum: **CHORDATA**
Subphylum: **Cephalochordata**
1. *Branchiostoma lanceolatum* (20x)
Subphylum: **Vertebrata**
Superclass: AGNATHA
Subclass: Myxinoidea
2. *Myxine glutinosa* (8x)
Subclass: Petromyzontoidea
3. *Lampetra fluviatilis* (8x)
Superclass: GNATHOSTOMA
Class: CHONDRICHTHYES
Subclass: Holocephali
4. *Hydrolagus collei* (1.3x)
Subclass: Elasmobranchii
5. *Heptanchrias perlo* (about 1.3x)
6. *Squalus acanthias* (1.3x)
7. *Carcharhinus sp* (1.3x)
8. *Raja clavata* (1.3x)
9. *Torpedo nobiliana* (1.3x)
Class: OSTEICHTHYES
Subclass: Actinopterygii
Superorder: Cladistia (Brachiopterygii)
10. *Erpetoichthys calabaricus* (8x)
Superorder: Chondrostei
11. *Scaphirhynchus platorynchus* (5x)
Superorder: Holostei
12. *Amia calva* (2.5x)
Superorder: Teleostei
13. *Salmo gairdneri* (3x)
14. *Trigla hirundo* (3x)
15. *Carassius carassius* (3x)
16. *Eigenmannia virescens* (8x)
17. *Gnathonemus petersii* (4x)
18. *Mormyrus kannume* (2x)
Subclass: Dipnoi
19. *Neoceratodus forsteri* (1.5x)
20. *Lepidosiren paradoxa* (3x)
Subclass: Crossopterygii
21. *Latimeria chalumnae* (1.5x)

Class: AMPHIBIA
Order: Urodela
22. *Ambystoma tigrinum* (5x)
Order: Gymnophiona (Apoda)
23. *Ichthyophis glutinosus* (8x)
23. *Ichthyophis kohtaoensis* (9x)
Order: Anura
24. *Rana esculenta* (6x)
Class: MAMMALIA
Subclass: Prototheria (Monotremata)
25. *Ornithorhynchus anatinus* (1.7x)
Subclass: Metatheria (Marsupialia)
26. *Didelphis virginiana* (1.6x)
Subclass: Eutheria (Placentalia)
27. *Erinaceus europaeus* (1.8x)
28. *Canis familiaris* (1.1x)
29. *Pan troglodytes* (0.7x)
30. *Balaenoptera physalus* (0.4x)
Class: REPTILIA
Order: Rhynchocephalia
31. *Sphenodon punctatus* (2.5x)
Order: Squamata
32. *Tupinambis teguixin* (2x)
33. *Varanus exanthematicus* (2x)
34. *Python regius* (3x)
Order: Crocodilia
35. *Alligator mississippiensis* (2.5x)
Order: Chelonia
42. *Testudo hermanni* (3.5x)
Class: AVES
Superorder: Palaeognathae
36. *Apteryx australis* (1.3x)
Superorder: Neognathae
37. *Phalacrocorax carbo* (1.5x)
38. *Columba livia* (1.8x)
39. *Corvus frugilegus* (?)
40. *Strix aluco* (?)
41. *Amazona ochrocephala* (1.2x)

Sources: *5*, Johnston (1906); *7*, Masai et al. (1973); *12*, Butler and Northcutt (1992); *16*, Heiligenberg et al. (1991, lateral view); *23*, Northcutt and Kicliter (1980: P2), Wicht and Himstedt (1990: P3), Kuhlenbeck et al. (1967: P3); *25*, Krubitzer et al. (1995); *28*, Singer (1962); *29*, Retzius (1906); *30*, Pilleri (1969); *31*, Christensen (1927), Haller von Hallerstein (1934), Platel (1989); *35*, Crosby (1917); *36*, Parker (1891), Craigie (1930), Durward (1932); *39*, *40*, *41*, Stingelin (1958)

labels can be applied in only some cases, and that, broadly speaking, the more rostral we get, the more difficult this labelling becomes.

The *rhombencephalon* (see Frame 2) constitutes the most caudal part of the brain. Rostrally, it borders on the mesencephalon and caudally it grades into the spinal cord (P3). The importance of this part of the brain stems from the facts that it harbours the chief centres of origin and termination of most of the cranial nerves (V through X and, where present, through XII) and that it also contains a number of relay centres including the larger part of the reticular formation. The cerebellum, although ontogenetically a derivative of the rostral part of the rhombencephalon, is generally considered a separate part of the brain (see below).

The rhombencephalon of fishes shows a number of notable specialised structures which relate directly to the behaviour and lifestyle of particular animals. Electric rays, such as *Torpedo*, possess enormous rhombencephalic electric lobes consisting virtually entirely of densely packed electromotoneurons which innervate the electric organs in the pectoral disc (P2,9). In some species this lobe accounts for almost 60% of the total brain weight (Roberts and Ryan 1975)! In electric rays, the discharge of the electric organ is both a means of protection and a device for stunning prey. Comparable electric organs are found in a group of teleosts known as stargazers (Fam: Uranoscopidae). In these animals the electric organs are formed of modified eye muscles and are innervated by the oculomotor nerve. Accordingly, this nerve and its centre are enormously enlarged in stargazers (Leonard and Willis 1979; Fig. 24.33).

Huge electric lobes (but quite different in origin and function from those in rays and stargazers) are present in the gymnotid and mormyrid groups of teleosts (P2,3: 16; P2). These vast lobes show a remarkable degree of internal differentiation and form the primary afferent centres of the elaborate electrosensory system of these fishes. Gymnotids and mormyrids are so-called active electrosensory teleosts: they use an array of electroreceptors on the body in conjunction with an electric organ that generates weak-pulsed electric current to communicate with conspecifics and locate objects in their environment on the basis of local distortions of the electric field (Chap. 15).

The large vagal lobe found in cyprinid teleosts is another example of a rhombencephalic specialisation (P2,3: 15). This lobe has no fewer than 16 layers and encompasses both sensory and motor components. The lobe innervates a complex and effective food-selecting organ, located in the roof of the oral cavity. In fact, in cyprinids and other teleosts, taste buds are not confined to the oral cavity but are scattered over the entire body. In cyprinids, the external, cutaneous gustatory system projects to an unpaired swelling, the facial lobe. In silurid teleosts, the gustatory system is also strongly developed, but here the 'exteriorisation' of the system is more advanced than in cyprinids. The silurid catfish *Ictalurus* has about 20 000 taste buds in the mouth but no fewer than 180 000 on the external body surface. The buds are particularly densely concentrated on the circumoral barbels, which play an important role in the search for food (Atema 1971). In silurids, the cutaneous and oral taste systems are centrally represented in separate facial and vagal lobes (Figs. 24.9, 24.70a), although the latter does not show the multitude of laminae seen in cyprinids.

Frame 2 – Rhombencephalon

Figure 24.3 shows a diagrammatic cross-section through the rhombencephalon. The wall of this brain part can be divided bilaterally into a basal plate, which contains numerous efferent or motor centres, and an alar plate, which harbours the various afferent or sensory areas. The motor basal plates and the sensory alar plates constitute the ventral and lateral walls, respectively, of the fourth ventricle. Dorsally, this ventricle is closed by a membranous roof. The boundary between the basal and alar plates is generally marked on the ventricular side by a distinct groove, the sulcus limitans of His. The basal and alar plates can both be subdivided into separate somatic and visceral zones, and so we obtain, from dorsal to ventral, a somatosensory, a viscerosensory, a visceromotor and a somatomotor zone (for details on this subdivision see Chap. 4, Sect. 4.6).

The *somatosensory zone* is occupied largely by centres related to the auditory and vestibular parts of the eighth cranial nerve and, in fishes and amphibians, to the lateral line nerves. The latter nerves carry impulses from the mechanoreceptive lateral line organs; in most anamniote groups the lateral line nerves are, in addition, peripherally related to electroreceptor organs. Many of the efferent fibres emanating from these centres, which are designated together as the octavolateral area, cross the median plane as fibrae arcuatae, and constitute an ascending pathway on the contralateral side, which is known as the lateral lemniscus. The *viscerosensory zone* represents the centre of termination of the afferent fibres of the seventh, ninth and tenth cranial nerves. Many of these

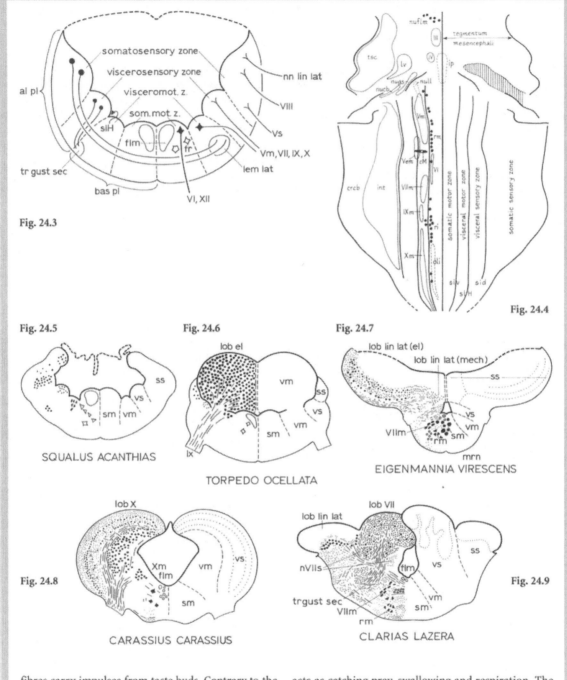

Fig. 24.3

Fig. 24.4

Fig. 24.5 SQUALUS ACANTHIAS

Fig. 24.6 TORPEDO OCELLATA

Fig. 24.7 EIGENMANNIA VIRESCENS

Fig. 24.8 CARASSIUS CARASSIUS

Fig. 24.9 CLARIAS LAZERA

fibres carry impulses from taste buds. Contrary to the projections from the somatosensory zone, those from the viscerosensory zone remain mostly on the ipsilateral side. They concentrate in a fibre system that ascends to the most rostral part of the rhombencephalon, where it terminates in a cell mass known as the nucleus gustatorius secundarius, or nucleus parabrachialis. The *visceromotor zone* contains the branchiomotor nuclei of the fifth, seventh, ninth and tenth cranial nerves, centres involved in such vital acts as catching prey, swallowing and respiration. The *somatomotor zone*, which may be considered as a direct continuation of the anterior horn of the spinal cord, passes rostrally over into the mesencephalon. It contains the nuclei which innervate the external eye muscles and, in amniotes, the nucleus of the hypoglossal nerve and harbours, in addition, the large cells of the reticular formation and a composite, mostly descending, fibre bundle, the fasciculus longitudinalis medialis (flm). The reticular cells subserve primarily a somatomotor coordinating function and their axons descend in the flm to the spinal cord. Vestibular centres situated in the alar plate also contribute fibres to the flm. Thus, throughout the length of the brain stem somatomotor centres of primary and higher order and a predominantly descending somatomotor fibre system are concentrated in a paramedian strip.

The four functional zones just discussed protrude into the fourth ventricle, constituting longitudinally arranged ridges, a feature which is particularly prominent in chondrichthyans (Fig. 24.5), chondrosteans, holosteans and in the crossopterygian *Latimeria chalumnae*. Figure 24.4 shows the zonal pattern in the rhombencephalon of the holostean fish, *Lepisosteus osseus*, as revealed by a topological analysis (cf. Chap. 4, Sect. 4.6.5; Chap. 6, Sect. 6.2.3). Although still discernable, the rhombencephalic zonal pattern is much less prominent in amniotes than in most fish.

The remarkable specialised structures discussed in the text can all be related to one of the functional zones. Thus, the large electromotor nucleus in stargazers (Frame 4, Fig. 24.33) forms part of the somatomotor zone, while the electromotor nucleus of electric rays (Fig. 24.6) comprises parts of the motor nuclei of VII, IX and X, and hence belongs to the visceromotor zone. The huge electrosensory lobes of gymnotids (Fig. 24.7) and mormyrids represent outgrowths of the somatosensory zone. The gustatory vagal lobes of cyprinids (Fig. 24.8) and the gustatory vagal and facial lobes of silurids (Fig. 24.9) are to be considered as local hypertrophies of the viscerosensory zone. In cyprinids, the visceromotor zone also participates in the formation of the vagal lobe (Fig. 24.8).

Abbreviations: *al pl*, alar plate; *bas pl*, basal plate; *cr cb*, crista cerebellaris; *flm*, fasciculus longitudinalis medialis; *fr*, formatio reticularis; *ip*, nucleus interpeduncularis; *lem lat*, lemniscus lateralis; *lob el*, lobus electricus; *lob lin lat (el)*, lobus lineae lateralis, electroreceptive portion; *lob lin lat (mech)*, lobus lineae lateralis, mechanoreceptive portion; *lob VII* (gustatory), lobus facialis; *lob X* (gustatory), lobus vagi; *lv*, nucleus lateralis valvulae; *mrn*, medullary relay nucleus of electromotor system; *nucb*, nucleus cerebelli; *nuflm*, nucleus of flm; *nugs*, nucleus gustatorius secundarius; *null*, nucleus lemnisci lateralis; *oli*, oliva inferior; *ri*, nucleus reticularis inferior; *rm*, nucleus reticularis medius; *sid*, sulcus intermedius dorsalis; *slH*, sulcus limitans of His; *sm*, somatic motor zone; *ss*, somatosensory zone; *tr gust sec*, tractus gustatorius secundarius; *Vem*, nucleus vestibularis magnocellularis; *vm*, visceromotor zone; *vs*, viscerosensory zone; *Vm*, motor part of trigeminal nerve, c.q. motor trigeminal nucleus; *Vs*, sensory part of trigeminal nerve; *VIIm*, motor facial nucleus; *IXm*, motor glossopharyngeal nucleus; *Xm*, motor vagal nucleus

In both shape and size, the *cerebellum* (see Frame 3) is by far the most variable part of the CNS. The cerebellum is entirely absent in hagfishes (P2,3: 2) and very small in lampreys (P2,3: 3), urodeles (P2,3: 22) and gymnophionans (P2,3: 23) but attains such enormous proportions in mormyrid teleosts that it covers all other parts of the brain. We do not think it likely that the highly different forms assumed by the cerebella of the various groups of vertebrates (Figs. 24.10–24.30) all represent specific adaptations to particular functional requirements. More plausibly, they simply represent different solutions to the simple issue of satisfying a given functional requirement in a certain prescribed volume of space. Traditionally, the size of the cerebellum has been correlated with locomotor performance, but Bullock (1984) pointed out that the evidence does not support this hypothesis, and even today we really have no clear idea about what this part of the brain does. Since the cerebellum poses many important questions in comparative studies, we shall return to this brain region several times in this chapter.

The dorsal part of the *mesencephalon* (see Frame 4) of all non-mammalian vertebrates is occupied by two correlation centres, the tectum mesencephali and the torus semicircularis. In the mammalian brain these structures are represented by the superior and inferior colliculi, respectively.

Frame 3 – Cerebellum

Embryological investigations have shown that the cerebellum develops from bilateral anlagen, which form the most rostral parts of the rhombencephalic alar plates. These anlagen become interconnected early in development by decussating fibres. They eventually fuse in the median plane to form a single cerebellar plate. The cerebellum is absent in myxinoids and very small in petromyzontoids, in which latter group a simple platelike configuration is maintained (Fig. 24.10). In most groups of vertebrates a further growth and elaboration of cerebellar tissue occurs. The modes in which the embryonic cerebellar plate is transformed into the adult organ differ widely among the various groups of gnathostomes.

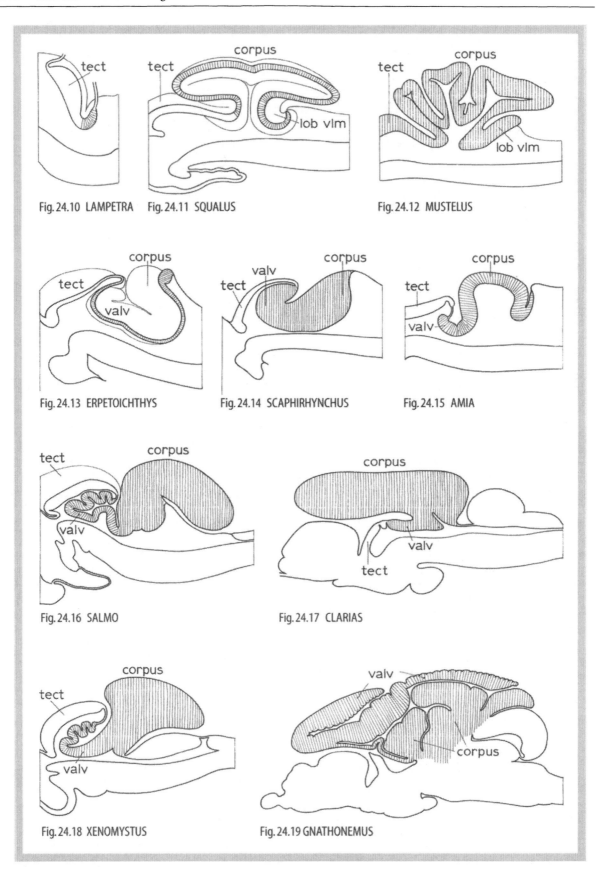

Fig. 24.10 LAMPETRA   Fig. 24.11 SQUALUS   Fig. 24.12 MUSTELUS

Fig. 24.13 ERPETOICHTHYS   Fig. 24.14 SCAPHIRHYNCHUS   Fig. 24.15 AMIA

Fig. 24.16 SALMO   Fig. 24.17 CLARIAS

Fig. 24.18 XENOMYSTUS   Fig. 24.19 GNATHONEMUS

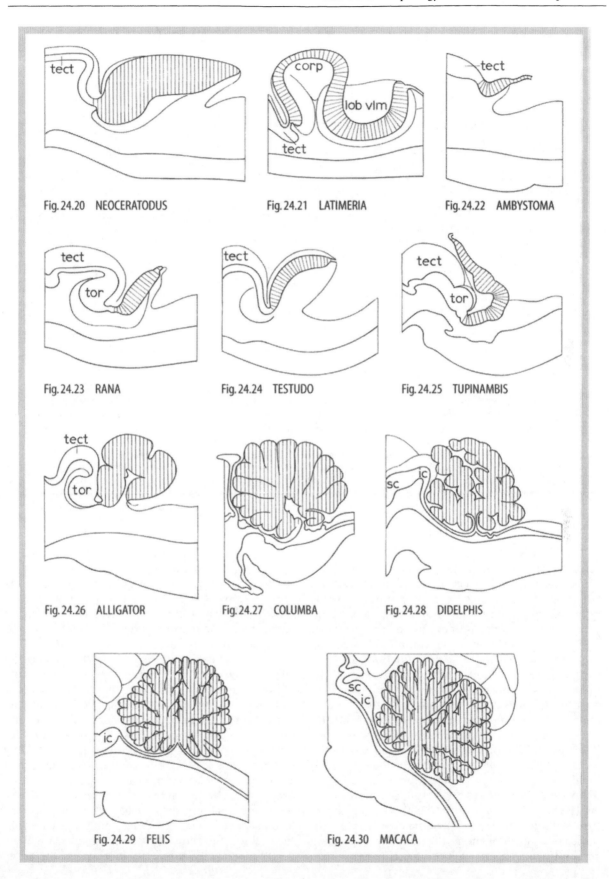

Fig. 24.20 NEOCERATODUS
Fig. 24.21 LATIMERIA
Fig. 24.22 AMBYSTOMA
Fig. 24.23 RANA
Fig. 24.24 TESTUDO
Fig. 24.25 TUPINAMBIS
Fig. 24.26 ALLIGATOR
Fig. 24.27 COLUMBA
Fig. 24.28 DIDELPHIS
Fig. 24.29 FELIS
Fig. 24.30 MACACA

The gross morphology of the cerebellum is shown in P2 (dorsal views), P3 (lateral views), and in Figs. 24.10–24.30, which present diagrammatic median sections through the cerebellar region in a number of representative vertebrates.

The studies of Larsell (summarised in Larsell 1967) have shown that in fishes, urodeles and larval anurans the cerebellum consists of two fundamental divisions, a caudobasal *lobus vestibulolateralis*, which, as its name indicates, receives vestibular and lateral line fibres, and a more rostrally situated *corpus cerebelli*, which is dominated by spinocerebellar, trigeminocerebellar, tectocerebellar and other systems. A similar subdivision can be made in adult anurans, reptiles, birds and mammals, but a lateral line system is absent in these groups and the caudobasal part of the cerebellum receives only primary and secondary vestibular fibres. The lobus vestibulolateralis, *casu quo*, vestibularis, consists of medial and lateral parts. In most groups of fish (P2: 4–14, 20–21), in urodeles (P2: 22), and in gymnophionans (P2: 23) the lateral parts are enlarged and form the *auriculae cerebelli*. The medial parts of the vestibulolateral lobe are strongly developed in chondrichthyans (Figs. 24.11, 24.12) and in the crossopterygian *Latimeria* (Fig. 24.21).

In actinopterygians, a rostral outgrowth of the corpus cerebelli forms the *valvula cerebelli*. This portion of the cerebellum usually occupies a subtectal position (Figs. 24.13–24.18), but in mormyrids the valvula is greatly enlarged and covers all other parts of the brain (P2: 17; P3: 18; Fig. 24.19). The valvula of mormyrid fishes is covered with fine ridges throughout. This can be clearly seen in P2: 17c, in which the valvula is unfolded. In P2: 17b the valvula is removed so as to reveal the other parts of the brain.

With regard to its shape, the *corpus cerebelli* is the most variable structure of the CNS. (a) In the dipnoans *Protopterus* and *Lepidosiren*, in amphibians (Figs. 24.22, 24.23), and in snakes the corpus cerebelli maintains a platelike configuration. (b) In chondrichthyans (Figs. 24.11, 24.12), holosteans (Fig. 24.15), teleosts (Figs. 24.16–24.18), the crossopterygian *Latimeria* (Fig. 24.21), crocodilians (Fig. 24.26), birds (Fig. 24.27) and mammals (Figs. 24.28–24.30), the corpus cerebelli is evaginated. (c) A condition intermediate between (a) and (b) is found in the dipnoan *Neoceratodus* (Fig. 24.20) and in the chelonian *Testudo* (Fig. 24.24). The corpus cerebelli is represented by a curved plate in these forms. (d) In cladistians (Fig. 24.13) and chondrosteans (Fig. 24.14) the cerebellum has invaginated into the fourth ventricle. (e) A clearly recurved cerebellum is found in many lizards (Fig. 24.25; P2: 33). (f) In many chondrichthyans (Fig. 24.12; P2: 7), some teleosts, and in crocodilians (Fig. 24.26), birds (Fig. 24.27; P3: 36–41) and mammals (Figs. 24.28–24.30; P2: 26–28; P3: 25–30) the corpus cerebelli shows transversely oriented external grooves. In chondrichthyans, the entire wall is involved in the folding, but only the external surface is convoluted in the others.

In mammals, the corpus cerebelli is differentiated into a median vermis and bilateral hemispheres (P2: 26–28). A small swelling, situated on either side of the base of the avian cerebellum, is homoplastic with the mammalian cerebellar hemispheres. For the structural organisation of the cerebellum, see Sect. 24.6.2).

Abbreviations: *corpus*, corpus cerebelli, *ic*, inferior colliculus; *lob vlm*, lobus vestibulolateralis, pars medialis; *sc*, superior colliculus; *tect*, tectum mesencephali; *tor*, torus semicircularis; *valv*, valvula cerebelli

The *tectum mesencephali* receives input from several sensory modalities, including auditory, somatosensory and, where present, electrosensory, but in most vertebrates the largest input by far is provided by primary visual fibres originating from ganglion cells of the retina. One of the principal functions of the tectum is the control of the orienting response, i.e. the rapid combination of movements of eyes, head and trunk towards external stimuli. The dominance of the visual input is frequently reflected in the size of the tectum. In animals with rudimentary eyes, such as hagfishes (P2,3:2), cave teleosts, cave salamanders, gymnophionans (P2,3:23), amphisbaenids and moles, the tectum/superior colliculus is small, whereas in such highly visual animals as the trout (P2,3:13) and many other teleosts, frogs (P2,3:24) and most reptiles (P2,3:31–35), and birds (P3:38–41), this structure is large and well differentiated internally. In birds, during ontogenesis the caudally expanding cerebral hemispheres push the tectal halves – here designated as optic lobes – laterally. The urodele tectum mesencephali represents the most simple condition among vertebrates with respect to its organisation. Roth and co-workers (1993) concluded that the tectum mesencephali and many sensory systems of salamanders, especially those of bolitoglossines, are secondarily simplified.

The *torus semicircularis/inferior colliculus* is the terminal station of the lateral lemniscus, a decussating pathway originating from the primary sensory auditory and, in appropriate species, lateral line centres in the rhombencephalon. The size and capacity of this mesencephalic complex vary with the degree of development of these specific sensory systems. Thus it is large in 'auditory' animals, such as frogs, bats and whales, and it attains amazing proportions in the active electrosensory teleosts, the gymnotids and mormyrids. In gymnotids, the torus semicircularis is a huge, multilayered structure which fills almost the entire mesencephalic ventricular cavity, while in mormyrids the torus semicircularis is differentiated into at least six separate cell masses which, having thrust aside the two halves of the tectum, bulge out of the mesencephalic ventricle (P2,17b). In most vertebrates, the torus semicircularis also receives somatosensory input.

Another macroscopically visible mesencephalic structure is the torus lateralis, which, as its name indicates, is an elevation of the lateral mesencephalic wall. This structure, which is confined to actinopterygians, clearly corresponds with a single mass of migrated cells. It is particularly strongly developed in the holostean *Amia calva* (P3,12). In contrast to the tectum mesencephali and the torus semicircularis, the function of the torus lateralis is unknown.

We come now to the diencephalon and telencephalon, which together constitute the forebrain or prosencephalon. The name *diencephalon* (see Frame 5), which means 'between-brain', is apt, because in many vertebrates this part of the brain is wedged in between its expanding neighbours, the tectum of the midbrain and the telencephalic cerebral hemispheres. In all vertebrates, the diencephalon is divisible into four major parts, epithalamus, dorsal thalamus, ventral thalamus and hypothalamus. These parts were originally thought of as horizontally arranged zones (e.g. Herrick 1910; Kuhlenbeck 1929a, 1973); however, thanks to the pioneering embryological studies of Bergquist (1932) and Bergquist and Källén (1954), which have been confirmed by recent studies on the expression patterns of certain regulatory genes (e.g. Bulfone et al. 1993; Puelles 1995), we know now that the dorsal and ventral thalami are both direct derivations of early

---

Frame 4 - Mesencephalon

The mesencephalon consists of the ventral tegmentum and the dorsal tectum (Fig. 24.31a). Schematically, it may be said that the rhombencephalic basal plate is rostrally continuous with the medial part of the tegmentum mesencephali, which has been designated on that account as the tegmentum motoricum. It contains the motor nuclei of the third and fourth cranial nerves, the rostral part of the reticular formation, the initial portion of the medial longitudinal fasciculus and, in many species, the nucleus ruber. The rhombencephalic alar plate passes over into the lateral part of the tegmentum and the tectum. Both of these are the recipients of important sensory pathways. The lateral tegmentum contains the end station of the lateral lemniscus, which represents the chief efferent pathway of the octavolateral area. In all non-mammalian vertebrates, the tectum is the principal centre of termination of the fibres which pass from the retina towards the brain. At the mesencephalic level these fibres constitute two pathways, the medial and lateral optic tracts, from which the optic fibres spread over the tectum. The general relations of the midbrain, just outlined, is exemplified in Fig. 24.31b for a shark.

In non-mammalian vertebrates, the nuclear formation which forms the end station of the lateral lemniscus is known as the torus semicircularis, a name which refers to the fact that this centre in many forms bulges into the mesencephalic ventricular cavity (Fig. 24.32a). In reptiles, the torus not only expands intraventricularly, but also produces an external bulge (Fig. 24.32b). With respect to the disposition of the torus semicircularis, birds and mammals show opposite tendencies. In birds (Fig. 24.32c), this formation shows a periventricular expansion and, contrary to reptiles, no external bulge. However, in mammals (Fig. 24.32d) the torus evaginates and becomes an externally visible elevation, situated behind the tectum. The mammalian homologues of tectum and torus are known as the superior and inferior colliculi, respectively.

Figures 24.33–24.41 illustrate the structural variability of the midbrain. Figure 24.33 shows that in the stargazer, *Astroscopus y-graecum*, the electromotor nucleus produces a marked expansion of the medial or motor tegmentum. The size of the torus semicircularis/inferior colliculus varies with the development of the auditory and/or lateral line systems. In gymnotid and mormyrid teleosts, the strong development of the electrosensory lateral line system has led to dramatic morphological changes at the mesencephalic level. In gymnotids (Fig. 24.34), the huge tori semicirculares show a distinct laminar differentiation and have fused entirely in the median plane. This massive toral body fills the mesencephalic ventricular cavity and separates both tectal halves. However, the gap between the latter is largely covered by the rostrally deflected corpus cerebelli (P2, P3: 16). In mormyrids (Fig. 24.35), the tori semicirculares are also strongly developed but show a nuclear, rather than a laminar, differentiation. They have pushed aside the relatively poorly developed tectal halves and are covered by the greatly enlarged valvula cerebelli (cf. P2: 17a and c). The tori semicirculares are also strongly developed (and partly fused) in ranid frogs (Fig. 24.37), and the same holds true for their homologues, the inferior colliculi in mammals with the capacity to echolocate (the microchiropteres and the cetaceans, e.g. the common porpoise, *Phocaena phocaena*: Fig. 24.40).

In birds (Figs. 24.32c, 24.38), the tectal halves, or optic lobes as they are called in this group, are displaced laterally and ventrally. Dorsally, they are partly or entirely covered by the caudally extending cerebral hemispheres (cf. P2, P3: 36–41).

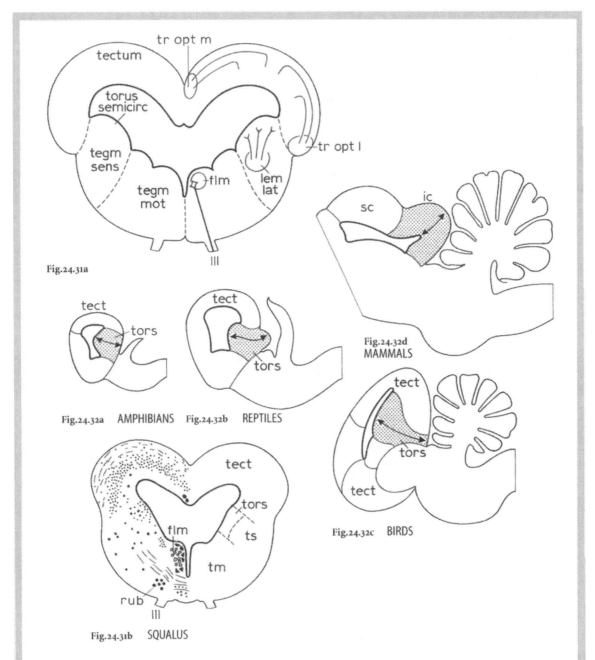

Fig.24.31a

Fig.24.32d MAMMALS

Fig.24.32a AMPHIBIANS Fig.24.32b REPTILES

Fig.24.32c BIRDS

Fig.24.31b SQUALUS

The basal portion of the mammalian mesencephalon is formed by an assembly of corticofugal (i.e. corticopontine and pyramidal) fibres, known as the pedunculus cerebri. This fibre compartment, while small in the opossum (Fig. 24.39) and other primitive mammals, is very large in primates, particularly in man (Fig. 24.41). Due to this addition and expansion of corticofugal projections, the configuration of the mesencephalon changes considerably. Thus, whereas in the midbrain of most non-mammalian vertebrates the dorsal, tectal parts are much wider than the basal parts (e.g. Figs. 24.33, 24.34, 24.38), in primates these relations are reversed. In actinopterygian fishes, a superficially situated tegmental cell mass produces a bulge, known as the torus lateralis, on the ventrolateral surface of the midbrain (P3, 12, 13). The functional significance of this cell mass, which attains amazing proportions in the holostean *Amia calva* (Fig. 24.36), is unknown.

Abbreviations: *bc*, brachium conjunctivum; *bp*, brachium pontis; *corp*, corpus cerebelli, *emn*, electromotor nucleus; *flm*, fasciculus longitudinalis medialis; *ic*, inferior colliculus; *lem lat (ll)*, lemniscus lateralis; *lih*, lobus inferior hypothalami; *lob opt*, lobus opticus; *nIII*, nervus oculomotorius; *ped*, pedunculus cerebri; *rub*, nucleus ruber; *sc*, superior colliculus; *sn*, substan-

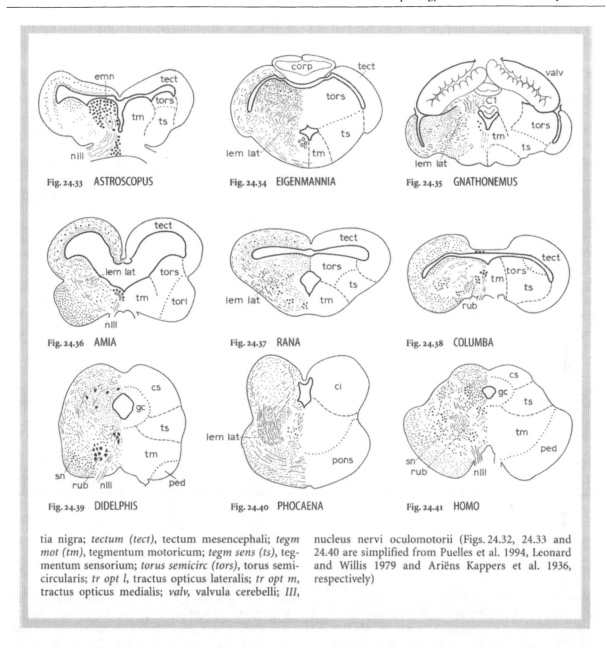

tia nigra; *tectum (tect)*, tectum mesencephali; *tegm mot (tm)*, tegmentum motoricum; *tegm sens (ts)*, tegmentum sensorium; *torus semicirc (tors)*, torus semicircularis; *tr opt l*, tractus opticus lateralis; *tr opt m*, tractus opticus medialis; *valv*, valvula cerebelli; *III*, nucleus nervi oculomotorii (Figs. 24.32, 24.33 and 24.40 are simplified from Puelles et al. 1994, Leonard and Willis 1979 and Ariëns Kappers et al. 1936, respectively)

embryonic neuromeres and so are essentially transversely oriented. The epithalamus stems from the same neuromere as does the dorsal thalamus, but there is no consensus about the neuromeres which participate in the formation of the hypothalamus. The tuberculum posterius, a structure which protrudes into the caudolateral part of the diencephalic ventricle, is probably a derivative of the most ventral parts of the 'dorsal' and 'ventral' thalamic neuromeres.

In most anamniotes, the *hypothalamus* forms the largest part of the diencephalon, and in chondrichthyan as well as actinopterygian fishes its walls are differentiated into bilateral inferior lobes. The hypothalamus is crucially involved in processes directly linked to the survival of the individual and the species. It plays a key role in: (a) the maintenance of homeostasis, a complex task accomplished by activation of visceral effector mechanisms and the initiation of feeding and drinking and related foraging behaviours; (b) agonistic (i.e. defence and attack) behaviours; (c) sexual and reproductive behaviour.

The *ventral thalamus* is formed by a narrow strip of tissue, intercalated between the hypothalamus and the dorsal thalamus. In mammals, the principal ventral thalamic cell mass, i.e. the nucleus subthalamicus, is reciprocally connected with the pallidal part of the basal ganglia and is known to play a role in motor control. In several anamniote species (for references and details see Chaps. 12, 15 and 19), however, the ventral thalamus has been shown to receive retinal, tectal and cerebellar projections and to innervate the telencephalon. In teleosts, the afferent projections to the ventral thalamus seem to be particularly multifarious and to include input from the telencephalon, the retina, the tectum, the torus semicircularis, the cerebellum, the main sensory trigeminal nucleus, the reticular formation and the spinal cord. Ventral thalamic projections have been traced in this group to a variety of telencephalic regions, the dorsal thalamus, the tectum, the torus semicircularis and the spinal cord. In anurans, the ventromedial thalamic nucleus receives substantial inputs from the spinal cord, from the main sensory trigeminal nucleus and from the dorsal column nucleus, and innervates the striatum. Judging from its afferent and efferent connections, no simple and generally valid functional label can be attached to the vertebrate ventral thalamus.

The *dorsal thalamus* may be broadly characterised as a complex of way-stations interposed between the contralateral sensory world and the telencephalon (Butler 1994a, 1995; Northcutt 1995). Unlike the ventral thalamus, the dorsal thalamus exhibits considerable differences in size and complexity among vertebrate species. In anamniote gnathostomes, the dorsal thalamus is relatively small and is made up of three nuclei, all of which project to the telencephalon. The component nuclei are: an anterior thalamic nucleus receiving retinal projections, a dorsal posterior or lateral thalamic nucleus involved in an indirect visual (and possibly multisensory) circuit from the tectum mesencephali to the telencephalon, and a central posterior or central thalamic nucleus which receives afferents conveying auditory, somatosensory and/or mechanoreceptive lateral line information from the torus semicircularis. The dorsal thalamus is larger and comprises many more individual cell masses in amniotes than in anamniotes.

Butler (1994a, 1995) recently enunciated what she called 'a dual elaboration hypothesis of dorsal thalamic evolution'. According to this hypothesis, the matrix or embryonic anlage of the dorsal thalamus comprises two parts that give rise to the two principal dorsal thalamic divisions, the collothalamus and the lemnothalamus (Fig. 24.42). The collothalamus receives its predominant input from the midbrain roof, while the lemnothalamus is supplied by lemniscal pathways, including the optic tract. The anuran ventromedial thalamic nucleus may therefore be included in the lemnothalamus. In anamniote vertebrates, the lemnothalamic matrix gives rise to the anterior nucleus and the collothalamic matrix gives rise to the dorsal posterior and central posterior nuclei, which all occupy a periventricular position. In amniotes, the collothalamic as well as the lemnothalamic matrices give rise to populations of neurons that migrate outward to form discrete nuclei. Thus, a dual elaboration of both divisions of the dorsal thalamus occurs which, according to Butler (1995, p. 49), "...is the defining difference that distinguishes the dorsal thalamus of amniotes from that of all other vertebrates." In amniotes, collothalamic nuclei relay visual, auditory and somatosensory-multisensory inputs to the telencephalon. Although the size and the topographical position of these nuclei differ considerably among the various amniote groups, evidence derived from their topology and connections strongly suggests that these midbrain-sensory relay nuclei are homologous to each other as discrete nuclei in all the vertebrate groups. For example, the dorsal posterior nucleus of cartilaginous and bony fishes, parts of the lateral thalamic nucleus of anurans, the nucleus rotundus of reptiles and birds, and the nucleus lateralis posterior-pulvinar complex of mammals are all strict homologues of each other. By contrast, the lemnothalamic matrix, which in anamniotes forms a simple periventricular cell mass, the nucleus anterior, gives rise in reptiles, birds and mammals to a whole series of nuclei. Most of these nuclei cannot be strictly homologised in a one-to-one fashion. According to Butler, however, the lemnothalamic nuclei in each amniote radiation are collectively homologous as a field to the nucleus anterior in anamniotes. A quite remarkable elaboration and differentiation of the lemnothalamus characterises the development of the mammalian thalamus. This development and the related unfolding of the neocortex endow mammals with a remarkable capacity for detailed analysis of their environment. A considerable, though somewhat more restricted and entirely independent, elaboration of the lemnothalamus is seen in birds.

The elaboration of the amniote dorsal thalamus is a prominent feature with clear macromorphological consequences; it is hidden from view, however, by another, related phenomenon: the caudal expansion of the cerebral hemispheres.

From the foregoing it appears that the dorsal thalamus is an important gateway to the telencephalon. It should be emphasised, though, that there are additional, as well as alternative, routes between more caudally situated parts of the brain and the telencephalon. Hypothalamo-telencephalic connections have been found in all vertebrate groups. In cartilaginous fishes and non-teleostean actinopte-

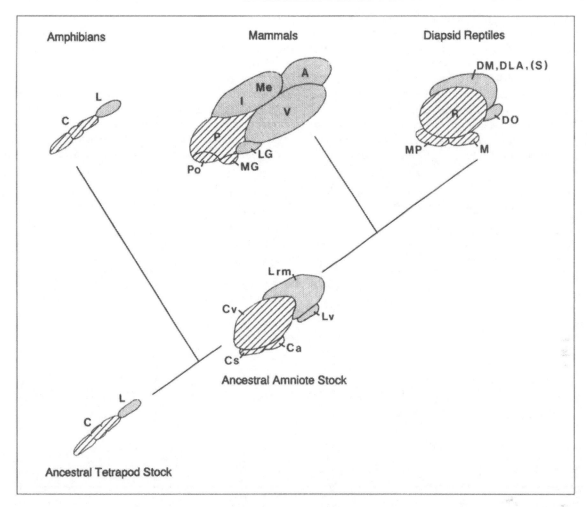

Fig. 24.42. Reconstruction of possible dorsal thalamic evolution in tetrapods: dendrogram illustrating the dorsal thalamus from an oblique dorsal view (rostral towards the *upper right*, caudal towards the *lower left*, medial towards the *upper left* and lateral towards the *lower right*) in ancestral tetrapods and amniotes and in extant amphibians, mammals and diapsid reptiles. Note the relatively greater expansion of the lemnothalamus in mammals as compared with diapsid reptiles. *A*, Anterior nuclear group; *C*, collothalamus; *Ca*, auditory-relay part of collothalamus; *Cs*, somatosensory-multisensory-relay part of collothalamus; *Cv*, visual-relay part of collothalamus; *DLA*, nucleus dorsolateralis anterior; *DM*, nucleus dorsomedialis; *DO*, dorsal lateral optic nucleus; *I*, intralaminar nuclear group; *L*, lemnothalamus; *LG*, dorsal lateral geniculate nucleus; *Lrm*, rostromedial part of lemnothalamus; *Lv*, lateral visual-relay part of lemnothalamus; *M*, nucleus medialis; *Me*, medial nuclear group; *MG*, medial geniculate body; *MP*, nucleus medialis posterior; *P*, lateral posterior-pulvinar complex; *Po*, posterior nuclear group; *(S)*, postulated lemniscal somatosensory relay; *V*, ventral nuclear group (from Butler 1995); reproduced by permission of S. Karger AG, Basel

rygians these fibres are concentrated in a massive tractus pallii (Johnston 1911a,b), which still awaits experimental exploration. In teleosts, the region of the tuberculum posterius gives rise to a number of migrated cell masses, collectively known as the preglomerulosus nuclear complex. Nuclei in this complex relay gustatory, mechanoreceptive lateral line and, in some species, also electrosensory information to the telencephalon (Fig. 24.47). The posterior tubercular region of cartilaginous fishes and amphibians is involved in comparable, though much less extensive, alternative routes to the telencephalon (see Fig. 24.46).

The *epithalamus* comprises the pineal complex and the habenular ganglion. The pineal complex is concerned primarily with the regulation of circadian rhythms. In several groups, including lampreys and lacertilians, the pineal complex is differentiated into separate pineal and parapineal organs. The habenular ganglion receives telencephalic afferent fibres by way of a pathway known as the stria medullaris, and its efferent fibres pass with the fasciculus retroflexus to the interpeduncular nucleus in the base of the midbrain. Edinger (1908, p. viii) remarked that the functional significance of this array of centres and pathways, which is a constant feature in all vertebrates from cyclostomes to man, is entirely unknown; little has changed in this respect during the intervening nine decades.

In all vertebrates, the *telencephalon* comprises three parts: the olfactory bulbs, the cerebral hemispheres and the telencephalon impar. In most vertebrates the *olfactory bulbs* are externally visible as separate structures, either directly apposed to the cerebral hemispheres or connected to them by stalks of variable length, called olfactory peduncles. Olfactory bulbs vary considerably in size, being entirely absent in certain cetaceans, minute in mormyrid teleosts, anolid lizards, most birds and primates, and very large, relative to the brain, in lampreys, hagfishes, many cartilaginous fishes, non-teleostean actinopterygians, turtles, and in many

Frame 5 – Diencephalon

The diencephalon can be divided into four principal regions, epithalamus, thalamus dorsalis, thalamus ventralis and hypothalamus (Fig. 24.43). Topographically, these four zones present themselves as dorsoventrally arranged, but embryological investigations have shown that two of these zones, the dorsal and ventral thalami, are direct derivatives of prosencephalic neuromeres, the prosomeres 2 and 3. Due to the strong curvature of the neural tube at diencephalic levels, these segments have become horizont-

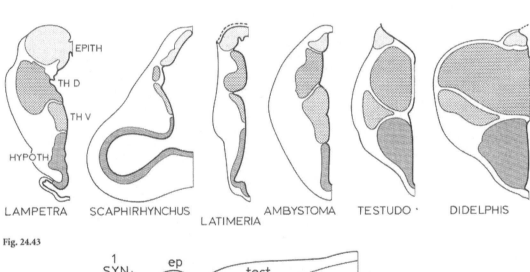

Fig. 24.43

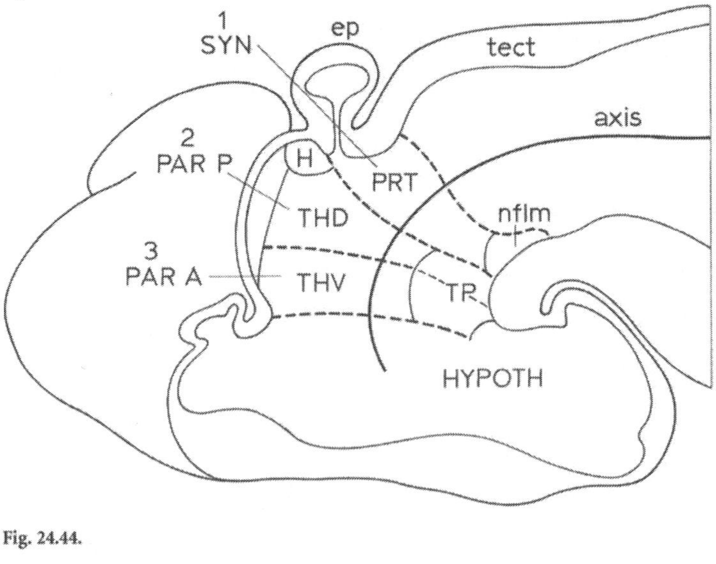

Fig. 24.44

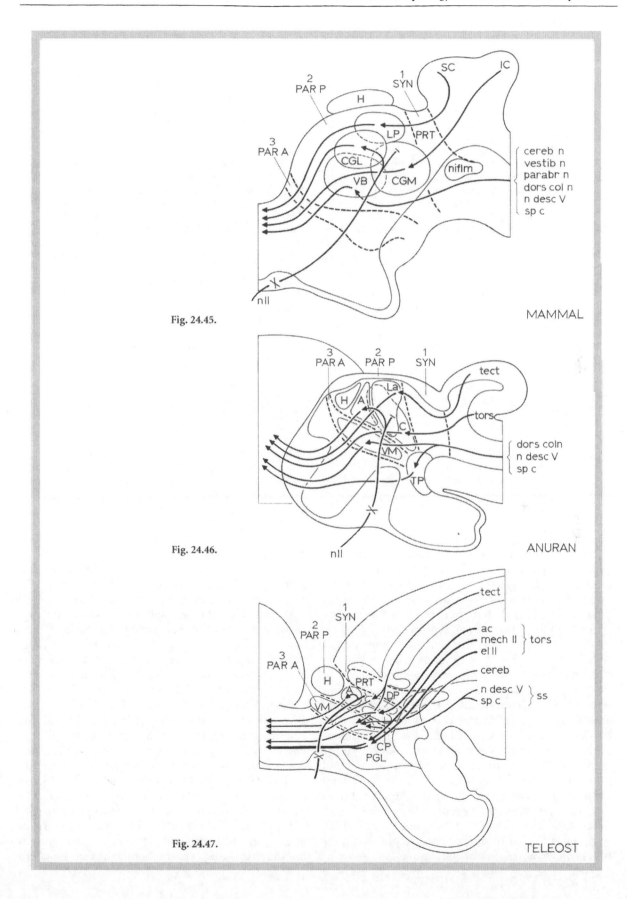

Fig. 24.45. MAMMAL

Fig. 24.46. ANURAN

Fig. 24.47. TELEOST

ally rather than vertically oriented. The second and third prosomeres are also known as the parencephalon posterius and anterius, respectively. Just like the dorsal thalamus, the epithalamic habenular ganglia arise from the parencephalon posterius. Caudally, the anterior and posterior parencephala converge into a cell mass situated in front of the tuberculum posterius, the nucleus tuberculi posterius. The first prosomere, or synencephalon, gives rise to the pretectal region and the nucleus of the fasciculus longitudinalis medialis. The hypothalamus is probably the product of two or more neuromeres which extend rostrally into the telencephalon, but at present there is no unanimity as to the number or nature of the segments involved. Figure 24.44, which is based on a graphical reconstruction of the diencephalon of a 16-mm embryo of the Australian lungfish, *Neoceratodus forsteri*, illustrates the relations discussed.

Thalamic nuclei relay sensory information to the telencephalon, but the cell masses involved in this function differ among the various groups of vertebrates. Thus in mammals, all of these nuclei are situated in the dorsal thalamus (Fig. 24.45), but in anurans the ventromedial thalamic nucleus, which forms part of the ventral thalamus, and the nucleus of the tuberculum posterius relay somatosensory information to the telencephalon (Fig. 24.46). In teleosts, finally, the dorsal and ventral thalami are both involved in relaying sensory information to the telencephalon. However, in this group the quantitatively most important ascending projections to the telencephalon arise from the so-called preglomerulosus complex (Fig. 24.47). The cells forming this complex have migrated rostroventrally from the region of the tuberculum posterius. The preglomerular nuclei receive ascending gustatory, acoustic, mechanosensory lateral line and, in some species also electrosensory lateral line inputs and project to the area dorsalis of the telencephalon (for references and details, see Chap. 15, Sect. 15.5.4).

Abbreviations: *A*, anterior thalamic nucleus; *ac*, acoustic part of torus; *C*, central thalamic nucleus; *cereb*, cerebellum; *cereb n*, cerebellar nuclei; *CGL*, corpus geniculatum laterale; *CGM*, corpus geniculatum mediale; *CP*, central posterior thalamic nucleus; *dors col n*, dorsal column nucleus; *DP*, dorsal posterior thalamic nucleus; *el l l*, electrosensory lateral line part of torus; *EPITH (ep)*, epithalamus; *H*, habenular nuclei; *HYPOTH*, hypothalamus; *IC*, inferior colliculus; *La*, nucleus lateralis thalami, anterior division; *LP*, lateralis posterior thalamic nucleus; *mech l l*, mechanosensory lateral line part of torus; *n desc V*, nucleus descendens nervi trigemini; *n gust sec*, nucleus gustatorius secundarius; *n (i) flm*, nucleus (interstitialis) of the fasciculus longitudinalis medialis; *nII*, nervus opticus; *PAR A*, parencephalon anterius; *parabr n*, parabrachial nuclei; *PAR P*, parencephalon posterius; *PGL*, preglomerulosus complex; *PRT*, pretectum; *SC*, superior colliculus; *sp c*, spinal cord; *ss*, somatosensory system; *SYN*, synencephalon; *tect*, tectum mesencephali; *TH D*, thalamus dorsalis; *TH V*, thalamus ventralis; *tors*, torus semicircularis; *TP*, nucleus tuberculi posterioris; *VB*, ventrobasal thalamic complex; *vestib n*, vestibular nuclei; *VM*, ventromedial thalamic nucleus; *1, 2, 3*, prosomeres (Fig. 24.44 is simplified from Bergquist 1932)

lizards, snakes, insectivores and rodents. As their name indicates, the olfactory bulbs receive their principal input from the olfactory epithelium in the nasal cavity. It is generally assumed that their size is directly proportional to the acuteness of the sense of smell and hints at the importance of this modality in the life of the animal. An *accessory olfactory bulb*, which is innervated by the vomeronasal organ, seems to be an adaptation to terrestrial life (Bertmar 1981). In caecilians, chemoreception in the vomeronasal organ is considered to be facilitated by a tentacle extending from the skull to the area of the orbit (Billo and Wake 1987).

The small *telencephalon impar* essentially maintains the early embryonic tubelike condition of the neuraxis. The structure consists of a thin roof and floor which, together with its thickened lateral walls, enclose the unpaired ventricular cavity. This structure is important because all ascending and descending fibre tracts which connect parts of the brain have to pass through it. In a single group, the holocephalean fishes, the telencephalon impar is drawn out into a remarkably long telencephalic peduncle (P2,3: 4).

The *telencephalic* or *cerebral hemispheres* (see Frames 6, 7) show an astonishing variability with respect to size, shape and structure. Together with the cerebellum, the hemispheres are by far the most variable parts of the vertebrate brain. As for size, in lampreys the telencephalic hemispheres are minuscule (though discernible; P2,3: 3), but in birds (P2,3: 36-41) and mammals (P2,3: 25-28) they are large and prominent, and in whales and primates the hemispheres grow out so enormously that they cover all other parts of the brain (P2,3: 29, 30), just like the cerebellum of mormyrid fishes (P2:17a; P3:18). With regard to shape, morphogenetic processes such as inversion, evagination, eversion, local thickening of their walls, narrowing and partial or even total obliteration of their ventricular cavities, coalescence in the median plane and enlargement of their external surface by formation of gyri, in different combinations, all contribute to the shaping of the cerebral hemispheres. The lateral and medial walls of the cerebral hemispheres consist in all vertebrates of dorsal, pallial and ventral, subpallial parts. In respect of the internal structure, in the telencephalic hemispheres of several vertebrate groups, including brachiopterygians, dipnoans and amphibians, the grey matter is confined mainly to a zone of periventricular grey, but in many other groups, including hagfishes, cartilaginous fishes, teleosts and amniotes, numerous telencephalic cells have migrated outward to form more peripheral neuronal aggregates. This tendency is particularly evident in the dorsal (pallial) parts of the mammalian cerebral hemispheres, where a superficially situated, multilayered neocortex is formed. Still of limited size in primitive mammals such as the opossum, the shrew and the hedgehog, this structure grows out in most mammals to become the single largest centre of their brains. A feature observed in the telencephalon of all amniotes is the formation of large cell masses in the lateral wall of the cerebral hemispheres, leading to the formation of thickenings and intraventricular protrusions of varying size. In mammals, the formation of such cell masses is confined to the ventral, subpallial parts of the hemispheres, but in reptiles and birds this process encroaches upon the dorsal, pallial parts of the hemispheric lateral walls. For this intraventricular protrusion, Johnston (1915) introduced the term dorsal ventricular ridge. In birds, almost the entire lateral wall of the hemispheres is involved in this expansion; consequently, the lateral ventricles in most of these animals are reduced to narrow, slitlike cavities.

Romer (1962, p. 544) stated: "The evolution of the cerebral hemispheres is the most spectacular story in comparative anatomy." This story, which features prominently in the works of such famous comparative neuroanatomists as Edinger (1908), Ariëns Kappers (1920/1921; 1929) and Herrick (1921, 1948), may be epitomised as follows.

1. The vertebrate telencephalic hemispheres began their history as purely olfactory centres. They received afferent fibres, originating from the olfactory bulbs, and their projections descended mainly to the epithalamic and hypothalamic regions of the diencephalon. Few fibres if any ascended from the more caudal parts of the brain to the telencephalon.
2. In most extant fish, the entire telencephalon is still dominated by the olfactory system, and even in amphibians no part of the cerebral hemisphere is wholly free from olfactory connections.
3. Gradually, the primitive rhinencephalon or 'smell brain' became invaded by fibres, ascending mainly from the dorsal thalamus and carrying non-olfactory sensory impulses to the telencephalon. This invasion led to the development of centres in which olfactory stimuli were correlated with impulses from other sources.
4. As often happens when invasions are involved, in later developments, the purely olfactory and olfactory correlation centres were pushed back by expanding non-olfactory centres, evolving in both the dorsal (pallial) as well as the ventral (subpallial) part of the cerebral hemisphere, until in man "the olfactory centers are crowded down into relatively obscure crannies of the hemisphere by the overgrown somatic systems" (Herrick 1921, p. 430).
5. The establishment of olfactory correlation centres and, notably, the formation and elaboration of non-olfactory areas led to an increasing expansion of the cerebral hemispheres, and this telencephalic expansion is the hallmark of the progressive evolution of the vertebrate brain.
6. It was recognised that the evolutionary development of the vertebrate brain, from a certain point onward, took a different course in mammals and birds. In mammals, the invasion of afferent fibres became associated with the development of the neocortex already alluded to, and this superimposed centre enabled these animals to make a detailed analysis of their environment and to attain remarkable motor skills and eventually became the seat of the highest mental faculties. In birds, on the other hand, structures thought to be equivalent to the mammalian corpus striatum expanded enormously, and these structures functioned in a way that was very different from that of the neocortex. "In birds we see a complex series of action patterns which may be called forth to meet a great variety of situations. Most are, however, stereotyped; the actions are innate, instinctive. The bird, its brain dominated by its basal nuclei, is essentially a highly complex mechanism with relatively little of the learning capacity which in mammals is associated with the development of the expanded neopallium" (Romer 1962, p. 597). And: "In the higher vertebrates, the birds and mammals, plasticity and flexibility are evident in all living species. Yet the birds developed along a different direction, perfecting to an unusual extent the fixed action pattern as the basic behavioral response to environmental requirements. It was really within the mammals that more flexible patterns of behavior have been the rule" (Jerison 1973, p. 433).

It is appropriate to comment on this story in the light of the new insights gained during the past several decades.

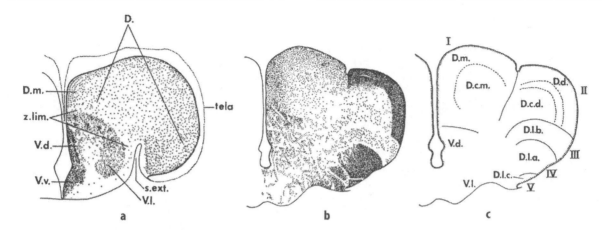

Fig. 24.48a–c. Transverse sections through the everted telencephalic hemispheres of actinopterygian fishes: **a** the bowfin, *Amia calva*; **b** the osteoglossomorph teleost, *Scleropagus formosus*; **c** the boundaries of the cell masses shown in **b**. The area dorsalis (*D*) or pallium is much further differentiated in *Scleropagus* than in *Amia* (reproduced from Nieuwenhuys 1962)

1. Its beginning is probably correct. In Amphioxus, which lacks an olfactory organ, there are no traces of any structures comparable to any parts of the vertebrate telencephalon. In hagfishes, the cerebral hemispheres have developed into a large and astonishingly highly differentiated secondary olfactory centre (including a multilayered cerebral cortex), and also in lampreys practically the entire telencephalon is dominated by the olfactory system. This is not to say, however, that the cyclostome's olfactory projections form the only afferent system to the cerebral hemispheres. Experimental studies (discussed in Chaps. 10 and 11) showed that the hemispheres in these animals receive ascending non-olfactory projections. In fact, this holds true even for the primary recipient centres of the olfactory system, the olfactory bulbs. Centres receiving their entire impulses from a single source only most probably do not exist at all in any CNS. This has implications for our concept of functional systems that will be elaborated later in this chapter.
2. Experimental neuroanatomical studies in gnathostome anamniotes have convincingly shown that the spread of secondary olfactory fibres over the cerebral hemispheres is much more restricted than was thought by previous investigators on the basis of non-experimental material. Thus, although there probably was an initial olfactory dominance, the 'de-olfactorisation' of the cerebral hemispheres has proceeded much more quickly than previously assumed.
3. In the classical view, the evolution of vertebrates has been envisioned as a linear process, proceeding from cyclostomes to mammals. Vertebrates have not evolved linearly, however. During vertebrate phylogeny four major radiations – agnathan, chondrichthyan, osteichthyan, tetrapod – have evolved and, remarkably, increases in telencephalic size and differentiation of cell masses occur in members of all of these radiations (Nieuwenhuys 1962; Northcutt 1977, 1981, 1984). Thus, the pallial parts of the cerebral hemispheres of osteoglossomorph teleosts are much further differentiated than those of chondrosteans and holosteans (Fig. 24.48), and these structural differences are strongly correlated with differences in the number of myelinated fibres connecting the telencephalon with the more caudally situated parts of the brain (Fig. 24.49). Since the ascending contingents of the fibre masses depicted do not originate from the dorsal thalamus, but rather from the tuberculum posterius region, i.e. an entirely different diencephalic province, and we know that similar or alternative routes to the telencephalon exist in chondrichthyans and in amphibians, the conclusion seems warranted that the evolution of the cerebral hemispheres is even a more spectacular story than our predecessors envisioned.
4. The idea that the avian cerebral hemispheres essentially consist of a greatly expanded corpus striatum came about because previous authors, knowing that the mammalian striatum develops from a thickening of the lateral walls of the hemispheres, felt justified in interpreting all laterally situated telencephalic structures that protrude into the lateral ventricle as corpora striata, or striatal complexes (Fig. 24.50), and then attached far-reaching functional conclu-

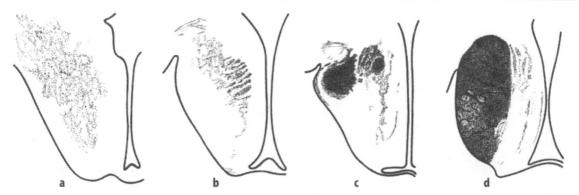

Fig. 24.49a–d. Transverse sections through the telencephalic peduncle in actinopterygian fishes, drawn from Weigert-Pal preparations: **a** the sturgeon, *Acipenser ruthenus*; **b** the bowfin, *Amia calva*; **c** the trout, *Salmo fario*; **d** the osteoglossomorph teleost, *Gnathonemus elephas*. In *Acipenser* and *Amia* the myelinated fibres connecting the telencephalon and the diencephalon constitute a diffuse bundle. In *Salmo* a medial and a lateral forebrain bundle can be distinguished. The lateral forebrain bundle is strongly developed in *Gnathonemus* (reproduced from Nieuwenhuys 1962)

sions to that interpretation. For example, Rabl-Rückhard (1883) interpreted the entire solid, telencephalic hemispheres of teleosts that expand into the ventricles as corpora striata and believed that the overlying membrane-like roof represented the pallium. On the basis of this interpretation, he arrived at the sweeping conclusion that the entire fish brain merely represents a dumb reflex mechanism: "... das Fischgehirn ist gewissermassen nur ein blödsinniger Reflexapparat" (Rabl-Rückhard 1883, p. 311). Similar, though somewhat less derogatory, conclusions have been attached, as we have seen, to the interpretations of the avian hemispheres as expanded corpora striata. The lateral walls of the caudal hemispheres of holocephalian, cartilaginous fishes are also greatly thickened and, hence, have also been interpreted as expanded corpora striata (Ariëns Kappers and Carpenter 1911), and all that has been said about birds holds in the broad outline also for reptiles, where the lateral hemispheric walls also participate in the formation of large intraventricular protrusions (i.e. the dorsal ventricular ridge).

Ontogenetic studies have clearly shown that in teleosts, reptiles and birds the dorsal parts of the intraventricular protrusions are pallial in nature and have nothing to do with the mammalian or indeed vertebrate striatum. Corresponding ontogenetic data are lacking for holocephalians; however, comparisons with the adult telencephalon of other cartilaginous fishes make it likely that in these forms also the dorsal parts of the thickened lateral hemispheric walls are of pallial origin.

Once it was accepted that the large dorsal parts of the reptilian and avian 'striatum' are pallial structures, the question arose as to whether neuronal aggregates present in the thickened pallia of these groups are homologous as individual structures to particular areas in the mammalian neocortex. The voluminous literature devoted to this problem has been extensively covered in Chaps. 4, 20 and 21, and briefly in Frame 7. It is noteworthy,

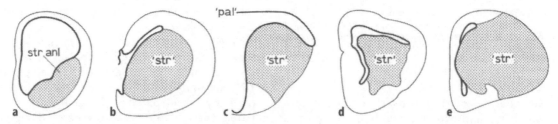

Fig. 24.50a–e. Diagrammatic transverse sections through the telencephalic hemispheres of (**a**) mammalian embryo, (**b**) holocephalian fish, (**c**) teleost (after Rabl-Rückhard 1883), (**d**) lizard, and (**e**) bird. In **a** the ganglionic eminences, from which the corpus striatum develops, are indicated. In **b**–**e** the structures which, on the basis of their intraventricular protrusion, were formerly interpreted as 'corpus striatum' are marked. '*pal*', Membranous pallium, according to the interpretation of Räbl-Rückhard (1883); '*str*', corpus striatum, according to previous interpretations; *str anl*, striatal anlage in the mammalian cerebral hemispheres

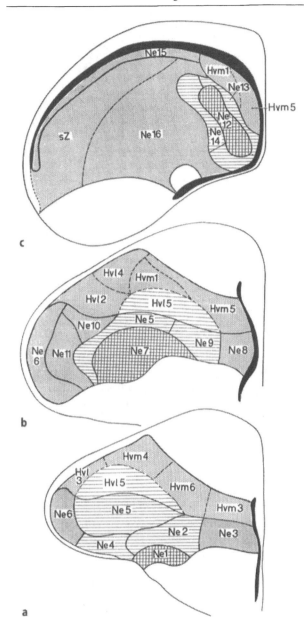

Fig. 24.51a–c. Diagrammatic transverse sections through the rostral (**a**), intermediate (**b**) and caudal parts (**c**) of the telencephalic hemispheres of the pigeon, showing the architectonic parcellation of the neostriatum and the ventral hyperstriatum. *Cross-hatching*, primary areas; *hatching*, secondary areas; *dots*, tertiary areas. *Hv*, Hyperstriatum ventrale; *l*, lateral; *m*, medial areae; *Ne*, neostriatum (reproduced from Rehkämper and Zilles 1991)

however, that recent studies using image analysis, histochemistry and experimental hodological data have led to a remarkable 'upgrading' of the avian 'striatum'. Thus, Rehkämper and Zilles (1991) reported that the complex in the avian pallium, constituted by the regions known as neostriatum and ventral hyperstriatum, and the mammalian neocortex, notwithstanding their profoundly different architecture ('nuclear' versus 'laminar'), show a strikingly similar functional organisation (Fig. 24.51). Specifically: (a) Both entities comprise primary target areas of the three major sensory systems (visual, auditory and somatosensory). (b) Both contain a primary output area. (c) In both, the primary sensory areas are surrounded by secondary areas, which are strongly interconnected with the respective sensory areas. (d) In both, a large region is occupied by tertiary areas; these are defined by massive intra-hemispherical connections with primary and secondary sensory areas on one side and output areas on the other side. (e) These tertiary areas represent in both classes regions of substantial integration, to be regarded as the morphological substrate of complex psychological functions, notably learning abilities, and consequently of a very flexible behavioural potential. According to Rehkämper and Zilles (1991), these similarities indicate that the same strategy of differentiation has been followed in the avian and mammalian telencephalon. They believed that similar parallel processes have occurred frequently in the evolution of the vertebrate brain, and they attached the following sweeping statement to their conclusions. "The task of contemporary comparative morphology is the analysis of the impact of function upon a given structure and the evaluation of its adaptive character and biological advantage. The question of homology (plesio-, apomorphy) has no priority" (Rehkämper and Zilles 1991, p. 24).

There can be no doubt that, whenever possible, structure and function should be studied conjointly, and obviously, the study of homoplastic structures such as the avian striatum and the mammalian neocortex is of paramount importance for our understanding of the evolution of the CNS. It should be pointed out, however, that the statement cited, in the light of our present-day knowledge of the structural and functional organisation of the brains of most vertebrate groups, is of a very limited operational validity, and that much work remains to be done in the realms of pure and phylogenetic morphology.

In the preceding pages a survey has been presented of the morphology of the vertebrate brain and spinal cord. This survey was confined primarily to the macromorphological level, though there were several micromorphological excursions. It was apparent that the brains of vertebrates, though built on a common plan, show amazing structural variety. It also became evident that differences in the development of particular sensory and motor systems may manifest themselves clearly at the macroscopic level, and that, hence, inspection of the gross morphology of brains may yield important information about the functional capacities, and possibly even the lifestyle of animals (though such information can be gathered more directly of course). Although these conclusions can have practical value, the functional significance of numerous other gross anatomical features evades us completely. It was pointed out that, contrary to the assumptions of many of our distinguished predecessors, the evolution of the telencephalon and its dependencies cannot be viewed as one linear series of increasing size and complexity. Rather, expansion and progressive differentiation of the telencephalon have occurred in all major vertebrate radiations. Finally, a brief summary of the work of Rehkämper and Zilles (1991) on the functional organisation of the avian 'striatum' served to emphasise the importance of the study of homoplastic structures for our understanding of neural adaptation.

Frame 6 – The telencephalic hemispheres. I. Anamniotes

The telencephalic hemispheres of anamniotes show considerable differences with regard to their development and general morphology. In hagfishes, paired anterior and posterior evaginations of the prosencephalic vesicle are formed (Wicht 1996). The caudal walls of the anterior vesicle and the rostral walls of the posterior vesicle join and, projecting rostromedially, fill the ventricular cavities of the anterior vesicles completely. Hence, in adult myxinoids, the telencephalic hemispheres are solid bodies (Fig. 24.52) in which the obliterated ventricular cavities are represented only by ependymal vestiges (*dashed curve* on the *right side* in Fig. 24.52). In lampreys (Fig. 24.53), evagination of the rostral parts of the prosencephalic walls leads to the formation of small but distinct lateral ventricles, but the evagination is incomplete, because the morphologically most dorsal parts of the pallium remain in the unevaginated portion of the prosencephalic vesicle. Although evagination is also the essential morphogenetic process in the telencephalon of cartilaginous fishes (Fig. 24.54), the forebrains of the various groups of this class show considerable structural differences. Thus, in holocephalians the ventrolateral hemisphere walls are greatly thickened, and in batoids, as well as in some groups of sharks, the lateral ventricles become reduced to short, horn-shaped rudiments. In actinopterygians, an everted forebrain is found, that is, the side walls of the telencephalon have recurved laterally during ontogenesis. This mode of development causes a considerable widening of the telencephalic roof plate. In cladistians (Fig. 24.55) a simple eversion is found, but in chondrosteans, holosteans (Fig. 24.56) and teleosts the recurvature is coupled with a considerable thickening of the telencephalic walls. The forebrain of dipnoans, like that of amphibians (Fig. 24.61), is evaginated into distinct cerebral hemispheres. However, within this group there exists an important structural difference between the forebrains of lepidosirenid lungfishes (Fig. 24.59) and that of the Australian lungfish, *Neoceratodus forsteri* (Fig. 24.58). In the former, the hemispheres consist entirely of nervous tissue, but in the latter a considerable part of the forebrain wall is formed by an ependymal membrane.

The telencephalon of the sole surviving crossopterygian, *Latimeria chalumnae*, shows a highly interesting structure intermediate between the forebrain of the actinopterygians on the one hand and the dipnoans and tetrapods on the other. In this fish (Fig. 24.57), the dorsal part of the forebrain wall is thickened and shows a slight tendency towards eversion, with widening of the ependymal roof plate, but the ventral region, in contrast, is thin-walled and clearly evaginated.

It is generally agreed that the telencephalon of the various groups of fish can be divided into a dorsal pallium and a ventral subpallium, and that within each of these two principal parts several (commonly three) separate cell masses or complexes can be distinguished. In Fig. 24.60, structural plans of the anamniote cerebral hemispheres, as constructed by Holmgren (1922), Kuhlenbeck (1929b) and Rudebeck (1945), as well as a morphotype of the forebrain of gnathostomes recently generated by Northcutt (1995), are shown, and in Table 24.1 the nomenclatures used by these authors are compared with that of Herrick (1948) for urodeles. In spite of the semantic differences, there is almost complete agreement among the authors cited concerning the basic subdivision of the telencephalic hemispheres of anamniotes. Nevertheless, for most groups different opinions have been expressed concerning the exact location of the palliosubpallial boundaries, as well as concerning the delineation and interpretation of the various cell masses. In what follows, some current controversies and recent reinterpretations will be highlighted.

1. Until recently, the interpretation of the parts of the hemispheres of myxinoid cyclostomes presented in Jansen's (1930) classical study of the brain of *Myxine glutinosa* was generally accepted. According to that interpretation, the superficial zone of the hemispheres consists of a highly differentiated, five-layered olfactory cortex, the central core is occupied by a primordium hippocampi, and a ventromedially situated area basalis represents, among other things, the corpus striatum (Fig. 24.52, *left side*). However, Wicht and Northcutt (1994) recently concluded, on mainly embryological and topological grounds, that the primordium hippocampi, designated by them as the cent-

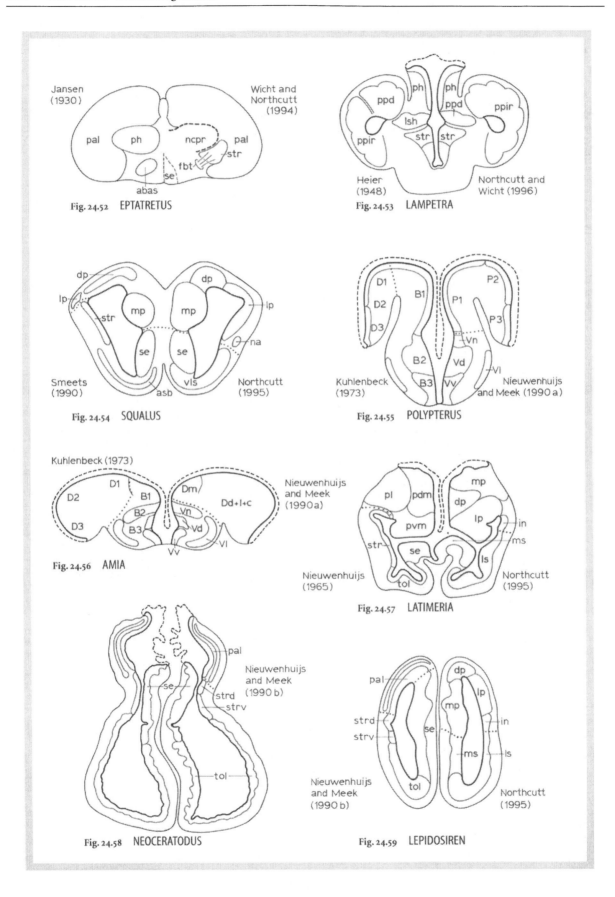

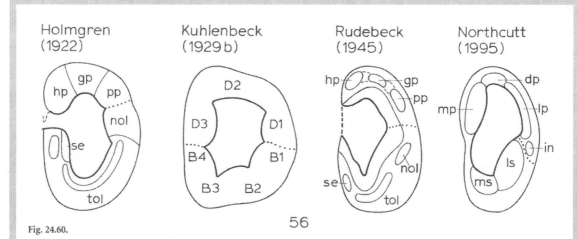

Fig. 24.60.

ral prosencephalic nucleus, is the product of a rostromedial projection of the lateral prosencephalic wall and most probably represents a derivative of the ventral thalamus. These authors also delineated a rostromedially situated septum and, using the basal forebrain bundle as a guide, they concluded that the striatum is located dorsolateral to Jansen's area basalis (Fig. 24.52, *right side*).

2. Another important reinterpretation concerns the pallium of petromyzontoids. According to Heier (1948), this pallium comprises an unevaginated primordium hippocampi and, in the hemisphere, the primordium pallii dorsalis and the primordium piriforme. A fourth area, the lobus subhippocampalis, represents the transitional field between the unevaginated and evaginated parts of the pallium (Fig. 24.53, *left*). However, Northcutt and Wicht (1997) interpreted the lobus subhippocampalis on positional grounds as the primordium pallii dorsalis and considered it likely that the primordium pallii dorsalis and the primordium piriforme of Heier's interpretation, both of which exhibit extensive reciprocal connections with the olfactory bulb, together represent the primordium piriforme or lateral pallium.

3. In his recent study on the forebrain of anamniotes, Northcutt (1995) positioned the lateral pallio-subpallial boundary more ventrally than most previous authors. Thus, he included a cell mass considered so far as being subpallial (nucleus A in selachians: Fig. 24.54; cell mass Vn in cladistians: Fig. 24.55; the dorsal part of the striatum in lepidosirenid lungfishes: Fig. 24.59, and in *Latimeria*: Fig. 24.57) as belonging to the pallium. He designated this cell mass as nucleus intercalatus and interpreted it as being a pallial amygdalar nucleus. Remarkably, the entire cell mass in the telencephalon of the spiny dogfish, interpreted as striatum by Smeets (1990; Chap. 12 of the present work; Fig. 24.54, *left*), was included in the lateral pallium by Northcutt (1995; Fig. 24.54, *right*).

4. The telencephalon of chondrosteans, holosteans and teleosts has been subdivided in many different ways. A discussion of all of these subdivisions is beyond the scope of the present synopsis. Suffice it to mention that Kuhlenbeck (1973; Fig. 24.56, *left*) positioned the pallio-subpallial boundary (indicated as a *dotted line* in Figs. 24.54–24.60) more dorsally than did Nieuwenhuys and Meek (1990a; Fig. 24.56, *right*) and most others, a decision which had considerable repercussions on the interpretation of the adjacent cell masses.

5. Nieuwenhuys (1965) and Nieuwenhuys and Meek (1990b) subdivided the thickened 'corpus pallii' of *Latimeria* into ventromedial, dorsomedial and lateral formations (Fig. 24.57, *left*). They were of

Table 24.1. Cell masses in the cerebral hemispheres of anamniotes: comparison of nomenclatures

Holmgren (1922) Rudebeck (1945)		Kuhlenbeck (1929b)	Herrick (1948) (urodeles)		Northcutt (1995)	
hp	hippocampal pallium	D3	ph	prim. hippocampi	mp	medial pallium
gp	general pallium	D2	ppd	prim. pallii dors.	dp	dorsal pallium
pp	piriform pallium	D1	pp	prim. piriforme	lp	lateral pallium
nol	n. olfact. lat.	B1	Strd	striatum dorsale	in	n. intercalatus
tol	tuberculum olfactorium	B2	Strv	striatum ventrale	ls	lateral subpallium
se	septal nuclei	B3+4	acc	n. accumbens	ms	medial subpallium
			se	septal nuclei		

prim, Primordium

the opinion that one of these, the ventromedial formation (*and not* the dorsomedial one, as was erroneously stated in Nieuwenhuys and Meek 1990b), which surrounds the central olfactory tract is directly comparable to the lateral, piriform or P1 pallial field of other gnathostomes. Northcutt (1995: Fig. 24.57, *right*) proposed a somewhat different subdivision of the pallium of *Latimeria* into (topologically) medial, dorsal and lateral areas.

6. There is a major disagreement concerning the location and extent of the pallium in lepidosirenid lungfishes. Northcutt (1986, 1995) and Reiner and Northcutt (1987) compared the telencephalon of this dipnoan group directly to that of amphibians. On the basis of topographical and histochemical criteria, they claimed that the lepidosirenid pallium can be divided into lateral, dorsal and medial formations, and that the latter occupies the dorsal part of the medial hemisphere wall (Fig. 24.59, *right*). Nieuwenhuys and Meek (1990b), who compared the telencephalon of lepidosirenids to that of their closest extant relative, the Australian lungfish (Fig. 24.58), arrived at a quite different interpretation. According to them, the dipnoan pallium is confined to the dorsolateral hemisphere wall and cannot be divided into separate fields (Figs. 24.58, 24.59, *left*).

7. Northcutt (1995) united the ventrolateral and ventral subpallial formations described by several previous authors (cf. Figs. 24.54, 24.57, 24.59, 24.60), on the basis of immunohistochemical data, into a single lateral zone.

This cursory survey may suffice to make clear that several parts of the forebrain of amniotes have recently been reinterpreted, and that many basic questions concerning the organisation of the cerebral hemispheres of these so-called lower vertebrates are still awaiting a solution.

Abbreviations: *abas*, area basalis; *asb*, area superficialis basalis; *B1,2,3*, subpallial areas; *D1,2,3*, pallial areas; *Dc, Dd, Dl, Dm*, central, dorsal, lateral and medial parts of the area dorsalis telencephali; *dp*, dorsal pallium; *fbt*, fasciculus basalis telencephali; *in*, intercalated nucleus; *lp*, lateral pallium; *ls*, lateral subpallium; *lsh*, lobus subhippocampalis; *mp*, medial pallium; *ms*, medial subpallium; *na*, nucleus A, *ncpr*, nucleus centralis prosencephali; *pal*, pallium; *ph*, primordium hippocampi; *ppd*, primordium pallii dorsalis; *ppir*, primordium piriforme; *se*, septum; *str*, striatum; *strd, strv*, dorsal and ventral parts of striatum; *tol*, tuberculum olfactorium; *Vd, Vl, Vn, Vv*, dorsal, lateral, dorsalmost and ventral parts of area ventralis telencephali

Frame 7 – The telencephalic hemispheres. II. Tetrapods

In all tetrapods the cerebral hemispheres are of the evaginated type (Figs. 24.61–24.65). In the amphibian hemispheres the grey matter forms a continuous periventricular zone (Fig. 24.61), in which homologues of the pallial and subpallial areas, present in most fish groups (Fig. 24.60), can be clearly recognised. The lateral pallio-subpallial boundary in the brains of reptiles and birds has been much debated in the literature, but thanks to the results of embryological, immunohistochemical and experimental hodological studies, their location can now be considered well-established (cf. *dotted lines* in Figs. 24.62 and 24.64). In general, the homologisation of the various subpallial cell masses in amniotes offers no major problems. It is important to note that during early development the amniote telencephalic hemispheres present themselves as thin-walled vesicles, but that during later development the ventrolateral hemispheric walls thicken to form elongated intraventricular protrusions. In mammals, two such protrusions, known as the medial and lateral ganglionic eminences (Fig. 24.65b: *a,b*), give rise to the caudate nucleus and the putamen.

Whereas in the amphibian pallium the cells are concentrated in the periventricular zone, the pallium of reptiles contains a well-differentiated cerebral cortex, extending from the medial to the lateral pallio-subpallial boundary (Fig. 24.62). This cortex is composed of three layers, and cytoarchitectonic differences allow a subdivision into four longitudinally arranged zones: the medial, dorsomedial, dorsal and lateral cortex. The reptilian pallium also comprises a large subcortical complex, which protrudes into the lateral ventricle and is known as the (anterior) dorsal ventricular ridge. In the rostral part of the hemispheres a cell mass is present between the dorsal cortex and the dorsal ventricular ridge. This cell mass, the pallial thickening, presents itself in some species as an infolding of the lateral part of the dorsal cortex.

Three-layered cortical formations are also present in the mammalian telencephalon, where they form the laterally situated olfactory, or prepiriform cortex and the medially situated hippocampal formation. However, in all mammals these two 'simple' areas are separated by a highly complex six-layered cortical formation, the neocortex or isocortex (Fig. 24.63). In primitive mammals such as the opossum or the hedgehog the extent of the neocortex is limited, but in some mammalian groups, particularly the primates and the cetaceans, this formation attains amazing dimensions and becomes by far the largest centre of the brain.

The avian pallium comprises a thin-walled medial and thick-walled lateral portion. The thin-walled portion contains two poorly differentiated cortical formations, known as the hippocampal and parahippocampal cortices. The latter is laterally continuous with a partially laminated structure, the eminentia sagittalis, or Wulst, which forms the most dorsal part of the avian hemispheres. The Wulst includes the hyperstriatum accessorium, an intercalated nucleus, the hyperstriatum intercalatum supremum and the hyperstriatum dorsale. The thickened lateral part of the pallium is homologous to the reptilian (anterior) dorsal ventricular ridge. It is composed of a number of cell masses, stacked one upon the other and known as the hyperstriatum ventrale, the neostriatum, the ectostriatum, and the nucleus basalis. Functionally, it is important to note that the reptilian and avian dorsal ventricular ridges and the mammalian neocortex all receive topographically organised somatosensory, visual and acoustic projections (see Frame 8), originating from discrete dorsal thalamic nuclei.

As regards the homology of the various parts of the reptilian, avian and mammalian pallium, the following hypotheses have been put forward:

1. Källén (1955, 1962) held that the reptilian and avian dorsal ventricular ridges, although of pallial origin, stem from a separate migration layer dIII, which produces no separately recognisable cell masses in the mammalian brain (cf. Chap. 4, Fig. 4.51).
2. Kuhlenbeck (1973) emphasised that the lateral pallial field D1 of anamniotes (Fig. 24.60) is not homologous to the lateral, olfactory cortex of reptiles and mammals. In his opinion, D1 becomes included into the basal nuclear complex of sauropsids and mammals by a thickening or 'introversion' of the ventrolateral hemisphere wall. Kuhlenbeck believed that in birds D1 expands enormously and gives rise to the entire dorsal striatal complex, and that in mammals D1 forms the lateral ganglionic eminence, which contributes to the caudate nucleus and the putamen (cf. Chap. 4, Fig. 4.50).
3. Northcutt (1981, 1987; Northcutt and Kaas 1995) conjectured that the mammalian isocortex and the reptilian and avian dorsal ventricular ridges evolved independently from different pallial fields of an amphibian-like ancestor. The isocortex arose in synapsid reptiles and mammals by a differentiation from the ancestral dorsal pallium, whereas the dorsal ventricular ridge in sauropsids (living reptiles and birds) arose by the enlargement of the ancestral lateral pallium. Thus, according to Northcutt, the isocortex and the dorsal ventricular ridge are homoplastic rather than homologous structures.
4. Karten (1969) speculated that the structural differences between the pallium of mammals on the one hand and that of reptiles and birds on the other reflect different patterns of alignment of pallial neurons rather than a fundamental difference in identity between the respective neuronal populations themselves. He believed that an embryonic structure comparable to the mammalian lateral ganglionic eminence is present in all amniotes

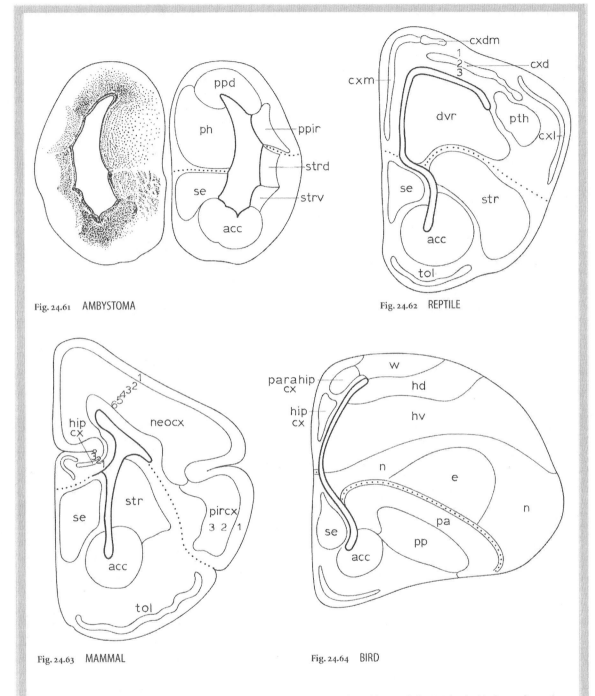

Fig. 24.61 AMBYSTOMA

Fig. 24.62 REPTILE

Fig. 24.63 MAMMAL

Fig. 24.64 BIRD

and argued that neurons, produced by the matrix zone of this eminence, accumulate in the dorsal ventricular ridge in reptiles and birds but contribute to the formation of the isocortex in mammals (*arrows 1* and *2*, respectively, in Fig. 24.65b). Butler (1994b) carried out a cladistic analysis of multiple features of the dorsal pallium in amniotes and arrived at the conclusion that the sets of neurons forming the dorsal ventricular ridge, the pallial thickening and the dorsal cortex in reptiles and those forming the dorsal ventricular ridge and the Wulst in birds are homologous to specified sets of cells in the mammalian isocortex. Figure 24.66 shows how the ontogenetic events postulated by Karten (1969) could account for Butler's (1994b) results.

5. Bruce and Neary (1995) recently questioned the homology between the dorsal ventricular ridge and isocortex and suggested that the reptilian dorsal ventricular ridge is related to the basolateral amygdalar complex of mammals.

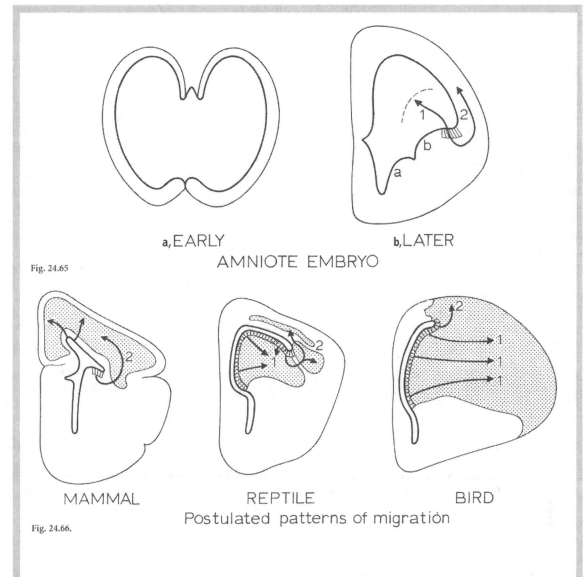

Fig. 24.65

Fig. 24.66. Postulated patterns of migration

Abbreviations: *a*, *b*, ganglionic eminences in the embryonic amniote telencephalon; *acc*, nucleus accumbens; *cxd*, cortex dorsalis; *cxdm*, cortex dorsomedialis; *cxl*, cortex lateralis; *cxm*, cortex medialis; *dvr*, dorsal ventricular ridge; *e*, ectostriatum; *hd*, hyperstriatum dorsale; *hipcx*, hippocampal cortex; *hv*, hyperstriatum ventrale; *n*, neostriatum; *neocx*, neocortex; *pa*, palaeostriatum augmentatum; *parahip cx*, parahippocampal cortex; *ph*, primordium hippocampi; *pircx*, piriform cortex; *pp*, palaeostriatum primitivum; *pd*, primordium pallii dorsalis; *ppir*, primordium piriforme; *pth*, pallial thickening; *se*, septum; *str*, striatum; *strd*, striatum dorsale; *strv*, striatum ventrale; *tol*, tuberculum olfactorium; *w*, Wulst; *1,2* etc., in Fig. 24.63: cortical layers; *1,2* in Figs. 24.65, 24.66: postulated patterns of migration of pallial cells; *dotted lines*, pallio-subpallial boundaries

24.3
Neuroembryology

It is obvious that ontogenetic studies are indispensable for a proper understanding of many macro- and micromorphological features of the CNS. During the past 150 years numerous neuroembryological studies have appeared, and the principal results of many of these have been incorporated in the present work. It is noteworthy, however, that during the past century only one group – formed by members of the Swedish or 'Holmgren' School of neuroembryology – thoroughly and systematically studied the ontogenetic development of the CNS in representatives of all major vertebrate radiations. The paucity of research of this type may be related to the effort required to prepare and examine material of this type.

The findings of the Swedish School have been incorporated in the general (Chaps. 4, 5) as well as in the specialised part (Chaps. 10, 12, 16, 18–21) of the present work. Amongst the major achievements of this school have been the demonstration of the segmental organisation of the vertebrate brain and the emphasis on its importance (e.g. Rendahl 1924; Bergquist 1932; Bergquist and Källén 1954, Fig. 24.67). It should be added that these accomplishments were made during a period when these structures were considered to be either of minor importance or non-existent.

During the past decade, however, the idea that the CNS is segmented has moved to the fore. We know that sets of genes which are highly conserved in evolution specify the boundaries of these segments, that many axonal tracts are formed at their boundaries, and that cells do not cross the borders of neuromeres (Chap. 4, Sect. 4.5). The neuromeric organisation of the spinal cord, and of most of the brain stem and diencephalon, has now been firmly established, but no consensus has yet been reached on the number and nature of the segments that participate in the formation of the hypothalamus and telencephalon (Chap. 4, Fig. 4.37). It will be clear that, if it could be proven that the brain in all vertebrate lineages develops from a set of corresponding neuromeres (as Bergquist and Källén 1954, anticipated), the resulting segmental paradigm would provide a new basis for comparative neuroanatomy. Much work remains to be done in order to achieve this goal, and it must be said that some investigators (e.g. Macdonald et al. 1994; Alvarez-Bolado et al. 1995; Guthrie 1995) doubt the existence of a segmental organisation in the forebrain.

The expression patterns of region-specific genes are likely to be of great importance for the interpretation of the parts of the CNS. For example, comparison of the anterior expression boundary of a particular homeobox gene (AmphiHox 3) in Amphioxus with that of a corresponding gene (Hox 2.7) in the mouse, the chicken and the frog strongly suggests that the vertebrate brain is homologous to an extensive region of the Amphioxus nerve cord, and not – as was previously thought – only to the region surrounding the tiny cerebral ventricle in the tapered rostral-most part of the cord in Amphioxus (Holland et al. 1992; Holland and Garcia-Fernández 1996).

From a morphological point of view, it is highly important to know where the various region-specific genes come to expression; biologically, it is even more important to know how these genes are 'translated' into morphology, that is, how the genes

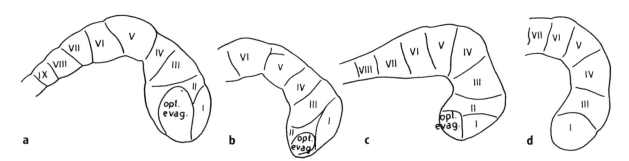

Fig. 24.67a–d. Neuromere in the salamander, *Ambystoma punctatum*. Drawings of wax-plate reconstructions of (**a**) the brain of a normal embryo and (**b–d**) operated embryos, in which the rostral part of the neural plate has been lifted up and the corresponding part of the substratum has been removed. In **b** and **c** small optic evaginations are present, while in **d** these structures are entirely lacking. Bergquist and Källén concluded from these experiments that the second neuromere is not a true neuromere, but only a secondary phenomenon, related to the evagination of the optic vesicles (reproduced form Bergquist and Källén 1955)

and their products act in concert with local epigenetic factors to shape the brain.

The modern neo-Darwinian theory of evolution states that gene mutations, acting in an environmental context, are the foundation of all evolutionary morphological changes. Within this framework, two questions have to be posed: (a) What has been the nature of the genomic changes which caused the changes in embryological development that led to morphological changes, and (b) how were these genomic changes channelled through, and structured by, the pathways and programs of development? The issue here is extremely complex since it involves, for instance, trying to decide how a sequence of unknown, random, gene mutations interacted with an environment that existed hundreds of millions of years in the past. Complete answers to such questions seem unattainable, but even partial answers will require the joint efforts of many specialists, including geneticists, molecular biologists, developmental neurobiologists and comparative neuroanatomists. This seems to be one of the major challenges and one of the major opportunities today.

24.4
Cell Masses and Fibre Connections

The cell masses within the gross morphological divisions and the connections between those masses form the next 'level of resolution'. Our goal will again be to distil down the numerous data presented in previous chapters to some general conclusions. This task is far from easy, and the results will not be as satisfying as could be wished; the origins of the difficulties will perhaps be as illuminating as our conclusions.

As an essential first step before presenting a synopsis of major results on cell masses and fibre connections, the limitations of the methods employed to obtain the data should be addressed. Much of this has already been discussed in Chaps. 2, 3 and 7, but it bears brief here, because without an awareness of the limitations of a technique the achievements cannot be fully comprehended.

The fundamental technique for revealing cell masses is the Nissl stain, which has been in use since its discovery in 1885 and is now known to stain the rough endoplasmic reticulum in the perikaryon of neurons and their nuclei (only the nuclei of glia stain, so they do not confuse the picture). Using this stain, cell aggregates are essentially revealed by the density and size of the cell bodies. In practice, analysis is usually carried out on a single specimen, judged to have been well fixed and stained, and details such as sexual dimorphism or, in the case of many animals collected in the wild, exact age or even species are not taken into account. Ideally, the cell masses revealed by the Nissl technique would have two identifying characteristics: (a) clear-cut boundaries of the mass and (b) well-defined location relative to brain landmarks, such as within a brain structure – tectum, brain stem and so on – and relative to, say, the ventricular surfaces. In practice, these criteria are often far from being satisfied. The subdivision of a mass of cells into different parts frequently involves subjective judgement. This holds in particular for secondarily simplified, cytoarchitectonically undifferentiated ('hidden nucleisation') brains such as those of urodeles and lepidoserinid lungfishes. Attempts have been made to remove some of the subjectivity by using image analysis (e.g. Zilles et al. 1978, 1982; Rehkämper et al. 1985), but this approach is limited, too, by the fidelity of the staining itself and by individual variation. Cell masses may also be far from convenient landmarks, which themselves are often quite variable. Thus, even at the first level of analysis – the taking of a simple inventory of what cell masses are there – uncertainties arise.

One way to improve on the Nissl method is to employ more recent immunohistochemical techniques that reveal the chemoarchitecture (Chap. 2, Sect. 2.9) of the brain. In this way, specific cell masses can be better delineated on the basis of, say, tyrosine hydroxylase reactivity or dopamine itself. This approach certainly has merit, and the data can be valuable in defining cell masses such as the striatum (see Smeets and Reiner 1994; Northcutt 1995), but it is also evident that each of the numerous antibodies now available is likely to reveal a different cell group, or to be of no help if a group of cells does not contain the substance in question. This approach, therefore, is at best an adjunct to the Nissl stain.

This, then, is the starting point assumed throughout this book: a collection of imperfectly delineated cell masses which nevertheless can be placed in some greater context, usually the macromorphology of the brain, within a given structure relative to a landmark. At this point some system of designations can be used to refer to the masses by convenient names, numbers or letters, and these can be chosen arbitrarily. The underlying assumption, however, is that if a similar specimen were prepared in the same way, similar masses could be identified and the same names ascribed to them.

From the perspective of the present work, the second, and crucial, phase of investigation is the construction of homologies between cell masses in different species. This may be more difficult than the first phase, and it is also clear that if the first phase has been relatively unsuccessful, the second phase will also have little hope of concluding well.

The complexities of establishing homologies has been dealt with in Chap. 6; here we just touch upon some major issues. One of these is that the relative or topological position of a given cell mass may be hard to establish, and consequently the first principle of homology is invalidated. Ontogenetic studies may be helpful in establishing the primary topological position of cell masses and, hence, may be of paramount importance for the solution of homology problems, but in many cases such material is unavailable. Another difficulty frequently encountered is that the number of cell masses delineated within a given region differs among the species studied and that the relationships of these cell masses are not clear. Here, again, ontogenetic material may be helpful. The central issue is that true homology, i.e. a one-to-one identification of cell masses, is probably never found. The masses may differ in size or shape between species. They may differ in location relative to landmarks. They may differ in the number of cell masses. That is to say that a single mass in one species may be replaced by two or more in another. This is a vexed issue that has to be resolved, when possible, by introducing the concept of a field.

It is evident from these comments alone that homology will not be established in many cases without additional criteria above and beyond the comparison of two inventories of cellular aggregates. The establishment of the fibre connections of a given cell aggregate can help to clarify its homology – this under the caveat that fibre connections are not entirely invariant. Data derived from immunohistochemical studies may also be helpful, and in some instances even decisive, but in other circumstances such data may actually confuse the issue.

As we have repeatedly stressed in this work, it is important to relate structure and function wherever possible. What can be said about the functional implications of a cell mass, or indeed a collection of cell masses, which comprise an identifiable macromorphological entity? There are several ideas that are implicit in our thinking but rarely stated openly, e.g. the notion that the larger a cell mass or centre, the greater the 'processing power'. In the case of the electric lobe of *Torpedo* this is evident in that the more electromotoneurons, the more modified muscle fibres that can be activated and the larger the electric current produced. We carry this thinking to the cerebral cortex, of course, regarding the vast expansion in man compared with all other species as a sign of superior intelligence. Yet, as the example of the electric lobe indicates, greater mass can simply mean that more motor units exist in a large animal

than a small one, and this is undoubtedly the basis for some of the cortical size found in large whales. On a comparative basis, the idea of size of a specific region relative to total brain size as a measure of functional capacity is born out by, for example, the olfactory systems found in several species.

Despite the extensive caveats described in the preceding paragraphs, it is evident that cell masses are identified and homologies are made; indeed, the existence of this book is a testament to the success of this approach. It is also evident that, presently, the task is not carried out in the ideal manner prescribed in Chap. 6, but rather through a heuristic process based on many lines of evidence, each of which has a certain weight in a given investigation.

The issues concerning the fidelity of fibre tract-tracing are similar to those involved in the delineation of cell masses. The first problem is that the extent of the fibres must be identified. A description of fibre-tracing methods is provided in Chap. 7, so only some major aspects will be considered here. In the classic studies of Ariëns Kappers et al. (1936) the approach of choice was to stain the myelin of fibre tracts with the Weigert method in normal material and follow along the pathways. This technique was widely employed but had the considerable shortcoming that it was easy to lose track of the pathway being followed, because of fibres entering and leaving the tract. Furthermore, the Weigert method gave no indication of the polarity of a pathway, i.e. in which direction information flowed. Because of such problems, numerous serious mistakes were made. Consequently, all the older material studied in this way should be treated with great caution. A major improvement in tract-tracing came about through the introduction of experimental methods, by which was meant the cutting of a pathway, or damaging of a cell mass, causing axon degeneration and the following of that degeneration with suitable methods, such as the Marchi method or the Nauta and the Fink-Heimer techniques. These methods were not without their drawbacks, however (see Chap. 7). For example, degeneration methods involve damage to tracts, and it is often difficult, if not impossible, to confine the damage to the tract of interest. Then there is the recurrent problem of sensitivity, the ability to see small degenerating fragments, and finally the not inconsiderable issue of determining the optimum time after damage to fix and process the material in order to observe the degeneration.

Today, these older methods have been replaced by techniques based on axonal transport of HRP, lectins or fluorescent molecules of various types (Fast Blue, Lucifer Yellow, dextran amines) or diffusion of carbocyanine dyes (DiI, DiO, etc.) combined with image-enhancement techniques. Nevertheless, these are relatively new techniques and they have yet to be applied in many systems, particularly in many non-mammalian species.

The above discussion implies, at least for modern experimental studies, that a lesion or injection is made at a certain location and a group of fibres followed along their course and to their targets. Another issue surfaces when fibre bundles are studied in isolation, for example, by taking a cross-section at the light- or electron-microscopic level. In almost all cases a spectrum of fibre sizes will be seen which may, in fact, have no relationship to each other, in the sense that the fibres merely travel together but do not originate or terminate in the same structure. This means that tracts can be described as entities in themselves or in terms of the centres they connect, but the two descriptions will not necessarily be the same. Classically, anatomists have preferred to describe fibre bundles as entities in themselves, but as the resolution of modern techniques reveals increasing numbers of components, many entering or leaving at different points along the way, the utility of this description may be questioned.

Finally, as noted in Chap. 3, fibres often make many lateral connections along their route, and so the bundle itself is not merely a conduit for impulses but also participates in the processing. A related aspect is that for some types of information processing, the time of arrival of the impulses via different pathways is likely to be important (the auditory system, for example) and so the length and calibre of fibres themselves play a role in information processing.

24.5
Macrocircuitry

Macrocircuitry and its representation as 'wiring diagrams' has always been, and will continue to be, an important 'level of resolution' in all branches of neuroanatomy. Looking at the similarities and differences in such diagrams across species is especially valuable in comparative anatomy.

In the past, wiring diagrams were based primarily on non-experimental material, often stained using the Weigert-Pal paracarmine technique (Fig. 24.68). Today, such diagrams are based on a variety of modern tract-tracing techniques, often supplemented with data from immunohistochemical analysis and other methods (see Posters 4 and 5).

In studies at the macrocircuitry level, cell masses, or even nuclear complexes, are usually considered as basic units. It is realised, of course, that the actual 'work', i.e. the processing of information, is done within these units, but this aspect is omitted from the macrocircuit level and treated at another level, that of the microcircuit (see Sect. 24.6). In macrocircuit analysis the fibre systems are also treated at a greatly simplified level, as conductors of impulses, and of their many attributes (Chap. 3, Sect. 3.6.1); usually, only their direction of conduction and size are represented. Given the extreme complexity of the CNS, this is a necessary simplification aimed at making the task of analysis tractable. Having said this, however, we are not absolved from critically examining the flaws inherent in this approach, so that we can realistically assess the reliability of a macrocircuit representation.

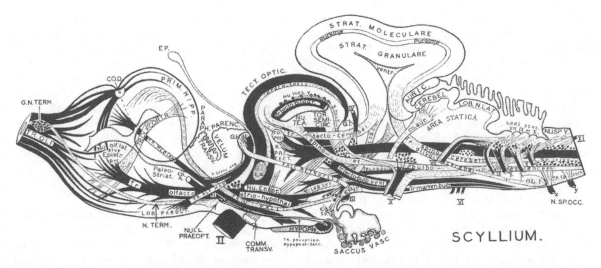

Fig. 24.68. Fibre connections in the brain of the shark *Scyliorhinus canicula* (reproduced from Ariëns Kappers 1947)

To begin, we note that there is not a single vertebrate brain the macrocircuitry of which is completely known. This is even true for the small and simple nerve cord of Amphioxus. Furthermore, the specialised chapters of this work reveal that our knowledge of the CNS differs greatly amongst the vertebrates. An enormous body of information is available on the rhesus monkey, the cat, the rat, pigeons and frogs, and much information has been accumulated on certain lizards and teleosts. But experimental hodological data on brachiopterygians, chondrosteans, holosteans and dipnoans are scant, while on the crossopterygian *Latimeria*, as well as on Amphioxus, they are entirely lacking. This implies that we must abandon any idea of a comprehensive cladistic analysis of phylogenetic history at the macrocircuit level. As noted in numerous papers on experimental hodology, much work remains to be done.

The first problem in all macrocircuit studies is deciding on the level of detail to be attempted. That is, the size of the neuronal aggregates that will form the 'nodes' in the wiring to be studied. An equally important question is whether the circuit will include all connections, however small, or whether it will be confined to only those above a certain size. It is not uncommon for studies from different investigators to begin from cytoarchitectonic divisions of differing resolution, or to be based on experimental tract-tracing techniques with differing selectivity or sensitivity. Such issues, often not taken into account, can seriously compromise efforts to compare results.

In Posters 4 and 5 the main sensory and motor circuitry is shown for representative species of the seven classes of vertebrates. These wiring diagrams are based on experimentally verified data. In Poster 4 somatosensory, auditory, visual, gustatory and olfactory projections to the CNS are shown, as well as the various ways in which these sensory systems reach the telencephalon. In Frame 8 the various telencephalic targets of visual, auditory and somatosensory projections are discussed. In Poster 5 the main motor systems are shown. In both posters a number of invariant centres and fibre systems are found, such as spinal projections to the reticular formation, a dorsal column nucleus, octavolateral projections via a torus semicircularis to the diencephalon, retinal projections to the tectum mesencephali, tectobulbar projections to the reticular formation and extensive reticulospinal projections to the spinal cord.

Before discussing some of the main differences and similarities in the sensory and motor circuitry of the CNS of the various vertebrate classes, we should note the following:

1. It is frequently seen that a sequence of centres are more strongly interconnected with each other than with other centres. Such a sequence is commonly designated as a *system*.
2. In many cases there is a polarisation associated with a given system, so that a regular and sequential decrease (Figs. 24.69a, 24.70a) or increase of the consecutive centres (Figs. 24.69b, 24.70b) can be observed. Such systems may be designated 'diminishing' or 'expanding', respectively.
3. Some sets of centres are strongly interconnected in a non-sequential manner. Such sets (e.g. the basal ganglia) may also be designated as systems, or the term complex may be preferred.

It is useful at this point to ask what meaning can be attached to a *system*, over and above a desire to bring some order to the endless complexity of the CNS. Neuroanatomists attempt to assign meaning to these configurations in three ways: (a) By starting their analysis at a functionally defined input, and then tracing the sequential pathways anterogradely. In this way sensory systems (somatosensory, visual, vestibular etc.) are delineated. (b) By starting their analysis at a functionally definable output, and then tracing the sequential pathways retrogradely. In this way, effector systems (i.e. in a gross sense, motor) are delineated. (c) By taking the results of experimental physiological, behavioural or clinical studies into consideration, and by attaching labels derived from these studies to the related centres and pathways (e.g. 'thermoregulatory', 'thirst' or 'limbic' system). It is interesting to note that at the beginning of this century, the members of the 'American School' of comparative anatomy, using techniques much less powerful than those available today, attempted to create an integral functional neuroanatomy (see Chap. 4, Sect. 4.6.2, and Chap. 6, Sect. 6.3.5)

There are at least two problems with this way of thinking, however. Firstly, the sequence of pathways called 'visual', for example, must end somewhere. Usually, this is at a level where another set of sequential pathways, with another designation, such as 'motor', begins. But the idea that a visual system has an abrupt end and is no longer to be judged visual after that point is disturbing. Secondly, as noted elsewhere, each cell mass is the recipient of other inputs, and furthermore, the cell masses always possess internal structure in the form of local circuits, so whatever the information is that enters the mass, we can be sure that different information comes out; this is the essence of information processing: if the information is unaltered, then no processing has occurred. To put it another way: there is no such thing as a pure 'relay' nucleus.

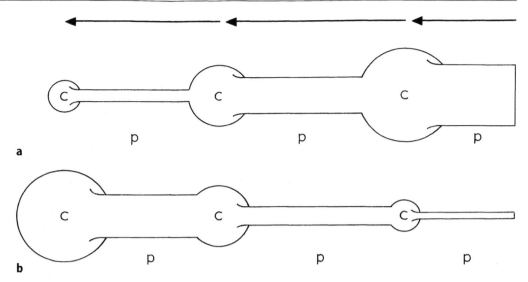

Fig. 24.69a,b. Sequences of centres (*c*) and pathways (*p*). **a** Diminishing system; **b** expanding system

To return to the visual system again, if an image were simply relayed to the cortex without substantial change, we would have to suppose that a cell or some other entity were present at that end point to 'view' the image; the infinite regression here is obvious. For this reason the label 'visual' attached to successive synapses from photoreceptor to primary visual cortex probably should be written in letters that become increasingly blurred.

Another issue is that there are structures so 'deep' in the nervous system, such as the cerebellum, that it seems futile to assign them to part of a system arising in the periphery, although stimulation of pathways regarded as visual or auditory near the periphery will give clear evoked responses in the cerebellum. The cerebellum, in fact, provides a recurring example of some of the major issues arising in the correlation of structure and function. One of the most enduring descriptions of cerebellar function is that it modulates and refines motor control, although this is obviously vague. This conclusion is based mainly on the results of ablation experiments, i.e. the removal of the cerebellum (to be precise, there are significant differences when cortex or cortex plus deep nuclei are removed, but that does not affect the present discussion). Ablation amounts to removal of a centre to see what function is lost and then ascribing the missing function to the brain area removed (Dow and Moruzzi 1958). This has obvious logical problems but nevertheless has so far proved more useful than a detailed analysis of the microcircuitry for defining overall function.

In a broader context it is probably true that, while the microelectrode is well-matched to the anatomical entity of the single cell, and even to some small microcircuits, there is no comparable technique for studying macrocircuits. Electroencephalographic (EEG) and magnetoencephalographic (MEG) recording approaches (see Chap. 7) give useful information under some circumstances, and both positron emission tomography (PET) and functional magnetic resonance imaging (fMRI) are promising, but all are limited in resolution to several millimetres, or even centimetres, under normal conditions, and so cannot be used to study the small centres of many of the species discussed in this work. Nevertheless, the rapid advances in functional imaging methods hold the distinct promise that we shall eventually be able to simultaneously follow the activation of various brain centres while the animal is performing a meaningful pattern of behaviour.

It is obviously impossible to discuss all the similarities and differences that have been seen at the macrocircuit level. Apart from purely quantitative differences in the *sensory circuitry* of the CNS correlated with the size or specialisation of a sense organ (Poster 4), the main differences between the sensory systems are found in their telencephalic representation (see Frame 8).

In the *motor circuitry* of the CNS of vertebrates, the main differences are found in the way in which the telencephalon influences motor control. A number of invariant motor centres and pathways are found throughout vertebrates. This is particularly evident for the descending supraspinal pathways and the basal ganglia. In Fig. 24.71 the cells of origin of descending supraspinal projections are shown for agnathans, cartilaginous fishes and a

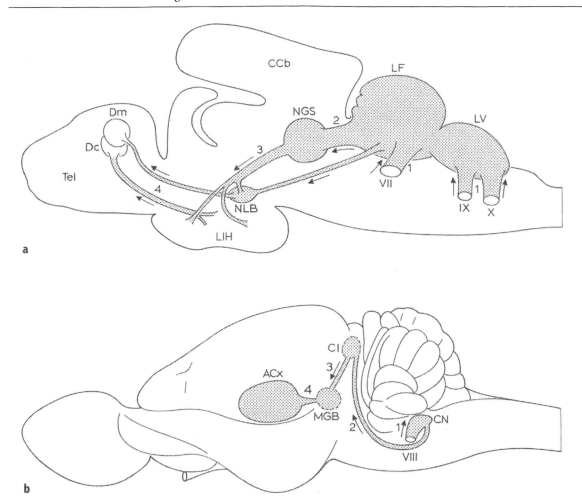

Fig. 24.70a,b. Examples of a diminishing (a) and an expanding (b) system. **a** Gustatory pathways in the brain of an ictalurid catfish (based partly on Kanwal and Finger 1992 and Lamb and Caprio 1993). **b** Auditory pathways in the North American opossum. *ACx*, Auditory cortex; *Ccb*, corpus cerebelli; *CI*, colliculus inferior (central nucleus); *CN*, cochlear nuclear complex; *Dc*, area dorsalis telencephali, pars centralis; *Dm*, area dorsalis telencephali, pars dorsalis; *LF*, lobus facialis; *LIH*, lobus inferior hypothalami; *LV*, lobus vagi; *NGS*, nucleus gustatorius secundarius; *NLB*, nucleus lobobulbaris; *Tel*, telencephalon; *VII*, nervus facialis; *VIII*, nervus octavus; *IX*, nervus glossopharyngeus; *X*, nervus vagus; *1, 2, 3, 4*, primary, secondary, tertiary and quaternary olfactory and auditory projections, respectively

lungfish. It is evident that there is a basic pattern in this series of anamniotes. Medial (reticulospinal) and lateral (octavomotor or vestibulospinal) columns are found in the brain stem. These columns give rise to the bulk of the descending motor pathways in anamniotes. Additionally, a rubrospinal projection was found in a ray (Fig. 24.71d) and a lungfish (Fig. 24.71e). Unfortunately, in these anamniotes the site of termination of these pathways is hardly known. Tract-tracing data on amphibians and amniotes (Fig. 24.72) show that throughout terrestrial vertebrates a classification can be made into medial and lateral systems as advocated in mammals by Kuypers (see Lawrence and Kuypers 1968a,b). The *medial· system* (reticulospinal and vestibulospinal projections) is functionally related to postural activities and progression and constitutes a basic system by which the brain stem exerts control over movements. The *lateral system* (the rubrospinal tract), at least in mammals, superimposes upon the general motor control by the medial system the capacity for the independent use of the extremities. A certain parallel can be drawn between the presence of limbs or limblike structures such as the large pectoral fins in rays and the presence of a rubrospinal tract (ten Donkelaar 1982, 1988). A rubrospinal tract is absent in apodans (Chap. 18) and snakes (Chap. 20).

The most notable difference between the motor circuitry of non-mammalian vertebrates and mammals is the apparent absence of somatomotor cortical areas giving rise to long descending projec-

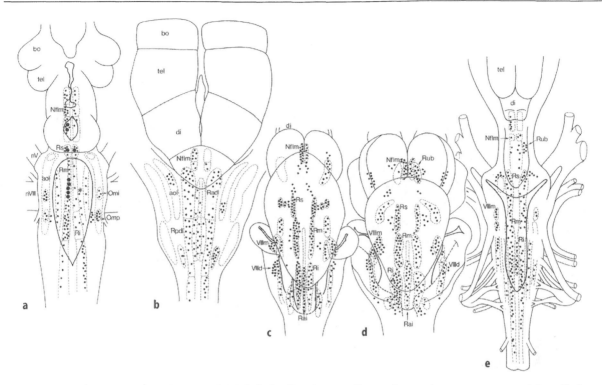

Fig. 24.71a-e. The distribution of retrogradely labelled cells in the brain stem after HRP injections into the left side of the spinal cord in (a) the silver lamprey, *Ichthyomyzon unicuspis*; (b) the Pacific hagfish, *Eptatretus stouti*; (c) the spiny dogfish, *Scyliorhinus canicula*; (d), a ray, *Raja clavata*; (e) the African lungfish (*Protopterus amphibians*). In a the *large dots* indicate Müller cells, the *asterisk* indicates the Mauthner cell. *aol*, Area octavolateralis; *bo*, bulbus olfactorius; *di*, diencephalon; *Nflm*, nucleus of the flm; *nVIII*, nervus octavus; *Omi, Omp*, intermediate and posterior octavomotor nuclei; *Radl*, dorsolateral part of nucleus reticularis anterior; *Rai*, nucleus raphes inferior; *Ri*, nucleus reticularis inferior *Rm*, nucleus reticularis medius; *Rpdl*, dorsolateral part of nucleus reticularis posterior; *Rs*, nucleus reticularis superior; *Rub*, nucleus ruber; *tel*, telencephalon; *VIIId, VIIIm*, descending and magnocellular vestibular nuclei (a, b, after Ronan 1989; c, d after Smeets and Timerick 1981; e, after Ronan and Northcutt 1985)

tions to the brain stem and spinal cord in nonmammalian vertebrates. The *corticospinal tract* is unique to mammals and its development appears to be associated with the acquisition of dextrous motor skills (see Kuypers 1981; Heffner and Masterton 1983; Porter and Lemon 1993). Instead, in anamniotes and reptiles, the striatum forms the main output centre of the telencephalon (see Poster 5). However, in birds, extensive descending pallial efferent systems arising from the anterior (somatosensory) part of the Wulst and from the anterior part of the archistriatum innervate the brain stem and possibly, at least in parrots and owls, also the rostral part of the spinal cord (see Frame 9). The evidence for an avian 'pyramidal tract' decussating into the spinal cord does not seem very convincing, however, since Webster et al. (1990) found no evidence for any direct spinal projections from the telencephalon in 'dextrous' parrots. A possible telencephalospinal projection was also noted in an advanced shark, the nurse shark, *Ginglymostoma cirratum* (Ebbesson and Schroeder 1971), after extensive lesions of the telencephalon. It would be of great interest to know where this pathway actually arises.

The *basal ganglia* in modern amniotes are very similar in connectivity and neurotransmitters (Medina and Reiner 1995; Veenman et al. 1995), suggesting that the evolution of the basal ganglia in amniotes has been very conservative. All amniotes possess a dorsal and a ventral striatum with major projections to a dorsal and a ventral pallidum, respectively. The chemoarchitecture and fibre connections of the basal ganglia in amphibians and fishes is basically similar to that of amniotes (Medina and Reiner 1995; Northcutt 1995; Marín et al. 1997b). A pallidal subdivision is 'hidden' within the striatal confines and needs further characterisation.

Despite the great similarities in chemoarchitecture and intrinsic connections of the basal ganglia in different species, the main differences are found in their extrinsic connections. Until recently, hardly any pallial projections were found to the striatum of

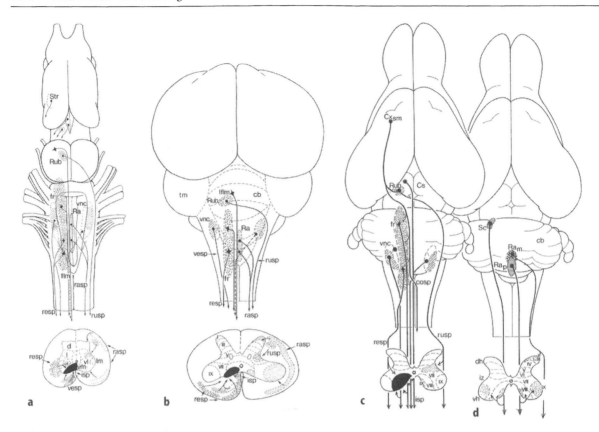

Fig. 24.72a–d. Diagrams summarising experimental data from anurans (a, the clawed toad, *Xenopus laevis*), birds (b, the pigeon *Columba livia*) and mammals (c, the North American opossum, *Didelphis virginiana*) on origin, course and site of termination of the medial (at the *left*) and lateral (at the *right*) systems of descending brain stem projections. In d, subcoeruleospinal and raphespinal projections are shown. *cb*, Cerebellum; *cosp*, corticospinal tract; *Cs*, colliculus superior; Cx_{sm}, sensorimotor cortex; *dh*, dorsal horn; *d, l, lm, vl, vm*, dorsal, lateral, lateral motoneuronal, ventrolateral and ventromedial fields of the spinal cord; *flm*, fasciculus longitudinalis medialis; *fr*, fasciculus retroflexus; *Iflm*, nucleus interstitialis of the flm; *isp*, interstitiospinal fibres; *iz*, intermediate zone; *Ra*, nucleus raphes; Ra_m, Ra_p, magnocellular and pallidal raphe nuclei; *rasp*, raphespinal fibres; *resp*, reticulospinal fibres; *Rub*, nucleus ruber; *rusp*, rubrospinal tract; *Sc*, subcoeruleus area; *Str*, striatum; *tm*, tectum mesencephali; *vesp*, vestibulospinal fibres; *vh*, ventral horn; *vnc*, vestibular nucleus complex; I–IX, layers I–IX of the spinal grey (after ten Donkelaar 1982, 1988)

amphibians and only restricted palliostriatal pathways in sauropsids, whereas the mammalian striatum is extensively innervated by most parts of the cerebral cortex. Recent tract-tracing studies (e.g. Veenman et al. 1995; Marín et al. 1997a), however, demonstrate that pallial projections to the striatum, although not as extensive as in mammals, are also present in anurans (from the medial pallium), in reptiles (from the dorsal cortex and ADVR), and especially in birds (extensive 'corticostriatal' projections from the Wulst, archistriatum and 'pallium externum'; see Chap. 21). Clear differences, however, are present in the output channels of the basal ganglia. In amphibians, reptiles and birds, the main output of the basal ganglia is via pretectal nuclei to the tectum mesencephali, whereas in mammals the main output channel reaches the premotor and motor cortex via the thalamus (see Poster 5).

Fibre systems that are clearly present in one species may be absent in another. As already noted, the corticospinal tract is unique to mammals. The extensive descending pallial efferent systems in birds arising from the anterior part of the Wulst and the anterior archistriatum are most likely comparable to the corticobulbar projections in mammals (see Frame 9). The Wulst and the archistriatum also innervate pontine nuclei, suggesting a (minor) cortico-ponto-cerebellar system in birds. Comparison at the macroscopic level between avian and mammalian brains reveals that in birds an externally visible pons and cerebellar hemispheres are lacking (P2: 36–41). Data on the organisation of the cerebellum in the pigeon suggest that a cortical zone comparable to the mammalian D-zone is absent in birds (Arends and Zeigler 1991a). It seems likely that in birds, an alternative pathway from the telencephalon to the cerebellum via the medial spiriform nucleus in the pretectum is more important.

These examples from the motor circuitry show that many similarities as well as important differences are found at the macrocircuit level.

In sharks, secondary olfactory projections reach the lateral pallium and the area superficialis basalis, a structure possibly homologous to both the globus pallidus and the olfactory tubercle (Northcutt et al. 1988). The thalamotelencephalic projections are mainly contralateral. Ascending thalamic projections converge on the medial pallium: visual input through the tectum via the anterior thalamic nucleus, electrosensory lateral line input via the lateral posterior nucleus, and mechanosensory lateral line and auditory input via the central posterior nucleus. In advanced, galeomorph sharks such as the nurse shark, *Ginglymostoma cirratum*, ascending thalamic input terminates in the pars centralis of the dorsal pallium (also shown in Poster 4). This proliferation (also known as the central nucleus) forms a telencephalic visual centre which receives visual information via the lateral geniculate nucleus and, furthermore, somatosensory input from a central thalamic nucleus (Schroeder and Ebbesson 1974). Gustatory information from a secondary gustatory nucleus (nucleus F), relayed via the lobus inferior (*in yellow*) of the hypothalamus, reaches the medial pallium in squalomorph sharks and the central nucleus in galeomorph sharks, as well as the area periventricularis ventrolateralis. The latter area is comparable to the striatum of amniotes (Northcutt et al. 1988).

The known sensory circuitry in teleosts is shown for the common goldfish, *Carassius auratus*. The goldfish exemplifies the 'typical' teleost (Nieuwenhuys and Meek 1990a). Secondary olfactory fibres terminate mainly in the posterior part of the area dorsalis. Visual, auditory and somatosensory information is relayed via the thalamus to various parts of the area dorsalis (Table 24.2): visual information via the anterior (retinal input) and dorsal posterior (tectal input) nuclei; auditory information via the central posterior nucleus, which receives auditory input from the central toral nucleus; and somatosensory input from the spinal cord and a lateral cuneate nucleus via the ventromedial thalamic nucleus. Auditory and lateral line input to the telencephalon is relayed predominantly via the lateral preglomerular nucleus, however (Striedter 1992). This large nucleus receives its afferent projections from the auditory medial pretoral nucleus and from the lateral line ventrolateral toral nucleus. Gustatory information from the facial lobe and particularly from the large vagal lobe (*in yellow*) passes to a secondary gustatory nucleus, which in its turn innervates the lobus inferior (in *yellow*) of the hypothalamus and the torus lateralis. The lobus inferior innervates the central part of the area dorsalis.

In amphibians, secondary olfactory projections terminate predominantly in the lateral pallium. The accessory olfactory bulb innervates the amygdala. Thalamotelencephalic projections (shown for *Rana esculenta*) are aimed mainly at the striatum: visual information via the anterior part of the lateral thalamic nucleus, which receives tectal input, auditory input (and, if present, lateral line input, as in *Xenopus laevis*) via the central thalamic nucleus, and somatosensory input via the ventromedial thalamic nucleus and presumably the nucleus of the posterior tubercle. These nuclei are innervated by spinothalamic fibres and projections from the dorsal column nucleus and a lateral cervical nucleus (Muñoz et al. 1997). The anterior thalamic nucleus is the only thalamic nucleus that innervates the medial pallium. The anterior nucleus receives retinal input, auditory input and somatosensory input via the dorsal column nucleus. Gustatory information passes via the nucleus of the solitary tract and the parabrachial nucleus (the secondary gustatory nucleus) to the hypothalamus and the basal forebrain.

In the red-eared turtle, *Pseudemys scripta elegans*, secondary olfactory projections terminate predominantly in the lateral cortex and in the amygdala. The reptilian thalamus gives rise to extensive thalamotelencephalic projections, terminating in the cortex as well as in the anterior part of the dorsal ventricular ridge (ADVR). Visual (via the dorsal lateral geniculate nucleus) and somatosensory (via the ventral nucleus) information reaches the dorsal or 'visual' cortex. Visual, auditory and somatosensory information to the ADVR is relayed via different thalamic nuclei. The visual (via the nucleus rotundus) and auditory (via the nucleus reuniens) channels to the ADVR remain anatomically separate in the midbrain, thalamus and telencephalon. There is a certain overlap between the auditory and somatosensory channels. In turtles, the core of the nucleus reuniens innervates the ventral part of the ADVR, whereas the belt area innervates a more dorsal part of the ADVR which overlaps with the somatosensory part of the ADVR (Belekhova et al. 1985). Moreover, the belt area of the chelonian nucleus reuniens innervates the amygdala (Belekhova 1994). Gustatory information via the nucleus of the solitary tract and the parabrachial nucleus reaches the hypothalamus, the basal forebrain and the ADVR. Künzle (1985) demonstrated an extensive projection from the parabrachial area/superior vestibular nucleus to a lateral part of the ADVR.

In birds, extensive thalamotelencephalic projections are aimed at the Wulst and the dorsal ventricular ridge. The Wulst includes an anterior, somatosensory part and a posterior, visual part which receive projections from the nucleus dorsalis intermedius ventralis anterior and the nucleus dorsolateralis anterior, respectively. In the dorsal ventricular ridge, four main sensory end-stations are found: (a) the rostral nucleus basalis, which receives quintofrontal fibres that convey trigeminal/tactile information via the main sensory trigeminal nucleus; (b) the ectostriatum, a visual end-station receiving tectal input via the nucleus rotundus; (c) a somatosensory zone in the intermediate part of the neostriatum which receives somatosensory and gustatory input via the nucleus dorsolateralis posterior, and (d) field L, an auditory centre in the caudal part of the neostriatum innervated by the nucleus ovoidalis. Field L includes three zones, L1–L3. The core of the nucleus ovoidalis innervates predominantly L2, whereas its shell area innervates L3 and the amygdalar part of the archistriatum.

In mammals, thalamotelencephalic sensory projections are aimed mainly at the cerebral cortex. In the North American opossum, *Didelphis virgiana*, the lateral geniculate body innervates the striate cortex, the lateral posterior nucleus the peristriate cortex, the medial geniculate body the auditory cortex, and the ventrobasal complex the sensorimotor cortex and a secondary somatosensory area that overlaps with the auditory cortex. Opossums have a complete overlap of the first somatosensory (S1) and the first motor (M1) areas of the cerebral cortex (Lende 1963a,b). In *D. virginiana*, secondary olfactory projections pass via the lateral olfactory tract to the olfactory tubercle, the piriform cortex, the lateral entorhinal cortex and the cortical amygdaloid nucleus.

This brief summary of the sensory circuitry in representative species of the seven classes of vertebrates shows remarkable similarities in the thalamotelencephalic projections, but clear differences in their telencephalic targets (Table 24.2).

Abbreviations used in Poster 4

A	anterior thalamic nucleus
Am	amygdala
aol (d,i,v)	dorsal, intermediate and ventral nuclei of the octavolateral area
Apvl	area periventricularis ventrolateralis
Asb	area superficialis basalis
Astr	archistriatum
aur	auricle of cerebellum
Bas	nucleus basalis
bo	bulbus olfactorius
boa	bulbus olfactorius accessorius
BON	basal optic nucleus
C	central thalamic nucleus
cb	cerebellum
ccb	corpus cerebelli
Cd	nucleus caudalis
Cgl	corpus geniculatum laterale
cho	chiasma opticum
CI	colliculus inferior
CN	central nucleus of dorsal pallium
Co	cochlear nuclear complex
Cp	central posterior thalamic nucleus
CS	colliculus superior
Cxd	dorsal cortex
Cxent	entorhinal cortex
Cxpir	piriform cortex
Cxpst	peristriate cortex
Cxst	striate cortex
DCN	dorsal column nucleus/nuclei
Dc, Dd, Dl, Dm, Dpo	central, dorsal, lateral, medial and posterior parts of area dorsalis (pallium)
Diva	nucleus dorsalis intermedius ventralis anterior
Dla, Dlp	nucleus dorsolateralis anterior, -posterior
Dm	nucleus dorsomedialis thalami (*Pseudemys*)
DMN	dorsal medullary nucleus
DP	dorsal pallium
Dp	dorsal posterior thalamic nucleus
Em	ectomamillary nucleus
Estr	ectostriatum
F	nucleus F (secondary gustatory nucleus)
fd	funiculus dorsalis
frh	fissura rhinalis
Gld	nucleus geniculatus lateralis pars dorsalis
Hb	habenula
HV	hyperstriatum ventrale
hyp	hypothalamus
Ico	nucleus intercollicularis
La	anterior division of lateral thalamic nucleus
lcb	tractus lobocerebellaris
LCN	lateral cervical nucleus
LF	lobus facialis
LGB	lateral geniculate body
LI	lobus inferior of hypothalamus
ll	lemniscus lateralis
lm	lemniscus medialis
LMN	lateral mesencephalic nucleus (torus in sharks)
LP	lateral pallium
Lp	lateral posterior thalamic nucleus
LRN	lateral reticular nucleus
lsp	lemniscus spinalis
LV	lobus vagalis
MGB	medial geniculate body
MON	medial octavolateral nucleus
MP	medial pallium
mrf	medullary reticular formation
MTN	medial pretoral nucleus
nlla, nllp	anterior and posterior lateral line nerves
Npo	pontine nuclei
Nstr	neostriatum
nvm	nervus vomeronasalis
nI	fila olfactoria
nII	nervus opticus
nVs	nervus trigeminus sensory part
nVIII	nervus octavus
nXII	nervus hypoglossus
Oli	oliva inferior
Ols	oliva superior
Ov	nucleus ovoidalis
Pb	nucleus parabrachialis
Pgl	lateral preglomerular complex
pin org	pineal organ
pret	pretectum
prf	pontine reticular formation

Reu	nucleus reuniens	Torc, Torl	central and laminar nuclei of (amniote) torus semicircularis
rf	reticular formation		
Ri	nucleus reticularis inferior	TP	tuberculum posterius
Rm	nucleus reticularis medius	V	ventral thalamic nucleus
Rmes	nucleus reticularis mesencephali	valv	valvula of cerebellum
Rot	nucleus rotundus	VB	ventrobasal thalamic complex
Rs	nucleus reticularis superior	Ve	area ventralis (subpallium)
S2	secondary somatosensory cortex	Visc	nucleus visceralis secundarius
SIN	superficial isthmic nucleus	VM	ventromedial thalamic nucleus
Sol	nucleus tractus solitarii	vnc	vestibular nuclear complex
Str	striatum	Vor	oral part of spinal trigeminal nucleus
thd	thalamus dorsalis	Vpr	principal sensory trigeminal nucleus
thv	thalamus ventralis	VIIIa, VIIId	ascending and descending octaval nuclei
Tl	torus lateralis		
tm	tectum mesencephali	IX	lobule IX of cerebellum
TO	tuberculum olfactorium		
tor (c, vl)	central and ventrolateral parts of (teleost) torus semicircularis	A	auditory part of ADVR
		S	somatosensory part of ADVR
tor (l, p, m)	laminar, principal and magnocellular nuclei of (anuran) torus semicircularis	S1	first somatosensory cortex
		V	visual part of ADVR

Frame 9 – Main motor circuitry in vertebrates

In Poster 5 the main motor circuitry found in vertebrates is summarised for representative species of the seven classes of vertebrates. Motor pathways are shown in contrasting colours: descending pathways to the spinal cord in *black*, tectobulbospinal projections in *orange*, basal ganglia projections in *dark blue* (the striatum, the substantia nigra and the nigrostriatal tract are shown in *green*), pallial projections and the fasciculus retroflexus mainly in *light blue*, cerebellar and vestibular projections in *red*.

In Table 24.3 the major sources of descending supraspinal pathways in vertebrates are summarised. A great number of invariant descending pathways are found throughout vertebrates; the main differences are found in the palliospinal projections. Spinal projections from the red nucleus, cerebellar nuclei and the dorsal column nucleus are apparently absent in agnathans and in many cartilaginous and bony fishes. A crossed rubrospinal tract, however, was demonstrated in rays, in a lungfish (see Fig. 24.71d,e), and in the common goldfish, *Carassius auratus* (Prasada Rao et al. 1987). Tectobulbar projections to the reticular formation are found throughout vertebrates (in Poster 5 only the crossed tectobulbo(spinal) projections are shown).

Basal ganglia projections show many similarities, but also major differences. The striatum, i.e. the part of the basal ganglia strongly innervated by dopaminergic fibres from the substantia nigra and related structures, is indicated in *green*. In lampreys, the striatum innervates the ventral thalamus, the mesencephalic tegmentum and the rostral part of the rhombencephalon. Palliostriatal projections have not been found so far. In the spiny dogfish, *Squalus acanthias*, the basal ganglia (or ventral subpallium) include the area periventricularis ventrolateralis (Apvl) and the area superficialis basalis (Asb). Both areas are innervated by dopaminergic fibres from the nucleus of the posterior tubercle, the substantia nigra and the ventral tegmental area, and also have other histochemical features that are most similar to those of the basal ganglia of other vertebrates (Northcutt et al. 1988): Apvl represents the striatal part, Asb may be comparable to the globus pallidus and the olfactory tubercle. Striatal projections pass via the basal forebrain bundle, decussate in the anterior commissure, and innervate the ventral thalamus, the posterior tubercle and the lobus inferior of the hypothalamus. Via the ventral thalamus and posterior tubercle the chondrichthyan basal ganglia may influence the tectum mesencephali. The strong lobobulbar projection (in *brown*) innervates the lateral reticular formation. In the common goldfish, *Carassius auratus*, the striatum is presumably represented by the dorsal part of the area ventralis (Nieuwenhuys and Meek 1990a). It receives a dopaminergic innervation, the origin of which has not yet been characterised. The efferent projections of this striatal part of the goldfish telencephalon are unknown.

In contrast to the basal ganglia of agnathans and cartilaginous and bony fishes, basal ganglia projections in amphibians and amniotes have been studied extensively. Marín et al. (1997a,b) distinguished dorsal, ventral and caudal subdivisions of the striatum of ranid frogs, each with its own complement of fibre connections. The dorsal and ventral parts of the striatum are innervated by the medial pallium and project to the posterior tubercle and the parabrachial area. The ventral striatum also has direct projections to the tectum mesencephali and the torus semicircularis.

24.5 Macrocircuitry

Table 24.3. Major sources of descending supraspinal pathways in vertebrates

Class	Reticular formation	Vestibular nuclear complex	Interstitial nucleus of flm	Red nucleus	Cerebellar nuclei	DCN	Spinal trigeminal nucleus	Nucleus of the solitary tract	Raphe complex	Locus coeruleus	Hypothalamus	Basal forebrain	Pallium
Agnatha	+	+	+	–	–	–	+	?	+	+	?	–	–
Chondrichthyes	+	+	+	±	–	+	+	+	+	+	+	–	?
Osteichthyes	+	+	+	±	+	–	+	+	+	+	+	–	–
Amphibia	+	+	+	±	+	+	+	+	+	+	+	+	–
Reptilia	+	+	+	+	+	+	+	+	+	+	+	+	+
Mammalia	+	+	+	+	+	+	+	+	+	+	+	+	?
Aves	+	+	+	+	+	+	+	+	+	+	+	±	?

DCN, Dorsal column nucleus/nuclei; *flm*, fasciculus longitudinalis medialis
+ present; – absent; ± present in certain species, absent in others; ? questionable

The caudal striatum is the source of striatal projections to the caudal brain stem and rostral spinal cord. Anuran accumbens projections are discussed in Chap. 19. Pallidal subdivisions are 'hidden' within the striatal confines. In the red-eared turtle, *Pseudemys scripta elegans*, the striatum is innervated by the dorsal cortex and the anterior dorsal ventricular ridge (ADVR). The striatum innervates the globus pallidus and, via the substantia nigra, the tectum mesencephali. Efferent fibres from the globus pallidus innervate the entopeduncular nuclei, the suprapeduncular nucleus, and the parabrachial area and continue into the rhombencephalic reticular formation. A main projection from the globus pallidus reaches the dorsal nucleus of the posterior commissure to influence the tectum mesencephali.

In birds, the basal ganglia are represented by two striatal subdivisions, i.e. the paleostriatum augmentatum (PA) and the lobus parolfactorius (LPO), and related pallidal subdivisions, the paleostriatum primitivum (PP) and a ventral pallidum. Extensive 'corticostriatal' projections are present arising in the Wulst, the archistriatum and the 'pallium externum', i.e. the outer rind of the pallium between the Wulst and the archistriatum (Veenman et al. 1995). The piriform cortex, hippocampal formation and posterior, amygdalar, part of the archistriatum innervate the LPO. The main efferent pathway from the basal ganglia is the pallidal (PP) projection to the lateral spiriform nucleus which innervates the tectum mesencephali. Although the connections of the basal ganglia have not been as extensively studied in opossums as in rodents (see Chap. 22), the organisation of the basal ganglia is comparable to that in placental mammals (Martin and Hamel 1967; Mickle 1976). Striatal projections from the caudate nucleus and the putamen innervate a small globus pallidus, the entopeduncular nucleus, the subthalamic nucleus and the substantia nigra (Mickle 1976). The substantia nigra is the only relay to the colliculus superior. The entopeduncular nucleus (comparable to the internal part of the globus pallidus of primates) innervates the ventromedial thalamic nucleus, which projects to the sensorimotor cortex (Donoghue and Ebner 1981). The nucleus accumbens, which is innervated by limbic forebrain structures such as the hippocampal formation and the amygdala, projects to the ventral pallidum [Loo's (1931) nucleus of the ansa lenticularis]. The ventral pallidum most likely innervates the mediodorsal thalamic nucleus, which projects to the frontal cortex (Tobias and Ebner 1973).

Pallial efferents in anamniotes arise mainly from the *medial pallium*. In lampreys, the medial pallium (the primordium hippocampi) innervates the piriform and dorsal pallial primordia and the olfactory bulb, but also sends a distinct fibre bundle via the dorsal thalamus to the pretectum, the tectum mesencephali, the torus semicircularis and the mesencephalic reticular area. The medial pallium also has access to the preoptic area and the hypothalamus. In the spiny dogfish, *Squalus acanthias*, the medial pallium and the dorsal pallium give rise to the tractus pallii, which decussates in the commissure of this tract to innervate the inferior lobe of the hypothalamus. It would be of great interest to know whether the central nucleus of galeomorph sharks, i.e. a hypertrophy of the dorsal pallium, gives rise to the telencephalospinal projection demonstrated by Ebbesson and Schroeder (1971). In the common goldfish, *Carassius auratus*, the medial part of the area dorsalis gives rise to pallial projections to the nucleus paracommissuralis in the pretectum, to the posterior tubercle and the midbrain tegmentum, and to the lobus inferior of the hypothalamus. The pretectal paracommissural nucleus is a relay nucleus for pallial information to the cerebellum. The central part of the area dorsalis also innervates the paracommissural nucleus, but, in addition, the torus semicircularis and, together with the lateral part of the area dorsalis, the tectum mesencephali. No telencephalic projections beyond the midbrain were observed. In amphibians, the medial pallium is the only pallial structure with projections beyond the telencephalon. It innervates the posterior tubercle and, both directly and indirectly (via the pretectum), the tectum mesencephali.

In *amniotes*, pallial efferent projections innervate the brain stem, in mammals, also the spinal cord. In turtles, the dorsal cortex gives rise to corticostriatal projections to the basal ganglia, both directly and indirectly via the ADVR. Moreover, the dorsal cortex innervates the dorsal lateral geniculate nucleus, the mesencephalon, the hypothalamus and, via the septum, the corpus mamillare. In birds, at least three descending pallial efferent systems have access to the brain stem (Ulinski and Margoliash 1990; Dubbeldam 1991). These corticobulbar projections arise from two parts of the Wulst and from the archistriatum. The Wulst gives rise to the septomesencephalic tract. Its caudal (*visual*) part (in *orange*) projects to the ventral lateral geniculate nucleus, the tectum and the basal optic nucleus. It plays a role in modulating the activity of tectal projections to premotor structures in the brain stem reticular formation. The anterior (*somatosensory*) part of the Wulst (in *green*) projects to the red nucleus, the medial pontine nucleus, the dorsal column nuclei, and possibly to the contralateral spinal cord. This is apparently a system related specifically to motor control of the talons in birds of prey (Ulinski and Margoliash 1990). The anterior part of the archistriatum gives rise to a major descending projection to the brain stem via the occipitomesencephalic tract. This tract (in *red*) terminates in the thalamus, tectum, tegmentum, lateral reticular formation, lateral pontine nucleus, hypoglossal nucleus, and possibly the rostral spinal cord. The occipitomesencephalic tract represents the telencephalic output channel for the trigeminal 'feeding circuit' and for the 'vocalisation circuit'. Via the frontoarchistriate tract the neostriatal nucleus basalis has access to the archistriatum. In songbirds, field L and related structures have access to the archistriatal nucleus robustus via the caudal part of the hyperstriatum ventrale. In mammals, restricted parts of the cerebral cortex (in *Didelphis virginiana*, the sensorimotor cortex) give rise to corticobulbar and corticospinal tracts. Moreover, large parts of the cerebral cortex have access to the cerebellum via their corticopontine projections.

Finally, *cerebellar* and *vestibular projections* (in *red*) will be briefly discussed. In lampreys, the small cerebellum has hardly been studied. The octavolateral area contains three octavomotor nuclei: the anterior and intermediate nuclei innervate the mesencephalon; the intermediate nucleus, moreover, gives rise to an ipsilateral octavospinal tract, whereas the posterior octavomotor nucleus gives rise to a crossed octavospinal projection. In *Squalus acanthias*, the large corpus cerebelli innervates the nucleus cerebelli which, via a brachium conjunctivum, projects to the contralateral red nucleus and ventral thalamus, and via a descending projection influences the reticular formation and presumably the vestibular nucleus complex. The valvula innervates the vestibular nuclei directly. The ascending vestibular nucleus (VIIIa) innervates the mesencephalon, the magnocellular (VIIIm) and descending (VIIId) vestibular nuclei the spinal cord. In teleosts, the eurydendroid cells form the output neurons of the cerebellum. They innervate the valvula, the tectum mesencephali and the torus semicircularis, the red nucleus, the ventromedial thalamic nucleus and the reticular formation. Magnocellular and posterior octaval nuclei innervate the spinal cord.

In amphibians, the cerebellar cortex innervates one cerebellar nucleus with ascending and descending projections. The anuran cerebellar nucleus gives rise to an ascending projection, the brachium conjunctivum, to the contralateral red nucleus (Fig. 24.73a), and to a descending projection to the contralateral vestibular nuclear complex. The latter bundle resembles the hook bundle, or fasciculus uncinatus, of amniotes. In amniotes, a longitudinal organisation pattern of the corticonuclear projections is evident. In reptiles, four longitudinal strips of Purkinje cells are found, each with a different target: a medial zone innervates the medial cerebellar nucleus, an intermediate zone the vestibular nuclear complex, a lateral zone the lateral cerebellar nucleus (Fig. 24.73b), and an outer zone the vestibular nuclear complex. The projections from the cerebellar nuclei are clearly different: the medial nucleus innervates mainly the contralateral vestibular nuclear complex, the lateral cerebellar nucleus, via the brachium conjunctivum, the contralateral red nucleus and the ventral thalamus. Such longitudinal zones are also evident in birds (Arends and Zeigler 1991a) and in the opossum (Klinkhachorn et al. 1984a,b). In birds, at least two cerebellar nuclei are found, giving rise to the bra-

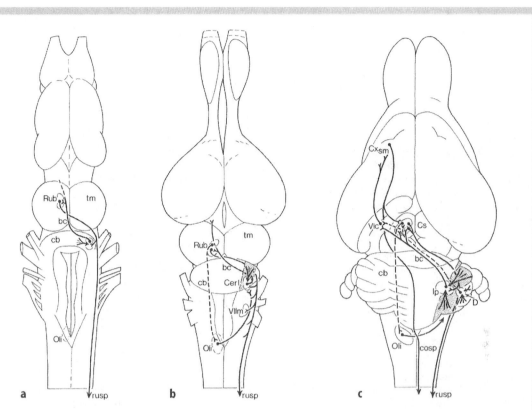

Fig. 24.73a-c. Diagrams showing the cerebellar corticonuclear zones innervating the cerebellar nuclei which project to the red nucleus in (**a**) the clawed toad, *Xenopus laevis*; (**b**) the savannah monitor lizard, *Varanus exanthematicus*; (**c**) the North American opossum, *Didelphis virginiana*. *bc*, Brachium conjunctivum; *cb*, cerebellum; *Cerl*, nucleus cerebellaris lateralis; *cosp*, corticospinal tract; *Cs*, colliculus superior; CX_{sm}, sensorimotor cortex; *D*, dentate nucleus; *Ip*, interposed nucleus; *Oli*, oliva inferior; *Rub*, nucleus ruber; *Rusp*, rubrospinal tract; *tm*, tectum mesencephali; *Vlc*, ventrolateral thalamic nuclear complex; *VIIm*, facial motor nucleus (from ten Donkelaar 1988)

chium conjunctivum and the fasciculus uncinatus, respectively (Arends and Zeigler 1991b). In mammals, in addition, a dentate nucleus is found which, together with the interposed nucleus, gives rise to the brachium conjunctivum (Fig. 24.73c). The mammalian medial or fastigial nucleus forms the fasciculus uncinatus (for further details and references see Chap. 22).

Abbreviations used in Poster 5

A-D	longitudinal zones of cerebellum
Acc	nucleus accumbens
ADVR	anterior dorsal ventricular ridge
Ala	nucleus ansae lenticularis anterior
Am	amygdala
Apvl	area periventricularis ventrolateralis
Asb	area superficialis basalis
aur	auricle of cerebellum
Bas	nucleus basalis
bc	brachium conjunctivum
bfb	basal forebrain bundle
bo	bulbus olfactorius
cb	cerebellum
ccb	corpus cerebelli
Cd	nucleus caudatus
Cer	nucleus cerebelli
Cerl	nucleus cerebellaris lateralis
Cerm	nucleus cerebellaris medialis
cosp	corticospinal tract
CS	colliculus superior
Cxd	dorsal cortex
Cxf	frontal cortex
Cxsm	sensorimotor cortex
Dc, Dl, Dm	central, lateral and medial parts of area dorsalis (pallium)
DCN	dorsal column nuclei
DP	dorsal pallium
Dt	nucleus dentatus
Em	ectomammillary nucleus
Ena	nucleus entopeduncularis anterior
Enp	nucleus entopeduncularis posterior
EP	nucleus entopeduncularis
F	nucleus fastigii
flm	fasciculus longitudinalis medialis
fr	fasciculus retroflexus

frat	fronto-archistriate tract	Rm	nucleus reticularis medius
Gld	nucleus geniculatus lateralis pars dorsalis	Rmes	nucleus reticularis mesencephali
		Rs	nucleus reticularis superior
Glv	nucleus geniculatus lateralis pars ventralis	Rub	nucleus ruber
		rusp	rubrospinal tract
GP	globus pallidus	SN	substantia nigra
Hb	habenula	Spd	nucleus suprapeduncularis
Hip	hippocampus	Spl, Spm	lateral and medial spiriform nuclei
hyp	hypothalamus	Sth	nucleus subthalamicus
HV	hyperstriatum ventrale	Str	striatum
ic	internal capsule	Strc, Strd, Strv	caudal, dorsal and ventral parts of (anuran) striatum
Iflm	interstitial nucleus of the flm		
Inp	nucleus interpeduncularis	tb	tectobulbar tract
Ip	nucleus interpositus	tbsp	tectobulbospinal tract
LF	lobus facialis	thd	thalamus dorsalis
LI	lobus inferior of hypothalamus	thv	thalamus ventralis
LP	lateral pallium	Tl	torus lateralis
LPO	lobus parolfactorius	tlb	tractus lobobulbaris
LV	lobus vagalis	tm	tectum mesencephali
Md	nucleus mediodorsalis thalami	tor	torus semicircularis
MP	medial pallium	TP	tuberculum posterius
mrf	medullary reticular formation	tpal	tractus pallii
Mth	Mauthner cell/fibre	valv	valvula of cerebellum
Nflm	nucleus of the flm	Vd	dorsal part of area ventralis
Nstr	neostriatum	vesp	vestibulospinal tract(s)
Npo	pontine nuclei	vespl, vespm	lateral and medial vestibulospinal tracts
nII	nervus opticus		
nVm	nervus trigeminus, motor part	VH	ventral hypothalamic nucleus
nVII	nervus facialis	VL	ventrolateral thalamic nucleus
nXII	nervus hypoglossus	VM	ventromedial thalamic nucleus
Oma, Omi, Omp	anterior, intermediate and posterior octavomotor nuclei	vnc	vestibular nuclear complex
		VP	ventral pallidum
PA	paleostriatum augmentatum	Vm	motor nucleus of trigeminal nerve
Pb	nucleus parabrachialis	Vsp	spinal nucleus of trigeminal nerve
Pgl	lateral preglomerular complex	VIIIa, VIIId VIIIm, VIIIp	ascending, descending, magnocellular and posterior nuclei of octaval nerve
Pom	nucleus preopticus magnocellularis		
PP	paleostriatum primitivum	VIIIc, VIIIv	caudal and ventral, vestibular, nuclei of octaval nerve
pret	pretectum		
prf	pontine reticular formation	XII	motor nucleus of hypoglossal nerve
Put	putamen		
resp	reticulospinal fibres	S, V	somatosensory and visual parts of Wulst
Ri	nucleus reticularis inferior		

24.6
Microcircuits

Analysis of the intrinsic organisation of grisea is an indispensable prerequisite for subsequent neurophysiological studies. During a long period of some 90 years (1873–1963), such knowledge was acquired almost exclusively from material impregnated by the Golgi technique. This method provides an inventory of the various types of neurons in a given griseum and an indication of their interconnections, revealing the microcircuitry. Since most neurons are polarised, that is, conduct in only one direction (see Chap. 1), such circuits can suggest how impulses traverse the neuronal microcircuit. Ramón y Cajal (1909, 1911) analysed numerous microcircuits, resulting in unequalled semidiagrammatic pictures, indicating the flow of impulses through a variety of grisea.

In the 1960s, new methods were introduced for revealing neuronal geometry, based on the injection of fluorescent and non-fluorescent markers from intracellular microelectrodes (Chap. 7). These techniques had the advantage that both electrical activity and cellular geometry could be correlated unambiguously, since the injecting electrode also recorded the potential of the cell. During that same decade it became feasible to use computers, connected to microscopes equipped with suitable video cameras, to construct three-dimensional representations of stained nerve cells. Finally, improved electrical recording techniques such as the whole-cell patch, together with significant advances in neuropharmacology, enabled greatly improved mapping of voltage-gated and ligand-gated channels in neurons and an understanding of their role in generating neuronal electrical activity. Such advances at the level of the single cell also assisted in the understanding of microcircuits. These were also known as local circuits and were the subject of an extensive review (Schmitt and Worden 1979). But the limitation has remained that only one or two cells can be studied at the same time. In a few cases, the use of dyes sensitive to membrane potential or intracellular ion concentration have helped to bridge this gap (see Chap. 7).

Within the mammalian CNS, a considerable number of centres have been analysed, both structurally and functionally. Since neurons represent a distinct, natural, structural and functional element in the CNS, it seems reasonable that analysis at the microcircuit level will reveal essential features of information processing in a given centre.

The results of integrated studies on many different structures in the mammalian CNS can be found in Shepherd (1990). It is possible to mention only briefly some of the issues arising from such studies.

We can look at microcircuits in two ways. We can compare the circuits found in different centres – olfactory bulb and reticular formation, for example – and we can compare the circuitry in one centre in different animals. Both of these approaches are useful but the second, perhaps the most relevant to the present work, is possible in only a few instances because detailed comparative studies of microcircuitry are generally lacking. This section concerns mainly the first topic, the similarities and the differences between brain regions. Underlying this issue is the understanding that all the brain masses described earlier are simply aggregates of microcircuits, and the macrocircuits are the interconnections between the aggregates of the microcircuits. In some sense, that hopefully becomes clearer as we proceed, the microcircuit is the brain – the most essential and distinctive element in a hierarchy of description.

What is a 'microcircuit'? Shepherd and Koch (1990) have defined three elements that constitute the triad of a microcircuit: the afferent, the projection neuron and the interneuron. These elements are linked by synaptic connections, predominantly the classical chemical synapse, with its inherent polarity. Shepherd and Koch (1990) have also provided a succinct survey of the architecture of chemical synapses, as well as discussion of the dendrite, the ramifying extension of the neuron that forms a receptive surface for synapses. These are the basic elements of microcircuits; they can be assembled in various ways to form the variety of circuits found in the different regions of the CNS.

An immediate and fundamental issue arises after we have accepted this definition of the microcircuit. As noted already, any brain region comprises a vast collection of microcircuits. The olfactory bulb or thalamus or cortex can be defined by saying it is an aggregate of similar microcircuits. But how do the individual circuits relate to the whole? Each circuit receives at least one input and has at least one output, but does it then function independently of its neighbouring circuits, or is the functional unit of a brain region an ensemble of microcircuits? If so, how are the ensembles delineated, and does the aggregate remain fixed, or does it vary with different tasks, in the way that a pool of people might supply different groups of workers for different tasks? And if microcircuits do function as units, how do individual circuits communicate with each other? Is it by conventional synaptic interconnections, or through other means, such as gap junctions or extrasynaptic communication?

The question of whether microcircuits form units rather than just functioning in parallel with little interaction is not resolved at this time, largely because of the lack of techniques for studying ensembles of microcircuits, though such techniques are beginning to appear (e.g. optical methods, PET and fMRI). Two examples of functional units can be mentioned. The first is the trivial observation that a microcircuit with but a single projection neuron obviously cannot form units, so in those animals that possess a Mauthner cell, this microcircuit does not act within a unit. The extreme rarity of the Mauthner cell, however, emphasises that virtually every centre has the capacity to form units because of the numbers of microcircuits present. The Mauthner cell participates in the escape reflex of some aquatic animals; at the other end of the hierarchy of brain function we find the neocortex. Here the question of units has been the subject of extensive debate. Mountcastle (1979) collected much evidence in favor of his argument that the whole cortex was organised into minicolumns or modules some 30 µm in diameter and containing about 110 neurons (260 in the primate striate cortex) extending across the width of the cortex. On this basis, Mountcastle (1979) estimated that there might be 400 million minicolumns in the human cortex and that they are usually aggregated into columns about 300 µm in diameter (see below). While such a view has received support from electrophysiological experiments, especially those on the visual cortex (e.g. ocular dominance columns), and from anatomical techniques (e.g. the staining of cytochrome oxidase), the contemporary view is that the cortex is not a vast ensemble of relatively isolated columns. The new techniques for tracing axons have revealed increasing numbers of lateral interconnections in the cortex, which imply that a more complex description of the microcircuitry is appropriate.

Our concepts about the nature of local circuits are being changed by the increasing evidence for modification by use or the ability of circuits to 'learn' to perform better by modifying their connections. The discussion of this elusive property has been focused on long-term potentiation (LTP) and long-term depression (LTD) (Bear and Malenka 1994; Maren and Baudry 1995) and the mechanisms remain controversial, but what is important is that the issue emphasises the disparity between anatomical structure and functional topology in neuronal networks.

Another issue that must be raised in connection with microcircuits is the role of glia. Glia were discussed briefly in Chap. 1 but otherwise have not featured much in this work. These cells are at least as numerous as neurons yet are almost always omitted from any representation of local circuits. The reason for their omission is the widely held view that glia are there primarily to support neurons, by providing metabolic substrates, synthesising structural proteins or removing excess transmitters. But there is increasing evidence that glia do mediate or facilitate some forms of signalling in the nervous system (Kettenmann and Ransom 1995). Along with other forms of extrasynaptic communication, this raises difficult problems for both the conceptualisation and representation of local circuits.

All of this discussion raises the issue of whether the anatomically defined 'triadic' microcircuit is the basic functional element. Since this argument cannot be taken further at present, the remainder of this discussion will continue to use the basic triadic microcircuit.

It is a truism that the microcircuit processes information; it is also a truism that if we cannot define the nature of the information we will have difficulty describing the function of the microcircuit. It follows that some of the best-understood microcircuits are found at the immediate sensory or motor interface of the CNS. Thus quite detailed descriptions exist of the circuitry in the retina and spinal cord. It is instructive to compare these structures.

24.6.1
Microcircuits at the Interface – Retina and Spinal Cord

The retina is a thin multilayered sheet of cells with a precise and complex organisation (Sterling 1990; Fig. 24.74a). To grossly oversimplify, the retina is an assembly of parallel processing circuits arranged at right angles to its surface with superimposed lateral interconnections between the parallel processing units. This organisation is obviously related to the retinal function, which is to form two-dimensional images (later synthesised into three-dimensional representations in the cortex by combining the two-dimensional images from each retina) by capturing photons originating in the visual world around the animal. The photoreceptors record the spatial location of the photons by virtue of the location of the receptor on the retina and encode the additional information about the intensity and colour by their responses. Subsequent layers of the retina, and particularly the lateral interconnections, work to improve the quality of information acquired by the retina through various forms of 'sharpening' of the image, controlling sensitivity and so on. This 'preprocessing' assumes importance in the retina because visual scenes contain enormous quantities of information and, from an engineering point of view, it is desirable to reduce the information leaving the retina and so reduce the size of the optic nerve and subsequent centres to manageable proportions. Much is known about how the photoreceptors function and how the circuitry of the retina operates (Sterling 1990). Some of the detailed mechanisms are unusual (amacrine cells lack axons, for example), but overall, the circuitry functions on the basis of chemically-mediated excitation and inhibition; glutamate is the predominant excitatory transmitter and GABA the predominant inhibitory transmitter. The retina also provides an example of extrasynaptic modulation by dopamine. This transmitter is released from a sheet of amacrine cells and diffuses throughout the retina to modify the behaviour of the microcircuits (Witkovsky et al. 1993). The input elements comprise the rods and cones (Fig. 24.74a) and the output elements are the ganglion cells, encompassing as many as 20 distinct types. The intrinsic elements (Fig. 24.74a) comprise bipolar cells, horizontal cells, amacrine cells and interplexiform cells; again, most of these elements can be further divided into several subtypes. One reason for the variety of cell types is that different microcircuits come into play under different conditions; for example, the circuits used in bright sunshine differ from those in starlight. Omitted from Fig. 24.74a, and from most

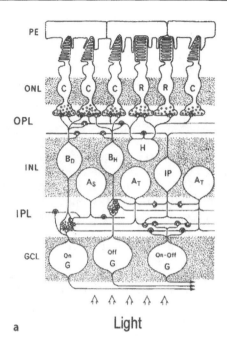

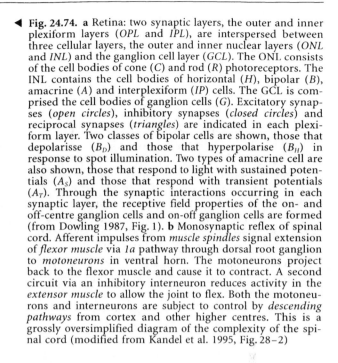

◀ **Fig. 24.74. a** Retina: two synaptic layers, the outer and inner plexiform layers (*OPL* and *IPL*), are interspersed between three cellular layers, the outer and inner nuclear layers (*ONL* and *INL*) and the ganglion cell layer (*GCL*). The ONL consists of the cell bodies of cone (*C*) and rod (*R*) photoreceptors. The INL contains the cell bodies of horizontal (*H*), bipolar (*B*), amacrine (*A*) and interplexiform (*IP*) cells. The GCL is comprised the cell bodies of ganglion cells (*G*). Excitatory synapses (*open circles*), inhibitory synapses (*closed circles*) and reciprocal synapses (*triangles*) are indicated in each plexiform layer. Two classes of bipolar cells are shown, those that depolarise (B_D) and those that hyperpolarise (B_H) in response to spot illumination. Two types of amacrine cell are also shown, those that respond to light with sustained potentials (A_S) and those that respond with transient potentials (A_T). Through the synaptic interactions occurring in each synaptic layer, the receptive field properties of the on- and off-centre ganglion cells and on-off ganglion cells are formed (from Dowling 1987, Fig. 1). **b** Monosynaptic reflex of spinal cord. Afferent impulses from *muscle spindles* signal extension of *flexor muscle* via *1a* pathway through dorsal root ganglion to *motoneurons* in ventral horn. The motoneurons project back to the flexor muscle and cause it to contract. A second circuit via an inhibitory interneuron reduces activity in the *extensor muscle* to allow the joint to flex. Both the motoneurons and interneurons are subject to control by *descending pathways* from cortex and other higher centres. This is a grossly oversimplified diagram of the complexity of the spinal cord (modified from Kandel et al. 1995, Fig. 28-2)

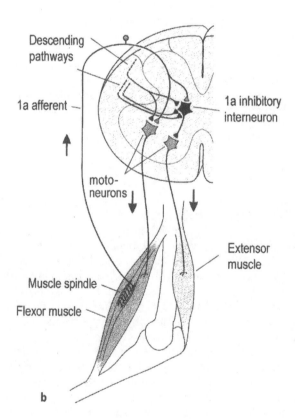

other figures showing the basic retinal circuitry, is the remarkable and prominent Müller cell. This glial cell extends from one surface of the retina to the other and is thought to play a major role in controlling extracellular potassium levels (Newman and Reichenbach 1996); by this means it may also modulate the function of retinal microcircuits.

From a comparative perspective, the circuitry of the retina shows differences. Interestingly, in animals with less central visual processing capability, more processing is done in the retina, so the frog has a more complex retina than the cat (Sterling 1990).

In stark contrast to the retina lies the spinal cord. The cord contains the motoneurons that form what Sherrington (1952) called the 'final common path' to the muscles of the extremities and the associated control circuitry. The cord has a long cylindrical organisation and, unlike the retina, is constituted not for parallel processing of spatial information, but for sequential processing of temporal information.

The intrinsic circuitry of the spinal cord is exceedingly complex. Ramón y Cajal (1909) devoted 300 pages to a description of the histology, and several thousand papers have been published since then on both histology and physiology. The basic circuit shown in Fig. 24.74b is therefore an enormous simplification, but it does show the hallmark feature of the cord: the monosynaptic reflex. Muscle spindles, in the flexor muscle associated with a limb joint, sense passive extension of the muscle and send a volley of impulses via the Ia afferent fibres and the dorsal root ganglion cells to excite motoneurons in the ventral horn of the cord. The motoneurons release acetylcholine at the neuromuscular junction of the muscle in question, causing contraction and thus opposing the original passive extension. This is, in fact, the only monosynaptic reflex known in the nervous system and so represents the simplest complete pathway from a sensory input to a motor output. The elegance and simplicity of this circuit have been one of the major stimuli for believing that an understanding of brain function is possible, on the grounds that all human behaviour is but an extension and elaboration of reflex pathways (Sherrington 1952; Swazey 1969). But the 'extension' is indeed 'elaborate', and it is far from clear that this supremely reductionist view is adequate to deal with the complexity of higher brain function. Even at the spinal level, the monosynaptic reflex is the exception. As shown in Fig. 24.74b, in order for the reflex to function, the Ia afferent fibres must also activate a second pathway via a Ia inhibitory neuron which suppresses activity in the motoneurons that innervate the extensor muscle associated with the limb joint in question. The whole ensemble is also subject to extensive descending influences from the cortex and elsewhere, so that the reflexes can be modulated, either to produce voluntary movement or to suppress inappropriate reflex activity. Furthermore, Fig. 24.74b hardly does justice to even the local circuitry, since a second set of local afferents from the Golgi tendon organ and joint receptors of the limb, along with cutaneous afferents, converge on the motoneurons illustrated in Fig. 24.74b via a population of inhibitory interneurons to maintain muscle tone; these circuits are also subject to descending control (Kandel et al. 1995).

The complexity of the local reflex pales, however, when it is appreciated that it is but a small component of a much more vast system that, in most animals, produces locomotion. Virtually all locomotor tasks involve coordinating different muscles to contract at different times in a pattern that produces linear motion. Our understanding of the spinal cord has been greatly enhanced by recent models of 'fictive locomotion' and central pattern generators (Arshavsky et al. 1993; Grillner et al. 1991) whereby entire motor patterns are generated in a cord isolated from the rest of the CNS. But this property of autonomy was already foreshadowed by the research of Sherrington on the spinal reflex arcs, such as the scratch reflex (Sherrington 1952). What has emerged is that the cord possesses the ability to generate intrinsic rhythms of activity, which are the basis of locomotion, be it swimming or walking, that such rhythms are subject to continual modification by sensory feedback reflecting the interaction of the limbs and body with the environment during the movement, and that there are intrinsic circuits that ensure the appropriate sequencing of muscles so that motion progresses in a single direction. A considerable degree of descending control can be exercised on these circuits, either to initiate movement in a certain direction or to accomplish complex locomotor tasks that require visual or vestibular information. Perhaps the best-understood example of how the coupling of intrinsic oscillators with local reflex circuits leads to motion is in the lamprey spinal cord, and the reader is encouraged to return to Chap. 10 for a detailed description of this system.

While the spinal cord differs in many ways from the retina, it does share two features. First, a great deal of information processing is done locally, thus relieving the higher centres of tasks and reducing the flow of necessary information from higher centres. This is exemplified by the fact that the whole basic paradigm of walking or swimming can be generated at the spinal level. Second, the intrinsic circuitry is again mediated through excitatory and inhibitory chemical transmission, again relying on glutamate and GABA for the majority of the synapses, though glycine is also found in the cord as an inhibitory transmitter.

It seems that little learning occurs in the adult retina and probably little in the spinal cord, though this is less certain. A case can be made that the interface with the outside world needs to function with dedicated and predictable behaviour and that plasticity is most usefully managed more centrally.

24.6.2
Central Microcircuits

We now jump at once to the deepest and most enigmatic parts of the nervous system: the cerebral cortex and the cerebellum, because by comparing and contrasting these two centres and by juxtaposing them with the brief description of the retina and spinal cord, some vision of the variety and complexities of microcircuits emerges.

We have chosen the cerebral cortex (hereafter simply called the cortex) and the cerebellar cortex (hereafter called the cerebellum) because these two centres have shown the greatest degree of expansion with evolution, because they are very deep in the CNS, and because they are very dissimilar in their microcircuitry.

Both cortex and cerebellum have a laminar structure, but there the similarity ends. The number of layers in the cortex is generally estimated at six (although the width of each layer varies with cortical region), while the cerebellum has three well-defined layers in most species. The cortex has been regarded from an anatomical perspective as a columnar structure, the basic unit being a cylinder (Fig. 24.75) extending through the layers (though as noted earlier the functional entity may not closely conform to this). Within each cylinder reside a great multiplicity of cells and along with them a multiplicity of transmitters. Glutamate, aspartate and GABA dominate, but the peptides: substance P, VIP, CCK, neuropeptide Y, SRIF, CRF and tachykinin, have been identified (Nieuwenhuys 1994). Columns are linked by interconnections which can be both excitatory and inhibitory.

The pyramidal cell predominates in the cortex, constituting about 70% of the total cell population (Nieuwenhuys 1994). There is a rich variety amongst these cells, but one important characteristic is that most axons remain within the cortex and some project several millimetres from their parent cell. Most axons project horizontally and contact both pyramidal and non-pyramidal cells. Cortical afferents arise from many sources but the majority come from two: other regions of the cortex and the thalamus. The interneurons of the cortex, making up 30% or less of the population, subdivide into excitatory and inhibitory. Spiny stellates form the majority of the excitatory interneurons, while a minority are bipolar cells. The inhibitory interneurons show a rich diversity, and among these are recognised: basket cells, chandelier cells, double bouquet cells, horizontal cells of Cajal, and Marinotti cells. Several other types of inhibitory interneurons are reported (Nieuwenhuys 1994). The overall picture, then, is of a rich collection of cell types with vast numbers of interconnections over varying distances. This emerging picture weakens the simple view of the cortex as an assembly of columnar aggregates of cells and suggests that we hardly know how to define the microcircuit. This has not prevented many imaginative attempts, however (Fig. 24.75).

In contrast to the cortex, the cerebellum (Fig. 24.76) is a uniform three-dimensional system arranged along rectangular geometrical lines. There are but five commonly identified cell types, and three of these are inhibitory interneurons. The inputs to the cerebellum arise from the vestibular nuclei, the spinal cord and, where appropriate, the cerebral cortex via the pontine nucleus. The mossy fibres constitute one of two major types of afferent endings and they excite an immense population of tiny granule cells which distribute their excitatory synapses to the remaining cells of the cerebellum.

The output of the cerebellar cortex originates from the most distinctive cell in the CNS, the Purkinje cell, which is known to inhibit its target neurons in the cerebellar nuclei (Eccles et al. 1967). The second afferent ending is the climbing fibre, originating in the inferior olive and terminating directly in a powerful excitatory synapse on the Purkinje cell. There are no intrinsic excitatory connections in the cerebellum, and the three interneurons, the stellate, basket and Golgi cells, are inhibitory. Finally there are only two major transmitters, an excitatory amino acid, probably glutamate, and GABA. From an electrophysiological perspective, the differences in intrinsic circuitry between cortex and cerebellum are manifested in many ways but we note here just one: the cortex readily sustains epileptic seizure while the cerebellum never does.

The cerebellum is one of the few brain regions in which the microcircuit has been subjected to detailed comparative analysis. In the late 1960s and early 1970s, a group of investigators under the direction of R. Llinás began a program of comparative anatomical and physiological study of the cerebellum, and their efforts were joined by others throughout the world. Most of their efforts are recorded in the book edited by Llinás (1969). As noted above, the basic circuit of the cerebellum is extremely simple (Fig. 24.76c,d) and in essence features one output, the Purkinje cell, which receives two inputs, climbing fibres which synapse directly and mossy fibres which have one neuron, the granule cell interposed. This basic circuit is found in frogs, alligators, elasmobranchs, teleosts, birds and all mammals so far studied (the lamprey cerebellum was too small for detailed analysis). In all these species the Purkinje cell maintains its distinctive geometry although the profusion of the dendritic tree varies, and the granule cells, along with their parallel fibres, are very similar. The elements that do vary among species are the inhibitory interneurons. Golgi cells, the interneurons of the granule cell layer, are sparse in the frog and become more numerous in the other species mentioned. The same holds true for the stellate cells of the molecular layer. But the most striking change is that basket cells, a specialised form of inhibitory interneurons that form pericellular baskets around Purkinje cells (Fig. 24.76a), occur only in birds and mammals; this appears to be a case of the same cell evolving twice. Thus one reaches two conclusions from this comparative study: (a) that the basic cerebellar circuit is highly invariant in all species, and (b) that as brain size and cerebellar mass increased, so did the inhibitory component of the cerebellum.

The differences between cortex and cerebellum are illustrated in the schematic diagrams of

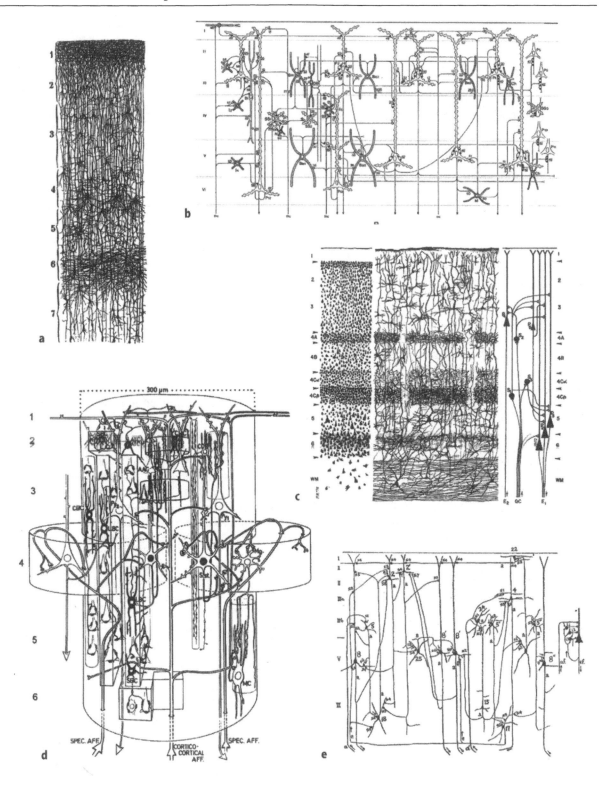

◀ **Fig. 24.75a–e.** Representations of cortical circuitry. **a** Ramón y Cajal's view of cortical architectonics. Section through the postcentral gyrus of a 1-month-old human infant, made with the Golgi method (Ramón y Cajal 1911, Fig. 405; DeFelipe and Jones 1988, Fig. 97). The *numbers* on the *left* indicate the cell layers: *1*, plexiform; *2*, small pyramidal cell; *3*, large and medium-sized pyramidal cells; *4*, external large pyramidal cells; *5*, granule cells or layer of small pyramidal cells and stellate cells; *6*, deep large pyramidal cells; *7*, medium-sized pyramidal cells. **b** Modern representation of the circuitry of the neocortex according to Nieuwenhuys (Nieuwenhuys 1994, Fig. 8). Although the detail cannot be seen in this figure, the diagram represents basket, bipolar chandelier, double bouquet cells, horizontal cells of Cajal, Martinotti cells, pyramidal cells and spiny stellate cells, as well as other elements. This diagram illustrates well the complexity of the cortical circuit. **c** Architectonics and circuitry according to Rakic (Rakic 1979, Fig. 5). Laminar and columnar organisation in the primary visual cortex the rhesus monkey. *Left panel* shows diagram of Nissl stained layers (note subdivision of *layer 4* in primate visual cortex). *Middle panel* is a representation of Golgi-stained material showing approximately one whole and two half columns, each 350–400 µm in diameter. *Right panel* shows simplified circuit, based largely on the work of Lund (1988) (see Fig. 24.77). **d** Neocortical circuitry according to Szentágothai (Szentágothai 1979, Fig. 6). This figure attempts to give a more three-dimensional representation of a cortical column 300 µm in diameter. The *left half* of the figure depicts inhibitory elements in black, while the *right half* shows mainly the excitatory elements; in reality, of course, both halves contain both cell types. Afferents are also shown and the bulges in *layer 4* indicate that these afferents are not necessarily in register with a single column. **e** Lorente de Nó's representation of neocortical circuitry. This figure first appeared in 1938 and was a major advance in understanding the cortex (based on Shepherd 1979, Fig. 15.5). The small *inset* to the *right* shows the internuncial chains of cells which were thought to be important in defining the vertical organisation of the circuitry; together with the vertical organisation of the afferents, this led to the idea of columnar organisation. Today this columnar organisation is thought to be less important because of the extensive lateral interconnections in the cortex

Fig. 24.77. Our knowledge of cortical circuitry (Fig. 24 77a–c) remains incomplete but the columnar organisation remains as a basis for approaching the function. A major feature of the cortex is undoubtedly the widespread distribution of excitatory interconnections. In contrast, the cerebellum is resolutely three-dimensional in its structure (Fig. 24.77d–f), with the parallel fibres stretching over several millimetres (Eccles et al. 1967); in species that lack folia, the parallel fibres cross the midline. The apparent simplicity of the cerebellar circuitry and the rigorous organisation of the structure have given rise to many hypotheses about the functional role of the microcircuitry which cannot be detailed here, though some brief remarks will be made in the next section.

24.6.3
Microcircuits in a Wider Context

If the premise is accepted that two-dimensional diagrams of connections capture the essentials of a microcircuit, then the way in which microcircuits are connected, i.e. the macrocircuit, is of great importance for deriving the next level of functionality. In this context the gross morphology can be seen as an externalisation of the internal organisation of a brain region. It would be exciting at this point to describe how a collection of microcircuits is combined to carry out some behaviour, but here we limit ourselves to certain aspects of that topic.

One of the most compelling examples emerging of a functional ensemble of microcircuits is the spinal cord of the lamprey; knowledge of the local circuits mediating swimming is being combined with an understanding of how the vestibular and visual circuits interact with the spinal ones (see Chap. 10). This work is driven largely by Grillner and associates, in cooperation with several other laboratories, and combines neuroanatomical, neurophysiological and modelling techniques, not only to solve the problem at hand but also to provide a paradigm that is likely to be paralleled in other contexts.

A second example of interest is recent work on the electrosensory system of fishes. This work brings to light some fundamental properties of cerebellar-like circuits. In a recent paper, Meek (1992) extended a former concept, based on the geometry and low conduction velocity of the parallel fibres, of the cerebellar microcircuitry as a timing device (Braitenberg 1967). It has been established that other centres, such as the electrosensory lateral line lobe (ELLL) of mormryrid electric fishes, have a cerebellar-like organisation (Bell and Szabo 1986). The ELLL is closer to the sensory surface of that animal that it serves, namely the electroreceptors located on the body surface, so that it is more amenable to research, using natural or controlled stimuli, than the cerebellum. These studies have resulted in the idea that the ELLL, and other non-electric centres, may be the site where information about the expected response to the electric organ is subtracted from the actual incoming data, which is a valuable technique for enhancing sensory capability (Bell et al. 1997). What is particularly interesting about this possibility is that it also seems to require a form of plasticity in the ELLL known as LTD, mentioned above (Bell et al. 1997). This type of 'learning' has been found to be present in the cerebellum (Ito 1993; Raymond et al. 1996) and appears to be an important aspect of the parallel fibre-climbing fibre systems. The notion that the

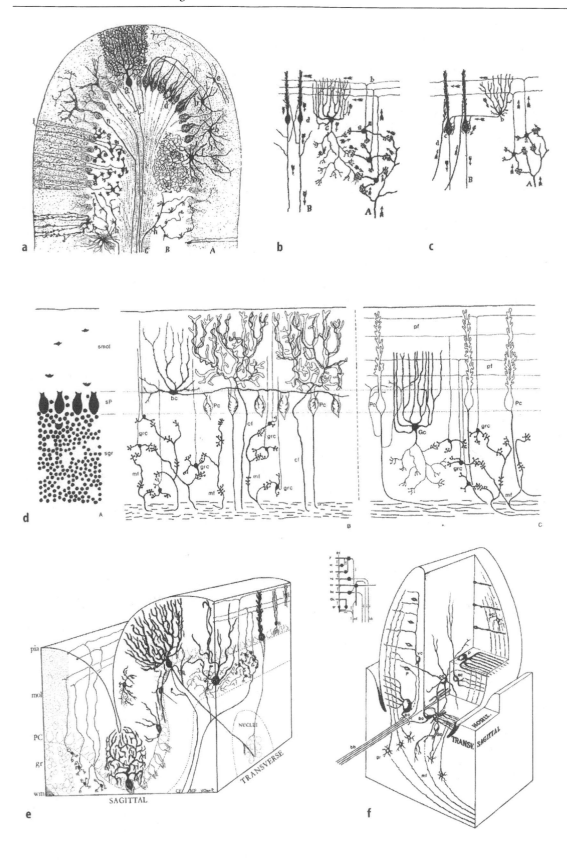

◀ **Fig. 24.76a–f.** Representations of the cerebellar cortex. **a** Ramón y Cajal's summary diagram of the mammalian cerebellar cortex in parasaggital section (Ramón y Cajal 1911). This orientation shows the spread of the Purkinje cell (*a*) dendrite. To the *right* of the figure the inhibitory stellate cells (*e*) and basket cells (*b*) make their characteristic pericellular axonal expansions around the bodies of the Purkinje cells in the molecular layer. The Golgi cell is also seen (*f*), with its vast axonal ramification in the granule cell layer. Beneath the Golgi cell is an entering mossy fibre afferent. On the *left* appear the granule cells with their vertical axons ascending into the molecular layer, where they form the characteristic T-shaped junctions and extend at right-angles to the plane of the figure to form the parallel fibres. Above ascending parallel fibres are the climbing fibre afferents and, finally, at the *base* of the *left-hand side* are the Bergmann glia (*j*) and astrocytes (*m*). **b** Diagram of circuit involving mossy fibres (*A*), granule cells (*a*), Golgi cells (*c*) and Purkinje cells (*d*) with their axons and collaterals (*B*) (Ramón y Cajal 1911). **c** Diagram of circuit involving mossy fibres, granule cells, basket cells (*b*), Purkinje cells and climbing fibres (*d*) (Ramón y Cajal 1991). In **b** and **c** Ramón y Cajal was able to indicate the probable flow of nerve impulses (*arrows*) based on his histology, but he did not know that the Golgi and basket cells were inhibitory. **d** Recent diagram of cerebellar circuitry (Chap. 2, this work). *Left panel* shows schematic of Nissl stain (*smol*, molecular layer; *sP*, Purkinje cell layer; *sgr*, granule cell layer), *centre panel* shows parasaggital view, *right* shows frontal section (*bc*, basket cell; *grc*, granule cell; *Pc*, Purkinje cell; *Gc*, Golgi cell; *pf*, parallel fibre; *mf*, mossy fibre; *cf*, climbing fibre). These circuits are based on panels *b* and *c* and other sources and show that no significant new results on cerebellar circuitry have emerged since Ramón y Cajal's studies. **e** An example from Palay and Palay (1974, Fig. 5) of a frequently used three-dimensional representation of the cerebellar cortex. **f** One of the few examples of a somewhat different arrangement of cells in the cerebellum, the valvula of the mormyrid fish (Nieuwenhuys and Nicholson 1969, Fig. 45). Here the axons of the granule cells enter the molecular layer at the base of the ridge but do not bifurcate. Furthermore, this part of the cerebellum also contains the intrinsic vertical cells (*vc*) and central cells (*cc*), as well as the basal cells (*bc*) (now called eurydendroid cells) that may constitute the homologue of the deep nuclei, since their axons project out of the cerebellum and the Purkinje cells project to them

ELLL is a sensory structure with a cerebellum-like circuitry, whereas the cerebellum is usually included in the motor sphere of activity, may be reconciled by the increasing evidence that the cerebellum is involved in cognitive functions (Baringa 1996).

Turning now very briefly to the cortex, there has been interest in the past few years in the long-range synchronisation of activity across the cortex by means of oscillations (Singer and Gray 1995). There is increasing evidence that, for example, disparate features of an object are 'bound' together by synchronising oscillations in different populations of cells. Local circuit properties are now being identified that may be the basis for such synchronisation (Jefferys et al. 1996).

What may be emerging from these various recent studies is that the microcircuitry of the cerebellum operates in a predominantely temporal domain by changing timing patterns into geometrical relationships. One might then be tempted to speculate that the cortex is concerned with spatial relationships, which it integrates by means of wavelike activity. The fundamental differences in these two modes of operation may be a reason for the dramatic and parallel expansion of the very different ensembles of microcircuits in cerebellum and cortex. Whatever the outcome of such speculation, it seems clear that the manifold new techniques at all levels of resolution are providing insights into both microcircuits and macrocircuits.

Some words of caution are needed in concluding this summary of microcircuitry. In looking at the various representations of circuitry we have to ask whether the two-dimensional representations exemplified by Fig. 24.77 are sufficient to describe the functional essence of brain microcircuitry. At least two problems exist. The first is how to capture the complex physiology of individual nerve cells (Segev 1992). We now know that neurons have repertoires of voltage-sensitive and ligand-gated channels that are distributed non-uniformly over the cell surface and can lead to different regions of the cell apparently functioning in different ways. This makes the old problem of representing the integrative properties of neurons within a circuit diagram more difficult. The second problem is that if circuits communicate to some degree by 'volume transmission' or 'extrasynaptic interaction', or if glial cells play a significant role in processing information, then diagrams of connectivity are likely to omit essential elements (Agnati et al. 1995).

A final caution may be appropriate. In their paper entitled: 'Neuronal circuits: an evolutionary perspective', Dumont and Robertson (1986) wrote: "Certain features of nervous systems may not have functional significance," and: "Evolution does not work logically, or with a long-term perspective, on the design of the neural circuits, but rather selects the most successful behavior from generation to generation. Thus there is no reason why the simplest solution to a problem should be the one actually used by the nervous system."

24.7
Conclusion

In this final chapter we have tried to encapsulate some of the wealth of detail that has been presented and to offer some comments. Looking back on this whole endeavour, spanning as it does more than two decades of work, we are struck by a combination of frustration and wonder.

We are frustrated because it would be satisfying to conclude by highlighting some clever and subtle principle that made sense of all that has gone

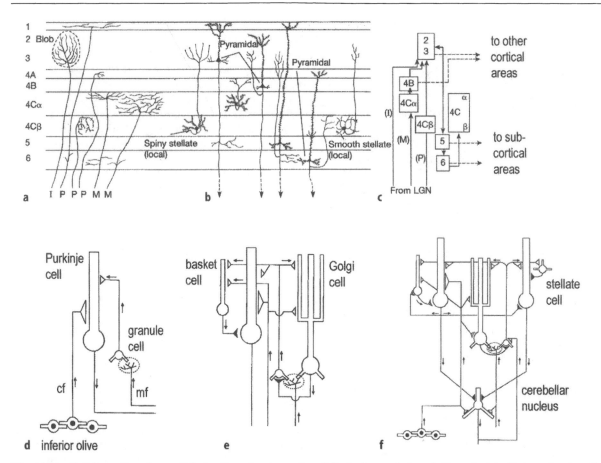

Fig. 24.77a-f. Schematics of essential circuitry of cortex and cerebellum. a-c Primary visual cortex of primate: a Most afferents (I, P, M) from the lateral geniculate (LGN) terminate in *layer 4C*. b Spiny stellate and pyramidal cells both have spiny dendrites and are excitatory. Smooth stellate cells are inhibitory. Only axons of pyramidal cells project out of the cortex. c Microcircuitry simplified. Afferents from M and P cells of LGN end on spiny stellates in *layer 4C*; these stellates project to *4B* and *layers 2* and *3*. Cells from interlaminar zone (*I*) in the LGN project directly to *layers 2* and *3*. From *2* and *3*, pyramidal cells project axon collaterals to *layer 5* pyramidal cells; the axon collaterals of these cells in turn project to *layer 6* and back to *layers 2* and *3*. Axon collaterals of *layer 6* pyramidal cells loop back to *layer 4C* to contact smooth stellate cells. Each layer, apart from *4C*, has different outputs. Cells in *layers 2, 3* and *4B* project to higher visual cortical areas. Cells in *layer 5* project to the superior colliculus, the pons, and the pulvinar. Cells in *layer 6* project back to the lateral nucleus and claustrum (modified from Kandel et al. 1995, Fig. 23-7). d-f Mammalian cerebellar circuitry: d Basic afferent circuit showing mossy fibre (*mf*)-granule cell-parallel fibre-Purkinje cell path and climbing fibre (*cf*)-Purkinje cell path from inferior olive. e Intrinsic neurons of the cerebellar cortex including the inhibitory Golgi cells and basket cells. f Complete cerebellar cortex and cerebellar nucleus. The axons of the Purkinje cells form inhibitory projections on the nuclear cells, which also receive collaterals from the mf and cf afferents. The cerebellar nuclei are outside the cerebellar cortex. Note the addition of the inhibitory stellate cells in panel f (from Llinás and Walton 1990)

before, to reveal the secret of brain structure and its organisation. Instead, we are left with a sense of awe at the myriad complexity of it all. Looking at the brains of these many animals, one has the feeling of being in a room, on the floor of which are closed boxes of many shapes and sizes. Opening any box, large or small, reveals something like an immensely intricate and exquisite assembly of components and circuits. The contents of each box resemble the contents of the others, yet each has a distinct identity and the assemblies are obviously functioning, but not according any known paradigm.

It is clear that our perception of the brain is governed by two related factors: the general state of our scientific knowledge and the availability of techniques for investigation. Thirty years or so ago, our thinking about brain function was dominated by the electrical and the computer metaphor. The dominant view of the brain was that neurons generated electrical impulses which were integrated by networks of cells in a positive or negative way via the synapse, until a threshold was reached and a further impulse generated. The view was that, by recording a sufficient amount of this electrical traffic and interpreting it on the basis of the anatomical wiring already revealed by studies such as those of Ramón y Cajal (1909, 1911), the brain could be understood. The computer provided a powerful metaphor since, while it was understood that the brain was not exactly like a computer, the basic idea of threshold logic and circuitry was seen as a common feature. These perspectives were captured in many monographs at that time, those by Arbib (1965) and Stevens (1966), for example. Today our perspective is different. Certainly, we have not abandoned the view that electrical impulses are crucial to brain activity, nor has the computer metaphor lost its value. Indeed, the breathtaking evolution and proliferation of the microprocessor holds valuable lessons for understanding the evolution and proliferation of the nervous system, and conversely, simplifed neural networks abstracted from ideas about neuronal microcircuits have begun to play useful roles in computation. What has really changed, however, is that the neuron has been recognised as a cell, with all the complex biological machinery that this entails. The revolution in cell biology, sparked by the understanding of DNA, took some time to permeate neuroscience, but it has now done so and our view of the brain has been altered forever. The revolution is most evident at the cellular level, where the neuron regarded as a simple threshold device is being replaced by a neuron capable of myriad forms of molecular transduction and modulation; exemplars here would be Shepherd (1990) and Kandel et al. (1995). It is noteworthy that the anatomical studies of Cajal retain their value, as shown by the appearance of the first English translation of the 1909 and 1911 volumes in 1995. As yet, the concepts of molecular biology have had limited impact on the larger scale of brain centres and macrocircuits that dominate this work. In a sense, therefore, this work comes at a very opportune time, gathering together and summarising our current structural knowledge as a prelude to the next phase of advancement. Just how this next phase will be we cannot say; that is, after all, the excitement of research, its unexpected discoveries and directions. We are probably safe in predicting, however, that the exploding knowledge of the genome and the process of development will greatly change our view of brain structure and homology.

Techniques are intimately bound to knowledge. A seemingly intractable problem that generates intense debate for years can be solved, and almost forgotten, overnight by the emergence of a new methodology. One thinks back to the great debates about the neuron doctrine – whether neurons were discrete entities or the nervous system a reticulum – a debate made irrelevant by the electron microscope. A more modest example occurred in 1971, when the existence of the Renshaw cell in the spinal cord was the subject of heated argument. The main evidence for its existence came from electrical recordings, but anatomists were unable to locate an appropriate neuronal candidate. Two extensive reviews appeared, one amassing the evidence for the existence of the cell (Willis 1971), the other against it (Scheibel and Scheibel 1971). Shortly afterwards, in a brief paper, Jankowska and Lindström (1971) used the new technique of simultaneously recording from a cell and revealing its morphology by injecting a fluorescent dye, were able to conclusively identify the elusive Renshaw cell and showed that it was not located where anatomists had sought it. The study of brain structure has been greatly helped by derivatives of the techniques used in the Renshaw cell experiments. Many new methods of introducing fluorescent and non-fluorescent components into nerve tracts, sometimes aided by immunocytochemistry, have revealed macro- and microcircuitry on an unprecedented scale, and surprising and unexpected connections are now reported almost daily.

This work is about the comparative anatomy of the brain. Despite the long history and distinguished achievements of this approach, it is not popular today. We hope it has become evident, though, that the study of different brains raises many crucial issues that would never be posed if attention were confined solely to the mouse or rat. Beyond this, the panorama of variety again and again prompts the question: How did that brain come to be the way it is? We are confronted with the mysteries of evolution and adaptation, not to mention the great debates that the topics engender. It transcends our powers of imagination to think that the processes so brilliantly conceived of by Darwin, working over enormous periods of time and under selective pressures that we can hardly begin to reconstruct, could arrive at the brains we see today. Yet we believe that is the case. Evolution and adaptation are among the most profound issues in biology and philosophy because, in contemplating the amazing variety among animal species, we are inevitably led to ponder our own nature.

References

Agnati LF, Zoli M, Stromberg I, Fuxe K (1995) Intercellular communication in the brain: wiring versus volume transmission. Neurosci 69:711–726

Alvarez-Bolado G, Rosenfeld MG, Swanson LW (1995) Model of forebrain regionalization based on spatiotemporal patterns of POU-III homeobox gene expression, birth dates, and morphological features. J Comp Neurol 355:237–295

Arbib MA (1965) Brains, machines and mathemathics. McGraw-Hill, New York

Arends JJA, Zeigler HP (1991a) Organization of the cerebellum in the pigeon (Columba livia). I. Corticonuclear and corticovestibular connections. J Comp Neurol 306:221-244

Arends JJA, Zeigler HP (1991b) Organization of the cerebellum in the pigeon (Columba livia). II. Projections of the cerebellar nuclei. J Comp Neurol 306:245-272

Ariëns Kappers CU (1920/1921) Die vergleichende Anatomie des Nervensystems der Wirbeltiere und des Menschen. Bohn, Haarlem

Ariëns Kappers CU (1929) The evolution of the nervous system in invertebrates, vertebrates and man. Bohn, Haarlem

Ariëns Kappers CU (1947) Anatomie comparée du système nerveux. Bohn, Haarlem

Ariëns Kappers CU, Carpenter FW (1911) Das Gehirn von Chimaera monstrosa. Folia Neurobiol 5:127-160

Ariëns Kappers CU, Huber GC, Crosby EC (1936) The comparative anatomy of the nervous system of vertebrates, including man, vol 1. MacMillan, New York

Arshavsky YuI, Orlovsky GN, Panchin YuV, Roberts A, Soffe SR (1993) Neuronal control of swimming locomotion: analysis of the pteropod mollusc Clione and embryos of the amphibian Xenopus. Trends Neurosci 16:227-233

Atema J (1971) Structures and functions of the sense of taste in the catfish (Ictalurus natalis). Brain Behav Evol 4:273-294

Barinaga M (1996) The cerebellum: movement coordinator or much more? Science 272:482-483

Bear MF, Malenka RC (1994) Synaptic plasticity: LTP and LTD. Curr Opin Neurobiol 4:389-399

Belekhova MG (1994) Thalamo-amygdalar auditory-somatic projections are ancient, conservative brain characters of amniotes. Zh Evol Biokhim Fiziol 30:454-473 (in Russian; English translation: J Evol Biochem Physiol 30:284-296)

Belekhova MG, Zharskaja VD, Khachunts AS, Gaidaenko GV, Tumanova NL (1985) Connections of the mesencephalic, thalamic and telencephalic auditory centers in turtles. Some structural bases for audiosomatic interrelations. J Hirnforsch 26:127-152

Bell CC, Szabo T (1986) Electroreception in mormyrid fish. Central anatomy. In: Bullock TH, Heiligenberg W (eds) Electroreception. Wiley, New York, pp 375-421

Bell C, Bodznick D, Montgomery J, Bastian J (1997) The generation and subtraction of sensory expectations within cerebellum-like structures. Brain Behav Evol 50, Suppl 1:17-31

Bergquist H (1932) Zur Morphologie des Zwischenhirns bei niederen Wirbeltieren. Acta Zool (Stockh) 13:57-303

Bergquist H, Källén B (1954) Notes on the early histogenesis and morphogenesis of the central nervous system in vertebrates. J Comp Neurol 100:627-660

Bergquist H, Källén B (1955) The archencephalic neuromery in Ambystoma punctatum. An experimental study. Acta Anat (Basel) 24:208-214

Bertmar G (1981) Evolution of vomeronasal organs in vertebrates. Evolution 35:359-366

Billo R, Wake MH (1987) Tentacle development in Dermopis mexicanus (Amphibia, Gymnophiona) with an hypothesis of tentacle origin. J Morphol 192:101-111

Braitenberg V (1967) Is the cerebellar cortex a biological clock in the millisecond range? Prog Brain Res 25:334-46

Breathnach AS (1960) The cetacean central nervous system. Biol Rev 35:187-230

Bruce LL, Neary TJ (1995) The limbic system of tetrapods: a comparative analysis of cortical and amygdalar populations. Brain Behav Evol 46:224-234

Bulfone A, Puelles L, Porteus MH, Frohman MA, Martin GR, Rubenstein JLR (1993) Spatially restricted expression of Dix-1, Dix-2 (Tes-1), Gbx-2, and Wnt-3 in the embryonic day 12.5 mouse forebrain defines potential transverse and longitudinal segmental boundaries. J Neurosci 13:3155-3172

Bullock TH (1984) The future of comparative neurology. Am Zool 24:693-700

Burr HS (1928) The central nervous system of Orthagoriscus mola. J Comp Neurol 45:33-128

Butler AB (1994a) The evolution of the dorsal thalamus of jawed vertebrates, including mammals: cladistic analysis and a new hypothesis. Brain Res Rev 19:29-65

Butler AB (1994b) The evolution of the dorsal pallium in the telencephalon of amniotes: cladistic analysis and a new hypothesis. Brain Res Rev 19:66-101

Butler AB (1995) The dorsal thalamus of jawed vertebrates: a comparative viewpoint. Brain Behav Evol 46:209-223

Butler AB, Northcutt RG (1992) Retinal projections in the bowfin, Amia calva: cytoarchitectonic and experimental analysis. Brain Behav Evol 39:169-194

Christensen K (1927) The morphology of the brain of Sphenodon. Univ Iowa Stud 12:1-29

Craigie EH (1930) Studies on the brain of the kiwi (Apteryx australis). J Comp Neurol 49:223-357

Crosby EC (1917) The forebrain of Alligator mississippiensis. J Comp Neurol 27:325-402

DeFelipe J, Jones EG (1988) Cajal on the cerebral cortex. Oxford University Press, New York

Dexler H, Eger O (1911) Beiträge zur Anatomie des Säugerrückenmarkes. I: Halicore dugong Erxl. Morphol Jahrb 43:107-207

Donoghue JP, Ebner FF (1981) The organization of thalamic projections to the parietal cortex of the Virginia opossum. J Comp Neurol 198:365-388

Dow RS, Moruzzi G (1958) The physiology and pathology of the cerebellum. University of Minnesota, Minneapolis

Dowling JE (1987) Retina, Vertebrate. In: Adelman G (ed) Encylopedia of neuroscience. vol II. Birkhäuser, Boston, pp 1061-1063

Dubbeldam JL (1991) The avian and mammalian forebrain: correspondences and differences. In: Andrew RJ (ed) Neural and behavioural plasticity. The use of the domestic chick as a model. Oxford University Press, Oxford, pp 65-91

Dumont JPC, Robertson RM (1986) Neuronal circuits: an evolutionary perspective. Science 233:849-853

Durward A (1932) Observations on the cell masses in the cerebral hemisphere of the New Zealand kiwi (Apteryx australis). J Anat (Lond) 66:437-466

Duvernoy H, Maillot Cl, Koritké JG (1970) La vascularisation de la moelle épinière chez le chat (Felis domestica). Les artères extramédullaires postérieures. J Hirnforsch 12:419-437

Ebbesson SOE, Schroeder DM (1971) Connections of the nurse shark's telencephalon. Science 173:254-256

Eccles JC, Ito M, Szentágothai J (1967) The cerebellum as a neuronal machine. Springer, Berlin Heidelberg New York

Edinger L (1908) Vorlesungen über den Bau der nervösen Zentralorgane. II. Vergleichende Anatomie des Gehirns, 7th edn. Vogel, Leipzig

Finger TE (1982) Somatotopy in the representation of the pectoral fin and free fin rays in the spinal cord of the sea robin, Prionotus carolinus. Biol Bull 163:154-161

Grillner S, Wallén P, Brodin L, Lansner A (1991) Neuronal networks generating behavior in lamprey: circuitry, transmitters, membrane properties, and simulation. Annu Rev Neurosci 14:169-199

Guthrie S (1995) The status of the neural segment. Trends Neurosci 18:74-79

Haller B (1891) Über das Centralnervensystem von Orthagoriscus mola. Morphol Jahrb 17:198-268

Haller von Hallerstein V (1934) Äußere Gliederung des Zentralnervensystems. In: Bolk L, Göppert E, Kallius E, Lubosch W (eds) Handbuch der vergleichenden Anatomie der Wirbeltiere, vol 2, part 1. Urban and Schwarzenberg, Berlin, pp 1-318

Heffner RS, Masterton RB (1983) The role of the corticospinal tract in the evolution of human digital dexterity. Brain Behav Evol 23:165-183

Heier P (1948) Fundamental principles in the structure of the brain. A study of the brain of Petromyzon fluviatilis. Acta Anat [Suppl] VI:1-213

Heiligenberg W, Keller CH, Metzner W, Kawasaki M (1991) Structure and function of neurons in the complex of the nucleus electrosensorius of the gymnotiform fish Eigenmannia: detection and processing of electric signals in social communication. J Comp Physiol [A] 169:151–164

Herrick CJ (1910) The morphology of the forebrain in Amphibia and Reptilia. J Comp Neurol 20:413–547

Herrick CJ (1921) A sketch of the origin of the cerebral hemispheres. J Comp Neurol 32:429–454

Herrick CJ (1948) The brain of the tiger salamander. University of Chicago Press, Chicago

Holland PWH, Garcia-Fernández J (1996) Hox genes and chordate evolution. Dev Biol 173:382–395

Holland PWH, Holland LZ, Williams NA, Holland ND (1992) An amphioxus homeobox gene: sequence conservation, spatial expression during development and insights into vertebrate evolution. Development 116:653–661

Holmgren N (1922) Points of view concerning forebrain morphology in lower vertebrates. J Comp Neurol 34:391–440

Ito M (1993) Synaptic plasticity in the cerebellar cortex and its role in motor learning. Can J Neurol Sci 20 [Suppl 3]:S70–S74

Jankowska E, Lindström S (1971) Morphological identification of Renshaw cells. Acta Physiol Scand 81:428–430

Jansen J (1930) The brain of Myxine glutinosa. J Comp Neurol 49:359–507

Jefferys JGR, Traub RD, Whittington MA (1996) Neuronal networks for induced '40 Hz' rhythms. Trends Neurosci 19:202–208

Jerison HJ (1973) Evolution of the brain and intelligence. Academic, New York

Johnston JB (1906) The nervous system of vertebrates. Blakiston, Philadelphia

Johnston JB (1911a) The telencephalon of selachians. J Comp Neurol 21:1–113

Johnston JB (1911b) The telencephalon of ganoids and teleosts. J Comp Neurol 21:489–591

Johnston JB (1915) Cell masses in the forebrain of the turtle, Cistudo carolina. J Comp Neurol 26:475–479

Källén B (1955) Notes on the mode of formation of brain nuclei during ontogenesis. CR Assoc Anat XLII:747–756

Källén B (1962) Embryogenesis of brain nuclei in the chick telencephalon. Ergeb Anat Entwgesch 36:61–82

Kandel ER, Schwartz JH, Jessell TM (1995) Essentials of neural science and behavior. Appleton and Lange, Norwalk

Kanwal JS, Finger TE (1992) Central representation and projections of gustatory systems. In: Hara TJ (ed) Fish chemoreception. Chapman and Hall, London, pp 79–102

Karten HJ (1969) The organization of the avian telencephalon and some speculations on the phylogeny of the amniote telencephalon. Ann N Y Acad Sci 167:164–179

Kettenmann H, Ransom BR (eds) (1995) Neuroglia. Oxford University Press, Oxford

Klinkhachorn PS, Haines DE, Culberson JL (1984a) Cerebellar cortical efferent fibers in the North American opossum, Didelphis virginiana. I. The anterior lobe. J Comp Neurol 227:424–438

Klinkhachorn PS, Haines DE, Culberson JL (1984b) Cerebellar cortical efferent fibers in the North American opossum, Didelphis virginiana. II. The posterior vermis. J Comp Neurol 227:439–451

Krubitzer L, Manger P, Pettigrew P, Calford M (1995) Organization of somatosensory cortex in monotremes: in search of the prototypical plan. J Comp Neurol 351:261–306

Kuhlenbeck H (1929a) Über die Grundbestandteile des Zwischenhirnbauplans der Anamnier. Morphol Jahrb 63:50–95

Kuhlenbeck H (1929b) Die Grundbestandteile des Endhirns im Lichte der Bauplanlehre. Anat Anz 67:1–51

Kuhlenbeck H (1973) The central nervous system of vertebrates, vol 3, part II: overall morphologic pattern. Karger, Basel

Kuhlenbeck H, Malewitz TD, Beasley AB (1967) Further observations on the morphology of the forebrain in Gymnophiona, with reference to the topologic vertebrate forebrain pattern. In: Hassler R, Stephan H (eds) Evolution of the forebrain. Plenum, New York, pp 9–19

Künzle H (1985) The cerebellar and vestibular nuclear complexes in the turtle. II. Projections to the prosencephalon. J Comp Neurol 242:122–133

Kusuma A, ten Donkelaar HJ, Nieuwenhuys R (1979) Intrinsic organization of the spinal cord. In: Gans C, Northcutt RG, Ulinski P (eds) Biology of the reptilia, vol 10: neurology B. Academic, London, pp 59–109

Kuypers HGJM (1981) Anatomy of the descending pathways. In: Brookhart JM, Mountcastle VB (eds) Handbook of physiology. The nervous system, vol II: motor control. American Physiological Society, Bethesda, pp 597–666

Lamb CF, Caprio J (1993) Diencephalic gustatory connections in the channel catfish. J Comp Neurol 337:400–418

Larsell O (1967) The comparative anatomy and histology of the cerebellum from myxinoids through birds. University of Minnesota Press, Minneapolis

Lawrence DG, Kuypers HGJM (1968a) The functional organization of the motor system in the monkey. I. The effects of bilateral pyramidal lesions. Brain 91:1–14

Lawrence DG, Kuypers HGJM (1968b) The functional organization of the motor system in the monkey. II. The effects of lesions of the descending brainstem pathways. Brain 91:15–36

Lende RA (1963a) Sensory representation in the cerebral cortex of the opossum (Didelphis virginiana). J Comp Neurol 121:395–414

Lende RA (1963b) Motor representation in the cerebral cortex of the opossum (Didelphis virginiana). J Comp Neurol 121:405–415

Leonard RB, Willis WD (1979) The organization of the electromotor nucleus and extraocular motor nuclei in the stargazer (Astroscopus y-graecum). J Comp Neurol 183:397–414

Llinás R (ed) (1969) Neurobiology of cerebellar evolution and development. American Medical Association, Chicago

Llinás RR, Walton KD (1990) Cerebellum. In: Shepherd GM (ed) The synaptic organization of the brain, 3rd edn. Oxford University Press, New York, pp 214–245

Loo YT (1931) The forebrain of the opossum Didelphys virginiana. Part II: histology. J Comp Neurol 52:1–148

Lund JS (1988) Anatomical organization of macaque monkey striate cortex. Annu Rev Neurosci 11:253–288

Macdonald R, Xu Q, Barth KA, Mikkola I, Holder N, Fjose A, Krauss S, Wilson SW (1994) Regulatory gene expression boundaries demarcate sites of neuronal differentiation in the embryonic zebrafish forebrain. Neuron 13:1039–1053

Maren S, Baudry M (1995) Properties and mechanisms of long-term synaptic plasticity in the mammalian brain: relationships to learning and memory. Neurobiol Learn Mem 63:1–18

Marín O, González A, Smeets WJAJ (1997a) Basal ganglia organization in amphibians: afferent connections to the striatum and the nucleus accumbens. J Comp Neurol 378:16–49

Marín O, González A, Smeets WJAJ (1997b) Basal ganglia organization in amphibians: efferent connections of the striatum and the nucleus accumbens. J Comp Neurol 380:23–50

Martin GF, Hamel EG (1967) The striatum of the opossum (Didelphis virginiana). J Comp Neurol 131:491–516

Masai H, Sato Y, Aoki M (1973) The brain of Mitsukurina owstoni. J Hirnforsch 14:493–500

Medina L, Reiner A (1995) Neurotransmitter organization and connectivity of the basal ganglia in vertebrates: implications for the evolution of basal ganglia. Brain Behav Evol 46:235–258

Meek J (1992) Why run parallel fibers parallel? – Teleostean Purkinje cells as possible coincidence detectors, in a timing device subserving spatial coding of temporal differences. Neuroscience 48:249–283

Mickle JP (1976) Efferent connections of the caudate nucleus in the Virginia opossum. J Comp Neurol 166:373–386

Millot J, Anthony J (1965) Anatomie de Latimeria chalumnae, vol II: système nerveux et organes de sens. Centre National des Rechèrches Scientifiques, Paris

Moodie RL (1915) A new fish brain from the coal measures of Kansas, with a review of other fossil brains. J Comp Neurol 25:135–181

Mountcastle VB (1979) An organizing principle for cerebral function: the unit module and the distributed system. In: Schmitt FO, Worden FG (eds) The neurosciences fourth study program. MIT Press, Cambridge, pp 21–42

Muñoz A, Muñoz M, González A, ten Donkelaar HJ (1997) Spinal ascending pathways in amphibians: cells of origin and main targets. J Comp Neurol 378:205–228

Newman EA (1995) Glial cell regulation of extracellular potassium. In: Kettenmann H, Ransom BR (eds) Neuroglia. Oxford University Press, New York, pp 717–731

Newman E, Reichenbach A (1996) The Müller cell: a functional element of the retina. Trends Neurosci 19:307–312

Nieuwenhuys R (1962) Trends in the evolution of the actinopterygian forebrain. J Morphol 111:65–88

Nieuwenhuys R (1965) The forebrain of the crossopterygian Latimeria chalumnae Smith. J Morphol 117:1–24

Nieuwenhuys R (1994) The neocortex. An overview of its evolutionary development, structural organization and synaptology. Anat Embryol (Berl) 190:307–337

Nieuwenhuys R, Nicholson C (1969) Aspects of the histology of the cerebellum of mormyrid fishes. In: Llinás R (ed) Neurobiology of cerebellar evolution and development. American Medical Association, Chicago, pp 135–169

Nieuwenhuys R, Meek J (1990a) The telencephalon of actinopterygian fishes. In: Jones, EG, Peters A (eds) Cerebral cortex, vol 8A. Plenum, New York, pp 31–73

Nieuwenhuys R, Meek J (1990b) The telencephalon of sarcopterygian fishes. In: Jones, EG, Peters A (eds) Cerebral cortex, vol 8A. Plenum, New York, pp 75–106

Nissl F (1885) Ueber die Untersuchungsmethoden der Großhirnrinde. Neurol Zentralbl 4:500–501

Northcutt RG (1977) Elasmobranch central nervous system organization and its possible evolutionary significance. Am Zool 17:411–429

Northcutt RG (1981) Evolution of the telencephalon in nonmammals. Annu Rev Neurosci 4:301–350

Northcutt RG (1984) Evolution of the vertebrate central nervous system: patterns and processes. Am Zool 24:701–716

Northcutt RG (1986) Lungfish neural characters and their bearing on sarcopterygian phylogeny. J Morphol, Suppl 1:277–297

Northcutt RG (1987) Evolution of the vertebrate brain. In: Adelman G (ed) Encyclopedia of neuroscience, vol I. Birkhäuser, Boston, pp 415–418

Northcutt RG (1995) The forebrain of gnathostomes: in search of a morphotype. Brain Behav Evol 46:275–318

Northcutt RG, Kaas JH (1995) The emergence and evolution of mammalian neocortex. Trends Neurosci 18:373–379

Northcutt RG, Kicliter E (1980) Organization of the amphibian telencephalon. In: Ebbesson SOE (ed) Comparative neurology of the telencephalon. Plenum, New York, pp 203–255

Northcutt RG, Wicht H (1997) Afferent and efferent connections of the lateral and medial pallia of the silver lamprey. Brain Behav Evol 49:1–19

Northcutt RG, Reiner A, Karten HJ (1988) Immunohistochemical study of the telencephalon of the spiny dogfish, Squalus acanthias. J Comp Neurol 277:250–267

Parker TJ (1891) Observations on the anatomy and development of Apteryx. Philos Trans R Soc [B] 182:25–134

Pilleri (1969) Das hirnanatomische Institut der psychiatrischen Universitätsklinik Bern. Hirnanatomisches Institut, Bern

Platel R (1989) L'encéphalisation chez le Tuatara de Nouvelle-Zélande Sphenodon punctatus Gray (Lepidosauria, Sphenodonta). Étude quantifiée des principales subdivisions encéphaliques. J Hirnforsch 30:325–337

Porter R, Lemon R (1993) Corticospinal function and voluntary movements. Clarendon, Oxford

Prasada Rao PD, Jadhao AG, Sharma SC (1987) Descending projection neurons to the spinal cord of the goldfish, Carassius auratus. J Comp Neurol 265:96–108

Puelles L (1995) A segmental morphological paradigm for understanding vertebrate forebrains. Brain Behav Evol 46:319–337

Puelles L, Robles C, Martínez-de-la-Torre M, Martínez S (1994) New subdivision schema for the avian torus semicircularis: neurochemical maps in the chick. J Comp Neurol 340:98–125

Rabl-Rückhard H (1883) Das Grosshirn der Knochenfische und seine Anhangsgebilde. Arch Anat Physiol Anat Abt (Lpz) 279–322

Rakic P (1979) Genetic and epigenetic determinants of local neuronal circuits in the mammalian central nervous system. In: Schmitt FO, Worden FG (eds) The neurosciences fourth study program. MIT Press, Cambridge, pp 109–127

Ramón y Cajal S (1909) Histologie du système nerveux de l'homme et des vertébrés. Tôme I. Malone, Paris (reprinted by CSIC, Madrid 1955; English translation published by Oxford University Press, New York, 1995)

Ramón y Cajal S (1911) Histologie du système nerveux de l'homme et des vertébrés. Tôme II. Malone, Paris (reprinted by CSIC, Madrid 1955; English translation published by Oxford University Press, New York, 1995)

Raymond JL, Lisberger SG, Mauk MD (1996) The cerebellum: a neuronal learning machine? Science 272:1126–1131

Rehkämper G, Zilles K (1991) Parallel evolution in mammalian and avian brains: comparative cytoarchitectonic and cytochemical analysis. Cell Tissue Res 263:3–28

Rehkämper G, Zilles K, Schleicher A (1985) A quantitative approach to cytoarchitectonics. X. The areal pattern of the neostriatum in the domestic pigeon, Columba livia f.d. A cyto- and myeloarchitectonical study. Anat Embryol (Berl) 171:345–355

Reiner A, Northcutt RG (1987) An immunohistochemical study of the telencephalon of the African lungfish. J Comp Neurol 256:463–481

Rendahl H (1924) Embryologische und morphologische Studien über das Zwischenhirn beim Huhn. Acta Zool (Stockh) 5:241–344

Retzius G (1906) Das Affenhirn in bildlicher Darstellung. Fischer, Jena

Roberts BL, Ryan KP (1975) Cytological features of the giant neurons controlling electric discharge in the ray Torpedo. J Mar Biol Assoc UK 55:123–131

Romer AS (1962) The vertebrate body. Saunders, Philadelphia

Ronan M (1989) Origins of the descending spinal projections in petromyzontid and myxinoid agnathans. J Comp Neurol 281:54–68

Ronan M, Northcutt RG (1985) The origins of descending spinal projections in lepidosirenid lungfishes. J Comp Neurol 241:435–444

Roth G, Nishikawa KC, Naujoks-Manteuffel C, Schmidt A, Wake DB (1993) Paedomorphosis and simplification in the nervous system of salamanders. Brain Behav Evol 42:137–170

Rudebeck B (1945) Contributions to forebrain morphology in Dipnoi. Acta Zool (Stockh) 26:9–156

Scheibel ME, Scheibel AB (1971) Inhibition and the Renshaw cell. A structural critique. Brain Behav Evol 4:53–93

Schmitt FO, Worden FG (1979) The neurosciences fourth study program. MIT Press, Cambridge

Schroeder DM, Ebbesson SOE (1974) Nonolfactory telencephalic afferents in the nurse shark (Ginglymostoma cirratum). Brain Behav Evol 9:121–155

Segev I (1992) Single neurone models: oversimple, complex and reduced. Trends Neurosci 15:414–421

Shepherd GM (1979) The synaptic organization of the brain, 2nd edn. Oxford University Press, New York

Shepherd GM (1990) (ed) The synaptic organization of the brain, 3rd edn. Oxford University Press, New York

Shepherd GM, Koch C (1990) Introduction to synaptic circuits. In: Shepherd GM (ed) The synaptic organization of the brain, 3rd edn. Oxford University Press, New York, pp 3-31

Sherrington C (1952) The integrative action of the nervous system. Cambridge University Press, Cambridge

Singer M (1962) The brain of the dog in section. Saunders, Philadelphia

Singer W, Gray CM (1995) Visual feature integration and the temporal correlation hypothesis. Annu Rev Neurosci 18:555-586

Smeets WJAJ (1990) The telencephalon of cartilaginous fishes. In: Jones EG, Peters A (eds) Cerebral cortex, vol 8A: comparative structure and evolution of cerebral cortex, part I. Plenum, New York, pp 3-30

Smeets WJAJ, Reiner A (1994) Phylogeny and development of catecholamine systems in the CNS of vertebrates. Cambridge University Press, Cambridge

Smeets WJAJ, Timerick SJB (1981) Cells of origin of pathways descending to the spinal cord in two chondrichthyans, the shark Scyliorhinus canicula and the ray Raja clavata. J Comp Neurol 202:473-491

Sterling P (1990) Retina. In: Shepherd GM (ed) The synaptic organization of the brain, 3rd edn. Oxford University Press, New York, pp 170-213

Stevens CF (1966) Neurophysiology: a primer. Wiley, New York

Stingelin W (1958) Vergleichend morphologische Untersuchungen am Vorderhirn der Vögel auf cytologischer und cytoarchitektonischer Grundlage. Helbing and Lichtenhahn, Basel

Striedter GF (1992) Phylogenetic changes in the connections of the lateral preglomerular nucleus in ostariophysan teleosts: a pluralistic view of brain evolution. Brain Behav Evol 39:329-357

Swazey JP (1969) Reflexes and motor integration: Sherrington's concept of integrative action. Harvard University Press, Cambridge

Szentágothai J (1979) Local neuron circuits of the cortex. In: Schmitt FO, Worden FG (eds) The neurosciences fourth study program. MIT Press, Cambridge, pp 399-415

ten Donkelaar HJ (1982) Organization of descending pathways to the spinal cord in amphibians and reptiles. Prog Brain Res 57:25-67

ten Donkelaar HJ (1988) Evolution of the red nucleus and rubrospinal tract. Behav Brain Res 28:9-20

Tensen J (1927) Einige Bemerkungen über das Nervensystem von Pipa pipa. Acta Zool 8:151-159

Tobias TJ, Ebner FF (1973) Thalamocortical projections from the mediodorsal nucleus in the Virginia opossum. Brain Res 52:79-96

Ulinski PS, Margoliash D (1990) Neurobiology of the reptile-bird transition. In: Jones EG, Peters A (eds) Cerebral cortex, vol 8A: comparative structure and evolution of cerebral cortex, part I. Plenum, New York, pp 217-265

Veenman CL, Wild JM, Reiner A (1995) Organization of the avian 'corticostriatal' projection system: a retrograde and anterograde pathway tracing study in pigeons. J Comp Neurol 354:87-126

Webster DMS, Rogers LJ, Pettigrew JD, Steeves JD (1990) Origins of descending spinal pathways in prehensile birds: do parrots have a homologue to a corticospinal tract of mammals? Brain Behav Evol 36:216-226

Wicht H (1996) The brains of lampreys and hagfishes: characteristics, characters, and comparisons. Brain Behav Evol 48:248-261

Wicht H, Himstedt W (1990) Brain stem projections to the telencephalon in two species of amphibians, Triturus alpestris (Urodela) and Ichthyophis kohtaoensis (Gymnophiona). Exp Brain Res Series 19:43-55

Wicht H, Northcutt RG (1994) An immunohistochemical study of the telencephalon and the diencephalon in a myxinoid jawless fish, the Pacific hagfish, Eptatretus stouti. Brain Behav Evol 43:140-161

Willis WD (1971) The case for the Renshaw cell. Brain Behav Evol 4:5-52

Witkovsky P, Nicholson C, Rice ME, Bohmaker K, Meller E (1993) Extracellular dopamine concentration in the retina of the clawed frog, xenopus-laevis. Proc Natl Acad Sci U S A 90:5667-5671

Zilles K, Schleicher A, Kretschman HJ (1978) A quantitative approach to cytoarchitectonics. I. The areal pattern of the cortex of Tupaia belangeri. Anat Embryol (Berl) 153:195-212

Zilles K, Stephan H, Schleicher A (1982) Quantitative cytoarchitectonics of the cerebral cortices of several prosimian species. In: Armstrong E, Falk D (eds) Primate brain evolution. Methods and concepts. Plenum, New York, pp 177-201

Subject Index

In this index the first number (in bold type) refers to the chapter's; the second number to the page. The synopsis at the top of this and of all the following odd numbered pages is intended to guide the reader to the structures, topics and animal groups in which he or she is interested. The larger parts of the brain, such as the telencephalon and the tectum, are not systematically indexed for the various groups of vertebrates. The pages on which these structures are dealt with can easily be found in the tables of contents on the first page of each chapter.

Synopsis
General introductory part: 1 Cellular elements; 2 Centres; 3 Fibre systems; 4 Morphogenesis; 5 Histogenesis; 6 Principles; 7 Techniques
Specialized part: 8 Introduction; 9 Amphioxus; 10 Lampreys; 11 Hagfishes; 12 Cartilaginous fishes; 13 Brachiopterygians; 14 Chondrosteans; 15 Teleosts; 16 Lungfishes; 17 Latimeria; 18 Urodeles; 19 Anurans; 20 Reptiles; 21 Birds; 22 Mammals
General concluding part: 23 Brain size; 24 Meaning

A

Acanthopterygii **15**: 766
accessory olfactory system **18**: 1128; **19**: 1275; **20**: 1468; **22**: 1894
accessory optic system **12**: 616; **18**: 1111; **19**: 1255; **20**: 1453; **22**: 1818
accessory spinal lobes **15**: 772, 794; **24**: 2137
Acipenser **14**: 702
Acipenseridae **14**: 701
– brain size **23**: 2106
acoustic communication **19**: 1241
acoustic nuclei **15**: 816
Acrania **8**: 357; **9**: 366
Actinistia **17**: 1007, 1041
Actinopterygii **8**: 358, 359
– brain size **23**: 2106
action potentials
– in axons **1**: 16
– in dendrites **1**: 18
adendritic cells **2**: 63
adenohypophysis **9**: 394; **10**: 463; **11**: 529; **12**: 625; **16**: 943, 986; **17**: 986, 1014; **18**: 1119; **19**: 1267; **20**: 1465; **22**: 1862
adhesio interthalamica **4**: 172
adhesion molecules **19**: 1162
adrenergic neurons **22**: 1759
adrenocorticotrophic hormone **11**: 530
aggression **22**: 1900
Agnatha **11**: 539
– brain size **23**: 2105
ala cinerea **22**: 1645
alar plate **4**: 194, 194; **10**: 406, 430, 446, 447; **13**: 657, 673; **14**: 703, 709, 713; **16**: 942; **17**: 1017; **18**: 1077, 1079; **19**: 1206; **24**: 2139
alar substrate pathway **5**: 258
alveus **22**: 1873, 1935, 1944
amacrine cells **10**: 420

Ambystoma tigrinum **8**: 360; **18**: 1048
Amia calva **8**: 359; **15**: 759, 786
Ammocoetes branchialis **10**: 398
Amniota **8**: 360
Amphibia **8**: 360
– brain size **23**: 2108
AmphiHox 3 gene **9**: 377, 393
ampulla caudalis **10**: 429; **15**: 790
ampulla, spinal terminal **9**: 387
ampullae of Lorenzini **12**: 583; **14**: 701, 721
ampullary electroreceptors **13**: 655; **14**: 701, 702; **16**: 940, 966; **17**: 1009, 1023
amygdala **13**: 693; **16**: 997; **17**: 1037; **18**: 1131; **19**: 1277, 1280; **20**: 1486; **22**: 1925
– connectivity **20**: 1486
– extended amygdala **22**: 1926
– olfactory amygdala **20**: 1486; **22**: 1925
– pars lateralis **18**: 1131; **19**: 1277
– pars medialis **18**: 1131; **19**: 1280
– subdivision **20**: 1486
– vomeronasal amygdala **20**: 1486; **22**: 1925
amygdaloid complex (mammals) **22**: 1872
– basal forebrain connections **22**: 1929
– brainstem connections **22**: 1928
– cortical connections **22**: 1931
– functional aspects **22**: 1931
– hippocampal connections **22**: 1930
– hypothalamic connections **22**: 1928
– intrinsic connections **22**: 1927
– olfactory connections **22**: 1928
– striatal connections **22**: 1901, 1916, 1929
– subdivision **22**: 1925
– thalamic connections **22**: 1929
analogy **6**: 279, 280
annelids **9**: 393
ansa lenticularis **21**: 1563, 1580, 1586, 1594, 1618; **22**: 1906, 1909

anterior dorsal ventricular ridge, see dorsal ventricular ridge
anterior neuropore   9: 368
anterior parencephalic segment   13: 680
anterograde degeneration   7: 329
anterograde degeneration techniques   7: 330
anterograde tracing techniques   7: 330
antidiuretic hormone   11: 529
antidromic stimulation   7: 340
Anura   8: 360
- brain size   23: 2108
apertura lateralis (Luschka)   22: 1645
apertura mediana (Magendie)   22: 1645
apomorphy   6: 311
Archaeopteryx   21: 1526
archencephalon   4: 161; 9: 375, 393; 11: 504; 21: 1533
archicortex   22: 1932
archipallium   16: 1002; 22: 1872
architriatum   21: 1607; 24: 2172
- amygdaloid region   21: 1588, 1607, 1611
- sensorimotor region   21: 1607, 1609, 1611
Archosauria   8: 360; 20: 1317
area X in LPO   21: 1596, 1604
area basalis (myxinoids)   11: 529; 24: 2159
area c   20: 1490
area corticoidea dorsolateralis   21: 1592, 1604, 1615
area d   20: 1490
area diagonalis   22: 1891
area dorsalis (rhombencephali)   4: 199, 202, 222; 13: 673; 16: 966; 17: 1017, 1021; 18: 1077; 19: 1206; 20: 1382
area dorsalis telencephali   14: 747, 753; 15: 906; 24: 2152
- pars centralis   14: 747; 15: 910
- pars dorsalis   14: 747; 15: 909
- pars lateralis   14: 747; 15: 910
- pars medialis   14: 747; 15: 909
- pars posterior   15: 907
area entorhinalis   21: 1614; 22: 1872, 1944
area hypoglossi   22: 1645
area hypothalamica lateralis   20: 1465; 22: 1865
area intermediodorsalis   4: 199, 202, 222; 13: 673; 14: 723; 16: 966; 17: 1017, 1023; 18: 1077; 19: 1206; 20: 1382
area intermedioventralis   4: 199, 202, 222; 13: 675; 14: 723; 16: 969; 17: 1017, 1024; 18: 1077; 19: 1206; 20: 1382
area intertrigeminalis   21: 1563
area lineae lateralis   10: 433
area magnocellularis tuberculi posterioris   14: 738
area octavolateralis   10: 430, 433, 485; 11: 517, 523; 12: 584; 14: 753; 16: 967; 17: 1024; 18: 1085; 19: 1216
- connectivity   18: 1089; 19: 1218
- cytoarchitecture   18: 1086; 19: 1217
- metamorphic changes   18: 1091; 19: 1228
area olfactoria dorsalis   14: 747
area olfactosomatica   14: 747
area optica accessoria   14: 733, 740
area pallialis (P1-P12)   12: 640
area parahippocampalis   21: 1614
area perforata anterior   22: 1891
area periventricularis dorsolateralis   12: 640
area periventricularis pallialis   12: 640
area periventricularis ventrolateralis   12: 642
area postrema   15: 808; 22: 1645, 1774
area preoptica   11: 529
area preoptica anterior   19: 1267
area preoptica lateralis   20: 1462
area preoptica posterior   19: 1267
area prepiriformis   21: 1616
area pretectalis   10: 453; 11: 524; 12: 611; 21: 1581
area retrobulbaris   12: 635; 14: 746; 17: 1039
area somatica   13: 693
area subpallialis (SP1-SP12)   12: 642
area superficialis basalis   12: 641; 17: 1037
area tegmentalis ventralis   11: 538; 20: 1411; 21: 1573; 22: 1907, 1916
area temporoparieto-occipitalis   21: 1592, 1604

area triangularis   20: 1457
area vagi   22: 1645
area ventralis (rhombencephali)   4: 199, 202, 222; 13: 676; 14: 709, 724; 16: 970; 17: 1017, 1024; 18: 1077; 19: 1206; 20: 1382
area ventralis telencephali   13: 688; 14: 740, 746, 753; 15: 906
- pars dorsalis   13: 688; 14: 746; 15: 906
- pars lateralis   13: 688
- pars postcommissuralis   15: 906
- pars supracommissuralis   15: 906
- pars ventralis   13: 688; 14: 746; 15: 906
area vestibularis   10: 434, 444
areae B1, B2, B3, B4 (Kuhlenbeck)   4: 214
areae D1, D2, D3 (Kuhlenbeck)   4: 214
arginin-vasopressin   9: 389
arginin-vasotocin   11: 531, 538; 16: 986, 988
ascending modulatory system   10: 481
ascending spinal pathways   6: 312; 10: 426, 444; 11: 516; 15: 793; 18: 1065, 1082; 19: 1188, 1210; 20: 1353, 1387
ascensus medullae   22: 1654, 1655; 24: 2135
association areas   22: 1955
- multimodal   22: 2025
- unimodal   22: 2025
association cortex   22: 1955
- parietotemporal   22: 1955
- prefrontal   22: 1955
association fibres   22: 1883
astrocytes   1: 9; 5: 235; 11: 507; 17: 1015
astroglia   11: 506
ATP   1: 13
atrial nervous system   9: 380, 388
auditory cortex   22: 1789
auditory system   12: 586, 590; 18: 1091; 19: 1223; 20: 1406; 22: 1777
auricula cerebelli   10: 478; 12: 601; 13: 657, 677; 14: 703, 726, 728, 753; 16: 943, 946, 966, 973; 17: 1011, 1027; 18: 1051; 19: 1229; 20: 1424; 21: 1564, 1565; 24: 2144
autapomorphy   6: 311
autonomic ganglia   11: 510; 19: 1187; 20: 1384
autonomic nervous system   10: 417; 21: 1549, 1560
autoradiographic tracing technique   7: 330
avidin-biotin-peroxidase technique   7: 334
axon   1: 7
axon guidance   19: 1245
axonal bundles   5: 257
axonal transport   7: 330

B

B cells of Bone   9: 385
Bagley's bundle   22: 1828
balance organ   9: 373
barbels   11: 497, 519
barrels   2: 51; 22: 1982, 2003
basal forebrain   20: 1477
basal forebrain bundle   3: 129
basal ganglia   18: 1131; 19: 1280; 20: 1487; 22: 1901
- chemoarchitecture   18: 1131; 19: 1284; 20: 1489
- connectivity   18: 1131; 19: 1281; 20: 1409, 1490, 1498; 22: 1908, 1911; 24: 2177
- cytoarchitecture   18: 1131; 19: 1281; 20: 1487
basal interstitial nucleus (Langer)   22: 1737
basal lamina   5: 235
basal plate   4: 161, 194, 196; 10: 406, 430, 436, 447; 13: 657, 673; 14: 703, 709; 16: 942; 18: 1077, 1079; 19: 1206; 24: 2139
basal substrate pathway   5: 258
basement membrane   1: 2
Bauplan   4: 220; 5: 251; 6: 285
Bdellostomatidae   11: 497
bed nucleus of pallial commissure   19: 1280
bed nucleus of stria terminalis   4: 217; 22: 1894, 1897, 1907
behaviour

Synopsis
General introductory part: 1 Cellular elements; 2 Centres; 3 Fibre systems; 4 Morphogenesis; 5 Histogenesis; 6 Principles; 7 Techniques
Specialized part: 8 Introduction; 9 Amphioxus; 10 Lampreys; 11 Hagfishes; 12 Cartilaginous fishes; 13 Brachiopterygians; 14 Chondrosteans; 15 Teleosts; 16 Lungfishes; 17 Latimeria; 18 Urodeles; 19 Anurans; 20 Reptiles; 21 Birds; 22 Mammals
General concluding part: 23 Brain size; 24 Meaning

- affective 21: 1616
- aggressive 22: 1900
- agonistic 21: 1596
- feeding 21: 1574, 1596, 1604
- goal-oriented 22: 2024
- reproductive 22: 1894, 1897, 1900
- sexual 22: 1869, 1900
- stereotyped 21: 1595
Bergmann glia 5: 239; 22: 1735
Betz cells 22: 1964
binocular vision 22: 1792
biochemical switching 3: 152; 6: 304
biocytin 7: 328, 331
biotic world 6: 307
biotinylated dextran amine 7: 331
birds
- brain size 23: 2110
blink reflex 22: 1718
blood vessels 5: 235; 11: 507
blood-brain barrier 1: 13; 11: 508
blueprint 5: 251
blueprint hypothesis 5: 258; 18: 1062
Bodian technique 7: 328
Bolk, subdivision of the cerebellum 22: 1724
Bone cells 11: 543
Brachiopterygii 8: 358; 13: 655, 697; 15: 766
brachium colliculi 21: 1578
brachium colliculi inferioris 22: 1646, 1788, 1720
brachium conjunctivum 10: 447; 12: 606; 14: 729; 16: 974; 17: 1028; 18: 1094; 19: 1234; 20: 1436; 22: 1712, 1720, 1752
- decussation 22: 1720, 1721
- uncrossed descending limb 22: 1752
brachium pontis 21: 1568; 22: 1712
brain atlases 7: 354
brain cell microenvironment 1: 11
brain size 23: 2099
- and allometry 23: 2099
- and convex polygon 23: 2099
- and ecology 23: 2107, 2111, 2116, 2122, 2124
- and ethology 23: 2107, 2112, 2116, 2127
- and feeding pattern 23: 2112, 2123, 2125, 2129, 2130
- and gradualism 23: 2127
- and group size 23: 2125
- and home range 23: 2125
- and K-selection 23: 2103, 2108, 2112, 2117, 2125
- and longevity 23: 2117, 2124
- and metabolic rate 23: 2116, 2123, 2127
- and monogamy 23: 2112
- and neonatal development 23: 2112, 2118
- and parental care 23: 2112, 2129
- and punctuated equilibria 23: 2127
- and r-selection 23: 2103, 2112
brain slice 7: 345
brain vesicles 4: 161, 164
branchiomotor nuclei 11: 521; 12: 593; 14: 723; 18: 1080; 19: 1210; 20: 1385
Branchiostoma lanceolatum 8: 357; 9: 365, 366
buccal cavity 11: 498
bulbothalamic fibres 10: 457
bulbus olfactorius 2: 61; 10: 464, 477; 11: 533; 12: 559, 632; 13: 657, 687; 18: 1124; 19: 1273; 20: 1468; 21: 1531, 1589, 1615; 22: 1885
bulbus olfactorius accessorius 18: 1124; 19: 1273; 20: 1468; 22: 1894; 24: 2152
bundle, see fasciculus

C

Cajal-Retzius cells 22: 1969, 1970
Calamoichthys 13: 655
calamus scriptorius 11: 521
calices of Held 22: 1783
camouflage 15: 748
campus Foreli 22: 1668
canalis centralis 11: 507; 12: 570
capillary networks 11: 507, 544
capsula externa 22: 1908
capsula extrema 22: 1908
capsula interna 22: 1883
capsula otica 11: 518
Captorhinomorpha 8: 360
caput nuclei caudati 22: 1903
carbocyanine tracers 7: 332, 344
Cartesian coordinate system 6: 275
cartilaginous fishes 12: 551
cauda equina 24: 2135
caudal neurosecretory system 15: 793, 794; 16: 965
Caudata 8: 360
caudate-putamen complex 22: 1903
cavitation 4: 160
cell culture 7: 343
cell membrane 1: 6
cells
- Golgi Type I 1: 7
- Golgi Type II 1: 7
- with ascending axon type 1 11: 515
- with ascending axon type 2 11: 515
central acoustic tract 22: 1786, 1787
central gray 10: 480
central layer (Senn) 5: 248
central pattern generators 12: 577, 600; 18: 1062; 19: 1198; 20: 1365
central prosencephalic complex 11: 532
- chemoarchitecture 11: 537
- connectivity 11: 532
- development 11: 504, 532
- homology 11: 532
central reticular fasciculus 22: 1721, 1760, 1761
centrum semiovale 3: 114; 22: 1883
cephalic flexure 4: 164
Cephalochordata 9: 366; 11: 539
cerebellar cortex 2: 56
cerebellar nuclei 22: 1735
- Cetacea 22: 1737
- connectivity 22: 1749, 1801, 1817
- dorsolateral hump 22: 1752
- interneurons 22: 1735
cerebellar plate 13: 660; 16: 946
cerebello-mesencephalic-olivary circuits 22: 1841
cerebellum 10: 446; 11: 523; 12: 601; 13: 657, 677; 18: 1093; 19: 1229; 20: 1422; 21: 1565; 22: 1645; 24: 2141, 2185
- A,X,B,C,D zones (compartments) (Voogd) 22: 1739–1745, 1751, 1842
- cell types 22: 1733, 1985
- central nuclei 21: 1567
- connectivity 10: 447; 12: 604; 14: 728; 15: 834; 18: 1093; 19: 1232; 20: 1428; 22: 1747, 1752
- corticonuclear projections 20: 1433; 22: 1739
- cytoarchitecture 12: 602
- development 12: 563, 601; 19: 1230; 20: 1423; 21: 1536

cerebellum
- development of zonal pattern 22: 1741
- effect of lesions 19: 1234; 20: 1439
- folial pattern 22: 1724
- function 12: 606
- gross anatomy 22: 1724
- hemispheres 22: 1645
- histochemistry 21: 1567
- longitudinal zones 22: 1739
- morphogenesis 22: 1731
- projections of cerebellar nuclei 20: 1435
- stimulation studies 19: 1235; 20: 1440; 21: 1565
- vermis 22: 1645
- zonal distribution of molecular markers 22: 1741
cerebral cortex 16: 990; 20: 1469; 24: 2161
- cellular physiology 20: 1473
- chemoarchitecture 20: 1474
- connectivity 20: 1475
- - commissural connections 20: 1478
- - intracortical connections 20: 1477
- - olfactory input 20: 1475
- - output 20: 1478
- - somatosensory input 20: 1477
- - visual input 20: 1476
- cytoarchitecture 20: 1469
- development 20: 1337
- evolution 20: 1497
- neuron types 20: 1471
- subdivision 20: 1469
cerebral hemispheres 10: 464; 22: 1647; 24: 2152
cerebralisation 21: 1528, 1620
cerebrospinal fluid-contacting cells 11: 528; 14: 743
cervical enlargement, see intumescentia cervicalis
cervical flexure 4: 164
channels
- anion 1: 16
- calcium 1: 15
- potassium 1: 15
- proton 1: 16
- sodium 1: 15
Chelonia 8: 361
chemical neurotransmission 2: 70
chemically defined neuronal populations
- dopamine 10: 460, 485; 11: 538; 12: 576, 598, 628, 644; 13: 684, 689, 693; 14: 747; 15: 887, 892, 896, 906; 16: 992; 18: 1096; 19: 1238; 20: 1339, 1410, 1487; 21: 1535, 1553, 1563, 1573; 22: 1760, 1907, 1913
- enkephalins 10: 463, 474; 11: 534, 539; 13: 689, 693; 15: 892, 895, 897; 16: 997; 18: 1133; 19: 1238, 1284; 20: 1489; 21: 1580, 1595
- neuropeptides 12: 575, 598, 628, 644
- noradrenaline 10: 472; 12: 576, 598, 628, 644; 15: 892, 895; 18: 1097; 19: 1239; 20: 1339, 1360, 1415; 21: 1535, 1563, 1573; 22: 1715, 1759
- serotonin 9: 388; 10: 419, 422, 472; 11: 510; 12: 575, 598, 628, 644; 13: 684, 689, 693; 14: 747, 751; 15: 892; 16: 993, 997; 18: 1097; 19: 1240; 20: 1360, 1417; 21: 1543, 1549, 1562; 22: 1715, 1757, 1911
- substance P 10: 422, 473; 11: 534, 539; 13: 684, 689, 693; 14: 747, 753; 15: 892, 895; 16: 993, 997; 18: 1133; 19: 1238, 1284; 20: 1411, 1489; 21: 1543, 1549, 1579, 1594
- vasotocin 11: 538; 14: 743; 15: 899; 16: 988; 18: 1118; 19: 1270; 20: 1465; 21: 1589
chemoarchitecture 1: 20; 2: 85
chemodifferentiation 2: 85
chiasma oculomotorii 10: 451
chiasma opticum 10: 449; 11: 501, 529, 543; 12: 613; 13: 679, 687; 14: 733, 739, 749; 16: 943, 989; 17: 1014; 22: 1646, 1792, 1862
Chiroptera 22: 1641
cholinergic basal forebrain 22: 1906
cholinergic cell groups 22: 1907
Chondrichthyes 8: 357; 17: 1008
- brain size 23: 2107

Chondrostei 8: 359
Chordata 8: 357; 11: 539
chromatolysis 7: 330
circadian pacemaker 11: 529
circadian rhythm 11: 525, 531; 21: 1583; 22: 1848
circumferential bundle 5: 246
circumventricular organs 11: 528, 541; 22: 1871, 2032
cladism 6: 314
Cladistia 13: 655; 15: 766
cladistic analysis 16: 940
cladistics 4: 221
cladograms 15: 766
claustrum 4: 217; 22: 1908
clear cells (amphioxus) 9: 388
Clupeomorpha 15: 766
coarctation 4: 171
Coelacanthini 17: 1007, 1041
cognitive brain 22: 2039
cognitive functions 24: 2190
colliculus inferior 21: 1579; 22: 1720, 1787; 24: 2145
- central nucleus 22: 1787
- commissure 22: 1720, 1787
- cortex dorsalis 22: 1787
- external nucleus 22: 1787
colliculus superior 22: 1646, 1720, 1752; 24: 2145
- commissure 22: 1720, 1812
- role in generation of saccades 22: 1792
collothalamus 19: 1266; 20: 1497; 24: 2148
Columba livia 8: 361
columna dorsalis (Clarke) 4: 177, 201; 21: 1543, 1549; 22: 1663, 1674
columna dorsolateralis 4: 177, 201
columna motoria spinalis 12: 572; 14: 724; 16: 970; 17: 1026
columna ventralis 4: 177, 201
columna ventrolateralis 4: 177, 201
columnar organisation 21: 1599, 1613
columns in neocortex 22: 2002
- macro columns 22: 2006
- mini columns 22: 2005
- motor 22: 2004
- ocular dominance columns 22: 2004
- orientation columns in neocortex 22: 2002
command systems 19: 1199
commissura ansulata 12: 611; 13: 679; 16: 977; 17: 1027
commissura anterior 12: 556; 13: 660, 686; 14: 749; 16: 943, 989; 17: 1014, 1036; 18: 1054; 19: 1160; 20: 1329; 21: 1609; 22: 1646, 1872
commissura cerebelli 4: 161; 10: 447; 11: 523; 14: 703, 729, 753; 16: 974
commissura dorsalis telencephali 10: 465, 467
commissura habenulae 4: 168; 10: 456, 470; 11: 525, 527, 533, 543; 12: 625; 13: 657, 681, 687; 14: 736, 750; 16: 943; 17: 1035; 18: 1115; 19: 1160; 20: 1412
commissura hippocampi 14: 750; 22: 1647, 1872
commissura infima (Haller) 11: 521; 15: 807; 16: 968; 21: 1559
commissura interbulbaris 11: 533
commissura octavolateralis 16: 974
commissura olfactoria 12: 636
commissura pallii 16: 997; 18: 1054; 19: 1160
commissura pallii anterior 20: 1478
commissura pallii posterior 16: 984; 20: 1478
commissura posterior 4: 161; 10: 453; 11: 524, 543; 12: 625; 14: 741; 16: 943, 979, 983, 986; 17: 1014, 1031; 18: 1051; 19: 1160; 20: 1456; 22: 1720, 1812
commissura postinfundibularis 10: 462
commissura postoptica 10: 449, 461, 462; 11: 527; 12: 625; 13: 686, 687; 14: 741; 16: 983, 985
commissura preinfundibularis 10: 462
commissura superior telencephali 12: 625
commissura supramamillaris 21: 1586
commissura supraopticus 22: 1801
commissura tecti 11: 527; 12: 611; 13: 679; 14: 730; 21: 1568

> **Synopsis**
> *General introductory part:* 1 Cellular elements; 2 Centres; 3 Fibre systems; 4 Morphogenesis; 5 Histogenesis; 6 Principles; 7 Techniques
> *Specialized part:* 8 Introduction; 9 Amphioxus; 10 Lampreys; 11 Hagfishes; 12 Cartilaginous fishes; 13 Brachiopterygians; 14 Chondrosteans; 15 Teleosts; 16 Lungfishes; 17 Latimeria; 18 Urodeles; 19 Anurans; 20 Reptiles; 21 Birds; 22 Mammals
> *General concluding part:* 23 Brain size; 24 Meaning

commissura transversa 12: 625
commissura tuberculi posterioris 11: 524, 543; 13: 686, 687; 14: 741; 17: 1033
commissura ventralis tegmentalis 11: 524
commissura vestibularis lateralis 22: 1726
commissura vestibulolateralis 10: 447
common mode rejection 12: 600
comparative embryology 4: 220
comparative neuroanatomy
- central aim 6: 302
- place 6: 273
- principles 6: 274
- programme 6: 302
confocal microscopy 7: 336
connexins 1: 21
conus medullaris 22: 1659
cornu Ammonis 22: 1647, 1872, 1944
cornu dorsale 14: 708, 722; 15: 791; 21: 1542; 22: 1661
cornu ventrale 12: 569; 14: 708; 15: 791; 21: 1542; 22: 1667
coronet cells 15: 895; 17: 1034
corpus amygdaloideum, see amygdaloid complex
corpus callosum 22: 1646, 1872
corpus cerebelli 12: 601; 14: 703, 726, 753; 15: 832; 16: 943, 946, 965, 973; 17: 1011, 1027; 18: 1093; 19: 1229; 20: 1424; 22: 1726
corpus commune posterius 12: 569
corpus gelatinosum 21: 1549, 1552
corpus geniculatum anterius 10: 449, 457
corpus geniculatum laterale 10: 457; 12: 616; 14: 753; 17: 1032
corpus geniculatum laterale (mammals), see nucleus geniculatus lateralis dorsalis
corpus geniculatum mediale (mammals), see nucleus geniculatus medialis thalami
corpus geniculatum posterius 10: 449, 457
corpus geniculatum thalamicum 18: 1117; 19: 1262
corpus habenulae 11: 525; 22: 1646
corpus juxtarestiforme 22: 1712
corpus pallii (*Latimeria*) 24: 2160
corpus restiforme 21: 1568; 22: 1712
corpus rostrale 17: 1014, 1035
corpus striatum 10: 465, 480; 13: 689, 695; 16: 991; 17: 1037; 21: 1589; 24: 2155
- pars dorsalis 16: 991
- pars ventralis 16: 991
corpus trapezoideum 21: 1557; 22: 1643, 1712, 1781
cortex dorsalis 20: 1469, 1478, 1480; 22: 1935; 24: 2161
cortex dorsolateralis 22: 1935
cortex dorsomedialis 20: 1469, 1478, 1480; 21: 1614; 24: 2161
cortex hippocampalis 24: 2161
cortex lateralis 20: 1470, 1478, 1479; 22: 1935; 24: 2161
cortex medialis 20: 1469, 1479; 22: 1935; 24: 2161
cortex parahippocampalis 24: 2161
cortex periamygdaloideus 22: 1873, 1891
cortex piriformis 21: 1615; 22: 1891; 24: 2161
cortex prepiriformis 22: 1873
cortex tuberculi olfactorii 17: 1037
cortico-cortical projections 22: 1960, 1964
corticobulbar connections 22: 1820, 1822, 1829
corticospinal connections 22: 1669-1672, 1820-1822, 1826, 1827
corticospinal tract, see pyramidal tract
corticostriate fibres 22: 1908
corticothalamic connections 22: 1858

corticotropin 14: 751
corticotropin-releasing factor 14: 751; 15: 896, 899
cranial nerves 18: 1079; 19: 1207; 20: 1367
Craniata 8: 357; 9: 366
crista cerebellaris 10: 433; 12: 582; 13: 673; 14: 713; 16: 966; 17: 1021
crossed caudal inhibitory neurons 10: 422, 428, 479
Crossopterygii 8: 359, 360; 13: 697; 16: 940, 1000; 17: 1007
- brain size 23: 2107
current-source density analysis 7: 339
Cyclostomata
- brain size 23: 2105
cytoarchitectonics 2: 47
cytogenetic sectors 5: 253
cytoskeleton 1: 5

D

Dahlgren cell 12: 569
decussatio supraoptica 21: 1578, 1579, 1584, 1585, 1611
decussatio tegmentalis dorsalis 22: 1721, 1794
decussatio tegmentalis ventralis 22: 1721
dendrites 1: 6
dendritic bundles 22: 1664, 1665
deoxyglucose technique 7: 333
descending octaval nucleus 16: 965, 967
descending pathways to the spinal cord 10: 435; 11: 515; 15: 793; 18: 1067; 19: 1190; 20: 1356; 21: 1553; 22: 1844
deuterencephalon 4: 161; 9: 375, 393; 11: 504; 21: 1532, 1534
dextran amines 7: 331
diagonal band of Broca 16: 997; 22: 1891
Didelphys virginiana 8: 361
diencephalic vesicle 11: 525
diencephalon 11: 500; 12: 618; 18: 1114; 19: 1258; 20: 1454; 22: 1646; 24: 2147
- chemoarchitecture 12: 628
- connectivity 12: 622, 623; 19: 1261; 18: 1116
- - auditory input 18: 1091; 19: 1263; 20: 1457
- - cerebellar input 20: 1458
- - gustatory input 18: 1085; 19: 1216
- - infrared input 20: 1396
- - lateral line input 19: 1264
- - pallidal input 20: 1459
- - retinal input 18: 1116; 19: 1263; 20: 1456
- - somatosensory input 18: 1116; 19: 1264; 20: 1457
- - tectal input 18: 1116; 19: 1263; 20: 1457
- - thalamotelencephalic projections 18: 1116; 19: 1266; 20: 1459
- cytoarchitecture 12: 560; 18: 1116; 19: 1260; 20: 1455
- development 11: 504, 532; 12: 561; 19: 1260; 20: 1335, 1455
- function 12: 630
- subdivision 18: 1114; 19: 1258; 20: 1455
- sulcal pattern 12: 618; 18: 1114; 19: 1160; 20: 1329
dinosaurs 8: 360; 20: 1317
- brain size 23: 2109
Dipnoi 8: 359; 17: 1007
- brain size 23: 2107
dissociated cell cultures 7: 343
divergence (of lateral plates) 4: 166
diversification 6: 306
DNA 1: 6
dopamine, see chemically defined neuronal populations

dopaminergic cell groups, see chemically defined neuronal populations
dorsal cells   10: 402, 418, 426, 432, 483; 11: 543; 16: 958
dorsal column nuclei   22: 1714, 1764, 1766
dorsal column nucleus, see nucleus funiculi dorsalis
dorsal column pathway   10: 411; 11: 516
dorsal commissural cells   9: 381; 16: 953
dorsal funiculus, see funiculus dorsalis
dorsal horn, see cornu dorsale
dorsal medullary nucleus   19: 1225
dorsal pallium   11: 543; 13: 694; 18: 1130; 19: 1277; 22: 1872
dorsal rhombencephalic zone   15: 815
dorsal root   9: 366; 10: 435, 457; 11: 515, 500, 511; 18: 1064; 19: 1180; 20: 1349; 22: 1661
dorsal thalamus, see thalamus dorsalis
dorsal ventricular ridge   4: 172; 21: 1539, 1591; 24: 2156, 2161
  - chemoarchitecture   20: 1484
  - cytoarchitecture   20: 1481
  - development   20: 1337
  - evolution   20: 1495
  - input
  - - auditory   20: 1460, 1484
  - - cortical   20: 1485
  - - gustatory   20: 1484
  - - infrared   20: 1396, 1484
  - - somatosensory   20: 1460, 1484
  - - visual   20: 1460, 1484
  - intrinsic organisation   20: 1481
  - output   20: 1485
  - subdivision   20: 1481
dorsolateral bundle   5: 259; 10: 403
dorsoventral diencephalic tract   5: 264
dual synapses   10: 426

E

echolocation   22: 1782
ectosomatic organs   6: 303
ectostriatum   4: 216; 21: 1584, 1597, 1603
edge cells   10: 421, 428; 11: 515; 20: 1347
Elasmobranchii   8: 358
electric organ discharge   12: 569
electric organs   15: 767
electrical junctions   1: 8
electrical synapses   1: 21
electrical transmission
  - in spinal cord   19: 1197
electroencephalogram   7: 337; 20: 1480
electromotor command nucleus   15: 801, 802
electromotor corollary circuits   15: 802
electromotor neurons   15: 792, 804, 849; 24: 2141
electromotor relay nucleus   15: 801, 802
electroreception   10: 433; 11: 499; 12: 600; 14: 701
electroreceptors
  - ampullary   15: 768
  - tuberous   15: 768
electrosensitivity   10: 477
electrosensory lateral line system   14: 754; 24: 1988, 2141
electrosensory lobes   15: 772, 820
Elepomorpha   15: 766
elevator movement (of matrix cells)   5: 232
embryos of hagfishes   11: 501
eminentia cerebelli ventralis   16: 974
eminentia facialis   22: 1645
eminentia granularis   14: 727; 15: 829; 17: 1027
eminentia medialis   22: 1645
eminentia mediana   11: 530; 12: 625; 14: 743; 17: 1034; 22: 1862
eminentia sagittalis   21: 1531, 1595
eminentia thalami   11: 528, 532; 12: 619; 14: 737, 752; 16: 948; 18: 1117

eminentia trigemini   10: 445
emotional behaviour   22: 2028
encephalisation   11: 500
encephalisation quotient   23: 2099
end bulbs   2: 63
endoplasmatic reticulum
  - rough   1: 5
  - smooth   1: 5
endosomatic organs   6: 303
entelechic tendency   6: 309
entorhinal cortex   22: 1955
ependymal cells   1: 9, 10; 11: 506; 10: 409; 17: 1034
ependymal gliocytes   13: 662; 16: 953; 17: 1015
ependymoglia   1: 11
epichordal zone   4: 208
Epigonichthys   9: 366
epiphysis cerebri   4: 168,169; 9: 378; 10: 399, 454, 477; 11: 525, 543; 12: 618; 13: 657, 581; 14: 736; 15: 884; 16: 943, 946, 978; 17: 1031; 18: 1115; 19: 1260; 20: 1456; 21: 1534; 22: 1846
epiphytic guidance   3: 147
epistriatum   16: 1001
epithalamus   4: 210, 212; 10: 45, 454; 11: 525; 12: 618; 15: 882; 16: 978; 17: 1031; 18: 1115; 19: 1260; 20: 1455; 22: 1845; 24: 2150
epithelium   1: 2
Eptatretidae   11: 497
Eptatretus stouti   8: 357
Erinaceidae   22: 1640
Erpetoichthys calabaricus   8: 358, 359; 13: 655
estivation   16: 983
eurydendroid cells   14: 728; 15: 829, 912
Euteleostei   15: 766
Eutheria   8: 361
  - brain size   23: 2119
evagination   4: 167; 6: 274; 10: 406; 13: 660; 14: 707, 744; 16: 946, 950; 17: 1036; 24: 2157
eversion   4: 170; 6: 274; 13: 660; 14: 703, 706, 744; 17: 1036; 24: 2157
evoked potentials   7: 337
evolutionary progress   6: 321
excitatory interneurons   10: 422, 428, 479
explant   7: 343
extended amygdala   22: 1907
external germinal layer   5: 237, 239
external granular layer   5: 239
extracellular field potentials   7: 337
extracellular marking techniques   7: 340
extracellular space   1: 12; 10: 409
extracellular volume fraction   7: 340
extramedullary neurons   19: 1178
extrapyramidal system   22: 1901, 1924
extrasynaptic interaction   24: 2190
extrasynaptic modulation   24: 2183
eye spots   9: 379

F

Falck-Hillarp technique   7: 333
fascia dentata   22: 1872, 1944
fasciculi Foreli   22: 1721
fasciculi proprii   22: 1668
fasciculi tegmentales   13: 680; 14: 725, 733; 16: 978; 17: 1027, 1031, 1040
fasciculus   3: 149
fasciculus anterolateralis (Gowers)   22: 1674, 1762, 1765
fasciculus basalis telencephali   11: 529, 533, 535; 12: 624, 642; 17: 1027, 1035, 1038, 1039
fasciculus communis   10: 430
fasciculus cuneatus   22: 1667
fasciculus dorsolateralis (Lissauer)   19: 1181; 20: 1349; 22: 1661, 1666
fasciculus gracilis   22: 1667

Synopsis
General introductory part: **1** Cellular elements; **2** Centres; **3** Fibre systems; **4** Morphogenesis; **5** Histogenesis; **6** Principles; **7** Techniques
Specialized part: **8** Introduction; **9** Amphioxus; **10** Lampreys; **11** Hagfishes; **12** Cartilaginous fishes; **13** Brachiopterygians; **14** Chondrosteans; **15** Teleosts; **16** Lungfishes; **17** Latimeria; **18** Urodeles; **19** Anurans; **20** Reptiles; **21** Birds; **22** Mammals
General concluding part: **23** Brain size; **24** Meaning

fasciculus lateralis telencephali   **3**: 129; **13**: 686; **14**: 741, 754; **15**: 910; **16**: 985; **18**: 1095; **19**: 1236, 1278; **20**: 1466; **21**: 1586, 1591, 1611, 1616
fasciculus lenticularis   **22**: 1909
fasciculus longitudinalis dorsalis (Schütz)   **3**: 142; **22**: 1720, 1866, 2030
fasciculus longitudinalis medialis   **3**: 126; **5**: 261; **10**: 423, 434, 460; **11**: 516; **12**: 570; **13**: 675, 676, 677; **14**: 703, 721, 724, 729, 733, 754; **15**: 805; **16**: 958, 965, 971, 973, 978; **17**: 1017, 1026, 1027, 1039; **18**: 1099; **19**: 1221; **20**: 1404; **21**: 1559, 1564, 1581; **22**: 1668, 1672, 1721
fasciculus mamillaris princeps   **22**: 1866
fasciculus medialis telencephali   **13**: 686; **14**: 740, 749, 750, 754; **15**: 910; **16**: 985; **17**: 1035; **18**: 1129; **19**: 1277; **20**: 1466; **21**: 1619; **22**: 1758, 1760, 1865, 2029, 2030
fasciculus medianus of Stieda   **12**: 575; **14**: 709
fasciculus predorsalis   **12**: 611; **22**: 1721, 1794, 1807, 1810
fasciculus prosencephali lateralis, see fasciculus lateralis telencephali
fasciculus retroflexus   **10**: 436, 456; **11**: 527; **12**: 619; **13**: 677, 680, 681; **14**: 733, 740, 754; **15**: 884; **16**: 983, 984, 985; **17**: 1031, 1032, 1035; **18**: 1094, 1115; **19**: 1236; **20**: 1412; **21**: 1583; **22**: 1720, 1845; **24**: 2150
fasciculus solitarius   **12**: 581; **13**: 673, 675, 677; **14**: 722; **16**: 973; **17**: 1023; **18**: 1084; **19**: 1216; **20**: 1397
fasciculus tectotegmentalis   **12**: 611
fasciculus thalamicus   **22**: 1909
fasciculus uncinatus   **18**: 1094; **19**: 1234; **20**: 1436; **21**: 1568; **22**: 1645
fate map   **18**: 1054; **19**: 1163
feeding   **14**: 742
Felis domestica   **8**: 361
fibrae arcuatae   **10**: 434, 446, 447, 449; **11**: 516, 519; **12**: 585; **13**: 673, 675; **14**: 717, 721; **16**: 966, 968; **17**: 1027; **22**: 1767
fibrae hypothalamotegmentales   **14**: 725, 741, 754
fibrae tectotegmentales dorsales   **13**: 679; **14**: 1029; **14**: 730
fibrae tectotegmentales ventrales   **13**: 679; **14**: 730; **17**: 1029
fibrae tectothalamicae   **16**: 985; **17**: 1029, 1040
fibre class   **3**: 150
fibre components   **3**: 113
fibre pattern   **3**: 133, 150
fibre spectrum   **3**: 133, 150
fibre systems
– chemodifferentiation   **3**: 146
– closed   **3**: 142
– compactness   **3**: 133
– composite   **3**: 143
– course   **3**: 129
– development   **5**: 256
– extent   **3**: 129
– functional aspects   **3**: 150
– open   **3**: 143
– origin   **3**: 125
– polarity   **3**: 125
– simple   **3**: 143
– size   **3**: 132
– somatotopic organisation   **3**: 144
– termination   **3**: 125
– topographical organisation   **3**: 144
– trajectory   **3**: 129
fictive locomotion   **24**: 2185
fictive scratch reflex   **20**: 1366
fictive swimming   **19**: 1171
field L   **21**: 1599, 1604
fields of Forel   **22**: 1861, 1909

fila olfactoria   **18**: 1124; **19**: 1273; **20**: 1468; **22**: 1647
filial imprinting   **21**: 1607
filum terminale   **9**: 383; **19**: 1170; **24**: 2135
Fink-Heimer technique   **7**: 330; **24**: 2166
fissura ansoparamedianus   **22**: 1727
fissura circularis   **10**: 405; **11**: 501
fissura endorhinalis   **22**: 1873
fissura hippocampi   **22**: 1647, 1944
fissura isthmi   **11**: 501, 517
fissura posterior superior   **22**: 1727
fissura rhinalis   **22**: 1873, 1891
fissura rhombo-mesencephalica   **13**: 657
fissura Sylvii   **22**: 1879
fissures, primitive   **22**: 2015
fixed action patterns   **24**: 2154
flexura cephalica   **22**: 1646
flexura pontis   **22**: 1646
flexural organ   **9**: 376, 393
flight   **21**: 1527
flocculus   **22**: 1728
floor plate   **4**: 161, 194; **9**: 369
fluid mosaic   **1**: 6
fluorescence histochemical techniques   **7**: 333
fluorescent probes   **7**: 336, 342
fluorescent tracers   **7**: 332
folding index, neocortex   **22**: 2012
follicle-stimulating hormone   **9**: 389
foramen interventriculare   **10**: 399, 464; **12**: 632; **16**: 946, 989; **22**: 1647, 1862, 1873
formatio hippocampi   **22**: 1647, 1872
formatio reticularis, see reticular formation
formatio vermicularis   **22**: 1725
fornix   **12**: 643; **18**: 1128; **19**: 1279; **20**: 1478; **22**: 1873
fossa interpeduncularis   **22**: 1643
fossa rhomboidea   **4**: 167
fossa Sylvii   **22**: 1879
fovea isthmi   **16**: 974
Freud'sche Hinterzellen   **16**: 958
frontal eye   **9**: 371, 377, 386
frontal organ   **4**: 169; **19**: 1260
Frontbildungstendenz   **21**: 1532
functional columns   **14**: 713, 752
functional mapping   **7**: 333, 335
fundamental components (Kuhlenbeck)   **5**: 247
funicular nuclei (of mormyrids)   **15**: 832
funiculus dorsalis   **10**: 420, 423; **11**: 510; **12**: 569; **13**: 666; **15**: 790; **16**: 958, 973; **17**: 1017; **22**: 1667
funiculus dorsolateralis   **22**: 1669
funiculus lateralis   **10**: 423; **11**: 510; **12**: 569; **13**: 666; **14**: 706; **15**: 790; **17**: 1017
funiculus ventralis   **10**: 423; **11**: 510; **12**: 569; **13**: 666; **14**: 706; **15**: 790; **16**: 958
funiculus ventralis   **17**: 1017

G

ganglion ciliare   **17**: 1029; **19**: 1210; **20**: 1367
ganglion habenulae   **10**: 456; **12**: 618; **13**: 657, 581; **14**: 736; **16**: 943, 978, 979
ganglion habenulae, see also nuclei habenulae
ganglion isthmi   **16**: 971
ganglionic eminences   **22**: 1872, 1903
gap junctions   **1**: 21
general pallium   **13**: 694; **16**: 990, 1002

general somatosensory system 14: 721
giant axon (of squid) 3: 118
giant interneurons 10: 420, 426, 483; 11: 515, 543
giant neurons, in hagfish brain 11: 506
Ginglymodi 8: 359; 15: 766
glia 1: 9; 9: 375; 11: 506, 507, 541; 24: 2182
glial fibrillary acidic protein 1: 9
glial index 1: 9
glioblasts 5: 234
gliogenesis 5: 234
Glires 22: 1642
globus pallidus 20: 1490, 1493, 1498; 21: 1594; 22: 1882, 1901, 1906
- pars externa 22: 1882
- pars interna 22: 1882
glomeruli (Held) 22: 1733
glucose 1: 13
Gnathostomata 11: 497
Golgi apparatus 1: 5, 8
Golgi technique 7: 328; 24: 2181
gonadal steriod-concentrating cells 22: 1897
gonadotropin-releasing hormone 11: 530, 538; 10: 463, 474; 14: 744, 746, 752; 15: 899, 905; 16: 996; 19: 1271
grand lobe limbique (Broca) 22: 2027
greater limbic system 22: 1865, 2024, 2029, 2030
grey matter 2: 25
griseum centrale rhombencephali 11: 523; 12: 579; 13: 677, 680; 14: 724; 16: 971, 973
griseum centrale mesencephali 22: 1720
griseum centrale metencephali 22: 1712
griseum superficiale isthmi et mesencephali 16: 953, 971, 978
Grundbestandteile (Kuhlenbeck) 4: 212, 214; 5: 247
Grundgebiete (Bergquist) 4: 175, 200; 6: 285, 287, 289; 16: 948
gustation 15: 767; 17: 1039
gustatory centre 21: 1560
gustatory pathways 22: 1774, 1777
gustatory system 11: 498, 516
Gymnophiona 8: 360
gyri of neocortex 22: 2013
gyrification 4: 172
gyrus ambiens 22: 1893
gyrus cinguli 22: 2015
gyrus dentatus 22: 1647
gyrus olfactorius lateralis 22: 1893
gyrus semilunaris 22: 1893

H

Häggqvist technique 7: 328
Halecomorphi 15: 766
Halecostomi 8: 359
Hatschek's pit 9: 394
hearing 18: 1091; 19: 1223; 20: 1406
Hemichordata 9: 366
hemispheres, diencephalic, of myxinoids 11: 504
hemispheres, telencephalic, of myxinoids 11: 504
Herring bodies 14: 743
Hinterzellen 11: 543
hippocampal cortex 22: 1932, 1944
hippocampal pallium 13: 694; 16: 990, 1002
hippocampal-pretectal-midbrain bundle 10: 470
hippocampus 21: 1614; 22: 1872
- microcircuitry 22: 1945
- postcommissural 22: 1873
- precommissural 22: 1873
- supracommissural 22: 1873
hodology
- comparative 3: 150
- general 3: 149
- special 3: 150
Holocephali 8: 358

Holostei 8: 359; 15: 759
- brain size 23: 2106
homeobox genes 4: 193; 5: 253; 6: 316; 19: 1160; 24: 2164
homeodomain proteins 6: 316
homeotic genes 9: 377
homing 21: 1528, 1616
Hominidae
- brain size 23: 2127
homology 6: 284; 24: 2166
- criteria 6: 284
- field 6: 280
- serial 6: 280
homoplasy 6: 282, 312
- convergent 6: 282
- parallel 6: 282
hook bundle, see fasciculus uncinatus
horseradish peroxidase 7: 328, 331
Hox genes 4: 193; 6: 316
HVC ('high vocalisation centre') 21: 1585, 1601, 1604
hydrodynamic mechanoreceptor organs 16: 940
hydrodynamic pressoreceptors 13: 655
hypercolumns (in visual cortex) 22: 2004
hyperinversion theory 6: 299
hyperstriatum accessorium 21: 1609, 1613; 24: 2161
hyperstriatum dorsale 4: 215; 21: 1609, 1612; 24: 2161
hyperstriatum intercalatum supremum 21: 1609; 24: 2161
hyperstriatum ventrale 4: 215; 21: 1595, 1604, 1607; 24: 2156
hypophyseal duct 17: 1011
hypophysis cerebri 11: 530; 12: 625; 13: 657, 584; 14: 739, 742; 18: 1119; 19: 1267; 20: 1465
hypothalamo-hypophysial relations 9: 376, 393; 11: 529; 12: 625; 13: 685; 17: 1034, 1040; 18: 1119; 19: 1270; 20: 1465; 22: 1868
hypothalamus 10: 461; 11: 524; 12: 619; 13: 684; 18: 1118; 19: 1267; 20: 1461; 21: 1534; 22: 1866; 24: 2150
- connectivity 10: 462; 11: 523; 18: 1118; 19: 1269; 20: 1463; 22: 1866
- cytoarchitecture 18: 1118; 19: 1267; 20: 1462
- pars dorsalis 13: 684; 14: 738; 16: 982; 17: 1033
- pars ventralis 13: 684; 14: 739; 16: 982; 17: 1033
Hyracoideae 22: 1642

I

Ichthyophis 18: 1046
idealistic morphology 6: 284
immediate-early genes 6: 304, 322; 7: 335
immune privilege (brain) 1: 13
immunocytochemical techniques 7: 333
immunohistochemistry 7: 333
in situ hybridisation 7: 335
in vitro preparations 7: 343
increase in complexity 6: 321
indusium griseum 22: 2019
inferior olive, see oliva inferior
infrared sensitivity 20: 1394
infundibular organ 9: 375, 384, 386, 393
infundibulum 4: 169; 12: 555; 13: 657; 14: 739, 742; 16: 983, 986; 17: 1014, 1040
inner ear 18: 1088; 19: 1219, 1223; 20: 1402, 1406
Insectivora 22: 1640
insula of Reil 22: 1879, 2019
insular cortex 22: 1882
intergeniculate leaflet, see nucleus geniculatus lateralis ventralis
interlimb coordination 20: 1351
intermediate zone (Boulder) 5: 237
intermediate zone (Herrick) 4: 198
intermediodorsal rhombencephalic zone 15: 807, 815
intermedioventral rhombencephalic zone 15: 806, 807
internuncial cells 9: 380; 11: 515
interstitiospinal projections 18: 1067; 19: 1193; 20: 1358

Synopsis
General introductory part: 1 Cellular elements; 2 Centres; 3 Fibre systems; 4 Morphogenesis; 5 Histogenesis; 6 Principles; 7 Techniques
Specialized part: 8 Introduction; 9 Amphioxus; 10 Lampreys; 11 Hagfishes; 12 Cartilaginous fishes; 13 Brachiopterygians; 14 Chondrosteans; 15 Teleosts; 16 Lungfishes; 17 Latimeria; 18 Urodeles; 19 Anurans; 20 Reptiles; 21 Birds; 22 Mammals
General concluding part: 23 Brain size; 24 Meaning

intracellular recording 7: 340
intracellular staining 7: 328
intracellular transport 1: 8
intralaminar nuclei 22: 1763; 22: 1855
intramedullary sensory cells 11: 519
intraspinal sensory cells 11: 515
intumescentia cervicalis 18: 1057; 19: 1170; 20: 1339; 21: 1529, 1542; 24: 2137
intumescentia lumbalis 18: 1057; 19: 1170; 20: 1339; 21: 1529, 1542, 1585
intumescentia lumbosacralis 24: 2137
intumescentia trunci 20: 1339; 24: 2137
invagination 4: 170
invasion hypothesis 6: 310
inversion 4: 167; 13: 660; 14: 744
ion-selective microelectrodes 7: 340
ionotropic (receptors) 1: 20
ipsilateral inhibitory interneurons 10: 422
islands of Calleja 22: 1906
isocortex 22: 1932; 24: 2161
isolated CNS preparations 7: 345
isthmus rhombencephali 4: 161; 10: 430, 445; 14: 724; 15: 840; 16: 943; 17: 1011, 1026

J

jamming avoidance response 15: 821, 861
Joseph cells 9: 371, 384, 393

K

karyoplasm 1: 5
Keimzellen (His) 5: 229
Klüver-Barrera technique 7: 328
Kölliker's pit 9: 369
Kolmer-Agduhr cells 18: 1061; 19: 1171
Kopfbeuge 16: 946

L

labyrinth 10: 434; 11: 498, 517; 12: 586
labyrinthectomy 19: 1221
labyrinthodonts 8: 360
Lagomorpha 22: 1642
lamellar body 9: 371, 378; 11: 543
lamellar cells 9: 386
lamellar complex 9: 393
lamina archistriatalis dorsalis 21: 1607
lamina frontalis superior 21: 1603, 1611, 1616, 1618
lamina frontalis suprema 21: 1618
lamina hyperstriatica 21: 1596, 1603
lamina medullaris dorsalis 21: 1591, 1596, 1618
lamina medullaris externa 22: 1906
lamina medullaris interna 22: 1906
lamina medullaris ventralis 21: 1591, 1616
lamina supraneuroporica 13; 16: 947, 989
lamina terminalis 4: 207; 10: 464; 16: 943, 946, 989; 17: 1036; 22: 1646, 1862, 1872
lamination 2: 72
Lampetra fluviatilis 8: 357; 10: 398
Lampetra planeri 10: 398
large dorsal cells 11: 515

large internuncial cells 10: 402
Larsell, subdivision of the cerebellum 22: 1726
lateral cervical nucleus, see nucleus cervicalis lateralis
lateral forebrain bundle 13: 686
lateral forebrain bundle, see fasciculus lateralis telencephali
lateral inhibitory interneurons 10: 421, 428, 479, 480, 483
lateral line efferent neurons 18: 1088; 19: 1218
lateral line system 11: 499, 517; 12: 583; 17: 1039; 18: 1086; 19: 1218
- electroreceptive component 12: 583; 14: 753; 18: 1086
- electrosensory component 15: 768
- mechanoreceptive component 12: 583; 14: 753; 18: 1086; 19: 1217; 10: 433
- mechanosensory component 15: 768
lateral pallium 11: 534, 543; 13: 694; 18: 1130; 19: 1277; 22: 1872
lateral reticular zone 16: 969; 18: 1097; 19: 1239; 20: 1415
lateral terminal nucleus of the accessory optic system 22: 1818
lateralisation 21: 1605, 1612
laterodorsal tegmental field 22: 1757
Latimeria chalumnae 8: 360; 13: 697; 16: 941, 971, 1000
laws of von Baer 4: 220; 6: 312, 316
learning 21: 1596, 1604, 1606, 1614; 22: 1900
lemniscus bulbaris 10: 431, 434, 449; 11: 518, 522, 528
lemniscus lateralis 10: 435; 12: 585; 13: 673, 675, 677; 14: 721, 725, 732, 754; 16: 966, 973, 978; 17: 1027; 18: 1089; 19: 1225; 20: 1406; 21: 1557, 1579; 22: 1646, 1721, 1786, 1787; 24: 2145
lemniscus medialis 10: 426; 18: 1083; 19: 1211; 20: 1387; 21: 1555, 1564; 22: 1720, 1721, 1764, 1767, 1770
lemniscus spinalis 10: 426, 427, 431, 444, 449, 452; 11: 516, 522, 528; 12: 570; 15: 805; 16: 973, 977; 17: 1017, 1027; 18: 1065, 1083; 19: 1213; 20: 1353, 1389; 21: 1554, 1564
lemniscus trigeminalis 11: 528; 12: 580; 18: 1084; 19: 1216; 20: 1392; 22: 1714
lemnothalamus 19: 1266; 20: 1497; 24: 2148
Lepidosauria 8: 360
Lepidosiren paradoxa 8: 359, 360; 16: 939; 17: 1039
Lepisosteus 8: 359; 15: 759, 786
ligand-gated channels 1: 18
limbic midbrain area 22: 2029
limbic system 22: 1901, 1922, 2024, 2026
line of Gennari 22: 1953
lineage tracing 19: 1162
liquor-contacting neurons 10: 466; 16: 982
lobi accessorii (Lachi) 21: 1549
lobi optici 21: 1531
lobule désert (Broca) 22: 1891
lobulus ansiformis 22: 1725
- crus I 22: 1727
- crus II 22: 1727
lobulus paramedianus 22: 1725, 1728
lobulus petrosus 22: 1728
lobulus simplex 22: 1724, 1727
lobulus VII (Larsell) 22: 1727
lobus anterior 22: 1724, 1727
lobus caudalis cerebelli 15: 829
lobus facialis 15: 772, 808, 912; 24: 2139
lobus flocculonodularis 22: 1726
lobus glossopharyngei 15: 811
lobus inferior hypothalami 12: 618; 14: 706, 7, 8; 15: 897
lobus lineae lateralis 13: 657; 16: 966
lobus lineae lateralis anterior 15: 818
lobus lineae lateralis posterior 15: 818

lobus olfactorius 22: 1891
lobus parolfactorius 21: 1591
lobus piriformis 22: 1647, 1873
lobus subhippocampalis 10: 460, 465
lobus subhippocampalis 24: 2160
lobus tuberis 14: 739
lobus vagi 2: 76; 10: 423, 430; 11: 521; 12: 581; 13: 657; 14: 722; 15: 772, 811, 912; 16: 968; 17: 1023, 1039; 24: 2141
lobus vestibulolateralis 12: 589; 14: 726; 15: 829; 17: 1011, 1027; 24: 2144
local circuit neurons 5: 233
locomotion 10: 443; 17: 1009; 18: 1062; 19: 1171; 22: 1918
– control 10: 443, 474; 18: 1062, 1071; 19: 1171
– fictive 10: 409, 418, 427, 429; 19: 1171
– neural network 18: 1062; 19: 1171; 10: 427
– spinal pattern generator 10: 428; 18: 1062; 19: 1171, 1198
locomotor circuitry 10: 485; 19: 1171, 1198
locomotor cycle 10: 477
locomotor networks 10: 481; 19: 1198
locomotor regions 10: 443; 18: 1072; 19: 1198; 20: 1365
locomotor responses 9: 373; 16: 953
locus coeruleus 11: 538; 13: 677; 15: 840; 16: 974; 18: 1068, 1097; 20: 1415; 21: 1563, 1573; 22: 1758, 1760
long-term depression 24: 2182
long-term memory 22: 1932
long-term potentiation 24: 2182
longitudinal zones 6: 316
'lower lip' 13: 657, 678
Lucifer Yellow 7: 328
lumbar enlargement, see intumescentia lumbalis
luteinizing hormone-releasing hormone 10: 468; 15: 899, 905; 18: 1121; 20: 1465; 22: 1897
lysosomes 1: 5

M

Macaca mulatta 8: 361
macrocircuitry 24: 2167
Macroscelidae 22: 1641
magnetic sense 21: 1528, 1618
magnetoencephalography 7: 337
Mammalia
– brain size 23: 2114
– phylogeny 23: 2115
mantle layer 5: 232; 10: 406
Marchi technique 7: 329
marginal cells 20: 1347; 21: 1549; 22: 1661
marginal layer 5: 231
Markgerüst (His) 5: 229
Marsupialia
– brain size 23: 2119
massa caudalis 15: 790
massa cellularis reuniens 16: 997
massa intermedia 4: 172; 22: 1646
matrisomes (striatum) 22: 1915
matrix 5: 232, 248
matrix (striatum) 22: 1905, 1914
matrix layer 10: 406
matrix zone 11: 504
Mauthner cell 2: 55; 3: 124, 129; 10: 403, 435, 441, 483, 485; 11: 506; 12: 592; 13: 676; 14: 709, 717, 723, 724, 725; 15: 787, 792, 799, 800, 801; 16: 958, 965, 967, 968, 969; 17: 1017, 1024, 1041; 18: 1091; 19: 1228
mechanoreceptive lateral line receptors 16: 966
mechanoreceptors 21: 1555
mechanosensory lateral line region 15: 818
medial forebrain bundle, see fasciculus medialis telencephali
medial lemniscus, see lemniscus medialis
medial pallium 11: 532, 534, 543; 13: 694; 18: 1128; 19: 1276; 22: 1872
medial reticular zone 13: 696; 15: 798; 16: 970; 17: 1026; 18: 1097; 19: 1238; 20: 1414

medial terminal nucleus of the accessory optic system 22: 1818, 1819
median cerebellar ridge 16: 974
median eminence 9: 394; 10: 463; 13: 685; 16: 983, 987; 18: 1119; 19: 1270; 20: 1463
median reticular zone ·13: 676; 16: 970; 17: 1026; 18: 1096; 19: 1238; 20: 1413
Megachiroptera 22: 1642
melanocyte-stimulating hormone 12: 627; 16: 997
melanophore-stimulating hormone 18: 1119; 19: 1270; 20: 1465
melatonin 10: 454, 455; 21: 1583; 22: 1848
melatonin-concentrating hormone 15: 896
membrana limitans externa 16: 953
membrana limitans gliae externae 10: 409
membrane
– lipid bilayer 1: 6
memory 22: 1900
Merkel cells 10: 431
mesencephalic command-associated nucleus 15: 850
mesencephalic locomotor region 22: 1908
mesencephalic reticular formation 15: 853, 854
mesocortex 22: 1955
mesomeres 21: 1535
messenger RNA 1: 8
metabotropic (receptors) 1: 20
metamorphosis 10: 450
Metatheria 8: 361
Meynert cells 22: 1964
Microchiroptera 22: 1641
microcircuit analysis 7: 335
microcircuitry 24: 2181
microelectrode techniques 7: 336
microfilaments 1: 5
microglia 1: 10; 11: 507; 17: 1015
microtubules 1: 5
midbrain extrapyramidal area 22: 1908, 1913
midline nuclei 22: 1855
migration
– radial 5: 240
– tangential 5: 244
– vectorial 5: 244, 253
migration area (Källén) 4: 175, 200; 5: 247; 6: 289
migration layers 5: 233
mitochondria 1: 5
mitral cells 10: 465
mixed synapses 10: 407, 485
modules 22: 2002; 24: 2182
molluscs 9: 373
monoaminergic systems, see chemically defined neuronal populations
monophyletic group 6: 311
Monotremata
– brain size 23: 2119
morphotype 4: 220; 6: 282, 284, 285, 314
mossy fibres 22: 1733, 1748
motivational pathway 21: 1608
motor circuitry in vertebrates 24: 2169, 2176
motor pool organisation, see spinal cord
motor systems
– (ventro)medial brain stem systems (Kuypers) 19: 1193; 20: 1362; 22: 1842, 2026
– emotional motor system (Holstege) 22: 1844, 2038, 2040
– lateral brain stem systems (Kuypers) 19: 1193; 20: 1362; 22: 1842, 2026
Müller cells 3: 115; 10: 403, 418, 423, 440, 465, 483; 11: 506, 522, 543; 16: 970
multineuromediator channel 3: 146, 150
multiple unit recording 7: 339
muscle spindles 19: 1182, 1197; 20: 1347, 1351
myelin 11: 506, 543
myelin sheath 1: 7
myelomeres 21: 1535
Myopterygii 8: 357; 11: 539

Subject Index   2207

Synopsis
General introductory part: 1 Cellular elements; 2 Centres; 3 Fibre systems; 4 Morphogenesis; 5 Histogenesis; 6 Principles; 7 Techniques
Specialized part: 8 Introduction; 9 Amphioxus; 10 Lampreys; 11 Hagfishes; 12 Cartilaginous fishes; 13 Brachiopterygians; 14 Chondrosteans; 15 Teleosts; 16 Lungfishes; 17 Latimeria; 18 Urodeles; 19 Anurans; 20 Reptiles; 21 Birds; 22 Mammals
General concluding part: 23 Brain size; 24 Meaning

myotomal motoneurons   **10:** 418, 428, 479, 481, 483; **11:** 510
mystacial vibrissae   **22:** 1772
Myxinidae   **11:** 497
Myxinoidea   **8:** 357

N

Na^+-K^+ pump   **1:** 14
natural coordinate system   **5:** 239; **6:** 274, 275
Nauta technique   **7:** 330; **24:** 2166
navigational map   **21:** 1616
neencephalon   **6:** 277, 309; **22:** 2024
Neoceratodus   **8:** 359; **16:** 939; **17:** 1039, 1041
neocortex   **22:** 1932; **24:** 2153, 2161
- asymmetry   **22:** 2019
- barrels   **22:** 1982, 2003
- cell types   **22:** 1734, 1938, 1947, 1968, 1969, 1986, 1988
- comparative aspects   **22:** 1990
- evolutionary development   **22:** 1957
- folding index   **22:** 2012
- lamination   **22:** 1952
- microcircuitry   **22:** 1989
- size   **22:** 2008
- subdivision   **22:** 1952
- surface area   **22:** 2012
- thickness   **22:** 2011
neocorticalisation   **22:** 2008
neodentate nucleus (parvocellular subdivision of the dentate nucleus)   **22:** 1737
neopallium   **22:** 1872, 1876; **24:** 2154
Neopterygii   **8:** 359; **15:** 766
neostriatum   **4:** 216; **21:** 1596; **24:** 2156
neostriatum auditivum   **21:** 1599
neostriatum posterodorsolaterale   **21:** 1603
Neoteleostei   **15:** 766
neoteny   **16:** 1000
nerve components   **4:** 196
nerve fibre   **3:** 115
nervus abducens   **10:** 430; **11:** 498, 543; **12:** 593; **16:** 970; **17:** 1011; **18:** 1079; **19:** 1207; **20:** 1367; **21:** 1555; **22:** 1645
nervus accessorius   **19:** 1209; **20:** 1383; **21:** 1554; **22:** 1645
nervus accessorius vagi s. spinalis   **18:** 1079
nervus anterior lineae lateralis   **11:** 517
nervus cochlearis   **22:** 1777
nervus facialis   **11:** 519, 541; **12:** 581; **17:** 1011; **18:** 1079; **19:** 1207; **20:** 1383; **21:** 1554; **22:** 1645
nervus glossopharyngeus   **11:** 519, 541; **12:** 581, 594; **18:** 1079; **19:** 1207; **20:** 1383; **21:** 1555, 1557; **22:** 1645, 1715, 1772
nervus hypoglossus   **18:** 1079; **19:** 1209; **20:** 1383; **21:** 1554; **22:** 1645
nervus intermedius   **22:** 1645
nervus lineae lateralis anterior   **10:** 433; **12:** 583; **13:** 673; **14:** 716; **16:** 943, 966; **17:** 1011, 1018; **18:** 1079, 1086; **19:** 1218
nervus lineae lateralis posterior   **10:** 433, 444; **12:** 583; **13:** 673; **14:** 716; **16:** 943, 966; **18:** 1079, 1086; **19:** 1218
nervus octavus   **10:** 434, 444; **11:** 517, 541; **12:** 586; **13:** 675; **16:** 967; **17:** 1011, 1024; **21:** 1554
nervus oculomotorius   **11:** 498, 543; **12:** 593; **18:** 1079; **19:** 1207; **20:** 1367; **21:** 1573; **22:** 1645
nervus olfactorius   **11:** 533, 538, 541; **16:** 946; **17:** 1035
nervus opticus   **10:** 444; **11:** 501, 528, 538, 541; **12:** 628; **13:** 686; **18:** 1104; **19:** 1250; **20:** 1445; **21:** 1581

nervus parietalis   **20:** 1456
nervus posterior lineae lateralis   **11:** 517, 518
nervus preopticus   **16:** 946, 996
nervus saccularis   **11:** 517
nervus stato-acusticus   **18:** 1079, 1088; **19:** 1219; **20:** 1383, 1402, 1406; **22:** 1645
nervus terminalis   **5:** 264; **10:** 467; **11:** 533, 538; **12:** 636; **13:** 688; **14:** 746; **16:** 946, 996; **17:** 1035; **18:** 1118; **19:** 1273; **21:** 1618; **22:** 1897
nervus trigeminus   **10:** 431; **11:** 518; **12:** 580; **13:** 675; **18:** 1079; **19:** 1207; **20:** 1367, 1391; **21:** 1555; **22:** 1643, 1645, 1714, 1770
nervus trochlearis   **11:** 498, 543; **12:** 593; **16:** 970; **18:** 1079; **19:** 1207; **20:** 1367; **21:** 1573
nervus vagus   **11:** 519, 541; **12:** 581, 594; **18:** 1079; **19:** 1209; **20:** 1383; **21:** 1554, 1557; **22:** 1645, 1716, 1772
nervus vestibularis   **22:** 1832, 1836
nervus vestibularis anterior   **10:** 444
nervus vestibulocochlearis   **18:** 1079; **19:** 1207, 1219; **20:** 1383, 1402, 1406
nervus vomeronasalis   **22:** 1647
neural arches   **11:** 497, 539
neural crest   **4:** 159
neural ectoderm   **4:** 159; **9:** 368
neural folds   **11:** 501, 544
neural plate   **4:** 159; **9:** 368; **11:** 501, 544; **18:** 1054; **19:** 1160
neural ridge   **19:** 1160
neural segmentation   **6:** 315
neural tube   **4:** 159; **6:** 274; **9:** 368; **11:** 501, 544; **19:** 1160; **20:** 1335
neurobiotaxis   **2:** 35; **11:** 521
neurochemistry (overview)   **1:** 13
neurocladistics   **6:** 311, 312
neuroepithelial cells   **5:** 232
neuroepithelial patches   **5:** 252
neuroepithelial zones   **5:** 253
neurofilaments   **1:** 5
neurogenesis   **6:** 274
neurohemal contact zone   **9:** 393; **11:** 531
neurohypophysis   **9:** 394; **10:** 461, 463; **11:** 529; **12:** 625; **14:** 743; **16:** 943, 986; **17:** 1014, 1033, 1034; **18:** 1119; **19:** 1267; **20:** 1465; **22:** 1862
neuromasts   **10:** 433
neuromasts   **11:** 499; **14:** 721; **15:** 768; **17:** 1009
neuromeres   **4:** 172; **5:** 233, 252, 259; **6:** 316; **11:** 505; **16:** 951; **20:** 1335; **21:** 1532-1534, 1581; **24:** 2147, 2164
neuromorphology
- evolutionary   **6:** 320
- functional   **6:** 318
- genetic   **6:** 315
- phylogenetic   **6:** 308
- pure   **6:** 308
neuron doctrine   **1:** 2, 21
neuronal cell culture   **7:** 343
neurons
- functional polarisation   **1:** 2
- structural features   **1:** 2
neuropil   **2:** 29
- deep   **2:** 30
- intermediate   **2:** 31
- superficial   **2:** 33
neuropil of Bellonci   **16:** 983; **18:** 1116
neuropil of the basal optic root   **18:** 1105
neuroporic recess, see recessus neuroporicus
neuroporus   **11:** 544

neuroporus anterior 4: 159
neuroporus posterior 4: 159
neurosecretory cells
- catecholaminergic 9: 393
- dorsal group 9: 387
- peptidergic 9: 393
- ventral group 9: 387
neurotensin 13: 694; 21: 1595
neurotransmission
- non-synaptic 2: 70; 3: 125
- synaptic 2: 70
neurulation 11: 501, 544
nigrostriatal projection 16: 992; 20: 1412; 22: 1911, 1913
nigrotectal projection 22: 1912
nigrotegmental projection 22: 1912
nigrothalamic fibres 22: 1909
Nissl substance 1: 5
Nissl technique 7: 328; 24: 2165
nitric oxide 1: 21
nodes of Ranvier 1: 7, 16; 3: 118, 124
nodulus 22: 1725, 1730
non-synaptic interneuronal communication 10: 408
non-synaptic neurotransmission 1: 13; 22: 1988, 2034
nonvesicular release (of neurotransmitters) 1: 20
noradrenergic cell groups, see chemically defined neuronal populations
notochord 9: 375, 382; 11: 497
nuclear membrane 1: 5
nuclei
- closed 2: 47
- intermediate 2: 47
- open 2: 47
- periventricular 2: 42
- superficial 2: 47
nuclei cochleares (mammals) 22: 1714, 1779
- cell types 22: 1781
- development 22: 1781
- nucleus dorsalis 22: 1779, 1781
- nucleus ventralis 22: 1779
nuclei pontis 22: 1750
nuclei raphes 22: 1715, 1757
nuclei vestibulares, see vestibular nuclei
nucleolus 1: 5
nucleus 'a' of Kusunoki 11: 522
nucleus 'glomerulosus' 15: 843, 887, 889, 891
nucleus (of cells) 1: 5
nucleus A 12: 595
nucleus accessorius medialis (Bechterew) 22: 1837
nucleus accessorius n. abducentis 18: 1080; 19: 1209; 20: 1385; 21: 1560, 1573; 22: 1715
nucleus accessorius n. facialis 22: 1718, 1719
nucleus accumbens 4: 217; 18: 1133; 19: 1281; 20: 1490; 21: 1595, 1596, 1615; 22: 1872, 1901, 1903, 1905, 1916
nucleus ambiguus 19: 1210; 20: 1386; 21: 1560; 22: 1715, 1716
nucleus angularis 21: 1557
nucleus ansae lenticularis 21: 1581
nucleus anterior areae octavolateralis 14: 713, 723
nucleus anterior n. octavi 19: 1217
nucleus anterior thalami 11: 532; 13: 681, 686; 14: 736, 741; 15: 885; 19: 1262
nucleus anterior tuberis 15: 896, 897
nucleus anterodorsalis (thalami) 22: 1857
nucleus anteromedialis (thalami) 22: 1857
nucleus anteroventralis (thalami) 22: 1857
nucleus arcuatus 20: 1462
nucleus B 12: 596
nucleus basalis 4: 217; 21: 1595, 1603
nucleus basalis (Kuhlenbeck) 21: 1539, 1589
nucleus basalis of Meynert 22: 1906
nucleus C1, C2 12: 590
nucleus caudalis areae octavolateralis 14: 713, 717; 17: 1018
nucleus caudalis n. octavi 19: 1217

nucleus caudalis thalami 20: 1458
nucleus caudatus 4: 217; 22: 1872, 1901
nucleus centralis hypothalami 15: 897
nucleus centralis posterior thalami 13: 681, 695; 14: 736, 741; 15: 885
nucleus centralis prosencephali (in myxinoids) 6: 299; 24: 2159
nucleus centralis superior 22: 1758
nucleus centralis thalami 19: 1262
nucleus centromedianus 22: 1856
nucleus cerebellaris lateralis 20: 1427
nucleus cerebellaris medialis 20: 1427
nucleus cerebelli 12: 597; 13: 678; 14: 722, 724, 728; 15: 834, 836; 16: 969, 974; 17: 1026; 18: 1094; 19: 1234
nucleus cervicalis centralis 22: 1661-1663, 1674, 1749
nucleus cervicalis lateralis 18: 1082; 19: 1211; 21: 1549; 22: 1666, 1766
nucleus cervicalis medialis 21: 1560
nucleus circularis 20: 1431
nucleus cochlearis angularis 20: 1406
nucleus cochlearis dorsalis magnocellularis 20: 1406
nucleus cochlearis laminaris 20: 1406
nucleus coeruleus, see locus coeruleus
nucleus commissurae anterioris 17: 1037
nucleus commissurae posterioris 10: 453; 11: 524; 12: 616; 15: 877; 16: 977; 17: 1032; 22: 1812
nucleus commissuralis of Cajal 11: 521; 15: 807, 808
nucleus corticalis 15: 880
nucleus cuneatus 21: 1557
nucleus cuneatus externus 21: 21: 1553, 1557, 1559; 22: 1674, 1766
nucleus cuneatus internus 22: 1764, 1766
nucleus cuneiformis 22: 1757
nucleus dentatus 22: 1735, 1752
nucleus descendens areae octavolateralis 14: 716, 723; 21: 1581
nucleus descendens lateralis n. trigemini 20: 1395
nucleus descendens n. trigemini 10: 444, 483; 11: 517, 519, 521, 544; 12: 581; 13: 675; 14: 708, 721, 723; 15: 814; 16: 965, 968; 17: 1024; 18: 1083; 19: 1214; 20: 1390; 22: 1714, 1770
nucleus diffusus lobi inferioris 14: 739
nucleus diffusus lobi lateralis hypothalami 15: 897
nucleus diffusus thalami 11: 528
nucleus dorsalis areae octavolateralis 10: 433; 11: 518, 543; 12: 582; 13: 673, 675; 14: 713, 717, 754; 16: 966, 973; 17: 1021, 1039; 18: 1086; 19: 1217
nucleus dorsalis intermedius ventralis 21: 1585
nucleus dorsalis of central prosencephalic complex 11: 532
nucleus dorsalis posterior thalami 13: 681, 695; 14: 736, 741; 15: 885
nucleus dorsointermedius posterior 21: 1583, 1595
nucleus dorsolateralis anterior 21: 1582, 1584
nucleus dorsolateralis, pars medialis 21: 1604
nucleus dorsolateralis posterior 21: 1583, 1585, 1603
nucleus dorsolateralis thalami 20: 1458
nucleus dorsomedialis anterior 21: 1585
nucleus dorsomedialis hypothalami 22: 1863
nucleus dorsomedialis thalami 20: 1458
nucleus dorsomedialis (thalami), see nucleus medialis dorsalis (thalami)
nucleus ectomamillaris 21: 1567, 1568, 1583
nucleus electrosensorius diencephalicus 15: 876, 880, 881
nucleus ellipticus 22: 1747, 1842
nucleus emboliformis, see nucleus interpositus anterior
nucleus endopiriformis 22: 1908
nucleus entopeduncularis 12: 641; 14: 737, 749; 22: 1882
nucleus entopeduncularis anterior 19: 1281; 20: 1461, 1492
nucleus entopeduncularis caudalis 13: 693
nucleus entopeduncularis dorsalis 13: 693
nucleus entopeduncularis posterior 19: 1267; 20: 1461, 1492
nucleus entopeduncularis ventralis 13: 693
nucleus epibasalis 21: 1539

> **Synopsis**
> *General introductory part:* 1 Cellular elements; 2 Centres; 3 Fibre systems; 4 Morphogenesis; 5 Histogenesis; 6 Principles; 7 Techniques
> *Specialized part:* 8 Introduction; 9 Amphioxus; 10 Lampreys; 11 Hagfishes; 12 Cartilaginous fishes; 13 Brachiopterygians; 14 Chondrosteans; 15 Teleosts; 16 Lungfishes; 17 Latimeria; 18 Urodeles; 19 Anurans; 20 Reptiles; 21 Birds; 22 Mammals
> *General concluding part:* 23 Brain size; 24 Meaning

nucleus externus thalami 11: 528
nucleus F 12: 597
nucleus fasciculi longitudinalis medialis 10: 453; 11: 516, 523; 12: 595; 13: 680; 14: 725, 733, 736; 15: 849, 850, 876; 16: 965, 975; 17: 1028, 1029; 18: 1097; 24: 2152
nucleus fasciculi solitarii, see nucleus tractus solitarii
nucleus fastigii 22: 1735, 1752, 1832
nucleus funiculi dorsalis 10: 483; 12: 570; 16: 973; 18: 1082; 19: 1211; 20: 1387
nucleus funiculi lateralis 12: 597; 14: 723
nucleus geniculatus lateralis 19: 1263; 20: 1456; 21: 1585
nucleus geniculatus lateralis dorsalis 22: 1853
 – binocular integration 22: 1859
 – binocular segment 22: 1854
 – lamination 22: 1854
 – monocular segment 22: 1854
 – receptive field properties 22: 1854, 1859
 – X-, Y- and W-cells 22: 1854
nucleus geniculatus lateralis ventralis 22: 1861
 – circadian rhythm 22: 1862
 – eye movements 22: 1862
 – pupillary light reflex 22: 1862
nucleus geniculatus medialis dorsalis (thalami) 22: 1852
nucleus geniculatus medialis medialis (thalami) 22: 1852
nucleus geniculatus medialis ventralis (thalami) 22: 1852
nucleus geniculatus pretectalis 20: 1453
nucleus gigantocellularis 22: 1756
nucleus globosus, see nucleus interpositus posterior
nucleus glomerulosus 15: 881
nucleus glomerulosus commissuralis 15: 891
nucleus gracilis 21: 1557; 22: 1764, 1766
nucleus gustatorius secundarius 15: 843
nucleus H 12: 595
nucleus habenulae 21: 1583
nucleus habenulae lateralis 22: 1846
nucleus habenulae medialis 22: 1846
nucleus habenularis dorsalis 14: 736; 18: 1115; 19: 1260
nucleus habenularis lateralis 20: 1455
nucleus habenularis medialis 20: 1455
nucleus habenularis ventralis 14: 736; 18: 1115; 19: 1260
nucleus hypothalamicus caudalis 15: 896
nucleus hypothalamicus dorsalis 19: 1267
nucleus hypothalamicus lateralis 15: 896; 19: 1267; 20: 1463
nucleus hypothalamicus medialis 15: 896; 20: 1463
nucleus hypothalamicus posterior 20: 1463
nucleus hypothalamicus ventralis 19: 1267
nucleus hypothalamicus ventromedialis 20: 1463, 1464
nucleus infundibularis 11: 528; 21: 1586; 22: 1863
nucleus intercalatus 24: 2160
nucleus intercalatus hyperstriati 21: 1611, 1612
nucleus intercalatus laminae supremae 21: 1611
nucleus intercollicularis 12: 570, 610; 20: 1389; 21: 1579, 1604
nucleus interfacialis 21: 1603, 1604
nucleus intermediolateralis 22: 1664
nucleus intermediomedialis 21: 1549; 22: 1664
nucleus intermedius 21: 1543
nucleus intermedius areae octavolateralis 10: 432, 433, 435, 446; 12: 582; 13: 675; 14: 713, 717; 16: 966; 17: 1021, 1039; 18: 1086; 19: 1217
nucleus intermedius dorsalis thalami 20: 1458
nucleus intermedius facialis (of Herrick) 15: 809, 810
nucleus intermedius thalami 13: 681; 14: 737, 740; 15: 886
nucleus internus thalami 11: 528

nucleus interpeduncularis 10: 451, 456; 11: 524; 12: 597; 13: 677, 680, 681; 14: 724, 741; 15: 853; 16: 971, 973, 985; 17: 1015, 1026; 18: 1094; 19: 1236; 20: 1413; 21: 1574; 22: 1720
nucleus interpositus anterior 22: 1735, 1752
nucleus interpositus posterior 22: 1735, 1752
nucleus interstitialis (Cajal) 21: 1573; 22: 1724
nucleus interstitialis commissurae anterioris 12: 641
nucleus interstitialis fasciculi basalis telencephali 12: 641
nucleus interstitialis of the flm 18: 1097; 19: 1239; 20: 1414; 21: 1580
nucleus interstitialis of the posterior commissure 20: 1431
nucleus interstitio-pretecto-subpretectalis 21: 1582
nucleus intracommissuralis thalami 11: 528
nucleus intrapeduncularis 21: 1591, 1594
nucleus isthmi 12: 597; 14: 722, 724; 15: 845; 18: 1110; 19: 1253; 20: 1448; 21: 1574
nucleus isthmo-opticus 21: 1574
nucleus laminaris 21: 1557
nucleus laminaris ventralis 14: 717
nucleus lateralis anterior 21: 1581
nucleus lateralis dorsalis (thalami) 22: 1854
nucleus lateralis habenulae 11: 525
nucleus lateralis n. octavi 19: 1217
nucleus lateralis posterior (thalami) 22: 1854
nucleus lateralis thalami 15: 891; 19: 1262
nucleus lateralis tuberis 12: 621; 14: 743; 15: 896
nucleus lateralis valvulae 13: 677, 678; 14: 732, 753; 15: 851, 852
nucleus lemnisci lateralis 14: 721, 723; 19: 1225; 20: 1408; 21: 1574, 1584; 22: 1721, 1786
nucleus lentiformis 22: 1883, 1906
nucleus lentiformis mesencephali 19: 1254; 20: 1453; 21: 1565, 1583
nucleus limitans (thalami) 22: 1855
nucleus lineae lateralis caudalis 19: 1217
nucleus lineae lateralis rostralis 19: 1217
nucleus linearis 21: 1562; 22: 1758
nucleus lobi lateralis 12: 624
nucleus lobobulbaris 15: 891
nucleus M 12: 642
nucleus magnocellularis areae octavolateralis 11: 516, 518; 14: 713, 723
nucleus magnocellularis cochlearis 21: 1557
nucleus magnocellularis neostriati 21: 1604
nucleus mamillaris 20: 1463
nucleus mamillaris 21: 1586
nucleus mamillaris dorsalis 22: 1863
nucleus mamillaris ventralis 22: 1863
nucleus marginalis (Hoffmann-Kölliker) 20: 1347; 21: 1549
nucleus medialis amygdalae 22: 1896
nucleus medialis areae octavolateralis 11: 517
nucleus medialis dorsalis (thalami) 22: 1857
nucleus medialis n. octavi 19: 1217
nucleus medialis of central prosencephalic complex 11: 532
nucleus medialis thalami 20: 1459
nucleus medianus magnocellularis 13: 676, 697
nucleus medianus tuberculi posterioris 13: 683, 684, 695; 14: 738, 741, 749
nucleus mediodorsalis (thalami), see nucleus medialis dorsalis (thalami)
nucleus medius hypothalami 12: 621
nucleus mesencephalicus lateralis 16: 971
nucleus mesencephalicus lateralis, pars dorsalis 21: 1557, 1579, 1584

nucleus mesencephalicus magnocellularis   15: 862
nucleus mesencephalicus n. trigemini   10: 432; 11: 519; 12: 580; 13: 679; 14: 723, 730; 15: 853; 16: 965, 975; 17: 1024, 1029; 18: 1083; 19: 1214; 20: 1389; 21: 1555, 1568; 22: 1714, 1719, 1770, 1772
nucleus mesencephalicus periventricularis   15: 850
nucleus motorius dorsalis n. vagi   19: 1210; 20: 1387; 22: 1715
nucleus motorius n. facialis   11: 521; 12: 593; 14: 723; 15: 806; 16: 969; 17: 1024; 18: 1080; 19: 1210; 20: 1386; 21: 1562; 22: 1717
nucleus motorius n. glossopharyngei   11: 521; 12: 593; 14: 723; 15: 807; 16: 969; 17: 1024; 18: 1080
nucleus motorius n. trigemini   10: 445; 11: 521; 12: 593; 13: 675, 696; 14: 723; 15: 806; 18: 1080; 19: 1210; 20: 1386; 21: 1562; 22: 1719
nucleus motorius n. vagi   12: 593; 14: 723; 15: 807; 16: 969; 17: 1024; 18: 1080
nucleus motorius tegmenti (of Jansen)   11: 522
nucleus N   12: 642
nucleus nervi abducentis   10: 436; 12: 592; 13: 676; 14: 724; 15: 803; 18: 1080; 19: 1209; 20: 1385
nucleus nervi hypoglossi   18: 1080; 19: 1210; 20: 1385; 21: 1543, 1560; 22: 1715, 1719
nucleus nervi oculomotorii   10: 436, 451; 12: 593; 13: 680; 14: 732; 15: 847, 848; 17: 1028, 1029; 18: 1080; 19: 1209; 20: 1385; 21: 1573; 22: 1723
nucleus nervi oculomotorii accessorius, see nucleus of Edinger-Westphal
nucleus nervi trochlearis   10: 436, 438; 12: 593; 13: 680; 14: 732; 15: 849; 17: 1026; 18: 1080; 19: 1209; 20: 1385; 21: 1573; 22: 1723
nucleus octavomotorius anterior   10: 485
nucleus octavomotorius intermedius   10: 423, 426, 444, 485
nucleus octavomotorius posterior   10: 423, 425, 426, 444, 485
nucleus octavus anterior   13: 675; 15: 816, 817; 17: 1024
nucleus octavus ascendens   12: 589
nucleus octavus descendens   12: 590; 15: 816, 817; 17: 1024
nucleus octavus magnocellularis   12: 589; 15: 816, 817; 16: 965
nucleus octavus posterior   15: 816, 817
nucleus octavus tangentialis   15: 816, 817
nucleus of Bellonci   18: 1115; 19: 1262
nucleus of Bischoff   22: 1766
nucleus of Darkschewitsch   21: 1581; 22: 1720, 1747, 1752, 1837
nucleus of Ebblonci   10: 449, 457
nucleus of Edinger-Westphal   12: 595; 15: 849; 17: 1029 18: 1081; 19: 1210; 20: 1386; 21: 1561, 1573; 22: 1724
nucleus of Onuf   22: 1665
nucleus of Perlia   22: 1724
nucleus of Roller   22: 1756
nucleus of the basal optic root   18: 1103, 1111; 19: 1255; 20: 1453
nucleus of the cerebellar crest   15: 846
nucleus of the commissure of Wallenberg   15: 846
nucleus of the diagonal band of Broca   22: 1899
nucleus of the fasciculus longitudinalis medialis   16: 965, 975, 977
nucleus of the periventricular organ   19: 1267
nucleus of the postoptic commissure   10: 449
nucleus olfactoretinalis   15: 884, 904
nucleus olfactorius anterior   22: 1890
nucleus olfactorius lateralis   4: 213; 16: 1003
nucleus optici accessorii   21: 1583
nucleus opticus accessorius dorsalis   15: 880
nucleus opticus accessorius ventralis   15: 880
nucleus opticus basalis   12: 616
nucleus opticus dorsolateralis   15: 885
nucleus opticus principalis   21: 1584
nucleus opticus tegmenti   16: 971
nucleus ovalis   20: 1457; 22: 1770
nucleus ovoidalis   21: 1575, 1579, 1585, 1599

nucleus P   12: 640
nucleus parabigeminus   22: 1721, 1761, 1797
nucleus parabrachialis   13: 673, 676; 14: 722, 724; 15: 843; 16: 969; 18: 1085; 19: 1216; 20: 1398; 21: 1562; 22: 1758, 1762, 1775
nucleus parabrachialis pigmentosus   22: 1907
nucleus paracommissuralis   11: 523, 528
nucleus paracommissuralis   15: 877
nucleus paramedianus   21: 1563
nucleus parasolitarius   22: 1774, 1775
nucleus parataenialis   22: 1856
nucleus paratubercularis posterior   10: 460
nucleus paraventricularis   20: 1462; 21: 1586; 22: 1863
nucleus pedunculopontinus   22: 1761, 1797, 1803
nucleus perivagalis   11: 516
nucleus periventricularis hypothalami   12: 621, 628
nucleus periventricularis magnocellularis   21: 1586
nucleus periventricularis tuberculi posterioris   13: 683, 684; 14: 736, 737, 747, 752
nucleus pontis   21: 1562, 1564; 22: 1750
nucleus postcommissuralis   13: 688, 693; 14: 747
nucleus posterior areae octavolateralis   14: 713, 716
nucleus posterior lateralis (thalami)   22: 1855
nucleus posterior medialis (thalami)   22: 1855
nucleus posterior thalami   15: 891; 19: 1262
nucleus posterior tuberis   15: 891, 892, 896
nucleus posterodorsalis   20: 1453
nucleus praethalamicus   15: 890
nucleus preglomerulosus anterior   15: 889
nucleus preglomerulosus dorsalis   15: 889
nucleus preglomerulosus gustatorius   15: 891
nucleus preglomerulosus lateralis   15: 888
nucleus preglomerulosus medialis   15: 889
nucleus preglomerulosus ventralis   15: 889
nucleus premamillaris dorsalis   22: 1863
nucleus premamillaris ventralis   22: 1863
nucleus preopticus   10: 461, 466; 12: 619; 21: 1589
nucleus preopticus anterior   19: 1242; 20: 1462
nucleus preopticus inferior   16: 993; 17: 1037
nucleus preopticus lateralis   22: 1865
nucleus preopticus magnocellularis   13: 693; 14: 743, 749; 15: 898; 16: 987, 993; 17: 1034, 1037; 18: 1118; 19: 1267
nucleus preopticus medialis   22: 1863
nucleus preopticus parvocellularis   13: 694; 14: 749; 15: 899; 16: 993; 17: 1037
nucleus preopticus periventricularis   20: 1462; 22: 1863
nucleus preopticus posterior   20: 1462
nucleus preopticus retinopetalis   15: 900
nucleus preopticus superior   16: 979, 993; 17: 1037
nucleus prepositus hypoglossi   19: 1233; 20: 1402; 22: 1761, 1804, 1834
nucleus pretectalis   19: 1253; 20: 1453; 21: 1581
nucleus pretectalis accessorius   15: 880
nucleus pretectalis anterior   22: 1812
nucleus pretectalis centralis   12: 611; 13: 683; 14: 737, 740, 753; 15: 880
nucleus pretectalis olivaris   22: 1812, 1817
nucleus pretectalis periventricularis   12: 611; 13: 683, 695; 14: 736, 751; 16: 979; 17: 1032
nucleus pretectalis posterior   22: 1812
nucleus pretectalis profundus   17: 1032
nucleus pretectalis superficialis   12: 616; 13: 683, 686; 14: 737, 739, 753; 16: 979; 17: 1032
nucleus pretectalis supracommissuralis   13: 683, 686
nucleus pretrigeminalis   19: 1242
nucleus princeps n. trigemini   10: 432; 11: 519; 12: 581; 16: 965, 968; 17: 1024; 18: 1083; 19: 1214; 20: 1390; 22: 1714, 1764, 1770
nucleus principalis precommissuralis   21: 1582
nucleus profundus mesencephali   13: 679, 680; 14: 732; 17: 1028, 1029
nucleus proprius of the dorsal horn   22: 1661
nucleus pulvinaris anterior   22: 1854
nucleus pulvinaris inferior   22: 1854

Synopsis
General introductory part: 1 Cellular elements; 2 Centres; 3 Fibre systems; 4 Morphogenesis; 5 Histogenesis; 6 Principles; 7 Techniques
Specialized part: 8 Introduction; 9 Amphioxus; 10 Lampreys; 11 Hagfishes; 12 Cartilaginous fishes; 13 Brachiopterygians; 14 Chondrosteans; 15 Teleosts; 16 Lungfishes; 17 Latimeria; 18 Urodeles; 19 Anurans; 20 Reptiles; 21 Birds; 22 Mammals
General concluding part: 23 Brain size; 24 Meaning

nucleus pulvinaris lateralis   22: 1854
nucleus pulvinaris medialis   22: 1854
nucleus Q   15: 846
nucleus raphes   10: 445; 11: 522; 15: 797; 18: 1097; 19: 1238; 21: 1553, 1562; 22: 1715, 1757
nucleus raphes dorsalis   22: 1720, 1758
nucleus raphes inferior   12: 595; 13: 676, 677; 14: 724; 16: 970; 17: 1026; 20: 1413, 1417
nucleus raphes magnus   22: 1757
nucleus raphes obscurus   22: 1757
nucleus raphes pallidus   22: 1757
nucleus raphes pontis   22: 1757
nucleus raphes superior   12: 595; 13: 676, 677; 14: 724; 16: 970; 17: 1026; 20: 1413, 1417
nucleus recessi posterior   15: 895
nucleus reticularis gigantocellularis   21: 1563
nucleus reticularis inferior   10: 440, 443, 481; 11: 522; 12: 595; 13: 676; 14: 724; 15: 798; 16: 970; 17: 1026; 18: 1097; 19: 1239; 20: 1414
nucleus reticularis isthmi   12: 596; 13: 677, 680; 14: 724; 18: 1097; 19: 1239; 20: 1414
nucleus reticularis lateralis   14: 723; 15: 799; 22: 1712, 1749, 1756
nucleus reticularis medius   10: 440, 443, 481; 11: 522; 12: 595; 13: 676; 14: 724; 15: 798; 16: 970; 17: 1026; 18: 1097; 19: 1239; 20: 1414
nucleus reticularis mesencephali   10: 440, 443, 451; 13: 680
nucleus reticularis paramedianus   22: 1712, 1756
nucleus reticularis parvocellularis   21: 1663
nucleus reticularis pontis caudalis   22: 1756
nucleus reticularis pontis oralis   22: 1756
nucleus reticularis superior   10: 440, 443, 481; 11: 522; 12: 595; 13: 676; 14: 724; 15: 798; 16: 970; 17: 1026; 18: 1097; 19: 1239; 20: 1414
nucleus reticularis tegmenti pontis   22: 1748, 1749, 1752
nucleus reticularis thalami   20: 1461; 22: 1860
nucleus retroambiguus   21: 1560; 22: 1668
nucleus retrobulbaris   13: 688
nucleus retrochiasmaticus   20: 1462
nucleus retrofacialis   21: 1560; 22: 1717
nucleus reuniens   20: 1459; 22: 1856
nucleus rhomboideus   22: 1856
nucleus robustus archistriati   21: 1580, 1604, 1609
nucleus rotundus   11: 532; 20: 1458; 21: 1584, 1599
nucleus ruber   11: 524, 543; 12: 598; 14: 733; 15: 854; 16: 965, 975, 977; 17: 1028, 1029, 1039; 18: 1094; 19: 1234; 20: 1437; 21: 1566, 1574; 22: 1721, 1837, 1840
nucleus sacci vasculosi   12: 527; 15: 896
nucleus sagulum   22: 1788
nucleus salivatorius   21: 1560
nucleus salivatorius inferior   19: 1210; 20: 1386; 22: 1715
nucleus salivatorius superior   19: 1210; 20: 1386; 22: 1715
nucleus semilunaris   21: 1574
nucleus semilunaris parovoidalis   21: 1575, 1584, 1599
nucleus sensorius n. vagi   11: 521
nucleus septalis   21: 1589, 1615
nucleus septalis lateralis   19: 1280; 20: 1488
nucleus septalis medialis   19: 1280; 20: 1488
nucleus septi lateralis   4: 214, 217; 13: 689
nucleus septi medialis   4: 214, 217; 13: 689; 16: 993; 17: 1037
nucleus sphericus   20: 1467
nucleus spinalis lateralis   22: 1666
nucleus spinalis n. accessorii   18: 1081; 19: 1210
nucleus spinalis n. trigemini   22: 1714, 1770

nucleus spiriformis lateralis   21: 1579, 1580, 1581, 1594
nucleus spiriformis medialis   21: 1566, 1583
nucleus striae terminalis   21: 1596
nucleus subcoeruleus   21: 1563; 22: 1758
nucleus subeminentialis   15: 846
nucleus subglomerulosus   15: 891
nucleus subhabenularis thalami   11: 528
nucleus subparafascicularis   22: 1827
nucleus subpretectalis   21: 1581, 1584
nucleus subrotundus   21: 1585
nucleus subthalamicus   22: 1882, 1901, 1907; 24: 2148
nucleus superficialis parvocellularis   21: 1585
nucleus superficialis synencephali   21: 1583
nucleus suprachiasmaticus   12: 620; 13: 686; 14: 740, 749; 15: 899; 18: 1118; 19: 1267; 20: 1462; 21: 1588; 22: 1848, 1862; 22: 1863, 1869
nucleus supracommissuralis   13: 688; 14: 747; 16: 993; 17: 1037
nucleus suprageniculatus   22: 1855
nucleus supraglomerulosus   15: 891
nucleus supraopticus   20: 1462; 21: 1589; 22: 1865
nucleus suprapeduncularis   20: 1458
nucleus supraspinalis   21: 1543, 1560
nucleus supratrigeminalis   21: 1555
nucleus tegmentalis dorsalis (Gudden)   22: 1712
nucleus tegmentalis dorsalis magnocellularis   15: 850
nucleus tegmentalis dorsalis posterior   15: 854
nucleus tegmentalis dorsolateralis   15: 854
nucleus tegmentalis lateralis   10: 451; 12: 598
nucleus tegmentalis medialis   14: 732
nucleus tegmentalis pedunculopontinus   22: 1757, 1913
nucleus tegmentalis ventralis   15: 854
nucleus tegmentalis ventralis (Gudden)   22: 1715
nucleus tegmenti dorsalis   21: 1563
nucleus tegmenti pedunculopontinus   21: 1573, 1581, 1594
nucleus thalamoretinalis   15: 887
nucleus tori lateralis   13: 662, 680, 683, 684
nucleus tori semicircularis   12: 598; 13: 679; 17: 102, 1029
nucleus tractus diagonalis   21: 1615
nucleus tractus olfactorii accessorii lateralis   20: 1486
nucleus tractus olfactorii lateralis   20: 1486
nucleus tractus olfactorius accessorius   22: 1895
nucleus tractus opticus   22: 1812
nucleus tractus septomesencephalicus   21: 1585
nucleus tractus solitarii   11: 519; 12: 581; 13: 673; 14: 722; 15: 808; 16: 965, 968; 18: 1084; 19: 1216; 20: 1398; 21: 1560; 22: 1715, 1772, 1774
nucleus triangularis thalami   11: 528; 21: 1599
nucleus tuberculi posterioris   10: 424, 460, 466; 11: 524; 12: 627; 13: 683, 684; 16: 979; 18: 1096; 19: 1238, 1266
nucleus tuberculi posterioris periventricularis   13: 695; 15: 891
nucleus tuberis   21: 1586
nucleus tuberis dorsalis   14: 739
nucleus tuberis lateralis   14: 739
nucleus tuberis posterior   14: 738, 739
nucleus tuberis ventralis   14: 739
nucleus uvaeformis   21: 1585, 1604
nucleus ventralis anterior (thalami)   22: 1851
nucleus ventralis areae octavolateralis   10: 426, 433, 434; 11: 517; 13: 675; 18: 1086; 19: 1217
nucleus ventralis habenulae   11: 525
nucleus ventralis lateralis (thalami)   22: 1851; 22: 1752
nucleus ventralis medialis (thalami)   22: 1850
nucleus ventralis posterior (thalami)   22: 1850

nucleus ventralis posterior lateralis (thalami) **22:** 1850
nucleus ventralis posterior medialis (thalami) **22:** 1850
nucleus ventralis posteromedialis (thalami) **22:** 1776
nucleus ventralis superficialis thalami **19:** 1263
nucleus ventralis tegmenti **21:** 1574
nucleus ventralis thalami **20:** 1458
nucleus ventrolateralis of central prosencephalic complex **11:** 532
nucleus ventrolateralis thalami **13:** 683, 686; **15:** 887; **17:** 1032; **19:** 1262; **20:** 1459
nucleus ventromedialis hypothalami **22:** 1863
nucleus ventromedialis thalami **13:** 683; **14:** 737, 739, 740; **15:** 887; **17:** 1032; **19:** 1262; **20:** 1461
nucleus vestibularis descendens **20:** 1402
nucleus vestibularis dorsolateralis **20:** 1402
nucleus vestibularis lateralis (Deiters) **21:** 1559; **22:** 1735, 1832
nucleus vestibularis magnocellularis **13:** 675; **14:** 713; **16:** 967, 969; **17:** 1024; **19:** 1218
nucleus vestibularis medialis **19:** 1218; **22:** 1832
nucleus vestibularis parvocellularis **13:** 675; **16:** 967, 969
nucleus vestibularis superior **20:** 1402; **22:** 1832
nucleus vestibularis tangentialis **20:** 1402
nucleus vestibularis ventrolateralis **20:** 1402; **20:** 1402
nucleus visceralis secundarius, see nucleus parabrachialis

O

obex **14:** 708
oblique bipolar cells **10:** 426
ocelli **9:** 371
octavocerebellar projections **11:** 523
octavolateral efferent nucleus **15:** 804, 805
octavospinal fibres **10:** 435; **11:** 516, 518
octavothalamic projections **11:** 518
olfactory amygdala **22:** 1897
olfactory system **10:** 468, 477; **11:** 498; **12:** 632; **15:** 767; **17:** 1009, 1039; **18:** 1124; **19:** 1273; **20:** 1468; **21:** 1528, 1615–1618; **22:** 1884
oligodendrocytes **1:** 7, 9; **5:** 235; **11:** 506; **17:** 1015
oliva inferior **12:** 597; **13:** 676; **14:** 707, 724; **15:** 840; **17:** 1026; **18:** 1094; **19:** 1232; **20:** 1430; **21:** 1562; **22:** 1643, 1712, 1733, 1741
– cortico-olivary connections **22:** 1745, 1747
– dorsal accessory olive **22:** 1712, 1741; **22:** 1744
– dorsal cap (Kooy) **22:** 1743, 1744
– dorsomedial subnucleus **22:** 1744
– group beta **22:** 1743, 1744
– medial accessory olive **22:** 1712.1741, 1745
– principal olive **22:** 1712, 1741
– subdivision **22:** 1712, 1743
– ventrolateral outgrowth **22:** 1744
oliva superior **15:** 817; **18:** 1091; **19:** 1225; **20:** 1406; **22:** 1782
– dorsomedial and the ventromedial nuclei in bats **22:** 1786
– oliva superior lateralis **22:** 1782
– oliva superior medialis **22:** 1782
olivocerebellar projections **14:** 729; **18:** 1094; **19:** 1232; **20:** 1430; **22:** 1743
olivocochlear bundle **22:** 1790
opercula **22:** 1879
optic system **12:** 617; **18:** 1105; **19:** 1250; **20:** 1445
optical recording **7:** 342
optokinetic nystagmus **21:** 1583, 1585, 1613
oral disc **10:** 431
organon paraventriculare **13:** 684; **14:** 738
organon subcommissurale **11:** 507, 541; **12:** 616; **13:** 681; **18:** 1057; **19:** 1260
organotypic culture **7:** 343
organs of Fahrenholz **13:** 656, 673; **16:** 940
organs of Hesse **9:** 371
organum vasculosum laminae terminalis **15:** 900
osmoregulation **11:** 539

osmotic stress **14:** 743
Osteichthyes **8:** 358; **15:** 759, 766
– brain size **23:** 2106
Osteoglossomorpha **15:** 766
Osteolepiformis **17:** 1041
oxygen **1:** 13

P

pacemaker nucleus **15:** 801, 802
paedomorphosis **16:** 939, 953
palaeocortex **22:** 1932
palaeoencephalon **6:** 277, 309; **22:** 2024
Palaeoniscoidei **8:** 359; **13:** 697
palaeopallium **16:** 1002; **22:** 1872
palatal organ **15:** 768
paleostriatum augmentatum **21:** 1591
paleostriatum primitivum **21:** 1591
paleostriatum ventrale **21:** 1596
pallial cortex **16:** 953
pallial fields D1, D2, D3 (Kuhlenbeck) **16:** 1002
pallial fields P1, P2, P3 **13:** 693; **17:** 1037
pallial region **10:** 465
pallial thickening **20:** 1470; **24:** 2161
pallidohabenular projection **22:** 1912
pallidostriatal projection **22:** 1910
pallidosubthalamic projection **22:** 1910
pallidotegmental projection **21:** 1595
pallidothalamic projection **22:** 1909
pallidum dorsale **22:** 1901, 1906
pallidum externum **22:** 1882
pallidum internum **22:** 1882
pallidum ventrale **22:** 1901, 1906
pallio-subpallial boundary **16:** 1001; **24:** 2161
palliostriatal pathways **24:** 2172
pallium **4:** 213; **11:** 527, 533; **13:** 693; **14:** 747, 753; **16:** 990, 1002; **17:** 1014, 1036; **21:** 1591; **22:** 1872
– chemoarchitecture **11:** 534
– connectivity **11:** 527, 534, 544
– cytoarchitecture **11:** 533
– dorsal **11:** 543; **12:** 640; **16:** 1002, 1003
– dorsomedial **17:** 1037
– external/internal **21:** 1594
– general **4:** 213
– hippocampal **4:** 213
– homology **11:** 534
– medial **11:** 532, 534, 543; **12:** 640; **16:** 1002, 1003
– lateral **11:** 534, 543; **12:** 640; **16:** 1002, 1003; **17:** 1037, 1039
– piriform **4:** 213
– ventromedial **17:** 1037
Paracanthopterygii **15:** 766
paracoerulear cell groups **15:** 843
paracore bundle, lateral **22:** 2031
paracore bundle, medial **22:** 2031
paracore, lateral **22:** 2029
paracore, medial **22:** 2029
paraflocculus **22:** 1725, 1728
paraflocculus accessorius (Henle) **22:** 1730
paralemniscal tegmental area, control of vocalisation in bats **22:** 1786
paraneurons **1:** 9
paraphyletic group **6:** 311
paraphysis **4:** 169; **11:** 525
parapineal nerve **10:** 399
parapineal organ **4:** 169; **10:** 399, 454, 477; **11:** 525, 543; **17:** 1031
parasubiculum **22:** 1872, 1944
parasympathetic fibres **19:** 1210; **20:** 1384
parasympathetic neurons **11:** 521
paraterminal body **16:** 1003
paraventricular organ **15:** 891, 892, 893; **16:** 982, 983; **18:** 1119; **19:** 1267; **20:** 1462

> **Synopsis**
> *General introductory part:* 1 Cellular elements; 2 Centres; 3 Fibre systems; 4 Morphogenesis; 5 Histogenesis; 6 Principles; 7 Techniques
> *Specialized part:* 8 Introduction; 9 Amphioxus; 10 Lampreys; 11 Hagfishes; 12 Cartilaginous fishes; 13 Brachiopterygians; 14 Chondrosteans; 15 Teleosts; 16 Lungfishes; 17 Latimeria; 18 Urodeles; 19 Anurans; 20 Reptiles; 21 Birds; 22 Mammals
> *General concluding part:* 23 Brain size; 24 Meaning

parcellation theory   2: 35; 6: 310, 313
parencephalon   21: 1534
parencephalon anterior   4: 194; 14: 734; 16: 951; 17: 1031; 19: 1260; 24: 2152
parencephalon posterior   4: 194; 14: 734; 16: 951; 17: 1031; 19: 1260; 24: 2152
parietal eye   20: 1456
parietal nerve   20: 1456
parietal organs   10: 453, 464; 11: 525
parsimony, principle of   6: 312
partial patterns (of activity)   5: 257
passive avoidance learning   21: 1595
passive electric properties   1: 14
patch clamp   7: 341
pathway   3: 150
pattern discrimination   21: 1614
pedunculi cerebelli   22: 1712
pedunculus cerebri   22: 1643, 1720, 1820, 1823; 24: 2146
pedunculus mamillaris   22: 1866
pedunculus olfactorius   17: 1035
Pelycosauria   8: 360
perihypoglossal nuclei   22: 1724
perikaryon   1: 5
periolivary cell groups   22: 1782
periventricular grey   2: 26
periventricular hypothalamus   15: 893
periventricular layer (Senn)   5: 248
periventricular organ   19: 1267
peroxidase anti-peroxidase technique   7: 334
Petromyzontoidea   8: 357
Phaseolus vulgaris leucoagglutinin   7: 331
pheromones   22: 1894
photoneuroendocrine cells   10: 455
photoreceptor cells   9: 379; 10: 454; 11: 498
phylogenetic relationships   4: 221
pineal complex   4: 168; 10: 456; 17: 1031
pineal eye   9: 393; 10: 397, 449
pineal gland, see epiphysis cerebri
pineal nerve   10: 399
pioneer neurons   5: 258
piriform cortex   22: 1932, 1940
piriform pallium   13: 694; 16: 990, 1002; 17: 1039
pit organs   20: 1394
pituitary gland, see hypophysis cerebri
Placodermi   8: 357
planum temporale   22: 2019
Platonic ideas   6: 305
Platypus   8: 361
plesiomorphic characters   11: 540
plexus brachialis   19: 1170; 20: 1343; 21: 1529
plexus lumbalis   21: 1529
plexus lumbosacralis   19: 1170; 20: 1343
plexus of Horsley   21: 1563
plica encephali ventralis   4: 161; 11: 501, 524; 12: 561; 14: 706, 733
plica valvulae   14: 703
Polistotrematidae   11: 497
Polyodontidae   14: 701
Polypteriformes   8: 358
Polypterini
- brain size   23: 2106
Polypterus   13: 655
polysomes   1: 5
pons   4: 171; 22: 1643
pontine flexure   4: 164; 21: 1534

pontine paramedian reticular formation (PPRF)   22: 1724, 1756
pontine taste area   22: 1776
pontocerebellar fibres   22: 1748, 1750
Porolepiformes   17: 1041
portal plexus   16: 988
posterior dorsal ventricular ridge, see amygdala
posterior parencephalic segment   13: 680
postoptic region   4: 194
postsynaptic dorsal column system   19: 1179; 20: 1387; 22: 1672, 1767
Potamogalidae   22: 1641
pouch of Rathke   13: 657, 685
pre-pacemaker nucleus (gymnotids)   15: 885
precerebellar nuclei   22: 1712
prefrontal cortex   22: 2025
prefrontal cortex (in birds)   21: 1603
preganglionic autonomic neurons   11: 516, 521; 18: 1081; 19: 1187, 1210; 20: 1343
preglomerular nuclei   15: 875, 887
preglomerulosus complex   13: 696; 14: 742; 15: 887; 24: 2149, 2152
premotor area   21: 1563, 1609
preoptic recess organ   19: 1267
preoptico-hypophysial neurosecretory system   10: 461, 466
preoptico-hypothalamic continuum   22: 1863
'pressure' cells   10: 419
presubiculum   22: 1872, 1944
pretectal region   24: 2152
pretectum   10: 463, 470; 11: 524; 12: 611; 15: 876; 18: 1111; 19: 1254; 20: 1453; 22: 1720, 1812
pretectum centrale   15: 879
pretectum periventriculare   15: 877
pretectum superficiale   15: 878
prey recognition   18: 1112; 19: 1257
prey-catching behaviour   18: 1101, 1111; 19: 1243, 1257; 20: 1320
primary cultures   7: 343
primary motor centre   9: 371
primary motor system   18: 1060; 19: 1171
primary neurons   18: 1060; 19: 1162
primary sensory system   18: 1060; 19: 1178
primates
- brain size   23: 2123
primitive motoneurons   10: 402, 403
primitive sensory tract   5: 259
primordium hippocampi   10: 406, 456, 460, 465, 480; 14: 747; 22: 1932; 24: 2159
primordium pallii dorsalis   10: 456, 466; 22: 1932; 24: 2160
primordium piriforme   10: 456, 466; 22: 1873, 1932; 24: 2160
principal commissure   9: 369
Proboscoida   22: 1642
Probst, commissure   22: 1721, 1786
processus neuroporicus   11: 501; 14: 706
progressive differentiation   6: 309
projection   3: 149
projection neurons   5: 233
prolactin   9: 389
proliferation zones   16: 949
prominentia granularis   14: 726
proprioceptive system   21: 1449, 1555
propriospinal fibres   20: 1351; 22: 1667
prosencephalic vesicle   11: 504, 543, 532
prosencephalon   9: 393; 11: 504; 19: 1260; 20: 1335

prosomeres   20: 1335; 21: 1591; 24: 2152
protein
 - synthesis   1: 8
 - trafficking   1: 8
Protopterus   8: 358, 359; 16: 939; 17: 1039
Prototheria   8: 361
prototype of the craniote brain   10: 480
psalterium   22: 1873
'pseudo-ventricle'   13: 660
pterosaurs   8: 360; 20: 1317
pure morphology   6: 284
putamen   4: 217; 22: 1883, 1901
pyramidal tract   3: 126, 129, 141; 22: 1645, 1668, 1720, 1823, 1830, 1832; 24: 2171
pyramis (lobulus VIII Larsell)   22: 1727

R

radial coupling   2: 79
radial glia   1: 10; 5: 234; 11: 544
radix descendens n. trigemini   13: 675, 677; 14: 721; 16: 968, 973
radix mesencephalicus n. trigemini   12: 580; 14: 721; 18: 1083; 19: 1213; 20: 1391; 22: 1770, 1772
Rana esculenta   8: 360; 19: 1153
raphespinal projections   11: 516, 522, 541; 18: 1068; 19: 1194; 20: 1362
receptive surface   1: 4; 2: 52
red nucleus, see nucleus ruber
reflex arc   1: 21
regio preoptica   11: 529; 14: 733, 749; 15: 898; 18: 1118; 19: 1267; 20: 1461; 21: 1589; 22: 1872
regulatory genes   4: 191; 6: 304
Reissner's fibre   9: 376, 384; 10: 429, 453; 11: 507, 541; 12: 616; 15: 790; 18: 1057; 19: 1261
Renshaw cell   22: 1663; 24: 2192
reproductive behaviour   22: 1894, 1897, 1900
reptile-bird transition   20: 1498
reptiles
 - brain size   23: 2109
respiration   10: 446; 21: 1560, 1579, 1604
resting potential   1: 14
reticular elements (lampreys)
 - bulbar group   10: 442
 - isthmic group   10: 442
 - Mauthner group   10: 442
 - mesencephalic group   10: 442
 - vagal group   10: 442
reticular formation   10: 423; 11: 522; 12: 595; 13: 676; 14: 724; 15: 798; 18: 1096; 19: 1238; 20: 1413; 22: 1715, 1753
 - connectivity   10: 443
 - chemoarchitecture   12: 598; 18: 1097; 19: 1239; 20: 1415
 - connectivity   12: 596; 18: 1099; 19: 1240; 20: 1418; 22: 1754, 1762
 - cytoarchitecture   9: 393; 10: 442; 12: 595; 18: 1096; 19: 1238; 20: 1413; 22: 1753
 - lateral part   13: 676; 14: 724
 - medial part   13: 676; 14: 724; 13: 676; 14: 724
reticulospinal projections   11: 516, 522, 524, 541; 18: 1067, 1099; 19: 1192; 20: 1358
retina   2: 58; 10: 450; 11: 531; 17: 1009; 22: 1791; 24: 2183
retinofugal fibres   11: 523; 13: 679; 14: 739; 16: 983
retinofugal projections   10: 444, 450; 11: 531; 15: 876
retinohypothalamic projections   18: 1105, 1118; 19: 1251; 20: 1447
retinopetal projections   10: 451, 452; 11: 525; 14: 736; 20: 1449
retinotectal projections   11: 523; 15: 787; 16: 977; 18: 1105; 19: 1250; 20: 1445, 1447; 22: 1797
retinothalamic projections   11: 528; 18: 1105, 1117; 19: 1251; 20: 1447
retrograde changes   7: 330
retrograde tracing techniques   7: 331

retrorubral area   20: 1410
Rexed's laminae   20: 1347
rhinencephalon   22: 1884, 2027; 24: 2153
Rhipidistia   17: 1007
rhipidistian crossopterygians   16: 941
rhombic lip   5: 239
rhombomeres   10: 406; 20: 1335; 21: 1535
ribosomes   1: 5
RNA   1: 8
Rodentia   22: 1642
Rohde cells   3: 115, 124; 9: 382, 390; 10: 483
Rohon-Beard cells   5: 259; 9: 390; 10: 403; 11: 543; 13: 662; 14: 706; 16: 953; 18: 1060; 19: 1171
roller-tube technique   7: 343
roof plate   4: 194
rostral interstitial nucleus of the medial longitudinal fasciculus (iMLF)   22: 1724
rostral organ   17: 1009, 1023, 1039
rough endoplasmic reticulum   1: 8
rubrospinal tract, see tractus rubrospinalis
Ruffini-type receptors   11: 498

S

saccus dorsalis   4: 168; 10: 399, 454, 464; 11: 525; 14: 703; 15: 883; 16: 943, 946; 17: 1014
saccus vasculosus   4: 169; 9: 375; 12: 627; 13: 657; 14: 739, 742; 15: 895; 16: 943; 17: 1034, 1041
Salmo gairdneri   8: 359
Sarcopterygii   8: 360; 15: 759; 16: 940; 17: 1007
 - brain size   23: 2107
scala naturae   6: 321
Scaphirhynchus platorhynchus   8: 359; 14: 702
Schreiner organs   11: 519
Schwann cells   1: 7, 9
seasonal changes   21: 1604
secondary simplified CNS   18: 1135
segmental organisation of the brain   24: 2164
Seitenwülste (Schaper)   4: 171
selective silver impregnation technique   7: 330
sense of taste   10: 431
sensory circuitry in vertebrates   24: 2169, 2173
septal region   16: 993; 22: 1898
 - connectivity   22: 1899
 - subdivision   22: 1898
septum   10: 465, 466, 480; 11: 535; 12: 642; 13: 688; 16: 950, 1003; 17: 1037; 18: 1131; 19: 1280; 20: 1477, 1479; 21: 1615
septum ependymale   4: 170; 16: 946, 948, 989; 17: 1014, 1036
septum pellucidum   22: 1647, 1898
septum precommissurale   22: 1899
septum verum   22: 1894
serotonergic cell populations, see chemically defined neuronal populations
sexual behaviour   22: 1869, 1900
sexual dimorphism   21: 1588, 1589, 1604; 22: 1897
simple reflexes (Windle)   5: 257
single-unit recording   7: 339
sinus rhomboidalis   21: 1549
Sirenia   22: 1642
sister groups   6: 311
skin photoreceptors   11: 498
slime   11: 498
slime gland plexus   11: 511
slime glands   11: 498, 510
smell   21: 1528, 1616
social interactions   22: 1932
soma   1: 4
somatic motor column   4: 197
somatic sensory column   4: 197
somatomotor bundle   9: 382
somatomotor neurons   10: 417
somatomotor nuclei   12: 592; 15: 802; 18: 1080; 19: 1209; 20: 1384

Synopsis
General introductory part: 1 Cellular elements; 2 Centres; 3 Fibre systems; 4 Morphogenesis; 5 Histogenesis; 6 Principles; 7 Techniques
Specialized part: 8 Introduction; 9 Amphioxus; 10 Lampreys; 11 Hagfishes; 12 Cartilaginous fishes; 13 Brachiopterygians; 14 Chondrosteans; 15 Teleosts; 16 Lungfishes; 17 Latimeria; 18 Urodeles; 19 Anurans; 20 Reptiles; 21 Birds; 22 Mammals
General concluding part: 23 Brain size; 24 Meaning

somatomotor zone   **9:** 380, 381; **10:** 436; **14:** 722, 724; **16:** 977; **24:** 2139
somatosensory zone   **10:** 430, 477; **16:** 977; **21:** 1612, 1613; **24:** 2139
sonic motor neurons   **15:** 803, 804
Soricidae   **22:** 1640
spatial memory, orientation   **21:** 1596, 1614
spawning   **14:** 743
speciation   **6:** 305
species   **6:** 304
Sphenodon   **8:** 360
spinal border cells   **22:** 1674, 1665
spinal cord   **3:** 115; **9:** 378; **11:** 510; **12:** 567; **18:** 1057; **19:** 1170; **20:** 1339; **22:** 1654; **24:** 2135
- chemoarchitecture   **12:** 575
- commissural neurons   **22:** 1663
- connectivity   **11:** 516; **12:** 570; **18:** 1065; **19:** 1188; **20:** 1349
- cytoarchitecture   **12:** 560, 567
- development   **12:** 564; **18:** 1061; **19:** 1170; **20:** 1335
- function   **12:** 576
- glial cells   **18:** 1062; **20:** 1364
- intermediate zone   **22:** 1662
- interneurons   **15:** 792; **18:** 1061; **19:** 1171; **20:** 1343; **22:** 1663, 1664
- laminae of Rexed   **22:** 1661, 1662
- laminar organisation   **20:** 1347
- motoneurons   **11:** 522; **15:** 791, 792; **18:** 1065, 1070; **19:** 1183; **20:** 1344; **22:** 1664, 1665
- regenerative capacity   **18:** 1069; **19:** 1195; **20:** 1364
- segmentation   **22:** 1659
- subdivision   **11:** 510; **18:** 1062; **19:** 1179; **20:** 1343
spinal fields   **19:** 1179
spinal lemniscus, see lemniscus spinalis
spinal motor columns   **13:** 676; **14:** 708; **18:** 1065; **19:** 1185; **20:** 1339
spinal nerves   **9:** 366
- general cutaneous component   **9:** 379
- visceral motor component   **9:** 379
- visceral sensory component   **9:** 379
spines (dendritic)   **1:** 7
spino-occipital nerves   **4:** 222; **13:** 676; **14:** 708, 724; **16:** 970; **17:** 1026
spinocerebellar fibres, see tractus spinocerebellaris
spinocervical system, see tractus spinocervicalis
spinoreticular connections, see tractus spinoreticularis
spinothalamic fibres, see tractus spinothalamicus
spongioblasts (His)   **5:** 229
Squalus acanthias   **8:** 358
Squamata   **8:** 360
staging of embryos   **11:** 497; **12:** 560; **18:** 1048; **19:** 1154; **20:** 1329
stem cells   **5:** 232
'Stirnorgan'   **19:** 1260
stratum cellulare externum   **13:** 662
stratum cellulare internum   **21:** 1586
stratum zonale   **5:** 248
stretch receptors
- excitatory   **10:** 421, 428, 480
- inhibitory   **10:** 421, 428, 480
stria acustica dorsalis   **22:** 1781
stria acustica intermedia   **22:** 1781
stria medullaris   **12:** 641; **13:** 681; **14:** 740, 750; **15:** 884; **16:** 984, 997; **17:** 1032, 1038; **18:** 1115; **19:** 1275; **20:** 1455; **21:** 1581, 1583, 1615; **24:** 2150
stria olfactoria lateralis   **22:** 1891

stria olfactoria medialis   **22:** 1891
stria terminalis   **20:** 1487; **22:** 1897, 1927
striato-amygdalar transition area   **20:** 1486
striatonigral projections   **20:** 1410; **22:** 1911
striatotegmental projection   **21:** 1594
striatum   **6:** 295; **11:** 535; **12:** 642; **13:** 695; **14:** 752; **16:** 950, 997; **18:** 1133; **19:** 1280; **20:** 1490; **22:** 1883; **24:** 2171
- cell types   **22:** 1913
- chemoarchitecture   **18:** 1133; **19:** 1284; **20:** 1490
- connectivity   **11:** 535; **18:** 1133; **19:** 1281; **20:** 1490, 1498
- dorsale   **22:** 1901
- subdivision   **18:** 1133; **19:** 1281; **20:** 1490
- ventrale   **22:** 1891, 1901
strio-amygdaloid complex   **16:** 993; **17:** 1037
striopallidonigral bundle   **22:** 1909
striosomes   **22:** 1905, 1914
structural plan   **6:** 282
subependymal layer   **5:** 237
subependymal zone   **16:** 953
subepithelial plexus   **9:** 380
subiculum   **22:** 1647, 1872, 1944
subpallium   **4:** 213; **11:** 535, 538; **12:** 641; **14:** 747, 753; **17:** 1014, 1036; **22:** 1872
substantia gelatinosa   **12:** 569; **22:** 1661, 1666
substantia gelatinosa dorsalis   **21:** 1542
substantia innominata   **22:** 1901, 1906, 1919
substantia nigra   **11:** 537; **19:** 1238; **20:** 1410; **21:** 1573, 1595; **22:** 1720, 1901
- dopaminergic neurons   **22:** 1760
- pars compacta   **22:** 1720, 1907
- pars reticulata   **22:** 1720, 1907
- projections to colliculus superior   **22:** 1801, 1803
substantia perforata anterior   **22:** 1891
substrate pathways   **5:** 257; **18:** 1062
subventricular zone   **5:** 237
sulci
- in brain stem   **18:** 1054; **19:** 1160; **20:** 1329
- in diencephalon   **18:** 1054,1114; **19:** 1160; **20:** 1329
- of neocortex   **22:** 2013
sulcus α   **22:** 2015, 2021
sulcus arcuatus   **22:** 2019
sulcus axial   **22:** 2014
sulcus calcarinus   **22:** 2019
sulcus callosus   **22:** 2015, 2018, 2019
sulcus centralis   **22:** 2019, 2021
sulcus cinguli   **22:** 2015, 2018, 2019, 2020
sulcus confinalis   **22:** 2020
sulcus coronalis   **22:** 2020
sulcus coronolateralis   **22:** 2019,2021
sulcus cruciatus   **22:** 2020
sulcus di-mesencephalicus   **11:** 501
sulcus diencephalicus dorsalis   **13:** 681; **14:** 736; **16:** 943, 978, 979; **17:** 1032
sulcus diencephalicus medius   **13:** 683; **16:** 943, 948, 978, 979; **17:** 1032
sulcus diencephalicus ventralis   **13:** 683; **16:** 943, 948, 949, 978, 979; **17:** 1032
sulcus dorsalis   **4:** 212
sulcus dorsomedialis   **22:** 2020, 2022
sulcus ectolateralis   **22:** 2019, 2020, 2023
sulcus ectosylvius   **22:** 2019, 2020
sulcus endorhinalis   **22:** 2015
sulcus hippocampalis   **22:** 1873, 2015
sulcus hypothalamicus   **16:** 948, 978
sulcus intermedius   **4:** 212

sulcus intermedius dorsalis 4: 199, 202; 10: 430; 13: 673; 14: 713; 16: 965; 17: 1017
sulcus intermedius ventralis 4: 199, 202; 10: 436, 445; 13: 673; 14: 723; 16: 965; 17: 1017
sulcus intraencephalicus anterior 10: 405; 16: 950
sulcus intraencephalicus posterior 10: 405
sulcus intrahabenularis 13: 681
sulcus intraparietalis 22: 2018, 2019, 2021
sulcus isthmi 16: 965; 17: 1017, 1028
sulcus lateralis 22: 1879, 2020, 2023
sulcus lateralis mesencephali 13: 678; 17: 1028
sulcus limitans (His) 4: 196, 199, 203, 222; 10: 406, 430, 445, 447, 466; 13: 674; 14: 709, 713, 724, 729; 15: 795; 16: 965, 975; 17: 1017, 1024, 1028; 22: 1645, 1714, 2014; 24: 2139
sulcus limitans pallii 16: 990
sulcus limitans telencephali 14: 747
sulcus longitudinalis 22: 2024
sulcus lunatus 22: 2019
sulcus marginalis 22: 2017, 2020, 2021
sulcus medianus inferior 10: 436
sulcus occipitalis ventralis 22: 2019
sulcus orbitalis 22: 2218
sulcus paramedianus 22: 1725, 1727
sulcus parieto-occipitalis 22: 2019
sulcus postsylvius 22: 2018, 2019, 2021
sulcus presylvius 22: 2019
sulcus pseudosylvius 22: 2017, 2020, 2021
sulcus rectus 22: 2218, 2019, 2021
sulcus rhinalis lateralis 22: 2015, 2022
sulcus splenialis 22: 2020, 2021
sulcus subhabenularis 10: 456; 13: 681; 16: 978
sulcus subpallialis 16: 949, 950
sulcus suprasylvius 22: 2019, 2020, 2022, 2023
sulcus Sylvius 22: 2018, 2019, 2021
sulcus tegmentalis mesencephali 17: 1028
sulcus telo-diencephalicus 11: 501
sulcus transversus telencephali 17: 1035
sulcus ventralis 4: 212
superficial layer (Senn) 5: 248
superior colliculus 22: 1793
- ascending connections 22: 1809
- efferent pathways 22: 1807, 1812
supracortical cells 16: 991
supramedullary neurons 15: 791
sympathetic fibres 19: 1187; 20: 1383; 21: 1554
symplesiomorphy 6: 312
synapomorphy 6: 312
synapses 1: 7
synaptic transmission
- excitatory 1: 20
- inhibitory 1: 20
synaptic vesicle 1: 8
synencephalic segment 13: 680
synencephalon 4: 194; 12: 611, 616; 14: 736; 15: 876; 16: 951; 18: 1115; 19: 1260; 21: 1534, 1581
system (the concept) 24: 2168

T

Tachyglossus 8: 361
Talpidae 22: 1640
taming effect 21: 1609
tanycytes 1: 10; 10: 463; 11: 529, 530; 14: 743
taste buds 11: 519; 14: 701; 15: 767; 21: 1528, 1560
taste organs 10: 430, 477
taste pathways 20: 1400
taxonomy 6: 277
tectum mesencephali 2: 59, 81; 9: 371; 10: 406, 448; 11: 523; 12: 606; 13: 678, 695; 18: 1101; 19: 1245; 20: 1441; 21: 1535, 1575, 1584
- connectivity 10: 449; 11: 523; 15: 863, 870; 18: 1104; 19: 1250; 20: 1445; 21: 1578
- cytoarchitecture 11: 523; 18: 1102; 19: 1247; 20: 1441
- development 18: 1102, 1107; 19: 1245; 20: 1335; 21: 1534
- effect of lesions 18: 1112; 19: 1257
- electrophysiological studies 20: 1447; 21: 1575, 1578
- neuron types 18: 1103; 19: 1248; 20: 1444
- stimulation studies 18: 1112; 19: 1257; 20: 1454
tegmentum dorsale 18: 1077, 1102
tegmentum isthmi 18: 1077
tegmentum mesencephali 11: 522, 523, 533; 12: 606; 18: 1107, 1102; 19: 1207
tegmentum motoricum 24: 2145
tegmentum motoricum mesencephali 10: 451
tegmentum pontis 22: 1712
tegmentum ventrale 18: 1107, 1102
tela diencephali 13: 657; 16: 943
tela telencephali 13: 660; 14: 703, 744; 16: 943; 17: 1014
telencephalon 11: 500, 501, 525, 541; 12: 632; 18: 1121; 19: 1273; 20: 1466; 22: 1871; 24: 2157
- chemoarchitecture 12: 644
- connectivity 11: 516, 524, 528, 534; 12: 637, 641; 18: 1128; 19: 1277
- cytoarchitecture 12: 637, 641
- development 11: 504, 532; 12: 559; 20: 1336
- effect of lesions 18: 1130; 19: 1286; 20: 1493
- frontal lobe 22: 1885
- function 12: 644
- occipital lobe 22: 1876
- piriform lobe 22: 1873
- stimulation studies 18: 1129; 19: 1286; 20: 1494
- subdivision 18: 1121; 19: 1273; 20: 1466
- temporal lobe 22: 1876
telencephalon impar 10: 480; 16: 943, 946, 988; 17: 1037
telencephalon medium 10: 464; 14: 744
Teleostei 8: 359; 15: 759
- brain size 23: 2107
telodendron 1: 4, 7
tendon organs 20: 1351
Tenrecidae 22: 1640
Terapsida 8: 360
terminal nerve ganglion 15: 904; 16: 996
terminals 1: 7
testosterone 21: 1579, 1604, 1612
Testudo hermanni 8: 361; 20: 13191
thalamic neuropil 16: 984
thalamocortical fibres 22: 1964
thalamostriate fibres 20: 1460, 1475, 1480; 22: 1910
thalamotelencephalic projections 10: 470; 11: 528; 18: 1116; 19: 1266; 20: 1459
thalamus 11: 527; 22: 1848
- association nuclei 22: 1848
- audition 22: 1852, 1855
- connections with cortex 22: 1858
- evolution 22: 1849
- nomenclature 22: 1850
- relay neurons 22: 1858
- sensory relay nuclei 22: 1848
- somatosensory 22: 1850, 1855
- subdivision 22: 1849
- vision 22: 1854, 1855
thalamus dorsalis 4: 212; 10: 422, 429, 453, 466, 472; 11: 527; 12: 619; 16: 951, 979; 17: 1040; 18: 1115; 19: 1261; 20: 1456; 21: 1584; 22: 1646; 24: 2150
- connectivity 11: 528; 18: 1116; 19: 1263; 20: 1456
- pars dorsalis 17: 1032
- pars ventralis 17: 1032
- subdivision 11: 528; 18: 1116; 19: 1261; 20: 1456
thalamus ventralis 4: 212; 10: 453; 11: 527; 12: 619; 16: 951, 979; 17: 1032; 18: 1117; 19: 1262, 1266; 20: 1461; 21: 1534, 1586; 22: 1860; 24: 2150
thirst 22: 1869
tissue print method 7: 343
tissue staining techniques 7: 328
tonotopic localisation 22: 1779, 1783, 1787, 1789
topographical position 6: 284
topological position 6: 285, 302

Synopsis
General introductory part: 1 Cellular elements; 2 Centres; 3 Fibre systems; 4 Morphogenesis; 5 Histogenesis; 6 Principles; 7 Techniques
Specialized part: 8 Introduction; 9 Amphioxus; 10 Lampreys; 11 Hagfishes; 12 Cartilaginous fishes; 13 Brachiopterygians; 14 Chondrosteans; 15 Teleosts; 16 Lungfishes; 17 Latimeria; 18 Urodeles; 19 Anurans; 20 Reptiles; 21 Birds; 22 Mammals
General concluding part: 23 Brain size; 24 Meaning

topology 6: 277
tortuosity (of extracellular space) 1: 12
torus externus 21: 1579
torus lateralis 4: 171; 13: 657; 15: 786, 855; 24: 2146
torus longitudinalis 13: 679; 14: 730; 15: 872, 912
torus semicircularis 2: 78; 10: 433, 435, 448; 12: 598; 13: 657, 673, 695; 14: 721, 730, 754; 15: 772, 855, 912; 16: 975, 977, 978; 18: 1091; 19: 1225; 20: 1407; 21: 1579, 1580; 24: 2145
– connectivity 19: 1226
– cytoarchitecture 19: 1225; 20: 1407
– nucleus centralis 20: 1408
– nucleus laminaris 19: 1225; 20: 1407
– nucleus magnocellularis 19: 1226
– nucleus principalis 19: 1225
total pattern (of activity) 5: 257
'touch' cells 10: 419
tracing of transmitter-specific pathways 7: 334
tract of posterior commissure 5: 261, 265
tractus 3: 149
tractus bulbospinalis 11: 518
tractus bulbospinalis cruciatus 11: 516, 522; 22: 1668, 1672
tractus bulbospinalis lateralis 22: 1669
tractus bulbospinalis rectus 11: 516, 522
tractus cerebellobulbaris 14: 729
tractus cerebellomotorius 10: 447; 14: 729
tractus cerebellovestibularis et cerebellobulbaris rectus 12: 605
tractus cervicothalamicus 22: 1766
tractus cortico-bulbospinalis 22: 1883
tractus corticotegmentalis 16: 985
tractus cuneocerebellaris 22: 1672, 1674, 1748
tractus descendens n. trigemini 10: 431; 11: 519; 12: 580; 18: 1084; 19: 1213; 20: 1391
tractus dorso-archistriaticus 21: 1618
tractus fronto-archistriaticus 21: 1597, 1608, 1618
tractus frontopontinus 22: 1720, 1821, 1883
tractus habenulo-interpeduncularis 10: 456; 21: 1583
tractus habenulothalamicus 10: 457
tractus hypothalamo-olfactorius 10: 462
tractus infundibuli 21: 1586
tractus interpedunculobulbaris 10: 436; 11: 523; 16: 973
tractus interpedunculotegmentalis 11: 523
tractus interstitiospinalis 18: 1067; 19: 1193; 20: 1358; 22: 1668, 1672
tractus isthmo-opticus 21: 1574
tractus lobobulbaris 10: 451; 12: 624; 14: 725
tractus lobocerebellaris 10: 447; 12: 624; 14: 729
tractus mamillobulbaris 14: 725
tractus mamillopeduncularis 10: 451
tractus mamillotegmentalis 10: 451
tractus occipitomesencephalicus 21: 1554, 1563, 1580, 1607, 1609, 1618
tractus octavomesencephalicus lateralis 10: 435
tractus octavomesencephalicus medialis 10: 435
tractus octavospinalis 10: 424, 435
tractus olfacto-dorsolateralis 16: 994
tractus olfactohabenularis 11: 527
tractus olfactorius 10: 467; 12: 635
tractus olfactorius accessorius 18: 1128; 19: 1275; 20: 1468; 22: 1895
tractus olfactorius centralis 17: 1038
tractus olfactorius dorsalis 17: 1038
tractus olfactorius externus 16: 989, 994
tractus olfactorius habenulae dorsalis 17: 1034, 1038
tractus olfactorius habenulae ventralis 17: 1034, 1038
tractus olfactorius lateralis 11: 533; 13: 688; 14: 749, 754; 15: 903; 16: 994; 18: 1127; 19: 1275; 20: 1468; 22: 1647, 1890, 1927
tractus olfactorius lateralis habenulae 16: 984, 994
tractus olfactorius medialis 11: 533, 535; 13: 688; 14: 749, 754; 15: 903; 16: 994; 18: 1127; 19: 1275; 20: 1468; 22: 1890
tractus olfactorius ventralis 11: 533; 16: 995; 17: 1038
tractus olfactothalamicus et -hypothalamicus 10: 467
tractus olivocerebellaris 12: 604; 17: 1028
tractus opticus 10: 457; 11: 527; 12: 610; 13: 679; 14: 730; 18: 1105; 19: 1250; 20: 1447; 22: 1720
tractus opticus accessorius 13: 680; 14: 733, 740
tractus opticus axialis 10: 449; 13: 686; 18: 1105; 19: 1250
tractus opticus basalis 12: 616; 17: 1035; 18: 1105; 19: 1250; 20: 1453
tractus opticus lateralis 10: 449; 13: 679, 686; 14: 730, 739; 17: 1029, 1035; 20: 1447
tractus opticus marginalis 13: 686; 14: 739; 16: 977; 18: 1105; 19: 1250; 13: 679
tractus opticus medialis 14: 730, 739; 16: 977; 17: 1029, 1035; 18: 1105; 19: 1250; 20: 1447
tractus pallialis septi 12: 643
tractus pallii 10: 462; 12: 624, 641; 13: 686; 17: 1038
tractus palliohabenularis 11: 534; 17: 1034, 1038
tractus palliothalamicus 11: 534
tractus parieto-temporo-pontinus 22: 1720, 1821, 1883
tractus pontospinalis cruciatus 22: 1668, 1672
tractus preoptico-hypophyseus 14: 743, 749; 16: 987; 10: 461; 12: 625
tractus pyramidalis, see pyramidal tract
tractus quintofrontalis 20: 1393; 21: 1555, 1563, 1586, 1618
tractus reticulocerebellaris 22: 1748
tractus reticulospinalis medialis 22: 1668, 1672
tractus rubrobulbaris 22: 1721, 1839; 12: 570; 17: 1039; 18: 1067; 19: 1190; 20: 1358, 1437; 21: 1553, 1564, 1574; 22: 1668, 1672, 1839, 1842, 1721
tractus sacci vasculosi 12: 627
tractus septo-tubercularis 17: 1033
tractus septomesencephalicus 21: 1554, 1564, 1580, 1612, 1618
tractus solitarius 22: 1715, 1772, 1774
tractus spino-olivaris 22: 1672, 1674
tractus spinobulbaris 11: 541
tractus spinocerebellaris 10: 477; 12: 570; 14: 729; 16: 974; 17: 1027; 18: 1083; 19: 1188, 1233; 20: 1356, 1428; 21: 1554, 1565; 22: 1672, 1674
tractus spinocerebellaris dorsalis 22: 1672–1674
tractus spinocerebellaris rostralis 22: 1674
tractus spinocerebellaris ventralis 22: 1672–1674, 1748
tractus spinocervicalis 18: 1083; 19: 1188, 1211; 22: 1672–1674
tractus spinomesencephalicus 17: 1027; 18: 1083; 19: 1213; 20: 1355
tractus spinoreticularis 11: 516, 522, 541; 18: 1083; 19: 1213; 20: 1353; 22: 1762
tractus spinotectalis 11: 516, 543; 22: 1672–1674
tractus spinothalamicus 10: 457; 11: 516, 528; 12: 570; 18: 1083; 19: 1213; 20: 1355; 21: 1554; 22: 1672, 1763
tractus spinothalamicus dorsalis 22: 1672, 1674, 1765
tractus spinothalamicus ventrolateralis 22: 1672, 1674, 1765
tractus spinovestibularis 22: 1672, 1674
tractus striomesencephalicus et -cerebellaris 21: 1619
tractus striopeduncularis 17: 1035; 18: 1095

tractus striotegmentalis 17: 1032, 1035; 18: 1095
tractus striothalamicus 17: 1035, 1040
tractus striothalamicus et -hypothalamicus 10: 467
tractus supraoptico-paraventriculohypophyseus 22: 1868
tractus tecto- et torotegmentalis 10: 451
tractus tecto-isthmicus 18: 1110; 19: 1254
tractus tectobulbaris 11: 522; 12: 611; 15: 805; 16: 977; 21: 1578
tractus tectobulbaris cruciatus 13: 679; 14: 732; 16: 971; 17: 1027, 1031; 19: 1254; 20: 1450
tractus tectobulbaris rectus 13: 679; 14: 732; 16: 971; 17: 1027, 1031; 19: 1254; 20: 1450
tractus tectobulbospinalis cruciatus 18: 1110
tractus tectobulbospinalis rectus 18: 1110
tractus tectocerebellaris 14: 729
tractus tectohabenularis 12: 622
tractus tectospinalis 11: 516, 523; 18: 1067; 19: 1190, 1254; 21: 1553; 22: 1668, 1672, 1809
tractus tectothalamicus 18: 1110; 19: 1253; 20: 1450; 21: 1579
tractus tectothalamicus et -hypothalamicus 10: 457, 461
tractus tegmentalis centralis 22: 1721, 1747, 1837, 1840
tractus tegmentalis medialis 22: 1741, 1747, 1837, 1840
tractus tegmentocerebellaris 14: 729, 732
tractus tegmentothalamicus 17: 1040
tractus thalamo-hypothalamicus caudalis 17: 1033
tractus thalamo-hypothalamicus et peduncularis cruciatus 16: 986
tractus thalamofrontalis 10: 458; 21: 1586, 1618
tractus thalamohypothalamicus 10: 461
tractus thalamopeduncularis 16: 986; 17: 1032
tractus thalamotectalis 12: 623
tractus thalamotegmentalis 12: 623
tractus thalamotelencephalicus 12: 623
tractus trigeminothalamicus 22: 1714
tractus trigeminothalamicus dorsalis (Wallenberg) 22: 1714, 1764, 1770
tractus tuberculo-infundibularis 17: 1033
tractus uncinatus 22: 1752
tractus vestibulospinalis 10: 426; 14: 721, 725; 15: 805; 18: 1067; 19: 1190; 20: 1358
tractus vestibulospinalis lateralis 19: 1193; 20: 1405; 22: 1668, 1672, 1835
tractus vestibulospinalis medialis 19: 1193; 20: 1405; 22: 1668, 1672, 1835
tractus visceralis secundarius 14: 722, 725; 16: 969, 973
trailing lag hypothesis 10: 429
transneuronal tracers 7: 332
transpeduncular tract 22: 1818
transport
- fast axonal 1: 9
- intracellular 1: 9
- slow anterograde 1: 9
triune brain (MacLean) 22: 2024
tuberculum acusticum 22: 1712
tuberculum olfactorium 4: 213, 217; 13: 689; 16: 950, 953, 984, 991, 1003; 17: 1037; 20: 1468; 22: 1647; 22: 1872, 1890, 1903, 1906; 18: 1131; 19: 1280
tuberculum posterius 9: 369; 12: 555; 13: 657, 695; 14: 706, 737; 15: 912; 16: 943; 17: 1014, 1031; 18: 1118; 19: 1266; 24: 2147
tuberculum quinti 22: 1643
Tunicata 9: 366
Tupaiidae 22: 1641
Tupinambis teguixin 8: 361; 20: 1319
type 4: 220
typology 6: 284

U

Übergangsganglion 15: 843
uncinate field 18: 1115; 19: 1254
Urodela

- brain size 23: 2108
urophysis 9: 384; 12: 569; 14: 709; 15: 793, 794
urotensin I 9: 389; 10: 474; 14: 709; 16: 965
urotensin II 13: 666; 14: 709; 16: 965
uvula 22: 1725, 1730

V

vallecula 21: 1531, 1593
valvula cerebelli 14: 703, 726; 15: 837, 912; 24: 2144
Varanus exanthematicus 20: 1322
vasotocin, see chemically defined neuronal populations
vector 6: 274
vectorial migration 5: 253
velum medullare anterius 4: 222; 12: 601; 13: 657; 14: 703, 726; 16: 943; 18: 1093; 19: 1157; 20: 1424; 22: 1645
velum medullare posterius 22: 1645
velum transversum 4: 162, 168; 12: 555; 13: 657; 14: 733, 744; 17: 1011, 1014
ventral amygdalofugal pathway 22: 1928
ventral cells of Bone 9: 385
ventral horn, see cornu ventrale
ventral nerves 9: 368
ventral pallidohabenular projection 22: 1918
ventral pallidonigral projection 22: 1918
ventral pallidum 21: 1596
ventral striatopallidal projection 22: 1917
ventral tegmental area, see area tegmentalis ventralis
ventral tegmental-substantia nigra complex 13: 684
ventral thalamus, see thalamus ventralis
ventricular system 11: 504; 12: 554; 18: 1051; 19: 1160; 20: 1329; 22: 1660
ventricular zone (Boulder) 5: 237
ventriculus cerebelli 17: 1011
ventriculus impar 17: 1037
ventrobasal complex 22: 1763
ventrolateral bundle 5: 261; 9: 371, 373; 10: 404
vertical cells 9: 381
vestibular apparatus 10: 435
vestibular nuclear complex 18: 1089; 19: 1221; 20: 1402
vestibular nuclei 22: 1714, 1724, 1735, 1752, 1832
vestibular system 10: 485
vestibulo-ocular reflex 19: 1221
vestibulo-oculomotor projections 18: 1089; 19: 1221; 20: 1405; 22: 1835, 1836
vestibulocerebellar fibres 18: 1093; 19: 1232; 20: 1430; 22: 1748
vestibulospinal projections, see tractus vestibulospinalis
viruses for transneuronal tracing 7: 333
visceromotor nuclei 12: 593, 595; 15: 806, 807; 18: 1081; 19: 1210; 20: 1386; 22: 1715, 1724
visceromotor zone 9: 380; 10: 436, 445; 14: 723; 24: 2139
viscerosensory nuclei 15: 807
viscerosensory zone 9: 382; 10: 430, 477; 14: 722, 723; 17: 1023; 24: 2139
vision 15: 768
- colour 21: 1585
- discrimination performance 21: 1584
visual cortex (primates) 22: 1958, 2003
visual field maps 22: 1797, 1801
visual imprinting 21: 1612
vocalisation 19: 1241; 21: 1526, 1560, 1579, 1580
- circuits 21: 1604
- programming network 21: 1604
voltage-gated channels 1: 15
voltage-sensitive dyes 7: 342
voltammetric microelectrodes 7: 340
volume transmission 1: 13
vomeronasal amygdala 22: 1895
vomeronasal organ 14: 746; 18: 1124; 19: 1274; 20: 1468; 22: 1871, 1894

> **Synopsis**
> *General introductory part:* 1 Cellular elements; 2 Centres; 3 Fibre systems; 4 Morphogenesis; 5 Histogenesis; 6 Principles; 7 Techniques
> *Specialized part:* 8 Introduction; 9 Amphioxus; 10 Lampreys; 11 Hagfishes; 12 Cartilaginous fishes; 13 Brachiopterygians; 14 Chondrosteans; 15 Teleosts; 16 Lungfishes; 17 Latimeria; 18 Urodeles; 19 Anurans; 20 Reptiles; 21 Birds; 22 Mammals
> *General concluding part:* 23 Brain size; 24 Meaning

W

Wallerian degeneration 7: 329
Weigert-Pal technique 7: 328
white matter 2: 25
Wulst 21: 1531, 1609

X

x (group × vestibular nuclei) 22: 1832
Xenopus laevis 19: 1154

Y

y (group y vestibular nuclei) 22: 1724, 1737, 1832, 1833

Z

zebrin 22: 1741
zona anteroventralis thalami 18: 1116
zona granularis marginalis 14: 726
zona incerta 22: 1804; 22: 1861
zona intermedia 14: 707
zona limitans 12: 632; 13: 683; 17: 1037
zona limitans diencephali 14: 736, 738
zona limitans intrathalamica 4: 194, 212
zona limitans telencephali 14: 747, 749
zona posterodorsalis thalami 18: 1116
zona subhabenularis thalami 18: 1116

Additional material from The *Central Nervous System of Vertebrates*, ISBN 978-3-642-62127-7, is available at http://extras.springer.com